5th European Conference on Neutron Scattering 2011

(ECNS 2011)

Journal of Physics: Conference Series Volume 340

Prague, Czech Republic
17-22 July 2011

ISBN: 978-1-62276-686-4
ISSN: 1742-6588

Printed from e-media with permission by:

Curran Associates, Inc.
57 Morehouse Lane
Red Hook, NY 12571

Some format issues inherent in the e-media version may also appear in this print version.

Printed by Curran Associates, Inc. (2013)

For permission requests, please contact the Institute of Physics
at the address below.

Institute of Physics
Dirac House, Temple Back
Bristol BS1 6BE UK

Phone: 44 1 17 929 7481
Fax: 44 1 17 920 0979

techtracking@iop.org

Additional copies of this publication are available from:

Curran Associates, Inc.
57 Morehouse Lane
Red Hook, NY 12571 USA
Phone: 845-758-0400
Fax: 845-758-2634
Email: curran@proceedings.com
Web: www.proceedings.com

TABLE OF CONTENTS

FUNDAMENTAL PHYSICS

STRUCTURE AND SPECTROSCOPY OF SOLIDS

MAGNETISM AND MAGNETIC EXCITATIONS

SUPERCONDUCTIVITY

GLASSES AND LIQUIDS

THIN FILMS AND INTERFACES

SOFT CONDENSED MATTER

BIOLOGY

FUNCTIONAL MATERIALS AND ENGINEERING

APPLIED SCIENCE (ENERGY, ENVIRONMENT, GEOSCIENCES, CULTURAL HERITAGE)

Opening of ECNS 2011; from the left: Dr. Petr Lukáš (Vice head of UCJF AVCR), Dr. Michael Steiner (Head of ENSA), Prof. Ivan Wilhelm (Deputy Minister of Education), Prof. Zdeněk Němeček (Dean of MFF UK), Ing. Josef Dobeš (Head of UCJF AVCR), Prof. Vladimír Sechovský (Vice-Dean of MFF UK).

Opening - Dr. Petr Lukáš (Vice head of UCJF AVCR and Co-Chairman of ECNS2011).

Poster session.

Concert in St. Nicolas Cathedral.

Dinner in the ZOOM restaurant (KCP praha).

Dinner in the ZOOM restaurant (KCP praha) - speaker Dr. Michel Steiner (Head of ENSA).

Walter Hälg Prize winner Gerry Lander (right).

Erwin Félix Lewy-Bertaut Prize winner Christian Ruegg (left).

Closing - Prof. Vladimír Sechovský (Vice-Dean of MFF UK and Co-Chairman of ECNS2011).

Introducing the next conference ECNS 2015 - Dr. Javier Campo (University of Zaragoza - Chairman).

List of participants

Surname	First Name	Company	City	Country
Abdel-Latif	Ihab	Najran University	Najran	Saudi Arabia
Abele	Hartmut	Technische Universität Wien	Wien	Austria
Abernathy	Douglas	Oak Ridge National Laboratory	Oak Ridge	USA
Abilama	Marc	Centronic Ltd	Croydon	U.K.
Ackermann	Ralf	Helmholtz-Zentrum Dresden-Rossendorf	Dresden	Germany
Ahlberg	Martina	Uppsala University	Uppsala	Sweden
Alba-Simionesco	Christiane	Laboratoire Léon Brillouin	Gif-sur-Yvette	France
Aliouane	Nadir	Paul Scherrer Institut	Villigen-PSI	Switzerland
Almisned	Ghada	Princess Nora Abdul Rahman University	RIYADH	Saudi Arabia
Alsmadi	Abdel	The Hashemite University	Zarqa	Jordan
Amoretti	Giuseppe	University of Parma	Parma	Italy
Andersen	Ken	ESS	Lund	Sweden
Andreani	Carla	UNIVERSITY OF ROME TOR VERGATA	ROME	Italy
Angelov	Borislav	Inst. Macromolecular Chemistry	Prague	Czech Republic
Arai	Masatoshi	JAEA	Tokai, Ibaraki	Japan
Argyriou	Dimitri	European Spallation Source ESS AB	Lund	Sweden
Audonnet	Fabrice	Laboratoire Léon Brillouin - CEA Saclay	Gif-sur-Yvette	France
Avdeev	Mikhail	Joint Institute for Nuclear Research	Dubna Moscow Reg.	Russia
Babcock	Earl	Juelich Centre for Neutron Science	Garching	Germany
Badurek	Gerald	TU Wien / Atominstitut	Wien	Austria
Bafile	Ubaldo	CNR-ISC	Sesto Fiorentino (FI)	Italy
Balasoiu	Maria	IFIN-HH	Bucharest - Magurele	Romania
Barchuk	Michal	Charles University	Prague 2	Czech Republic
Barroso-Bujans	Fabienne	Centro de Fisica de Materiales, CSIC-UPV/EHU	San Sebastian	Spain
Bartkowiak	Maciej	ANSTO	Lucas Heights	Australia
Bartkowiak	Marek	Paul Scherrer Institut	Villigen PSI	Switzerland
Bataille	Alexandre	CEA	Gif sur Yvette	France
Bauer	Guenther	Charles University in Prague	Prague	Czech Republic
Bażela	Wiesława	Cracow University of Technology	Cracow	Poland
Beil	Johannes	Niels Bohr Institute	Copenhagen	Denmark
Bellissent-Funel	Marie-Claire	CNRS	Gif-sur-Yvette	France
Belushkin	Alexander	Joint Inst. Nuclear Res.	Dubna	Russia
Benedetto	Antonio	Messina University	Messina	Italy
Beran	Přemysl	Nuclear Physics Institute ASCR	Řež	Czech Republic
Bert	Nickel	Ludwig-Maximilians-Universität	München	Germany
Bertelsen	Mads	Niels Bohr Institute	Copenhagen	Denmark
Berts	Ida	Institut Laue-Langevin	Grenoble	France
Bestel	Martina	Paul Scherrer Institute	Villigen PSI	Switzerland
Bick	Jens-Peter	University of Luxembourg	Luxembourg	Luxembourg
Bigault	Thierry	Institut Laue Langevin	Grenoble	France
Birk	Jonas Okkels	Copenhagen University	Copenhagen E.	Denmark
Birke	Michael	DESY	Hamburg	Germany
Blackburn	Elizabeth	University of Birmingham	Birmingham	U.K.
Blanco	Jesus A.	University of Oviedo	Oviedo	Spain
Boag	Stephen	STFC	Didcot, Oxfordshire	U.K.
Boehm	Andreas	University of Leoben	Leoben	Austria

Boehm	Martin	Institut Laue Langevin	Grenoble	France
Boin	Mirko	Helmholtz Centre Berlin	Berlin	Germany
Böni	Peter	Technische Universität München	Garching	Germany
Bordallo	Heloisa	University of Copenhagen	Copenhagen	Denmark
Botegard	Sofie	European Spallation Source ESS AB	Lund	Sweden
Bottyán	László	KFKI Res Inst Particle and Nucl Physics	Budapest	Hungary
Bourgeois	Julie	CEA	Gif-sur-Yvette Cedex	France
Bousige	Colin	ILL	Grenoble	France
Bouwman	Wim	Delft University of Technology	Delft	The Netherlands
Bove	Livia Eleonora	CNRS & Universitè P. et M. Curie	Paris	France
Bowden	Zoë	Science & Technology Facilities Council	Didcot	U.K.
Bowron	Daniel	Science and Technology Facilities Council	Chilton, Didcot	U.K.
Bramwell	Steven			U.K.
Brandl	Georg	TU Munich	Garching	Germany
Brok	Erik	Technical University of Denmark	Kgs. Lyngby	Denmark
Bruchhaus	Rainer	Platinum partner - JCNS		Germany
Brüning	Beate-Annette	Helmholtz Zentrum Berlin	Berlin	Germany
Bulavin	Maxim	Joint institute for nuclear research	Dubna	Russia
Bull	Martyn	Science and Technology Facilities Council	Didcot	U.K.
Burgoyne	John	Oxford Instruments	Abingdon	U.K.
Burgoyne	John	Oxford Instruments	Abingdon	U.K.
Burchard	Walther	University of Freiburg	Freiburg i(Breisgau)	Germany
Cabral	Joao	Imperial College London	London	U.K.
Cabrillo	Carlos	Instituto de Estructura de la Materia, IEM-CSIC	Madrid	Spain
Cadogan	Sean	University of Manitoba	Winnipeg	Canada
Campbell	Richard	Institut Laue-Langevin	Grenoble	France
Campo	Javier	ICMA	Zaragoza	Spain
Canadillas-Delgado	Laura	ICMA-CSIC-Universidad de Zaragoza	Zaragoza	Spain
Cardenas	Marite	University of Copenhagen	København ø	Denmark
Carlile	Colin	ESS AB	Lund	Sweden
Carlsson	Patrik	European Spallation Source ESS AB	Lund	Sweden
Carpenter	John M	Oak Ridge & Argonne National Laboratories	Lisle	USA
Carva	Karel	Charles University	Prague	Czech Republic
Caspi	El'ad	NRCN	Beer-Sheva	Israel
Cicognani	Giovanna	ILL		France
Clausen	Henrik	Centre for Materials Crystallography	Aarhus	Denmark
Clausen	Kurt	Paul Scherrer Institut	Villigen PSI	Switzerland
Coduri	Mauro	Uniersita' degli Studi di Milano	Romano di Lombardia (BG)	Italy
Comsan	MNH	Afaq Scientific	Cairo	Egypt
Correa	Jonathan	ILL	Grenoble	France
Courtois	Pierre	Institut Laue Langevin	Grenoble	France
Cousland	Geoffrey	ANSTO	Kirrawee DC	Australia
Cronenberg	Gunther	Vienna University of Technology	Vienna	Austria
Cuello	Gabriel	Institut Laue Langevin	Grenoble	France
Culicov	Otilia Ana	JINR	Dubna	Russia
Cura	Harald	Siemens AG	Erlangen	Germany

Custers	Jeroen	Charles University	Prague	Czech Republic
Čermák	Petr	Charles University	Praha	Czech Republic
Danilkin	Sergey	ANSTO	Lucas Hieghts	Australia
Daniš	Stanislav	Charles University Prague	Prague	Czech Republic
Davydov	Vadim	Paul Scherrer Institute	Villigen	Switzerland
Daymond	Mark	Queen's University	Kingston, Ontario	Canada
De Francesco	Alessio	CNR-IOM	Grenoble	France
de Groot	Joost	Forschungszentrum Jülich GmbH	Jülich	Germany
de Pedro	Manuel	Universidad de Cantabria	Santander	Spain
Deen	Pascale	ESS - AB	Lund	Sweden
Deriu	Antonio	UNIVERSITA' DEGLI STUDI DI PARMA	PARMA	Italy
Devishvili	Anton	Ruhr Unvirsitaet Bochum	Bochum	Germany
Dewhurst	Charles	Institut Laue Langevin	Grenoble	France
Dimeo	Robert	NIST	Gaithersburg	USA
Divis	Martin	Charles Univerzity Prague	Prague	Czech Republic
Dlouhá	Maja	Czech Technical University, Prague	Praha 1	Czech Republic
Dobeš	Jan	Academy of Sciences of the Czech Republic	Prague	Czech Republic
Döbrich	Frank	University of Luxembourg	Luxembourg	Luxembourg
Dos Santos-García	Antonio J.	Universidad Complutense de Madrid	Madrid	Spain
Down	Richard	STFC RAL	Didcot	U.K.
Dráb	Martin	FNSPE CTU in Prague	Praha 1	Czech Republic
Drahoš	Jiří	Academy of Sciences of the Czech Republic	Prague	Czech Republic
Dubbers	Dirk	Universität Heidelberg		Germany
Dunstan	Matthew	University of Sydney	University of Sydney	Australia
Dyadkin	Vadim	Petersburg Nuclear Physics Institute	Gatchina, Saint-Petersburg	Russia
Dyadkina	Ekaterina	Petersburg Nuclear Physics Institute	Gatchina	Russia
Ebrahimi	Omid	Beuth Hochschule für Technik Berlin	Berlin	Germany
Eckerlebe	Helmut	Helmholtz-Zentrum Geesthacht	Geesthacht	Germany
Eckold	Götz	Georg-August-Universität Göttingen	Göttingen	Germany
Egelhaaf	Stefan	Heinrich-Heine University	Duesseldorf	Germany
Egetenmeyer	Nikola	Paul Scherrer Institut	Villigen PSI	Switzerland
Em - STORNO	Vyacheslav	Korea Atomic Energy Research Institute	Daejeon	South Korea
Embs	Jan Peter	Paul Scherrer Institut	Villigen PSI	Switzerland
Enderle	Mechthild	Institut Laue-Langevin	Grenoble	France
Eriksson	Roger	ESSS		Sweden
Fabelo	Oscar	1Instituto de Ciencia de Materiales de Aragón	Zaragoza	Spain
Fabian	Margit	Res. Inst. for Solid State Physics and Optics	Budapest	Hungary
Fak	Bjorn	CEA	Grenobloe	France
Farhi	Emmanuel	ILL	Grenoble Cedex 9	France
Fedorov	Valery	Petersburg Nuclear Physics Institute	Gatchina, St. Petersburg reg.	Russia
Fenske	Jochen	Helmholtz-Zentrum Geesthacht	Geesthahct	Germany
Ferenc	Mezei	ESS AB	Lund	Sweden
Fernandez-Alonso	Felix	Science and Technology Facilities Council	Chilton,Didcot	U.K.
Fikáček	Jan	MFF UK	Praha	Czech Republic
Filges	Uwe	Paul Scherrer Institut	Villigen PSI	Switzerland
Forster	Miriam	ILL	Grenoble	France

Forsyth	Trevor	Institut Laue-Langevin/Keele University	Grenoble	France
Fouquet	Peter	Institut Laue-Langevin	Grenoble Cedex 9	France
Fragneto	Giovanna	ILL	Grenoble	France
Frank	Alexander	JINR	Dubna	Russia
Franklyn	Chris	Necsa	Pretoria	South Africa
Freeman	Paul	Helmholtz Zentrum Berlin	Berlin	Germany
Freund	Andreas	Via Cordis	Bordeaux	France
Frielinghaus	Henrich	Forschungszentrum Jülich GmbH	Garching	Germany
Frontera	Carlos	Institut de Ciencia de Materials de Barcelona-CSIC	Cerdanyola del Vallès	Spain
Frontzek	Matthias	ORNL	Oak Ridge	USA
Fujita	Masaki	Tohoku university	Sendai	Japan
Fukunaga	Toshiharu	Kyoto University	Osaka	Japan
Furrer	Albert	SwissNeutronics AG	Klingnau	Switzerland
Furusaka	Michihiro	Hokkaido University	Sapporo Hokkaido	Japan
Füzi	János	Res. Inst. of Solid State Physics and Optics	Budapest	Hungary
Gabel	Frank	IBS Grenoble	Grenoble	France
Gahl	Thomas	Paul Scherrer Institut	Villigen PSI	Switzerland
Galan	Isabel	Eduardo Torroja Institute IETcc-CSIC	Madrid	Spain
Gallova	Jana	Faculty of Pharmacy, Comenius University	Bratislava	Slovak Republic
Gan	Weimin	Helmholtz-Centre Geesthacht	Garching	Germany
García-Muñoz	José Luis	Institut de Ciència de Materials de Barcelona-CSIC	Bellaterra (Barcelona)	Spain
Gaspard	Jean-Pierre	University of Liege	Sart-Tilman	Belgium
Gast	Heike	Helmholtz-Zentrum Berlin		Germany
Gavilano	Jorge	Paul Scherrer Institut	Villigen PSI	Switzerland
Georgiev	Peter	University of MIlano	Milano	Italy
Gerber	Simon	Paul Scherrer Institut	Villigen PSI	Switzerland
Gerelli	Yuri	Institut Laue-Langevin	Grenoble	France
Gerth	Stefan	University of Erlangen-Nuremberg	Erlangen	Germany
Gibhardt	Holger	Georg-August-Universität	Göttingen	Germany
Gilles	Ralph	TU Munich	Garching	Germany
Glavatskyi	Illia	Helmholtz Centre Berlin for Materials and Energy	Berlin	Germany
Glyde	Henry	University of Delaware	Newark, Delaware	USA
Godfrin	Henri	CNRS	Grenoble	France
Goerigk	Guenter	Helmholtz-Zentrum Berlin	Berlin	Germany
Golosovsky	Igor	Petersburg Nuclear Physics Institute	Gatchina, Leningrad district	Russia
Gonthier	Julien	ILL	GRENOBLE	France
Gonzalez	Miguel	Institut Laue-Langevin	Grenoble	France
Gonzalez Larrea	Lander	Scientifica Internacional, S.L.	Elgoibar	Spain
Gonzalez Larrea	Lander	Scientifica Internacional, S.L.	Elgoibar	Spain
Gorria	Pedro	Universidad de Oviedo	Oviedo	Spain
Gösselsberger	Christoph	Atominstitut, TU Wien	Wien	Austria
Gottschlich	Michael	Research Center Juelich	Juelich	Germany
Grazzi	Francesco	Consiglio Nazionale delle Ricerche	Sesto Fiorentino	Italy
Grigoriev	Sergey	Petersburg Nuclear Physics Institute	St-Petersburg	Russia
Grigoryeva	Natalia	Saint-Petersburg State University	Saint-Petersburg,	Russia
Groitl	Felix	HZB	Berlin	Germany
Grosse	Mirco	Karlsruhe Institute of Technology	Karlsruhe	Germany
Grünwald	Alexander	Helmholtz-Zentrum Geesthacht	Geesthacht	Germany
Gubanova	Nade	Petersburg Nuclear Physics Institute	Gatchina	Russia
Guerard	Bruno	ILL	Grenoble	France

Gutberlet	Thomas	Helmholtz-Zentrum Berlin GmbH	Berlin	Germany
Habicht	Klaus	Helmholtz Zentrum Berlin	Berlin	Germany
Haese	Harald	S-DH		Germany
Hall-Wilton	Richard	European Spallation Source ESS AB	Lund	Sweden
Hampl	Václav	Charles University in Prague	Prague	Czech Republic
Han	Young-Soo	Korea Atomic Energy Research Institute	Daejeon	South Korea
Hansen	Britt Rosendahl	Risø DTU	Roskilde	Denmark
Hansen	Ursula Bengaard	Niels Bohr institute	Broenshoej	Denmark
Hanslik	Romuald	Forchungszentrum Jülich	Jülich	Germany
Haramus	Vasyl	Helmholtz-Zentrum Geesthacht	Geesthacht	Germany
Harmat	Péter	Mirroton Ltd	Budapest	Hungary
Harrison	Andrew	Institut Laue Langevin	Grenoble Cedex 09	France
Hase	Masashi	National Institute for Materials Science	Tsukuba	Japan
Hasegawa	Yuji	Atominstitut, TU-Wien	Vienna	Austria
Hatnean	Monica	CEA Saclay	Gif-sur-Yvette	France
Hauß	Thomas	Helmholtz-Zentrum Berlin für Materialien und Energ	Berlin	Germany
Häußler	Wolfgang	Technische Universität München	Garching	Germany
Havela	Ladislav	Charles University	Prague 2	Czech Republic
Havlíček	Josef	Charles University in Prague		Czech Republic
Havlíčková	Alena	Charles University in Prague	Prague	Czech Republic
Hayden	Stephen	University of Bristol	Bristol	U.K.
Heinemann	Andre	Helmholtz-Zentrum Geestacht	Geesthacht	Germany
Helgesen	Geir	Institute for Energy Technology	Kjeller	Norway
Henry	Paul	European Spallation Source ESS AB	Lund	Sweden
Hermes	Helen E.	Heinrich-Heine-University Duesseldorf	Duesseldorf	Germany
Hernandez-Velasco	Jorge	Inst.Ciencia Materiales Madrid, CSIC	Madrid	Spain
Hetmańczyk	Łukasz	Jagiellonian University	Kraków	Poland
Hiess	Arno	Institut Laue Langevin	Grenoble	France
Hinderer	Joerg	Bruker Biospin AG	Faellanden	Switzerland
Hippert	Françoise	LMGP	Grenoble	France
Hirschner	Jan	ČVUT FJFI	Praha	Czech Republic
Hlinka	Jiří	IOP AS CR	Prague 8	Czech Republic
Hoelzel	Markus	FRM-II	Garching near Munich	Germany
Hofmann	Michael	TU München	Garching	Germany
Hofmann	Tommy	Helmholtz-Zentrum Berlin	Berlin	Germany
Holderna-Natkaniec	Krystyna	Adam Mickiewicz University	POZNAN	Poland
Holm	Sonja	University of Copenhagen	København Ø	Denmark
Holy	Vaclav	Charles University in Prague	Praha	Czech Republic
Honecker	Dirk	University of Luxembourg	Luxembourg	Luxembourg
Horak	Lukas	Faculty of Mathematics and Physics	Prague	Czech Republic
Horigane	Kazumasa	Tohoku University	Sendai	Japan
Horsewill	Tony	University of Nottingham	Nottingham	U.K.
Hoser	Andreas	Helmhlotz-Zentrum Berlin	Berlin	Germany
Houben	Andreas	RWTH Aachen University	Aachen	Germany
Hrušák	Jan	MŠMT		Czech Republic
Huber	Klaus	University of Paderborn	Paderborn	Germany
Huber	Norman	Huber Diffraktionstechnik		Germany

Humbert	Christophe	MAATEL	VOREPPE	France
Hundal	Amandeep	ISIS		U.K.
Hutanu	Vladimir	RWTH Aachen	Garching	Germany
Chaboussant	Gregory	Laboratoire Leon Brillouin	Gif sur Yvette	France
Chang	Lieh-Jeng	National Cheng Kung University	Tainan	Taiwan
Chatterji	Tapan	Institut Laue-Langevin	Grenoble	France
Chen	Wangchun	NIST	Gaithersburg	USA
Chernenkov	Yury	Petersburg Nuclear Physics Institute	Gatchina	Russia
Chetverikov	Iurii	PNPI	Gatchina	Russia
Chlebeckova	Marie	Cryogenic Ltd		U.K.
Christensen	Mogens	Aarhus University	Aarhus C	Denmark
Christensen	Sebastian	Aarhus University	Aarhus C	Denmark
Chumakov	Andrey	Petersburg Nuclear Physics Institute	Gatchina	Russia
Igartua	Josu M	Basque Country University	Leioa (Bizkaia)	Spain
Ino	Takashi	KEK	Tsukuba, Ibaraki	Japan
Inosov	Dmytro	MPI Stuttgart	Stuttgart	Germany
Ioffe	Alexander	Jülich Centre for Neutron Science	Garching	Germany
Ionita	Ion	Institute for Nuclear Research Pitesti Romania	Mioveni	Romania
Ishii	Takahiro	Kuraray Co.,LTD.	Kurashiki	Japan
Ishikado	Motoyuki	Japan Atomic Energy Agency	Tokai, Naka, Ibaraki	Japan
Isnard	Olivier	Institut Néel, CNRS/University J Fourier	GRENOBLE	France
Ivanov	Alexandre	Institut Laue-Langevin	Grenoble	France
Jacobsen	Henrik	Niels Bohr Institute, University of Copenhagen	København Ø	Denmark
Jacobsen	Johan	Univesity of Copenhagen	Copenhagen	Denmark
Jalarvo	Niina	Juelich research centre	Oak Ridge TN	USA
Javorský	Pavel	Charles University in Prague		Czech Republic
Jensen	Pia	University of Copenhagen	Copenhagen NV	Denmark
Jericha	Erwin	Vienna University of Technology	Wien	Austria
Jirak	Zdenek	Institute of Physics ASCR	Prague 8	Czech Republic
Jobic	Herve	IRCELYON	Villeurbanne	France
Káli	György	MTA SZFKI	Budapest	Hungary
Kalvoda	Ladislav	CTU in Prague, FNSPE	Prague 2	Czech Republic
Kamarad	Jiri	Institute of Physics, ASCR, v.v.i.	Praha 8	Czech Republic
Kamata	Yohei	University of Sheffield	Sheffield	U.K.
Kamenev	Konstantin	University of Edinburgh	Edinburgh	U.K.
Kampmann	Reinhard	Helmholtz-Zentrum Geestgacht	Geesthacht	Germany
Kandemir	Timur	Fritz-Haber-Institut der MPG	Berlin	Germany
Kargl	Florian	German Aerospace Center (DLR)	Köln	Germany
Karlsson	Maths	Chalmers University of Technology	Göteborg	Sweden
Kaszas	György	MIRROTRON, Ltd.	H-1121 Budapest	Hungary
Kaštil	Jiří	Charles University in Prague, MFF	Prague 2	Czech Republic
Katsaros	Fotios	NCSR Demokritos	Athens	Greece
Kawakita	Yukinobu	Japan Atomic Energy Agency	Tokai, Naka, Ibaraki	Japan
Kawasaki	Takuro	Japan Atomic Energy Agency	Tokai, Ibaraki	Japan
Kazimirov	Viacheslav	Joint Institute for Nuclear Research	Dubna	Russia
Keiderling	Uwe	Helmholtz-Zentrum Berlin	Berlin	Germany
Keller	Thomas	MPI for solid state Physics		Germany
Kempa	Martin	Institute of Physics, ASCR	Prague	Czech Republic
Kenzelmann	Michel	Paul Scherrer Institut	Villigen	Switzerland
Keum	Jong	Oak Ridge National Lab	Oak Ridge	USA
Khaplanov	Anton	ESS/ILL	Grenoble	France

Khaydukov	Yury	Max Planck Institute for Solid State Research	Stuttgart	Germany
Kibalin	Yury	Petersburg Nuclear Physics Institute RAS, Gatchina	Gatchina	Russia
Kiefer	Klaus	Helmholtz-Zentrum Berlin	Berlin	Germany
Kichanov	Sergey	Joint Institite for Nuclear Research	Dubna, Moscow reg.,	Russia
Kim	Tae-Hwan	Korea Atomic Energy Research Institute	Daejeon	South Korea
Kimura	Hiroyuki	Tohoku University	Sendai	Japan
Kindervater	Jonas	Technische Universität München	Garching	Germany
King	Philip	STFC	Didcot	U.K.
Kitazawa	Hideaki	National Institute for Materials Research	Tsukuba	Japan
Kiyanagi	Yoshiaki	Hokkaido University	Sapporo	Japan
Klauser	Christine	Institut Laue Langevin	Grenoble	France
Klenø	Kaspar Hewitt	University of Copenhagen	Copenhagen	Denmark
Klepp	Juergen	University of Vienna	Vienna	Austria
Klicpera	Milan	Charles University, MFF	Prague	Czech Republic
Klíma	Jan	MFF KU	Prague	Czech Republic
Knaapila	Matti	Institute for Energy Technology	Kjeller	Norway
Knoll	Wiebke	Institut Laue-Langevin, Grenoble	Grenoble	France
Knudsen	Kenneth D.	Institute for Energy Technology	Kjeller	Norway
Kohlbrecher	Joachim	LNS	Villigen PSI	Switzerland
Köhler	Ralf	Technical University Berlin	Berlin	Germany
Kolomiets	Alexandre	Charles University	Praha 2	Czech Republic
Konrad	Gertrud	Vienna University of Technology	Vienna	Austria
Kopeček	Jaromír	Institute of Physics of the AS CR	Praha	Czech Republic
Kopitsa	Gennady	Petersburg Nuclear Physics Institute	Gatchina	Russia
Korolkov	Denis	Forschungszentrum Jülich GmbH	Garching	Germany
Kovács-Mezei	Rita	Mirrotron Ltd.	Budapest	Hungary
Kozhevnikov	Sergey	Joint Institute for Nuclear Research	Dubna, Moscow Region	Russia
Kozlenko	Denis	Joint Institute for Nuclear Research	Dubna Moscow Reg.	Russia
Krakovsky	Ivan	Faculty of Mathematics & Physics	Praha	Czech Republic
Kratochvilova	Marie	MFF, Charles University	Prague	Czech Republic
Kreuzer	Martin	University of Heidelberg	Heidelberg	Germany
Krezhov	Kiril	Institute for Nuclear Research and Nuclear Energy	Sofia	Bulgaria
Kriska	Martin	K.U.Leuven	Heverlee (Leuven)	Belgium
Krist	Thomas	Helmholtz-Zentrum Berlin	Berlin	Germany
Krotscheck	Eckhard	JKU Linz	Linz	Austria
Krůželová	Monika	FNSPE CTU Prague	Prague	Czech Republic
Krzytyniak	Matthew	NOTTINGHAM TRENT UNIVERSITY	NOTTINGHAM	U.K.
Kulda	Jiri	Institut Laue-Langevin	Grenoble	France
Kulikov	Sergey	Joint Institute for Nuclear Research	Dubna	Russia
Kunzmann	Marc	Quantum Design (LOT-Oriel)		Germany
Kurbakov	Alexander	1Petersburg Nuclear Physics Institute	Gatchina	Russia
Kurt	Walther	Karlsruhe Institute for Technology	Karlsruhe	Germany
Kuznetsov	Sergey	P.N. Lebedev Physical Institute	Moscow	Russia
Kužel	Radomír	UK MFF	Prague 2	Czech Republic
Lacerda	Alex	LANL - LANSCE		USA
Lahoubi	Mahieddine	Department of Physics, University of Annaba	Annaba	Algeria
Lake	Bella	Helmholtz Zentrum Berlin	Berlin	Germany

Lander	Gerry			France
Lapčák	Ladislav	Charles University in Prague	Prague 2	Czech Republic
Lappas	Alexandros	Foundation for Research & Technology-Hellas (FORTH	Heraklion	Greece
Larsen	Jacob	Risoe DTU	Roskilde	Denmark
László	Zoltán	MTA-SZFKI	Budapest	Hungary
Le	Manh Duc	Helmholtz-Zentrum Berlin für Materialen und Energi	Berlin	Germany
Le Dréau	Loïc	Paul scherrer institute	Villigen	Switzerland
Lee	Chul-Ho	National Institute of Advanced Industrial Science	Tsukuba, Ibaraki	Japan
Lee	Seongsu	KAERI	Daejeon	South Korea
Lee	Wai Tung	Australian Nuclear Science and Technology Organisa	Lucas Heights	Australia
Lefmann	Kim	University of Copenhagen	Copenhagen	Denmark
Lehmann	Eberhard	Paul Scherrer Institut	Villigen	Switzerland
Leist	Jeannis	Georg August Universität Göttingen	Göttingen	Germany
Lelièvre-Berna	Eddy	Institut Laue Langevin	Grenoble	France
Lerbret	Adrien	CEA Saclay	Gif-sur-Yvette	France
Lieutenant	Klaus	Helmholtz-Zentrum Berlin	Berlin	Germany
Lin	Tsang-Lang	National Tsing Hua University	Hsinchu	Taiwan
Lindenau	Bernd	Forschungszentrum Juelich GmbH	Juelich	Germany
Lindner	Peter	Institut Laue-Langevin	Grenoble	France
Ling	Chris	The University of Sydney	Sydney	Australia
Litvin	Vasily	Institute for Nuclear Research RAS	Moscow	Russia
Loewenhaupt	Michael	TU Dresden	Dresden	Germany
Lokitz	Brad	Oak Ridge National Laboratory	Oak Ridge	USA
Lommatzsch	Ina	FRM II		Germany
Lott	Dieter	Helmholtz-Zentrum Geesthacht	Geesthacht	Germany
Ludanyi	Zsolt	MIRROTRON, Ltd.	H-1121 Budapest	Hungary
Lukas	Petr	Nuclear Physics Institute	Rez	Czech Republic
Maccarini	Marco	Institut Laue Langevin	Grenoble	France
Magerl	Andreas	University Erlangen-Nuernberg	Erlangen	Germany
Magi	Ádám	Mirroton Ltd	Budapest	Hungary
Malikova	Natalie	Laboratoire Leon Brillouin (CEA-CNRS)	Gif-sur-Yvette	France
Mancinelli	Rosaria	Università di Roma Tre	Roma	Italy
Mantlíková	Alice	Charles University in Prague		Czech Republic
Manuel	Pascal	ISIS Facility	Chilton	U.K.
Manzin	Giuliana	Institut Laue Langevin	Grenoble	France
Marko	Márton	RISP	Budapest	Hungary
Marti	Xavier	Charles University in Prague	Prague	Czech Republic
Martin	Nicolas	CEA Grenoble INAC-SPSMS-MDN	Grenoble	France
Martin Rodriguez	Damian	Forschungszentrum Juelich GmbH	Juelich	Germany
Masayasu	Takeda	Japan Atomic Energy Agency	Tokai-mura	Japan
Maškova	Silvie	Charles University in Prague	Prague 2	Czech Republic
Matej	Zdenek	Charles University	Prague 2	Czech Republic
Matejova	Jana	MFF UK, Charles University, Prague	Praha 2	Czech Republic
Matveev	Vasily	Petersburg Nuclear Physics Institute	Gatchina	Russia
Mazumder	Subhasish	Bhabha Atomic Research Centre	Mumbai	India
McEwen	Keith	University College London	London	U.K.
Medarde	Marisa	Paul Scherrer Institut	Villigen PSI	Switzerland
Meier	Gerhard	FZ-Jülich	Jülich	Germany
Mergia	Konstantina	NCSR "DEMOKRITOS"	ATHENS	Greece
Merkel	Dániel	KFKI Research Inst. for Part. and Nucl. Physics	Budapest	Hungary
Messoloras	Spyros	NCSR "DEMOKRITOS"	ATHENS	Greece

Meven	Martin	TU München	Garching	Germany
Middendorf	Dieter	University of Oxford	Oxford	U.K.
Michels	Andreas	University of Luxembourg	Luxembourg	Luxembourg
Michor	Herwig	Vienna University of Technology	Wien	Austria
Mikula	Pavol	Nuclear Physics Institute ASCR, v.v.i.	Řež	Czech Republic
Mistonov	Alexander	Saint Petersburg State University	Saint Petersburg, Petrodvorets	Russia
Míšek	Martin	Institute of Physics, ASCR, v.v.i.	Prague 8	Czech Republic
Mitchinson	Simon	Oxford Instruments NanoScience	Abingdon	U.K.
Mitschler	Andre	IGBMC	Illkirch	France
Mitsuda	Setsuo	Tokyo University of Science	Tokyo	Japan
Mityukhlyaev	Victor	PNPI	Gatchina	Russia
Mixa	Martin	Charles University in Prague	Prague	Czech Republic
Molchanov	Vyacheslav	Moscow State University	Moscow	Russia
Mondelli	Claudia	CNR-IOM ILL	Grenoble cedex 9	France
Monkenbusch	Michael	Forschungszentrum Juelich	Juelich	Germany
Mori	Kazuhiro	Kyoto Univ.	Osaka	Japan
Mortensen	Kell	University of Copenhagen	Frederiksberg	Denmark
Moskvin	Evgeny	Petersburg Nuclear Physics Institute	St.Petersburg, Gatchina	Russia
Moudřín	Jan	Charles University in Prague		Czech Republic
Moulin	Jean-Francois	HZG	Geesthacht	Germany
Mukhin	Konstantin	Joint Institute For Nuclear Research	Dubna Moscow region	Russia
Müller	Martin	Helmholtz-Zentrum Geesthacht	Geesthacht	Germany
Náfrádi	Bálint	Max Planck Institute for Solid State Research	Stuttgart	Germany
Nagano	Mikinori	Osaka University	Suita	Japan
Nagy	Bela	KFKI Res Inst Particle and Nuclear Physic		Hungary
Nagy	Gergely	Res. Inst. for Solid-state Physics and Optics	Budapest	Hungary
Natali	Francesca	CNR	Grenoble	France
Natkaniec	Ireneusz	Joint Institute for Nuclear Research	Dubna	Russia
Natour	Ghaleb	Forschungszentrum Jülich GmbH	Jülich	Germany
Nawroth	Thomas	Gutenberg-University	Mainz	Germany
Němeček	Zdeněk	Charles University in Prague	Prague	Czech Republic
Nemkovskiy	Kirill	Forschungszentrum Jülich	Jülich	Germany
Nesvizhevsky	Valery	Institut Laue-Langevin	Grenoble	France
Neuhaus	Jürgen	Technische Universität München	Garching	Germany
Niedermayer	Christof	Paul Scherrer Institut	Villigen	Switzerland
Nichtova	Lea	MFF Charles University	Prague 2	Czech Republic
Nilsson	Monica	ESSS		Sweden
Nishiki	Naomi	Panasonic		Japan
Noirez	Laurence	Laboratoire Léon Brillouin (CEA-CNRS)	Gif-sur-Yvette	France
Novotny	Tomas	MFF UK Praha	Praha 2	Czech Republic
Nowak	Dorota Marta	Joint Institute for Nuclear Research	Dubna	Russia
Nowak	Sebastian	Atominstitut, TU Wien	Wien	Austria
Nylander	Tommy	Lund University	Lund	Sweden
Ohl	Michael	Forschungszentrum Juelich GmbH	Oak Ridge	USA
Ohoyama	Kenji	Tohoku University	Sendai	Japan
Oku	Takayuki	Japan Atomic Energy Agency	Tokai	Japan
Oliver	Kirstein	European Spallation Source ESS AB	Lund	Sweden
Ollivier	Jacques	Institut Laue-Langevin	Grenoble	France
Ondrejkovic	Petr	Institut Laue-Langevin	Grenoble	France

Onodera	Yohei	Kyoto University Resarch Reactor Institute	Sennan	Japan
Orban	Janos	RESEARCH INSTITUTE FOR SOLID STATE PHYSICS AND OPT	Budapest	Hungary
Orecchini	Andrea	Institut Laue Langevin	Grenoble	France
Ostermann	Andreas	Technische Universität München	Garching	Germany
Otake	Yoshie	RIKEN	Wako-shi, Saitama	Japan
Otomo	Toshiya	KEK	Tsukuba	Japan
Padilla	Jessica	ICMAB (CSIC)	Bellaterra	Spain
Pajskr	Karel	Charles University in Prague		Czech Republic
Pajzderska	Aleksandra	A.Mickiewicz Univeristy	Poznan	Poland
Palacio	Fernando	CSIC	Zaragoza	Spain
Palacios Gómez	Jesús	Instituto Politécnico Nacional	Mexico City	Mexico
Panarin	Justin	SPSMS, UMR-E CEA / UJF-Grenoble 1	Grenoble	France
Pantalei	Claudia	C.E.A Saclay	Gif-sur-Yvette	France
Panzner	Tobias	Paul Scherrer Institut	Villigen PSI	Switzerland
Paolasini	Luigi	ESRF	Grenoble	France
Pappas	Catherine	TU Delft	Delft	The Netherlands
Parker	Dinah	Oxford Instruments NanoScience		U.K.
Paul-Boncour	Valérie	CNRS	THIAIS	France
Pawlukojc	Andrzej	Frank Laboratory of Neutron Physics, JINR	Dubna	Russia
Perelli Cippo	Enrico	Università degli Studi di Milano-BICOCCA	Milano	Italy
Peters	Judith	Institut de Biologie Structurale	Grenoble	France
Peterson	Joe			USA
Petry	Winfried	TUM	Garching	Germany
Pfleiderer	Christian	Technische Universität München	Garching	Germany
Philipp	Schmakat	Technische Universität München	Garching	Germany
Piarristeguy	Andrea	Institut Charles Gerhardt Montpellier	Montpellier	France
Piegsa	Florian	ETH Zürich	Zürich	Switzerland
Pierret	Stephane	Paul Scherrer Institute	Villigen	Switzerland
Pilgrim	Wolf-Christian	Philipps-University of Marburg	Marburg	Germany
Pilch	Jan	Institute of Physics AS CR, v. v. i.	Prague	Czech Republic
Pipich	Vitaliy	Forschungszentrum Jülich GmbH	GARCHING	Germany
Platz	Martin	Institut Laue Langevin	Grenoble	France
Plewinski	François	ESS	Lund	Sweden
Podjarny	Alberto	IGBMC	Illkirch	France
Podlesnyak	Andrey	Oak Ridge National Laboratory	Oak Ridge	USA
Poltierova Vejpravova	Jana	Institute of Physics of the AS CR, v.v.i.	Prague 8	Czech Republic
Pomjakushin	Vladimir	Paul Scherrer Institute	Villigen	Switzerland
Pomjakushina	Ekaterina	Paul Scherrer Institute	Villigen	Switzerland
Poole	Amy	Paul Scherrer Institut	Villigen	Switzerland
Porter	Daniel	Royal Holloway, University of London	Egham	U.K.
Pospisil	Jiri	Charles University	Praha	Czech Republic
Potapova	Nadya	PNPI	Gatchina	Russia
Pracharova	Marcela	Charles University	Prague 2	Czech Republic
Pranzas	P. Klaus	Helmholtz-Zentrum Geesthacht	Geesthacht	Germany
Pregelj	Matej	Paul Scherrer Institue	Villigen	Switzerland
Prévost	Sylvain	TU Berlin	Berlin	Germany
Prchal	Jiri	Charles University in Prague	Praha 2	Czech Republic
Prokes	Karel	Helmholtz Zentrum Berlin	Berlin	Germany
Prokhnenko	Oleksandr	Helmholtz Zentrum Berlin	Berlin	Germany
Prokleska	Jan	Charles University	Prague	Czech Republic

Prsa	Krunoslav	ETH Zurich	Zurich	Switzerland
Radaelli	Paolo G.	Oxford University	Oxford	U.K.
Radulescu	Aurel	Forschungszentrum Jülich GmbH	Garching	Germany
Ramirez-Cuesta	Anibal	ISIS Facility, Rutuerford		U.K.
Randau	Christian	HZG	Geesthacht	Germany
Rauch	Helmut	Vienna University of Technology	Wien	Austria
Rebelo Kornmeier	Joana	FRM II - TUM	Garching	Germany
Recarte	Vicente	Universidad Pública de Navarra	Pamplona	Spain
Rećko	Katarzyna	University of Białystok	Białystok	Poland
Rennie	Adrian	Uppsala University	Uppsala	Sweden
Repper	Julia	TU Muenchen	Garching	Germany
Reviakine	Ilya	CIC biomaGUNE	San Sebastian	Spain
Richter	Dieter	Forschungszentrum Jülich	Jülich	Germany
Ritter	Clemens	Institut Laue Langevin	Grenoble	France
Rodriguez-Velamazan	J. Alberto	Inst. de Ciencia de Materiales de Aragon at ILL	Grenoble	France
Rolfs	Katharina	Helmholtz-Centre Berlin for materials and energy	Berlin	Germany
Rols	Stéphane	ILL	grenoble	France
Romanos	George	NCSR Demokritos	Athens	Greece
Rømer	Astrid Tranum	University of Copenhagen	København Ø	Denmark
Rondelli	Valeria	Università degli Studi di Milano	Segrate	Italy
Rosenfeldt	Sabine	University of Bayreuth	Bayreuth	Germany
Rosta	Laszlo	Budapest Neutron Centre	Budapest	Hungary
Roth	Georg	RWTH Aachen University	Aachen	Germany
Roux	Stephane	ILL		France
Rovira-Esteva	Muriel	Universitat Politècnica de Catalunya	Barcelona	Spain
Rudajevova	Alexandra	Charles University, MFF	Prague	Czech Republic
Rüegg	Christian	Paul Scherrer Institute	Villigen PSI	Switzerland
Runov	Vladimir	Petersburg Nuclear Physics Institute RAS	Gatchina	Russia
Russina	Margarita	Helmholtz Zentrum Berlin	Berlin	Germany
Russo	Daniela	CNR/IOM	grenoble	France
Ryan	Dominic	McGill University	University St	Canada
Ryukhtin	Vasyl	Hemholtz Zentrum Berlin	Berlin	Germany
Sadakane	Koichiro	High Energy Accelerator Research Organization	Ibaraki	Japan
Sadykov	Ravil	Institute for Nuclear Research RAS	Moscow	Russia
Saez Puche	Regino	Universidad Complutense Madrid	Madrid	Spain
Saito	Hiroshi	Toshiba Electron Tubes & Devices Co., Lta.		Japan
Sakai	Kenji	Japan Atomic Energy Agency	Tokai-mura	Japan
Sales	Morten	University of Copenhagen	Copenhagen	Denmark
Sánta	Zsombor	Mirrotron		Hungary
Sapalidis	Andreas	NCSR "Demokritos"	Aghia Paraskevi, Athens	Greece
Savey-Bennett	Marc	Cryogenic Ltd		U.K.
Savin	Juliette	NMI3		France
Sechovsky	Stepan	MFF UK	Prague	Czech Republic
Sechovsky	Vladimir	Charles University	Prague	Czech Republic
Seki	Yoshichika	RIKEN	Wako, Saitama	Japan
Senyshyn	Anatoliy	Technische Universität München	Garching b. München	Germany
Seto	Hideki	High Energy Accelerator Research Organization	Tsukuba	Japan
Seydel	Tilo	Institut Max von Laue - Paul Langevin	Grenoble	France

Shamoto	Shin-ichi	Japan Atomic Energy Agency	Tokai	Japan
Sharp	Melissa	ESS AB	Lund	Sweden
Sheptyakov	Denis	Laboratory for Neutron Scattering	Villigen PSI	Switzerland
Shimakura	Hironori	Kyushu University	Fukuoka	Japan
Shvetsov	Alexey	PNPI	Gatchina	Russia
Schäffer	Turi	University of Copenhagen	Copenhagen	Denmark
Schaniel	Dominik	Nancy Universite	Nancy	France
Schanzer	Christian	SwissNeutronics		Switzerland
Schefer	Jürg	Paul Scherrer Institute	Villigen PSI	Switzerland
Scheffzük	Christian	University Karlsruhe KIT	Karlsruhe	Germany
Scherzinger	Christine	RWTH-Aachen	Aachen	Germany
Schmalzl	Karin	JCNS at ILL	Grenbole	France
Schmidt	Christian	S-DH		Germany
Schmidt	Wolfgang	JCNS @ ILL	Grenoble	France
Schneider	Michael	SwissNeutronics		Switzerland
Schneider	Rudolf	Huber Diffraktionstechnik GmbH & Co. KG	Rimsting	Germany
Schöbel	Michael	TU Vienna	Vienna	Austria
Schober	Helmut	ILL	Grenoble	France
Schrader	Tobias	Forschungszentrum Jülich	Garching	Germany
Schreiber	Frank	University of Tübingen	Tübingen	Germany
Schulz	Christian	Helmholtz-Zentrum Berlin	Berlin	Germany
Schulz	Jennifer	HZB Berlin	Berlin	Germany
Schulz	Michael	FRM II, TUM	Garching	Germany
Schurtenberger	Peter	Lund University	Lund	Sweden
Schwahn	Dietmar	Forschungszentrum Jülich	Jülich	Germany
Schweika	Werner	Forschungszentrum Jülich	Jülich	Germany
Schweins	Ralf	Institut Laue - Langevin	GRENOBLE CEDEX 9	France
Siebrecht	Ralf	MAATEL	VOREPPE	France
Simonet	Virginie	Institut Néel	Grenoble	France
Simonsen	Marius	Niels Bohr Institute, University of Copenhagen	Copenhagen o	Denmark
Siouris	Ioannis	Democritu Univ. Thrace	Xanthi	Greece
Sittner	Petr	Fyzikální ústav AV ČR,v.v.i.	Praha	Czech Republic
Skelboe	Stig	ESS/University of Copenhagen	Copenhagen	Denmark
Skipper	Neal	University College London	London	U.K.
Skoulatos	Markos	Helmholtz Center Berlin	Berlin	Germany
Skripov	Alexander	Institute of Metal Physics	Ekaterinburg	Russia
Slawinski	Wojciech	University of Warsaw	Warsaw	Poland
Souza Dos Santos	Edgar	Institut Neel, CNRS/UJF	Grenoble cedex9	France
Spehr	Tinka Luise	Technische Universität Darmstadt	Darmstadt	Germany
Spiegelhalder	Berno	Astrium GmbH	Immenstaad	Germany
Sponar	Stephan	TU Wien - Atominstitut	Vienna	Austria
Stadler	Andreas	Forschungszentrum Juelich GmbH	Juelich	Germany
Stahn	Jochen	Paul Scherrer Institut	Villigen PSI	Switzerland
Stampfl	Anton	ANSTO	Lucas Heights	Australia
Staron	Peter	Helmholtz-Zentrum Geesthacht	Geesthacht	Germany
Stefanopoulos	Konstantinos	NCSR "Demokritos"	Athens	Greece
Steigenberger	Uschi	STFC	Chilton	U.K.
Stein	Wolf-Dieter	Helmholtz-Zentrum Berlin	Berlin	Germany
Steiner	Michael	ENSA	Berlin	Germany
Steitz	Roland	Helmholtz-Zentrum Berlin	Berlin	Germany
Stelzer	Hermann	Forschungszentrum Jülich	Jülich	Germany
Steuwer	Axel	ESS AB	Lund	Sweden
Stewart	Ross	ISIS	Didcot	U.K.
Stieper	Gitte	Copenhagen university	Copenhagen Ø	Denmark

Stockert	Oliver	Max-Planck-Institut CPfS	Dresden	Germany
Störmer	Michael	Helmholtz-Zentrum Geesthacht	Geesthacht	Germany
Strobl	Markus	Helmholtz Zentrum Berlin	Berlin	Germany
Strunz	Pavel	Nuclear Physics Institute	Rez near Prague	Czech Republic
Stunault	Anne	ILL	Grenoble	France
Stüßer	Norbert	Helmholtz-Zentrum Berlin	Berlin	Germany
Su	Yixi	Forschungszentrum Juelich GmbH	Garching	Germany
Sugiyama	Masaaki	Kyoto University	Osaka	Japan
Sultan	Ahmad	Institut Néel, CNRS	Grenoble, cedex 9	France
Sura	Ravichandra	PAUL SCHERRER INSTITUT	VILLIGEN PSI	Switzerland
Svoboda	Pavel	Charles University	Prague 2	Czech Republic
Swiderska	Karolina	Institute of Atomic Energy POLATOM	Otwock	Poland
Szakal	Alex	HAS RISP	Budapest	Hungary
Szanto	Peter	Kurt J. Lesker	Budapest	Hungary
Száraz	Zoltán	Joint Research Centre, Institute for Energy	Petten	The Netherlands
Székely	Noemi Kinga	HAS Res. Inst. Solid State Phys.	Budapest	Hungary
Szubrin	Daniel	Philipps-Universität Marburg	Marburg	Germany
Szytula	Andrzej	Jagiellonian University	Kraków	Poland
Šaroun	Jan	Nuclear Physics Institute, ASCR	Řež	Czech Republic
Taketani	Kaoru	KEK	Tsukuba	Japan
Tasaki	Seiji	Kyoto University	Kyoto	Japan
Taylor	Andrew	STFC	Didcot	U.K.
Tehei	Moeava	UOW & AINSE	Wollongong	Australia
Tennant	David	Helmholtz Zentrum Berlin	Berlin	Germany
Terada	Noriki	National Institute for Materials Science	Ibaraki	Japan
Theis-Bröhl	Katharina	University of Applied Sciences Bremerhaven	Bremerhaven	Germany
Tietze	Ursula	Helmholtz-Zentrum Geesthacht	Geesthacht	Germany
Tjebane	Tjatji	Necsa	Pretoria	South Africa
Tkach	Ilya	Charles University	Prague	Czech Republic
Toft-Petersen	Rasmus	Risø National Laboratory, DTU	Roskilde	Denmark
Tomeno	Izumi	Akita University	Akita	Japan
Tomchuk	Oleksandr	Joint Institute for Nuclear Research	Dubna	Russia
Torikai	Naoya	Mie University	Tsu	Japan
Torralvo Fernandez	Maria Jose	Universidad Complutense Madrid	Madrid	Spain
Toshihiro	Ctotoh	Panasonic Corporation		Japan
Toth	Sandor	Helmholtz Zentrum Berlin	Berlin	Germany
Trapp	Marcus	Helmholtz-Zentrum Berlin für Materialien und Energie GmbH	Berlin	Germany
Treimer	Wolfgang	Beuth Hochschule für Technik Berlin	Berlin	Germany
Tsapatsaris	Nikolaos	Helmholtz-Zentrum Berlin	Berlin	Germany
Turek	Ilja	Charles University, Faculty of Math. and Physics	Prague 2	Czech Republic
Udalov	Oleg	Institute for physics of microstructures RAS	Nizhny Novgorod	Russia
Udby	Linda	University of Copenhagen	2100 København Ø	Denmark
Ueno	Hiroki	Kyushu University	Fukuoka	Japan
Uhlirova	Klara	Charles University in Prague	Prague	Czech Republic
Uhrikova	Daniela	Faculty of Pharmacy, Comenius University	Bratislava	Slovak Republic
Ulbricht	Andreas	Helmholtz-Zentrum Dresden-Rossendorf	Dresden	Germany

Ulivi	Lorenzo	CNR, Consiglio Nazionale delle Ricerche	Sesto Fiorentino	Italy
Unnep	Renata	MTA-SZFKI	Budapest	Hungary
Unruh	Tobias	Friedrich-Alexander-Universität Erlangen-Nürnberg	Erlangen	Germany
Urcelay	Irene	Institut Laue Langevin	Grenoble	France
Uribe Laverde	Miguel Angel	Fribourg University	Fribourg	Switzerland
Valenta	Jaroslav	Charles University in Prague	Praha 2	Czech Republic
Vales	Vaclav	Charles University	Praha	Czech Republic
Vališka	Michal	Charles University in Prague	Praha 8	Czech Republic
Valvoda	Václav	UK, Faculty of Mathematics and Physics	Prague 2	Czech Republic
van Eijck	Lambert	Delft University of Technology	Delft	The Netherlands
van Well	Ad	Delft University of Technology	Deflt	The Netherlands
Varro	Sandor	Hungarian Academy of Sciences, RISSPO	BUDAPEST	Hungary
Vasilev	Nikolay	Lancaster University	Lancaster	U.K.
Velický	Bedřich	Faculty of Mathematics and Physics, Charles U.	Praha 2	Czech Republic
Venter	Andrew	Necsa Limited	Pretoria	South Africa
Ventluka	Petr	Ministry of Education, Youth and Sports		Czech Republic
Veres	Tamás	Research Institut for Solid State Physics and Opt.	Budapest	Hungary
Vettier - STORNO	Christian	ESRF	GRENOBLE	France
Vickery	Anette	Niels Bohr Institute, ESS	Copenhagen	Denmark
Vidal Gallego	Samuel	University of the Basque Country	Bilbao	Spain
Violini	Nicolò	Università di Perugia	Perugia	Italy
Virginie	Marry	PECSA-UMR7195, Université Pierre et Marie Curie	Paris	France
Voigt	Jörg	Forschungszentrum Jülich	Jülich	Germany
von Klitzing	Regine	TU Berlin	Berlin	Germany
Voronin	Vladimir	Petersburg nuclear physics institute	Gatchina	Russia
Vrána	Miroslav	Nuclear Physics Institute ASCR		Czech Republic
Vratislav	Stanislav	CTU in Prague, Faculty of Nuclear Sciences and	Prague	Czech Republic
Wacklin	Hanna	European Spallation Source ESS AB	Lund	Sweden
Wagner	Richard	Institut Laue Langevin (ILL)	Grenoble	France
Wagner	Werner	Paul Scherrer Institut	Villigen	Switzerland
Wahyudi	Olivia	CNRS UMR 6226	Rennes	France
Wallace	Warren	Paul Scherrer Institut	Villigen PSI	Switzerland
Wallacher	Dirk	Helmholtz-Zentrum Berlin	Berlin	Germany
Wardecki	Dariusz	University of Warsaw	Warsaw	Poland
Wasicki	Jan	A.Mickiewicz University	Poznan	Poland
Weber	Frank	Karlsruhe Institute of Technology	Eggenstein-Leopoldshafen	Germany
Webster	John	STFC	Didcot	U.K.
Wellert	Stefan	TU Berlin	Berlin	Germany
White	John	Australian National University	Canberra	Australia
White	Jonathan	Paul Scherrer Institut	Villigen	Switzerland
Wiedenmann	Albrecht	ILL	Grenoble	France
Wilhelm	Ivan	Ministry of Education, Youth and Sports	Prague	Czech Republic
Willendrup	Peter	Risø DTU	Roskilde	Denmark
Witzanyova	Nadezda	Ministry of Education, Youth and Sports	Prague	Czech Republic
Wuttke	Joachim	Forschungszentrum Juelich GmbH	Juelich	Germany
Yamada	Norifumi	high energy accelerator research	Naka	Japan

		organization		
Yamagata	Yutaka	RIKEN	Wako	Japan
Yamaguchi	Akinori	Mirrotron Ltd.		Japan
Yamamura	Kazuya	Osaka University	Suita, Osaka	Japan
Yanez	Marianna	Lund University	Lund	Sweden
Yonemura	Masao	High Energy Accelerator Research Organization(KEK)	Tokai	Japan
Yoshio	Furuya	Panasonic Production Technology Co., Ltd.	Osaka	Japan
Yutaka	Tanaka	Toshiba Electron Tube & Divices Co.,Ltd	Otawara-shi,Tochigi	Japan
Zabel	Hartmut	Ruhr-University Bochum	Bochum	Germany
Zaharko	Oksana	Laboratory for Neutron Scattering	Villigen	Switzerland
Zamponi	Michaela	Forschungszentrum Juelich GmbH	Garching	Germany
Zeilinger	Anton	University of Vienna	Wien	Austria
Zhang	Junrong	High Energy Accelerator Research Organization	Naka	Japan
Zheludev	Andrey	ETH Zurich	Zurich	Switzerland
Zoeckler	Bill	Quantum Design (LOT-Oriel)		Germany

Suppressed Meissner-effect in Niobium: Visualized with polarized neutron radiography

Journal of Physics: Conference Series

S Aull [1], O Ebrahimi [2,3], N Karakas [2,3], J Knobloch [1], O Kugeler [1], W Treimer [2,3]

[1]Helmholtz Zentrum für Materialien und Energie, G- I4, D - 14109 Berlin, Germany
[2]Helmholtz Zentrum für Materialien und Energie, G-G1, D - 14109 Berlin, Germany
[3]Beuth Hochschule für Technik, FB II, D - 13353 Berlin, Germany

treimer@helmholtz-berlin.de

Abstract. Low temperature superconductors ideally exhibit the Meissner-effect, i.e. magnetic flux is expelled from the material during superconducting transition. We report about the complete suppression of the Meissner-effect in two differently surface-treated niobium samples by means of polarized neutron radiography. Both samples were studied in the Meissner phase, $T < T_c = 9.25[K]$ with an external magnetic field $B_{ext} = 6.3mT$ and for $T < T_c$ and $B_{ext} = 0$. Neutron radiographs of both samples were recorded, imaging the depolarization of the neutron spin for $T > T_c$ and $B_{ext} = 6.4mT$, $T < T_c$ and $B_{ext} = 6.3mT$, and for $T < T_c$ and $B_{ext} = 0$. After turning off B_{ext} at a temperature below T_c strong position dependent flux pinning was observed in the untreated sample and more uniform flux pinning in the case of the surface - BCP treatment.

1. Introduction

Flux pinning in type II superconductors has been extensively studied and a large number of publications discuss pinning centres, interactions between flux lines, and the investigations on the geometry of pinning centres. Already very early papers deal with temperature and field dependence of dislocation pinning in Niobium alloys, attributing the pinning process to a second-order elastic interaction between dislocations and the fluxoid lattice [1]. Several mechanisms have been suggested to explain currents in rough surfaces of type II SC such as Meissner currents, the surface sheath, the surface barrier and flux pinning. Meissner currents and flux pinning were connected with a loss of the Meissner state [2], however, the change in flux density between the outside and inside of the superconductor, the greatest contribution is suggested to come from the Meissner currents [2]. In order to understand flux pinning in type II SC a concept of flux-lattice elasticity was developed predicting pinning functions, which explained measured Lorentz force curves [3]. Another aspect was promoted by the observation of the paramagnetic Meissner effect (PME) [4], which was explained by the persistence of the giant vortex state with the fixed orbital quantum number L. The numerical solution of the Ginzburg-Landau equations showed that the compression of the flux trapped inside the giant vortex state results in the PME. The PME is suppressed, and the normal diamagnetic response is recovered, by increasing the applied field as was found out in [5,6]. Corresponding vortex phase diagrams are given in [7], however, they could not observe a multi-critical point in Nb crystals. Some works were performed to investigate Meissner and shielding effects recording zero field cooling (ZFC) and FC on polycrystalline Niobium samples with different pinning strengths [8]. There the Meissner fraction was found almost unity for the annealed sample from which pinning centers were removed and the effect was much less for the unannealed one.

Niobium is commonly used for superconducting radio frequency resonators. The material must fulfil purity criteria concerning dissolved gases like hydrogen, oxygen and nitrogen that decrease the heat conductivity at liquid helium temperature and degrade the cooling of the rf surface. Moreover, any deviation from the ideal Niobium lattice, like contaminations by foreign metals, grain boundaries, dislocations, dissolved oxygen or hydrogen can lead to magnetic flux pinning. Flux-pinning contributes to the total surface resistance of the cavity, and thus leads to increased heat dissipation in rf fields [9]. It can even dominate over the (unavoidable) baseline BCS-resistance making it the cost driver of the cryo-plant and thus the whole superconducting accelerator – a solid reason to put more effort in understanding and possibly avoiding flux pinning in SRF applications.

By means of radiography with polarized neutrons [10]-[15] one can image magnetic fields in the bulk of massive samples and thus get information about the magnetic field distribution inside and outside of a sample. Low temperature superconductors ideally exhibit the Meissner-effect, i.e. magnetic flux is expelled from the material during the superconducting transition. However, a marginal type II superconductor like niobium can behave like a type II superconductor and trap up to 100% of the ambient magnetic flux during the superconducting transition.

Thus, by means of polarized neutron radiography it can be distinguished if the sample is in the Meissner state or the intermediate state where magnetic flux pinning occurs. We have measured the nearly complete suppression of the Meissner-effect in two differently surface-treated Niobium samples by means of polarized neutron radiography. It turned out that the magnetic distribution of the trapped field depended strongly on the surface treatment.

2. Polarized neutron radiography

The neutron radiographies of these Niobium samples were recorded with PONTO, an instrument of the University of Applied Sciences "Beuth Hochschule für Technik" Berlin, dedicated to radiography and tomography with polarized neutrons. It is located in neutron guide hall of the BER II reactor of the Helmholtz Zentrum Berlin. Neutrons are perfect to study bulk materials and magnetic fields, the latter due to the spin of the neutron which interacts with a magnetic field B as

$$\dot{s}_j(t) = \frac{\mu_N}{\hbar} \cdot g \cdot \left[\vec{S}(t) \times B(t)\right]_j \quad j = x,y,z \tag{1}$$

with $\mu_N = 5.05078343*10^{-27}$ [J/T] is the nuclear magneton of the neutron and $g = -3.826085$ the Landé-factor for neutrons, $h = 6.6260755*10^{-34}$[J.s], $\hbar = h/2\pi$. The spin rotation can be treated like a classical dipole having the Larmor frequency $\omega_L = \gamma_L.B$. γ_L is the gyromagnetic ratio of the neutron, $\gamma_L = g.\mu_N/\hbar = -1.83247\mathrm{x}10^8 \mathrm{rad}\cdot \mathrm{s}^{-1}\cdot\mathrm{T}^{-1}$. Entering a magnetic field the neutron spin starts Larmor precessions, whereas the number of rotations depend the strength of magnetic field and on the path length trough the field, as

$$\phi = \omega_L \cdot t = \gamma_L \cdot B \cdot t = \frac{\gamma_L}{v} \int B \cdot ds = \frac{\gamma_L \cdot m}{h} B \cdot s \cdot \lambda \tag{2}$$

ϕ = rotation angle, t = time the neutron traverses the field, v = velocity of the neutron, B = magnetic field, m = neutron mass, s path length in the field, λ = wavelength and h= Planck constant. With a mean neutron wavelength $0.39.10^{-9}$[m], $v = h/m.\lambda$ is 1014[ms^{-1}]. Leaving the sample, i.e. the magnetic volume, the neutron spin has made position –and path-length dependent Larmor precessions that can be analyzed with a spin-analyzer in front of a two-dimensional detector. The intensity is two-dimensionally detected as described in [11] as

$$I(x,y) = I_0(x,y)\exp[-\int \alpha(s)ds]\frac{1}{2}(1+\cos\Theta(x,y)) \tag{3}$$

I(x,y) = measured intensity in the (x,y,z) co-ordinate system, the detector plane (x,y) is perpendicular to the neutron flight direction (z-axis), I_0= incident intensity, α = linear attenuation coefficient of the sample, s is the path through the sample. In order to perform radiography or tomography with polarized neutrons, the neutron beam has to be polarized, of course, in front and analyzed behind the sample which was done with so-called benders [16], polarizing units consisting of Silicon wafers which were coated with an FeSi substrate that were magnetized. The polarization P for the following experiments was determined with two different methods, once with the shim method, measuring the depolarization of polarized neutrons when a demagnetizing shim plate is put in the polarized beam as and by measuring the spin rotation with a spin flipper.

$$P = \left\{ \frac{I_{pol} - I_{background}}{I_{depol} - I_{background}} - 1 \right\} \cdot 100 \tag{4}$$

Niobium is a type II superconductor and the superconductivity occurs immediately after reaching the critical Temperature T_c which is in the case of Niobium 9.25K. The critical magnetic field Bc of Niobium at 0K allows an external field of 150mT to be expelled, which depends on the temperature T as

$$B_c(T) = B_c(0K) \cdot \left[1 - \left(\frac{T}{T_c} \right)^2 \right] \cdot (1 - N) \tag{5}$$

One has to add a factor (1-N) to the formula taking into account the demagnetisation factor N which is given by

$$N = 1 - \frac{1}{\left(1 + \frac{q.a}{b} \right)} \tag{6}$$

For a = 22.5mm (radius of the sample), b = 1.4mm (half height of the sample), q is yielded as

$$q = \frac{4}{3\pi} + \frac{2}{3\pi} \cdot \tan\left(1,27 \cdot b / a \cdot \ln\left(1 + a / b\right)\right) \tag{7}$$

Thus the demagnetisation factor N was for the samples 0.8837. In these experiments the external field B_{ext} = 6.4mT was for T = 5.5K well below the critical $B_c(5.5K)$ = 11mT (see Fig.1).

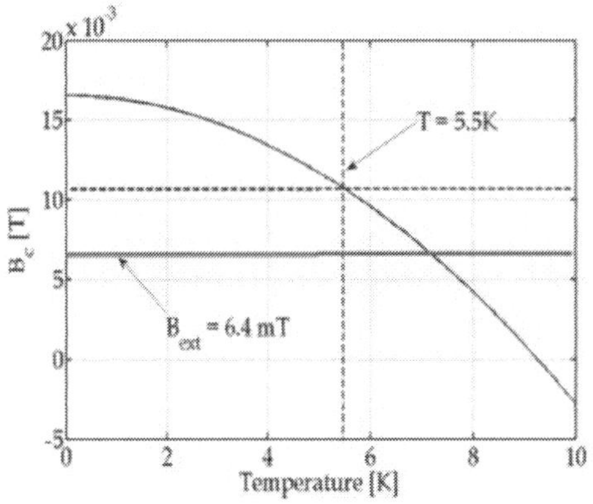

Fig.1 The applied external magnetic field B_{ext} = 6.4mT was at T = 5.5K well below 10.7mT.

3. Experiments and results

The experimental setup is shown in Fig.2. A graphite monochromator in neutron guide reflected neutrons with a mean wave length = 0.39nm to an optical bench. The neutron beam was horizontally and vertically collimated (0.1° and 0.2°) and polarized with the bender. The overall polarization of the beam P = 85% was measured with both methods, mentioned above. The samples were 2.8mm thick and had a diameter of 45mm. The influence of the surface on magnetic flux trapping was studied by means of a special surface treatment. One of the samples got a BCP = Buffered Chemical Polishing: Hydrogen fluoride (HF) 48%, Nitric acid (HNO_3) 65%, Phosphoric acid (H_3PO_4) 85% at the ratio of 1:1:2. And therefore the treated Niobium sample was 100µm thinner than the other one which was left untreated. The samples were kept in special mountings that fitted in the cryostat (Fig.2). The samples themselves were larger than the field of view of the spin analyzer and then they had to be adjusted in the cryostat properly in order to realize images for both orientations (0° and 90°) with respect to the incident neutron beam and magnetic field, respectively.

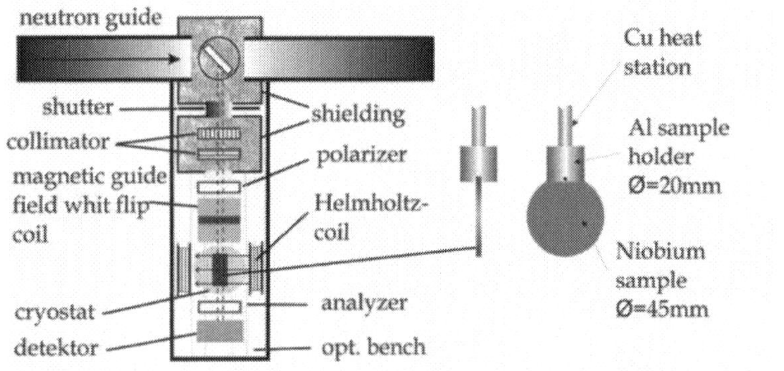

Fig.2 Experimental layout of the instrument PONTO (polarized neutron tomography) and sample mounting for the cryostat.

At first the flux pinning was investigated for the untreated sample, i.e., it was cooled down with a conventional cryostat below $T_c = 9.25K$ in the presence of an external homogenous magnetic field $B_{ext} = 6.4mT$, generated with two Helmholtz coils (Fig.3). At each particular cooling temperature (11K, 9K and 5.5K) T was kept constant for more than three hours to ensure a homogeneous temperature all over the sample. The exposure time for each radiograph was again three hours, flat and dark field images were recorded at the beginning and end of each series. In the "Flux-pinning-experiments" (B_{ext} = on) the samples were oriented with the surface first perpendicular (0°-orientation) and then parallel (90°-orientation) to B_{ext}, These experiments were repeated with zero external magnetic field $B_{ext} = 0$ for both orientations.

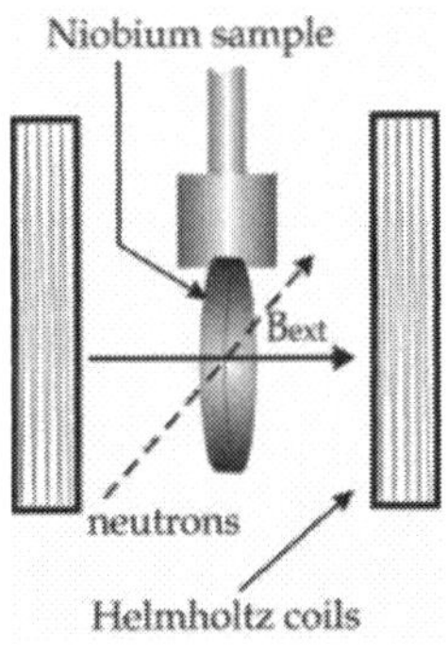

Fig.3 The Niobium sample was kept at cryogenic temperatures in an external magnetic field B_{ext} = 6.4mT, neutrons traversed the sample parallel (0°) (this case is shown in the figure) and perpendicular (90°) the surface.

If the sample was in the Meissner phase in the 0°- position the expelled magnetic field should have caused Larmor precessions of the neutron spins that should have been observed as intensity variations on the 2D detector. The path lengths were long enough to cause spin rotations of more than 2π if the expelled magnetic field would be app. 1mT (at a path length of 45mm) and less for other paths. However, this is not what we observed: The results are shown in Fig.4.

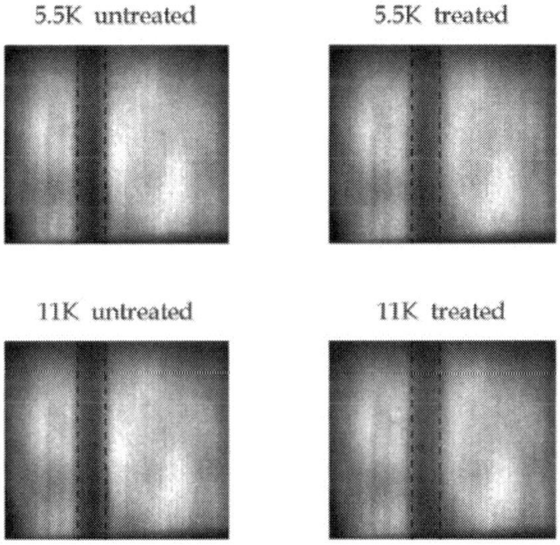

Fig.4 Nb sample, untreated and surface treated, orientation of surface perpendicular to B_{ext} = 6.4mT, sample being in the Meissner phase for T = 5.5K, black dotted lines indicate the surfaces of the sample, black area between dotted lines shows the pure absorption contrast due to the sample; there is no difference to the 11K – images.

No intensity changes could be detected in the area outside the sample. Inside the sample (black area between the dotted lines) no typical fringe pattern due to spin rotations, as observed in lead [13] were measured. The whole systems behaved in the same way as it would be above T_c. Inhomogeneities were due to the inhomogeneous field of view (FOV) of the spin analyzer. Thus (almost) all of the offered flux must have been pinned into the sample.

The sample was then rotated by 90° such that its surface was perpendicular to the incident neutron beam and the external magnetic field was switched off to measure possible flux pinning. Fig.5 shows the measured spin orientations for untreated and treated Niobium samples below and above T_c (at T = 9K and 11 K, respectively).

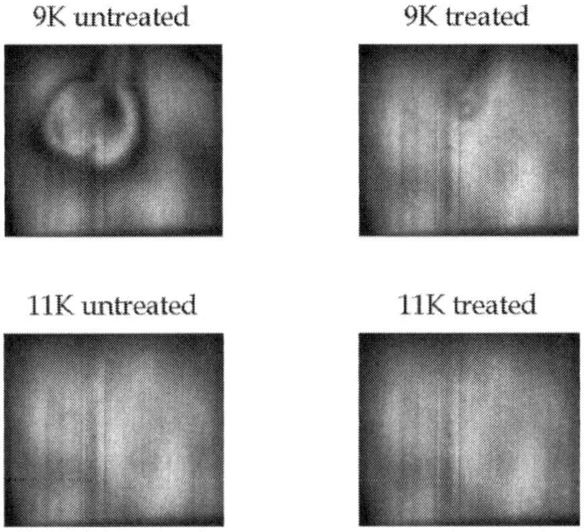

Fig.5: Flux pinning in an untreated and surface treated Nb sample; images show the position dependent depolarisation of the neutron spin due to pinned magnetic fields in the bulk of the sample. Bright areas and dark areas show spin rotations of whole-number multiples of 2π. Note the rather homogeneous field distribution in the case of the surface treated sample.
The feature in the 9K – image was caused by a slit in the sample due to the preparation process. BCP smoothens the edges of the slit which lead to less flux pinning. In the case of T = 11K no differences could be observed.

The distinct feature in the pictures of the low temperature measurements hints at the existence of frozen flux. It is due to a cut in the sample and becomes invisible to the neutrons in the normal conducting state. The feature is much less pronounced in the treated sample, meaning that either less flux is frozen and/or the flux is distributed more homogenously.

A more thorough quantitative analysis of the results is necessary, but has not been performed yet. Qualitatively it can be stated that the BCP had a significant influence on the captured magnetic flux. It must be emphasized that the magnetic flux is measured unobtrusively as a bulk effect in these experiments. Other methods like SQUID or flux gate magnetometers can only measure the part of the flux lines that is outside the bulk.

4. Summary

In this work the suppression of the Meissner effect in superconducting niobium was investigated by using polarized neutron radiography. Below the critical temperature and the lower critical magnetic field, niobium should be in the Meissner state and an ambient magnetic field should be expelled. However, defects in the crystal lattice suppress the expulsion of the magnetic field during the superconducting transition so that the magnetic field is trapped inside the material. Radiography via polarized neutrons allows the study of the imperfect Meissner effect and has been performed on two niobium samples with different surface treatment history. Since the neutron radiography allows studying magnetic fields in bulk materials, information of the local field distribution in the samples can be derived. It is shown that both studied samples suppress the Meissner effect and that the surface has a great influence on the local field distribution of the trapped field.

This work was part of the BMBF project 05KN7KF1

References

[1] Fietz W A and Webb W W 1969 *Phys. Rev.* 178 2 657
[2] Melville P H 1972 *J. Phys. C, Solid State Phys.* 5 147
[3] Dew-Hughes D 1974, 30, 2 , 293
[4] Thompson D J , Minhaj M S M, Wenger, L E , Chen J T 1995 , *Phys. Rev. Lett.* 75 3 529
[5] Moshchalkov V V, Qiu X G, Bruyndoncx V 1996 *Journ. Low Temp Physics*, 105 Nos. 3/4 515
[6] Moshchalkov V V, Qiu X G, Bruyndoncx V 1997 *Phys. Rev* 55 17 11793
[7] Das P, Tomy C V , Takeya H, Ramakrishnan S, Grover A K 2009 *Journal of Physics: Conference Series* 150 052041 1
[8] Tomioka Y , Naito M , Kitazawa K 1993 *Physica C* 215 297
[9] Padamsee H, Knobloch J, Hays T 1998 *RF Superconductivity for Accelerators*, Wiley VCH
[10] Treimer W, Hilger A, Kardjilov N , Strobl M 2005 *Nucl. Instr. & Meth.* A 542 367
[11] Kardjilov N, Manke I, Strobl M, Hilger A, Treimer W, Meissner M , Krist T, Banhart J 2008 *Nature Physics* 4, 399
[12] Strobl, M, Treimer W, Walter P, Keil S, Manke I 2007 *Appl. Phys. Lett.* 91 254104
[13] Treimer W , Ebrahimi O, Karakas N , Seidel S O, 2011 *Nucl. Instr. & Meth.* A 2011 (in press)
[14] Schulz M, Neubauer A, Masalovich S, Mühlbauer M, Calzada E, Schillinger B, Pfleiderer C, Boeni P 2010 *Journ. of Phys., Conference Series* 211 012025
[15] Schulz M, Neubauer A, Mühlbauer M, Calzada E, Schillinger B, Pfleiderer C, Böni P 2010 *Journal of Physics: Conference Series* 200 112009
[16] Krist T, Kennedy, S J, Hick T J, Mezei F 1998 *Physica* B 241–243 82

PA20 : A new SANS and GISANS project for soft matter, materials and magnetism

Grégory Chaboussant[1], Sylvain Désert[1], Pascal Lavie[1] and Annie Brûlet[1]

[1] Laboratoire Léon Brillouin, UMR12 CEA-CNRS, CEA Saclay, 91190 Gif sur Yvette, France

E-mail: gregory.chaboussant@cea.fr

Abstract. This article presents the new Small Angle Neutron Scattering (SANS) instrument PA20 which will replace the PAXE instrument at LLB-Orphée. SANS is well-known to be especially well adapted to research in soft matter, materials and nanosciences and SANS is particularly powerful in the studies of complex systems, with isotopic labeling and contrast variation method, but also for large-scale structures (magnetic or not). PA20 is part of the LLB instrumental upgrade program "CAP2015". PA20 will not only maintain LLB's capabilities in SANS, but also considerably extend them in terms of SANS for magnetism with a polarized neutron option and Grazing Incidence SANS (GISANS), with an improved dynamical Q-range. The total length of PA20 will be 40 m, including a 19 m collimation length, a 20 m detector tank containing high-resolution/high-efficiency XY detectors, and a casemate containing a monochromator (velocity selector $\lambda = 0.3 - 2$ nm), a chopper system for Time-of-Flight (TOF) mode, a polarizer and an RF spin flipper. PA20 will allow faster measurements, with "single-shot" access to a wider range of scattering vectors, on possibly small samples (few mm in size). In addition, polarized neutrons will enable magnetic studies in both SANS and GISANS configurations. Studies of nanostructured surfaces and interfaces (deposited or embedded nano-objects), magnetic domain formation, multilayered materials or magnetic thin films through specular and off-specular signals will be possible through GISANS setups. The versatility of PA20 should contribute to both enlarge the neutron user community, especially in expanding areas like nanosciences, and offer improved services for users in the years to come.

1. Motivations and Objectives

At the heart of the LLB instrumental upgrade program CAP2015 lies the need to improve and expand the instrumental suite to meet users demands in the next 10-15 years. The LLB is equipped with 4 SANS instruments (PAXE, PAXY, PACE and TPA [1]) and 2 reflectometers (PRISM and EROS) [2]. This instrumental suite covers small and very small neutron scattering, polarized neutron reflectometry and liquid reflectometry. As SANS instruments represent over a third of the total beam time requests at LLB, it is imperative to maintain top-level instruments in this area. On the other hand, increased interests in nanostructured materials with surface or interface issues [3, 4, 5] have brought about the need for LLB to offer a complete new GISANS instrument. In this context, a new Small Angle Neutron Scattering (SANS) and Grazing Incidence Small Angle Neutron Scattering (GISANS) instrument, PA20, is under construction in place of the PAXE spectrometer located at the end of the G5 neutron guide. PA20 aims at providing a better service to users, with more versatility, and at answering the needs of the expanding nanosciences community. The length of the instrument, approximately 40 meters, will make it possible to exploit both the flux and the resolution. PA20 will cover a broad dynamical range, from very small scattering vectors, lower than 0.01 nm^{-1}(at $\lambda = 8$ Å and 19 m) with reasonable flux, up to 7 nm^{-1} (at $\lambda = 4$ Å and 1.0 m). The increased performances (flux, resolution and Q-range, versatility, polarized neutrons) will also broaden the span of experiments that can be performed with SANS and GISANS experiments. The global progress compared to existing SANS at LLB will be particularly appreciable to study the objects from a few nanometers up to several hundreds nanometers [24].

1.1. SANS

SANS is a particularly relevant technique to investigate inhomogeneities of size in the range between 1 and 500 nm [6, 7], presenting a neutron scattering contrast being structural, chemical or magnetic in origin. Information like density or size distributions, shape or interface effects, and magnetic structurations [8, 9] can be addressed using SANS instruments. The fact that several SANS instruments are currently under construction in Europe highlights the importance and relevance of SANS studies in these wide range of topics [10, 11, 12]. This is especially true for soft matter (polymers, polyelectrolytes, micelles, vesicles, biological systems like proteins) [13, 14, 15, 16], materials science (chemical demixing, precipitates, porous materials, alloys of industrial interest) [17] and magnetic materials (critical magnetic fluctuations near transitions, magnetic particles of ferrofluids [18], vortices in superconductors, nanomagnets, long-range helicoïdal orders).

SANS is without doubt one of the main experimental technique for soft matter and, to a lesser extent, biosystems. Its success originates obviously from H/D isotopic substitution of hydrogeneous materials allowing labeling and contrast matching experiments but also to the capacities of neutron scattering to probe form and structure factors of large objects. Increased performances in terms of flux and efficiencies will help to investigate small volume samples like thin polymer films, confined systems, grafted objects, finely tuned molecular assemblies where isotopic labeling is possible on small fraction of the materials. This is particularly relevant to biological systems [25, 26, 27] which, until now, are mostly studied with X-ray diffraction, but do not profit to the powerful insight offered by contrast variation methods. Other examples are the studies of materials based on nano-particles inserted in polymeric matrices [28, 29, 30], which could be done at lower concentration of particles, and rheological studies of complex fluids under strain, which could be done in-situ with greater ease. The polarized neutron mode is also of prime interest in the case of magnetic nanostructured materials like inserted magnetic objects [31]. Polarization will also be extremely helpful to eliminate the incoherent background of hydrogeneous systems, from small molecules to proteins.

PA20 will improve SANS experiments in a variety of ways in terms of neutron flux, scattering angle coverage, Q-resolution and neutron polarization. For instance, faster data acquisition rates enables kinetics measurements or systematic studies as a function of temperature near phase transitions where better resolution are critical to differentiate between models or theories. The increased Q-range coverage allowed by the 2D front and rear detectors will permit a more rigorous analysis of interface effects, their fractal aspects (Porod regime) compared to volume effects (Guinier regime). Pushing the limits in terms of flux and resolution is also critical when dealing with samples that are difficult to synthesize or which can be obtained only in modest quantity. The new instrument PA20 will make it possible to tackle studies of such samples.

Regarding magnetic studies, PA20 will be the first SANS with multi-wavelength neutron polarization at LLB, fulfilling the needs to study magnetization processes in nano-objects and thin films. Another range of topics which will benefit from SANS with polarized neutron option relates to magnetic systems and correlated electron systems [32, 33]. The existence of a magnetic contrast in SANS and GISANS allows to study magnetic structures. Cluster formation in spin glasses, long-range helicoidal orders, vortices in superconductors [34], phase segregation in CMR materials like manganites [35] have been widely and successfuly investigated with SANS . This technique has provided decisive contributions because the response function can be measured across a broad range of scattering vectors ($10^{-2} - 5$ nm^{-1}). Weak Bragg peaks, even with some disorder, are accessible providing sufficient neutron flux. This also applies to study the role played by long-range modulations (several tens of nm) in multiferroic materials. With PA20, it will also be possible to study thin magnetic films in both SANS and GISANS modes to address questions related to surfaces and interfaces effects, alteration of the properties when going from bulk to layers, etc. The prospect of having a dedicated upgraded SANS instrument to magnetic studies at the nanometer scale opens the way to more detailed approaches.

Finally, PA20 will be especially helpful in materials sciences [17, 36] as SANS proved to be an indispensable tool to study microstructural heterogeneities on a nanometric scale, in particular precipitation or segregation in metal alloys. In comparison with the usual techniques used in metallurgy (TEM, SAXS), SANS provides quantitative data at the sub-micron scale on large volumes (several mm^3). This is a key asset for industrial materials which present heterogeneities or multiples phases. Irradiated materials, exposed to high-energy neutrons and/or elevated temperatures over very long periods of time, often present microstructural changes like swelling, cavities, hardening, etc. SANS can address these issues and provides invaluable information to engineers and scientists in the nuclear industry.

1.2. GISANS

GISANS is less developed worldwide but, as a surface and interface probe, complements SANS and neutron reflectivity techniques [19] as well as Xrays GISAXS. The 2D detectors allow measurements of off-specular signal, characteristic of structural and magnetic surface inhomogeneities. PA20 will have the capacity to probe structural or magnetic inhomogeneities in the range of 1 to 500 nm on surfaces and buried interfaces along both orientations - vertical or horizontal. Combining high-precision slits collimations and high angular coverage by both rear and front detectors will permit off-specular scattering studies in the most versatile possible way. Particularly relevant issues are related to magnetic layered materials, magnetic domain formation, organic or polymeric layered films, etc. [20, 21, 22, 23]. If GISANS is known to be complementary to GISAXS X-Rays experiments, neutron sources cannot compete with modern synchrotrons in terms of flux. Therefore, GISANS within PA20 will focus essentially to the study of networks of magnetic nano-objects (wires, ribbons), magnetic correlations or domains formed at the surface or at interfaces. Important issues related to emergent materials such as thermoelectrics or multiferroics could be addressed.

2. Instrument description

This section describes the technical key components of PA20 whose general layout is shown in Figure 1.

2.1. Casemate: monochromator, chopper, polarizer, RF flipper

The 3 m long casemate, composed of lead and heavy concrete, will house a Dornier-type velocity selector that can be replaced by a chopper when PA20 is operating in Time of Flight (TOF) mode, an escapable polarizer placed in vacuum, and a RF spin flipper (see Figure 2). The multi-blade velocity selector (Astrium-EADS) has a transmission larger than 95% (for a theoretical

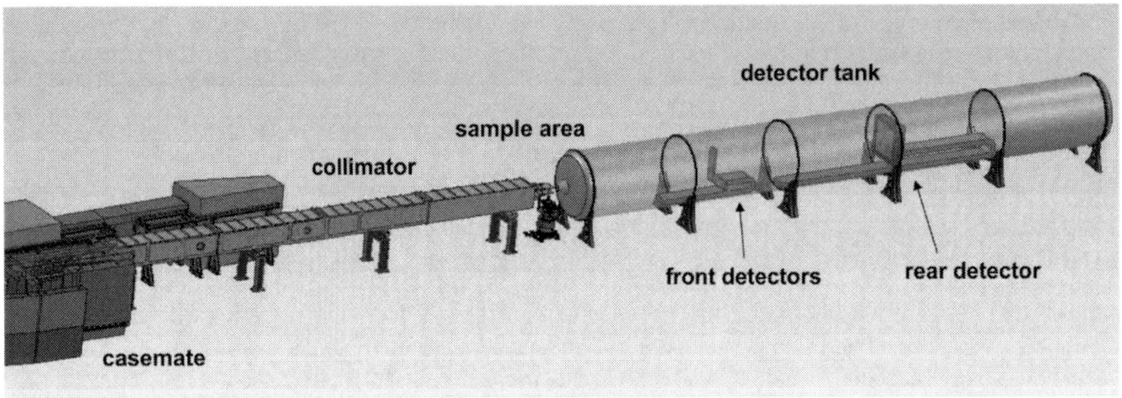

Figure 1. General layout of the PA20 spectrometer (in orange) in LLB-Orphée Guide Hall. From left to right: Casemate housing monochromator, polarizer and flipper; collimation line; sample area; detectors tank. The 20 m evacuated tube contains a 64x64 cm^2 ^{3}He rear detector (5 mm vertical and horizontal pixel size) which can be moved from 1 m to 19 m from the sample position and a front detector consisting of two sets of 10 mm diameter tubes placed perpendicular one to another. Each set will be 20 cm wide and 64 cm high. The front detectors can be located between 1 and 10 m from the sample.

non diverging neutron beam) and a wavelength distribution of about 10% .

Beam polarization will be ensured by a 1.4 m long cavity polarizer [43] with a tapering angle of 1.6 degrees and with a double-V configuration (with transmission $\approx$ 80 %) [44, 37, 38, 39] using 0.9 m long silicon substrates, double-side coated with $m = 3.4$. A 450 Gauss magnetic field will be generated to ensure magnetization. This setup will provide a polarization ratio superior to 98 % from 4 Å upwards. A classical radio-frequency spin flipper located after the polarizer can reverse the polarization direction. PA20 will also offer the possibility to perfom high Q-resolution experiments with high-Q dynamical range (typically $50 \times \frac{\lambda_{max}}{\lambda_{min}}$). This is best done using a chopper in TOF mode where wavelength spread could be limited to less than 1 % . This setup complements the velocity selector mode and will find applications in various areas where very high resolution is needed.

2.2. Collimation System

The collimation system has been designed to offer maximum flexibility in terms of collimated beam shapes and accuracy. The maximum collimation length, in polarized mode, will be 16 m from the exit point of the casemate to the sample position. In non polarized mode an additional 3 m of collimation is available with an extra slit system located at the entrance of the casemate. From the casemate exit point to the end-point of the collimation, there will be 7 collimation slits separating 5 super-mirror guide elements (Ni/Ti, m = 2) of 85 × 25 mm^2 internal section. The available collimation lengths will therefore approximately be 19, 16, 12, 8, 4, 2.5 and 1 m. A free space is booked inside the collimator and over all its length for future implementation of modern optics such as new guide shapes or multi-beam mode [45, 47]. At the end of the collimator, an under vacuum variable-length extension (50-80 cm) will optimize neutron transmission to the sample.

A system of 4 independent slits, 2 horizontal and 2 vertical, has been chosen to provide square shaped collimation for SANS and vertical or horizontal rectangular shape for GISANS. These slits have a positioning accuracy of 50 μm and offer openings up to 80 mm along the vertical

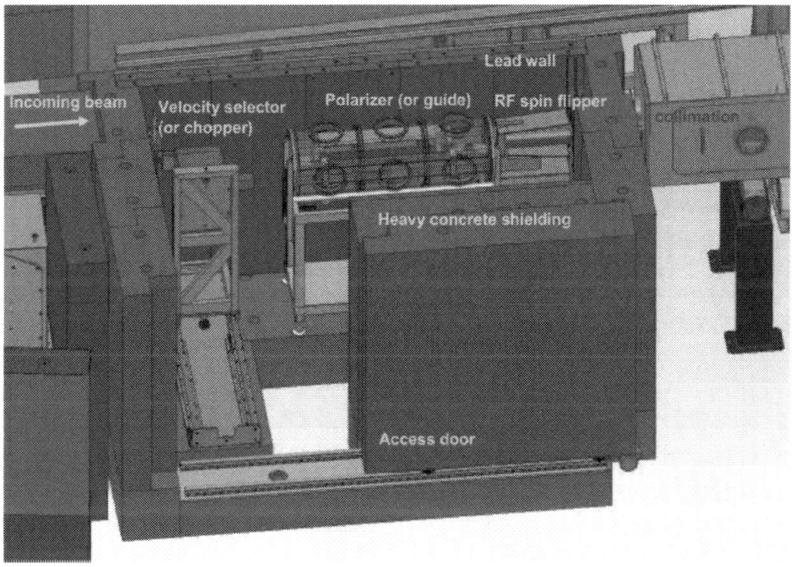

Figure 2. Casemate plan (roof removed): Velocity selector monochromator (left) and polarizing setup including transmission polarizer and RF spin flipper (right).

direction and 30 mm along the horizontal direction. To increase the scattered intensity at small scattering vectors or to be able to get even smaller scattering vectors, two motorized systems of MgF_2 focusing lenses [40, 46, 41] can be inserted inside the collimator, just before the sample, with a 25 mm diameter, a 25 mm radius of curvature and a 1 mm thickness at the centre. The number of lenses for each device will be 19 and 37 respectively.

This option is well adapted to long wavelengths and large sample-to-detector distances, a situation where counting times are usually the longest. Neutron polarization after the casemate is maintained by a vertical 30 Gauss guiding magnetic field, generated by vertical columns of permanent cylindrical magnets (NbFeB, Ø = 25 mm, 1 Tesla) and two iron horizontal plates situated below and above the neutron guides and separated by approximately 200 mm. At the very end of the collimation, right before the extension, the polarization is switched to horizontal to ease the use of polarized neutrons with horizontal magnetic fields at the sample position (2 Tesla electromagnets and 10 Tesla cryomagnets currently available at LLB). Both the extension of the collimator and the entrance part of the detectors tank will be non magnetic.

2.3. Detectors

In order to achieve high angular resolution and large angular coverage, two sets of 2D XY ^3He detectors will be placed inside the detector tank (see Figure 1). The rear detector is a 64x64 cm^2 ^3He single-bloc Aluminum multi-tube detector, with 5 mm vertical and horizontal pixel size and square section tubes, which can move 30 cm sideways from its central position to access larger scattering vectors still with a good pixel definition. Its expected efficiency of 85% at 5 Å and 15 bar pressure is among the best in the world. As high resolution at large scattering angle is not absolutely necessary, the front detector, also with ^3He, with tube sections in the order of 10 mm, consists of two sets of tubes placed perpendicular one to another. Each set will be 20 cm wide and 64 cm high. The front detector will allow covering an extended Q-range and its distance to sample (between 1 and 10 m) will be adapted according to the position of the rear detector [42] (between 1 and 19 m). The simultaneous use of the detectors will allow the collection, in a single measurement, of the scattered intensity over a dynamic Q-range of $Q_{max}/Q_{min} \approx 50$ and could even be pushed further by shifting the rear detector and the vertical front detector sideways. These high-resolution and high-efficiency XY detectors will greatly increase both the counting rate and the Q-range covered in a single measurement compared to our current SANS. A total gain factor of 15 is anticipated in a standard configuration, which will go at least to 50 with the focusing lenses in place (for $Q_{min} = 10^{-2}$ nm^{-1}, $\lambda = 8$ Å , $L = 12$ m detector distance).

3. Conclusion

The new Small Angle Neutron Scattering instrument, PA20, has been presented. PA20 will be located at the LLB-Orphée reactor in Saclay and will replace the actual spectrometer PAXE. The total length of PA20 will provide a 19 m collimation length and a 20 m detector tank. Among its specifications, PA20 will offer an extended dynamical Q-range of 50, and an intensity gain of 15 compared to PAXE, in monochromatic mode through the simultaneous use of high resolution and high efficiency rear and front 2D detectors. The collimator will offer the possibility to use either square or slit shape apertures for GISANS or specific SANS measurements, two sets of MgF_2 neutron lenses, and also leave room for future implementation of modern techniques in neutron optics such as new guide shapes or multibeam devices. PA20 will also enable the use of polarized neutrons with a polarization ratio larger than 98 % from 4 Å upwards. Finally, TOF option is envisaged in the future with a resolution close to 1 % and a dynamical scattering range of 500. Therefore, due to its design and versatility, PA20 will contribute to enlarge the neutron user community, in particular in nanosciences.

4. Acknowledgments
The authors thank Sébastien Gautrot and Marc Detrez for their technical support and Patrice Permingeat for fruitful discussions. The authors acknowledge financial support from RTRA "Triangle de la Physique" and CNANO-IDF 2009-2010 funding bodies.

5. References

[1] For more details on TPA, see: Désert S., Thevenot V., Gabriel A., Permingeat P., Oberdisse J. and Brûlet A., 2007, *J. Appl. Cryst.* **40**, 945949; Désert S., Thevenot V., Oberdisse J. and Brûlet A., 2007, *J. Appl. Cryst.*, **40**, s471s473.
[2] see LLB spectrometers web site at : www-llb.cea.fr
[3] Muller-Buschbaum P., Gutmann J.S., Stamm M., Cubitt R., et al, 2000, *Physica B* **283**, 53-59.
[4] Muller-Buschbaum P., Gutmann J.S., Cubitt R., Petry W., 2004, Physica B **350**, 207-210.
[5] Cousin F, Jestin J, Chaboussant G, et al., 2009, *Eur. Phys. J. - Special Topics* **167**, 177-183.
[6] Shibayama M, 2011 *Polymer Journal* **43** 18-34.
[7] Karino T, Okumura Y, Ito K, et al., 2004 *Macromolecules* **37** 6177-6182
[8] Michels A, and Weissmuller J, 2008, *Rep. Prog. Phys*, **71** 066501
[9] Sankar S., et al, 2000, *J. Mag. and Magn. Mat.*, **221**, 1-9.
[10] Gilles R., Ostermann A., Schanzer C., Krimmer B., Petry W., 2006, *Physica B* **385386**, 11741176.
[11] Gilles R., Ostermann A. and Petry W., 2007, *J. Appl. Cryst.*, doi:10.1107/S0021889807006310.
[12] Dewhurst C.D., 2008, *Meas. Sci. Technol.* **19**, 034007.
[13] Dubois E, Boue F, 2001 *Macromolecules*, **34**, 3684-3697.
[14] Cousin F, Gummel J, Ung D, et al. 2005 *Langmuir* **21**, 9675-9688.
[15] Gayet F., C. El Kalamouni, P. Lavedan, J.-D. Marty, A. Brûlet, and N. Lauth de Viguerie, 2009, *Langmuir* **25**, 97419750.
[16] Agut W., A. Brûlet, C. Schatz, D. Taton, and S. Lecommandoux, *Langmuir* 2010, **26**, 1054610554.
[17] Melnichenko Y.B. and Wignall G.D., 2007, *J. Appl. Phys.* **102**, 021101.
[18] Wiedenmann A., U. Keiderling, R.P. May, and C. Dewhurst, 2006 *Physica B*, **385-386**, 453456.
[19] Hamilton W.A., 2005, *Current Opinion in Colloid and Interface Science*, **9**, 390-395.
[20] Steitz R., P. Muller-Buschbaum, S. Schemmel, R. Cubitt and G. H. Findenegg, 2004, *Europhys. Lett.*, **67**, 962968.
[21] Leiner V. and H. Zabel, 2004, *Phys. Rev. Lett.*, **92**, 255501-1.
[22] Wolff M., A. Magerl, and H. Zabel, 2005, *Eur. Phys. J. E*, **16**, 141145.
[23] Lauter-Pasyuk V., H. J. Lauter,G. P. Gordeev, P. Muller-Buschbaum, B. P. Toperverg, M. Jernenkov, and W. Petry, 2003, *Langmuir* **19**, 7783-7788.
[24] Brûlet A., V. Thevenot, D. Lairez, S. Lecommandoux, W. Agut, S. P. Armes, J. Duc and S. Désert, 2008, *J. Appl. Cryst.* **41**, 161166.
[25] Schmidt I, Cousin F, Huchon C, et al., 2009, *Biomacromolecules*, **10**, 1346-1357.
[26] Sanson C., C. Schatz, J.-F. Le Meins, A. Brûlet, A. Soum, and S. Lecommandoux, 2010, *Langmuir*, **26**, 27512760.
[27] Sanson C., Diou O., Thevenot J., Ibarboure E., et al, 2011, *ACS NANO*, **5**, 11221140.
[28] Jestin J., Cousin F., Dubois I., et al. 2008, *Advanced Materials* **20**, 2533.
[29] Oberdisse J, Deme B, 2002, *Macromolecules*, **35**.
[30] Jouault N, Dalmas F, Said S, et al., 2010, *Physical Review E*, **82** 031801.
[31] Wiedenmann A., et al, 2008, *Physical Review B* **77**, 184417.
[32] Wiedenmann A., 2000, *J. Appl. Cryst.*, **33**, 428-432.
[33] Wiedenmann A., U. Keiderling, K. Habicht, M. Russina, and R. Gahler, 2006, *Phys. Rev. Lett.*, **97**, 057202.
[34] Pautrat A, Aburas M, Simon C, et al., 2009, *Physical Review B*, **79**, 184511.
[35] Saurel D, Simon C, Pautrat A, et al., 2010, *Physical Review B*, **82**, 054427.
[36] Mathon MH, de Carlan Y, Geoffroy G, et al. 2003, *Journal Of Nuclear Materials*, **312**, 236-248.
[37] Stahn J and Clemens D., 2002, *Appl. Phys. A: Mat. Sci. Proc.*, **74**, s1532-s1534.
[38] James M., Nelson A. and Klose F., 2009, *Neutron News*, **20**, 21-23.
[39] Böni P, Clemens D., Senthil Kumar M. and Tixier S., 1997, *Physica B: Cond. Matt.*, **241-243**, 1060-1067.
[40] Choi S.M., J. G. Barker, C. J. Glinka, Y. T. Cheng and P. L. Gammel 2000, *J. Appl. Cryst.* **33**, 793-796.
[41] Ott. F, 2008, *Modern Developments in X-Ray and Neutron Optics, Springer Series in Optical Sciences*, **137/2008**, 113-135.
[42] Kuklin A.I., A. Kh. Islamov, and V. I. Gordeliy, 2005, *Neutron News*, **16**, 16-18.
[43] Mezei F., in "Neutron Optical Devices", SPIE conferences proceedings, ed. C.F. Majkrzak (SPIE, 1988), vol 983, 10.
[44] Böni P., Munzer W., Ostermann A., 2009, *Physica B*, **404**, 2620-2623.
[45] Glinka C.J., Rowe J.M., La Rock J.G., 1986, *J. Appl. Cryst.* **19**, 427-439.
[46] Eskildsen M.R., Gammel P.L. and Isaacs E.D., 1998, *Nature*, **391** 563-566.
[47] Len A, Pépy G, Rosta L, 2004, *Physica B*, **350** e771-e773.

Journal of Physics: Conference Series **340** (2012) 012003 doi:10.1088/1742-6596/340/1/012003

TAIPAN: First Results from the Thermal Triple-axis Spectrometer at OPAL Research Reactor

S.A. Danilkin[1], M. Yethiraj[1], T. Saerbeck[1,2], F. Klose[1], C. Ulrich[1,3], J. Fujioka[4], S. Miyasaka[4,5], Y. Tokura[4] and B. Keimer[6]

[1] Bragg Institute, Australian Nuclear Science and Technology Organization, New Illawarra Road, Lucas Heights, NSW 2234, Australia

[2] School of Physics, University of Western Australia, WA 6009, Australia

[3] The University of New South Wales, Sydney, NSW 2052, Australia

[4] Departments of Applied Physics, University of Tokyo, Tokyo 113-8656, Japan

[5] Department of Physics, Graduate School of Science, Osaka University, Osaka 560-0043, Japan

[6] Max – Planck -Institute for Solid State Research, D-70569 Stuttgart, Germany

E-mail: Sergey.Danilkin@ansto.gov.au

Abstract. The thermal triple-axis spectrometer TAIPAN is the first instrument for inelastic neutron scattering at the new Australian research reactor OPAL. TAIPAN started operation in February 2009 and is in full user service since November 2010. Conceptually, it is similar to the triple-axis spectrometers IN8 (ILL) and PANDA (FRM-II) with variable incident and final energies and a secondary spectrometer with a single detector. The instrument can be operated either in a high flux mode with a double-focusing monochromator and analyser, or with Soller collimators - gaining resolution at the expense of intensity. Presently the PG (002) double-focusing monochromator and analyser are in use. The incident energy range on the TAIPAN TAS is from ~5 meV up to ~100 meV with neutron flux at sample position of $2.4 \leq 10^7$ n/cm²/s

at incident energy of 14.8 meV. First experiments were performed with superionic conductor $Cu_{2-\delta}Se$. The measurements reveal the presence of a soft mode related to ordering of Cu atoms followed by α - β phase transition at a lower temperature. The evolution of the magnetic structure with temperature in a magnetically modulated $FePt_3$ thin film was investigated in the diffraction mode of TAIPAN. The results show that the film fabricated by modulation of the chemical order parameter consists of a magnetic FM/AFM superlattice in single-crystalline $FePt_3$. The spin wave and phonon dispersion was recently investigated in $TbVO_3$ single crystal. The acoustic and optical magnon branches were observed in the same energy range. This indicates that the 'orbital Peiers state' also exists in $TbVO_3$.

1. Introduction

Triple axis spectroscopy is one of the most commonly used and powerful techniques for inelastic neutron scattering studies at steady-state reactor sources. The triple-axis spectrometer (TAS) is a very flexible instrument permitting the scattered intensity to be measured as a function of energy or momentum transfer in such a way that a scan can be performed in any desired direction in the Brillouin zone. Depending upon the scientific goals of a particular experiment the TAS can be optimized for either high neutron flux at the sample (double focusing mode) or high resolution (conventional mode). Currently the TAIPAN TAS (figure 1) is the only instrument for inelastic scattering amongst the initial suite of seven instruments at the Australian research reactor OPAL. It is installed on a thermal beam guide at the face of the reactor shielding in order to maximize the neutron flux. Conceptually, TAIPAN TAS is similar to the triple-axis spectrometers IN8 (ILL) [1] and PANDA (FRM-II) [2] with variable incident and final energies and a secondary spectrometer with a single detector. The monochromator and analyzer on TAIPAN can each be independently focused in both vertical and horizontal directions and TAIPAN can be operated in either a high flux, relaxed momentum resolution, double focusing mode or in the conventional high resolution, flat or vertically focused monochromator and analyzer, mode of operation with the option to insert Soller collimators. Polarization analysis capabilities will be provided by the use of polarized ^{3}He spin filters and "Magic box" magnetostatic cavities developed at the ILL [3]. These are expected to be commissioned late in 2011. TAIPAN started operation in February 2009 and is in full user service since November 2010.

Figure 1. The triple-axis spectrometer TAIPAN

2. Instrument characteristics and performance.

The TAIPAN TAS is located on the thermal tangential beamtube TG4 at OPAL reactor. Outside of the heavy water moderator vessel the sides of the beam tube are lined with m = 3 supermirror guide and filled with helium heat exchange gas. The size of the exit beam at the reactor face (4.5 m distance from the reactor core) is 50×175 mm^2. The input optics assembly for TAIPAN is located between the exit of the beam at reactor face and the monochromator. It consists of a horizontal slit at a 1.8 m distance from the monochromator ("virtual source"), a secondary beam shutter, a sapphire filter and Soller collimators. The width of the virtual source slits can be tuned to optimize the signal-to-noise ratio. It also optimizes the conditions for monochromatic horizontal focusing [4].

Table 1. TAS TAIPAN main characteristics

Size of beam at reactor face	50 × 175 mm^2 (W×H)
Horizontal virtual source aperture	(0 - 65) × 200 mm^2
Monochromator	HOPG (002) 24′ and Cu (200) 20′, Double-focusing; 9 rows × 7 columns
Monochromator, take-off angle	$16° \leq 2\theta_M \leq 85°$
HOPG // Cu monochromator, incident momentum range	~1.5 $Å^{-1} \leq k_i < 6.5\ Å^{-1}$ // ~3.8 $Å^{-1} < k_i <$ ~10 $Å^{-1}$
HOPG // Cu monochromator, incident energy range	~ 5 meV < E_i < 88 meV // ~30 meV < E_i < ~200 meV
Sample table	Non-magnetic double goniometer, on air-pads, maximum central weight 5 kN,
Sample scattering angle	$-145° \leq 2\theta_S \leq 115°$
Analyser	HOPG (002) 24′ Double-focusing; 7 rows × 5 columns, 160 × 140 mm^2
Analyser scattering angle	$-110° \leq 2\theta_A \leq 110°$
Detector	∅25 mm ^{3}He detector (focused analyser);
	∅50 mm ^{3}He detector (collimator)
Distance, Source – Monochromator	6500 mm
Distance, Reactor Face – Monochromator	2000 mm
Distance, Virtual Source – Monochromator	1740 mm
Distance Monochromator – Sample	1750 - 2000 mm
Distance Sample – Analyser	810 - 1125 mm
Distance Analyser – Detector	810 - 1125 mm
Pre-monochromator collimators	15′, 30′, Open; 90 × 185 mm^2(W×H)
Post-monochromator collimators	10′, 20′, 40′, 60′, 80′; 50 × 130 mm^2 (W×H)
Pre-analyser and pre-detector, collimators	20′, 40′, 60′, 150′, Open; 50 × 130 mm^2 (W×H)
Polarisation	Polarized 3He spin filters

The sapphire filter reduces the neutron background due to epithermal and fast neutrons and the pre-monochromator collimators define the resolution. The virtual source and the collimators can be remotely adjusted. TAIPAN is designed to have two double-focusing monochromators: highly oriented pyrolitic graphite HOPG (002) and copper Cu (200). The two monochromators are to be mounted back to back, on a goniometer with linear translation stage (currently, we have the HOPG (002) monochromator installed). The filtering of the high – order monochromator contaminations is achieved by 5 cm thick HOPG filter for selected incident or final neutron energies, depending on the experimental setup.

The shielding of the monochromator drum, which has an outer diameter of 2140 mm and a 600 mm diameter inner space for monochromators, is constructed entirely of non-magnetic materials. The variable monochromator take-off angle $2\theta_M$ can be changed from 16° to 85°. For this purpose, the shielding has a mobile central block with a 62° open segment for the incoming beam from the reactor face and three movable wedges (each of which subtends 14°) are moved to the other side through the incident beam at specific values of the takeoff angle.

The secondary spectrometer consists of separate sample, analyzer and detector stages mounted on air pads moving on a polished granite floor. TAIPAN has a conventional secondary spectrometer with polyethylene shielding installed around the analyzer and detector elements. The double focusing analyzer consists of 35 HOPG (002) crystals with an active area of 160 × 140 mm^2. As is usual on a TAS, Söller collimators can be installed in the monochromator shielding-drum exit beam tube, in front of the analyzer and in front of the detector. The basic geometrical dimensions and characteristics of TAIPAN TAS are given in table 1 and in references [5-6].

The instrument is controlled via the LabView-based data acquisition program, SPICE [7] which allows an easy-to-use graphical user interface for novice users, but also allows customizations required by expert users. Notably for TAIPAN, it uses a UB matrix formulation which makes reaching out-of-plane areas of q-space available to a triple-axis machine in both elastic and inelastic modes of operation.

For investigations at very low temperatures and high magnetic fields a 12 T vertical cryomagnet provided with a recondensation system and dilution insert (20 mK – 100K) is available. Moreover the standard OPAL neutron scattering facility sample environment can be used at TAIPAN; including a closed cycle refrigerator (2.0 K < T < 300 K); closed cycle cryo-furnaces (5 K < T < 400 K and 20 K < T < 750 K with the high temperature insert); a 70 mm "orange" cryostat with low temperature insert (1.8 K < T < 300 K) and a 7.4 T vertical magnet [8].

The Q - ωrange of spectrometer depends on the experimental setup. Figure 2 shows the accessible areas for HOPG and Cu monochromators in the most often used configuration with fixed final energy. At high Q values the Q - ωrange is limited by the dance floor size as indicated by the dashed line.

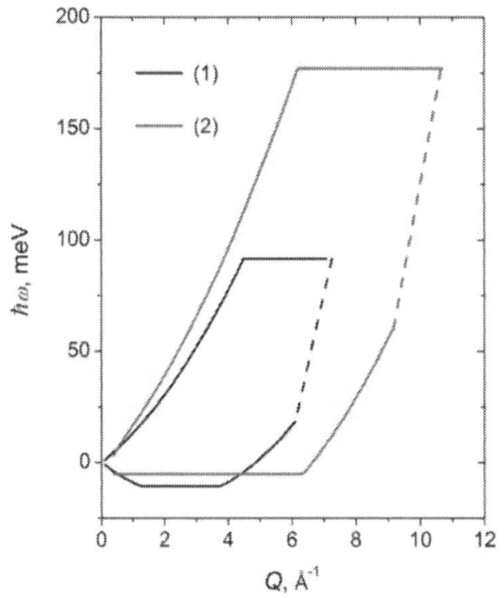

Figure 2. The $Q - \hbar\omega$ range of the TAIPAN spectrometer for measurements with fixed energy of scattered neutrons (E_f). Positive $\hbar\omega$ values correspond to the neutron energy loss.

(1) HOPG (002) monochromator; E_f = 14.7 meV.
(2) Cu (002), monochromator; E_f = 30.5 meV.

As with all TAS instruments, the resolution depends upon the values of the collimation, the initial and final energies and the focusing conditions. The full-width of the resolution function, as measured with a vanadium standard, using the HOPG (002) vertically focused monochromator and 40′

collimators before and after sample, varies from 6% at an incident energy of 14.8 meV to about 10% at E_i = 42 meV (figure 3). The flux of the monochromatic neutrons was measured with gold foils and

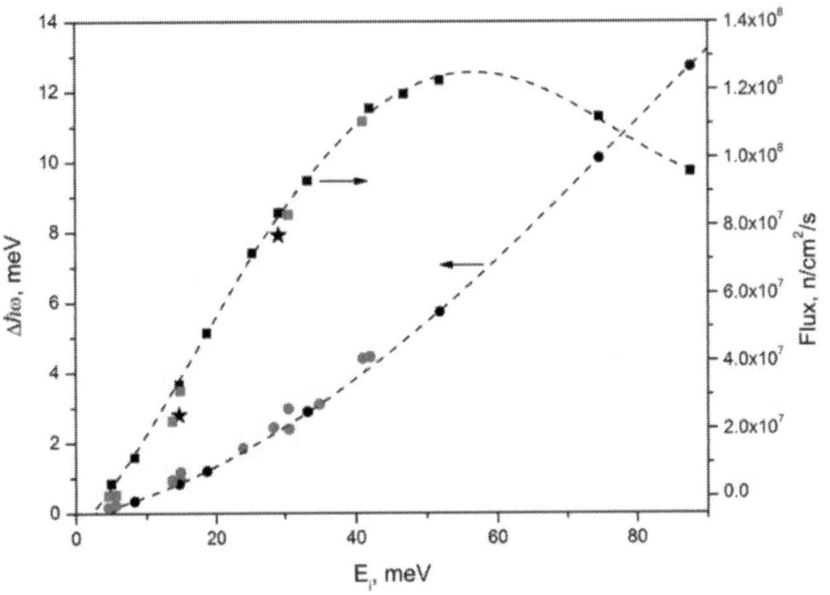

Figure 3. The dependence of monochromatic neutron flux and resolution of TAIPAN dependent on incident neutron energy.

■ - calculated neutron flux at the sample position from HOPG (002) monochromator (McSTAS);

★ - neutron flux measured with gold foils from vertically focused HOPG (002) monochromator; with PG filter for E_i=14.8 meV and corrected for second order contamination for E_i= 30.6 meV;

■ - neutron flux measured with vanadium of monochromatic neutrons from vertically focused HOPG (002) monochromator and with PG filter;

●- calculated FWHM of resolution function at sample position for HOPG monochromator (McSTAS);

● - measured FWHM of vanadium elastic peak for vertically focussed HOPG (002) monochromator and collimation 40′ before and after sample.

with a vanadium rod at the sample position. For the vertically focused HOPG monochromator it ranges from $2.4 \cdot 10^7$ n/cm^2/s at E_i = 14.7 meV to $1.1 \cdot 10^8$ n/cm^2/s at E_i = 42 meV (figure 3). These fluxes are quite competitive with similar high-flux instruments at ILL, FRM-II, NIST and ORNL in similar instrument configurations. Both resolution and neutron flux data agree well with results of previous McSTAS [9-10] simulations.

3. First experiments on TAIPAN

Several scientific problems were studied during the commissioning of the instrument in 2009. We started with the previously studied compound $Cu_{1.85}Se$, a mixed ionic-electronic conductor which is in a superionic α-phase at room temperature. Other experiments were performed in a diffraction mode of TAIPAN to explore the evolution of the magnetic structure with temperature in magnetically modulated $FePt_3$ thin film. The third example described in this paper is from very recent experiment on the search of the orbital Peierls state in $TbVO_3$ performed in the frame of the normal user proposal program.

3.1. Soft phonon modes in superionic conductor $Cu_{1.85}Se$

$Cu_{2-\delta}Se$ is a mixed ionic-electronic conductor with a superionic transition at 414K for the stoichiometric composition. At room temperature the superionic α-phase exists in the concentration range from $\delta = 0.15$ to 0.25 [11]. The characteristic structural features of copper selenide are the ordering of Cu atoms in the low-temperature phase and a random distribution of Cu over interstitial sites in high-temperature superionic phase [12-13]. However we observed that superstructure reflections in the $Cu_{1.8}Se$ compound appear at a temperature ~ 40 K above the α - β transition temperature; which is slightly below 300K for this composition [14]. Intense superstructural reflections were observed at the wave vectors $\boldsymbol{q}$ = 1/2 (±1, 1, 1); 1/3 (0, 2, 2) and 1/2 (2, 0, 0). We also found the lower intensity peaks at $\boldsymbol{q}$ = 1/4 (±1, 1, 1) and $\boldsymbol{q}$ = 1/2 (2, 0, 0). These satellites are in general agreement with observation of $\boldsymbol{q}$ = 1/8 (1, 1, 1) and $\boldsymbol{q}$ = 1/3 (0, 2, 2) reflections by Kashida and Akai in the low temperature β-phase of Cu_2Se at ambient temperature [12-13]. Note that the strongest superstructural reflections are located at the BZ boundaries or close to them. This causes effects similar to folding of the BZ due to the doubling of the lattice parameter.

The lattice dynamics of $Cu_{2-\delta}Se$ is similar to other Cu and Ag fast ionic conductors showing the presence of low-energy excitations. The importance of low-lying modes that give the major contribution to thermal motion due to a high density of states and low activation energy is widely recognized. In our previous experiments we found that the transverse acoustic (TA) phonon branches show drastic change in the slope and flatten out at phonon wave vectors $q/q_m > 0.4$ with energy of about 4 meV [15]. At the same time, phonon peaks show a strong broadening at these wave vectors.

Measurements of phonon dispersion in $Cu_{1.8}Se$ at TAIPAN were performed with fixed final neutron energy of 14.87 meV, a vertically focused HOPG (002) monochromator and 40′ collimation before and after the sample. The dispersion curves were measured in [100], [110] and [111] directions [16]. Due to a strong broadening of phonon peaks data are limited to low energy modes. The onset of the broadening as a function of q is rapid and takes place at phonon wave vectors $q/q_m \approx 0.5$ for all measured phonon branches. This is clearly seen in figure 4, where the typical q-dependence of the phonon peaks is shown for the TA_1 [110] and TA [111] branches. Phonons with wave vectors $q/q_m \leq 0.5$ are narrow and show a fast decrease in intensity with increasing wave vector which is characteristic for acoustic phonons. At wave vectors $q/q_m \geq 0.5$ a change-over from strong q-dependence to broad phonon peaks with minor modifications in peak shape and intensity takes place. The elastic peaks ($\hbar\omega = 0$) at $q/q_m \approx 0.66$ and 1.0 in [110] direction and at $q/q_m \approx 0.25$ and 0.5 in [111] direction are from the superstructure mentioned above. In general the present data agree well with our previous results measured with spectrometer UNIDAS (FZ Jülich) [15]. However since the new data cover a wider q-range and have been measured with higher accuracy we found that the TA_1 [110] branch with the polarization vector along [1-10] direction and TA phonons in [111] direction (figure 4) demonstrate a presence of soft mode condensing at the X and L points of BZ in addition to flat optic-like modes observed previously. In contrast, the TA_2 [110] phonons polarized in [001] direction and longitudinal modes show the almost linear q-dependence at $q/q_m < 0.5$. At higher wavevectors these phonon peaks spread out and are no longer visible [16].

Experimental results were compared with dispersion curves calculated with density functional theoretical (DFT) approach [17] for stoichiometric composition and antifluorite crystal structure. Results are shown in figure 4 together with experimental data. The most remarkable features of calculated acoustic modes are the low phonon frequencies and the instability over a large area of reciprocal space. Acoustic modes TA [100], TA [111], TA_1 [110] and LA [110] are unstable and go to negative eigenvalues in vicinity of BZ centre and at $q/q_m \geq 0.3 - 0.4$ in case of transverse modes and at K – point for LA [110] mode. This indicates that the stoichiometric compound is dynamically unstable and the antifluorite structure is not a true low-temperature one. Indeed, a phase transition to low-temperature β-phase takes place at 414 K for stoichiometric Cu_2Se [11].

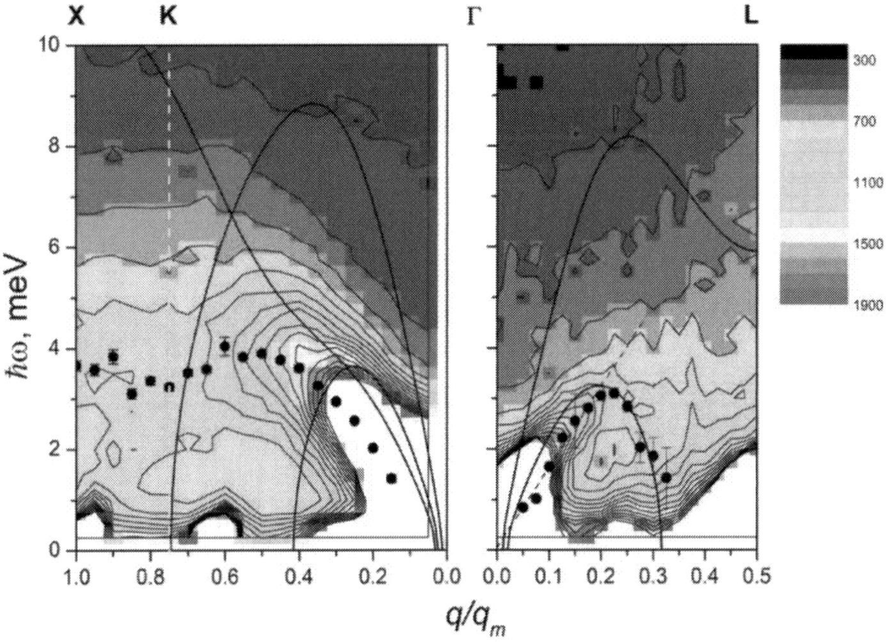

Figure 4. Contour plot of inelastic scattering intensity in $Cu_{1.8}Se$ determined from series of constant–Q scans. Transverse acoustic TA_1 [110] (left) and TA [111] branch (right). Dots correspond to the centres of phonon peaks; lines show calculated longitudinal and transverse acoustic dispersion curves.

The instability of acoustic modes is directly related to the ordering of Cu atoms observed in copper-selenide. The intensity of the superstructure reflections is quite large. As noted above the appearance of strong reflections at the edge of the BZ in this direction can cause effects similar to the folding of the Brillouin zone. Indeed, figure 4 show that TA phonon branches in [110] and [111] direction have tendency to soften at the boundary of BZ, although phonon intensities at "new" BZ centres are weak. These superstructural points are in the area of pronounced instability of acoustic modes indicating that ordering process is actually driven by a soft mode. The effect of folding of the BZ has a direct relationship to the suggestion of Wakamura who considers that the low-energy mode in β-AgI originates from the zone-edge acoustic phonons in γ-AgI because of BZ folding [18]. However, the important difference in the case of the $Cu_{1.8}Se$ compound is that the ordering process and the folding of the BZ are driven by a soft mode, that is a second order transition compared to a first order transition as found in AgI.

3.2. Investigation of a modulated antiferromagnetic/ferromagnetic structure in a quasi - single crystal $FePt_3$ thin film

Stoichiometric $FePt_3$ bulk crystals in perfectly chemically ordered fcc structure show antiferromagnetic (AFM) order below T_N=160 K [19]. In a chemically disordered phase (i.e. random distribution of the Fe and Pt atoms on lattice sites) $FePt_3$ exhibits ferromagnetic (FM) ordering. In thin

Journal of Physics: Conference Series **340** (2012) 012003 doi:10.1088/1742-6596/340/1/012003

films of $FePt_3$ the degree of chemical order can be controlled by temperature modulation during growth, thus creating a single-crystalline $FePt_3$ film of homogeneous composition throughout consisting layers of alternating chemically ordered AFM and chemically disordered FM structures. Using this method the magnetically modulated thin films of a superlattice of five repetitions 200 Å $FePt_3$ (T_s=873 K) / 100 Å $FePt_3$ (T_s=673 K) were grown via magnetron sputtering on MgO (001) substrates [20].

(a) (b)

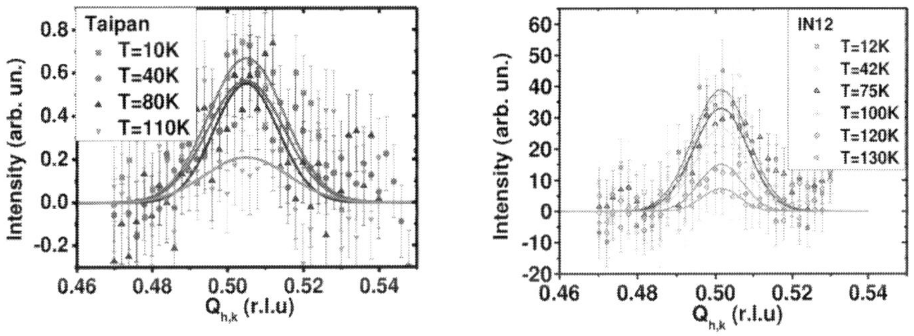

Figure 5. Antiferromagnetic Bragg peak as a function of temperature measured at TAIPAN and IN12 spectrometers.

In order to elucidate the AFM ordering in the superlattice, with only ~220 µg of AFM material, neutron diffraction using a triple-axis spectrometer represents an excellent tool due to very high signal to noise ratios. The experiments were performed using the triple-axis spectrometers IN12 (Institute Laue Langevin, Grenoble) and TAIPAN at ANSTO. In both experiments the incoming neutron wavelength was fixed (IN12: λ= 3.141 Å; TAIPAN: λ= 2.35 Å) and the spectrometers were setup to detect elastic scattering. Diffraction peaks of the $FePt_3$ superlattice were recorded by setting the spectrometer to particular combinations of incident and scattered angles and subsequently measuring rocking curve profiles. Similar to chemically ordered $FePt_3$ bulk crystals with T_N =160 K, the superlattice exhibits the onset of a (½ ½ 0) AFM Bragg peak below a temperature of 140 K. Rocking curve profiles recorded at IN12 and TAIPAN are shown in figure 5(a) and 5(b) respectively. The results from both instruments agree very well with each other [20]. This magnetic Bragg peak demonstrates the existence of antiferromagnetism within the thin chemically ordered $FePt_3$ layers of the superlattice and confirms the successful creation of single composite AFM/FM interfaces. Figure 6 shows the temperature dependency of the integrated AFM peak intensity in comparison to peak intensities obtained from a 2800 Å $FePt_3$ thin film without chemical order modulation. The rocking curve profiles, recorded with TAIPAN, are shown in figure 7. The Nèel temperature of the film with an order parameter of S = 83% is shifted towards 160 K showing a more bulk - like attribute. Otherwise, the integrated Bragg peak intensities follow a similar behavior. Otherwise, the integrated Bragg peak intensities follow a similar temperature behavior. Furthermore, no (½ 0 0) reflection could be observed at any temperature, implying the chemically ordered phase of the sample fully orients in the antiferromagnetic phase. Using complementary polarized neutron reflectometry, the existence of layered FM phases in the single-crystalline $FePt_3$ superlattice was confirmed [20]. The results represent a new approach to the creation of AFM and FM interfaces without chemical modulation in single composite epitaxial thin films [20].

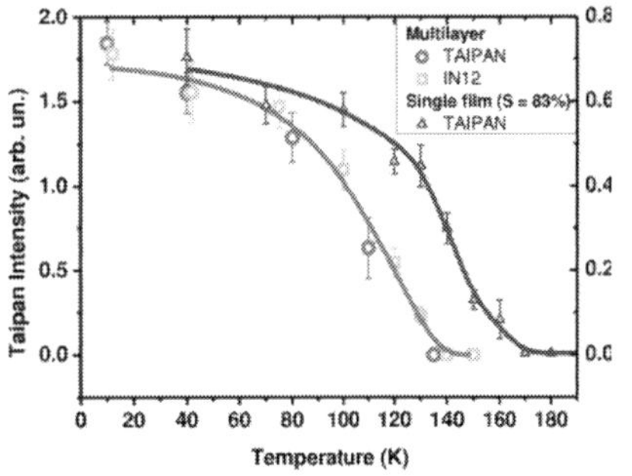

Figure 6. Integrated AFM Bragg peak intensities of multilayered and single thin films as a function of temperature.

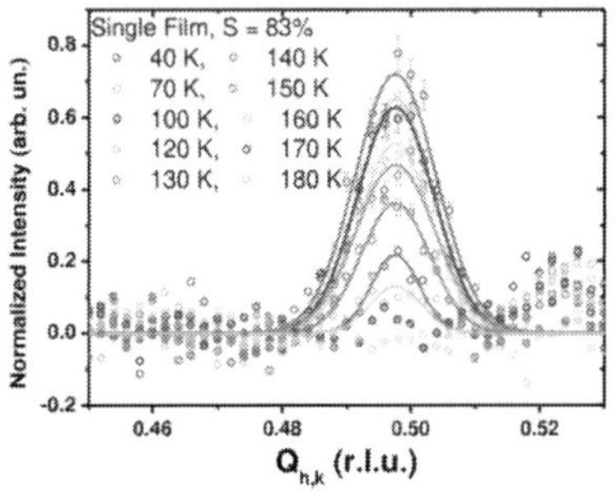

Figure 7. AFM Bragg peak of single 2800 Å film as a function of temperature measured at TAIPAN.

3.3. Search for the Orbital Peierls State in the Vanadates

Orbital degrees of freedom play an important role in the physics of strongly correlated electron systems. Our early studies on the insulating titanates $LaTiO_3$ and $YTiO_3$, with only one electron in the t_{2g} levels, yielded unusual static and dynamic properties of the magnetic spins, which are difficult to reconcile with 'standard' theoretical models of the standard lattice driven orbital - order, but point towards an explanation using a novel spin - orbit liquid state as a consequence of strong orbital quantum fluctuations [21-24].

Strong orbital quantum fluctuations were also expected in the vanadates, which possess two t_{2g} electrons. In the family of the vanadates, YVO_3 is of particular interest since it undergoes a series of temperature induced magnetization reversals as an evidence for a spontaneous redistribution of the valence electron density far below the magnetic ordering temperature [25]. YVO_3 undergoes two magnetic phase transitions: in the "high temperature" phase between 116 K and 77 K, YVO_3 exhibits a *C*-type spin structure. Below 77 K the spin structure is *G*-type, i.e. antiferromagnetic in all directions. Our recent neutron scattering experiments on YVO_3 have revealed a highly unusual magnetic ground state as consequence of orbital fluctuations [26-27]. While the static spin structure and the spin wave dispersion relation of the low temperature *G*-type structure is consistent with standard theoretical descriptions, the *C*-type phase exhibits quite unusual magnetic properties. The magnetic moment is strongly reduced (1 μ_B as opposed to ~2 μ_B expected for a d^2-system) and the spin structure is highly noncollinear. Furthermore, the spin wave dispersion relation exhibits an unusual hierarchy of exchange interactions, i.e. large ferromagnetic and weak antiferromagnetic superexchange, opposite to the prediction of the Goodenough - Kanamori rules. Further, the magnon spectrum is split into optical and acoustic branches, contrary to conventional expectations. Based on a combination of analytical and numerical results [28-29], this phase was identified as a theoretically predicted but hitherto unobserved collective singlet state, the "orbital Peierls state", an insulating analogue of the well-known Peierls phase in low-dimensional metals.

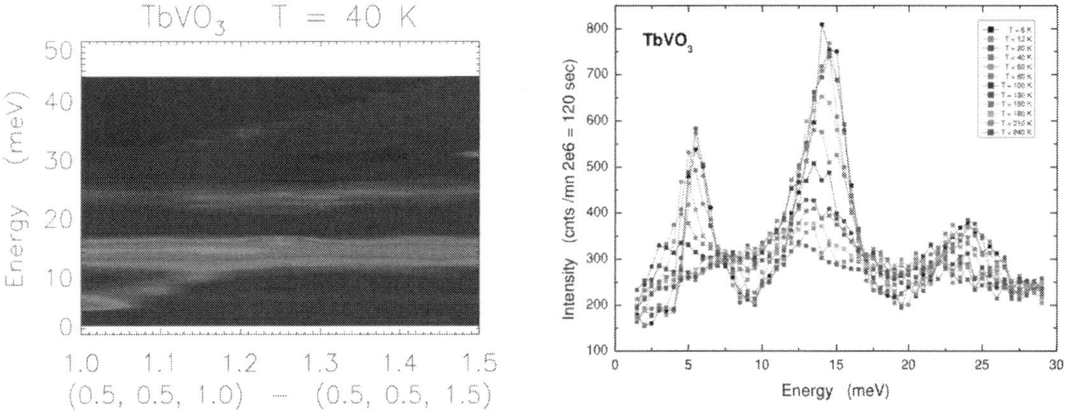

Figure 8. Inelastic neutron scattering data of $TbVO_3$ taken at TAIPAN (left). Temperature dependence of the spectrum of $TbVO_3$ at the centre of the magnetic Brillouin zone at (½ ½ 1) (right).

When replacing Yttrium with a larger lanthanide, the *G*-type antiferromagnetic phase disappears whereas the *C*-type magnetic structure remains almost unchanged [30-32]. $TbVO_3$ is the first member of the RVO_3 family where the vanadium moments order solely the *C*-type structure. In addition, the magnetic moments of the Tb-ions order below 11.5 K [27]. Large single crystals of $TbVO_3$ were synthesized in the group of Prof. Y. Tokura at the University of Tokyo. Inelastic neutron scattering experiments were performed at the triple axis spectrometer TAIPAN at ANSTO. Figure 8 shows the contour plot of the excitation spectrum of $TbVO_3$ along the Q_L direction. In order to assign the observed features, the temperature dependence of the spectrum at the centre of the *C*-type antiferromagnetic Brillouin zone (½, ½, 1) was measured (see figure 8). The peaks at 5.5 meV and at 24.0 meV can be attributed to spin wave excitations. The mode at 14.5 meV is perfectly dispersionless in both energy and intensity. In contrast to the expected behaviour of a phonon mode, this peak decreases and disappears with increasing temperature above the magnetic phase transition temperature. Therefore, this mode can be assigned to crystal field excitations arising form the Tb-ions. However, the optical phonon modes arising from the vibration of the Tb-ions should appear in the same energy range. The resulting spin wave dispersion relation is shown in figure 9. The blue symbols can be associated with spin wave excitations, whereas the black symbols correspond to phonon modes. The red symbols originate from either an optical phonon or crystal field excitations.

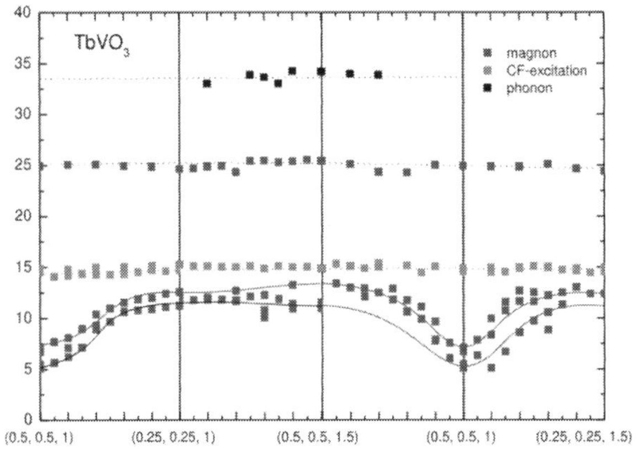

Figure 9. Magnon dispersion relation of $TbVO_3$ taken at a temperature of 40 K. The blue symbols can be associated with spin wave excitations, whereas the black symbols correspond to phonons. The red symbols originate from either an optical phonon or crystal field excitations

As in YVO_3 an acoustic and an optical magnon branch is observed in the same energy range. This indicates that the 'orbital Peiers state' also exists in $TbVO_3$. While the *G*-type antiferromagnetic phase disappears with changing the tilt of the VO_6 octahedra in the RVO_3 series, the 'orbital Peierls state' is robust concerning changes in the width of the valence band.

The TAIPAN characteristics and the experiments described above demonstrate the high performance of new thermal triple axis spectrometer at OPAL research reactor. Instrument is open for users through the research proposals system of the Bragg Institute, ANSTO [33]

4. Acknowledgements

The authors acknowledge the early design and conceptual work for TAIPAN carried out by L. Cussen, M. Hagen and G. Horton. We have also greatly benefited from the freely given help and advice of T. Finlayson (Monash University), J. Kulda (ILL) and L. Robertson (ORNL). At ANSTO, the ongoing input from R. Moore, E. Imamovic, G. Braoudakis, M. Deura, P. Hansen, F. Darmann, A. McGregor and all of the members of the Neutron Beams Instrument Project is gratefully acknowledged. This development was fully funded by the Australian Federal Government.

References

[1] A. Hiess, M. Jimenez-Ruiz, P. Courtois, R. Currat, J. Kulda, F.J. Bermejo, ILL's renewed thermal three-axis spectrometer IN8: A review of its first three years on duty, Physica B-Condensed Matter, 385-86 (2006) 1077-1079.

[2] A. Schneidewind, P. Link, D. Etzdorf, R. Schedler, M. Rotter, M. Loewenhaupt, PANDA - first results from the cold three-axes spectrometer at FRM-II, Physica B-Condensed Matter, 385-86 (2006) 1089-1091.

[3] http://www.ill.eu/industry/neutron-technology/neutron-optics/polarised-3he-spin-filters/.

[4] L. Pintschovius, Performance of a three-axis neutron spectrometer using horizontally and vertically focussing monochromators, Nuclear Instruments and Methods in Physics Research Section A: Accelerators, Spectrometers, Detectors and Associated Equipment, 338 (1994) 136-143.

[5] S.A. Danilkin, G. Horton, R. Moore, G. Braoudakis, M. Hagen, The TAIPAN thermal triple-axis spectrometer at the OPAL reactor, Journal of Neutron Research, 15 (2007) 55 - 60.

[6] S.A. Danilkin, M. Yethiraj, TAIPAN: Thermal Triple-Axis Spectrometer, Neutron News, 20 (2009) 37 - 39.

[7] M.D. Lumsden, J.L. Robertson, M. Yethiraj, SPICE - Spectrometer and Instrument Control Environment, Physica B-Condensed Matter, 385-86 (2006) 1336-1339.

[8] http://www.ansto.gov.au/research/bragg_institute/facilities/sample_enviros.

[9] K. Lefmann, K. Nielsen, McStas, a general software package for neutron ray-tracing simulations, Neutron News, 10 (1999) 20 - 23.

[10] P. Willendrup, E. Farhi, K. Lefmann, McStas 1.7 - a new version of the flexible Monte Carlo neutron scattering package, Physica B: Condensed Matter, 350 (2004) E735-E737.

[11] N.H. Abrikosov, M.A. Korzhuev, G.K. Demenski and O.A. Teplov, Sov. Phys. - Solid State, 25 (1983) 1678.

[12] S. Kashida, J. Akai, X-Ray-Diffraction and Electron-Microscopy Studies of the Room-Temperature Structure of Cu_2Se, J Phys C Solid State, 21 (1988) 5329-5336.

[13] K. Yamamoto, S. Kashida, X-Ray Study of the Cation Distribution in Cu_2Se, $Cu_{1.8}Se$ and $Cu_{1.8}S$ - Analysis by the Maximum-Entropy Method, Solid State Ionics, 48 (1991) 241-248.

[14] S.A. Danilkin, Diffuse scattering and lattice dynamics of superionic copper chalcogenides, Solid State Ionics, 180 (2009) 483-487.

[15] S.A. Danilkin, A. Skomorokhov, A. Hoser, H. Fuess, V. Rajevac, N.N. Bickulova, Crystal structure and lattice dynamics of superionic copper selenide $Cu_{2-\delta}Se$, J Alloy Compd, 361 (2003) 57-61.

[16] S.A. Danilkin, M. Yethiraj, G.J. Kearley, Phonon Dispersion in Superionic Copper Selenide: Observation of Soft Phonon Modes in Superionic Phase Transition, J Phys Soc Jpn, 79SA (2010) 25.

[17] G. Kresse, J. Furthmueller, Efficient iterative schemes for ab initio total-energy calculations using a plane-wave basis set, Phys Rev B, 54 (1996) 11169.

[18] K. Wakamura, Origin of the low-energy mode in superionic conductors, Phys Rev B, 59 (1999) 3560.
[19] S. Maat, O. Hellwig, G. Zeltzer, E.E. Fullerton, G.J. Mankey, M.L. Crow, J.L. Robertson, Antiferromagnetic structure of $FePt_3$ films studied by neutron scattering, Phys Rev B, 63 (2001) 134426.
[20] T. Saerbeck, F. Klose, D. Lott, G.J. Mankey, Z. Lu, P.R. LeClair, W. Schmidt, A.P.J. Stampfl, S. Danilkin, M. Yethiraj, A. Schreyer, Artificially modulated chemical order in thin films: A different approach to create ferro/antiferromagnetic interfaces, Phys Rev B, 82 (2010) 134409.
[21] B. Keimer, D. Casa, A. Ivanov, J.W. Lynn, M.v. Zimmermann, J.P. Hill, D. Gibbs, Y. Taguchi, Y. Tokura, Spin Dynamics and Orbital State in $LaTiO_3$, Physical Review Letters, 85 (2000) 3946.
[22] G. Khaliullin, S. Maekawa, Orbital Liquid in Three-Dimensional Mott Insulator: $LaTiO_3$, Physical Review Letters, 85 (2000) 3950.
[23] C. Ulrich, G. Khaliullin, S. Okamoto, M. Reehuis, A. Ivanov, H. He, Y. Taguchi, Y. Tokura, B. Keimer, Magnetic Order and Dynamics in an Orbitally Degenerate Ferromagnetic Insulator, Physical Review Letters, 89 (2002) 167202.
[24] G. Khaliullin, S. Okamoto, Quantum Behavior of Orbitals in Ferromagnetic Titanates: Novel Orderings and Excitations, Physical Review Letters, 89 (2002) 167201.
[25] Y. Ren, T.T.M. Palstra, D.I. Khomskii, E. Pellegrin, A.A. Nugroho, A.A. Menovsky, G.A. Sawatzky, Temperature-induced magnetization reversal in a YVO_3 single crystal, Nature, 396 (1998) 441-444.
[26] C. Ulrich, G. Khaliullin, J. Sirker, M. Reehuis, M. Ohl, S. Miyasaka, Y. Tokura, B. Keimer, Magnetic Neutron Scattering Study of YVO_3: Evidence for an Orbital Peierls State, Physical Review Letters, 91 (2003) 257202.
[27] M. Reehuis, C. Ulrich, P. Pattison, B. Ouladdiaf, M.C. Rheinstaedter, M. Ohl, L.P. Regnault, M. Miyasaka, Y. Tokura, B. Keimer, Neutron diffraction study of YVO_3 , $NdVO_3$, and $TbVO_3$, Phys Rev B, 73 (2006) 094440.
[28] G. Khaliullin, P. Horsch, A.M. Olesacute, Spin Order due to Orbital Fluctuations: Cubic Vanadates, Physical Review Letters, 86 (2001) 3879.
[29] P. Horsch, G. Khaliullin, A.M. Olesacute, Dimerization versus Orbital-Moment Ordering in a Mott Insulator YVO_3, Physical Review Letters, 91 (2003) 257203.
[30] S. Miyasaka, Y. Okimoto, M. Iwama, Y. Tokura, Spin-orbital phase diagram of perovskite-type RVO_3 (R=rare-earth ion or Y), Phys Rev B, 68 (2003) 100406.
[31] J. Fujioka, S. Miyasaka, Y. Tokura, Orbital disordering and the metal-insulator transition with hole doping in perovskite-type vanadium oxides, Phys Rev B, 72 (2005) 024460.
[32] J. Fujioka, T. Yasue, S. Miyasaka, Y. Yamasaki, T. Arima, H. Sagayama, T. Inami, K. Ishii, Y. Tokura, Critical competition between two distinct orbital-spin ordered states in perovskite vanadates, Phys Rev B, 82 (2010) 144425.
[33] http://www.ansto.gov.au/research/bragg_institute/users/call_for_proposals.

Journal of Physics: Conference Series **340** (2012) 012004 doi:10.1088/1742-6596/340/1/012004

Experimental corrections in neutron diffraction of ambient water using H/D isotopic substitution

J. Dawidowski [1] **and G. J. Cuello** [2].

[1] Centro Atómico Bariloche and Instituto Balseiro, Bustillo 9500 (R8402AGP) Bariloche, Argentina

[2] Institut Laue Langevin, 6, rue Jules Horowitz (38042) Grenoble, France

E-mail: [1]`javier@cab.cnea.gov.ar`

Abstract. We report on the procedure to treat neutron diffraction data, in order to obtain the differential cross sections, a step that is essential in the analysis of structural information for liquids and amorphous materials. The treatment is applied to new measurements that have been made on the D4C diffractometer for eight different isotopic compositions of H_2O/D_2O mixtures at ambient temperature. The procedure is based on Monte Carlo simulations to perform multiple scattering, attenuation and detector efficiency corrections, applied iteratively. The normalization process allows the direct comparison of measured data from different samples, and requires the knowledge of the total cross-section at the incident neutron energy. They high contrast between the cross sections (coherent and incoherent) for heavy and light water allows to test the procedure under very different conditions of multiple scattering effects.

1. Introduction

The availability of improved neutron spectrometers, makes it possible to reach a high degree of refinement in the experimental data. However these improvements have not reached yet the customarily employed analysis techniques for diffraction, and outdated correction procedures borrowed from different techniques are still being employed, due to the lack of customary software specific for neutron diffraction.

It is accepted nowadays that the most reliable procedure to correct for multiple scattering and beam attenuation effects in neutron scattering experiments is achieved through Monte Carlo simulations. Since the great variety of neutron techniques pose very different experimental conditions, the numerical simulations must be devised for each specific situation. Prior to the present work, a correction tool had been previously developed and benchmarked for neutron diffraction in liquid and amorphous materials [1]. In the particular case of light and heavy water, different models were explored, and it was determined that in the case of deuterated samples, the multiple scattering signal showed a structure very different from the oversimplified calculations borrowed from X-rays techniques.

The structure of water has been a subject of intensive research during the last decades. The efforts aim at a better knowledge of the the structural characteristics of liquid water in terms of the partial pair correlation functions and the orientational correlations between neighbouring molecules. The purpose of this work is to show the data processing procedure we apply to subtract the multiple scattering effects, attenuation of the neutron beam and detector efficiency to obtain $d\sigma(\theta)/d\Omega$ in an absolute scale in a series of light water/heavy water mixtures measured at diffractometer D4C (ILL, Grenoble)[2], thus allowing the direct comparison of measured data from different samples. We show the theoretical foundations of the process, as well as the procedure of data normalizations to achieve differential cross sections expressed in a physically meaningful scale.

2. Theoretical background

In this section we will develop the mathematical expressions of the magnitudes calculated by the Monte Carlo program employed to process the data.

2.1. Overview

The magnitude measured in a neutron diffraction experiment is the macroscopic differential cross section, *i.e.* the angular distribution of the neutrons emerging from the sample. The goal of the experiment is to deduce the microscopic differential cross section, that is directly related with the structural and dynamical properties of the sample.

If the initial and final neutron wave vectors are $\mathbf{k}_0$ and $\mathbf{k}$ respectively, the expression for the macroscopic cross section as a function of the scattering angle θ measured in diffraction experiments [3], is obtained after integration on the final energies, and results

$$\frac{1}{A(\hat{\mathbf{k}}_0)}\left(\frac{\mathrm{d}\Sigma}{\mathrm{d}\Omega}\right)_{\mathrm{scatt}} = \frac{N\sigma_{\mathrm{scatt}}}{4\pi A(\hat{\mathbf{k}}_0)}\int \mathrm{d}E\ \frac{k}{k_0}\varepsilon(k)s(\mathbf{k}_0,\mathbf{k}) \quad , \tag{1}$$

where $\hat{\mathbf{k}}_0$ is the direction of the incident beam, $A(\hat{\mathbf{k}}_0)$ is the cross sectional area exposed at the incident beam, N is the number of atoms, σ_{scatt} the bound-atom scattering cross section of the sample, $\varepsilon(k)$ is the detector efficiency and $s(\mathbf{k}_0,\mathbf{k})$ is the effective scattering function defined by Sears [3]. This function can be decomposed in contributions according to the number of collisions. For systems consisting in a sample placed into a container, the function $s(\mathbf{k}_0,\mathbf{k})$ contains the information of the whole set, and it can be conveniently decomposed into a part due to singly scattered neutrons by the sample, another due to single scattering from the container, and a third due to multiply scattered neutrons with any combination of sample-can scattering events,

$$s(\mathbf{k}_0,\mathbf{k}) = s_1(\mathbf{k}_0,\mathbf{k}) + s_{\mathrm{C}}(\mathbf{k}_0,\mathbf{k}) + s_{\mathrm{M}}(\mathbf{k}_0,\mathbf{k}) \quad . \tag{2}$$

The Monte Carlo code samples the density of neutrons in the steady state regime. To this end Copley's scheme [4, 5] provides an efficient method, that consists in generating neutron histories biased in such way that the neutron never leaves the sample nor is absorbed. To compensate this bias, a weight (that is initially 1) is computed for the history, that decreases at every path traversed and at every scattering step as described in Ref. [1]. Neutron histories are originated in a random point of the sample surface that faces the incoming beam. The distance traveled by the neutron between two scattering steps is determined by the mean free paths of the materials that the neutron has to traverse at the current flight direction and energy E, to emerge from the sample. After the scattering event, new energies and directions of flight are randomly sampled from the joint probability distribution

$$P(E_0,E,\theta) = \frac{1}{\sigma_{\mathrm{scatt}}(E_0)}\frac{\mathrm{d}^2\sigma}{\mathrm{d}\Omega\mathrm{d}E} \quad , \tag{3}$$

where $\sigma_{\mathrm{scatt}}(E_0)$ is the total scattering cross section at energy E_0 (before the scattering interaction) and $\mathrm{d}^2\sigma/\mathrm{d}\Omega\mathrm{d}E$ is the double differential cross section. For molecular systems a suitable model for these magnitudes is provided by Granada's Synthetic Model [6]. Although this is an incoherent model, it proved to be sufficient in the calculation of multiple scattering events, even when coherent scattering dominates.

The scored magnitudes at each step are the contributions of the history to the detectors. The most useful estimators to evaluate the corrections needed in a diffraction experiment are enumerated in the following paragraphs.

2.1.1. Singly scattered neutrons The basic information that provides a diffraction experiment is contained the *angular distribution of neutrons after the first scattering.* In the Monte Carlo procedure this value is calculated by sampling the first collision point from the probability distribution

$$P(x) = \frac{\Sigma_{\rm tot}(\mathbf{k}_0)\exp[-\Sigma_{\rm tot}(\mathbf{k}_0)x]}{w_1} \quad , \tag{4}$$

with

$$w_1 = 1 - \exp[-\Sigma_{\rm tot}(\mathbf{k}_0)d(\mathbf{r}_0, -\hat{\mathbf{k}}_0)] \quad , \tag{5}$$

where $d(\mathbf{r}_0, -\hat{\mathbf{k}}_0)$ is the total thickness of the sample measured from the point of incidence $\mathbf{r}_0$ in the direction $\hat{\mathbf{k}}_0$, and $\Sigma_{\rm tot}(\mathbf{k}_0)$ is the macroscopic total cross section. The term w_1 is a bias in the probability so the neutron does not leave the sample.

w_1 is the probability that the neutron incident at $\mathbf{r}_0$ will interact within the sample, and is the weight of the history at the first step. When a container is present, the factor $\exp[-\Sigma_{\rm tot}(\mathbf{k}_0)d(\mathbf{r}_0, -\hat{\mathbf{k}}_0)]$ (called *attenuation factor*) must also include the path traversed in the container and its cross section.

The sought distribution is calculated by scoring over a large number of histories the estimator

$$\chi_1(\mathbf{r}, \mathbf{k}_0, \mathbf{k}) = w_1 \frac{\Sigma_{\rm scatt}(\mathbf{k}_0)}{\Sigma_{\rm tot}(\mathbf{k}_0)} P(\mathbf{k}_0, \mathbf{k}) \quad , \tag{6}$$

which is the probability that the incoming neutron is scattered, times the probability distribution in $\mathbf{k}$ of the scattered neutrons. $P(\mathbf{k}_0, \mathbf{k})$ is the same function introduced in Eq. (3). As described in Ref. [1], this probability can be evaluated making resource the experimental data and implementing an iterative correction scheme as described below.

In the sampling process the estimator (6) must be weighed by the density of first collisions [7]

$$\widetilde{\Psi}_1(\mathbf{r}, \mathbf{k}_0) = \frac{1}{A(\hat{\mathbf{k}}_0)} \frac{\exp[-\Sigma_{\rm tot}(\mathbf{k}_0)L(\mathbf{r}, -\hat{\mathbf{k}}_0)]}{w_1} \quad , \tag{7}$$

where the tilde over Ψ indicates that the function has been modified (dividing by w_1) to compensate the bias and $L(\mathbf{r}, -\hat{\mathbf{k}}_0)$ is the distance from point $\mathbf{r}$ (inside the sample), to the sample surface in the direction $\hat{\mathbf{k}}_0$. The estimator (6) weighed over the collision density (7) in the Monte Carlo process, is the integral over the sample volume (see Ref. [3])

$$z_1(\mathbf{k}_0, \mathbf{k}) = \int \mathrm{d}^3r\, \widetilde{\Psi}_1(\mathbf{r}, \mathbf{k}_0)\chi_1(\mathbf{r}, \mathbf{k}_0, \mathbf{k}) = \frac{\Sigma_{\rm scatt}(\mathbf{k}_0)}{\Sigma_{\rm tot}(\mathbf{k}_0)}(1 - t(E_0))P(\mathbf{k}_0, \mathbf{k}) \quad , \tag{8}$$

where $t(E_0)$ is the total fraction of transmitted neutrons. Equation (8) expresses that the intensity of singly scattered neutrons is equal to the ratio of the scattering and the total cross sections, times the fraction of neutrons interacting in the sample $(1-t(E_0))$, and it is distributed according to the microscopic double differential cross section. The integration of Eq. (8) over final energies is the magnitude of interest in diffraction experiments

$$z_1(\mathbf{k}_0, \theta) = \frac{\Sigma_{\rm scatt}(\mathbf{k}_0)}{\Sigma_{\rm tot}(\mathbf{k}_0)}(1 - t(E_0))\frac{1}{\sigma_{\rm scatt}(E_0)}\frac{\mathrm{d}\sigma}{\mathrm{d}\Omega}(E_0, \theta) \quad , \tag{9}$$

i.e. the intensity of the *angular distribution of neutrons after the first scattering*. This equation shows that the sought magnitude ($d\sigma/d\Omega$) is proportional to $z_1(\mathbf{k}_0, \theta)$, and shows the proportionality constant. This magnitude is not directly accessible from the experiment due to the presence of multiple scattering, and even when we can subtract this component (by calculation) the single scattering component is still affected by attenuation and detector efficiency, as will be shown in the next section. When all the corrections have been performed, the resulting component is expressed by Eq. (9), which is the expression that must be employed to compare the diffractograms from different samples.

2.1.2. Detected singly scattered neutrons The mentioned contribution of the singly scattered neutrons (at position $\mathbf{r}$) to the detectors (placed in the direction $\mathbf{k}$) is scored by the estimator

$$\chi_1'(\mathbf{r}, \mathbf{k}_0, \mathbf{k}) = w_1 \frac{\Sigma_{\mathrm{scatt}}(\mathbf{k}_0)}{\Sigma_{\mathrm{tot}}(\mathbf{k}_0)} P(\mathbf{k}_0, \mathbf{k})\, \mathrm{e}^{-\Sigma_{\mathrm{tot}}(\mathbf{k}) L(\mathbf{r}, \hat{\mathbf{k}})} \varepsilon(k) \quad , \tag{10}$$

where $\mathrm{e}^{-\Sigma_{\mathrm{tot}}(\mathbf{k}) L(\mathbf{r}, \hat{\mathbf{k}})}$ is the attenuation factor from $\mathbf{r}$ to the detector position and $\varepsilon(k)$ is the detector efficiency.

When a container is present, the attenuation factor must also include the path traversed in the container and its cross section. In the Monte Carlo code this estimator is scored over the neutron histories as in Eq. (8). The result converges to the integral over the sample volume of the estimator (10) weighed over the collision density (7), resulting

$$\tilde{z}_1(\mathbf{k}_0, \mathbf{k}) = \frac{V}{A(\hat{\mathbf{k}}_0)} \Sigma_{\mathrm{scatt}}(\mathbf{k}_0)\, P(\mathbf{k}_0, \mathbf{k})\, \varepsilon(k) H_1(\mathbf{k}_0, \mathbf{k}) \quad , \tag{11}$$

where $H_1(\mathbf{k}_0, \mathbf{k})$ is the primary attenuation factor described by Sears [3]. This expression is directly related with the single scattering component of Eq. (1) through

$$\tilde{z}_1(\mathbf{k}_0, \mathbf{k}) = \frac{N\sigma_{\mathrm{s}}}{4\pi A(\hat{\mathbf{k}}_0)} \frac{k}{k_0} \varepsilon(k) s_1(\mathbf{k}_0, \mathbf{k}) \quad . \tag{12}$$

As in Sect. 2.1.1 the integration of (11) over energies gives the specific result for diffraction

$$\tilde{z}_1(\mathbf{k}_0, \theta) = \frac{V}{A(\hat{\mathbf{k}}_0)} \Sigma_{\mathrm{scatt}}(\mathbf{k}_0) \int \mathrm{d}E\, \frac{1}{\sigma_{\mathrm{scatt}}(E_0)} \sigma(E_0, E, \theta)\, \varepsilon(E) H_1(E_0, E, \theta) \quad . \tag{13}$$

2.1.3. Container singly scattered neutrons Similar expressions as (11) and (13) hold for the singly scattered neutrons in the container. The transmission factors must include the traversed paths in different media and their cross sections. The calculated component $\tilde{z}_C(\mathbf{k}_0, \theta)$ is related with an expression similar as (12) to obtain the corresponding component in Eq. (1).

2.1.4. Multiply scattered neutrons The estimator for the detected multiply scattered neutrons, scored in the Monte Carlo procedure is evaluated by the recurrence relation

$$\chi_n'(\mathbf{r}, \mathbf{k}_{n-1}, \mathbf{k}) = w_n \frac{\Sigma_{\mathrm{scatt}}(\mathbf{k}_{n-1})}{\Sigma_{\mathrm{tot}}(\mathbf{k}_{n-1})} P(\mathbf{k}_{n-1}, \mathbf{k})\, \mathrm{e}^{-\Sigma_{\mathrm{tot}}(\mathbf{k}) L(\mathbf{r}, \hat{\mathbf{k}})} \varepsilon(k) \quad , \tag{14}$$

where the weight at the n-th step is determined from the weight at step $n-1$ as

$$w_n = w_{n-1} \left\{1 - \exp[-\Sigma_{\mathrm{tot}}(\mathbf{k}_{n-1}) L(\mathbf{r}_{n-1}, -\hat{\mathbf{k}}_{n-1})]\right\} \frac{\Sigma_{\mathrm{scatt}}(\mathbf{k}_{n-1})}{\Sigma_{\mathrm{tot}}(\mathbf{k}_{n-1})} \quad . \tag{15}$$

Similarly as in Eq. (4) the path length from the $(n-1)$-th to the n-th step is sampled from a biased probability so the neutron never leaves the sample, changing $d(\mathbf{r}_0, -\hat{\mathbf{k}}_0)$ by $L(\mathbf{r}_{n-1}, -\hat{\mathbf{k}}_n)$, *i.e.* the distance from position $\mathbf{r}_{n-1}$ to the sample surface in the direction of $\mathbf{k}_n$.

The contribution of the n-th scattered neutrons is obtained after sampling the estimator (14) with the modified density of n-collisions. The resulting expression similarly as in Eq. (8), is

$$\tilde{z}_n(\mathbf{k}_0, \mathbf{k}) = \int \mathrm{d}^3r\, \widetilde{\Psi}_n(\mathbf{r}, \mathbf{k}_{n-1}) \chi'_n(\mathbf{r}, \mathbf{k}_{n-1}, \mathbf{k}) \quad , \tag{16}$$

where $\widetilde{\Psi}_n$ results from the recursive relation

$$\widetilde{\Psi}_n(\mathbf{r}, \mathbf{k}_{n-1}) = \int \mathrm{d}^3k' P(\mathbf{k}', \mathbf{k}_{n-1}) \int_0^{L(\mathbf{r}, \hat{\mathbf{k}}_{n-1})} \mathrm{d}s\, \exp[-\Sigma_{\mathrm{tot}}(\mathbf{k}_{n-1})s]\, \widetilde{\Psi}_{n-1}(\mathbf{r} - s\hat{\mathbf{k}}_{n-1}, \mathbf{k}') \quad . \tag{17}$$

As in Eq. (12) the multiple scattering component is related with Eq. (1) through

$$\tilde{z}_n(\mathbf{k}_0, \mathbf{k}) = \frac{N\sigma_{\mathrm{scatt}}}{4\pi A(\hat{\mathbf{k}}_0)} \frac{k}{k_0} \varepsilon(k) s_n(\mathbf{k}_0, \mathbf{k}) \tag{18}$$

and its corresponding magnitude for diffraction is

$$\tilde{z}_n(\mathbf{k}_0, \theta) = \frac{N\sigma_{\mathrm{scatt}}}{4\pi A(\hat{\mathbf{k}}_0)} \int \mathrm{d}E \frac{k}{k_0} \varepsilon(E) s_n(E_0, E, \theta) \quad . \tag{19}$$

The total contribution of multiple scattering is

$$\tilde{z}_{\mathrm{M}}(\mathbf{k}_0, \theta) = \sum_{n=1}^{\infty} \tilde{z}_n(\mathbf{k}_0, \theta) \quad . \tag{20}$$

In practice the sum is truncated by the Russian roulette method, when the neutron weight drops below significant values [7].

2.2. Correction Method

The Monte Carlo code calculates the measured macroscopic cross section as well as its components as shown in the former sections. The total magnitude calculated by Monte Carlo is

$$\frac{1}{A(\hat{\mathbf{k}}_0)} \left(\frac{\mathrm{d}\Sigma}{\mathrm{d}\Omega}\right)_{\mathrm{scatt}} = \tilde{z}_1(\mathbf{k}_0, \theta) + \tilde{z}_{\mathrm{C}}(\mathbf{k}_0, \theta) + \tilde{z}_{\mathrm{M}}(\mathbf{k}_0, \theta) \quad , \tag{21}$$

which is a description of the measured diffractogram and provides the means to separate $\tilde{z}_1$ from the rest of the contributions that must be evaluated. The sought magnitude z_1 (Eq. (9)), is also calculated in the Monte Carlo procedure.

The proposed procedure to process the experimental data is performed through correction factors following an iterative procedure. We define the multiple scattering factor at the iteration i as

$$f_{\mathrm{MS}}^{(i)}(\mathbf{k}_0, \theta) = \frac{\tilde{z}_1^{(i)}(\mathbf{k}_0, \theta)}{\tilde{z}_1^{(i)}(\mathbf{k}_0, \theta) + \tilde{z}_{\mathrm{C}}^{(i)}(\mathbf{k}_0, \theta) + \tilde{z}_{\mathrm{M}}^{(i)}(\mathbf{k}_0, \theta)} \quad , \tag{22}$$

and the attenuation factor

$$\mathcal{A}^{(i)}(\mathbf{k}_0, \theta) = \frac{\tilde{z}_1^{(i)}(\mathbf{k}_0, \theta)}{z_1^{(i)}(\mathbf{k}_0, \theta)} \quad . \tag{23}$$

In the first step the uncorrected experimental data function $\mathcal{E}^{(0)}(\mathbf{k}_0, \theta)$ is employed to calculate the estimators given in Eqs. (6), (10) and (14) in the Monte Carlo code, and $\tilde{z}_1$, $\tilde{z}_C$, and $\tilde{z}_M$ are obtained as outputs. After the convergence of the Monte Carlo runs, the factors (22) and (23) are defined and applied to $\mathcal{E}^{(0)}(\mathbf{k}_0, \theta)$, thus defining $\mathcal{E}^{(1)}(\mathbf{k}_0, \theta) = \mathcal{E}^{(0)}(\mathbf{k}_0, \theta) f_{\mathrm{MS}}^{(1)}/\mathcal{A}^{(1)}$. The corrected data are employed as input of the next Monte Carlo run, thus defining an iterative process. In general the corrected spectrum at run i-th is defined as

$$\mathcal{E}^{(i+1)}(\mathbf{k}_0, \theta) = \frac{f_{\mathrm{MS}}^{(i)}(\mathbf{k}_0, \theta)\, \mathcal{E}^{(0)}(\mathbf{k}_0, \theta)}{\mathcal{A}^{(i)}(\mathbf{k}_0, \theta)} \quad . \tag{24}$$

The iterative process converges typically after 8 runs. We thus obtain the experimental angular distribution of singly scattered neutrons, $\mathcal{E}_1(\mathbf{k}_0, \theta)$. Its expression is proportional to $z_1(\mathbf{k}_0, \theta)$ (Eq. (9)), with a proportionality constant that links the experimental data with the theory

$$\mathcal{E}_1(\mathbf{k}_0, \theta) = K z_1(\mathbf{k}_0, \theta) \quad . \tag{25}$$

Integrating over all directions of solid angle in Eq. (25) (or equivalently form 0 to $2k_0$ in Q), from Eq. (9) we get the value of K

$$K = \frac{\Sigma_{\mathrm{tot}}(\mathbf{k}_0)}{\Sigma_{\mathrm{scatt}}(\mathbf{k}_0)} \frac{1}{1 - t(E_0)} \frac{2\pi}{k_0^2} \int_0^{2k_0} \mathrm{d}Q\, \mathcal{E}_1(\mathbf{k}_0, Q) Q \quad . \tag{26}$$

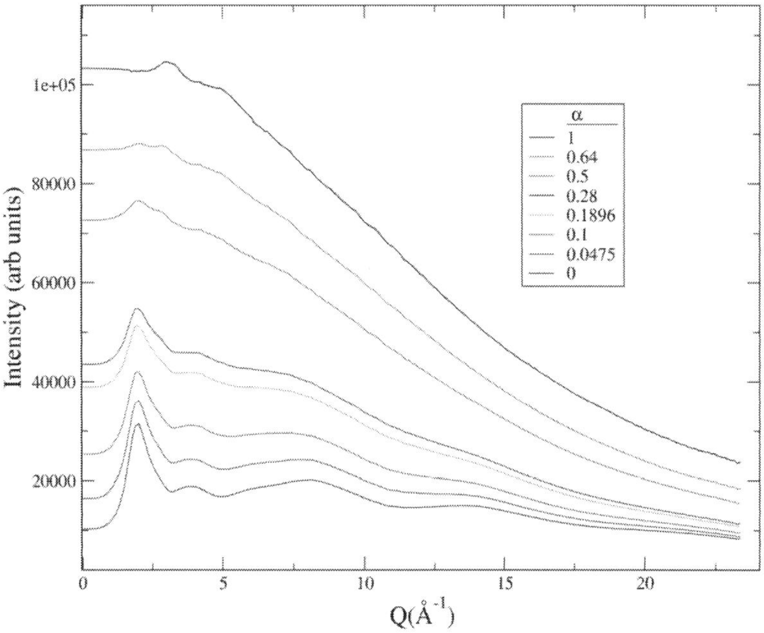

Figure 1. Experimental results of diffraction on the eight samples, classified according their light water concentration (α).

3. Experimental procedure

The samples employed in the present work were the water mixtures. Eight different isotopic mixtures ($H_{2\alpha}O_\alpha$ / $D_{2(1-\alpha)}O_{(1-\alpha)}$, α being the concentration of light water) were used in this

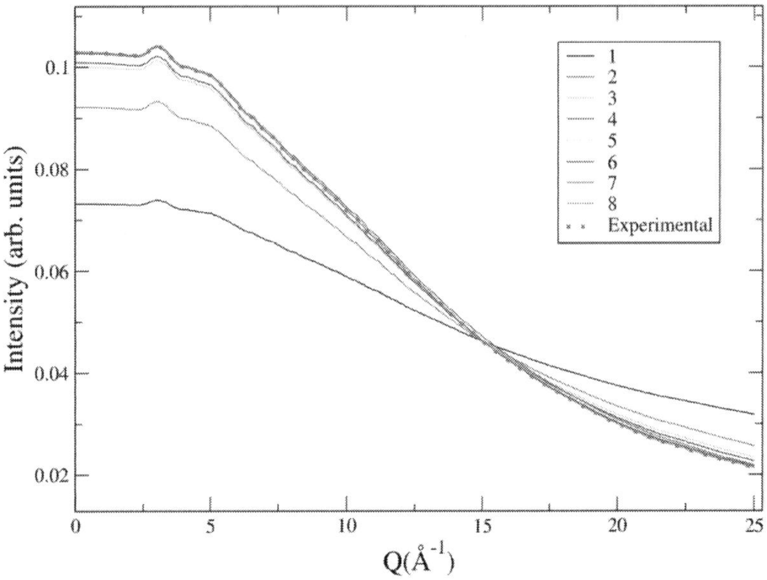

Figure 2. Convergence of the iterative method as a function of the iteration number, for the case of light water. The dotted line represents the experimental data. Starting from iteration 8 no significant changes are observed.

experiment and were prepared by careful weighing inside a moisture-controlled glove box. The mixtures were prepared with $\alpha = 1$, 0.64, 0.5, 0.28, 0.19, 0.1, 0.05 and 0. The sample holder was a thin-walled cylindric vanadium can, 6 mm inner diameter and 60 mm height, situated at the centre of an evacuated bell jar.

The diffraction experiments were carried out at room temperature, (25 $\pm$ 2)°C, at the diffractometer D4C [2]. The incident neutron beam wavelength was $\lambda_0 = 0.5$ Å (energy 0.324 eV). The beam size was chosen to bath the central part of the sample. Its width was chosen to cover exactly the width of the sample. The scattered intensity was recorded moving the multidetectors over the available angular range, which effectively covered a Q-range from 0.05 to 24 Å^{-1}. Similar measurements were made for the empty cell, background and a standard vanadium rod of 6 mm diameter. The detector efficiency profile was determined from measurements using a vanadium sample.

4. Data treatment

In Fig. 1 we show the experimental results for all the samples as a function of the elastic Q_e value defined as $Q_e = 4\pi/\lambda_0 \sin(\theta/2)$. In the results, the background (measured by removing the sample from its position), was subtracted from the experimental data.

The experimental results serve as starting point for the above described iterative correction method. In Fig. 2 we show the behavior of the iterative process described in Sect. 2.2, in the case of the light water sample ($\alpha = 1$). The criterion of convergence is that the total scattering as calculated by Monte Carlo, should reproduce the experimentally observed spectra. A good convergence to the experimental data is achieved starting from the eighth iteration, and no significant changes are observed in the subsequent iterations.

Figure 3 displays the contribution of single, multiple and container scattering (barely visible) to the total observed spectra as calculated from the last Monte Carlo iteration, for each sample. We also show the calculated distributions of neutrons after the first scattering (Eq. (8)).

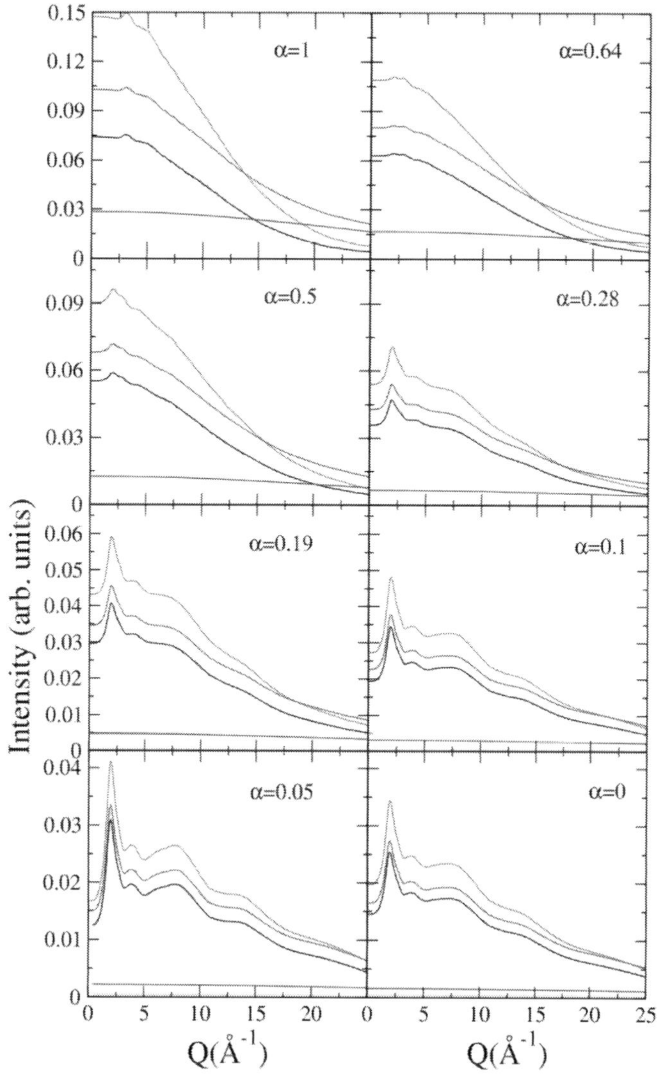

Figure 3. Detail for the contribution of the detected single scattering (black) $\tilde{z}_1$, multiple scattering (blue) $\tilde{z}_M$, scattering from container (green) $\tilde{z}_C$ and the total (violet) $\tilde{z}_{tot}$, from the Monte Carlo simulations for each sample. In red, the sought distribution of neutrons after the first scattering z_1 (Eq. (9)).

The results shown in Fig. 3 provide the necessary functions to apply the correction procedure described in Sect. 2.2, obtaining the functions $\mathcal{E}_1(\mathbf{k}_0, \theta)$ for each sample.

To obtain the functions $d\sigma/d\Omega$, we firstly need the total cross sections $\sigma(E_0)$, at the incident neutron energy, which were calculated by means of Granada's Synthetic Model [6]. The cross sections are shown in Fig. 4. We observe that the values of $\sigma(E_0)$ differ from the free atom limit, due to the chemical bonding effects. Thus, the values can not be obtained directly from tables. Instead, either a model or a measurement are necessary to determine them.

It is also necessary to know the ratio $\Sigma_{tot}(\mathbf{k}_0)/\Sigma_{scatt}(\mathbf{k}_0)$, that can be calculated from Fig. 4 and the absorption cross sections, from tables [8]. With all the collected information now we can calculate the values of K from Eq. (26). To calculate the integral of $\mathcal{E}_1(\mathbf{k}_0, \theta)$ we have to

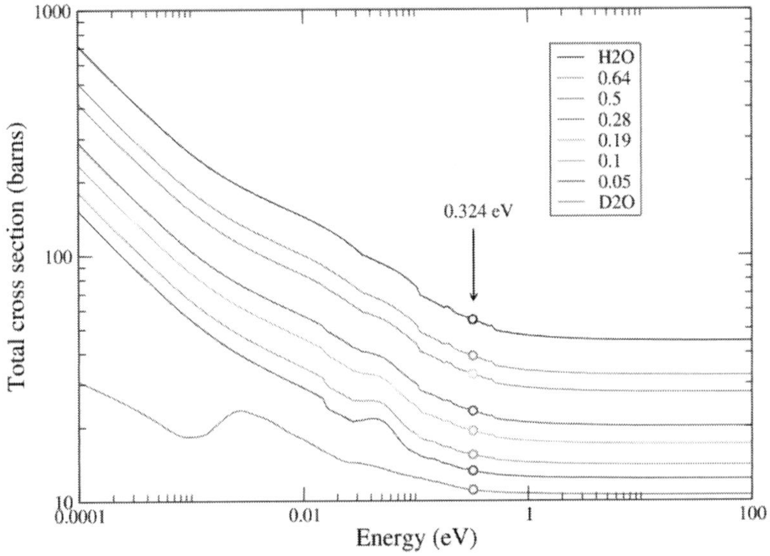

Figure 4. Total cross sections as a function of neutron energy for the different water mixtures (values of α indicated). The total cross sections at the incident energy (0.324 eV) are shown.

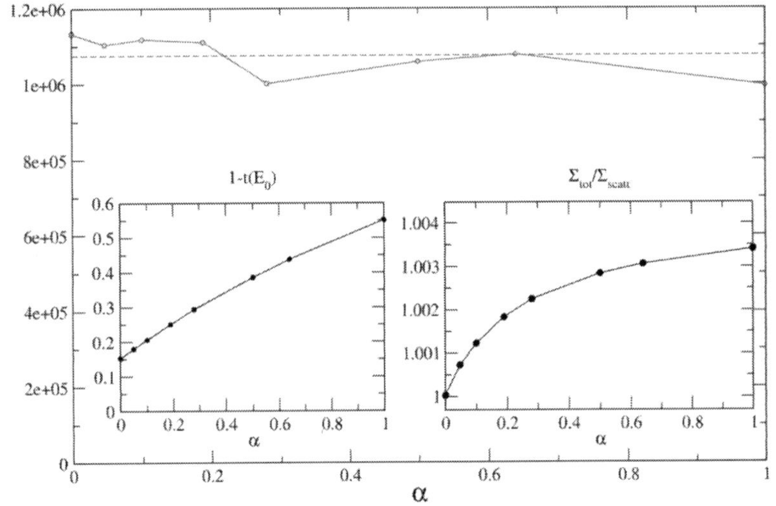

Figure 5. The normalization constant K defined in Eq. (26) for the different water mixtures. Left inset: $1 - t(E_0)$ (scattering power of the samples as a function of α). Right inset: the ratio $\Sigma_{\rm tot}(\mathbf{k}_0)/\Sigma_{\rm scatt}(\mathbf{k}_0)$ at the incident neutron energy.

extrapolate the experimental data to $\theta = 180°$ (the experimental data not going beyond 140°), and the transmission coefficient can be calculated as shown in Ref. [1].

Figure 5 shows the resulting K factors for all samples, with a mean value of 1.074×10^6 and a spread of about 7%. A difference between the samples of lower and higher hydrogen contents is also observed. With the values of K obtained in this process we can calculate $z_1(\mathbf{k}_0, \theta)$ from Eq. (9), and from them $d\sigma/d\Omega$ employing $\sigma_{\rm scatt}(E_0)$ shown in Fig. 4. In Fig. 6 we show the final results of $d\sigma/d\Omega$ as a function of Q.

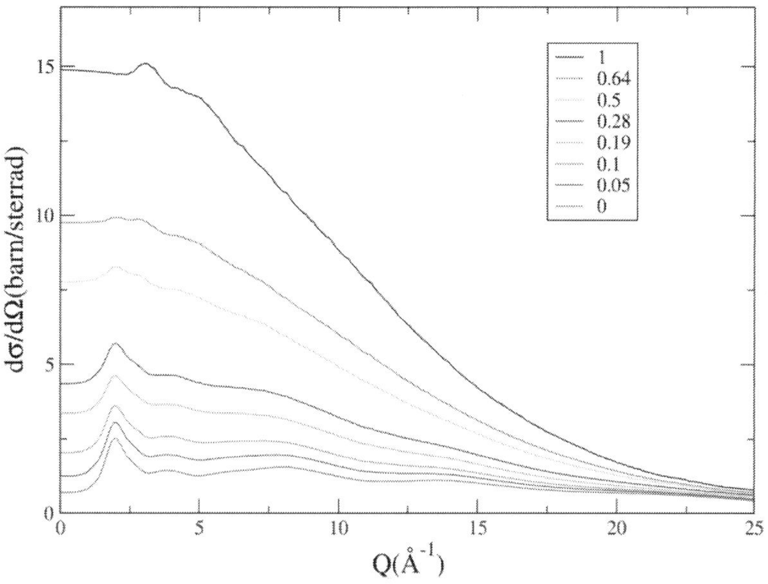

Figure 6. The differential cross sections of the water mixtures analyzed in this paper.

5. Summary and conclusions

In this paper we showed the procedure to obtain the link between the arbitrary experimental scale in neutron diffraction experiments (number of recorded counts per monitor counts) with the corresponding cross sections. The process involves multiple scattering, attenuation and detector efficiency corrections, that are calculated by a Monte Carlo method. We have developed the mathematical expressions of the magnitudes calculated by Monte Carlo and their experimental counterparts.

It is worth to comment about the coefficient K, the constant linking the scattering power of the samples with the number of registered counts. This coefficient should be an instrumental constant independent of the sample. However, we observe a decreasing trend with the higher Hydrogen-containing samples, *i. e.* the samples with the higher scattering power. A similar effect was observed in Ref. [1]. The causes there analyzed were a possible improper description of the detector bank efficiency, and the presence of a sample-dependent background, that increases with the higher Hydrogen-containing samples. Regarding the detector efficiency, it was exhaustively checked during the process of calibration of D4C [2]. However, the sample-dependent background is a subject that remains to be treated, both from the numerical and the experimental points of view, and is not only a problem in diffraction experiments but a general issue in neutron scattering experiments.

Acknowledgements

We are indebted to John Dore and Imre Bakó for providing us the experimental data and valuable suggestions about the procedures developed in this paper. We acknowledge ANPCyT (Projects No. PICT Raíces 2006-574), and Universidad de Cuyo (Project No. 06/C288) for financial support.

References

[1] Rodríguez Palomino L A, Dawidowski J, Blostein J J and Cuello G J 2007 *Nuclear Instruments and Methods in Physics Research Section B: Beam Interactions with Materials and Atoms* **258** 453 – 470 ISSN 0168-583X

[2] Fischer H E, Cuello G J, Palleau P, Feltin D, Barnes A C, Badyal Y S and Simonson J M 2002 *Appl. Phys. A* **74**(Suppl.) S160–S162
[3] Sears V 1975 *Advances in Physics* **24** 1–45
[4] Copley J R D 1974 *Computer Physics Communications* **7** 289–317
[5] Copley J R D, Verkerk P, van Well A A and Fredrikze H 1986 *Computer Physics Communications* **40** 337–357 ISSN 0010-4655
[6] Granada J R 1985 *Phys. Rev. B* **31**(7) 4167–4177
[7] Bischoff F G, Yeater M L and Moore W E 1972 *Nuclear Science and Engineering* **48** 266–280
[8] Mughabghab S, Divadeenam M and Holden N 1984 *Neutron Cross Sections: Neutron Resonance Parameters and Thermal Cross Sections* Neutron Cross Sections Vol. 1 (Academic Press) ISBN 9780125097116

New Data Collection Subsystem for Powder Neutron Diffractometer with Position Sensitive Detectors

Martin Drab, Ladislav Kalvoda, Stanislav Vratislav, Maja Dlouha

Czech Technical University in Prague, Faculty of Nuclear Sciences and Physical Engineering, Dept. of Solid State Engineering, Trojanova 13, 120 00 Prague 2, Czech Republic

E-mail: martin.drab@fjfi.cvut.cz

Abstract. Project INDECS (Integrated Neutron Diffraction Experiment Control System) is a newly developed software system created for the purpose of data acquisition from, and controlling of the upgraded version of the KSN-2 powder neutron diffractometer equipped with Position Sensitive Detectors. The KSN-2 neutron diffractometer (belonging to LND, DSSE, FNSPE, CTU in Prague, CZ, placed at the NRI, Řež near Prague, CZ) was recently upgraded with the possibility to contain up to three parallel linear PSDs which can be mounted on the detector arm instead of the original single counter detector. That was the motivation for the project INDECS to be started. For the actual data acquisition and initial data analysis of the raw sampled signals within the project INDECS a special modular structure called the PSD Acquisition Path (or PSDAP) was designed. Raw sampled signals from either end of the PSD are taken as the main input to the subsystem. The signals are then split into multiple events, analyzed for the position of the obtained events on the PSD, and a histogram of the diffraction events is created as the main output of this subsystem. Moreover, the PSDAP is capable of storing the raw signals either before event splitting or after, so that later the whole process of acquisition can be replayed in software, perhaps with different settings of the processing parameters or perhaps even algorithms. It can also be switched into a mode where under special conditions a correction curve specific for the given PSD can be constructed and later in normal operation it can be used to do the corrections of detected event positions to enhance the results on that particular PSD. This whole subsystem then acts as a single modular command unit (External Execution Module or EEM) within the entire system of project INDECS.

1. Introduction

The KSN-2 neutron diffractometer belongs to the Laboratory of Neutron Diffraction at the Dept. of Solid State Engineering of the Faculty of Nuclear Sciences and Physical Engineering of the Czech Technical University in Prague, CZ, and it is located at one of the horizontal channels of the LVR-15 nuclear reactor at the Nuclear Research Institute in Řež near Prague, CZ. The KSN-2 has been upgraded with the possibility to allow it to switch the single counting detector head with a head containing up to three parallelly placed linear position sensitive detectors (PSDs) on the detector arm of the diffractometer (currently, the diffractometer can be changed into any of these two basic configurations: single counter or linear PSD(s)).

Introducing the linear PSDs into the KSN-2 system also required to introduce new supportive analog electronics, computer hardware and of course software as well. For this purpose the development of a new modular controlling and data collection software system began. The project is called INDECS (Integrated Neutron Diffractometer Experiment Control System) [1],

[2] and it is entirely based on stream communication mechanisms provided by the stream caching and multimedia stream routing libraries which were both created partially with project INDECS in mind (see [2]).

One of the main parts of project INDECS with respect to the diffractometer is the subsystem that acquires data from the detectors and does their initial processing. We call this subsystem the PSD Acquisition Path (PSDAP).

The PSDAP subsytem opens the possibility to experiment with the algorithms used to decipher the position of the event from the sampled detector signals. This algorithm can be easily replaced by a completely different one or perhaps with just slightly modified version of that algorithm, or you can just play with its various parameter settings that this algorithm may possibly have, in order to achieve the best results. This is possible due to the modular structure of the PSDAP subsystem itself.

Another feature of the PSDAP is that the raw sampled data (digitized signals coming out of either end of the linear PSDs) can be stored and read and reprocessed (or you may say "replayed") again by the INDECS system later. Of course by doing so we gain no new information about the sample, that is understandable and not really intended. But what we do gain is an option to do a comparison of the processing of the exact same raw sampled data processed by either different algorithms or different processing parameter settings within the INDECS system. That allows us to actually compare the processing algorithms or processing parameter settings with one another without the influence of the semi-random nature of the diffraction pattern caused by the semi-random nature of the neutron flux and the nature. Because even when measuring the same sample twice in a row in exactly the same experiment, you never get exactly the same raw sampled data from the detectors and so comparing these two sets of data processed with different algorithms or their parameter settings does not give us only the comparison of the algorithms or their settings themselves, but is also distorted by the randomness of the neutron events, while when comparing with the replayed data, we only get the information about the changes in processing. Of course, this feature may perhaps be more usable for the development of the data processing algorithms and their optimization, than for the actual routine measurement of the samples, but even there it may become handy from time to time. As for instance replaying of the experiments may also possibly give us the opportunity to process the previously acquired data with new algorithms developed or implemented within the INDECS project later in time.

2. PSD acquisition path

The schematic of the modular structure of the PSDAP can be seen on figure 1. There are several modules dealing with data storage, two data processing modules, one data source and one module which collects the result. The entire subsystem then hovers above the External Execution Module (EEM), which is an interface block into the rest of the INDECS system as described in [2]. All the communication with the PSDAP subsystem is done through the EEM interface via commands that this interface implements from the external point of view of system INDECS. During the initialization of the subsystem, it is literally assembled from the various modules. For some of them there can be more alternatives to choose from, depending on how many are implemented at the time.

Briefly, it works in the following way. At first the raw sampled signals are acquired by the Data Source module. Afterwards they can be sliced into multiple events in the Multi Event Separator module. From there the sampled event signals go to the Peak Analyzer module, where the position of the neutron event along the PSD is decoded. The decoded position then goes to the final Histogrammer module, where it is stored into the building histogram. The events are accompanied by their timestamps along the way, so that it would be possible to do the reconstruction in proportional time later or perhaps some statistics of the neutron distribution

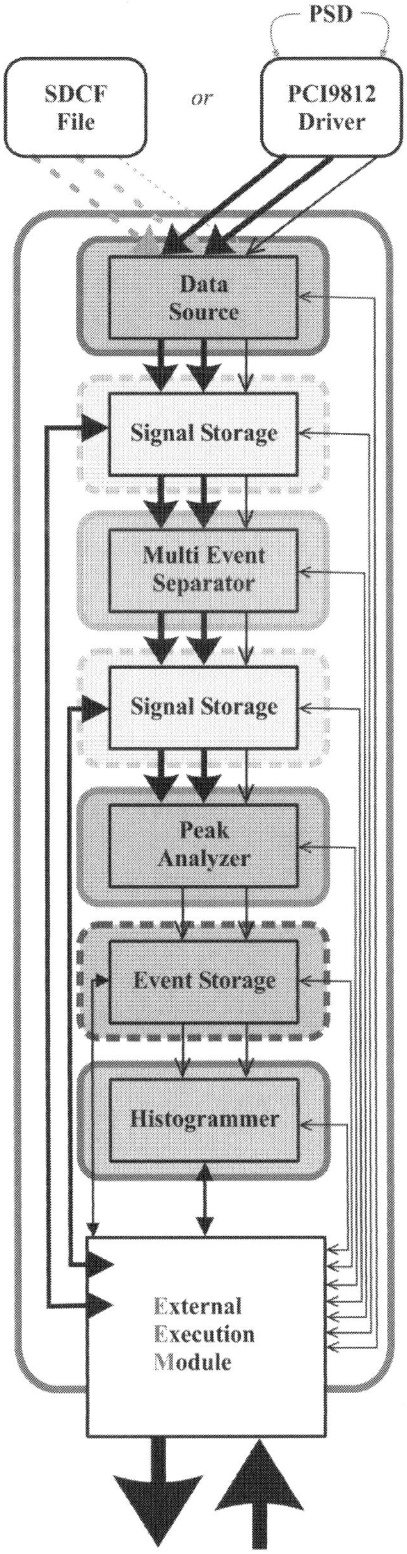

Figure 1. The schematic of the modular structure of the PSDAP with the data flow routes depicted by arrows.

in time can be procured from that if necessary.

In between these functional blocks there are optional storage blocks, which can store the signals at the adequate point of view or perhaps divert it (via the EEM interface) into another similar subsystem somewhere else if necessary.

2.1. Data source module

The data source is the interface between the PSDAP and the actual source of sampled event signals, whether these are acquired from the actual data acquisition (DAQ) card or from a file, where the signals were stored previously. The ultimate goal of this module is to prepare one sampled signal burst from each end of the PSD detector containing at least one neutron event in it, and a timestamp for that burst.

At the present time, there are two modules that fit into this slot. Since the PSD head of the KSN-2 diffractometer uses the ADLink PCI-9812 A/D converter card to obtain these signals, one of the implemented modules acquires the data from that card. The card can sample a fixed number of samples for each channel, based on a preset hardware trigger. For performance reasons a special RTLinux [3], [4] (hard-real-time sublayer of the ordinary Linux system) driver for this card was written, which is running as a real-time process and the module communicates with it via the RT-FIFO mechanism of the RTLinux. The real-time approach has its advantages, but it also has quite some disadvantages. We have no space to discuss either of them in this article, but there is a another module and driver work in progress, which would allow to use the device safely on an ordinary Linux, hopefully eliminating all the negative aspects (or at least the most severe ones) that this approach would normally have.

The second existing Data Source module does the same work, but instead of acquiring the sampled signals from the DAQ card, it rather reads them from a file, where they were stored previously during the actual experiment and, thus, allowing replaying the experiment off- line. The file to read should be of a special SDCF (Streaming Data Container Format) format. This format was specifically designed to carry streaming data, including those produced by the PSDAP.

2.2. Signal storage module

The Signal Storage module has several purposes. By far, the most important one being to store the data of raw sampled PSD signals that pass through it. From the data flow point of view within the PSDAP subsystem, the Signal Storage module is completely transparent. What is passed on to it is also passed on to the next module along the path. The storage can be done either directly to the SDCF file (the Streaming Data Container Format was specially developed for quick, yet relatively robust storage of the sampled signals and metadata from various devices, and was also partially developed with the scope of the project INDECS in mind, see [2]), or the data can be sent into the INDECS system via a virtual instruction using the EEM module interface, and then possibly stored or processed by some other part of the INDECS system.

Aside from that, the Signal Storage module can divert the entire data stream into another PSDAP for further processing, or opposite to that, it can receive such a stream from the Signal Storage module in another PSDAP and let this PSDAP continue in the processing. The communication is done via other special virtual instructions carrying the data from one PSDAP to another (usually on another computer) within the INDECS system.

Reason for this other function of the Signal Storage module is that the data acquisition may require quite a lot of CPU processing power to react quickly enough, so that some slower computers may have difficulties to catch up with the processing of the signals, when some more time-consuming algorithms are used. This allows diverting the signal processing to another computer, which may not be bound by the burden of data acquisition.

The rough data rate estimation that we are talking about here can be done based on the basic specifications of the connected devices used on the KSN-2. Given that for instance one of the used PSDs, Reuter Stokes RS-P4-1224-201 has a given averaged rate of 4200 counts per second, and that the pulse width in 0.1 % of height of the semi-gaussian shape pulses produced by the used EG&G ORTEC 570 shaping amplifier is up to 63.8 μs, and that the ADLink PCI-9812 can sample at 20 MHz per channel with data stride of 16-bits per sample (though the A/D converter is actually 12-bit), that gives us a rough average estimate of about 20 MB/s data rate per one PSD. Of course, this is also very much dependent on the neutron flux, measured sample, and setting of the detector electronics, and the flow is varying in time. The actual amount of data collected, however, also depends on the length of the measurement, thus, giving about 1.2 GB/min or 70 GB/h.

Again, to be clear, the function of the Signal Storage module depends on which one of its implementations is used.

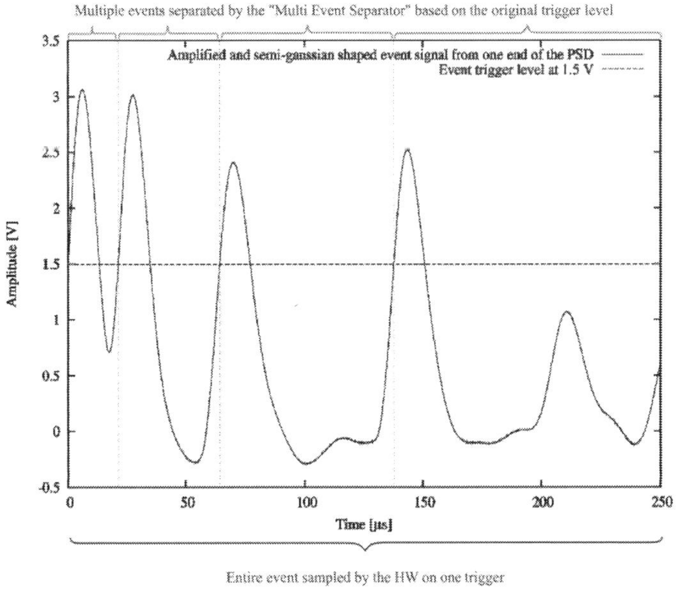

Figure 2. The example of a shaped signal from one end of a PSD, sampled on one hardware trigger event, separated into four shorter events using the same trigger level (1.5 V at this case) that was used to obtain this entire sample.

2.3. Multi event separator module

Sampling signals via a trigger based DAQ card, such as the PCI-9812, has a problem of dead times. While the internal buffer of the card is filled with a sampled event, no other events are accepted until the data have been transferred away to the computer. That may take some time (though perhaps relatively short) during which we lose whatever events the detector throws at us. Neutron events generated by the avalanche effect in a ^{3}He gas filled PSD detector produce very quick and sharp peaks (with a bit longer decaying tail of course). To make them easier to detect we use the shaping amplifier (in our case it is the EG&G ORTEC 570), which shapes and broadens the peak a bit to have about 3 to 64 μs. You can see an example of such a shaped signal on figure 2.

Since the card can sample at least few hundreds of s, we can use it to our advantage and sample as much as possible on one event and subdivide the sample into multiple shorter events (as seen on figure 2) to improve the number of detected neutrons. This is the job of the Multi Event Separator module. So far, we have only one instance of this module implemented that uses a simple trigger level separation.

2.4. Peak analyzer and histogrammer modules
The Peak Analyzer module evaluates the sampled event signals and decodes the position of appropriate neutron events along the PSD, while using the preloaded compensation curve to compensate for nonlinearities of the detector. This operation produces one simple number determining the linear position and that is then passed to the Histogrammer module, which stores it into a histogram of such events and allows retrieving the resulting histogram by other parts of the INDECS system as a result of the diffraction experiment for further evaluation.

Currently there is just one implementation of the Peak Analyzer determining the position from the height of the peaks on either side of the PSD, with the use of given PSD specifications and its current position with respect to the sample and a compensation curve for the specific PSD profile corrections.

There is currently also just one simple implementation of the Histogrammer, which has a number of bins determined by the experiment setup (length and spatial resolution of the PSD and the area to be covered by the PSD arm movement). The events are counted based on the posion of the individual events produced by the peak analyzer module and current detector arm position as obtained by the axis control subsystem of the INDECS project.

The events positions can be stored or inspected by the Event Storage module in between these two modules, but at the present time it is only used for debugging purposes.

3. Conclusions
While most of the essential parts of the PSDAP subsystem mentioned in this text are more or less finished and tested with artificial testing inputs, the current development of the project INDECS for the KSN-2 neutron diffractometer is mainly focused on creating the ordinary Linux port of the PCI-9812 driver based on the Q-Buf Kernel Stream Engine technology (see [2]) to get it as close to the behaviour of the now hard to operate on modern computers real-time (RTLinux) variant of the driver written in the past and used for the initial tests of the PSDs data output, before the core of the project INDECS was implemented. When this driver is finished the PCI-9812 data source module can then be used to actually obtain the data directly from the PSD detectors into the PSDAP subsystem. Currently the modules of the subsystem can only be fed by artificial or presampled data, which is good enough for the initial tests of the PSDAP modules, to determine their functionality and for their debugging purposes, but not enough for testing the actual properties that the PSDAP subsystem of the INDECS system shall have when really connected and fed from the PSDs of the KSN-2 in real time to show some real experimental results. That is why the development of the new PCI-9812 driver and several other supportive parts of the INDECS project are so important now.

Acknowledgments
This work was supported by the Ministry of Education, Youth, and Sports of the Czech Republic grants MSM6840770040 and MSM6840770021.

References
[1] Dráb M 2003 *Projekt: INDECS, The Design of a Superior System for Data Collection and Analysis* Masters thesis Faculty of Nuclear Science and Physical Engineering, Czech Technical University Prague, CZ

[2] Drab M 2011 *Project: INDECS – Integrated Neutron Diffractometer Experiment Control System* Ph.D. thesis Faculty of Nuclear Sciences and Physical Engineering, Czech Technical University in Prague

[3] Yodaiken V 1999 *The RTLinux Manifesto* (Raleigh, NC) URL `http://www.yodaiken.com/papers/rtlmanifesto.pdf`

[4] Yodaiken V 2009 The RTLinux Manifesto Tech. rep. annotated version URL `http://www.yodaiken.com/papers/rtlmanifesto_annotated.pdf`

A compact SEOP ^{3}He neutron spin filter with AFP NMR

Takashi Ino[1], Yasushi Arimoto[1], Hirohiko M. Shimizu[1], Yoshifumi Sakaguchi[2], Kenji Sakai[3], Hiroshi Kira[3], Takenao Shinohara[3], Takayuki Oku[3], Jun-ichi Suzuki[3], Kazuhisa Kakurai[3], Lieh-Jeng Chang[4]

[1]KEK, Tsukuba, Ibaraki 305-0801, Japan

[2]Research Center for Neutron Science and Technology, Comprehensive Research Organization for Science and Society (CROSS), Tokai, Ibaraki 319-1106, Japan

[3]JAEA, Tokai, Ibaraki 319-1195, Japan

[4]Department of Physics, National Cheng Kung University, Tainan 70101, Taiwan

takashi.ino@kek.jp

Abstract. We developed AFP NMR in an aluminum container for polarized noble gas nuclei. The radio frequency magnetic field inside the aluminum container was designed from computer simulations. The polarization loss by the AFP spin flip of ^{3}He was measured to be as low as 3.8×10^{-4}. With this technique, a compact *in-situ* polarizing ^{3}He neutron spin filter with AFP NMR is demonstrated.

1. ^{3}He neutron spin filter and AFP NMR

The ^{3}He neutron spin filter (NSF) attracts wide attention in neutron scattering because of its broad energy range and wide solid angle coverage. Some of such ^{3}He neutron spin filters are equipped with adiabatic fast-passage (AFP) NMR [1] that allows one to measure the ^{3}He nuclear polarization as well as to flip the spin of ^{3}He nuclei and, in consequence, the neutron spin direction is reversed [2-6]. An AFP NMR system consists of static and radio-frequency (RF) magnetic fields. Both fields are perpendicularly aligned, and by slowly sweeping either the strength of the static field or the frequency of the RF field through the Larmor resonance, the ^{3}He spin direction can be reversed. During the sweep, the ^{3}He nuclear spin adiabatically follows the vector sum of the static and the RF magnetic fields in the rotating frame at the Larmor frequency, and the NMR signal from the processing nuclei can be detected by a pick-up coil. It is required to keep the AFP condition in the following equations [1] to minimize the polarization losses during the spin flip process:

$$\frac{1}{T_2}\frac{B_1}{2} << \left|\frac{dB_0}{dt}\right| << \gamma\left(\frac{B_1}{2}\right)^2 \quad \text{or} \quad \frac{1}{T_2}\frac{B_1}{2} << \frac{1}{\gamma}\left|\frac{d\omega}{dt}\right| << \gamma\left(\frac{B_1}{2}\right)^2,$$

where γ is the gyromagnetic ratio of ^{3}He nuclei, T_2 is the transverse relaxation time, and B_0 is the flux density of the static magnetic field. The amplitude and the angular frequency of the RF magnetic field are B_1 and ω, respectively. Large inhomogeneity of B_1 does not fulfill the AFP condition and will increase depolarization in the AFP spin flip for a given sweeping rate.

2. Designing a compact ^{3}He spin filter with AFP NMR

Many compact ^{3}He neutron spin filters with solenoid magnetic cavities have already been developed successfully [2, 5, 6]. A solenoid magnetostatic cavity may reduce radiation shielding void space compared with those using permanent magnets [7], and it is essential to minimize such void spaces especially at a spallation neutron source since high energy neutrons and gammas will increase the background level at instruments without proper shielding.

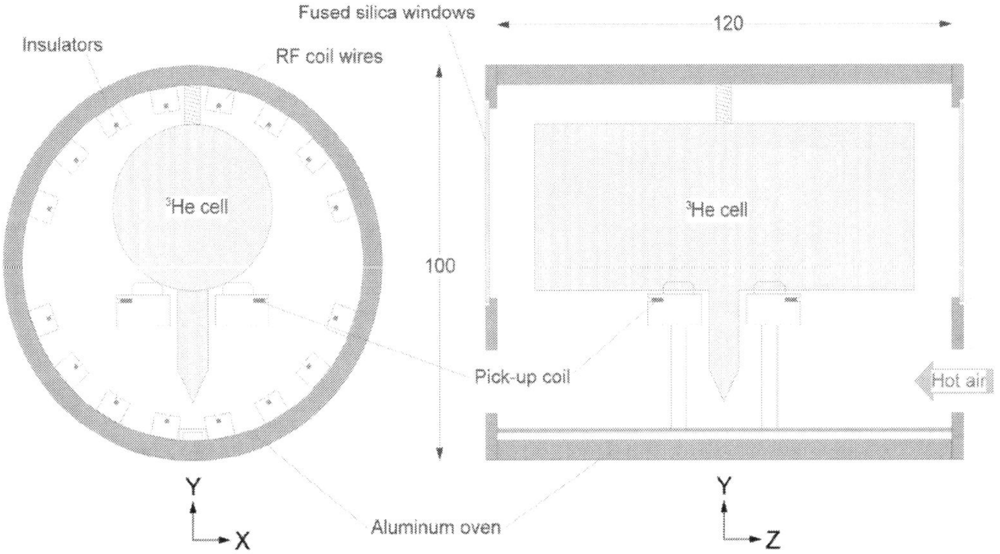

Figure 1. Cross sectional views of the oven, the RF coil, the pick-up coil, and the ^{3}He cell. The lengths are in mm.

An additional demand is the *in-situ* polarization of ^{3}He nuclei at neutron instruments or on neutron beamlines in the caves. It will provide a neutron beam with a steady polarization, otherwise the beam polarization decreases with time as ^{3}He nuclei depolarize. Previously, we demonstrated such an *in-situ* polarizing compact NSF but without AFP NMR [8]. In the process of polarizing ^{3}He nuclei by the spin-exchange optical pumping (SEOP), the ^{3}He cell has to be kept at an elevated temperature of 150°C to 200°C in an oven, and the choice of materials for the oven always becomes an issue when one designs the *in-situ* polarizing NSF with AFP NMR because the RF magnetic field interferes with conductive materials. In the case of Ref. 5, the authors used Ultem (polyetherimide) for the oven, Teflon in Ref. 6, and calcium silicate in Ref. 4. Unlike these insulating materials, the oven of our previous *in-situ* polarizing NSF was made of aluminum for easy and cheap fabrication. It did not seem to allow AFP NMR at a first sight, but we tried RF electromagnetic field simulations to introduce a uniform RF magnetic field over a ^{3}He cell inside the aluminum oven for AFP NMR. A special RF coil was finally designed to be installed inside the aluminum oven. The cross sectional views of the aluminum oven and the RF coil inside are shown in Fig. 1. A cosine winding is employed to produce a uniform RF magnetic field in the horizontal direction of the cylindrical oven. A circular pick-up coil is aligned perpendicular to the RF magnetic field under the ^{3}He cell. Polyimide insulated copper wires are used for both the RF and pick-up coils for better thermal properties. The oven is placed in a homogeneous static magnetic field produced by a solenoid with a diameter of 200 mm and a length of 300 mm, additional compensation coils, and permalloy magnetic shielding (Figs. 2 and 3). A picture of the modified ^{3}He NSF is shown in Fig. 4.

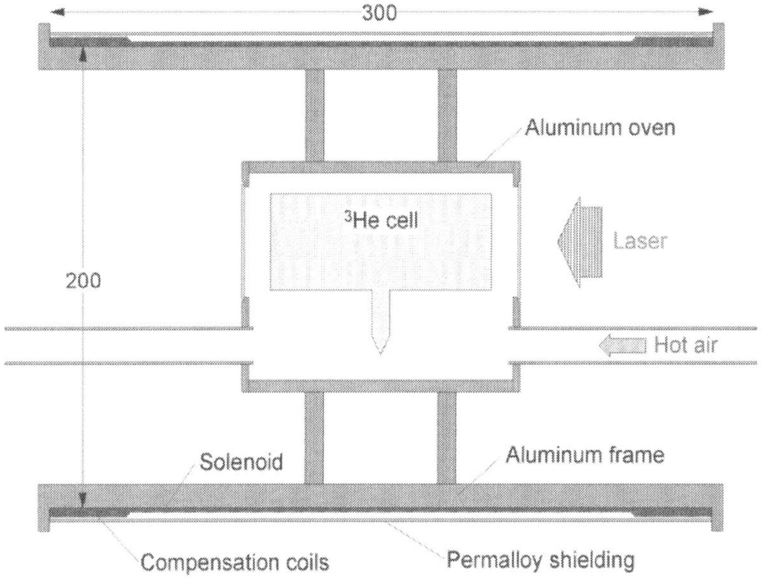

Figure 2. A cross sectional view of the solenoid magnetic cavity and the oven.

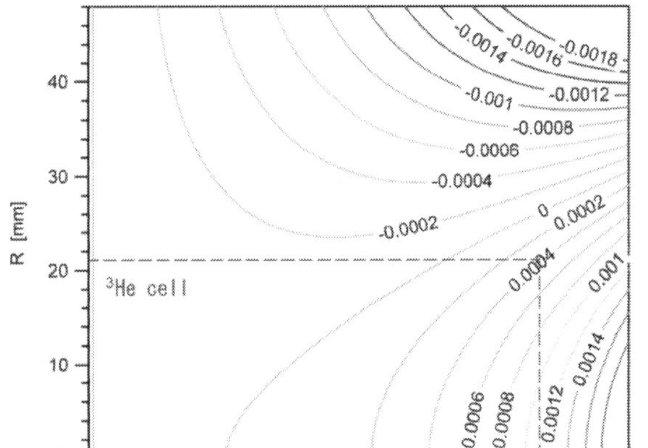

Figure 3. Field gradient ($dB_0 / dR / B_0$ in [/cm]) of the solenoid magnetostatic cavity (calculation). The center of the solenoid is at R = 0 (radial direction) and Z = 0 (solenoid axis).

Figure 4. The modified ^{3}He NSF. The fused silica windows are removed to show the inside of the aluminum oven.

3. Radio-frequency electromagnetic field simulation for the ^{3}He AFP NMR

There are many methods developed for computational calculations of the radio frequency (RF) electromagnetic field. We used the finite-difference time-domain (FDTD) method to simulate the RF field for the ^{3}He AFP NMR. The FDTD method consists in computing the finite-differences of electromagnetic field in a meshed space with time according to the Maxwell's equations [9]. In the FDTD, the simulated space is segmented into small unit cells, and the electric and magnetic field components are calculated in every cell from those in the adjacent cells using the Maxwell's partial differential equations. The time evolution of the electromagnetic field can easily be computed by the leapfrog scheme in this simulation method.

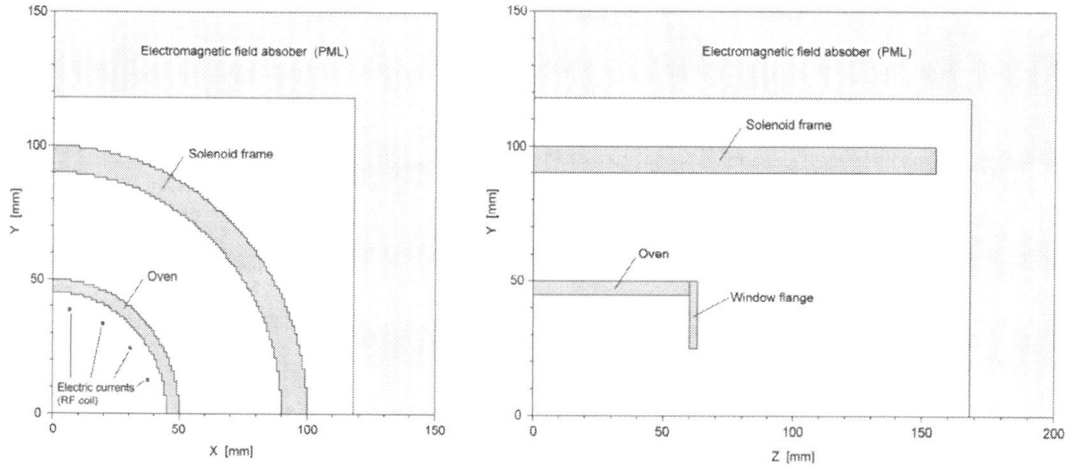

Figure 5. The configuration of the ^{3}He AFP NMR for the FDTD simulation. The cross sectional views on the x-y plane at z =0 (left) and on the y-z plane at x = 0 (right) are shown.

Three dimensional FDTD simulations were performed for the ^{3}He AFP NMR. The volume was segmented into cubic cells of 1 mm side, and approximate representations of the 1/8 parts of the RF coil, the oven, the window flange, and the solenoid frame were configured by considering the symmetry (Fig. 5). Aluminum parts (the oven, the window flange, the solenoid frame) were simulated as perfect conductors, which have no resistivity or magnetization. The outer edges of the simulated space were filled with electromagnetic field absorber (perfectly matched layers or PML) [10]. The FDTD calculations were performed for the following five configurations to see the effects of the solenoid frame, the window flange, and the end of the winding of the RF coil:

1 : The oven and only the linear parts of the RF coil wires were configured; no solenoid frame, no window flange, no end of the winding of the RF coil wires (the RF coil wires were omitted at the edges and outside of the oven).
2 : The oven, the solenoid frame, only the linear parts of the RF coil wires.
3 : The oven, no solenoid frame, the linear parts and end of the winding of the RF coil wires.
4 : The oven, the solenoid frame, the linear parts and end of the winding of the RF coil wires.
5 : The oven, the solenoid frame, the linear parts and end of the winding of the RF coil wires, the window flanges (a 3 mm thick aluminum disc with a circular hole at the center approximated the window flange; the real flange has two holes that are not at the center) .

The configuration #1 is ideal and the simplest; #4 and #5 are rather realistic. The current in the RF coil was powered at a frequency of 100 kHz. The results of the FDTD simulations are presented in Figs. 6 and 7 for the root mean square (RMS) amplitudes of the RF magnetic fields. The two dimensional RF magnetic field maps are shown for the configuration #5, and the RF fields along the x, y, and z axes for all the configurations. The contribution of the end of the winding of the RF coil is rather significant while the reflection effect by the solenoid frame is almost negligible. The static magnetic field produced by the RF coil (the linear parts only) was also calculated by the Biot-Savart equation. It was found that the static field was roughly three times larger than that by the alternating current of 100 kHz because of the reflected RF wave from the oven. The RF magnetic field simulated for the configuration #5 was found almost uniform in the area for the ^{3}He cell, a cylindrical area with a diameter of 40 mm and a length of 100 mm, and the polarization loss by the AFP spin flip is expected to be quite limited.

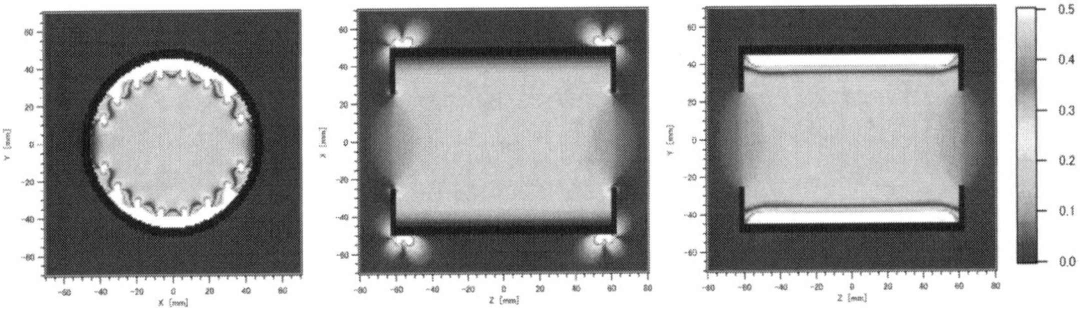

Figure 6. Two dimensional images of the RF magnetic field by the FDTD simulation for the configuration #5. The cross sectional mappings are at z = 0, y = 0, and x = 0, respectively. The cylindrical oven is shown in dark blue. The scale is in Gauss.

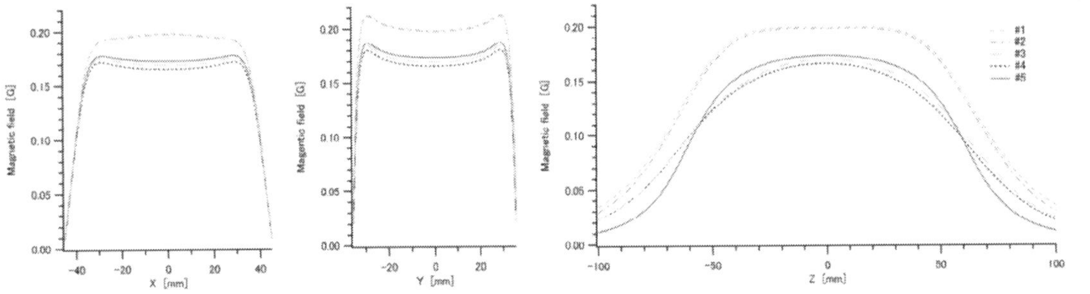

Figure 7. RF magnetic field along the x, y, and z axes by the FDTD simulations.

4. RF magnetic field measurement

The RF field strength inside the oven was measured with a small search coil magnetometer. A small circular coil with a diameter of 5 mm was wound with a fine insulated copper wire. The voltage induced in the coil is proportional to the RF field strength. The positions of the coil were scanned, and the RF field mapping was measured without the window flanges. The RF magnetic field strength along the x, y, and z axes as well as the FDTD calculation for the configuration #4 are represented in Fig. 8. A slight difference is observed in the field strength between the measurement and the simulation especially around the edges of the cylindrical oven ($z = \pm 50$ mm in the right plot). It may be due to the fact that the structural arrangement in the FDTD simulation is limited to the size of the meshed cell while the real oven and the RF coil are more complex arrangements. Nevertheless, the simulations show excellent agreements with the measurements.

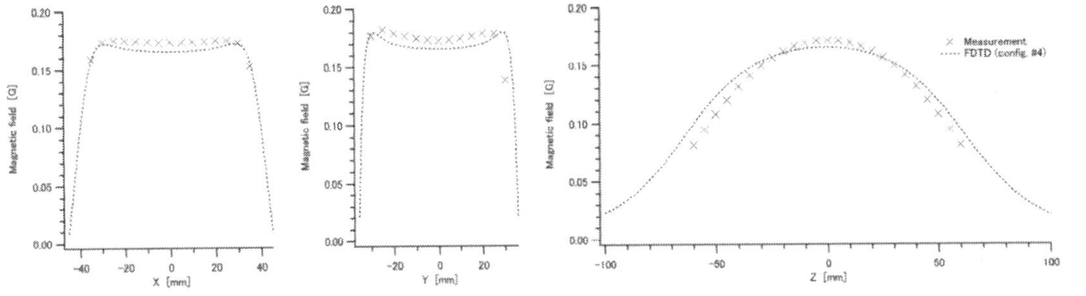

Figure 8. RF magnetic field strength along the x, y, and z axes deduced from the measurements and the FDTD simulations for the configuration #4 (without the window flanges).

Let us compare the RF magnetic field with the AFP condition mentioned in *Section 1*. The RF magnetic field B_1 was measured to be ~0.17 G near the center of the oven. The RF frequency was 100 kHz, and the corresponding static magnetic field B_0 was 30.8 G for the ^{3}He nuclei ($\gamma = -2.038 \times 10^8$ /s). In the AFP spin flip process, B_0 was swept at a speed of 11 G/s. Thus,

$$\frac{1}{T_2}\frac{B_1}{2} \approx 0.85\,\mathrm{G/s}, \quad \left|\frac{dB_0}{dt}\right| \approx 11\,\mathrm{G/s}, \text{ and } \gamma\left(\frac{B_1}{2}\right)^2 \approx 1.5x10^6 \text{ G/s}.$$

Here, the transverse relaxation time T_2 is assumed ~0.1 s [11]. The above numbers satisfy the AFP condition in *Section 1*, and the polarization losses in the AFP spin flip process is expected to be minimal.

5. AFP NMR test

The polarization loss by the AFP spin flip was measured for a cylindrical ^{3}He cell with a diameter of 3 cm and a length of 5 cm. The ^{3}He cell was polarized in a different laser pumping station and moved into the *in-situ* AFP NMR system. In this test, the static magnetic field was swept over the Larmor resonance while the frequency and the amplitude of the RF magnetic field were fixed. One AFP sweep took 0.5 second, and the AFP NMR measurement was repeated 60 times (Fig. 9). The polarization loss in one AFP spin flip was found only to be 3.8×10^{-4}. Note that the RF coil was operated by a common function synthesizer with an output power of ~0.3 W in this measurement; no RF power amplifier was used.

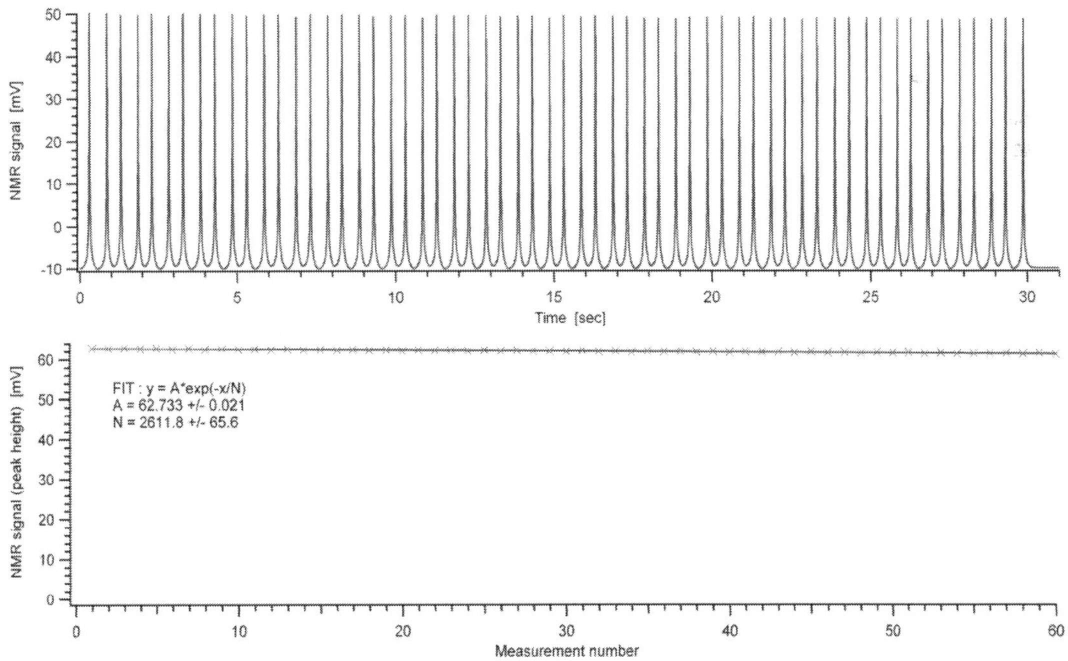

Figure 9. The AFP NMR signals (raw data and the peak heights) are plotted for 60 continuous measurements. The loss of ^{3}He polarization in one AFP spin flip was calculated from the exponential fit to the data. Note that the spin relaxation of ^{3}He (T_1) was 187 hours and enough long compared to the time duration of the measurements.

6. Conclusions

We have developed an *in-situ* compact ^{3}He neutron spin filter with AFP NMR. The oven was made of aluminum for easy and low-cost fabrication. The RF magnetic field inside the aluminum oven was designed with the FDTD simulation. The FDTD calculation of the RF field was in good agreement with the measurement, and one may be able to rely on the FDTD simulation to design this kind of AFP devices. The polarization loss of ^{3}He in one AFP spin flip was measured to be very little, and it will further be reduced with the amplitude modulation AFP technique [12]. A neutron beam test of this ^{3}He NSF was performed at J-PARC just before the disastrous earthquake in Japan, and the results are presented in a separate article [13].

References

[1] Abragam A 1961 *The Principles of Nuclear Magnetism* (Oxford: Oxford University Press)
[2] Ino T and Muto S 2005 A compact solenoid ^{3}He neutron spin filter with a fast spin flip capability *Nucl. Instrum. Methods* A 548 pp 499-506
Ino T and Muto S 2006 A tiny neutron spin filter *Physica* B 385-386 pp 1138-1141
[3] Petoukhov A K *et al* 2006 Recent advances in polarised ^{3}He spin filters at the ILL *Physica* B 385-386 pp 1146-1148
Chupp T E *et al* 2007 A large area polarized ^{3}He neutron spin filter *Nucl. Instrum. Methods* A 574 pp 500-509
Babcock E *et al* 2007 AFP flipper devices: Polarized ^{3}He spin flipper and shorter wavelength neutron flipper *Physica* B 397 pp 172-175
Fu C B *et al* 2011 A wide angle neutron spin filter system using polarized ^{3}He *Physica* B 406 pp 2419-2423
McKetterick T J *et al* 2011 Optimised adiabatic fast passage spin flipping for ^{3}He neutron spin filters *Physica* B 406 pp 2436-2438
[4] Babcock E *et al* 2009 In-situ SEOP polarizer and initial tests on a high flux neutron beam *Physica* B 404 pp 2655-2658
S. Boag *et al* 2009 In situ SEOP polarised ^{3}He neutron spin filter for incident beam polarisation and polarisation analysis on neutron scattering instruments *Physica* B 404 pp 2659–2662
[5] Jones G L *et al* 2005 Continuously operating compact ^{3}He-based neutron spin filter *Physica* B 356 pp 86-90
Jones G L *et al* 2006 Test of a continuously polarized ^{3}He neutron spin filter with NMR-based polarization inversion on a single-crystal diffractometer *Physica* B 385-386 pp 1131-1133
[6] Lee W T *et al* 2010 In-situ Polarized ^{3}He-Based Neutron Polarization Analyzer for SNS Magnetism Reflectometer *J. Phys.:Conf. Ser.* 251 p 012086
[7] Petoukhov A K *et al* 2006 Compact magnetostatic cavity for polarised ^{3}He neutron spin filter cells *Nucl. Instrum. Methods* A 560 pp 480–484
Hutanu V *et al* 2008 Development of compact magnetostatic cavities for ^{3}He spin filter cells *Meas. Sci. Technol.* 19 p 034010
McIver J W *et al* 2009 End-compensated magnetostatic cavity for polarized ^{3}He neutron spin filters *Rev. Sci. Instrum.* 80 p 063905
[8] Ino T *et al* 2009 Development of a compact on-beam SEOP neutron spin filter *Physica* B 404 pp 2667-2669
Kira H *et al* 2011 Development and test of SEOP neutron spin filter in Japan *Physica* B 406 pp 2433-2435
[9] Yee K S 1966 Numerical solution of initial boundary value problems involving Maxwell's equations in isotropic media *IEEE Trans. Ant. and Prop.* 14 pp 302-307 (there are a lot of textbooks available for electromagnetic field simulation by the FDTD method.)
[10] Berenger J P 1994 A perfectly matched layer for the absorption of electromagnetic waves *J. Comp. Phys.* 114 pp 185-200

[11] Cates G D, Schaefer S R, and Happer W 1988 Relaxation of spin due to field inhomogeneities in gaseous samples at low magnetic fields and low pressures *Phys. Rev.* A 37 pp 2877-2885
McGregor Douglas D 1990 Transverse relaxation of spin-polarized ^{3}He gas due to a magnetic field gradient *Phys. Rev.* A 41 pp 2631-2635
[12] Garwood M and DelaBarre L 2001 The Return of the Frequency Sweep: Designing Adiabatic Pulses for Contemporary NMR *J. Magn. Reson.* 153 pp 155-177
[13] Sakai K *et al* Development of polarized Xe gas target for neutron experiment at J-PARC in these proceedings

Experimental and methodic progress in ultra-small-angle polarised neutron scattering on novel magnetic materials

E. Jericha[1], G. Badurek[1], C. Gösselsberger[1], D. Süss[2]

[1]Atominstitut, Vienna University of Technology, Stadionallee 2, A-1020 Wien, Austria
[2]Institute of Solid-State Physics, Vienna University of Technology, Wiedner Hauptstraße 8, A-1040 Wien, Austria

E-mail: jericha@ati.ac.at

Abstract. Ultra-small-angle scattering of polarised neutrons (USANSPOL) allows for the study of magnetic structure in the micrometre range. The technique takes advantage from the narrow angular width of the Bragg reflection curve of perfect crystals and is employed in a double-crystal configuration of channel-cut perfect silicon crystals. Polarisation of the neutron beam is obtained by placing magnetic prisms, which act as birefringent regions, between the monochromator and analyser crystal. Samples are placed between the polariser prisms and the analyser crystal. Scattering of spin-up and spin-down neutrons is recorded in a single measurement and identified by an angular shift of their respective scattering curves. We have developed a prototype sample environment and handling system by which anisotropic samples may be aligned in different orientations and be subjected to external magnetic fields and stresses. Here, we present new experimental results on iron-boron ribbons of varying composition with remarkable magnetostriction properties, highly promising for technological application, and methodic development.

1. Introduction

Ultra-small-angle neutron scattering (USANS) allows to resolve scattering patterns in the µrad range which gives access to structures with sizes in the range of a few tenths of a micrometre up to a few tens of micrometres, correspondingly. The potential of ultra-small-angle scattering of polarised neutrons (USANSPOL) for the study of magnetic microstructure of matter was then recognised a few years ago [1, 2]. First tests were performed with double perfect-crystal diffractometers in Bonse-Hart configuration where the neutrons were polarised by inserting magnetic prisms between the monochromator and the analyser crystals [2, 3]. In recent years we have extended the scope of these exemplary measurements to the study of novel amorphous magnetic ribbons with high potential of technological application [4, 5]. USANSPOL is particularly attractive for these samples since their domain sizes are found to be in the appropriate range for this technique. Moreover USANSPOL has also allowed for the study of their structural evolution when subjected to external magnetic fields or mechanically induced stress. At the same time we have realised that a proper sample environment is essential for thorough application of the method, specifically with anisotropic structure. Requirements in this context comprise a continuous variation of sample orientation, external magnetic field axis and strength of the induced mechanical stress. In this work we test the opportunities and limitations of such a prototype sample environment under given instrumental conditions and put in the underlying methodical context.

2. The instrument

The USANS technique relies on the narrow angular reflection width for neutrons reflected by perfect crystals. Therefore, the main components of a corresponding instrument are channel-cut perfect silicon crystals acting as monochromator and analyser with the sample placed in between. For this purpose two instruments are regularly at our disposal, the CRG-B instrument S18 at the Institut Laue Langevin, Grenoble [6] and the KWS instrument at the Atominstitut TRIGA reactor of the Vienna University of Technology [7]. For the USANSPOL setup two NdFeB-based prisms are introduced between the monochromator and the analyser crystal which produce a strong static magnetic field of about 0.9 T and act as birefringent media for neutrons of different spin state which pass the magnetic prisms through an air gap [1, 8]. Spin-up and spin-down neutrons are deflected by these prisms in opposite directions about an angle which is larger than the full width of the total reflecting range of the perfect crystal reflectivity curve. This criterion is necessary for application in the USANSPOL scheme. Then scattering of spin-up and spin-down neutrons may be observed in sufficiently separated angular regions of the measured intensity curve which allows to discern magnetic and nonmagnetic scattering [1, 2]. Implementation of the USANSPOL option on both instruments was presented in [3] and the current setup described in detail in [4].

A USANSPOL pattern is produced by rotating the analyser crystal with respect to the monochromator about their common parallel orientation while the reflected neutron intensity is recorded for each rotation angle. As a result we obtain the so-called rocking curve which exhibits two sharp peaks belonging to the reflection of spin-up and spin-down neutrons that were separated about their deflection angles by the magnetic prisms. The instrument curve is additionally influenced by diffraction effects encountered by the neutrons during their passage through the slit system of the instrument which defines the neutron beam cross section and the prism-shaped magnetic field of the polariser prisms. These effects may be seen as shoulder-like structures in the wings of the rocking curve and are often asymmetrically enhanced on one side of the peaks as seen in Fig. 1 for instrument curves recorded on S18. Thereby, proper alignment of the components is often a tradeoff between higher intensity, optimum shape and the amount of setup time for the USANS option of the combined neutron interferometry/USANS instrument S18. In principle, both aspects are tremendously important for a USANSPOL instrument. Due to the employment of perfect crystal reflections the technique is inherently intensity limited and a high ratio of peak intensity to instrument background is essential for the scattered intensity at larger scattering angles which defines the shape of the scattering structure. This, e.g., is the main reason why USANSPOL studies at the high-resolution KWS instrument in Vienna are limited to special model structures only [3]. On the other hand a narrow shape in the peak region of the instrument curve is crucial to determine the size of large scatterers when the scattered intensity is concentrated at very low scattering angles [4].

Therefore, good knowledge of the instrument curve is a requirement for later analyses of the scattered intensity. The rocking curve is to be measured with statistical significance throughout the complete measurement range to determine the properties of the peak (peak position and peak separation, peak height, and peak shape and peak width), the characteristics of the wings and the background level. Two points apply in this context: the discrimination of the scattering signal from the instrument curve when both curves are close to each other and the fact that the measured intensity is a convolution of the actual scattering with the rocking curve.

We model the instrument curve for the USANSPOL setup by a sum of elementary functions

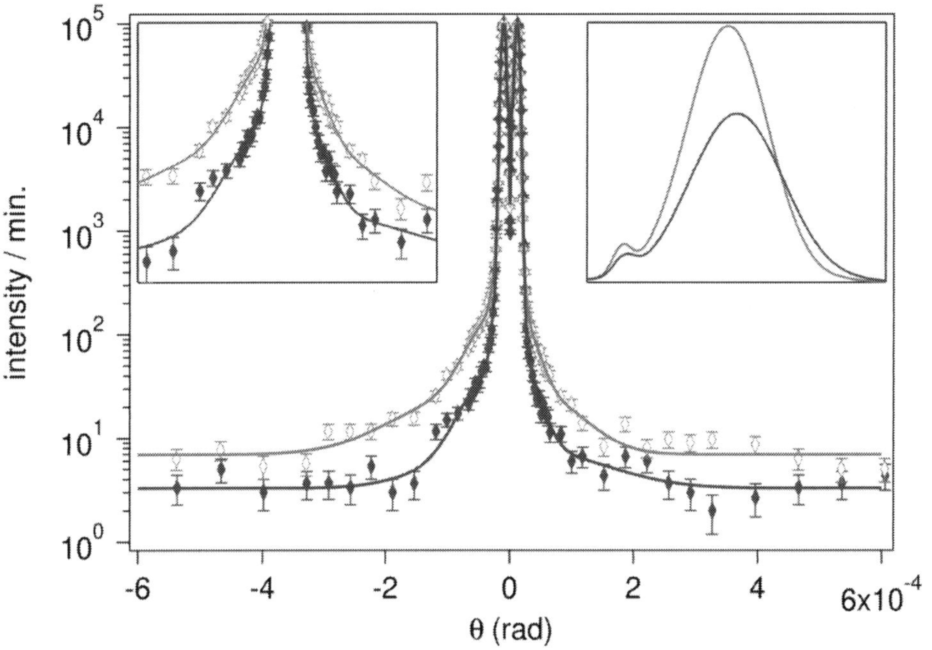

Figure 1. Instrument curves for two different alignment settings of the monochromator crystal and the polariser prisms. The red setting (open symbols) was taken with a Cd square aperture of 1 cm^2 and the blue setting (full symbols) with a circular Cd aperture of 1 cm diameter at the sample position.

which are fitted to the experimental data,

$$R(\theta) = I_0 \sum_{i=1}^{2} \left[A_t \Lambda \left(\frac{\theta - \theta_i}{\tau} \right) + A_0 \mathcal{N}(\theta_i, \sigma_0^2) + \sum_j A_j \mathcal{N}_a(\theta_i, \sigma_{aj}^2, \sigma_{bj}^2) \right] , \tag{1}$$

where Λ is the triangular function centred at θ_i with width τ, $\mathcal{N}(\theta_i, \sigma_0^2)$ a Gaussian centred at θ_i with width σ_0 and $\mathcal{N}_a(\theta_i, \sigma_a^2, \sigma_b^2)$ an asymmetric Gaussian centred at θ_i with width σ_a on the left side and width σ_b on the right side of a peak. The triangle is motivated by the fact that the rocking curve is the convolution of the reflectivities of the monochromator and the analyser crystal which are essentially rectangular in the centre, [9], while the wings of the ideal rocking curve are modelled by the Gaussian and the asymmetric diffraction effects are taken into account by the asymmetric Gaussians. Their number is adapted to the degree of asymmetry and may vary for different instrumental settings. The sum has to be taken over both peaks located at θ_1 and θ_2. We assume the same shape for spin-up and spin-down neutrons since their angular separation is macroscopically spoken extremely small and both spin states should undergo the same effects during their passage through the instrument without sample for all practical purposes. It was shown experimentally that already the unpolarised setup exhibits the main features of the instrument curve which is then shifted by the separation angle and superposed in the polarised setup [5]. The amplitudes A are relative weights for the individual contributions and the peak intensity is denoted by I_0.

Repeated measurement of the instrument curve during a measurement series is likewise important to monitor possible changes in the describing parameters. Applying the wrong

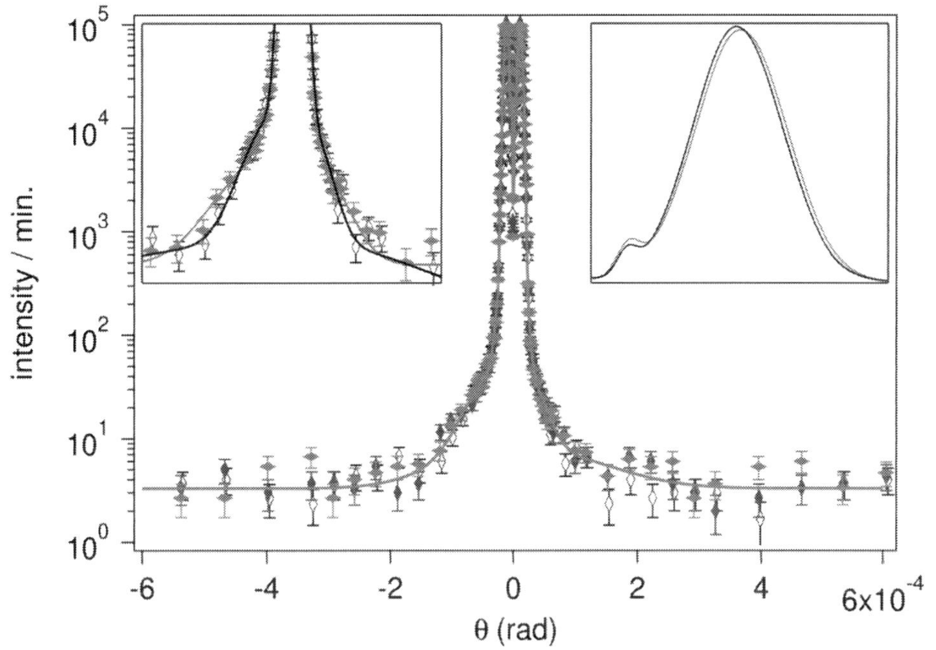

Figure 2. Several rocking curves from the same measurement series (blue setting of Fig. 1). The overall shape of the curves is essentially the same. The inset on the right shows the right peaks of the two different instrument curves. Slight deviations of the peak positions and peak heights are attributed to minor changes in the wavelength distribution during the series. The left inset illustrates possible uncertainties in the wings. Both instrument curves coincide within the error bars, however.

rocking curve to a given scattering measurement could easily lead to misinterpretation and wrong conclusions about the scattering structure. Changes in the overall intensity or the background level might diminish or overemphasise the scattering signal and changes in the shape of the instrument curve might create or remove structural details artificially. Fig. 2 displays several instrument curves taken during a measurement series at S18. Apart from some minor differences which are attributed to slight changes of the incident neutron spectrum, the overall shape of these instrument curves is the same. For data analysis, changes in the peak positions are taken into account at any case by treating them as fit parameters in the course of parameter estimation.

For analysis of the scattered neutron intensity the independent quantity has to be transformed from the rotation angle of the analyser crystal θ to the scattering vector $\mathbf{Q}$. USANS is a one-dimensional technique with high resolution only in the direction perpendicular to the Bragg-reflecting lattice planes of the silicon crystals. Rotating the analyser crystal then spans a plane of high resolution which in current instrument design coincides with the horizontal plane. Therefore, only one component of $\mathbf{Q}$ is accessible for ultra-small-angle scattering which we denote by Q_y. The transformation is simply linear and reads as

$$Q_y = 2\pi\theta/\lambda \tag{2}$$

with the neutron wavelength λ as constant of proportionality. It was proposed in [10] to use artificial one-dimensional line gratings, e.g. fabricated from silicon wafers, as neutron-optical components for wavelength determination. The Q_y values of the maxima produced by lattice

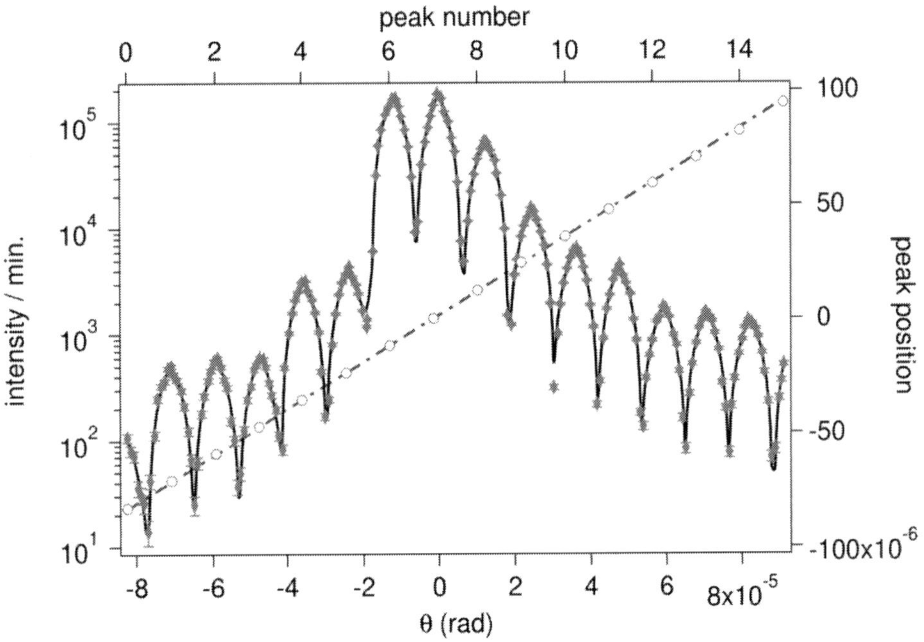

Figure 3. Lattice diffraction pattern of a silicon line grating with 16 μm lattice constant used for wavelength determination; neutron intensity and diffraction angle are shown on the left and bottom axes. The peak positions of the diffraction maxima are located and the angular increment between them determined by a linear fit; diffraction angle of the maxima and the respective peak number are displayed on the right and top axes. The slope of the line is directly proportional to the effective wavelength λ.

diffraction is analogous to light optics and easily calculated to give

$$Q_y = \frac{2\pi n}{a} = Q_y(\theta_n), \quad \lambda = \frac{a\theta_n}{n}, \tag{3}$$

with a the lattice constant of the grating and n the number of the diffraction maximum at diffraction angle θ_n. The wavelength may be obtained by applying a linear fit through the positions of the diffraction maxima as a function of the peak number and scaling the slope with the lattice constant.

An advantage of this method is the fact that λ which then enters into (2) is already an effective wavelength, characteristic for the wavelength distribution at the sample position. Since the wavelength is determined by a diffraction process which is effectively identical to the scattering process of the unknown sample structure all averaging effects are taken inherently into account by this procedure.

During our measurement series (blue setting of Fig. 1) we have performed 4 wavelength determinations with a 16 μm silicon grating. An example is displayed in Fig. 3 where the diffraction pattern is shown as function of the diffraction angle, and the peak positions together with the linear fit as a function of the peak number. The wavelengths were calculated from the slope of this line by relation (3). As first result we obtained $\lambda_1 = 1.87853(18)$ Å. The accuracy given is the statistical error of the linear fit and its value corresponds to the order of magnitude of 10^{-4} as envisaged in [10]. The other measurements yielded values for the wavelength with a similar precision but their average value $\bar{\lambda} = 1.9082(15)$ Å differed considerably from the previous

Setting	red	blue
I_{max} ($\mathrm{min}^{-1}\,\mathrm{cm}^{-2}$)	$1.510(2) \times 10^5$	$1.261(3) \times 10^5$
I_B($\mathrm{min}^{-1}\,\mathrm{cm}^{-2}$)	6.5(4)	4.7(4)
$\Delta\Theta_{sep}$ (µrad)	20.92(1)	22.02(1)
$\Delta\Theta$(FWHM) (µrad)	7.2(2)	7.97(3)
$\bar{\lambda}$ (Å)	1.912(11)	1.9082(15)

Table 1. Compilation of characteristic quantities of the instrument curves shown in Fig. 1.

on this level of accuracy. The error quoted for this average value is the standard deviation of the three measured values and represents a reasonable estimate for the actual level of accuracy for an instrument under operating conditions with data taking periods of a few hours for a single scattering pattern.

The difference of the first value and the following average correspond well to the observed shift of the peak positions of the instrument curve as shown in Fig. 2. The peak positions of the two curves presented in the right inset of the figure are located at $\theta_b = 10.70$ µrad and $\theta_r = 11.01$ µrad. Since the deflection angle of the polariser prisms is proportional to λ^2, [8], the ratio $\theta_b/\theta_r = \lambda_b^2/\lambda_r^2 = 1.029$ gives the ratio of the two wavelengths involved squared. To confirm that this shift may indeed be attributed to the change of the effective wavelength we take the ratio of the two wavelengths involved in the preceding paragraph $\bar{\lambda}^2/\lambda_1^2 = 1.032$ an find that the associated values are perfectly consistent.

For comparison of the parameters involved we compile characteristic quantities in Table 1. These include the maximum intensity of the peaks I_{max}, the background intensity I_B, the peak separation $\Delta\Theta_{sep}$ and the full width at half maximum $\Delta\Theta$(FWHM).

3. Investigation of magnetic samples

As in a typical SANS experiment USANSPOL measures variations of the neutron scattering length density $\rho(\mathbf{r}) = N\langle b\rangle(\mathbf{r})$ across the sample. Here, N is the local particle density (atoms or molecules per unit volume) and $\langle b\rangle$ the average neutron scattering length which may vary with position. The scattered intensity is determined by its Fourier transform $\rho(\mathbf{Q})$,

$$I(\mathbf{Q}) \propto |\rho(\mathbf{Q})|^2 = (N\langle b(\mathbf{Q})\rangle V)^2 S(\mathbf{Q}) \tag{4}$$

with the scattering function $S(\mathbf{Q})$ and V the volume of the scatterer. In the case of magnetic scattering $b(\mathbf{Q})$ contains contributions from nuclear and magnetic scattering, see e.g. [11],

$$\langle b(\mathbf{Q})\rangle = b_c + b_m\langle\sigma\rangle \cdot \hat{\mathbf{M}}_\perp(\mathbf{Q}) f_m(\mathbf{Q})\,, \tag{5}$$

where b_c denotes the coherent nuclear scattering length, $b_m = -\gamma_n r_e S$ the magnetic scattering length with $\gamma_n = -1.913$ the gyromagnetic ratio, $r_e = 2.818$ fm the classical electron radius and S the effective atomic spin, σ the Pauli spin operator, $\hat{\mathbf{M}}_\perp$ a unit vector in the direction of the component of the average atomic magnetisation perpendicular to the scattering plane, i.e. perpendicular to the scattering vector $\mathbf{Q}$ and $f_m(\mathbf{Q})$ the atomic magnetic form factor which depends on the magnitude of $\mathbf{Q}$. In USANS studies of magnetic structures $\mathbf{Q}$ is sufficiently close to zero that the magnetic form factor can be set to unity. In the USANSPOL setup, neutrons are polarised in the vertical direction which we will denote as the z-direction from which $\langle\sigma\rangle = \sigma_z = \pm 1$ follows. Eq. (5) then reduces to

$$\langle b(\mathbf{Q})\rangle = b_c \pm b_m \hat{M}_{\perp z}(\mathbf{Q})\,, \tag{6}$$

where the + sign refers to spin-up and the − sign to spin-down neutrons. It is worth noting that in an USANSPOL measurement, as in any magnetic scattering experiment, the intensity depends on the relative orientation of the local magnetisation, the scattering vector and the neutron polarisation. In contrast to magnetic Bragg scattering however, where the scattering vector $\mathbf{Q}$ is fixed for a given Bragg reflection, or in the case of 2-dimensionally resolved SANS where the scattering vector is fixed for a given detector pixel, in a USANSPOL measurement at any horizontal angular position θ (equivalently $Q_y = 2\pi\theta/\lambda$) the scattered intensity is integrated over all all vertical scattering angles θ_{vert} ($Q_z = 2\pi\theta_{vert}/\lambda$), [5]. Now, $\hat{M}_{\perp z}(\mathbf{Q})$ may be expressed as

$$\hat{M}_{\perp z}(\mathbf{Q}) = \frac{Q_y}{Q^2}(Q_y\hat{M}_z - Q_z\hat{M}_y) = \frac{\theta}{\Theta^2}(\theta\hat{M}_z - \theta_{vert}\hat{M}_y) \tag{7}$$

where $\Theta^2 = \theta^2 + \theta_{vert}^2$. At any angle θ there is a continuous contribution from scattering into all vertical angles θ_{vert} and thus a contribution of continuously varying values of $\hat{M}_{\perp z}(\mathbf{Q})$ scaled with the corresponding value of the scattering function $S(\mathbf{Q}) = S(Q_y, Q_z) = S(\theta, \theta_{vert})$. As a consequence, the magnetic contribution of structures magnetised in the horizontal plane will vanish for scattering into the horizontal plane ($Q_z = 0$) only. In case the scattering vector has a vertical component, however, $\hat{M}_{\perp z}(\mathbf{Q}) \neq 0$ and magnetic scattering results. When we turn the neutron polarisation into the horizontal plane along the y-axis, $\hat{M}_{\perp y}$ takes the role of $\hat{M}_{\perp z}$ in the preceding relations and is given by

$$\hat{M}_{\perp y}(\mathbf{Q}) = -\frac{Q_z}{Q^2}(Q_y\hat{M}_z - Q_z\hat{M}_y) = -\frac{\theta_{vert}}{\Theta^2}(\theta\hat{M}_z - \theta_{vert}\hat{M}_y) = -\frac{\theta_{vert}}{\theta}\hat{M}_z\,. \tag{8}$$

Proper linear combinations of (7) and (8) may be used for arbitrary neutron polarisation in the (y, z) plane.

We adopt the notation of [5] and denote by $P_{\uparrow\downarrow scat}(\theta)$ the probability for a scattered neutron to be scattered about an angle θ. We obtain for neutron polarisation in z-direction

$$P_{\uparrow\downarrow scat}(\theta) = \frac{1}{C_N}\int d\theta_{vert}(N[b_c \pm b_m\hat{M}_{\perp z}(\theta, \theta_{vert})]V)^2 S(\theta, \theta_{vert})\,. \tag{9}$$

with the normalisation constant C_N, give by

$$C_N = \int d\theta P_{\uparrow\downarrow scat}(\theta)\,. \tag{10}$$

$P_{\uparrow\downarrow scat}(\theta)$ describes the conditional probability that a neutron is scattered about an angle θ in the horizontal plane given that it is scattered at all. The integral probabilities for spin-up and spin-down scattering are denoted by $\alpha_{\uparrow\downarrow}$ and the scattered intensity as a function of the horizontal scattering angle θ is obtained by convolution with the rocking curve,

$$I_{\uparrow\downarrow scat}(\theta) = \alpha_{\uparrow\downarrow}\int d\theta' R_{\uparrow\downarrow}(\theta' - \theta)P_{\uparrow\downarrow scat}(\theta)\,. \tag{11}$$

It is quite clear from the preceding discussion that the analysis of USANSPOL data is quite complex and the information from a single orientation of the sample with fixed neutron polarisation suffice in special cases only. In general we will have to gather information from different sample orientations combined with different directions of the neutron polarisation. If we keep the sample environment constant during these operations, i.e. keeping an external magnetic field or the direction of mechanically induced stress constant with respect to the sample, we will have the constraint for all measuring conditions that the shape of the scatterer and its magnetisation stay the same. For this purpose we have built a prototype sample environment as shown in Fig. 4 having in the first instance ribbon samples in mind.

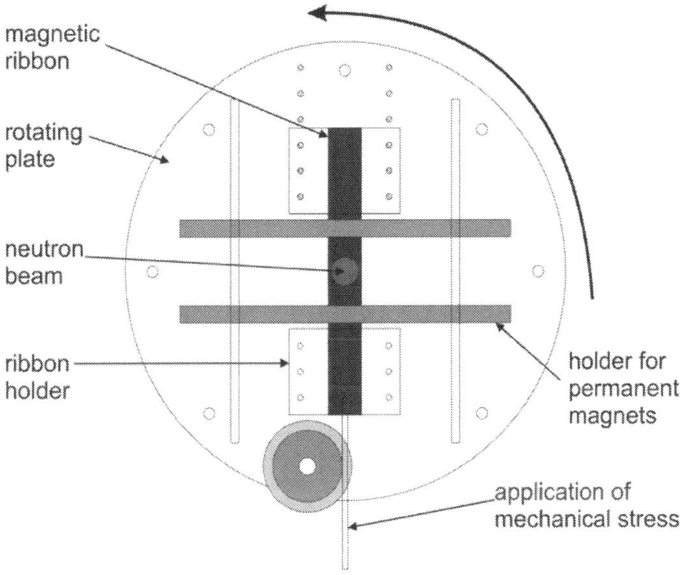

Figure 4. Prototype sample environment for USANSPOL studies of magnetic ribbons. A rotating plate allows for continuous variation of the ribbon axis orientation. An external magnetic field may be applied by permanent magnets mounted on separate holders onto the rotating plate. The field axis is then turned together with the ribbon to maintain the sample in a defined configuration. Mechanical stress may be induced in direction of the ribbon axis by adding some weight to the ribbon holder. Owing to the deflexion pulley the force is always exerted along the ribbon axis while the sample is rotated.

The ribbons are clamped by two ribbon holders. Then the upper holder is screwed to a manually rotating plate which contains a central opening through which the neutron beam passes. The lower holder is held by a guide rail mounted to the rotating plate where it may run freely to ensure that no extra force is exerted on the ribbon except its own weight. An external magnetic field can be produced by small permanent magnets which are fixed to the rotating plate and rotate together with the ribbon. Mechanical stress is induced by adding a container with weight to the ribbon holder. This container is connected to the ribbon holder via a string supported by a deflexion pulley when rotated out of the vertical direction to ensure that the stress is always directed along the ribbon axis. The plate allows for rotation about 90 degrees between the vertical and horizontal direction of the ribbon axis as indicated by the curved arrow in Fig. 4. For symmetry reasons we have attached a circular cadmium screen of 1 cm diameter to the plate to define the neutron beam cross section.

The following features may be achieved with this type of sample environment: By rotating the sample with respect to the high resolution y-axis anisotropies in the shape of the scatterers are identified. The magnetisation vector $\hat{\mathbf{M}}$ is affected by this rotation as well. The evolution of the microstructure under the influence of external stress can be monitored in the same positions with constant neutron polarisation in z-direction. An external magnetic field may be applied with fixed direction relative to the ribbon. The neutron polarisation is affected by the external magnetic field and will rotate with the field concurrently. Thereby different projections of $\hat{\mathbf{M}}_\perp$ are obtained. Environmental conditions combining external magnetic field and mechanically induced stress can be applied.

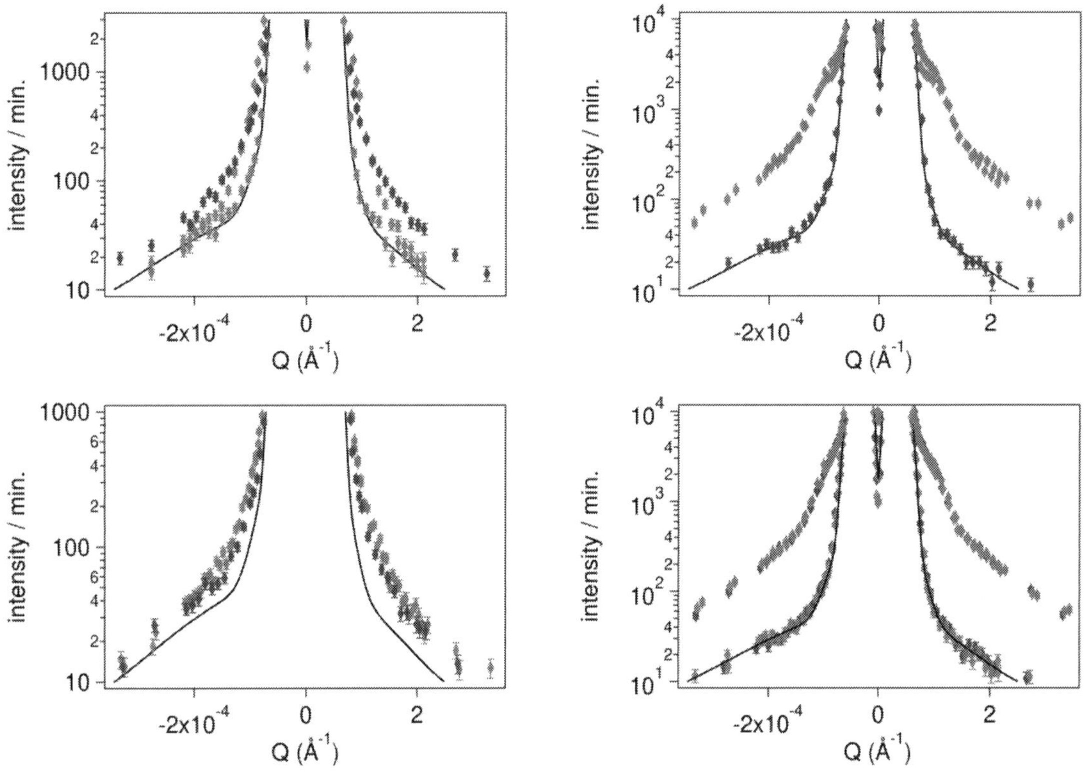

Figure 5. USANSPOL measurements on iron-boron based amorphous magnetic ribbons. The upper left graph shows data for an $Fe_{65}Co_{18}Si_1B_{16}$ compound, the other graphs show experiments with an $Fe_{78}Mo_2B_{20}$ ribbon. Upper left panel: sample placed in a small ambient vertical magnetic field of $B_z = 5$ G with the ribbon oriented at an angle of 0° (red symbols), 45° (green) and 90° (blue) with respect to the vertical z-direction. Upper right panel: ribbon axis fixed in the vertical direction, measurements were performed at varying external magnetic field of $B_z = 5$ G (red symbols), 12 G (green) and 1 kG (blue). Lower right panel: ribbon rotated to 0° (red symbols), 45° (green) and 90° degrees (blue) with respect to the vertical direction, the upper curve triple shows corresponding measurements with $B_z = 5$ G, the lower curve triple at $B = 1$ kG. Lower left panel: the ribbon was subjected to an external stress of 70 MPa at an ambient magnetic field of $B_z = 5$ G and the ribbon rotated to 0° (red symbols), 45° (green) and 90° (blue) with respect to the vertical direction.

4. Experiments

For test experiments we chose to follow the route taken in [4, 5] and studied the scattering of amorphous magnetic ribbons. Special iron-boron or iron-cobalt-boron compounds with certain admixtures may exhibit strong saturation magnetostriction which makes them attractive candidates as novel technological materials to be applied in magnetic sensors or actuators. We selected several compounds which are available as large enough ribbons with typical sizes of 10×1.2 cm^2 and about 20–30 µm thickness.

Fig. 5 shows an illustrative selection of measurements performed with these samples mounted to the sample environment shown in Fig 4. The sample area at S18 is surrounded with a Helmholtz coil that can produce a variable magnetic field of −12 to +12 G at the position of the sample to compensate for possible stray fields of the surroundings and to provide for zero field reference conditions. Unfortunately, this compensation can only be applied for the vertical direction and we found out that the conditions for our experiments included stray field components in the horizontal plane. Tests with a Magnaprobe, a small alnico bar magnet that is gimbal mounted, exhibited a wobbling behaviour in the magnetic stray field when the vertical field was adjusted to zero. The same behaviour could be expected for the neutron polarisation which would have not been unambiguously defined under such conditions. At an ambient vertical field of $B_z = 5$ G the test magnet remained stable in the vertical orientation. Therefore, we chose this setting as reference condition for our low-field measurements.

As a common characteristics for the subsequent discussion we point out that the USANSPOL pattern is predominantly determined by the structure dimension along the y-axis and the magnetisation in z when the neutrons are polarised and the external field points in that direction, although, according to the smearing integral (9), not independent of the structure size in z direction and the magnetisation in the horizontal plane.

Experiments with $Fe_{65}Co_{18}Si_1B_{16}$ are shown in the upper left graph of Fig. 5. Data show experimental results when the sample was placed in the small ambient vertical field of $B_z = 5$ G with the ribbon oriented at an angle of 0° (vertical, red symbols), 45° (diagonal, green) and 90° (horizontal, blue) with respect to the vertical z-direction. When the ribbon is oriented vertically the scattering pattern practically coincides with the instrument curve which represents the resolution function for the experiments. The scatterers are apparently large perpendicular to the axis and exceed the size range of the USANS technique. The main part of the scattering is directed into the forward direction and hidden by the resolution function. Careful inspection of the scattered intensity may indicate that the wings of the scattering curve are slightly broader and sophisticated parameter estimation, similar to the one applied in [4], may still yield a realistic size estimate in that direction. Rotating the sample raises the scattering pattern from the level of the resolution function and indicates smaller structure size in the direction of the band axis. The slope at the diagonal position at 45° is steeper than in the horizontal position at 90°. This evidence is consistent with the picture of a stripe structure of the ribbons with the broad stripe dimension perpendicular and the narrow stripe dimension parallel to the ribbon axis. It can be further seen that the 5 G field seems to induce magnetic ordering already. An asymmetry in the scattering in favour of spin-up neutrons on the right side of the scattering pattern can be seen for the ribbon positions at 45° and 90°. This example highlights the importance of sample rotation in case of anisotropic scatterers and the necessity for a precise determination of the instrument curve to exploit the Q_y range down to smallest accessible values with clear distinction of the scattering and the resolution curve.

The data shown in the remaining panels of Fig. 5 were taken with an $Fe_{78}Mo_2B_{20}$ ribbon. In the upper right graph measurements with varying external field are presented. The ribbon axis was kept in vertical orientation and experiments performed at a varying external magnetic field of $B_z = 5$ G (red symbols), 12 G (green) and 1 kG (blue). Due to the applied magnetic field the neutron polarisation was kept in the z-direction. We see slow structural development from $B_z = 5$ G to 12 G, at 1 kG the ribbon is saturated and the scattering curve runs along the resolution function. No residual nuclear scattering is visible in the latter case from which the purely magnetic origin of the scattering observed at low external magnetic field may be concluded.

The extremal values of the external magnetic field used above were investigated for the lower right graph of Fig. 5. Under these conditions the ribbon was rotated to 0° (vertical, red symbols), 45° (diagonal, green) and 90° degrees (horizontal, blue). The upper curve triple belongs to $B_z = 5$ G, the lower curve triple to $B = 1$ kG. The strong field was generated by small permanent magnets which were attached to the rotating plate as indicated in Fig. 4 and its axis rotated together with the ribbon axis. In the 0° position the field pointed in vertical direction, $B_z = 1$ kG, in the diagonal position at 45° we obtain $B_x = B_y = 1/\sqrt{2}$ kG $= 0.7$ kG and in the horizontal position at 90° $B_y = 1$ kG. Due to this external magnetic field the direction of the neutron polarisation is rotated together with the ribbon axis. As can be seen in the graph the scattering structures appear rather isotropic, also in the case of saturation at 1 kG. Due to the absence of visible nuclear scattering the ribbon then becomes optically transparent for the neutrons and represents an ideally amorphous neutron optical medium.

As pointed out above the USANSPOL pattern is predominantly determined by the structure dimension along the y-axis, i.e. perpendicular to the ribbon axis, and the magnetisation in z-direction, parallel to the ribbon axis, in the 0° position. In the 90° position the scattering pattern is still predominantly determined by the structure dimension along the y-axis, i.e. now parallel to the ribbon axis, and differently by the magnetisation in y-direction, still parallel to the ribbon axis but now the projection of the total magnetisation according to Eqs. (7) and (8), due to the dot product in Eq. (5). Of course, in the latter case magnetic scattering takes place with a finite vertical component of the scattering vector only.

Finally, in the lower left graph measurements with applied mechanical stress are presented. Lead weights were put into the container attached to the ribbon holder to produce a stress of 70 MPa on the ribbon. The samples were placed in the ambient vertical magnetic field of $B_z = 5$ G and the ribbons again rotated to the 0° (vertical, red symbols), 45° (diagonal, green) and 90° (horizontal, blue) positions. The scattering pattern has to be compared with upper curve triple in the lower right panel of Fig. 5 which were taken under the same environmental conditions at zero induced stress. We observe that the scattering structures have grown considerably when compared to the unstressed ribbon. Since the scattering is of practically pure magnetic nature this is equivalent to a growth of the magnetic domain size with externally induced stress and mirrors the magnetostriction properties of the sample. Again, also magnetic ordering is induced as can be seen from the enhanced spin-up neutron scattering in the right part of the scattering curves.

These experiments illustrate the particular value of a proper sample environment for USANSPOL measurements on magnetic samples. At the same time we also encountered the limitations of our prototype sample environment. Specifically the lack of compensation for the stray fields in the horizontal plane prohibited the measurement in zero field environment under the given conditions. Especially for soft-magnetic samples as studied in [4] zero field measuring conditions are absolutely essential. A second limitation was encountered with the external magnetic field generated by the small permanent magnets. Favourable measuring conditions were created with magnetic fields of the order of 1 kG but smaller field values, as would have been needed to study the structure evolution in the upper right panel of Fig. 5, exhibited considerable field inhomogeneities along the neutron trajectories. A third limitation can be identified regarding the generation of mechanical stress by the application of lead weights. The presence of a free hanging load in the vicinity of the high resolution analyser axis and the perfect silicon analyser crystal causes considerable tension for the experimenter during sample positioning and measurement. Automatisation of the sample rotation which was performed manually in the current design is a final issue for improvement. A concept to overcome these limitations already exists and awaits realisation for upcoming USANSPOL tests and measurements.

5. Conclusions

As seen by the examples compiled in Fig. 5, USANSPOL is a potentially powerful technique for the study of magnetic microstructure of matter. Due to the complexity of novel magnetic materials relevant for technological application and the smearing effects for scattering, see Eqs. (9) and (11), inherently involved in the technique data modelling and routine analysis are still in the developing phase. The role of the instrument curve as resolution function and stability measure was highlighted. As in many occasions, the measurement time spent for determining the instrument curve and its variations should compare to the measurement time of the samples themselves. With the application of artificial line gratings and the observation of the peak separation in the instrument curve the effective wavelength is determined to a high precision and its possible shifts well under control. This is not so important for determination of the Q_y scale where tiny scaling errors are outshone by the size distribution of scatterers in novel materials but for a correct comparison of the scattered intensity with the resolution function of the instrument. The theoretical basis for analysis exists, and its implementation into model calculations is underway. A prototype sample environment was tested which opens up a wide variety of possibilities for the study of magnetic materials, with an initial emphasis put on magnetic ribbon samples. In parallel, limitations of the first design were identified and an improved new prototype is currently under development. With these ingredients the study of magnetic microstructure in modern materials has become feasible and its evolution under different environmental conditions accessible.

Acknowledgement
The authors thank H. Lemmel for invaluable support during the measurements at S18. This project was supported by the European Integrated Infrastructure Initiative for Neutron Scattering and Muon Spectroscopy NMI3, FP7 Grant Agreement No. 226507-NMI3.

References

[1] Littrell K C and Lee W T 2004 *Physica B* **345** 246–249

[2] Wagh A, Rakhecha V, Strobl M and Treimer W 2004 *Pramana J. Phys.* **63** 369–373

[3] Jericha E, Badurek G and Trinker M 2007 *Physica B* **397** 88–90

[4] Badurek G, Jericha E, Grössinger R and Sato-Turtelli R 2010 *J. Phys.: Conf. Ser.* **211** 012027

[5] Jericha E, Badurek G and Grössinger R 2011 *Physica B* **406** 2401–2404

[6] Hainbuchner M, Baron M, Lo Celso F, Triolo A, Triolo R and Rauch H 2002 *Physica A* **304** 220–229

[7] Jericha E, Baron M, Hainbuchner M, Loidl R, Villa M and Rauch H 2003 *J. Appl. Cryst.* **36** 88–90

[8] Badurek G, Buchelt R J, Kroupa G, Baron M and Villa M 2000 *Physica B* **283** 389–392

[9] Rauch H and Petraschek D 1978 *Neutron Diffraction* (*Topics in Current Physics* vol 6) (Berlin: Springer) chap Dynamical Neutron Diffraction and Its Application, pp 303–351

[10] Trinker M, Jericha E, Loidl R and Rauch H 2008 *Nucl. Instrum. Methods A* **586** 124–128

[11] Sears V F 1989 *Neutron Optics* (Oxford: Oxford University Press)

Journal of Physics: Conference Series **340** (2012) 012008 doi:10.1088/1742-6596/340/1/012008

Development of high pressure gas cells at ISIS

O Kirichek, R Done, C M Goodway, M G Kibble, B Evans, Z A Bowden

ISIS Facility, STFC, Rutherford Appleton Laboratory, Harwell, Didcot, UK

oleg.kirichek@stfc.ac.uk

Abstract. High-pressure research is one of the fastest-growing areas of natural science, and one that attracts as diverse communities as those of physics, bio-physics, chemistry, materials science and earth science. In condensed matter physics there are a number of highly topical areas, such as quantum criticality, pressure-induced superconductivity or non-Fermi liquid behaviour, where pressure is a fundamental parameter. Reliable, safe and user-friendly high pressure gas handling systems with gas pressures up to 1GPa should make a significant impact on the range of science possible. The ISIS facility is participating in the NMI3 FP7 sample environment project supported by the European Commission which includes high pressure gas cell development. In this paper the progress in designing, manufacturing and testing a new generation of high pressure gas cells for neutron scattering experiments is discussed.

1. Introduction

The combination of high pressure techniques with neutron scattering proves to be a powerful tool for studying the phase transitions and physical properties of solids in terms of inter-atomic distances. In condensed matter physics there are a number of highly topical areas, such as quantum criticality, pressure-induced superconductivity or non-Fermi liquid behaviour, where pressure is a fundamental parameter. The high pressure technique based on a gas medium compression is one of the most popular in neutron scattering experiments [1, 2]. Currently this technique covers the pressure range up to ~0.7 GPa and typically uses compressed helium gas as the pressure medium. An increase in the range of available gas pressures up to 1 GPa (10 kbar) should make a significant impact on the range of science possible. However, high pressure gas sample cells require thick cell walls which may lead to an unacceptable neutron background and thus the choice of materials and geometries is critical to improving the quality of the data. Reliable, safe and user-friendly high pressure gas handling systems are also an essential part of the development. ISIS facility participates in the NMI3 FP7 sample environment project supported by the European Commission which includes a high pressure gas cell development task.

This paper will discuss the sample chamber seal design of a pressure vessel intended to operate up to 1 GPa (10 kbar) with inert gas as well as 0.8 GPa (8 kbar) hydrogen gas on ISIS neutron instruments. The working temperature will be from 300K down to 4K. The material of construction of the pressure vessel should be neutron compatible and able to retain the necessary strength to endure the enormous stresses likely to occur at the intended high pressures. In addition, it should be completely resistant to hydrogen embrittlement.

2. Seal test arrangements

A standard high-pressure seal for this type of pressure regime would be an unsupported area seal arrangement also known as a Bridgman seal [3]. Any such seal arrangement would generally consist of three components: two soft sacrificial seals, normally made from lead, and a slightly harder principle seal, normally made from annealed copper (Fig. 1 and Fig. 2).

Fig. 1 Typical Bridgman seal arrangement showing a three part seal assembly and the supporting backing nut.

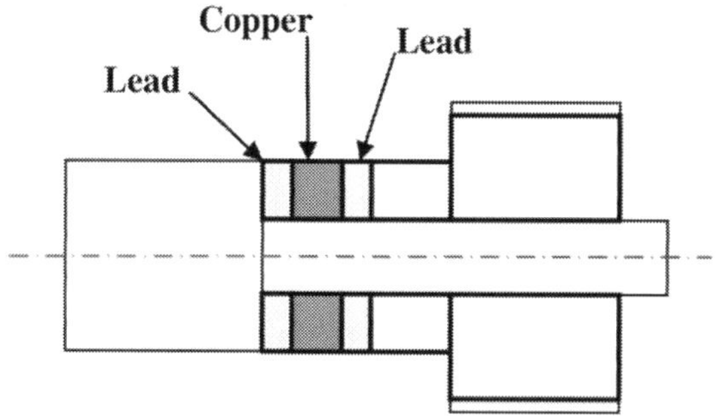

Fig. 2 Typical Bridgman seal arrangement.

All pressure testing at ISIS is carried out in a special purpose test bunker made from concrete, which has camera viewing only during the test procedure. All valve operations are made externally from an intensifier control panel thus protecting personnel in the event of a failure of the equipment inside the test bunker. The extremely high pressures needed for the test were generated using a multi-stage hydraulic oil intensifier which is rated to 1.38 GPa (13.8 kbar) and is operated by a combination of manual pumps taking an initial low inlet pressure up to the system maximum. The seal test vessel is shown in Fig. 3 and is registered as an ISIS pressure vessel.

Fig. 3 Seal test vessel.

3. Material selection

The material selection for the proposed seal combinations was based on both the specification requirements of pressure and temperature, along with previous experience of high-pressure Bridgman seal designs. Materials were chosen that are able to withstand the enormous loadings created at the working pressures, along with the ability to retain ductility at cryogenic temperatures. The materials also had to be inert to the effects of hydrogen gas [4]. The materials selected include pure aluminium, pure copper and pure lead, with backing-ring supports made from either stainless steel or beryllium copper. Two examples of seal material arrangements are shown: an aluminium/copper seal combination and a lead/copper seal combination in Fig. 4.

Fig. 4 Aluminium/copper seal combination (left); lead/copper seal combination (right).

4. Seal test results and discussion

The double Bridgman seal combination shown in Fig. 5 and Fig. 6, using five seal components of lead and lead-coated copper, was chosen as a seal likely to be leak tight at these pressures. By adding an additional sacrificial and principle seal, it was thought that any failure of the active combinations during pressurisation would always have a supporting backup seal to prevent overall breakdown. It did seal successfully, with the lead seals extruding in the process. As a sacrificial seal this is to be expected due to the high ductility of the metal and the presence of a slight clearance between the vessel bore and the seal piston. The 1 GPa (10 kbar) pressure was held for 60 minutes with no apparent leakage. An interesting finding is that a short while after release of the pressure the remaining lead seals expand, causing a clear rippling effect in the assembly (Fig. 7). It is thought that this is possibly caused by residual stresses in the lead relaxing over time creating a reverse-yielding situation allowing the seal to expand. It could also be caused by retained pressure within the structure of the metal forcing the weaker external surface outwards.

The aim of the seal test programme is to examine the quality of the seals at both room temperature and cryogenic conditions. A purpose-built cryostat centre-stick is required for the cryogenic tests and all high-pressure fittings used need to withstand both the 1 GPa (10 kbar) pressures as well as hydrogen corrosion. In many cases such fittings are not commercially available and so have been designed in-house to suit. The longer term requirements of the test programme will require an intensifier to reach over 0.8 GPa (8 kbar) with hydrogen. Room temperature tests have begun using an existing helium intensifier.

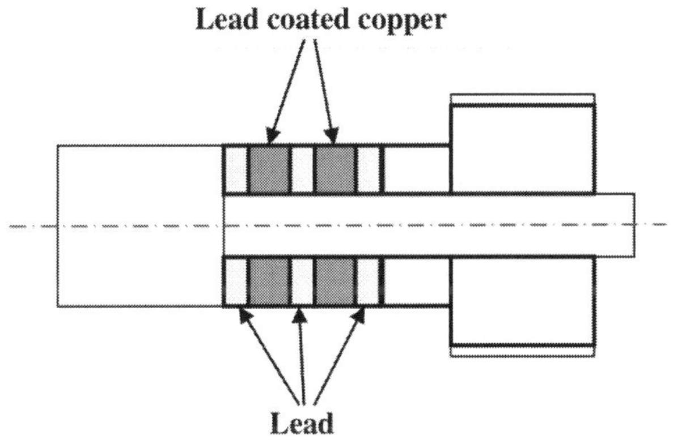

Fig. 5 Double Bridgman seal arrangement.

Fig. 6 Double Bridgman seal prior to pressurisation.

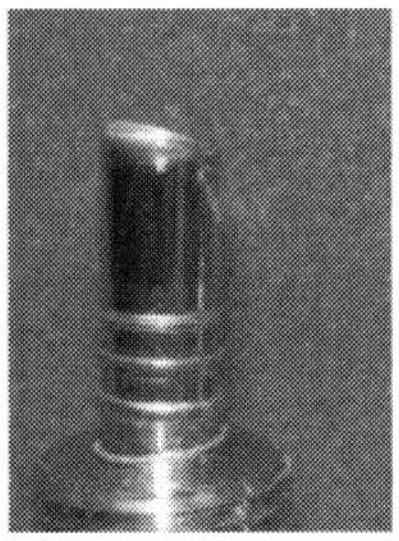

Fig. 7 Double Bridgman after pressurisation, showing the ripple effect of the lead seals.

5. Conclusion and further work

To date, the double Bridgman seal combinations of lead/lead-coated copper, and lead/aluminium, have been successfully tested up to 1 GPa (10 kbar). Seal materials were chosen that are able to withstand the corrosive effects of hydrogen gas and the enormous loadings caused by the high working pressures.

The need to optimise neutron transmission through pressure cells restricts the choice of construction materials available. This limited choice is reduced further when the gas medium required is hydrogen. An important part of this project has thus been the investigation of cell design, hydrogen embrittlement and material suitability.

A literature review of hydrogen embrittlement and material resistance has been undertaken to increase understanding of the issue. Tensile and fatigue testing has been carried out by Imperial College on behalf of ISIS on material samples, with and without hydrogen charging. The fatigue test results are particularly important as the pressure cells will experience repeated pressurisation cycles during experimental use and it is thought that hydrogen embrittlement has a greater effect on the fatigue characteristics of materials due to the nature of the corrosion mechanisms occurring.

The technical aspects of cell design have been investigated, in particular the technique of autofrettaging (a mechanical strengthening process by which pressurised cells are deliberately subjected to an internal force greater than the limiting elastic strength of the cell wall material). Experimental investigations using both neutron (ISIS) and x-ray (Diamond) instruments have been undertaken to confirm, or allow revision of, empirical and FEA modelling of pressure cell designs.

A more detailed account of this work is beyond the scope of this paper and will be covered in future publications.

6. Acknowledgments

We are grateful to Jean-Michel Mignot, Burkhard Annighofer and Michael Meißner for their contribution and valuable discussions.

This research project has been supported by the European Commission under the 7th Framework Program through the 'Research Infrastructures' action of the 'Capacities' Programme, Contract No: CP-CSA_INFRA-2008-1.1.1 Number 226507-NMI3.

References

[1] NMI3, FP7, JRA: Report on current inert gas pressure cell technology 2010
[2] Done R, Kirichek O, Evans B E and Bowden Z A 2010 *arXiv* 1007.3135
[3] Bridgman P W The physics of high pressure G. Bell and Sons Ltd., London, 1931
[4] Jewett R P, Walter R J, Chandler W T and Frohmberg R P Hydrogen environment embrittlement of metals NASA-CR-21, Rocketdyne, 1973

Journal of Physics: Conference Series **340** (2012) 012009 doi:10.1088/1742-6596/340/1/012009

Cryogen free sample environment for neutron scattering experiments at ISIS

O Kirichek, R B E Down, J Keeping, B Evans, Z A Bowden

ISIS Facility, STFC, Rutherford Appleton Laboratory, Harwell, Didcot, UK

oleg.kirichek@stfc.ac.uk

Abstract. Most neutron facilities have a fleet of cryostats providing low temperature and high magnetic fields for sample environment. This large scale usage of cryogenic equipment requires significant resources and can create a number of problems including health and safety issues and the considerable cost of the required cryogens. The last problem has become more significant due to the increasing costs of liquid helium caused by global helium supply problems. The ISIS facility has an internal development programme intended to gradually substitute all conventional cryogenic systems with cryogen free systems preferably based on the pulse tube refrigerator. The programme includes a number of development projects which are aiming to deliver a range of cryogen free equipment including a top-loading 1.5 K cryostat, superconducting magnets in re-condensing cryostats and cryogen free dilution refrigerators. Here we are going to describe the design of these systems and discuss the results of prototype testing.

1. Introduction

The increasing costs of liquid helium caused by global helium supply problems raises significant concerns about the affordability of conventional cryogenic equipment based on the evaporation of liquid cryogens. This problem poses a serious threat to the future experimental programmes of large scientific facilities. However the progress in closed cycle refrigerator (CCR) technology offers a new generation of cryogenic systems with significantly reduced consumption and, in some cases, near complete elimination of cryogens. This breakthrough became possible due to a new generation of commercial cryo-coolers developed during the last decade. The most successful representative is the pulse tube refrigerator (PTR). The unique feature of the PTR is the absence of cold moving parts, which considerably reduces the noise and vibrations generated by the cooler. It also increases the reliability of the cold head because no expensive high-precision seals are required and the cold head can be operated without any service inspection.

There are two options available in order to reduce or, in the ultimate case, eliminate the usage of cryogens: so-called "dry" systems that do not contain liquid cryogens at all, built around a cryo-cooler that utilizes the cooling power from the cold head [1, 2] and re-condensing systems in which evaporating helium is cooled and returned back to the cryostat by a CCR [3]. The ISIS facility is carrying out an internal development programme intended to gradually substitute all conventional cryogenic systems with cryogen-free systems preferably based on the PTR. This paper will describe the design of these systems and discuss the results of prototype testing.

2. Top-loading cryostat based on a CCR

Since the late 1970s the Orange Cryostat designed at the ILL [4] has remained the workhorse of all neutron scattering facilities. This is a top-loading liquid helium bath cryostat with a base temperature of ~1.5K. However today the Orange Cryostat is losing its popularity due to its high helium consumption and the significant resources required for routine operation.

Almost a decade ago ISIS started a development programme aimed at developing a cryogen-free top-loading system based on the CCR as an ideal substitute for the Orange Cryostat. This project consists of two phases. During the first phase, top-loading systems based on the Gifford McMahon (G-M) cold head and the PTR have been designed and tested [2, 5]. These top-loading systems are fitted with rotation stages which allow the rotation of a sample around its vertical axis with precision better than ±0.5°. The photos of both systems are presented in Fig. 1 and comparisons of technical specifications achieved in tests are presented in Table 1. It can be seen that the top-loading system based on the PTR, reaching a base temperature of 3.5K in under three hours, cooled faster than the G-M CCR based system. In addition, the vibration level measured in the PTR cryostat was also an order of magnitude less than that measured in a similar system based on a standard G-M cold head. Low level of mechanical vibrations is essential for high resolution neutron reflectometry and SANS experiments. For a rough estimation of the CCR-based top loading system's cost efficiency one can compare the cost of liquid helium required to run a standard Orange Cryostat (~30 £/day) with the cost of the electric energy consumed by a CCR (~8 £/day). This difference is only likely to increase in the foreseeable future.

Table 1 Comparison between the basic technical specification parameters of top-loading systems based on the G-M and PTR CCRs.

Type of system	Base sample temperature	Cool down time	Sample change time	Number of systems available
PTR Based	3.5K	<3h	<1.2h	2
GM Based	4.4K	~4h	<1.25h	8
Orange cryostat	≤1.5K	~ 24 h	~ 2 h	21

During the second phase of the project a 1.5K cryogen-free system based on a PTR, with a 50 mm diameter top-loading sample access for neutron scattering experiments has been developed and successfully tested [6]. The system design is based on the combination of a top-loading system [2] with a helium condensation loop, which consists of three heat-exchangers mounted on the cryocooler, stainless steel impedance and 1.5K heat-exchanger on variable temperature inserts (VTI). The sample temperature range of this system is 1.45 – 300K in the continuous flow regime. A sustainable cooling power of 55 mW at 1.82K and temperature stability of ± 0.1K in temperature range 5 - 200K, and ±0.01K below 5K has been achieved. From a user perspective, the system offers operating parameters very similar to those of an Orange cryostat, but without the complication of cryogens. The system may also be used with ultra-low temperature inserts such as the Kelvinox VT dilution refrigerator. Recent preliminary tests have successfully demonstrated the operation of a Kelvinox VT insert at 75 mK.

3. Cryogen free dilution refrigerators

The ISIS facility uses a number of dilution and ^{3}He refrigerator inserts which can be used with the VTI of a conventional cryostat or superconducting magnet. These systems consume cooling power produced at the VTI heat exchanger and can easily be used with cryogen-free, or reduced cryogen, cryostats. This can be achieved by inserting the refrigerator either in the top-loading system based on the PTR discussed above or in a re-condensing cryostat with a VTI. However the range of applications of conventional ultra-low temperature inserts is limited by their relatively low cooling power (~30 μW at 100mK) and small sample space (OD ≤ 37 mm and height ≤80 mm).

A growing number of neutron scattering experiments now require a combination of extreme conditions such as high pressure, high magnetic field or large sample size, with ultra-low temperatures. In order to satisfy these emerging requirements one needs a powerful dilution refrigerator with a sample space big enough to accommodate a high pressure cell or a small superconducting magnet. Oxford Instruments, in collaboration with ISIS, have developed a powerful cryogen-free dilution refrigerator (E-18) for neutron scattering experiments. The design of the refrigerator is based on a prototype of a "dry" dilution unit with PTR pre-cooling [7]. The refrigerator is capable of cooling large (OD 200 mm; height 250 mm) and heavy (up to 20 kg) samples and provides access for the neutron beam through a set of thin aluminium alloy windows. The base temperature of the refrigerator is 15mK and the cooling power is approximately 370 μW at 100mK. The cool down time of E-18 is ~36 hours, which means using only one sample during single experiment. Currently Oxford Instruments, again in collaboration with ISIS, is developing a split pair superconducting magnet for neutron scattering experiments; the magnet is going to be attached to the 4K shield of the E-18 refrigerator. The design of the magnet is based on a prototype successfully tested by Oxford Instruments [8].

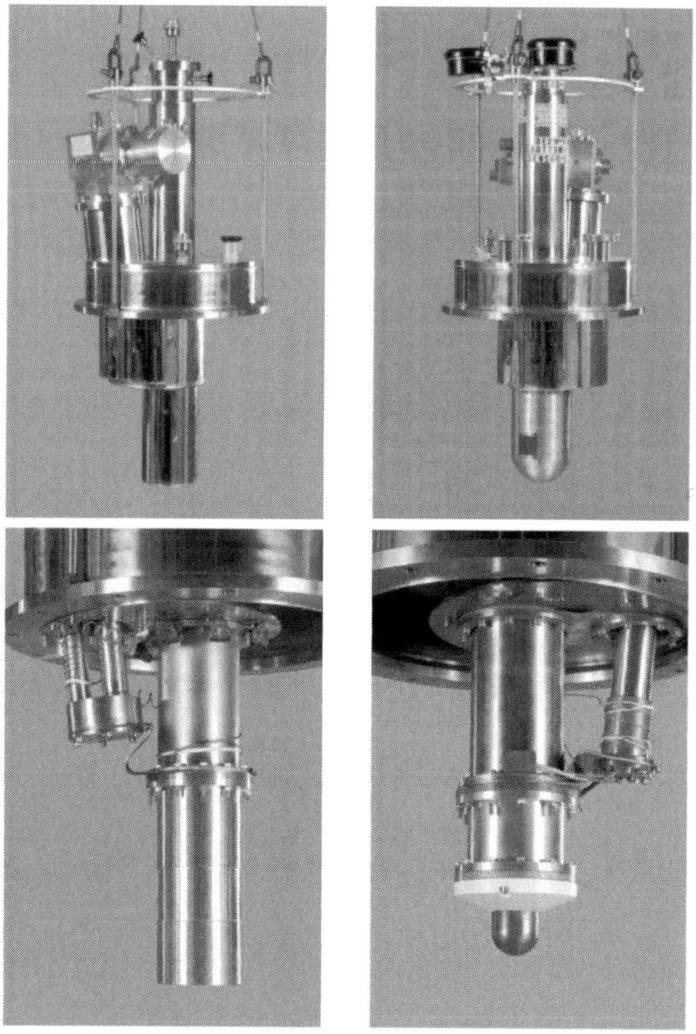

Fig. 1 Top loading CCRs based on PTR (left); and on GM cryo-cooler (right)

4. Superconducting magnets actively cooled by a PTR

Neutron scattering experiments at high magnetic field and ultra low temperatures consume significant amount of the total cryogens spent by ISIS user program. In 2006 ISIS commissioned three superconducting magnets for use in neutron scattering experiments [5, 9]. All three magnets have used re-condensing technology based on the PTR. The design of re-condensing magnet cryostats is usually based on similar helium bath cryostats [3]. The superconducting magnet is immersed in the liquid helium. The radiation shield is cooled by the cooler's first stage and the second stage re-condenses helium directly in the helium vessel. Thus the re-condensing magnet cryostat does not consume any liquid helium in normal operation. The main advantage of this system is that all magnet operating procedures; for example cooling, running up to the field and quenching are identical to those of a standard magnet in a bath cryostat. All standard magnet system accessories like current leads, power supply, helium transfer siphon etc. can be used. This system also provides a homogeneous temperature distribution, which is crucial for optimum magnet performance. Another important advantage is the ability of the magnet, in the case of a cryo-cooler power failure, to stay at field for more than 48 hours before quenching. This feature also makes it possible to move the cold magnet between different neutron scattering instruments and/or the cryogenic preparation laboratory.

Zero helium boil-off operation of the re-condensing 9T superconducting magnet cryostat with a Kelvinox VT dilution refrigerator insert running at a base temperature of 63mK during a 24 hour period has recently been demonstrated at ISIS [10]. This combined system allows a cryogen-free high magnetic field and ultra-low temperature sample environment for neutron scattering experiments. The system offers a significant reduction in liquid helium consumption as well as the advantage of operational simplicity and a significant improvement in user safety. Estimations based on half a year of 14T magnet cryostat operation statistics show that one re-condensing magnet cryostat should save us ~7500 £/year.

5. Conclusion

The ISIS facility is carrying out an internal development programme with the intention of gradually substituting all conventional cryogenic systems with cryogen-free systems preferably based on the PTR. This paper has described the design of ISIS cryogen-free and re-condensing systems and discussed the results of prototype testing and operational experiences of these systems in the ISIS user programme. This approach offers a significant reduction in liquid helium consumption as well as the advantage of operational simplicity and a significant improvement in user safety. The ISIS facility plans a gradual substitution of all conventional cryostats with cryogen free systems in the next decade.

6. Acknowledgements

The authors are grateful to Oxford Instruments for fruitful collaboration and ISIS colleagues for valuable discussions and contributions to the development programme.

References

[1] Oliver E, Evans B E, Chowdhury M A H, Major R A, Kirichek O and Bowden Z A 2008 *Meas. Sci. Technol.* **19** 034019
[2] Evans B E, Down R B E, Keeping J, Kirichek O and Bowden Z A 2008 *Meas. Sci. Technol.* **19** 034018
[3] Kirichek O, Philip C, Johnson C and Atrey M 2005 *Review of Scientific Instruments* **76** 055104
[4] Brochier D 1977 *ILL Technical Report* 77/74
[5] Kirichek O, Evans B E, Down R B E and Bowden Z A 2009 *J. Phys.: Conf. Ser.* **150** 012022
[6] Chapman C R, Evans B E, Dudman M P, Keeping J, Down R B E, Kirichek O and Bowden Z A 2011 *Cryogenics* **51** 146
[7] Uhlig K 2002 *Cryogenics* **42** 73
[8] Batey G, Buehler M, Cuthbert M, Foster T, Matthews A J, Teleberg G and Twin A 2009 *Cryogenics* **49** 727
[9] Down R B E, Kouzmenko G, Kirichek O, Wotherspoon R, Brown J and Bowden Z A 2010 *J. Phys.: Conf. Ser.* **251** 012092
[10] Kirichek O, Down R B E, Kouzmenko G, Keeping J, Bunce D, Wotherspoon R and Bowden Z A 2010 *Cryogenics* **50** 666 - 669

A new imaging method using pulsed neutron sources for visualizing structural and dynamical information

Y Kiyanagi[1], **H Sato**[1,2], **T Kamiyama**[1] **and T Shinohara**[2]
[1] Faculty of Engineering, Hokkaido University, Kita 13, Nishi 8, Kita-ku, Sapporo, Hokkaido 060-8628, Japan
[2] J-PARC Center, Japan Atomic Energy Agency, 2-4 Shirane Shirakata, Tokai-mura, Naka-gun, Ibaraki 319-1195, Japan

E-mail: kiyanagi@qe.eng.hokudai.ac.jp

Abstract. Neutron imaging using pulsed neutron sources coupled with a 2-dimensional position sensitive detector applicable to the time-of-flight method can give information on the crystal texture of coherently scattering materials, dynamical information of incoherently scattering materials such as hydrogen, and magnetic field information. Bragg edges appeared at cold neutron region reflect the preferred orientation, crystallite size, and lattice spacing. To deduce such information from the neutron transmission data depending on the position we have developed a data analysis code, and applied this code to data of a welded iron sample. Furthermore, as examples of more realistic materials we have investigated quenched iron rods. The quenched region was clearly demonstrated by the lattice space distribution. Furthermore, difference in the bound state of water or hydrogen in wet and dry cement pastes have been observed by analyzing the gradient of the neutron transmission cross section at the cold neutron region. The magnetic field has been also measured by using the polarized neutrons, and the strength of the field was estimated easily by analyzing the wave length dependent data.

1. Introduction

Neutron transmission imaging based on steady neutron sources has been used in various fields as a method of non-destructive inspection. On the other hand, imaging using pulsed neutron sources has been developing a new field of the imaging with the aid of its energy resolving nature, since the time-of-flight method for energy analysis adopted in the pulsed neutron source allows us easily to obtain the neutron transmission depending on the neutron wavelength at each pixel of a two dimensional position sensitive detector. The transmission spectrum, in principle, depends on the total neutron cross section of the object material. The total neutron cross section generally consists of coherent elastic scattering, coherent inelastic scattering, incoherent elastic scattering, incoherent inelastic scattering, and absorption cross section. Furthermore, magnetic scattering should be included when the polarized neutron transmits a magnetic field. Coherent scattering is major cross section for almost all materials other than materials such as hydrogen, vanadium and so on. The neutron cross section exhibits change of its value at the cold neutron region; for the coherent scatterer there are Bragg edges corresponding to the Bragg scattering of each lattice plane, and for incoherent scatterer the cross section increases in proportion to the wavelength. Furthermore, absorption cross sections of some materials changes drastically at a specific energy, for example, 0.4 eV for cadmium. At early stage of the time-of-flight imaging we tried to identify each alloy in an object[1], and then we got different contrast images at once by using the time-of-flight imaging[2]. Furthermore, a spatial depending image of crystal structural information was also obtained by analyzing the transmission spectra. Spatial distribution of strain, phase transition, texture, crystallite size were observed by using the pulsed neutron sources[3-11]. For deducing quantitative information on texture, crystallite size and strain, we have to analyze wavelength dependent transmission over wide wavelength region. Therefore, we have developed an analysis code 'RITS' [12] including similar functions of the analysis code of the neutron diffraction.

We have performed pulsed neutron transmission experiments at Hokkaido linac facility and also at J-PARC neutron source to study the crystallographic structure of a welded iron plate with a thickness of 6 mm and a quenched iron rod with a length of 20 mm. Furthermore, as an example of hydrogenous material we have studied water and hydrogen dynamical property in cement pastes to see difference between a dry cement paste and a wet cement paste. Finally, by using polarizing neutron we have observed image of a magnetic field. Here, we first explain the new analysis code 'RITS' and present the experiments performed at pulsed neutron source which indicating usefulness of the pulsed neutron imaging.

2. Crystallographic information imaging

2.1. Data analysis method

A Rietveld-type spectral analysis code for Bragg-edge transmission imaging, RITS (Rietveld Imaging of Transmission Spectra), was developed for quantitative evaluation of crystallographic information of a material, e.g., crystal lattice strain, preferred orientation, degree of crystallographic anisotropy and crystallite size [12-14]. This code calculates position-dependent transmission spectra based on an effective total cross section model, and then refines the structural parameters at each position by fitting the calculation data to the experimental data. It is very important to develop a theoretical model being able toanalyze the experimental data. Hereafter, we present a new theoretical expression of a Bragg-edge transmission spectrum to deduce the crystallographic and metallographic information.

The neutron transmission $Tr(\lambda)$ as a function of wavelength λ is represented by

$$Tr(\lambda) = \exp\left(-\sigma_{\text{tot}}(\lambda)\rho t\right). \tag{1.1}$$

$\sigma_{\text{tot}}(\lambda)$ is the neutron total cross section, ρ is the number density, and t is the thickness of a material. In the low energy region, the total cross section consists of elastic coherent scattering, elastic incoherent scattering, inelastic coherent scattering, inelastic incoherent scattering and absorption parts, as follows:

$$\sigma_{\text{tot}}(\lambda) = \sigma_{\text{coh}}^{\text{ela}}(\lambda) + \sigma_{\text{incoh}}^{\text{ela}}(\lambda) + \sigma_{\text{coh}}^{\text{inela}}(\lambda) + \sigma_{\text{incoh}}^{\text{inela}}(\lambda) + \sigma_{\text{abs}}(\lambda). \tag{1.2}$$

The elastic coherent scattering cross section is the most important for crystallographic information analysis because this component uniquely relates to Bragg-edges, and reflects the crystal structure. To describe an actual Bragg-edge transmission spectrum of a polycrystalline material, we proposed a modified expression of a Bragg-edge transmission cross section that is combined by the kinematical diffraction theory [15] with three new factors, $R_{hkl}(\lambda - 2d_{hkl})$, $P_{hkl}(\lambda, 2d_{hkl})$ and $E_{hkl}(\lambda, 2d_{hkl}, F_{hkl})$, as follows:

$$\sigma_{\text{coh}}^{\text{ela}}(\lambda) = \frac{\lambda^2}{2V_0}\sum_{hkl}\left|F_{hkl}\right|^2 d_{hkl} R_{hkl}(\lambda - 2d_{hkl}) P_{hkl}(\lambda, 2d_{hkl}) E_{hkl}(\lambda, 2d_{hkl}, F_{hkl}), \tag{1.3}$$

where V_0 is the unit cell volume, d_{hkl} is the crystal lattice plane spacing, and F_{hkl} is the crystal structure factor including the Debye-Waller factor.

The first new factor $R_{hkl}(\lambda - 2d_{hkl})$ is the edge profile function including the d_{hkl} information. This function describes the edge broadening due to the neutron pulse shape, the strain and the microstructure. The Dreele-Jorgensen-Windsor model [16] was implemented for strain imaging.

$$R_{hkl}(\lambda - 2d_{hkl}) = \frac{1}{2}\mathrm{erfc}(w) - \frac{\beta_{hkl}\exp(u)\mathrm{erfc}(y) - \alpha_{hkl}\exp(v)\mathrm{erfc}(z)}{2(\alpha_{hkl} + \beta_{hkl})}, \quad (1.4)$$

where

$$w = \frac{\lambda - 2d_{hkl}}{\sqrt{2}\sigma_{hkl}}, \quad (1.5)$$

$$u = \frac{\alpha_{hkl}}{2}\left[\alpha_{hkl}\sigma_{hkl}^2 + 2(\lambda - 2d_{hkl})\right], \quad (1.6) \qquad v = \frac{\beta_{hkl}}{2}\left[\beta_{hkl}\sigma_{hkl}^2 - 2(\lambda - 2d_{hkl})\right], \quad (1.7)$$

$$y = \frac{\alpha_{hkl}\sigma_{hkl}^2 + (\lambda - 2d_{hkl})}{\sqrt{2}\sigma_{hkl}} \quad \text{and} \quad (1.8) \qquad z = \frac{\beta_{hkl}\sigma_{hkl}^2 - (\lambda - 2d_{hkl})}{\sqrt{2}\sigma_{hkl}}. \quad (1.9)$$

σ_{hkl}, α_{hkl} and β_{hkl} are the broadening parameters of the Gaussian, the rise exponential and the decay exponential of the $\{hkl\}$ edge, respectively.

The crystallographic anisotropy due to the preferred orientation in a polycrystalline material changes the whole shape of a Bragg-edge transmission spectrum. Therefore, we implemented the modified March-Dollase preferred orientation distribution function [17] as the second new factor $P_{hkl}(\lambda,2d_{hkl})$ in the RITS code. The formulation is

$$P_{hkl}(\lambda,2d_{hkl}) = \frac{1}{2\pi}\int_0^{2\pi}\left(R^2B_{hkl}^2 + \frac{1-B_{hkl}^2}{R}\right)^{-\frac{3}{2}}\mathrm{d}\phi, \quad (1.10)$$

where

$$B_{hkl} = \cos A_{hkl}\sin\theta_{hkl} + \sin A_{hkl}\cos\theta_{hkl}\sin\phi. \quad (1.11)$$

θ_{hkl} is the scattering angle as a function of λ and $2d_{hkl}$ from the Bragg's law. φ represents the angle on a certain Debye-Scherrer ring of the scattering angle 2θ. A_{hkl} is the angle between the Bragg reflection vector <*hkl*> and the preferred orientation <*HKL*> that is parallel or perpendicular to the beam direction. The March-Dollase coefficient R provides the information of the degree of crystallographic anisotropy. If there is no anisotropy, both R and $P_{hkl}(\lambda,2d_{hkl})$ are unity. When the texture grew, R departs from one. $R = 0$ or ∞ means a single crystal specimen. Thus, we can get the information of the preferred orientation <*HKL*> and the degree of crystallographic anisotropy R from the shape change of a Bragg-edge transmission pattern.

The primary extinction effect is caused by the re-diffraction phenomenon inside one perfect crystal block (crystallite). This effect reduces the diffraction intensity by returning the diffracted neutrons in the direction of the transmitted beam. Hence, this phenomenon increases the transmitted neutron intensity, and decreases the Bragg-edge transmission cross section. Therefore, we implemented the Sabine's primary extinction function [18] as the third new factor $E_{hkl}(\lambda,2d_{hkl},F_{hkl})$ in the RITS code. This formulation is expressed by combining the backward-scattering (Bragg) component E_B and the forward-scattering (Laue) component E_L, as follows:

$$E_{hkl}(\lambda,2d_{hkl},F_{hkl}) = E_B\sin^2\theta_{hkl} + E_L\cos^2\theta_{hkl}, \quad (1.12)$$

where

$$E_B = \frac{1}{\sqrt{1+x}}, \quad (1.13)$$

$$E_L = 1 - \frac{x}{2} + \frac{x^2}{4} - \frac{5x^3}{48} + \cdots \quad \text{for} \quad x \le 1, \quad (1.14)$$

$$E_{\mathrm{L}} = \sqrt{\frac{2}{\pi x}}\left[1 - \frac{1}{8x} - \frac{3}{128x^2} - \frac{15}{1024x^3} - \cdots\right] \quad \text{for} \quad x > 1, \tag{1.15}$$

and

$$x = S^2\left(\frac{\lambda F_{hkl}}{V_0}\right)^2. \tag{1.16}$$

S is the crystallite size along the beam direction. Thus, we can get the crystallite size information from the intensity increase of transmission data.

Figure 1 shows an example of the spectral fitting using the RITS code to Bragg-edge transmission data along the normal direction of an α -iron plate of 5 mm thickness (cold rolled steel: 0.03%C, 0.04%Si, 0.5%Mn, 0.03%P and 0.03%S). The experiment was performed at the electron linac facility at Hokkaido University. The RITS model is more consistent with the experimental data than the traditional kinematical diffraction theory. Moreover, we can simultaneously obtain various crystallographic parameters. In the previous studies, it has been already indicated that the RITS code gave valid results of crystallographic information, by using various methods, e.g., the preferred orientation theories, the grains observations by an optical microscope, and the neutron diffraction experiments analyzed with their own Rietveld analysis code [12, 14].

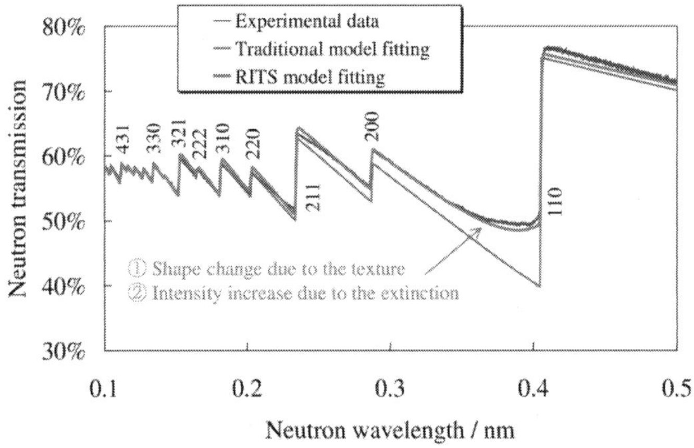

a = 0.28630 nm, <HKL> = <530>, R = 1.90, S = 3.44 μm & $\rho \times t = 4.42 \times 10^{22}$ cm^{-2}

Figure 1. Result of the Rietveld-type analysis of a Bragg-edge transmission spectrum of an α-iron plate of 5 mm thickness. The crystal lattice parameter a, the preferred orientation <HKL>, the degree of crystallographic anisotropy R, the crystallite size S and projected density $\rho \times t$ can be simultaneously extracted by the RITS code

2.2 Lattice space distribution of quenched steel rods

Here we refer one of examples of the spectroscopic imaging and its analysis. The lattice space analysis of quenched steel rods processed with the induction hardening. By the quenching process we obtain a toughened piece of steel. Such phenomenon is caused by the diffusionless transformation, and forms the martensitic microstructure. Its crystal structure expands a little from the original, and then the quenched piece has a strain inside and shows hardening. Normally, the hardening area is checked by method such as a hardness tester, but the information is limited in a local area. On the other hand, because the neutron has a large penetration depth, we can obtain the mean lattice space correlated with strain through the bulk depth on a wide area. For a realistic sample used in industry we need to measure thick sample. In this measurement feasibility for the thick bulk material is also tested. So we applied the Bragg edge lattice space analysis to the quenched steel measurement.

The experiment was carried at BL10 in J-PARC facility, Japan. It views the decoupled liquid hydrogen moderator, and the proton beam power was around 200 kW during the experimental period. The samples were cut out to 20 mm axial length from long quenched steel rods (JIS S45C, 20 mm diameter) which controlled the quenching depth as 3 mm, 5 mm and 7 mm. with 10 kHz induction hardening and tempering at 423 K. The used detector was the micro-channel plate (MCP) type, which has the 0.055 mm pixel resolution with 14 mm x 14 mm detecting area and is driven under the time event position recording [19]. The samples were set just before the detector window as their center axis parallel to the neutron beam. The obtained transmission spectra were analyzed with the RITS code and deduced the distributions of parameters such as the lattice space.
Figures 2(a), 2(b) and 2(c) are the energy sliced radiographies for the quenching depth 3mm, 5mm and 7 mm. We can recognize the brightness difference between the sample center and their rim. The resultant lattice space images after the RITS code analysis are shown in Figures 3(a), 3(b) and 3(c). From these figures we understand that the rim areas are expanded in comparison with the sample

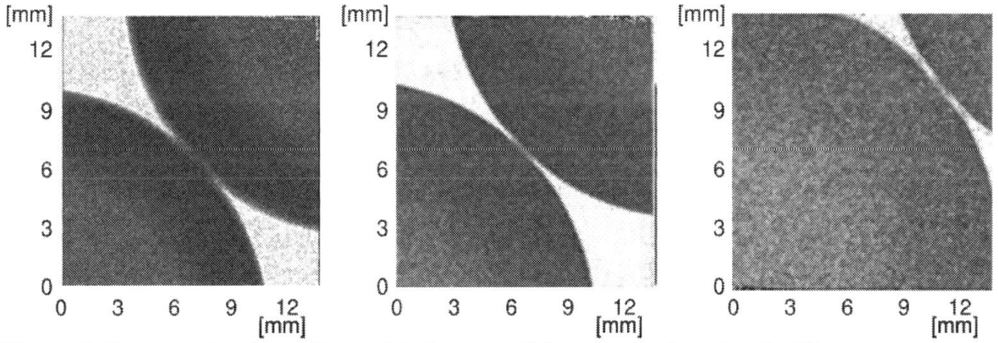

Figure 2. Energy selective radiographic images of the quenched steel rods. The neutron wave length region is from 4.100 to 4.111 Å. The productive quenching depths are (a) 3 mm, (b) 5 mm and (c) 7 mm, respectively.

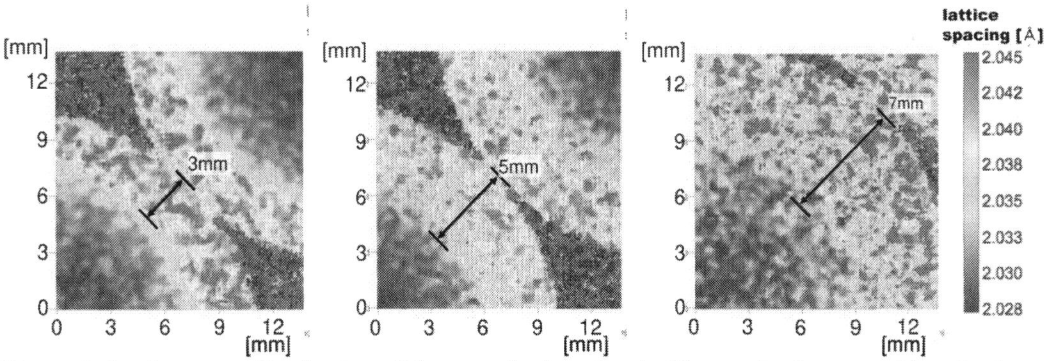

Figure 3. Lattice space distribution of the quenched steel rods. The productive quenching depths are (a) 3 mm, (b) 5 mm and (c) 7 mm, respectively.

centers, and that the effect is caused by the martensite phase formed in the quenching process. The hardening depth can be confirmed for each sample as is successively formed in the planned depth. Such information is important for industrial production, the RITS code analysis can offer the quantitative microstructural image piles upon the real image.

3. Imaging of dynamical information of hydrogen in cement pastes

Hydrogen is a typical incoherent scatterer and is included in many materials. Plant is one example of the typical material including water and the information of not only the water content but also the bound state of water will be sometime important. The freeze of water in a fuel cell is one of the issues for developing the fuel cell. Cement is also important as structure material, and the content of hydrogen is observed by using the traditional neutron radiography. In the cement paste there exist various types of hydrogen; free water, crystal water and hydrides. It is useful to obtain the information on the hydrogen distribution in the cement paste with the information of its bound state relating to the dynamical motion of hydrogen. It is well known that the total neutron cross section of the water molecule increases with the neutron wavelength [20, 21] and at long wavelength region the total cross section is expressed by $\sigma = a\lambda + b$[21]. Here, σ is the total neutron cross section, λ the neutron wavelength, a and b are coefficients. It was also indicated that 'a', the gradient of the total cross section, become larger with increasing the water temperature [21]. The total cross section of bound water and bound hydrogen in a cement paste will be similar to that of ice since the quasi-elastic neutron scattering spectra of a cement paste were re-produced by superposition of free water and ice spectra [22]. The difference of the gradients of the total cross section between ice and water will be used for distinguish the free water and the bound water or the hydride in the cement paste. In order to study this feature, we performed experiments on two kinds of the cements pastes, dry and wet cement pastes, since in the wet sample free water exists in vacancies and in the dry cement the rate of the free water is much smaller than that of the wet sample. The experiments were performed at BL10 beam line in J-PARC/MLF, and the detector used was the MCP detector [19]. The flight path length was 14.5m from the moderator surface to the detector. We put the cement paste sample in front of the detector. The resolution and the active area was written in previous chapter. The cement pastes were made by gradually mixing water and Portland cement powder. The weight percent of the water and the cement powder was 50%, respectively. After the mixing the paste was casted in a mold with a diameter of 50 mm and length of 100 mm, and after about 72 hours several pieces of cement paste were cut out. The thickness of the samples was 0.5 cm and the diameter 50 mm. The thickness was almost the thinnest one we could make. To prepare the wet sample and the dry sample the pieces were cured in water and then dried.

We could obtain the data up to about 12 Å due to frame overlap of the neutron beam. Figure 6 shows the transmission images obtained by whole energy range of neutrons measured. There is difference of the contrast in each sample corresponding to mainly hydrogen content at each pixel. It is also recognized that the wet sample includes the larger number of hydrogen atoms compared with the dry one. At each pixel we got a wavelength dependent transmission and deduced a micro-cross section.

Figure 6. Transmission images of a dry cement paste (left) and a wet cement paste (right). The darker image of the wet sample indicates higher content of hydrogen.

By fitting a linear function to the transmission cross section in a range of 7-11.3 Å, the gradient a[barns/Å/hydrogen] was determined. In this process we summed up 16x16 pixel data to improve the statistics. Figure 7 shows spatial distributions of the gradient of the total cross section obtained by the transmission data. In the results, the gradient changes dependent on position in the same sample. In the dry cement case the gradients are generally larger than those of the wet one. This is inconsistent with the expected result that bound water would give smaller gradient. This is considered to be due to a reason that since in the wet sample there are many free water molecules and sample thickness was not sufficiently thin to avoid the up-scattering of the long wavelength neutrons. The temperature of the sample was much higher than the neutron energy used for analysis. Therefore, the long wavelength neutrons incident on the sample have high possibility for scattering obtaining the energy, namely up-scattering. The energy transferred to the neutron is much higher than the incident neutron energy. Therefore, the energy spectrum of the up-scattered neutrons will not depend very much on the incident energy. As a result the transmitted neutron has similar energy spectrum. This trend makes the total cross section gentler dependence of the wavelength. Therefore, the effective cross section obtained by the transmission data was in some sense an averaged cross section over the thermal energy. On the other hand, in the case of the dry cement paste the scattering was mainly elastic by the bound hydrogen. Therefore, the energy of the neutron changed not so much, and the effective cross section obtained by the transmission data reflects the incident energy. We have confirmed this effect by simulation calculation. The result indicate that for deducing the gradient we should use thinner sample or use a collimator to remove the scattered neutrons. However, by adopting suitable experimental method we will get information on the bound state of hydrogen in materials. Anyway, we need more complicated analysis for getting the quantitative information.

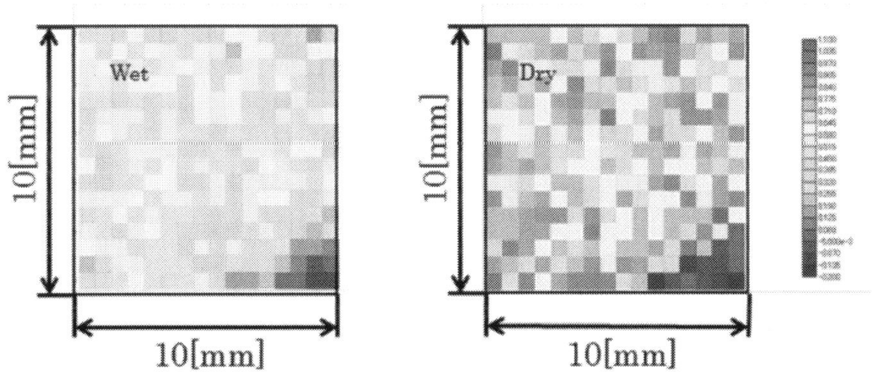

Figure. 7. Gradient distribution of the cross section of the wet (left) and the dry (right) samples. The pixcel size was about 0.65 mm square. The gradient becomes larger from blue to red. The blue zones at the right left corners in both figures are the areas where the samples do not exist, namely, which appear due to the processing of the colouring.

4. Magnetic field imaging using polarized neutrons

The neutron imaging technique using neutron polarization analysis method has been developed to visualize the spatial distribution of magnetic field [23]. This technique is based on the motion of a neutron spin passing through a magnetic field, and the rotation of a neutron spin is measured as the change in the neutron polarization with spatial resolution using a two-dimensional neutron detector.
The neutron spin rotation angle φ due to the Larmor precession in a magnetic field B is expressed by following equation,

$$\varphi = \frac{\gamma}{v}\int_{path} B ds \tag{4.1}$$

where γ is the gyromagnetic ratio of the neutron and v the neutron velocity. Therefore, we can get the cumulated magnetic field strength along the neutron beam path from the rotation angle using polarized monochromatic neutron beam. However, as measurements of neutron polarization with a monochromatic neutron beam results in an uncertainty of the rotation angle due to the periodicity of 2π, the study of wavelength dependence of the polarization is necessary to determine the rotation angle. Then, integral of magnetic field along the flight path of polarized neutrons can be quantified.

The usage of the pulsed neutron beam, whose energy can be recorded precisely by the time-of-flight (TOF) method, possesses an advantage on the measurements of energy-dependent phenomena, so the polarized pulsed neutron beam is suitable for the magnetic field imaging experiments. Recently, we have performed the first proof-of-principle experiment of polarized pulsed neutron imaging and successfully confirmed the possibility of quantification of magnetic field strength inside the solenoid coil and the direction from the quantized axis of the incident neutron spin by analysing the wavelength dependence of polarization after transmission of the sample [24].

On the other hand, to analyse the magnetic field vector, 3D polarization analysis is required, for which each polarization vector of incident and transmitted neutron beams was controlled three dimensionally [25]. The polarization change due to the spin rotation in a magnetic field is described by a (3 × 3) matrix ***D***, and its elements are expressed by direction cosines of the magnetic field vector [26]. Thus, measuring and analysing the wavelength dependence of each element of matrix ***D*** at each position allow determination of the spatial distribution of the magnetic field vector. Preliminary imaging experiments using 3D polarization analysis were performed at the beam line of BL10 of MLF at J-PARC. The schematic illustration of our experimental setup was shown in Figure 8. Electric currents, which changed synchronized with the flight time of the pulsed neutron to give the same spin rotation angle to wide wavelength range neutron beam, were applied to the spin rotators.

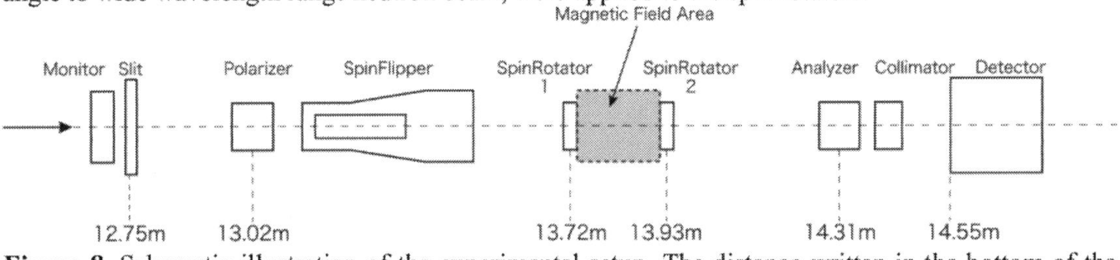

Figure 8. Schematic illustration of the experimental setup. The distance written in the bottom of the figure is from the source of pulsed neutron beam. Both polarizer and analyser consist of magnetic super mirrors. 2D detector was composed of a position-sensitive photomultiplier and a ZnS scintillator [27]. The beam size was 12 mm x 12 mm and the channel width of this detector was 0.25 mm.

After deciding the applied current condition of each spin rotator, test measurements were performed by superposing a small guide field at the centre of two spin rotators to check whether they work properly. Results of the wavelength dependence of matrix elements D_{ij} ($i, j = x, y, z$, where z axis is taken parallel to the beam direction) were shown in Figure 9. In this figure, matrix elements regarding x and z axis showed oscillating behaviour apparently. The diagonal elements D_{xx} and D_{zz} could be fitted with cosine curves though the off-diagonal elements D_{xz} and D_{zx} could be done with sine curves. This indicates that the magnetic field was along y axis and its orientation was a positive direction. In addition, the averaged magnetic field strength was evaluated to be 0.175 mT from the oscillation frequency. These results were in good agreement with the results measured by a magnetic field detector.

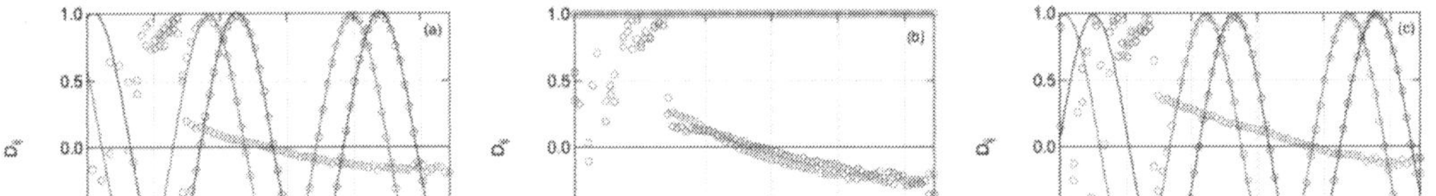

Figure 9. Wavelength dependence of matrix element D_{ij} ($i, j = x, y, z$). Incident neutron beam polarization was parallel to (a) x axis, (b) y axis and (c) z axis. Solid lines indicated the results of fitting using a sinusoidal function.

Then, we placed a soft magnetic metal foil sample with a thickness of 0.03 mm, in the midway of the spin rotators and performed imaging measurements. This soft magnetic foil was a commercial amorphous Fe alloy used for the core of an electric transformer, with almost isotropic magnetic properties. The sample size was 50mm in height, 30 mm in width and the thickness of 0.03 mm. The neutron beam passed across the central 12 mm by 12 mm region. The obtained images of diagonal elements were shown in Figure 10. The images of D_{xx} and D_{zz} showed stripe patterns, whose width were 2 ~ 3 mm, while that of D_{yy} was almost homogeneous. These results suggest that the component of magnetic field vector along *y* axis is homogeneous inside the foil, but *x* and *z* axis components were different at each position. The stripe structures shown in the images of D_{xx} and D_{zz} can be regarded as the evidences of the magnetic domain structure in the soft magnetic foil. Moreover, the features of polarization distribution images shown in Fig. 10 can be explained by a picture, that the small magnetic field applied along y axis magnetized the sample but partially there remained some magnetization component which could not align with the applied field. In other words, the magnetization vector inside the foil almost directed along the *y* axis, but it should canted toward *x* or *z* direction.

Because of insufficiently optimized set up, quantifying each component of the magnetic field vector inside the sample using off-diagonal matrix elements was difficult. By removing the stray field and tuning the spin rotators more precisely, polarized pulsed neutron imaging using 3D polarization analysis method would enable the quantitative imaging of arbitral magnetic field vector.

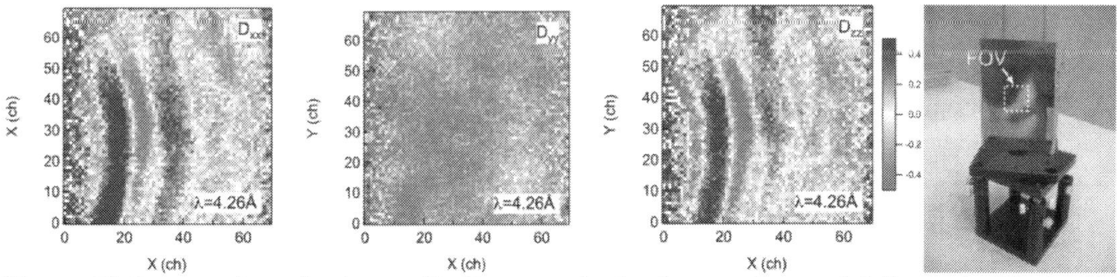

Figure 10. Images of matrix element D_{ii} (i = x, y, z) of soft magnetic metal foil sample. The incident neutron wavelength was 4.26 Å. The right figure is the photograph of the sample. FOV means field of View.

5. Conclusions

It has been demonstrated that the pulsed neutron imaging can give image of the crystallographic information by one measurement. Such information for thick materials has not been obtained by other method such as X-ray and electron beam scattering. The imaging of the hydrogen dynamical (bound state) information was tried. The difference of the gradient of the total neutron cross section was indicated in two sample, which suggests the possibility to obtain the information on the hydrogen bound state. However, for the detailed data analysis we need a simulation calculation using a precise neutron cross section. Furthermore, it has been indicated that the pulsed neutron measurement can give easily an image of the strength of the magnetic field. The pulsed neutron imaging is still at developing stage and further improvement is required to deduce more precise quantitative information and also to deduce the new information that has not been obtained so far.

.

Acknowledgement

The authors thanks Dr. Tremsin very much for giving us opportunities to use his MCP detector and joining with our experiments. The discussion with him was very useful for promoting the research. This work was supported by KAKENHI (JSPS) 23226018.

References

[1] Kiyanagi Y and Iwasa H, 1993 *JAERI-M 93-228* **2** 796
[2] Sakamoto N, Kiyanagi Y, Sato S, Sagehashi H, Furusaka M, Suzuki J, Littrell K C, Loong C K, Gorin A, Manuilov I, Ryazantsev A, Kuroda K, Sakai K, Tokanai F, Adachi T, Oku T,

Ikeda K, Miyasaka H, Suzuki S, Morimoto K and Shimizu H M 2003 *J. Applied Crystallography* **36** 820
[3] Santisteban J R, Edwards L, Steuwer A and Withers P J, 2001 *J. Appl. Crystallogr.* **34**, 289
[4] Santisteban J R, Edwards L, Fizpatrick M E, Steuwer A, Withers P J, 2002 *Appl. Phys. A* **74** S1433
[5] Mizukami K, Sato S, Sagehashi H, Ohnuma S, Ooi M, Iwasa H, Hiraga F, Kamiyama T and Kiyanagi Y, 2004 *Nucl. Instr. and Meth. Phys. Res.* **A529** 310
[6] Kiyanagi Y, Kamiyama T, Iwasa H and F. Hiraga F, 2004 *Key Engineering Materials* **270** 1371
[7] Kiyanagi Y, Sakamoto N, Iwasa H, Kamiyama T, Hiraga F, Sato S, Sagehashi H, Ino T, Furusaka M, Suzuki J, Gorin A, Manuilov I, Ryazantsev A, Kuroda K, Sakai K, Tokanai F, Miyasaka H, Adachi T, Oku T, Ikeda K, Suzuki S, Morimoto K, and Shimizu H M 2005 *IEEE Tran.Nuclear Science* **52** 371
[8] Kiyanagi Y, Mizukami K, Kamiyama T, Hiraga F and Iwasa H 2005 *Nucl. Instr. and Meth., Phys. Res.* **A542** 316
[9] Kiyanagi Y, Kamiyama T, Nagata T and Hiraga F 2006 *Key Eng. Mat.* **321-323** 1663
[10] Iwase K, Sakuma K, Kamiyama T and Kiyanagi Y, 2009 *Nucl. Instr.. and Meth., Phy. Res.* **A605** 1
[11] Takada O, Kamiyama T and Kiyanagi Y 2010 *J. Nucl. Mat.* **398** 129
[12] Sato H, Kamiyama T and Kiyanagi Y, 2011 *Materials Transactions*, **52** 1294
[13] Sato H, Takada O, Iwase K, Kamiyama T and Kiyanagi Y 2010 *J. Phys. Conf. Ser.* **251** 012070
[14] Sato H, Kamiyama T, Iwase K, Ishigaki T and Kiyanagi Y *Nucl. Instr. Meth. A* (in press)
[15] Fermi E, Sturm W J and Sachs R G 1947 *Phys. Rev.* **71** 589
[16] Vogel S 2000 *Ph. D. Thesis* Christian Albrechts Universität
[17] Larson A C and Von Dreele R B 2004 *LAUR 86-748* Los Alamos National Laboratory
[18] Sabine T M, Von Dreele R B and Jørgensen J -E 1988 *Acta Crystallogr. Sec. A* **44** 374
[19] Tremsin A S, McPhate J B, Kockelmann W A, Vallerga J V, Siegmund O H W and Feller W B 2009 *IEEE Trans. Nucl. Sci.* **56** 2931
[20] Heinloth K 1961 *Zeitschrift fuer Physik* **163** 218
[21] Zaitsev K N et al. 1991 *Soviet Journal of Atomic Energy*, **70**, 238
[22] Inoue K, Kiyanagi Y and Sakamoto Y 1980 *J. At. Energy Soc. Japan* **22** 189
[23] Kardjilov N, Manke I, Strobl M, Treimer W, Meissner M, Krist T and Banhart J 2008 *Nature Physics* **4** 399
[24] Shinohara T, Sakai K, Ohi M, Kai T, Harada M, Oikawa K, Maekawa F, Suzuki J, Oku T, Takata S, Aizawa K, Arai M and Kiyanagi Y *Nucl. Instrum. Methods* A (in press)
[25] Janoschek M, Klimko S, Gähler R, Roessli B and Böni P 2007 *Phys.* B **397** 125
[26] Rekveldt M Th 1973 *Z. Phys.* **259** 391
[27] Hirota K, Shinohara T, Ikeda K, Mishima K, Adachi T, Morishima T, Satoh S, Oku T, Yamada S, Sasao H, Suzuki J and Shimizu H M 2005 *Phys. Chem. Chem. Phys.* **7** 1836

Journal of Physics: Conference Series **340** (2012) 012011 doi:10.1088/1742-6596/340/1/012011

High precision depolarisation measurements with an opaque test bench

C Klauser [1,2], J Chastagnier[1], D Jullien[1], A Petoukhov[1] and T Soldner[1]

[1] Institut Laue-Langevin, 38042 Grenoble Cedex 9, France

[2] Excellence Cluster Universe, Technische Universitaet Muenchen, 85748 Garching, Germany

E-mail: klauser@ill.fr

Abstract. We present a tool to investigate neutron depolarisation effects with 10^{-4} precision. The test bench consists of two opaque ^{3}He cells with in-situ adiabatic fast passage flipping of the helium spin. The cells polarise a neutron beam to more than 99.99 % and analyse its polarisation with high accuracy. For depolarisation studies, a sample can be inserted between the two cells and its effect on a primarily highly polarised beam is analysed. The test bench has been validated at the cold neutron beam PF1B at Institut Laue-Langevin, France. We present here preliminary results for the wavelength range from 5Å to 7Å. Polarisation with super mirrors is limited to about 99.7%. Direct evidence of depolarisation in the order of 10^{-3} in polarising super mirrors was found by the test bench.

1. Introduction

High precision in polarisation of large neutron beams to this day has been of the order of 10^{-3}, mostly produced by high performance polarising super mirror benders. The best polarisation obtained with super mirrors is of 99.7 % [1], using the crossed X-SM geometry. We present a tool to measure depolarising effects in samples to an accuracy level of 10^{-4}. On our test bench, we use ^{3}He for both polarisation and polarisation analysis. The transmission T through a cell of polarised helium to a degree of P_{He} for neutrons of wavelength λ with spins parallel (+) and antiparallel (-) to the helium's spin can be described by:

$$T_{\pm} = \frac{1}{2} \exp^{-(1 \mp P_{\mathrm{He}})O} \quad (1)$$

$$O = \frac{0.0733}{\mathrm{bar} \cdot \mathrm{cm} \cdot \text{Å}} (p \cdot l \cdot \lambda) \quad (2)$$

where p is the helium pressure and l refers to the cell's length. The letter O is called the cell's opacity. The resulting polarisation can hence be given as:

$$P = \frac{T_+ - T_-}{T_+ + T_-} = \tanh(P_{\mathrm{He}} O) \quad (3)$$

For details, see [2]. An adequate choice of the cell opacity by driving its parameters to appropriate values results thus in high neutron spin polarisation in the theoretical description of the cell, see Eqn. (3). As an example, neutrons of 5 Å wavelength are polarised to 99.99 % after crossing a cell of 15 cm of length, He polarised to 75 % at a pressure of 1.4 bar.

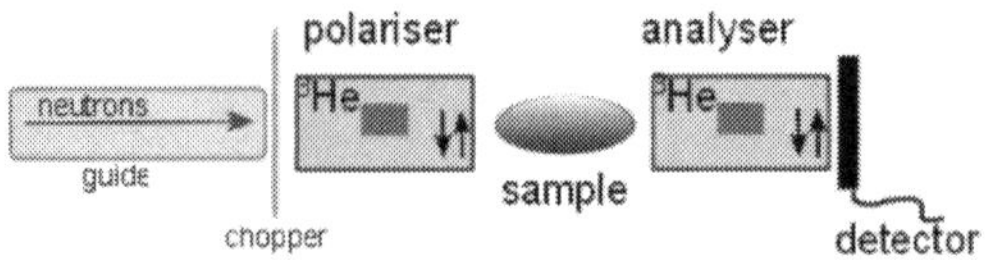

Figure 1. Set-up of the test bench as used on PF1B in November 2010.

2. Set-up

A sketch of the test bench is shown in Fig. (1). The test bench consists of a polarising and an analysing helium cell contained in so-called magic boxes [3]. A neutron detector follows directly after the second magic box. The sample is placed between the two cells. Wavelength resolution is provided by a chopper in front of the first magic box and time of flight. Measurements are taken in two different cell configurations: The ^{3}He polarisations in two cells are either parallel (countrate N_+) or anti-parallel (countrate N_-) alignment to the ^{3}He spin. Thus, the product of the polarising power P and the analysing power A is:

$$A \cdot P = \frac{N_+ - N_-}{N_+ + N_-} \tag{4}$$

As both polariser and analyser cell do have the same shape and were filled with ^{3}He of the same pressure and initial polarisation, the product $A \cdot P$ can be estimated via Eqn. 3 as :

$$A \cdot P = \tanh(P_{\text{He}}O) \cdot \tanh(P_{\text{He}}O) = \tanh^2(P_{\text{He}}O) \tag{5}$$

The experiment took place at the high flux instrument PF1B of Institut Laue-Langevin, Grenoble [4]. Measurements were taken for cells with 1.7 bar pressure and an initial Helium spin polarisation of 75 %. The cells' relaxation times were found to be longer than 140 hours and measurements were performed within the first five hours after the filling of the cells. Flipping of the ^{3}He polarisation took place in situ; polarisation loss in ^{3}He was less than 10^{-5} per flip so that we could safely neglect the number of spin flips for the cells' performance. Between the Helium cells, a guiding field was installed. With no sample present, the resulting polarisation was higher than 99.99 % (see red triangular data points in Fig. 2) for wavelengths longer than 5 Å. For wavelengths longer than 7Å, the transmission (see also Eqn. (1)) was too low to provide good statistics. Data were treated assuming a constant background and were dead-time corrected. Data points were fitted to Eqn. (5) using the helium polarisation as free parameter and taking into account the finite wavelength resolution (dominated by the chopper).

3. Depolarisation In Polarising Supermirror Bender

We introduced on the sample site a Co-Ti Schaerpf Bender [5] in a magnetizing field of 0.04 T. In Fig. 2, the polarisation of the two ^{3}He cells only is compared to the final polarisation obtained in a set-up with cell-bender-cell. Clearly, polarisation of the latter is several times 10^{-3} smaller than for the other configuration. Hence, the highly polarised neutrons coming out of the polariser cell undergo depolarisation when subject to a super mirror bender. Depolarisaton effects by the bender's housing were excluded.

4. Conclusion

We have shown a possibility to measure depolarisation effects of neutrons with an accuracy of 10^{-4} by the use of an opaque test bench. The sample is placed between two helium

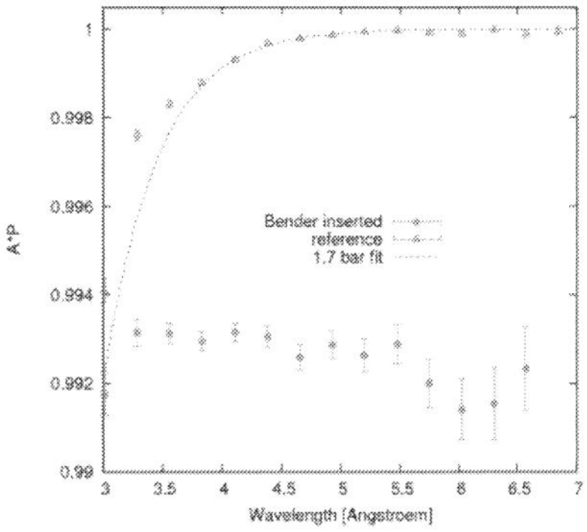

Figure 2. Product of polarisation and analysis power versus wavelength. Plot shows both a reference case without a sample (red triangles, with fit in dark red performed) and the case with the sample of a super mirror bender (green dots).

spin filters. Under ideal conditions, i.e. a sufficiently homogeneous high flux neutron beam, the cells' parameters can be chosen in a way that polarisation of 99.99 % and the required experimental sensitivity are achieved. When applied to a supermirror bender, depolarising effects were observed. The test bench may find application in neutron scattering processes to closely investigate depolarisation effects as well as in particle physics to develop highly performant polarisation devices.

5. Acknowledgement

This work is supported by the Deutsche Forschungsgemeinschaft, SSP 1491 project SO 1058/1-1.

6. References

[1] Kreuz M, Nesvizhevsky V, Petoukhov A and Soldner T 2005 *Nucl. Instr. and Meth.* A **547** 583
[2] Coulter KP, A B, Happer W, Chupp T E and Wagshul M E 1988 *Nucl. Instr. and Meth.* A **270** 90-4
[3] Petoukhov AK 2006 *Nucl. Instr. and Meth.* A**560** 480
[4] Abele H et al 2006 *Nucl. Instr. and Meth.* A **562** 407
[5] Schaerpf O 1989 *Physica B* **639** 156-7

Journal of Physics: Conference Series **340** (2012) 012012 doi:10.1088/1742-6596/340/1/012012

Neutron Larmor Diffraction study of the $BaM_2(XO_4)_2$ (M = Co, Ni; X = As, P) compounds

Nicolas Martin[1], Louis-Pierre Regnault[1] and Serguei Klimko[2]

[1] SPSMS, UMR-E CEA / UJF-Grenoble 1, INAC, Grenoble, F-38054, France

[2] Laboratoire Léon Brillouin, CEA-CNRS, CEA/Saclay, 91191 Gif sur Yvette, France

E-mail: `regnault@ill.fr`

Abstract. Neutron Larmor Diffraction is a relatively new concept in neutron scattering which allows high resolution studies of subtle structural features in condensed matter. Several setups exist in various neutron facilities around the world. One of them is the Zero-field spin-Echo for Triple-Axis option (ZETA), available on the thermal triple-axis spectrometer IN22 (CEA-CRG beamline at Institut Laue Langevin). We have used ZETA to study thermal evolution of lattice constants in serveral systems of physical interest. Among them, the $BaM_2(XO_4)_2$ family is a serie of compounds which have been intensively studied for three decades, mainly because of their fundamental though puzzling magnetic properties. In this paper, we aim to shed a new light on the physical understanding of those systems by investigating the link between structural and magnetic ordering. The magnetoelastic effect is found to be weak but can be resolved thanks to the high resolution provided by NLD.

1. Introduction

Materials with strong magnetoelastic couplings are the subject of a large number of studies because of their certain proximity to a multiferroic state. For instance, this point has been stressed in the case of hexagonal manganites $RMnO_3$ ([1]). However, magnetostriction is a more general property of magnetic materials and is observed in many systems. To probe the existence of such a coupling between spin and lattice degrees of freedom in the $BaM_2(XO_4)_2$ family, we chose to measure the thermal evolution of lattice constants in three systems belonging to this serie by using Neutron Larmor Diffraction (NLD). Interestingly, all of them cristallize within the same space group but display different magnetic structures when cooled below their respective Néel temperature. In the following, we will introduce the main features and interesting facts about the studied compounds (Sec. 2). As already reported elsewhere (*e.g.* in [2]), a state of the art Zero-field spin-Echo for Triple-Axis (ZETA) option has been recently commisionned and is now available on the thermal TAS spectrometer IN22. Initially designed for studying lifetimes of elementary excitations in different class of single-crystalline samples, this setup can easily be operated into a high resolution diffraction mode using NLD. This elegant method allows to circumvent the usual tight compromise between intensity and resolution. A handwaving explanation of its principle and the specificities of the ZETA option will be given (Sec. 3). Finally, we will present the results obtained on the thermal lattice response of $BaM_2(XO_4)_2$ (Sec. 4).

2. Basic properties of the $BaM_2(XO_4)_2$-family

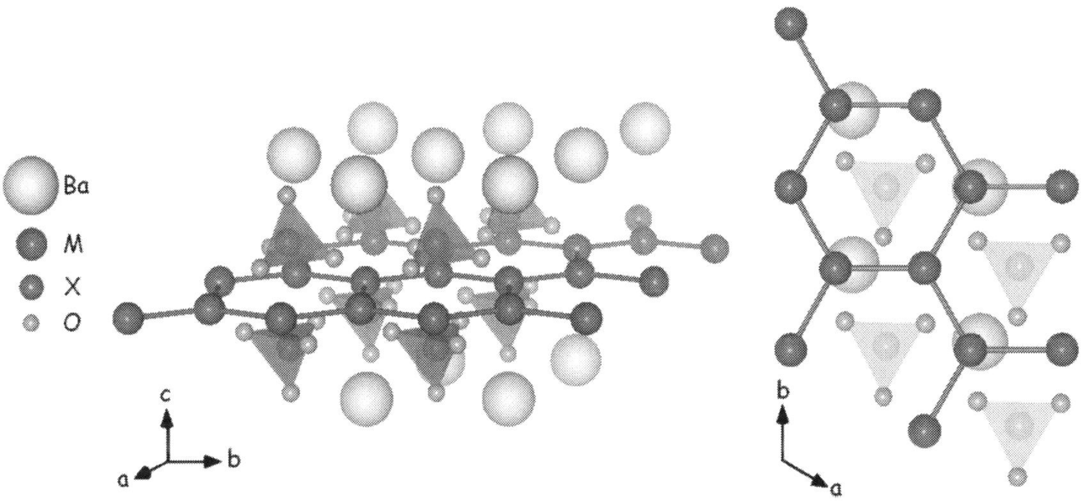

Figure 1. General structure of $BaM_2(XO_4)_2$ compounds. Each plane is monoatomic. Magnetic ions (in red) form a honeycomb lattice.

Table 1. Basic properties of the $BaM_2(XO_4)_2$ compounds studied in this paper. Lattice parameters are given at T = 4.2 K and d_2/d_1 is the ratio between the in-plane and interplane M-M distances.

Formula	a (Å)	c (Å)	d_2/d_1
$BaNi_2(PO_4)_2$	4.81	23.22	2.78
$BaNi_2(AsO_4)_2$	4.94	23.43	2.85
$BaCo_2(AsO_4)_2$	5.00	23.25	2.89

$BaM_2(XO_4)_2$, with M being a transition metal (Co or Ni) and X a pnictogen (As or P) is a family of compounds with interesting magnetic properties. All of them crystallize within the rhombohedral space group $R\bar{3}$ and no strucural phase transition has ever been measured down to the lowest temperature. Their layered structure consists in a stacking of atomic planes where magnetic ions sit on a honeycomb lattice. Theses planes are well separated from each other which naturally makes $BaM_2(XO_4)_2$ a very good playground to study two-dimensionnal magnetism (see Fig. 1 and Tab. 1). For instance, this 2-d character is evidenced by inelastic neutron scattering and the presence of a T^2-term in the low temperature specific heat ([3]). On the other hand, magnetization and neutron diffraction measurements reveal a strong planar spin anisotropy. From the microscopic magnetism point of view, each compound behave differently upon cooling. All of them order at finite temperatures but, even if their nuclear structure stays the same, different structures, *i.e.* described by different $\vec{k}$-vectors, arise. In Tab.2, we give the different exchange integrals determined from the analaysis of spin waves dispersion laws. We note that in each case, the third neighbour exchange integral is negative. In $BaNi_2(PO_4)_2$, $\mathcal{J}_1$ and $\mathcal{J}_2$ are also negative but in arsenides, they are positive. In fact, it is helpful to define an effective exchange integral for each compound like $\tilde{\mathcal{J}} = -E(\vec{k})/3\mathcal{S}^2$, where $E(\vec{k})$ is the classical energy of the ground state. $\tilde{\mathcal{J}}$ is actually a measure of the interactions present in these systems. Measurement of the thermal evolution of the magnetic moment yields different behaviours for each sample. If we define a critical scaled temperature like $t_c = T_N/|\tilde{\mathcal{J}}|\mathcal{S}(\mathcal{S}+1)$, it follows that the lower t_c is, the steeper the moment variation will be (see Fig. 2). Those differences motivated us to lead a comparative study of their thermal expansion properties along both in-plane and out-of-plane axis to probe the existence of a hypothetical magnetoelastic coupling and the differences which may appear.

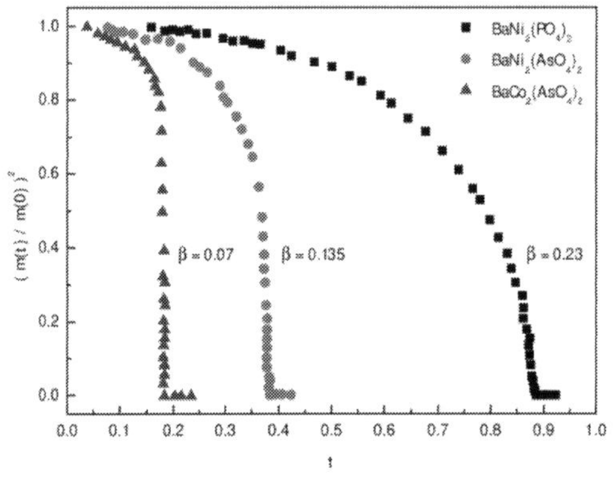

Figure 2. Square of the magnetic moment obtained by neutron diffraction plotted as a function of the reduced temperature $t = T/|\tilde{\mathcal{J}}|\mathcal{S}(\mathcal{S}+1)$. Fitting the data to a power law of the form $I(T)/I(0) = A(1 - T/T_N)^{2\beta}$ yields dramatically different critical exponents β.

Table 2. Magnetic properties of the $BaM_2(XO_4)_2$ compounds studied in this paper. $\mathcal{S}$ is the spin born by magnetic ions, $\mathcal{J}_1$, $\mathcal{J}_2$ and $\mathcal{J}_3$ are respectively the first, second and third nearest neighbour exchange integrals, T_N is the Neel temperature, β the critical exponent and $\vec{k}$ is the propagation vector of the magnetic structure in each compound (from [4]).

Formula	$\mathcal{S}$	$\mathcal{J}_1/k_b$ (K)	$\mathcal{J}_2/k_b$ (K)	$\mathcal{J}_3/k_b$ (K)	T_N (K)	β	$\vec{k}$
$BaNi_2(PO_4)_2$	1	-2.2	-0.3	-8.8	24	0.23	(0, 0, 0)
$BaNi_2(AsO_4)_2$	1	3.4	0.2	-10.0	18.5	0.135	(0.5, 0, 0.5)
$BaCo_2(AsO_4)_2$	1/2	38	1.5	-10.0	5.35	0.07	(0.27, 0, -1.33)

3. Neutron Larmor Diffraction with ZETA

Neutron Larmor Diffraction (NLD) is now well established as a powerful derivative of Neutron Spin Echo for structural studies in condensed matter physics. To understand how this technique works, let us first consider a general feature of elastic neutron scattering. Bragg condition selects the component of the neutron wavevector perpendicular to the lattice planes involved in the very process of diffraction according to the following vector relation:

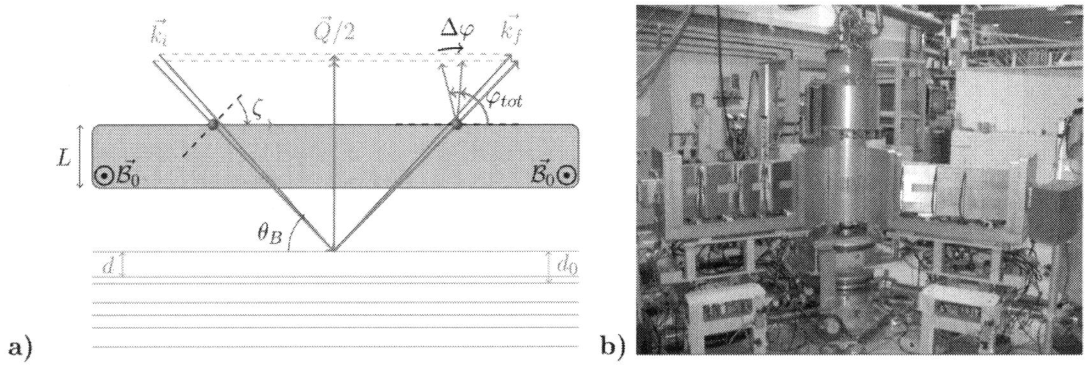

Figure 3. *a)* Layout depicting the principle of NLD. *b)* ZETA option as installed on IN22. The upper part of the μ-metal shielding is removed to show the four pairs of radio-frequency spin flippers (see text).

$$\vec{k}\sin\theta_B = \vec{Q}/2 \tag{1}$$

where $|\vec{k}| = m_n v/\hbar$ and $|\vec{Q}| = 2\pi/d$ with m_n the neutron mass, v its velocity and d the lattice spacing being measured. The latter equation means that the scattering angle is completely controlled by d (or its reciprocal space counterpart, $|\vec{Q}|$) for a given neutron velocity v (or wavevector $|\vec{k}|$). In general, a neutron beam contains a distribution of $\vec{k}$ and the evaluation of d is consequently vitiated by an error calculated as follows:

$$\frac{\delta d}{d} = \left(\left(\frac{\delta k}{k}\right)^2 + \left(\frac{\delta\theta_B}{\theta_B}\right)^2 \cot^2\theta_B\right)^{1/2} \tag{2}$$

Thus, measuring d with high accuracy can be done by conventionnal neutron diffraction (CND) but this requires a tight collimation of the beam. This naturally results in a loss of intensity as while constraining neutrons trajectories, the ones which can travel through the apparatus become more and more scarce. To circumvent this limitation, it has been proposed to measure neutrons velocity, and hence d, by means of Larmor precession using a setup such as the one which is sketched in Fig. 3 *a)* ([5]). The neutron beam is first polarized. After passing through the apparatus, the total phase, *i.e.* the rotation angle of the neutron magnetic moment around the magnetic field $\vec{B}_0$, is simply proportional to the time spent by each neutron within this field region and hence to sample dependent quantities:

$$\varphi_{tot}(d) = \omega_L t = \frac{\omega_L L}{v\cos\zeta} = 4\frac{m_n\omega_L L}{h}\cdot\frac{\sin\theta_B}{\cos\zeta}\cdot d \tag{3}$$

where $\omega_L = \gamma_n|\vec{B}_0|$ is the Larmor pulsation associated with the field $\vec{B}_0$ and $\gamma_n = 2\pi\cdot 2.916$ rad.s^{-1}.G^{-1} is the neutron gyromagnetic ratio. Aligning the field boundaries parallel to the lattice planes, *i.e.* setting $\zeta = \pi/2 - \theta_B$ (*NLD condition*), cancels the velocity dependence of the phase (Eq. 3) and we find that this quantity is simply proportionnal to d on the one hand, and to user-defined quantities such as ω_L and L on the other hand. Then, it follows that any change in d will affect φ_{tot} accordingly:

$$\Delta\varphi_{tot} = \varphi_{tot}(d) - \varphi_{tot}(d_0) = \varphi_{tot}(d_0)\cdot\frac{\Delta d}{d_0} \tag{4}$$

where d_0 is to be taken as a reference. The latter equality obviously implies that relative evolution of the total phase is identical to the relative evolution of lattice constants under pressure, stress or temperature variation. Thus, the measurement of the relative evolution of d is done by slightly changing the field integral $\omega_L L$ and the intensity is modulated because the beam is finally analyzed. Another implication of Eq. 4 is that NLD allows to measure the lattice parameters distribution function width which can be retrieved by measuring the final polarization $\mathcal{P}_f$ of the beam as a function of total phase. Using Eq. 4, the latter quantity turns out to be the cosine Fourier transform of the lattice spacing distribution which is generally assumed to have a gaussian shape:

$$\mathcal{P}_f(\varphi_{tot}, \delta) = \mathcal{P}_0 \langle \cos \Delta\varphi_{tot} \rangle = \mathcal{P}_0 \langle \cos \left[\varphi_{tot}(d_0) \cdot \frac{\Delta d}{d_0} \right] \rangle = \mathcal{P}_0 \sum_i \exp\left(-\frac{\delta_i^2 \varphi_{tot}^2(d_0)}{16 \ln 2} \right) \quad (5)$$

where $\mathcal{P}_0$ is the natural beam polarization, δ_i is the full width at half maximum of the individual lattice spacings distribution functions while i has to be equal to 1 for a true single crystal. Thus, in theory, increasing φ_{tot} provides a better resolution but also leads to a substantial depolarization of the beam and, in the extreme cases, it becomes more and more difficult to maintain high accuracy. We shall come back to this point later on.

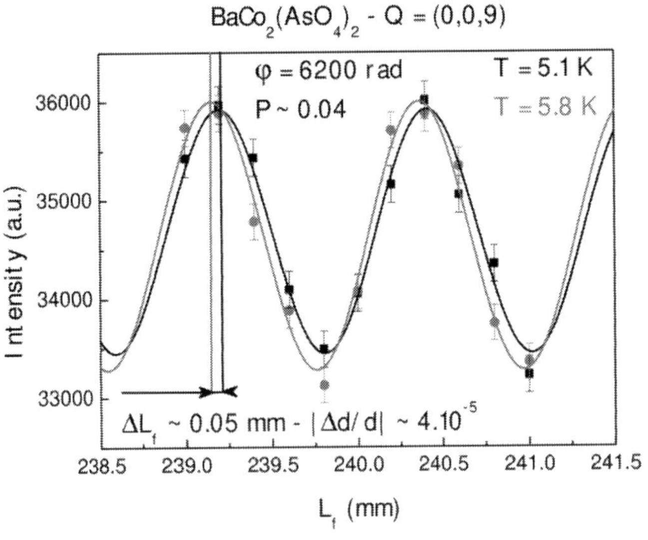

Figure 4. Typical signal measured with Neutron Larmor Diffraction. When d changes as a consequence of a temperature change, the signal is slightly shifted and the phase is recorded as the position of a local maximum. This example shows that a positionning accuracy of the translation stage of the order of 10 μm is needed to achieve a resolution $\Delta d/d \sim 10^{-5}$ at $\varphi_{tot} \sim 6200$ rad.

On the technical point of view, field regions are simulated by pairs of radio-frequency spin flippers (RFSFs). A RFSF consists of a flat solenoid which produces a homogeneous vertical static field $\vec{\mathcal{B}_0}$ and encloses radio-frequency coils producing a small oscillating field with frequency ω_{rf} and amplitude $\vec{\mathcal{B}_{rf}}$ in the horizontal scattering plane. When the RFSFs are set to resonance, *i.e.* when $|\vec{\mathcal{B}_0}| = \omega_{rf}/\gamma_n$, one pair simulates a field of magnitude $\mathcal{B}_0$ over a length

L, where L is the distance between coils ([6]). ZETA option is composed of 4 pairs of RFSFs which are located inside a double μ-metal shielding to avoid spurious magnectic fields which would induce unwanted beam depolarization (Fig. 3 *b)*). L can reach 172 cm in each arm (using the *Bootstrap* technique, [7]) and ω_{rf} can be varied up to 700 kHz which leads to a maximum Larmor phase of $\varphi_{tot}^{max} \sim 9 \cdot 10^3$ rad for the most commonly used wevelength, namely $\lambda = 2.36$ Å. This high limit is imposed in practice by the maximum current which can pass through the static coils wires which are water and air cooled and can produce at best a magnetic field of 240 G. Furthermore, using those flippers allows to modify the shape of the precession area as they can be easily rotated by an angle ζ with respect to the beam axis. In the case of ZETA, ζ can reach 70° which means that the *NLD condition* can be met in many cases.

With such a setup, the phase is measured by slightly changing the distances between coils in one of the two spectrometer arms[1]. The signal is modulated because this slight change in the field integral results in a small rotation of the final polarization. As the beam is analyzed by reflexion on a Heusler monochromator, the beam intensity can be fitted to a cosine law of the form $I_0 \left(1 + \mathcal{P}_f \cdot \cos\left(2\pi \frac{L-L_0}{w}\right)\right)$ where I_0 is the average intensity of the scattered beam, $\mathcal{P}_f$ its final polarization, w is the period of the signal and L_0 a reference length. Then, when d changes, the figure will be slightly shifted (for instance, see Fig. 4). To refine the phase change properly, one needs to record at least one period of the signal and to work with a sufficiently polarized beam. So things are simple if we refer to the formalism but we see that some difficulties may arise *e.g.* in the case of poor quality single crystals. There are in addition many sources of systematic errors which can not be avoided like thermal dilatation of the pieces composing the spectrometer or positionning errors of the translation stage. Those effect are on the other hand very small. Anyway, NLD potentially offers an appreciable gain in resolution in neutron diffraction whithout working with tight collimations, as it would be needed in CND, thanks to the aforementionned *NLD condition*.

4. Thermal expansion and magnetoelastic effect in $BaM_2(XO_4)_2$

Thermal evolution of the lattice parameters a (in-plane) and c (out-of-plane) in $BaM_2(XO_4)_2$ have been measured in each compounds using NLD. We used a neutron wavelength $\lambda = 2.36$ Å and worked with natural collimations $30' - 40' - 40' - 110'$[2]. The high temperature parts of each dilatation curves have been fitted to a simple polynomial model and extrapolated to T = 0. On the next step, this is considered as a "background" which is not linked with the magnetic ordering and is substracted from the raw curves. This is a mean to extract the "excess" lattice strain due to the magnetic ordering only.

4.1. $BaNi_2(PO_4)_2$

Thermal evolutions of a and c parameters have been inferred from the (2,0,2)[3] and (0,0,9) Bragg peaks respectively in a temperature range going from T = 1.5 up to 200 K. For both reflections, we worked at $\varphi_{tot} \simeq 6200$ rad which corresponds to a theoritical resolution of $\Delta d/d \simeq 8 \times 10^{-6}$. At the magnetic transition, a and c lattice parameters experience a strain which is respectively negative and positive (Fig. 5 a). The volumic magnetoelastic effect is

[1] This is possible because the last RFSFs pair is mounted on a translation stage.
[2] Collimations are given here in the usual order *i.e.* before monochromator, sample, analyzer and detector.
[3] In $BaM_2(XO_4)_2$ compounds, we have $c \gg a$ thus reflexions of the form (h,0,l) with h = 2 and $l = \bar{1}, 2$ mainly reflect a^*. In the case of $BaNi_2(PO_4)_2$ and $BaNi_2(AsO_4)_2$, the angle betwen $\vec{Q} = (2,0,2)$ and $\vec{a^*}$ is $\sim 10°$ while in $BaCo_2(AsO_4)_2$ we get, between $\vec{Q} = (2,0,\bar{1})$ and $\vec{a^*}$, an even smaller angle of $\simeq 5°$.

obtained through $\Delta V/V = 2\Delta a/a + \Delta c/c$[4] and is slightly negative *i.e.* the unit cell undergo a weak spontaneous contraction at T_N (Fig. 5 c). At T = 0, we find $\Delta V/V \sim -4 \cdot 10^{-5}$. On Fig. 5 *b)* and *c)*, we report the relative lattice parameters and volume variations as a function of the intensity of the (2,0,2) Bragg peak, itself proportional to the square of the magnetic moment. We note that this peak is both structural and magnetic. This implies that it may be subjected to extinction at the transtion but this effect is in this case negligible because the nuclear structure factor is weak. Direct scaling to the square of the magnetic moment (Fig. 2) indeed gives a similar result. Correlation between both quantities is quite good which indicates that the unit cell volume is a quadratic function of the order parameter. This result is expected for single lattice ferromagnets but has also been observed in the two sublattices antiferromagnets transition metals difluorides MF_2 ([8]). This spontaneous strain is very weak in magnitude but can be resolved thanks to the high accuracy provided by NLD.

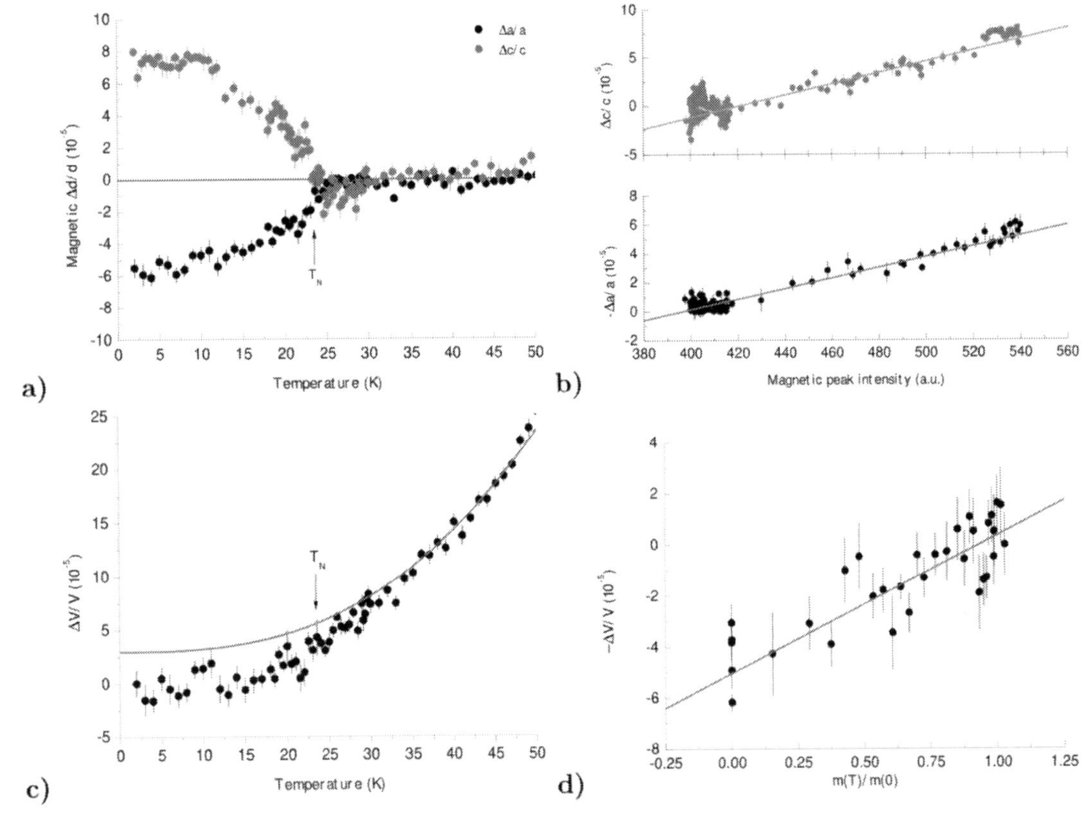

Figure 5. Magnetoelastic effect in $BaNi_2(PO_4)_2$. *a)* Thermal variation of the "excess" axial strain induced in the ordered phase. *b)* Axial strains as a function of (2,0,2) magnetic peak intensity (see text). *c)* Relative variation of the unit cell volume. Solide line is the sum of polynomial fits performed on the high temperature axial expansion data (see text). *d)* Volumic strain as a function of the ordered magnetic moment.

[4] Volume dilatation coefficient β is equal to the trace of the linear thermal dilatation tensor. In the case of hexagonal unit cell, we simply get $\beta = \frac{dV}{VdT} = 2\frac{da}{adT} + \frac{dc}{cdT}$ which leads to the above relation.

4.2. $BaCo_2(AsO_4)_2$ and $BaNi_2(AsO_4)_2$

The situation is rather different in baryum arsenides. For instance, in $BaCo_2(AsO_4)_2$, the magnetic strain goes in opposite sense as compared with the previous case, being respectively positive and negative for a and c lattice constants. The measurement has been made at $\varphi_{tot} = 4300$ and 6200 rad for the $(2,0,\bar{1})$ and (0,0,9) respectively between 2 and 60 K. Expected resolution is therefore of the order of 8×10^{-6} and 1×10^{-5} which is enough to resolve the very steep lattice constant variation along both directions (Fig. 6 a). The volumic effect at T = 0 is positive with $\Delta V/V \simeq +10^{-4}$. The situation is then clearly different as compared with the observations made in $BaNi_2(PO_4)_2$ even though the steeper moment variation gives a steeper lattice constant evolution at the magnetic transition.

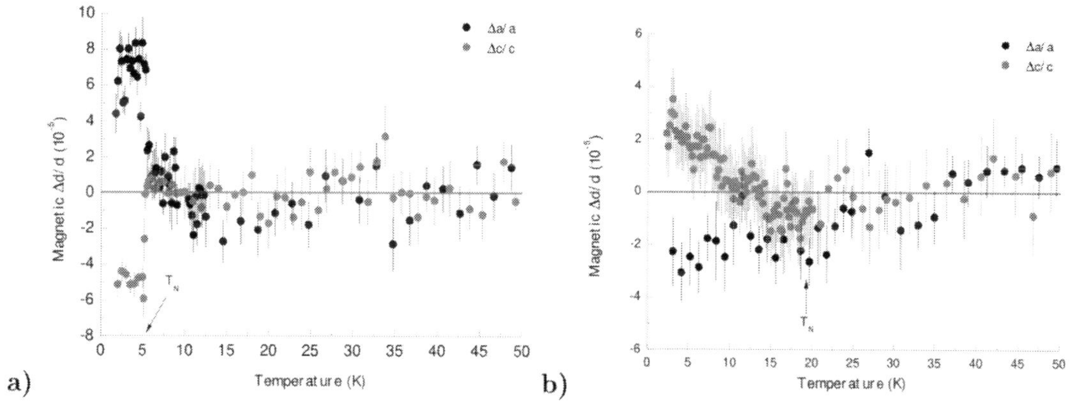

Figure 6. *a)* Axial magnetoelastic effect in $BaCo_2(AsO_4)_2$. The magnetic transition is marked by a step-like variation in lattice constants. *b)* Axial magnetoelastic effect in $BaNi_2(AsO_4)_2$. In this case, it is barely resolved but still visible (see text).

The experiment with $BaNi_2(AsO_4)_2$ has been more difficult to conduct as the polarization decay was is in this case rather fast when increasing φ_{tot} (a consequence of the large lattice spacing distribution, from Eq. 5.). For this measurement, we could only achieve a modest resolution of about 10^{-4} and 1.7×10^{-5} for the (2,0,2) and (0,0,9) reflection respectively. This is not good engouh to clearly resolve the volumic magnetoelastic effect but it nevertheless seen. From the data, we obtain an estimation of $\Delta V/V \simeq -3 \times 10^{-5}$ (Fig. 6 b).

5. Conclusion and outlook

We have investigated magnetoelastic properties of $BaM_2(XO_4)_2$ compounds using Neutron Larmor Diffraction. Our results show that in the case of $BaNi_2(PO_4)_2$ and $BaCo_2(AsO_4)_2$, magnetostriction is strongly correlated to the square of the order parameter, namely the value of the ordered magnetic moment. Strains are found to be weak but we were able to resolve them thanks to the high resolution provided by NLD. For the first sample, there is a volume contraction at T_N while the second one experience an expansion. The case of $BaNi_2(AsO_4)_2$ is more difficult because the relatively large lattice spacing distribution led to a substantial depolarization of the beam. Consequently, the measurement has been conducted at a rather poor resolution and the magnetoelastic effect is only barely resolved but seems to exist anyway. A rough estimation gives a negative value for the relative volume variation. In those three cases, we indeed observe an isostrucural transition at T_N. The way the lattice responds to magnetic ordering depends in a non-trivial way on the value of the spin or the magnetic structure propagation vector. Now, the next step in the understanding of the magnetostructural phase transitions in $BaM_2(XO_4)_2$ would require theoritical work to be done to link our thermal expansion result with a physical model. We hope that those results will trigger new studies aiming to understand the magnetic and structural behaviours in this interesting family with 2-d topology and honeycomb magnetic network.

Acknowledgments

We would like to aknowledge technical support by B. Geffray, B. Longuet and F. Mantegazza during the course of the experiments. This work has been supported by the FRENCH ANR project NEMSICOM and the European NMI3 network under Contract HII3-CT-2003-505925.

References

[1] Lee S., Pirogov A., Kang M., Kwang-Hyun J., Yonemura M., Kamiyama T., Cheong S.-W., Gozzo F., Namsoo S., Kimura H., Noda Y. and Park J.-G. 2008 *Nature* **451** 805-8
[2] Martin N., Regnault L.-P., Klimko S., Lorenzo J.E. and Gaehler R. 2011 *Physica B: Condensed Matter* **406** 2333-36
[3] Regnault L.-P. and Rossat-Mignod J. 1990 *Magnetic Properties of Layered Transition Metal Compounds* ed De Jongh L.J. (Kluwer Academic Press) 271-321
[4] Regnault L.-P. 1981 *PhD thesis*
[5] Rekveldt M.T.H. and Kraan W.H 1999 *J. Neutron Research* **8** 53-70
[6] Golub R. and Gaehler R. 1987 *Phys. Lett. A* **123** 43-48
[7] Gaehler R. and Golub R. 1988 *J. Phys. France* **49** 1195-1202
[8] Chatterji T., Iles G.N., Ouladdiaf B. and Hansen T.C. 2010 *J. Phys. Condens. Matter* **22** 316001

Use of parabolic mirrors as guide splitters

D. Martin Rodriguez[1], S. J. Kennedy[2], P. M. Bentley[2] and U. Rücker[3]

[1] Jülich Centre for Neutron Science, Forschungszentrum Jülich GmbH, 52425 Jülich, Germany

[2] Bragg Institute, ANSTO, Locked Bag 2001, Kirrawee DC NSW 2232, Australia

[3] Jülich Centre for Neutron Science and Peter Grünberg Institut, Forschungszentrum Jülich GmbH, 52425 Jülich, Germany

d.martin-rodriguez@fz-juelich.de

Abstract. The use of non-linear tapered neutron guides has been extended over the last years and there are many examples of their performance and potential in several neutron scattering facilities around the world. However, the potential of these geometries is not fully explored. On the other hand, the increasing demand of neutron scattering instruments creates the need of more space for more instruments. One way to obtain it is by guide splitting. In this report we propose a guide splitter using a parabolic mirror. We make an analytical study and Monte Carlo simulations in order to observe its properties, performance and potential. The advantages of this configuration are clear: the filtering of unwanted neutrons in a shorter space than the traditional curved guide sections, the low divergence of the obtained beam and the lack of making larger moderators. However, the main disadvantage is the non-uniform divergence distribution of the beam.

1. Introduction

The use of non-linear tapered neutron guides has been extended over the last years and it has been used in neutron sources all over the world [1-4]. Elliptical guides have shown outstanding neutron transport properties [5] and they also have been used as focusing devices. On the other hand, Parabolic guides have been also used as focusing devices in order to use them in small samples [6].

However, the potential of non-linear geometries for neutron transport has not been completely explored. One of the main issues when designing the neutron beam transport system is the space needed for instruments. The demand for using neutron scattering instruments has increased with time, and new powerful instruments are needed to meet this demand. One possible solution is splitting the guides. However, if straight guides are used, the splitting demands larger moderators in order to have decent neutron transport or smaller guides to fit the size of the moderator. The first solution would increase the maintenance costs of the moderator and the second one would increase the reflection losses.

In this report, we propose a novel design to split guides without using larger moderators. For this purpose we want to develop the potentials of parabolic geometry for neutron transportation. We will study the suitability of this design analytically and by means of Monte Carlo (MC) simulations.

2. Description of the beam splitter and coupling of a point source

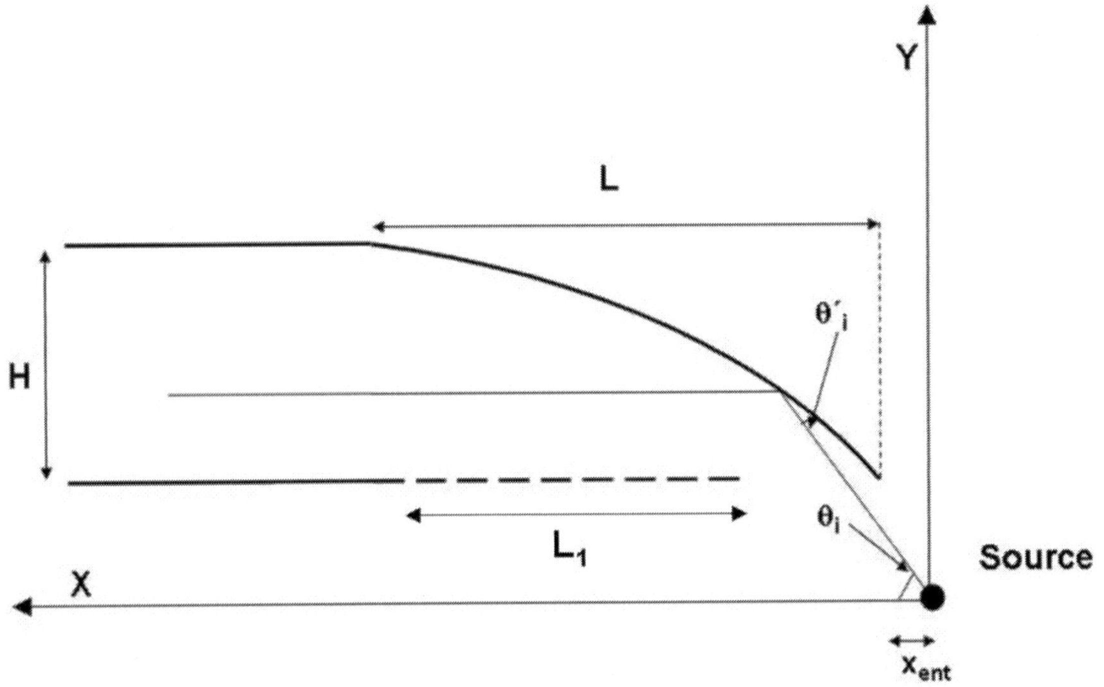

Fig. 1: Sketch of the parabolic neutron beam splitter for a point source.

Figure 1 shows the proposed beam splitter for a point source. The source will be at the focus of the parabolic mirror and will convert the most divergent neutrons coming from the source into a new neutron beam (called secondary guide thereafter), which, in the case of a point source, has zero divergence. The length of the parabolic mirror (*L*) will depend on its shape and the dimensions of the secondary guide, as it will be discussed later. On the other hand, in order to filter the unwanted neutrons, it is necessary to extend the lower part of the new guide (L_1 is the length of that extension) avoiding the direct view of the source in the guide. The rest of the neutrons will go to the main guide (it is not sketched in the figure).

The focus of the parabolic mirror will be at the origin, and therefore, the equation of the mirror will be defined with the following equation

$$y = \sqrt{4p(x+p)} \quad (1)$$

where p is the distance between the focus and the edge of the parabola and it is a parameter related with the shape of the parabola. The different divergences of the neutrons before and after the reflection as a function of the point they reflect (x) will be defined as follows.

$$\theta_i(x) = \frac{\sqrt{4p(x+p)}}{x} \quad (2)$$

$$\theta_i'(x) = \sqrt{\frac{p}{x+p}} \quad (3)$$

$$\theta_r(x) = 0 \quad (4)$$

Where θ_i is the angle of incidence of the neutron against the main guide direction at the source, θ_i' is the angle of incidence of the neutron at the surface of the parabolic mirror and θ_r, which does not appear in fig. 1, is the angle of the reflected neutron against the direction of the secondary guide.

The limit in the reflection will be determined by the coating following the condition $\theta_i'(x) \leq \theta_c(\lambda)$. As θ_i' decreases with increasing *x*, that also means that the coating in the mirror has to change accordingly in order to fix a maximum wavelength of the filtered neutrons. The *m* value of the coating as a function of the position will be defined by the following equation.

$$m(x) = \frac{\theta_i'(x)}{\theta_{c,Ni}(\lambda)} = \frac{1}{\theta_{c,Ni}(\lambda)}\sqrt{\frac{p}{x+p}} \quad (5)$$

where $\theta_{c,Ni}(\lambda)$is the critical angle for the Ni/vacuum interface. Taking this into account, it is important to note that *p*, which is related to the shape of the parabolic mirror, will be set according to the shortest desired wavelength (λ_{min}) and the maximum *m* value of the coating that can be used. In this case, the maximum *m* value will be at the entrance.

$$p = \frac{\lambda_{\min}^2 m^2(x_{ent})x_{ent}\theta_{c,Ni}^2(1A)}{1-\lambda_{\min}^2 m^2(x_{ent})\theta_{c,Ni}^2(1A)} \quad (6)$$

$\theta_{c,Ni}(1A)$is the critical angle for the Ni/vacuum interface and for 1Å neutrons. Taking into account the usual values for all the magnitudes, *p* will be in the range of 10^{-4}-10^{-5} m.

Once the shape of the mirror is determined, it is necessary to determine the length of the mirror (L) which will depend on the shape of the mirror and by the height of the secondary guide (H) using the following equation:

$$H = 2\left(\sqrt{p(x_{ent}+L+p)} - \sqrt{p(x_{ent}+p)}\right) \quad (7)$$

From this equation we can obtain L as a function of the shape of the mirror and the height of the guide:

$$L = \frac{H\left(H+4\sqrt{p(x_{ent}+p)}\right)}{4p} \quad (8)$$

Figure 2 shows an example of eq. (8) for x_{ent}=0.3 m. and two different values of the height of the guide. As shown in the figure, the length of the parabolic mirror has a strong dependence on *p* and on the height of the guide (*H*). Taking into account that usually the curved guide sections are around 20-40 m. long, there is a practical limit of *p*, in which the length of the mirror is larger than the usual curved guide sections, and therefore, considering mirrors with a lower *p* has no sense.

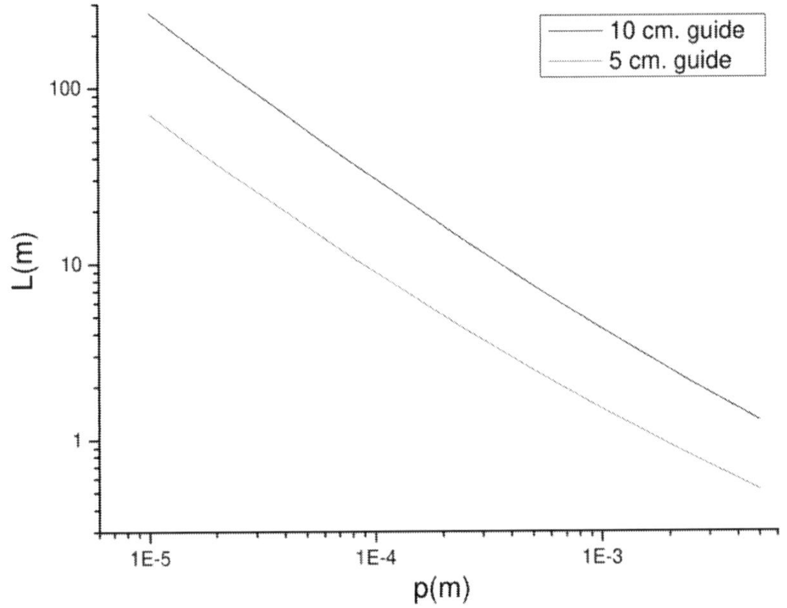

Fig. 2: Relationship between the length of the parabolic mirror (L) and the shape of the parabola (see eq. (1)) using eq. (8) for x_{ent}=0.3 m. and two different values of the height of the secondary guide (H).

The existence of a practical minimum p imposes a limit in the transport, although the size of the secondary guide is also important to determine this practical limit. The smaller the H, the smaller the minimum p of this practical limit. Taking into account that in the ideal case of a point source the divergence is zero, there is no problem on making H smaller as there will be no reflection loses in the guide.

Finally, as explained in fig. 1, the lower part of the guide has to be extended in order to avoid the direct view. The following equation will determine the length needed to avoid that:

$$L_1 = \frac{(x_{ent} + L)\sqrt{p(x_{ent} + p)}}{\sqrt{p(x_{ent} + L + p)}} \quad (9)$$

Taking into account all the analytical study of the ideal case, the advantages of the proposed parabolic mirror guide splitter are obvious. First of all, it is not necessary to build a larger moderator to obtain another neutron beam. On the other hand, the divergence of the beam is zero in the ideal case, which means it can be ideally transported to long distances with no reflection losses.

3. Coupling of an extended source with the guide splitter

As shown in the former section, the ideal case shows the advantages of the use of a parabolic mirror as a guide splitter. However, this is not realistic as the sources are extended. On the other hand, it is convenient to have an extended source if the main guide is a straight guide. For this reason, we must study analytically the effect of extended sources in the proposed guide splitter.

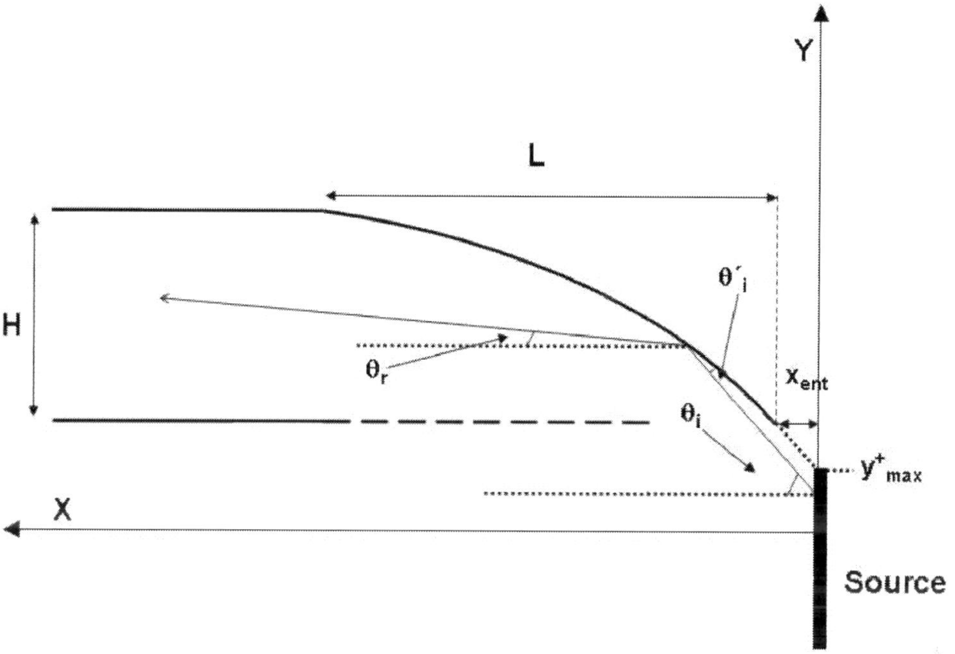

Fig. 3: Sketch of the parabolic neutron beam splitter for an extended source.

Figure 3 shows the sketch of the guide splitter for an extended source. In this case, the reflected neutron beam is not going to have zero divergence as most of the neutrons will not come from the focus of the parabolic mirror. The equations for the incident and reflected angles will be as follows:

$$\theta_i = \frac{\sqrt{4p(x+p)} - y}{x} \quad (10)$$

$$\theta_i'(x,y) = \sqrt{\frac{p}{x+p}} - \frac{y}{x} \quad (11)$$

$$\theta_r(x,y) = \frac{y}{x} \quad (12)$$

Where x will be the reflection point and y the origin of the neutrons at the source. As the origin of the system is in the focus of the parabola, there will be different situations depending on where from the source are the neutrons coming from.

In the case of the neutrons coming from the positive side of the source, we will find that $\theta_i' < \theta_c(\lambda)$ and $\theta_r > 0$. For these neutrons multiple reflections in the parabolic mirror are likely to happen. When that happens, the divergence of the neutron will become smaller with every reflection in the mirror. Therefore, this imposes an upper limit in the divergence of the neutron beam, which is also the limit where the multiple reflections start.

$$\theta_{r,\max} = \frac{2\left(\sqrt{p(x_{ent}+L+p)} - \sqrt{p(x_{ent}+p)}\right)}{L} \quad (13)$$

Equalising eqs. (12) and (13) for $x=x_{ent}$ we will obtain the upper limit on y for multiple reflections.

$$y^{+}_{mult} = \frac{2x_{ent}\left(\sqrt{p(x_{ent}+L+p)}-\sqrt{p(x_{ent}+p)}\right)}{L} \quad (14)$$

For neutrons coming from the positive side of the source there is also another limit which will impose the maximum extent of the source in the upper part. If the neutron were coming from an upper position, those would not reflect on the mirror as one of the conditions for being reflected by the mirror is that $\theta'_i \geq 0$. If $\theta'_i < 0$, then the neutron would not reflect and would contribute to the direct view, which is something that must be avoided. The maximum extent of the source will be:

$$y^{+}_{\max} = \sqrt{\frac{p}{x_{ent}+p}(x_{ent}+2p)} \quad (15)$$

For neutrons coming from the negative part of the source ($y<0$) $\theta_r < 0$ and there will be no multiple reflections. While in the positive part, the divergence of the neutrons is limited by the shape (p) and the position (x_{ent}) where the parabolic mirror starts, in the negative part it will be limited not only by the shape, but also by the coating, and will be wavelength dependent, as shown in the following equation:

$$y^{-}_{\max}(x) = x\left(\sqrt{\frac{p}{x+p}} - \theta_c(\lambda)\right) \quad (16)$$

This means that the upper and lower part of the divergence of the reflected neutron beam will be different and therefore, the divergence distribution of the reflected beam will be asymmetric.

In summary, the study of the more realistic case of an extended source shows the main disadvantage of this concept: an asymmetric divergence distribution for the outgoing neutron beam.

4. MC simulations on real systems

MC simulations will give us a way to study the applicability of this concept in real systems. For that purpose we used the software VITESS for neutron scattering instrument simulation [7] and the sketch of the simulated system can be found in fig. 4. An important fact to take into account is that the parabolic mirror in real systems has not a perfect parabolic shape, but it will have a polygonal shape with 50 cm. long linear sections as an approximation a parabolic surface.

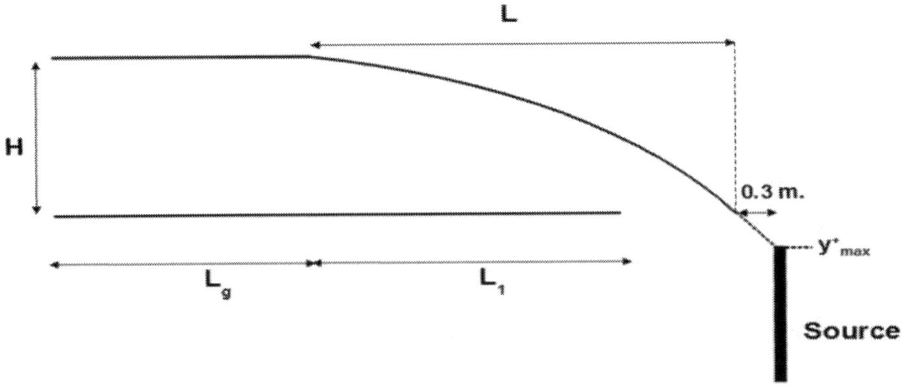

Fig. 4: Sketch of simulated system.

The width of the mirror and the secondary guide is 10 cm., and H=10 cm.. The coating of the parabolic mirror has been calculated using eq. (5), while the coating of the secondary guide is m=1. As both, the parabolic mirror and the extension of the lower part of the secondary guide (L_1) are built with 50 cm. long sections (and therefore L_l has to be a multiple of 50 cm), the avoidance of the direct view is not perfect and the secondary guide must be extended by L_g.

We use two values of p in order to study the performance of this configuration as a function of the shape. Table 1 will show the parameters used in the simulation:

Name	p (m)	L (m)	L_l (m)	y^+_{max} (cm)	L_g (m)
P1	3.47 x 10^{-4}	10.0	9.0	1.0	11.5
P2	5.43 x 10^{-4}	7.0	6.5	1.3	4.0

These parabolic mirror configurations will be compared with a system composed by the source and a slit system which will give a similar angular extent for the divergence as the parabolic mirror configuration (1° x 1°). This is achieved with a two 10 x 10 cm. slits which are 5.75 m. from each other. In this way, we would like to compare the results in terms of source brightness losses.

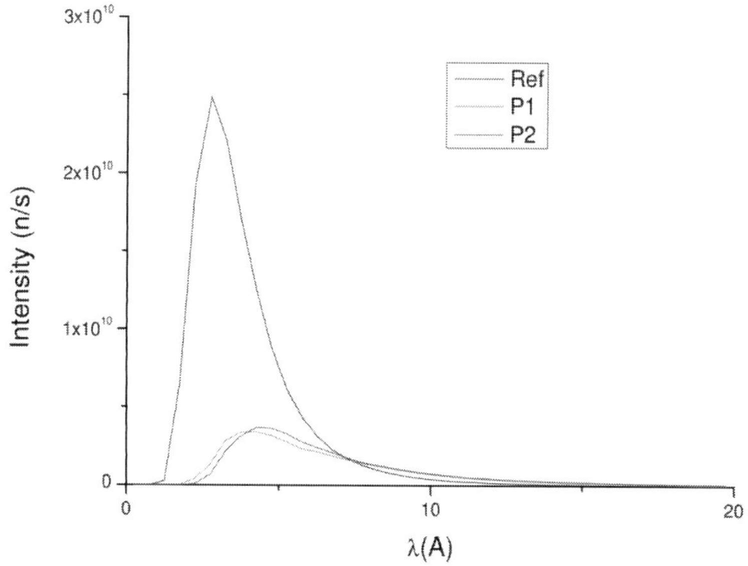

Fig. 5: Neutron spectra of the different parabolic mirror configurations compared with the reference

Figure 5 shows the spectrum for the different parabolic mirrors and for the reference system. The capability of the parabolic mirror to filter unwanted neutrons depends on its shape, when p increases the parabolic mirror filters neutrons with longer wavelength.

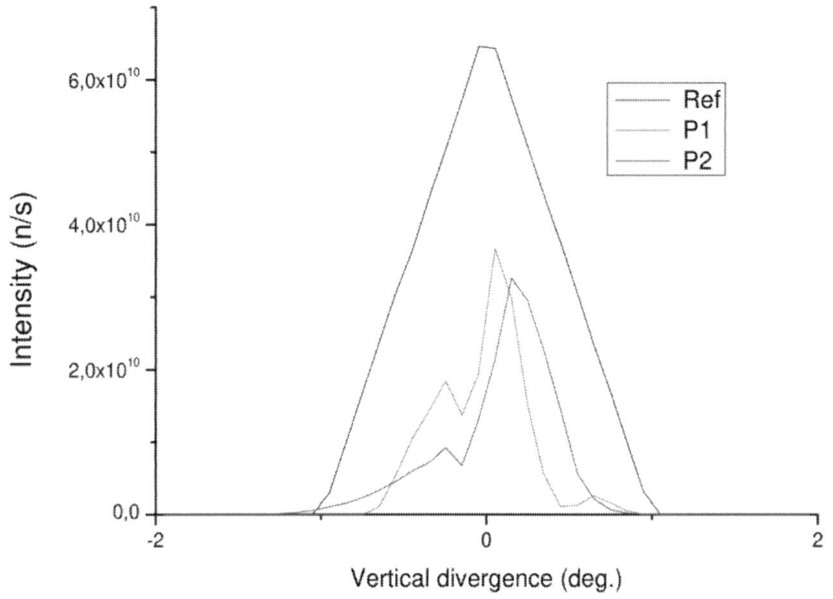

Fig. 6: Divergence distribution of the neutron beam at the end of the guide (see fig. 4) for both parabolic mirror configurations and for the reference.

Figure 6 shows the vertical divergence distribution of the different parabolic mirror configurations and the reference. The divergence distribution of the parabolic mirror configurations is non-uniform while the reference is completely uniform. It is also evident that the upper and lower limits in the divergence distribution are different, as it was predicted in the former section.

Simulations confirm the advantages and disadvantages deducted from analytical calculations. With the proposed configuration it is possible to filter unwanted neutrons in a shorter space, and it is possible to split a guide without using a larger moderator. However, the divergence distribution of the upcoming beam is non uniform, and therefore, unattractive for neutron scattering instrument designers.

One of the possible solutions is to place a collimator after the guide. This would uniformise the divergence distribution of the beam at the cost of losing flux. Another possible solution to explore is the use of hybrid geometries [8, 9] which could improve the quality of the beam, i. e., the uniformity of the divergence distribution. This last solution will be subject of further studies.

5. Conclusions

We present a guide splitter using a parabolic mirror. From analytical calculations and Monte Carlo simulations it is deduced that this guide splitter shows several advantages:

- It is not necessary to build a larger moderator
- It filters unwanted neutrons in a shorter space compared to curved guide sections
- The beam is collimated in the reflection plane

However, the main disadvantage is:

- Non-uniform divergence distribution

Solutions to avoid this disadvantage are under study.

References

[1] P. Boni 2008 *Nucl. Inst. Methods Phys. Res. Sect. A* **586** 1.
[2] S. Muhlbauer, M. Stadlbauer, P. Böni, C. Schanzer, J. Stahn and U. Filges 2006 *Physica B* **385-386** 1247
[3] H. Arima, K. Komatsu, K. Ikeda, K. Hirota and H. Kagi 2009 *Inst. Methods Phys. Res. Sect. A* **600** 71
[4] J. C. Cook 2009 *Rev. Sci. Inst.* **80** 023101
[5] D. Martin Rodriguez, S. J. Kennedy and P. M. Bentley 2011 *J. Appl. Cryst* **44** 727
[6] S. Yamada, T. Shinohara, H. Sasao, T. Oku, J. I. Suzuki, H. Matsue and H. M. Shimizu 2006 *Physica B* **385-386** 1243
[7] D. Wechsler, G. Zsigmond, F. Streffer and F. Mezei 2000 *Neutron News* **11** 25
[8] S. Khaykovic, M. V. Gubarev, Y. Bagdasarova, B. D. Ramsey and D. E. Moncton 2011 *Inst. Methods Phys. Res. Sect. A* **631** 98
[9] P. M. Bentley, S. J. Kennedy, K. H. Andersen, D. Martin Rodriguez and D. F. R. Mildner, *unpublished.*

Double bent crystal dispersive arrangement for high resolution diffractometry

P Mikula[1], M Vrána[1], J Šaroun[1], V Em[2], B S Seong[2], W Woo[2]

[1]Nuclear Physics Institute ASCR, v.v.i., 25068 Rez, Czech Republic
[2]Neutron Science Division, Korea atomic Energy Research Institute, 1045 Daaedeok-daero, Daejeon 305-353, Republic of Korea

E-mail: mikula@ujf.cas.cz

Abstract. In this paper some properties of the dispersive double-bent-crystal setting are described. Thanks to an easy manipulation with focusing of the monochromatic beam, high reflection probability of both bent crystals and very narrow (1-3 mm) obtained monochromatic beam such an arrangement could be attractive for an employment. We tested the properties of the double-bent-crystal setting of the bent Si(111) + bent Si(220) slabs on the diffractometer at the neutron wavelength of 0.163 nm for various curvatures of the Si(220) slabs. By using a standard polycrystalline sample of Fe it has been found that besides an excellent resolution the neutron current is sufficient even for powder diffraction experiments.

1. Introduction

It is commonly accepted meaning that a dispersive double-crystal (DDC) setting is not convenient for neutron monochromatisation due to a low monochromatic neutron current and thus low luminosity of the scattering instrument. This consideration is true in the case of mosaic crystals. However, in the case of bent perfect crystals (e.g. Si or Ge) some advantages of the DDC setting can be found. First, one should consider that when the curvature of the crystals is not drastic, the reflection probability of both bent perfect crystals can be close to 1 and there is a small loss of neutrons due to the double reflection process. The general formula for the reflection probability (often called as peak reflectivity) of neutron passing through a homogeneously deformed perfect crystal has been derived in a form as [1]

$$r(R_M) = \{1 - \exp[-Q_{hkl}(\partial\Delta\theta/\partial s_0)^{-1}]\}, \quad (1)$$

where $(\partial\Delta\theta/\partial s_0)$ is the rate of change of the Bragg angle on the flight path s_0 in the crystal in the incident beam direction and Q_{hkl} is the kinematical reflectivity of the crystal unit volume. For cylindrically curved monochromator it has been derived in the general form [2] when including also asymmetric diffraction geometry given by the angle ψ of the asymmetry cut of the crystal

$$\partial\Delta\theta/\partial s_0 = (1/R_M)(\cos\psi/\cos\theta_M)[1-(1+\sigma)\sin(\theta_M+\psi)\ \sin(\theta_M-\psi)], \quad (2)$$

where R_M is the bending radius, θ_M the Bragg angle and σ the Poisson constant. Furthermore, curved crystal instruments are known by the fact that they are working with the open beams, they do not use Soller collimators and their resolution can be controlled through bending the crystals. Possible

[1] To whom any correspondence should be addressed.

employment of the focusing does not necessarily imply only focusing in real space, but it can cover also focusing in scattering. In the case of a single bent perfect crystal monochromator focusing in scattering has been widely used in powder diffractometry, namely, for strain/stress measurements in polycrystalline materials [3-8] and can also be exploited even in three axis spectrometry [9-12]. A special category of instruments employing two bent perfect crystals are high-resolution small-angle neutron scattering diffractometers [13-18]. With two bent crystals in successive reflection, the double crystal rocking curve is generally much broader in comparison with the flat perfect crystals. However, when the phase space windows of both bent crystals matches, the double crystal rocking curve is narrow because it is related only to the reflectivity curves and simultaneously the intensity of the double reflected beam is high. This matching of the phase space windows can be even achieved with the bent crystals using different *d*-spacings [19]. All this is interest of SANS instruments using two bent crystals which are in fact in the nondispersive settings.

The situation is rather different in the case of really dispersive double-crystal setting. Due to an extremely high angular and $\Delta\lambda/\lambda$ resolution and rather low flux of neutron sources in comparison with e.g. synchrotron ones, the use of perfect crystals in neutron diffractometry is not practical. However, the use of mosaic crystals, or bent perfect crystals, or their combination can appear useful. Namely, it could be useful in high resolution experiments carried out on small samples when one can benefit from real space mikrofocusing and also from focusing in scattering. First, this possibility was pointed out in ref. 5, where the properties of the combination of a flat Si(111) mosaic crystal with a Cu(220) bent mosaic crystal in (+,+) setting are described. Then, very promising results of an extensive investigation of dispersive (+,-) combination of a vertically focusing mosaic Ge(331) crystal with a horizontally bent perfect Si(111) crystal can also be found in ref. [20,21]. Here, it was clearly demonstrated both real space focusing and focusing in scattering on a standard polycrystalline γ-Fe sample (austenite pin of the diameter of 2 mm) at λ=0.183 nm. Furthermore, it has been found an excellent property that for an optimum curvature of the bent Si(111) crystal, *FWHM*s of the γ-Fe(hkl) diffraction profiles on the $2\theta_S$-angular scale are practically the same for both parallel and antiparallel settings. This fact can have enormous impact in powder diffractometry and namely for residual strain/stress measurements when permitting to measure two strain components simultaneously.

In this paper we deal with the results of investigation of properties of the dispersive Si(111)+Si(220) double bent crystal (+,-) setting for neutron monochromatisation and its possible to employ it in neutron diffractometry.

2. Experimental arrangement

For this experiment we used test neutron optics diffractometer installed at the 10 MW medium power research reactor LWR-15. The diffractometer operates at fixed neutron wave length of λ=0.163 nm which is providing cylindrically bent perfect Si(111) crystal of the dimensions of 200x40x4 mm^3 (length x width x thickness). Schematic sketch is drawn in figure 1. Depending on the experimental requirements, the diffractometer can operate in two or three axis mode. It should be pointed out that the first Si(111) crystal of the DDC setting is bent with a fixed radius (R≈10 m) which was not changed during the experiment. The second bent perfect Si(220) monochromator was in fact a sandwich of two slabs each of the dimension of 200x40x1.3 mm^3 (length x width x thickness). By using a sandwich instead of a single piece it could be curved in a large range of curvatures without a danger of breaking. For neutron detection either a point detector or Neutron Imaging Plate (IP) were used. Sample could be put on the third axis of the instrument which is at

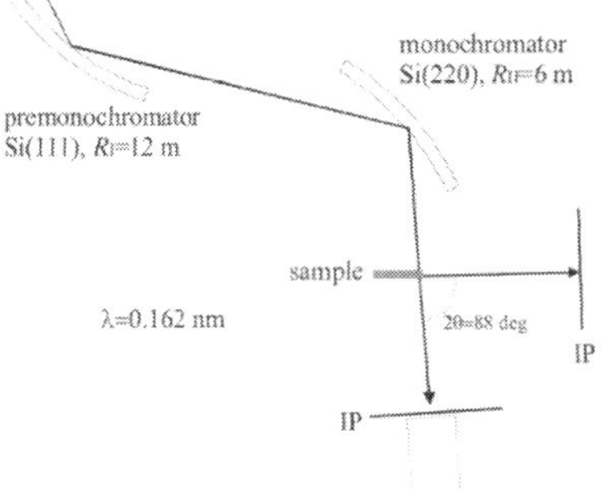

Figure 1. Experimental performance as used for the experimental investigations.

the distance of 50 cm from the Si(220) crystal. The distance between the Si(111) and Si(220) crystals was fixed and equal to 180 cm.

3. Experimental results

First of all, the profile of the beam diffracted from the bent Si(220) sandwich was investigated. For this purpose Imaging Plate was situated either just behind the sandwich (distance about 10 cm) or in

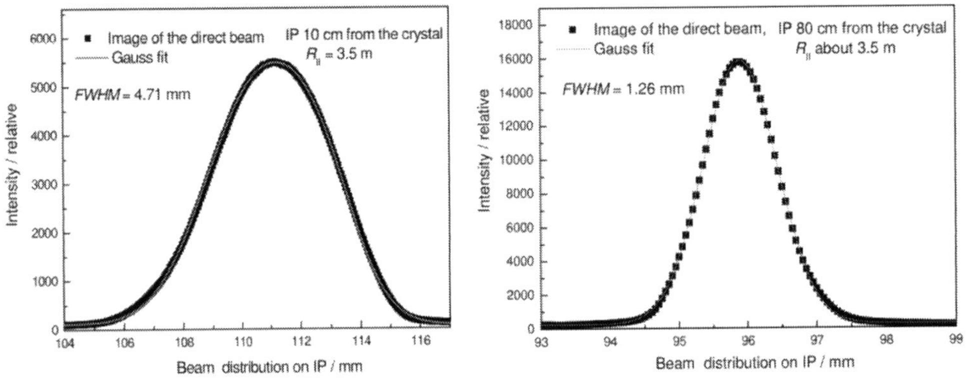

Figure 2. Profiles of the double diffracted beam as taken by IP situated at 10 cm and 80 cm distance from the second crystal for the radius of 3.5 m.

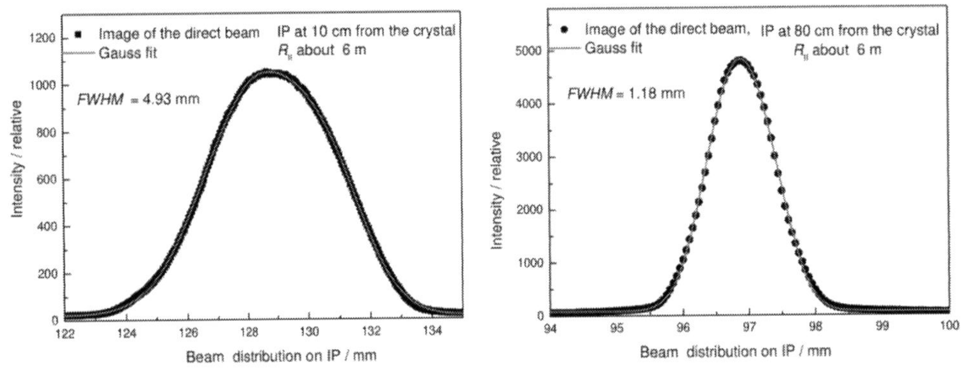

Figure 3. Profiles of the double diffracted beam as taken by IP situated at 10 cm and 80 cm distance from the second crystal for the radius of 6 m.

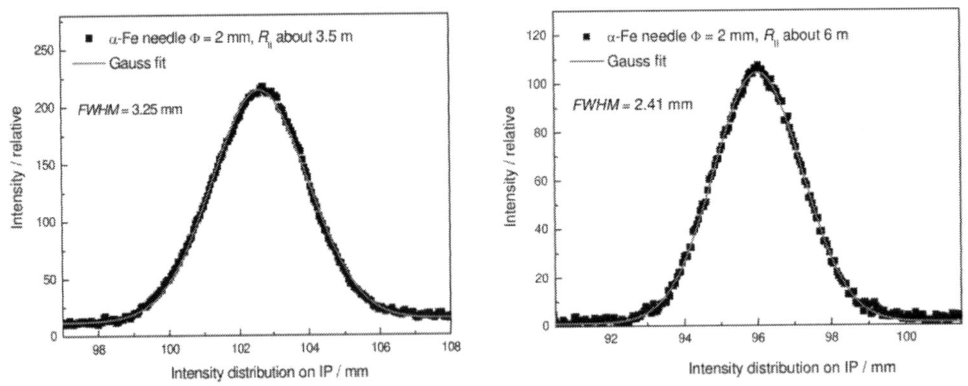

Figure 4. Neutron diffraction profiles from the α-Fe(211) polycrystalline standard sample situated at 50 cm from the Si(220) sandwich and with IP at 45 cm from the sample for two different curvatures of the sandwich.

the front of the point detector at the distance of 80 cm. The images of the beam were taken for different radii of curvature of the Si(220) sandwich. Figures 2 and 3 show some of the related results. Inspection of these results reveals that the dispersion double bent crystal setting provides, as expected, an intensive monochromatic beam of very small width, but also a strong real space focusing when comparing the width of the beam profile just behind the Si(220) sandwich with the profile taken at some distance e.g. 80 cm. It should be pointed out that no collimators were used. Of course, even in this case the focusing is slightly imperfect mainly due to an uncertainties coming from the effective mosaicity of the bent Si(220) sandwich dependent on its nonnegligible thickness (2x1.3 mm). Therefore, when using thinner crystals, the *FWHM* of the beam profile can be still narrower. In this way it should be mentioned that successful studies of a possible employment of a 1.3 mm thin bent perfect crystal monochromator in a strain/stress diffractometer in order to improve its resolution have been already carried out [8]. In the next step diffraction profiles on a well annealed standard α-Fe solid polycrystalline pin of 2 mm diameter have been taken for different curvatures of the Si(220) sandwich. The sample was installed on the third axis of the instrument at the distance of 50 cm from the Si(220) crystal and the Imaging Plate was at 45 cm from the sample. Scattering angle on the sample was $2\theta_S=88^{\circ}$. Figure 4 shows two of several obtained results including the one providing a minimum *FWHM*=2.41 mm on the Imaging Plate. Finally, by using the incident monochromatic beam related to the bending radius of about 6 m (with *FWHM* = 2.41 mm), a test with steel samples having the form of plates with the thickness of 2 mm and the width of 30 mm was carried out. Two steel samples were at a disposal: virgin one as well as the one ruptured after mechanical loading in a tension machine (see Figure 5). In the case of the ruptured sample two positions in the beam, marked as A and B, were used. Fig. 6 shows the obtained data. On the basis of these results, it could be seen that the virgin sample itself was already deformed and the related *FWHM* is much larger than that of the standard sample. Further deformation in the tensile rig makes additional plastic deformation resulted in a slight widening of *FWHM*. It means that the effect of plastic deformation by peak profile analysis can be easily investigated by this technique. As IP was used for imaging of the diffraction profiles and each profile had to be imaged and scanned individually, there are no results on effect of elastic

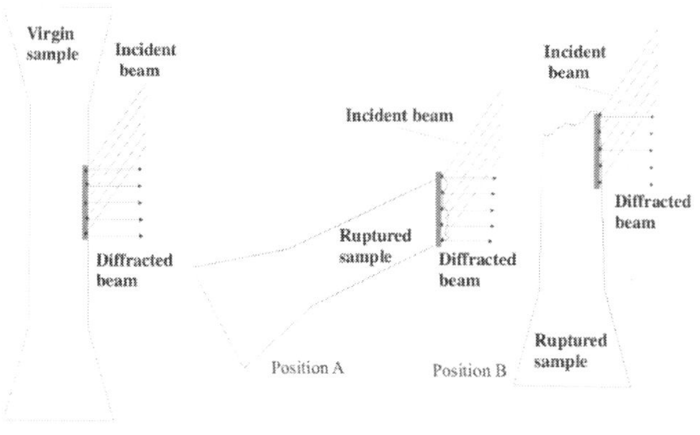

Figure 5. Schematic sketch of the diffraction geometry of the virgin as well as the ruptured sample and the positions in which the measurement of the diffraction profiles was carried out.

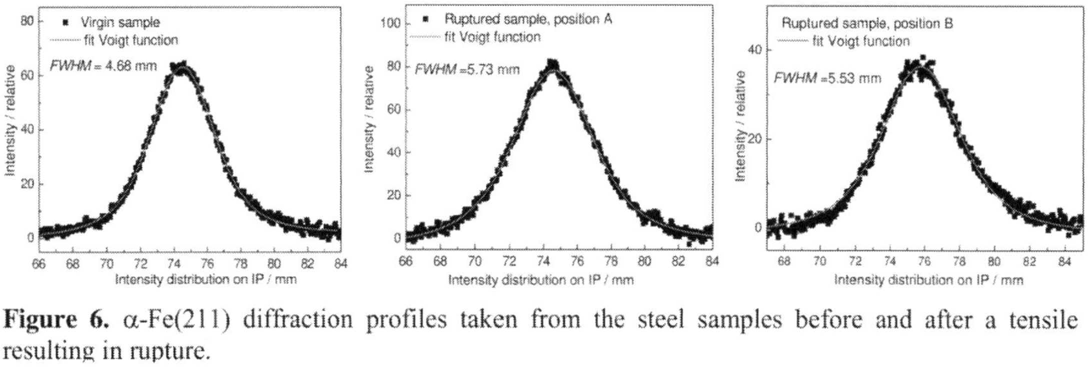

Figure 6. α-Fe(211) diffraction profiles taken from the steel samples before and after a tensile load resulting in rupture.

deformation from the residual stress which brings about the angular shift of the diffraction profile with respect to the angular position of the one related to the standard stress free sample. For such measurements a high resolution position sensitive detector is required. It should be pointed out that it is a common view that the measurement of angular peak shift is easier to observe than a change of the width of the diffraction profile with a required accuracy. In our case, 1 mm on the IP (in the scattering plane) represents $\Delta(2\theta_S)\approx 2.2\times10^{-3}$ rad.

4. Conclusion

Inspection of figures 2-6 reveals that contrary to a common view, the presented double crystal monochromator with crystals in the dispersive setting can provide a sufficient neutron current at the sample position which can be used in powder diffractometry, of course, at a reasonably high flux neutron source. One should keep in mind that the dispersive setting behaves as a self collimation assembly producing a highly collimated and highly monochromatized beam. Using the dispersive double crystal monochromator, one works with open beams without any Soller collimators. In the present case, it is necessary to point out that the instrument could not be fully optimized when using a fixed curvature of the Si(111) crystal. Furthermore, due to the given experimental conditions, the distance between the bent crystals was rather large (1.8 m). Another parameter which was fixed it was the thickness of the crystals. By including all these parameters as well as possibility of vertical focusing, we believe that after an optimization the luminosity of the presented experimental setup could be one order of magnitude higher. Even with this performance, when using a high resolution position sensitive detector instead of IP having a rather low efficiency, the time expected for collection of the data related to individual diffraction profile is expected to be less than 20 minutes at our medium power neutron source. For design and full optimization of such DDC setting, Monte Carlo simulations are desirable. For this purpose one can use a code which has been already developed and permits an evaluation of the resolution function and intensity properties of neutron diffractometer taking into account a spatial configuration of the experimental setup and the monochromator curvature [22]. Some help may provide also an analytical approximation of instrumental line shapes [23] and matrix computation technique (program DAX [24]).

Acknowledgements
Bragg diffraction optics investigations are in the Czech Republic supported by the research projects AVOZ104805505, ASCR-M100480901, GACR P204/10/0654 and also by FP7-NMI3 "Integrated Infrastructure Initiative for Neutron Scattering and Muon Spectroscopy", 2009-2013.

References
[1] Kulda J, 1984 *Acta Cryst.* **A40** 120-126; 1088 *Acta Cryst.* **A44** 283-285.
[2] Mikula P, Kulda J, Lukáš P, Vrána M and Wagner V 1994 *Nucl. Instrum. Methods in Phys. Res.* A **338** 18-26.
[3] Wagner V, Mikula P and Lukáš P 1994 *Nucl. Instrum. Methods in Phys. Research,* A **338** 53-59.
[4] Vrána M, Lukáš P, Mikula P and Kulda J 1994 *Nucl. Instrum. Methods in Phys. Research,* A **338** 125-131.
[5] Popovici M and Yelon W B 1995 *J. Neutron Research* **3** 1-25; 1997 *J. Neutron Research* **5** 227-239.
[6] Wimpory R C, Mikula P, Šaroun J, Poeste T, Li J, Hoffman M and Schneider R 2008 *Neutron News* **19** 16-19.
[7] Seong B S, Em V, Mikula P, Šaroun J and Kang M H 2010 *J. Appl. Cryst.* **43** 654-658.
[8] Seong B S, Em V, Mikula P, Šaroun J and Kang M H 2011 *Mater. Sci.Forum* **681** 426-430.
[9] Popovici M and Yelon W B *Nucl. Instrum. Methods* 1994 A **338** 132-135.
[10] Kulda J and Šaroun J 1995 *Nucl. Instrum. Methods* A **379** 155-156.
[11] Kulda J 2006 *Nuclear Engineering and Technology* **38** No.5 433-436.
[12] Stoica A D, Popovici M, Yelon W B and Berliner R 2000 *J. Appl. Cryst.* **33** 147-155.

[13] Kulda J and Mikula P 1983 *J. Appl. Cryst.* **16** 498-504.
[14] Mikula P, Wagner V and Scherm R 1991 *J. Appl. Cryst.* **24** 298-303.
[15] Šaroun J, Lukáš P, Mikula P and Strunz P 1993 *Journal de Physique*, Coll. C7, **3** 439-442.
[16] Popovici M, Yelon W B, Berliner R and Stoica A D 1993 Proc. 41st Annual Midwest Solid State Conference, Columbia, USA.
[17] Šaroun J, Lukáš P, Mikula P and Alefeld B 1994 *J. Appl. Cryst.* **27** 80-88.
[18] Hempel A, Eichhorn F, Reichel P and Boede W 1996 *Nucl. Instrum. Methods in Phys. Res.* A **381** 466-471.
[19] Popovici M, Stoica A D, Chalupa B and Mikula P 1988 *J. Appl. Cryst.* **21** 258-265.
[20] Mikula P, Vrána M, Choi Y N, Kim S A, Oh H S, Sung K H and Lee C H 2002 *Mater.Sci. Forum* **404-407** 299-302.
[21] Mikula P, Vrána M, Wagner V, Choi Y N, Moon M K, Lee C H and Inoue K 2004 Proc. of 12th Int. Conf. on Experimental Mechanics - ICEM12, 29 August - 2 September, Bari, Italy, McGraw-Hill, ed. C Pappalettere, full paper No. 038 on CD-ROM.
[22] Šaroun J, *http://neutron.ujf.cas.cz/restrax/*.
[23] Stoica A D, Popovici M and Yelon W B 2000 *J. Appl. Cryst.* **33** 137-146.
[24] Ionita I and Stoica A D 2000 *J. Appl. Cryst.* **33** 1067-1074.

Journal of Physics: Conference Series **340** (2012) 012015 doi:10.1088/1742-6596/340/1/012015

Investigation of multiple Bragg reflections at a constant neutron wavelength and their possible separation

P Mikula[1,2], M Vrána[1], J Šaroun[1,2], V Em[3] and B S Seong[3]

[1]Nuclear Physics Institute, v.v.i., ASCR, 25068 Řež, Czech Republic
[2]Research Centre Řež, Ltd., 25068 Řež, Czech Republic
[3]Neutron Science Division, Korea Atomic Energy Research Institute, 1045 Daedok daero, Yuseong-gu, Daejeon, 305-353, Republic of Korea

E-mail: mikula@ujf.cas.cz

Abstract. Multiple Bragg reflections (MBR) realized in one bent-perfect crystal (BPC) slab by sets of different lattice planes behave differently in comparison to the case of perfect nondeformed or mosaic crystal. Individual sets of lattice planes are mutually in dispersive diffraction geometry and the kinematical approach can be applied on this MBR process. It has been found that contrary to the perfect nondeformed or mosaic crystal, individual reflections participating in the MBR process can be spatially separated.

1. Introduction

Multiple reflections in a single crystal occur when two or more than one set of reflection planes fulfil the Bragg condition, i.e. when two or more reciprocal lattice points defined by the vectors $\boldsymbol{g}_i$ appear on the Ewald sphere (n-beam case, n≥2) [1-4]. In the case of the perfect crystals a dynamical neutron wave field is created where all momentum components $\boldsymbol{k}_o+\boldsymbol{g}_i$ are present and cannot be mutually separated. Similar kinematical mixture of the $\boldsymbol{k}_o+\boldsymbol{g}_i$ components is created also in mosaic crystals, namely in the case of very small mosaic blocks. Though that multiple reflections are mostly studied by X-rays because of sufficiently high beam current, several papers dealing with neutron diffraction studies on mosaic crystals have already appeared [5-7]. Contrary to the case of perfect or mosaic crystal, multiple Bragg reflections realized in a bent-perfect crystal slab behave differently. It is given by the homogeneous deformation of the crystal. It means that neutron reflected by a set of lattice planes in a local point A of the bent crystal cannot meet again the Bragg condition on the same set of planes on the path within the crystal. However, it can be reflected by a set of another lattice planes at a local point B resulting (in the simplest case) in a double reflection process and the local points A and B (see figure 1) are spatially separated [8]. Of course, that the situation is

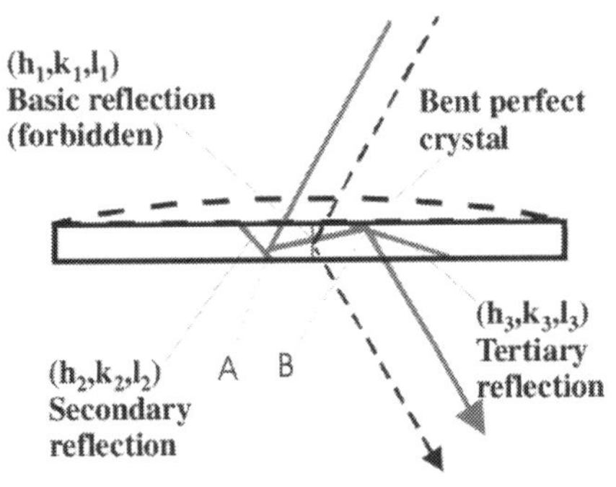

$$\vec{g}_1 = \vec{g}_2 + \vec{g}_3$$

Figure 1. Schematic diagram of a simple double reflection process accompanying allowed or forbidden primary reflection.

much more complicated when n≥3. Then, several multiple reflections with the momentum components $\boldsymbol{k}_o+\boldsymbol{g}_i-\boldsymbol{g}_j$ ($\boldsymbol{g}_i-\boldsymbol{g}_j=\boldsymbol{g}_1$) accompanying a primary reflection with the momentum component $\boldsymbol{k}_o+\boldsymbol{g}_1$ (for which the crystal sample is set) can occur. The purpose of this paper is to show that in a rather simple case of two double reflection processes accompanying primary reflection (n=3) they can be spatially separated just by the elastic deformation.

2. Experiment

The experiment was carried out on the multipurpose diffractometer installed at the research reactor LWR-15 in Řež. Contrary to our earlier investigations of MBR effects carried out by setting the forbidden reflections of Si or Ge single crystals in polychromatic beam and then by θ-2θ scanning the MBR effects were searched (see e.g. ref. 9-12), in this case the method of azimuthal rotation around the scattering vector of a forbidden reflection at the fixed neutron wavelength was used. The

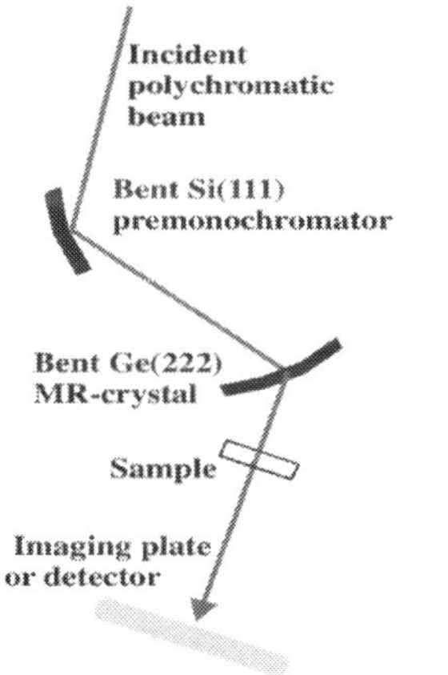

Figure 2. Schematic diagram of the diffractometer performance.

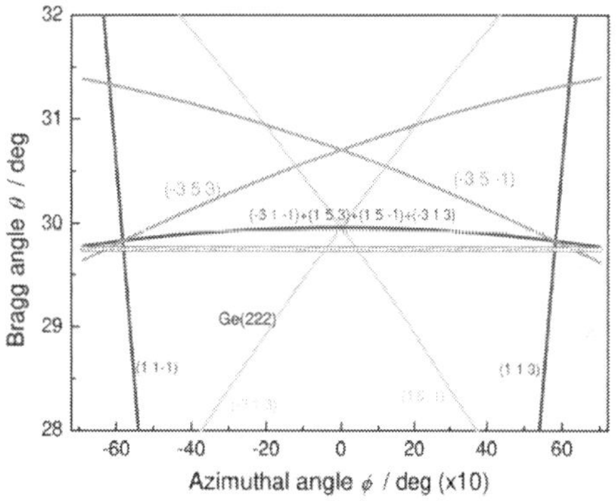

Figure 3. Part of the azimuth-Bragg angle relationship for 222 primary reflection of the diamond structure at the vicinity of the Bragg angle of 30°. Indexes are related only to secondary reflections.

motivation of this experiment consists in obtaining a luminous high resolution monochromatic beam of a large cross section with a possible exploitation in neutron diffractometry or radiography. The experimental performance is schematically shown in figure 2. The premonochromator is set for fixed Bragg angle of θ_M=15.01° and provides monochromatic neutrons of λ=0.162 nm. For searching the MBR effects, the method of azimuthal rotation of the crystal around the scattering vector $\boldsymbol{g}_1$ of the Ge(222) basic (primary) reflection was used. The beam coming from the bent Si(111) crystal was limited by 8x20 mm^2 Cd slit. As the Ge(222) reflection is forbidden, the MBR effects can be thus easily observed. In fact, they simulate the forbidden reflection. Figure 3 shows the azimuth-Bragg angle relationship for occurence of the multiple Bragg reflections of the diamond structure in the vicinity of the Bragg angle of 30° where they can be expected and accompany the Ge(222) primary reflection. The horizontal line correspods to the azimuthal rotation of the crystal slab around the 222 scattering vector at the mean Bragg angle of 29.75°. In the experiment we carried out several θ-rocking curves with respect to the Si(111) premonochromator for different azimuthal angles ϕ and the Figure 4 shows the properties of the strong MBR effect coming from the simultaneous action of 2 pairs (-31-1)/(513) and (153)/(1-3-1) of the secondary/tertiary reflections. It should be pointed out that in this favourite case, during the MBR process the neutrons diffracted by the secondary planes (at some point A) move in the bent crystal slab along its longest edge before they are reflected again by tertiary set of planes (in some point B). In the next step, the spatial distribution and collimation properties of the beam obtained by MBR effect was studied by using Neutron Imaging Plate (IP). Figure 5 shows the image of holes of different diameter in Cd-plate but taken by IP at the distance of

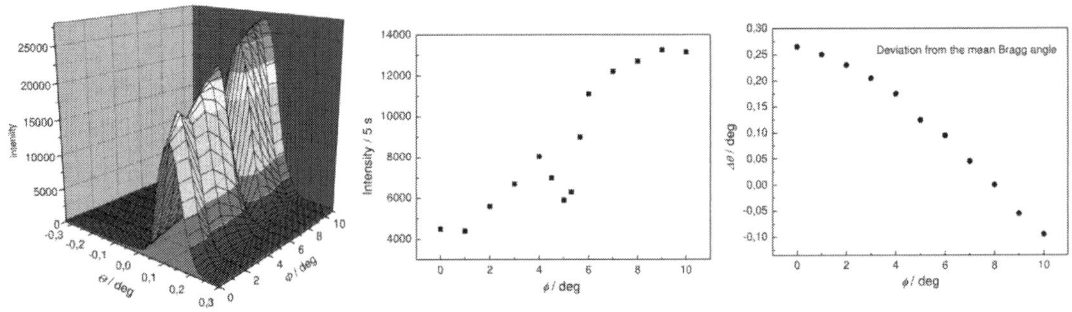

Figure 4. The 3-d intensity distribution constructed from the individual rocking curves for different azimuthal angles ϕ - (a), the peak intensities related to the rocking curves - (b) and the deviation of the rocking curve maxima from the mean Bragg angle 29.75° - (c).

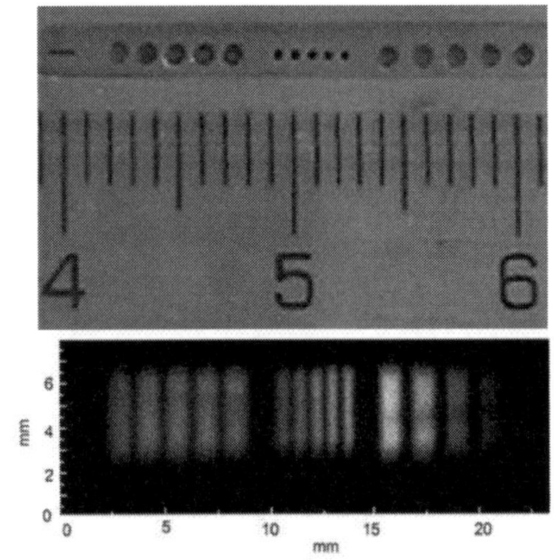

Figure 5. Photo of the series of the holes of different diameters and their image taken by IP at the distance of 70 cm from the bent Ge crystal.

70 cm from the Ge crystal. On the other hand, figure 6 displays the spatial distribution of the beam taken by IP at the distance of about 10 cm from the bent Ge crystal. Inspection of figure 6 reveals that thanks to the favourite geometry of the secondary and tertiary diffraction when diffracted neutrons feel the change of the Bragg angle due to crystal curvature at rather long distance, two individual MBR-processes (-31-1)/(513) and (153)/(1-3-1) are separated. The strong vertical part of the image corresponds to the (153)/(1-3-1) process and the "tail" of rather

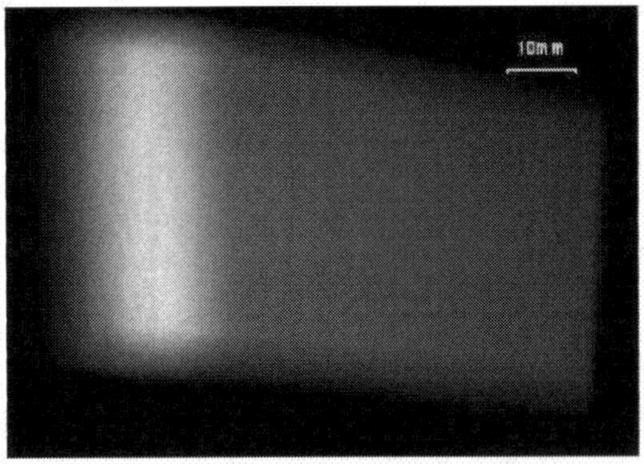

Figure 6. Image of the monochromatic beam obtained on the basis of the MBR process.

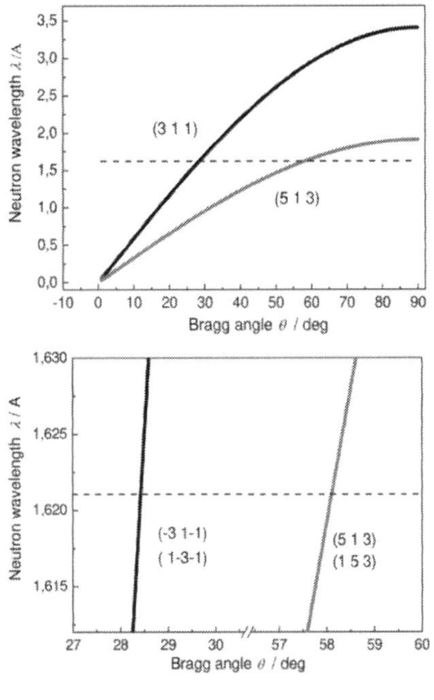

Figure 7. Bragg condition law for 311 and 513 reflections and enlarged part related to the described MBR process.

large cross-section corresponds to the (-31-1)/(513) one. The explanation of this effect can be derived from figure 7 where Bragg condition dependences related to the reflections participating in the MBR process as well as the enlarged parts in the vicinity of λ=0.162 nm are displayed. It is clear that if secondary reflection is of the 311 type then the related phase space element ($\Delta\lambda$ x $\Delta\theta_{311}$) is spread over much larger $\Delta\theta_{513}$ of the tertiary reflection of the 153 type and *vice versa*. On the other hand, inspection of figure 6 reveals that the monochromatic beam obtained by MBR process is highly collimated in the scattering plane, but not in the perpendicular direction. This is brought about by a fact that all planes participated in the MBR process are perpendicular to the scattering plane and the secondary and tertiary reflection planes are mutually in dispersive diffraction geometry. However, thanks to this dispersive geometry, also the monochromacity of the beam is very high and depending on crystal radius of curvature, $\Delta\lambda/\lambda$ spread is less than 10^{-3} [13,14].

3. Conclusion

Multiple Bragg reflections realized in bent perfect crystals can provide very intensive highly monochromatic and highly collimated beams of rather large cross section. There is as good chance to exploit such monochromatic beam in high (or extremely high) resolution neutron diffractometry or spectrometry and also for high resolution radiography [14,15,16], especially at high-flux neutron sources. As there are still many opened questions which should be solved, Monte Carlo simulations of the MBR processes would be desirable. First attempts of MC simulations have been already done [17,18].

Acknowledgements

Bragg diffraction optics investigations are in the Czech Republic supported by the research projects AVOZ104805505, ASCR-M100480901, GACR P204/10/0654 and also by FP7-NMI3 "Integrated Infrastructure Initiative for Neutron Scattering and Muon Spectroscopy", 2009-2013.

References

[1] Ewald P P 1937 *Z. Kristallographie* **A97** 1-27.
[2] Renninger M 1960 *Z. Kristallographie* **113** 99-103.
[3] Pinsker Z G 1982 *X-ray Crystal Optics*, Nauka, Moscow.
[4] Chang S H 1984 *Multiple Diffraction of X-rays in Crystals*, Springer Verlag, Berlin.
[5] Moon R and Shull C G 1964 *Acta Cryst.* **17** 805-812.
[6] Kuich G and Rauch H 1965 *Acta Physica Austriaca* **20** 7-16.
[7] Seong B S, Shin E, Lee C H, Em V T and Shim H S 2001 *J. Neutron Research* **9** 363-371.
[8] Kulda J and Mikula P, *Phys. Stat. Sol.* (a), **92** (1985) 95-100.
[9] Mikula P, Vrána M and Wagner V 2006 *Z. Kristallographie*, Suppl. **23** 205-210.
[10] P. Mikula, M. Vrána, D. Lott, V. Wagner 2006 *Physica B* **385-386** 1274-1276.
[11] P. Mikula, M. Vrána, V. Wagner, M. Furusaka 2008 *Nucl. Instrum. Methods in Phys. Research* A**586** 18-22.
[12] Mikula P, Vrána M, Šaroun J, Seong B S, Em V and Moon M K 2011 *Nucl. Instrum. Methods in Phys. Res.* A, **634** S108-S111.
[13] Mikula P, Vrána M, Wagner V, Choi Y N, Moon M K, Lee C H and Inoue K 2004 Proc. of 12th Int. Conf. on Experimental Mechanics August 29 – September 2, Bari, Italy, McGraw-Hill, ed. C. Pappalettere, full paper No. 038 on CD-ROM.
[14] Mikula P and Vrána M 2007 Proc. of 13th Int. Conf. on Experimental Mechanics July 1-6, Alexandropolis, Greece, Springer, ed. E Gdoutos, chapter 2T25–*Structural Testing*, paper 18_Mik on CD.
[15] Kardjilov N 2003 *Further Developments and Applications of Radiography and Tomography with Thermal and Cold Neutrons*, PhD-Thesis, Technische Universität München.
[16] Kardjilov N, Lee S V, Lehmann E, Lim I C, Sim C M and Vontobel P 2005 *Nucl. Instrum. Methods in Phys. Research* A**542** 100-105.

[17] Šaroun J and Kulda J 2009 *Monte Carlo simulation package RESTRAX 5.2.0 and SIMRES 6.0.6*, http://neutron.ujf.cas.cz/restrax/index.php?chap=3 /
[18] Šaroun J, Mikula P and Kulda J 2011 *Nucl. Instrum. Methods in Phys. Research* A**634** S50-S54.

High-precision figured thin supermirror substrates for multiple neutron focusing device

M. Nagano[1], F. Yamaga[1], D. Yamazaki[2], R. Maruyama[2], H. Hayashida[2], K. Soyama[2] and K. Yamamura[1]

1 Research Center for Ultra-precision Science and Technology, Graduate School of Engineering, Osaka University, 2-1, Yamadaoka, Suita, Osaka 565-0871, Japan

2 J-PARC Center, Japan Atomic Energy Agency, 2-4 Tokai, 319-1195, Japan

yamamura@upst.eng.osaka-u.ac.jp

Abstract. An aspherical supermirror is one of the most useful neutron-focusing optics. We aim to develop multiple aspherical supermirror devices using high-precision figured aspherical focusing supermirrors to focus neutron beams with high intensities, because multiple mirrors collect a very large beam divergence. Thin mirrors with a millimeter thickness are required to minimize the absorption loss of incident neutron beams since the thickness of a mirror shadows the reflective area of the other mirrors. However, it is difficult to fabricate thin mirror substrates with a form accuracy of sub-micrometer level by conventional machining. Conventional machining deforms a substrate by machining force and spring back after machining causes figure error. Furthermore the deposition of supermirrors deforms the mirror substrate by film stress. Thus, we developed a new process of fabricating a precise millimeter-thick elliptical supermirror. This process consists of noncontact figuring by the numerically controlled local wet etching technique and the ion beam sputter deposition of NiC/Ti multilayers on both sides of the mirror substrate to compensate for film stress. In this paper, we report on the fabrication results and focusing performance of elliptical supermirrors with a thickness of 1.5 mm.

1. Introduction

Recently, various measurements using neutrons have been of considerable practical interest because of their ability to identify hydrogen atoms and water molecules precisely, sensitivity to magnetism and penetrating power. High-intensity proton accelerator facilities, such as J-PARC in Japan, ISIS-TS2 in UK and SNS in USA, are constructed for neutron research. However, neutron intensity is very low compared with that of X-ray in synchrotron orbital radiations, even in these latest neutron sources. Therefore, high-performance optical devices for focusing neutron beams are required to realize practical measurements. Reflective aspherical supermirror optics would be advantageous for focusing neutron beams with a wide wavelength range since it has no chromatic aberration. Neutrons can be focused within a small area of less than 1 mm^2 using a high-performance elliptical supermirror with a high form accuracy and a low roughness at both the substrate surface and the multilayer interface. Therefore, reflective optics using an elliptical supermirror improves S/N ratio, reduces measurement time and can be applied to a wide range of neutron experiments, such as the diffraction-based measurement of small biomolecules and position-sensitive prompt gamma analysis, at neutron reactors and pulsed neutron sources.

We developed a numerically controlled local wet etching (NC-LWE) technique[1] to figure aspherical mirror substrates and an ion beam sputter deposition technique [2] to deposit a NiC/Ti multilayer for the realization of usable aspherical supermirror optics. Previously, we fabricated a plano-elliptical supermirror with a clear aperture size of 90 x 40 mm^2 by applying these techniques, and obtained a focusing gain of 6 and a reflectivity of 0.64-0.7 in the near-critical-angle region for m=4 [3,4]. The m-value indicates the ratio of the effective critical angle of the supermirror to that of natural nickel.

To increase the focusing gain and reflectivity, a decrease in surface roughness and the enlargement and multiplexing of supermirrors are essential. Previously, we succeeded in developing a fabrication process for an aspherical mirror substrate with a large clear aperture size [5]. In the case of the multiplexing of supermirrors, mirror substrates must be thin to reduce the absorption loss of incident neutron beams. Previously, a Kirkpatrick-Baez mirror consisting of orthogonal stacks of reflectors for an X-ray telescope was reported [6]. However, the figure accuracy of the mirror has not been referred to because the weight of the mirror is more important than its shape in space X-ray optics. On the other hand, the figure accuracy of the mirror is most significant for the beam width in neutron focusing optics. It is difficult to fabricate thin mirror substrates with a form accuracy of sub-micrometric level and a surface roughness of sub-nanometric level because conventional precision grinding deforms the workpiece by machining force and spring back after grinding causes figure error. In addition, conventional polishing for reducing surface roughness increases figure error. Furthermore, the deposition of supermirors deforms the mirror substrate by film stress. As a fabrication technique for SiC mirrors with a millimeter thickness, ELID grinding with a feedback compensation method has been reported. In this method, the tool path is determined from numerical calculation results of the profile deformation analyzed by FEM [7].

On the other hand, the NC-LWE technique is very suitable for fabricating an ultraprecise millimeter-thick mirror substrate, because such a technique enables us to fabricate an aspherical shape without processing pressure to the workpiece and wear of machine tools because of its noncontact removal mechanism utilizing a chemical reaction. Moreover, a high reproducibility is also achieved owing to the insensitivity of such a technique to external disturbances, such as vibration or thermal deformation. Thus, the NC-LWE technique is very feasible for fabricating an ultraprecise millimeter-thick mirror substrate with a high reproducibility. We developed a new process of fabricating a precise millimeter-thick elliptical supermirror. This process consists of noncontact figuring by the NC-LWE technique, the minimization of surface roughness without degrading form accuracy by low-pressure polishing with a polishing pressure less than about 7 kPa (1 psi) and the ion beam sputter deposition of NiC/Ti multilayers on both sides of mirror substrate surfaces to compensate for film stress.

In this work, we fabricated and evaluated a plano-elliptical neutron focusing supermirror with a thickness of 1.5 mm.

2. Multiple neutron focusing device

An aspherical supermirror is one of the most useful optics for focusing neutrons with a wide wavelength range. Multiple aspherical supermirror devices using high-precision figured aspherical focusing supermirrors have received considerable attention with the aim of focusing neutron beams with high intensities, because multiple mirrors collect a very large beam divergence. Thin mirrors with a millimeter thickness are required to minimize the absorption loss of incident neutron beams since the thickness of a mirror shadows the reflective area of the other mirrors.

2.1. Optical design of multiple mirror system

To increase the focusing gain, it is effective to multiplex the supermirror. We designed a multiple mirror neutron focusing system to be applied to reflective small angle scattering measurements. Figure

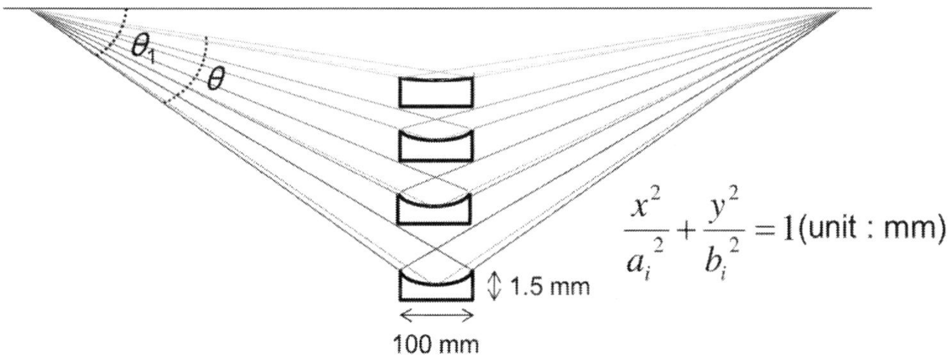

Figure 1. Design of multiple mirror system.

1 shows the optics design of the multiple mirror neutron focusing system. This focusing system consists of 4 mirrors with elliptical shapes represented by

$$\frac{x^2}{a^2}+\frac{y^2}{b^2}=1 \text{ (unit:mm)}. \quad (1)$$

The a and b values of each mirror are shown in Table 1. The thickness of each mirror substrate is 1.5 mm and the clear aperture size is 100 x 50 mm^2. The magnification of the mirrors and the focal length and incident angle of the outmost mirror center are 1:1, 1050 mm and 1.4°, respectively.

2.2. High-precision figured thin elliptical supermirror

Thin mirrors with a millimeter thickness are required to minimize the absorption loss of incident neutron beams since the thickness of a mirror shadows the reflective area of the other mirrors. However, it is difficult to fabricate thin mirror substrates with a form accuracy of sub-micrometer level by conventional machining. Conventional machining deforms the substrate by machining force and spring back after machining causes figure error. Furthermore, the deposition of supermirors deforms the mirror substrate by film stress.

We developed a new process of fabricating a precise millimeter-thick elliptical supermirror. This process consists of noncontact figuring by the NC-LWE technique and the ion beam sputter deposition of NiC/Ti multilayers on both sides of mirror substrate surfaces to compensate for film stress.

In this work, we fabricated the mirror #4 shown in Table 1 by the NC-LWE technique. The material and outer size are synthesized quartz glass and 150 x 150 x 1.5 mm^3, respectively. In the NC-LWE technique, figuring is performed by controlling the dwelling time of the nozzle head. Because of its noncontact chemical removal mechanism, the NC-LWE technique enables us to fabricate a millimeter-thick mirror substrate without generating subsurface damage and deformation. As the processing parameters in NC-LWE figuring, we used 37 wt% HF aqueous solution with a temperature of 42°C, a circular nozzle with a diameter of 15 mm and a feed pith of 0.5 mm.

In NC-LWE figuring, raster-type numerically controlled scanning is applied to correct the residual figure error. After raster scanning, shallow periodic grooves, which are similar to the tool marks formed in conventional lathe turning, are inevitably generated on the figured surface by the overlapping of the scanned etching path. We have reported a method of decreasing the height of the tool marks by a two-stage NC-LWE process [3]. However, in NC-LWE figuring with raster scanning, the tool marks cannot be completely eliminated. Low-spatial frequency roughness (LSFR: λ_s >1 mm), such as that caused by tool marks, degrades the focusing beam profile. To elliminate LSFR, we applied a first polishing process with a hard polishing pad.

Additionally, the mid-spatial frequency roughness (MSFR: 1 mm > λ_s > 1 μm) and high-spatial frequency roughness (HSFR: 1 μm > λ_s > 1 nm) components of the processed surface tend to increase after NC-LWE figuring. The MSFR and HSFR components are one of the causes of the decrease in the reflectivity of the supermirror. Subsurface damage is considered as one of the causes of the increases in MSFR and HSFR because the etching rate of the damaged layer is different from that of an undamaged layer. Thus, it is essential to minimize the surface roughness to obtain a higher reflectivity and thus realize the practical application of the neutron focusing mirror. Therefore, a second polishing process using a soft polishing pad was applied to improve the MSFR and HSFR of the figured mirror substrate.

The deposition of supermirrors deforms the mirror substrate by film stress. In particular, the film stress with a GPa level generated by ion beam sputter deposition deforms the mirror substrate with a millimeter level thickness. We performed the ion beam sputter deposition of NiC/Ti multilayers on both sides of mirror substrate surfaces to compensate for film stress. A NiC/Ti supermirror with m=3 was fabricated. The layer sequence of the deposited supermirror was designed using the algorithm proposed by Hayter and Mook. The thickness, total number of multilayers and size of deposition were 3 μm, 400 and 110 x 60 mm^2, respectively.

2.3. Evaluation of focusing performance and reflectivity

We evaluated the focusing performance of the fabricated mirror at pulsed neutron instrument with disc chopper (CHOP) in Japan Research Reactor No. 3 Modified (JRR-3M) of Japan Atomic Energy Agency. Figure 2 shows a schematic of the setup for measuring the neutron beam profile. The wavelength of the neutron beam ranged from 3 to 10 +. The width of the slit in this experiment was 0.25 mm. The magnification of the optical design was 1. The beam profiles were measured by using an imaging plate.

The reflectivity of the fabricated supermirror was evaluated at the Neutron Multilayer Interferometer Reflectometer 2 (MINE2) of JRR-3M. The wavelength of the monochromatized neutron beam was 8.8 +. The meridional direction of the mirror under test was set vertically, and the neutron beam defined by two slits was irradiated in the sagittal direction, in which the test mirror has no curvature.

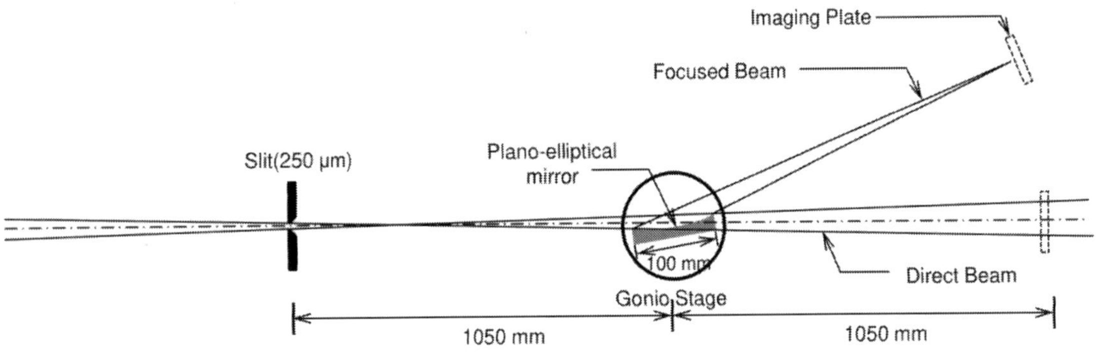

Figure 2. Schematic of the setup for measuring the neutron beam profile at CHOP.

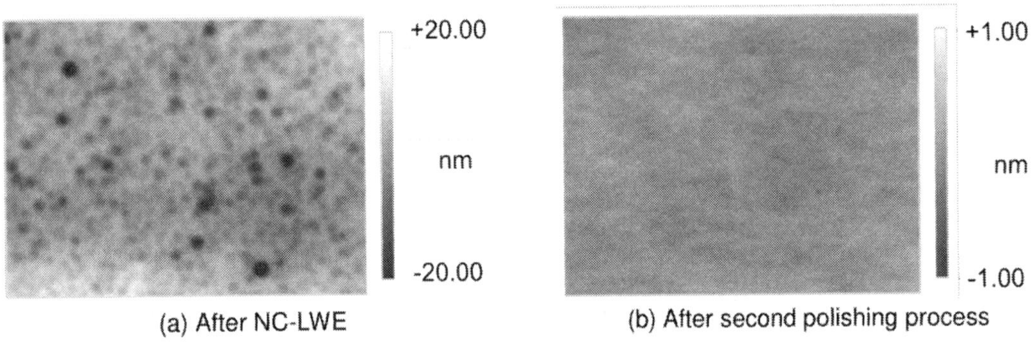

Figure 3. The surface roughness of the mirror substrate. (64 x 48 μm^2)

3. Results and discussion

The residual figure error after NC-LWE was about 0.2 μm. The shape of the mirror substrate was measured with a laser autofocus measuring machine (Mitaka Kohki NH-3SP, z resolution: 1 nm). This result indicates that the noncontact removal by controlling the dwelling time of the etchant nozzle deterministically figures the aspherical shape with a form accuracy of sub-micrometric level.

Figure 3 (a) and (b) show the surface roughnesses of the figured mirror substrate after NC-LWE correction and second polishing, respectively. The surface roughness was measured using a scanning white-light interferometer (Zygo NewView 200CHR). Many pits were observed on the surface after NC-LWE correction, as shown in Figure 3 (a), and the rms surface roughness was 4.6 nm. To remove these undesirable surface structures, a soft polishing pad was used in the second polishing process with a removal depth of about 40 nm. Figure 3 (b) indicates that the surface morphology was improved and an rms roughness of less than 0.15 nm was obtained by applying the second polishing process. It also indicates that the second polishing process is very effective for improving the overall surface roughness of the mirror substrate in the MSFR regions.

In general, the application of a polishing process to improve the surface roughness degrades the form accuracy because it is difficult to maintain a constant removal rate over the area of the clear aperture of the mirror. The residual figure error after the second polishing process was about 0.2 μm p-v. The low-pressure polishing process has the ability to improve the surface roughness satisfactorily while maintaining a sub-micrometer form accuracy. These results indicate that low-pressure polishing is very effective for improving the surface roughness without degrading the shape of the mirror substrate.

We deposited a NiC/Ti supermirror with m=3 on both sides of the figured mirror substrate by ion beam sputter deposition. Figure 4 shows the residual figure error observed after deposition. If one side of the mirror substrate is deposited, film stress deforms the mirror substrate with a deformation amount of 50-60 μm. However, we succeeded in inhibiting deformation less than 1 μm p-v by depositing NiC supermirror on both sides of the figured mirror substrate. Figure 5 shows the surface roughness observed after deposition. The rms roughness of the supermirror was 0.2 nm, thus maintaining the surface roughness of the mirror substrate.

Table 1. Optical parameters of neutron focusing mirrors

Mirror #1	θ_1=1.40(dgree),	a_1=1050.31,	b_1=25.66 (mm)
Mirror #2	θ_2=1.19(dgree),	a_2=1050.23,	b_2=21.83 (mm)
Mirror #3	θ_3=1.00(dgree),	a_3=1050.16,	b_3=18.34 (mm)
Mirror #4	θ_4=0.83(dgree),	a_4=1050.11,	b_4=15.17 (mm)
Focal length	1050 mm		

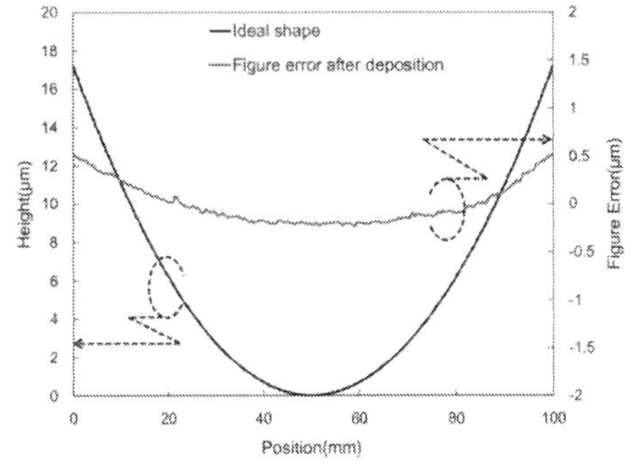

Figure 4. Figure errors of the fabricated mirror in meridional direction.

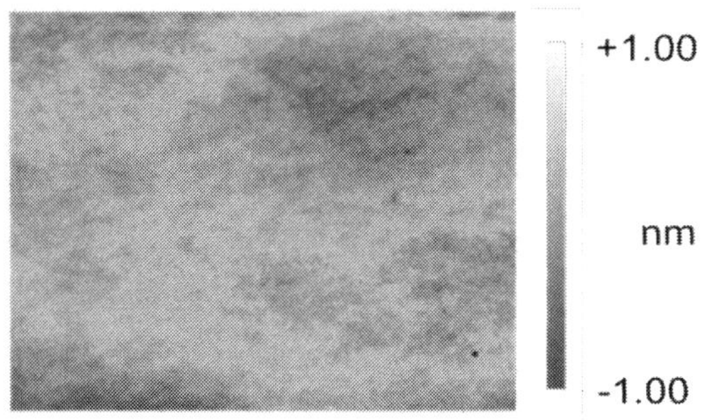

Figure 5. The surface roughness of the supermirror. (64 x 48 μm^2)

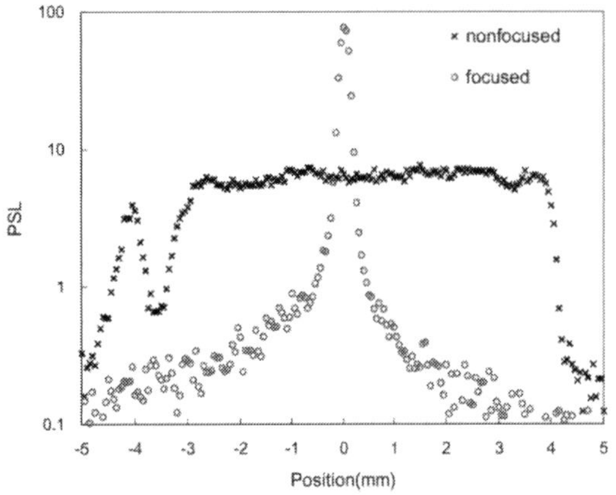

Figure 6. Focused beam profile.

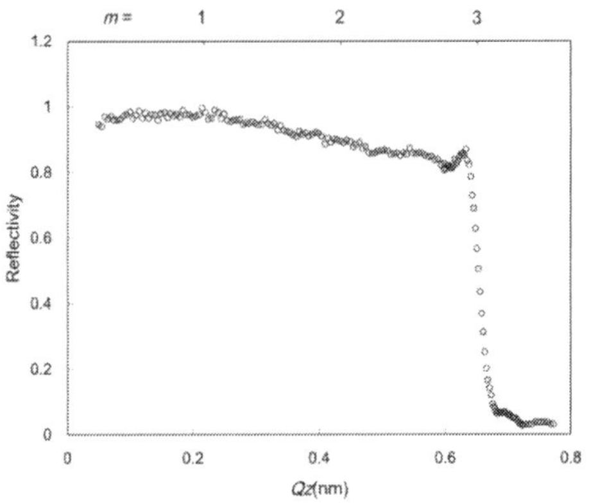

Figure 7. Reflectivity of figured elliptical supermirror.

These results suggest that the application of the combined fabrication process consisting of NC-LWE figuring, low-pressure polishing and ion beam sputter deposition is a highly effective method for fabricating a high-precision millimeter-thick aspherical supermirror with a form accuracy of sub-micrometric level and a surface roughness of sub-nanometric level.

Figure 6 shows the beam profiles of the focused and non-focused beams. The full width at half maximum (FWHM) of the focused beam was 0.21 mm. The focusing gain of 12 in peak intensity was achieved compared with that in the case of the nonfocused directly irradiated beam, and the peak intensity was more than 100-fold higher than background intensity. These results indicate that our thin elliptical supermirror has a high figure accuracy for focusing neutron beams with sub-millimeter focusing widths.

The result of the $\theta 2\theta$ scan is shown in Figure 7. For the performance of neutron reflection, the critical angle for m=3 (q_z=0.65 nm^{-1}) was achieved. This result shows that the designed d-space distribution of the NiC/Ti multilayer was precisely formed on the elliptically figured substrate. The reflectivity in the near-critical-angle region was more than 0.8. This result suggests that the fabricated supermirror has an interface roughness of sub-nanometer level.

The above results indicate that the fabricated millimeter-thick aspherical supermirror has the ability to focus neutron beams with sub-millimeter widths and high reflectivities. In our future study, we will attempt to fabricate and evaluate a multiple mirror system using a millimeter-thick aspherical supermirror.

4. Conclusions

We fabricated a plano-elliptical neutron focusing supermirror with a thickness of 1.5 mm by using a combined fabrication process consisting of NC-LWE figuring, low-pressure polishing and ion beam sputter deposition. We succeeded in obtaining a figure error of 0.82 μm p-v with a surface roughness of less than 0.15 nm rms, a FWHM of 0.21 mm and a focusing gain of 12 in peak intensity with a reflectivity of more than 0.8 in the near-critical-angle region.

Acknowledgements

This work was partially supported by a grant for "Development of High-Precision Aspherical Supermirror for Neutron Focusing", "Development of System and Technology for Advanced Measurement and Analysis (SENTAN)" from Japan Science and Technology Agency (JST), the "Industrial Technology Research Grant Program in 2005" from the New Energy and Industrial Technology Development Organization (NEDO) of Japan, a Grant-in-Aid for the Global COE Program "Atomically Controlled Fabrication Technology" from the Ministry of Education, Culture,

Sports, Science and Technology, Japan, and Grant-in-Aid for JSPS Fellows. The authors would like to thank Masahiro Hino, PhD for assistance with experiments at MINE2 and Kazuya Aizawa, PhD and Shinichi Takata, PhD for assistance with experiments at CHOP.

References

[1] Yamamura K 2007 Science and Technology of Advanced Materials. **8** 158
[2] Maruyama R, Yamazaki D, Ebisawa T, Soyama K 2009 Nucl. Instr. and Meth. A. **600** 68
[3] Yamamura K, Nagano M, Takai H, Zettsu N, Yamazaki D, Maruyama R, Soyama K and Shimada S 2009 Optics Express **17** 6414
[4] Yamamura K, Nagano M, Zettsu N, Yamazaki D, Maruyama R and Soyama K 2010 Nucl. Instr. and Meth. A, **616** 193
[5] Nagano M, Yamaga F, Zettsu N, Yamazaki D, Maruyama R, Soyama K, Yamamura K 2010 Nucl. Instr. and Meth. A. **634** S112
[6] Hudec R, Marsikova V, Pina L, Inneman A and Skulinova M 2010 International Conference on Space Optics.
[7] Ohmori H, Dai Y, Lin W, Suzuki T, Katahira K, Itoh N, Makinouchi A and Tashiro H 2003 ADVANCES IN ABRASIVE TECHNOLOGY V **238-2** 65

Journal of Physics: Conference Series **340** (2012) 012017 doi:10.1088/1742-6596/340/1/012017

Analysis of Neutron Polarimetry data using MuFit

Amy Poole, Bertrand Roessli

Laboratory for Neutron Scattering, Paul Scherrer Institut, CH-5232 Villigen, PSI, Switzerland

E-mail: `amy.poole@psi.ch`, `bertrand.roessli@psi.ch`

Abstract. This paper presents a new single crystal neutron scattering data analysis program, Mufit. The program can be used to refine spherical neutron polarimetry (SNP) data as well as nuclear and magnetic single crystal diffraction data. The article presents a brief overview of the SNP process and the core mathematics used to calculate the SNP matrices and gives an example of data that has been fitted using the MuFit package, which is available on request from the author: Bertrand.Roessli@psi.ch.

1. Neutron Polarimetry

Spherical neutron polarimetry (SNP) is the measurement of the both the longitudinal and transverse components of a polarized neutron beam after the scattering event. It is the experimental realization of the independent, theoretical work of Blume and Maleev who put forward a general description of the scattering cross section for a neutron beam that includes an arbitrary initial and final polarization, [1, 2, 3]. Their work was an extension of that by Moon, Riste and Koehler who developed longitudinal polarization analysis (LPA) and a set of equations to describe the observed experimental intensities [4]. LPA type experiments require a guide field to maintain the polarization of the neutron beam. Any transverse components of the polarization vector precess around this field and the information contained in these components is lost. CRYOPAD (CRYOgenic Polarization Analysis Device) and, subsequently, MuPAD (MU-metal Polarization Analysis Device) prevent this information loss by shielding the polarized beam from stray magnetic fields, which negates the requirement for a guide field and allows SNP-type measurements [6, 7]. The power of the technique lies in the accuracy with which one may determine the orientation of $\mathbf{M}_{\perp}(\mathbf{Q})$ and the ability to determine the change of intensity when the structure has out of phase components [8, 9, 10].

2. Experimental Process

The neutron polarimetry process may be considered in three key stages. The first step is to orient a polarized neutron beam, usually in one of the directions, x, y or z where x is in the direction of the scattering vector $\mathbf{Q}$, z is along the instrument axis, perpendicular to the scattering plane and y makes the right hand set [11]. The oriented, polarized beam then enters the 'zero-field' chamber where it meets the sample. On scattering, the neutron spins are rotated by the magnetic scattering vector, $\mathbf{M}_{\perp}(\mathbf{Q})$, that lie in the direction of polarization and flipped by components that are perpendicular. The third step is then to measure the intensity of the spin-flip and non spin-flip scattering in each of the directions x, y and z. In the case of purely magnetic scattering the process may be visualized as a rotation about $\mathbf{M}_{\perp}(\mathbf{Q})$, but in which polarization may be created along, **Q**, when $\mathbf{M}_\perp(\mathbf{Q})$ has both real and imaginary components, i.e. for helical and cycloidal structures [12].

3. Mathematical Summary

The spin-flip and non-spin flip scattering intensities, I^- and I^+ respectively, are measured for each of the initial X, Y and Z and final x, y and z directions. The 3×3 polarization matrix,

$$\begin{pmatrix} P_{Xx} & P_{Xy} & P_{Xz} \\ P_{Yx} & P_{Yy} & P_{Yz} \\ P_{Zx} & P_{Zy} & P_{Zz} \end{pmatrix},$$

is found from these 18 measurements as each element of the matrix is given by,

$$P_{If} = \frac{I_{If^+} - I_{If^-}}{I_{If^+} + I_{If^-}} \tag{1}$$

The measured intensity, I_{If}, is a sum of the scattering from all of the n magnetic domains, weighted by the volume fraction δ^n, that are present in the sample:

$$I_{If} = \sum_n \delta^n I^n_{If} \tag{2}$$

To determine the intensity of the polarized scattering in the purely elastic case the general scattering cross section has the form,

$$\left(\frac{\mathrm{d}\sigma}{\mathrm{d}\Omega}\right) = \sum_{\sigma_I \sigma_f} p_{\sigma_I} p_{\sigma_f} \times |\langle \sigma_f | \hat{V}(\mathbf{Q}) | \sigma_I \rangle|^2, \tag{3}$$

where σ_I and σ_f are spinors that describe the incident and final polarization respectively and p_{σ_I} and $p_{\sigma f}$ describe the efficiency of that polarization. The term $\hat{V}(\mathbf{Q})$ is the Fourier transform of the interaction potential between the incident neutron and the target system multiplied by $(m/2\pi\hbar)$, [13]. When the polarization of the beam is not unitary, both before and after the sample, a sum has to be made over all the combinations of σ_I and σ_f which are weighted according to the polarization efficiency. If we assume that there is no contribution to the magnetism from the nuclear spin the term $\hat{V}(\mathbf{Q})$ may be decomposed into parts that describe the nuclear interaction $\hat{V}_N(\mathbf{Q})$ and the magnetic interaction $\hat{V}_M(\mathbf{Q})$,

$$\hat{V}(\mathbf{Q}) = \hat{V}_N(\mathbf{Q})\,\mathbf{1} + \hat{V}_M(\mathbf{Q}) \cdot \boldsymbol{\sigma}. \tag{4}$$

The term $\hat{V}_N(\mathbf{Q})$ is multiplied by a 2×2 unitary matrix, represented by $\mathbf{1}$, and $\hat{V}_M(\mathbf{Q})$ is multiplied by a vector of Pauli matrices, $\boldsymbol{\sigma}$. In this form the potential can be considered as a rotational matrix that operates on the initial spin state σ_I to give the final spin state σ_f. The measurement of the intensity of the neutrons with final spin states σ_{f^+} and σ_{f^-} reveals the rotational operation that has occurred and from this the potential may be deduced.

If we consider the final polarization vector $\mathbf{P}'_I$, with initial polarization $\mathbf{P}_I$, where $I = X, Y$ or Z we have the polarization vector described in the Blume-Maleev equations:

$$\mathbf{P}'_I = (P_{Ix}, P_{Iy}, P_{Iz}) = \frac{\mathrm{Tr} \cdot \hat{\rho} \langle \hat{V}^\dagger | \sigma | \hat{V} \rangle}{\mathrm{Tr} \cdot \hat{\rho} |\hat{V}|^2} \tag{5}$$

where $\hat{V}^\dagger$ is the complex conjugate transpose of $\hat{V}$, Tr$\cdot$ denotes the trace over neutron spin variables and $\hat{\rho}$ is a probability density matrix, described by,

$$2\hat{\rho} = \mathbf{1} + \mathbf{P}_I \cdot \hat{\boldsymbol{\sigma}}. \tag{6}$$

From the expansion of equation 5 it is possible to derive the Blume-Maleev equations, [1, 2, 3],

$$\begin{aligned}\mathbf{P}'_I = & [(\mathbf{P}_I NN^*) \\ & + (-\mathbf{P}_I(\mathbf{M}_\perp \cdot \mathbf{M}^*_\perp) + (\mathbf{M}_\perp(\mathbf{P}_I \cdot \mathbf{M}^*_\perp) + \mathbf{M}^*_\perp(\mathbf{P}_I \cdot \mathbf{M}_\perp)) \\ & + (-i(\mathbf{M}^*_\perp \times \mathbf{M}_\perp)) \\ & + (N\mathbf{M}^*_\perp + N^*\mathbf{M}_\perp - i(N\mathbf{M}^*_\perp - N^*\mathbf{M}_\perp) \times \mathbf{P}_i)]/\sigma, \end{aligned} \tag{7}$$

$$\begin{aligned}\sigma = & (NN^*) + (\mathbf{M}_\perp \cdot \mathbf{M}^*_\perp) \\ & + (i\mathbf{P}_I \cdot (\mathbf{M}^*_\perp \times \mathbf{M}_\perp) + (\mathbf{P}_I \cdot (\mathbf{M}_\perp N^* + \mathbf{M}^*_\perp N)). \end{aligned} \tag{8}$$

Where $\mathbf{M}_\perp$ is the magnetic interaction vector, $\mathbf{M}^*_\perp$ is the complex conjugate of the magnetic interaction vector and N is the nuclear structure factor. The $\mathbf{Q}$ dependence of $\mathbf{M}_\perp$ and N has been omitted for clarity. The first and second rows of equation 7 give the nuclear and magnetic intensities respectively. The third row describes the contribution to the scattering from the real and imaginary components of $\mathbf{M}_\perp$ and is zero for all magnetic structures that are not either helices or cycloids. The final row gives the interference effects from coincident nuclear and magnetic scattering and is zero in all cases where $\mathbf{k} \neq 0$, [14].

4. Data Analysis

To analyze and fit the polarimetry data, dedicated software was developed in the interpreted language Octave, which is an open source software developed to perform numerical and mathematical operations [15]. The Octave language has the advantage that it is mostly compatible with Matlab and hence is easily accessible to most users. Furthermore, it has the option to dynamically load C++ functions to reduce long computation times.

Currently MuFit can be used to calculate the nuclear and magnetic structure factors including anisotropic thermal factors, and extinction corrections, however, absorption corrections cannot be applied. Furthermore, a simultaneous refinement of nuclear and magnetic structures cannot be performed. The magnetic structure factors are calculated by making use of the Fourier components of the atomic moment, analogous to the approach taken in FullProf [16]. The weighted population of the symmetry related twins and rotationally related magnetic domains is summed and the weighting may be refined, as described in equation 2 to give the correct final polarization. Structures that are described by more than one $\mathbf{k}$ vector cannot currently be calculated within the program. For comparison the Blume-Maleev equations have been programmed and can be used to calculate the matrices, the equations are not used to fit the data as the depolarization corrections and domain populations cannot be applied within the vector formalism adopted by the equations. To correct the polarization matrices for the finite polarization of the beam the MuFit software calculates the neutron cross sections weighted by the beam polarization, $p_{\sigma I}$ and $p_{\sigma f}$, and sums to find the final intensity in each channel, equation 3. The step described in equation 3 describes the 'leakage' of intensity from one polarization channel to another and is not explicitly included in the Blume-Maleev polarization equations.

In order to speed up the calculations, the mathematical operations were vectorized as much as possible and C++ was used for the calculation of the elements of the polarization matrix to avoid loops that are costly in terms of computing time as no Just-In-Time compiler is included in Octave. To perform least-square refinements with MuFit, the user has the choice between three algorithms [17]: Levenberg-Marquardt, Simplex and Simulated Annealing. These routines have been implemented to fit nuclear and magnetic intensities from single crystals and polarimetry data. The program has been used to determine the magnetic structure of $ErMn_2O_5$ [18] and $BaMnF_4$ [10] and were found to be in agreement with the CCSL crystallographic library [19]. A plot of the goodness of fit of the nuclear and magnetic calculations for $BaMnF_4$ is shown

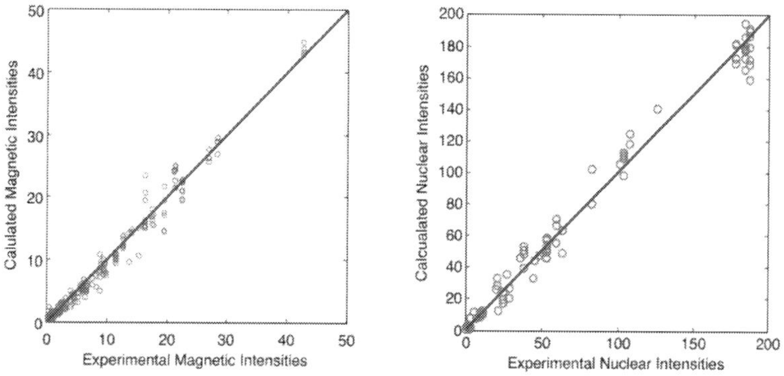

Figure 1. A comparison of the calculated and experimental magnetic and nuclear intensities for $BaMnF_4$, the line indicates $x = y$ and is a guide for the eye. [10]

in figure 1, where a description of the experiment and an example polarization matrix is given in reference [10]. The package also contains routines to draw nuclear and magnetic structures based on the Octave-Forge package 'vrml' [17]. MuFit is issued under the terms of the GNU General Public License and is freely available. It can be obtained by contacting the author at Bertrand.Roessli@psi.ch.

5. Conclusion

A summary of the essential equations for the description of the intensity of the polarised neutron scattering have been given and a description of how these are used in the MuFit computer package is described. The MuFit package has been successfully used in the analysis of experimental data and has been compared to CCSL. The results from this comparison indicates that the routines give very similar results.

6. Acknowledgements

The authors would like to thank P J Brown, P Fischer and M Janoschek for useful discussions.

7. References

[1] Izyumov Y. and Maleev S., *Soviet Phys. - JETP* **14** 1668 (1962)
[2] Blume M *Phys. Rev.* **130** 1670 (1963)
[3] Blume M., *Phys. Rev.* **133** A1366 (1963)
[4] Moon R M, Riste T and Koehler W C *Phys. Rev.* **181**, 920 (1969)
[5] Tasset F, *Physica B*, **156**, 627, (1989)
[6] Lelièvre-Berna E, Bourgeat-Lami E, Fouilloux, *et al*, *Physica B* **356** 131 (2005)
[7] Janoschek M, Klimko S, Gähler R, Roessli B, Böni P *Physica B* **397** 125 (2007)
[8] Brown P J, Forsyth B and Tasset F *J. Phys.: Condens. Matter* **10** 663 (1998)
[9] Brown P J and Chatterji T, *J. Phys.: Condens. Matter*, **18**, 10085 (2006)
[10] Poole A, Roessli B, Zaharko O and Kraemer K, *J. Phys.: Condens. Matter*, **23** 266004 (2011)
[11] Brown P J, Forsyth B and Tasset F *Proc. R. Soc. Lond. A* **442** 147 (1993)
[12] Brown P J *Physica B*, **297**, 198 (2001)
[13] Lovesey S W *Theory of Neutron Scattering from Condensed Matter: Polarization Effects and Magnetic Scattering II* , Oxford University Press (1989)
[14] Tasset F *Physica B* **297** (2001)
[15] http://www.octave.org
[16] Rodríguez-Carjaval J, *Physica B*, **192**, 55 (1993).
[17] These packages can be downloaded from http://octave.sourceforge.net/
[18] Roessli B *et al.*, *J. Phys.: Condens. Matter*, **20**, 48521 (2008).
[19] Brown P J and Matthewman J C, computer code CCSL, (2008).

Journal of Physics: Conference Series **340** (2012) 012018 doi:10.1088/1742-6596/340/1/012018

First Implementation of Novel Multiplexing Techniques for Advanced Instruments at Pulsed Neutron Sources

M. Russina[1] , **F. Mezei**[2,3] , **G. Kali**[3]

[1]Helmholtz Zentrum Berlin, Hahn-Meitner Platz 1, 14109 Berlin, Germany

[2]European Spallation Source ESS AB, PO BOX 176, 22100 Lund, Sweden,

[3]SzFKI, Hungarian Academy of Sciences, Pf. 49, 15125 Budapest, Hungary

Margarita.russina@helmholtz-berlin.de

Abstract. Novel multiplexing techniques, such as Repetition Rate Multiplication and Wavelength Frame Multiplication imply the use of a set of monochromatic wavelengths or a set of wavelength bands coming from the same source pulse by using novel combinations of standard mechanical neutron choppers. In this case the instrumental parameters, such as wavelength resolution, wavelength band, repetition rate are not any more determined by the source parameters, but can be flexibly defined by the chopper frequencies, speeds and slits. Here we report about the first experimental implementation of Repetition Rate Multiplication (RRM) and Wavelength Frame Multiplication (WFM). For this purpose the TOF spectrometer NEAT at HZB, Berlin and TOF diffractometer at BNC, Budapest have been used in non-standard modes of operation. Our results provide full proof-of-principle of the RRM and WFM methods and clearly show the extensive capability of these methods to achieve multiply enhanced data collection rates by individually tuning for each experiment the pulse length and/or pulse repetition rate within broad limits and independently from the actual source pulse parameters.

1. Principles of the multiplexing techniques.

The major limitation of the instruments operating at the existing pulsed spallation sources is that most of the instrumental parameters such as accessible $S(Q,\omega)$ domain, resolution and repetition rate are permanently fixed by the characteristics of the source pulses. For instance, in neutron spectroscopy the source pulse length and the length of the instrument define the instrumental resolution, while the source frequency is setting the instrumental repetition rate. Low repetition frequency of the source is often the reason why most of the TOF instruments at pulsed sources suffer from modest data collection efficiency. Compared to the "ideal" pulse repetition rate usually in the 50-300Hz range data collection can be realized only in 10 – 20 % of the time due to the low source frequencies varying in the 10 – 50 Hz range. The situation is becoming more complex in the case of the future European Spallation Source. According to the current project design ESS will deliver pulses with unprecedented peak flux, but at relatively long pulses of about 2-2.8 ms pulse length and low frequency in 10-20 Hz range. Using traditional design approach one needs an instrument length of several hundred meters to match the resolution of the instruments at ESS to those at the short pulses sources, which would increase the cost of the instruments substantially. Shortening the pulse can be achieved by introducing pulse shaping choppers; however, the low pulse repetition rate set by the source frequency would still make the use of the source peak intensity less efficient.

The main challenge in the instrument design at the future European Spallation Source is, therefore, to develop new kind of techniques to take full advantage of the high neutron peak flux and at the same time to take control over the other pulse parameters to best suit the instruments and individual experiments.

Novel multiplexing techniques, such as Wavelength Frame Multiplication [1] and Repetition Rate Multiplication [2] allow us to create instead of one long pulse a number of mini-pulses with variable frequencies and pulse lengths but with the same peak flux as the original pulse. For instance, Wavelength Frame Multiplication allows us to shape a moderator pulse to desired pulse length without the reduction of the width of the wavelength band [1]. It is realized with the help of a fast rotating chopper system that defines several wavelength bands emitted from the same source pulse. The wavelength bands arrive at the detector one after another and combined together they continuously cover the wide wavelength band sought for (Fig.1). Instead of polychromatic wavelength bands, Repetition Rate Multiplication implies using a set of several monochromatic wavelengths, which are emitted from the same source pulse. In this case the resulting instrument repetition rate increases proportionally to the number of the wavelengths used from one pulse. Furthermore, combination of several wavelength bands or monochromatic wavelengths allows us simultaneous access to larger S(Q,ω) space domain in a single experiment. Both methods can be implemented by means of a fast rotating set of mechanical choppers. In this case the instrumental parameters, such as wavelength resolution, wavelength band, and repetition rate will *not anymore be determined by the source pulse parameters, but can be flexibly defined by the chopper geometry, pulse frequency and speed.*

The design of the chopper systems for realization of multiplexing methods is different from those at reactors or currently operating at short pulse sources. Similarly to the reactor based instruments the main requirement for multiplexing chopper system is to assure that a neutron detected at any given time can only come from a single pulse shaping chopper pulse. However, multiplexing entails "quantized" chopper configurations in terms of the choice of the chopper positions L_i and chopper pulse frequencies f_i, respectively relative to the source and source frequency f. In order to make sure that all multiplexing chopper system pulses originate at the same time at the source, i.e. at proper chopper phasing all multiplexing frames look at the source within the same time period of the same duration, the time between the pulses of multiplexing chopper #i, $1/f_i$ must be proportional to the distance of the chopper from the source L_i. Furthermore, all frequencies f_i shall be multiples of the source frequency f, which lead to two RRM relations [3]

$$f_i L_i = \text{const}, \qquad f_i/f = \text{integer} \qquad (1)$$

It is obvious that the data collection and evaluation using multiplexing techniques is more complex. For proper data collection it is very important that multiplexing choppers create signals which are well separated in time, but allow us at the same time to continuously cover the S(Q,ω) space. There are many issues to be addressed, most importantly how to match the information provided by different wavelengths or different wavelengths bands. We performed extended experimental studies of the multiplexing methods using reactor based instruments: the Time-of-Flight (TOF) spectrometer NEAT at Helmholtz Zentrum Berlin and the Time-of-Flight diffractometer at Budapest Neutron Scattering Center. Here we overview our published work and also present some new results regarding the implementation of Wavelength Frame Multiplication.

2. Implementation of the Wavelength Frame Multiplication.

As mentioned above, we used the TOF diffractometer at Budapest Neutron Scattering Center (BNC) for the implementation of the Wavelength Frame Multiplication (WFM). In conventional, non-multiplexing operation the chopper system of the TOF spectrometer at BNC defines a single wavelength frame repeating itself with each pulse shaping chopper pulse [4]. In our most recent study of WFM we used the multiplexing setup illustrated in Figure 1. The first chopper was emulating the source and at 10.61 m from the "source" the second chopper was used for pulse shaping. The sample was at 13.07 m from the pulse shaping chopper, and a single detector of 12 x 30 cm^2 area at 169.5° scattering angle at 3.57 m from the sample. For better calibration to the incoming wavelength spectrum we placed a monitor at 13.79 m from the pulse shaping chopper.

Since the chopper control system available on the BNC TOF diffractometer was not designed for synchronizing choppers with very different rotation speeds (8 Hz and 200 Hz, respectively in this case), the data collection was realized by selecting events in the event recording [5] (list mode) data sets, when the source emulation chopper was in phase with the pulse shaping chopper within a prescribed error (10 % of the source pulse length). The choppers basically ran asynchronously, and at data evaluation events were retained only if the phases of the choppers were within desired limits from each other. Using the event recorded data (which consist of all time stamped timing signals of all choppers, in addition to the time stamped monitor and detector counting events) we were also able to verify the fully uniform distribution of the relative phases between source and pulse shaping chopper pulses.

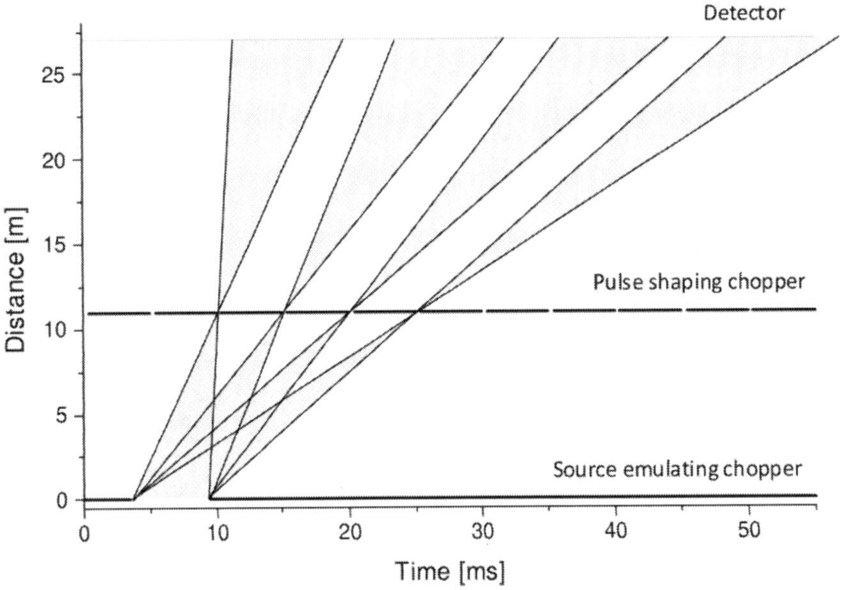

Figure 1 Principle of Wavelength Frame Multiplication diffractometry represented by the lay-out used for its emulation using the TOF Diffractometer at the Budapest Neutron Center. The chopper emulated long pulse source parameters are: 5.2 ms FWHM pulse length, 8 Hz repetition rate. The pulse shaping chopper parameters are: 0.139 ms FWHM pulse length, 199.97 Hz repetition rate, 10.61 m distance from source. The RRM frames (i.e. the distinct parts of the spectrum arriving at the detector in intervals separated by gaps around 20, 33 and 46 ms on the time axis) correspond to neutrons originating from subsequent pulse shaping chopper pulses

For the present experimental emulation of pulsed source WFM diffraction we have used an Al_2O_3 ceramic sample (which was found to also contain small amounts of other phases). As mentioned above, the multiplexing chopper system must create signals which are well separated in time, but cover continuously the wavelength range. Figure 2 shows the distribution of neutron detection times measured with respect to the specific pulse shaping chopper timing signal which was synchronized to the source pulse (i.e. found simultaneous within error with the source emulating chopper timing signal emitted at the center of its 5.2 ms long pulse). Background counts, that arrived to the detector at times when they could not originate from any of the chopper pulses, are naturally discarded in the event recording data evaluation based on the unambiguous assignment of each count to a chopper pulse. One can clearly observe the distinct multiplexed RRM frames originating from the same source pulse and subsequent pulse shaping chopper pulses, with each covering about 10 ms time frame with 2.9 ms black-out periods in-between (cf. also Figure 1). It is important to observe that the wavelength bands in subsequent multiplexing frames overlap and form a continuous broad spectrum. For example there is only a 2.9 ms black-out time gap between the wavelength frames from 7.3 ms to 17.3 ms and from 20.2 ms to 30.2 ms, respectively, while they originate from pulse shaping chopper pulses separated by 5 ms.

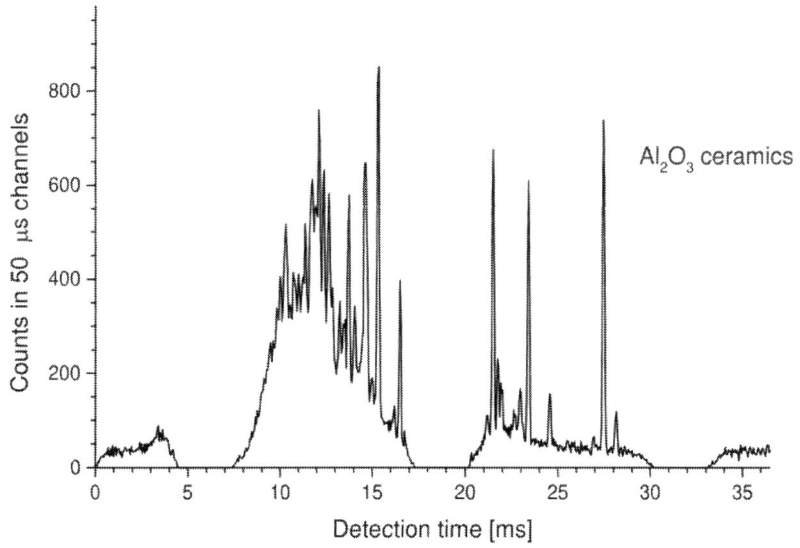

Figure 2 Time distribution of the neutron counts at the detector in synchronous source – pulse shaping chopper operation. Due to the overlap of the wavelength bands of the in time separated RRM frames, some Bragg peaks appear two times i.e. on both sides of a black-out gap.

Thus e.g. the Bragg peak observed at 16.5 ms in one RRM frame is the same as the one at 21.5 ms in the next frame, exactly by the 5 ms pulse shaping chopper period later. The event recording data evaluation consists of determining on the basis of the time stamps of all chopper and neutron detection events the one and only one pulse shaping chopper pulse a given neutron could come from and calculate the neutron distribution as a function wavelength (i.e. the time-of-flight measured for each neutron count from the corresponding pulse shaping chopper pulse). Therefore these two peaks will combine into a single one at 11.50 ms flight time from the pulse shaping chopper, i.e. 2.733 Å in the wavelength histogram (cf. Figure 4 below).

The correction of the spectra in the area of the overlap between two neighboring wavelength bands should be handled with great care. If the choppers rotate at fixed phase difference relative to each other (synchronous operation), then the intensity at the overlap area will depend on the details of this overlap and its handling in the data evaluation. This is illustrated by our results on the first implementation of WFM described in Ref. [6], as shown in Figure 3. Here the joining together of the subsequent RRM wavelength frames are marked by well visible dips around 0.7, 1.8 and 2.9 Å d spacings.

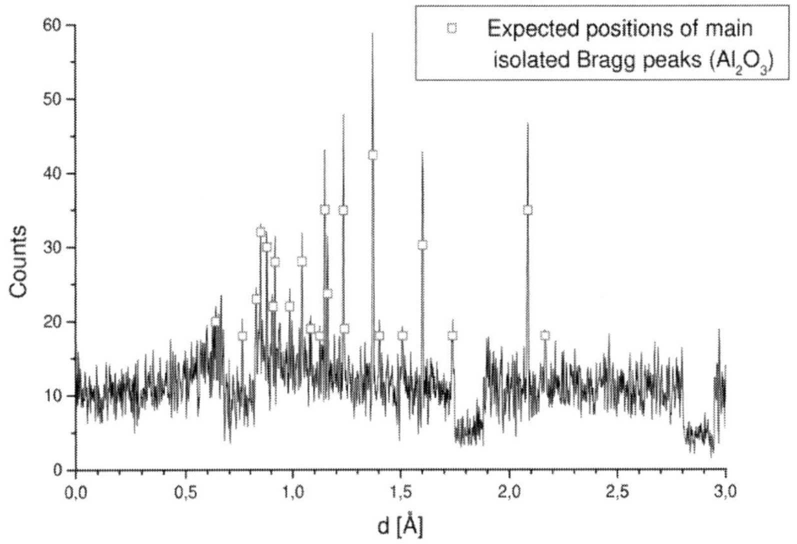

Figure 3. The first neutron diffraction pattern obtained by Wavelength Frame Multiplication [6]. The dips in the background line show the regions of switching from one wavelength frame to the next. The results also illustrate the good resolution capability accessible with the pulse shaping disc choppers (25 μs FWHM).

In our more recent study we have successfully tested two methods for perfect "stitching" between the different multiplexed wavelength frames. The first approach is based on asynchronous operation of the pulse shaping chopper with respect to the source pulse frequency. In addition, we have avoided accidental synchronization by selecting 199.97 Hz for the pulse shaping chopper in order to avoid 200 Hz, a simple multiple of the 8 Hz source frequency (cf. Figure 1). In this case the RRM wavelength frames are uniformly slewed over the whole spectrum, so after a sufficient number of cycles (10^4 source pulses or more) the transition/overlap regions between these frames (such as the dips illustrated in Figure 3) will affect every wavelength precisely the same way. The uniformity of the distribution of the chopper phases over the whole wavelength range could be easily verified in our complete event recording data set. The so observed asynchronous, phase slewing diffraction pattern observed with good statistical accuracy is shown in Figure 4.

The phase slewing mode makes multiplexing WFM exactly equivalent to any TOF diffraction method operating with a single wavelength frame, both in principle and in practicalities. The normalization to the incoming neutron spectrum remains in both cases the only calibration to perform. For this purpose it suffices to determine this spectrum under identical data collection conditions, which is in practice either achieved with the help of a calibration run (e.g. with vanadium as sample) or by the use of monitor calibrated with high precision.

The other method we have tested is based on high precision and reliable normalization to the actual combined incoming neutron wavelength spectrum over all RRM wavelength frames in synchronous, fixed phase data collection mode. This approach has the advantage that it can also be used for short scans, only consisting of a few (or even a single) source pulse. In the present work this was achieved by simultaneously using a high counting rate monitor in the incoming neutron beam. This calibration/normalization also takes care of the parameter settings used in the event recording data evaluation process, e.g. TOF safety margins used in assigning the correct pulse shaping chopper pulse to each retained neutron count. Our results obtained by this method are identical within statistical error to the ones shown in Figure 4 for the phase slewing data evaluation, as illustrated in Figure 5, which focuses to a wavelength domain affected by the overlap of neighbouring wavelength frames.

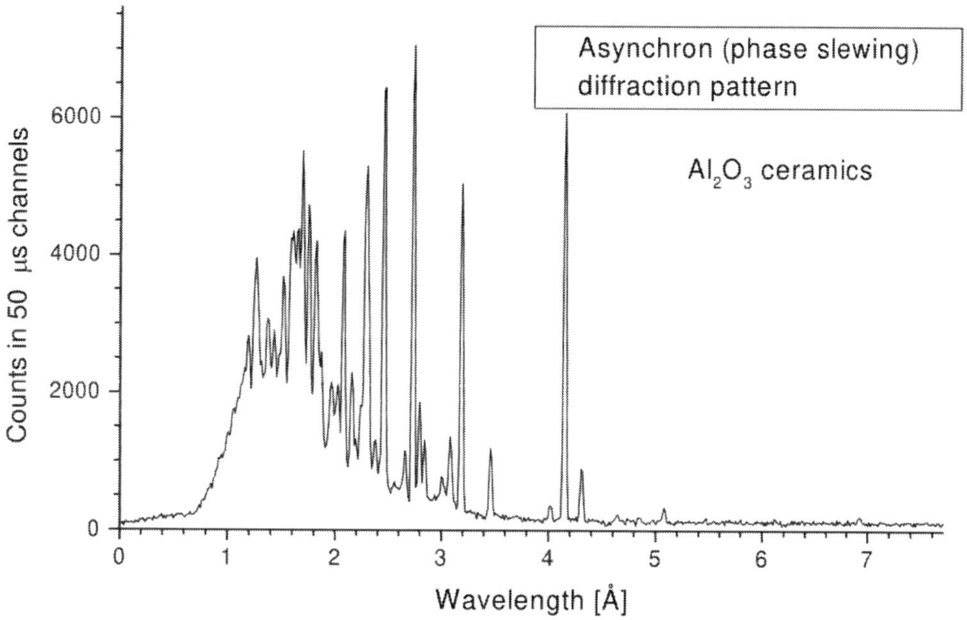

Figure 4. Diffraction pattern obtained by asynchron (phase slewing) Wavelength Frame Multiplication data collection mode. The chopper system parameters are given in the caption of Figure 1. The wavelength range shown was fully covered by 4 multiplexed RRM wavelength frames for each source pulse.

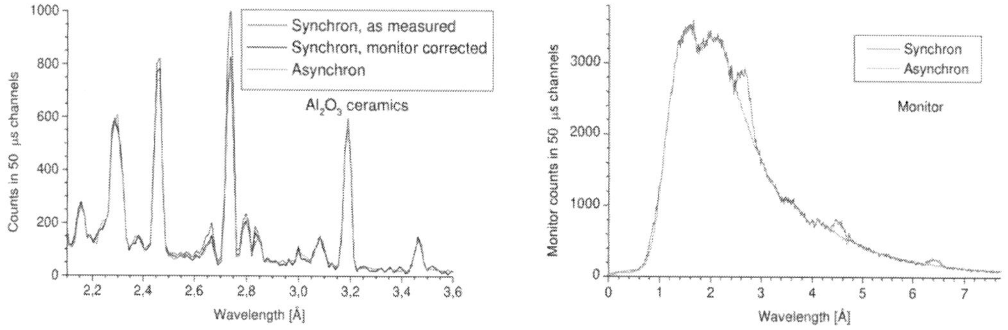

Figure 5. Comparison of the WFM diffraction patterns measured in the range of an overlap of neighbouring multiplexed RRM wavelength frames. The enhanced intensities in the as measured synchron (fixed phase) spectrum due to RRM wavelength frame overlap between 2.45 and 2.9 Å can be fully corrected for with the help of the variation of the corresponding monitor spectrum (blue line in the right hand side) with respect to the asynchron (slewed phase) monitor spectrum (red line).

The chopper system used for the proof of principle of WFM in [6] was actually more complex than the one shown in Figure 1. Instead of emulating the pulsed source we have used the WFM multiplexing chopper configuration best adapted for long pulse sources, due to its property of defining "virtual source pulse" periods where the neutrons are exclusively accepted from. These "virtual source pulses" can be tuned to overlap with the most intense flat top part of the real long source pulses, by which the WFM chopper system extracts its continuous neutron spectrum exclusively from the brightest intensity part of the source pulses. This means that neither the rising initial portion nor, most importantly, the extended decaying tail of the source pulse can dilute or contaminate the peak pulse intensity. Such a multiplexing chopper system consists – in addition to the pulse shaping chopper – a set of source frame overlap and RRM frame overlap choppers, as shown in [6]. (For the set-up shown in Figure 1. frame overlap was anyway negligible due to the low source repetition rate and the short source to detector distance.)

3. Repetition Rate Multiplication

The first experimental implementation of the repetition rate multiplication was realized using the reactor based TOF spectrometer NEAT in Berlin and is reported in Ref. [3] in details. In this study the pulse of the source has been emulated by the first NEAT chopper (Ch1-2), operating at low speed and delivering short (about 200 μs FWHM) "source" pulses at 40 and 23.67 Hz in two separate experimental runs. Sets of monochromatic neutron pulses have been cut out from the pulsed white beam by the last chopper (Ch6-7) at frequencies that were equal to an integer multiple of the frequency of the first chopper pulses $\nu_{Ch6\text{-}7} = m\ \nu_{Ch1\text{-}2}$. The three choppers, placed between the first and last chopper and used for regular operation of NEAT, have been stopped in open position to enable the arrival of the full cold neutron beam spectrum to the last chopper. The integer m was chosen to allow for about 6 ms time between the different wavelength pulses selected by the last chopper. This was necessary to keep at minimum the overlap between detected inelastic spectra from adjacent pulses at the sample. The data collection has been triggered by the pulse of the first chopper, thus spectra for all monochromatic wavelengths have been recorded one after the other in one data frame.

The left side of the Figure 6 shows the signal at the monitor taken at 23.67 Hz "source" chopper pulse frequency, with the monitor placed shortly before the sample. One can see a set of 7 clear well distinguished monochromatic wavelengths emitted from the same "source" pulse. The increase of the "source" chopper frequency to 40 Hz led to observation of neutrons with longer incoming wavelengths $\lambda_{inc} > 9.2$ Å emitted by the previous "source" pulse in addition to the neutrons emitted from the same "source" chopper pulse. The elimination of this frame overlap neutrons can be done by specially designed frame overlap choppers. Missing these choppers in the available instrument set-up we emulated the frame overlap correction by measuring each wavelength separately and subtracting them from the multiplexed spectra [6]. The resulting repetition rate frequency at the detector was in both cases close to 120 Hz, demonstrating very clearly that the pulse repetition rate in a multiplexing instrument is independent from the source frequency.

The right hand side in Figure 6 shows the spectra collected at the detector which was placed at 2.5 m from the sample. The shape of the polychromatic scattered spectra reproduces those taken with monochromatic wavelengths. The data demonstrate very well that there is enough time between pulses to record spectra for different wavelengths coming from one source pulse. Almost no disturbing overlap is observed between the data sets for different wavelength RRM frames even at high temperatures and wavelengths, except for very long ones, i.e. here 13.04 Å. However, the time required to record each individual scattered spectrum at different incoming wavelengths is different and need to be individually adjusted for optimal data collection efficiency. For instance in our study 6 ms was sufficient for the spectra at 3.09 Å, while 9.2 ms were required for the spectra at 11.14 Å. At a conventional instruments operating with a single wavelength the time between pulses needs to be a constant and this is commonly realized by a slowly rotating chopper letting only through every second, every third, etc.... pulse of the monochromating chopper.

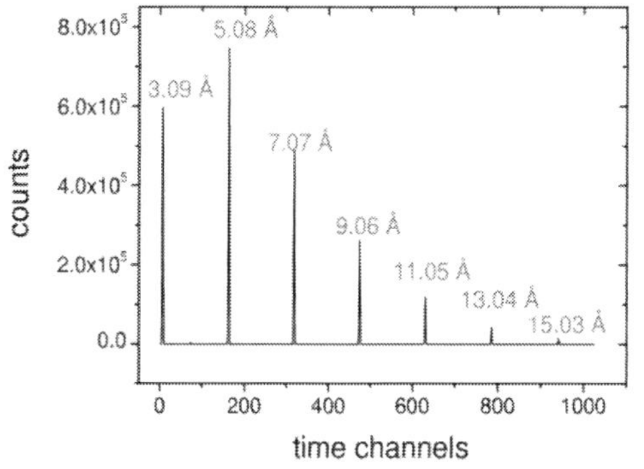

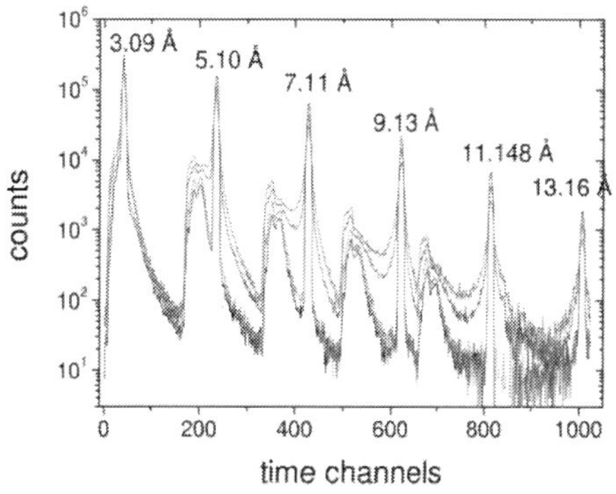

Figure 6. The monitor signal and polychromatic scattering spectra taken in Repetition Rate Multiplication mode on NEAT. The sample is water confined in porous silica (MCM-41)

However, in polychromatic mode of operation we need variable times between pulses, which can be achieved by selective pulse suppression [7]. Such suppression can be provided by the chopper shown in Figure 7. The selective suppression chopper rotates at the same frequency as the frequency of the source. The special configuration of the chopper slits allow us to transmit the first set of pulses without suppression, and then transmit only one in two, then only one in three (and so on) pulses. The application of selective pulse suppression will allow us using Repetition Rate Multiplication more efficiently for recording both thermal and cold neutrons in a single experiment.

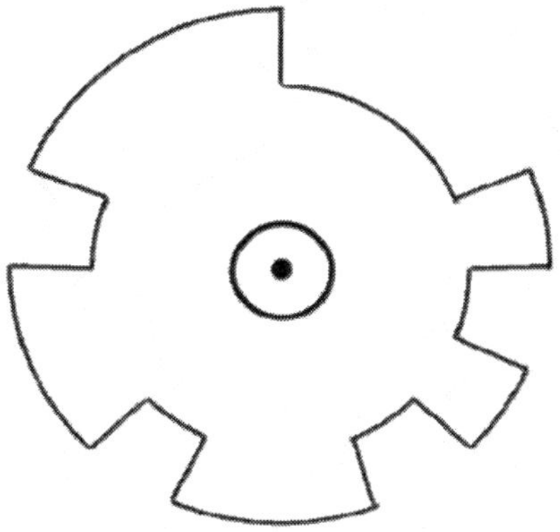

Figure 7. Example of a selective pulse suppression chopper disc for Repetition Rate Multiplication combining thermal and cold neutrons. At 14 Hz rotation frequency it would select a neutron pulse sequence of 3 pulses at 3.57 ms after the previous one, followed by 2 pulses at 7.14 ms and 2 pulses at 10.71 ms.

The spectra collected in the Repetition Rate Multiplication mode can be treated exactly in the same way as spectra taken at one monochromatic wavelength only. Figure 8 shows the energy dependent spectra for 4 individual incoming wavelengths, extracted from the time-dependent polychromatic data set in the Figure 6. No artificial shifting and deformation of the line shape is observed for the vibrational peaks at the higher frequencies (5 -10 meV). Observation of the quasielastic data obviously depends on the resolution at individual wavelengths. For the analysis of the data one can advantageously combine the information provided by different wavelengths: the short wavelengths can be used for the information at higher momentum transfer, while the long wavelengths can be used for the high resolution data. In fact, in some cases using long wavelength neutrons one can relax the resolution requirements for the data at shorter wavelengths, which in turn can provide for higher intensity. In the example of Figure 8, the vibrational peaks have nearly 3 orders of magnitude higher intensity at 30 times lower quasielastic resolution at 3.09 Å incoming wavelength than at 9.13 Å,

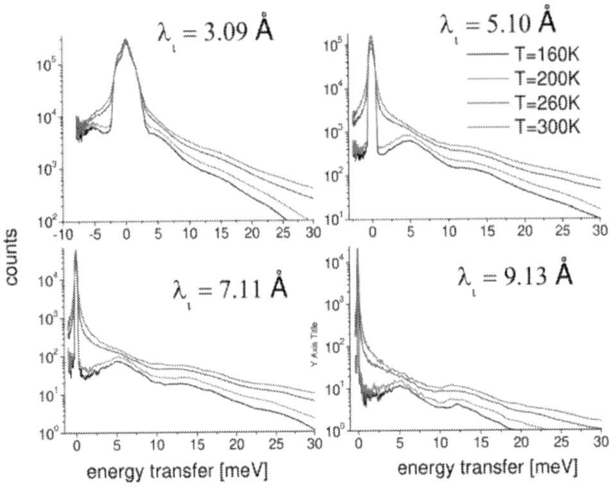

Figure 8. Energy converted spectra taken using Repetition Rate Multiplication (RRM) mode on water confined in MCM-41. It can be clearly seen; that combining the information contents from spectra at different wavelengths, all dynamical features, vibrations as well the quasielastic broadening can be defined with higher accuracy and more details.

4. Conclusions

With the help of unconventional use of existing time-of-flight instruments at continuous reactor sources we have experimentally implemented and tested for the first time various multiplexing chopper systems earlier proposed [1] as novel, high power instrumentation approaches for pulsed spallation sources. These experiments practically established the basic "quantized" design rules of multiplexing chopper systems and provide full proof of principle of these methods. Repetition Rate Multiplication (RRM) was found [3] to enhance data collection rates by up to an order of magnitude, including the quality and breadth of the information collected in a single RRM experimental run. It allows us to select the pulse repetition rate at the sample essentially independently of the frequency of a pulsed source. The Wavelength Frame Multiplication (WFM) variant of the RRM approach was found to provide unrestricted capability for flexible pulse shaping for high and variable resolution diffractometry at long pulse sources, overcoming the conventional wavelength band limitations due the source pulse parameters. In a first experimental test [6] we were able to demonstrate in particular the important "virtual source" feature of the method for selecting the period of highest source pulse brightness for beam extraction at all wavelengths. The latest implementation of WFM on an actual long pulse source emulated by a beam chopper has provided evidence for the capability to continuously cover any desired wavelength band by a set of multiplexed wavelength frames. It has shown constant TOF resolution and unperturbed spectral distributions, fully comparable to the common single wavelength frame operation of pulsed source diffractometers, even for every single source pulse. This study also included the first successful experimental realization of phase slewing, asynchronous pulse shaping chopper operation for diffraction work on a pulsed source, which makes a long pulse source to fully behave like an ideal short pulse source with flexibly adjustable and ideally shaped sharp pulses.

References

[1] Mezei F and Russina M 2002 *in: Advances in Neutron Scattering Instrumentation, I.S. Anderson, B. Guerard (Eds.), Proc. of SPIE* **4785** 24
[2] Mezei F 1997 *J. Neutron Research* **6** 3 and references therein
[3] Russina M, Mezei F 2009 *J. Nuclear Instr. and Methods*, **A604** 624
[4] Kali Gy, Santa Zs, Bleif H.-J., Mezei F, Rosta L, Szalok M, 2007 *Z.Kristallogr. Suppl.* **26** 165
[5] Russina M, Mezei F, Kozlowski T, Lewis P, Pentilla S, Fuzi S, David E, Messing G, Jülich 2003 *in:G. Mank, H. Conrad (Eds.): Proceedings of ICANS-XVI*, 667
[6] Russina M, Kali Gy, Santa Zs and Mezei F 2011 *J. Nucl. Instr. and Methods.* **A654** 383
[7] Russina M, Mezei F 2010 *Journal of Physics: Conference Series* **251** 012079

Journal of Physics: Conference Series **340** (2012) 012019 doi:10.1088/1742-6596/340/1/012019

Improving energy resolution on neutron monochromator arrays

Markos Skoulatos, Klaus Habicht and Klaus Lieutenant

Helmholtz Zentrum Berlin, Berlin, 14109, Germany

markos.skoulatos@helmholtz-berlin.de

Abstract. Horizontal focusing properties of a new neutron monochromator geometry are investigated by Monte Carlo simulations. The geometry is based on the Rowland circle, which satisfies ideal focusing conditions, as opposed to standard planar monochromator devices. For asymmetric geometries one can get optimised conditions for both intensity and energy resolution for such a "Rowland monochromator", in contrast to standard planar arrays. Energy resolution gains are also found for symmetric geometries, in particular for cold neutrons. Additional advantages include a symmetric beam well centered in energy. No changes are observed in integrated intensities.

Since many decades, neutron monochromators have been used for both diffraction and inelastic experiments in neutron reactors as well as spallation sources, in order to focus in real and *k*-space [1]. In particular, focusing properties have been employed exhaustively in different experimental setups by means of focusing with a planar array of crystals onto a spot (sample for a monochromator or detector for an analyser).

However, there is a further potential in using neutron monochromators which has not been exploited so far, due to its complexity in design or lack of Monte Carlo (MC) simulation results. Let us describe this arrangement, which is based on the so-called Rowland circle geometry [2] and a Johansson type curved crystal [3], most common for x-ray setups. In order to satisfy the ideal focusing condition, the source (A), monochromator (B) and sample (C) must lie on the Rowland circle, as shown in fig. 1. We stress that in this ideal case, the monochromator surface is not planar, but instead lies on the Rowland circle, with center R. Furthermore, the orientation of the lattice planes of the reflecting crystallites are determined by a different circle which we call the focusing circle, with center F. For the general asymmetric Rowland case, in which $AB \neq BC$, points B, R and F are not collinear. This introduces the angle $\delta\theta$, which equals 0 for the symmetric Rowland geometry ($AB=BC$). This general layout forms the basis of our motivation to study neutron monochromator properties, since the focusing circle is being used extensively, unlike the Rowland circle which has never been reported on neutron monochromators.

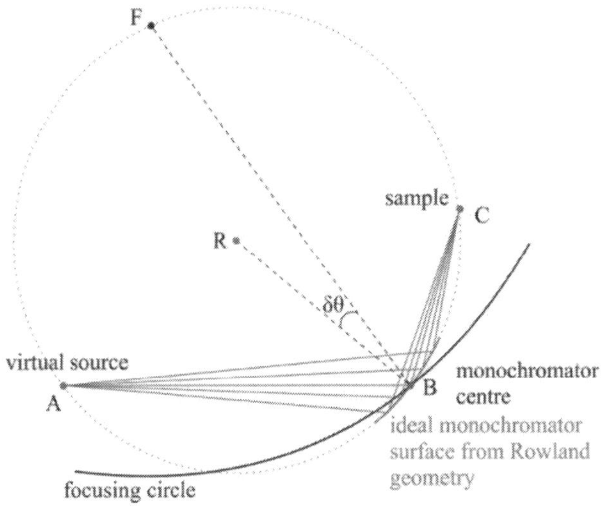

Figure 1 Rowland circle geometry indicating the Rowland and focusing circles. Distances here are AB=2m and BC=1m.

In order to investigate this model, we construct a generic instrument within the McStas [4,5] MC simulation package, which consists of a neutron source, followed by a slit (which plays the role of a virtual source, read more on [6]), a monochromator and a monitor, as shown in fig. 2a). The distances slit-monochromator and monochromator-monitor are equal to L_1 and L_2 respectively and can be varied. For a practical application of such a neutron monochromator it is necessary to be able to change the selected wavelength (and hence the radii of curvature and Rowland circle simultaneously). We approximate the ideal case described in fig. 1, by that shown in Model B of fig. 2b), where individual crystal pieces are arranged with their centres on the Rowland circle. We compare this model to a standard planar monochromator (Model A) as well as a hybrid case (Model C). This last model is a mixture of models A and B as it is a planar device, however it is still a good approximation of the Rowland circle for the loci of the monochromator pieces. The way this is achieved is by including the extra $\delta\theta$ in the take-off angle of the whole array, while turning back every individual crystal by $-\delta\theta$ in order to satisfy the elastic condition. It is essentially a linear approximation of a small arc of the Rowland circle, and only requires individually turning each piece, however not shifting them on the exact Rowland circle as is required by model B. A similar setup to this engineering design (Model C) is in use at the MACS [7,8] monochromator and as a multiplexing analyser-detector system in various triple-axis spectrometers, including the RITA-II [9,10], SPINS [11], IMPS [12] option at IN8, SIKA [13], MACS [14], as well as PUMA [15] and UFO [16,17] option of IN12, with the last two having more degrees of freedom. Note however that, with the exception of the MACS monochromator, their usage is completely different, focusing primarily on simultaneous data collection in (Q,ω) space (multiplexing). We emphasize on the use of such a design for the monochromator (elastic case), which in addition spatially focuses onto the sample (and does not disperse on a position sensitive detector). Furthermore, the shifts we are interested in are those dictated purely by the Rowland geometry, unlike shifts of the PUMA and UFO option, which are on a kinematic basis for following particular paths in (Q,ω).

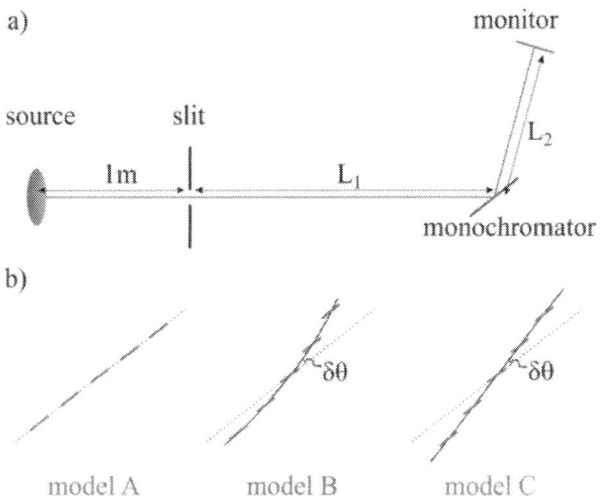

Figure 2 Monte Carlo simulation model (L_1=2m, L_2=1m and λ=4.05 Å). Part a) shows the whole instrument layout, and b) the 3 different monochromator modules simulated: A: standard planar, B: Rowland, C: hybrid geometry. In the monochromator models B and C, the dashed line is taken from model A for comparison. For simplicity we only show 5 monochromator crystals which correspond to an unfocused device in this figure.

For the MC simulations we assume a 31cm array in the horizontal plane, consisting of 1cm pyrolytic graphite (PG) crystals in the (002) orientation, with mosaic η=30'. Simulations were performed for both cold and thermal neutrons (λ=4.05 and 1.8 Å), while the sample has a width of 1 cm in the horizontal plane. Three distinct cases are modelled: the symmetric Rowland geometry (L_1=L_2) and two asymmetric cases (L_1>L_2 and L_1<L_2). Note that optimal focusing for model A is determined analytically by the standard formula

$$R^{standard}_{Model\ A} = \frac{2L_1L_2}{L_1+L_2}\frac{1}{sin\theta} \qquad \text{eq. (1),}$$

whilst for models B/C the following can be shown [18]:

$$R^{hor.\ shifts}_{Models\ B/C} = \frac{L_1+L_2}{2sin\theta} \qquad \text{eq. (2).}$$

Apart from the analytical derivations, both of these equations have been also verified during these M.C. simulations. Optimum focusing is used throughout.

The results, summarized in fig. 3 are striking. In all cases, model B gives a better energy resolution (FWHM throughout) and overall beam properties (symmetry and nominal center). In particular, for the asymmetric cases the energy resolution gets drastically better (a factor of 3-8) as a result of following the right crystal geometry. We can understand this qualitatively by inspecting model B of fig. 2b. This is a drawing (in scale) of how far away the crystal elements are in standard monochromators (dashed line) compared to the ideal case (solid red lines). Note that fig. 2 is for an asymmetric case (for the symmetric one the shifts are much smaller, hence smaller gains there).

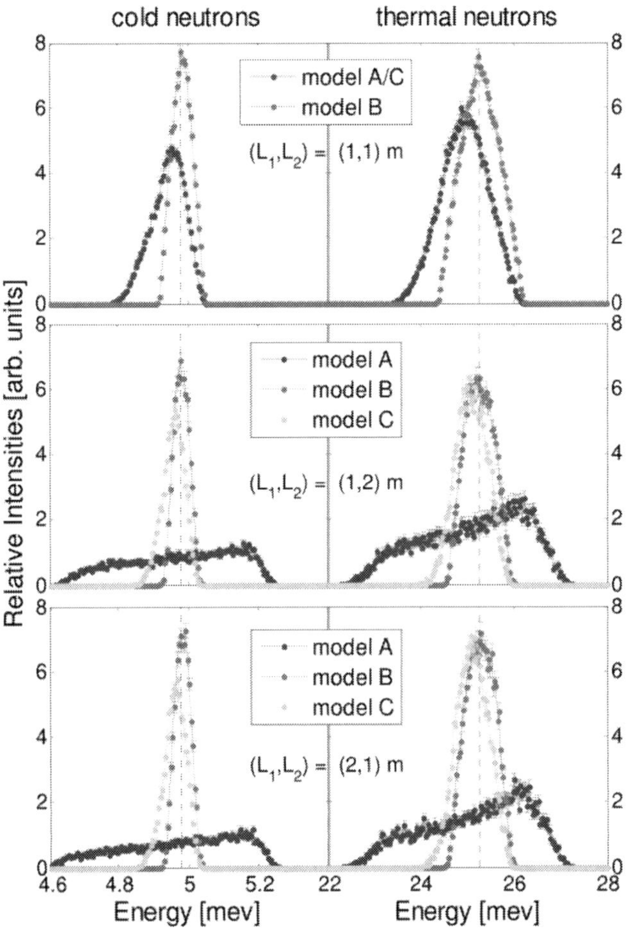

Figure 3 McStas MC simulation results for cold (left) and thermal (right) neutrons. The top part shows a symmetric geometry while the middle and bottom plots are for asymmetric cases. Large energy resolution gains are obtained by using models B/C, while in particular model B ("Rowland monochromator") benefits on top from symmetric energy lineshapes which are well-centered w.r.t. the nominal energy.

For cold neutrons and for the symmetric Rowland geometry we get a 43% increase in energy resolution (and a 5% increase in integrated intensity) when comparing model B to A (equivalent to C). Furthermore, the energy lineshape forms a very symmetric peak and in addition is well centered to the nominal energy for the given monochromator take-off angle, indicated by the dashed line in fig. 3 (model A deviates by 25μeV, a significant fraction of the energy resolution that can be important when studying low-gapped excited states). For L_1=1m, L_2=2m the energy resolution gets drastically better with ratios 1:7.6:6.3 for models A:B:C respectively. Similarly for L_1=2m, L_2=1m the energy resolution ratios are 1:7.1:6.0 for models A:B:C. In all cases, the integrated intensities are similar with small deviations around 3-5%. This is a real effect and comes from the size of the monochromator as viewed by the source, since the whole device is tilted when changing configurations and is a side effect.

For thermal neutrons and the symmetric Rowland scenario with $L_1=L_2=1$m we have a small gain of 13% in the energy resolution. This happens because the deviations away from the ideal Rowland geometry are smaller at these wavelengths. However, the peak is more symmetric and centered exactly at the nominal wavelength for the given take-off angle as shown in fig. 3. For the asymmetric cases we find that models B and C give similar results in width, however model B is indeed better centered. The energy resolution gains are large for thermal wavelengths too: for $L_1=1$m, $L_2=2$m ($L_1=2$m, $L_2=1$m) we get ratios 1:3.5 (1:3.1) when comparing models A:B.

In general, when designing an instrument the distance from the real neutron source to the monochromator is not equal to the distance from monochromator to sample. Under these conditions, one cannot achieve real and k-space focusing simultaneously. The standard solution is to introduce a virtual source by means of a slit so that $L_1=L_2$. The other way around it is to use a "Rowland" monochromator, model B (or an approximation to it as used on MACS, model C).

Summarizing, we have comprehensively shown that this Rowland monochromator gives in all cases good intensity and energy resolution simultaneously. Furthermore, the beam one gets is symmetric in energy and centred at the correct energy corresponding to the nominal monochromator take-off angle. These gains in energy resolution as well as energy lineshapes are also observed for symmetric Rowland cases and are larger for the colder wavelengths. The multiplexing secondary spectrometers existing to date are a clear indication that the technology for such a device exists. What is needed on top of the horizontal shifts and focusing is to combine independent vertical focusing in order to create such beneficial instruments.

Acknowledgements MS would like to thank E. Farhi, P. Willendrup, P. Freeman and N. Tsapatsaris for useful information and discussions.

References

[1] Proc. Workshop on Focusing Bragg Optics, Braunschweig Germany 1993 *Nucl. Instr. and Meth. A* **338**

[2] Born M and Wolf E 2002 Principles of optics, Cambridge University Press, 7th edition

[3] Johansson T 1933 *Zeitschrift für Physik A Hadrons and Nuclei* **82** 7 507

[4] Lefmann K and Nielsen K 1999 *Neutron News* **10** 20

[5] Willendrup P, Farhi E and Lefmann K 2004 *Physica B* **350** 735

[6] Habicht K and Skoulatos M Optimization of virtual source parameters in neutron scattering instrumentation, *these proceedings*

[7] Broholm C 1996 *Nucl. Instr. and Meth. A* **369** 169

[8] Smee S A, Orndorff J D, Scharfstein G A, Qiu Y, Brand P C, Broholm C L and Anand D K 2002 *Appl. Phys. A* **74** S255

[9] Lefmann K, McMorrow D F, Rønnow H M, Nielsen K, Clausen K N, Lake B and Aeppli G 2000 *Physica B* **283** 343

[10] Bahl C R H, Andersen P, Klausen S N and Lefmann K 2004 *Nucl. Instr. and Meth. B* **226** 667

[11] Zaliznyak I A and Lee S-H 2004 "Magnetic Neutron Scattering and Recent Developments in Triple Axis Spectroscopy" in Modern Techniques for Characterizing Magnetic Materials ed. by Z. Yimei, (Kluwer)

[12] Jiménez-Ruiz M, Hiess A, Currat R, Kulda J and Bermejo F J 2006 *Physica B* **385-386** 1086

[13] Vorderwisch P and Wu C-M, SIKA cold TAS, http://www.ansto.gov.au/__data/assets/pdf_file/0016/37042/Sika_Fact_Sheet.pdf

[14] Rodriguez J A, Adler D M, Brand P C, Broholm C, Cook J C, Brocker C, Hammond R, Huang Z, Hundertmark P, Lynn J W, Maliszewskyj N C, Moyer J, Orndorff J, Pierce D, Pike T D, Scharfstein G, Smee S A and Vilaseca R 2008 *Meas. Sci. Technol.* **19**

[15] Eckold G December 2009 The multianalyzer/Multidetector Design of PUMA, *FRM II news* No **3**
[16] Schmidt W, Rheinstädter M C, Raymond S and Ohl M 2004 *Physica B* **350** e849
[17] Schmidt W and Ohl M 2006 *Physica B* **385** 1073
[18] Habicht K and Skoulatos M Phase-space tailoring using the virtual source concept, *review paper to be published*

Journal of Physics: Conference Series **340** (2012) 012020 doi:10.1088/1742-6596/340/1/012020

A Note on Observation and Theoretical Description of the Neutron Whispering Gallery Effect

V.V. Nesvizhevsky[1], **A.Yu. Voronin**[2], **R. Cubitt**[3], **K.V. Protasov**[4],
Production Editor, *Journal of Physics: Conference Series*,
IOP Publishing, Dirac House, Temple Back, Bristol BS1 6BE, UK

E-mail: nesvizhevsky@ill.eu

Abstract. The "whispering gallery" effect has been known since ancient times for sound waves in air, later in water and more recently for a broad range of electromagnetic waves from the radiofrequency region, through visible light to X-rays. It consists of wave localization in the vicinity of concave surfaces. For matter waves, it would include a new feature: a massive particle would be settled in quantum states, with parameters depending on its mass. Here, we present the observation and theoretical description of such an effect.

1. Introduction

Owing to whispering gallery effect, the sound can reach a person on the opposite side of a building, or even complete a circle, imitating an "echo". Lord Rayleigh explained and described quantitatively this phenomenon in his "Theory of sound" (1), thus giving birth to the science of acoustics. Whales are believed to communicate over long distances, profiting from a similar effect in surface layers of sea water. The electromagnetic whispering gallery waves from radio to light ("glory" or "heiligenschein") and X-ray frequencies are of ever-growing interest (2), (3), owing to their multiple applications. Prior to the work described here (4), an analogous phenomenon has never been measured with matter waves. At certain condition, the problem can be reduced to a quantum particle above a mirror in a linear potential – the so-called quantum bouncer, in analogy with the neutron quantum motion in the Earth's gravitational field above a flat mirror (5). On the other hand, an optical analogue of a quantum particle bouncing on a hard surface under the influence of gravity is experimentally demonstrated using a circularly curved optical waveguide (6). Quantum revivals of a quantum bouncer (7) could be applied to experiments with the GRANIT spectrometer (8), (9), and with the neutron whispering gallery quantum states (10), (11), (4), (12), (20). Two kinds of initial localization of neutrons could be considered: in position (easier applied to gravitational quantum states (5), (13), (14), (15), (16)) and in energy (easier applied to centrifugal quantum states). Revivals is an extremely sensitive tool to study

[1] nesvizhevsky@ill.eu, Institut Laue-Langevin, 6 rue Jules Horowitz, Grenoble, France, F-38042.
[2] dr.a.voronin@gmail.com, Lebedev Physical Institute, 53 Leninsky pr., Moscow, R-119991.
[3] cubitt@ill.eu, Institut Laue-Langevin, 6 rue Jules Horowitz, Grenoble, France, F-38042.
[4] protasov@lpsc.in2p3.fr, Laboratoire de Physique Subatomique et de Cosmologie, IN2P3/UJF/INPG, 53 rue des Martyrs, Grenoble, France, F-38026.

surface potentials, however, their extreme sensitivity to various experimental parameters might lead to various false effects, to be studied experimentally.

2. Centrifugal quantum states of neutrons.

Consider scattering of a cold neutron by a perfect cylindrical mirror with a radius of a few centimetres as shown in figure 1.

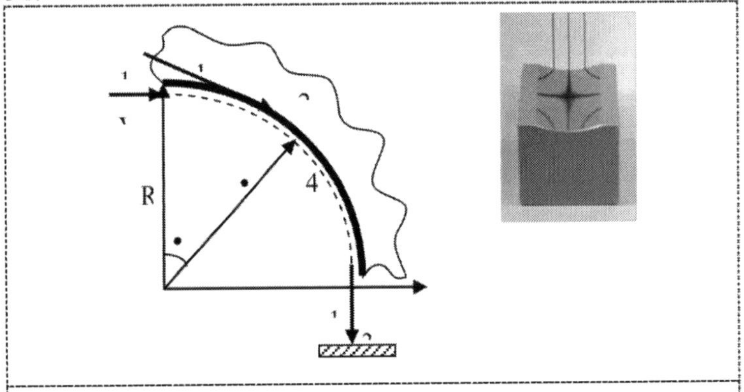

Figure 1. A scheme of the neutron centrifugal experiment. 1: classical trajectories of incoming and outcoming neutrons, 2: cylindrical mirror, 3: neutron detector, 4: quantum motion along the mirror surface. Insert: A photo of the single-crystal silicon mirror used for the presented experiments, with an optical reflection of black stripes for illustrative purposes.

The mirror is described by a uniform neutron-nuclear optical potential, which reflects (classically) neutrons with the radial velocity components lower than the critical velocity corresponding to this potential. On the other hand, the neutron is affected by huge centrifugal acceleration of the order of a million times the Earth's gravitational acceleration. The tangential neutron motion is essentially classical, while for the radial motion quantum effects are dominant. The quantum states are settled in a bounding well of nearly triangular shape formed by the centrifugal potential and the mirror potential. These are quasi-stationary states, as the probability of tunneling through the trapping potential barrier is never zero, although it could be negligible for deeply bound states.

The method to study the quantum states is based on a continuous variation of the bounding triangle barrier width. The width, the energy and the number of states depend strongly on the neutron velocity. In particular, the transmitted neutron flux increases sharply from zero when the neutron velocity (wavelength) approaches a characteristic cut-off value corresponding to the appearance of the lowest quasi-stationary state.

An alternative method for studying such centrifugal states consists of measuring the radial velocity distribution using a position-sensitive neutron detector, placed at a distance from the mirror. In particular, if a single long-living state is populated, we could measure directly the distribution of radial velocity components in this quantum state. Evidently, the most powerful method consists of combining the two options that is a simultaneous measurement of the scattering angle and the neutron velocity.

A typical scattering pattern is shown in figure 2.

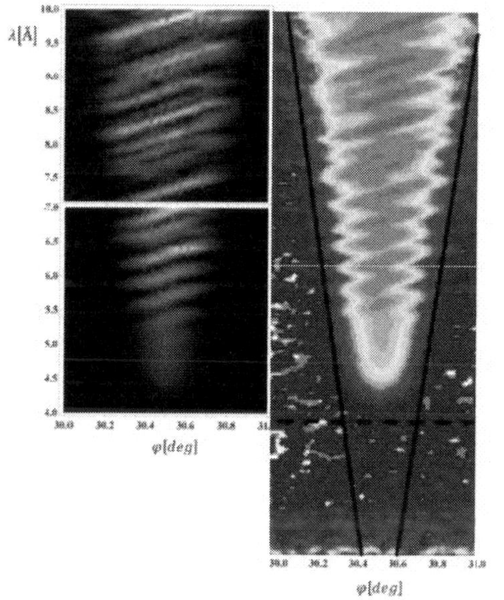

Figure 2. a) The scattering probability as a function of neutron wavelength (vertical axis) and deviation angle (horizontal axis). The geometrical angular size of the mirror is 30.5°. The inclined solid lines show the signal shape for the classical Garland trajectories. The dashed line illustrates a characteristic wavelength cutoff. b) Theoretical simulation of the data.

Neutrons enter from the entrance edge of a truncated cylindrical mirror (figure 1). The "fingerprints" of the quantum states in figure 2 have a "V" shape. This can be understood as follows. The average deviation angle is equal to the angular mirror size; the "fingerprint" is centered on this value; thus, the radial velocity distribution is symmetric relative to the zero value. For classical consecutive Garland trajectories of neutrons, the width of the "V" letter would be exactly proportional to the neutron wavelength (figure 2). A manifestation of the observed centrifugal quasi-stationary states consists of the sharp cutoff in the neutron flux wavelength, corresponding to the appearance of the lowest quantum state. Another manifestation is the stripe structure inside the "V" shape. It is explained by interference of a few transmitted quantum states. A theoretical simulation of the data, shown in figure 2, reproduces the measurement in detail. Some of the difference between the experimental data and the simulation is probably due to the thin oxide layer on the mirror surface.

At certain condition, the problem can be reduced to a quantum particle above a mirror in a linear potential – the so-called quantum bouncer, in analogy with the neutron quantum motion in the Earth's gravitational field above a flat mirror (5). These two phenomena provide a first direct demonstration of the weak equivalence principle for an object in a quantum state: although the independence of a free fall on mass does not hold in the quantum limit, quantum states of a massive body in a locally uniform gravitational field and those in a system moving with equal acceleration are equivalent. Both problems, the centrifugal and gravitational ones, provide an excellent laboratory for studying neutron quantum optics phenomena. Deeply bound whispering-gallery states are long-living and weakly sensitive to surface potentials thus could be used for searches for extra short-range forces (17), (18), (19); highly excited states are short-living and very sensitive to the wall potential shape. Therefore, they are a promising tool for studying fundamental neutron-matter interactions, quantum neutron optics and surface physics.

3. Theoretical description of centrifugal quantum states.

The problem of neutron scattering on a cylindrical mirror can be solved in a closed form, free from fitting phenomenological parameters (10), (12). This clear mathematical background makes the problem attractive for studying physical effects, responsible for deviations from the benchmark theory of neutron scattering on an ideal cylinder, which is characterized by neutron-optical potential. Such additional effects include neutron scattering by hypothetical extra-forces, suggested in extensions of the Standard Model. So far the constraints on the strength and the characteristic range of such extra-forces could be derived by comparing experimental data to theoretical predictions. The main result of our theoretical study of the whispering gallery scattering of neutrons on a cylindrical mirror consists in establishing a relation between the centrifugal quasi-stationary states of neutrons, localized due to superposition of centrifugal and optical potentials, and the neutron scattering amplitude. This relation is given by the following equation:

$$f(\varphi) = -2i\left(\frac{\mu_0}{2}\right)^{1/3} \exp(i\mu_0\varphi)\sqrt{\frac{2\pi\hbar}{p}}\sum_{n=1}^{\infty}\text{Res}\, S(\lambda_n)\exp\left(-i\lambda_n \frac{\mu_0}{2}^{2/3}\varphi\right) \quad (1)$$

Here $\mu_0 = \frac{MvR}{\hbar}$, M is the neutron mass, v is the neutron velocity, R is the radius of a cylindrical mirror, $p = Mv$, S is the scattering S-matrix and λ_n is the pole of the S-matrix in the complex momentum plane (10), (12).

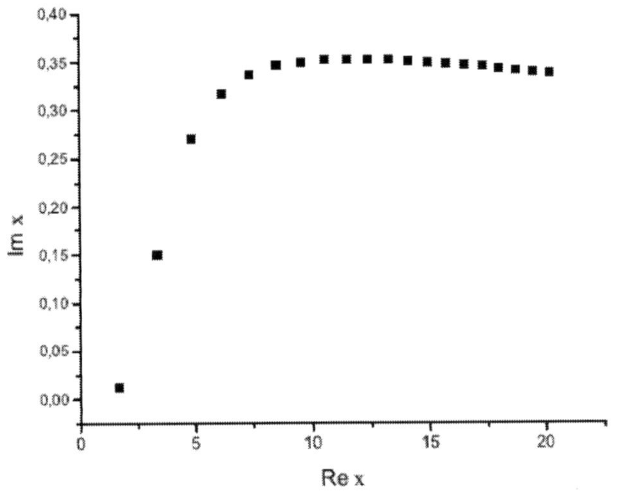

Figure 2. Values of the S-matrix poles in the first quadrant of the complex momentum plane.

The physical sense of the above expression is transparent. For large scattering angles (which we are interested in) the main contribution to the scattering amplitude is given by the S-matrix poles with the smallest imaginary parts. These poles correspond to the mentioned above narrow quasi-stationary centrifugal states. The energy and width of such states can be exactly calculated. Indeed, such states are the well-known solution of the Schrodinger equation in the linear potential with the emission boundary condition:

$$\chi \sim \begin{cases} Bi(u_0 - x - \lambda_n) + iAi(u_0 - x - \lambda_n),\ z \geq 0 \\ Ai(-x - \lambda_n),\ z < 0 \end{cases} \quad (2)$$

where $u_0 = U\left(\frac{2R^2}{\hbar^2 M v^4}\right)^{1/3}$, $x = z\left(\frac{2M^2 v^2}{\hbar^2 R}\right)^{1/3}$, U is the depth of optical potential.

4. Conclusion

We described a new phenomenon of the neutron whispering gallery and a theoretical formalism to describe it. It is a promising tool for studying fundamental neutron-matter interactions, quantum neutron optics and surface physics effects. In particular, we estimate that it provides the best neutron method for constraining extra short-range forces in a broad distance range, with large potential for further improvement in sensitivity.

Acknowledgments

This work was supported by BLANC ANR-05-BLAN-0098-01 (France).

References

[1] Strutt, J.W., The Theory of Sound, v. 2, s.1: Macmillan, 1878
[2] Mie, G., 1908 *Ann. Phys.* **25**, 377
[3] Vahala, K.J., 2003 *Nature* **424**, 839
[4] Nesvizhevsky, V.V. et al, 2010 *Nature Phys.* **6**, 114
[5] Nesvizhevsky, V.V. et al, 2002 *Nature* **415**, 297
[6] Della Valle, G. et al, 2009 *Phys. Rev. Lett.* **102**, 180402
[7] Robinett, R.W., 2004 *Phys. Rep.* **392**, 1
[8] Baessler, S. et al, 2011 *Compt. Rend. Acad. Sciences* **12**, 707
[9] Kreuz, M. et al, 2009 *NIM A* **611**, 326
[10] Nesvizhevsky, V.V. et al, 2008 *Phys. Rev. A* **78**, 033616
[11] Cubitt, R. et al, 2009 *NIM A* **611**, 322
[12] Nesvizhevsky, V.V. et al, 2010 *New J. Phys.* **12**, 113050
[13] Nesvizhevsky, V.V. et al, 2003 *Phys. Rev. D* **67**, 102002
[14] Nesvizhevsky, V.V. et al, 2005 *Europ. Phys. J. C* **40**, 479
[15] Baessler, S., 2009 *Phys. G Nucl. Part. Phys.* **36**, 104005
[16] Nesvizhevsky, V.V., 2010 *Physics-Uspekhi* **53**, 645
[17] Antoniadis, I. et al, 2011 *Compt. Rend. Acad. Sciences* **12**, 755
[18] Baessler, S. et al, 2007 *Phys. Rev. D* **75**, 075006
[19] Nesvizhevsky, V.V. et al, 2008 *Phys. Rev. D* **77**, 034020
[20] Nesvizhevsky, V.V. et al, 2011 *Compt. Rend. Acad. Sciences* **12**, 791

Neutron optics in cryogenic sample environment

Marek Bartkowiak, Uwe Filges and Tobias Panzner
LDM, Paul Scherrer Institut, 5232 Villigen PSI, Switzerland
E-mail: marek.bartkowiak@psi.ch

Abstract. Neutron focusing devices can be used to enhance neutron instruments when measuring small samples. However, for an optimum performance they need to be as close as possible to the sample which, in most situations, conflicts with the need for sample environment such as pressure cells, cryostats or magnets. In this paper we explore the potential and feasibility to incorporate a neutron lens based on supermirror technology into a cryostat containing a Paris-Edinburgh pressure cell. We present experimental results on the performance of super-mirrors between 5 and 300 K. Based on the experimental results we estimate the expected gain factors for a setup with a parabolic lens and a pressure cell in a cryogenic environment using Monte Carlo simulations.

1. Introduction

The use of supermirror (SM) technology for neutron guides and beam focusing has ensured an increasing performance of neutron instruments (see for example [1], [2] and [3]). In particular the ability to produce supermirrors with large m-values on curved surfaces provide huge potential for improved instrument performance and new instrument concepts [4], [5].

Since this technology relies on the total reflexion of neutrons at a surface it is crucial to extend the neutron lens as close as possible to the sample to obtain maximum intensity gain. In many situations this conflicts with the requirements on the sample environment such as the need of low or high temperatures, high pressures or magnetic fields by utilising cryostats, furnaces, pressure cells or magnets. Moreover, the ever accelerating development of new materials and the growing need to study their properties requires the ability to measure even small quantities of these new materials with sample sizes below 1 mm^3.

A logical step to resolve this dilemma is to incorporate neutron lenses into the sample environment. In this paper we would like to explore the feasibility and the potential to use SM-lenses in a cryogenic environment.

2. Experiments

To check the feasibility of our idea we have conducted several cryogenic tests. The SM which were used for these tests were prepared at PSI. Float glass substrates with an area of 50x50 mm^2 were coated with Ni-Ti SM-layers ($m = 3.6$). In first tests, these samples were repeatedly thermally cycled between 300 K and 77 K. After visual inspection, we did not observe any degradation of the film itself nor did we see any indications for lift-off of the SM-layers from the substrate.

Encouraged by our first tests we went on to study the neutron reflectivity properties of our SM-film as a function of temperature. We performed θ-2θ scans on the Morpheus reflectometer at the SINQ (PSI) to measure the q dependence of the reflectivity at 5.012 Å. The sample was

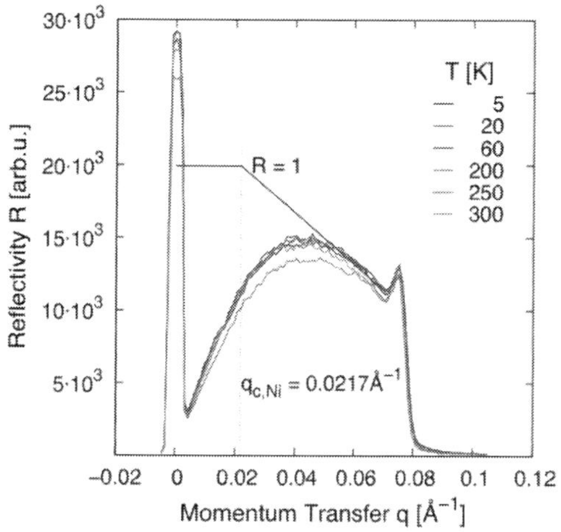

Figure 1. Temperature dependence of the reflectivity of a supermirror with $m = 3.6$. The wavelength for the experiment was chosen to be 5.012 Å. The deviation of the curve at 250 K is a result of the miss-alignment of the sample due to the thermal contraction of the sample stick. The peak at $q = 0$ is the observed direct beam. It was always measured to ensure that the geometrical configuration had not changed during the temperature scan.

installed in an Orange Cryostat, a cryostat with static exchange gas, to vary the temperature between 5 and 300 K and to ensure good thermalisation of the glass. The resulting data for a full temperature scan upon warming are shown in Fig. 1 omitting some intermediate temperature steps for clarity.

The measured reflectivity curves at low temperatures are unchanged as compared to the curve measured at room temperature. For the curve at 250 K small corrections to the sample alignment had to be made but the expected shape could not be fully recovered. However, the observed discrepancy is solely caused by a slight miss-alignment of the sample due to thermal contraction of the cryostat sample stick and not an effect of the SM. This can be concluded since we were able to almost recover the same shape and level of reflectivity by a realignment of the instrument.

3. Monte Carlo Simulations

To estimate the potential of a cryogenic lens we have run Monte Carlo simulations using McStas [7], [8]. As mentioned before, one promising application would be the combination of a neutron lens with a pressure cell where samples are generally small and surrounded by the pressure medium and the cell material. Previous work [6] reporting on the combination of a pressure cell and a neutron lens at ambient temperature achieved gain factors of 3 using a horizontal and vertical focussing unit. Concerning pressure cells available at SINQ, the largest potential for improvement is held by the Paris-Edinbugh pressure cell (PE-cell) in conjunction with its closed cycle refrigerator (CCR). Details about this device can be found in [9] and [10]. Figure 2 shows a sketch of the important components and dimensions.

Here, we considered a parabolic lens. The lens has a length of 75 mm which is limited by the diameter of the innermost radiation shield of the CCR. In this model we only consider vertical focusing. We set the focal length to 14 mm and the vertical height of the lens to 1.5 mm at the exit which results in a 5 mm opening at the beam entrance. The supermirrors were chosen to have $m = 7$ with a reflectivity of 64 %. That is the highest m-value which is currently available. We allowed for a 1 mm clearance between the lens and the anvil which would be largely occupied by the mirror substrate in a real setup.

The simulations were performed assuming a neutron source with predefined divergence. Figure 3 shows the dependence of the neutron intensity gain on the vertical beam divergence at the focal point with a chosen sample height of 1 mm. The intensity gain is defined as the count ratio at the sample position with and without lens. The wavelength of the beam was set

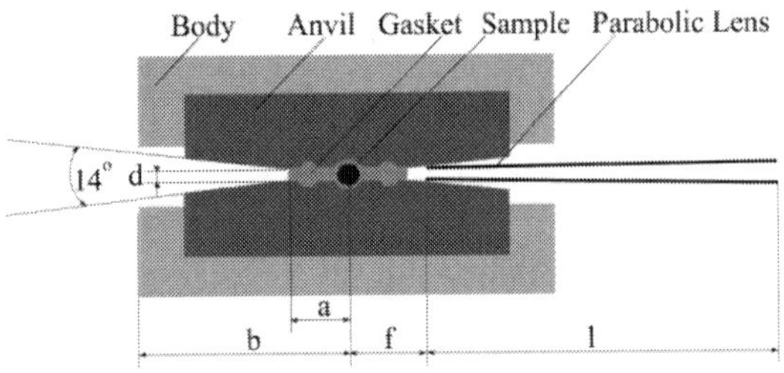

Figure 2. Schematics of the inner components of a PE pressure cell. The sample is enclosed in a pressure gasket together with the pressure medium. The gasket has a height (d) of 2 mm and a radius (a) of 10 mm. The anvils allow an opening angle of 14° with a total radius (b) for the anvil together with its holder of 37 mm. The focal length of the lens (f) was set to 14 mm and its length (l) of 75 mm is determined by the diameter of the inner radiation shield which has a diameter of 180 mm.

to 5.0(±0.2) Å. As one would expect the gain increases with decreasing divergence since a single sharp focal point can only be observed for a parallel incoming beam.

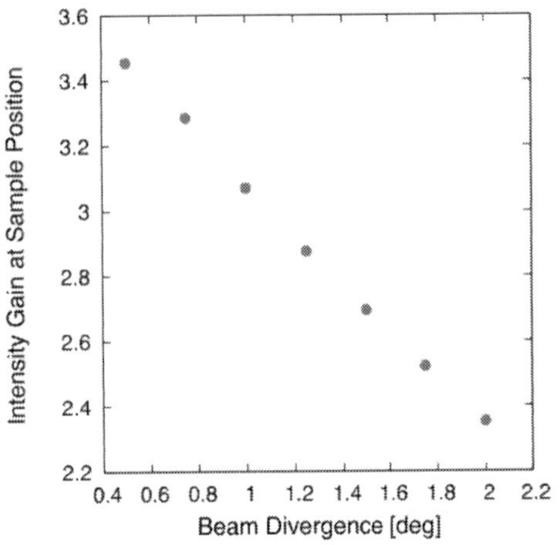

Figure 3. Intensity gain as a function of vertical beam divergence. The calculations were done for an assumed sample height of 1 mm and a neutron wavelength of 5 Å.

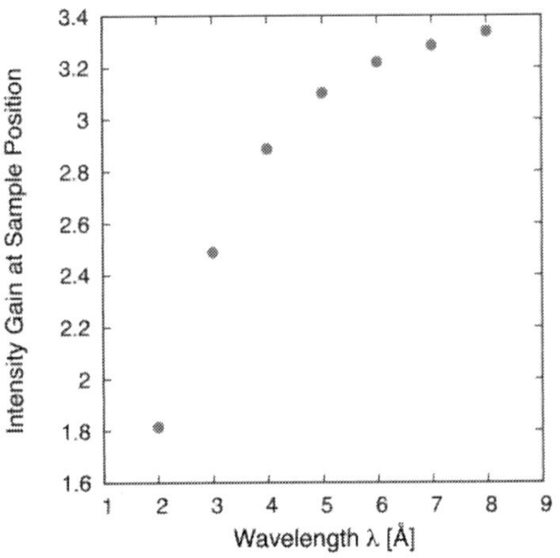

Figure 4. Intensity gain as a function of wave length. The vertical divergence of the beam was set to 1°. Samples height was 1 mm.

For a selected divergence of 1 % the wavelength dependence for the investigated setup is shown in Fig. 4. Depending on the incoming wavelength gain factors between 1.8 and 3.4 can be achieved.

In Fig. 5 we examine the influence of the SM-coating reflectivity on the intensity gain. The wavelength was set again to 5 Å and curves for maximum and minimum beam divergence were considered. For both cases we find a saturation of the gain for $m > 5$. We think this observation is a consequence of the geometrical constrains of the setup. The m-value is proportional to the critical angle of the total reflection. In the example considered, the beam divergence as well as the curvature of the lens are rather small and total reflexion is always fulfilled for $m > 5$. In case of the small divergence data set, this saturation starts a bit earlier than for the large divergence data set.

The effect of the sample size on the intensity gain is also very important. As one can clearly see in Fig. 6 small samples benefit most from this kind of intensity enhancement. This becomes clear if one recalls that the intensity gain is basically the ratio of the total beam to the direct

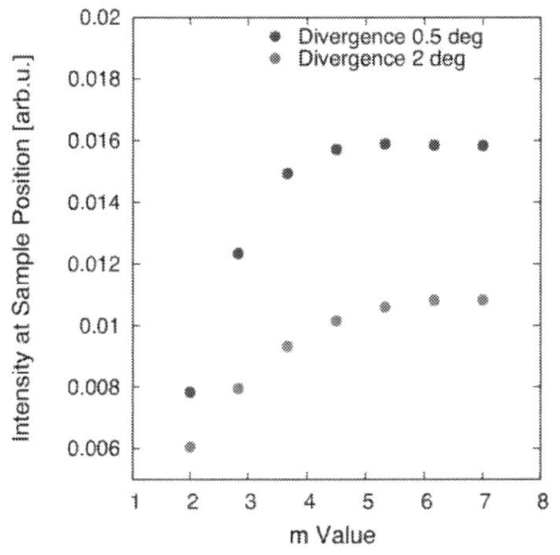

Figure 5. Intensity at the sample position as a function m-value. For smaller divergences the choice of m-value becomes less important.

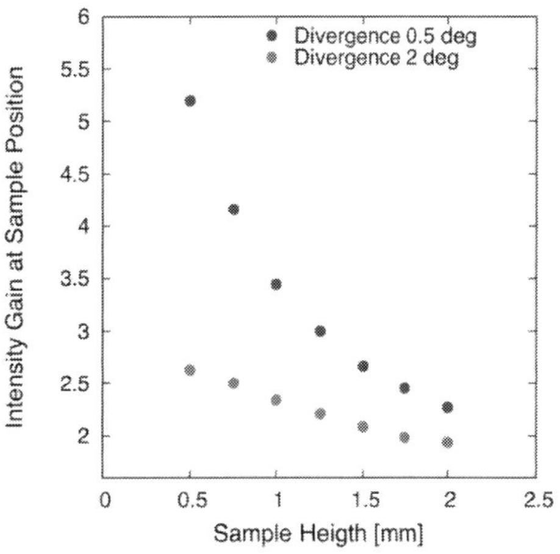

Figure 6. Intensity gain as a function of sample size at $\lambda = 5$ Å and $m = 7$.

beam at the sample . The intensity of the direct beam scales with the sample size (i.e. here, it scales with the height) while the reflected beam does not loose intensity assuming that the sample is perfectly in focus. This, however, is only true if the beam divergence is exactly zero. For finite divergences the reflected beam is also loosing intensity due to the blurring of the focal spot.

4. Conclusion

We have shown that it is technically feasible to envision neutron optics inside a cryogenic sample environment. Moreover, we have shown that such a lens would yield sizeable intensity gains for small samples. Furthermore, we have shown that the finite beam divergence has the largest influence on the performance of the lens. The gain could be further enhanced by increasing the length of the lens which in case of the PE-cell and its CCR would only require a new set of modified radiation shield to increase the length by a factor two. Ultimately, intensity gains of the order of 10 and higher could be achieved by combining a cryogenic and an external neutron lens.

We would like to thank M. Zolliker and J. Stahn for their support with the sample environment and the measurements on Morpheus as well as the fruitful discussions.

References

[1] Maier-Leibnitz H and Springer T 1963 *Reactor Sci. Technol* **17** 217
[2] Mezei F 1976 *Commun. Phys.* **1** 81
[3] Mezei F 1977 *Commun. Phys.* **2** 41
[4] Schanzer C, Böni P, Filges U and Hils T 2004 *Nuc. Instr. and Meth.* A **529** 63
[5] Böni P 2008 *Nuc. Instr. and Meth.* A **586** 1
[6] Goncharenko I N, Mirebeau I, Molina P and Böni P 1997 *Physica* B **234-236** 1047
[7] Lefmann K and Nielsen K 1999 *Neutron News* **10** 20
[8] Willendrup P, Farhi E and Lefmann K 2004 *Physica* B **350** 735
[9] Besson J M, Nelmes R J, Hamel G, Loveday J S, Weill G and Hull S 1992 *Physica* B **180&181** 907
[10] Bailey I F 2003 *Z. Kristallogr.* **218** 84

Monte Carlo simulations for the analysis of texture and strain measured with Bragg edge neutron transmission

M Boin[1,2], R C Wimpory[1], A Hilger[1], N Kardjilov[1], S Y Zhang[3] and M Strobl[1]

[1]Helmholtz Centre Berlin for Materials and Energy, Hahn-Meitner-Platz 1, 14109 Berlin, Germany
[2]The Open University, Walton Hall, Milton Keynes, MK7 6AA, England
[3]ISIS Facility, Rutherford Appleton Laboratory, Chilton, Didcot, OX11 OQX, England

boin@helmholtz-berlin.de

Abstract. The transmission spectrum of cold and thermal neutrons displays sudden intensity changes, known as Bragg edges, whose appearance is connected to the crystallographic structure of the investigated sample. However, many difficulties in interpreting and understanding the detected signal exist, in particular when results are evaluated quantitatively. We measured strain and texture properties of different samples utilizing the Bragg edge method on the CONRAD and VSANS instruments at Helmholtz Centre Berlin as well as on ENGIN-X at ISIS, UK. The different results will be compared and validated with conventional diffraction measurements and theoretical considerations. According to this, a Bragg edge neutron transmission simulation was developed to include instrument effects together with sample characteristics. Therefore, a neutron ray-tracing simulation package was applied to emulate the employed instruments in combination with the Bragg edge transmission technique. A new sample module has been created to realize energy-dependent transmission simulations based on neutron cross sections calculations. Towards the analysis of the conducted experiments the virtual sample features the expected strain and texture parameters. The results of the simulation are in good agreement with the experiments and will be discussed with regard to the prospects and limitations of this measurement method and the utilized instruments.

1. Introduction

Monte Carlo simulations have become more and more an important tool for the development of neutron instruments and experimental data analysis [1, 2]. Due to the complexity of a neutron instrument, a large number of parameters must be considered when instrument optimizations and new beamlines are realized. The valuation of the instrument's influences on the measured detector signal is useful when more flux or higher resolution, for example, are aimed at and extra effects to the detector signal must be quantified. A Monte Carlo neutron instrument simulation therefore enables the testing of varying parameters while their influences on the sample position and the detector are monitored. Furthermore, a simulation could be utilized to study the feasibility of alternative measurement methods. Then, the value of the measured results could be validated by a comparison with simulated experiments taking the instrument characteristics and well-known sample properties into account.

In the present work, we want to bring forward to development of the so-called Bragg edge transmission technique as part of the neutron imaging applications. In the past, many advances have been achieved in neutron imaging [3]. Especially energy-resolved neutron transmission has enabled the possibility to investigate and visualize the effects of crystallographic structures and their properties. The basic principle of such investigations is the analysis of patterns occurring in the transmission spectrum – the Bragg edges. According to Bragg's law, these sudden intensity changes appear due to coherent elastic scattering of neutrons with wavelength λ at selected *hkl* lattice planes with interplanar distance d.

$$n\lambda = 2d_{hkl} \sin\theta \tag{1}$$

The scattering angle is denoted with θ and n is the order of the reflection, which accounts for the fact that also higher orders of the wavelength might be reflected under the conditions of Bragg's law. Applying the transmission geometry, the Bragg edge phenomenon becomes visible when $\lambda = 2d$, i.e. $\theta = 90°$. For wavelengths longer than this position no Bragg scattering will be possible for the selected lattice plane. The reason for the sudden intensity increase is that the fraction of neutrons to be scattered by this lattice plane disappears at this point – more neutrons can contribute to the transmitted intensity. As a result, the step-like Bragg edges occur in the transmission spectrum. A simulation of this effect and the prediction of Bragg edges are of considerable benefit for any beamline concerned with such measurements and for anybody who plans or performs such experiments. Thus, appropriate measurement positions will be known beforehand. Moreover, the performance of future imaging instruments [4] using the Bragg edge mode could be estimated, if a calculation of this effect is available.

Since, coherent elastic scattering is responsible for the Bragg transmission edges it seems obvious that crystal structure information can be derived from such measurements. Here, we refer to the analysis of crystallographic preferred orientations, known as texture [5], and lattice strains, induced by residual stresses for example [6]. Therefore, different samples featuring these properties have been investigated on instruments utilizing different methods to enable wavelength-dispersive neutron transmission experiments. A simulation of these approaches including the sample features is demonstrated in this paper, which allows conclusions to be drawn about the quantitative analysis of the crystallographic properties using the Bragg edge transmission method.

2. Neutron energy selection and detection

In order to realize Bragg edge experiments, different approaches exist. In principle, wavelength-sensitive neutron transmission is performed either by selecting a single wavelength prior to the sample – scanning through a series of wavelengths and measuring the transmission signal for each step – or by analyzing the complete spectrum of a transmitted beam at the detector position, for example in terms of the neutron time-of-flight (ToF) method. Both methods have been utilized within this work, namely on the CONRAD tomography machine and the (very) small angle neutron scattering instrument VSANS, both at the BER II continuous source of the Helmholtz Centre Berlin (HZB), and the ENGIN-X diffractometer at the ISIS pulse spallation source, UK.

2.1. CONRAD tomography instrument

The CONRAD instrument is dedicated to radiography and tomography experiments using a spectrum of cold neutrons with wavelengths between 2 Å and 12 Å [7]. The instrument setup includes an option to select a monochromatic incident beam (see figure 1). Therefore, a double-monochromator device [8] is installed to scan through individual wavelengths by rotating and shifting two highly-orientated pyrolytic graphite (HOPG) single crystals according to the Bragg equation (1). With this setup a wavelength resolution of $\Delta\lambda/\lambda \approx 3\%$ is achieved. The specimen is mounted on a translation and rotation table 5 m away from the monochromator position, followed by the detector stage. A $ZnS/^{6}Li$ scintillator screen converts the neutron intensity to visible light captured by a CCD camera, which is reflected upwards for shielding reasons.

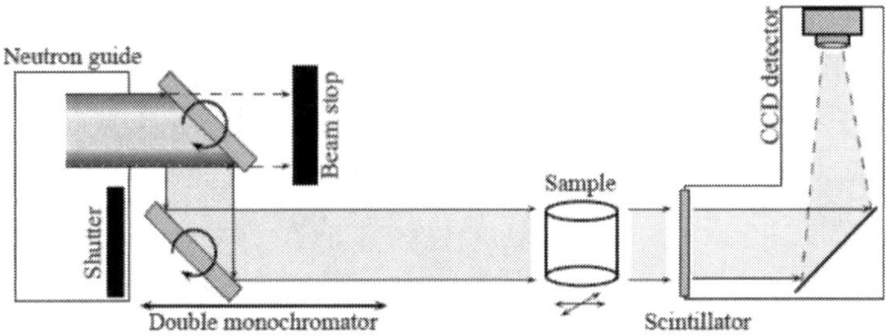

Figure 1. Basic principle of the CONRAD tomography machine using a double-monochromator device: The monochromator crystals are aligned according to Bragg's equation (1) to select single wavelength from the incident cold neutron spectrum.

2.2. VSANS instrument

The VSANS instrument is located at the neutron guide NL4C of the BER II providing a cold neutron spectrum with a lower limit of around 3.5 Å. The principle setup is visualized in figure 2, which shows the workflow of the chopper system along the neutron guide. Initially, a supermirror coated multichannel bender extracts neutrons to the neutron guide which has a window of 4×4 cm^2 at the beginning and 8×8 cm^2 at the end. In between, four chopper discs create a pulsed beam from the incident continuous beam. After that, a collimation chamber, consisting of variable neutron guide sections, can be utilized to measure with different divergence settings. The geometry and the wavelength selection of the new VSANS instrument were used to measure neutron transmission in the Bragg edge region, such that the original sample position was modified. Within a proof-of-principle study [9], the position at the end of the detection chamber was used for the sample position and the detection system. The latter was a CCD camera to take radiographic images triggered with a MCP amplifier to measure the transmitted neutron intensity versus the neutron ToF. The total distance of the detector system was L = 31.85 m away from the master chopper. In order to resolve the neutron wavelength λ, the measured ToF signal t is converted using the de Broglie relation:

$$\lambda = \frac{ht}{mL} \qquad (2)$$

Where h is Planck's constant and m is the neutron mass. The original sample position applied for small angle scattering experiments is located 20 m downstream from the first (master) chopper – the place between both chambers. Here, a pinhole with a diameter of 3 cm has been used for the imaging experiment presented in this work.

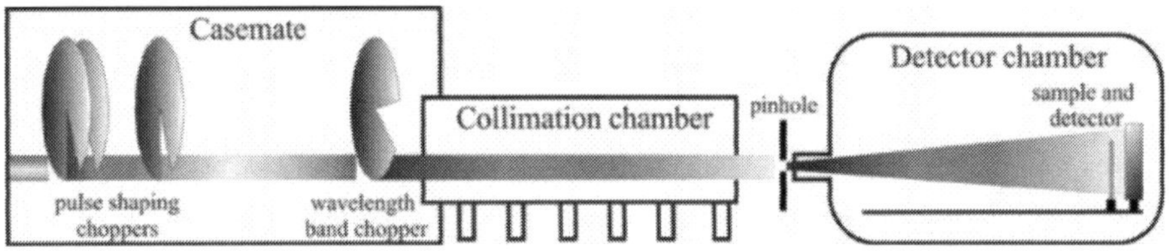

Figure 2. Schematic drawing of the VSANS instrument: The continuous neutron signal is transformed into a pulsed beam using rotating disc choppers. For the experiment presented in this paper, the sample and detector stage was placed at the end of the detector chamber.

2.3. ENGIN-X diffractometer

ENGIN-X is a time-of-flight diffractometer designed to study residual stresses of engineering materials with neutrons [10]. Viewing a 100K liquid methane moderator, the instrument provides a spectrum of approx. 0.5 Å to 6 Å with a high-flux range over 1 Å to 3 Å [11] at a moderator-sample distance of L = 50 m. Therefore, two detector banks located 1.5 m away from the sample table and ±90° from the incident beam direction are installed to measure the ToF signal. In addition to the diffraction mode, the instrument is able to measure the neutron intensity in the transmission direction. A third detector system, consisting of a 10 × 10 array of 2 × 2 mm^2 scintillating ToF detector elements on a vertical and horizontal pitch of 0.5 mm, is placed behind the sample table to measure the neutron transmission signal. Figure 3, visualizes the instrument setup. Each detector element is connected to a Hamamatsu H6568 photo-multiplier tube via fiber optic light guides [12] and triggered with the neutron pulse frequency to count the neutron intensity against the neutron time-of-flight. Thus, transmission spectra in the Bragg edge region can be recorded.

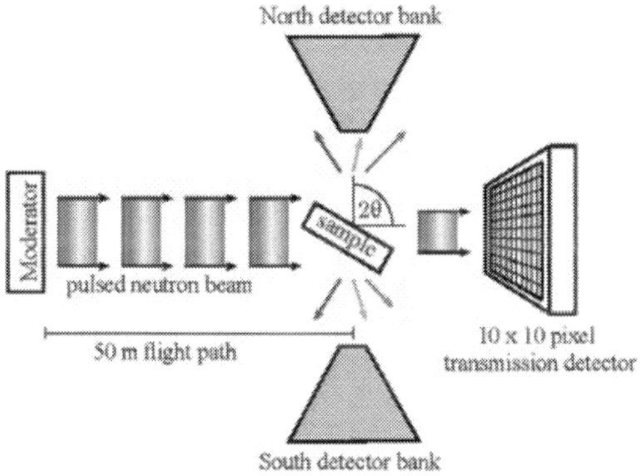

Figure 3. The ENGIN-X residual stress diffractometer at the ISIS pulsed spallation source measures the diffracted neutron intensity versus the time-of-flight. For Bragg edge measurements, the instrument is equipped with a detector array placed behind the sample.

3. Samples and reference measurements

Within this study a number of samples provided with different characteristics, namely strain and texture properties, have been investigated. All samples have also been investigated with neutron diffraction techniques in order to provide reference values from a conventional and well-established measurement method. Thus, a comparison with the Bragg edge transmission results will be possible. Further, the reference values will be used as input parameters for the developed simulations.

3.1. Residual stress samples

For the analysis of strain parameters an Al shrink-fit ring and plug cylinder has been used as a reference sample. The sample was part of the VAMAS TWA20 round-robin exercise and its strain values have been determined many times [13]. With neutron diffraction, the lattice strain was measured by means of the lattice spacing d along the sample diameter. Using the value for the unstrained lattice spacing d_0 of the same material, the strain ε is then calculated by:

$$\varepsilon = \frac{d - d_0}{d_0} \tag{3}$$

Moreover, an analytical solution to the stress and strain gradients is available due to the knowledge of its manufacturing parameters. The height and diameter of the cylinder are each 5 cm. Furthermore, a reference plug with a diameter of 2.5 cm was available to provide a specimen with the same material for measurements of the stress-free lattice parameter d_0. A photograph of the ring and plug samples and the theoretical values for the three principle strain directions (hoop, radial and axial) are shown in figure 4a and 4d respectively.

Furthermore, dieless-drawn steel rods [14] shown in figure 4b have been investigated on the E3 residual stress diffractometer at the HZB [15] to obtain the strain values along the sample diameter from the three principle directions. The sample symmetry is similar to that from the ring and plug sample, i.e. the hoop, radial and axial strains have been measured using the angular-dispersive (constant wavelength) mode on E3. The steel rods have an initial diameter of 5 mm which was drawn down to 3.5 mm. However, for a comparison with the strain data obtained with the Bragg edge method (see section 6), we refer to the 'untreated' 5-mm-region. Here, the strain values obtained on the E3 diffractometer are plotted in figure 4e.

3.2. Texture sample

The crystallographic orientation distribution of a drawn aluminium wire sample was determined by means of pole figure measurements conducted on E3 at HZB and the STRESS-SPEC diffractometer at FRM-II [16, 17]. Therefore, an Eulerian cradle was employed to tilt and rotate the sample, such that every possible grain orientation was measured by means of the reflection intensities. A picture of the sample is shown in figure 4c, where RD indicates the rolling or reference (drawing) direction of the cylinder piece. The height of the cylinder is 8 mm and the diameter is 11.8 mm. The resulting pole figures reveal a circular symmetry around the sample axis RD which is known as fiber texture. Exemplarily, the pole figure of the (111) reflection is shown in figure 4f. The texture data is used as a reference for the comparison with the Bragg edge method.

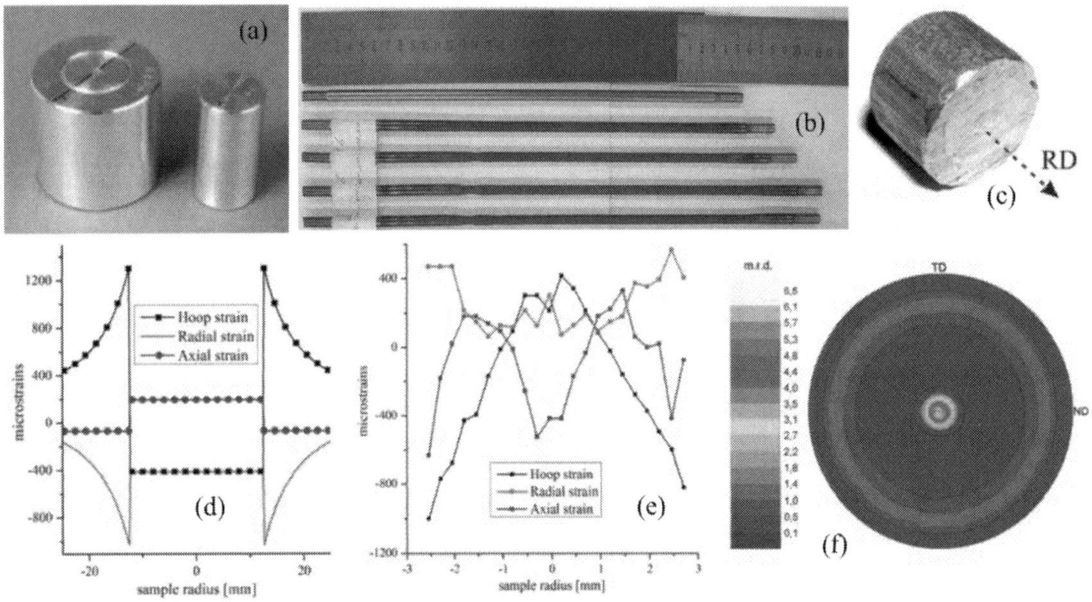

Figure 4. Investigated samples and reference data: The ring and plug Al samples (a) from the VAMAS TWA20 round-robin exercise [13] and dieless-drawn steel rods (b) [14] for residual stress investigations and a fiber-textured Al cylinder (c). The dedicated reference data is shown below: the theoretical strain values of the ring and plug (d) from the analytical solution, the strain data of the rod specimens measured on E3 (e) and a selected pole figure of the textured Al cylinder (f).

4. Development of a neutron experiment simulation

The McStas neutron ray-tracing software package [18, 19] has been utilized for the present study. McStas provides virtual neutrons propagating through an instrument emulating the conditions of a real neutron instrument. Therefore, the McStas library already includes modules that represent the individual assembly of a real instrument. These modules are ready to use and cover devices such as neutron-producing sources, guide components, slit systems, samples and detectors. Thus, each of the previously described CONRAD, VSANS and ENGIN-X machines has been mapped to a simulation taking the experimental conditions into account. Although, a few sample modules for different purposes are available, a new virtual sample simulating the Bragg edge neutron transmission effect has been developed.

4.1. Bragg edge sample module

The basic principle in McStas of a neutron propagating from the source to the end of the instrument definition (e.g. to a detector position), is to change its state parameters, i.e. its position, its flight direction and its velocity, according to the physics implemented in each of the instrument's components. In the case of the wavelength-dispersive neutron transmission method a sample module must therefore handle the neutron-matter interaction with respect to the neutron wavelength λ and the sample's crystal structure. A new sample component for McStas was created to account for these parameters and to realize Bragg edge experiment simulations.

The probability for a neutron to be transmitted is based on the computation of neutron cross sections, which describe quantitatively the interaction between neutrons and matter. The total microscopic neutron cross section σ_{total} of an isotope is the cross sectional area "presented" to an incoming neutron to interact with the material. The sum of its individual coherent, incoherent and absorption contributions (σ_{coh}, σ_{inc}, σ_{abs}) is used to determine the transmission probability Tr:

$$\sigma_{total}(\lambda) = \sigma_{coh}(\lambda) + \sigma_{inc}(\lambda) + \sigma_{abs}(\lambda)$$
$$Tr = \frac{I}{I_0} = \exp(-\sigma_{total}(\lambda) N l) \quad (4)$$

The transmitted intensity I is given by an exponential attenuation of the incident intensity I_0. N is the number of scattering centers per unit volume and l is the transmission path length through the sample. With λ being in the range of interatomic distances, i.e. for thermal and cold neutrons, the transmission spectrum of crystalline samples displays Bragg edges, which occur due to the discontinuities in the coherent elastic part of the neutron cross section:

$$\sigma_{coh}^{el}(\lambda) = \frac{\lambda^2}{2V_0} \sum_{d_{hkl}=0}^{2d_{hkl}<\lambda} |F_{hkl}|^2 d_{hkl} \quad (5)$$

The summation runs over all set of hkl lattice planes as long as d_{hkl} is smaller than half the size of the selected neutron wavelength. F_{hkl} is the structure factor and V_0 is the unit cell volume. Thus, sudden intensity changes at $\lambda = 2d_{hkl}$ in the transmission spectrum can be quantified.

The calculation of the coherent elastic scattering cross section can be applied to single atoms as well as to material compositions. However, the knowledge of the sample's crystallographic structure, its lattice parameters and the position of atoms included inside the unit cell is required. For the new sample module all parameters will be provided as input parameters to the McStas simulation. Thus, different materials and sample characteristics can be tested with the same module. The choice for a single neutron intersecting the sample volume to be transmitted is then based on the probability Tr, calculated by equation (4). The incoherent and absorption contributions are described elsewhere [20] and have been implemented as well.

4.2. Texture modification

Unlike powders, most engineering materials and components reveal a crystallographic preferred orientation due to the process of their manufacturing. In the case of diffraction experiments, the intensities captured with a detector vary with the number of orientated grains inside a polycrystalline sample. If the sample is textured, the probability of detecting grains via selected lattice planes differs with the sample orientation, such as shown with the pole figure in figure 4f, where the intensities, normalized to multiples of a random distribution (m.r.d.), are plotted against the sample orientation.

In order to correct for the influence of crystallographic preferred orientations, the March-Dollase model was used to apply changes to the diffraction intensities. Following the descriptions from [21], the texture model was implemented with respect to the Bragg edge transmission geometry. Hence, any correction factor will also affect the transmission spectrum. The basic principle of this model is to correct the diffraction intensities using the March-Dollase equation:

$$P_{\vec{H}}\left(\alpha_{\vec{H}}\right)=\left(r^{2}\cos\alpha_{\vec{H}}+\frac{1}{r}\sin^{2}\alpha_{\vec{H}}\right)^{-\frac{3}{2}} \tag{6}$$

where $\alpha_{\vec{H}}$ is the angle between the preferred orientation direction $\vec{H}$ and orientation axis, for example the rolling direction, and r is a variable describing the degree of the texture. Within the computation of the neutron cross sections inside the Bragg edge simulation module, the correction is applied to the calculation of the individual coherent elastic scattering cross section for each contributing lattice plane with the normal vector $\vec{h}$ and $\alpha_{\vec{h}}$ being the angle between $\vec{h}$ and the preferred orientation axis:

$$\sigma_{coh}^{el}(\lambda)=\sum_{\vec{h}}\sigma_{coh,\vec{h}}^{el}(\lambda)P_{\vec{h}}\left(\alpha_{\vec{h}}(\lambda)\right) \tag{7}$$

Basically, the procedure is to derive the correction functions $P_{\vec{h}}$ for the individual reflections $\vec{h}$ from the master correction function $P_{\vec{H}}$ [21]. Thus, a simulation of texture influences parameterized by only two values, $\vec{H}$ and r, is realized using McStas and the developed Bragg edge module.

4.3. Strain implementation

A second modified version of the Bragg edge sample module was created to include local variations of lattice spacings, i.e. strain. The basic principle of the calculation implemented in the sample module is shown in figure 5 and is explained based on the example of the ring and plug cylinder sample.

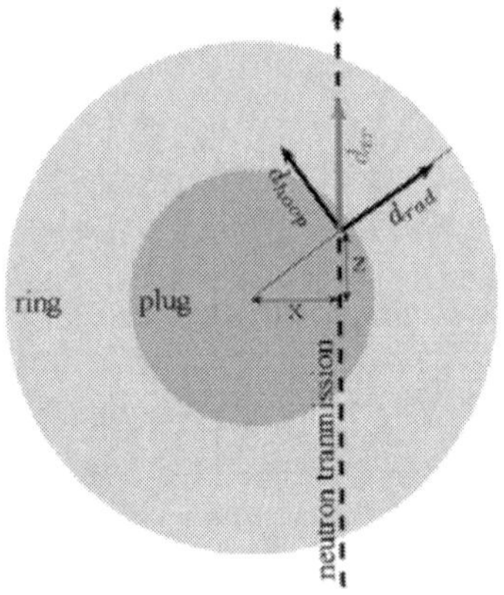

Figure 5. The calculation of the effective lattice spacing d_{tr} considers the neutron position within the sample as well as the flight direction. The lattice spacing values in the principle strain directions are chosen according to the position on sample radius. In this figure, the calculation is simplified to the *xz*-plane (hoop and radial strain).

In order to ensure a realistic calculation, the computation of the lattice spacing in the transmission direction not only depends on the spatial position inside the sample but also on the neutron flight direction. Hence, the three strain components (radial, hoop, axial) are utilized to calculate the effective d-value in the neutron flight direction:

$$\begin{aligned} d_{xz} &= \left(d_{hoop}-d_{rad}\right)\sin^{2}\psi_{1}+d_{rad} \\ d_{tr} &= \left(d_{xz}-d_{axial}\right)\sin^{2}\psi_{2}+d_{axial} \end{aligned} \tag{8}$$

In a first instance, the lattice spacing d_{xz} in the *xz*-plane (see figure 5) is determined based on the neutron flight direction in that plane, where ψ_1 is the angle between d_{xz} and the radial component d_{rad}. In the second step the axial strain is applied to account for any neutron direction outside the *xz*-plane. Hence, ψ_2 is the angle between the neutron flight direction and the y-axis (height) of the sample cylinder. For both steps, the $sin^2\psi$ approach is utilized, which assumes that the lattice spacing is linear with the square of sine of the angle between two principle directions [22]. Thus, the calculation of the effective d-value d_{tr} along the neutron transmission path also considers any degree of beam divergence and therefore accounts for the influences of the instrument parameters. Since every neutron is considered individually during the simulation, the computation of the appropriate d-value is performed for every neutron passing through the sample. The decision whether a virtual neutron is transmitted depends again on the probability calculated with equation (4) – but this time using the value of d_{tr} computed for each individual neutron. As a result the position of a Bragg edge in the transmission spectrum is affected by the lattice strain in the sample. As the experiments follow a radiographic approach, the detected signal is a convolution of all effective lattice spacings along the transmission path through the sample. Hence, the strain values evaluated from such measurements will be referred to as the "projected" strain.

5. Experiments and simulations

Utilizing the presented instruments, different methods have been applied to measure Bragg edges and the properties of the investigated samples.

On CONRAD, an iterative alignment of the double-monochromator device has been conducted to perform a wavelength scan from 3 Å to 5 Å in steps of 0.01 Å and 0.02 Å for the ring and plug residual stress specimen and the textured Al cylinder respectively. The radiographic images taken by the CCD camera have been normalized by flatfield measurements to correct for the incident intensity variation and the inhomogeneous screen illumination. The images in figure 6 show the corrected radiographs for selected wavelength positions for both experiments. The resulting grey values have been used to analyze the transmission spectrum. Thus, the Bragg edge positions could be identified. In order to improve the statistics, the evaluated regions along the diameter of the ring and plug sample have been selected with regions of interest of 4.5×9 mm^2 each, to determine a single grey value for each wavelength step. These regions are indicated in figure 6a. For the textured sample (figure 6b), the region of the circular base area was analysed.

The same samples have been investigated on ENGIN-X measuring the ToF signal with the transmission detectors, i.e. the complete transmission spectrum from 2 Å to 5 Å has been captured at once. The samples were aligned such that the middle of the array detector was faced with the sample center. Because of the limited detector size and the field of view, the ring and plug sample was driven sideways from -25 mm to +25 mm in 5 mm steps to scan through the sample diameter with the center of the detector array. For each step, transmission spectra have been recorded by each of the detector elements. Initially, the 6th detector column was aligned to the sample center (radius r = 0 mm), such that the 7th column faced the position at r = 2.5 mm.

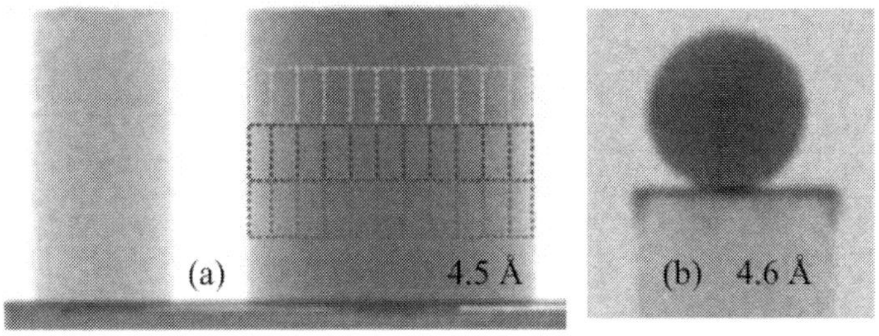

Figure 6. A normalised radiograph of the reference plug (left) and the ring and plug sample (right) are shown in (a). The coloured regions mark the areas along the sample diameter that have been evaluated. In (b) the texture cylinder, placed on a sample table, is shown.

The parameters used to run the presented experiments have been transferred to the developed Bragg edge simulation using McStas, i.e. the program has been executed using the same arrangements as the real experiment. In particular, this refers to the wavelength scans and the additional flatfield measurements performed with CONRAD. Although, it is possible with McStas to record the neutron state parameters (including the neutron wavelength) at any time during the simulation and thus a distinction between individual wavelengths in the transmission spectrum would be enabled, the study was focused on the influence of the double-monochromator. Hence, the device was included in the simulation and aligned stepwise such that the wavelength scans mentioned above could be emulated. The ring and plug sample has been included in terms of the developed Bragg edge sample module. The residual stress properties have been inherited from the theoretical strain values from figure 4d such that a calculation of the effective lattice spacings along the neutron transmission direction has been enabled as described in section 4.3.

The dieless-drawn steel rods have been measured on the VSANS machine. Therefore, the incident neutron beam has been transformed into a pulsed beam such that the neutron ToF signal could be triggered with the CCD camera in order to map the transmitted intensity to individual wavelengths. Details of the experimental setup and the chopper settings to realize a constant wavelength resolution of about $\Delta\lambda/\lambda$ = 1.5 % for a wavelength band from 3 Å to approximately 5 Å have already been reported in [9]. A simulation of the complete VSANS machine is currently under development. However, the experimental parameters have been applied to the Bragg edge simulation using the strain parameters shown in figure 4c for the dieless-drawn steel specimens.

6. Results and Comparison

6.1. Texture Al cylinder

In figure 7, the measured and simulated transmission spectra of the textured cylinder sample are shown. Using the incident and transmitted neutron intensities, the attenuation coefficient was used to compare the results with the theoretical texture-free Bragg edges. It is apparent, that the resulting spectra show large deviations due to the influence of crystallographic preferred orientations. Especially the (111) Bragg edge at 4.6 Å reveals a large increase of the macroscopic cross section which coincides with the intensity in the center (RD) of the pole figure shown in 4f. Furthermore, a difference between the CONRAD and ENGIN-X results can be observed. In particular, the height of the Bragg edge varies by a factor of around 2.5, which is a result of the wavelength-selection with the double-monochromator. The wavelength resolution of 3% smears the evaluated attenuation coefficient due to wavelengths before and after the selected position.

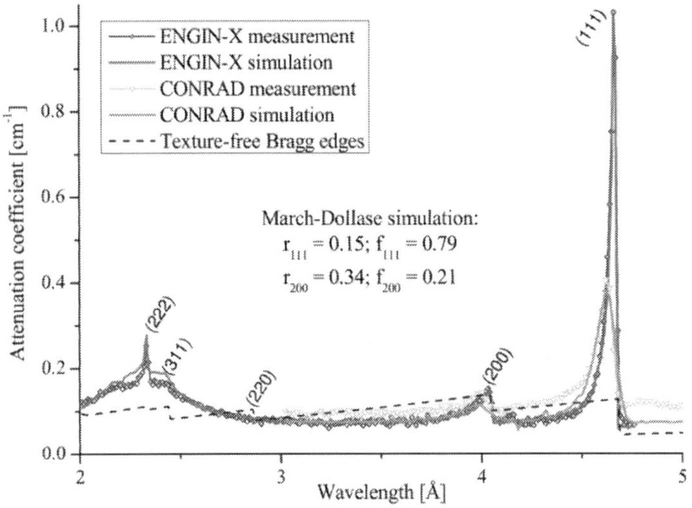

Figure 7. Comparison of the transmission spectra of the textured Al sample measured and simulated with the CONRAD and ENGIN-X instruments. Compared to the texture-free curve (dashed) the Bragg edges captured in RD show large deviations due to preferred crystallographic orientations.

Nevertheless, the simulations, executed with the March-Dollase modification, agree with the individual experiments. Therefore, the same March-Dollase parameters have been applied for both, the CONRAD and ENGIN-X simulations. As shown in the figure, two preferred orientation directions (111 and 200) were used for the simulation. Here a further parameter f was established to account for the fraction of grains exhibiting one of the given directions. Thus, a good agreement with the ENGIN-X data could be achieved. Only a small misfit in the region of the (311) and (222) Bragg edge can be found. The CONRAD results also show the same trend. The difference between simulation and experiment, appearing like a vertical shift, occurs due to the uncertainties of the monochromator description. For the simulation, a single value has been used to describe the mosaic spread of the two applied single crystals. Moreover, the strength of higher-order reflections from the monochromator contributing to the measured intensity is unknown, which could lead to the deviations shown.

6.2. Ring and plug

The strain values of the ring and plug sample have been compared similarly. Therefore, the Bragg edge positions of the measured and simulated transmission spectra from ENGIN-X have been fitted to a function proposed in [23], which accounts for the asymmetric and symmetric instrument broadenings affecting the Bragg edge. For the CONRAD results, the same function with a modified kernel has been used to fit the data. Here, a single parameter σ was used to describe the instrumental broadening.

$$Tr(\lambda) = \exp(-n\omega\sigma_0)\left\{\exp(-n\omega\sigma_{hkl}) + (1-\exp(-n\omega\sigma_{hkl}))\frac{1}{2}erfc\left(-\frac{\lambda-\lambda_0}{\sqrt{2}\sigma}\right)\right\} \tag{9}$$

The atomic density n, the sample thickness ω and the total neutron cross sections σ_{hkl} and σ_0 at the short and long wavelength side of the selected *hkl* edge can be defined before the fitting. The fit result is the Bragg edge position at λ_0. The resulting projected strains, calculated with equation (3), are shown in figure 8.

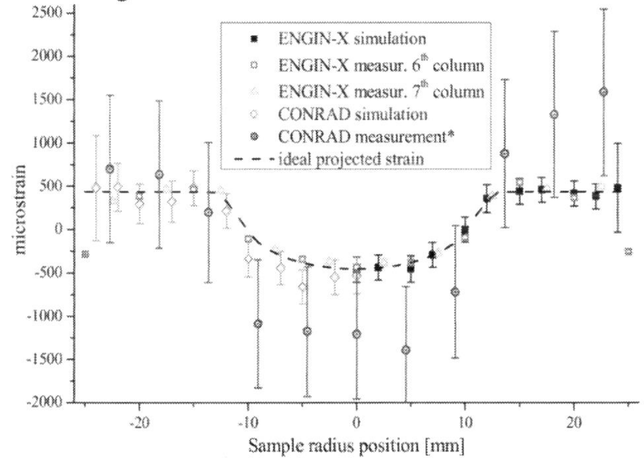

Figure 8. The projected strains in the transmission direction evaluated from measurements and simulations of the CONRAD and ENGIN-X instruments are compared with the ideal strain curve (dashed line). The asterisk at the CONRAD measurement curve indicates that the error bars have been scaled down by a factor of 5 for reasons of clarity.

The plotted results follow the same trend of the ideal projected strain (dashed curve) that was obtained from the calculations presented in section 4.3. The best fit to that reference results from the ENGIN-X experiment and simulation data. This is also true for both evaluated detector columns (6th and 7th. The only mismatch occurs at the two sample edge positions at ±25 mm, because for these positions the 2 × 2 mm^2 detector element is covered by only half of the sample area. The rest already faces neutrons outside the sample volume, which affects the fitting of Bragg edge position systematically. Moreover, it is visible that the error bars become larger with an increasing radius position due to the fact that the sample gets thinner which downgrades the statistics.

The strains obtained on CONRAD exhibit the largest deviation from the reference. The standard deviation of the data set about the ideal projected strain suggests a random error of about 750 microstrain (the size of the error bars depicted here). However, the actual fits from the fitting routine used returned much larger uncertainties (an average of about 2500 microstrain if one does not include the uncertainty of the reference specimen and 4000 microstrain if one does include the reference uncertainty). These fitting uncertainties are clearly unrealistic. The comparison of the data sets depicted above presents the opportunity to evaluate the approximate actual random uncertainty. Nevertheless, the simulation of the CONRAD experiment reveals an acceptable agreement with the dashed curve. Consequently, for the case of the strain and stress analysis the description of the simulated CONRAD instrument is not sufficiently accurate. The above mentioned uncertainties of the monochromator parameterization and the estimated wavelength resolution manifest the difficulties of Bragg edge position evaluations from CONRAD experiments.

6.3. Dieless-drawn steel rods

The untreated region of the dieless-drawn steel rods has been investigated in this study. The observed strain values measured with conventional neutron diffraction have been applied to the developed

Bragg edge simulation to obtain the theoretical projected strain for these samples. Thus, the transmission measurements on VSANS could be validated, although a model of the instrument is not yet available. In figure 9, the ideal projected strain (dashed curve) is compared with the measured curves evaluated based on a 5 × 5 binning of the captured radiographic images [9].

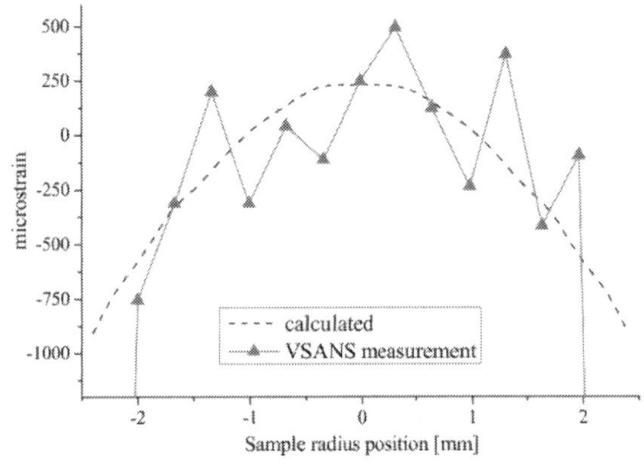

Figure 9. Comparison of the measured and calculated projected strain along the sample diameter of a dieless-drawn steel rod.

The diagram shows that the evaluated strain along the sample diameter follows the expected trend of the calculated projected strain. The variations occur due to the limited neutron statistics and the thin samples, in particular at the sample edge.

7. Conclusion

We have created new program routines for the calculation of neutron cross sections. These routines have been used with the McStas simulation environment to represent a sample module that is able to perform realistic Bragg edge transmission simulations. The presented work has shown, that the simulation is capable to emulate different materials based on their crystallographic description. Furthermore, the McStas module was modified for two cases to simulate texture and strain properties. The former was achieved by the application of the March-Dollase model, which enables to adjust the Bragg edge spectrum according preferred crystallographic orientations based on two parameters. The latter has been realized using the strain values measured with conventional diffraction methods. The calculation of the effective strain in the transmission direction was implemented and applied to determine the measureable projected strain.

Regarding the analysis of strain properties applied the Bragg edge method, the ENGIN-X instrument showed the best results. The main beneficial aspect here is the wavelength resolution of the instrument, which was given in [24] in terms of the lattice spacing precision of $\Delta d/d = 10^{-4}$. Hence, the ToF diffractometer suggests the reliability for further experiments of this kind. In contrast, the wavelength-selective option on CONRAD could not provide such results. However, the simulation showed, that the lower resolution could also deliver comparable results. As a consequence, the parameterization of the double-monochromator for the simulation requires further investigations. In between, the 1.5% wavelength resolution of the VSANS setup has proven the feasibility of the strain measurements. However, the agreement to the calculated reference data may be improved.

From the analysis of the textured Al sample cylinder on the CONRAD and ENGIN-X instruments several conclusions can be drawn. Both the ENGIN-X measurement and simulation showed very good agreement and the strength of the texture coincides with the reference pole figure measurements obtained with neutron diffraction. The CONRAD results show a considerably smaller degree of texture due to wavelength spread of the monochromator mosaicity. However, using the simulation and applying the same March-Dollase parameters from the ENGIN-X, the same texture influence on the Bragg edge spectrum was reproduced with McStas. Nevertheless, more powerful texture models may be needed if more complex samples shall be investigated.

In summary, a Monte Carlo application was developed to validate measured strain and texture data with Bragg edge neutron transmission and to provide test cases for further investigation regarding instrument developments.

Acknowledgements
We would like to thank Javier Santisteban, Jon James, Brian Abbey, Daniel Clemens, Christian Randau and Michael Hofmann for their support with the instruments and the helpful discussions on the experimental methods. We thank Peter Tiernan for permission to reprint figure 4b.

References
[1] Šaroun J, Pirling T and Rogge R B 2002 *Appl. Phys. A* **74** s1489-s91
[2] Santisteban J R, Daymond M R, James J A and Edwards L 2006 *J. Appl. Cryst.* **39** 812-25
[3] Strobl M, Manke I, Kardjilov N, Hilger A, Dawson M and Banhart J 2009 *J. Phys. D: Appl. Phys.* **42** 243001
[4] Kockelmann W, Zhang S Y, James J A, Burca G, Paradowska A and Chapon L 2009 IMAT: An Imaging and Engineering Facility for ISIS TS-II *Proc. NEUWAVE2 (Abingdon, UK, July 12-15, 2009)*
[5] Bunge H-J 1993 *Texture Analysis in Materials Science: Mathematical Methods* (Göttingen: Cuvillier Verlag)
[6] Hauk V 1997 *Structural and Residual Stress Analysis by Nondestructive Methods: Evaluation - Application - Assessment* (Amsterdam: Elsevier Sience)
[7] Hilger A, Kardjilov N, Strobl M, Treimer W and Banhart J 2006 *Physica B* **385** 1213-5
[8] Treimer W, Strobl M, Kardjilov N, Hilger A and Manke I 2006 *Appl. Phys. Lett.* **89** 203504
[9] Strobl M *et al.* 2010 *Nucl. Instrum. Methods Phys. Res., Sect. A* in press
[10] Dann J A, Daymond M R, James J A, Santisteban J R and Edwards L 2003 ENGIN-X: A New Diffractometer Optimised for Stress Measurement *Proc. ICANS XVI (Düsseldorf-Neuss, Germany, May 12 - 15, 2003)* eds G Mank and H Conrad pp 231-8
[11] Kockelmann W, Frei G, Lehmann E H, Vontobel P and Santisteban J R 2007 *Nucl. Instrum. Methods Phys. Res., Sect. A* **578** 421-34
[12] Santisteban J R, Steuwer A, Edwards L, Withers P J and Daymond M R 2003 Phase analysis and imaging by TOF neutron transmission *Proc. ICANS XVI (Düsseldorf-Neuss, Germany, May 12 - 15, 2003)* eds G Mank and H Conrad pp 239-46
[13] Webster G A 2000 *Neutron Diffraction Measurements of Residual Stress in a Shrink-fit Ring and Plug* (Teddington, UK: VAMAS Report No. 38, National Physics Laboratory)
[14] Carolan R and Tiernan P 2009 *J. Mater. Process. Technol.* **209** 3335-42
[15] Wimpory R C, Mikula P, Šaroun J, Poeste T, Li J, Hofmann M and Schneider R 2008 *Neutron News* **19** 16-9
[16] Hofmann M, Rebelo-Kornmeier J, Garbe U, Wimpory R C, Repper J, Seidl G A, Brokmeier H-G and Schneider R 2007 *Neutron News* **18** 27-30
[17] Brokmeier H-G, Gan W M, Randau C, Völler M, Rebelo-Kornmeier J and Hofmann M 2011 *Nucl. Instrum. Methods Phys. Res., Sect. A* **642** (1) 87-92
[18] Lefmann K and Nielsen K 1999 *Neutron News* **10** 20-3
[19] Willendrup P K a r 2004 *Physica B* **350** E735-E7
[20] Granada J R 1984 *Z. Naturforsch. A* **39** 1160-7
[21] Vogel S C 2000 *A Rietveld-Approach for the Analysis of Neutron Time-Of-Flight Transmission Data* PhD thesis Christian-Albrecht-Univ. Kiel
[22] Fitzpatrick M E and Lodini A 2003 *Analysis of Residual Stress by Diffraction using Neutron and Synchrotron Radiation* (London: Taylor & Francis)
[23] Santisteban J R, Edwards L, Steuwer A and Withers P J 2001 *J. Appl. Cryst.* **34** 289-97
[24] Steuwer A, Santisteban J R, Withers P J, Edwards L and Fitzpatrick M E 2003 *J. Appl. Cryst.* **36** 1159-68

Effects of the sample environment and collimation in the background measurement

N Cuello [1], **G J Cuello** [2]

[1] Università degli studi di Torino, Via Pierto Giuria 1, 10125, Torino, Italy

[2] Institut Laue Langevin, 6, rue Jules Horowitz, 38042 Grenoble, France

E-mail: cuello@ill.eu

Abstract. Most of diffractometers have collimators in front of detectors either to reduce the background contribution or to increase the angular resolution. These collimation devices are designed to collect only the neutrons coming from the sample position. When a sample environment is required, its walls are close to the sample position and produce some unwanted effects on the diffractograms. A simple geometrical model allows a description of the observed patterns in the two-axis diffractometer D4 (ILL, Grenoble, France). The calculated effects are compared with experimental data for a vanadium bar as measured in the standard cryostat of that instrument. This kind of measurements allows the experimental determination of the sample transmission.

1. Introduction

In a scattering experiments the design of the instrument is conceived in a way allowing the detector to see the whole sample. In order to reduce the background signal, the majority of instruments has a collimation system focused at the sample position, as shown in Fig. 1. Because the instruments must adapt to different sample sizes, usually the collimators are able to see a volume larger than the sample size, which is not a problem for ambient condition experiments, *i.e.* without a sample environment around the sample position. When a sample environment is required, if it is large enough to be far from the sample, its signal will stopped by the collimator, as happens with the signal of the vacuum chamber in Fig. 1. But there is a region near the sample position where any material present will produce a scattered beam reaching partially the detector, producing a shadow effect on the observed diffractogram.

The cylindrical symmetry of a two axis diffractometer advice the use of cylindrical samples and environments. Thus the sample environment (a cryostat, a furnace or others) can be represented by a cylindrical screen around and centered at the sample position (Fig. 1).

We will consider the problem of an incoming neutron beam diffracted at three points, namely the sample and the two boundaries of the sample environment (Fig. 1). The scattered beam is registered by the detector placed at the end of the collimator. The shadow effects observed in the diffractograms are determined by the following set of parameters (Fig. 2): (i) the scattering angle (2θ), (ii) the window length ($2w$), (iii) the detector length (2ℓ), (iv) the sample-detector distance (D), (v) the sample-window distance (d), (vi) the sample diameter ($2s$), and (vii) the screen diameter ($2f$).

The geometrical relationship of all these parameters will describe the shadow effects. In particular, we will consider the case of neutron scattering and the D4 instrument at ILL [1],

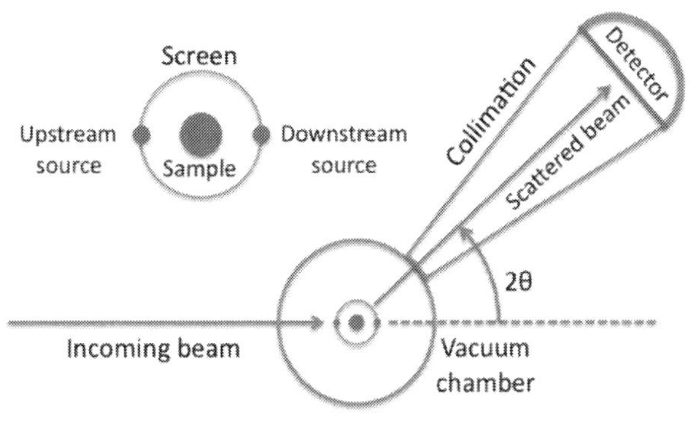

Figure 1. Sketch of the diffractometer device.

but it is worth noticing that these effects could also be observed for other kinds of radiation or instruments with collimation tubes in front of the detectors. The aim of this work is to provide a geometrical interpretation to the effects observed in diffraction experiments performed at D4.

2. Geometrical calculation

We will consider the problem of an incoming beam of neutrons onto a cylindrical sample surrounded by a cylindrical screen. At this point, for the sake of simplicity, we make two assumptions: (i) we consider just only one screen around the sample, and (ii) we assume the three hitting points are punctual sources of scattered radiation. The former assumption is reasonably because considering a more complex situation with several concentric screens could hinder the understanding of a quite simple effect; the generalization to a more complex case can be easily made. The latter is also a reasonable hypothesis, because the non-punctual character of these sources will only contribute to produce smooth shapes instead of sharp features, but the rational behind is exactly the same.

In Fig. 2 we show the notation adopted for the description of the problem. The sample is located at the origin and the segment $\overline{\mathrm{AB}}$ represents the detection surface; the segment $\overline{\mathrm{PQ}}$ is the window of the collimator, which in fact defines the view of any point on the detector. Apart the sample itself, two other sources (s_{u} and s_{d}) are defined by the intersection of the incoming beam (the $x-$axis) with the cylindrical sample environment. Only using some elementary geometrical considerations the coordinates of the points lying on the detector can be calculated as follows:

$$\left\{ \begin{array}{lcl} x & = & D\cos 2\theta + (\ell - z)\sin 2\theta \\ y & = & D\sin 2\theta - (\ell - z)\cos 2\theta \end{array} \right. , \tag{1}$$

where z is a variable increasing from A to B: $z = 0$ is A, $z = \ell$ is C and $z = 2\ell$ is B. The same geometrical considerations are valid for the points lying on the window, whose coordinates can be obtained just replacing D by $D - d$, ℓ by w and z by u in Eq. (1):

$$\left\{ \begin{array}{lcl} x & = & (D-d)\cos 2\theta + (w - u)\sin 2\theta \\ y & = & (D-d)\sin 2\theta - (w - u)\cos 2\theta \end{array} \right. , \tag{2}$$

where u is a variable increasing from P to Q: $u = 0$ is P, $u = w$ is R and $u = 2w$ is Q.

Tracing the straight lines from the upstream and the downstream sources located a $(-f, 0)$ and $(+f, 0)$ to the window edges Q and P, we obtain the *light* cone of both sources. At this point we identify two critical points (α and α') defining two special segments $\overline{\alpha\mathrm{B}}$ and $\overline{\alpha'\mathrm{A}}$. These

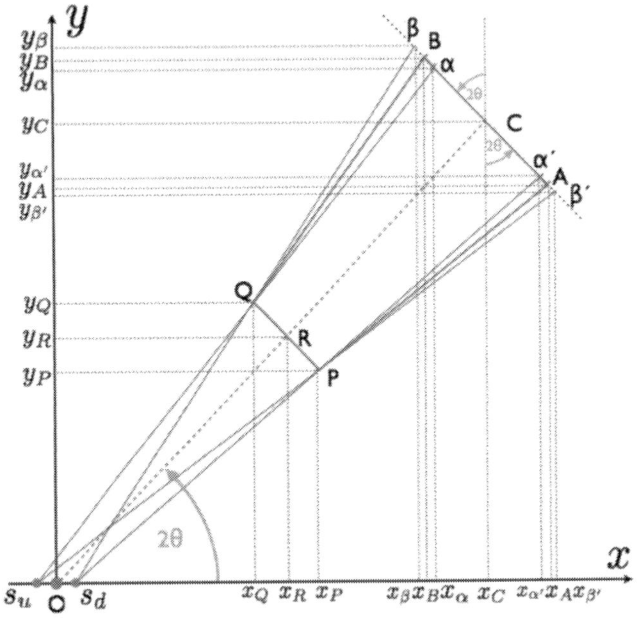

Figure 2. Sketch of the experimental device. We call $\overline{\mathrm{AB}} = 2\ell, \overline{\mathrm{PQ}} = 2w, \overline{\mathrm{OC}} = D, \overline{\mathrm{OR}} = d, 2f =$ furnace diameter, $2s =$ sample diameter, $2\theta =$ scattering angle. We have $\overline{\mathrm{QR}} = \overline{\mathrm{PR}} = w, \overline{\mathrm{AC}} = \overline{\mathrm{BC}} = \ell$

regions see only two of the three sources: s_d and O, s_u and O, respectively. Using Eqs. (1) and (2) we can compute the coordinates of α

$$\begin{cases} x_\alpha &= \dfrac{y_\mathrm{P}\, f\,(x_\mathrm{B}-x_\mathrm{A}) + (x_\mathrm{P}-f)\,(x_\mathrm{B}\, y_\mathrm{A}-x_\mathrm{A}\, y_\mathrm{B})}{y_\mathrm{P}\,(x_\mathrm{B}-x_\mathrm{A}) - (x_\mathrm{P}-f)\,(y_\mathrm{B}-y_\mathrm{A})} \\ y_\alpha &= \dfrac{x_\alpha-f}{x_\mathrm{P}-f}\, y_\mathrm{P} \end{cases} , \tag{3}$$

and α'

$$\begin{cases} x_{\alpha'} &= \dfrac{y_\mathrm{Q}\, f\,(x_\mathrm{B}-x_\mathrm{A}) + (x_\mathrm{Q}-f)\,(x_\mathrm{B}\, y_\mathrm{A}-x_\mathrm{A}\, y_\mathrm{B})}{y_\mathrm{Q}\,(x_\mathrm{B}-x_\mathrm{A}) - (x_\mathrm{Q}-f)\,(y_\mathrm{B}-y_\mathrm{A})} \\ y_{\alpha'} &= \dfrac{x_{\alpha'}-f}{x_\mathrm{Q}-f}\, y_\mathrm{Q} \end{cases} . \tag{4}$$

Using the explicit expressions for points A, B, P and Q we obtain the following formulae showing the explicit dependence of the coordinates with the instrumental parameters:

$$\begin{cases} x_\alpha &= \dfrac{fD - (D-d)\,[f\,(\sin 2\theta)^2 + D\,\cos 2\theta] - w\,[f\,\cos 2\theta\,\sin 2\theta + D\,\sin 2\theta]}{f\,\cos 2\theta - (D-d)} \\ y_\alpha &= \dfrac{(x_\alpha-f)\,[(D-d)\,\sin 2\theta - w\,\cos 2\theta]}{(D-d)\,\cos 2\theta + w\,\sin 2\theta - f} \end{cases} , \tag{5}$$

and

$$\begin{cases} x_{\alpha'} &= \dfrac{fD - (D-d)\,[f\,(\sin 2\theta)^2 + D\,\cos 2\theta] - w\,[f\,\cos 2\theta\,\sin 2\theta - D\,\sin 2\theta]}{f\,\cos 2\theta - (D-d)} \\ y_{\alpha'} &= \dfrac{(x_{\alpha'}-f)\,[(D-d)\,\sin 2\theta + w\,\cos 2\theta]}{(D-d)\,\cos 2\theta - w\,\sin 2\theta - f} \end{cases} . \tag{6}$$

These equations allow the calculation of the geometrical pattern shown in Fig. 3, where the instrumental parameters of D4 were used as example. As expected, because of the symmetry of the problem, the effect is maximum at 90° and null at 0° or 180°.

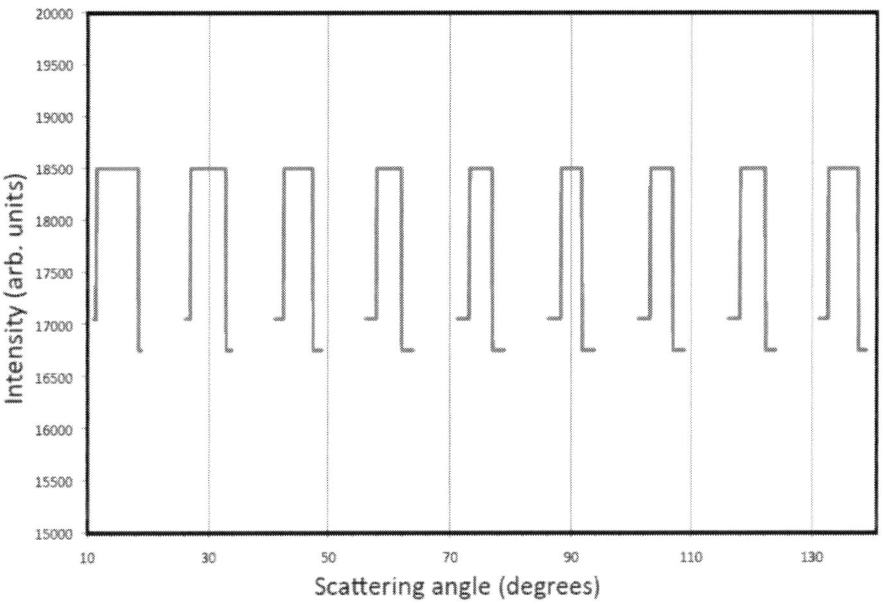

Figure 3. Calculation of the shadow effect for a given position of the detector ensemble, where we observe its angular dependence. The maximum effect is observed around 90°.

3. Comparison with D4 data

In previous section we have identified three different regions on the detector surface. A central region ($\overline{\alpha\alpha'}$) that can see the three sources, and two side regions ($\overline{\alpha' \mathrm{A}}$, $\overline{\alpha \mathrm{B}}$). For a flat diffraction pattern we should expect to observe step features on the detector with more intensity in the middle and less near the borders. In fact, taking into account the beam attenuation produced by the sample we should expect the low-angle step having more intensity than the high-angle one.

Figure 4 shows a comparison of our calculation with a real experiment performed at the two-axis neutron diffractometer D4. The experiment was performed on a vanadium sample (6.37 mm diameter), which is ideal to compare with our calculation because of its almost incoherent flat diffractogram [2]. The sample was contained in a standard orange cryostat, which here is represented by a vanadium cylinder of 50 mm of diameter surrounding the sample position (the screen in Fig. (Fig. 2)). The sample-detector and sample-window distances are 1146 mm and 280 mm, respectively. The example shown in Fig. 4 corresponds to two scattering angles near 90 degrees, where the effect is expected to be more important; the sample-window distance was left as a free parameter in the model. The results shown in Fig. 4 correspond to a final value of 770 mm.

In Fig. 4 we observe three levels of intensities for a given detector: at low-angles (I_1), at central-angles (I_2) and at high-angles (I_3). These intensities are related to the scattered intensities at the three sources (I_u, I_s and I_d, for upstream, sample and downstream, respectively) by the following expressions: $I_1 = I_\mathrm{s} + I_\mathrm{u}$, $I_2 = I_\mathrm{u} + I_\mathrm{s} + I_\mathrm{d}$, and $I_3 = I_\mathrm{s} + I_\mathrm{d}$. The three levels were fitted with constants and from them the intensities of the sources were obtained using the following relationships: $I_\mathrm{u} = I_2 - I_3$, $I_\mathrm{d} = I_2 - I_1$ and $I_\mathrm{s} = I_1 + I_3 - I_2$.

In this case, for the cylindrical vanadium sample the observed intensities around $2\theta = 90°$ (in counts per million of monitor counts) are $I_1 = 16860 \pm 30$, $I_2 = 18090 \pm 30$ and $I_3 = 16420 \pm 30$ (see Fig. 4), which yield to $I_\mathrm{s} = 15190 \pm 50$, $I_\mathrm{u} = 1670 \pm 40$ and $I_\mathrm{d} = 1230 \pm 40$. Now we are able to calculate the transmission of the sample as the ratio $T = I_d/I_u$, which gives for our vanadium cylinder $T = 0.74 \pm 0.03$. This transmission can also be calculated independently from tabulated values using the expression

$$T = \exp(-\sigma n x) \quad , \tag{7}$$

where σ is the total cross section (absorption plus scattering), n is the atomic density and x is the thickness of the sample. For vanadium the scattering and absorption (for neutrons of 0.5 Å) cross sections are 5.1 b and 1.41 b, respectively [3]; the density is 6.11 g/cm^3 [4], which represents an atomic density of 0.07223 Å^{-3}. The thickness, in the punctual approximation, is actually the diameter, *i.e.* 6.37 mm. Using these values the calculated transmission is 0.741 in a very good agreement with the measured value.

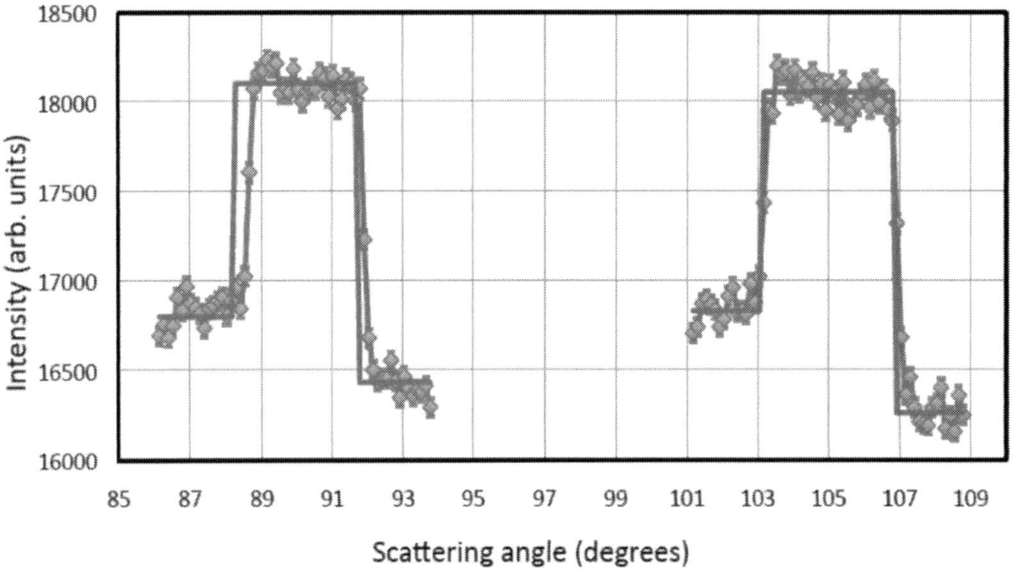

Figure 4. Comparison of experimental data (points with error bars) and the calculation.

4. Conclusions

The effect of the sample environment can be qualitatively described with a very simple geometrical model. The small differences with the experimental data likely stem from the rough assumptions made in the model, mainly the fact of considering only punctual sources. The main objective of this article is to show that this geometrical effect exists and to give an interpretation. A more detailed analysis could provide a quantitative description without changing the essence of the observed effect. It is worth noticing that this *a priori* unwanted effect opens the possibility of measuring the sample transmission as the ratio between the downstream and upstream intensities.

References

[1] Fischer H E, Cuello G J, Palleau P, Feltin D, Barnes A C, Badyal Y S and Simonson J M 2002 *Appl. Phys. A: Material Sci. Process.* **74** S160.
[2] Cuello G J, Dawidowski J and Granda J R 1997 *Physica B* **234-236** 11/4 .
[3] Sears V F 1992 *Neutron News* **3** 26.
[4] Winter M *WebElements* (Univ. of Sheffield) http://webelements.com/.

Journal of Physics: Conference Series **340** (2012) 012024 doi:10.1088/1742-6596/340/1/012024

Efficient implementation of multiple scattering Monte Carlo estimates in time-of-flight neutron spectrometry exploiting wide-area detectors

Alessio De Francesco[1], **Ubaldo Bafile**[2], **Ferdinando Formisano**[1] **and Eleonora Guarini**[3]

[1] Consiglio Nazionale delle Ricerche, Istituto Officina dei Materiali, Operative Group in Grenoble (OGG), c/o Institut Laue Langevin, 6 rue Jules Horowitz, BP 156 F-38042 Grenoble Cedex, France

[2] Consiglio Nazionale delle Ricerche, Istituto dei Sistemi Complessi, via Madonna del Piano 10, I-50019 Sesto Fiorentino, Italy

[3] Dipartimento di Fisica e Astronomia, Università di Firenze, via G. Sansone 1, I-50019 Sesto Fiorentino, Italy

E-mail: `defrance@ill.fr`

Abstract. The substantial upgrade in new-generation reactor-based time-of-flight spectrometers lies in their hugely increased detection area ensuring high neutron-collection power and remarkably good count statistics in relatively short times. Dealing with thousands of time channels and several tens of thousands of detection pixels is, however, quite punishing for data handling and correction. Real-geometry multiple scattering evaluation, even in an approximate way, is often the most demanding step in the treatment of inelastic neutron data, and becomes a very hard task in widely-extended detection geometries, as those of spectrometers like BRISP, IN4 or IN5 at the Institut Laue Langevin in Grenoble. We refreshed our approach to multiple scattering calculations, in order to obtain reasonably accurate real-geometry results in nearly real-time conditions. Our new code, originating from a long standing experience in the application of Monte Carlo (MC) integration techniques to multiple scattering calculations, is now made particularly efficient in computing time both by a careful application of the MC importance sampling method, and by the use of programming languages allowing for an efficient use of matrix algebra to avoid the far slower nested-loop logic of more traditional languages. The concepts at the basis of the algorithm and several implementation details are presented, together with the application to a real experimental test case.

1. Introduction

Multiple scattering is an unavoidable phenomenon in all situations where a radiation or particle beam is scattered upon interaction with a matter specimen. In the case of neutron diffraction or spectroscopy, multiple scattering plays a particularly important role due to weakness of the probe, which imposes the use of larger samples (volumes in the cm^3 range) and beam cross-sections (of the order of cm^2) than with other scattering techniques.

The common experimental practice looks for a compromise between the need of a sufficiently strong scattering signal making experiments feasible in a reasonable time, and the requirement to keep multiple scattering and other phenomena, such as attenuation, resolution loss, and others, at an acceptable level. It is generally considered as a reasonable compromise to make the sample scatter a fraction of the order of 10 to 20% of the incoming neutrons, which is also the order of magnitude of the expected multiple-to-single signal ratio. However, in particular cases, for instance when a high-scattering or thick-walled container is needed, it can be even larger.

Excluding the lucky and few cases when it can just be neglected, or treated simply as a flat background, multiple scattering is a typical effect that needs to be corrected for to access the sought-for structural or dynamical quantities (related to the single-scattering component only) from the experimental raw data. In no way multiple scattering contributions to the total signal can be measured independently, therefore they need to be estimated otherwise.

Much effort has been devoted in the past to multiple scattering evaluation in neutron diffraction [1, 2, 3, 4, 5]. In the case of inelastic scattering, with which we are concerned here, an essential point is that the simple knowledge of the multiple-to-single scattering ratio, in terms of just one number telling us the strength of this effect, is not at all sufficient for an appropriate correction of the experimental spectra, since the spectral distributions of multiple and single scattering signals are substantially different. This can be intuitively understood considering that, roughly speaking, the energy transferred from the neutron to the sample after p scattering events covers a range at least as broad as the p-fold convolution with itself of the one-scattering energy transfer probability distribution.

Therefore, multiple scattering evaluation in inelastic experiments is quite harder than in diffraction measurements, requiring the consideration of each energy change along the neutron zig-zag paths in the sample, ruled by the (unknown) sample-specific, anisotropic, and inelastic scattering law, and giving rise to different attenuation effects. It is then clear that a reasonable estimate of multiple inelastic scattering requires to follow, in some way, the "history" of a neutron from its arrival at the sample to its absorption into a detector, during which a number of scattering events may take place.

Two main routes of approach have been traditionally followed to tackle this problem from the computational point of view, both relying on the application of random Monte Carlo (MC) statistical methods. One, the most frequently applied [6, 7, 8, 9, 10], treats the problem as one of "neutron histories simulation" which, in a very recent development [11], becomes a part of an overall simulation of the spectrometer-plus-sample system in a virtual experiment. In this type of algorithms, a scattering event is forced to take place along a straight path of the neutron into the material system, and the new energy and direction of the emerging neutron are chosen according to a joint probability distribution function determined by the scattering law assumed as a model. Along the new path a second scattering is considered and so on, until the neutron is forced to fly to the detector carrying with itself as a label the product of all the attenuation factors along the travelled path and all the scattering strengths due to the various events. This approach is very efficient but rather heavy in terms of computer time and of programming for different sample geometries. In particular the choice of energy and vector momentum transfer at every scattering point requires a sampling from a four-dimensional probability distribution which can only be done in a numerically approximate and slow way.

Much less attention [12, 13, 14] has so far received an alternative route based on using MC techniques to calculate multiple scattering intensities as multidimensional integrals of appropriate functions. Though in principle less efficient because a certain number of samplings of the integration variables has to be discarded, this approach is characterised by a much simpler implementation and turns out to be more easily optimized as far as the statistical accuracy is concerned, thus allowing for a fast computation.

In this paper we show that this second way of facing the problem of the multiple inelastic scattering correction can be successfully applied to new-generation instruments which, with their extremely high space and time resolution and their very wide detection areas, still challenge neutronists aiming at a quick, statistically accurate, and reasonable estimate of the quoted effects in real-geometry conditions. In particular, we will consider time-of-flight (ToF) experiments, where each detection element records a full inelastic scattering spectrum.

We show that good estimates of multiple scattering can be obtained in time of the order of minutes with standard "off-the-shelf" computers. This renders an algorithm as the one described here a tool not only useful for the analysis of neutron data but also suitable to a real-time monitoring of the experiment. On ToF instruments this strongly suggests that the multiple scattering signal can be usefully evaluated as a function of flight time, so that it can be directly subtracted without converting it into a more usual energy distribution.

The intensity of inelastic scattering in a neutron experiment is, in fact, customarily described in terms of an energy distribution for a given angle θ_D defined by the position P_D of a detector element with respect to the sample position. (In this paper an energy E or the corresponding frequency $\omega = E/\hbar$, where $\hbar$ is the Planck constant, will be used equivalently.) As explained before, however, calculating scattering intensities directly on the flight time scale of the measurements is quite a natural choice for a ToF experiment. We therefore select a suitable set of time values, and for each of them we calculate the intensity due to scattering sequences where the neutron arrives at the detector at time τ having been scattered a fixed number of times. We assume a common start time for all neutrons, neglecting the neutron pulse duration.

In this paper we limit ourselves to double scattering only, since higher-order scattering sequences add to the total signal a contribution which is usually little, if not negligible at all, and increasingly structureless and similar to a flat background that can be more easily modelled and subtracted in other ways, while, on the other hand, its computation becomes more time-consuming. If necessary, however, the computation of higher-order processes is a lengthy but in principle straightforward task.

2. Single and double scattering

A set of orthogonal cartesian axes having its origin O at the centre of the sample position is used to define the geometry of the system (see figures 1 and 2), with the y-axis in the direction of the incident beam. To simplify the description, we consider here a somewhat idealized situation. The actual implementation, still in progress, takes care of further details of a realistic case. We assume an incident non-diverging neutron beam which illuminates the sample with a uniform intensity. Moreover, we neglect the presence of a container and consider only the bare sample in the form of a slab perpendicular to the beam, with its vertical axis along the z direction, and a (x, z)-plane two-dimensional position-sensitive detector facing the beam at a distance L_0 from the sample center. The distance between O and the detector element is $L_\mathrm{D} = L_0/\cos\theta_\mathrm{D}$. The flight time start $\tau = 0$ is taken here to coincide with the instant when the neutron pulse arrives at the beam cross-section plane $y = 0$, although an instrument-dependent offset may be present in the definition of the ToF frame. In the following, we will make use of the well-known relationships among energy, wave vector k, and speed v of a neutron, namely

$$\omega = \frac{\hbar k^2}{2m}, \tag{1}$$

and

$$v = \frac{\hbar k}{m}, \tag{2}$$

where m is the neutron mass.

For single-scattering calculations, we consider an event where a neutron with fixed energy ω_0 enters the sample material crossing its surface at a point A, is inelastically scattered in P at an angle θ, and travels in the direction of P_D crossing again the sample surface at point B. The path travelled inside the material is made of two line segments of lengths ℓ_{AP} and ℓ_{PB}. The vector positions of P and $P_{\rm D}$ are $\boldsymbol{r}$ and $\boldsymbol{r}_{\rm D}$ with $|\boldsymbol{r}_{\rm D}| = L_{\rm D}$. The situation is depicted in figure 1. Upon scattering, the neutron energy changes to ω_1, with an energy transfer $\omega = \omega_0 - \omega_1$.

The quantity to be calculated is the scattering intensity distribution per unit time of flight

$$\frac{dI_1}{d\tau}(\theta_{\rm D},\tau) = \Phi_0 n \int_V d\boldsymbol{r} \exp[-\mu_{\rm tot}(k_0)\ell_{AP}] \frac{d^2\sigma}{d\Omega d\omega}(\theta,\omega) \exp[-\mu_{\rm tot}(k_1)\ell_{PB}] \frac{\Delta_{\rm D}\cos\theta}{L^2}\eta(k_1)\frac{d\omega}{d\tau} \quad (3)$$

where Φ_0 is the incident flux, n is the number density of scattering units, the exponentials are the attenuation factors along the incident- and scattered-neutron path, $\mu_{\rm tot} = n(\sigma_{\rm scat} + \sigma_{\rm abs})$ is the attenuation coefficient due to both scattering and absorption cross sections, and η is the detector efficiency. Both $\mu_{\rm tot}$ and η depend, in general, on the neutron speed. ℓ_{AP} is a function of $\boldsymbol{r}$, while ℓ_{PB}, the distance L between P and $P_{\rm D}$, and the actual scattering angle θ depend on both $\boldsymbol{r}$ and $\boldsymbol{r}_{\rm D}$. Moreover, in equation (3) $\frac{d^2\sigma}{d\Omega d\omega}$ is the double-differential cross section, $\Delta_{\rm D}$ is a surface element on the detector around $P_{\rm D}$, and the integration is performed over the whole illuminated sample. The function $\omega(\tau) = \omega_0 - \omega_1(\tau)$ is obtained from (1) with

$$k_1(\tau) = \frac{mL}{\hbar\left(\tau - \dfrac{\boldsymbol{r}\cdot\hat{\boldsymbol{y}}}{v_0}\right)}, \quad (4)$$

where $\hat{\boldsymbol{y}}$ is the unit vector in the y direction. Equation (4), in turn, is derived from the imposed total flight time

$$\tau = \frac{\boldsymbol{r}\cdot\hat{\boldsymbol{y}}}{v_0} + \frac{L}{v_1}. \quad (5)$$

In (3) is also used the derivative

$$\frac{d\omega}{d\tau} = \frac{\hbar^2}{m^2 L}k_1^3(\tau). \quad (6)$$

In the case of double scattering, the neutron is inelastically scattered twice, first in P_1 at an angle θ_1 and then in P_2 at an angle θ_2. P_1 and P_2 have vector positions $\boldsymbol{r}_1$ and $\boldsymbol{r}_2$, and $\boldsymbol{r}_{12}$ is the vector distance between the two points. The path travelled inside the material is now made of three straight parts of lengths ℓ_{AP_1}, $\ell_{P_1P_2} = r_{12}$, and ℓ_{P_2B}. The energy transfers are $\omega' = \omega_0 - \omega_1$ and $\omega'' = \omega_1 - \omega_2$. The geometry of a double-scattering process is schematically represented in figure 2.

Now the specified (i.e. pertaining to a given time channel of the acquisition time frame) flight time τ and the positions of the scattering centres do not determine ω' in a unique way, as this quantity can assume any value such that the time of flight up to the second scattering point does not exceed the value prescribed for the whole path to the detector. Then ω' can vary between $-\infty$ and $\omega'_{\rm max} = \omega_0 - mr_{12}^2/\left[2\hbar(\tau - \boldsymbol{r}_1\cdot\hat{\boldsymbol{y}}/v_0)^2\right]$. The calculation has therefore to include the integration over one energy variable, in our case ω' between the given limits. However, when ω' is chosen, ω'' is fixed by the constraint on the total flight time, so that $\omega'' = \omega''(\omega',\tau)$.

The double-scattering ToF intensity can be written as

$$\frac{dI_2}{d\tau}(\theta_{\rm D},\tau) = \Phi_0 n^2 \int_{V_1} d\boldsymbol{r}_1 \int_{V_2} d\boldsymbol{r}_2 \exp[-\mu_{\rm tot}(k_0)\ell_{AP_1}] \int_{-\infty}^{\omega'_{\rm max}} d\omega' \frac{d^2\sigma}{d\Omega d\omega'}(\theta_1,\omega')$$
$$\times \frac{\exp[-\mu_{\rm tot}(k_1)r_{12}]}{r_{12}^2} \frac{d^2\sigma}{d\Omega d\omega''}(\theta_2,\omega'') \exp[-\mu_{\rm tot}(k_2)\ell_{P_2B}] \times \frac{\Delta_{\rm D}\cos\psi}{L^2}\eta(k_2)\frac{d\omega''}{d\tau} \quad (7)$$

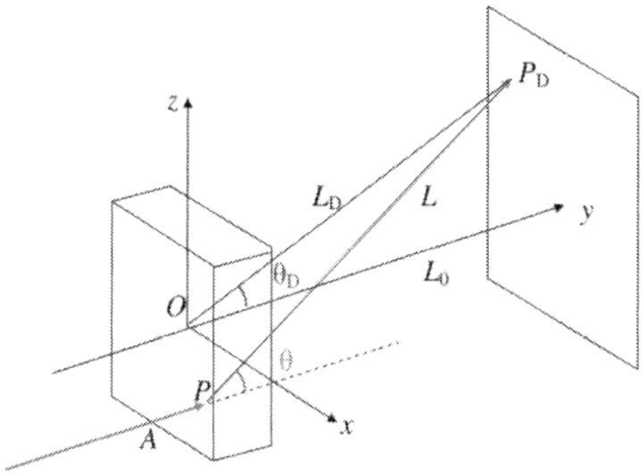

Figure 1. Schematic representation of single-scattering geometry. The position vector $\boldsymbol{r}$ and the point B, located on the exit slab face, are not drawn.

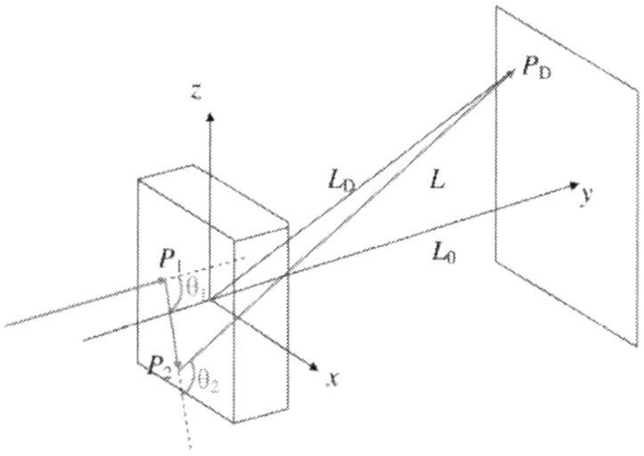

Figure 2. Schematic representation of double-scattering geometry. The position vectors $\boldsymbol{r}_1$ and $\boldsymbol{r}_2$ are not drawn. For clarity's sake, the points O, A, and B, and the angles θ_D and ψ are also not indicated.

where now $k_1 = k_1(\omega')$ and $k_2 = k_2(\omega', \tau)$. In (7) there appear three attenuation exponential factors, two double-differential cross sections, and the six-dimensional integration over the coordinates of the two scattering points. ℓ_{AP_1} is a function of $\boldsymbol{r}_1$, while r_{12} and θ_1 depend on $\boldsymbol{r}_1$ and $\boldsymbol{r}_2$. The path lengths ℓ_{P_2B} and L are functions of $\boldsymbol{r}_2$ and $\boldsymbol{r}_D$, and θ_2 depends on $\boldsymbol{r}_1$, $\boldsymbol{r}_2$, and $\boldsymbol{r}_D$. $\psi(\boldsymbol{r}_2, \boldsymbol{r}_D)$ is the angle between the directions of P_2P_D and the y axis. The integration volumes V_1 and V_2 do not coincide if the second scattering can occur in a region of the sample not directly illuminated by the incident beam but from which the detector can be reached. The calculation of (7) is performed with the help of the following relationships. For a given ω' one has

$$k_1(\omega') = \sqrt{\frac{2m}{\hbar}(\omega_0 - \omega')} \tag{8}$$

and the corresponding $v_1(\omega')$. From the total time of flight

$$\tau = \frac{\boldsymbol{r}_1 \cdot \hat{\boldsymbol{y}}}{v_0} + \frac{r_{12}}{v_1} + \frac{L}{v_2} \tag{9}$$

one has

$$k_2(\omega', \tau) = \frac{mL}{\hbar\left[\tau - \dfrac{\boldsymbol{r}_1 \cdot \hat{\boldsymbol{y}}}{v_0} - \dfrac{r_{12}}{v_1(\omega')}\right]}, \tag{10}$$

$$\omega''(\omega', \tau) = \omega_0 - \omega' - \omega_2(\omega', \tau), \tag{11}$$

and

$$\frac{d\omega''}{d\tau}(\omega', \tau) = \frac{\hbar^2}{m^2 L} k_2^3(\omega', \tau). \tag{12}$$

In our case the multiple scattering evaluation is essentially given by the calculation of (7), since, as said before, we are neglecting higher-order scattering contributions. However, the calculated $\frac{dI_2}{d\tau}(\theta_D, \tau)$ cannot be usefully applied to the correction of experimental data unless first normalized in a consistent way. This problem, unsolvable in practice, can be circumvented by calculating, with a negligible extra cost in computing time, the single-scattering intensity $\frac{dI_1}{d\tau}(\theta_D, \tau)$ and taking the double-to-single scattering intensities ratio, which is independent of any instrumental or experimental normalization, and of undetermined arbitrary quantities such as Φ_0 and Δ_D as well. In this way, to each point in the experimental ToF spectrum the correct single-scattering intensity can be assigned, resulting in an effective subtraction of double-scattering contributions.

An essential ingredient for these calculations is the scattering law expressed through the double-differential cross section. For simplicity, we discuss this issue with reference to a situation typical of a small-angle spectrometer like BRISP [15] having as one of the main fields of application the investigation of liquid-state dynamics. Assuming for simplicity a monatomic sample characterised by both coherent and incoherent scattering, with the respective bound-atom cross-sections denoted as $\sigma_{\rm coh}$ and $\sigma_{\rm incoh}$, the scattering process is ruled by

$$\frac{d^2\sigma}{d\Omega d\omega}(\theta, \omega) = \frac{k'}{4\pi k}[\sigma_{\rm coh} S(Q, \omega) + \sigma_{\rm incoh} S_{\rm s}(Q, \omega)] \tag{13}$$

where $Q = |\boldsymbol{k} - \boldsymbol{k}'|$ is the momentum transfer, in units of $\hbar$, in a generic event where the neutron wave vector changes from k to k', $S(Q, \omega)$ is the dynamic structure factor describing the spectrum of density fluctuations with wave vector Q, and $S_{\rm s}(Q, \omega)$ is its self part.

A typical Q range of interest for such experiments, relevant for the study of collective dynamics, is around $1 \lesssim Q/\text{nm}^{-1} \lesssim 10$, where $S(Q, \omega)$ is well described by model line shapes, based on hydrodynamic or viscoelastic concepts [16], which usually display a central peak plus more or less broad inelastic contributions related to the presence of excitation modes. On the other hand, $S_{\rm s}(Q, \omega)$ describes the single-particle motions reflecting some kind of diffusive-like dynamics and is usually given by a quasielastic spectral line, often modelled as a Lorentzian curve [17, 18]. These models provide suitable inputs for a scattering law to be used in the MC calculation of single scattering, at least as an initial guess to be refined later.

On the other hand, the same models do not, generally, describe correctly the multiple scattering situation. To see why, we first recall that the total, i.e. frequency-integrated, coherent scattering intensity reflects, although is not exactly proportional to, the static structure factor

$$S(Q) = \int_{-\infty}^{+\infty} d\omega S(Q, \omega) \tag{14}$$

while the frequency integral of $S_{\rm s}(Q, \omega)$ is just unity.

At a small angle θ_D, the detector collects neutron scattered once at actual angles $\theta \simeq \theta_D$, so that the low-Q dynamic structure factor is to be modelled. Multiple-scattering processes, instead, can end with the neutron travelling in a direction close to that of the incoming beam after having been scattered more than once at any angle. Indeed, the most common situation for double scattering is that a small θ_D is obtained with two large-angle, and therefore large-Q, processes, as the scattering sample usually extends much more transversally than in the beam direction, the sample thickness being typically of the order of a few mm. In figure 3 we show that in the MC evaluation of (7) most double-scattering sequences are characterised by angles distributed around the value $\pi/2$. As a consequence, a reasonable model for double-scattering calculations should be one that represents the dynamical properties at a rather large Q, in a region, typically beyond the position of the main peak of the static structure factor, where the differences between $S(Q, \omega)$ and $S_{\rm s}(Q, \omega)$ are greatly reduced and $S(Q)$ oscillates around unity.

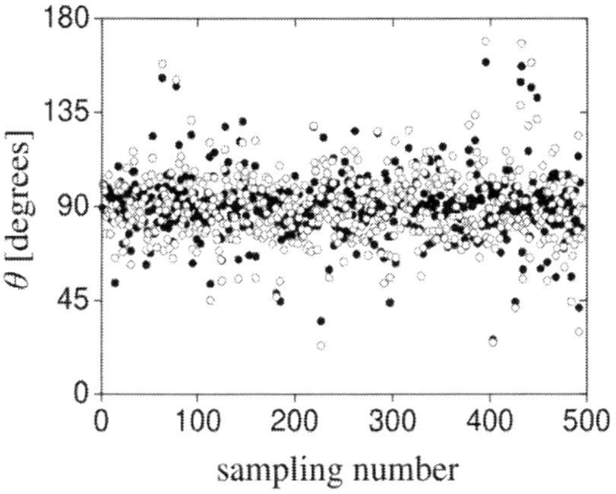

Figure 3. Angles at first (dots) and second (circles) scattering in 500 random samplings of P_1 and P_2 for the calculation of double-scattering intensity at $\theta_D = 11.9°$ with $L_0 = 4$ m.

The above considerations suggest that even in condensed samples, a reasonable, easy-to-use and very general scattering law to be applied in the evaluation of $\frac{dI_2}{d\tau}(\theta_D, \tau)$ is provided by the well-known free-particle dynamic structure factor

$$S(Q,\omega) = S_s(Q,\omega) = \frac{1}{s\sqrt{2\pi}} \exp\left[-\frac{(\omega-\omega_r)^2}{2s^2}\right] \quad (15)$$

where $s = Q\sqrt{k_B T/M}$, k_B is the Boltzmann constant, T is the temperature of the sample, M is the atomic mass, $\omega_r = \hbar Q^2/(2M)$ is the so-called recoil frequency, and $S(Q) = 1$. Equation (15) gives the correct limit form of $S(Q,\omega)$ for very large Q. In this paper we refer to calculation of $\frac{dI_2}{d\tau}(\theta_D, \tau)$ performed with a double-differential cross-section defined by (13) in the 'self approximation' $S(Q,\omega) = S_s(Q,\omega)$ defined by (15), which turns out to be very convenient for an efficient application of MC integration, as shown below.

3. Monte Carlo integration

The MC method of integration [19, 20] is based on the fact that the integral

$$I = \int_V dx\, dy \cdots dz\, f(x,y,\cdots,z) \quad (16)$$

can be estimated as

$$I_{MC}(N) = \frac{V}{N}\sum_{i=1}^{N} f(x_i, y_i, \cdots, z_i) \quad (17)$$

if a set $(x_i, y_i, \cdots, z_i)$ is randomly and uniformly sampled N times in the integration domain V of volume V. $I_{MC}(N)$ is a random variable with mean I and variance

$$\mathrm{var}\, I_{MC}(N) = \frac{V^2}{N}\mathrm{var} f = \frac{1}{N}\left[\left(V\int_V dx\, dy \cdots dz\, f^2(x,y,\cdots,z)\right) - I^2\right] \quad (18)$$

where $\mathrm{var} f = \langle f^2\rangle - \langle f\rangle^2$ is the variance of the function f in V, required to be finite.

The deviations of $I_{MC}(N)$ from the true value I are of order of the standard deviation given by the square root of (18), which implies that, with increasing N, the rate of convergence of $I_{MC}(N)$ is $1/\sqrt{N}$, a fundamental property of MC integration. This very slow convergence makes the method inefficient for integrating functions of one variable, if compared with common deterministic quadrature algorithms [19], but the power of the method appears clearly when applied to the integration in more dimensions, since the $1/\sqrt{N}$ convergence rate is independent of the number of dimensions. Therefore, MC integration is the best choice for the evaluation of quantities as those in (3) and (7) where a three- and a seven-dimensional integral needs to be computed, respectively.

Equation (18) shows that the statistical accuracy of MC integration depends on two factors, the variance of the integrand function over the integration domain and the number of samplings (i.e. of function evaluations). This has led to the development of variance-reduction techniques, in order to allow for the desired accuracy while keeping N, and therefore the computation load, at a minimum. A general approach to this problem is the so-called method of *importance sampling* (IS) based on the use of non-uniform distributions of random numbers and aiming at the application of the intuitive idea that a greater efficiency is achieved by evaluating the function f the more often in ranges where it contributes the more to the integral I. Limiting ourselves to only quoting the results, the method consists of replacing (17) with [19, 20]

$$I_{\rm MC}(N) = \frac{1}{N}\sum_{i=1}^{N}\frac{f(x_i, y_i, \cdots, z_i)}{w(x_i, y_i, \cdots, z_i)} \tag{19}$$

where w is a weight function required to be strictly positive and normalized to unity in V, and the set $(x_i, y_i, \cdots, z_i)$ is randomly sampled according to a joint probability distribution given by w. In the case of interest here, only one variable, say z, is sampled non-uniformly, and $w(x, y, \cdots, z)$ splits into the product $w_{x,y,\cdots}(x, y, \cdots)w_z(x, y, \cdots|z)$ where the first term is a constant in the volume spanned by $x, y, \cdots$ and the second rules the IS of z according to the conditioned probability distribution of z given $x, y, \cdots$. The method is most easily implemented when a generator of random numbers distributed according to w is available. The purpose of using (19) lies in the fact that a variance reduction may be achieved if w can be chosen to be similar in shape to f, so that the ratio f/w has a smaller variance than f alone, improving the convergence of the MC estimate of I. Since IS is exploited in the calculation of the double-scattering intensity, we report here for later reference the corresponding expression of the variance, given by

$$\mathrm{var}\, I_{\rm MC}(N) = \frac{1}{N}\left[\left(\int_{\rm V} dx\, dy \cdots dz\, \frac{f^2(x, y, \cdots, z)}{w(x, y, \cdots, z)}\right) - I^2\right] \tag{20}$$

which replaces the one given in (18).

For the MC calculation of the integral in (3) we use (17) where the integration variables are the coordinates of the scattering point P. For a slab sample having its sides parallel to the coordinate axes, the obvious choice is to use cartesian coordinates x, y, and z, obtained from uniform random sampling in the ranges corresponding to the volume directly exposed to the beam. From the coordinates of P and those, known, of $P_{\rm D}$ at the detector, it is immediate to derive the positions of A and B, the distances ℓ_{AP}, ℓ_{PB}, and L, and the angle θ. From equations (4)-(6), the known scattering properties of the material, and the detector efficiency, the values of the integrand function are readily calculated for each of the N iterations and finally averaged.

The MC calculation of (7) presents particular features to be dealt with, concerning the integration over both space and energy variables. As far as the spatial integration is concerned, one has to consider the presence of the term $1/r_{12}^2$ which makes the function to be integrated to have an infinite variance, rendering a direct calculation subject to wild fluctuations depending on whether the points P_1 and P_2 happen to be sampled very close to each other or not. We are in fact dealing with a function presenting a singularity in $r_{12} = 0$, though its volume integral remains finite (obviously, since otherwise the multiple scattering intensity would be infinite).

The standard approach to this problem is to look for a change of variables that removes the singularity. For example, a procedure based on this idea has been mentioned in Ref. [13].

In this case an effective solution is found passing from the set of variables $(\boldsymbol{r}_1, \boldsymbol{r}_2)$ to the set $(\boldsymbol{r}_1, \boldsymbol{r}_{12})$ where $\boldsymbol{r}_{12}$ is defined in spherical coordinates $(r_{12}, \theta_{12}, \phi_{12})$ with $d\boldsymbol{r}_{12} = dr_{12} r_{12}^2 d\theta_{12} \sin\theta_{12} d\phi_{12}$. In this way, the factor $r_{12}^2 \sin\theta_{12}$ is included into the integrand function compensating $1/r_{12}^2$ and removing the singularity and the diverging variance. It is convenient to choose the polar axis along the y direction so that θ_{12} coincides with the first scattering angle θ_1. The integral over $\boldsymbol{r}_{12}$ can be written in the form $\int d\boldsymbol{r}_{12}\, f(\boldsymbol{r}_{12}) = \int_0^{r_{12}^*} dr_{12} \int_0^{\pi} d\theta_{12} \int_0^{2\pi} d\phi_{12} [r_{12}^2 \sin\theta_{12} f(\boldsymbol{r}_{12})]$. The meaning of the upper limit r_{12}^* for the variable r_{12} is understood by observing that, once $\boldsymbol{r}_1$ is determined as in the calculation of single scattering, the sampling of $(r_{12}, \theta_{12}, \phi_{12})$ selects a random point within a sphere of radius r_{12}^* around P_1. Such a point however can be a valid choice for P_2 only if it lies inside the scattering material, otherwise it needs to be discarded.

On the other hand, if the distance between P_1 and P_2 is kept above a certain minimum the singularity does not show up and the variance of the integrand function is greatly reduced even when the choice of the two scattering points is made by a standard uniform sampling of $\boldsymbol{r}_1$ and $\boldsymbol{r}_2$ in their respective volumes.

One is then led to find an optimum strategy in the splitting of the integral in (7) into the sum of two parts: one, I_{short}, 'at short distances', is performed as described above using as variables $\boldsymbol{r}_1$ and $\boldsymbol{r}_{12}$ with spherical coordinates for the latter, and one, I_{long}, 'at long distances', is carried out with uniform sampling of $\boldsymbol{r}_1$ and $\boldsymbol{r}_2$. The first computation requires to discard the whole sampling when P_2 falls outside V_2, the second one requires discarding pairs which are at distances less than r_{12}^*. Therefore, r_{12}^* is a parameter to be optimized with the criterion of obtaining the smallest variance for the total $I_{MC} = I_{\text{short}} + I_{\text{long}}$. In fact, with a larger r_{12}^* more samplings are to be discarded in the computation of both I_{short} and I_{long}, but this loss of efficiency is more than compensated by the fact that a larger r_{12}^* increases the integration domain of a function with an intrinsically lower variance due to the compensation of the factor $1/r_{12}^2$, and also reduces the variance of the function to be integrated in I_{long}. On the other hand too large an r_{12}^* makes the loss of samplings too great to be compensated.

The integration volume V_1 is the same as for single scattering. As for V_2, in the calculation of I_{short} it is given by the volume of $[0, r_{12}^*] \times [0, \pi] \times [0, 2\pi]$, that is $V_2 = 2\pi^2 r_{12}$, while it equals the real sample volume when I_{long} is also computed with cartesian coordinates for P_2.

We next discuss the integration over the energy transfer ω' contained in (7). The integrand function, denoted here for short as $F(\omega')$, depends strongly on τ and on the actual Q values involved in the two scattering events. With a dynamic structure factor in the form (15), one can write both double-differential cross sections as functions of ω' as

$$\frac{d^2\sigma}{d\Omega d\omega'}(\omega') = \frac{k_1(\omega')}{k_0} \frac{\sigma_{\text{scat}}}{4\pi} \frac{1}{s(Q_1)\sqrt{2\pi}} \exp\left[-\frac{(\omega' - \omega_{\text{r}}(Q_1))^2}{2s^2(Q_1)} \right] \tag{21}$$

and

$$\frac{d^2\sigma}{d\Omega d\omega''}(\omega') = \frac{k_2(\omega')}{k_1(\omega')} \frac{\sigma_{\text{scat}}}{4\pi} \frac{1}{s(Q_2)\sqrt{2\pi}} \exp\left[-\frac{(\omega_0 - \omega' - \omega_2(\omega') - \omega_{\text{r}}(Q_2))^2}{2s^2(Q_2)} \right], \tag{22}$$

where the dependence of the various quantities on τ, $\boldsymbol{r}_1$, $\boldsymbol{r}_2$, and $\boldsymbol{r}_{\text{D}}$ is not indicated to simplify the notation, and $\sigma_{\text{scat}} = \sigma_{\text{coh}} + \sigma_{\text{incoh}}$. Here,

$$Q_1(\omega') = k_0^2 + k_1(\omega')^2 - 2k_0 k_1(\omega') \cos\theta_1, \tag{23}$$

$$Q_2(\omega') = k_1(\omega')^2 + k_2(\omega')^2 - 2k_1(\omega')k_2(\omega')\cos\theta_2, \tag{24}$$

and (11) has been used. Expressions (21) and (22) are bell-shaped curves centred at two, in general different, values of ω' and having different widths. Although, due to the ω' dependence of Q_1, Q_2, k_1/k_0, and k_2/k_1, they do not have a Gaussian shape, they do not differ strongly from it and so neither does their product. Adding to this that all the other ω'-depending terms in (7) are very smooth functions of ω' leads one to conclude that what is to be performed is the integration of a function $F(\omega')$ rather similar to a Gaussian one.

This situation, together with the property of the integration domain of being unbounded on one side, lends itself naturally to the application of the IS technique using a suitable Gaussian weight function $w(\boldsymbol{r}_1, \boldsymbol{r}_2|\omega')$, or $w(\omega')$ for short. This choice exploits the availability in many software programs of generators of normally-distributed random numbers.

The actual implementation of the IS integration over ω' proceeds as follows. Once the scattering points are sampled as described before, while retaining the explicit dependence on ω' in the exponentials of (21) and (22), we choose a value of ω' only for the purpose of fixing Q_1, Q_2, and the derived quantities. A convenient choice turns out to be $\omega' = 0$. Then, the product of the two exponentials, so modified into true Gaussian curves, is also a Gaussian function of ω' with peak position and width that are immediately calculated. However, for the width it is advisable to take a slightly larger value, in order to ensure that the tails of $w(\omega')$ always decay to zero more slowly than those of $F(\omega')$, preventing problems of divergence in the ratios $F(\omega')/w(\omega')$ of (19) or $F^2(\omega')/w(\omega')$ of (20). The actual sampling of ω' is then obtained according to the so-defined $w(\omega')$ by means of the normally-distributed random number generator and, together with the spatial variables already determined, allows for the calculation of $F(\omega')$ and the entire function to be integrated in (7). The rare case in which the value of ω' is larger than $\omega'_{\max}$ requires the sampling of all variables to be discarded.

The effectiveness of this IS implementation has been estimated in a simplified situation where we consider only that part of $F(\omega')$ that contributes most to its variance, namely the product $\frac{d^2\sigma}{d\Omega d\omega'}(\omega')\frac{d^2\sigma}{d\Omega d\omega''}(\omega')$, and using (18) and (20) to calculate the expected variances in the ω' integration alone, for some random choices of the other spatial variables. This has been done with the actual instrumental setup and the same sample properties of an experiment described later as a test case for the comparison with calculated scattering intensities. The typical variance reduction factor, shown in figure 4, is of the order of a few tens, which translates immediately into the same factor of computer time saving for a given required statistical accuracy.

4. Calculations and results

Turning to the real case, where the above MC integrations have to be performed efficiently for ToF instruments using thousands of detection pixels and time-channels, it is necessary to adopt programming methods ensuring an extremely fast computation of the integrands in (3) and (7) for all samplings and, ideally, all pixels and all time-channels. However, it is immediate to realize that, even by exploiting the IS method and other tools, such an ambitious project is still out of reach of medium-performance 'normal' personal computers, unless weeks or months of computing time are accepted.

To obtain real-time double scattering results, it is thus unavoidable to reduce the number of evaluations by reasonably decreasing, in the calculations, the spatial and time resolution with respect to that of the real instrument. As common practice also in the gathering of experimental data, spatial resolution is reduced by grouping 'equivalent' pixels, i.e. those pixels corresponding to the same nominal scattering angle θ_D within an appropriate step $\Delta\theta_D$. In the case of flat two-dimensional detectors as the BRISP one, this corresponds to grouping the detector pixels into a limited number of rings, each delimited by the intersections of the detector surface with

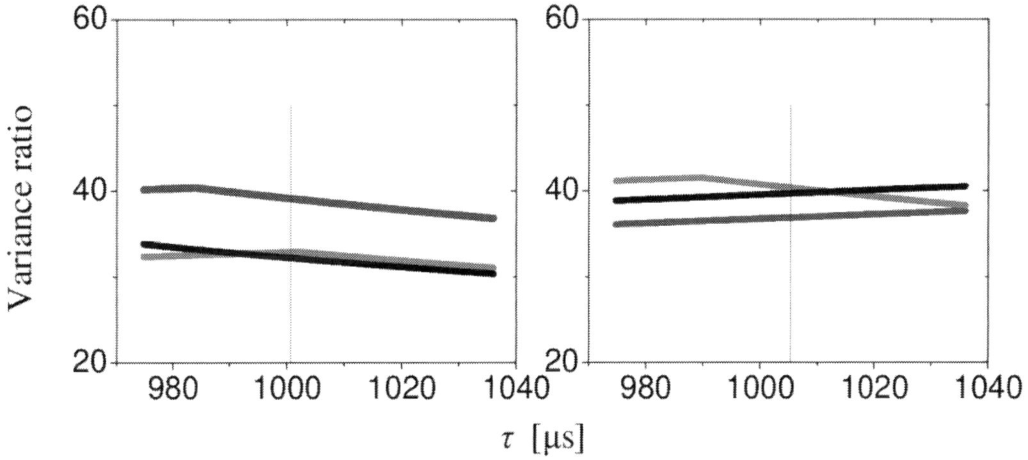

Figure 4. Ratio of the standard variance (18) to the IS variance (20) in the integration over ω' of the product $\frac{d^2\sigma}{d\Omega d\omega'}(\omega')\frac{d^2\sigma}{d\Omega d\omega''}(\omega')$, for three random samplings of the spatial coordinates performed in the 'short distance' (left) and 'long distance' (right) modes. The horizontal scale gives the flight time τ, the vertical line marks the position of the 'elastic channel', i.e. the flight time of a neutron undergoing one elastic scattering at the origin O. The data refer to a detector element at $\theta_D = 6.5°$ with $L_0 = 4$ m.

the Debye-Scherrer cones [18] of semiaperture $\theta_D - \Delta\theta_D/2$ and $\theta_D + \Delta\theta_D/2$, respectively. In this way the integrand evaluation is performed not for all available pixels, but only for few tens of scattering angle values.

Moreover a further reduction of computing time is achieved by limiting the double scattering calculation to far less and more widely spaced time channels than those set through the detector electronics in the real-instrument acquisition configuration. Such a choice is well justified by the rather flat and structureless profile typical of double scattering intensities (see below), which does not require a high time resolution.

This is not the case for single scattering evaluations, due to the much narrower and sometimes structured corresponding ToF spectrum. However, computing time is saved in this case not at the expenses of the time resolution, but by the fact that the spectrum is concentrated in a much narrower time window than that actually provided by the whole ToF frame. In any case, the evaluation of (3) for one time value is much faster than that of (7).

The above reasonable approximations of the detection geometry and time-acquisition properties of the instrument are enough to ensure the wanted performances in terms of computing time, if of course used in connection with a modern programming approach, based on row-by-column matrix products in place of nested loops. Suitable commercial software efficiently allows for large-size matrix handling and multiplication, and therefore for single and double scattering evaluations in real-time conditions.

In particular, for each nominal scattering angle θ_D, the random coordinates of P_1 and P_2, within their scattering volumes, are initially stored in $N \times 3$ matrices, and the whole MC computation, including the IS of ω', is carried out simultaneously for all n_{ch} time channels and all samplings, thanks to the possibility of calculating the needed quantities in the form of $N \times n_{ch}$ matrices, which can be properly combined in an extremely fast way. As a final output we arrive at an $N \times n_{ch}$ matrix containing the integrand of (3) or (7) for each sampling (rows of the final matrix) and for each time channel (columns of the final matrix). To obtain the MC integral at

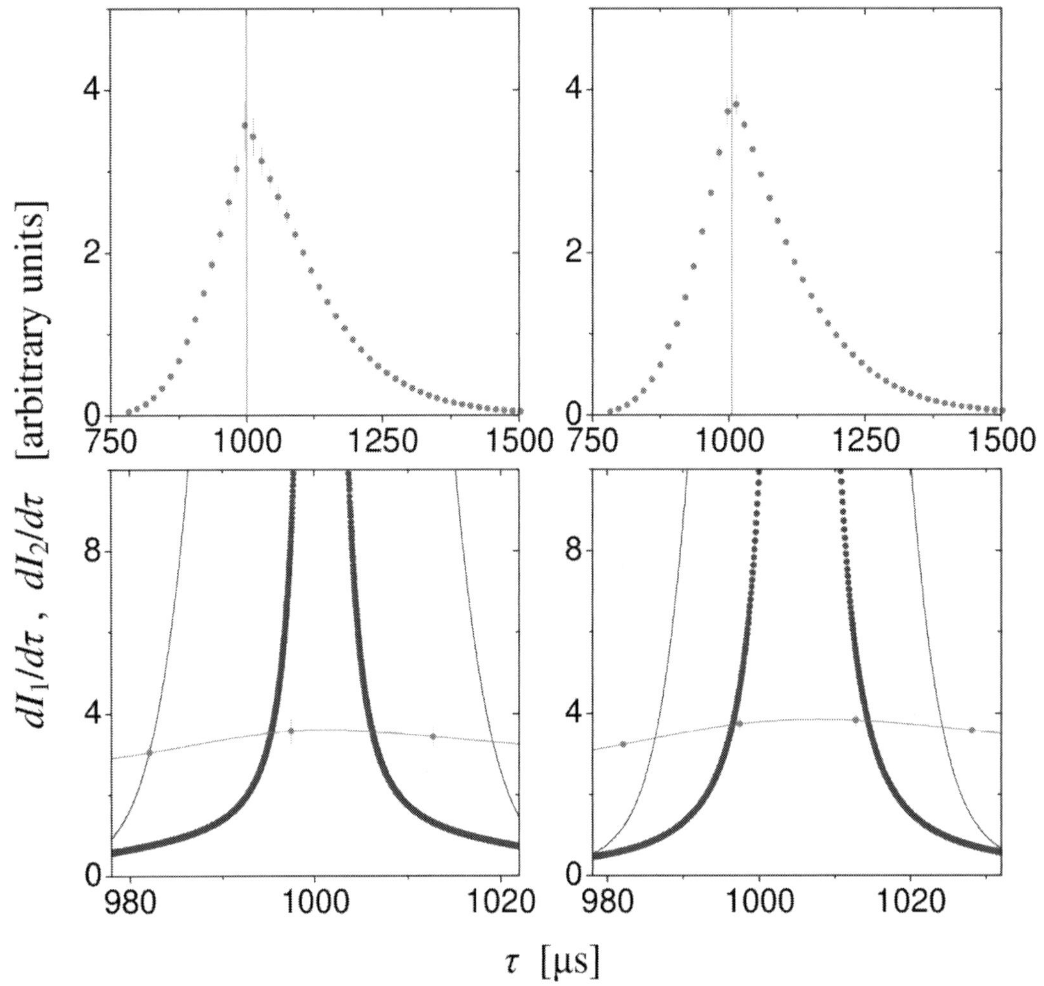

Figure 5. Calculated ToF spectra. The double-scattering intensity (red dots with error bars and red line interpolated through the calculated points) is shown in the top frames and compared with single scattering in the bottom ones. The blue dots and the black line show the tails of single scattering spectrum before and after the convolution with the experimental resolution. The data refer to $\theta_D = 3.3°$ (left) and 6.5° (right). The vertical lines mark the positions of the 'elastic channel' as in figure 4.

a given angle θ_D as a function of time one needs only to sum up the elements of each column and properly normalize to the MC volumes. To obtain the statistical spread in the calculation, we repeat the MC cycle several times (at each scattering angle), and finally take the average and the spread over the performed cycles.

As an example of the application of this computer program to the correction of multiple scattering effects, we refer to measurements recently performed [21] on BRISP with an incident neutron wavelength of 0.988 Å and $L_0 = 4$ m. The sample was liquid D_2S at a temperature of 205 K in a slab-shaped container, with a sample thickness of 6 mm. The calculation of single scattering has been carried out assuming a scattering law of the type shown in (13), with $S(Q, \omega)$ given by the Rayleigh-Brillouin three-line spectrum [16] and the simple Lorentzian line

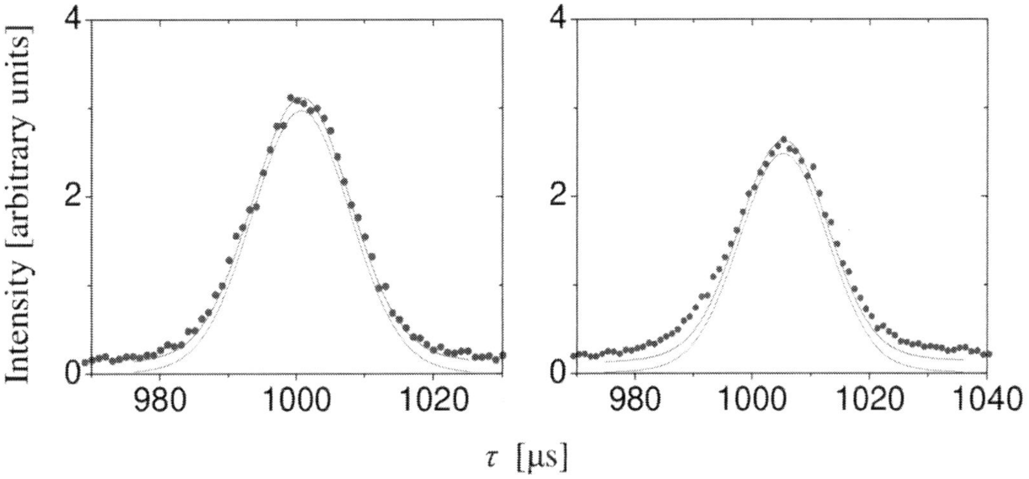

Figure 6. BRISP spectra of D_2S (dots) at the same angles as in figure 5. The lower black line is the single scattering intensity after convolution with the instrumental resolution. The upper red line also includes the double scattering contribution. The calculations have been arbitrarily scaled to allow for a comparison of the spectral shapes.

of the Fick's law [17] describing the self diffusion dynamics. Both models are derived within hydrodynamics concepts. However, the presence of incoherent scattering cross-sections and the fact that $S(Q)$ is much smaller than unity due to the low compressibility of the liquid [17] make incoherent scattering dominate.

As already said, double scattering was calculated with the free-particle $S(Q,\omega)$ of (15). The calculations were obtained with rather small values of N. The double scattering was evaluated at 128 time channels with 500 samplings for I_{long} and only 70 for I_{short}, where $r^*_{12} = 3$ mm, i.e. half the sample thickness. The single scattering was computed at 800 time channels with $N = 500$. At each scattering angle, 5 and 3 cycles were performed for double and single scattering, respectively.

We present preliminary results at two scattering angles, namely $\theta_D = 3.3°$ and $6.5°$. In figure 5 we show the intensity of double scattering, which appears to be distributed over a very large time range. In fact, in the range where the experimental spectrum is appreciably different from zero, the double scattering is rather flat and low, as seen in the bottom frames of the figure. In order to allow for a meaningful comparison with the experiment, the calculated single scattering intensity must be convoluted with the experimental resolution, which, on the other hand, has a negligible effect on the double scattering spectrum. The comparison is displayed in figure 6, where the lower line gives $\frac{dI_1}{d\tau}$ and the upper line is the sum of $\frac{dI_1}{d\tau}$ and $\frac{dI_2}{d\tau}$. Both calculated lines are arbitrarily but consistently scaled to be plotted together with the measured spectra, which are, in turn, not yet normalized in absolute units. The experimental data shown here are obtained after subtraction of the container scattering, where the attenuation due to the presence of the sample has been properly accounted for.

From figure 6 it appears that the calculated total spectra are in rather good agreement with the experiment. However, we remind that it is not the purpose of this algorithm to simulate the correct scattering law. Instead, the use of very simple, though reasonably realistic, models for $\frac{d^2\sigma}{d\Omega d\omega}$ is aimed at an estimate of the multiple-to-single scattering ratio which is believed to be

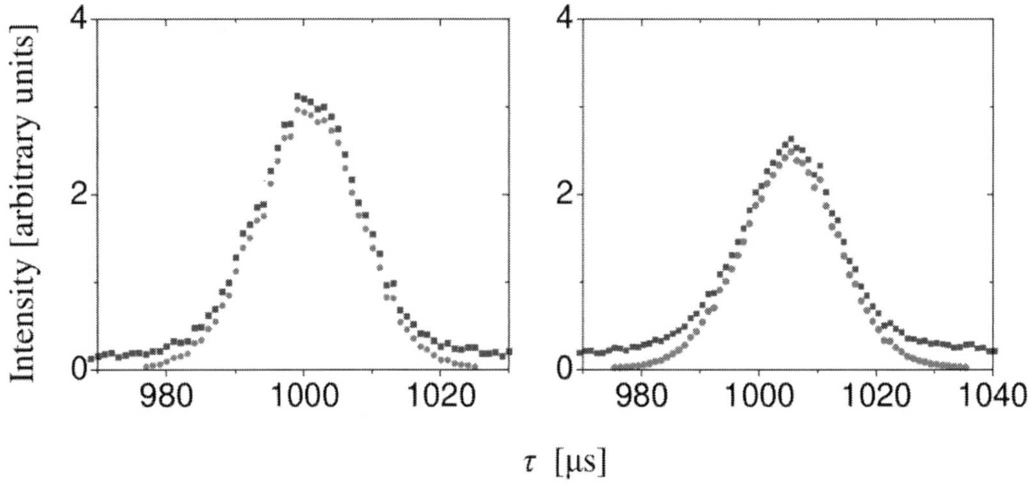

Figure 7. BRISP spectra of D_2S (blue dots) at the same angles as in figure 5, together with the results of the multiple scattering subtraction (red dots) performed as described in the text.

a good approximation to the real one, only weakly dependent on the adopted $S(Q, \omega)$ model. Indeed, the agreement between the calculation and the measurement is better at the lower angle, according to the fact that hydrodynamic models are known to be valid only at very low Q.

The subtraction of multiple scattering from the experimental signal proceeds then from the assumption that

$$\left(\frac{dI_1/d\tau}{dI_1/d\tau + dI_2/d\tau}\right)_{\rm exp} \simeq \left(\frac{dI_1/d\tau}{dI_1/d\tau + dI_2/d\tau}\right)_{\rm calc}, \tag{25}$$

so that, to a good approximation, the single-scattering experimental spectrum is obtained from the measured one by multiplication for the right-hand-side quantity, which is close to one in the peak region and decreases to zero in the tails. The application of this procedure leads to the results shown in figure 7.

We conclude by noting that, once the feasibility of such computations is demonstrated, further improvements can be easily envisaged in order to bring a program of this kind to the level of a reliable real-time tool for both planning and checking experimental work in the field of inelastic neutron scattering, besides of course helping in the analysis of data already collected. In this direction, foreseeable steps are, for example, the extension to the more realistic case of a scattering system inside a container, and the availability of a variety of models for the scattering law of different types of samples, starting from the schematization of molecular systems based on effective-mass approaches [22].

Acknowledgments

Useful discussions with R. Mancinelli on the application of the IS method are gratefully acknowledged.

References

[1] Vineyard G H 1954 *Phys. Rev.* **96** 93
[2] Sears V F 1975 *Adv. Phys.* **24** 1
[3] Blech I A and Averbach B L 1965 *Phys. Rev.* **137** A1113
[4] Soper A K and Egelstaff P A 1980 *Nucl. Instrum. and Meth.* **178** 415

[5] Bausenwein T, Bertagnolli H, Tödheide K and Chieux P 1991 *Nucl. Instrum. and Meth.* B**61** 527
[6] Copley J R D 1974 *Comput. Phys. Commun.* **7** 289
[7] Copley J R D, Verkerk P, van Well A A and Fredrikze H 1986 *Comput. Phys. Commun.* **40** 337
[8] Mayers J and Cywinski R 1985 *Nucl. Instrum. Meth. Phys. Res.* A **241** 519
[9] Dawidowski J, Cuello G J and Granada J R 1993 *Nucl. Instrum. Meth. Phys. Res.* B **82** 459
[10] Mayers J, Fielding A L and Senesi R 2002 *Nucl. Instrum. Meth. Phys. Res.* A **481** 454
[11] Farhi E, Hugouvieux V, Johnson M R and Kob W 2009 *J. Comput. Phys.* **228** 5251
[12] Johnson E and Robinson L 1989 *Rev. Sci. Instrum.* **60** 3447
[13] Petrillo C and Sacchetti F 1992 *Acta Cryst.* A**48** 508
[14] Guarini E, Orecchini A, Formisano F, Demmel F, Petrillo C, Sacchetti F, Bafile U and Barocchi F 2005 *J. Phys.: Condens. Matter* **17** 7895
[15] Aisa D *et al* 2005 *Nucl. Instrum. Meth. Phys. Res.* A**544** 620
[16] Bafile U, Guarini E and Barocchi F 2006 *Phys. Rev.* E **73** 061203
[17] Balucani U and Zoppi M 1994 *Dynamics of the Liquid State* (Oxford: Clarendon)
[18] Furrer A, Mesot J and Strässle T 2009 *Neutron Scattering in Condensed Matter Physics* (Singapore: World Scientific Publishing Company)
[19] James F 1980 *Rep. Prog. Phys.* **43** 1145
[20] Frenkel D and Smit B 1996 *Understanding Molecular Simulation from Algorithms to Applications* (San Diego: Academic Press)
[21] Sacchetti F et al, unpublished
[22] Krieger T J and Nelkin M S 1957 *Phys. Rev.* **106** 290

Journal of Physics: Conference Series **340** (2012) 012025 doi:10.1088/1742-6596/340/1/012025

Residual stress determination in thick welded steel plates

Vyacheslav Em[1], Wanchuck Woo[1], Baek-Seok Seong[1], Pavol Mikula[2], Jongdae Joo[1], Mi-Hyun Kang[1], Kyu Hong Lee[1]

[1]Neutron Science Division, Korea atomic Energy Research Institute, 1045 Daedeok-daero, Daejeon 305-353, Republic of Korea
[2]Nuclear Physics Institute ASCR, v.v.i., and Research center Rez Ltd, Rez, Czech Republic

E-mail: vtem@kaeri.re.kr

Abstract. Through thickness strain distribution of 50 mm thick welded ferritic steel (bcc) plate was studied using diffraction reflections 211 and 110 and neutron wavelengths of 1.55 and 2.39Å, respectively. Experimental results showed that for stress measurements in a possibly maximum thick weld, the different strain components should be measured with different reflections 211 and 110. The strains measured with these reflections for the same component are close. Since planes (211) and (110) of bcc ferrite have the same diffraction elastic constants the appropriate values of stresses could be derived from strains measured with reflections 211 and 110.

1. Introduction

Neutron diffraction method is widely used for strain measurements deep inside engineering structures and components [1]. Limitation of the path length is known to be about 60 mm in iron [2, 3], which corresponds to the plate thickness ~ 40mm. However, the ship building industry demands stress measurement in welded ferritic steel plates with thickness up to 80 mm now, and even more in future. The maximum available path length can be increased by using larger sampling (gauge) volume and/or longer measurement time. However, maximum size of the gauge volume is limited by the required spatial resolution. Increasing measurement time is not effective because intensity related to the diffracted peak exponentially decreases with neutrons path length through the material [1].

The more effective way for increasing available path length has been suggested and tested in ref. 4 based on the selection of the neutron wavelength corresponding to the local minimum of neutron total cross-section near the Bragg edges. The results showed that the available depth in steels can be remarkably increased by the use of focusing bent perfect crystal Si monochromators [5-8] and as well as selection of the proper neutron wavelength λ and *hkl* reflection for the measurements of the particular strain component. In ferritic steel (α-Fe, bcc structure) with a low texture, an instrument configuration using wavelength $\lambda = 1.55$Å and reflection 211 ($2\theta_S = 82.5^{\circ}$) is preferable for measuring normal (ND) strain components (reflection geometry, Figure 1c), while another configuration using wavelength $\lambda = 2.39$ Å and reflection 110 ($2\theta_S = 72.4^{0}$) is advantageous for measuring transverse (TD) and longitudinal (LD) strain components (transmission geometry, Figure 1d). Thus, it is necessary to measure different strain components with different reflections 110 and 211 for inferring stresses at the maximum possible depth in low textured ferritic steels.

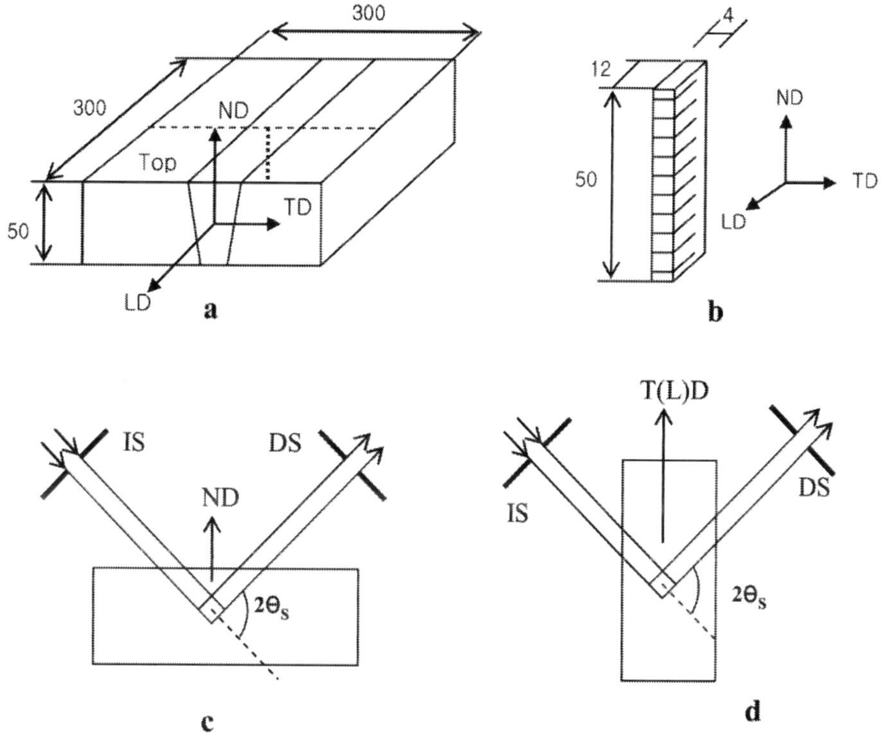

Figure 1. a) Longitudinal (LD), transverse (TD), and normal (ND) directions in the welded plate; b) Schematic of the stress-free comb taken along the LD direction; c) Orientation of the plate for the measurement of the ND component in reflection geometry; d) Orientation of the plate for measurement of TD and LD components in transmission geometry.

In general, the calculation of stresses from strains measured with different *hkl* reflections can cause a misleading because the diffraction elastic constants and inter-granular strains are *hkl*-dependent. However, planes (211) and (110) of α-Fe (bcc) are known to have the same diffraction elastic constants calculated by the Kroner model: $E_{211,110} = 225.5$ GPa, $\nu_{211,110} = 0.28$ [1]. Furthermore, planes (211) and (110) are less sensitive to the inter-granular strains and evident inter-granular effects in ferritic steel weld could not be observed [9]. Therefore, one can expect that appropriate stresses can be derived from the strains measured with reflections 211 and 110.

In thick weld joints the microstructure and texture between the weld and base metal are different. A large grain size and strong texture are often formed in the weld [1]. The strong texture can considerably change intensity of the diffraction peak in a certain sample orientation and make a difficulty to measure the strain component along the particular principal direction. Moreover, texture differently affects intensity of different *hkl* reflections as well.

The instrument configuration with the wavelength $\lambda = 2.39$ Å and reflection 110 enabled us to measure all principal strain components in a 50 mm thick welded ferritic steel plate as reported in ref. 10. The TD component was measured much faster (~1 h) than ND component (~3 h) in the mid-thickness of the plate. Because of the strong texture in the weld a notable difference in the intensity of the 110 diffraction peak from the weld and base metal was observed for the same path length. In this work we performed the through-thickness scan of strains using the instrument configuration for the wavelength $\lambda = 1.55$ Å and 211 reflection in the weld of the same 50 mm thick welded ferritic steel plate. In low textured sample this configuration enables measurements of all principal strain components in 50 mm thick plate but it is particularly advantageous for the ND component [4].

The aim of this paper is to develop a guideline for the stress measurements in possibly maximum thick welded ferritic steel plate by comparing two instrument configurations, i.e., 110 and 211 reflections, for the measurements of the different strain components. Thus, the results will compare strain/stress components measured with both 110 and 211 reflections in the weld region.

2. Experiment

The two plates (600 mm long x 150 mm wide x 50 mm thick) of high-strength low-carbon steel plates with grain size of about 20 μm have been joined vertically using 1-pass 2-pole electro-gas welding technique. The details of base material and welding process are reported in [11]. The 600 mm long welded plate was cut across the welding direction and two pieces of dimension 300 mm x 300 mm x 50 mm were prepared. One piece (Plate 1, Fig. 1a) was used for the residual strain scanning and another one (Plate 2) was used for preparing reference "stress free" sample.

For getting the stress free reference spacing (d_0) a 4 mm thick through-thickness slice with dimensions 12 (LD) x 4 (TD) x 50 (ND) mm^3 was machined out at the middle of the weld length from the Plate 2 by electro-discharge machining. A comb-like sample (Fig. 1b) was carefully machined from this slice with 4.7 mm wide "teeth" along the depth (ND) at the same locations where strains had been measured. The long (10 mm) axis of the "teeth" is parallel to the welding direction (LD) because no steep variation of d_0 is expected in this direction [1]. It is supposed that "teeth" of the comb are free from the stress field.

The experiments were carried out by using the Residual Stress Instrument at HANARO reactor of Korea Atomic Energy Research Institute [8]. The bent perfect crystal Si(111) and Si(220) monochromators at take-off angles $2\theta_M = 45^\circ$ and 48° were used to provide neutrons with wavelength $\lambda = 2.39$ Å and 1.55Å for measurement of diffraction peak (110) and (211) at scattering angle $2\theta_S = 72.4^\circ$, 82.5°, respectively. The section of the gauge volume in the diffraction plane was defined by 2 mm wide cadmium slits on incident and diffracted beams as shown in Figs. 1 b and 1 c. The height of the gauge volume (GV) was defined by the height of the incident beam as well. It was 20 mm for measurement of the ND or TD components (GV ~ 2 x 2 x 20 mm^3) and 8 mm for the LD component (GV ~ 2 x 2 x 8 mm^3). Through-thickness scan (at 9 depths from 5 mm from the top surfaces with 5 mm step) of diffraction peaks was conducted in the Specimen 1 at the middle of the weld centerline.

For the measurement of the comb the height of the incident slit was reduced to 2mm (GV ~ 2 x 2 x 2 mm^3). For each component, in order to increase the number of reflecting grains, the comb was oscillated in the interval (±3°) along the diffractometer axes with 1° step and measured in seven rotations. The obtained seven diffraction peaks were summed for the peak analysis.

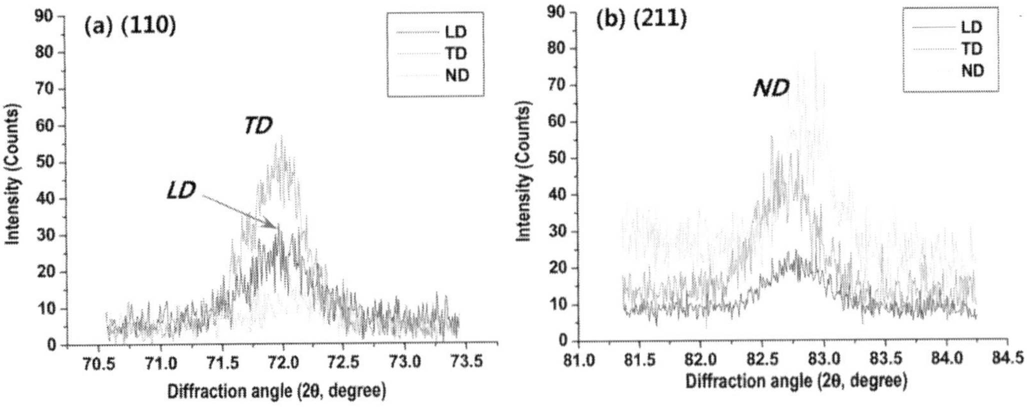

Figure 2. Diffraction peak measured at the mid-thickness (25 mm) and weld centerline by using a) (110) reflection and; b) (211) reflection in 50 mm thick ferritic steel weld plate.

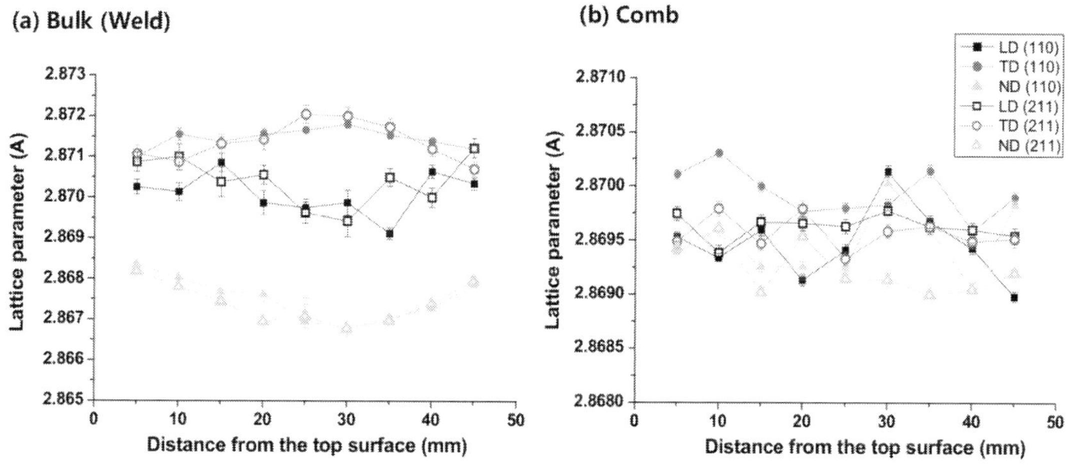

Figure 3. Through thickness variation of the lattice parameter measured a) in bulk weld sample and b) in "stress free" comb.

3. Results and discussion

Figure 2 shows the diffraction peaks of the two configurations of the reflections, i.e., (110) and (211). Note that it was measured at the mid-thickness of the weld (25 mm depth) for different orientations of the plate. The result shows that the intensity of the 110 diffraction peak in the LD and TD sample orientations is higher than that of the ND orientation (see Fig. 2a). Meanwhile, the intensity of the 211 peak is higher for ND sample orientation than those of the LD and TD orientations (see Fig. 2b). Thus, for the investigation of the LD and TD strain components in a welded plate thicker than 50 mm using the 110 reflection is advantageous and on the other hand for ND components it is more suitable using 211 reflection.

The lattice spacing (d_{hkl}) for different reflections can be compared on a common scale by comparing corresponding lattice parameters (a_{hkl}):

$$a_{hkl} = \sqrt{(h^2 + k^2 + l^2)}d_{hkl} \ .$$

The macro-strain in particular sample direction, corresponding to the macroscopic stresses field is given by

$$\varepsilon_{hkl} = \frac{d_{hkl} - d^0_{hkl}}{d^0_{hkl}} = \frac{a_{hkl} - a^0_{hkl}}{a^0_{hkl}} \ ,$$

where d^0_{hkl} / a^0_{hkl} is lattice spacing/parameter of reference "stress free" comb for the particular direction. It is assumed that each "tooth" with dimensions 4 x 4.7 x 10 mm^3 does not hold macro-strains associated with the elastic macro-stresses. It may have micro-strains and effects of chemistry but they should be the same as in the bulk plate and will cancel out in the macro-strain calculation.

Figure 3a shows the variation of lattice parameter measured with reflections 110 (a_{110}) and 211 (a_{211}) as a function of depth from the top surface in the weld. It shows that the lattice parameters measured for ND and TD components smoothly change with the depth and discrepancy between parameters measured with 211 and 110 reflections in the most points is within the experimental error. For LD component the discrepancy is larger and there is a notable scatter in the measurement values. The micrographs showed that the average grain size in the weld is about 180 μm and this may be the source of the scatter for the LD component, which was measured with the smaller gauge volume. However, if consider the insignificant dependence of the lattice parameter with the depth the discrepancy between the smooth lines is smaller (~0.0005 Å) which corresponds to strain error $\pm10^{-4}$ and it is acceptable for the most stress measurements. Close values of the lattice parameters measured with reflections 110 and 211 suggested that most probably the macro-strains and micro-strains (if latter exist) measured with 211 and 110 reflections are close. Therefore, appropriate stresses could be derived from the strain components measured with different reflections 211 and 110, if corresponding values of the reference lattice spacing $d^0_{211/110}$ have been measured.

Figure 3b shows the variation of the lattice parameters with depth in the comb measured with reflections 211 (a^0_{211}) and 110 (a^0_{110}). There is a fair amount of scatter in the measurements, much larger than experimental error. Relatively large grain size, possibly remained macro-stresses in "teeth" [10], and inter-granular micro-stresses can be the reason of the discrepancy (~0.0005Å) between the lattice parameters for the different directions and reflections. Further experiments with "stress free" samples are necessary.

Figure 4 compares the stresses derived from the strain components measured with i) one reflection 110 and ii) two reflections 110 (TD, LD) and 211 (ND). It shows that there is a good agreement between stresses measured with one and two reflections in most points. Notable (~50 MPa) difference in several points is mainly caused by difference between the two reflections in reference lattice spacing/parameter d_0/a_0 measured in the normal component. Obviously the agreement would be better if the ND component is also measured with the 110 reflection up to available depth and only in few points when the gauge volume is deep inside the sample near the mid-thickness should be measured with the reflection 211. Thus, for stress measurements in 50 mm thick (and more) welded ferritic steel plates the following procedure can be recommended: measurement of all TD and LD components along all depths, and ND component up to about 20 mm depth by using wavelength λ =2.39 Å and reflection 110; measurement of ND component in the rest deeper locations by using the wavelength λ =1.55Å and reflection 211.

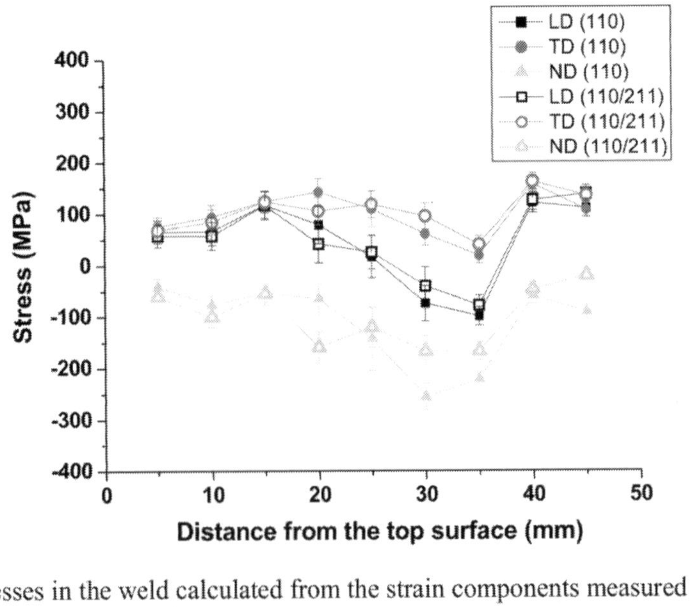

Figure 4. The stresses in the weld calculated from the strain components measured with one reflection (110) and two reflections (110/211).

Through-thickness variations of residual stresses along the weld centerline can be discussed. Figure 4 shows that the three profiles of components similarly fluctuate with a sine-wave like distribution. It can be observed in the previous results of residual stress measurements in thick components [12]. Compared to the previous results it shows a similar distribution. For example, the minimum longitudinal residual stress (σ_{LD}) located at the depth of 0.5~0.7, while the higher tensile σ_{LD} (up to 180 MPa) were observed close to the surfaces at the depth of 0.2~0.3 of the thickness from both sides. Since such a tensile σ_x can often initialize the transverse cracks in thick welds the location of the maximum tensile residual stress can be critical to the integrity of the final weld components.

4. Conclusions

The presented experiments showed that available thickness of welded ferritic steel plate for stress measurement can be increased by using two instrument configurations with the neutron wavelengths 2.39 and 1.55Å and corresponding reflections 110 and 211, respectively. Then, appropriate residual stresses can be obtained from the transformation of these strain components by using appropriate diffraction elastic constants.

Acknowledgements

This research was carried out under the Nuclear Research and development program of the Ministry of Education, Science and Technology of the Republic of Korea. In the Czech Republic it is supported by the projects GACR P204/10/0654 and ASCR-M100480901

References

[1] Hutchings M T, Withers P J, Holden T M and Lorentzen T 2005 *Introduction to the Characterization of Residual Stress by Neutron Diffractiont* (Boca Raton, FL: CRS Press)
[2] Withers P J 2004 *J. Appl. Cryst.* **37** 607 -612
[3] Santistaban J R, Daymond M R, James J A and Edvards L 2006 *J. Appl. Cryst.* **39** 812-825
[4] Woo W, Em V, Seong B, Shin E, Mikula P, Joo J and Kang M 2011 *J. Appl. Cryst.* **44** 747-754
[5] Mikula P, Kulda J, Lukas P, Vrana M and Wagner V 1994 *Nucl. Instrum. Methods Phys. Res.* A **338**, 18-26
[6] Popovici M, Yelon W B, Berliner R, Stoica A D, Ionita I and Law R 1994 *Nucl.Instrum. Methods Phys. Res.* A **338**, 99-110
[7] Moon M K, Em V, Lee C H, Mikula P, Hong K P, Choi Y N, Cheon J K, Nam U W, Kong K N and Jin K C 2005 *Physica* B **368** 70-75
[8] Seong B S, Em V, Mikula P, Saroun J and Kang M, *J. Appl. Cryst.* **43** 654-658
[9] Holden T M, Suzuki H, Carr D G, Ripley M I and Clausen B 2006 *Mater. Sci. Eng.* A **185** 33-37
[10] Woo W, Em V, Mikula P, An G B and Seong B 2011 *Mater. Sci. Eng.* A **528** 4120-4124
[11] Ganguly S, Edwards L and Fitzpatrick M E 2011 *Mater. Sci. Eng.* A **528** 1226-1232
[12] Smith D J, Bouchard P J, George D, 2000 *J. Strain Analysis* 2000 **35** 287-305.

Journal of Physics: Conference Series **340** (2012) 012026 doi:10.1088/1742-6596/340/1/012026

New Trends in Thermal Neutron Spectrometry Getting High-Resolution Performances in Thermal Neutrons Spectrometry by Using Inverse Space Focusing

I. Ionita, C. Paunoiu, V. Florescu
Institute for Nuclear Research Pitesti
ICN Pitesti, Campului Street No. 1, Pitesti – Mioveni, Arges, Romania

E-mail: ionionita@lycos.com, cpaunoiu@email.com, valeriu.florescu@nuclear.ro

Abstract. The characteristics of a Q-space focusing configuration are given. The procedure to get inverse space conditions is given also. Both the diffraction and inelastic scattering are considered. For the neutron diffractometry both instruments with one crystal monochromator and with two crystals monochromators are taken into consideration. The time of flight diffractometers both for a pulsed neutron source and a steady state one is considered also. For inelastic scattering the inverse space focusing conditions for triple axis spectrometers are givene line-width.

Introduction

In neutron spectrometry the optimal use of the available neutrons is of the greatest importance. The general philosophy to do it is to maximize the neutron flux at sample by using neutron guides, supermirrors [1], or spatial focusing effects involving flexible configurations and curved crystals. A different approach [2], [3], is to obtain the required optimum experimental conditions not by getting focused beams at sample or anywhere else, but only by decreasing as much as possible the scan variable variances.

A Q-space focussing configuration is characterized by high-resolution even when no spatial focussing exist at sample position or anywhere else or is used quite open beam without any Soller collimators; only coarse collimators should be used to reduce the background level.

This is possible just by decreasing the scan variable variances.

The following steps should be followed:

- to define the scan variable
- to define the significant spatial variables as are the monochromator and sample width for example
- to express the scan variable, using specific geometry characteristics and the existing correlation between variables, through the spatial variables
- to cancel the significant contributions to the scans variable variances by cancelling the important variables coefficients, from the scan variable expression; *when correlation between variable exists*, as is the case for these kind of experimental configurations, such a coefficient has more then one term, not all of the same sign, appearing the possibility to be cancelled and therefore to cancel entirely the corresponding contribution of this spatial variable to the line-width.

Examples for neutron diffractometers and triple axis spectrometer are given bellow.

Neutron Diffractometers

2.1. The General Method

In the neutron diffractometry the scan variable is X_1, the X vector component along the Q_0 direction, where $X=Q-Q_0$ and $Q = k_f-k_i$; k_f, k_i are the wave vectors for the incoming respectively the diffracted neutrons and "0" refers to the corresponding most probable value. The scan variable is given by:

$$X_1 = 2 \cdot \Delta k \cdot \sin\theta_s + k \cdot (\gamma_f - \gamma_i) \tag{1}$$

where γ_i, γ_f are the horizontal divergence angles from the most probable trajectories, Δk is the wave vector spread, $2\theta_s$ is the scattering angle; the indices i,f refer to the incoming respectively the scattered neutrons. The variables defining the neutron trajectories in phase space are all the spatial relevant coordinates to which has to be added, for the time-of-flight diffractometry, the temporal variables defining its emission respectively detection. All the spatial and temporal variables are measured from the corresponding most probable values. The relation defining the temporal variables is:

$$\frac{h}{2\pi m} \cdot (t_f - t_0) = \frac{\Delta L}{k} - \frac{L}{k} \cdot \frac{\Delta k}{k} \tag{2}$$

where L is the neutron flight path from source (from chopper centre for a steady state source) to detector, h is the Plank constant and m is the neutron mass.
For the crystal diffractometry the Bragg constraints have to be considered:

$$\left.\begin{aligned} \gamma_i - \gamma_f &= 2 \cdot \frac{l_m}{R_m} \cdot sign(\theta_m + \chi_m) \\ \frac{\Delta k}{k} &= \cot\theta_m \cdot \frac{\gamma_i - \gamma_f}{2} \end{aligned}\right\} \tag{3}$$

The relation (2) and (3) defines the Δk in (1) for the time-of-flight respectively the crystal diffractometry. To these relations must be added the relations giving the horizontal angular variables expressed through the spatial variables l_0, l_s, l_d, l_m the source, sample, detector and monochromator lengths; R_m is the crystal radius of curvature. The general procedure is to express the scan variable X_1 through spatial and, for time-of-flight diffractometry, temporal variables and to cancel the major contributions, i.e. the l_0, l_s, l_d, l_m coefficients.

2.2. The Crystal Diffractometry

2.2.1. One Crystal Monochromator

For this geometry (see Figure 1), following the above-described procedure and cancelling the l_m coefficient, one obtains:

$$R_m = \frac{2a}{2a-1} \cdot \frac{L \cdot sign(\theta_m + \chi_m)}{\sin(\theta_m - \chi_m)} \tag{4}$$

where

$$a = -\tan\theta_s / \tan\theta_m$$

$$\text{sign}\alpha = \text{abs}\alpha / \alpha$$

Cancelling the "l_s" coefficient one obtains:

$$\left.\begin{aligned} \tan\chi_s &= \cot\theta_s\cdot[1-\frac{2}{1+(2a-1)\cdot\frac{L_2}{L_1}}] \\ \cot\alpha_s &= \frac{\cos(2\theta_s)-(2a-1)\cdot\frac{L_2}{L_1}}{\sin(2\theta_s)} \\ \alpha_s &= \chi_s+\theta_s+\pi/2 \end{aligned}\right\} \qquad (5)$$

where α_s as the inclination angle measured from the monochromatic beam direction.

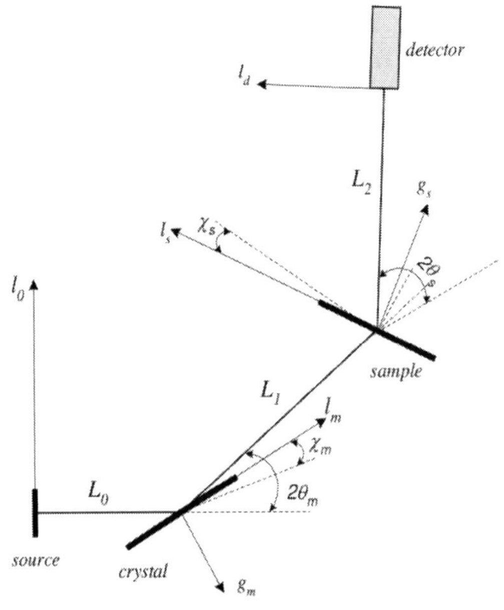

Figure 1

The (5) condition is fulfilled if the sample is rotated during the experimental diffraction pattern raising; for each value of the scattering angle the sample has to be positioned accordingly to the (5) relation. In this case the contribution to the line-width of the sample width is cancelled. But, as in this situation, the sample effective volume where the neutrons are scattered, is different for each value of the scattering angle, the normal Rietveld experimental data processing procedure should be modified accordingly.

The (4) relation is the condition to cancel the monochromator width contribution to the line-width. But as the crystal reflectivity is strongly dependent on the radius of curvature, is not possible to be changed during the diffraction pattern raising; this would lead to a significant variation of the incident beam intensity and no data processing is possible in this situation. It is convenient to keep the monochromator radius of curvature constant, at a value corresponding to that given by (4) for a scattering angle $2\theta_s$ in the range 90-110 degrees, where the overlapping problems are the most significant and therefore the line-width should be minimum. A value of around 10 m for the monochromator radius of curvature is quite convenient, giving a minimum line-width of 12-15 minutes and not more than 25 minutes for the rest of scattering angle values.

An important problem appears. The sample position given by (5) correspond to a certain value of the scattering angle $2\theta_s$. The question is: if we keep the same sample position corresponding to a certain scattering angle value, how much is decreased the resolution if we measure the pattern for a certain

angular range, around this value? Can be used a position sensitive detector in the case of Q-space focussing configurations?
To find an answer to this important problem, computations were made, using the DAX program [4], for the configuration existing at INR Pitesti and a sample of 2.5 cm width, 3 cm height and .5 cm thickness, for 5 values of the scattering angle, 30^0, 50^0, 70^0, 90^0 and 110^0, for which the optimum sample orientation angles are 29^0, 37^0, 90^0, 121^0, 144^0 respectively; were computed the line widths for a range of 20^0 around the above mentioned scattering angle values. The results are given in Figure 2.

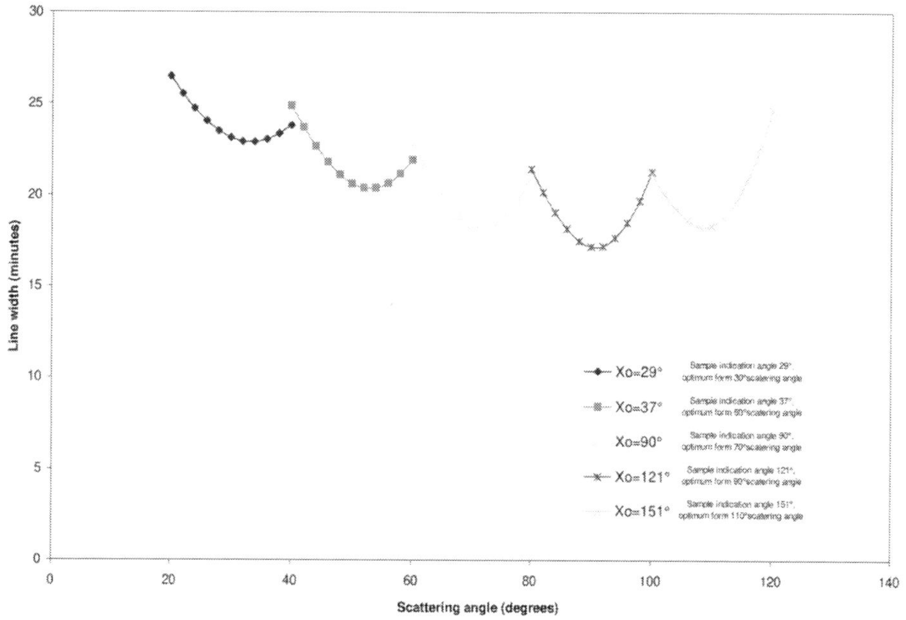

Figure 2

The conclusion is quite obvious: the resolution is still very good even for a range of 20^0 around the scattering angle for which the sample is in the optimum position; if still a better resolution is desired, the detector angular width can be lowered at 15^0; that means that the diffraction pattern can be got from 5 -7 steps.
Considering the advantages offered by a Q-space focussing configuration in comparison with the high resolution conventional ones, we have to mention that the neutrons use efficiency is better with a factor of $10^3/5$, were 10^3 is roughly the transmission of the Soller collimators system defining the conventional instrument considered; that means that with a lower flux of two order of magnitude can be obtained the same luminosity (the same measure time); for comparable neutrons flux the measure time can be reduced correspondingly.

2.2.2. Two Crystals Monochromator
For this geometry (see Figure 3), following the above-described procedure, tacking account that the Bragg constraints, for the two crystals, are:

$$\left.\begin{aligned} \gamma_0 + \gamma_1 &= 2 \cdot sign(\theta_{m1} + \chi_{m1}) \cdot \frac{l_{m1}}{R_{m1}} \\ \gamma_2 + \gamma_1 &= 2 \cdot sign(\theta_{m2} + \chi_{m2}) \cdot \frac{l_{m2}}{R_{m2}} \end{aligned}\right\} \tag{6}$$

$$\left.\begin{aligned} \frac{\Delta k_1}{k} &= (\gamma_0 - \gamma_1)\cdot\frac{\cot\theta_{m1}}{2} \\ \frac{\Delta k_2}{k} &= (\gamma_1 - \gamma_2)\cdot\frac{\cot\theta_{m21}}{2} \end{aligned}\right\} \quad (7)$$

to which it has to be added the condition that the neutrons reflected by the first crystal must have a proper wave-length to be reflected by the second crystal, i.e.

$$\left.\begin{aligned} &\Delta k_i = \Delta k_f \\ &or \\ &\gamma_2 = \gamma_1 \cdot\ 1 - \mathrm{a_m}\ + \mathrm{a_m}\cdot\gamma_0 \end{aligned}\right\} \quad (8)$$

with $a_m = \tan\theta_{m2} / \tan\theta_{m1}$.

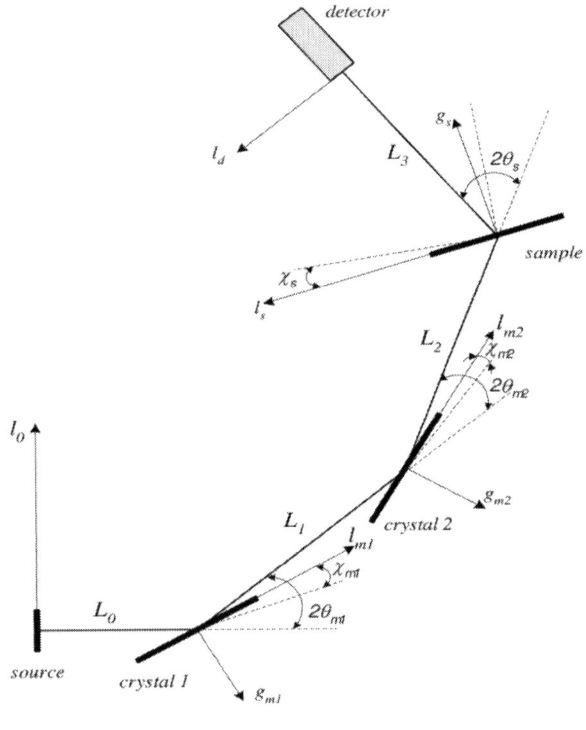

Figure 3

One obtains the following focusing conditions, giving the crystals radii of curvature:

$$\left.\begin{aligned} \frac{1}{R_1} &= \frac{a_m - 1}{a_m \cdot L_1}\cdot sign(\theta_{m1} + \chi_{m1})\cdot\sin(\theta_{m1} - \chi_{m1}) \\ \frac{1}{R_2} &= -\frac{a_m - 1}{L_1}\cdot sign(\theta_{m2} + \chi_{m2})\cdot\sin(\theta_{m2} + \chi_{m2}) \end{aligned}\right\} \quad (9)$$

By cancelling the "l_0" and "l_s" coefficients, one obtains respectively

$$\left.\begin{aligned} \frac{L_2}{L_1} &= \frac{1+\frac{a_m}{2a}}{1-a_m}\cdot\frac{\sin(\theta_{m2}-\chi_{m2})}{\sin(\theta_{m1}+\chi_{m1})} \\ \tan\chi_s &= \cot\theta_s\cdot\frac{\xi-1}{\xi+1} \end{aligned}\right\} \tag{10}$$

with

$$\left.\begin{aligned} \xi &= -\frac{L_3}{L_2}\cdot\frac{2\cdot(a+a_m)-1-2\cdot\alpha_1\cdot(1-\frac{1}{a_m})\cdot(a+a_m-1)}{2a_m-1-2\cdot\alpha_1\cdot\frac{(1-a_m)^2}{a_m}+\alpha_2\cdot(1+\alpha_1\cdot\frac{2-a_m}{a_m})} \\ \alpha_1 &= \frac{L_0}{L_1}\cdot\frac{\sin(\theta_{m1}-\chi_{m1})}{\sin(\theta_{m1}+\chi_{m1})},\quad \alpha_2=\frac{L_2}{L_1}\cdot\frac{\sin(\theta_{m2}-\chi_{m2})}{\sin(\theta_{m2}+\chi_{m2})} \end{aligned}\right\} \tag{11}$$

2.3. The Time-of-Flight Diffractometry

2.3.1. Pulsed Source

For this geometry (see Figure 4) following the above-described procedure, tacking account that

$$L = L_0 + L_1$$

$$\Delta L = -l_d\cdot\sin\chi_d - 2\cdot l_s\cdot\sin\theta_s\cdot\cos\chi_s + l_0\cdot\sin\chi_0,$$

by cancelling the "l_d", "l_s" and "l_0" coefficients, one obtains, respectively:

$$\left.\begin{aligned} \tan\chi_d &= \cot\theta_s\cdot\frac{L}{2\cdot L_1} \\ \tan\chi_s &= \frac{\frac{4\cdot\tan^2\theta_s}{L}+\frac{1}{L_1}+\frac{1}{L_0}}{\tan\theta_s\cdot(\frac{1}{L_0}-\frac{1}{L_1})} \\ \tan\chi_0 &= -\cot\theta_s\cdot\frac{L}{2\cdot L_0} \end{aligned}\right\} \tag{12}$$

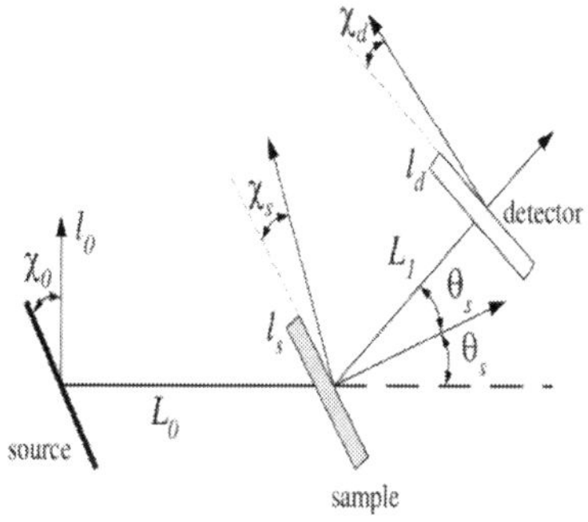

Figure 4

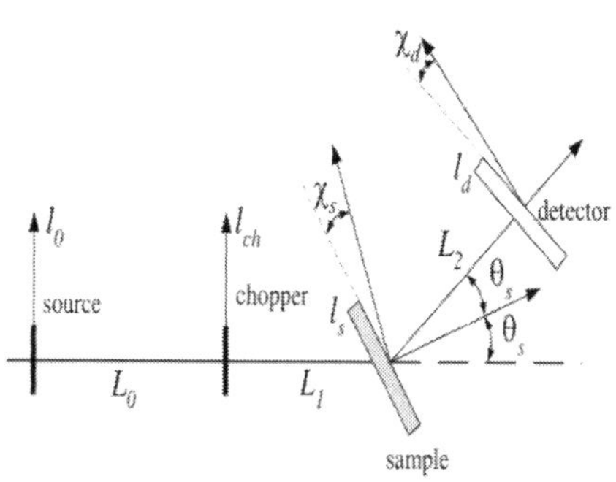

Figure 5

2.3.2. Steady State Source

For this geometry, (see Figure 5), following the above-described procedure, tacking account that in this case t_0 is the moment when the neutron reaches the chopper centre, $L = L_1 + L_2$,

$\Delta L = l_d \cdot \sin\chi_d - 2\cdot l_s \cdot \cos\chi_d \cdot \sin\theta$, that the constraint introduced by the chopper presence is given by $t_0 = \gamma_2/\omega$, cancelling the "l_0", the "l_d" and "l_s" coefficients, one obtains respectively:

$$\left.\begin{aligned} \tan\theta_s &= -\frac{\pi m\omega}{hk}\cdot(L_1+L_2) \\ \tan\chi_d &= -\frac{L\cdot\tan\theta_s}{2\cdot L_2} \\ \tan\chi_d &= -\frac{L\cdot\tan\theta_s}{2\cdot L_2} \end{aligned}\right\} \qquad (13)$$

Three Axis Spectrometers

The experimental setup is given in Figure 6. As the real focusing effects are possible only when no Soller collimators are present between the spectrometer components this situation is considered below. Let us consider the idealized case of a neutron spectrometer with ideal perfect crystals and sample, monochromator and analyzer of negligible thickness.

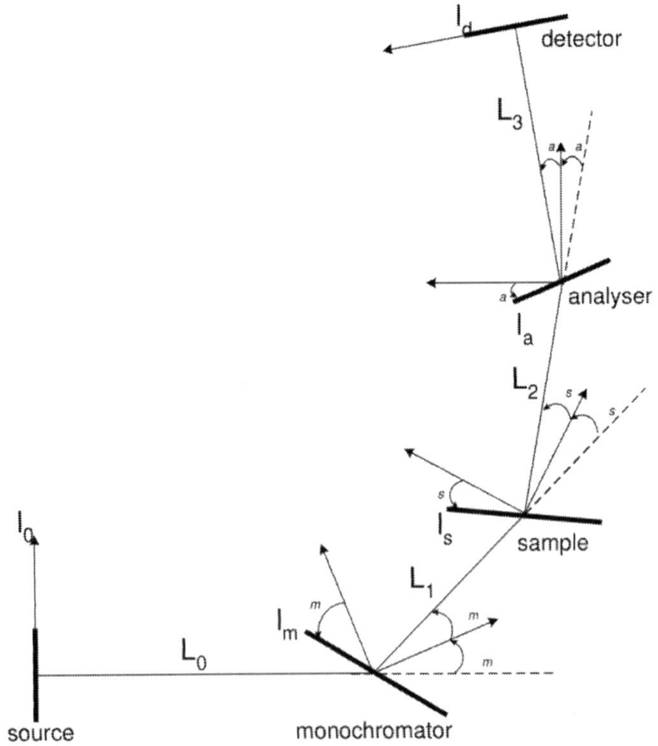

Figure 6

The spread of the scan variable can be written as a linear combinations between X vector components

$$\Delta = X_4 + C_1\cdot X_1 + C_2\cdot X_2 + C_3\cdot X_3 \qquad (14)$$

where X is the vector of components ($Q-Q_0$, $\omega-\omega_0$), $Q = k_i - k_f$ and $h\omega = E_i - E_f$; k_i, k_f, E_i, E_f are the wave vectors respectively the energy for the incoming and the diffracted neutrons, while the "0" index refers to the corresponding most probable values. The X_1 and X_2 components are along respectively normal to Q_0 direction, all situated in the scattering plane. In the first approximation the X_3 component, normal to the scattering plane, should not be taken into consideration as there is no correlation

between horizontal and vertical variables. The variables defining the neutron trajectories in phase space are all the spatial relevant coordinates.
The expression giving the X components are:

$$X_1 = \cos\phi \cdot \Delta k_i + k_{i0} \cdot \sin\phi \cdot \gamma_i - \cos\phi' \cdot \Delta k \cdot - k_{f0} \cdot \sin\phi' \cdot \gamma_f \tag{15}$$

$$X_2 = -\sin\phi \cdot \Delta k_i + k_{i0} \cdot \cos\phi \cdot \gamma_i + \sin\phi' \cdot \Delta k_f - k_{f0} \cdot \cos\phi' \cdot \gamma_f \tag{16}$$

$$X_4 = \frac{h}{m} \cdot (k_{i0} \cdot \Delta k_i - k_{f0} \cdot \Delta k_f) \tag{17}$$

where

$$\left.\begin{aligned} \tan\phi &= \frac{k_{f0} \cdot \sin 2\theta_s}{k_{i0} - k_{f0} \cdot \cos 2\theta_s} \\ \varphi' &= \varphi - 2 \cdot \theta_s \end{aligned}\right\}, \tag{18}$$

and γ_i, γ_f are the horizontal divergence angles from the most probable trajectories, Δk_i, Δk_f are the corresponding wave vector spread, $2\theta_s$ is the scattering angle; the index i, f refers to the incoming respectively scattered neutrons; φ is the angle between Q_0 and k_{i0}.
To these relations must be added the Bragg constraints related to the monochromator and analyzer respectively:

$$\left.\begin{aligned} &\gamma_0 + \gamma_1 = 2 \cdot l_m \cdot \rho_m \\ &\frac{\Delta k_i}{k_{i0}} = \frac{\gamma_0 - \gamma_1}{2} \cdot \cot\theta_m \\ &\gamma_2 + \gamma_3 = 2 \cdot l_a \cdot \rho_a \\ &\frac{\Delta k_f}{k_{f0}} = \frac{\gamma_2 - \gamma_3}{2} \cdot \cot\theta_a \end{aligned}\right\} \tag{19}$$

where $\rho_m = 1/R_m$, $\rho_a = 1/R_a$ and R_m, R_a are the radius of curvature for the monochromator and analyzer respectively.
The geometry of the experimental configuration gives the following expressions for γ_0, γ_1, γ_3, γ_4:

$$\left.\begin{aligned} L_0 \cdot \gamma_0 &= -l_0 + l_m \cdot \sin(\theta_m + \chi_m) \\ L_1 \cdot \gamma_1 &= l_s \cdot \cos(\theta_s + \chi_s) + l_m \cdot \sin(\theta_m - \chi_m) \\ L_2 \cdot \gamma_2 &= -l_s \cdot \cos(\theta_s - \chi_s) + l_a \cdot \sin(\theta_a + \chi_a) \\ L_3 \cdot \gamma_3 &= l_d + l_a \cdot \sin(\theta_a - \chi_a) \end{aligned}\right\} \tag{20}$$

The angles χ_m, χ_a, all measured in trigonometric sense, represent the angle between to the lattice plane and the normal to the crystal surface, for monochromator and analyzer respectively.
Eliminating the angular variables γ_0 and γ_3 one obtains the following expressions for Δk_i, Δk_f:

$$\left.\begin{aligned} \Delta k_i &= k_{i0} \cdot \cot\theta_m \cdot (l_m \cdot \rho_m - \gamma_1) \\ \Delta k_f &= k_{f0} \cdot \cot\theta_a \cdot (l_a \cdot \rho_a - \gamma_2) \end{aligned}\right\} \tag{21}$$

Using (15), (16), (17) in (14) on obtains for the scan variable Δ:

$$\Delta = \Delta k_i \cdot A_i - A_f \cdot \Delta k_f + k_{i0} \cdot B_i \cdot \gamma_1 - k_{f0} \cdot B_f \cdot \gamma_2 \tag{22}$$

where:

$$\left.\begin{aligned} A_i &= k_{i0}\cdot\frac{h}{m} + C_1\cdot\cos\phi - C_2\cdot\sin\phi \\ A_f &= k_{f0}\cdot\frac{h}{m} + C_1\cdot\cos\phi' - C_2\cdot\sin\phi' \\ B_i &= C_1\cdot\sin\phi + C_2\cdot\cos\phi \\ B_f &= C_1\cdot\sin\phi' + C_2\cdot\cos\phi' \end{aligned}\right\} \tag{23}$$

Using (19) and (20) the relation giving the scan variable Δ becomes:

$$\Delta = k_{i0}\cdot l_m\cdot[\cot\theta_m\cdot\rho_m\cdot A_i + D_i\cdot\frac{\sin(\theta_m-\chi_m)}{L_1}] + l_s\cdot[D_i\cdot k_{i0}\cdot\frac{\cos(\theta_s+\chi_s)}{L_1} - D_f\cdot k_{f0}\cdot\frac{\cos(\theta_s-\chi_s)}{L_2}] + k_{f0}\cdot l_a\cdot[\frac{\sin(\theta_a+\chi_a)}{L_2} - \cot\theta_a\cdot\rho_a\cdot A_f] \tag{24}$$

with

$$D_i = B_i - A_i\cdot\cot\theta_m \qquad D_f = -(B_f + A_f\cdot\cot\theta_a) \tag{25}$$

Cancelling the coefficients of l_m, l_a and l_s one obtains:

$$\frac{L_1}{2\cdot f_1} = 1 - \frac{B_i}{A_i}\cdot\tan\theta_m \qquad \frac{L_2}{2\cdot f_2} = 1 + \frac{B_f}{A_f}\cdot\tan\theta_a \tag{26}$$

$$\frac{\cos(\theta_s+\chi_s)}{\cos(\theta_s-\chi_s)} = \frac{L_1\cdot k_{f0}}{L_2\cdot k_{i0}}\cdot\frac{A_f\cdot\cot\theta_a + B_f}{A_i\cdot\cot\theta_m - B_i} \tag{27}$$

with:

$$f_1 = \frac{\sin(\theta_m-\chi_m)}{2\cdot\rho_m} \qquad f_2 = \frac{\sin(\theta_a+\chi_a)}{2\cdot\rho_a} \tag{28}$$

The first two conditions can be fulfilled by properly choosing the radius of curvature R_m, R_a for monochromator and analyzer respectively when varying k_i and k_f if L_1 and L_2 are fixed while the third one can be fulfilled by a proper sample orientation. The focusing conditions given by (26) assure the compensation of the contributions given by monochromator and analyzer respectively to the line width while the focusing condition (27) assures the sample contribution compensation.

The focusing conditions deduced above for an idealized configuration can serve as a "0"-th order approximation in a numerical optimization process.

Conclusions

Very high resolution performances can be obtained through inverse-space focusing in thermal neutrons spectrometry if focusing conditions involving monochromator and analysis radius of curvature and the sample orientation are fulfilled.

For time-of-flight instruments the inverse space focusing conditions given in [5] should be considered.

Inverse space focusing represents a very promising way to get high-resolution instruments in neutron spectrometry. A new concept of high-resolution instrument based on inverse space focusing effects could be developed.

This work was partially supported by IAEA Vienna under the contract 16049/RO.

References

[1] P. Boni 1996 *J. Neutron Research* **5**, 63
[2] M. Popovici, A. D. Stoica and I. Ionita 1987 *J.Appl. Cryst.* **20**, 90
[3] I. Ionita, A. D. Stoica, M. Popovici, N. C. Popa 1999 *Nucl. Instr. & Meth. in Physics Research*

A431 509-520
[4] I. Ionita, A. D. Stoica 2000 The Crystal Neutron Diffractometer Resolution Function Spatial Effects Included, *J. Applied Cryst.* **33**, 1067
[5] I. Ionita 2002 Focusing Conditions in Neutron Spectrometry, *Nucl. Instr.& Meth. in Physics Research* 313-336.

Journal of Physics: Conference Series **340** (2012) 012027 doi:10.1088/1742-6596/340/1/012027

Multi-Incident Beam Compact Time-Of-Flight SANS Instrument with 1 Millimeter Diameter beams

M Furusaka[1], S Takeda[1], A Homma[1], F Fujita[1], T Miyata[1], Y Kiyanagi[1] and M Ohnuma[2]

[1] Faculty of Engineering, Hokkaido University, Kita-13 Nishi-8, Kita-ku, Sapporo, Hokkaido 060-8628, Japan
[2] Quantum Beam Unit, National Institute for Materials Science, 1-2-1 Sengen, 305-Tsukuba 0047, Japan

E-mail: furusaka@eng.hokudai.ac.jp

Abstract. Although scanning small-angle x-ray scattering (SAXS) instruments are available, almost no attempt has been made to realize scanning small-angle neutron scattering (SANS) instruments because of rather low flux available in the case of neutron scattering. We propose the use of a multi-pinhole collimator–pinhole combination to reduce the beam size to about 1–2-mm diameter immediately in front of a sample. With such small pinholes, we can produce a rather short instrument, on the order of 1–2 m, to have access to conventional minimum Q of about 0.03 nm^{-1}. Because multiple beams hit the same detector at different parts of the detector, the obtained SANS patterns mutually overlap. However, using a wavelength-dependent SANS pattern, we were able to resolve the overlap in theory. We conducted a proof-of-principle type of experiment at the Hokkaido University electron linear accelerator based pulsed cold neutron source (HUNS) facility. We made several multi-pinhole plates with thin beam holes arranged in hexagonal patterns. By changing the moderator and detector pixel size and the instrument overall size, we can optimize the pinhole size and their layout. We conducted experiments using welded steel samples and seek the difference in SANS pattern in the welded and non-welded parts, as well as the heat-affected zone between them. We demonstrated the principle of resolving overlapped SANS patterns from the adjacent beams using wavelength-dependent scattering.

1. Introduction

Scanning X-ray SAS [1–5] is a standard method used for analyzing position-dependent nanostructures in objects that have hierarchical structures such as bone, but people usually avoid performing scanning "NEUTRON" SAS because of the low flux available to neutron instruments in general. In fact, SAS measurements necessitate tight angular collimation of about a few millirads, resulting in low flux at the sample position. The only available means of overcoming this is to enlarge the sample size, i.e. just opposite to scanning concept.

We are proposing the use of multiple pinholes of approximately 1 mm diameter for collimation. Because of the small pinhole size, the use of rather short path lengths of about 1–2 m can provide the necessary collimation. Minimum Q that are typically available using conventional SANS instruments, such as several times 0.01 nm^{-1} to 0.1 nm^{-1}, can be realized by the configuration. The intensity drops rapidly and concomitantly with the (pinhole diameter)$^{-2}$. To compensate for this, we propose the use of

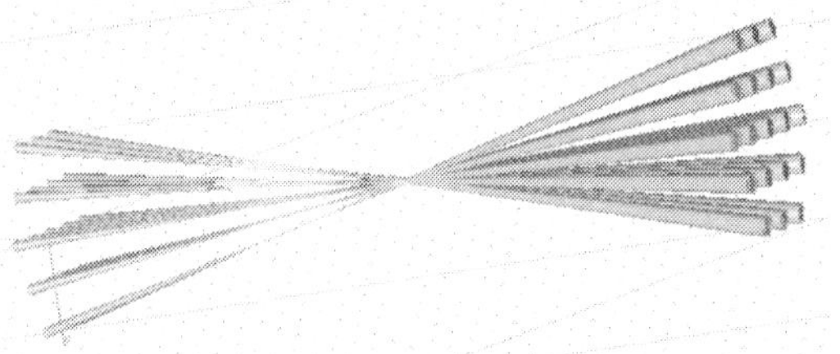

Figure 1. "Intensity gain type (type 1)" multi-pinhole arrangement. A multi-pinhole plate is placed at the entrance (left-hand side) of the collimator section, with a single pinhole at the middle immediately in front of the sample position, and a detector at the right end.

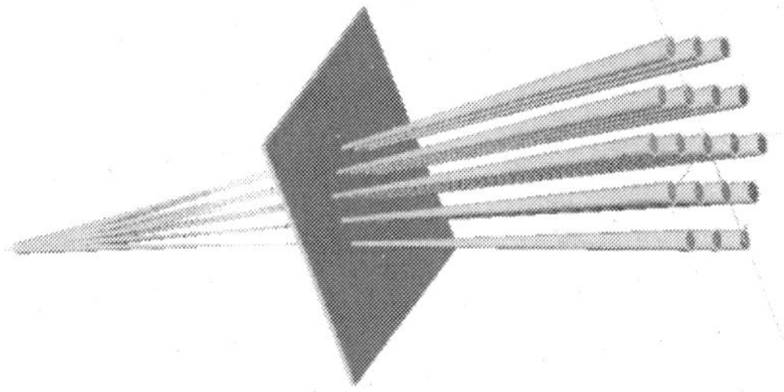

Figure 2. "Scanning SANS type (type 2)" multi-pinhole arrangement. A single pinhole is placed at the entrance (left-hand side) of the collimator section, with a multi-pinhole plate placed at the middle just in front of the sample position, and a detector at the right end.

many small pinholes for the entrance of the collimator section to gain intensity and use a single pinhole immediately in front of the sample position to define small area. The possibility also exists of using a multi-pinhole aperture plate immediately in front of the sample.

We then confront overlapping small-angle scattering from different pinholes. Separating them is expected to be difficult in the case of a reactor instrument, but in the case of a time-of-flight instrument, it is possible to resolve the overlap because we can obtain multiple $I(q)$ information for the same q from different wavelengths and position regions.

We therefore made a proof-of-principle mockup instrument to study whether such a concept actually works at the Hokkaido University electron linear accelerator based pulsed cold neutron source facility (HUNS).

2. Principle of multi-incident beam instrument

The Q resolution ΔQ of a pinhole-geometry SANS instrument can be expressed as

$$\Delta Q \approx \frac{4\pi}{\lambda}\cdot\Delta\theta = \frac{2\pi}{\lambda}\cdot\sqrt{\frac{2\ln 2}{3}\left(\frac{A_0^2+A_S^2}{L_1^2}+\frac{A_S^2+A_D^2}{L_2^2}\right)}\cdot\frac{\Delta Q}{Q}, \qquad (1)$$

where λ is the wavelength, $\Delta\theta$ the angular resolution, A_0 is the first aperture, A_S is the aperture immediately before the sample, A_D is the detector position resolution, L_1 the length between A_0 and A_S, and the L_2 sample to detector distance. In the case of time of flight (TOF) instrument, $\Delta\lambda$ is negligible in general.

Typically, A_S is chosen as nearly 10 mm or slightly less in diameter. When $L_2 \approx 5$ m and $\lambda \approx 0.6$ nm is assumed, ΔQ becomes about 0.01–0.02 nm^{-1} and Q_{min} of about 0.03–0.06 nm^{-1} can be measured. If we were able to reduce A_s to about 1 mm, then we would be able to obtain the same Q-range using only $L_2 \approx 0.5$ m, with the sacrifice of intensity to 1/100.

To regain the intensity, we can use a multi-pinhole aperture for the first or second aperture, as presented in figures 1 and 2. For the first case (intensity gain type, (type-1)), all neutrons passed through the pinholes converge onto the single pinhole immediately before the sample position that is placed exactly halfway between the incident pinholes and the detector position at the end. Every single beam generates SANS and on the detector, all the SANS patterns from every beam overlap. A method of unwrapping the degenerated SANS patterns will be presented later in this paper. We are using the TOF method. Consequently, we can make use of wavelength dependence to unwrap the pattern. For example, with short wavelength neutrons, the SANS patterns are expected to be concentrated near each direct beam and are expected to accommodate the unwrapping fairly easily.

In the second case (scanning SANS or imaging type, (type-2)), the beam goes through a single pinhole at the entrance of the collimator section. Because the multiple pinholes are immediately in front of the sample, neutrons hit different positions of the sample. It is like scanning SANS experiment, meaning that we observe different patterns from each beam. Again the SANS patterns overlap, but we can unwrap them as described later.

Both types of configuration can be mixed. Some examples are related to the first aperture: we can use a seven-pinhole plate and put three pinholes immediately before the sample to produce 21 different SANS patterns from three different positions on the sample.

Because all SANS become the same, it is far easier to resolve the SANS scattering if we use type-1 geometry. The time required to scan is the same for both type-1 and type-2 measurements.

3. Instrument design principle

One example of multi-pinhole patterns is presented in figure 3.

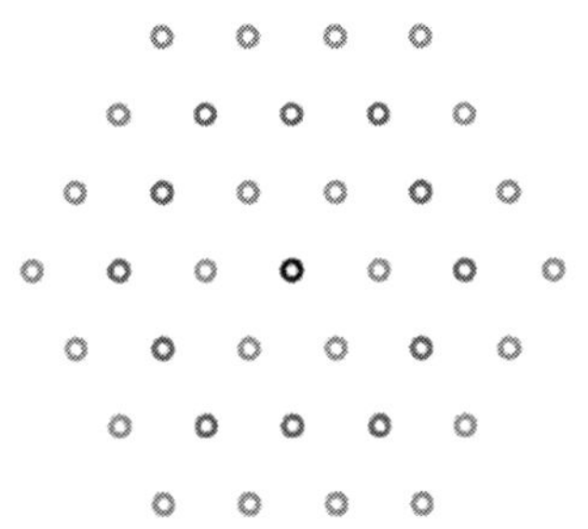

Figure 3. Example of multi-pinhole patterns.

The number of pinholes $n_{pinhole}$ is calculated as

$$n_{\mathrm{pinhole}} = 1 + \sum_{i=1}^{m-1} 6i \ , \tag{2}$$

where m is an integer parameter related to the approximate distance between the center to the position of pinholes specifying the number of layers, starting from 1. For m=1, there is only one pinhole at the center. For m=2, six pinholes are added, making the total number 7. The total numbers of pinholes therefore become 1, 7, 19, 37, 61, ... as parameter m increases.

When designing the pinhole pattern, a constraint arises from the detector pixel size d_{pix}, pixel number n_{pix}, and the moderator size d_{mod}. We first assumed the direct beam size d_{dir} as r_1 times larger than d_{pix}. Here, r_1 should be a number greater than 2.

$$d_{\mathrm{dir}} = r_1 \, d_{\mathrm{pix}} \tag{3}$$

The distance between the direct beam images on the detector can be expressed as

$$d_{patt} = r_2\, d_{dir}\,. \tag{4}$$

When the detector accommodates m layers of the pinhole pattern, the overall detector size d_{det} can be expressed as

$$d_{det} = (2\,m - 1)\, d_{patt}\,, \tag{5}$$

allowing a 1/2 d_{patt} margin outside of the pinholes.

Then, the sample to detector distance L_{sd} and the collimator length L_{cs} can be related to the moderator to sample distance L_{ms} and the moderator size d_{mod}, using d_{det} as

$$d_{mod} \approx \frac{L_{ms}}{L_{sd}}\, d_{det}\,, \qquad L_{sd} \approx \frac{d_{det}}{d_{mod}}\, L_{ms} \qquad \text{(type 1 case)}, \tag{6}$$

$$d_{mod} \approx \frac{L_{mc}}{L_{cs} + L_{sd}}\, d_{det}\,, \qquad L_{cs} + L_{sd} \approx \frac{d_{det}}{d_{mod}}\, L_{mc} \qquad \text{(type-2 case)}, \tag{7}$$

respectively corresponding to type-1 and type-2 cases. L_{ms} , which is not a free parameter, is usually of the order of 5–10 m. When the pinhole size at the sample position d_{patt} is fixed, L_{cs} or L_{sd} is determined according to the requirement from the Q resolution ΔQ.

ΔQ is a function of the pinhole size and L_{sd}. When we choose collimator length L_{cs} as equal to L_{sd}, the optimum ratios of the pinholes become

$$A_0 : A_s : A_d = 2:1:4\,. \tag{8}$$

Actually, ΔQ of about 0.01 nm^{-1} can be attained using long-wave neutrons of 0.3–1.2 nm if we choose A_s = 2 mm and L_{cs}=L_{sd}=1.5 m.

We can use 7–19 or more pinholes for the incident aperture that helps in recovering the intensity. Of course we face overlapping of small-angle scattering from different pinholes. Separating them is expected to be difficult in the case of a reactor instrument, but in case of time-of-flight instrument, it will be possible because we can get multiple $I(q)$ information for the same q from different wavelengths and position regions. The possibility also exists of using a multi-pinhole aperture plate immediately in front of the sample.

4. Experiment

We tested the principle of the instrument type-2 at HUNS. It has an extremely efficient solid methane cold moderator, but the time-integrated neutron flux at sample position is rather low: a few times 10^3 n/cm^2/s.

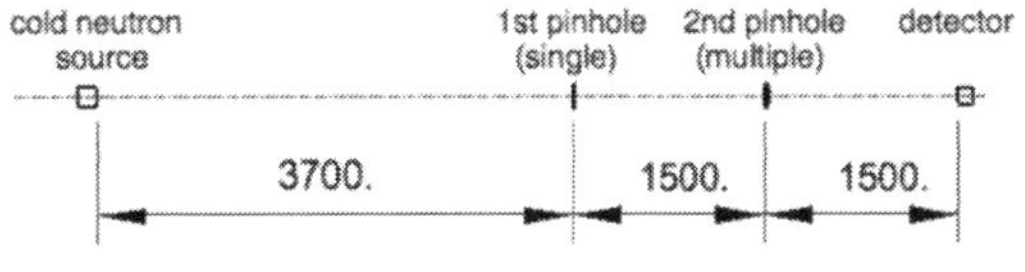

Figure 4. Instrument layout of the test experiment.

The instrument layout is shown briefly in figure 4, and the multiple pinhole plates we used are presented in figure 5. Cadmium plates of 0.5-mm thickness backed with borated resin plates are used and multiple pinholes were opened. The pinhole size of the first was 4 mm in diameter, and that for multiple pinholes on the plate

Figure 5. Photographs of the first plate with a pinhole of 4 mm diameter and the second one with multi-pinholes of 2 mm diameter placed immediately in front of the sample.

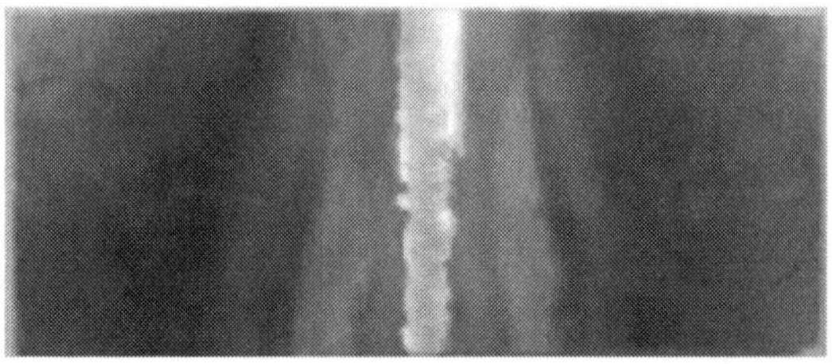

Figure 6. Welded steel sample SS400. The weld is at the center.

immediately before the sample was 2 mm. The sample was a welded steel (SS400) sample as shown in figure 6, and the central part of the multi-pinhole pattern is placed at the welded position.

For the detector, we used a Li-6 loaded ZnS scintillator coupled with a position-sensitive photomultiplier tube (R3292; Hamamatsu Photonics K.K.)[6]. The detector resolution is less than 1 mm and the effective area is about 100 mm diameter. The whole system together with a data acquisition electronics based on VME modules and control software (LabView; National Instruments Corp.) was obtained from Japan Neutron Optics Inc.

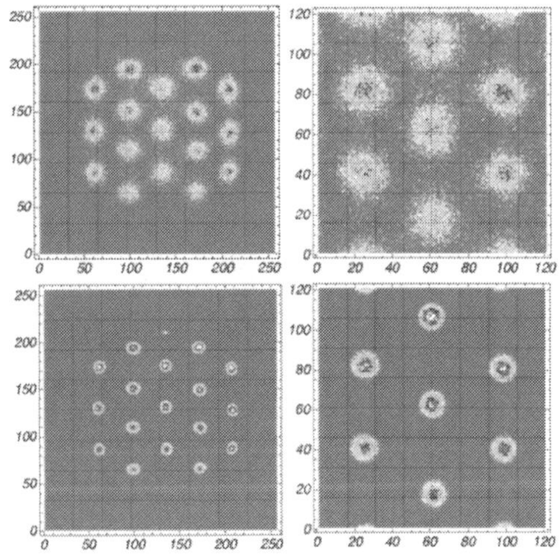

Figure 7. Measured direct beam patterns on the bottom and SANS patterns in the welded steel sample on the top. Enlarged images of the central parts of the images on the left columns are portrayed in the right columns. The scattering is integrated over wavelengths of 0.35–0.62 nm.

Figure 7 shows measured direct beam patterns and SANS patterns for the welded steel sample. Enlarged images of the central parts of the images on the left columns are also shown in the right columns. The detector image is oriented as though one is looking from the back of the detector: it has 256 × 256 position channels with 0.45 mm/channel. Unfortunately, some image distortion is apparent towards the periphery of the detector image. Data are integrated for wavelengths of 0.35–0.62 nm.

Despite the lack of a vacuum chamber and despite the exposure to atmosphere in between the first pinhole to the detector, the direct beam profile shows little indication of diffuse tailing down to about 3.5 orders from the peak. Because of the pinhole camera geometry, the intensity distribution on the moderator surface directly reflects the intensity distribution of the pinhole images, but it is upside down and flipped left to right on a vertical axis. Although there were 19 pinholes, the top and bottom images at the central column are missing because the beams corresponding to the pinholes are at the outside edges of the moderator.

5. Position-dependent SANS using 1-mm-diameter pinholes

We also conducted a 1-mm-diameter pinhole experiment. The sample was also welded steel, but this time there was a heat-affected zone (HAZ) of about 1 mm width between the welded and non-welded parts. We put a multi-pinhole plate upstream that had two sets of nine pinholes and two pinholes

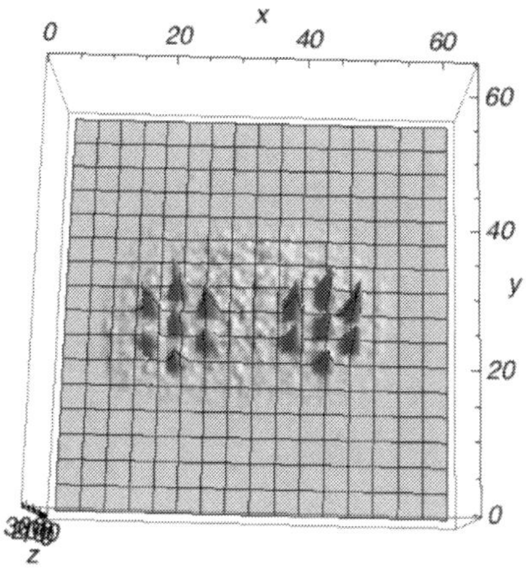

Figure 8. Direct beam pattern from a couple of nine pinholes upstream and two 1-mm-diameter pinholes immediately in front of the sample.

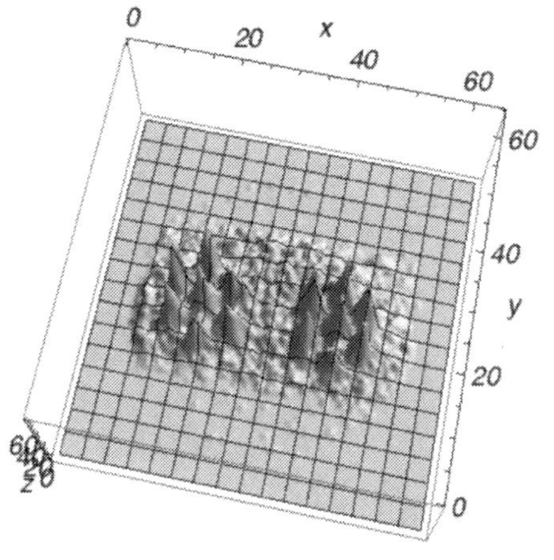

Figure 9. SANS pattern from a couple of the nine pinholes upstream and two 1-mm-diameter pinholes immediately before the sample.

immediately before the sample. The direct-beam patterns are shown in figure 8. In figure 9, the pattern with the sample is shown. A strong SANS contribution is apparent. We are conducting analyses now.

6. Resolving overlapped scattering

We are now developing a method of resolving overlapping SANS from multiple pinholes, although it is in a preliminary stage. Figure 10 shows wavelength-dependent circular-averaged SANS patterns from a pinhole with wavelength regions of 0.28–0.35 nm, 0.31–0.39 nm, and 0.35–0.44 nm. In each figure, thin upper curves show SANS with the sample and lower ones without the sample. The red thick curves the background subtracted SANS patterns. Because of scattering from adjacent pinholes, upturn behavior is shown towards the higher-Q region from about 0.025 nm^{-1}. With shorter wavelength regions, we assume that the data are reliable for higher-Q regions. With longer wavelength regions, lower-Q regions would be reliable. We combined data from good regions and obtained the SANS curve shown in figure 11. The upper three curves show scattering from welded parts and the lower three show that from non-welded parts. Although the analysis remains in a preliminary stage, it is apparent that welded parts produce higher scattering intensities than non-welded parts do.

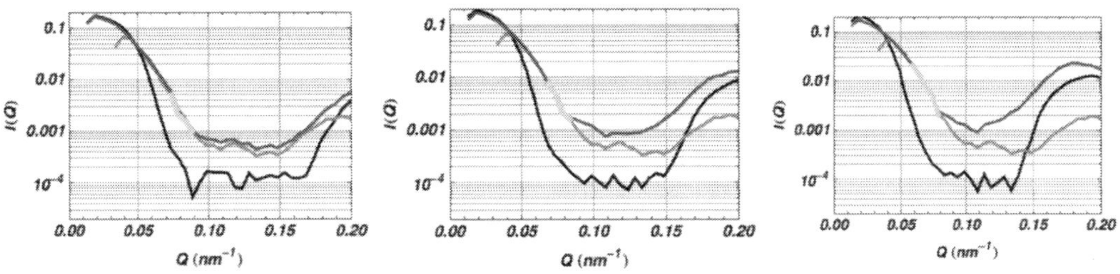

Figure 10. Wavelength-dependent SANS patterns from a pinhole with the wavelength regions of 0.28–0.35 nm, 0.31–0.39 nm, and 0.35–0.42 nm from left to right. The upper curves show that of the scattering with sample and the lower ones without the sample. The red curves show scattering corrected for the background. The thick green parts on the red curves show Q-ranges where the data are reliable with the selected wavelength ranges.

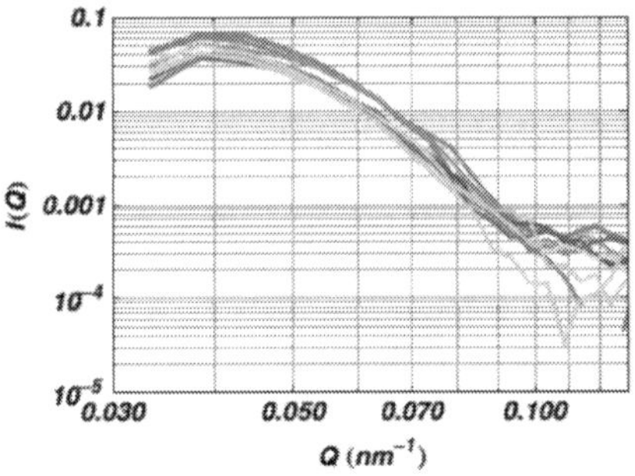

Figure 11. Reconstructed SANS patterns in welded parts (upper three curves) and in non-welded parts (lower six curves).

Acknowledgments
This work was partially supported by the Quantum Beam Technology Program of JST and (JSPS) KAKENHI (23226018).

References

[1] Bunk O, Bech M, Jensen TH, Feidenhans'l R, Binderup T, Menzel A, and Pfeiffer F 2009 *New Journal of Physics* **11** 123016

[2] Fratzl P, Weinkamer R, 2007 Nature's hierarchical materials, *Progress in Materials Science*, **52** 1263-1334

[3] Cedola A, Mastrogiacomo M, Lagomarsino S, Cancedda R, Giannini C, Guagliardi A, Ladisa M, Burghammer M, Rustichelli F, and Komlev V 2007 *Spectrochimica Acta Part B: Atomic Spectroscopy*, **62** 642-647

[4] Dinapoli R, Bergamaschi A, Henrich B, Horisberger R, Johnson I, Mozzanica A, Schmid E, Schmitt B, Schreiber A, and Shi X 2010, *Nucl. Instrum. and Meth.* A, doi:10.1016/j.nima.2010.12.005 in press.

[5] Cancedda R, Cedola A, Giuliani A, Komlev V, Lagomarsino S, Mastrogiacomo M, and Peyrin F, Rustichelli F 2007 *Biomaterials* **28** 2505-2524

[6] Hirota K, Shinohara T, Ikeda K, Mishima K, Adachi T, Morishima T, Satoh S, Oku T, Yamada S, Sasao H, Suzuki J-i, and Shimizu HM 2005 *Phys. Chem. Chem. Phys.* **7** 1836-1838

Journal of Physics: Conference Series **340** (2012) 012028 doi:10.1088/1742-6596/340/1/012028

Neutron beam tailoring by means of a novel pulsed spatial magnetic spin resonator

C. Gösselsberger, H. Abele, G. Badurek, E. Jericha, W. Mach, S. Nowak, T. Rechberger

Atominstitut, Vienna University of Technology, Stadionallee 2, 1020 Vienna, Austria

E-mail: goesselsberger@ati.ac.at

Abstract. Since long neutron spin resonance in an arrangement of crossed homogeneous and spatially alternating transversal magnetic fields is known to allow for wavelength-selective polarisation manipulation. Combined with a pair of highly efficient polarising supermirrors it is an elegant method to single out a specific wavelength from an initially polychromatic polarised neutron beam, based upon the fact that in its rest frame each neutron creates its individual frequency. A spin flip process can then take place only if this individual frequency equals the neutron Larmor frequency in the homogeneous field. Here, we present the first experimental results we have obtained with a conceptually novel type of such spatial spin resonator, consisting of a series of separate modules which can be controlled independently from each other. Thus it becomes possible to vary e.g. the amplitude distribution of the spatially alternating transversal field without any geometric modification of the setup. Moreover this device fulfils the requirements for fast electronic switching of each module, which will be an important asset for the possible decoupling of the minimal neutron pulse duration and the achievable wavelength resolution. We compare the actual performance of this prototype resonator system with the theoretically expected behaviour.

1. Introduction

Spatial magnetic neutron spin resonance, first proposed more than four decades ago by Drabkin [1], is a meanwhile well understood method to achieve wavelength-selective spin flipping of polarised thermal and cold neutron beams [2, 3]. Combined with a pair of polarising supermirrors it has found since then a series of applications, as e.g. a neutron monochromator with variable output wavelength at fixed take-off angle [4–6], as an electronically tunable energy analyser in inverted geometry neutron time-of-flight spectrometers [7], or as neutron time-of-flight focusing device [8]. Essentially stimulated by the implementation of new generations of pulsed high-flux neutron sources, there is a still ongoing development [9, 10] and an increasing need [11] of high performance supermirror arrangements and of increasingly sophisticated polarisation manipulation devices. Probably the most appealing feature inherent to a spatial magnetic spin resonator becomes evident if it is operating in a chopped mode, namely the potential feasibility of decoupling its wavelength resolution from the minimally achievable duration of the transmitted neutron pulses. Following this goal we have realised a novel design of such a resonator, based upon a sequence of individually switchable stages, each consisting of a low-inductance single-turn aluminium coil. Although it has not yet been optimised with respect to fast beam chopping [12], this prototype resonator allows for a wavelength-selective creation of neutron pulses of virtually arbitrary length. The first instalment of this novel type of resonator in a real experiment is envisaged for the setup of the PERC (Proton Electron Radiation Channel) project [13, 14], which searches for new physics beyond the Standard Model of particle physics via ultra-high precision measurements of the beta-decay of free neutrons. Here we present the essential conceptual details of the magnetic field design of this resonator prototype as well as first experimental results, both in static and pulsed mode of operation.

2. Design of the resonator prototype

Without going into detail, we just remind that spatial neutron spin resonance is based upon the nuclear magnetic resonance-like superposition of a static field $\mathbf{B_0}$ and a crossed time-dependent magnetic field $\mathbf{B_1}(t)$, with the conceptual difference that here - in the most simple arrangement - the transversally oscillating field component is just produced by a zigzag-folded conducting aluminium foil, so that according to its speed (*viz.* wavelength) each neutron in its own rest frame creates its individual frequency $\omega(\lambda)$, governed only by the spatial period $2a$ of the conducting foil. A complete polarisation reversal takes then place if $\omega(\lambda)$ equals the LARMOR frequency $\omega_0 = |\gamma|B_0$, which is controlled by the strength of the static field ($\gamma = -1.83 \times 10^8\,\mathrm{s^{-1}T^{-1}}$), and if the amplitude of $\mathbf{B_1}(t)$ is properly tuned so that upon passage through the oscillating field region a total spin rotation angle of π (or odd multiples of π) around $\mathbf{B_1}$ is accumulated. Disturbing subsidiary maxima of the spinflip probability, which can be attributed to cut-off effects in the Fourier spectrum of $\mathbf{B_1}(t)$, can be suppressed quite effectively by appropriate amplitude modulation of the transversal field oscillations [15].

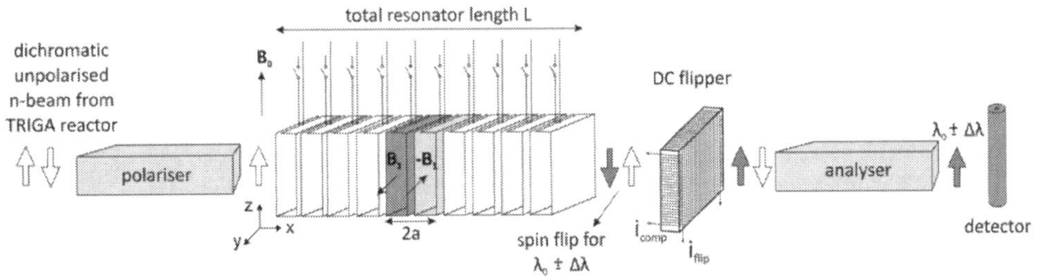

Figure 1. Schematic sketch of the experimental setup at the polarised neutron beamline of the 250 kW research reactor of the Vienna University of Technology. 1st and 2nd order Bragg reflection at the (002) lattice planes of a highly oriented pyrolithic graphite single crystal delivers a dichromatic incident neutron beam ($\lambda_{1,2} = 2.6 \pm 0.04$ and 1.3 ± 0.02 Å, respectively). The resonator prototype, composed of 10 separate single-turn coils of $a = 1$ cm thickness (each made of a 0.3 mm thin aluminium foil) is placed between two polarising supermirrors acting as polariser and analyser, respectively. A rectangular Helmoltz-type coil pair (not shown here) produces a variable vertical homogeneous field $\mathbf{B_0}$ whose strength determines the 'resonance' wavelength λ_0, where the spinflip probability of the resonator reaches its (sharp) maximum. A standard DC spinflipper is used only to determine the polarisation degree of the beam as well as the spinflip efficiency of the resonator, but is otherwise not activated (see Subsection 3.1).

The design of a prototype of a spatial magnetic spin resonator requires hence two major steps: *i*) to calculate for each specific resonator configuration the three-dimensional magnetic field distribution both for the static field that is produced by a Helmholtz-type coil pair and for the spatially alternating field generated by the sequence of resonator elements, and *ii*) then to solve the spin equation of motion along many possible particle trajectories. In a previous work [11] magnetic field simulations for a neutron spin resonator for the above mentioned PERC project have revealed that a thickness $a = 1$ cm of each resonator element should be quite appropriate

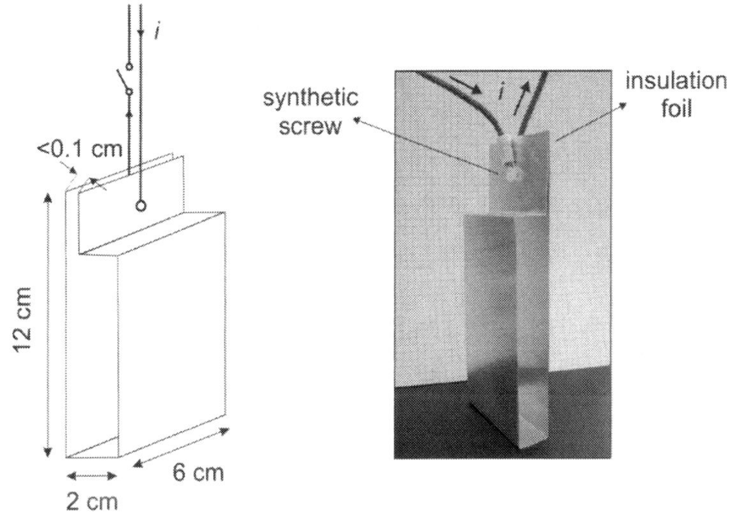

Figure 2. Detailed sketch and photo of a single-turn coil. For the present prototype setup the coil current is supplied just via point contacts.

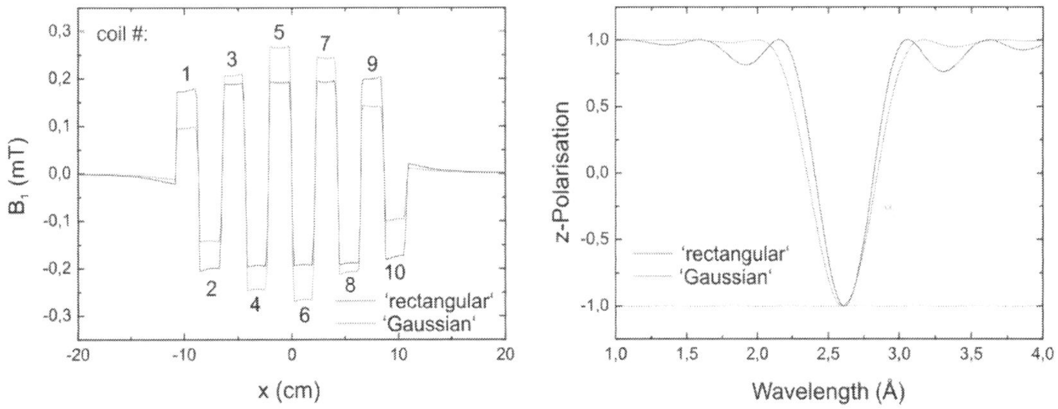

Figure 3. (*Left*) Simulated transversal magnetic field distribution of the resonator prototype without ('rectangular') amplitude modulation and with 'Gaussian' shaped amplitude. The two other field components are small and hence not explicitly shown, but they were taken into account in the derivation of the final polarisation vector. Notice that each resonator period requires two single-turn coils producing oppositely oriented transversal fields. The coils are numbered consecutively from 1 to 10. (*Right*) Numerical integration of the spin equation of motion yields the expected polarisation of the neutron beam after its passage through the respective field configuration.

to flip cold neutrons with a central wavelength of 5 Å. However, at the tangential beamline of the 250 kW research reactor of the Vienna University of Technology no cold source is available. Hence, we use a dichromatic 2.6 Å and 1.3 Å thermal neutron beam produced by 1st and 2nd order Bragg reflection at the (002) lattice planes of a highly oriented pyrolithic graphite single crystal. To achieve similar conditions as within PERC, we selected the 2.6 Å component of the

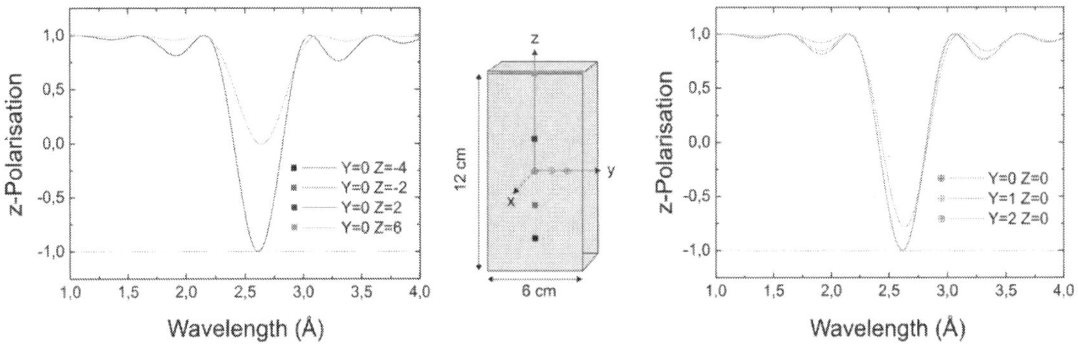

Figure 4. Simulated exit polarisation of a beam intially fully polarised in vertical direction after its passage through the resonator along different neutron trajectories. While the field is obviously very homogeneous in z-direction, it decreases quite strongly towards the borders in y-direction [12].

spectrum to evaluate the performance of a prototype resonator which consists of 10 individual single-turn coils, each with a thickness $a = 2\,\mathrm{cm}$ (see Fig. 1). In Fig. 2 the geometry of such a single-turn coil is sketched and illustrated by a photo.

According to the chosen geometry of the resonator elements as well as their total number, the wavelength resolution would be given as $\Delta\lambda/\lambda = 0.68\,/n \simeq 13.6 \times 10^{-2}$, where $n = L/2a$ is the number of resonator periods, provided a polychromatic incident beam is available. In our case, however, the actual wavelength resolution is defined by the spectral width $\Delta\lambda_{\mathrm{M}}$ delivered by the monochromator crystal. The main purpose of the simulations presented here was to find a resonator configuration which is just sufficient to invert the polarisation of the 1st order neutrons with a wavelength of 2.6 Å, and to obtain a detailed knowledge of the homogeneity of the magnetic fields following from the specific geometry of this resonator. Using the CST-Studio Suite^{TM} the three-dimensional magnetic field distribution was simulated both for'rectangular' and 'Gaussian' shaped amplitude modulation of the transversal field oscillations. Subsequently, assuming an incident beam fully polarised in z-direction, all three components of the polarisation vector behind the resonator were calculated for a band of wavelengths ranging from 1 to 4 Å by means of numerical integration of the spin equation of motion $d\boldsymbol{P}/dt = \gamma \boldsymbol{P} \times \boldsymbol{B}$. In Fig. 3 both the transversal field component $B_1(x)$ along the beam axis and the resulting polarisations are shown for these two different resonator configurations. In order to quantify the homogeneity of the magnetic field, a series of simulations for different neutron trajectories through the resonator were performed. Their results are plotted in Fig. 4. It is clearly seen that the simulated results for neutron trajectories at distances of up to 4 cm off the center in z-direction and of up to 1 cm in y-direction are completely indistinguishable from each other. Since in our experimental setup the dimensions of the neutron beam are 2 cm in z-direction and 1 cm in y-direction, one can thus expect that for both field configurations the flipping efficiency of the resonator will be definitely homogeneous over the the whole beam cross-section.

3. First experimental results

The resonator prototype was built as exactly as possible according to the underlying design simulations: i.e. 10 individually switchable single-turn aluminium coils with a foil thickness of 0.3 mm, a height of 120 mm, a lateral dimension of 60 mm and a downstream thickness of 20 mm. To avoid electric short-circuits a gap of 1 mm between neighbouring coils was established by thin insulating frames. As sketched in Fig. 1 the resonator is placed between two between two polarising supermirrors acting as polariser and analyser, respectively. A rectangular Helmholtz-coil pair is used to define the resonator's static magnetic field $\mathbf{B_0}$, which is required to guide the neutron spin as well as to define the resonance wavelength λ_0. The strength of the spatially alternating field $\mathbf{B_1}$ is controlled by the (individually tunable) current that is fed to each sub-coil of the resonator. Each coil can be powered with a current of up to 15 A in order to create a magnetic field of up to 0.25 mT. Due to the already mentioned lack of a 'white' neutron beam we tuned the static field $\mathbf{B_0}$ to match the resonance condition for neutrons with 2.6 Å wavelength. In Fig. 5 the transmitted neutron intensity is plotted as a function of the DC current in the

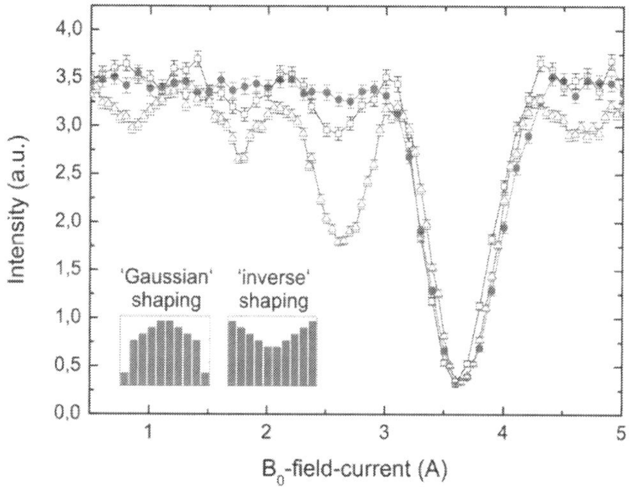

Figure 5. The amplitude of the alternating field $\boldsymbol{B_1}$ of the resonator prototype was modulated according to the different shaping functions shown as insets. A 'Gaussian-shaped' field indeed minimises the disturbing subsidiary maxima (*viz.* intensity minima) in the spinflip efficiency compared to the case of no amplitude modulation (open squares). However, as expected, an 'inverse-Gauss-shaping' leads to a drastic enhancement of these disturbing maxima.

Helmholtz-type coil pair both without (i.e. 'rectangular') and with 'Gaussian' shaping of the alternating field amplitude. For purpose of demonstration the effect of 'inverse' Gaussian shaping is shown too.

Additionally, in order to resolve the two discrete spectral components of the incident dichromatic neutron beam, a mechanical Fermi-type neutron chopper was used to perform neutron time-of-flight measurements. As can be seen explicitly in Fig. 6, tuning the resonator to one of the two available neutron wavelengths strongly suppresses the respective spectral contribution. The inset shows the operational simplicity and hence reliability of the corresponding tuning procedure.

3.1. Resonator efficiency

To analyse the efficiency of the resonator a second spin flipper was used, in this case a simple standard DC coil flipper. From the combination of four different neutron intensities accumulated during successive experimental runs with *i*) both the resonator and the DC flipper turned off (I_0), *ii*) with inactive resonator but activated DC flipper (I_F), *iii*) with activated resonator but inactive flipper (I_R), and *iv*) finally with both devices activated (I_{RF}) one can use the following

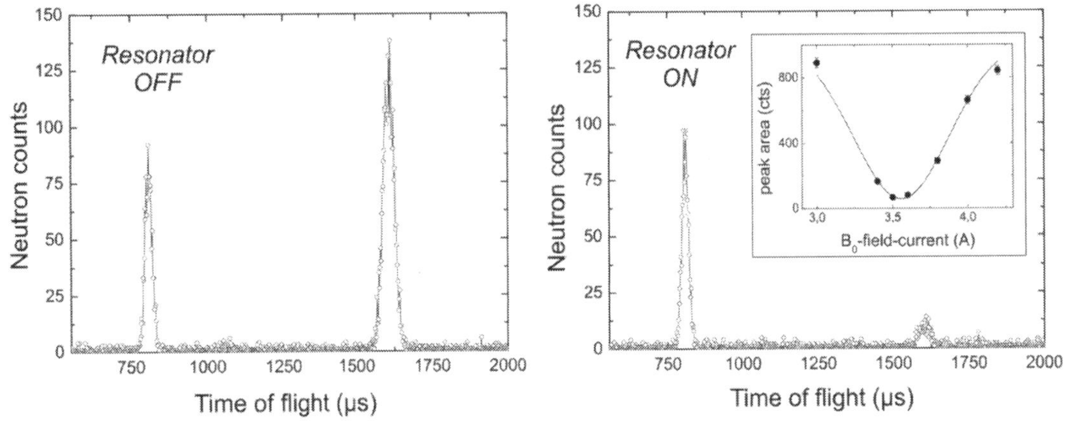

Figure 6. (*Left*) Results of time-of-flight measurements using a mechanical chopper when the resonator is not activated. The two available spectral contributions are clearly separated according to their different arrival times at the detector. (*Right*) As expected, activation of the resonator and tuning it to a resonance wavelength of 2.6 Å minimises the contribution of the corresponding spectral component. In the inset the detected count rate is plotted against the DC current that is fed to the Helmholtz coil pair which generates the static resonator field $\mathbf{B_0}$ and thus defines the respective resonance wavelength.

relations to determine the efficiencies $e_{\rm R}$ and $e_{\rm F}$, respectively, of both the resonator and the DC flipper, as well as the product P^2 of the (identical) polarisation powers of polarising and analysing supermirror:

$$\begin{aligned} \frac{I_{\rm F}}{I_0} &= \frac{1 - e_{\rm F}P^2}{1 + P^2} \\ \frac{I_{\rm R}}{I_0} &= \frac{1 - e_{\rm R}P^2}{1 + P^2} \\ \frac{I_{\rm RF}}{I_0} &= \frac{1 + e_{\rm R}e_{\rm F}P^2}{1 + P^2} \end{aligned}$$

Since an initially fully polarised beam would leave any flipping device which exhibits a spinflip probability of 50% like an unpolarised one with respect to its spin 'up' and spin 'down' contributions, the simple relation $e = 2k - 1$ holds between the efficiency e and the spinflip probability k. Averaging of the results of a series of repeated measurements the following values for the polarisation degree product and the spinflip probabilities of the resonator and the DC flipper were obtained: $P^2 = 0.838 \pm 0.044$, $k_{\rm R} = 0.990 \pm 0.024$, $k_{\rm F} = 1.018 \pm 0.030$. This means that even this rather simple prototype resonator exhibits almost perfect performance. The rather poor polarisation product is very likely due to an admixture of unpolarised neutrons which have passed the mirrors without reflection. An improved mirror alignment will be required in order to suppress this contribution.

4. Pulsed mode of operation

The creation of wavelength-selected neutron pulses of arbitrary time structure will be the ultimate goal for this new type of neutron spin resonator. Here, we present first results of a prototype driven in 'classic' pulse mode, i.e. by turning the whole resonator on and off again.

Thereby the neutron passage time through the resonator defines the shortest possible pulse length, which in our case is roughly about 140 μs. In Fig. 7 neutron pulses with a duration of 140 μs and 500 μs, respectively are shown. A 'travelling wave' mode of operation, where at any moment of time only one section of the resonator after the other is activated and deactivated, should allow to produce much shorter pulses and to decouple the minimal neutron pulse width from the achievable wavelength resolution, which for given resonator period depends upon its total length, as has been proposed for the first time in [16]. To realise this ambitious goal we have developed a quite sophisticated computer controlled pulsed power supply. Its details as well as the results of the corresponding experiments, which meanwhile have begun, will soon be presented in forthcoming publications.

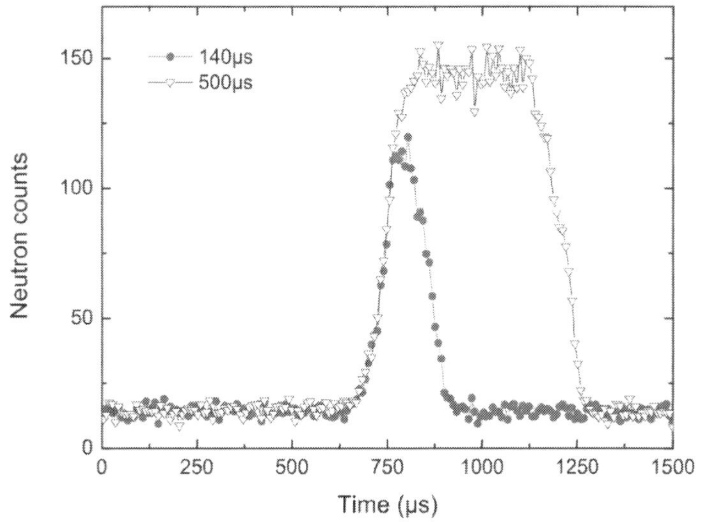

Figure 7. Time-of-flight spectrum obtained by operating the resonator prototype in a 'classical' way by pulsing it as a whole in order to create neutron pulses of $140\mu s$ and $500\mu s$, respectively.

It is worth to mention that the achievable 'switching ratio', i.e. the ratio between maximal and minimal intensity, becomes nearly completely independent of the efficiency of any type of pulsed flipping device, provided an additional static spin flipper - which can easily be designed to yield virtually 100% efficiency - is inserted into the beam path [17]. Thus the quality of the pair of polarisers remains as the only crucial quantity which has to be pushed to its limits in order to achieve a satisfactory overall performance of the setup and to minimise the unmodulated wavelength independent background intensity. Meanwhile switching ratios well beyond 10^2 have been realised by using crossed pairs of supermirrors [9], and neutron polarisation degrees of 99.99% have been achieved by means of ^3He spinfilters [10], so that switching ratios of the order of 10^4 are seemingly feasible. However, with current technologies such an extremely high degree of polarisation can be attained only on the cost of an enormous loss of beam intensity. This is a severe problem which has to be solved before the implementation of our device as integrated monochromator & chopper for neutron time-of-flight spectroscopy may become a really useful option. However, we are quite sure that for the specific configuration of the already mentioned PERC project - where our new spin resonator has to undergo its first test under realistic experimental conditions - it will be possible to minimise the disturbing influence of the unmodulated background by gating the neutron decay product detectors according to the pulse structure of the spatial spin resonator. But this implementation is a complex task still requiring considerable technological efforts and its treatment is beyond the scope of the present article.

5. Conclusion and outlook

We have simulated and designed a novel pulsed spatial magnetic neutron spin resonator. We could show that the experimental results of our fully working prototype fit very well to the data we got from our simulations. The next step will be to enhance the performance of this resonator in order to create shorter and sharper pulses utilising a 'travelling-wave' mode operation. Based upon the experience gained by studying the performance of the present prototype, we will then begin to design a more elaborate resonator system with at least one order of magnitude larger number of elements and bigger dimensions to allow for the passage of beams with larger cross-sectional area, provided, of course, that we can raise the required monetary funding for this project.

Acknowledgments

Financial support by the Austrian Fonds zur Förderung der wissenschaftlichen Forschung (Project No. I528-N20, MONOPOL) and by the FP7 Grant Agreement No.226507-NMI3 (Integrated Infrastructure Initiative for Neutron Scattering and Muon Spectroscopy) is gratefully acknowledged.

References

[1] Drabkin G M 1963 *Sov. Phys. JETP* **16** 282

[2] Drabkin G M, Trunov V A and Runov V B 1968 *Sov. Phys. JETP* **27** 194

[3] Agamalyan M M, Drabkin G M and Sbitnev V I 1988 *Phys. Rep.* **168** 265–303 ISSN 0370-1573

[4] Majkrzak C F and Shirane G 1982 *J. Phys.* **43** Coll. C7215

[5] Parizzi A, Lee W T and Klose F 2002 *Appl. Phys. A* **74**(0) s1498–s1501 ISSN 0947-8396

[6] Yamazaki D, Soyama T, Aizawa K and Tasaki S 2004 *Nucl. Instr. Meth. Phys. Res. A* **529** 204–208

[7] Badurek G, Kollmar A, Seeger A and Schalt W 1991 *Nucl. Instrum. Meth. A* **309** 275–283 ISSN 0168-9002

[8] Alefeld B, Kollmar A, Badurek G and Drabkin G 1991 *Nucl. Instrum. Meth. A* **306** 300–304 ISSN 0168-9002

[9] Kreuz M, Nesvizhevsky V, Petoukhov A and Soldner T 2005 *Nucl. Instrum. Meth. A* **547** 583–591 ISSN 0168-9002

[10] Klauser C, Chastgnier J, Jullien D, Petoukhov A and Soldner T 2011 *J. Phys.: Conf. Ser.* **this issue**

[11] Gösselsberger C, Abele H, Badurek G, Jericha E, Nowak S, Wautischer G and Welzl A 2011 *Physics Procedia* Accepted,
see also: Badurek G, Gösselsberger C and Jericha E 2011 *Physica B* **406** 2458

[12] Badurek G, Gösselsberger C and Jericha E 2011 *Physica B* **406** 2458–2462 ISSN 0921-4526

[13] Dubbers D, Abele H, Baessler S, Märkisch B, Schumann M, Soldner T and Zimmer O 2008 *Nucl. Instrum. Meth. A* **596** 238 – 247 ISSN 0168-9002

[14] Abele H 2008 *Prog. Part. Nucl. Phys.* **60** 1–81 ISSN 0146-6410

[15] Agamalyan M M, Drabkin G M and Lebedev V T 1977 *Sov. Phys. JETP* **73** 382

[16] Badurek G and Jericha E 2003 *Physica B* **335** 215–218 ISSN 0921-4526

[17] Badurek G 1981 *Nucl. Instrum. Meth.* **189** 543

Optimization of virtual source parameters in neutron scattering instrumentation

K Habicht and M Skoulatos
Helmholtz-Zentrum Berlin für Materialien und Energie,
Hahn-Meitner-Platz 1, D-14109 Berlin, Germany

E-mail: habicht@helmholtz-berlin.de

Abstract. We report on phase-space optimizations for neutron scattering instruments employing horizontal focussing crystal optics. Defining a figure of merit for a generic virtual source configuration we identify a set of optimum instrumental parameters. In order to assess the quality of the instrumental configuration we combine an evolutionary optimization algorithm with the analytical Popovici description using multidimensional Gaussian distributions. The optimum phase-space element which needs to be delivered to the virtual source by preceding neutron optics may be obtained using the same algorithm which is of general interest in instrument design.

1. Introduction

Inelastic neutron scattering provides a versatile method for investigations of single crystal excitations. To overcome the intensity limits of the technique focusing by either supermirror neutron optics, crystal optics or a combination of both is applied in state-of-the-art neutron scattering instrumentation. The latest generation three-axis spectrometers (TAS) consequently seeks to employ monochromators which do not only focus in the vertical direction but also horizontally [1-4]. The basic idea is to increase intensity per area and per unit energy by imaging a well defined aperture onto the sample, known as the virtual source concept. Although the principle is quite general and can be used in either diffractometers or spectrometers, here we consider the situation in the particular context of TAS where energy resolution is crucial.

Since beam monochromaticity, wave-vector resolution and intensity depend on a sizable number of parameters, namely on the absolute wave vector as well as on the virtual-source size, the horizontal curvature, the crystal mosaic, the geometrical dimensions of the monochromator setup and the sample size, it is advantageous to have an optimization routine at hand suggesting best performance to the experimenter. Here we report on the implementation of an evolutionary algorithm applied to a generic virtual source setup and present selected optimization results.

In particular we consider three different optimization tasks: (1) Given a built TAS with a limited number of variable parameters, what is the optimum choice of instrument parameters for given constraints on energy or wave-vector resolution? (section 4), (2) Given a defined divergence at the virtual source, fixed sample size and given wave vector, what is an optimum set of parameters for instrument design? (section 5) and (3) Given a typical sample size and fixed wave vector, what is the optimum phase-space element which needs to be delivered to the virtual-source setup? (section 6).

2. Virtual source model

We reduce the primary spectrometer part of a TAS to a generic virtual source setup consisting only of: a slit, a focussing PG(002) monochromator and a sample. As a major simplification we consider the horizontal dimension only. This is legitimate since vertical and horizontal resolution entirely decouple for practical means.

Strict monochromatic and simultaneously spatial focusing using a crystal monochromator can only be achieved for either a divergent neutron beam emerging from a point-like source or a non-divergent beam emerging from a spatially extended source. The virtual-source concept is closest to the case where a point source is imaged onto a point-like sample although in practice finite spatial dimensions apply to the source and sample. The geometry of the virtual-source setup and the ideal radii of curvature are introduced in figure 1 for the symmetric Rowland geometry. More details on this model can be found in [5].

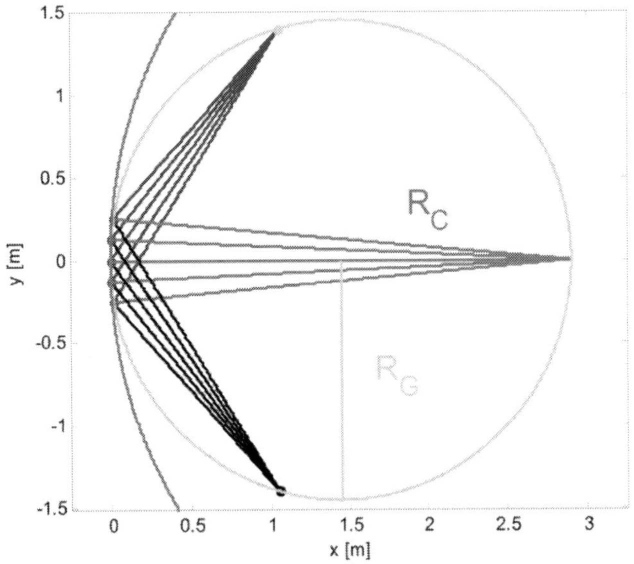

Figure 1. Numerically calculated neutron trajectories for the symmetric Rowland geometry. Parameters are k_I = 1.55 Å^{-1} and L_0 = 1.75 m. The red circle corresponds to the curvature of the lattice planes with R_C = 2.897 m. The green circle is the Rowland circle with radius R_G=R_C/2 which defines the ideal loci of reflection for the individual monochromator crystallites. Black lines are neutron trajectories incident on the PG(002) monochromator while blue lines are neutron trajectories Bragg reflected by the monochromator. Red lines are normal to the reflecting lattice planes of the local crystallites. The angular range covered is ±5°.

Stoica [6] and Popovici [7] have given the normal approximation of the resolution function for a conventional TAS in a covariance matrix formalism as derived earlier by Cooper and Nathans [8]. By comparing to Monte Carlo simulations we have shown that the Popovici model is completely adequate to investigate the virtual source problem. This allows looking into optimization issues using the much faster analytical path which reduces computational time compared to Monte Carlo simulation.

To prove the adequacy of the Gaussian model of a virtual source primary spectrometer we have cross-checked intensity, spatial beam width, width in incident energy, width in transverse incident wave vector and correlation in wavelength and divergence as a function of horizontal curvature and incident wave vector k_I with the Monte Carlo simulation package McStas [9]. There are of course differences between the McStas model and the Popovici description, e.g. the Monte Carlo model is more accurate since it does not rely on the Gaussian approximation except for the monochromator mosaicity. Slits have rectangular openings. Spatial monochromator parameters have rectangular dimensions with a sharp cutoff rather than Gaussian shapes. Also the McStas model is three-dimensional but we do not consider the vertical dimensions and effectively integrate over the vertical components. In the McStas component the thin crystal approximation is used while the Popovici description adopted allows for finite monochromator thickness. For both parameterizations of the monochromator the crystal surface is parallel to the lattice planes.

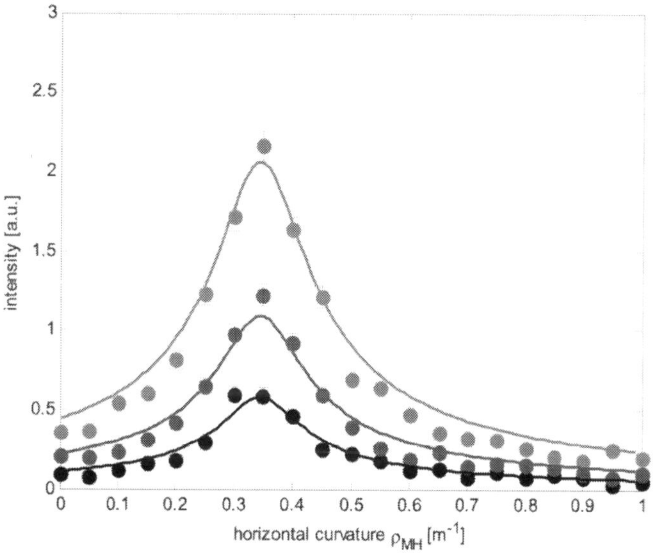

Figure 2. The intensity at the sample position as a function of the curvature parameter ρ_{MH} for different virtual source width w_0 and sample width w_S. Red: w_0 = 2 cm, w_S = 2 cm; blue: w_0 = 1 cm, w_S = 2 cm; black: w_0 = 1 cm, w_S = 1 cm. Note that the intensity is not normalized to area! The lines are results of the numerical evaluation of the analytical Popovici expression.

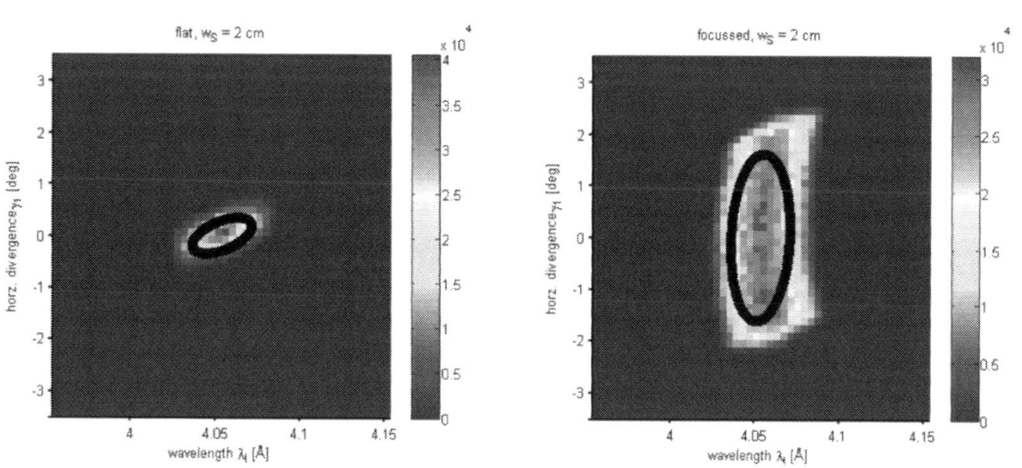

Figure 3. Wavelength-divergence correlation plots for flat (*left*) and horizontally focussed monochromator (right) integrated over a sample width w_S = 2 cm. The color map data are simulation results using McStas. The black ellipses are derived from the Popovici model and are lines of constant probability 50 % .

Despite the differences in the models, there is in general an excellent agreement between the simulated and the calculated results. An example for the intensity as a function of monochromator curvature is shown in figure 2. The agreement between the second moments of distributions obtained from the Monte Carlo model and the same quantities obtained from the Popovici description is found to be excellent. Naturally deviations from the Gaussian distributions are not covered by the Popovici description as is seen from the phase space elements shown in figure 3.

3. Evolutionary algorithm

Here we apply an evolutionary optimization routine, known as genetic algorithm [10, 11] which is sufficiently efficient for the 10-dimensional parameter space considered. We calculate the intensity at the sample position I_S, the energy width ΔE_S and the transverse wave vector width $\Delta k_\perp$ perpendicular to k_I with Popovici's approach. All optimization runs discussed here use genetic evolution parameters: 95% cross-over (the percentage of reproduction events), 2% mutation (the percentage of genes in next-generation instruments that mutate) and exclusivity 1 which is the number of best instruments kept unchanged for the next generation. These parameters are explained in detail in reference [11]. We run 20 generations with a population of 30 instrument configurations per generation which samples only about 10^{-4} of the total number of possible instrument configurations. The best instrument is chosen on the basis of best fitness, i.e. the maximum figure of merit.

4. Experiment parameter optimization

For practical TAS applications the simplest optimization task is met at a given spectrometer with only a reduced number of variable parameters. We ask for the optimum virtual source width and monochromator curvature giving maximum intensity under the boundary condition of a defined energy width or a defined wave-vector width. We assume that k_I is given by other boundary conditions e.g. the total energy transfer and the accessibility of the total wave-vector transfer **Q**.

As an example we consider k_I = 1.55 Å^{-1} and a set of neutron optical components which effectively provides $m = 5$ divergence at the entrance of a virtual source slit which can be varied in the range 0.1 cm $\leq w_0 \leq$ 3 cm (in steps of 0.1 cm). The monochromator width is w_M = 30 cm, monochromator mosaic is η_M = 45'. The distances from the virtual source to the monochromator L_0 =1.75 m and from the monochromator to the sample L_1 =1.75 m are chosen to be equal for the symmetric Rowland configuration. A standard sample size w_S = 2 cm is considered. Limiting the energy or wave-vector width to defined values is achieved by setting the figure of merit to zero for those instrument configurations which give unacceptable resolution. We grade instrument configurations with the figure of merit

$$f_1 = \begin{cases} 0 & \text{if} \quad \Delta E_S, \Delta k_\perp > \Delta E_{\text{limit}}, \Delta k_{\perp\text{limit}} \\ 10^{10} \times I_S & \text{else} \end{cases} \tag{1}$$

thus maximizing intensity.

Table 1 and figure 4 (left) show the results of the optimization when imposing a limit on the energy width. While the curvature stays at its optimum value for a divergent point source, the virtual source width is reduced to meet the energy width required. The worst energy width achievable with the virtual source width limited to 3 cm is 0.091 meV. Energy widths below 0.055 meV cannot be reached and thus the optimization breaks down if $\Delta E_{\text{limit}} < 0.055$ meV. The result of our series of optimization runs is that the energy resolution may be improved by a factor of 1.7 just by adapting the virtual source width while the intensity is reduced by a factor of 4.3. We note that with a flat monochromator $\rho_{MH} = 0$ and fully open virtual source width w_0 = 3 cm the intensity is $I_S = 11.23 \times 10^{-10}$ giving an energy width ΔE_S = 0.091 meV. This energy width cannot be further increased since it is essentially limited by the mosaic of the monochromator crystals and w_0.

Table 2 and figure 4 (right) give optimized parameters w_0 and ρ_{MH} for a given limit on the transverse wave-vector width $\Delta k_\perp$. Clearly the virtual source width stays always at the maximum constraint while the curvature changes to adapt for the wave-vector width requested. Indeed, for $\Delta k_{\perp \mathrm{limit}} = 0.060$ Å^{-1} and 0.055 Å^{-1} the monochromator is overfocused. The worst wave-vector width achievable with the optimum focusing ρ_{MH} = 0.34 m^{-1} is 0.1 Å^{-1}. For wave-vector widths below 0.055 Å^{-1} the optimization breaks down. As a result of our optimization the wave-vector resolution may be improved by a factor of 1.83 just by adapting the monochromator curvature ρ_{MH} while the intensity is reduced by a factor of 2.

Table 1. Results of the instrument parameter optimization maximizing intensity given an energy resolution limit. k_I = 1.55° Å^{-1}, m=5, w_M = 30 cm, η_M = 45′, w_S = 2 cm.

$\Delta E_{\mathrm{limit}}$ *(meV)*	w_0 *(cm)*	ρ_{MH} *(m^{-1})*	I_S $\times 10^{10}$	ΔE_S *(meV)*	$\Delta k_\perp$ *(Å^{-1})*
0.100	3.0	0.34	17.9	0.091	0.090
0.080	2.3	0.32	13.8	0.078	0.098
0.060	1.2	0.34	7.3	0.060	0.097
0.055	0.7	0.30	4.2	0.055	0.095

Table 2. Results of the instrument parameter optimization maximizing intensity given a transverse wave vector resolution limit $\Delta k_\perp$. k_I = 1.55° Å^{-1}, m=5, w_M = 30 cm, η_M = 45′, w_S = 2 cm.

$\Delta k_{\perp \mathrm{limit}}$ *(Å^{-1})*	w_0 *(cm)*	ρ_{MH} *(m^{-1})*	$I_S \times 10^{10}$ *(a.u.)*	ΔE_S *(meV)*	$\Delta k_\perp$ *(Å^{-1})*
0.120	3.0	0.34	17.9	0.091	0.099
0.080	3.0	0.14	14.5	0.091	0.079
0.060	3.0	0.73	10.1	0.090	0.059
0.055	3.0	0.80	9.0	0.090	0.054

5. Instrument design parameter optimization

We have also investigated the situation where the available phase space before the virtual source is known and consider instrument design optimization with fixed parameters: k_I = 1.00 Å^{-1}, 1.55 Å^{-1}, 3.70 Å^{-1}, $L_0 = L_1$ = 1.75 m, w_S = 2 cm, monochromator crystal thickness t_M = 2 mm and variables to be optimized: virtual source width w_0, monochromator width w_M, crystal mosaic η_{MH} and curvature ρ_{MH}. Each variable can take discrete values with 0.1 cm $\leq w_0 \leq$ 3 cm (in steps of 0.1 cm), 1 cm $\leq w_M \leq$ 50 cm (in steps of 1 cm), 5' $\leq \eta_M \leq$ 60' (in steps of 1'), 0 m^{-1} $\leq \rho_{MH} \leq$ 1.0 m^{-1} (in steps of 0.01 m^{-1}). Thus the parameter space consists of 8.5×10^6 instrument configurations. We have considered four different cases for the divergence in front of the virtual source slit: small divergence (^{58}Ni neutron guide), medium (*m*=3 neutron guide) or larger *m*=5, *m*=7 (as is the case for e.g. elliptically tapered guide sections).

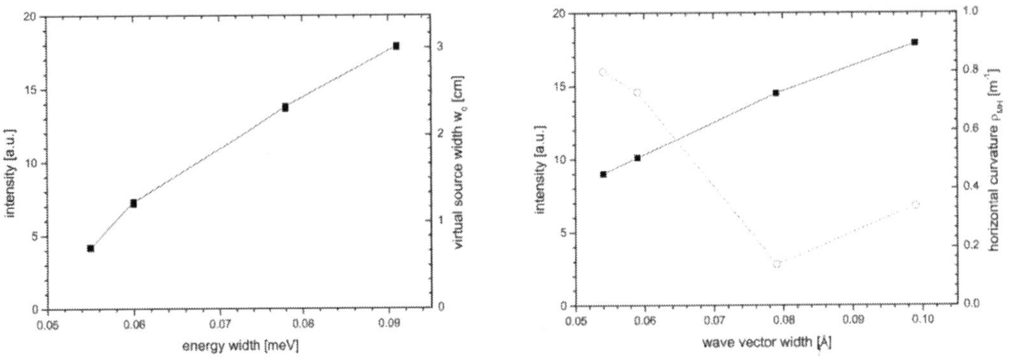

Figure 4. Intensity optimizations when imposing a limit on the energy width (left) or a limit on the transverse wave-vector width (right). Since the intensity scales linearly with the virtual source width, the curves for intensity and virtual source width fall on top of each other for the chosen scales in the left hand panel.

We wish to maximize the intensity at the sample position I_S and minimize the energy width ΔE_S of the neutron beam at the sample and define the dimensionless figure of merit

$$f_2 = 10^{10} \times I_S + \frac{1}{100} \times \frac{meV}{\Delta E_S}. \tag{2}$$

Since I_S is on the order of $10^{-10} - 10^{-9}$ and ΔE_S is on the order of 0.01 meV (at $k_I = 1.0$ Å^{-1}) this choice of prefactors ensures that the individual terms are roughly the same order of magnitude. Results are summarized in Table 3.

The optimum mosaic η_M ranges between 51 and 60 arcmin throughout the whole wave-vector range considered indicating little sensitivity to this parameter in agreement with our analytical results on the intensity as a function of η_M. The optimum mosaic value is also in agreement with extended Monte Carlo simulations performed for the upgrade of the cold triple axis spectrometer FLEX [4]. The optimized monochromator curvature is found to be close to the theoretical curvature appropriate for a divergent point-source ρ_{MH} = 0.53 m^{-1}, 0.35 m^{-1}, 0.14 m^{-1} for k_I = 1.00 Å^{-1}, 1.55 Å^{-1}, 3.70 Å^{-1} respectively. Interestingly for the smallest divergences incident on the virtual source the optimum monochromator curvature is found slightly smaller. This is expected as the optimum monochromator curvature ρ_{MH} depends on the divergence and is reduced by a factor 2 in the limit of a non-divergent beam as can be derived easily from geometrical optics.

For all wave vectors and independent of the incident divergence the virtual source width w_0 is found to be the maximum allowed (constrained to 3 cm). The monochromator width w_M is also found to be the maximum allowed (constrained to 50 cm). One cannot expect the optimization to find balanced values for monotonic parameters where the maximum figure of merit is at the end of the allowed parameter range. For such parameters we suggest to "optimize" by defining an acceptance level for the figure of merit.

The intensity at the sample varies within a factor of 50 (I_S =1.5 × 10^{-10} … 7.6 × 10^{-9}) while the energy width at the sample varies by a factor of ~150 (ΔE_S = 10 μeV … 1.5 meV). As a result of the optimization the energy resolution is essentially independent of the incident divergence while the intensity scales linearly with the incident divergence at the expense of transverse wave vector resolution $\Delta k_\perp$.

Table 3. Results of the instrument design optimization based on the figure of merit f_2. Incident divergence and k_I are fixed input parameters, while η_M and ρ_{MH} are optimized parameters.

Incident divergence	k_I ($Å^{-1}$)	η_M (*arcmin*)	ρ_{MH} (m^{-1})	$I_S \times 10^{10}$ (*a.u.*)	ΔE_S (*meV*)	$\Delta k_\perp$ ($Å^{-1}$)	f_2
^{58}Ni	1.0	60	0.42	1.47	0.011	0.022	2.4
m=3	1.0	56	0.47	3.62	0.011	0.045	4.5
m=5	1.0	53	0.53	5.76	0.011	0.070	6.7
m=7	1.0	57	0.51	7.53	0.011	0.090	8.5
^{58}Ni	1.55	60	0.26	5.19	0.092	0.043	5.3
m=3	1.55	60	0.33	12.83	0.092	0.074	12.9
m=5	1.55	51	0.34	20.25	0.092	0.111	20.4
m=7	1.55	60	0.34	26.72	0.092	0.142	26.8
^{58}Ni	3.70	55	0.08	14.75	1.500	0.198	14.8
m=3	3.70	54	0.13	36.41	1.500	0.242	36.4
m=5	3.70	52	0.16	57.33	1.500	0.305	57.3
m=7	3.70	59	0.16	75.84	1.500	0.368	75.8

6. Optimum phase space tailoring

For instrument design the most important question is: what is the optimum phase-space element which needs to be delivered to the virtual source? Modern neutron optics provides an elegant solution to manipulate the phase-space area at the virtual-source section while increasing intensity in the center of the slit. Combining a neutron guide with a converging section in the horizontal dimension allows spatial compression at the expense of increase in divergence, e.g. elliptical or parabolic compression [12,13]. The basic idea here is to ensure that the phase space element transported through the neutron guide system can actually be used at the sample position. We define an extended figure of merit

$$f_3 = 10^{10} \times I_S + \frac{1}{100} \times \frac{meV}{\Delta E_S} + \frac{1}{100} \times \frac{rad}{\gamma_0} \tag{3}$$

aimed at maximizing the intensity at the sample I_S, minimizing the width of the energy spectrum at the sample ΔE_S and keeping the divergence γ_0 before the virtual source as minimal as possible. In addition Δk_0 can be minimized by using a velocity selector which pre-limits the width of the spectrum incident on the virtual source to reduce background.

We assume that the area of the phase space element $k_I\gamma_0 \times w_0$ before the converging section is constant and is equal to the area equivalent to an m = 2.5 guide with width w_0 = 6 cm. Thus, assuming a square distribution the divergence has a variance

$$\gamma_0 = \frac{1.1296 \times 10^{-3}}{w_0[m]k_I[Å^{-1}]}. \tag{4}$$

Any decrease in spatial width requires an increase in divergence due to Liouville's theorem. We have allowed the virtual source width to vary in the range of 0.1 cm $\leq w_0 \leq$ 12 cm (in steps of 0.1 cm). Note that a source width above 6 cm would actually mean a spatially expanding beam with reduced divergence relative to m=2.5. The monochromator mosaic and curvature are allowed to vary within 5' $\leq \eta_M <$ 60' (in steps of 1') and 0 m^{-1} $\leq \rho_{MH} \leq$ 1.0 m^{-1} (in steps of 0.01 m^{-1}). The monochromator width is kept fixed at w_M = 30 cm. Under these conditions we find a virtual source width which is 4 cm at the highest wave vector considered (k_I = 3.70 Å^{-1}), i.e. moderate beam compression is required corresponding to m = 3.75 after the converging neutron optical section.

Our optimizations find the minimum virtual source width allowed (w_0 = 1 mm) to be optimal at small wave vector and small sample size, however leading to a large, technically not achievable effective supermirror m. The explanation can be found when considering the figure of merit as a function of virtual source width. At the smallest virtual source width the figure of merit has a global maximum since it is dominated by the $1/\Delta E_S$-term. We emphasize that great care is needed in defining the figure of merit. We note that a threshold on intensity or a limit on the effective m-coating gives the solution w_0 = 4 cm for all wave vectors suggesting rather modest beam compression to be optimal.

7. Conclusions

We have investigated optimization of instrument parameters for a generic virtual source setup employing an evolutionary algorithm. As the combination of virtual source width and monochromator curvature gives additional flexibility to tune the energy and wave-vector resolution in order to match the instrument properties to a particular experiment it is desirable to have an optimization routine at hand suggesting best performance.

We have adopted Popovici's analytical approach for a fast evaluation of the quantities of interest: intensity and resolution. Comparison to Monte Carlo simulation proves adequate accuracy. Most useful for the experimenter is the routine to identify instrument parameters required to achieve a particular resolution either in energy or in wave vector.

We have also applied the optimization routine to instrument design tasks. Independent of incident wave length we find the optimum PG002 mosaic at values which are conventionally in practical use for horizontally flat monochromators. In addition we have applied the algorithm to identify the optimum phase-space element to be delivered to the virtual source and found moderate compression ratios of 1.5 optimal for medium size samples. Smaller sample sizes and higher neutron energies require higher compression ratios.

Our study of the optimization problem has drawn our attention to a few general insights applicable to any optimization problem. The genetic algorithm works best on parameters which show a maximum within the parameter range as is seen with the monochromator curvature in our case. One cannot expect the optimization to find balanced values for monotonic parameters where the maximum figure of merit is at the end of the allowed parameter range. For such parameters it is best to"optimize" by defining an acceptance level for the figure of merit. In general, the definition of the figure of merit is crucial and the optimum parameters depend sensitively on the weight that a particular instrument property is given.

Acknowledgments

We would like to thank P.M. Bentley for his generous agreement to use part of his computational C++ code for the optimizations based on the genetic algorithm.

References

[1] Böhm M *et al.* 2007 *J. Magn. Magn. Mater.* **310** e965
[2] Janoschek M, Böni P and Braden M 2010 *Nucl. Instrum. Methods Phys Res A* **613** 119
[3] Schmalzl K *et al.* 2008 FZ Jülich Annual Report
[4] Skoulatos M, Habicht K 2011 *Nucl. Instr. and Meth. A Preprint* doi:10.1016/j.nima.2011.05.037
[5] Skoulatos M, Habicht K, Lieutenant K *'Improving energy resolution on neutron monochromators'*, these proceedings
[6] Stoica A 1975 *Acta Cryst. Sec. A* **31** 189
[7] Popovici M 1975 *Acta Cryst. Sec. A* **31** 507

[8] Cooper M, Nathans R 1967 *Acta Cryst.* **23** 357
[9] Lefmann K, Nielsen K 1999 *Neutron News* **10** 20; McStas CVS-080208 Version 0.2 2009
[10] Bentley P M, Pappas C, Habicht K, Lelièvre-Berna E 2006 *Physica B* **385-386** 1349–1351
[11] Hölzel R, Bentley P M, Fouquet P 2006 *Rev. Sci. Instrum.* **77** 105107
[12] Böni P 2008 *Nucl. Instrum. Methods Phys Res A* **586** 1–8
[13] Mühlbauer S, Niklowitz P G, Stadlbauer M, Georgii R, Link P, Stahn J and Böni P 2008 *Nucl. Instrum. Methods Phys Res A* **586** 77–80

Neutron-spin-echo from polarizing samples

J. Kindervater[1], W. Häußler[2,1], A. Tischendorf[1] and P. Böni[1]

[1]Technische Universität München, Physik Departement E21, 85748 Garching, Germany
[2]Forschungs-Neutronenquelle Heinz Maier-Leibnitz (FRM II), 85748 Garching, Germany

E-mail: `jonas.kindervater@frm2.tum.de`

Abstract. The dependence of the neutron spin polarization on the difference of the magnetic field integrals along the precession fields before and after the sample gives rise to a spin-echo signal. In neutron spin-echo (NSE) experiments with non-polarizing samples, a single spin-echo group is observed (or two in case of some magnetic sample systems). However, in NSE experiments using polarizing samples, three different groups appear: i) the spin rotation (SR) group, ii) the antiparallel echo (APE) group and iii) the parallel echo (PE) group. The echo groups contain information on the life time of the excitations from which the intermediate scattering function $S(Q,t)$ can be derived. In this article, we discuss the basic strategies for NSE measurements using the 2-, 4-, and 20-point method. Finally, we report on quasi elastic NSE experiments on a polarizing magnetic sample MnSi, which stabilizes a single-handed spin spiral. Here, the overlap of the spin rotation and spin-echo group for small spin-echo times requires a measurement of the NSE signal over a wide range of phase angles.

1. Introduction

Neutron Spin Echo (NSE) spectroscopy [1] as a quasi elastic neutron scattering technique uses the neutron spin to detect small changes in kinetic energy of the neutrons before and after the sample. The NSE-technique yields directly the intermediate scattering function $S(Q,t)$ being the Fourier transform of the scattering function $S(Q,\omega)$. NSE allows to measure the dynamics of excitations with sub-μeV resolution. A measure for the resolution is given by the spin echo time $\tau = (\gamma m_n{}^2 \bar{\lambda}_0{}^3 BL)/(2\pi h)$, where γ is the gyromagnetic ratio, m_n the mass and $\bar{\lambda}_0$ the average wavelength of the neutrons. τ is proportional to the magnetic field integral BL. In this paper we report on the effects of polarizing samples on the NSE-signal at small τ when the spin echo groups interfere with a spin rotation (SR) group. Therefore, to determine $S(Q,t)$ from a NSE-measurement, the influence of the sample on the polarization of the beam has to be taken into account.

Section 2 gives a short introduction to the NSE-technique. In section 3, three standard methods for analysing NSE-data are discussed. In part 4 we show how to treat NSE-data from polarizing samples in the limit of small τ.

2. Neutron Spin Echo Spectroscopy

Fig. 1 displays a schematic of the NSE spectrometer RESEDA (Ref. [2]) located at the end position of neutron guide NL5 at the Forschungs-Neutronenquelle Heinz Maier-Leibnitz (FRM II). The wavelength band of the incident polarized beam is determined by a velocity selector. While passing the precession region in the first coil, the neutron spin accumulates a spin phase

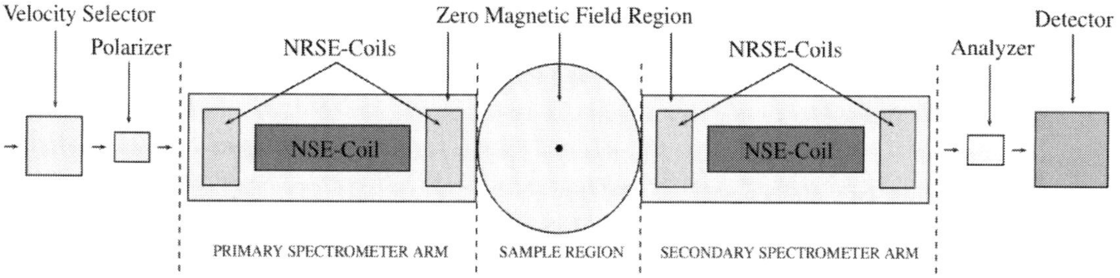

Figure 1: Principal components of RESEDA. The zero field regions are produced by means of double μ-metal shielding.

ϕ_1 that is given by $\phi_1(\lambda) = \gamma B_1 L_1 \lambda/(hm_n)$. In the second spectrometer arm the field B_2 is anti parallel to B_1. Therefore the spin of the neutrons precesses backwards and the net spin phase is $\phi = \phi_2 - \phi_1$. The final polarization P of the neutron beam is directly proportional to $S(Q,t)$. The constant magnetic fields as used in NSE can be replaced in the Neutron Resonance Spin Echo (NRSE) technique by RF-spin flippers that define the boundaries of regions with zero field as shown in Fig. 1 (Ref. [3, 4]). The dependence of the neutron spin polarization on the difference of the magnetic field integrals along the precession fields before and after the sample gives rise to a spin echo signal. For both, NSE and NRSE in case of a monochromatic beam, the intensity at the detector, i.e. the so called spin echo group, is given by

$$I(\lambda) = I_{av}(P\cos(\phi) + 1) = A(\cos(\phi) + 1) + I_{BG}, \tag{1}$$

where I_{av} is the average intensity, P the polarization, $\phi(\lambda)$ the spin phase and I_{BG} the background. In the ideal case all neutrons of the beam contribute to the spin echo signal and precess the same way along the magnetic field. Commonly this is described by a fully polarized echo i.e. $P = 1$. A real spin echo measurement is influenced by both depolarization effects and unpolarised background, which can not be distinguished. Therefore a real measurement can be described either by a reduced polarization $P < 1$ or introducing a background $I_{BG} > 0$. In spin echo the normalized intermediate scattering function $S(Q,\tau)$ is measured. This can either be determined from the polarization P or the normalized amplitude A/A_0, where A_0 is the amplitude for elastic scattering.

3. Standard data analysis strategies

There are various methods used to calculate the polarization of neutrons at the detector. In the following, we discuss the statistical errors that are involved in conducting three different strategies.

For our further discussion we consider an ideal spin echo instrument for which the intensity and therefore the polarization are given by Eq. (1). For the velocity selector we adopt a Gaussian wavelength distribution with $\Delta\lambda/\lambda_0 = 15\%$ corresponding to a relative standard deviation $\sigma_r = \sigma/\lambda_0 = 0.0637$ and with a mean wavelength $\lambda_0 = 5$Å:

$$f(\lambda, \lambda_0) = \frac{1}{\sigma\sqrt{2\pi}} \exp\left[-\frac{(\lambda - \lambda_0)^2}{2\sigma^2}\right] \tag{2}$$

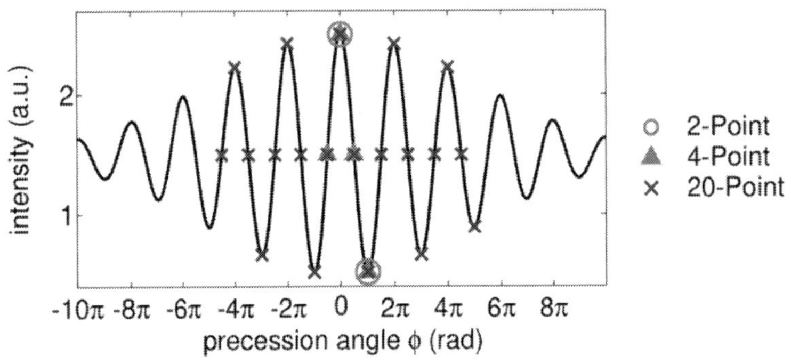

Figure 2: black curve shows the intensity of a spin echo signal from Eq. (3), the different markers shows the points used for the different methods discussed below

The intensity of the spin echo group is given by folding Eq. (1) with the wavelength distribution Eq. (2), i.e.

$$I(\phi(\lambda_0)) = \int_{-\infty}^{\infty} f(\lambda, \lambda_0)(A\cos(\phi(\lambda)) + A + I_{BG})\mathrm{d}\lambda = A\left(\cos(\phi)\exp\left[-\frac{\phi^2}{2\frac{\lambda_0^2}{\sigma^2}}\right] + 1\right) + I_{BG}, \tag{3}$$

Here, ϕ designates the net spin phase for the average wavelength λ_0. Fig. (2) shows the spin echo signal given by Eq. (3). It is a cosine function that is damped by a Gaussian with a relative standard deviation σ^{-1}. In the following we discuss two models for each strategy based on the assumption that i) a total number of 1000 neutrons per unit time is scattered by the sample into the polariser before the detector and ii) a background of 500 neutrons per unit time is added to the signal. For example, if a n-point echo is measured, for each of the n measured points of the echo signal an intensity of $1000/n$ reaches the analyser before the detector.

3.1. 2-Point-Echo method

The polarization of a neutron beam can be determined by measuring the intensities of the spin up I_{up} and spin down I_{down} neutrons at the detector, i.e. by conducting two measurements:

$$P_1 = \frac{I_{\mathrm{up}} - I_{\mathrm{down}}}{I_{\mathrm{up}} + I_{\mathrm{down}}}. \tag{4}$$

I_{up} and I_{down} can also be interpreted as two points of the cosine function of the spin echo group shown as green circles in Fig. 2 that are separated by a phase of $\phi = \pi$. Therefore, we call this measurement a *2-point-echo*. From Eq. (1) the amplitude is given by $A = P \cdot I_{\mathrm{av}}$. Introducing the average counts $I_{\mathrm{av}} = (I_{\mathrm{up}} + I_{\mathrm{down}})/2$ there are altogether three possibilities to determine the amplitude A.

- $A_1 = (I_{\mathrm{up}} - I_{\mathrm{down}})/2$
- $A_2 = I_{\mathrm{av}} - I_{\mathrm{down}}$
- $A_3 = I_{\mathrm{up}} - I_{\mathrm{av}}$.

For general NSE measurements, where for example 16 spin echo times are measured, one can consider I_{av} as independent from τ, therefore the statistical error of I_{av} is decreased by a factor of 4. The statistical errors for A_i as determined by using Gaussian error-propagation are given in Tab. 1. A major disadvantage of the *2-point-echo Method* is the loss of the phase information leading to large errors in P if the spin echo group accidentally shifts.

3.2. 4-point-echo Method

An alternative strategy is to take four points in steps of $\pi/2$ starting at $-\pi/2$ (red crosses in Fig. 2) allowing to analytically calculate the amplitude of the cosine function. This strategy is used by various groups at the FRM II and the ILL, because it has the advantage of determining A and the phase of the spin echo group. The four points are given by

$$I_n = A(\cos(\phi + \frac{\pi}{2}(n-1)) + 1) + I_{BG} \tag{5}$$

with $n = 0 \ldots 3$. From these points, the amplitude is given by

$$A = \frac{1}{2}\sqrt{(I_0 - I_2)^2 + (I_1 - I_3)^2}. \tag{6}$$

The statistical error of the amplitude as calculated by Gaussian error-propagation is given in Tab. 1. The uncertainty is increased from 0.032 to 0.045 when compared with the optimum *2-point-echo Method.*

3.3. 20-point-echo Method

The 20 points of the spin echo group are selected in the interval $-9/2\pi \leq \phi \leq 5\pi$ in steps of $\pi/2$ around the spin echo point (blue triangles in Fig. 2). The amplitude of the oscillation decreases with increasing distance from the spin echo point $\phi = 0$. Therefore the statistical error is strongly affected by the width of the wavelength band that defines the width of the Gaussian envelope of the signal. To calculate the statistical error of A we divide the 20-points of the echo into 5 groups of 4 echo points. Each of the groups is analysed according to Eq. 6. The total statistical error is then determined by quadratic summation.

$$\frac{\Delta A}{\bar{A}} = \frac{1}{5\bar{A}}\sqrt{\sum_{i=1}^{4} \Delta A_i} \tag{7}$$

The statistical error of A is slightly increased from 0.045 to 0.053 when compared with the *4-point-echo Method* (Tab. 1).

3.4. Comparison of the methods

An overview of the statistical errors of the different methods is given in Tab 1. Considering only the statistical error and the measuring time the best way to measure the polarization is the *2-point-echo Method.* Considering that the phase may vary, the *4-point-echo Method* becomes the best choice. The benefit of a well known envelope in the *20-point-echo Method* is accompanied by a larger statistical error.

If a background of 500 neutrons per unit time is assumed, the statistical error increases by approximately 25% which is a similar effect as a loss of polarization. To compensate the effect of the background, the measurement time has to be increased by a factor of 1.5.

Following the discussion from section 3.1, our standard method for data analysis is as follows: First, all echos are fitted within the *4-point-echo Method* with free fitting parameters, which are A, I_{av}, and ϕ. If all I_{av} are identical within the errors, the average of I_{av} is calculated and fixed. In a second data treatment the data are now fitted with A and ϕ as free parameters. If ϕ turns out to be constant too, it can also be fixed in a third data treatment.

4. Neutron spin echo from polarizing samples

Magnetic samples without a centre of symmetry like the B20 compound MnSi may act as a polariser [5] leading to three spin echo like groups. Fig. 3 shows the spin echo signal versus

Table 1: Statistical error of the amplitude (described in section 3). I_{BG} are the counts added as background to the NSE-signal of the sample. No background: $I_{\mathrm{BG}} = 0$, Background: $I_{\mathrm{BG}} = 500$.

		2-point		4-point	20-point
	A_1	A_2	A_3	A	A
$I_{\mathrm{BG}} = 0$					
$\Delta A/A$	0.032	0.046	0.077	0.045	0.053
$I_{\mathrm{BG}} = 500$					
$\Delta A/A$	0.039	0.065	0.089	0.056	0.064

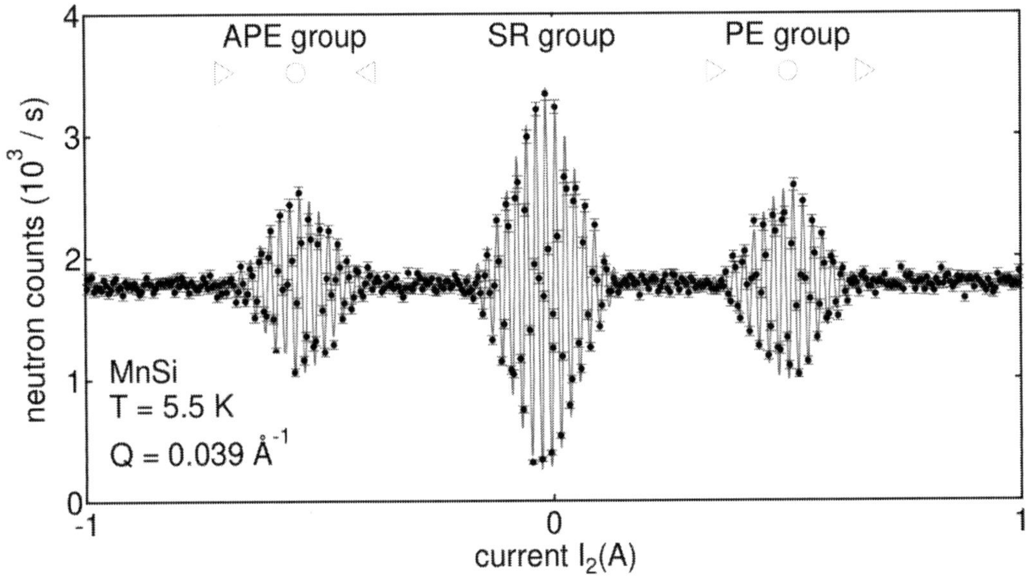

Figure 3: The single-handed magnetic spiral in non-centrosymmetric MnSi polarizes the neutron beam during diffraction from the magnetic satellites. A complete NSE-scan exhibits two spin echo groups (APE and PE) and a spin rotation group (SR). I_2 designates the current in the second spectrometer arm.

the current I_2 in the second spectrometer arm for a fixed current $I_1 = 0.519\,\mathrm{A}$ in the first arm. In the centre there is the SR group and on the left and right hand side the anti parallel echo (APE) group and the parallel echo (PE) group, respectively. The APE group is observed in classical NSE-experiments on non-magnetic materials where the magnetic precession fields are directed anti parallel. The PE group is typically found in magnetically scattering samples where the sample magnetization component transverse to the polarization leads to spin-flip scattering and therefore occurs at parallel field adjustment. The rotation group occurs because the sample polarizes the beam. The spins of the repolarised beam start to precess in the the second NSE coil after scattering. After a few rotations the polarization of the beam is lost due to the broad wavelength distribution. In contrast, the polarization of the APE and PE groups recurs due to the fact that the absolute value of the magnetic field integrals in both spectrometer arms become equal. In order to record the whole NSE signal, the current of the second NSE coil has to be scanned over a wide range. In magnetic systems, NSE measurements are ideally performed on the PE group where nuclear background scattering does not influence the results. The SR group is independent of the spin dynamics of the sample and does not depend on τ.

4.1. Theoretical spin echo from a polarizing sample

From now on we focus on a special case: We assume that the sample system acts as a polariser, polarizing along an axis in the precession plane. The intensity from such a sample system is given by

$$I_{\text{final}} = I_S \left(\frac{1+\cos(\phi_1(\lambda))}{2}\right)\left(\frac{1+\cos(\phi_2(\lambda))}{2}\right). \tag{8}$$

Here, I_S is the mean intensity scattered from the sample into the detector. ϕ_1 and ϕ_2 describe the phase of the spin precession after the primary and secondary spectrometer arm, respectively. The intensity at the detector is given by integrating Eq. (8) over the Gaussian wavelength distribution yielding

$$\begin{aligned} I =& \frac{I_S}{4} + \frac{I_S}{4}\cos(\phi_1)\exp\left[-\frac{\phi_1^2}{2\frac{\lambda_0{}^2}{\sigma_\lambda^2}}\right] \\ &+ \frac{I_S}{4}\cos(\phi_2)\exp\left[-\frac{\phi_2^2}{2\frac{\lambda_0{}^2}{\sigma_\lambda^2}}\right] \\ &+ \frac{I_S}{4}\cdot\frac{1}{2}\left[\cos(\phi_1-\phi_2)\exp\left[-\frac{(\phi_1-\phi_2)^2}{2\frac{\lambda_0{}^2}{\sigma_\lambda^2}}\right] + \cos(\phi_1+\phi_2)\exp\left[-\frac{(\phi_1+\phi_2)^2}{2\frac{\lambda_0{}^2}{\sigma_\lambda^2}}\right]\right]. \end{aligned} \tag{9}$$

Here, ϕ_i $(i = 1, 2)$ designate the spin phase of neutrons with the average wavelength λ_0 after the primary and secondary arms. As discussed above the magnetic field is only changed in the secondary arm, i.e. $\phi_1 = \text{const}$. Therefore, the first line of equation (9) is a constant contribution to the intensity. There are three echo like groups: The SR group (line 2) and the two spin echo groups (line 3). The factor of $\frac{1}{2}$ in line 3 shows that the polarization of the APE and PE group limit the maximum available polarization to 50% of the incident polarization.

4.2. Analysis of data from polarizing samples

For long spin echo times, the SR and spin echo groups are well separated (Fig. 4). In this case, the *4-point-echo method* is the most appropriate technique to extract the amplitude A from the spin echo groups. Note that the SR group provides the intrinsic polarization of the NSE-spectrometer in combination with the sample. Therefore, depolarization effects by the sample may be directly accounted for. However, with decreasing τ, the SR and spin echo groups start overlapping and the *4-point-echo method* cannot be applied any more. Therefore, more spin echo points are required to disentangle the SR and spin echo groups. As discussed above, the SR group is independent of τ. Hence, the data can be analysed in the following way: First, the data sets for all τ are fitted using Eq. (9) with all parameters free primarily the amplitude and the average intensity of the SR, PE and APE group. In a second step, the average amplitude A^{SR} of the SR group and $I_{\text{av}}^{\text{SR}}$ are determined. If A^{SR} and $I_{\text{av}}^{\text{SR}}$ for all measurements are equal within the statistical error, these two parameters are fixed and all data sets are re-analysed and the amplitude of the spin echo groups is extracted.

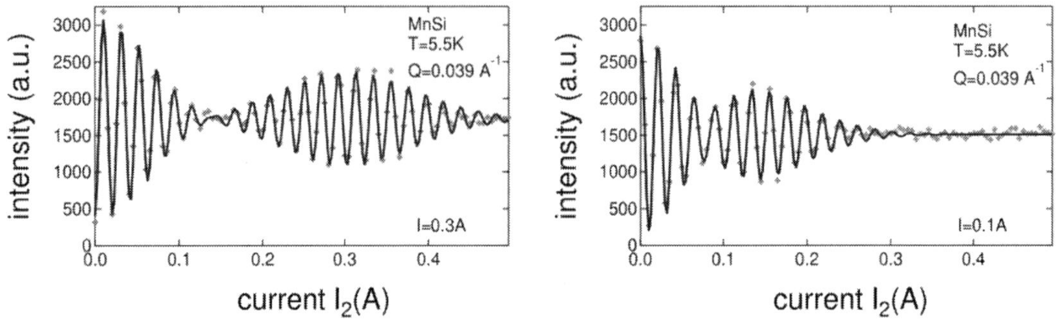

Figure 4: Spin-Rotation and PE group from MnSi as measured for long (left) and short (right) spin echo times. Short τ lead to a strong overlap of the SR and PE groups. I_2 designates the current in the second spectrometer arm.

5. Summary

By means of the spin echo method, the intermediate scattering function $S(Q,t)$ can be directly determined from the dependence of the polarization of the scattered neutrons on the spin echo time. Depending on the sample under investigation, various strategies for analysing the data can be applied. If the spin echo groups are clearly separated from the SR group, the *4-point-echo method* may be the most efficient method to extract the amplitude of the polarization. The statistical error can be minimized by fitting multiple data sets thus fixing the value of the average intensity more accurately. In addition an excellent phase stability is obtained. NSE-experiments from polarizing samples lead to the appearance of a spin rotation group that may overlap at short spin echo times with the spin echo groups. Here, the number of the measured echo points has to be significantly increased. We have developed an appropriate strategy to analyse such data and successfully applied the technique to analyse the quasielastic scattering from polarizing MnSi.

References

[1] Mezei F 1972 *Zeitschrift für Physik A Hadrons and Nuclei* **255**(2) 146–160 ISSN 0939-7922 10.1007/BF01394523 URL `http://dx.doi.org/10.1007/BF01394523`

[2] Häußler W, Gohla-Neudecker B, Schwikowski R, Streibl D and Böni P 2007 *Physica B: Condensed Matter* **397** 112–114 ISSN 0921-4526 URL `http://www.sciencedirect.com/science/article/pii/S0921452607001408`

[3] Gähler R and Golub R 1987 *Zeitschrift fr Physik B Condensed Matter* **65**(3) 269–273 ISSN 0722-3277 10.1007/BF01303712 URL `http://dx.doi.org/10.1007/BF01303712`

[4] Golub R and Gähler R 1987 *Physics Letters A* **123** 43 – 48 ISSN 0375-9601 URL `http://www.sciencedirect.com/science/article/pii/0375960187907602`

[5] Shirane G, Cowley R, Majkrzak C, Sokoloff J B, Pagonis B, Perry C H and Ishikawa Y 1983 *Phys. Rev. B* **28** 6251–6255

Towards polarizing beam splitters for cold neutrons using superparamagnetic diffraction gratings

J Klepp[1], I Drevenšek-Olenik[2,3,4], S Gyergyek[4], C Pruner[5], R A Rupp[1] and M Fally[1]

[1]University of Vienna, Faculty of Physics, A-1090 Vienna, Austria
[2]University of Ljubljana, Faculty of Mathematics and Physics, SI 1000, Ljubljana, Slovenia
[3]Center of Excellence for Polymer Materials and Technologies, SI 1000, Ljubljana, Slovenia
[4]Jožef Stefan Institute, Materials Synthesis Department, SI 1001 Ljubljana, Slovenia
[5]University of Salzburg, Department of Materials Science and Physics, A-5020 Salzburg, Austria

E-mail: juergen.klepp@univie.ac.at

Abstract. For holographic gratings recorded in superparamagnetic nanoparticle-polymer composites the diffraction efficiency should – next to grating spacing, nanoparticle concentration and grating thickness – depend on the strength of an external magnetic field and the incident neutron spin state. As a consequence, diffraction gratings should be tunable to act as mirrors for one spin state, while being essentially transparent for the other. Thus, polarizing beam splitters for cold neutrons become feasible.

1. Introduction

Polarized neutron experiments are a key technique for materials science [1, 2] as well as fundamental physics [3, 4]. The most common methods to prepare neutrons in a well-defined spin state parallel or antiparallel to an external magnetic field, i.e. to produce polarized neutron beams, are: *i)* Bragg reflection from magnetic single crystals such as Cu_2MnAl Heusler crystals, *ii)* total reflection from magnetic mirrors or supermirrors and *iii)* spin filters such as polarized ^{3}He-gas filters that have high absorption for neutrons with spin antiparallel to the nuclear spins of the gas. The decision for or against one of these three methods depends on the application one has in mind, since all of them have certain advantages and drawbacks. For instance, high degrees of polarization can be obtained with bent polarizing supermirrors, but these devices change the direction of the incident beam and exhibit small-angle scattering. Both properties cause severe problems for some neutron scattering applications. For polarized-SANS, cavity transmission polarizers [5, 6] can be a good option if their length of $\sim 2\,\mathrm{m}$ is suitable in a particular setup. In turn, the above problems do not occur when employing ^{3}He spin filters. Beam polarizations up to 80% have been reached [1, 2] with this technique. However, such devices can become a costly alternative due to shortages in ^{3}He supplies. In the present paper, design of low-cost, business-card size polarizers for cold and very-cold neutrons is proposed.

We employ light-sensitive materials combined with holographic techniques to produce diffraction gratings for neutron-optics. The materials exhibit a neutron refractive-index pattern, arising from a light-induced redistribution of nanoparticles in a polymer matrix, an effect referred to as the photo (neutron)- refractive effect [7, 8]: Upon illumination with a spatial light pattern, a neutron refractive-index change occurs in the recording material that is proportional to the change of the coherent scattering length density $b_c\rho$. Here, b_c is the coherent scattering length of a specific isotope and ρ is its corresponding number density. If superposed light beams generate a periodic modulation of $b_c\rho$, neutron diffraction gratings are recorded. One needs to adjust accessable parameters such as b_c, the grating spacing Λ, the grating thickness d_0 and ρ, so that peak-values of the diffraction efficiency (or reflectivity) η are suitable for a particular application one has in mind. Depending on η, diffraction gratings for neutrons can be used as wavelength filters, guides, mirrors or beam splitters for interferometry, for instance.

Nanoparticle-polymer composites [9] allow efficient tuning of the refractive-index modulation by suitable choice of the species of possible nanoparticles, i.e. choice of b_c via the nuclides contained in the nanoparticles. Recently, we have demonstrated that holographic SiO_2 nanoparticle-polymer composite gratings can be used as 50:50 beam-splitters [10] or three-port beam splitters (30:30:30) for cold neutrons [11]. Exploiting the Pendellösung interference effect [12] by tilting of the gratings [13], diffraction efficiencies of 83% have been reached for very-cold neutrons [14].

One question appears naturally in view of the above statements: Is it possible to produce holographic gratings whose diffraction properties can be controlled via external means – apart from tilting – such as cooling or heating, mechanical strain or electromagnetic fields? The answer is affirmative at least for the latter, as is suggested in the following.

2. Magnetic contribution to the neutron refractive index modulation in nanoparticle-polymer gratings

A one-dimensional sinusoidal grating can be described by the periodically modulated refractive index

$$n(x) = \overline{n} + \Delta n \cos\left(\frac{2\pi}{\Lambda}x\right), \tag{1}$$

where $\overline{n}$ and $\Delta n \simeq \lambda^2 b_c \Delta\rho/(2\pi)$ are the overall refractive-index change and the refractive-index modulation amplitude, respectively, arising from photon-induced reactions and the resulting redistribution of nanoparticles due to the recording light-pattern.

It is well known (see, for instance, Refs. [3] or [12]) that for a magnetic material the neutron-optical potential V_n – describing neutron-nucleus interactions – must be modified by adding a magnetic term $V_m = -\vec{\mu}\cdot\vec{B}(\vec{r})$ to account for the interaction of the neutron magnetic dipole moment $\vec{\mu}$ with the magnetic field $\vec{B}$ inside the material. V_m is related to the magnetic scattering length $b_m(\vec{Q})$ that contains the magnetic form factor. The neutron refractive index becomes

$$n = \sqrt{1 - \frac{V_n + V_m}{E}} \simeq 1 - \lambda^2 \frac{[b_c \pm b_m(0)]\rho}{2\pi}, \tag{2}$$

where $b_m(0)$ refers to

$$b_m(\vec{Q}) = -\frac{m_N}{2\pi\hbar^2\rho}\vec{\mu}\cdot\vec{B}(\vec{Q}) \tag{3}$$

at $\vec{Q} = 0$. Here, E and m_N are the energy of the incident neutron and the neutron mass, respectively. Since the structure of holographic nanoparticle-polymer gratings is based upon redistribution of nanoparticles, i.e. a modulation of ρ, we get

$$\Delta n \simeq \lambda^2 \frac{[b_c \pm b_m(0)]\Delta\rho}{2\pi} \tag{4}$$

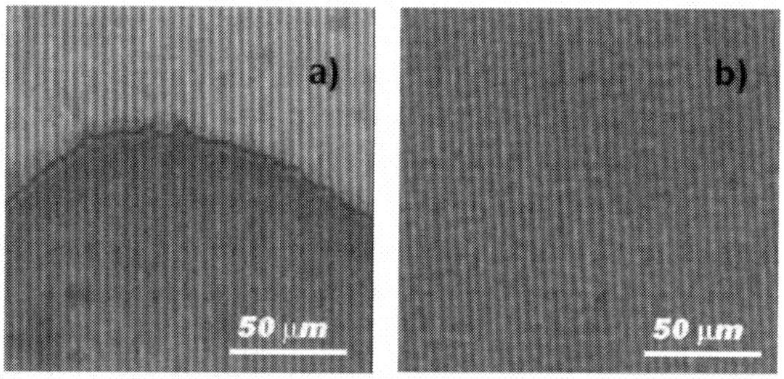

Figure 1. Optical polarization microscopy image of grating structures with maghemite nanoparticles. Dark regions are rich with maghemite. a) Structure before secondary polymerization, b) after secondary polymerization.

for the neutron refractive index modulation amplitude in gratings with *magnetic* nanoparticles. One can see from Eqs. (2) and (4) that for holographic gratings containing sufficiently high concentrations of magnetic nanoparticles the modulation in Eq. (1) can be switched off for one spin component just by adjusting the strength of an external field, i.e. the grating becomes essentially invisible for neutrons in this spin state. Furthermore, by carefully choosing the grating parameters [14], the grating can at the same time act as a mirror for the other spin component. That means polarizing beam splitters for cold and very-cold neutrons should be feasible.

3. Superparamagnetic nanoparticles

The superparamagnetic state of noninteracting single-domain ferro- or ferrimagnetic nanoparticles is governed by the thermal energy and the potential energy of the particles within an external magnetic field [15]. Typical relaxation times (the so-called Néel relaxation time) are very short compared to measurement times and so the net magnetization appears to be zero in the absence of an external magnetic field. On application of a field, the nanoparticles get magnetized as it is the case for paramagnets. However, the magnetization can become very large, depending on the type of nanoparticles and their preparation.

Next to the micro- or nanostructure of the nanocomposite, mainly the content of nanoparticles in the matrix determines the magnetic properties of the material. As in most applications, superparamagnetic nanocomposites with high magnetization are desired in our case. It is therefore necessary to prepare a nanocomposite containing large amounts of dispersed superparamagnetic nanoparticles. However, to lower their surface energy, nanoparticles show a strong tendency to agglomerate. In order to prevent such agglomeration and to provide for compatibility of nanoparticles with the polymer matrix it is essential to coat the nanoparticles with a suitable surfactant (see, for instance, [16, 17]).

Development of optical methods to structurize superparamagnetic nanocomposite materials as described in the latter two references are currently under way. Preliminary results for maghemite nanoparticles are shown in Fig. 1. Many of the superparamagnetic nanoparticle species absorb optical radiation. This brings about some problems to usual methods of holographic structuring. Therefore, a two-step structuring method is considered. First, a template grating structure is recorded in transparent materials. Second, one of the contituents of the template is removed by a suitable solvent and the remaining voids are refilled with the solution of magnetic nanoparticles in a photosensitive monomer. After refilling, secondary photopolymerization by homogeneous illumination ensures that the final composite structure remains stable. Note that estimations of nuclear and magnetic (saturation) scattering length densities of maghemite nanoparticles are $b_c\rho \simeq 6.7 \times 10^{10}\text{cm}^{-2}$ and $b_m\rho \simeq 1 \times 10^{10}\text{cm}^{-2}$, respectively (see, for instance, [18] and references therein). Materials with larger b_m/b_c will be investigated in the near future.

4. Conclusion

In this paper, small sized and low-cost polarizing beam splitters for cold and very cold neutrons – based upon diffraction gratings recorded in superparamagnetic nanoparticle-polymer composites by holographic means – have been proposed. Because of the spin-dependent magnetic contribution to the coherent scattering length, the properties of such diffraction gratings can be tuned such that for one spin component the grating acts as a mirror, while for the other it is transparent. Due to high absorption of maghemite nanoparticles in the visible wavelength range, a new method to structurize such materials using optical holography is currently developed.

Acknowledgments

Financial support by the Austrian Science Fund (FWF): P-20265 and the ÖAD in the frame of the bilateral project SI 07-2011 is greatly acknowledged.

References

[1] Willis B T M and Carlile C J 2009 *Experimental Neutron Scattering* (New York-Oxford: Oxford University Press)
[2] Furrer A, Mesot J and Strässle T 2009 *Neutron Scattering in Condensed Matter Physics* (Singapore: World Scientific)
[3] Rauch H and Werner S A 2000 *Neutron Interferometry* (*Neutron scattering in condensed matter* vol 12) (New York–Oxford: Oxford University Press)
[4] Abele H 2008 *Prog. Part. Nuclear Physics* **60** 1
[5] Keller T, Krist T, Danzig A, Keiderling U, Mezei F and Wiedenmann A 2000 *Nucl. Instrum. Methods A* **451** 474
[6] Aswal V K, van den Brandt B, Hautle P, Kohlbrecher J, Konter J A, Michels A, Piegsa F M, Stahn J, Van Petegem S and Zimmer O 2008 *Nucl. Instrum. Methods A* **586** 86
[7] Rupp R A, Hehmann J, Matull R and Ibel K 1990 *Phys. Rev. Lett.* **64** 301
[8] Fally M 2002 *Appl. Phys. B* **75** 405–426
[9] Suzuki N, Tomita Y and Kojima T 2002 *Appl. Phys. Lett.* **81** 4121
[10] Fally M, Klepp J, Tomita Y, Nakamura T, Pruner C, Ellabban M A, Rupp R A, Bichler M, Drevenšek-Olenik I, Kohlbrecher J, Eckerlebe H, Lemmel H and Rauch H 2010 *Phys. Rev. Lett.* **105** 123904
[11] Klepp J, Pruner C, Tomita Y, Kohlbrecher J and Fally M 2011 In preparation
[12] Sears V F 1989 *Neutron Optics* (*Neutron scattering in condensed matter* vol 3) (New York–Oxford: Oxford University Press)
[13] Somenkov V A, Shilstein S Sh, Belova N E and Utemisov K 1978 *Solid State Commun.* **25** 593
[14] Klepp J, Pruner C, Tomita Y, Plonka-Spehr C, Geltenbort P, Ivanov S, Manzin G, Andersen K H, Kohlbrecher J, Ellabban M A and Fally M 2011 *Phys. Rev. A* **84** 013621
[15] Wagner W and Kohlbrecher J 2005 *Small-Angle Neutron Scattering (SANS)* (Boston – Dortrecht – London: Kluwer Academic Publishers) chap 2
[16] Gyergyek S, Huskic M, Makovec D and Drofenik M 2008 *Colloids and Surfaces A: Physicochem. Eng. Aspects* **317** 49
[17] Makovec D, Čampelj S, Bele M, Maver U, Zorko M, Drofenik M, Jamnik J and Gaberšček M 2009 *Colloids and Surfaces A: Physicochem. Eng. Aspects* **334** 74
[18] Avdeev M V, Dubois E, Meriguet G, Wandersman E, Garamus V M, Feoktystov A V and Perzynski R 2009 *J. Appl. Cryst.* **42** 1009

A new time-of-flight neutron reflectometer and SANS instrument GORIZONT at IN-06 spallation neutron source

V S Litvin[1], V A Trounov[2], V A Ulyanov[2], A P Boulkine[2], S I Kalinin[2], A A Alekseev[1], R A Sadykov[1,3] and E A Koptelov[1]

[1]Institute for Nuclear Research RAS, 60-letiya Oktyabrya 7a, Moscow, 117312, Russia
[2]Petersburg Nuclear Physics Institute, Gatchina, 188350, Russia
[3]Institute for High Pressure Physics RAS, Troitsk, 142190, Moscow Region, Russia

E-mail: vlitvin@inr.ru

Abstract. A new time-of-flight neutron reflectometer GORIZONT with vertical scattering plane was constructed and tested at the IN-06 spallation pulsed neutron source at the Institute for Nuclear Research RAS. The instrument can be used also as a small angle neutron instrument. Due to the vertical scattering plane it is well-suited for the studies of free-liquid surfaces. A Monte-Carlo simulation and experimental testing of instrument have been performed. The neutron spectra at the guide exit and at the sample position and beam profiles were measured. Based on the experimental data the momentum transfer range and the instrument resolution were estimated.

1.Introduction

The neutron reflectometry (NR) is a powerful experimental technique to study surfaces and interfaces of materials such as thin films and multilayers in many fields of condensed matter physics, materials sciences and life sciences. This method provides the coherent scattering length density and the magnetization depth profile. The off-specular NR can give valuable information about the roughness of surfaces or interfaces as well as domain structure [1]. Reflectometers with a vertical scattering plane can investigate free liquids surfaces and interfaces, thin films on liquid surfaces, for example: polymers, proteins, lipid membranes, ferrofluids, microemulsions and surface-active substances. The small angle neutron scattering (SANS) option is well-suited for the investigation of the nm-size nanostructures and inhomogeneities in solid and liquid samples [2]. The SANS method is very useful for studying nanoparticles, polimeres, alloys, proteins and other life objects. As a rule the time-of-flight instruments have a rather broad momentum transfer range compared to the constant wavelength instruments. Such instruments are very efficient for in-situ studies.

2.Design of instrument

A new time-of-flight reflectometer GORIZONT (HORIZON) with a vertical scattering plane was designed and constructed at the pulsed spallation neutron source IN-06 at the Institute for Nuclear Research RAS (INR RAS) in Troitsk near Moscow [3]. The neutron beamline of the instrument is provided by neutrons from the 5 cm thick water neutron moderator. The maximum of neutron spectra is at about 1.3 Å, the primary proton pulse duration can be chosen from the range 0.25 to 200 μs and the repetition frequency range is 1 to 50 Hz. The typical pulse duration for the IN-06 source is 60 μs [4]. Fig. 1 shows the layout of the instrument. The instrument consist of the 7 meter long curved neutron guide, two slit packages, the deflecting supermirror, the sample stage and the hand with a neutron detector [5]. A shutter is installed before the guide. The guide has a mirror channel with a cross-section 10x70 mm^2 (height and width, respectively) and a curvature radius in the vertical plane of 520 m. It consists of 14 sections of flat supermirrors. The high and both side mirrors are based on multilayer NiMo/Ti coatings with m=2, and the bottom mirrors on Ni coating (m=1). The guide is mounted in a vacuum tube filled with polyethylene grains for neutron capture. The slit packages can provide neutron beams with sizes 0.1 - 10 mm in the vertical plane and 0.1-70 mm in the horizontal plane. A deflecting supermirror is based on the multilayer NiMo/TiZr coating with m=2. The neutron beam after the guide is inclined down at angle 0.77°. The supermirror is designed to increase the incident angle up to 1.5° for a horizontal sample surface (some samples can not be inclined, for example liquid). The sample stage stands on a shake-proof pedestal. It can provide samples translations and incline them in the scattering plane. The moderator to sample distance is 14.4 m and the sample to detector distance is 4.0 m. The hand with a detector can be inclined in the vertical plane by maximum angle from the incident neutron beam 12°. Before the detector the vacuum vessel is mounted. For the NR mode the neutron beam is vertically well collimated and horizontally broad. In the SANS and GISANS mode the square shaped neutron beam is used. For the polarised neutrons scattering measurements the supermirror can be replaced by the neutron polarisation device. Now we use single ^{3}He-filled neutron counters for specular NR. Two dimension position-sensitive detector with thin ^{10}B converters is developing in INR RAS. It will allow off-specular NR, GISANS and SANS measurements. The detector sensitive area will be 200x200 mm^2. The expected resolution is about 2 mm. We have successfully tested the prototype of this detector [6].

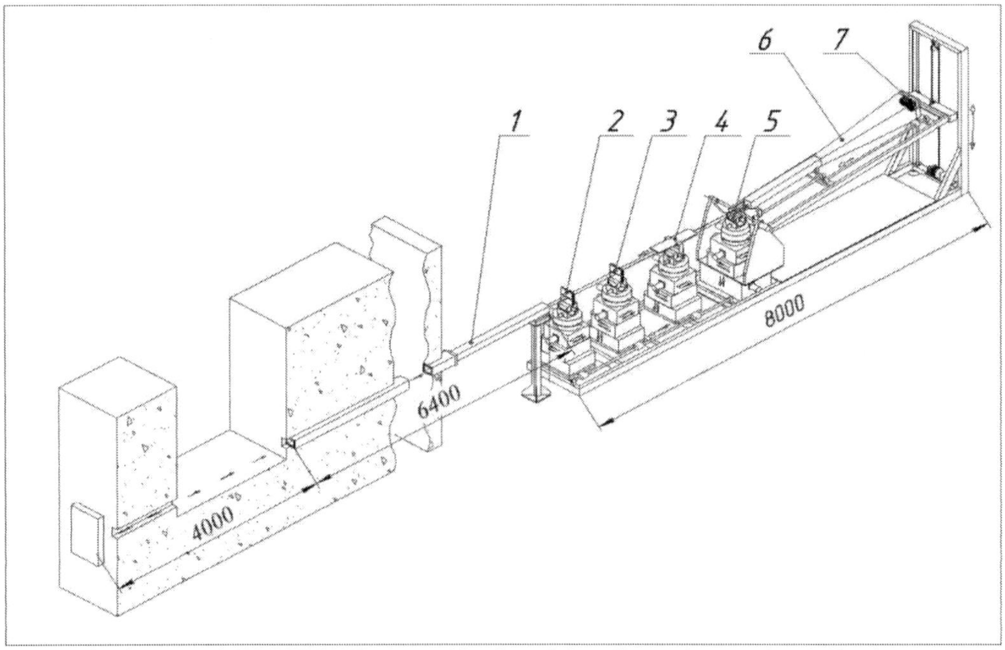

Figure 1. The layout of GORIZONT: 1 - guide, 2 and 3 - slit packages, 4 - deflecting supermirror, 5 - sample stage, 6 - hand with vacuum vessel, 7 - neutron detector.

3.Instrument testing results

In November 2010 we have performed a successful test of the instrument. The neutron spectra (see fig. 2) were measured using the time-of-flight method and ^{3}He-filled neutron counters. Fig. 2 demonstrates the experimental spectrum at the guide exit. For comparison the neutron spectrum of another IN-06 beamline without any neutron optic device is shown. The shaped pulses from counters were registered by the time to digital converter with CAMAC interface developed in INR RAS. For the beam profile measurement the neutron counter with a 1 mm-cadmium slit was used. This counter was translated in the vertical direction by a special step motor. The neutron beam divergence at the sample position was simulated using the VITESS package [7] as well as the neutron spectra after the guide [5]. The divergence estimated from this beam profile in the vertical plane is 0.2° at 10 mm slits, similar to the results of our Monte-Carlo simulation. The divergence can be decreased by using smaller slits. According to our simulation at 0.2 mm slits the divergence will be 0.005°. Because the wavelength resolution is 1,5% (at neutron pulse duration 130 μs) the momentum transfer resolution can be down to 2 % [5]. As seen in fig. 2 the guide absorbs all the neutrons with a wavelength less than 1 Å. The maximum of the neutron spectrum at the sample position was observed at about 2 Å. We have observed that the guide provides a substantial increase of the neutron flux for the wavelength >2 Å . The effective neutron wavelength range is 2 to 9 Å and the momentum transfer range 0.005-0.7 $Å^{-1}$ at the pulse repetition frequency 25 Hz. The main parameters of the instrument are shown in the table 1.

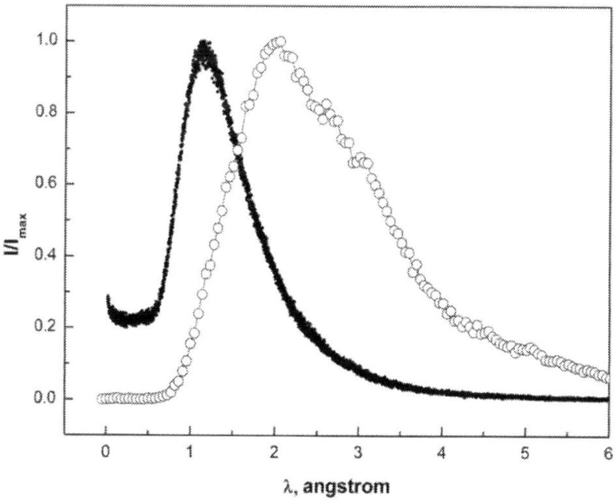

Figure 2. Experimental neutron spectra at guide exit (empty circles) and at another IN-06 beamline without any guide (points).

4.Conclusions

We have constructed and tested a new time-of-flight neutron reflectometer with a small angle neutron scattering option. Due to the vertical scattering plane the instrument is well-suited for the studies of liquid surfaces and interfaces. In the future we plan to add to the instrument extra option including the neutron polarizers, spin-flippers for studies of magnetic liquids, multilayers and nanostructures.

Table **1**. The main parameters of GORIZONT.

Parameter	Value
Pulse repetition frequency, Hz	25 and 50
Pulse duration time, μs	60÷200
Wavelength range, Å	2÷9
Momentum transfer range, $Å^{-1}$	0.005÷0.7
Maximum beam size, mm	10 (height)/70 (width)
Moderator to sample distance, m	14.4
Sample to detector distance, m	4
Maximum scattering angle (2θ), deg.	12

References

[1] Daillant J and Gibaud A 2009 *X ray and Neutron Reflectivity: Principles and Applications, Lect. Notes Phys.* (Berlin Heidelberg: Springer) p 770

[2] Zhu Y. 1999 *Modern techniques for characterizing magnetic materials* (Berlin: Springer Verlag) p 600

[3] E A Koptelov, Yu V Ryabov, A A Alekseev, S F Sidorkin, M I Grachev, V A Fedchenko, R A Sadykov, Yu B Lebed, V S Litvin, S P Kuznetsov, E S Clementyev, P A Alekseev, V A Trunov, A P Bulkin, V A Ulianov and S N Axenov 2011 *J. Phys.: Conf. Ser.* **291** 012012

[4] Koptelov E A., Fedchenko V A., Goncharenko O N. et al. 2005 (*Proc. XVII Meet. Int. Collaboration on Advanced Neutron Sources, Santa Fe, New Mexico, USA April 25-29 2005)* P. 22.

[5] Litvin V S, Trounov V A, Bulkine A P, Ulyanov V A, Solovei V A, Kolkhidashvili M R, Kalinin S I, Sadykov R A and Koptelov E A 2010 *J. of Surface Investigations. X-Ray, Synchrotron and Neutron Techniques* **4** 873

[6] Litvin V S, Potashev S I, Razin V I, and Sadykov R A Bulletin of the Russian Academy of Sciences: Physics **75** p. 229

[7] Zsigmound G and Lieutenant K 2002 *Neutron News* **13** 11

Multiple neutron scattering corrections. Some general equations to do fast evaluations.

Rosaria Mancinelli

Dipartimento di Fisica, Universita' degli Studi di Roma Tre, via della Vasca Navale, Roma, Italy

E-mail: mancinelli@fis.uniroma3.it

Abstract. An efficient Monte Carlo algorithm has been developed aimed at simulating neutron tracing in a liquid or amorphous sample, during a diffraction experiment. At odds with approximated analytical estimates, such a method allows to easily take into account some physical and geometrical features, which play an important role in determining the measured scattered neutron intensity.

Some general laws giving the multiple scattering intensity, in terms of few adimensional parameters, are provided. In particular, it results that the most relevant quantity involved in the neutron transport is the mean free path (defined as the inverse of the attenuation factor), which can be helpfully assumed as *natural* length unit. This yields to some important consequences: for istance, in the case of absorbing or structured samples, the mean free path can strongly depend on the neutron energy and this should be considered in the choice of the experimental apparatus.

1. Introduction

The diffraction of thermal neutrons is widely used to study the interatomic structure of isotropic systems [1]. As a matter of fact, the general theory predicts that, in case of neutrons scattered at most once, the intensity of detected radiation is directly proportional to the differential cross section as a function of momentum transferred, giving a measure of the atomic ordering within the system. This implies that, in reality, the measured data should be corrected for neutrons scattered more than once, in order to get the static structure factor.

Analytical estimates of multiple (more than one) scattering have been provided since 1954 [2, 3, 4]: all of these are based on the assumption that the angular distribution of primary scattered neutrons is nearly isotropic so that it makes sense to assume that the ratio of next scattering probabilities is constant. More recently, an analytical algorithm to extract single scattering from measured profiles in small angle experiment has been proposed [5]: this is based on the formalism developed for investigating large scale structures [6] and the main issue is the estimation of the scattering mean free path, which depends on the size of scatterers and the density contrast. Here we consider a quite different case: the scatterers are point, and no inhomogeneities are taken into account so that the small angle scattering (SAS) is null and the mean free path depends only on the strenght of interaction between the nuclei of the sample and the probing radiation. While the last feature is a simplification *vis-a-vis* the SAS, to our best knowledge the need of considering an interatomic structure factor does not lead any inversion method, to get the single scattering from the measured detected beam.

In particular, in the following, we'll point out that when considering a real experiment, consisting of a finite-size neutron beam, incident on a finite-size specimen, which scatters differently at different angles, the analytical approach carried out for diffraction data analysis may break down. Instead of it, numerical integrations of Boltzmann neutron transport equation, have traditionally been used to tackle real cases. The more general numerical approach is given by Monte Carlo simulations, which have been widely used to include all the features of the experiment (from [7] to [8]). But even nowadays, running simulations to get a rough but reliable real-time correction could be too slow. Here we propose to recover an analytical approach, based on equations fitting a plenty of simulated data. As shown in the following for elastic scattering, some universal scaling laws hold. This is due to the fact that a *natural* length unit, depending on the energy, can be introduced, so that many scattering signals coming from different samples collapse into few general behaviours.

Even if this research has been assigned to a specific task, *i.e.* to improve and speed up the analysis on data collected at diffractometers at ISIS [9], its results could be easily generalized, or give suggestions for further developments. Moreover, it's noteworthy that modern diffractometers cover wider and wider Q-ranges, detecting both large-scale and microscopic structure, so in next future an integrated analytic method shall hopefully be developed.

2. From neutron transport theory to numerical evaluation of scattered neutron intensity

A brief review on the neutron transport theory could be helpful to introduce the implemented algorithm, which is similar, but not identical, to some previous code [10].

Consider a neutron, which enters at $\vec{r_0}$ and goes out of the sample at $\vec{r_{out}}$. It may have n collisions in the sample, at positions $\vec{r_1}$, $\vec{r_2}$, $\vec{r_3}, \cdots \vec{r_n}$.

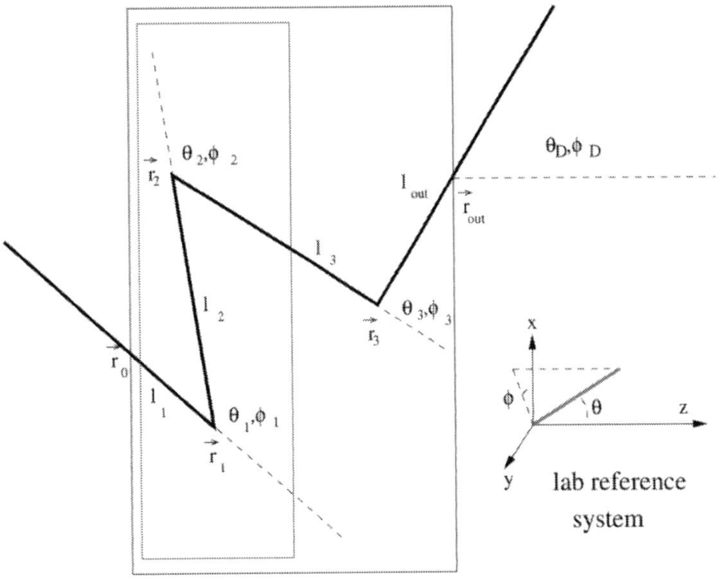

Figure 1. Skeme of a three collisions scattering event. The tracing in the inner rectangle corresponds to the product of the sequence in (1) for $n = 3$. For each collision the free path between two next collisions and the scattering solid angle should be chosen.

After each collision, the neutron changes flight direction (see Fig.1) so that

$$\vec{r}_j = \vec{r}_{j-1} + \vec{l_j} \quad \text{being} \quad \vec{l_j} = l_j \begin{pmatrix} \sin\theta_{j-1}\cos\phi_{j-1} \\ \sin\theta_{j-1}\sin\phi_{j-1} \\ \cos\theta_{j-1} \end{pmatrix}$$

While (θ_j, ϕ_j) gives the direction after the j-collision in the *lab reference frame*, the direction after the j-collision in the *neutron reference frame* are (θ_j^s, ϕ_j^s). In particular, the angle formed by the neutron flight directions after and before the j-collision, $\theta_j^s = \arccos(\hat{r_j} \cdot \hat{r}_{j-1})$, is called **scattering angle**.

In the following, we'll tackle only **elastic scattering** events, i.e. we'll suppose that neutron energy doesn't change after each collision in the sample.

Under this hypothesis, the neutron beam detected at θ_D, due to n collisions, can be written as [4]:

$$\begin{aligned} T_n(\theta_D|\lambda) = \left(\tfrac{\mu_s}{4\pi}\right)^n \prod_{i=1}^{n-1} \int_0^{l_i^{max}} dl_i e^{-\mu_T l_i} \int_0^{4\pi} d\Omega_i^s \left[\tfrac{\mu_{coh}}{\mu_s} S(Q_i) + \tfrac{\mu_{inc}}{\mu_s}\right] \\ \int_0^{l_n^{max}} dl_n e^{-\mu_T(l_n + l_{out})} \left[\tfrac{\mu_{coh}}{\mu_s} S(Q_n) + \tfrac{\mu_{inc}}{\mu_s}\right] I_0(\vec{r_0}, \lambda) \end{aligned} \tag{1}$$

where $Q_i = \frac{4\pi}{\lambda} \sin\frac{\theta_i^s}{2}$ is the momentum transferred in the i-th collision, μ_T, μ_{coh} and μ_{inc} are the attenuation, coherent and incoherent coefficients ($\mu_{inc} + \mu_{coh} = \mu_s$), l_i^{max} is the longest distance a neutron can travel along the $\Omega_{i-1} = (\theta_{i-1}, \phi_{i-1})$ direction, depending on its own on $\overrightarrow{r_{i-1}}$ and Ω_{i-1}^s. This integral means that all the possible histories $(l_1, \Omega_1, l_2, \Omega_2, \ldots, l_{n-1}, \Omega_{n-1}, l_n)$, corresponding to n-th elastic collisions of neutrons having a wavelength λ and going out at a detected angle θ_D should be considered. For $n = 1$, Eq. (1) becomes:

$$T_1(\theta_D|\lambda) = \frac{\mu_s}{4\pi} \int_0^{l_1^{max}} dl_1 e^{-\mu_T(l_1 + l_{out})} \left[\frac{\mu_{coh}}{\mu_s} S(Q_1) + \frac{\mu_{inc}}{\mu_s}\right] I_0(\vec{r_0}, \lambda)$$

which, in the case of extremely thin sample, reduces to $T_1(\theta_D|\lambda) = \left[\frac{\mu_{coh}}{4\pi} S(Q_1) + \frac{\mu_{inc}}{4\pi}\right] I_0(r_0, \lambda)$. Dividing $T_1(\theta_D|\lambda)$ by the incident flux, $I_0(r_0, \lambda)$, and the sample density, ρ, one gets the differential cross section: $\frac{d\sigma}{d\Omega}(\theta_D|\lambda) = \frac{\sigma_{coh}}{4\pi} S(Q_1) + \frac{\sigma_{inc}}{4\pi}$.

In (1) μ_T indicates the total attenuation coefficient; this is given by the product of the sample density, ρ, and the total cross section σ_T: $\mu_T = \rho \cdot \sigma_T$.

The total cross section can be calculated by tables [11] as $\sigma_T(\lambda) = \sigma_s + A_{abs}\lambda$ being σ_s the scattering cross section, A_{abs} the ratio between the thermal absorption cross section at $\lambda^* = 1.7979\mathring{A}$ and λ^*, and λ the neutron wavelength. Otherwise, the total cross section can be evaluated from the intensity measured by the transmission monitor, T_0, as: $\sigma_T(\lambda) = -\frac{1}{\rho \cdot th} \ln T_0(\lambda)$, being th is the thickness of the sample.

Similarly, μ_s can be calculated from tables, being $\mu_s = \rho\sigma_s$, or from the experimental value of μ_T: $\mu_s = \mu_T - \rho \cdot A_{abs}\lambda$.

When the sample is contained in a container, an "average effective" sample should be simulated, i.e. the total attenuation coefficient of the sample in the container is calculated as a weighted average of the attenuation coefficients of the sample (^{s}) and the container (^{c}):

$$\mu_T = \frac{\sigma_T^s \rho^s th^s + \sigma_T^c \rho^c th^c}{th^s + th^c}$$

and, analogously, the absorption coefficient is given by

$$\rho A_{abs} = \frac{A_{abs}^s \rho^s th^s + A_{abs}^c \rho^c th^c}{th^s + th^c}.$$

The integral (1) can be evaluated by running a **Monte Carlo** (MC) simulation of the neutron tracing: this consists in following the history of a plenty of neutrons in the sample. For each neutron, for each collision, but the last,

- the free path before collision l
- the new direction of flight (θ, ϕ)

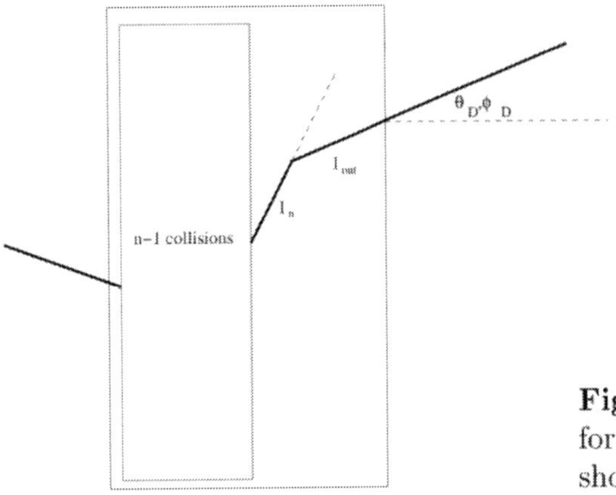

Figure 2. Given the position of the detector D, for the last collision only the free path before it should be chosen.

should be chosen. For the last collision, the final flight direction, that pointing the detector, is given, as sketched in Fig.2; so only the free path, l_n, before the last collision should be chosen, according to (1), where the integral runs over the last free path, given the final direction.

The evaluation of the generic term of the product of the sequence in (1)

$$\underbrace{\int_0^{l_j^{max}} dl_j e^{-\mu_T l_j}}_{1.} \underbrace{\int_0^{2\pi} d\phi_j^s}_{2.} \underbrace{\int_0^{\pi} d\theta_j^s \sin\theta_j^s \left[\frac{\mu_{coh}}{\mu_s} S(Q_j) + \frac{\mu_{inc}}{\mu_s}\right]}_{3.} \tag{2}$$

can be done through an **importance-sampling** [12] of the neutron free path before the j-th collision, l_j, and the scattering angles θ_j^s, ϕ_j^s. This consists in forcing successive scattering events to occur within the sample and by calculating the score at each scattering point. The optimization of the numerical evaluation of (2) can be achieved by changing integration variables. Be:

$$t_1 = \frac{1 - e^{-\mu_T l_j}}{1 - e^{-\mu_T l_j^{max}}}, \tag{3}$$

$$t_2 = \frac{\phi_j^s}{2\pi}, \tag{4}$$

$$t_3 = \frac{1 - \cos\theta_j^s}{2} \tag{5}$$

$$t_4 = P(Q_j) \text{ where } P(Q_j) = \frac{I(Q_j)}{I(2k)} \text{ and } I(x) = \int_0^x dQ_j Q_j S(Q_j) \tag{6}$$

In terms of t_1, t_2, t_3, t_4, Eq.(2) can be written as

$$\underbrace{\left[\frac{1 - e^{-\mu_T l_j^{max}}}{\mu_T} \int_0^1 dt_1\right]}_{(i)} \underbrace{\left[2\pi \int_0^1 dt_2\right]}_{(ii)} \underbrace{\left[2\frac{\sigma_{inc}}{\sigma_s} \int_0^1 dt_3 + 2\frac{\sigma_{coh}}{\sigma_s} \frac{I(2k)}{k^2} \int_0^1 dt_4\right]}_{(iii)} \tag{7}$$

This suggests how the corresponding Monte Carlo algorithm works.

(i) Given a random number t_1 uniformly distributed in $[0, 1]$, from (3) the free path before the j-th collision, l_j, is found:

$$l_j = -\frac{1}{\mu_T} \ln\left(1 - (1 - e^{-\mu_T l_j^{max}}) t_1\right).$$

Journal of Physics: Conference Series **340** (2012) 012033 doi:10.1088/1742-6596/340/1/012033

Eq. (7) requires to scale the neutron weight:

$$weight = weight \cdot \frac{1 - e^{-\mu_T l_j^{max}}}{\mu_T}.$$

(ii) Given a random number t_2 uniformly distributed in $[0, 1]$, ϕ_j^s can be found from (4):

$$\phi_j^s = 2\pi t_2.$$

According to Eq.(7) the neutron weight should be scaled:

$$weight = weight \cdot 2\pi.$$

(iii) Two possible scattering events may occur: according to Eq.(5), given a random number t_5 uniformly distributed in $[0, 1]$,

- if $t_5 < \frac{\sigma_{inc}}{\sigma_s}$, an incoherent elastic scattering is scored, by rescaling the neutron weight as

$$weight = weight \cdot 2$$

and selecting the scattering angle as

$$\theta_j^s = \arccos(1 - 2t_3),$$

being t_3 a random number uniformly distributed in $[0, 1]$.

- Otherwise, a random number t_4 uniformly distributed in $[0, 1]$ is chosen, a **coherent** elastic scattering is scored, the scattering angle is calculated as

$$\theta_j^s = P^{-1}(t_4)$$

and the weight is updated as

$$weight = weight \cdot 2\,\frac{I(2k)}{k^2}$$

The maximum scattering order is not decided *a priori* but according to its weight in the total multiple scattering (MS) intensity, given a certain level of accurancy.

To avoid correlations between events, each neutron is used once (i.e. for one scattering order, for one wavelength). A faster algorithm could be implemented to exploit a neutron history for all the scattering orders.

3. Natural units and relevant parameters

The neutron scattering depends on both the physical properties of the sample (the density, ρ, the total and the absorption cross section, σ_t, σ_{abs}) and the geometrical features of the source (the beam size, b, and the incident neutron wavelength, λ), the sample itself (shape and size) and the detectors (size and positions).

The general expression of the neutron beam, having wavelength λ, detected at θ_D, emerging after n collisions, is given by Eq. (1). It's worth noting that the integrations on the free paths l_i can be substituted with the corresponding integration on the adimensional variable $\xi_i = \mu_t l_i$ so that

$$T_n(\theta_D|\lambda) = \left(\frac{\mu_s}{\mu_t 4\pi}\right)^n \prod_{i=1}^{n-1} \int_0^{\xi_i^{max}} d\xi_i e^{-\xi_i} \int_0^{4\pi} d\Omega_i^s \left[\frac{\mu_{coh}}{\mu_s} S(Q_i) + \frac{\mu_{inc}}{\mu_s}\right]$$
$$\int_0^{\xi_n^{max}} d\xi_n e^{-(\xi_n + \xi_{out})} \left[\frac{\mu_{coh}}{\mu_s} S(Q_n) + \frac{\mu_{inc}}{\mu_s}\right] I_0(\vec{r_0}, \lambda).$$

This means that, at given λ, the absolute linear dimensions are not relevant but their products with the total attenuation $\mu_t = \mu_t(\lambda)$. In other words, at given wavelenght λ, the mean free path mfp defined as

$$mfp(\lambda) = \frac{1}{\rho\sigma_t(\lambda)},$$

can be assumed as "natural" length unit. As a matter of fact, the distinction between thin and thick samples contained in a slab can be given [4, 13] using the adimensional parameter

$$\tau = \frac{thickness}{mfp},$$

so that a sample is said thin if $\tau \ll 1$, thick is $\tau \geq 1$. Note that, in general, the mean free path is a function of λ, as the fastest neutrons have the longest typical path. So, in principle, a sample could be thin for high-energy neutrons and thick for low-energy ones.

Generalizing the normalization done for the thickness, all the linear geometrical features can be usefully expressed in mfp units. Moreover, to take into account the finite size of the beam we introduce the adimensional beam size

$$\beta = \frac{beam\ size}{mfp}.$$

Regarding the geometry of the detecting apparatus, the detector size is usually negligible in comparison with the sample-detector distance and, if the sample size is small with respect to the detecting apparaturs, the detector position can be taken into account considering only the detecting angle, θ_D, defined with respect to the incident beam direction.

Another adimensional parameter to take into account is the percentage of absorption

$$\alpha = \frac{\sigma_{abs}}{\sigma_t}.$$

To evaluate the single and multiple scattering, it's important to recall that these intensities can be easily calculated if we consider a sample for which
- the structure factor is flat: $S(Q) = 1$;
- the trasversal size is infinite;
- incident neutron beam is extremely collimated ($\beta = 0$);
- absorption is negligible ($\alpha = 0$).

From now on, this case will be assumed as reference, for both the slab and the cylinder geometry. Some formulas will be provided to get the multiple scattering intensity from the single scattering intensity In the following, we'll see how to correct the multiple scattering calculated for the reference case, to get a prediction for the general case: by processing a plenty of data coming from Monte Carlo simulations, some general laws, reported in the next pages, have been found.

4. Slab geometry

In case of sample contained in a slab, having a square as trasversal face, the linear dimensions are the thickness and the width, which are, respectively, the size along the direction of the incident beam and the size along the trasversal directions. In mfp units, these correspond to the adimensional parameters:

$$\tau = \frac{thickness}{mfp} \qquad \omega = \frac{width}{mfp}.$$

The meaning of adimensional parameters α, β, ω, τ, θ_D is sketched in Fig.3.

For the real cases I've taken into account (vanadium, empty slab container of silica or titanium-zirconium, water in a slab, measured on ISIS diffractomet [9]), the variability ranges of the relevant parameters are reported in Tab.1.

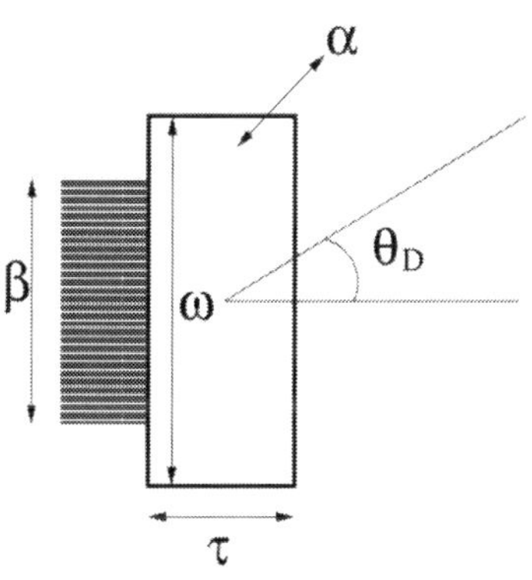

Figure 3. Sketch of relevant adimensional parameters in slab geometry.

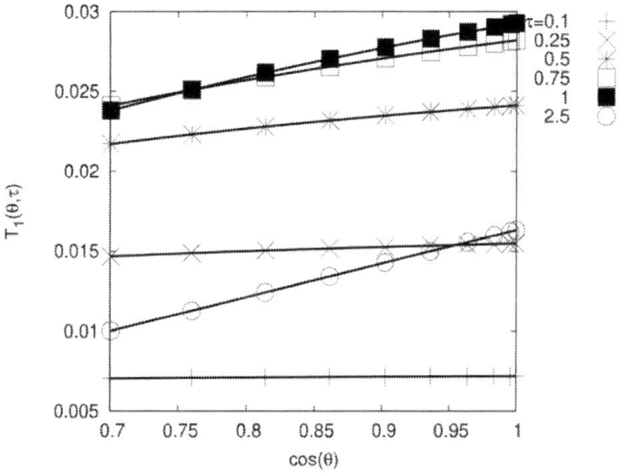

Figure 4. T_1 from Monte Carlo simulations (lines) and from analytical calculation (marks).

Table 1. Variability ranges of the relevant parameters.

τ	ω	β	θ_D (rad)	α
[0.02, 5]	[0.8,50]	[0,40]	[0,0.8]	[0,0.75]

4.1. Angular and thickness dependence for the reference case

In the reference case we can analytically calculate the intensity of neutrons scattered once, i.e.

$$T_1(\theta_D|\lambda) = (1-\alpha)\frac{e^{-\tau \sec\theta_D} - e^{-\tau}}{4\pi(1-\sec\theta_D)} I_0(\lambda)$$

where θ_D is the scattering angle, $\mu_t = \rho\sigma_t$ the total attenuation coefficient, $\mu_s = \mu_t - A_{abs}\lambda$, th is the thickness, $l_{out} = (th - z)sec(\theta_D)$ the distance to exit from z at θ_D angle, I_0 the incident flux (put $I_0 = 1$). This can be used to test to goodness of the Monte Carlo simulations (Fig.4).

For the smallest detecting angles, the single intensity is the highest as the attenuation is the lowest. Increasing θ_D, the single intensity decreases. Moreover, by increasing τ, the angular dependence gets stronger, as the range between the shortest and the longest path through the sample is enhanced. Note that while T_1 increases for $\tau \leq 1$ and decreases for $\tau > 1$ (Fig.5), the beam of multiple scattered neutrons, T_m, defined as:

$$T_m \equiv \sum_{i=2}^{n} T_i \quad \text{(for } n \text{ as large as required by the accurancy level)}$$

increases for all τ and has a weaker angular dependence for high τ values, as the incident direction is not relevant for high order scattering events (Fig.6).

Instead of considering T_1 and T_m separately, look at the ratio $\frac{T_m}{T_1}$ (Fig.7). As a function of θ, this value is almost constant for low τ values, while it's slightly increasing with θ_D for high τ values. Incidentally, in SAS experiment, the functional dependence of the scattered intensity in the Porod region of momentum transfer has been found to remain invariant under multiple

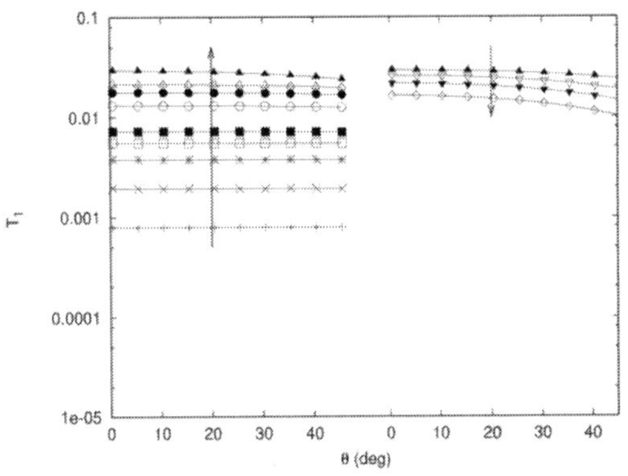

Figure 5. Angular dependence of single intensity for thin and thick slabs.

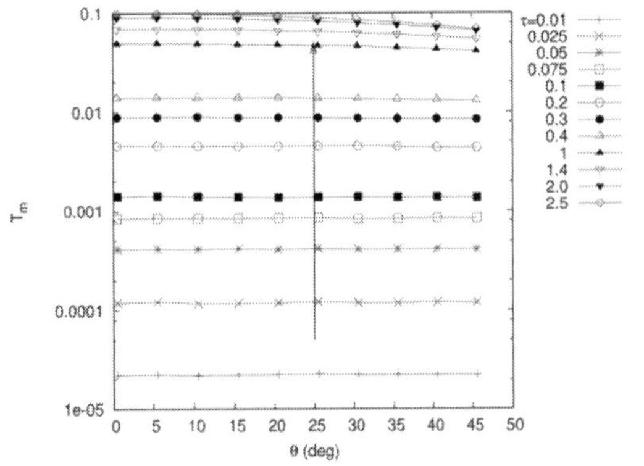

Figure 6. Angular dependence of multiple intensity for thin and thick slabs.

scattering [13], that is to say that the scattering profiles of two samples of the same substance but different thickness differ from one another only for a scale factor.

We propose to fit $\frac{T_m}{T_1}$ by a linear function of θ_D: $\frac{T_m}{T_1} \simeq C_\tau + B_\tau \theta$.

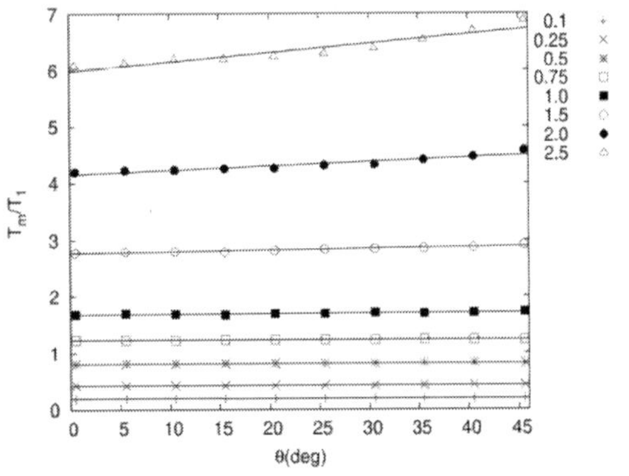

Figure 7. For the cases here examined, the ratio $\frac{T_m}{T_1}$ may be fitted with a linear function of θ_D, where the coefficients depend on τ.

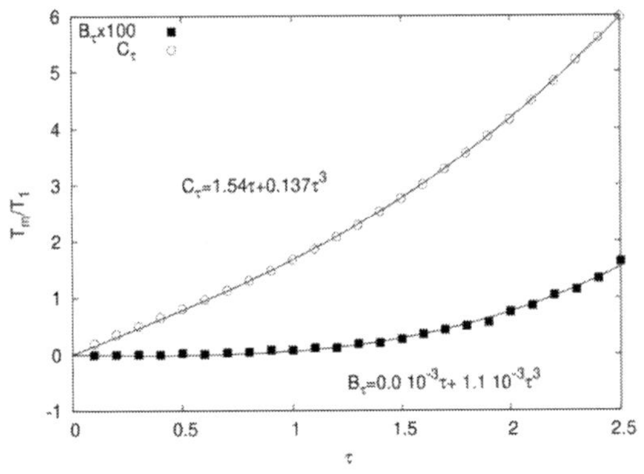

Figure 8. The fitting curves for C_τ and B_τ coefficients.

By fitting a plenty of data in the examined range (see Fig.7), we get

$$C_\tau = 1.54\tau + 0.137\tau^3 \quad B_\tau = 0.0 \cdot 10^{-3}\tau + 1.1 \cdot 10^{-3}\tau^3,$$

as showed in Fig.8. In conclusion, the θ-dependence is very weak, with respect to the τ-dependence, so the multiple scattering can be estimated by:

$$T_m^{ref}(\lambda) \simeq (1.54\ \tau(\lambda) + 0.137\tau(\lambda)^3)\ T_1^{ref}(\lambda) \tag{8}$$

This result is consistent with previous analytic estimates by Sears and Vineyard (see Fig.9).

Furthermore, the expansion of $\frac{T_m}{T_1}$ in odd powers of τ can be justified by considering that the most probable scattering angle is $\frac{\pi}{2}$; this means that at small θ_D are detected neutrons which have had two or four collisions at about $\frac{\pi}{2}$ in the sample. As $\frac{T_n}{T_1} \sim \tau^{n-1}$, the formula (8) is a reasonable approximation.

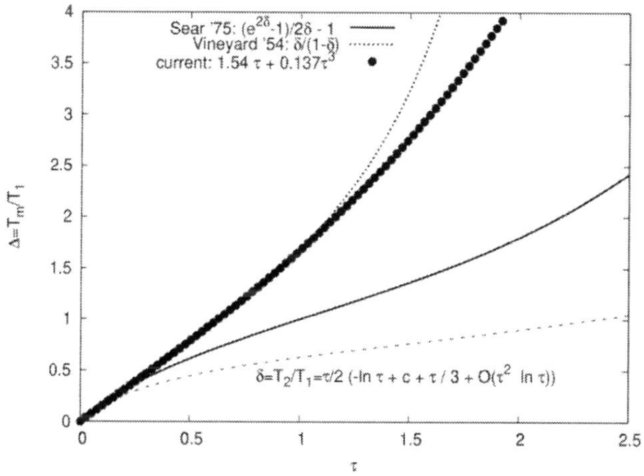

Figure 9. Current result in comparison with estimates in [2, 4]. According to [4], the expansion of δ in terms of τ is valid for $\tau \ll 1$, indeed.

4.2. Dependence on the finite size of the sample

4.2.1. Squat samples Let us consider a finite slab containing a not absorbing sample, wet by an extremely collimated neutron beam ($\alpha = 0$, $\beta = 0$). In such a case, the ratio $\frac{T_m}{T_1}$ is a function of τ, θ_D, ω. In particular, the single is not sensitive on the finite size if θ_D is so small that $tg\theta_D < \frac{\omega}{2\tau}$. On the contrary, the angular dependence is relevant for thick samples with a small width. In Figg.10,11,12, the single, the multiple scattering intensities and their ratio are

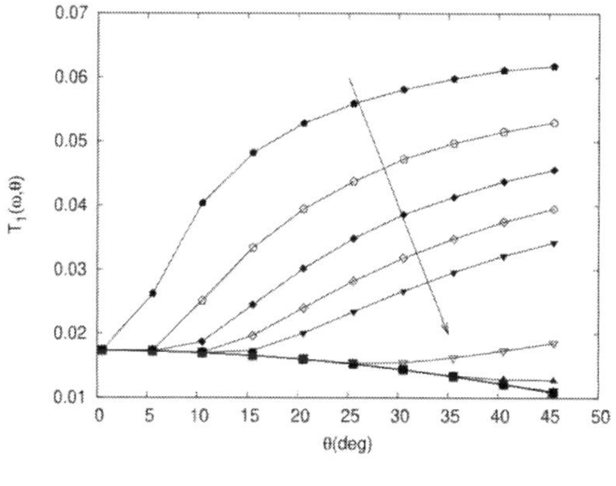

Figure 10. Single intensity taking into account the finite size effect for a thick slab ($\tau = 2.4$), for many ω values.

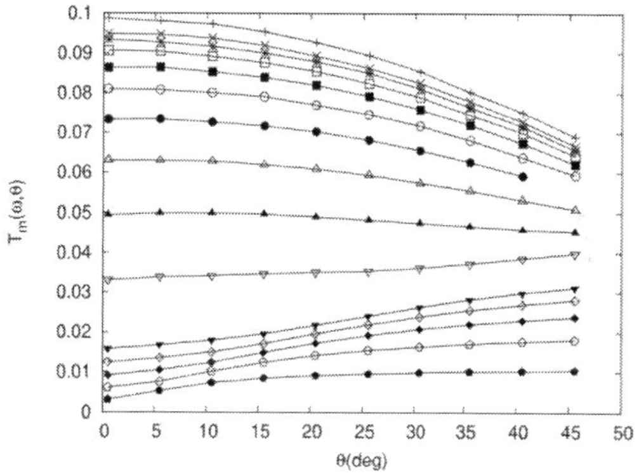

Figure 11. Multiple intensity in the same condition of Fig.10.

reported for $\tau = 2.4$. Note that in the cases under study (Tab.1), $\frac{\omega}{2\tau} = \frac{width}{2thickness} > 5$ so the angular dependence of the single scattering intensity for a finite slab does not depend on the width. In any case, it results that the angular dependence of the ratio $\frac{T_m}{T_1}$ is less important than the dependence on the width, so it can be neglected, also in this case, at least in a first approximation:

$$\frac{T_m}{T_1} \simeq C(\tau, \omega).$$

As usual, we'll refer to the ratio between multiple and single scattering $C = \frac{T_m}{T_1}$, as it's more flat vs θ_D than both the single intensity and the multiple intensity.

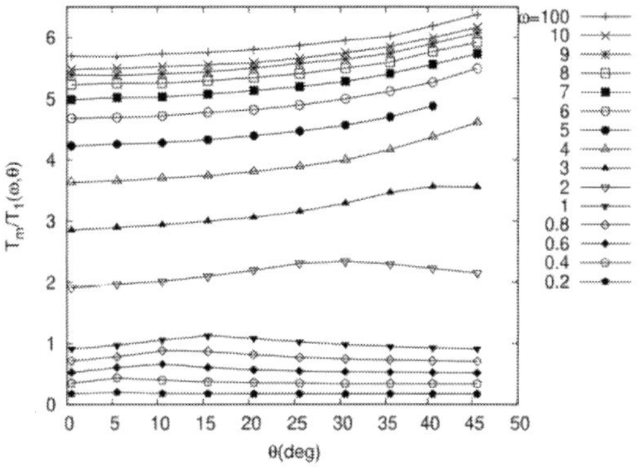

Figure 12. Ratio between multiple and single intensities in the same condition of Fig.10.

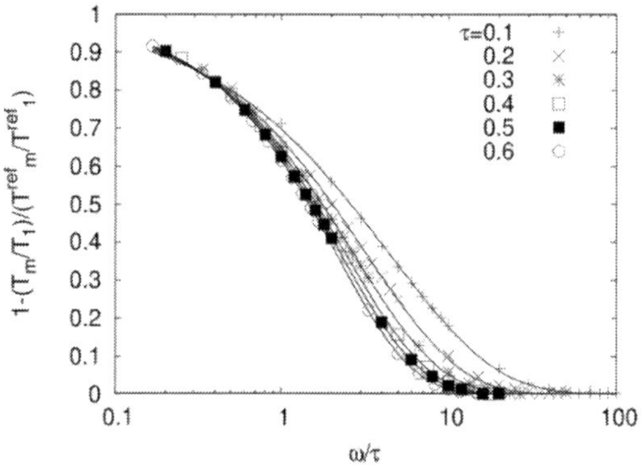

Figure 13. The correction for finite-size slab can be achieved using the stretched exponential function in Eq.(9).

Note that if $\theta_{D,max} < tg\frac{\omega}{2\tau}$ (as usually happens), the single scattering is not ω-dependent, i.e. $T_1(\omega,\tau) = T_1(\infty,\tau)$ so $\frac{T_m(\tau,\omega)}{T_m(\tau,\infty)} = \frac{C(\tau,\omega)}{C(\tau,\infty)}$. A description of the τ and ω behaviour of T_m can be achieved by considering the quantity:

$$1 - \frac{C(\tau,\omega)}{C(\tau,\infty)} = f(\tau,\omega),$$

where f is a decreasing function going from 1 (for $\omega = 0$) to 0 (for $\omega = \infty$). A good fit of a plenty of numerical data produced by the Monte Carlo simulation can be achieved using a stretched exponential function in $\frac{\omega}{\tau}$ (Fig.13). In particular,

$$1 - \frac{C(\tau,\omega)}{C(\tau,\infty)} \simeq \exp\left\{-A_\tau \left(\frac{\omega}{\tau}\right)^{d_\tau}\right\} \tag{9}$$

where A_τ and d_τ are power laws in τ. Fitting d_τ and A_τ in the current range, it results :

$$d_\tau = 1.0\tau^{0.19} \quad A_\tau = 0.52\tau^{0.15}.$$

It's not surprising that the main parameter to be considered in the case of a finite slab should be the ratio between the lateral and the longitudinal sizes $\frac{\omega}{\tau} = \frac{width}{thickness}$: so that, in first approximation, a slab can be considered infinite if $width \gg thickness$. Summarizing the helpful formula is

$$T_m(\tau,\omega) = \left[1 - \exp\left(-0.52\tau^{0.15}\left(\frac{width}{thickness}\right)^{\tau^{0.19}}\right)\right] T_m^{ref}(\tau) \tag{10}$$

4.3. Absorption dependence

Come back to an infinite slab ($\omega = \infty$). So far we have considered a not absorbing sample. Be $\alpha = \frac{\sigma_{abs}}{\sigma_t}$ (between 0 and 1) the absorption parameter.

In the case of $\beta = 0$, the ratio $\frac{T_m}{T_1}$ can be compared with the corresponding not-absorbing case, and approximated with a polynomial in α with degree 4 (Fig.14). It results:

$$1 - \frac{\frac{T_m}{T_1}(\alpha)}{\left[\frac{T_m}{T_1}\right](\alpha = 0)} = A_\tau\alpha + B_\tau\alpha^2 + C_\tau\alpha^3 + D_\tau\alpha^4 \tag{11}$$

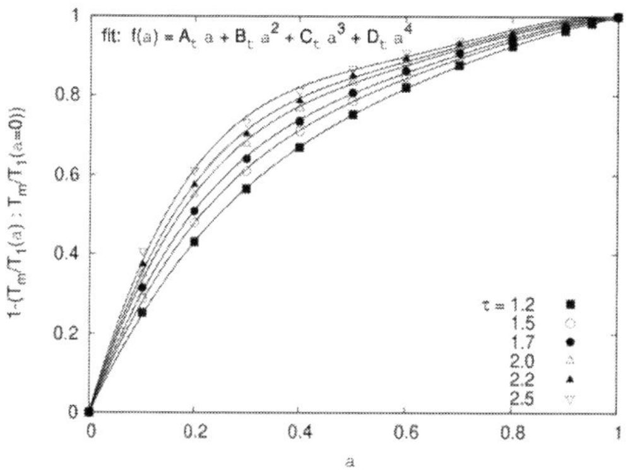

Figure 14. Fitting polynomial curves to correct for absorption.

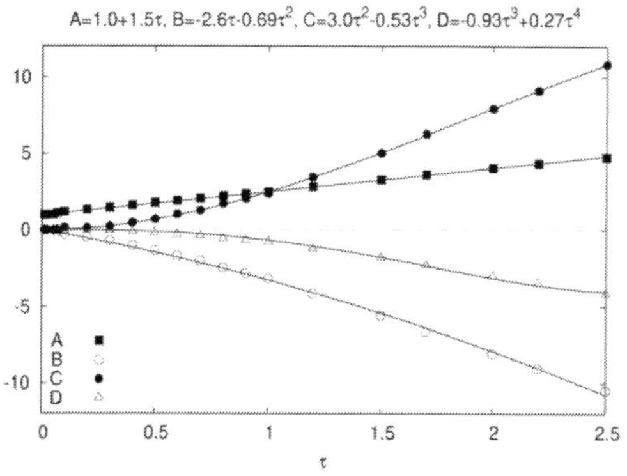

Figure 15. Coefficients of the polynomial series, Eq.(11).

The coefficients of the polynomial series as function of τ are shown in Fig.27.

In conclusion, the absorption can be taken into account by using:

$$T_m(\alpha) = T_m^{ref} \cdot \left[1 - \alpha(1 + 1.5 \cdot \tau) + 2.6\tau\alpha^2 + O(\tau^2, \alpha^3)\right] \tag{12}$$

4.4. Dependence on the beam size

Further calculations involve a real beam, i.e. having a finite size. Introducing

$$\gamma = \frac{\beta}{\omega} = \frac{beam\ \ size}{width}$$

we can estimate how important is the effect of the finite beam size. Put:

$$D(\gamma, \omega, \tau, \theta) = \frac{C(\gamma, \omega, \tau, \theta)}{C(0, \omega, \tau, \theta)},$$

being $C(\gamma, \omega, \tau, \theta) = \frac{T_m(\gamma,\omega,\tau,\theta)}{T_1(\gamma,\omega,\tau,\theta)}$. $D(\gamma, \omega, \tau, \theta)$ is almost independent on θ, so, from now on, we assume $D(\gamma, \omega, \tau, \theta) = D(\gamma, \omega, \tau)$.

The quantity $1 - D(\gamma, \omega, \tau)$ may be fitted with an even polynomial in γ:

$$1 - D(\gamma, \omega, \tau) = A_{\omega,\tau}\gamma^2 + 0.05\gamma^4. \tag{13}$$

In Fig.16, the results for $\tau = 2$ are shown. From these, it results that

- $D(\gamma, \omega, \tau)$ is always decreasing with γ, meaning that a less collimated and wider beam makes the ratio between multiple and single scattering lower;
- at γ fixed, $D(\gamma, \omega, \tau)$ is not monotonic with ω: it decreases for $\omega < \tau$ while it increases for $\omega > \tau$.
- at γ and ω fixed, $D(\gamma, \omega, \tau)$ variation is enhanced for τ increasing.

The coefficient $A_{\omega,\tau}$ can be fitted as a function of ω, which is 0 for $\omega = 0$ and $\omega \to \infty$:

$$A_{\omega,\tau} = A_\tau(\omega) = A_\tau \omega e^{-C_\tau \omega^{d_\tau}}. \tag{14}$$

The fits are reported in Fig.17.

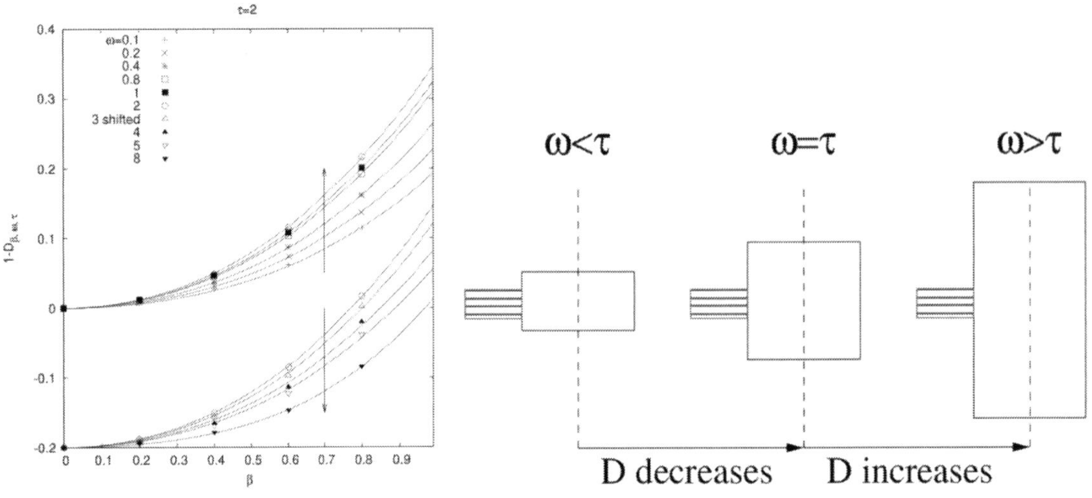

Figure 16. Left: $1 - D(\gamma, \omega, \tau = 2)$ for many ω-values and the corresponding polynomial fits. Data for $\omega \geq \tau$ have been shifted for clarity. Right: what happens to D by increasing ω.

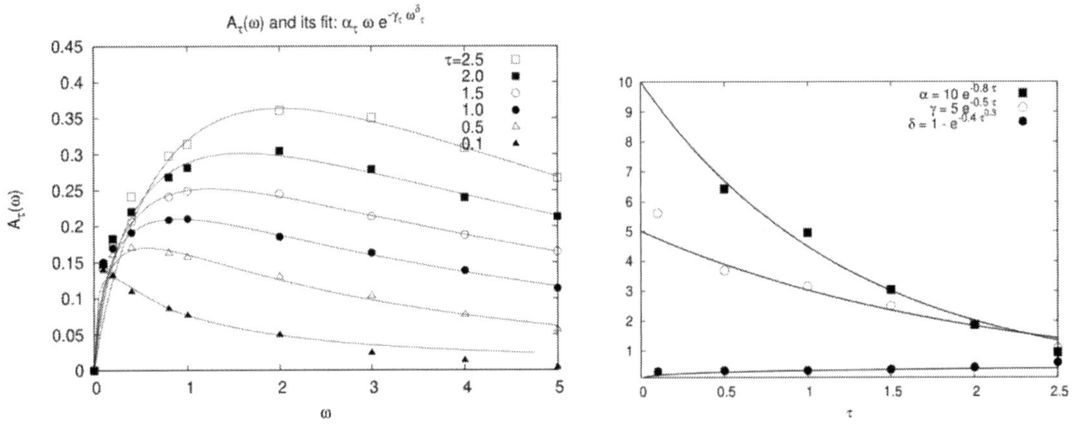

Figure 17. $A_{\omega,\tau}$ (marks) and the fitting functions (lines).

5. Cylinder geometry

Consider now a sample contained in a cylinder, the axis of which is transversal to the incident beam. In this case we introduce the adimensional diameter given by

$$2\rho = \frac{diameter}{mfp},$$

and the adimensional height, defined by:

$$\eta = \frac{height}{mfp}.$$

In this geometry, the reference case is obtained for $\eta = \infty$, $\beta = 0$, $\alpha = 0$.

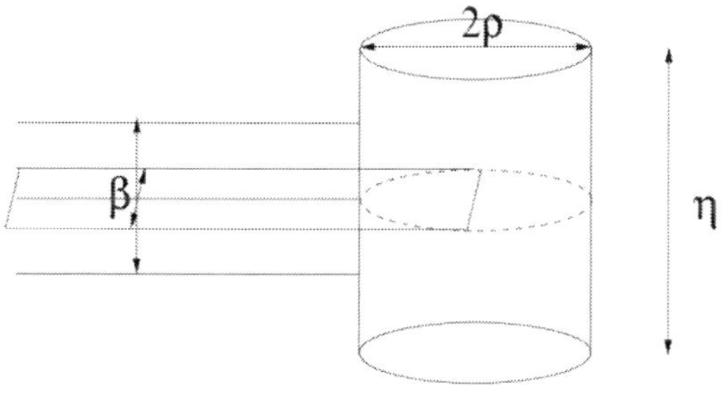

Figure 18. Parameters considered in the case of a cylinder.

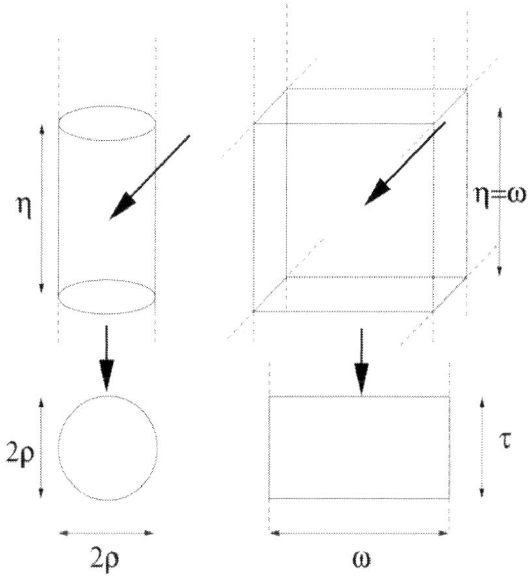

Figure 19. Comparison between the linear sizes in the two geometries.

5.1. Some results on the reference case

Many remarks done for the slab are still valid for cylinder: while T_1 and T_m may depend on the detected angle θ_D, the ratio $\frac{T_m}{T_1}$ is much less dependent on θ_D (Fig.20). So $\frac{T_m}{T_1}$ may be well

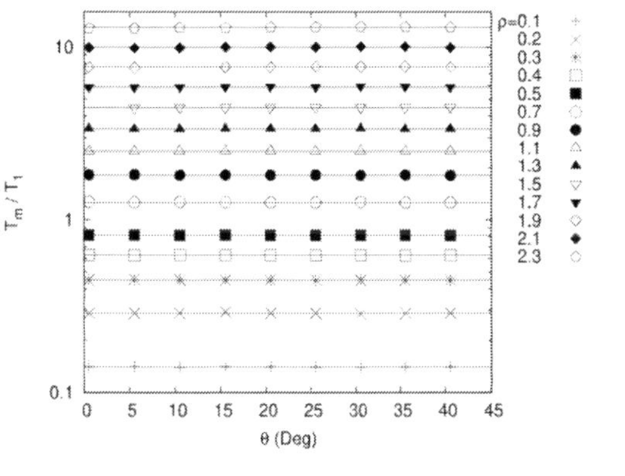

Figure 20. Angular dependence of single and multiple intensity for thin and thick samples.

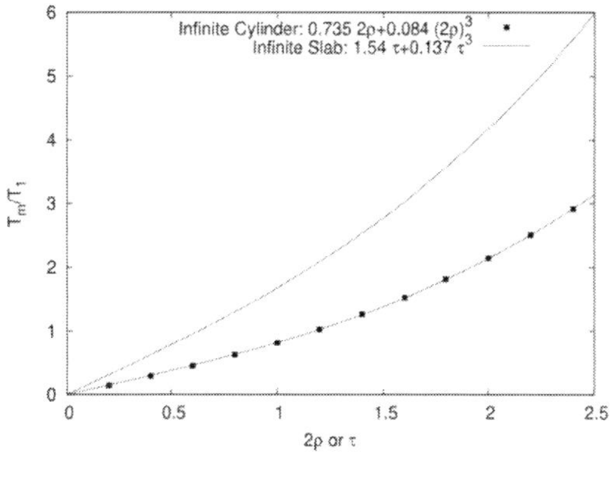

Figure 21. $\frac{T_m}{T_1}$ versus τ (for slab) or 2ρ (for cylinder).

fitted with a function of ρ, as shown in Fig.21.

5.1.1. Comparison with the reference slab Note that the intensities of scattered neutrons should be quite different in the two geometries as, by definition, the transversal sizes are equal in case of a square slab while are different in the case of a cylinder: one is the diameter, the other one is the height (see Fig.19). In other words, by comparing the two reference cases, an infinite slab is a ∞^2 object while an infinite cylinder is ∞^1 object.

It follows that the ratio $\frac{T_m}{T_1}$ is lower for an infinite cylinder of radius R than for a slab having thickness equal to $2R$ (Fig.21).

5.2. Dependence on the finite size of the sample

Let us consider cylinders which have finite height, say $\eta = 0.1, 1, 10$. While the single and multiple scattering intensities are dependent on the detected angle (Fig.22), it results that the ratios $\delta = \frac{T_2}{T_1}$ and $\Delta = \frac{T_m}{T_1}$ are almost independent on the position of the detectors (Fig.23).

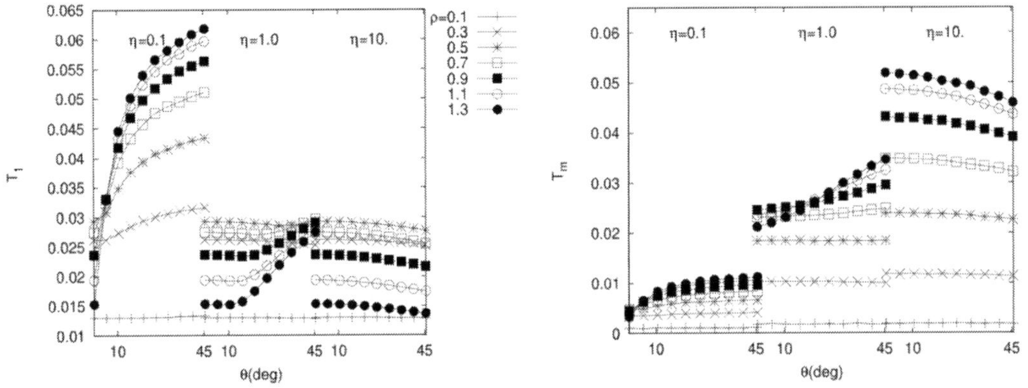

Figure 22. Angular dependence of single and multiple intensity for finite cylindrical samples.

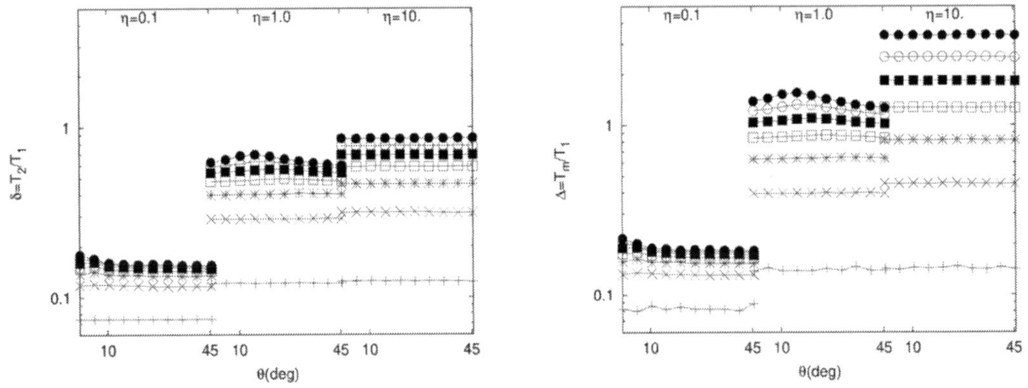

Figure 23. Angular dependence of the ratios $\delta = \frac{T_2}{T_1}$ and $\Delta = \frac{T_m}{T_1}$ for finite cylindrical samples. The symbols are the same of Fig.22.

Moreover, Δ can be fitted as an odd polynomial function of δ, being the coefficients almost, but not exactly, independent on η, as shown in Fig.24. Here we propose an alternative approach which follows just what we've done for specimens contained in slabs.

So, introducing

$$C(\rho, \eta) = \frac{T_m(\rho, \eta)}{T_1(\rho, \eta)},$$

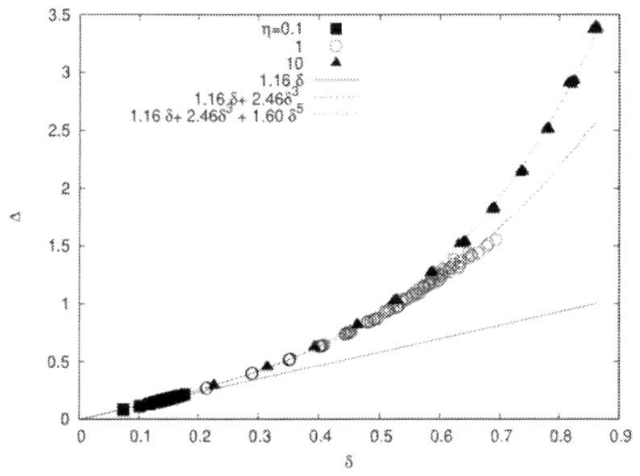

Figure 24. $\Delta = \frac{T_m}{T_1}$ as a function of $\delta = \frac{T_2}{T_1}$ for three different heights ($\eta = 0.1, 1, 1$).

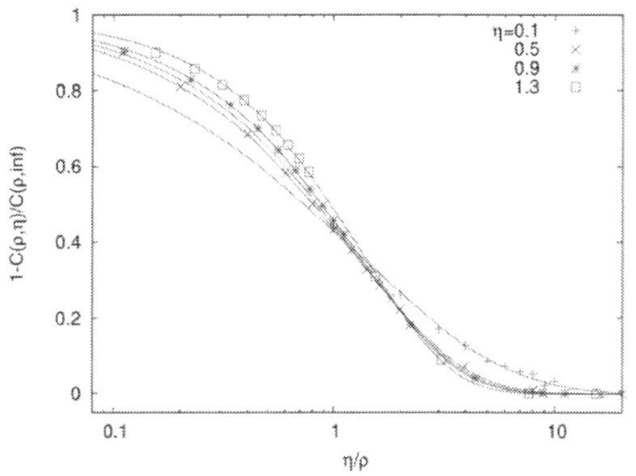

Figure 25. The correction for finite-size cylinder can be achieved using the stretched exponential function in Eq.(15).

and processing many numerical data, a fit can be performed looking at

$$1 - \frac{C(\rho,\eta)}{C(\rho,\infty)} = f(\rho,\eta),$$

where f is a decreasing function going from 1 (for $\eta = 0$) to 0 (for $\eta = \infty$). A good fit can be achieved using a stretched exponential function of $\frac{\eta}{\rho}$ (Fig.25). In particular,

$$1 - \frac{C(\rho,\eta)}{C(\rho,\infty)} \simeq \exp\left\{-A_\rho \left(\frac{\eta}{\rho}\right)^{d_\rho}\right\} \tag{15}$$

where A_ρ and d_ρ are power laws in ρ. Fitting d_ρ and A_ρ in the current range, it results:

$$d_\rho = 0.99\rho^{0.008} \quad A_\rho = -0.079\rho^2 + 0.85.$$

5.3. Absorption dependence

Come back to an infinite cylinder ($\eta = \infty$). So far we have considered a not absorbing sample.

Be $\alpha = \frac{\sigma_{abs}}{\sigma_t}$ (between 0 and 1) the absorption parameter.

In the case of $\beta = 0$, the ratio $\frac{T_m}{T_1}$ can be compared with the corresponding not-absorbing case, and approximated with a polynomial in α with degree 4 (Fig.26).

It results:

$$1 - \frac{\frac{T_m}{T_1}(\alpha)}{\left[\frac{T_m}{T_1}\right](\alpha = 0)} = A_\rho\alpha + B_\rho\alpha^2 + C_\rho\alpha^3 + D_\rho\alpha^4 \tag{16}$$

The coefficients of the polynomial series, as functions of ρ, are shown in Fig.27.

In conclusion, the absorption can be taken into account by using:

$$T_m(\alpha) = T_m^{ref} \cdot \left[1 - \alpha(1 + 1.5\cdot\rho) - 1.5\rho\alpha^2 + O(\rho^2, \alpha^3)\right] \tag{17}$$

Note that Eq.(17) is very similar to Eq.(12); this implies that if the sample is not very absorbent ($\alpha \ll 1$) and quite thin ($\tau \ll 1$ or $2\rho \ll 1$) the absorption is almost independent on the shape of the sample.

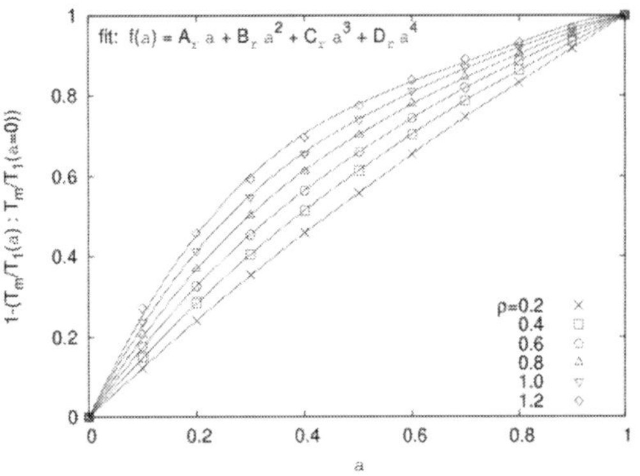

Figure 26. Fitting polynomial curves to correct for the absorption.

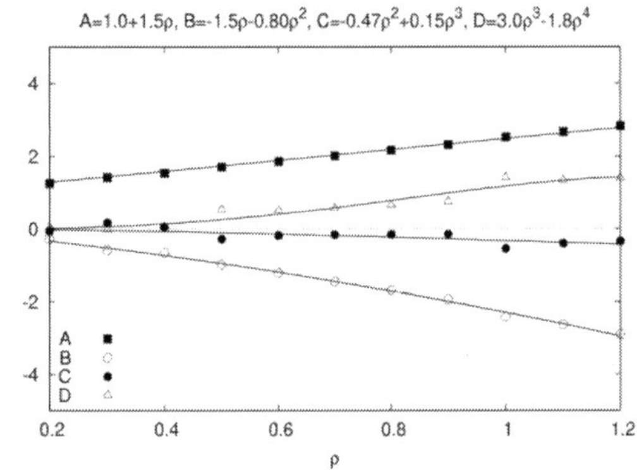

Figure 27. Coefficients of the polynomial series, Eq.(16).

5.4. Comparison with previous analytical results

In early studies [3, 4] approximated results were found for the cylinder geometry. In particular, $\delta = \frac{T_2}{T_1}$ was tabulated as a function of ρ and η. The present estimates give quite similar results, showing some limits of the previous.

First of all when a squat cylinder is considered ($\rho \sim \eta$) the angular dependence is not negligible (see Fig.28).

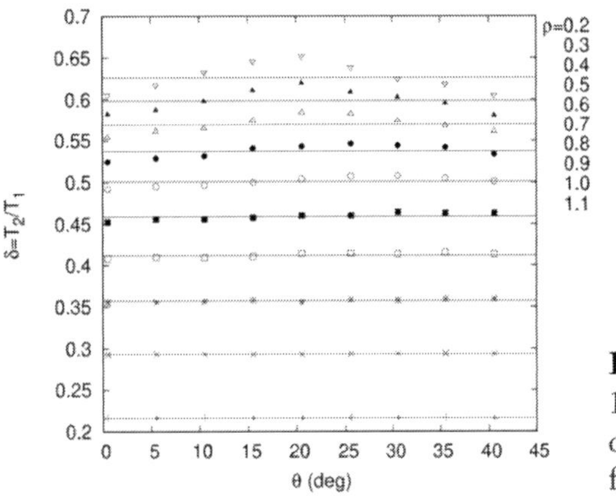

Figure 28. Monte Carlo results for $\eta = 1.1$. While for a rod ($\eta \gg \rho$) the angular dependence is negligible, this doesn't happen for a squat sample ($\eta \sim \rho$).

Secondly, approximating with a constant mean value, I get the values shown in Fig.29, which are slightly different from those obtained from Sears. Here, the analytical estimates for rods and discs are shown and the result tabulated by Sears[4] for the same case is shown for comparison.

I've also done a comparison with the results given by Blech-Averbach [3] for rod having $\rho = 0.1, 0.2, 0.3, 0.4$ and $\eta = 10 \cdot \rho$ (Fig.30).

6. How to use the fitting curves to predict the multiple scattering

In the following I'll show an example illustrating how to calculate quickly the multiple scattering corrections: this is the case of a typical silica slab, as measured on NIMROD diffractometer at ISIS-TS2 [9].

The input data needed for evaluation, as reported in the previous pages, are:

(i) the sample density, $\rho = 0.0663 \frac{atoms}{\AA^3}$;

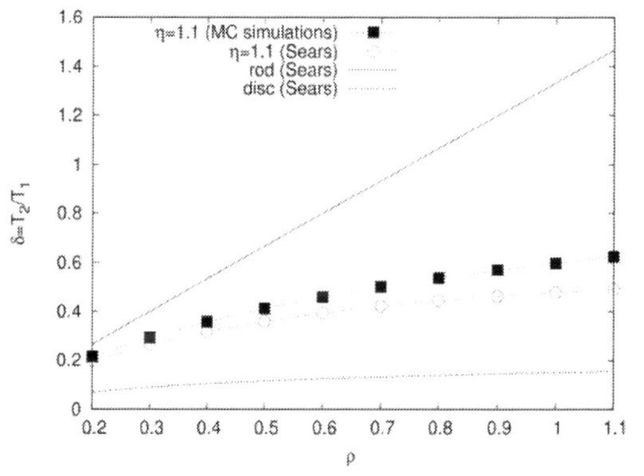

Figure 29. Monte Carlo simulation results and analytical estimates by Sears.

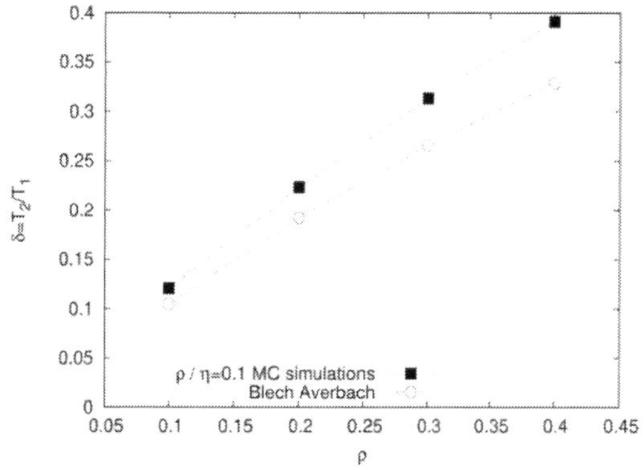

Figure 30. Monte Carlo simulation results and analytical estimates by Blech and Averbach.

(ii) the thickness, 0.4 cm;
(iii) the width, 4 cm;
(iv) the data collected by the transmission monitor, giving $\sigma_t(\lambda)$;
(v) the absorption $\sigma_{abs}(1.7979\mathring{A}) = 0.057\ barn$;
(vi) the beam size, 3 cm.

The relevant parameters, τ, ω, α and β, as functions of λ, are shown in Fig.31. The not-trivial λ-behaviour is due to the fact that $\sigma_t(\lambda)$, and so the mean free path, are not flat. The sample can be considered thin for almost all wavelenghts and the absorption is at most 10% of the scattering power. The adimensional width ω is not greater than 1 (which should be for being the infinite slab approximation valid) so a large correction is due to the finite size of the slab. Also the beam size is not negligible in comparison with the sample width ($\beta \simeq 0.75$).

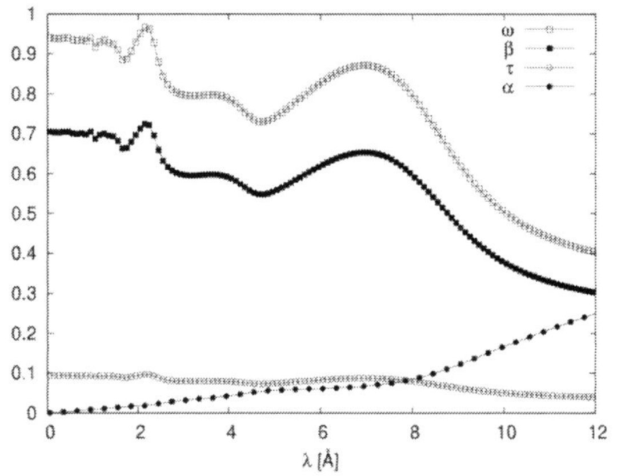

Figure 31. λ-behaviour of the relevant parameters in the real case of a silica slab.

The evaluation of the multiple scattering can be done according to the following steps, performed for each λ value, at $\theta_D = 0$.

(i) Calculation of the single and multiple scattering intensities for the reference case (according to Eq.(8)):

$$\tau(\lambda) = \rho \cdot thickness \cdot \sigma_t(\lambda) \qquad T_1^{ref}(\lambda) = \frac{\tau e^{-\tau}}{4\pi} \qquad T_m^{ref}(\lambda) = 1.52\tau T_1^{ref}(\lambda).$$

(ii) Correction for the absorption (according to Eq.(12)):

$$\sigma_{abs}(\lambda) = \frac{\sigma_{abs}(1.7979\mathring{A})}{1.7979}\lambda \qquad \alpha(\lambda) = \frac{\sigma_{abs}(\lambda)}{\sigma_t(\lambda)}$$

$$f(\alpha) = \alpha(1 + 1.5 \cdot \tau) - 2.6\tau\alpha^2 + .. \qquad T_m^{\alpha}(\lambda) = (1 - f(\alpha))T_m^{ref}(\lambda)$$

(iii) Correction for the finite width (according to Eq.10):

$$T_m^{\omega,\alpha}(\lambda) = \left[1 - \exp\left(-0.52\tau^{0.15}(10)^{\tau^{0.19}}\right)\right] T_m^{\alpha}(\lambda).$$

(iv) Correction for the finite size of the beam (according to Eq.13 e 14):

$$T_m^{\gamma,\omega,\alpha}(\lambda) = T_m^{\omega,\alpha}(\lambda)\left[1 - \left(10e^{-0.8\tau}\omega \cdot e^{-5e^{-0.5\tau}\omega^{1-e^{-0.4\tau^{0.3}}}}\right)\gamma^2 + 0.05\gamma^4\right]$$

The results of these next steps are reported in Fig.33.

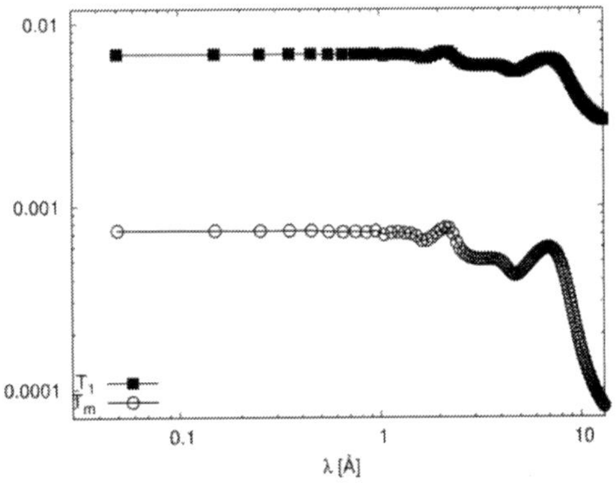

Figure 32. Single and multiple scattering intensities calculated by the equations here proposed at $\theta_D = 0$.

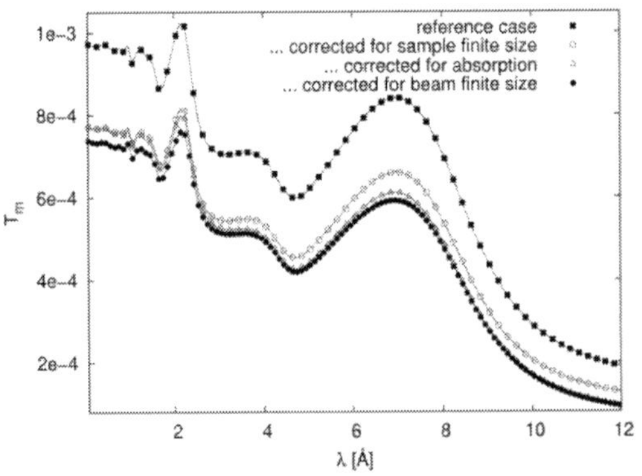

Figure 33. The effect of the next corrections on the multiple scattering intensity at $\theta_D = 0$.

7. Conclusions

A method to get an approximated analytical estimate of multiple scattering intensity of thermal neutron beam in a diffraction experiment has been presented. Even if the formulas provided are valid only for the specimens and instruments here studied, namely in the variability range given in Tab.1, the general approach could be followed aimed at implementing a fast real-time data analysis. It's worth noting that each geometrical feature can be expressed in *mfp* (standing for mean free path) units, so that a sample is said thin or thick, short or long, according to its physical features, in particular depending on its total cross section. Given an experimental setup and a scientific case (source, detectors, sample containers and specimens of interest), Monte Carlo simulations can be run in order to get fitting curves dependent on the adimensional parameters. As a matter of fact, these scaling laws can provide a quick and reliable evaluation of the multiple scattering correction, at least when the sample is quite disordered.

Interestingly, an analytical inversion algorithm for SAS experiment has been found elsewhere, by Mazumder *et al* [5]: this shall be considered in next future when further developments will be required to get an unique method for analysing data, collected on wide Q range diffractomers.

Acknowledgments
I'm grateful to U. Bafile, E. Guarini for clarifying discussion on Monte Carlo methods; W.S. Howells for giving me the code he developed many years ago. I also wish to acknowledge assistance and encouragement from A.K. Soper, M.A. Ricci and the financial support from CNR - Istituto Sistemi Complessi and Ministero dell'Universita' e Ricerca - PRIN 2008. At last, I'm greatly indebted to the reviewer to make me aware of last studies about multiple corrections in SAS experiments.

References

[1] Van Hove L 1954 *Phys. Rev.* **95** 249
[2] Vineyard G H 1954 *Phys. Rev.* **96** 93
[3] Blech I A 1965 *Phys. Rev.* **137** 1113
[4] Sears V F 1975 *Adv. in Phys.* **24** 1
[5] Mazumder S, Sen D, Patra A K 2003 *J. Applied Crystall.* **36** 840; Sen D, Bahadur J, Mazumder S 2009 *Eur. Phys. J. B* **71** 75
[6] Mazumder S, Sequeira A 1990 *Phys. Rev. B* **41** 6272; Mazumder S, Sequeira A 1990 *Physica B* **174** 200
[7] Bischoff F G, Yeater M L, Moore 1972 *Nucl. Sci. Eng.* **48** 266
[8] Lefmann K, Nielsen K *Neutron News* 1999 **10** 20
[9] http://www.isis.stfc.ac.uk/instruments/instruments2105.html
[10] Copley J R D, Verkerk P, Rowe J M *Comp. Phys. Commun.* 1986 **40** 337
[11] Sears V F 1992 *Neutron News* **3** 26
[12] James F 1980 *Rep. Prog. Phys.* **43** 1146
[13] Mazumder S, Sen D, Roy S K, Hainbuchner M, Baron M, Rauch H 2001 *J. Phys: Condens. Matter* **13** 5089

Journal of Physics: Conference Series **340** (2012) 012034 doi:10.1088/1742-6596/340/1/012034

One-dimensional neutron focusing with large beam divergence by 400mm-long elliptical supermirror

M. Nagano[1], F. Yamaga[1], D. Yamazaki[2], R. Maruyama[2], H. Hayashida[2], K. Soyama[2] and K. Yamamura[1]

1 Research Center for Ultra-precision Science and Technology, Graduate School of Engineering, Osaka University, 2-1, Yamadaoka, Suita, Osaka 565-0871, Japan

2 J-PARC Center, Japan Atomic Energy Agency, 2-4 Tokai, 319-1195, Japan

yamamura@upst.eng.osaka-u.ac.jp

Abstract. Reflective optics is one of the most useful techniques for focusing a neutron beam with a wide wavelength range since there is no chromatic aberration. Neutrons can be focused within a small area of less than 1 mm^2 by high-performance aspherical supermirrors with high figure accuracy and a low smooth substrate surface and a multilayer interface. Increasing the mirror size is essential for increasing the focusing gain. We have developed a fabrication process that combines conventional precision grinding, HF dip etching, numerically controlled local wet etching (NC-LWE) figuring, low-pressure polishing and ion beam sputtering deposition of the supermirror coating to fabricate a large aspherical supermirror. We designed and fabricated an plano-elliptical mirror with large clear aperture size using the developed fabrication process. We obtained a figure error of 0.43 μm p-v and an rms roughness of less than 0.2 nm within an effective reflective length of 370 mm. A NiC/Ti supermirror with m=4 was deposited on the substrate using ion beam sputtering equipment. The results of focusing experiments show that a focusing gain of 52 at the peak intensity was achieved compared with the case without focusing. Furthermore, the result of imaging plate measurements indicated that the FWHM focusing width of the fabricated mirror is 0.128 mm.

1. Introduction

In recent years, high-intensity proton accelerator facilities have been constructed for neutron research such as J-PARC in Japan, ISIS-TS2 in UK and SNS in USA. High-intensity pulsed neutron beams with a wide wavelength range are produced from these facilities. These beams are used extensively in combination with the time-of-flight (TOF) method to reduce the measurement time in various experiments using neutron beams. A neutron-focusing device is required to improve the S/N ratio and resolution and reduce the measurement time. Solid-state lenses and magnetic lenses have been developed and used for focusing cold neutron beams [1,2]. However, it is difficult to focus a neutron beam including thermal neutrons onto a small point without chromatic aberration. On the other hand, reflective optics is one of the most useful techniques for focusing a neutron beam with a wide wavelength range since there is no chromatic aberration. Neutrons can be focused within a small area of less than 1 mm^2 using high-performance aspherical supermirrors with high figure accuracy and a smooth substrate surface and a multilayer interface. We have developed a numerically controlled local wet etching (NC-LWE) technique [3] to figure aspherical mirror substrates and an ion beam sputtering deposition technique [4] to deposit a NiC/Ti multilayer to realize practically applicable aspherical supermirror optics. Previously, we fabricated a plano-elliptical supermirror with a clear aperture of size of 90 × 40 mm^2 by applying these techniques and obtained a focusing gain of 6 and a reflectivity of 0.64-0.7 in the near-critical-angle region for m=4 [5,6], where m is the ratio of the effective critical angle of the supermirror to that of nickel.

Increasing the mirror size is essential for increasing the focusing gain. We have developed a fabrication process that combines conventional precision grinding, HF dip etching, NC-LWE figuring, low-pressure polishing and ion beam sputtering deposition of the supermirror coating to fabricate a large aspherical supermirror [7]. Here, we report fabrication result of an elliptical supermirror with large clear aperture size and the focusing performance of the neutron beam.

2. Elliptical supermirror with large clear aperture

2.1. Optical design

It is effective to increase the length of the mirror to increase the focusing gain. We designed a plano-elliptical neutron-focusing mirror that can be applied to reflective small-angle scattering measurement in SUIREN (Apparatus for Surface and Interface Investigations with Reflection of Neutrons) of JRR-3M (Japan Research Reactor No.3 (Modified) of Japan Atomic Energy Agency) and BL17 at J-PARC. The effective reflective area, magnification, focal length and incident angle at the mirror center were 370 mm x 30 mm, 1:1, 1050 mm and 1.4° assuming that an m=4 supermirror was deposited, respectively. The designed shape of the mirror surface is given by

$$\frac{x^2}{1050.31^2} + \frac{y^2}{25.66^2} = 1 \text{ (unit : mm)}. \quad (1)$$

For the designed mirror with a length of 370 mm, the maximum depth of the mirror is 420 μm.

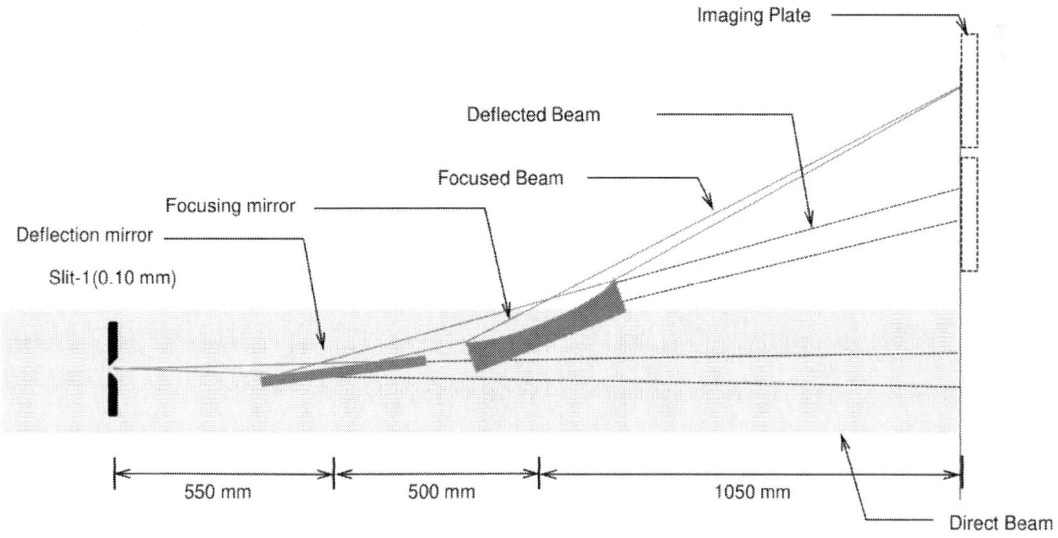

Figure 1. Schematic of the setup for measuring the neutron beam profile (top view).

2.2. Fabrication of neutron-focusing elliptical supermirror with large clear aperture size

We have developed a fabrication process that combines conventional precision grinding, NC-LWE figuring, low-pressure polishing and ion beam sputtering deposition of the supermirror coating to fabricate a large aspherical supermirror.

For the efficient figuring of the mirror shape, precision grinding was conducted with a computer-based numerically controlled grinding machine using resin-bond diamond wheels of #325, #1200 and #3000 in turn, whose feed pitches were 3 mm, 1 mm and 0.5 mm, respectively. The mirror material was synthetic quartz glass with dimensions of 458 mm × 100 mm × 35 mm, and the figured and evaluated area was 370 mm × 50 mm. Grinding techniques make it possible to obtain micrometer form accuracy in a short time. We performed HF dip etching and a first polishing process using CeO2 slurry before figure correction by NC-LWE to remove the subsurface damage after figuring by grinding.

The NC-LWE technique enables an aspherical shape to be figured deterministically with sub-micrometer form accuracy because it utilizes a noncontact chemical removal process that is insensitive to external disturbances such as vibration and/or thermal deformation. As the processing parameters in NC-LWE figuring, we applied 37 wt% HF solution with a temperature of 41.6°C, a circular nozzle with a diameter of 15 mm and a feed pitch of 0.5 mm.

In NC-LWE figuring, raster-type numerically controlled scanning is applied to correct the residual figure error. After raster scanning, shallow periodic grooves are inevitably generated on the figured surface by the overlapping of the scanned etching path. We applied a second polishing process with a hard polishing pad to remove low-spatial-frequency roughness (LSFR: λs >1 mm) such as tool marks.

The mid-spatial-frequency roughness (MSFR: 1 mm > λs > 1 μm) and high-spatial-frequency roughness (HSFR: 1 μm > λs > 1 nm) components of the figured surface increase after NC-LWE. A possible interpretation of the increases in MSFR and HSFR is that subsurface damage appears because the etching rate of the damaged layer is different from that of the undamaged layer. It is essential to minimize the surface roughness to obtain higher reflectivity and thus realize the practical application of the neutron-focusing mirror. Therefore, a third polishing process using a soft polishing pad was applied to improve the MSFR and HSFR of the figured mirror substrate.

The surface of the plano-elliptical substrate was deposited with NiC/Ti multilayers by ion beam sputtering (IBS) to fabricate a supermirror for increasing the critical angle of neutron reflection. We used the IBS system installed at JAEA, which has an effective deposition area of 0.2 m^2 (diameter of 500 mm) [8]. The NiC layer was obtained by Ni-C cosputtering. The carbon atoms added to the nickel layer suppressed interfacial roughness by reducing the crystal grain size of the nickel layer [9-12]. Majkrzak and Wood have reported achieving greater than 90% reflectivity at $m = 3$ [9, 11]. Maruyama *et al.* have succeeded decreasing diffuse intensity by more than one order of magnitude by adopting NiC/Ti multilayers instead of conventional Ni/Ti multilayers and achieving 80% reflectivity at $m = 4$ [13]. A NiC/Ti supermirror with m=4 was fabricated on the figured substrate. The thickness and total number of multilayers were 6 μm and 1200, respectively.

2.3. Evaluation of focusing performance and reflectivity

The focusing performance of the fabricated elliptical supermirror was evaluated at NOBORU at J-PARC of JAEA. We show a schematic of the setup for measuring the neutron beam profile in Figure 1. A flat supermirror with $m = 4$ is used to reflect the neutrons into the focusing device to reduce the background and the gamma radiation from the direct beam. The wavelength of the neutron beam was varied from 3.5 to 10 Å by calibration using a chopper. The sagittal plane of the mirror was installed horizontally on a goniostage to focus the neutron beam vertically. The mean incident angle of the neutron beam was 1.40 °. The horizontal divergence of the neutron beam was 0.60 °. Two-dimensional images of focused and nonfocused direct neutron beams were measured using an imaging plate (Fujifilm BAS-ND; resolution: 50 μm). The width of the slit in this experiment was 0.10 mm. The magnification of the optical design was 1.

3. Results and discussion

We designed and fabricated a plano-elliptical mirror with large clear aperture size using the proposed fabrication process. The shape of the mirror substrate was measured with a laser autofocus measuring machine (Mitaka Kohki NH-5N; z resolution: 10 nm). We were able to reduce the figuring time of NC-LWE to 1 day by applying grinding and obtained a damage-free mirror substrate by applying HF dip etching and the first polishing process. The maximum difference from the designed shape was 0.43 μm after the third polishing process. This result indicates that the residual figure error was corrected deterministically with sub-micrometer form accuracy when NC-LWE was used to figure the mirror substrate with a large clear aperture (370 mm) and high mirror depth (ca. 420 μm) and that the second and third polishing processes that we developed and applied were effective for improving the surface roughness while maintaining sub-micrometer form accuracy. Figure 2 shows the surface roughness of the figured mirror substrate after the third polishing process. The surface roughness was measured using a scanning white-light interferometer (Zygo NewView 200CHR). To remove the undesirable surface structures, a soft polishing pad was applied in the third polishing process to a removal depth of about 40 nm. This result indicates that the surface morphology was improved and that an rms roughness of less than 0.2 nm was obtained by applying the third polishing process.

From these results, it is concluded that the mirror fabrication process combining precision grinding, NC-LWE and low-pressure polishing is a highly effective method of fabricating an ultraprecision aspherical mirror substrate with a large clear aperture size, sub-micrometer form accuracy and sub-nanometer surface roughness.

Figure 3 shows the surface roughness of the mirror surface after NiC/Ti supermirror deposition. There was no difference between the surface roughness after deposition and the surface roughness of the mirror substrate, indicating that the surface of the mirror substrate has sub-nanometer surface roughness dominated by the crystal grain aggregate of NiC.

Figure 4 shows two-dimensional images of the neutron beam detected using the imaging plate. A neutron beams with a wide wavelength range was uniformly focused on the mirror where it had an effective width of 30 mm. The neutrons detected beyond the mirror were due to the horizontal divergence of the neutron beam. The shape of nonfocused beam is asymmetric. This is because the center of the slit could be out of alignment from the center of the upstream collimators or slits. Figure 5 shows detailed profiles of both focused and nonfocused direct neutron beams along cross section A-A'. The full width at half maximum (FWHM) of the focused beam was 0.128 mm, and a focusing gain of 52 in terms of the peak intensity was achieved compared with the nonfocused direct beam. The intensity of the focused neutron beam was 16.7 times greater than that of the nonfocused beam at a width of 0.5 mm. Figure 6 shows the result of TOF measurement. The neutron beam with a wavelength range of 3.5-10 Å was focused effectively. These results suggest that the fabricated elliptical supermirror enables to focus neutron beams with wide wavelength range to be focused onto samples of sub-millimeter size with high intensity.

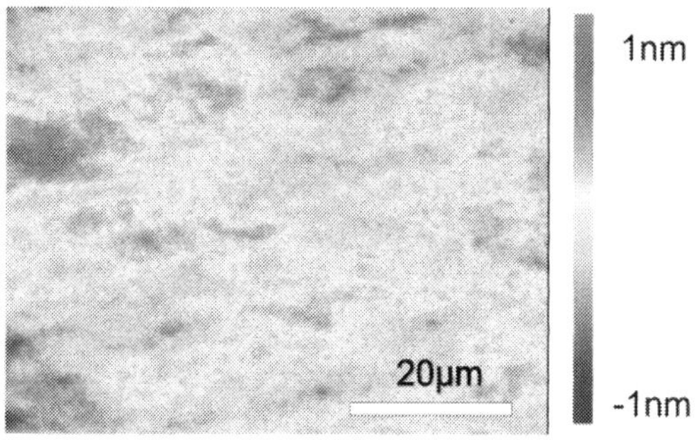

Figure 2. Surface roughness after 3rd polishing (0.211 nm rms in 64 x 48 μm^2).

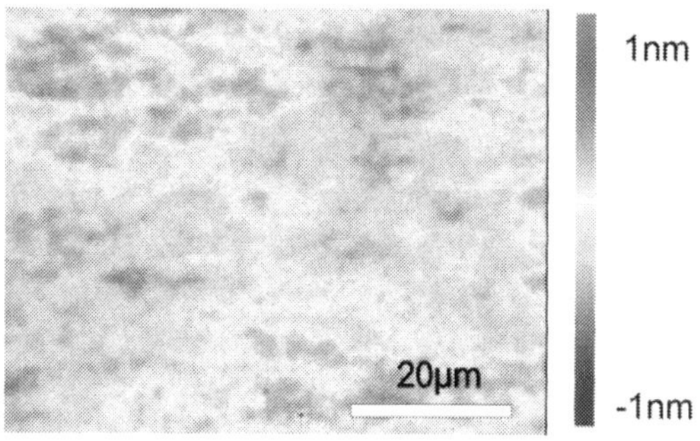

Figure 3. Surface roughness after supermirror deposition (0.202 nm rms in 64 x 48 μm^2).

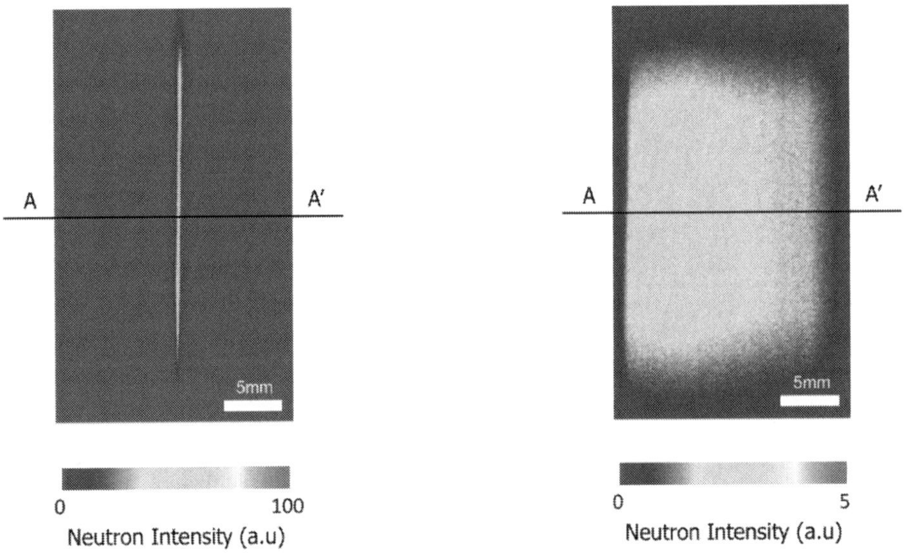

(a)focused beam (b)nonfocused beam

Figure 4. Neutron beam images detected by the imaging plate.

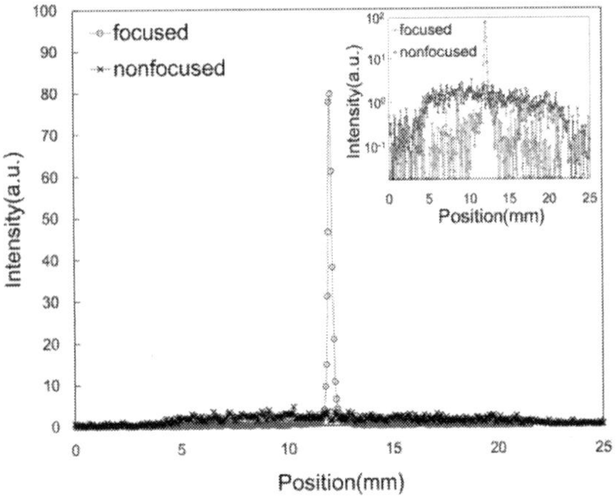

Figure 5. Focused beam profile.

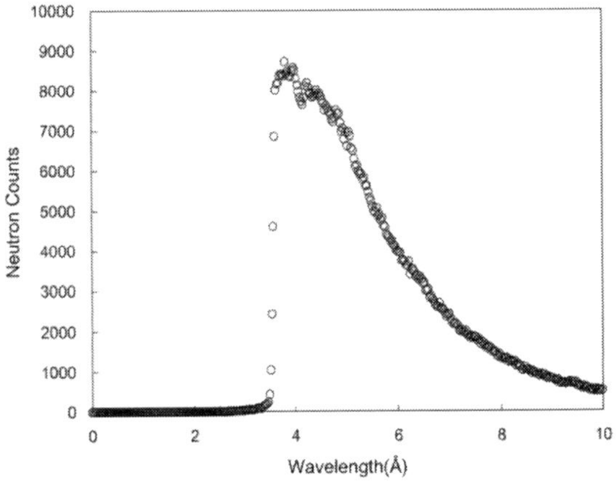

Figure 6. The result of TOF measurement.

4. Conclusions

We fabricated a 370-mm-long elliptical neutron-focusing supermirror by a combined fabrication process consisting of precision grinding, HF dip etching, NC-LWE figuring, low-pressure polishing and ion beam sputtering deposition. We succeeded in obtaining a figure error of 0.43 μm p-v with a surface roughness of less than 0.2 nm rms, an FWHM of 0.128 mm and a focusing gain of 52 at the peak intensity for a neutron beam with a wavelength range of 3.5-10 Å.

NC-LWE figuring technique has higher material removal rate than other chemical figuring techniques, such as ion beam figuring (IBF) [14], plasma chemical vaporization machining (PCVM) [15], and elastic emission machining (EEM) [16]. Therefore our process can be applied to fabrication of elliptical supermirror with a length of more than 1 m efficiently.

The magnification of the optical system we designed was 1 because the fabricated mirror is a prototype for evaluation of the fabrication process, the mirror shape and the surface roughness. We can achieve a spot size as small as a previous results using bending mirror system [17, 18] and a focusing gain higher than a previous results [19] by designing an optical system with high magnification. The figured mirror is more cost-effective than the bending mirror system and needs no time to align a mirror shape.

We aim to develop a multiple aspherical supermirror devices using high-precision figured aspherical focusing supermirror to focus neutron beam with high intensity, because multiple mirror collect very large beam divergence. In the case of the multiplexing of supermirrors, mirror substrates must be thin to reduce the absorption loss of incident neutron beams. Our process can be applied to fabrication of precise millimeter-thick elliptical supermirror since NC-LWE technique is noncontact removal process without machining pressure. In our future study, we will attempt to fabricate and evaluate a multiple mirror system using a millimeter-thick large aspherical supermirror.

Acknowledgements

This work was partially supported by a grant for "Development of System and Technology for Advanced Measurement and Analysis (SENTAN)" from Japan Science and Technology Agency (JST), the "Industrial Technology Research Grant Program in 2005" from the New Energy and Industrial Technology Development Organization (NEDO) of Japan and Grant-in-Aid for JSPS Fellows.

References

[1] Oku T, Kira H, Shinohara T, Takata S, Arai M, Suzuki J, Shimizu H 2010 J. Phys. Conf. **251** 012078
[2] Mildner D F R, Hammouda B, Kline S R 2005 J. Appl. Crystallogr. **38** 979
[3] Yamamura K, 2007 Ann. of the CIRP **56/1** 541
[4] Maruyama R, Yamazaki D, Ebisawa T, Soyama K 2009 Nucl. Instr. and Meth. A. **600** 68
[5] Yamamura K, Nagano M, Takai H, Zettsu N, Yamazaki D, Maruyama R, Soyama K and Shimada S 2009 Opt. Express **17** 6414
[6] Yamamura K, Nagano M, Zettsu N, Yamazaki D, Maruyama R, Soyama K 2010 Nucl. Instr. and Meth. A, **616** 193
[7] Nagano M, Yamaga F, Zettsu N, Yamazaki D, Maruyama R, Soyama K, Yamamura K 2010 Nucl. Instr. and Meth. A. **634** S112
[8] Maruyama R, Yamazaki D, Ebisawa T, Hino M, Soyama K 2006 Physica B **385–386** 1256.
[9] Wood J SPIE 1992 **1738** 22
[10] Vidal B, Jiang Z, Samuel F 1992 SPIE**1738** 141
[11] Majkrzak C, Ankner J 1992 SPIE**1738** 150
[12] Hino M, Sunohara H, Yoshimura Y, Maruyama R, Tasaki S, Yoshino H, Kawabata Y 2004 Nucl. Instr. and Meth. A **529** 54
[13] Maruyama R, Yamazaki D, Ebisawa T, Soyama K 2009 J. Appl. Phys. **105** 083527
[14] Martin W 2009 Nucl Instr and Meth. B **267** 1930
[15] Yamamura K, Shimada S, Mori Y, Ann. of the CIRP, **57** (2008) 567
[16] Yamauchi K, Mimura H, Inagaki K, Mori Y 2002 Rev. Sci. Instrum. **73** 4028
[17] Ice G E, Hubbard C R, Larson B C, Pang J W L, Budai J D, Spooner S, Vogel S C 2005 Nucl. Instr. and Meth. A **539** 312
[18] Ice G E, Hubbard C R, Larson B C, Pang J W L, Budai J D, Spooner S, Vogel S C, Rogge R B, Fox J H, Donaberger R L 2006 Mat. Sci. Eng. A **437** 120
[19] Kardjilov N, Hilger A, Dawson M, Manke I, Banhart J, Strobl M, Böni P 2010 J. Appl. Phys.**108** 034905

Journal of Physics: Conference Series **340** (2012) 012035 doi:10.1088/1742-6596/340/1/012035

Research toward the development of compact neutron interference imaging instrument with gratings

Yoshié OTAKE[1], ***Margie OLBINADO***[6], ***Yoshichika SEKI***[1], ***Katsuya HIROTA***[1], ***Yutaka YAMAGATA***[1], ***Jungmyoung JU***[1], ***Tomohiro ADACHI***[1], ***Shinya MORITA***[1], ***Yoshihisa IWASHITA***[2] ***Masahiro HINO***[3], ***Masahiro ICHIKAWA***[2], ***Masaaki KITAGUCHI***[3], ***Toshio TAKAHASHI***[4], ***Hideki YOSHIZAWA***[4], ***Seung Wook LEE***[5], ***Wataru YASHIRO***[6], ***Atsushi MOMOSE***[6]

[1]**RIKEN,** Wako, Saitama, 351-0198, Japan

[2]UKUICR, Uji, Kyoto, 611-0011, Japan

[3]KURRI, Kumatori, Osaka, 590-0490, Japan

[4]ISSP U-Tokyo, Kashiwanoha, Kashiwa, Chiba, 277-8581, Japan

[5]KAERI, 1045 Daedeok-daero, Daejeon, Korea

[6]Advanced Material Science FS U-Tokyo Kashiwanoha, Kashiwa, Chiba, 277-8562, Japan

yotake@riken.jp

Abstract. The neutron interference imaging experiment with two absorption gratings has been done. We develop new way to fabricate absorption gratings for neutron with the pitch of 150 μ m, 180 μ m, and 200 μ m. Compact imaging detector system is developed, whose weight is 2kg. Small break in an acrylic plate, the welded place of Nb plate were observed, it reveals that the neutron interference imaging method can become new method of a nondestructive inspection for reinforced plastics and composite material such as CFRP and so on.

1. Introduction

The composite material consisting of light elements becomes widely used such as a body of the airplane, cars, even space rockets. It is the serious problem that there is no established way for non-destructive inspection of the composite material until now. Here we show one potential method to observe the crack, internal delamination of them.

There are several methods of phase contrast imaging in the x-ray field, such projection imaging method as phase contrast imaging with perfect crystal interferometer, with prism interferometer, with grating interferometer, by detecting refraction angle, and such technique of microscope as Zernike phase difference, differential phase, holography, and so on. The phase contrast imaging using Talbot-Lau interferometer for x-ray [1] [2] [3] [4] is now widely used. The neutron phase contrast imaging with grating [5] [6] [7] becoming more popular now a days. However, in neutron case, to fabricate absorption grating is much more difficult than in X-ray case, because Gd is unworkable material. We, here, show the result of neutron interference image with two absorbing gratings. We have fabricated neutron absorbing gratings whose pitches are as large as 150 μ m, 180 μ m, and 200 μ m with gadolinium oxide powder, while neutron absorption gratings with wide pitch is normally made with Gd sputtering [8], or electroforming process. We also introduce a portable imaging detector system as an imaging detector. Three types of the images, absorption image, differential phase image, visibility contrast image, are obtained with them, as in the case of Talbot interferometer imaging method. A small crack in an acrylic plate, and a weld part of niobium plates were observed. We discuss future possibility in last chapter shortly.

2. Neutron Absorbing Grating

We have made new type of the neutron absorption gratings with large pitch. 100 μ m width grooves with 200 μ m depth are cut with the pitch of 150 μ m, 180 μ m, 200 μ m on the surface of silicon substrate as is shown in Fig.1 (a). The silicon gum which contains Gd_2O_3 powder is filled in the channels, as is shown in Fig.1 (b).

(a)

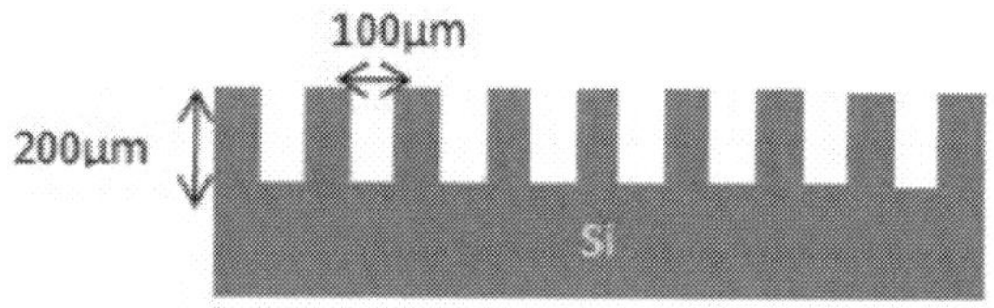

(b)

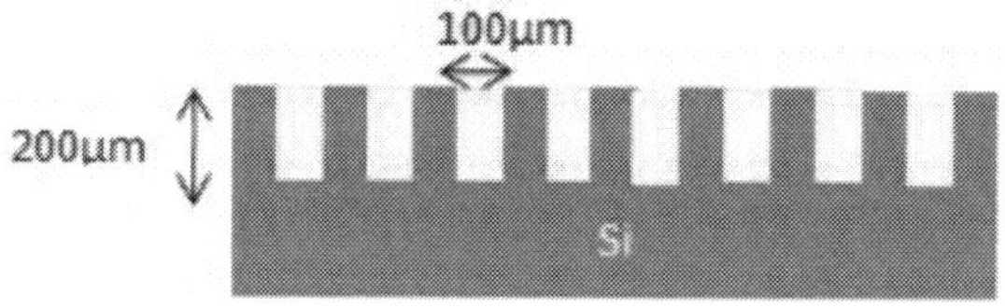

Fig.1 (a) Channels of the width of 100 μ m and the depth of 200 μ m are cut on the surface of silicon substrates. (b) The silicon gum which contains gadolinium oxide powder is compacted there.

The aspect of this absorbing grating is shown in Fig.2. The diameter is 150mm.

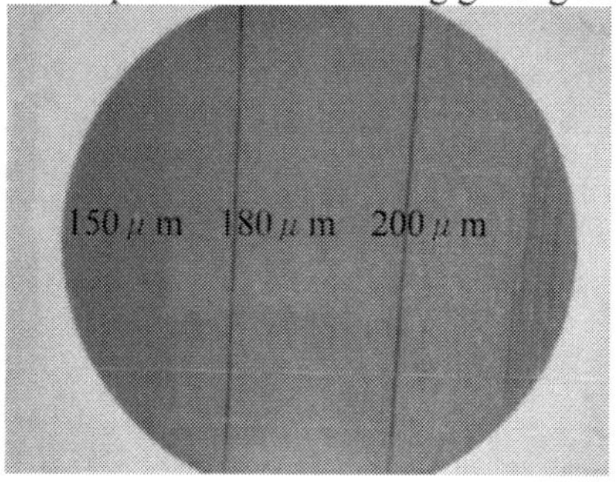

Fig.2 The photograph of absorption grating is shown. There are three parts with different pitches, as the pitches of 150 μ m, 180 μ m and 200 μ m with the width of absorption material line is 100 μ m.

The surface observation with ten times magnified is shown in Fig.3. The white lines are the silicon lines and the dark lines are the mixture of gadolinium oxide powder and silicon gum.

(a) (b) (c)

Fig3. The surface observation of absorbing grating with NH3.. (a)Pitch 150 μ m, the absorbing line which consists of Gd_2O_3 powder and silicon gum is shown as gray line. (b) 180 μ m pitch, (c) 200 μ m pitch, the same width of the Si and Gd_2O_3, The width of dark colored line are all the same.

We made two of the same absorption grating by changing the weight ration of the Gd2O3 powder in Si-gum with 50%, and 60%. The neutron moiré fringes created by them are shown in Fig.4. It ensures their completeness as absorption gratings.

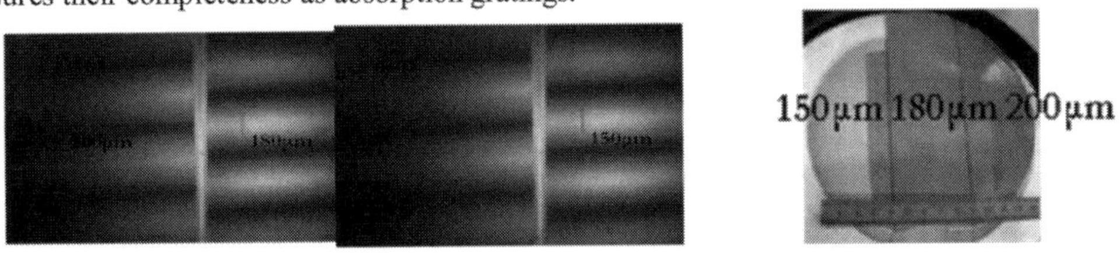

Fig.4 Images of moiré fringes, left side. The two gratings are piled together, right side, and put just in front of the detector system.

We can find the boundary line in the picture of the moiré fringes, the largest pitch of the moiré fringes is created by the superposition of the part of 150 μ m pitch grating, and so on as shown in Fig.4.

3. Interference imaging with two absorbing gratings

The neutron interference imaging experiment has been done at ULS, C1-3 beam port of JRR3 in JAEA (Japan Atomic Energy Agency), Tokai, Japan. Neutron wave length is 0.44nm.

The interference imaging experiment has been done with two absorption gratings, and portable 2D detector system. Two neutron absorbing gratings are set behind the sample; the distance between two gratings is 14.5mm as shown in Fig. 5(a). Portable compact neutron camera system is developed with $Zn/S/^6LiF$ scintillator of 0.25mm thickness and CCD (IMB-7012G) together with camera box. The total weight of this system is less than 2kg. The effective area of imaging is $53 \times 40mm^2$, and the pixel number is 656×484.

The experiment has performed by scanning one of two absorption gratings. We use two 180 μ m pitch gratings.

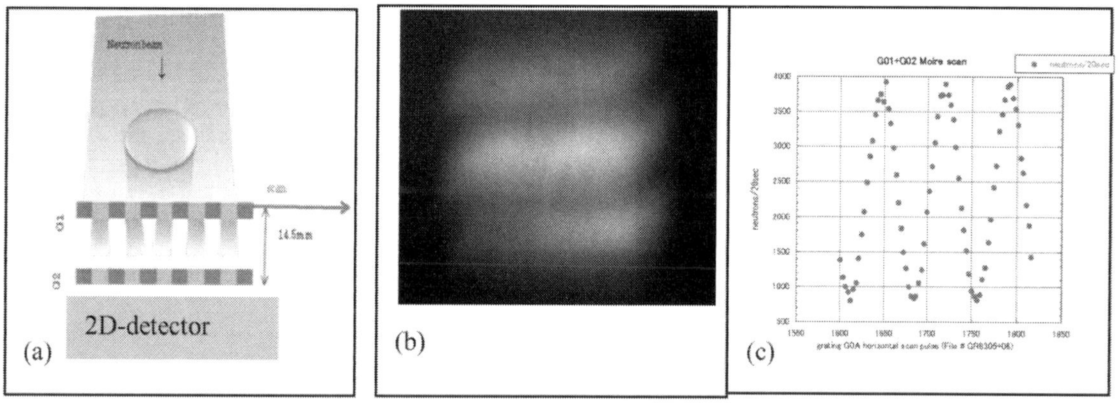

Fig.5 Experimental setting and the measured moiré fringes without sample
(a) Two absorption gratings are set behind the sample.
(b) The moiré fringes observed behind the second gratings with camera system
(c) The intensity measured with 3He detector.

The distance between two absorption gratings which is 14.5mm is determined by the estimated source size. To take the interference imaging, five shots are taken with and without sample by scanning the second gratings. The samples are acrylic plate with small crack as shown in Fig. 6 (c) and the weld part of Nb plate as shown in Fig. 6 (g).

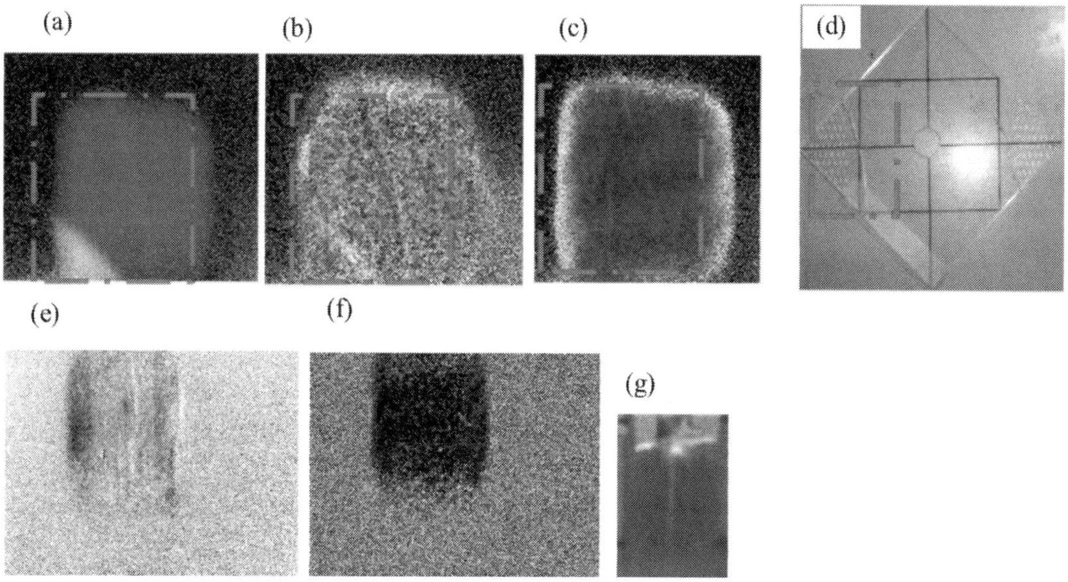

Fig.6 Images of a small crack in an acrylic plate, (a) absorption image (b) differential image (c) visibility contrast image with neutron, and (e) its photo, small crack part is surrounded by red dotted line Images of welded part of Nb plate, (e) absorption image (f) differential image with neutrons, (g) its photo

Small break and crack of an acrylic plate which is in the area of red square in the picture, Fig.6 (d), is observed in absorption image and differential phase image clearly with two absorption gratings. The weld part of niobium plates shown in Fig.6 (g) is observed with absorption image, Fig.6 (e), small spots are observed in differential phase image, Fig.6 (f).

As is shown in Fig. 6 (a)-(c) and (e)-(g), with two absorption gratings, we could get phase contrast imaging. It means that we use the deformation of the shadow of the first absorption grating which is created by phase shift through the sample, in other words, the deformation of the wave front, caused by the sample, and draw out the phase information of the sample by scanning the second absorption grating. In Fig.6 (c) shows the deformation of the shadow of the first grating schematically. The deformed shadow is scanned by the second absorption grating show as in Fig. 6(d). Actually the special coherence length at the second grating is 50.1 μ m. With this method, we don't need the longer special coherence length than the pitch of the second grating.

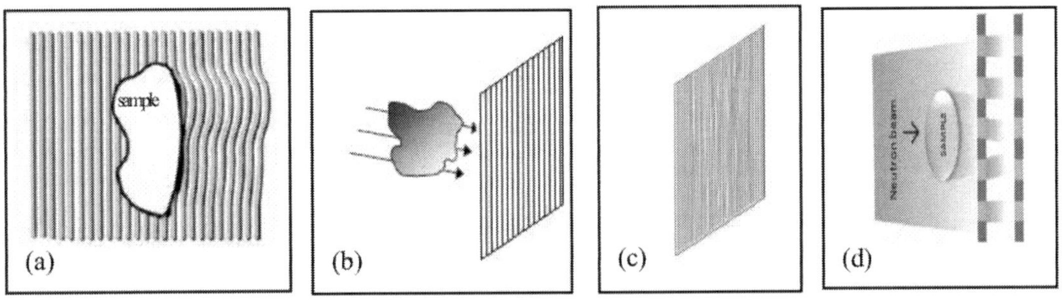

Fig.7 Schematic view of neutron wave front and the experimental set up
(a) Wave front deformation caused by sample (b) the first absorption grating is placed after the sample (c) the shadow of the absorption grating deformed by the phase shift through the sample

One of our important aims of the useful neutron interference imaging system is following. We like to develop non-destructive inspection system for such composite materials as resin material, CFRP, FRP, reinforced plastic and so on. These materials become widely used as a body of car, airplane or important parts in aeronautical field because the light weight and strength although until now there is no way to detect their internal deletion, internal water penetration and so on. The main elements of CFRP or reinforced plastic are H, C, and O. The neutron scattering cross section of C and O are more than thousands times larger than absorption cross section. So, we evaluate neutron phase imaging method be one of the best way to detect their internal small deletion. The crack in the acrylic plate is clearly shown in absorption imaging as shown in Fig. 6(a), and differential phase imaging as shown in Fig.6 (b). This method is different from the experiment of two absorption gratings in [9].

4. Conclusion

The small break in an acrylic plate was successfully observed using two neutron absorption gratings. This result ensures us that the interference neutron imaging method using gratings can be one method of nondestructive test for the deletion or small crack or internal delamination inside reinforced plastics, composite material such as CFRP, and so on.

References

[1] A.Momose et al. Jpn.J.Appl. Phys.**42**(2003)L866
[2] T.Weitkamp et al. Appl.Phys.Lett.**86**(2005) 054101
[3] F.Pfeiffer et al. Nat.ure letters **2**(2006)258

[4] A.Momose et al. Jpn.J.Appl.Phys **45**(2006)5254
[5] Pffeifer, et al. Phys.Rev.Lett. **96** (2006)215505
[6] C.Grunzweig, et al Phys., Rev. Lett. **101** (2008) 025504
[7] S.W.LEE et al. Nucl. Instr. and Meth. **A605** (2009) 16
[8] C. Grunzweng, et al. Rev. of Sci. Instr. **79** (2008) 053703
[9] Zhi-Feg Huan, et al. Phys. Rev. **A 79** (2009) 013815

The new polarizer devices at RESEDA

J. Repper[1,2,*], W. Häußler[1], A. Ostermann[1], L. Kredler[1,2], A. Chacón[2] and P. Böni[2]

[1] Forschungs-Neutronenquelle Heinz Maier-Leibnitz (FRM II), Lichtenbergstr. 1, 85748 Garching, Germany

[2] Lehrstuhl für Experimentalphysik E21, TU München, James-Franck-Str. 1, 85748 Garching, Germany

* now at: Paul Scherrer Institut, Materials Science and Simulation, NUM/ASQ, 5232 Villigen PSI, Switzerland

E-mail: julia.repper@psi.ch

Abstract. In the neutron resonance spin echo method the information about sample dynamics is encoded in the neutron beam polarization measured in the analyzer-detector unit. Thus, the method is not applicable for sample systems and environments, which depolarize the neutron beam strongly. To over come this draw back a neutron analyzer directly before the sample position may be installed to perform MIEZE-I experiments. We compared the performance of a transmission polarizer and a solid-state bender at this position for the neutron resonance spin echo spectrometer RESEDA by Monte Carlo simulations. It turned out, that the polarization as well as the intensity transmitted to the sample position is more advantageous for the transmission polarizer as for the bender. In addition, we present measurements of the polarization and intensity performance of the transmission polarizer already installed at RESEDA to polarize the neutron beam coming from the reactor FRM II. The measurements are in good agreement with Monte Carlo simulations.

1. Introduction

Neutron spin echo (NSE) is a well known method to measure dynamics with high energy resolution [1]. Based on NSE several neutron resonance spin echo (NRSE) instruments were developed in the last decades [2, 3, 4]. At the neutron resonance spin echo spectrometer RESEDA at FRM II, studies on magnetic fluctuations in ferro- and helimagnetic compounds and on the diffusion of water and hydrogen in porous samples were performed up to now [5]. In NSE and NRSE, the energy resolution is decoupled from the wavelength spread of the primary neutron beam. Thus, NRSE and NSE can provide high primary intensities, combined with a high momentum and energy resolution required for studies of relatively slow dynamics. The information about the dynamics in the sample is encoded in the beam polarization, which is determined by the analyzer-detector unit.

As the beam polarization is the measured quantity the NSE and NRSE techniques suffer from a serious draw back: the beam polarization for spin-incoherent scattering systems, i.e. hydrogen containing materials, is limited to 1/3, so that the statistical accuracy of these measurements is decreased. Moreover, measurements in ferro magnetic materials are not possible due to the depolarization of the beam by the magnetic domains. The MIEZE-I technique, a variant of the NRSE method, measures the intermediate scattering function independently of the beam polarization. Thus, the depolarization neither due to the sample nor due to the sample environment does affect the measurement. The MIEZE-I unit consists of a polarizer, two radio-frequency spin flipper coils (NRSE coils) and an additional analyzer. All components are located before the sample position. At the detector, the neutron beam may be unpolarized and the oscillating amplitude of the intensity yields the information about the sample dynamics. A comprehensive introduction to MIEZE-I is given e.g by Keller et al. in [6] and in a recent realization by Georgii et al. [7]. The MIEZE-I unit is well suited as an additional option for the NRSE instrument RESEDA as the polarizer as well as the NRSE devices are already installed, and only an additional spin analyzer in front of the sample is needed.

In this paper, we compare the results of Monte Carlo simulations for the performance of RESEDA equipped with a solid-state bender and a transmission polarizer (V-cavity), both located behind the NRSE devices, directly in front of the sample position. In addition, the experimentally observed performance of the V-cavity, which is installed in front of the NRSE devices is compared with Monte Carlo simulations.

2. Polarization devices

Polarizer in front of NRSE devices The transmission polarizer, a single V-cavity (Mezei, 1988), between the reactor and the first NRSE devices has an entrance window of 36.6 x 36.6 mm^2 and a length of 2 m. The wafers for the V are coated with Fe-Si supermirrors $m = 3$ and the outer walls of the neutron guide consist of glass plates coated with Ni-Ti layers with m=1.2. The cavity is installed at RESEDA since 2006.

Polarizer behind NRSE devices The simulated solid-state bender consists of a stack of 270 silicon wafers. The wafer thickness is 0.15 mm. Each wafer is coated with Fe-Si supermirrors $m = 3$ at both sides. The wafers are bent with a radius of 0.50 m, leading to a critical wavelength of $\lambda^* \approx 2.3$Å. The neutron entrance window of the bender has dimensions of 40 x 40 mm^2 and the length of the bender is 50 mm. As the bender emits the spin up and spin down neutrons within different angles relative to the incident beam direction a collimator with a length of 30 mm has to be used to absorb the spin down component.

The simulated transmission polarizer consists of five V-cavities, which are arranged in parallel channels vertical and perpendicular to the beam direction. The silicon wafers used to build the Vs have a thickness of 0.3 mm each and are coated with Fe-Si supermirrors to reach an m-value of 4. The cavity has a length of 260 mm and in total a neutron entrance window of 40 x 40mm^2.

3. Experimental details

3.1. Measurements

The set up to determine the polarization performance of the V-cavity in front of the NRSE devices consists of two spin flippers at a distance of 20 cm to each other and the first one 30 cm away from the exit of the cavity, a ^{3}He cell for spin analysis and a CCD detector camera with a cross section of 65 x 96 mm^2, which is located 100 cm behind the second spin flipper (compare figure 1). To avoid depolarization between the cavity and the analyzer appropriate guide fields were installed. Four measurement steps are necessary to determine the spin flip efficiency of the flippers and the polarization of the neutron beam after passing through the polarizer, the analyzer and the guide fields. At first, the spin flippers are switched off, and the neutron beam polarization is given by the transmission polarizer. In the second step, one of the spin flippers is switched on, the other remains switched off, and in the third step vice versa. Finally, the two spin flippers are operated simultaneously. Here, the analyzer efficiency and the guide field efficiency is assumed to be 100 %. For the measurements a wavelength of $\lambda = 5.3$ Å was chosen. A slit of 33 x 33 mm^2 was positioned behind the exit of the V-cavity to avoid the detection of neutrons, which did not pass the cavity.

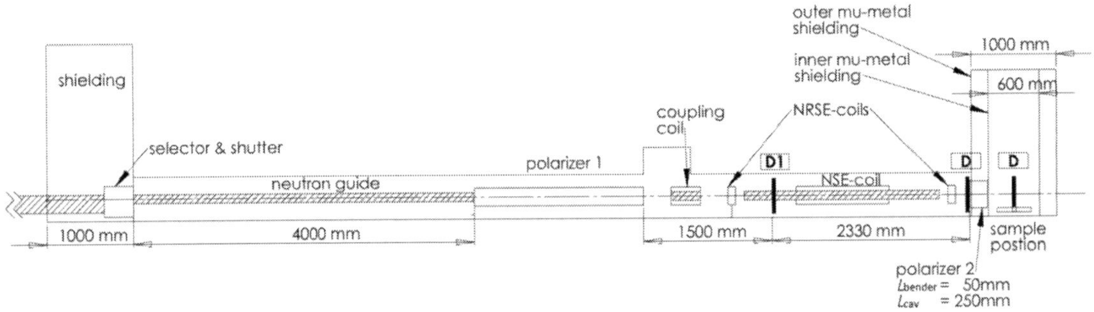

Figure 1. Schematic view of the neutron resonance spin echo instrument RESEDA as used for the McStas simulations presented. The detector D1 has dimension of 65 x 96 mm^2 and was used for gaining data for the simulation of the transmission polarizer in front of the NRSE devices. The detectors D have cross sections of 40 x 40 mm^2 and were used for the simulation of the polarizer devices behind the NRSE devices. This second polarizer is either a solid state bender (bender) or a V-cavity (cavity).

3.2. Monte Carlo simulations

All Monte Carlo simulations were performed using McStas [8]. Here, self-written components were used to simulate the polarization devices. Figure 1 shows a schematic view of the neutron resonance spin echo spectrometer RESEDA as used for the simulations.

Polarizer in front of NRSE devices A position sensitive detector (PSD) with a detector area of 65 x 96 mm^2 was located 150 cm behind the end of the V-cavity. A slit of 33 x 33 mm^2 was positioned directly behind the V-cavity. To simulate the neutron background, detected during the experiment, an uniform counting rate of 0.2 Hz/pixel was added to the intensities simulated. The polarization P was determined by using $P = (I_{up} - I_{down})/(I_{up} + I_{down})$, where I_{up} and I_{down} give the intensity of the spin up and spin down neutrons as measured in the PSD detector D1 (figure 1).

Polarizer behind NRSE devices The polarizers behind the NRSE devices were simulated with an unpolarized incident neutron beam, as e.g. in MIEZE the beam may become depolarized by the NRSE coils. Therefore, the transmission polarizer in front of the NRSE devices was replaced in the simulations by a neutron guide with equal dimensions with a coating of $m = 2$. The wavelength distribution in the incident neutron beam was chosen to 1 to 13 Å. The used detectors had cross sections of 40 x 40 mm^2 and were positioned directly in front of the second polarizer and at the sample position (figure 1). We used these two detectors (D in figure 1) to calculate the transmission of the beam through the polarizer device. The polarization P was determined by using $P = (I_{up} - I_{down})/(I_{up} + I_{down})$ as measured in the detectors D.

4. Results

Polarizer in front of NRSE devices Figures 2 and 3 show the intensity distribution and the polarization distribution over the 65 x 96 mm^2 position sensitive detector area at a distance of 150 cm from the exit of the polarizer. In both figures the experimentally determined data are compared to simulated ones.

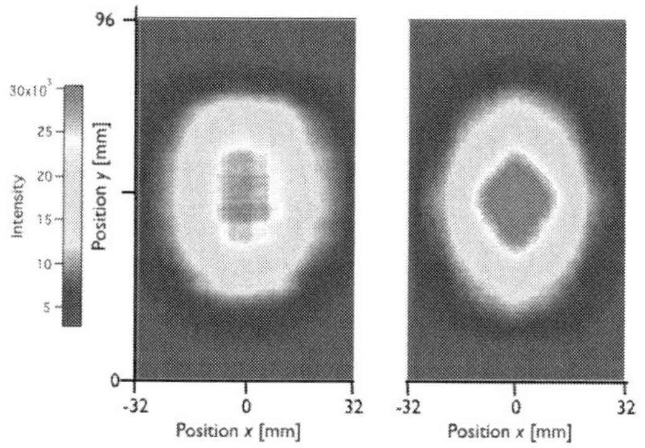

Figure 2. Intensity distribution over the PSD detector D1 (compare figure 1) as experimentally determined by a measurement (left) and simulated by Monte Carlo simulations (right).

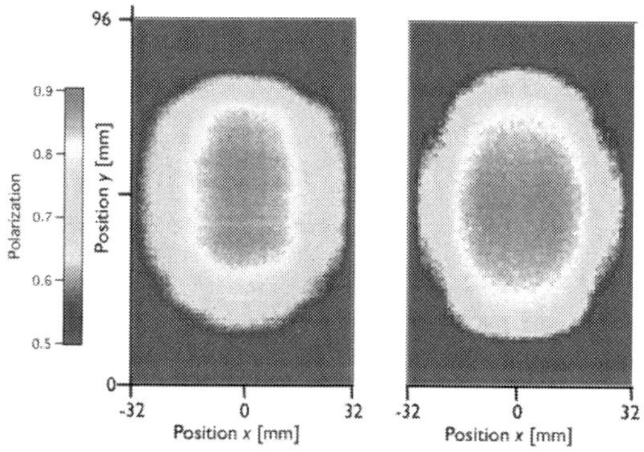

Figure 3. Polarization distribution over the PSD detector D1 (compare figure 1) as experimentally determined by a measurement (left) and simulated by Monte Carlo simulations (right).

Polarizer behind NRSE devices Figure 4 compares the polarizations of the neutron beam caused by the solid-state bender and the V-cavity in dependence of the wavelength. For both detectors with cross sections of 40 x 40 mm^2 at the sample position were used. The figure of merit $T^* = T \cdot P^2$ of the solid-state bender (T^*_{bender}) and of the V-cavity (T^*_{cavity}) at the sample position is shown in figure 5 as determined by detectors with an area of 40 x 40 mm^2.

5. Discussion

Polarizer in front of NRSE devices The shape of the neutron beam measured by the CCD camera shows a good agreement with the shape of the simulated neutron beam (figure 2). The observed intensity distribution over the cross section of the neutron beam is consistent with the distribution found in the McStas simulations. Slight differences are caused by mechanical imperfections of the polarizer itself as well as of the neutron guides, which guide the neutron beam from the reactor tank to RESEDA. The polarization distribution over the neutron beam as measured by the PSD detector is in good agreement with the compared simulation data (figure 3). The averaged polarization is 86 and 75% in the simulation compared to 84 and 72% in the measurement analyzed for detector areas of 33x33 mm^2 and of 65 x 96 mm^2, respectively. The small deviations between the simulated and the measured values may arise from a non-perfect analyzer and guide field efficiency in the real experiment in contrast to the simulations, where the efficiencies are assumed to be 100%.

Polarizer behind NRSE devices The beam polarization directly behind the solid-state bender for a wavelength of 3.8 Å, which is the minimum wavelength typically used at RESEDA, is 40 %, compared to 80 % behind the V-cavity. In contrast, the polarization of both devices is almost equal for the upper wavelength range used at RESEDA (around 6 Å). Whereas, the Bender causes beam polarizations of 94 %, the polarization with the V-cavity reaches 91 %.

In general, the figure of merit $T^* = T * P^2$ curve is larger for the cavity as for the bender. This reflects the large absorption of the neutron beam due to the 50 mm stack of silicon wafers in the bender. In contrast to the beam polarization P_{bender} the values of T^*_{bender} decrease for wavelengths larger than 7 Å. This shows the even increasing neutron absorption effects with an increasing wavelength.

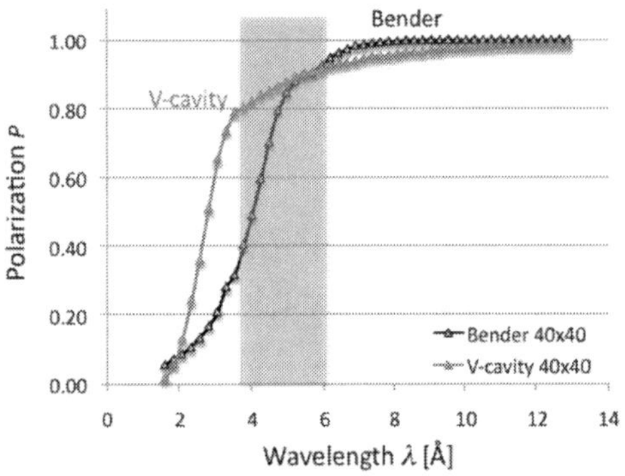

Figure 4. Polarization of the bender and the V-cavity both located behind the NRSE devices, measured with detectors with cross sections of $40 \times 40\,\text{mm}^2$ positioned at the sample position. The highlighted area marks the wavelength range usually used at RESEDA (3.8 - 6Å).

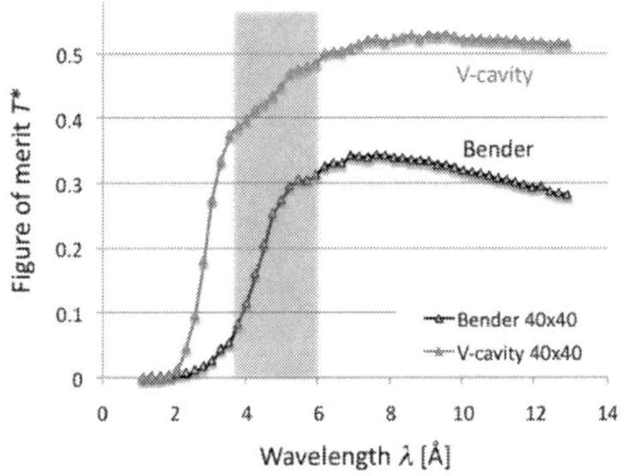

Figure 5. Figure of merit $T^* = T * P^2$ of the bender and the V-cavity both located behind the NRSE devices, measured at the sample position. Detectors with cross sections of $40 \times 40\,\text{mm}^2$ are used. The highlighted area marks the wavelength range usually used at RESEDA (3.8 - 6Å).

6. Conclusion

We could show that the transmission V-cavity polarizer already installed at RESEDA delivers the performance, which was expected by Monte Carlo simulations with respect to intensity and polarization of the neutron beam. In addition, we decided, based on the here presented data to install an additional V-cavity, which consists of five parallel channels just before the sample position in order to use a MIEZE-I unit as optional instrument set up. This decision is based on the simulated polarization and intensity performance for the wavelength range needed. Also the given geometrical restrictions at RESEDA will allow us to use the cavity with a total length of 270 mm (including magnetic housing). However, in cases the space is more limited the performance of the solid state bender with its short length of only 90 mm is convincing.

7. References

[1] F. Mezei: Neutron spin echo: A new concept in polarized thermal neutron techniques, Z. Phys. A, **255**(2), 1972, 146-160.
[2] T. Keller; R. Gähler; H. Kunze; R. Golub: Features and performance of an NRSE spectrometer at BENSC, Neutron News, **6**(3), 1995, 1 - 17.
[3] M. Kitaguchi, M. Hino, Y. Kawabata, H. Hayashida, S. Tasaki, R. Maruyama, T. Ebisawa: Correction for beam divergence effect in a NRSE spectrometer with high resolution, Meas. Sci. Technol., **19**, 2008, 034014.
[4] M. Köppe, P. Hank, J. Wuttke, W. Petry, R. Gähler, R. Kahn: Performance and Future of a Neutron Resonance Spinecho Spectrometer, J. Neutron Res., **4**(1 - 4), 1996, 261 - 273.
[5] W. Häussler, P. Böni, M. Klein, C. J. Schmidt, U. Schmidt, F. Groitl, and J. Kindervater: Detection of high frequency intensity oscillations at RESEDA using the CASCADE detector, Rev. Sci. Instrum., **82**, 2011, 045101.
[6] T. Keller, R. Golub, R. Gähler: Neutron spin echo - a technique for high-resolution neutron scattering, in: Scattering and Inverse Scattering in Pure and Applied Science, R. Pike and P. Sabatier (eds.), Academic Press, San Diego, 2002.
[7] R. Georgii, G. Brandl, N. Arend, W. Häußler, A. Tischendorf, C. Pfleiderer, P. Böni, J. Lal: Turn-key module for neutron scattering with sub-micro-eV resolution, Appl. Phys. Lett., **98**(7), 2011, 073505.
[8] K. Lefmann and K. Nielsen: McStas, a General Software Package for Neutron Ray-tracing Simulations, Neutron News, **10**(3), 1999, 20 - 23.

Development of polarized Xe gas target for neutron experiment at J-PARC

K. Sakai[1], T. Oku[1], T. Shinohara[1], H. Kira[1], M. Ooi[1], F. Maekawa[1], K. Kakurai[1], T. Ino[2], Y. Arimoto[2], H.M. Shimizu[2], Y. Sakaguchi[3], J. Suzuki[3], K. Ohoyama[4], L.J. Chang[5]

[1]JAEA, Tokai, Ibaraki 319-1195, Japan
[2]KEK, Tsukuba, Ibaraki 305-0801, Japan
[3]Research Center for Neutron Science and Technology, Comprehensive Research Organization for Science and Society (CROSS), Tokai, Ibaraki 319-1106, Japan
[4]Institute for Materials Research, Tohoku University, Sendai, Miyagi 980-8577, Japan
[5]Department of Physics, National Cheng Kung University, Tainan 70101, Taiwan

E-mail: kenji.sakai@j-parc.jp

Abstract. At the Materials and Life science experimental Facility (MLF) in J-PARC, an experiment of detecting a neutron polarizing ability caused by a neutron-nuclear spin correlation at a resonant peak of ^{129}Xe is planned. We evaluated measurable quantities based on a neutron optical theorem, developed a polarized Xe gas system, and carried out a feasibility test of our apparatus.

1. Introduction

Recent developments of spin exchange optical pumping (SEOP) technique for polarizing noble gases have enabled to obtain the gases of $10^{-2} - 10^{-1}$ nuclear spin polarization at pressures of $10^{-1} - 10^{0}$ atm [1]. The polarized gases have been applied extensively in various fields. For example, the polarized Xe gas has been utilized as a probe for investigating nuclear electric dipole moment (EDM) by measuring Larmor precession precisely in a fundamental physics [2], a standard sample for nuclear magnetic resonance (NMR) method in a material science, and a blood flow tracer in a medical industry [3]. For low energy neutron experiments, the Xe gas and solid are also expected to be utilized as high polarized targets and samples under a low magnetic field. For example, an NMR-modulated neutron scattering is proposed in order to study slow conformational changes in liquids by measuring a neutron scattering from the polarized Xe nuclei which are solved in water solutions [4]. It is also interesting to measure a correlation term $\boldsymbol{s} \cdot \boldsymbol{I}$ of a neutron spin $\boldsymbol{s}$ and a target nuclear spin $\boldsymbol{I}$ with a high polarized Xe target for a verification of a neutron optical theorem (NOPT). However, there seem to be few experimental data on the $\boldsymbol{s} \cdot \boldsymbol{I}$ term of Xe at the present [5]. At a Materials and Life science experimental Facility (MLF) in J-PARC, an experiment of detecting a neutron polarizing ability caused by the $\boldsymbol{s} \cdot \boldsymbol{I}$ term at a resonant peak of ^{129}Xe is planned, and a polarized Xe gas target based on the SEOP has been developed. This paper reports on the plan and preparative status of our experiment.

2. Evaluation of neutron-nuclear spin correlation based on the NOPT

According to the NOPT, a propagation of low energy neutron through target material is described by a forward scattering amplitude f,

$$f = A + B\hat{\boldsymbol{s}} \cdot \hat{\boldsymbol{I}} + C\hat{\boldsymbol{s}} \cdot \hat{\boldsymbol{k}} + D\hat{\boldsymbol{s}} \cdot (\hat{\boldsymbol{k}} \times \hat{\boldsymbol{I}}), \tag{1}$$

where $\hat{\boldsymbol{s}}$, $\hat{\boldsymbol{I}}$ and $\hat{\boldsymbol{k}}$ denote unit vectors representing directions of a neutron spin, a target nuclear spin and a incident neutron momentum, respectively [6, 7]. The complex coefficients A, B, C, D are described as a function of a neutron energy E_n though they are considered to be constant below a thermal neutron energy. A and B represent parity conserving (PC) amplitudes caused by strong interactions, and C and D represent parity non-conserving (PNC) amplitudes caused by weak interactions. In Eq. (1), imaginary parts of the $\boldsymbol{s} \cdot \boldsymbol{I}$, $\boldsymbol{s} \cdot \boldsymbol{k}$ and $\boldsymbol{s} \cdot (\boldsymbol{k} \times \boldsymbol{I})$ terms induce neutron spin $\boldsymbol{s}$ dependent cross sections $\Delta\sigma_I$, $\Delta\sigma_k$ and $\Delta\sigma_{k\times I}$ relative to directions of $\boldsymbol{I}$, $\boldsymbol{k}$ and $\boldsymbol{k} \times \boldsymbol{I}$, while their real parts induce neutron spin rotations ϕ_I, ϕ_k and $\phi_{k\times I}$ around axes of $\boldsymbol{I}$, $\boldsymbol{k}$ and $\boldsymbol{k} \times \boldsymbol{I}$.

Observations of the PNC amplitudes are important for study of neutron fundamental physics. The PNC effects caused by the $\boldsymbol{s} \cdot \boldsymbol{k}$ term have been estimated about 10^{-7}, and verified by observing helicity asymmetries A_k of the spin dependent cross sections $\Delta\sigma_k$ with proton-proton scatterings [8]. However, they have been found to be enhanced in neutron-nucleus scatterings, especially, values of A_k at p-wave resonances of ^{139}La, ^{131}Xe and so on, have reached up to $10^{-2} - 10^{-1}$ [9, 10, 11]. These enhancements are explained by a model based on a interference between the p-wave and its neighboring s-wave resonances [12]. According to the model, A_k and ϕ_k are described as a function of E_n with a parity mixing matrix element xW and Briet-Wigner (B-W) resonance formula. In particular, both A_k and ϕ_k of ^{139}La have been observed with cold and epithermal neutrons, and they were consistent with predictions deduced from the model [11, 13, 14, 15]. The result suggests that the NOPT on the $\boldsymbol{s} \cdot \boldsymbol{k}$ term is valid up to an epithermal neutron energy region. But for verification of the NOPT based on Eq. (1), it becomes necessary to measure the quantities caused by the $\boldsymbol{s} \cdot \boldsymbol{I}$ and $\boldsymbol{s} \cdot (\boldsymbol{k} \times \boldsymbol{I})$ terms with polarized targets.

The enhancement of the PNC effect at the p-wave resonance is considered to be useful for detecting a time reversal non-conserving (TRNC) effect caused by the $\boldsymbol{s} \cdot (\boldsymbol{k} \times \boldsymbol{I})$ term [7, 16]. For realizing to measure the TRNC effect, it becomes a key technique to reduce an affect of neutron spin rotation frequency ω_I and ω_H caused by the $\boldsymbol{s} \cdot \boldsymbol{I}$ term and an external magnetic field $\boldsymbol{H}$ for holding target polarizations, where $\omega_I(= d\phi_I/dt)$ and ω_H are known as a pseudo-magnetic frequency and a Larmor precession frequency. Especially it is important to comprehend ω_I as a function of E_n because ω_I was predicted to increase in vicinities of resonant peaks [17, 18].

We attempted to evaluate measurable quantities of ^{129}Xe($I = 1/2$) and ^{131}Xe($I = 3/2$) caused by the $\boldsymbol{s} \cdot \boldsymbol{I}$ term. As the PNC effects are normally much smaller than the PC effects, Eq. (1) is approximately described as $f \approx A + B\hat{\boldsymbol{s}} \cdot \hat{\boldsymbol{I}}$, and $\Delta\sigma_I$ and ω_I are given as

$$\Delta\sigma_I = \frac{4\pi}{k} Im(f_+ - f_-) \qquad \omega_I = -\frac{2\pi N_I \hbar}{m_n} Re(f_+ - f_-), \tag{2}$$

where N_I and m_n denote a number density of target and a neutron mass, and $f_+(f_-)$ is the amplitude f with the neutron spin $\boldsymbol{s}$ parallel (anti-parallel) to the target spin $\boldsymbol{I}$ [18]. It is considered to be difficult to estimate neutron scattering amplitudes caused by the $\boldsymbol{s} \cdot \boldsymbol{I}$ term because a complete set of resonance parameters is not known in most case, at least not for bound-state resonances ($E_n < 0$) [19]. But the amplitudes caused by the $\boldsymbol{s} \cdot \boldsymbol{I}$ term at resonant peaks with strong spin selectivity could be estimated by using Eq. (2) and the B-W formula [17, 18]. Fig. 1 represents the estimated $\Delta\sigma_I$ and ω_I of ^{129}Xe as a function of E_n around a 9.6 eV s-wave resonant peak (total angular momentum $J = 1$). This resonance selects neutrons

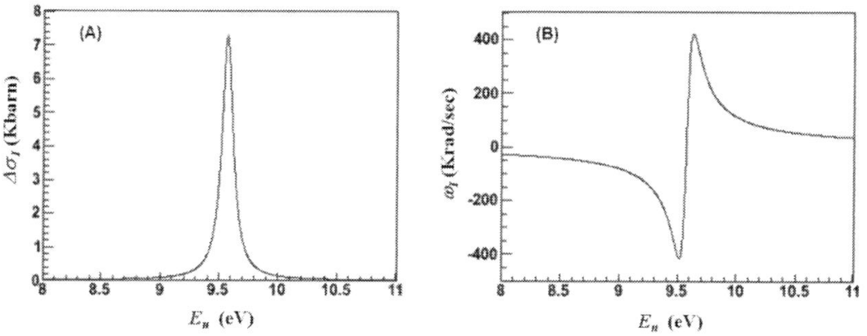

Figure 1. Estimated values of (A) spin dependent cross section $\Delta\sigma_I$ and (B) spin rotation frequency ω_I at 9.6 eV s-wave resonant peak of ^{129}Xe

with spin $s = +1/2$ based on a conservation of an angular momentum. Fig. 1 was calculated by using resonance parameters of ^{129}Xe based on the evaluated data in ENDF/B-VII.0 [20] and by assuming a ^{129}Xe gas of 1 atm pressure. It predicts that $\Delta\sigma_I$ and ω_I depend on E_n strongly and increase at the resonant peak. Especially ω_I exhibits a dispersion curve as a function of E_n, and its maximum value is 4×10^5 rad/sec which corresponds to a pseudo-magnetic field $\boldsymbol{H}^* \approx 2.2$ mTesla. We also calculated ω_I of ^{131}Xe at a 14.4 eV s-wave resonant peak ($J = 2$), and obtained the maximum value of $\omega_I \approx 1 \times 10^7$ rad/sec which corresponds to $\boldsymbol{H}^* \approx 57$ mTesla.

3. Plan to measure neutron-nuclear spin correlation at BL10 in J-PARC

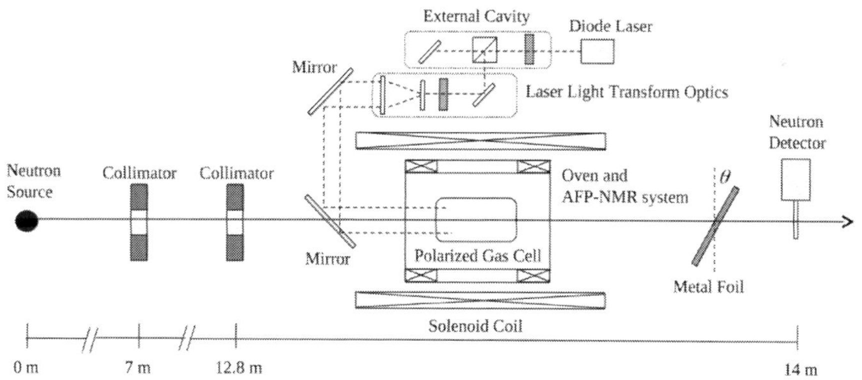

Figure 2. Experimental apparatus for detecting neutron polarizing ability of ^{129}Xe

As the first step to detect the $\boldsymbol{s} \cdot \boldsymbol{I}$ term, we will measure a neutron polarizing ability at the 9.6 eV s-wave resonant peak of ^{129}Xe when unpolarized neutrons transmit through a polarized Xe target. It has an advantage of realizing the measurement without a neutron spin polarizer though its predicted value is smaller than that of a spin dependent cross section asymmetry $A_I = (\sigma_{I+} - \sigma_{I-})/(\sigma_{I+} + \sigma_{I-})$, where σ_{I+} and σ_{I-} denote the cross section with $\boldsymbol{s}$ parallel and anti-parallel to $\boldsymbol{I}$.

Our experimental apparatus is schematically shown in Fig. 2. Our experiment will be carried out at a neutron beam line 10 (BL10) of the MLF in J-PARC [21]. Pulsed neutron beams supplied from a spallation neutron source transmit through collimators, a polarized Xe gas target, a referential metal foil, and are counted by a neutron detector placed at 14 m position apart from the source. The movable collimators placed at 7 m and 12.8 m positions are used for adjusting the neutron beam size and intensity. The metal foil mounted on a goniometer is used in order to compensate time fluctuations of the detector by comparing resonant peak amplitudes between the foil and ^{129}Xe measured with the same detector. A ratio R_x between neutron transmissions with the target polarized and unpolarized is expressed as

$$R_x = \cosh(N_x \sigma_x l_x \rho P_x), \tag{3}$$

where N_x, σ_x, l_x, P_x denote a number density, cross section, thickness and polarization of target nuclei x, respectively. Spin dependent parameters ρ of the target become 1/3 of ^{129}Xe and 1 of ^{3}He. Its neutron polarization is given as $P_n = \sqrt{1 - R_x^{-2}}$ with R_x in Eq. (3).

The Xe gas target is polarized by utilizing an in-situ SEOP system developed for a polarized ^{3}He neutron spin filter at J-PARC [22, 23]. It is designed compactly for mounting on experimental tables in narrow areas of beam lines. It consists of a diode laser (nLight VSA-100-796 DL) driven at an output power of ~47 W, an external cavity for narrowing wave length of laser, laser light transform optics, and a solenoid coil of 20 cm in diameter and 30 cm long for holding the target polarization. A Xe gas of ~1 atm is contained in a cylindrical glass cell about 3 cm in diameter and 5 – 10 cm long with a N_2 gas and a small amount of Rb. Atomic spins of Rb are polarized by a resonance absorption of a circularly polarized laser light of a 795 nm wave length, and Xe nuclear spins are polarized by the spin exchange with the polarized Rb atoms. The cell is mounted in an oven of 10 cm in diameter and 12 cm long. The oven is made of aluminum (Al) and placed at a center of the solenoid coil. A temperature inside the oven is kept to 350 – 450 Kelvin for obtaining a Rb vapor. Around the cell, a pickup coil and a drive coil are mounted for the NMR measurement by an adiabatic fast passage (AFP) method. The AFP-NMR system is very useful for evaluating and monitoring a polarization P_{Xe} of the Xe cell by periodic measurements of NMR signals during beam experiments. These components of the SEOP system are set in a box of $40 \times 40 \times 60$ cm^3 size for shielding laser light.

We guess that practical values of $\Delta R_{Xe} = 1 - R_{Xe}$ in Eq. (3) will be $10^{-3} - 10^{-2}$. Actually ΔR_{Xe} is estimated about 0.003 with a natural Xe gas of 15 atm·cm thickness and 0.012 with an enriched ^{129}Xe gas of 10 atm·cm thickness if a realizable $P_{Xe} \sim 0.2$ is assumed. For obtaining the Xe gas cell with a practical thickness l_{Xe} and P_{Xe}, it is important to know the pressure dependence of P_{Xe} and the effect of the relaxation due to collisions between Xe atom and the cell wall, which is expressed by the wall relaxation time t_w [24]. Time evolution of a noble gas x polarization $P_x(t)$ is described as

$$P_x(t) = P_{Rb} \frac{\gamma_{se}}{\gamma_{se} + \Gamma_w} (1 - e^{-(\gamma_{se} + \Gamma_w)t}) = \bar{P}_x (1 - e^{-\gamma_{gr} t}), \tag{4}$$

where P_{Rb}, γ_{se}, Γ_w and γ_{gr} denote Rb atom polarization, Rb-x spin exchange rate, wall relaxation rate and growth rate of P_x. $\bar{P}_x$ represents an attainable value of $P_x(t)$, and a inverse of Γ_w corresponds to t_w. In the case of Xe, it is known that P_{Xe} decreases with increasing a Xe atom pressure because of a decrement of P_{Rb} by a increment of a Rb spin destruction rate due to the effect of Rb–Xe binary collisions. While, for reducing the wall relaxation, some papers have reported to obtain long values of t_w more than 20 minutes by coating inner surfaces of borosilicate glass cells (Pyrex) with SurfaSil [24, 25]. For neutron experiments, however, it is required to select cell materials with their neutron cross sections as small as possible. We attempted to measure t_w of the cell filled with a natural Xe gas of 2.6 atm pressure in a spherical quartz glass of 3 cm diameter, and obtained long t_w more than 50 minutes. This result suggests that quartz glass becomes a suitable material for the polarized Xe cell.

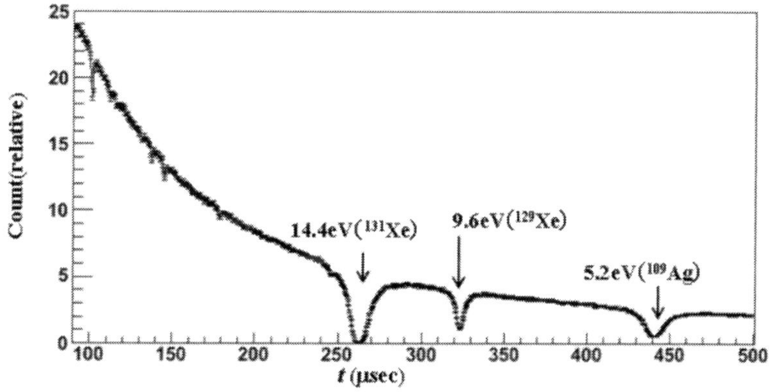

Figure 3. Time of flight (TOF) spectrum of neutrons transmitting through unpolarized Xe cell

4. Feasibility test for measuring neutron polarizing ability of ^{129}Xe

According to the apparatus in Fig. 2, we have attempted to check the compensation method of time fluctuations by comparing resonant peak amplitudes between the Xe cell and the referential metal foil. In the test, a cylindrical quartz cell of 5 cm long filled with a natural Xe gas of 3 atm pressure, a silver (Ag) foil of 10 μm thickness mounted on a goniometer, and a Li glass scintillator of 1 mm thickness, were used as the cell, metal foil and neutron detector. Fig. 3 represents a time of flight (TOF) spectrum of neutrons transmitting through the unpolarized Xe cell. It suggests that the Xe cell and Ag foil are suitable thickness because the both resonant peaks at 9.6 eV of ^{129}Xe and 5.2 eV of ^{109}Ag could be distinguished with enough peak amplitudes. In Fig. 2, an effective thickness l_{Ag} of the Ag foil depends on a inclination angle θ of the foil relative to the neutron beam direction, and changes by rotating the goniometer. We attempted to demonstrate a sensitivity of our system by detecting a slight change of the ratio $\Delta R(\theta) = 1 - r_{res}(\theta)/r_{res}(0^\circ)$ as a function of θ, where $r_{res}(\theta)$ represents the amplitude ratio between the 9.6 eV resonant peak of ^{129}Xe and the 5.2 eV resonant peak of ^{109}Ag. In the beam test during a few hours, $\Delta R(30^\circ) = 0.060 \pm 0.014$ and $\Delta R(20^\circ) = 0.013 \pm 0.011$ have been obtained. This preliminary result suggests that the sensitivity corresponding to $\Delta R_{Xe} = 1 - R_{Xe}$ in Eq. (3) reaches to $\sim 10^{-2}$ in statistical errors.

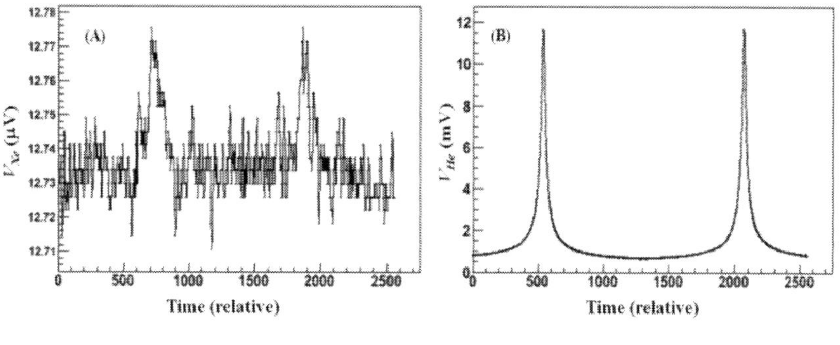

Figure 4. Typical NMR amplitudes (A) V_{Xe} of ^{129}Xe and (B) V_{He} of ^{3}He observed at ν_{res} =50 kHz

For monitoring the polarization of the cell during beam experiments, an AFP-NMR system has been developed [26, 27]. It can work inside of the Al oven in our SEOP system, and detect NMR signals both of ^{3}He and ^{129}Xe at a same resonant frequency ν_{res}. For demonstrating an ability of our NMR system, the measurement of P_{He} of a polarized ^{3}He filter with cold neutrons has been carried out with observing NMR signal amplitudes V_{He} of ^{3}He periodically. According to Eq. (3), P_{He} was determined from the ratio R_{He} between neutron transmissions with the ^{3}He filter polarized and unpolarized. In the test, a cylindrical glass cell of 5 cm long containing in a ^{3}He gas of 3.2 atm was used as the ^{3}He filter. Fig. 4 represents the typical NMR amplitudes V_{Xe} of ^{129}Xe and V_{He} of ^{3}He observed at ν_{res} =50 kHz. Fig. 5 represents time evolution process of P_{He} deduced from R_{He} and V_{He} from a start time ($t = 0$) of the optical pumping. By fitting the observed values with Eq. (4), $\bar{P}_{He} = 0.26 \pm 0.02$ was obtained from Fig. 5 (A). It corresponds to $V_{He} \approx 18.3$ mV deduced from Fig. 5 (B). While, $\gamma_{gr}^{-1} = 9.54 \pm 0.19$ hours was also obtained as a growth time of Fig. 5 (B). It was consistent with the γ_{gr}^{-1} of Fig. 5 (A) in statistical errors. We guess that the observed P_{He} might be small due to the lack of adjustment of the laser light transform optics because $P_{He} \geq 0.6$ is obtained by using our SEOP system usually. By using the correlation between P_{He} and V_{He} obtained from this test, the polarization P_{Xe} of the Xe cell can be deduced from the NMR amplitude V_{Xe}. Actually we could obtain about 0.01 as P_{Xe} corresponding to the amplitude V_{Xe} in Fig. 4 (A).

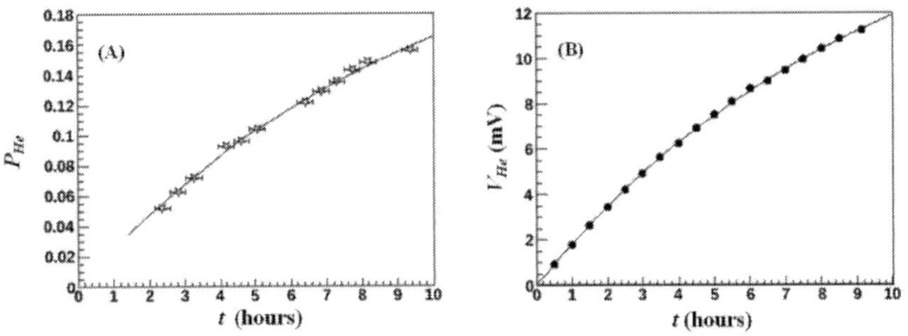

Figure 5. Time evolutions of (A) P_{He} deduced from R_{He} and (B) V_{He} of NMR as a function of time t from start time ($t = 0$) of the optical pumping

5. Summary

We mentioned interest in the measurement of the $\boldsymbol{s} \cdot \boldsymbol{I}$ term as a function of E_n for verification of the NOPT, and reported the experimental plan and feasibility tests for detecting the neutron polarizing ability at the 9.6 eV resonant peak of ^{129}Xe. Although the statistical accuracy of ΔR_{Xe} is dominated by the counting capability of our system at the present, it will be achieved to $\sim 10^{-3}$ by improving the detector and data acquisition system. For obtaining the practical l_{Xe} and P_{Xe}, we will attempt to optimize conditions of the SEOP system, and prepare a Xe cell filled with an enriched ^{129}Xe gas of a low pressure in order to avoid a decrement of P_{Xe} with increasing a Xe pressure. For the study on the $\boldsymbol{s} \cdot \boldsymbol{I}$ term in detail, however, the measurements of A_I and ω_I with polarized neutron beams are indispensable. Actually, by using $P_n = 0.5$ polarized neutron beams, $A_I \approx 0.038$ is expected with a natural Xe cell of $P_{Xe} = 0.2$ and 15 atm·cm thickness though $\Delta R_{Xe} \approx 0.003$ is estimated with the same cell. For realizing these measurements, developments of a polarized proton filter and a high pressure ^{3}He filter will be desirable as neutron spin filters at the MLF in J-PARC [28].

Acknowledgments

We are grateful to the staffs of the MLF at J-PARC for their useful help and suggestions. This research has been partly supported by Grants-in-Aid for Scientific Research (No.22604005) and the Quantum Beam Fundamental Development Program, MEXT.

References

[1] B. Larson et al., *Phys. Rev.* A **44** (1991) 3108
[2] A. Yoshimi et al., *Phys. Lett.* A **304** (2002) 13.
[3] I.C. Ruset et al., *Phys. Rev. Lett.* **96** (2006) 053002.
[4] A.D. Buckingham, *Chem. Phys. Lett.* **371** (2003) 517.
[5] V.R. Skoy, *J. Res. Natl. Inst. Stand. Technol.* **110** (2005) 471.
[6] L. Stodolsky, *Phys. Lett.* B **172** (1986) 5.
[7] P.K. Kabir et al., *Phys. Rev.* D **25** (1982) 25.
[8] R. Balzer et al., *Phys. Rev.* C **30** (1984) 1409.
[9] Y. Masuda et al., *Nucl. Phys.* A **478** (1988) 737.
[10] V.W. Yuan et al., *Phys. Rev.* A **44** (1991) 2178.
[11] H.M. Shimizu et al., *Nucl. Phys.* A **552** (1993) 293.
[12] O.P. Sushkov et al., *JETP Lett.* A **32** (1980) 352.
[13] T. Haseyama et al., *Phys. Lett.* B **534**/1-4 (2002) 39.
[14] E.A. Kolomensky et al., *Phys. Lett.* B **107** (1981) 272.
[15] B. Heckel et al., *Phys. Rev.* C **29** (1984) 2389.
[16] L. Stodolsky, *Nucl. Phys.* B **197** (1982) 213.
[17] V.E. Bunakov et al., *Phys. Lett.* B **429** (1998) 7.
[18] V.P. Gudkov, *Physics Repors* **212** No.2, (1992) 77.
[19] L. Koester et al., *Atomic Data and Nuclear Data Tables* **49** (1991) 65.
[20] M.L. Chadwick et al., *Nuclear Data Sheets* **107** (2006) 2931. <http://t2.lanl.gov/data/neutron7.html>
[21] K. Oikawa, et al., *Nuclear Instr. and methods,* A **589** (2008) 310
[22] H. Kira, et al., *Physica* B **406** (2011) 2433.
[23] T. Oku et al., "Developments of In-Situ SEOP Polarized ^{3}He Neutron Spin Filter at J-PARC", in these proceedings.
[24] H. Sato et al., *Nuclear Instr. and methods* A **402** (1998) 241.
[25] X. Zeng et al., *Phys. Lett.* A **96** (1983) 191.
[26] T. Ino et al., "A compact SEOP ^{3}He neutron spin filter with AFP NMR", in these proceedings.
[27] T. Ino et al., *Physica* B **404** (2009) 2667.
[28] K. Sakai et al., *Nuclear Instr. and methods* A **402** (1998) 244.

A device for sample rotation under external load for the simultaneous measurement of strain and orientation dependent material properties by means of TOF neutron diffraction

Ch. Scheffzük[1,2], H. Hempel[3,4], A. Frischbutter[5], K. Walther[1] & F.R. Schilling[1]
[1] Karlsruhe Institute for Technology KIT, Kaiserstr. 12, Dept. of Appl. Geosciences, 76131 Karlsruhe, Germany
[2] Frank Laboratory of Neutron Physics, Joint Institute for Nuclear Research, ul. Joliot-Curie 6, 141980 Dubna, Russia
[3] TU Bergakademie Freiberg, Institute for machine elements, construction and fabrication, Akademiestr. 6, 09599 Freiberg, Germany
[4] SKF Competence Center Leipzig, Hauptstr. 101, 04416 Markkleeberg, Germany
[5] Am Feldgraben 25, 14548 Schwielowsee, OT Geltow, Germany

E-mail: scheff@gfz-potsdam.de

Abstract. The time-of-flight diffractometer EPSILON-MDS at the modernised pulsed reactor IBR-2M at Dubna is designed for the detection of residual and applied strains in polycrystalline materials. The instrument allows the recording of diffraction patterns with high resolution ($\Delta d/d = 5\text{x}10^{-3}$ to the best) over a wide wavelength band (λ_{max} = 7.8 Å; d_{max} = 5.6 Å). Complicated polyphase materials like rocks consisting of minerals with large unit cells and low crystal symmetry can be investigated. The diffractometer is equipped with 81 detectors positioned behind nine radial collimators at a unique Bragg angle of $2\theta = 90°$ for the central detector and at different radial angles allowing the simultaneous measurement of nine sample directions. A pressure device for the *in situ* investigation of intra-crystalline strain under uniaxial load up to 150 MPa on cylindrical samples is available. The macroscopic strain on the sample surface can be determined simultaneously by means of a laser extensometer. The sample can be rotated around its axis, *i.e.* limited information on texture can be obtained. The simultaneous strain and CPO measurements are helpful for a better understanding of the deformational behaviour of polycrystalline materials, *e.g.* to investigate elastic processes in rocks before, during and after rock failure.

1. Introduction

The moving plate boundaries of the Earth's crust are marked by the concentration of volcanism and earthquakes with focus depths almost between 2 and 10 km, occasionally up to 60 km. Earthquakes may have enormous consequences for humanity as well as economics. That's why countries like Japan, China, and Russia make large effort to understand the mechanisms of earthquake generation in order to predict earthquakes. Nevertheless, up to now, geoscientists agree that it will be impossible to predict location, time, magnitude, focus depth or any other parameter of seismic events, because the processes of earthquake generation are poorly understood.

By the end of 2011, the reconstructed pulsed reactor IBR-2M of the Frank Laboratory of Neutron Physics at JINR (Dubna, Russia) will be available for physical experiments. The neutron time-of-flight (TOF) diffractometers EPSILON and SKAT are especially designed for strain and texture investigations on geological samples, which may open new possibilities for the investigation of the mechanical properties of rocks. In addition to the simultaneous investigation of the textural and strain development with time, it will be possible to determine the strain storage capacity and strain release characteristics of rocks. Information on the loading capacity of rocks in the deep crust is expected, as well as information on strain-stress redistribution processes acting within the crust during seismic activity.

The mechanical properties of rocks like the deformational behaviour are strongly influenced by the composition, the grain size, the grain size distribution, the grain orientation, the internal strain, the strain distribution and other parameters. Non-destructive methods are favoured to characterize the internal structure of materials. TOF neutron diffraction offers the possibility to measure the crystallographic preferred orientation and the internal strain on bulk samples with sample volumes up to several cm^3, in addition, all diffraction peaks in the available range of lattice spacings can be measured at the same time. This is of interest for *in situ* load experiments in order to investigate the strain development and the strain relaxation, because the diffraction geometry has not to be changed in the course of the experiment. The TOF neutron diffraction technique has been successfully used to characterize the deformational behaviour and the structure of polyphase rocks [1, 2]. A pressure device for *in situ* texture investigations by conventional neutron diffraction method has been presented by [3].

2. TOF neutron diffraction as a tool for strain and texture determinations

The TOF neutron diffractometer EPSILON-MDS at beamline 7A of the pulsed neutron source IBR-2M at JINR (Dubna) has been designed for high resolution measurements of applied and residual differential strains $\varepsilon = (d-d_0)/d_0$, including *in situ* strain/stress experiments as well [4]. The research reactor IBR-2M generates neutron pulses with a frequency of 5 Hz, corresponding to a maximal accessible neutron wavelength of $\lambda_{max} = 7.8$ Å. This allows the investigation of large unit cells up to a maximal lattice spacing of $d_{max} = 5.6$ Å. Furthermore, a good spectral resolution of $\Delta d/d = 5\text{x}10^{-3}$ at $d=$ 2Å is achieved, mainly due to collimation of the diffracted beam. The high resolution is advantageous for the investigation of polyphase materials with many Bragg reflections, whereas the wide *d*-range allows the investigation of materials with large unit cells. Both these properties apply to geological materials.

The diffractometer EPSILON-MDS (Fig. 1) is a multi-detector system equipped with 81 ^{3}He-detectors. The detectors are arranged in nine banks, each bank locates behind a radial collimator. The detectors of a bank cover the 2θ-range from $82° \leq 2\theta \leq 98°$, with the central detector at $2\theta = 90°$. The detector banks are arranged around the incident beam at radial angles of –21°, 0°, 21°, 69°, 90°, 111°, 159°, 180°, and 201°, *i.e.*, the angular coverage is 222°. This arrangement allows the simultaneous detection of Bragg diffraction patterns for nine different sample directions. Because the individual detectors of a detector bank are arranged at slightly different Bragg angles and at different secondary flight paths, the data of the nine detectors will be summarized applying 'on-the-fly' time focusing method.

3. Pressure device and rotatable sample

For the investigation of applied load sample states, an electrically operated uniaxial pressure device is mounted on the sample goniometer (Figs. 1a, 1b). The pressure device allows the *in situ* investigation of intracrystalline strains to a force of up to $F = 100$ kN, corresponding to a pressure of $p_{max} = 150$ MPa [5]. The corresponding cylindrical sample dimensions are $d=$ 30 mm and $l=$ 60 mm. The deformation axis is oriented at 45° with respect to the incident neutron beam. This allows the detection of the intracrystalline strain in different sample directions at the same time, *e.g.*, in axial (detector 2) and radial direction (detector 8, Fig. 2). Poisson's ratio, ν, can be determined, which is strongly

depending on the sample composition and the crystallographic preferred orientations (CPOs = textures) of the constituents. An arrangement of incident beam diaphragms determines the gauge volume. Mostly, a gauge volume of 2 cm^3 is used.
The macroscopic strain on the sample surface is measured simultaneously. Using a laser extensometer, any sample-sensor contact can be avoided and the bulk strain can be determined with an accuracy of 0.5 μm. The combined measurement of macro stresses and micro stresses allows conclusions on the sample fabric, *e.g.*, the pore volume. Acoustic emissions, generated by microcracks during deformation, can be detected in the course of the *in situ* deformation experiment (Fig. 1b). The intensity as well as the local and chronological distribution of acoustic emissions characterises the formation of microcracks.

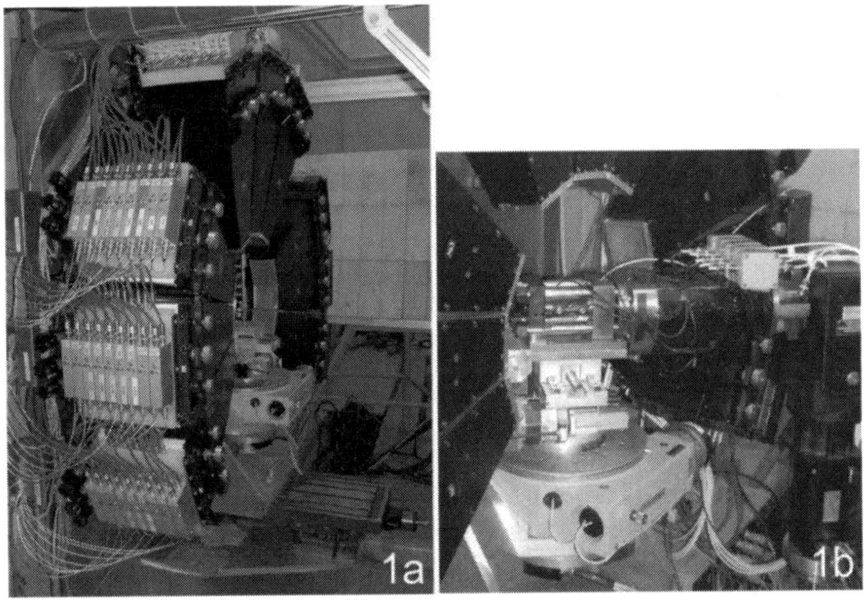

Fig. 1a: The neutron time-of-flight diffractometer EPSILON-MDS for the in situ investigation of residual and applied strains, mainly of geological materials. 1b: Unixaial pressure device (p_{max} = 150 MPa) with the deformation axis oriented at 45° with respect to the incident beam.

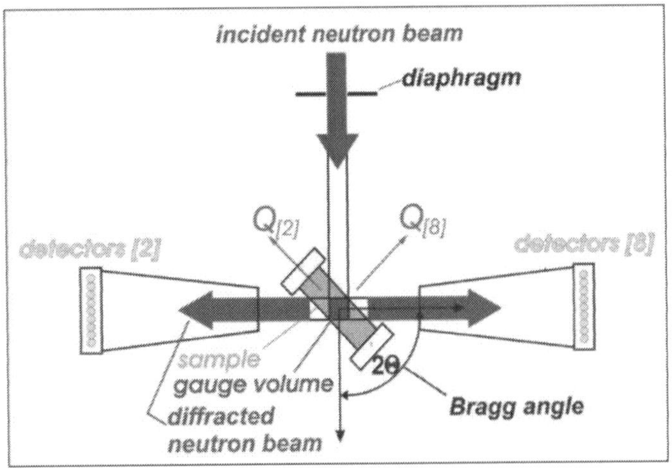

Fig. 2: Schematic sketch of the experimental geometry at the EPSILON-MDS.

A sufficient number of recorded sample directions is required to get information on the orientation dependence of applied and/or residual strain. For the investigation of residual strain and its distribution within the sample, the sample can be translated and/or rotated using a four-axis goniometer. The range of horizontal x- and y- translation is 120 mm, vertical translations up to 40 mm are possible (z-axis). The rotation axis Φ covers the whole 2π range. The accuracy of sample movements is 0.0025 mm and 0.0025°, respectively. For the investigation of applied strain, the pressure device has been equipped with a mechanical gear to rotate the sample under load in order to measure the strain at different orientations (Figs. 3, 4). Sample rotation can be done in arbitrary steps. This allows access to further orientation dependent information, which one can get by analysis of the neutron diffraction pattern relating the Bragg diffraction peak position for strain characterization and the Bragg diffraction peak intensity for the characterization of crystallographic preferred orientation. High precision of the rotation mechanics is required to avoid eccentric moments of force.

As the diffractometer EPSILON-MDS is optimized for strain/stress investigations, the suitability for crystallographic preferred orientation measurements is limited. The better CPO information can be obtained applying specialized equipment like the SKAT texture diffractometer [6], which is operated at the neighbouring beamline.

Fig. 3: Schematic sketch of the rotatable sample chamber inside the uniaxial pressure device.

Fig. 4: Photograph of the rotatable pressure device.

Tab. 1: Technical parameters of the pressure device for sample rotation.

Motor	- High torque step motor with a micro step-by-step control - maximal moment of a torque: 39.6 Ncm - resolution: < 0.36°
gear mechanism	- 3 stage gear - overall multiplication factor: 100.3:1 - stage 1: planetary gear with gear ratio 100:1, max. outgoing unit 2000 Ncm - stage 2: synchronically load distribution on the intermediate shaft, gear ratio 1:1 - stage 3: gear ratio 1.3:1
encoder	- optical - directly flanged on the metering shaft - resolution: 500 pulses per rotation
Max. possible axial load of the rotation plane	F = 265 kN, corresponding to p = 400 MPa

4. Conclusions

Nondestructive intracrystalline strain investigations on geological materials can be carried out by means of neutron diffraction method. The description of the mechanical behaviour of polycrystalline materials requires the analysis of the crystallographic structure, especially the crystallographic texture and the intracrystalline strain. The combined analysis of applied and/or residual strain and texture can be done only by a rotatable sample in the deformation device. The strain storage capacity of mineral lattice planes, which have an effect on the crustal stability, can be determined by time-resolved experiments. Applied and/or residual strains under variable load states, *e.g.*, cyclic loading of the sample, can be recorded.

The uniaxial pressure device at the neutron time-of-flight strain diffractometer EPSILON-MDS has been extended by a rotatable sample chamber. The deformation axis is oriented at 45° with respect to the incident beam, allowing access to a maximal number of sample directions by means of sample rotation, and to achieve minimal geometrical restrictions for the incident and backscattered beam paths. By this way, information on crystallographic preferred orientation of the rock constituents is obtained simultaneously to the time-depending variation of applied strain.

High precision of the rotation mechanics is required to avoid the eccentric moment of force. The simultaneous strain and CPO measurements are helpful for a better understanding of the deformational behaviour of polycrystalline materials, in particular rocks.

Acknowledgements

The project has been supported by the German Federal Ministry of Education and Research (BMBF) under contract number 05K09CPA. The authors thank K. Ullemeyer for helpful comments.

References

[1] Frischbutter A, Neov D, Scheffzük Ch, Vrána M. and Walther K 2000 Lattice strain measurements on sandstones under load using neutron diffraction. *J. Struct. Geol.* **22** (11/12), 1587-1600.
[2] Scheffzük Ch, Walther K & Frischbutter A 2001 Applied and residual strain/stress measurements on a dolomite rock sample using neutron time-of-flight diffraction. *J. Neutron Res.* **9** (2-4), 187-192.
[3] Brokmeier H-G, Zink U, Reinert T and Murach W 1996 The new loading device for texture measurement on the neutron diffractometer TEX-2 *J. Appl. Cryst.* **29**, 501-502.

[4] Walther K, Scheffzük C and Frischbutter A 2000 Neutron time-of-flight diffractometer epsilon for strain measurements: layout and first results. *Physica B, Condensed Matter* **276-278**, 130-131.
[5] Walther K, Frischbutter A, Lieckefett R and Scheffzük C 1999 A pressure device for strain measurements of geomaterials at the neutron diffractometer "EPSILON". In: Szpunar, J.A. (Ed.): Proceedings of the *Twelfth Interational Conference on Textures of Materials (ICOTOM-12).* NRC Research Press, Ottawa, Ontario, Canada. Vol. 2, 1605-1609.
[6] Ullemeyer K, Spalthoff P, Heinitz J, Isakov NN, Nikitin AN and Weber K 1998 The SKAT texture diffractometer at the pulsed reactor IBR-2 at Dubna: Experimental layout and first measurements. *Nucl. Instr. Meth. Phys. Res.* **A 412** (1), 80-88.

Demonstration of optical thickness measurement using multilayer cold neutron interferometer

Y Seki[1], J Uda[2], H Funahashi[2], M Kitaguchi[3], M Hino[3], Y Otake[1], K Taketani[4] and H M Shimizu[4]

[1] RIKEN, Wako, Saitama 351-0198, Japan
[2] Institute for the Promotion of Excellence in Higher Education, Kyoto University, Kyoto 606-8501, Japan
[3] Research Reactor Institute, Kyoto University, Kumatori, Osaka 590-0494, Japan
[4] KEK, Tsukuba, Ibaraki 305-0801, Japan

E-mail: yseki@riken.jp

Abstract. We have measured the optical thickness of a phase object for the first time using multilayer cold neutron interferometer. The measured phase shift of 15.1 ± 1.9 wavelength agreed with the expected value of 17.4 ± 0.7 wavelength due to an about 600-μm-thick silicon plate. This demonstration reconfirmed that two paths in our new interferometer were completely separate, and showed its applicability into various precise measurements.

1. Introduction

The multilayer mirror, which can reflect cold and very cold neutrons, is one of powerful neutron optical devices for interferometry. In addition to utilization of the long-wavelength neutrons, accurate alignment of multilayer mirrors with a large distance makes it possible to increase interaction time of neutrons with external potentials in the interferometer. Such a multilayer interferometer will be more sensitive in phase detection than the conventional silicon single-crystal interferometers, and contribute to precise measurements of small interactions.

We have recently developed a new type of multilayer neutron interferometer [1]. The multilayer interferometer consisted of a pair of beam splitting etalons (BSEs) (Figure 1) [2] with an air gap of 189 μm in spacing. Two paths of the interferometer were completely separated with a center-to-center distance of 328 ± 9 μm at an incident angle of 0.99 ± 0.05 degrees, while the previous multilayer interferometers had almost overlapped paths. Clear interferograms with maximum contrast of 67 % were also successfully observed.

This progress enables us to carry out experiments in more various configurations than before. As one of the configurations, we demonstrated the optical thickness measurement of a phase object inserted into the one-side path using the multilayer interferometer.

2. Experiment and discussion

The experimental setup is shown in Figure 2. A Jamin-type interferometer consisted of a pair of BSEs. In the interferometer, the incident neutron waves were firstly converted to superposition of the spin-up and spin-down states by the polarizer and the first $\pi/2$ RF spin flipper: $\frac{1}{\sqrt{2}}(|\uparrow\rangle + |\downarrow\rangle)$. The superposition state was then split spatially by the first BSE. For the symmetric geometry,

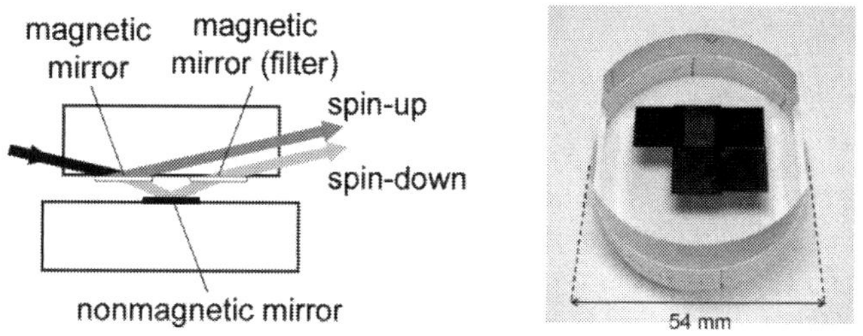

Figure 1. Schematic view (left) and photograph (right) of a beam splitting etalon (BSE). The BSE separates the incident spin superposition state into two parallel paths with the spin-up and the spin-down states.

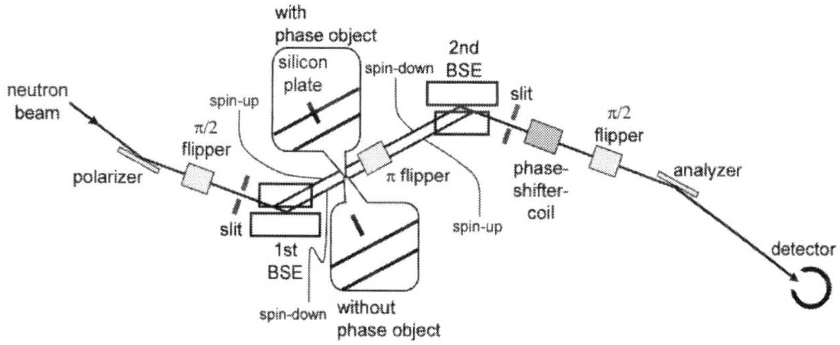

Figure 2. Experimental setup. A weak magnetic field of about 10 G was generated vertically by a guide coil all over the beam (omitted in the figure). The distance between two BSEs was 300 mm.

each spin state was reversed by the π RF spin flipper: $\frac{1}{\sqrt{2}}(|\uparrow\rangle + |\downarrow\rangle) \rightarrow \frac{1}{\sqrt{2}}(|\downarrow\rangle + e^{2i\chi_\pi}|\uparrow\rangle)$, where χ_π is the phase of oscillating magnetic field in the π RF spin flipper relative to the other $\pi/2$ RF spin flippers, and the phase factor $e^{2i\chi_\pi}$ is due to the spin state transition [3]. After the spin reversal, the spatially separate paths were recombined by the second BSE. Finally, the spin state from each path was converted to the spin-up state by the second $\pi/2$ RF spin flipper and the analyzer: $\frac{1}{2}(|\uparrow\rangle + e^{2i\chi_\pi}|\uparrow\rangle)$. The interferograms were obtained by scanning the relative phase χ_π. The change of χ_π of 180 degrees gave a interferograms of one cycle. For the present measurement, a solenoid as a phase-shifter-coil was arranged behind the second BSE. The experiment was carried out at the monochromatic cold neutron beamline MINE2 [4] of Japan Research Reactor No.3 (JRR-3) in Japan Atomic Energy Agency (JAEA). The mean wavelength of the beam was $\lambda_0 = 0.88$ nm with a bandwidth of 2.7% in full width at half-maximum.

As verified in the previous studies [5, 6], the contrast of interferograms in the multilayer interferometer is proportional to the coherence function $\Gamma(\sigma_k, L_0)$ as

$$\Gamma(\sigma_k, L_0) = \exp\left[-\frac{1}{2}(\sigma_k L_0)^2\right] \quad , \tag{1}$$

when we use semi-monochromatic beam with a mean wavenumber k_0 and the standard deviation σ_k of Gaussian distribution. L_0 is the optical path difference between the two paths and L_0 of zero makes an "echo point", that is, the maximum contrast point.

In the case of the present demonstration, L_0 is expressed as

$$L_0 = L_B + L_b + L_i \quad . \tag{2}$$

L_B was provided by the phase-shifter-coil. The phase object inserted into the one-side path caused L_b. L_i arose from the shift of relative angle between two BSEs. L_B was proportional to the current j supplied to the phase-shifter-coil as

$$L_B = -\alpha_B j \quad . \tag{3}$$

The coefficient α_B of the coil, which was also established in the previous studies [6], was $9.1\lambda_0$/A. The insert was a silicon plate with the thickness D_{Si} of 600 ± 25 μm. L_b is therefore expressed as

$$L_b = -(n(k_0) - 1)\, D_{Si} \tag{4}$$
$$\simeq \frac{2\pi\nu b}{k_0^2} D_{Si} \quad , \tag{5}$$

where n is the refractive index of the material for slow neutrons, b is the neutron scattering length of the nucleus, ν is the average number density, and k_0 is the neutron wavenumber.

The phase-shifter-coil induced the optical path differences L_B of 0, $9.1\lambda_0$, $18\lambda_0$, $27\lambda_0$, and $36\lambda_0$, corresponding to the currents of 0, 1, 2, 3, and 4 A, respectively. These optical path differences affected the contrast of interferograms according to Equation 1. Four times measurements of interferograms at each current were carried out iteratively with/without the silicon plate. The sequential change of the contrast are shown in Figure 3 and 4. The transition of average contrast for each data set is shown in Figure 5. These transitions were fitted to Gaussian curves with a common σ_k and maximum contrast. The fitting gave the common σ_k of 0.079 ± 0.008 nm^{-1}, which reproduced well the wavelength dispersion of 0.082 nm^{-1} at the MINE2 beamline. The optical path difference L_b of the inserted silicon plate caused an echo-point-shift. The observed echo-point-shift was $15.1 \pm 1.9\lambda_0$. Within the margin of error, this shift was consistent with the expected value of $17.4 \pm 0.7\lambda_0$ calculated by using the number of density ν of 5.00×10^{22}, and the scattering length b of 4.15 fm.

3. Conclusions

We have succeeded in measuring the optical thickness of the phase object inserted into the one-side path for the first time using the multilayer neutron interferometer. This measurement demonstrated the functional feature of completely separate paths in our new interferometer. With improved phase stability we can advance this pilot experiment into precise measurements, for example, measurement of neutron scattering lengths [7] for the study of many-body-force between nucleons. Cold neutron interferometers have a great advantage in measuring scattering lengths because the phase shift due to the nuclear interaction is proportional to the square of neutron wavelength as shown in Equation 5.

Acknowledgments

This work was primarily supported by the interuniversity program for the common use of JAEA and KURRI. This work was also assisted financially by a Grant-in-Aid for Scientific Research (B) (No. 19340065) of JSPS, a Grant-in-Aid for Creative Scientific Research (No. 19gs0210) of MEXT, and a 47th Toray Science and Technology Grant. One of the authors (Y. S.) was supported by the JSPS Research Fellowships for Young Scientists and the Junior Research Associates Program of RIKEN.

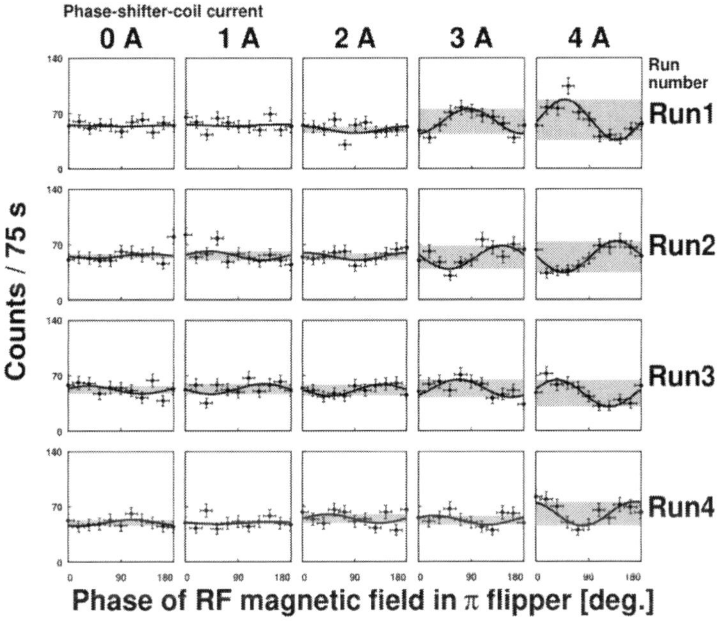

Figure 3. Transition of interferograms due to additional optical path differences induced by magnetic field of the phase-shifter-coil. No phase object was inserted into any paths of the interferometer. All the data were fitted by sine curves with the cycle fixed at 180 degrees.

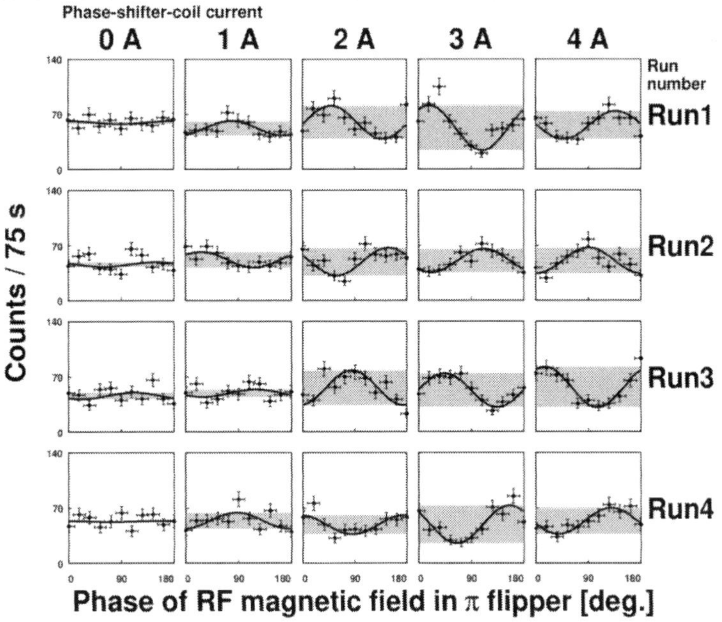

Figure 4. Transition of interferograms due to additional optical path differences induced by magnetic field of the phase-shifter-coil. A silicon plate was inserted into the one-side path of the interferometer. All the data were also fitted by sine curves with the cycle fixed at 180 degrees.

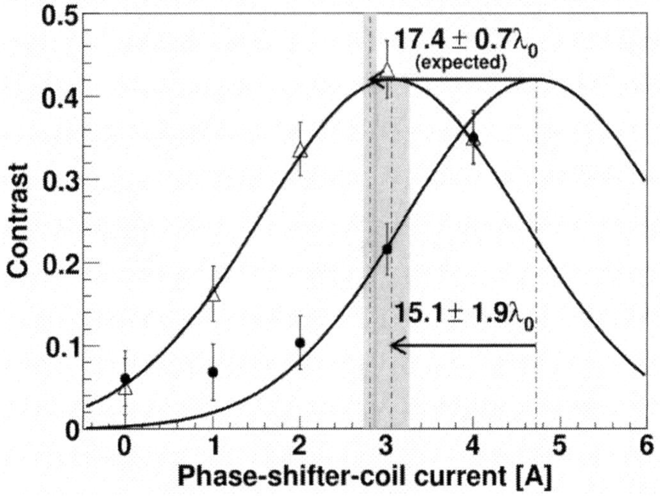

Figure 5. Variation in average contrast of four times measurements. Solid circles and open triangles indicate the case of Figure 3 and 4, respectively. Observed shift of echo point corresponds to optical thickness of about 600-μm-thick silicon plate for 0.88-nm neutrons.

References

[1] Seki Y, Funahashi H, Kitaguchi M, Hino M, Otake Y, Taketani K and Shimizu M H 2010 *J. Phys. Soc. Jpn.* **79** 124201.
[2] Kitaguchi M, Funahashi H, Nakura T, Hino M and Shimizu M H 2003 *Phys. Rev.* A **67** 033609.
[3] Yamazaki D 2002 *Nucl. Instrum. Methods Phys. Res., Sect.* A **488** 623.
[4] Suzuki J, Soyama K, Tasaki S and Ebisawa T 2001 *Proc. 15th Meet. Int. Collaboration on Advanced Neutron Sources* (Tsukuba: JAERI) p 547.
[5] Morikawa Y and Otake Y 1990 *Nuovo Cimento* B **105** 507.
[6] Kitaguchi M, Funahashi H, Nakura T, Taketani K, Hino M, Otake Y and Shimizu M H 2003 *J. Phys. Soc. Jpn.* **72** 3079.
[7] Huber M G, Arif M, Black T C, Chen W C, Gentile T R, Hussey D S, Pushin D A, Wietfeldt F E and Yang L 2009 *Phys. Rev. Lett.* **102** 200401.

Current status of a time-of-flight single crystal neutron diffractometer SENJU at J-PARC

I Tamura[1], K Oikawa[1], T Kawasaki[1], T Ohhara[2], K Kaneko[1], R Kiyanagi[1], H Kimura[3], M Takahashi[4], T Kiyotani[5], M Arai[1], Y Noda[3] and K Ohshima[4]

[1] Japan Atomic Energy Agency, Tokai, Ibaraki, Japan
[2] Comprehensive Research Organization for Science and Society, Tsuchiura, Ibaraki, Japan
[3] Institute of Multidisciplinary Research for Advanced Materials, Tohoku University, Sendai, Miyagi, Japan
[4] Institute of Materials Science, University of Tsukuba, Tsukuba, Ibaraki, Japan
[5] Showa Pharmaceutical University, Machida, Tokyo, Japan

E-mail: kawasaki.takuro@jaea.go.jp

Abstract. SENJU is a state-of-the-art single crystal time-of-flight Laue diffractometer in Materials and Life Science Experimental Facility at Japan Proton Accelerator Research Complex (J-PARC). The diffractometer is designed for precise crystal and magnetic structure analyses under multiple extreme conditions, such as low temperature, high-pressure, high-magnetic field. Measurements using small sample less than 1.0 mm^3 will be also realized, which allows us to study wide variety of materials. SENJU is using a poisoned decoupled moderator to obtain peak profiles of Bragg reflection, and intensity distributions of superlattice reflections and diffuse scatterings with good accuracy. At the beginning, the diffractometer will have 31 two-dimensional scintillator detectors to cover wide area of reciprocal lattice space by a single measurement. The instrument is currently under construction and is scheduled to start on-beam commissioning in February 2012.

1. Introduction

Crystal structure is one of the most fundamental properties of materials and closely relates with physical characteristics. Detailed information on the structure and its external-field response are indispensable to understand and to control physical properties of the materials. Diffraction experiments are a basic but an essential method for the structural study of the materials.

Time-of-Flight (TOF) single crystal neutron diffraction gives us an efficient way to precisely analyse the crystal and magnetic structures and structural fluctuation of the materials. SXD at ISIS [1] has been a leading instrument in this field for decades, and TOPAZ at SNS [2] has been just launched. At Materials and Life Science Experimental Facility at Japan Proton Accelerator Research Complex (MLF/J-PARC), construction of a TOF single crystal neutron diffractometer SENJU has been started in 2010. The primary aim of SENJU is precise crystal and magnetic structure analyses under multiple extreme conditions. SENJU also sheds light on superlattice reflections and diffuse scatterings in order to analyse local structures as well as phase transitions of the materials. Measuring a small sample, less than 1.0 mm^3, is another important feature of this diffractometer. SENJU will be able to extend a range of materials that can be measured by neutron diffraction method in the field such as science and engineering. In this paper, overview and current status of SENJU are presented.

2. Instrument design

2.1. General design of diffractometer

Choice of the moderator is one of the most crucial decisions made in the first stage. To meet the targeted science at SENJU, a poisoned decupled moderator was selected since it provides a symmetrical pulse shape and resultant better S/N ratio at the pulse tail compared with other types of moderators. The sample position is set to be 34.8 m from the moderator. As for the high-resolution capability, $\Delta d/d < 0.2$ % will be easily achieved by the high-angle detectors at this position. A bandwidth of 4.4 Å should be adequate for the most of the experiments at SENJU. A straight beamline was adopted instead of a curved one due to the necessity of short wavelength neutrons, up to 0.4 Å (500 meV). A uniform beam divergence of $\pm$ 0.30° at the sample position will be realized by an elliptical shape neutron guide, as briefly described in the next paragraph. The distance between the sample and the detector banks is set to be 0.8 m in minimum, which can be extended up to 1.3 m to incorporate various accessories and to realize better spatial resolution. An exchangeable vacuum sample chamber is employed to accept a superconducting magnet and to simplify a beam alignment process. Figure 1 shows the illustration of the chamber and detectors array of SENJU. General specifications of SENJU are listed in table 1.

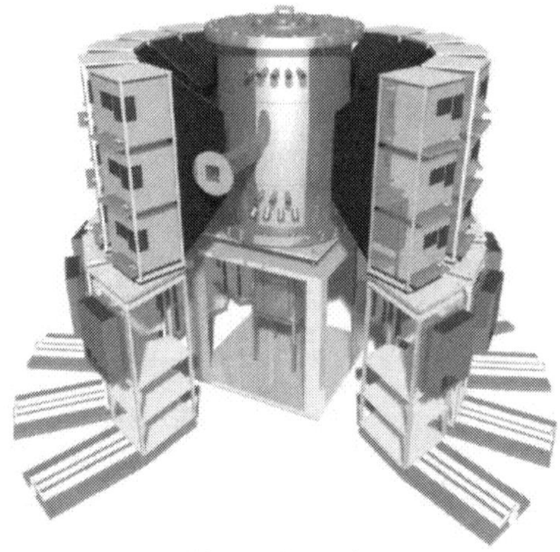

Figure 1. 3D illustration of SENJU

Table 1. General specifications of SENJU

Beamline	BL18
Moderator	Decoupled poisoned H_2 (20 K)
Neutron wavelength	0.4 ~ 4.4 Å (1st frame)
	4.6 ~ 8.8 Å (2nd frame)
Sample position	34.8 m from moderator
Guide tube	Straight elliptic supermirror
Detector	WLSF-type scintillator detector
	4×4 mm^2 / pixel
	0.8 m or 1.3 m from sample
Range of 2θ	-13° ~ -167°, +58° ~ +167°
Beam divergence	± 0.30° (Standard mode)
	± 0.45° (High-intensity mode)

2.2. Beamline layout

The beamline of SENJU is designed to transport neutrons of $\lambda = 0.4$ Å at the minimum with a uniform divergence to the sample position. A square cross-sectional elliptic straight neutron guide is adopted where focal points of the guide are 9.8 and 34.8 m from the moderator. The start and the end positions of fixed section of the guide are 15.2 and 31.8 m, respectively. The optical components to control beam divergence and size consist of a subsequent interchangeable section of the guide (supermirror coating $m = 5$) and two B_4C slits as shown in figure 2. The beam divergence will be increased from $\pm$ 0.30° to $\pm$ 0.45° by the use of this section with a little extension of the minimum wavelength. A better collimation, *i.e.*, the divergence smaller than $\pm$ 0.30°, will be realized by use of the slits with a compensation of the neutron flux. Bandwidth choppers are necessary to select a single frame of 4.4 Å and eliminate other frames. Instead of the use of the second frame, a double frame mode with a half repetition will be effective in some experiments. So-called T_0 chopper is also necessary to stop fast neutrons and prompt γ-ray from the moderator for experiments using a sensitive sample. To optimize parameters of the beamline components, such as the guide, choppers and slits, Monte Carlo simulations were carried out using the program McStas [3]. Note that the simulated beam profile showed that the beam halo does not exceed 30 × 30 mm^2 at the sample position without the slits. A schematic drawing of the beamline layout of SENJU is shown in figure 2.

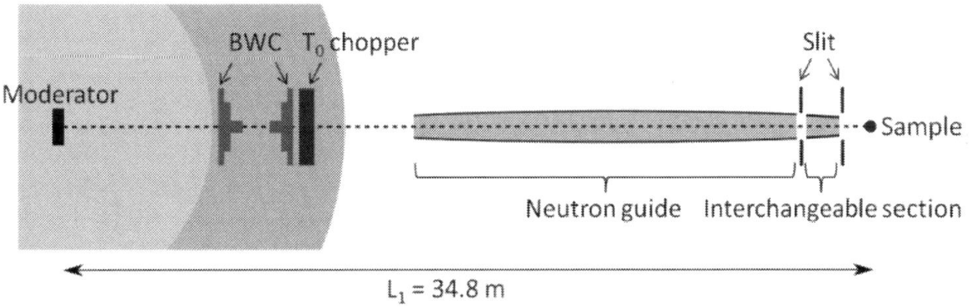

Figure 2. Schematic drawing of the beamline layout of SENJU

2.3. Detector system

The success of SENJU depends on the reliability and stability of the detector system. Recently, a scintillator detector has been an only choice for the diffractometer because of shortage of ^{3}He gas. A wavelength-shifting fiber (WLSF) -type scintillator detector developed by Japan Atomic Energy Agency [4] is already in operation at the biological TOF single crystal neutron diffractometer iBIX [5] in MLF/J-PARC. This type of detector satisfies the required performance for SENJU, such as spatial resolution and count capability. Due to the experimental requirements, the pixel size of the detector was relaxed for SENJU to 4 × 4 mm^2 from 0.52 × 0.52 mm^2 of the iBIX detector. A sensitive area of the detector was widened for SENJU to 256 × 256 mm^2 from 133 × 133 mm^2 of iBIX. A chassis size of the detector was also changed to 300 × 300 mm^2 from 160 × 160 mm^2. Reliability and stability of the system were improved and the relative dead area was reduced by these modifications. The Monte Carlo simulation using McStas showed that at least 3 × 3 pixels were illuminated for a Bragg spot from an ideal single crystal, and in reality 5 × 5 or more pixels will detect the spot.

At the maximum, 36 detectors will be installed around the sample in a cylindrical arrangement to minimize the dead area. In addition, one detector will be installed underneath the sample to cover qualitatively different reciprocal lattice space from the others. This makes UB matrix more precisely and gives crucial information especially for magnetic structure analyses. The total covered solid angle around the sample will be 4 sr, which enables the efficient measurements and improves the completeness of the data. Schematic drawings of the detectors arrangement are shown in figure 3.

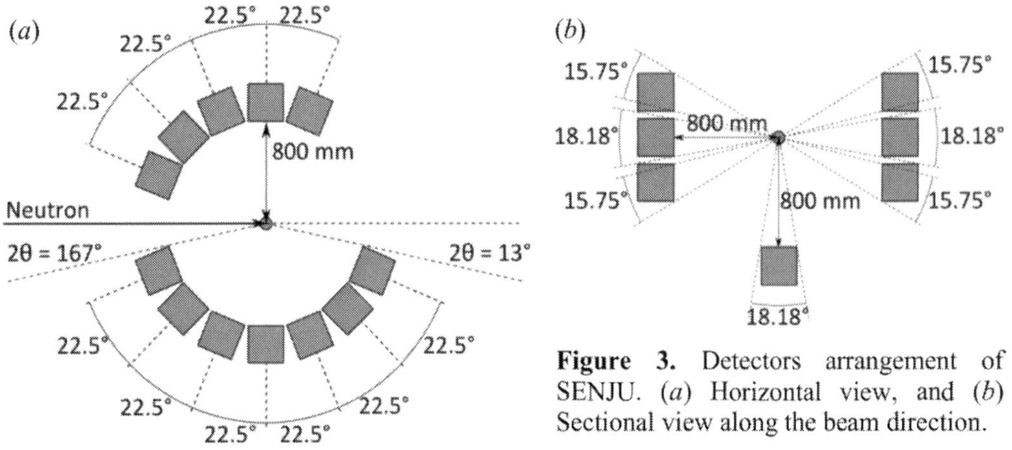

Figure 3. Detectors arrangement of SENJU. (*a*) Horizontal view, and (*b*) Sectional view along the beam direction.

2.4. Sample environment

One of the main aims of SENJU is to pursue precise crystal and magnetic structure analyses under multiple extreme conditions. SENJU is designed to accept various sample environment apparatuses for this purpose. A room temperature apparatus, a standard closed-cycle refrigerator, a 7 T superconducting magnet and a dilution refrigerator will be introduced in the initial stage; one typical "multiple extreme" condition at the beginning is under the magnetic field of 7 T below 50 mK. It is of great importance to reduce background from such apparatuses for the precise structural study. In this respect, unique designs, such as a piezo motor based goniometer inside the cryostat, is employed. In order to realize distinct experimental conditions, installation of an incident neutron polarization device, a high-pressure cell, a high-temperature furnace and a photo-irradiation device is planned as a future upgrade.

3. Construction schedule and current status

The designs of all the devices and the floor layout have been already finished. The devices are manufactured in factories and almost completed. The software to operate the devices from GUI and to perform the data reduction and visualization is also being developed. The data reduction and visualization components of the software are based on STARGazer [6]. Although the construction schedule has been delayed because of the devastating disaster, SENJU will be in commission in 2012.

References

[1] Keen D A, Gutmann M J, and Wilson C C 2006 *J. Appl. Cryst.* **39** 714

[2] Frost M J, M. Austin D, Viola R, Thomison C, Carmen P J, Hoffmann C, Miller E M, Mosier L B, and Overbay M A 2010 *J. Phys.: Conf. Ser.* **251** 012084

[3] Lefmann K, and Nielsen K 1999 *Neutron News* **10** 20

[4] Hosoya T, Nakamura T, Katagiri M, Birumachi A, Ebine M, and Soyama K 2009 *Nucl. Instr. Methods Phys. Res.* A **600** 217

[5] Tanaka I, Kusaka K, Hosoya T, Niimura N, Ohhara T, Kurihara K, Yamada T, Ohnishi Y, Tomoyori K, and Yokoyama T 2010 *Acta Cryst.* D **66** 1194

[6] Ohhara T, Kusaka K, Hosoya T, Kurihara K, Tomoyori K, Niimura N, Tanaka I, Suzuki J, Nakatani T, Otomo, Matsuoka S, Tomita K, Nishimaki Y, Ajima T, and Ryufuku S 2009 *Nucl. Instr. Methods Phys. Res.* A **600** 195

Neutron Powder Diffraction Measurements of the Spinel $MgGa_2O_4:Cr^{3+}$ - A Comparative Study between the High Flux Diffractometer D2B at the ILL and the High Resolution Powder Diffractometer Aurora at IPEN

M A F M da Silva[1], L P Sosman[1] , F Yokaichiya[2,3], V L Mazzocchi[4], C B R Parente[4], J Mestnik-Filho[4], P F Henry[2, 5] and H N Bordallo[2, 5, 6]
[1] Instituto de Física, Universidade do Estado do Rio de Janeiro, Rio de Janeiro, Brazil
[2] Helmholtz-Zentrum Berlin, Hahn-Meitner Platz, Berlin, D-14109, Germany
[3] Laboratório Nacional de Luz Síncrotron, P.O. Box 6192, Campinas 13084-971, SP, Brazil
[4] Instituto de Pesquisas Energéticas e Nucleares, P.O. Box 11049, São Paulo 05422-970, SP, Brazil
[5] European Spallation Source ESS AB, P.O. Box 176, SE-22100 Lund, Sweden
[6] Niels Bohr Institute, University of Copenhagen, Universitetsparken 5, DK-2100 Copenhagen, Denmark

E-mail: sosman@uerj.br; fabiano.yokaichiya@lnls.br

Abstract. Optical materials that emit from the visible to the near-infrared spectral region are of great interest due to their possible application as tunable radiation sources, as signal transmission, display, optoelectronics signal storage, cellulose industry as well as in dosimetry. One important family of such systems are the spinel compounds doped with Cr^{3+}, in which the physical the properties are related to the insertion of punctual defects in the crystalline structure. The purpose of our work is two fold. First, we compare the luminescence of the $MgGa_2O_4$-Ga_2O_3 system with the single phase Ga_2O_3 and $MgGa_2O_4$ and relate structural changes observed in $MgGa_2O_4$-Ga_2O_3 system to the optical properties, and secondly, to compare the neutron powder diffraction results obtained using two diffractometers: D2B located at the ILL (Grenoble, France) and Aurora located at IPEN (São Paulo, Brazil). In the configuration chosen, Aurora shows an improved resolution, which is related to the design of its silicon focusing monochromator.

1. Introduction

Solid-state lasers are widely used in various scientific areas as material processing, spectroscopy, medicine, communications, displays, photonics, etc. For all of these applications two types of laser are particularly useful: tunable lasers and micro-lasers adequate for "integration" in an optical circuit. Both types of laser consist on a single crystal host having good optical and thermal characteristics doped with one or more laser-active impurities. The impurities are transition metal ions as Cr^{3+}, V^{2+}, Ti^{3+}, etc. or rare earth ions as Nd^{3+}, Er^{3+}, Dy^{3+}, Yb^{3+}. The hosts are double oxides or halide compounds. In order to facilitate the substitution, the host ion must have the same valence and atomic radius value quite close to the impurity one. However, it is not unusual a host which does not satisfy these requirements. In this case the difference of valence and ionic radius are compensated by a new position of the host ions during the synthesis process. When the impurity ion is inserted in the host lattice, the energy levels of the host can be split, in a trend that depends strongly on the potential generated by their neighbourhood (crystal-field). This dependence can be directly associated to small structural differences in the occupation site of the doping ion. However, in many cases, the precise structure of the emitting centres is not well known, even if the determination of such structures is of capital interest for the optimization and application of new lasers systems. For example, Cr ion is a typical laser-active impurity in solid-state lasers and it is its valence and local symmetry in the lattice that completely determine the emitting properties. So, Cr^{3+} coordinated with six O^{2-} ions gives laser emission in the visible range but Cr^{4+} coordinated with four O^{2-} ions gives laser emission in the infrared range. [1, 2]

Following this idea we have combined neutron powder diffraction (NPD) to X-ray powder diffraction (XPD) and optical measurements to get insight on the structure of $MgGa_2O_4$: Cr^{3+}. The interest in this compound comes from the fact that nowadays the most efficient solid-state lasers emit between 720 and 1010 nm.[3,4] For instance, YAG: Mn^{3+} lasers emit between 600 and 800 nm, with maximum of intensity at 640 nm, but with very low efficiency (<1%)[5], when compared to Cr^{3+} systems. Moreover, Cr^{3+} ions are stable in the crystal synthesis but Mn^{3+} ions are not, originating compounds with Mn^{2+} and Mn^{4+}, making its spectral interpretation difficult and more importantly contributing to the laser efficiency decrease. Therefore obtaining a new material with emission located under 700 nm, i.e. in the visible range, having Cr^{3+} as impurity is of great interest.

Magnesium gallate, $MgGa_2O_4$, has a partially inverted spinel structure belonging to the AB_2O_4 class of materials in which there are two positions for the A^{2+} and B^{3+} cations. It crystallizes in the Fd3m cubic space group with cell parameter a = 8.286 Å, with 16±2% of the Mg^{2+} population in tetrahedral and 81±1% in octahedral sites.[6,7] Phase diagram studies[8] have indicated the possibility of some amount of beta-gallium, $\beta-Ga_2O_3$, together with the principal compound $MgGa_2O_4$, which was indeed confirmed by XPD and optical measurements.[9, 10] Interestingly, the presence of the $\beta-Ga_2O_3$ compound in the $MgGa_2O_4$ samples certainly introduces an additional perturbation on the observed spectra since Cr^{3+} optical spectroscopic properties depend on the environment of the dopant ion. In fact the $\beta-Ga_2O_3$ presents a monoclinic structure[11, 12] crystallizing in the C2/m space group with lattice parameters a = 12.23 Å, b = 3.04 Å, c = 5.80 Å and β = 103.7° , with distorted octahedral Ga^{3+} sites that can be occupied by Cr^{3+} ions. It has been shown[9] that the $MgGa_2O_4$: Cr^{3+} systems presents a broad emission band in red and near infrared spectral region between 670 nm and 850 nm, which is attributed to Cr^{3+} ions in octahedrally coordinated sites, and in reality produced by variations in the crystal field due to the inverse nature of this spinel.

In this work we demonstrate that the optical characteristics of $MgGa_2O_4$: Cr^{3+} are enhanced when $MgGa_2O_4$ - β-Ga_2O_3: $Cr^{3+}$is formed. This fact is directly related to the structural characterization using the combination of x-rays and neutron diffraction data. The diffraction pattern shows that the compound is a mixture of the inverted spinel and the monoclinic phase. The calculated percentages show that the dominant phase is the spinel arrangement. Finally we confirm that for the determination of structures with lower symmetry and larger unit cells, the resolution of the spectra has great importance.

2. Experimental details

2.1.1. Sample preparation
Powder samples of the system $MgGa_2O_4$ - β-Ga_2O_3 with 0.1, 0.5, 1 and 5% (in mass) Cr^{3+} substituting the Ga^{3+} ions were produced by a solid-state reaction of PA oxides, MgO, Ga_2O_3 and a desired amount of Cr_2O_3, by sintering at 1300°C using a Jung 5713 oven. After the thermal treatment the samples presented the characteristic green variation obtained for similar doped Cr^{3+} samples.

2.1.2. Optical measurements
Photoluminescence measurements were performed at ambient condition at the Optical Spectroscopy Laboratory at Pontificia Universidade Católica (PUC-RIO) using a QuantaMaster™ UV VIS, with a 3 nm resolution.

2.1.3. Short description of the neutron powder diffractometers (NPD): D2B and Aurora
The neutron powder diffractometer D2B is located at the Institut Laue Langevin (ILL, Grenoble) and is characterised by the very high take-off angle (135°) for the monochromator, which is 300 mm high, focusing vertically onto about 50 mm and has a relatively large mosaic spread of 20'. In addition, 200 mm high detectors and collimators match the large incident vertical divergence. [13] In this particular study a wavelength of 1.2307 Å was used. Such a configuration allows for fast measurements time due to higher neutron flux on the sample.
Aurora is a high-resolution neutron powder diffractometer installed on the IEA-R1 research reactor located at Instituto de Pesquisas Energéticas e Nucleares (IPEN-CNEN/SP, São Paulo). [14] This diffractometer is equipped with a position sensitive detector (PSD) and a double-focusing perfect single crystal silicon monochromator. Installed in a take-off angle of 84°, the monochromator can be positioned to produce 4 different wavelengths, namely 1.111, 1.399, 1.667 and 2.191 Å (nominal values). In this study an adjusted value for the wavelength equal to 1.4119 Å was used, allowing for improved resolution in the q-range of interest.

2.1.4. X-ray powder diffractometer - XPD
Room temperature XPD measurements were performed using a X'Pert PRO Panalytical Diffractometer located at the Crystallography and X-ray Diffraction Laboratory at the Centro Brasileiro de Pesquisa Físicas (CBPF, Rio de Janeiro), with 40 KV and 40 mA and CuKα radiation, λ = 1.5418Å in Bragg-Brentano geometry. A 2θ range from 10° to 100° was covered using a step of 0.05 ° and a count time of 1 sec. NPD and XPD data were analyzed using the program FullProf.[15]

3. Structural results based on X-rays and neutron powder diffraction
Both XPD and NPD experiments were used to take advantage of the different scattering lengths ratios, which should facilitate the determination of the different site occupancies. Moreover, by sampling small amounts of the different pellets by XPD, it was possible to check for sample inhomogeneities in the larger samples produced for the NPD measurements.
The powder patterns were initially modeled for all samples (Cr^{3+} concentration = 0.1%, 0.5%, 1% and 5%) employing the cubic spinel structure described by the space group Fd3m with Ga or Mg in the Wyckoff position 16d (0.5, 0.5, 0.5) (center of the octahedra) and Ga or Mg ions at the Wyckoff position 8a (0,125, 0.125, 0.125) (center of the tetrahedra). The O atoms reside on Wyckoff position 32e (x, x, x), with the x-value being the only variable parameter to be determined. The initial value x=0.256(5) obtained for $MgGa_2O_4$ was used as initial parameter.[6] However all attempts of indexing the diffraction peaks in this structure failed indicating that the system under study is a mixed phase. Therefore, a second phase described by the β-Ga_2O_3 structural model was introduced into the refinement. In β-Ga_2O_3 the atoms are in five sets of 4i special positions, (0, 0, 0, 0.5, 0.5, 0) ± (x, 0, z), with two crystallographically different Ga atoms in the asymmetric unit, one with tetrahedral and the

other with octahedral coordination geometry. The results of the Rietveld refinements using NPD and XPD of the crystalline structure of the system $MgGa_2O_4$-β-Ga_2O_3:0.1Cr^{3+} are shown in Figure 1.

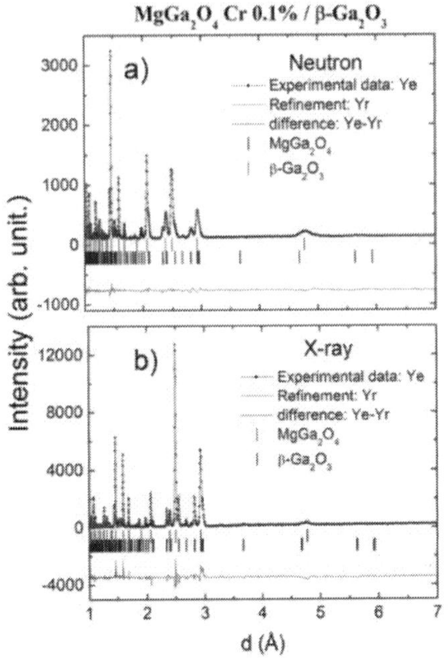

Figure 1. Comparison of NPD and XPD diffraction patterns at RT room temperature for $MgGa_2O_4$+β-Ga_2O_3 with 0.1% Cr^{+3} measured using D2B-ILL, λ = 1.2307 Å (a) and a X'Pert PRO Panalytical Diffractomter with CuKα radiation, λ = 1.5418 Å (b). The red line is the Rietveld refinement, the blue line is the difference between the experimental data and the calculated one; the marks are (i) the upper mark is related to $MgGa_2O_4$ and (ii) the lower mark is related to the β-Ga_2O_3.

The calculated inverted spinel phase percentages are 72%, 77%, 72% and 79% for the 0.1%, 0.5%, 1% and 5% Cr^{3+} concentrations respectively. Figure 2 shows the behavior of the lattice parameter, *a*, and the occupation of Ga and Mg in the A and B sites in the spinel structure as a function of Cr^{3+} concentration. We observed a distinct influence of the Cr^{3+} cation in the structure, and in fact when the concentration of Cr^{3+} cation increases there is a reversion of the occupancy between the Ga and Mg atoms in the B site (the octahedral site) as shown in the Figure 2 (b). These findings corroborate with luminescence results showing a distinct difference between the lines as a function of concentration.[16]
As the crystalline arrangement of the spinel system studied here contains highly symmetric atomic arrangement the resolution of the powder pattern is of great interest. Therefore, the experiments for the crystal structure were also carried out using the NPD Aurora for selected compositions. An example is given in Figure 3 showing the measurement for the $MgGa_2O_4$-β-Ga_2O_3:0.1Cr^{3+} at both neutron powder diffractometers. It is observed that the average separation between adjacent Bragg reflections is improved using Aurora, for instance the shoulder around 1.45Å, attributed to the β-Ga_2O_3 phase related to the (-2 0 4) reflection, can be clearly distinguished due to the instrumental higher resolution. Moreover, an asymmetric peak shape can noticeably be observed and explained by the fact that overlapping peaks are discriminated. For instance, the peak around 1.6 Å related to the reflections (3 3 3) from $MgGa_2O_4$ and the peak (-3 1 3) from β-Ga_2O_3,, follow in this category see Figures 3 (c) and 3 (d).

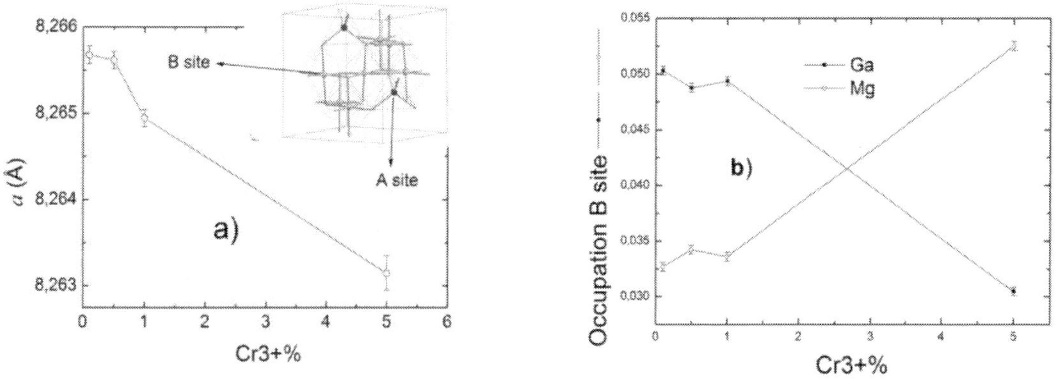

Figure 2. a) Lattice parameter *a* and b) the occupation of the Ga and Mg atoms in the B sites of the spinel structure (as shown in the insert figure) as a function of Cr^{3+} concentration.

Table 1 gives the values for the typical Caglioti resolution parameters accounting for the angular dependence of the width of the Bragg reflections obtained from the Rietveld analysis for ILL-D2B, conventional XPD and the IPEN-Aurora diffractometers.

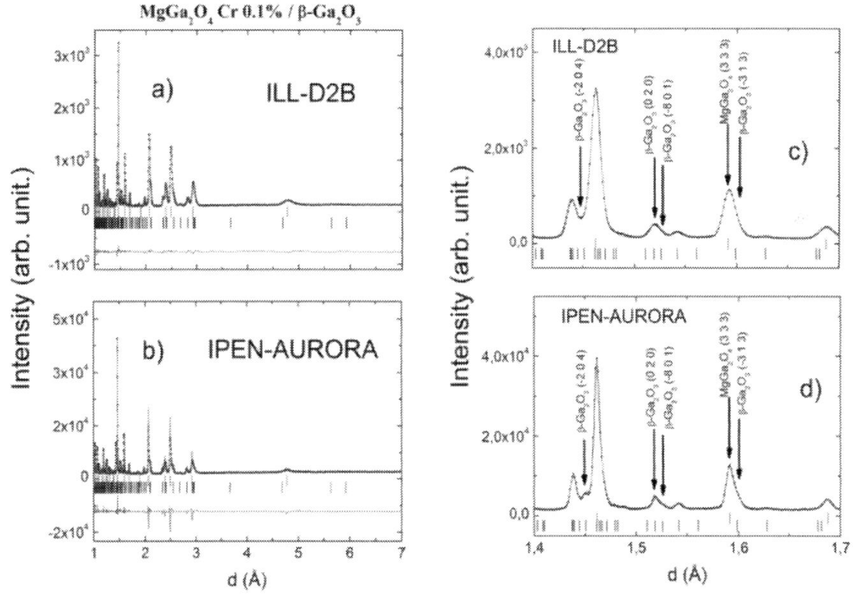

Figure 3. (a) Data obtained at ILL-D2B, λ=1.2307 Å and (b) data obtained at IPEN-AURORA, λ=1.4119 Å. (c) and (d) show details of the resolution of both instruments.

Figure 4 shows a resolution curves for both diffractometers D2B and Aurora. The optimized conditions of Aurora allows for a better resolution than D2B, for angles below 70°. These results are quite stimulating as they show that despite the lower flux Aurora suits well studies related to magnetic structures, as well as small structural distortions.

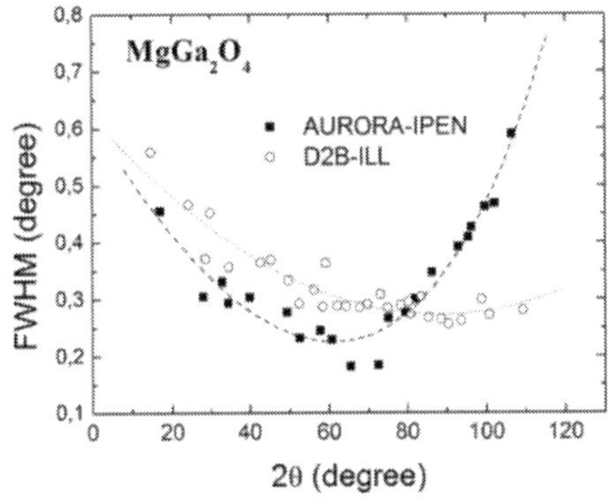

Figure 4. Resolution curves (FWHM) vs. scattering angles (2θ) for ILL-D2B and IPEN-AURORA for $MgGa_2O_4$ obtained from the Rietveld refinement. The lines are only guide for the eyes.

Table 1. Typical Caglioti resolution parameters.

	ILL-D2B λ=1.2307 Å (°)	XPD λ=1.5418 Å (°)	IPEN-AURORA λ=1.4119 Å (°)
U	0.066832	0.060727	0.254962
V	-0.202147	-0.098025	-0.384117
W	0.181029	0.085859	0.006255

4. Photoluminescence of the Cr^{3+} ion in Ga_2O_3, $MgGa_2O_4$+Ga_2O_3 and $MgGa_2O_4$

Figure 5 exhibits the room temperature photoluminescence spectra for samples doped with 0.1% of Cr^{3+} (in mass) excited with 450 nm. The luminescence of the impurity ion in $MgGa_2O_4$ (dotted line) presents a maximum intensity at 708 nm and is assigned to the spin forbidden $^2E(^2G) \rightarrow {}^4A_2\ (^4F)$ electronic transition of Cr^{3+} octahedral. The luminescence of β-Ga_2O_3 presents a peak at 689 nm, due to the emission associated with the $^4T_2(^4F) \rightarrow {}^4A_2(^4F)$ spin-allowed transition of Cr^{3+} octrahedrally coordinated to oxygen ions. Finally, we can see the spectrum of the system $MgGa_2O_4$+β-Ga_2O_3 showing the broader band, with barycentre at 697 nm (full line). Most likely, a host distortion was generated with the presence of β-Ga_2O_3, leading the Cr^{3+} ions to an intermediate crystal field. Therefore, the $^4T_2\ (^4F) \rightarrow {}^4A_2(^4F)$ energy transition originates the broad observed band.

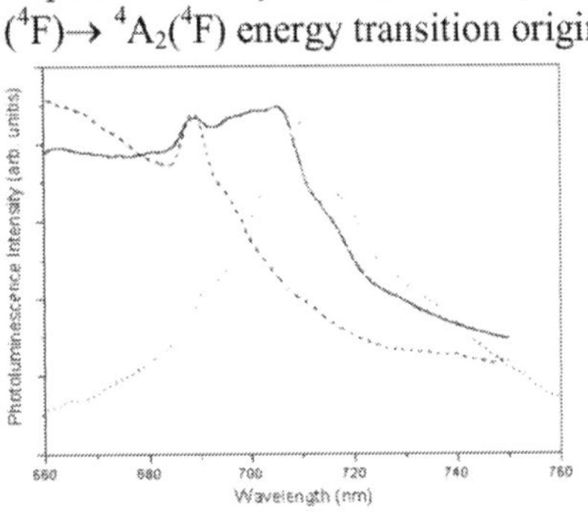

Figure 5. Room temperature photoluminescence of the Cr^{3+} ion in Ga_2O_3 (dashed line), $MgGa_2O_4$ (dotted line) and $MgGa_2O_4$+Ga_2O_3 (solid line).

5. Conclusion

XPD and NPD Rietveld refinements were used for quantitative phase analysis to determine the amount of $MgGa_2O_4$ and β-Ga_2O_3 in the system $MgGa_2O_4$-β-Ga_2O_3:Cr^{3+} at room temperature. The precursor phases were found to present similar unit cells in all composition studied. Moreover, to assess the quality of the powder diffraction method, the bulk properties of system analyzed by an optical study are also presented. From the luminescence data, we observe the broadest band emission for the mixed system in comparison with those individual phases. As tunable materials should offer broad emission and intense bands this finding is extremely important. Besides, the samples synthesis process is easy and cheap, leading us to conclude that the system is a promising material for technological applications.

High-resolution NPD obtained using the instrument Aurora allowed to confirm the presence of β-Ga_2O_3. In the particular configuration chosen for this study, the higher resolution of Aurora can be attributed to the combination of the focusing silica monochromator to the ensemble of collimation set-up in addition to the placement of a sapphire filter. This result places Aurora as a potential NPD diffractometer to be used to determine magnetic structures as well as to perform phase analysis.

Acknowledgments
The results presented in this work were obtained during the development of the PhD thesis of MAFMS under the supervision of LPS at the Universidade do Estado do Rio de Janeiro, and financially supported by FAPERJ. MAFMS acknowledges the partial financial support given by the Berlin Neutron Scattering Center (BENSC) for a visit to discuss on his XPD and NPD results. Support given by the Institut Laue-Langevin (ILL) for the NPD experiments is also acknowledged.

References

[1] Martines M A U, Davolos M R, Jafelicci Jr M, Souza D F, Nunes L A O 2008 *J. Lumin.* **128** 1787
[2] Laroche M, Gilles H, Girard S, Passilly N and Aït-Ameur, K 2006 *IEEE Photonics Technology Letters* **18** 764.
[3] Shand M L and Lai S T 1984 J. *Opt. Soc. Am. B* **20** 105
[4] Stalder, M, Chai, B H T and Bass M 1991 *Appl. Phys. Lett.* **58** 216
[5] Kück S 2001 *Appl. Phys. B* **72** 515
[6] Weidenborner J E, Stemple N R and Okaya, Y 1966 Acta Cryst. 20 761
[7] Hoffman C W W and Brown J J 1968 J. Inorg. Nuclear Chem. 30 63
[8] Brown J J 1967 J. Electrochem. Soc. 3 245
[9] Sosman L P, Abritta T, Nakamura O and D'Aguiar Neto M F 1995 J. Mat. Sci. Lett.14 19
[10] Garapon C, Brenier A and Moncorgé R 1998 Optical Materials 10 177
[11] S. Geller S J. 1960 Chem. Phys. 33 676
[12] Åhman, J, Svensson, G and Albertsson, J 1996 Acta Cryst. C52, 1336
[13] http://www.ill.eu/instruments-support/instruments-groups/instruments/d2b/characteristics/
[14] Parente C B R, Mazzocchi V L, Mestnik-Filho J, Mascarenhas Y P and Berliner R 2010 Nuclear Instruments and Methods in Physics Research A 622, 678
[15] Rodriguez-Carvajal J 1990 *FULLPROF: A Program for Rietveld Refinement and Pattern Matching Analysis.* In: Abstracts of the Satellite Meeting on Powder Diffraction of the XV Congress of the IUCr
[16] M A F M da Silva, PhD Thesis, Universidade Estadual do Rio de Janeiro, 2010

Accelerating medium effect as a general wave phenomenon

A I Frank[1, 5], P Geltenbort[2], M Jentschel[2], G V Kulin[1], D V Kustov[1,3] and A N Strepetov[4]

[1]Frank Laboratory of Neutron Physics, Joint Institute for Nuclear Research, Dubna, Russia
[2]Institut Laue Langevin, Grenoble, France
[3]Institute for Nuclear Research, Kiev, Ukraine
[4]National Research Centre "Kurchatov Institute", Moscow, Russia

Abstract. Already at the end of the last century theory predicted that the wave number and frequency of any wave will change when passing an accelerating refractive medium. The effect was calculated both for electromagnetic and neutron waves. As a refractive index may be introduced for waves of any nature one can speak about a very general Accelerating Medium Effect. As far as we know this effect has not yet been observed for light. Here we report on a neutron-optics experiments performed with ultra-cold neutrons where this effect has been demonstrated for the first time ever. The maximum energy transform in the experiment was $\pm$ $(2 \div 6) \times 10^{-10}$ eV which agrees with theory within less than 10%. Possibilities for future investigations of the Accelerating Medium effect will be discussed.

1. Introduction

Similarities in the description of light optics and quantum mechanics have given raise to particle optics as an independent field of research in the middle of the last century. In the forties E. Fermi introduced the index of refraction for neutron waves [1]. Later the work of Foldy [2] and Lax [3,4] allowed to understand that an index of refraction can be introduced for any material and waves, if the material contains scattering centers for the considered wave. Of course that is also true for matter-waves. In this case the index of refraction is given as

$$n^2 = 1 + 4\pi\rho k_0^{-2} f(0 \to 0) , \qquad (1)$$

where k_0 – is the wave number of the incoming wave , ρ - the spatial density of the scattering centers and $f(0 \to 0)$ - the forward scattering amplitude.

In neutron optics the majority of phenomena studied so far were such previously investigated and known from light optics. This was also the case, when during the seventies of the 20th century Horn and Zeilinger [5] started to investigate the interaction of neutron waves with moving matter. Due to

[5] frank@nf.jinr.ru

their analogies with light optics the experiments were called neutron Fizeau experiments. However, regardless the large number of similarities, there are substantial differences in the passage of electromagnetic and neutron waves through moving mater [6].

The phase shift of electromagnetic waves in Fizeau experiments can be explained by relativistic transformations of the velocity terms into a moving coordinate system. Neutron optics considers non-relativistic particles and the transformation into a coordinate system moving with the particle is a Galilean transformation. Further, in the most frequent cases, the forward scattering amplitude for neutrons on atomic nuclei can be considered as constant $f(0 \rightarrow 0) = -b$ and the interaction of the neutron with matter is described by means of an effective potential

$$U = 2\pi\hbar^2 m^{-1} \rho b,$$

where b – is the volume average of the coherent scattering length. Any movement of the matter itself which does not change its effective potential has no effect on the neutron and a phase shift only occurs when the sample boundaries move.

However, if a layer of material moves together with its boundaries, it will cause the wave frequency in the matter, measured in the laboratory system, to be different from the one in vacuum. This is valid –in light [7] as well as in neutron optics [8]. In both cases the frequency shift is caused by the Doppler effect

$$\Delta\omega = (n-1) k_0 \mathrm{v} \ ,$$

where n – is the index of refraction in the moving reference system, in which the matter is at rest, k_0 – is the wave number of the incoming wave, v – is the speed of the moving matter.

For a constant motion of the matter the Doppler shifts during the passage of the two boundaries have the same amplitudes but different signs. The overall effect is zero and the frequency of the wave transferred through uniformly moving matter does not change. For a long time the condition that the matter moves with constant velocity was underestimated, although it is decisive. In the case of non-constant motion, however, the frequency shifts at the sample boundary do not cancel and hence, the frequency of the transmitted wave differs from the incoming one. This was firstly shown by Tanaka [9], who solved the theoretical problem of electromagnetic wave transmission through linearly accelerated dielectric matter on the basis of covariant generalization of the Maxwell equations. Corresponding experiments in light optics have never been carried out although different possibilities have been discussed [10].

The transmission of neutrons through a layer of matter moving with linear acceleration was considered by Kowalski [11] in the context of a new type of experiment for the verification of the equivalence principle. The author concludes that the energy of the neutrons must change after transmission through such a linearly accelerated layer. This was confirmed later by Nosov and Frank [12] who calculated the velocity of neutrons transmitted through the accelerated boundaries of a sample.

The first experimental evidence of a neutron energy change after transmission through accelerated matter can be found in [13], while a more detailed investigation of the accelerated matter effect (AME) is given in [14]. Recently we observed the acceleration and deceleration of neutrons after transmission through an oscillating sample for the first time. The results of these experiments are given in section 3.

The AME is a general optical effect, which has only been demonstrated so far in neutron-optical experiments. Some new and more general aspects of this effect will be discussed in section.

2. The accelerating medium effect and the equivalence principle

As shown in ref. [14] the energy change due to the AME can be derived in good approximation from the equivalence principle without too detailed calculations. We will follow this derivation here, as it demonstrates the universality of the effect. According to Kowalski [11] the following Gedanken experiment is considered (fig. 1):

Assume the wave moves from the source towards the detector in the field of gravity such that its frequency increases during flight (fig. 1a). As there are no dissipative processes involved, the introduction of a refracting sample between source and detector cannot change the wave frequency when its detected. Any frequency change can be excluded as one imagines the wave to be elastically back reflected (without passage through the sample). In case of a frequency change the energy conservation law would be violated. Now one assumes (see figure 1b) the source, the detector and the refracting sample move all together with an acceleration $w=g$. Here g – is the free fall acceleration in the system of inertia and the wave, measured in the laboratory system, has a constant frequency. According to the equivalence principle the frequencies measured in both systems have to be the same.

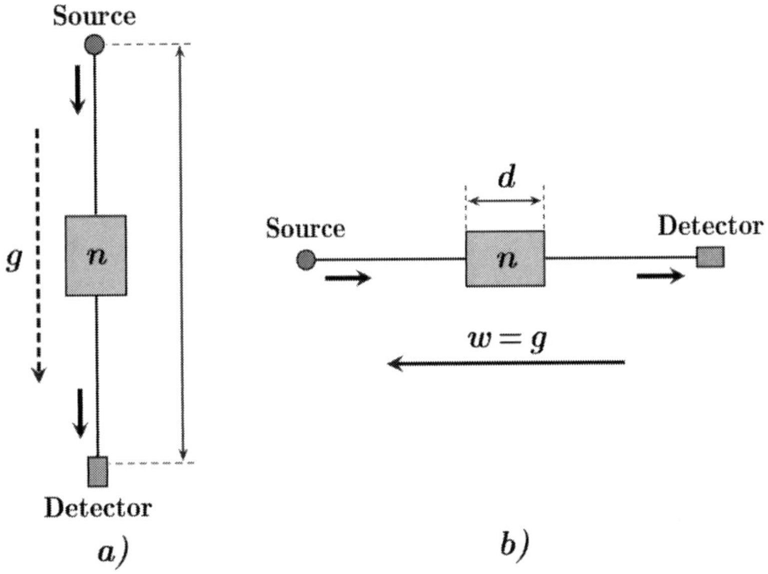

Figure 1. Neutron propagation through a refractive sample (a) in an inertial reference frame in the presence of gravity and (b) in a noninertial reference frame

However, the refracting sample in the flight path of the wave acts on its propagation time as the phase velocity inside the sample differs from the velocity in vacuum. For simplicity we assume that the inserted sample increases the propagation time by Δt. During this delay time the detector continues to accelerate and the additional (relative to the situation without sample) change of its velocity is $\Delta \mathrm{v} = w\Delta t$. Therefore, if the refracting sample would only cause a time delay, the speed of the detector at the time when the wave is detected would be different in the laboratory system with and without refracting sample. Therefore, due to the Doppler-effect, the registered wave frequency would also depend on the delay time and therefore on the sample properties. As this contradicts the equivalence principle we have to conclude that during the transmission through the refracting sample also the frequency of the wave chnages. This frequency change should compensate the Doppler-shift.

For neutrons and other non-relativistic particles the delay time is

$$\Delta t_{(n)} = \frac{d}{\mathrm{v}}\left(\frac{1}{n} - 1\right), \tag{2}$$

where t – is the time of flight, v – is the particle velocity and d – is the sample thickness. Accordingly, the energy, necessary for to compensate the Doppler-shift is $\Delta E = m\mathrm{v} \cdot w\Delta t$. Inserting equation (2) into this expression, we obtain

$$\Delta E \cong mwd\left(\frac{1}{n}-1\right), \tag{3}$$

which is in complete agreement with the results in [11,12]. For photons and ultra-relativistic particles the delay time is

$$\Delta t_{(\gamma)} = \frac{d}{c}(n-1), \tag{4}$$

and the Doppler-shift of the frequency (in first order of v/c) is $\Delta\omega \approx \omega \Delta \mathrm{v}/c = \omega w \Delta t/c$. Inserting equation (4) into this expression we get the same expressions as Tanaka [9] for light

$$\Delta\omega = \frac{\omega}{c^2}(n-1)wd. \tag{5}$$

It is worth mentioning that for the derivation above we were only exploring the fact that the wave number inside the accelerated matter is different to the one in vacuum. Therefore the effect should occur not only for accelerating matter containing scattering centers but much more generally for any volume containing force fields.

3. Observation of the accelerating matter effect in experiments with ultracold neutrons.

3.1. Spectroscopy with UCN

The change of energy by transmission through an accelerating sample was firstly measured several years ago in experiments with ultracold neutrons. Neutrons passed through a silicon sample, which osillates. Accordingly the energy change of the quasi-monochromatic neutrons is given as

$$\Delta E \cong -mA\Omega^2 d\frac{1-n}{n}\sin\Omega t, \qquad \Omega << \frac{\mathrm{v}_i}{d}, \tag{6}$$

where A and Ω - denote the oscillation amplitude and frequency of the sample and v_i – their velocity inside the sample material. The layout of the experimental setup, a modified spectrometer of [15,16], is shown in Figure 2.

Ultra-cold neutrons arrive at the main part of the device, passing through an annular corridor. At the exit of the corridor, there is a monochromator *1*, realized via a neutron interference filter – a Fabry-Perrot interferometer (FPI). It transmits UCNs with a narrow spectrum of vertical velocities. Sample *2*, a silicon plate of 0.6 or 1.85 mm thickness, can execute harmonic motion by means of the special driver *3* and is placed just below the monochromator. Sample oscillation frequencies of 40 or 60 Hz were used. The variable acceleration of the sample reached 90m/s^2.

Passing the monochromator and sample, neutrons arrive at a vertical mirror neutron guide, where the second FPI, analyzer *4*, is located and whose position can be varied in height. A scintillation detector for UCNs *5* is placed below the analyzer. The transmission line maxima of the monochromator and analyzer correspond to 107 and 127neV, respectively. The dependence of the count rate on the distance between the filters is qualitatively presented in Figure 3. The sign-alternating change in the neutron energy caused by passage through a sample moving with variable acceleration leads to the periodic variation in the count rate, as shown by the dashed straight lines in Figure 3.

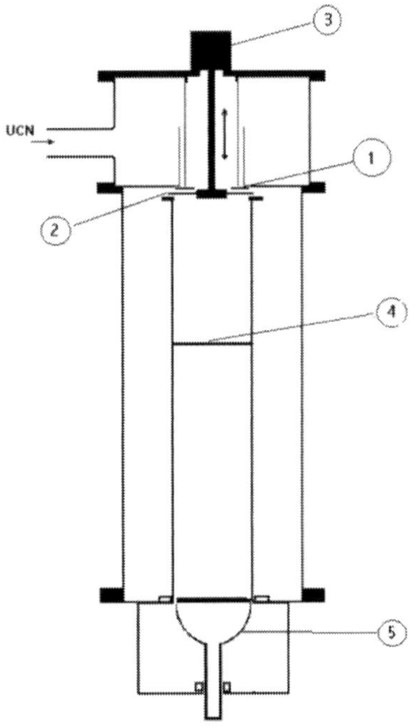

Figure2. Layout of the experimental setup

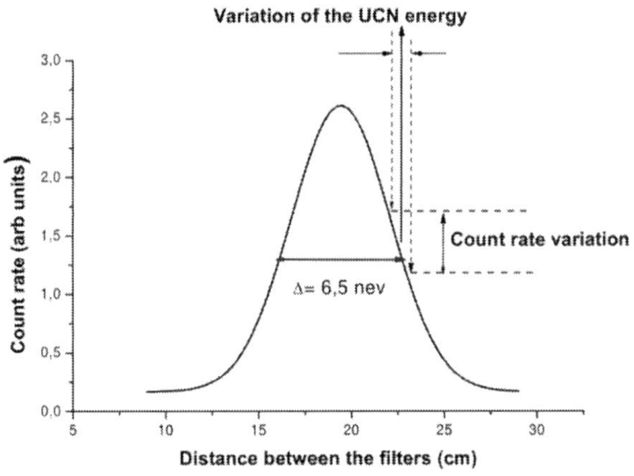

Figure 3. Count rate vs. the distance between the filters (scanning curve) and the detection principle of periodical variation of energy

The maximum change in the energy of UCNs, which is determined by Eq. (6), reaches approximately 0.2 neV in a first experiment [13] and 0.6neV in the second experiment [14].

Unfortunately, the effect of accelerating matter is not the only origin of count rate modulation. Obviously the velocity $V(t) = A\Omega \cdot Cos(\Omega t)$, as well as the acceleration of the plate $w(t) = -A\Omega^2 Sin(\Omega t)$, change simultaneously when the plate was moved. The periodic variation in the sample transparency modulates the intensity with a relative amplitude of about 8%. Therefore the relative change in the intensity is given by the expression

$$P(t) = \frac{I(t)}{I_0} = 1 - B \cdot cos\left(\Omega t + \alpha\right), \ B = \sqrt{a^2 + b^2}, \ \alpha = arcsin\left(\frac{a}{B}\right), \qquad (7)$$

Here, the amplitude a determined by the acceleration of the sample is proportional to the derivative of the scanning curve and depends on the analyzer position. At the maximum of this curve, $a = 0$ and the count rate oscillation is determined only by the velocity effect in the transmission with amplitude b. Formula (7) determines the experimental procedure. Measuring the modulation phase for various analyzer positions and using the known scanning curve, one can determine the amplitude of the periodic change in the energy of the neutron passing through the accelerating sample.

The phase and amplitude of the count rate modulation were measured in the experiment for various positions of the analyzer filter. The time dependence of the count rate was determined in a time interval equal to the oscillation period. The origin of the scale was specified by a generator controlling the sample motion. The data were normalized to the averaged count rate and fitted to the function $f(t) = 1 + B\, sin(\Omega t - \varphi)$. The amplitude B and phase φ of the count rate oscillation were determined by such a way in each value of the position of the analyzer. Some results from the phase measurements are shown in Figure 4.

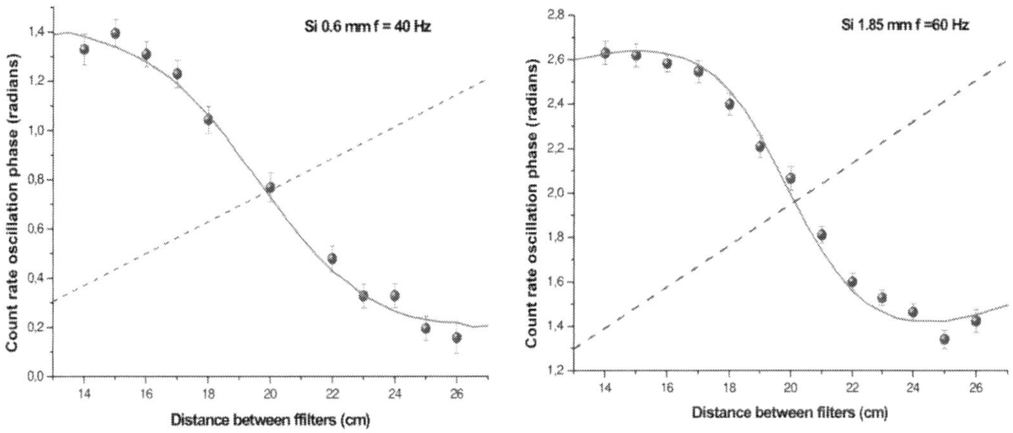

Figure 4. Measured phase of the counting-rate oscillation versus the analyzer position. Solid (red) curve – Monte-Carlo simulation. Dashed (blue) line – theoretical prediction for the phase in the absence of the Accelerating medium effect

Monte Carlo simulation of the oscillation phases were repeated using the relation $\Delta E \cong -KmA\Omega^2 L\left[(1-n)/n\right]\sin\Omega t$ which differs from equation (7) by the correction factor K. For the latter one a value of $K = 0.94 \pm 0.06$ was found. Thus, the results of the measurements of the counting-rate oscillation phase are in reasonable agreement with the existing theory of the accelerating medium effect.

3.2. The Accelerating Medium Effect and time focusing of neutrons.

3.2.1. Experimental Principle

Recently a new type of experiment for the observation of the AME was carried out [18]. Here a plate, vibrating in space, was used as periodic modulator to change the velocity of ultra-cold neutrons. The main idea of this experiment is rather similar to experiments on time focusing [19] and its main principle is illustrated in Figure 5. Monochromatic neutrons are transmitted through a modulator – acting as a time lens. The neutron velocity was changed periodically such that in the ideal case all neutrons arrive simultaneously in the detector L. The lens is working in a cyclic regime and time focuses those neutrons passing through it in one cycle. In reference [18] an aperiodic diffraction grating moving across the neutron beam was used as a lens. In the current experiment we used a plate of silicon oscillating along the propagation direction of the ultra-cold neutron beam. Moving with periodic acceleration it was periodically accelerating or slowing down the neutrons due to the AME. Here the momentum transfer to the neutrons was insufficient for an efficient focusing and the focal plane was significantly away (to the right side) from the plane L. However, in the detection plane there occurs a substantial concentration in time of flight of the incoming neutrons. This leads to a time modulation of the count rate at the detector. It can be shown that in this case the time dependence of the detector count rate is determined by the derivative of the modulation function $\Delta v(t)$. The change of velocity by passing the oscillating plate is according to equation (6) given by

$$\Delta v_n = -A\Omega^2 \frac{d}{v_0}\left(\frac{1-n}{n}\right) \sin \Omega t$$

and therefore the weak focusing leads to the following time dependence of the count rate

$$N_a(t) = N_0 + C_a A\Omega^3 \left(\frac{1-n}{n}\right) \cos \Omega(t+\tau) \qquad (8)$$

where N_0 – is the mean count rate, τ - is the average time of flight and C_a – is a constant coefficient.

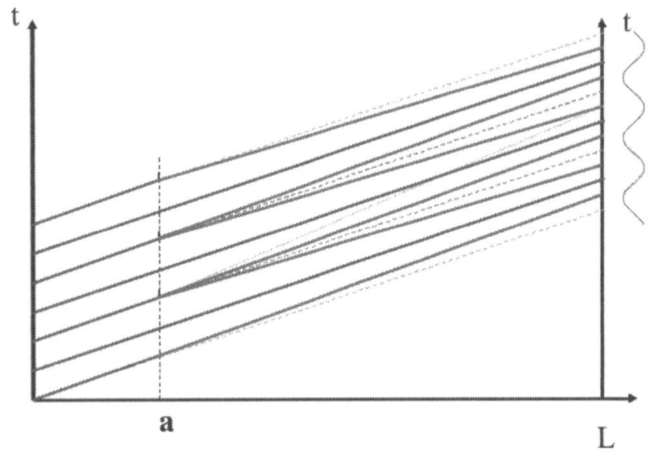

Figure 5. A coordinate versus time diagram illustrating the idea of time focusing.

Obviously the harmonic movement of the sample leads to a periodic variation of the neutron velocity with respect to the sample. As the dependence of all absorbing processes in the case of ultra-cold neutrons is proportional to 1/v the transmission of the sample is also changing accordingly. Therefore, additionally to the AME given by equation (8) there is another systematic effect described by the following equation

$$N_v(t) = N_0 + C_v A\Omega \cos(t+\tau)$$

As both effects are synchronous, the harmonic modulation of the count rate is obtained by summing the amplitudes of both

$$\Delta N(\Omega) = A\Omega^2 \left(C_a \Omega + \frac{C_v}{\Omega} \right) \qquad (9)$$

Expression (9) is written such, that it underlines the main principle of the given experiment: the measurement is carried out such that the quantity $A\Omega^2$ stays constant and the AME is growing proportionally while the systematic velocity effect if inversely proportional to the frequency of modulation.

The systematic velocity effect was encountered in former experiments [13,14]. In these measurements it was already found that if the monochromatisation happens after transmission through the sample the AME is excluded. All experimentally observed modulations are due to the systematic velocity effect. This finding has been explored for the current measurement.

The experimental strategy consisted in the measurement of the modulation amplitude of the ultra-cold neutron count rate transmitted through an oscillating sample for a large set of frequencies Ω and fixed values of $A\Omega^2$. The measurement was carried out in two geometries. In the first case the monochromatic neutrons were transmitted through the sample and the amplitude modulation was given by equation (9). In the second case the neutrons were transmitted through the oscillating sample and only after through the mononchromator. In the second case the modulation amplitude is defined by the systematic velocity effect only. The frequency dependence of the difference of both effects is defined exclusively by the AME and should depend linearly on the frequency of modulation.

3.2.2. Experimental realisation and results.
For the experiment the same spectrometer as in reference [14] was used. A schematic illustration is given in Figure 2. The only difference consists in the absence of the analyzer 4. As in earlier works we used a Fabri-Perot interferometer as monochromator. It transmits a single wave length at about *107 neV* with a relative width of $\Delta E/E \approx 0.04$. As a sample we used a wafer with *1.85 mm* thickness, which was put into oscillation by the same driving stage as in [14]. The time depending acceleration was permanently measured by a piezoelectric sensor, the sinusoidal signal of which was also used for the stabilization of the amplitude of the driving stage. This allowed to stabilize the quantity $w_{max} = A\Omega^2$ on the level of *2%*. The measurement was carried out for two values w_{max} = 57 m/s^2 and w_{max} = 72 m/s^2 and for a frequency range of f = 20 ÷100Hz. The obtained results are shown in Figure 6.

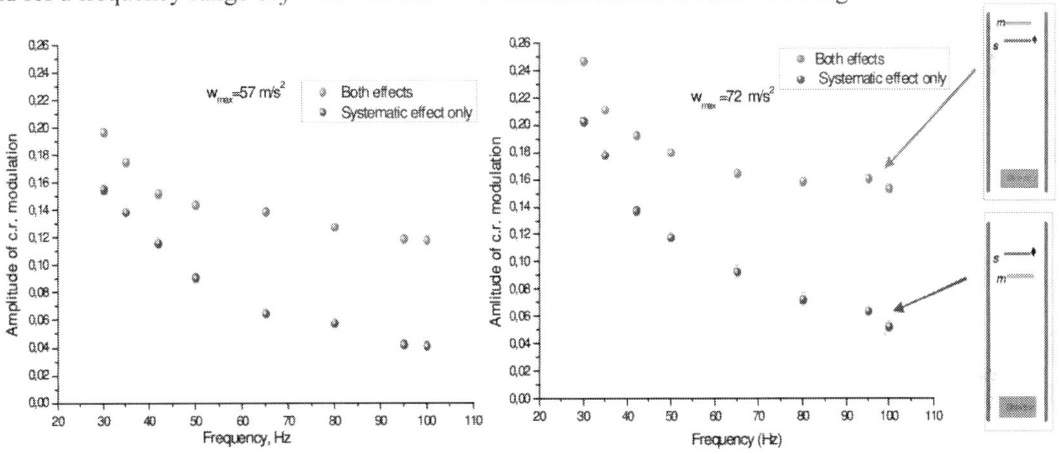

Figure 6. The count rate oscillation amplitude as a function of frequency for a fixed value of $w_{max} = A\Omega^2$. On the left hand side w_{max} = 57m/s^2, while on the right hand side w_{max} = 72m/s^2. The measurements were done in two geometries, which symbolically are indicated on the very right. Here M – is the monochromator and S – is the sample.

The results show without any doubt the presence of weak time focusing, i.e. the acceleration/slowing down of neutrons when passing the oscillating sample. The modulation amplitude in the geometry excluding the AME effect (lower blue points) is substantially lower than the amplitude of modulation sensitive to both effects (upper red points). Further, it can be seen that the difference between both effects is growing with an increase of the modulation frequency.

Unfortunately it is difficult to directly compare these data with calculations. The reason is that the modulation amplitude is also depending on the dispersion of the time of flight, i.e. on the width of the spectrum transmitted by the monochromator. Additionally it turned out that the background of the spectrometer is slightly changing, when the monochromator is changed from one position into another. Therefore a number of calibration measurements were added, in which the modulation of the ultra-cold neutron flux was realized via a mechanical chopper. In these measurements the non moving silicon sample was also present and the monochromator was changed between different positions.

Such calibrations allowed to normalize the data of Figure 6. The difference of the two curves should be a straight line, corresponding to the first term of equation (9). These results are shown in Figure 7 and it is possible to compare them directly to theoretical calculations. The main parameter here is the inclination angle of the straight line. The calculations were done by assuming several origins of the background. The dispersion of the calculated background values was included into a systematic error.

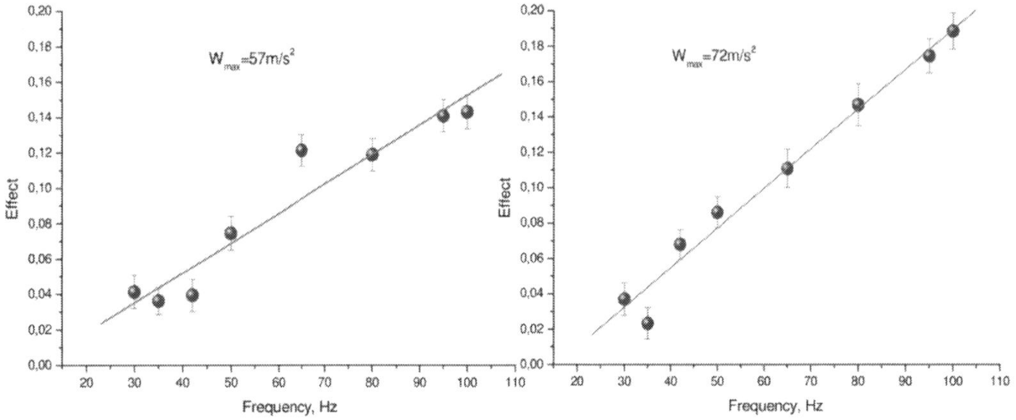

Figure 7. The difference between normalized amplitudes of the count rate oscillation versus frequency. The value $w_{max} = A\Omega^2$ was fixed. On the left $w_{max} = 57m/s^2$, on the right $w_{max} = 72m/s^2$.

The error of the linear fit of the data in Figure 7 was considered to be the statistical error. From the obtained results it was possible to estimate the agreement of the calculated velocity change when neutrons are transmitted through an oscillating refractive sample.

For the factor $K = \Delta v_{exp}/\Delta v_{exp}$ and values of $w_{max} = A\Omega^2$ as mentioned above the agreement of experiment and calculation was: $K_1 = 0.95 \pm 0.10_{stat} \pm 0.05_{syst}$ and $K_2 = 0.95 \pm 0.05_{stat} \pm 0.05_{sys}$

4. The Accelerating Matter Effect in the case of double refraction

In the present works concerning the AME the polarization of waves was not considered so far. In order to partially fill this gap we will consider briefly the case of matter with double refraction, which is characterized by two indices of refraction $n_{\pm}$, according to two polarizations of the incoming wave.

In light optics double refraction (birefringence) is a commonly known phenomenon. Obviously, when light is moving through double refracting matter, the phase shift given by the Tanaka equation (5) becomes depending on the direction of the E-vector with respect to the sample. It is possible that a comparably low frequency of rotation of the polarization can be easier detected than the small shift of frequency.

In neutron optics the quantities $n_{\pm}$ correspond to two different projection of the neutron spin on a physical axis. Accordingly we will rewrite equation (3) as

$$\Delta E_{\pm} \cong mwd\left(1-n_{\pm}\right)/n_{\pm}, \quad \Delta\omega_{\pm} = \Delta E_{\pm}/\hbar$$

After transmission through an accelerated birefringent sample the two spin components of the neutron wave function differ by a frequency and form a non-stationary superposition. In the case of an arbitrary polarization of the original wave function $\Psi_0(x,t)$ the final state will have the form

$$\Psi(x,t) = \left\{exp\left[-i\left(\Delta k_{+}x + \Delta\omega_{+}t + \chi_{+}\right)\right]\left|\uparrow\right\rangle + exp\left[-i\left(\Delta k_{-}x + \Delta\omega_{-}t + \chi_{-}\right)\right]\left|\downarrow\right\rangle\right\} \Psi_0(x,t) \quad (10)$$

where

$$\Delta k_{\pm} = \frac{m}{\hbar^2} \frac{\Delta E_{\pm}}{k_0}.$$

The constant phase angles, $\chi_{\pm}$, that are irrelevant to what follows, determine spin directions after transmission through a moving sample [19]. The wave function in equation (10) describes the state with spin precession. The precession angle is obtained from the difference between the phase angles of two spin components:

$$\varphi(x,t) = (\Delta k_{+} - \Delta k_{-})x - (\Delta\omega_{+} - \Delta\omega_{-})t + \chi_{+} - \chi_{+}$$

Assuming here the simplifying condition $\Delta E_{\pm}/E << 1$ it is possible to write that $(\Delta k_{+} - \Delta k_{-})x = (\Delta\omega_{+} - \Delta\omega_{-})x/v$, from which follows that in the reference system moving with the velocity of the neutron the direction of the spin vector is unchanged. However, in a fixed point of observation, $x=L$, the spin direction changes periodically in time with the frequency $\Omega = \Delta\omega_{+} - \Delta\omega_{-}$. This periodic change of the polarization direction can be measured. The beat frequency Ω and respective energy transfers $\hbar\Omega$ can be quite small.

In neutron optics there can be several physical reasons for double refraction. First of all one should mention the rather simple case of interaction of the magnetic momentum of the neutron μ with a magnetic field B. Obviously any space in which the magnetic field has a nonzero value is acting with two refraction indexes $n_{\pm} = (1 \mp \mu B/E)^{1/2}$. If this space moves together with its field boundaries it will generate according to what was mentioned above a non-stationary state (10). Sample material, put into a magnetic field can also act as double refracting material for neutron waves [20,21]. This will be caused by different neutron wave numbers due to the presence of the magnetic field and the dispersion of the material itself. An accelerated motion of the sample in a constant magnetic field will also lead to states with different frequencies for the different spin components of the wave function. Double refraction might also occur in the absence of a magnetic field. First of all one might focus here on nuclear pseudomagnetism, which takes place when a neutron wave is propagating in matter with polarized nuclei. [22,23]. Due to the spin dependence of the nuclear interaction the coherent scattering length $b_{\pm}$ is different for the two values of the total spin. As a result, and in accordance with (1) the medium has two refractive indices.

Finally, double refraction might be caused by parity violation in neutron-nucleon interaction. The forward scatter length and consequently the refractive index will be depending on the neutron spin orientation [24,25]

5. Conclusion

It was shown, that the Accelerating Medium Effect is closely related to the equivalence principle (EP). Consequently, the equations which describe the frequency change after their passage through accelerated refractive samples, may be derived not only from first principles but from the EP too. This can be interpreted as additional evidence for the general nature of this effect which exists for waves and particles of different nature. Two experiments detecting the AME in neutron optics were described. The measured energy variations were equal or less than 5×10^{-10} eV while the velocities changed by about 1 cm/s. The measurements agree with theoretical predictions better than 10%. New possibilities for the detection and for applications of the AME may be possible by the use of birefringent material.

This work is partly supported by the Russian Foundation for Basic research (project 11-02-00271).

References

[1] Anderson H L Comment to the papers 217 and 220. In: E. Fermi. Note e memorie (collected papers). Accademia Nazionale dei Lincei -The University of Chicago Press. Vol.2, p.425
[2] Foldy L J 1945 *Phys. Rev.* **67** 107
[3] Lax M 1951 *Rev. Mod. Phys*. **23** 287
[4] Lax M 1952 *Rev. Phys.* **85** 621
[5] Horn M A and Zeilinger A 1979 *Neutron Interferometry*, Eds. Bonse U and Rauch H (Clarendon Press, Oxford) 350
[6] Peres A 1983 *Am. J. Phys*. **51** 947
[7] Yeh J 1965 J. Appl. Phys. **36** 3513
[8] Arif M, Kaiser H, Werner S, Cimmino A A, Hamilton W A, Klein A G, and Opat G I 1985 *Phys. Rev*. **A 31** 1203.
[9] Tanaka K, 1982 *Phys. Rev. A* **25**, 385
[10] Neutze R and Stedman G E 1998 *Phys. Rev. A* **58** 82
[11] Kowalski F V 1993 *Phys. Lett. A* **182**, 335
[12] Nosov V G and Frank A I 1998 *Phys. At. Nucl.* **61** 613
[13] Frank A I, Geltenbort P, Kulin G V, Kustov D V, Nosov V G and Strepetov A N 2006 *JETP Letters*, **84** 363
[14] Frank A I, Geltenbort P, Jentschel M, Kustov D V, Kulin G V, Nosov V G and Strepetov A N 2008 *Phys. At. Nucl.* **71** 1656
[15] Bondarenko I V, Bodnarchuk V I, Balashov S N, Geltenbort P, Klein A G, Kozlov A V, Korneev D A, Masalovich S V, Nosov V G, Frank A I, Høghøj P, and Cimmino A 1999 *Phys. At. Nucl.* **62** 721
[16] Bondarenko I V, Frank A I, Balashov S N, Masalovich S V, Nosov V G, Geltenbort P, Høghøj P Cimmino A and Klein A G 2000 *NIM A* **440** 591
[17] Frank A I, Geltenbort P Jentschel M, Kustov D V, Kulin G V and Strepetov A N 2011 *JETP Letters* **93** 361
[18] Frank A I, Geltenbort P, Kulin G V and Strepetov A N 2003 *JETP Letters* **78** 188
[19] Littrell K C, Werner S A and Allman B E 1996 *J. Phys. Soc. Jpn*. **65**, Suppl. A, 98
[20] Baryshevskii V G, Cherepitsa S V and Frank A I 1991 *Phys.Lett*. A **153** 299
[21] Frank A I, Bondarenko I V, Kozlov A V, Høghøj P and Ehlers G 2001 *Physica B* **297** 307
[22] Baryshevkii V G, Podgoretsky M I 1964 *JETP* **47** 1050
[23] Abragam A, Bacchella G L, Glätti H, Meriel P, Pinot M, and Piesvaux J 1973 *Phys.Rev.Lett* **31** 776
[24] Forte M, Heckel B R, Ramsey N F, Green K, Greene G L, Byrne J and Pendlebury J M 1980 *Phys.Rev.Lett*. **45** 2088
[25] Sushkov O P and Flambaum V V 1982 Sov. Phys. Usp. **25** 1

A proposed search for new light bosons using a table-top neutron Ramsey apparatus

F.M. Piegsa[1], **G. Pignol**[2]

[1] ETH Zürich, Institute for Particle Physics, 8093 Zürich, Switzerland

[2] LPSC, Université Joseph Fourier Grenoble 1, CNRS/IN2P3, INPG, 38000 Grenoble, France

E-mail: florian.piegsa@phys.ethz.ch

Abstract. If a new light boson existed, it would mediate a new force between ordinary fermions, like neutrons. In general such a new force is described by the Compton wavelength λ_c of the associated boson and a set of dimensionless coupling constants. For light boson masses of about 10^{-4} eV/c^2, λ_c is of the order millimeters. Here, we propose a table-top particle physics experiment which provides the possibility to set limits on the strength of the coupling constants of light bosons with spin-velocity coupling. It utilises Ramsey's technique of separated oscillating fields to measure the pseudo-magnetic effect on neutron spins passing by a massive sample.

1. Introduction

In particle physics, the interaction between matter particles (fermions) is understood as an exchange of vector particles (bosons). The Standard Model of particle physics explains electromagnetism, nuclear weak and strong interactions in terms of exchange of photons, weak heavy bosons and gluons. It is believed that the Standard Model is a low energy limit of some more fundamental theory, that will ultimately also describe gravity in the same framework. Various extensions of the Standard Model based on new ideas, such as Supersymmetry or Superstring theory are being developed in this direction. As a general feature of such theoretical investigations, new particles, and thus new interactions beyond the standard model, are expected. These hypothetical new particles have escaped detection up to now, either because they are too heavy, thus mediating an interaction which is too short in range, or because they are too weakly coupled with ordinary matter, thus mediating an interaction which is too weak. The latter case is of interest for this proposal: we are looking for a weak, long range new interaction. This new interaction between ordinary fermions would be mediated by a new light boson with an interaction range $\lambda_c = \hbar/Mc$, where M is the mass of the new boson, $\hbar$ is Planck's constant and c is the speed of light. For a light mass of 10^{-4} eV/c^2, the range is about 2 mm. Depending on the details of the fermion-boson coupling and whether the new boson is scalar (spin 0) or vectorial (spin 1), such a force between two fermions could depend on the spins of the fermions or not. It could also depend on the relative velocity between the fermions. For a scalar boson ϕ the interaction with a fermion ψ, described by the Lagrangian function $\mathcal{L} = \overline{\psi}(g_S + ig_P\gamma^5)\psi\phi$ is parametrised by two constants: the scalar coupling g_S and the pseudoscalar coupling g_P. For a vector boson X_μ the interaction $\mathcal{L} = \overline{\psi}(g_V\gamma^\mu + g_A\gamma^\mu\gamma^5)\psi X_\mu$ is parametrised by the vector coupling g_V and the axialvector coupling g_A. Hence, in general a new force is described by the

range λ_c and a set of dimensionless coupling constants.
Most current laboratory experiments regarding new light bosons do not search for the boson itself, but for the additional interaction the boson would mediate between ordinary particles. We restrict our discussion to the coupling with nucleons, although sensitive experiments probing the coupling with electrons have also been performed. The non-relativistic interaction potential between two fermions contains several terms, bearing diverse phenomenological consequences [1], calling for diverse experimental techniques to probe them. In the case of a spin 0 boson, several experiments investigate the gravity-like spin-independent scalar-scalar coupling, i.e. the so-called 5th force [2, 3, 4, 5, 6]. Furthermore, the scalar-pseudoscalar coupling also called axion-like interaction is studied by measuring the influence of an unpolarised source on a polarised probe, e.g. on neutron spins or the nuclear spins of a polarised ^{3}He gas etc. [7, 8, 9, 10]. Finally, for experiments investigating the pseudoscalar-pseudoscalar coupling the source as well as the probe have to be polarised. A recent experiment performed at Princeton probed this coupling at a meter distance, using a ^{3}He spin source and a K-^{3}He co-magnetometer probe [12]. Similarly, a spin 1 boson can mediate vector-vector, vector-axial and axial-axial interactions. The first type corresponds to a spin-independent repulsive new force, which is probed in the millimeter region by the Seattle torsion pendulum experiment [4]. The most recent measurement yields the upper limit:

$$g_\mathrm{V}^2 < 5 \times 10^{-40} \tag{1}$$

Next, there are terms depending on the spins of the two interacting particles (spin-spin interaction). Ref. [12] reports the limits, in terms of vector exchange parameters:

$$g_\mathrm{V} g_\mathrm{A} \left(1 + \frac{r}{\lambda_\mathrm{c}}\right) e^{-r/\lambda_\mathrm{c}} < 5 \times 10^{-25} \tag{2}$$

and

$$g_\mathrm{A}^2 e^{-r/\lambda_\mathrm{c}} < 1.5 \times 10^{-40} \tag{3}$$

where in this experiment $r = 0.5$ m is the distance from the ^{3}He spin source with the K-^{3}He co-magnetometer. In addition to the spin-spin interaction, a vector boson would mediate two kinds of spin-velocity interactions between a source particle and a probe particle:

$$V_\mathrm{VA}^\mathrm{point}(r) = \frac{g_\mathrm{V} g_\mathrm{A}}{2\pi}\, \hbar c\, \boldsymbol{\sigma} \cdot \frac{\boldsymbol{v}}{c}\, \frac{e^{-r/\lambda_\mathrm{c}}}{r} \tag{4}$$

and

$$V_\mathrm{AA}^\mathrm{point}(r) = \frac{g_\mathrm{A}^2}{16\pi} \frac{(\hbar c)^2}{mc^2}\, \boldsymbol{\sigma} \cdot \left(\frac{\boldsymbol{v}}{c} \times \frac{\boldsymbol{r}}{r}\right) \left(\frac{1}{\lambda_\mathrm{c}} + \frac{1}{r}\right) \frac{e^{-r/\lambda_\mathrm{c}}}{r} \tag{5}$$

where $\boldsymbol{\sigma}$ is the spin and m is the mass of the probe particle, r is the distance between the source and the probe and v is the relative velocity between the source and the probe. These potentials corresponds to f_v and $f_\perp$ in the notations of [1].The fact that the interaction does not depend on the source particle's spin allows to use an unpolarised macroscopic body as a source, the potential of each particles in the body adding coherently.
Here, we propose to probe these exotic interactions, in the millimeter range, with a beam of polarised slow neutrons as a probe and macroscopic bulk matter as a source. Since the new interaction potentials are linear proportional to the neutron spin, it is equivalent to a interaction with a magnetic field which can be described by the relation $V = -\frac{1}{2}\gamma_\mathrm{n}\hbar\, \boldsymbol{\sigma} \cdot \boldsymbol{B}$, where γ_n is the gyromagnetic ratio of the neutron. Hence, in case of a semi-infinite plane of bulk material as a source, one can perform the integration over the source volume analytically to obtain the corresponding *pseudo-magnetic fields*:

$$B_\mathrm{VA}(\Delta y) = \frac{2}{\gamma_\mathrm{n}} g_\mathrm{V} g_\mathrm{A}\, N \lambda_\mathrm{c}^2\, \boldsymbol{v}\, e^{-\Delta y/\lambda_\mathrm{c}} \tag{6}$$

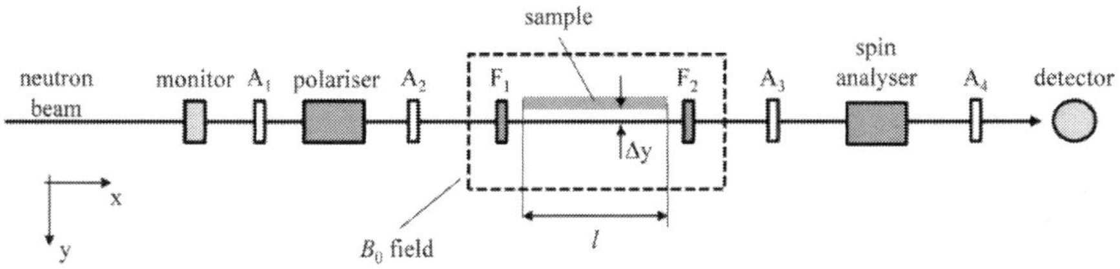

Figure 1. Topview of the proposed setup for the measurement of the axial-axial interaction between polarised neutrons and a macroscopic sample. The cold neutron beam is shaped by several apertures A_{1-4}. The sample is placed parallel in a distance Δy with respect to the beam and is located between the two $\pi/2$-spin flippers F_1 and F_2. They produce a radio frequency field perpendicular to the static magnetic field B_0 which points into z-direction, i.e. perpendicular to the drawing plane.

$$B_{\rm AA}(\Delta y) = \frac{1}{\gamma_{\rm n}} \frac{g_{\rm A}^2}{4} \; N \; \frac{\hbar c}{m_{\rm n} c^2} \; \lambda_{\rm c} \; (\boldsymbol{v} \times \boldsymbol{e}_y) \; e^{-\Delta y/\lambda_{\rm c}} \tag{7}$$

where $\Delta y > 0$ is the distance from the probe particle, the neutron, to the surface, N is the nucleon number density of the source and $m_{\rm n}$ is the neutron mass. For the vector-axial case, the pseudo-magnetic field is directed along the neutron beam, whereas for the axial-axial case, the direction is orthogonal to both the neutron beam and the plane normal. These pseudo-magnetic fields induce corresponding pseudo-magnetic neutron precessions $\varphi_{\rm VA} = -\gamma_{\rm n} B_{\rm VA} \tau$ and $\varphi_{\rm AA} = -\gamma_{\rm n} B_{\rm AA} \tau$, where τ is the time the neutron spin experiences the pseudo-magnetic potential. A method to measure these precession angles very accurately is Ramsey's technique of separated oscillating fields adapted to neutrons [13].

2. Proposed experimental setup

In Fig. 1 a schematic drawing of the proposed Ramsey setup is depicted, which can be used to probe the axial-axial pseudo-magnetic field $B_{\rm AA}$ as a function of the distance Δy between the neutron beam and the sample surface. Firstly, a cold monochromatic neutron beam is polarised by means of a polarising supermirror in z-direction and collimated by several apertures to a width, which is much smaller than the beam-sample distance.[1] Along the beam path two phase-locked radio frequency $\pi/2$-spin flippers are placed in a magnet which provides a static field B_0 in z-direction. The investigated sample with a length l is placed between these $\pi/2$-spin flippers, which produce linear oscillating fields in y-direction. The spins of the neutrons are analysed with a second polarising supermirror and are finally detected employing a ^{3}He gas detector. The number of detected neutrons is normalised using a monitor detector placed in the incidenting neutron beam. The additional appertures behind the sample (A_3 and A_4) avoid detecting neutrons which are accidentially scattered in air or on the sample surface.
A so-called Ramsey oscillation pattern is obtained by measuring the detector count rate as a function of the spin flipper frequency close to the neutron Larmor resonance. Any additional phase precession due to the sample will cause a corresponding phase-shift of the sinusoidal-shaped Ramsey pattern. In order to avoid false effects the sample must be chosen to be non-magnetic and any phase drifts, e.g. caused by thermal expansion, phase drifts of the radio frequency fields, instabilities of the external magnetic field etc. must be suppressed as much as possible. Besides monitoring and stabilising the magnetic field, the implementation of a second reference neutron beam would be advantageous to correct for occurring phase deviations [14, 15]. In similar neu-

[1] The neutron wavelength distribution should have a width of less than 10% to avoid smearing of the Ramsey pattern due to magnetic field inhomogeneities of B_0.

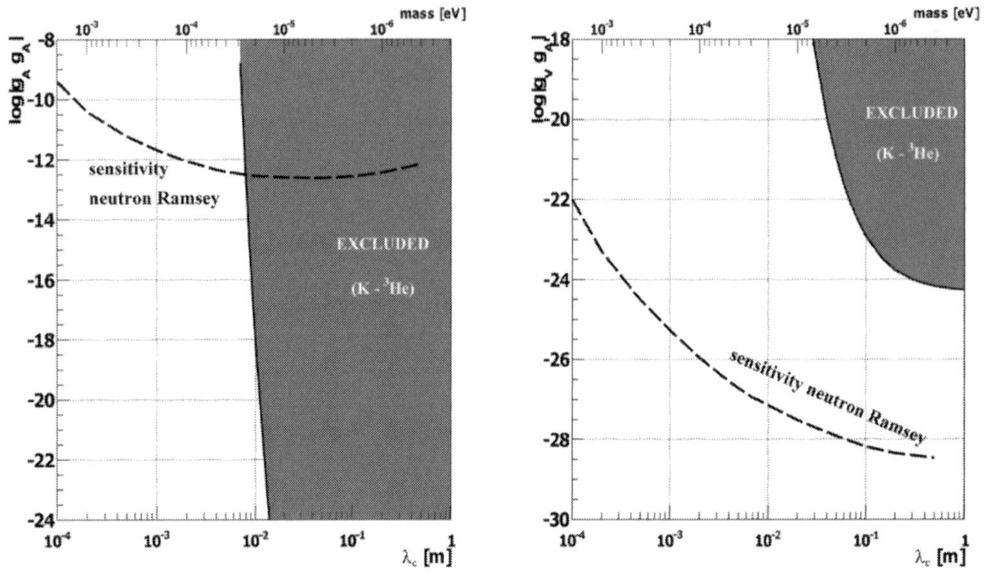

Figure 2. Expected sensitivity of the neutron Ramsey setup (dashed lines) for the axial-axial interaction (left) and the vector-axial interaction (right) as a function of the Compton wavelength λ_c. The grey shaded areas represent the exclusion by the Princeton experiment [12].

tron Ramsey apparatuses a phase stability of better than 1° could be established [15, 16, 17]. Such a Ramsey setup could be easily installed at an already existing polarised neutron reflectometer, which would have the advantage of usually very precise alignment possibilities at such beam lines. A time-of-flight reflectometer would even allow measuring the velocity dependence of the exotic interaction for different neutron de-Broglie wavelengths. Moreover, a similar setup with a longitudinal static magnetic field, or equally a neutron spin-echo setup, can be used to investigate the vector-axial coupling.

3. Sensitivity of the neutron Ramsey technique

For the consideration of the achievable sensitivity to probe for the exotic interactions, we assume a phase stability of the setup of $\delta\varphi = 1°$, a neutron wavelength of 0.5 nm, i.e. a velocity of $v \approx 800$ m/s, and a 500 mm long aluminium sample with a plane surface. In order to minimise systematic errors due to the paramagnetic property of the aluminium ($\chi_{\rm para} = 2.1 \times 10^{-5}$) the applied field B_0 should not exceed a few millitesla. For example, a field of 3 mT corresponds to a neutron Larmor resonance frequency of about 87.5 kHz and would lead to a paramagnetic phase-shift for a neutron beam crossing the entire sample, which corresponds to the worst case, of about +0.4° [18]. Hence, applying the following formula:

$$\delta B = \frac{v}{\gamma_{\rm n}\, l}\, \delta\varphi \tag{8}$$

one finds that the sensitivity of such an apparatus would be about 150 nT. Measuring with this sensitivity at a beam-sample distance $\Delta y = 0.3$ mm, yields with Eqn. (6) and (7) the exclusion plots for the axial-axial and vector-axial interactions presented in Fig. 2. Note, using a copper sample instead of aluminium would even increase the sensitivity by a factor of about 3, due to the higher density of the material.

4. Conclusion

We have proposed a table-top neutron Ramsey experiment to probe the strength of the coupling constants of light spin 1 bosons with spin-velocity coupling. The method has a sensitivity of about 10^{-11} for the axial-axial interaction and 10^{-25} for the vector-axial interaction at $\lambda_c = 1$ mm and represents, to our knowledge, so far the only method to probe spin-velocity interactions in the millimeter range.

References

[1] B.A. Dubrescu and I. Mocioiu, JHEP **11** (2006) 005.
[2] E. Fischbach, C. Talmadge, Nature **356** (1992) 207.
[3] S. Schlamminger et al., Phys. Rev. Lett. **100** (2008) 041101.
[4] E.G. Adelberger et al., Prog. Part. Nucl. Phys. **62** (2009) 102.
[5] D.J. Kapuet et al., Phys. Rev. Lett. **98** (2007) 021101.
[6] A.A. Geraci et al., Phys. Rev. D **78** (2008) 022002.
[7] S. Baeßler, V.V. Nesvizhevsky, K.V. Protasov, A. Yu. Voronin, Phys. Rev. D **75** (2007) 075006.
[8] A.N. Youdin, D. Krause, K. Jagannathan, L.R. Hunter, S.K. Lamoreaux, Phys. Rev. Lett. **77** (1996) 2170.
[9] A.K. Petukhov, G. Pignol, D. Jullien, K.H. Andersen, Phys. Rev. Lett. **105** (2010) 170401.
[10] A.P. Serebrov et al., JETP Lett. **91** (2010) 6.
[12] G. Vasilakis, J.M. Brown, T.W. Kornack, M.V. Romalis, Phys. Rev. Lett. **103** (2009) 261801.
[13] N. Ramsey, Phys. Rev. **78** (1950) 695.
[14] O. Zimmer, G. Ehlers, B. Farago, H. Humblot, W. Ketter, R. Scherm, EPJ direct A **1** (2002) 1.
[15] F.M. Piegsa, B. van den Brandt, H. Glättli, P. Hautle, J. Kohlbrecher, J.A. Konter, B.S. Schlimme, O. Zimmer, Nucl. Instr. and Meth. A **589** (2008) 318.
[16] F.M. Piegsa, B. van den Brandt, P. Hautle, J.A. Konter, Nucl. Instr. and Meth. A **605** (2009) 5.
[17] F.M. Piegsa, B. van den Brandt, P. Hautle, J. Kohlbrecher, J.A. Konter, Phys. Rev. Lett. **102** (2009) 145501.
[18] F.M. Piegsa, B. van den Brandt, P. Hautle, J.A. Konter, Physica B **406** (2011) 2409.

Journal of Physics: Conference Series **340** (2012) 012044 doi:10.1088/1742-6596/340/1/012044

Triple Entanglement in Neutron Interferometric and Polarimetric Experiments

S. Sponar[1], J. Klepp[2] R. Loidl[1], K. Durstberger-Rennhofer[1], H. Geppert[1], G. Badurek[1], H. Rauch[1,3], and Y. Hasegawa[1]

[1]Atominstitut, Vienna University of Technology, Stadionallee 2, 1020 Vienna, Austria
[2] Faculty of Physics, University of Vienna, Boltzmanngasse 5, 1090 Wien, Austria
[3] Institut Laue-Langevin, B.P. 156, 38042 Grenoble Cedex 9, France

E-mail: sponar@ati.ac.at

Abstract. Entanglement is a remarkable peculiarity in quantum mechanics. It occurs in quantum systems that consist of space-like separated parts, or in systems whose observables belong to disjoint Hilbert spaces. The latter is the case in single-neutron systems. Entangled states are renowned for exhibiting non-classical correlations between observables of individual sub-systems. In a perfect Si-crystal interferometer experiment entanglement between three degrees of freedom in a single-neutron system is created. The prepared entanglement of spin, path and energy is induced by interaction with an oscillating magnetic field. The generated Greenberger-Horne-Zeilinger (GHZ) state is analyzed with an inequality derived by Mermin, yielding a value $M = 2.558(4) \nleq 2$, which exhibits a clear violation of the classical assumption. In addition observation of a GHZ entanglement, consisting of spin, momentum and total energy, in a neutron polarimetric experiment is presented. Here the advantages of neutron polarimetry, such as high contrast or insensitivity to ambient disturbances, are utilized resulting in final value of $M = 3.936(2) \nleq 2$.

1. Introduction

Quantum mechanics is probably the most successful physical theory ever. No other theory has yet given more accurate predictions from elementary-particle physics to the early stages of our universe. However, quantum mechanics only gives probabilistic predictions for individual events. Consequently, according to Einstein, a more complete, deterministic, hidden theory must underlie quantum mechanics [1]. In 1951 Bohm reformulated the EPR argument for spin observables of two spatially separated entangled particles to illuminate the essential features of the EPR paradox [2]. After this, Bell proved in his celebrated theorem that all hidden variable theories, which are based on the joint assumption of locality and realism, conflict with the predictions of quantum mechanics [3]. Local hidden variable theories (LHVTs) assume that the outcome of a measurement is predetermined and independent of spacelike separated measurements. Therefore, Bell introduced inequalities which hold for the predictions of any LHVT considered, but are violated by quantum mechanics. Violation of a Bell inequality proves the presence of entanglement and thus, according to Bells theorem, non-local characteristics of the quantum systems. From this, one can conclude that quantum mechanics cannot be reproduced by local hidden variable theories. Only five years later Clauser, Horne, Shimony and Holt (CHSH) reformulated Bells inequalities pertinent for the first practical test of quantum non-locality [4]. Polarization measurements with correlated photon pairs [5], produced by atomic cascade [6, 7] and parametric down-conversion of lasers [8, 9, 10], demonstrated violation of the CHSH inequality. Up to date several systems [11, 12, 13, 14] have been examined, including neutrons [15].

Not a statistical violation, but a contradiction between quantum mechanics and local hidden variable theories was found by Greenberger, Horne and Zeilinger (GHZ) in 1989 for tripartite entanglement [16, 17]. The GHZ argument is independent of the Bell approach, thereby demonstrating in a nonstatistic manner that quantum mechanics and local realism are mutually incompatible. Several experimental realizations, using multipartite entanglement have been achieved: among them are photon polarization experiments [18, 19, 20, 21], atoms [22] and trapped ions [23]. The GHZ argument was analyzed in detail by Mermin [24], where a inequality is derived for a state of n spin-1/2 particles that is violated by quantum mechanic by an amount that increases exponentially with n.

LHVTs are a subset of a more general class of hidden variable theories, namely the noncontextual hidden variable theories. Noncontextuality implies that the value of a dynamical variable is determined and independent of the experimental context, i.e. of previous or simultaneous measurements of a commuting observable [25, 26]. Noncontextuality is a more stringent demand than locality because it requires mutual independence of the results for commuting observables even if there is no spacelike separation [27]. First tests of quantum contextuality, based on the Kochen–Specker theorem [28], have been proposed [29, 30] and performed successfully using trapped ions [31], photons [32] and neutrons [33, 34].

In the case of neutrons entanglement is not achieved between particles but between different degrees of freedom. Since the observables of one Hilbert spaces (describing a certain degree of freedom) commute with observables of a different Hilbert space, the single-neutron system is suitable for studying noncontextual hidden variable theories with multiple degrees of freedom.

2. Interferometric Experiment

Neutron interferometry [35, 36] has become a unique tool for studies on the quantum-mechanical nature of matter waves. A beam of massive particles is coherently split by amplitude division and recombined after passing through different regions of space. During these space-like separation of typically a few centimeters, the neutron's wavefunction is modified in phases (and amplitude), due to various possible interactions. This interaction may be of nuclear, magnetic, electric or gravitational type, depending on which neutron optical devices are inserted. The topic of neutron interferometry belongs to the field of self-interference, since at a given instant at most one neutron propagates through the interferometer. Using neutron interferometry (IFM) single-particle entanglement, between the spinor and the spatial part of the neutron wave function [15], as well as full tomographic state analysis [37], have already been accomplished.

2.1. Coherent Energy Manipulation

The neutron's total energy degree of freedom is an almost ideal candidate for a degree of freedom to be manipulated, due to its experimental accessibility within a magnetic resonance field [38]. For this purpose the time evolution of the system is described by a photon-neutron state vector, which is an eigenvector of the corresponding modified Jaynes-Cummings (J-C) Hamiltonian [39]. The J-C Hamiltonian can be adopted for a system consisting of a neutron coupled to a quantized rf-field [40]. Manipulating the total energy provides realization of triple-entanglement between the neutron's path, spin and energy degrees of freedom.

2.1.1. The corresponding neutron Jaynes-Cummings Hamiltonian

Since two RF-fields, operating at frequencies ω and $\omega/2$, are involved in the actual experiment, the modified J-C Hamiltonian is denoted as

$$\begin{aligned}\mathcal{H}_{\text{J-C}} &= -\frac{\hbar^2}{2m}\nabla^2 - \mu B_0(\vec{r})\sigma_z + \hbar(\omega\hat{a}^\dagger_\omega\hat{a}_\omega + \frac{\omega}{2}\hat{a}^\dagger_{\omega/2}\hat{a}_{\omega/2}) \\ &+ \mu\left(\frac{B_1^{(\omega)}(\vec{r})}{\sqrt{N_\omega}}(\hat{a}^\dagger_\omega\sigma_+ + \hat{a}_\omega\sigma_+) + \frac{B_1^{(\omega/2)}(\vec{r})}{\sqrt{N_{\omega/2}}}(\hat{a}^\dagger_{\omega/2}\sigma_+ + \hat{a}_{\omega/2}\sigma_+)\right), \end{aligned} \quad (1)$$

where m and μ are the mass ($1.6749\ 10^{-27}$ kg) and the magnetic moment ($-1.913\mu_\text{N}$, with $\mu_\text{N} = 5.051\ 10^{-27}$ J/T) of the neutron, respectively and $\sigma_\pm = \frac{1}{2}(\sigma_x \pm i\sigma_y)$. The first term accounts for the kinetic energy of the neutron. The second term leads to the usual Zeeman splitting of $2|\mu|B_0$. The third term adds the photon energy of the oscillating fields of frequencies ω and $\omega/2$, by use of the creation and annihilation operators $a^\dagger$ and a. Finally, the last term represents the coupling between photons and the neutron, where $N_{\omega_j} = \langle\hat{a}^\dagger_{\omega_j}\hat{a}_{\omega_j}\rangle$ represents the mean number of photons with frequencies ω_j in the RF-field. Note that the first two and the last terms concern the spatial $|\psi(\vec{r})\rangle$ and the (time-dependent) energy $|E(t)\rangle$ subspaces of neutrons, respectively.

The state vectors of the oscillating fields are represented by coherent states $|\alpha\rangle$, which are eigenstates of $a^\dagger$ and a. The eigenvalues of coherent states are complex numbers, so one can write

$$\hat{a}|\alpha\rangle = \alpha|\alpha\rangle = |\alpha|e^{i\phi}|\alpha\rangle \text{ with } |\alpha| = \sqrt{N}. \quad (2)$$

Using Eq. (1) one can define a total state vector including not only the neutron system $|\Psi_\text{N}\rangle$, but also the two quantized oscillating magnetic fields:

$$|\Psi_\text{i}\rangle = |\alpha_\omega\rangle \otimes |\alpha_{\omega/2}\rangle \otimes |\Psi_\text{N}\rangle. \quad (3)$$

In a perfect Si-crystal neutron interferometer the wavefunction behind the first plate (acting as a beam splitter) is a linear superposition of the sub-beams belonging to the right ($|\text{I}\rangle$) and the left path ($|\text{II}\rangle$), which are laterally separated by several centimeters. The sub-beams are recombined at the third crystal plate and the wave function in forward direction then reads as $|\Psi_\text{N}\rangle \propto |\Psi_\text{N}^{(\text{I})}\rangle + |\Psi_\text{N}^{(\text{II})}\rangle$, where $|\Psi_\text{N}^{(\text{I})}\rangle$ and $|\Psi_\text{N}^{(\text{II})}\rangle$ only differ by an adjustable phase factor $e^{i\chi}$ ($\chi = N_\text{ps}b_c\lambda D$, with the thickness of the phase shifter plate D, the neutron wavelength λ, the coherent scattering length b_c and the atom number density N_ps in the phase shifter plate). By rotating the plate, χ can be varied systematically. This yields the well known intensity oscillations of the two beams emerging behind the interferometer, usually denoted as O- and H-beam [36]. A sketch of the setup is depicted in Fig. 1.

In our experiment, only the beam in path II is exposed to the rf-field of frequency ω, resulting in a spin flip process. The spin flip configuration of the first rf-field ensures entanglement of spin and spatial degree of freedom of the neutron state [15]. Interacting with a time-dependent magnetic field, the total energy of the neutron is no longer conserved after the spin-flip [41, 42, 43, 44, 45]. Photons of energy $\hbar\omega$ are exchanged with the rf-field. This particular behavior of the neutron is described by the 'dressed-particle' formalism [40, 46]. Consequently the two sub-beams $|\text{I}\rangle$ and $|\text{II}\rangle$ now differ in total energy (see Fig. 1(b)). Therefore the neutron state can be considered to consist of the three subsystems, namely the total energy, path and spin degree of freedom. A coherent superposition of $|\text{I}\rangle$ and $|\text{II}\rangle$ results in the multiply entangled state vector, expressed as

$$|\Psi(t)\rangle \propto |\alpha_\omega\rangle \otimes |\alpha_{\omega/2}\rangle \otimes \frac{1}{\sqrt{2}}\Big(|\text{I}\rangle \otimes |\Uparrow\rangle + e^{i\omega t}e^{i\chi}|\text{II}\rangle \otimes e^{i\phi_\omega}|\Downarrow\rangle\Big). \quad (4)$$

According to the arguments above Eq.(4) can be rewritten as

$$|\Psi(t)\rangle \propto |\alpha_\omega\rangle \otimes |\alpha_{\omega/2}\rangle \otimes \frac{1}{\sqrt{2}}\Big(|\mathrm{I}\rangle \otimes |E_0\rangle \otimes |\Uparrow\rangle + e^{i\chi}|\mathrm{II}\rangle \otimes |E_0 - \hbar\omega\rangle \otimes e^{i\phi_\omega}|\Downarrow\rangle\Big) \tag{5}$$

The spin in path $|\mathrm{II}\rangle$ is flipped by a RF-flipper, which requires two magnetic fields: A static field $B_0 \cdot \hat{\mathbf{z}}$ and a perpendicular oscillating field $B^{(1)} = B_{\mathrm{rf}}^{(\omega)} \cos(\omega t + \phi_\omega) \cdot \hat{\mathbf{y}}$ satisfying the amplitude

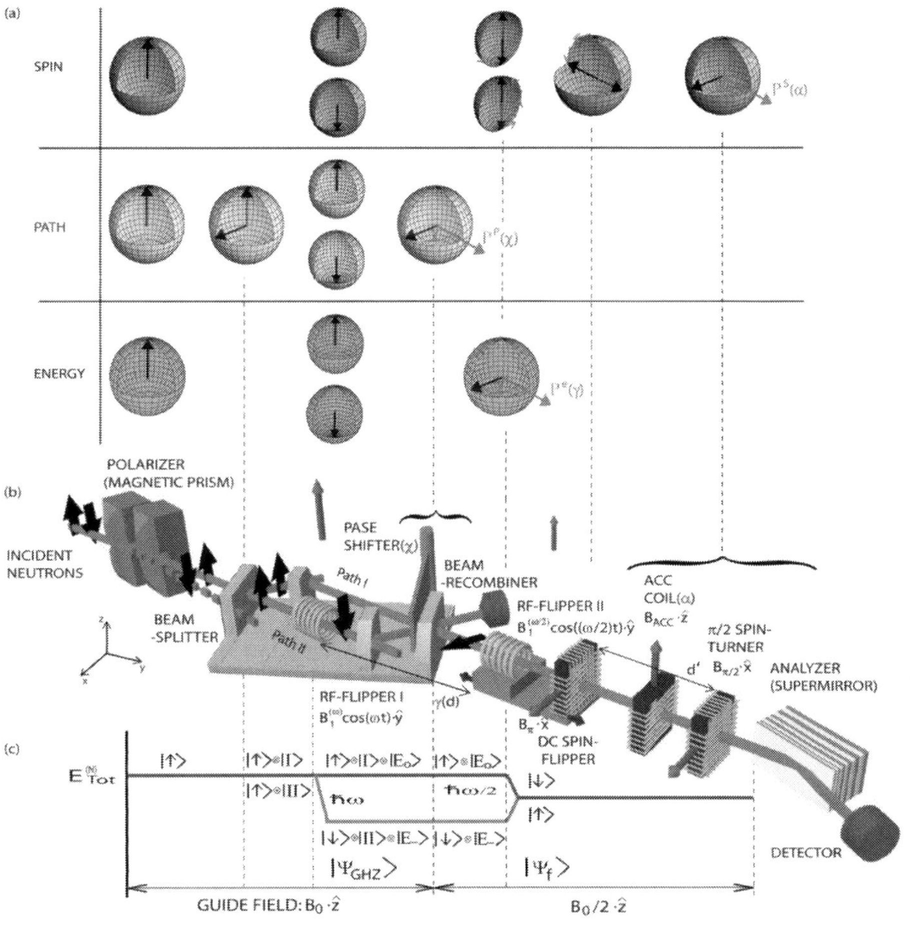

Figure 1. (b) Schematic view of the experimental setup for stationary observation of interference between two rf-fields. Showing the arrangement of two radio-frequency flip coils (the first within one path of the skew-symmetric Mach-Zehnder-type neutron interferometer and the other driven by half the frequency behind the interferometer), accelerator coil and $\pi/2$ spin-turner (the purpose of the DC spin-flipper is explained in Sec. 2.2.1). (c) Energy level diagram of the two interfering sub-beams $|\mathrm{I}\rangle$, $|\mathrm{II}\rangle$ during their passage through the different static field regions (B_0,$\mathrm{B}_0/2$), including corresponding energy eigenstates $|E_0\rangle$, $|E_-\rangle := |E_0 - \hbar\omega\rangle$ and spin states $|\uparrow\rangle$, $|\downarrow\rangle$ and taking into account the spin flips at rf-frequencies ω and $\omega/2$. (a) Bloch-sphere descriptions to depict evolutions of each quantum state (i.e., spin, path and energy degrees of freedom) in the GHZ measurement. The directions of the projective measurements P^j are depicted by thick red arrows in Bloch spheres.

and frequency resonance condition

$$B_1^{(\omega)} = \frac{\pi\hbar}{\tau|\mu|} \text{ and } \omega = \frac{2|\mu|B_0}{\hbar}\Big(1 + \frac{B_1^2}{16B_0^2}\Big), \tag{6}$$

where μ is the magnetic moment of the neutron and τ denotes the time the neutron is exposed to the RF-field. The second term in ω is due to the Bloch-Siegert shift [47]. The oscillating field is produced by water-cooled RF-coil, operating at a frequency of $\omega/2\pi = 58\,\mathrm{kHz}$. The static field is provided by a uniform magnetic guide field $B_0^{(\omega)} \sim 2\,\mathrm{mT}$, produced by a pair of water-cooled Helmholtz coils. The state vector of the neutron acquires a phase $\pm\phi_\omega$ during the interaction with the oscillating field ($B(t) = B_1\cos(\omega t + \phi_\omega)$), induced by the action of the operators $\hat{a}_\omega$ and $\hat{a}_\omega^\dagger$ in the last term of Eq. (1). The neutron part of the total state vector in Eq.(5) represents a path-energy-spin entanglement within a single-neutron system.

The two sub-beams are recombined at the third plate, which is described by the projection operator $\hat{O}^{(\mathrm{P})} = \frac{1}{2}(|\mathrm{I}\rangle + |\mathrm{II}\rangle)(\langle\mathrm{I}| + \langle\mathrm{II}|)$, resulting in a time-dependent state vector due to the different energies of the two partial wavefunctions. Due to the orthogonality of the energy and spin eigenstates the measured polarization is zero and no intensity modulations are observed in the H-beam:

$$\vec{P}_{\mathrm{H}}(t) = \Big(\cos(\chi + \pi - \omega t - \phi_\omega), \sin(\chi + \pi - \omega t - \phi_\omega), 0\Big). \tag{7}$$

This behavior [38] is related to the spinor precession known from zero-field spin-echo experiments [42, 43].

The beam recombination of the O-beam is followed by an interaction with the second RF-field, with half frequency $\omega/2$. Here the operator

$$\hat{O}^{(\mathrm{E})} = \frac{1}{\sqrt{2}}|E_0 - \hbar\omega/2\rangle(\langle E_0| + (\langle E_0 - \hbar\omega|), \tag{8}$$

represents mathematically the energy transfer and the total state vector is given by

$$\begin{aligned} |\Psi_{\mathrm{f}}\rangle \;\propto\; & |\alpha_\omega\rangle \otimes |\alpha_{\omega/2}\rangle \otimes (|\mathrm{I}\rangle + |\mathrm{II}\rangle) \otimes |E_0 - \hbar\omega/2\rangle \\ \otimes\; & \frac{1}{\sqrt{2}}\Big(e^{i\phi_{\omega/2}}|\Downarrow\rangle + e^{-i\omega T}e^{i\chi}e^{i(\phi_\omega - \phi_{\omega/2})}|\Uparrow\rangle\Big). \end{aligned} \tag{9}$$

where ϕ_ω and $\phi_{\omega/2}$ are the phases induced by the two RF-fields. The second RF-flipper is operating at $\omega/2\pi = 29\,\mathrm{kHz}$, which is half the frequency of the first RF-flipper. The oscillating field is denoted as $B_1^{(\omega/2)}\cos((\omega/2)t + \phi_{\omega/2})\cdot\hat{\mathbf{y}}$, and the strength of the guide field was tuned to $B_0^{(\omega/2)} \sim 1\,\mathrm{mT}$ in order to satisfy the frequency resonance condition. By choosing a frequency of $\omega/2$ for the second RF-flipper, the time-dependence of the state vector is eliminated since both components acquire a phase $e^{\pm i\omega/2(t+T)}$, depending on the spin orientation. Only a constant phase offset of $e^{-i\omega T}$ remains, which is referred to as *zero-field phase*, with T being the neutron's propagation time between the centre of the first and second RF-flipper coil, remains in the stationary state vector [48]. Hence the second RF -flipper compensates the energy difference between the two spin components, by absorption and emission of photons of energy $E = \hbar\omega/2$. Thus, the energy difference between the orthogonal spin states is compensated by choosing a frequency of $\omega/2$ for the second RF-flipper, resulting in a stationary state vector, which is graphically illustrated in (see Fig. 1(c)). Hence also the time dependence of the polarization vector is eliminated:

$$\vec{P}_{\mathrm{O}}^{\mathrm{fin}} = (\cos\Delta_{\mathrm{tot}}, \sin\Delta_{\mathrm{tot}}, 0), \tag{10}$$

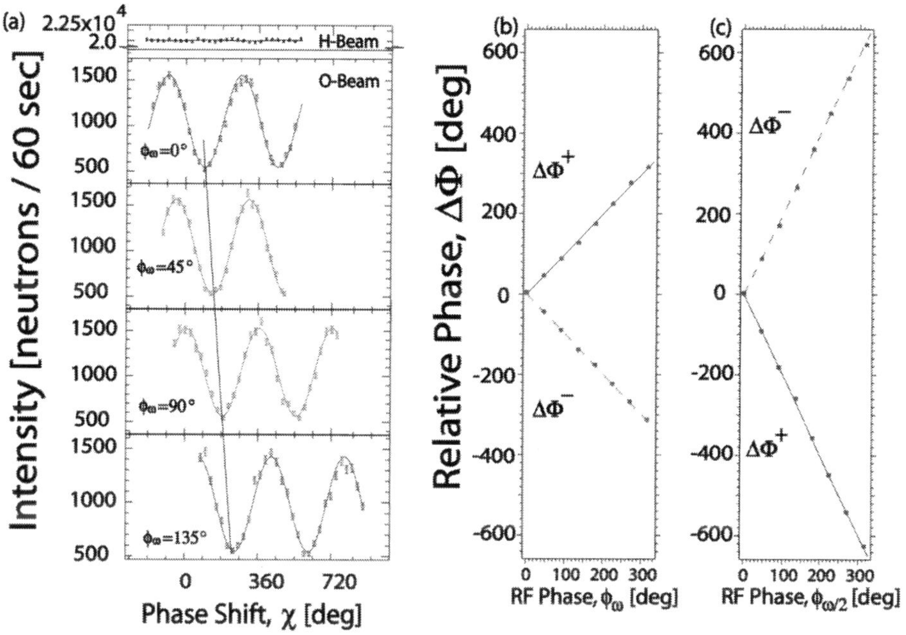

Figure 2. (a) Typical interference patterns of the H- and the O-beam. In the H-beam no interference fringes are observed due to orthogonal spin states in the interfering sub-beams, whereas the O-beam exhibits time-independent sinusoidal intensity oscillations, when rotating the phase shifter plate (χ). A phase shift occurs by varying ϕ_ω. Here $\phi_{\omega/2}$ is keept constant at $\phi_{\omega/2} = 0$. (b) Relative phase $\Delta\Phi^\pm$ vs. ϕ_ω, and (c) $\Delta\Phi^\pm$ vs. $\phi_{\omega/2}$. The sign of the phase depends on the chosen initial polarization. An additional tunable accelerator coil compensates the phase contributions of the constant zero field phase ωT and the dynamical phase accumulated by Larmor precession.

with

$$\Delta_{\text{tot}} = (\chi - 2\phi_{\omega/2} + \phi_\omega - \omega T), \tag{11}$$

consisting of the phases induced the path (phase shifter χ), spin (phases of the two RF-fields $\phi_\omega, \phi_{\omega/2}$), and energy manipulation (zero-field phase ωT).

Finally, the spin is rotated back by an angle $\delta = \pi/2$ (in the $\hat{\mathbf{x}}, \hat{\mathbf{z}}$ plane) to the $\hat{\mathbf{z}}$-direction by use of a $\pi/2$ static field spin-turner. Due to a spin dependent reflection within a Co-Ti multilayer supermirror, the spin is analyzed along the $\hat{\mathbf{z}}$-direction, denoted as a projection operator $\hat{O}^{(S)} = |\Uparrow\rangle\langle\Uparrow|$.

The interference oscillations can be written in form of

$$I_0 \propto 1 + C\cos(\chi + \Phi - \omega T), \tag{12}$$

introducing the contrast C and the relative (spin) phase $\Phi = \phi_\omega - 2\phi_{\omega/2}$. Typical interference patterns, when scanning χ, for different settings of ϕ_ω, are depicted in Fig. 2 (a). In the O-beam an average fringe contrast of $C = 52.4(2)\,\%$ is achieved, whereas no interference fringes are observed in the H-detector, where no further manipulations are applied. However, even for $\phi_\omega = \phi_{\omega/2} = 0$ the relative phase Φ exhibits an offset, as seen in Fig. 2 (a). This is due to the zero-field phase contribution ωT, but also due to dynamical phase contribution, resulting from

Larmor precession within the guide field regions $B_0^{(\omega)} \cdot \hat{\mathbf{z}}$ and $B_0^{(\omega/2)} \cdot \hat{\mathbf{z}}$, denoted as $\delta_{\mathrm{GF\,I}}$ and $\delta_{\mathrm{GF\,II}}$, which have not been mention yet. These contributions are compensated by an additional Larmor precession within a tunable accelerator coil with a static field, pointing in the $\hat{\mathbf{z}}$-direction. Thus by choosing

$$\delta_{\mathrm{ACC}} = -(\delta_{\mathrm{GF\,I}} + \delta_{\mathrm{GF\,II}} - \omega T) \tag{13}$$

no more offset is observed for $\phi_\omega = \phi_{\omega/2} = 0$, as shown in Fig. 2 (b) and (c).

2.1.2. Phase measurement results

The experiment was carried out at the neutron interferometer instrument S18 at the high-flux reactor of the Institute Laue-Langevin in Grenoble, France. Here a silicon perfect-crystal monochromator is installed permanently in the neutron guide to monochromatize the incident neutron beam to a mean wave length of $\lambda_0 = 1.91$ Å with the monochromaticity$\Delta\lambda/\lambda_0 \sim 0.015$. The cross section of the beam is roughly 5x5 mm^2. Birefringent magnetic field prisms are used to polarize the incident beam in $\hat{\mathbf{z}}$-direction, before the beam enters the interferometric setup.The angular separation can be used such that only the spin-up (or spin-down) component fulfills the Bragg-condition at the first interferometer plate (beam splitter).

It is possible to invert the initial polarization simply by rotating the interferometer by a few seconds of arc, thereby selecting the spin-down component to enter the interferometer, which is expected to lead to an inversion of the relative phase. Thus a relative phase difference $\Delta\Phi^\pm = \pm\phi_\omega \mp 2\phi_{\omega/2}$, where $\pm$ denotes the respective initial spin orientation is observed. Figure 2 (a) shows a plot of the relative phase $\Delta\Phi^\pm$ versus ϕ_ω, with $\phi_{\omega/2} = 0$, and a phase shift $\Delta\Phi^\pm$ caused by a variation of ϕ_ω. As expected, the slope is positive for initial spin up orientation (1.007(8)), and negative for the spin down case (-0.997(5)). In Fig. 2 (b) $\phi_{\omega/2}$ is varied, while ϕ_ω is kept constant, yielding slopes of -1.995(8) and 1.985(7) (since $\Delta\Phi$ is shifted by $-2\phi_{\omega/2}$,), depending again on the initial beam polarization.

The arrangement of two RF-flippers of frequencies ω and $\omega/2$ can be interpreted as an interferometer-scheme for the neutron's total energy. Due to energy splitting the first RF-flipper generates a superposition of two coherent energy states, similar to the action of the first beam-splitter of a Mach-Zehnder interferometer, where a single beam is split spatially into two coherent sub-beams. The second flipper compensates the energy difference and therefore acts as a beam analyzer equivalent to the last beam-splitter of the interferometer. The results of the experiment discussed in this Section can be found in more detail in [48].

2.2. GHZ-States in Neutron Interferometry

The neutron part of the multi entangled statevector given by

$$|\Psi_{\mathrm{N}}\rangle \propto \frac{1}{\sqrt{2}}\Big(|\mathrm{I}\rangle \otimes |E_0\rangle \otimes |\Uparrow\rangle + |\mathrm{II}\rangle \otimes |E_0 - \hbar\omega\rangle \otimes |\Downarrow\rangle\Big), \tag{14}$$

denotes a Greenberger-Horne-Zeilinger (GHZ)-like state [16, 17]. A contradiction between quantum mechanics and local hidden variable theories for the GHZ state is found only for perfect situations (which cannot be realized experimentally), an inequality is used to demonstrate the peculiarities of the GHZ state. Mermin analyzed the GHZ argument in detail and derived an inequality pertinent for an experimental test of local hidden variable theories [24]. Since our experiment, GHZ-like entanglement exists between degrees of freedom in a single-neutron system we test noncontextual hidden variable theories using a sum of four expectation values defined as

$$M = E(\sigma_x^{(\mathrm{P})}\sigma_x^{(\mathrm{E})}\sigma_x^{(\mathrm{S})}) - E(\sigma_x^{(\mathrm{P})}\sigma_y^{(\mathrm{E})}\sigma_y^{(\mathrm{S})}) - E(\sigma_y^{(\mathrm{P})}\sigma_x^{(\mathrm{E})}\sigma_y^{(\mathrm{S})}) - E(\sigma_y^{(\mathrm{P})}\sigma_y^{(\mathrm{E})}\sigma_x^{(\mathrm{S})}), \tag{15}$$

where $\sigma_{x,y}^{(i)}$ are the Pauli operators in path, total energy and spin subspace. The value of M is bounded by 2 for any noncontextual hidden variable theorie, whereas quantum mechanics predicts an upper limit of 4 for the GHZ state. The Pauli operators used in Eq.(15) can be decomposed as

$$\begin{aligned} \sigma_x^{(i)} &= \hat{P}^{(i)}(0) - \hat{P}^{(i)}(\pi) \\ \sigma_y^{(i)} &= \hat{P}^{(i)}(\pi/2) - \hat{P}^{(i)}(3\pi/2), \end{aligned} \tag{16}$$

with $\hat{P}(\alpha)^{(\mathrm{S})}$, $\hat{P}(\beta)^{(\mathrm{k})}$and $\hat{P}(\gamma)^{(\mathrm{E})}$ being the projection operators onto an up-down superposition on the equatorial plane in path, total energy and spin Bloch sphere. The azimuthal angle is given by an angle parameter χ, γ and α, respectively. The operators are defined as

$$\begin{aligned} \hat{P}^{(\mathrm{P})}(\chi) &= \frac{1}{\sqrt{2}}(|\mathrm{I}\rangle + e^{-i\chi}|\mathrm{II}\rangle)(\langle\mathrm{I}| + e^{i\chi}\langle\mathrm{II}|) \\ \hat{P}^{(\mathrm{E})}(\gamma) &= \frac{1}{\sqrt{2}}(|E_0\rangle + e^{-i\gamma}|E_0 - \hbar\omega\rangle)(\langle E_0| + e^{i\gamma}\langle E_0 - \hbar\omega|) \\ \hat{P}^{(\mathrm{S})}(\alpha) &= \frac{1}{\sqrt{2}}(|\Uparrow\rangle + e^{-i\alpha}|\Downarrow\rangle)(\langle\Uparrow| + e^{i\alpha}\langle\Downarrow|), \end{aligned} \tag{17}$$

where χ, γ and α are the azimuthal angles on the Bloch spheres depicted, having only the values 0 and π or $\pi/2$ and $3\pi/2$, sine only σ_x and σ_y occur in Eq.(15) (see Fig. 1). Each expectation Value $E(\sigma_{x,y}^{(\mathrm{P})}\sigma_{x,y}^{(\mathrm{E})}\sigma_{x,y}^{(\mathrm{S})})$ is experimentally determined by a combination of normalized count rates, using appropriate setting of χ, γ and α. For instance

$$\begin{aligned} E(\sigma_x^{(\mathrm{P})}\sigma_y^{(\mathrm{E})}\sigma_y^{(\mathrm{S})}) = E(\alpha:(0;\pi), \beta:(\tfrac{\pi}{2};\tfrac{3\pi}{2}), \gamma:(\tfrac{\pi}{2};\tfrac{3\pi}{2})) \\ = \langle\Psi_{\mathrm{GHZ}}|(\hat{P}(0)^{(\mathrm{P})} - \hat{P}(\pi)^{(\mathrm{P})}) \otimes (\hat{P}(\tfrac{\pi}{2})^{(\mathrm{E})} - \hat{P}(\tfrac{3\pi}{2})^{(\mathrm{E})}) \otimes (\hat{P}(\tfrac{\pi}{2})^{(\mathrm{S})} - \hat{P}(\tfrac{3\pi}{2})^{(\mathrm{S})})|\Psi_{\mathrm{GHZ}}\rangle = \tfrac{A}{B} \end{aligned} \tag{18}$$

with

$$\begin{aligned} A &= \Big(N(0,\frac{\pi}{2},\frac{\pi}{2}) - N(\pi,\frac{\pi}{2},\frac{\pi}{2}) - N(0,\frac{\pi}{2},\frac{3\pi}{2}) + N(\pi,\frac{\pi}{2},\frac{3\pi}{2}) \\ &- (N(0,\frac{3\pi}{2},\frac{\pi}{2}) - N(\pi,\frac{3\pi}{2},\frac{\pi}{2}) + N(0,\frac{3\pi}{2},\frac{3\pi}{2}) + N(\pi,\frac{3\pi}{2},\frac{3\pi}{2})\Big) \end{aligned} \tag{19}$$

and

$$\begin{aligned} B &= \Big(N(0,\frac{\pi}{2},\frac{\pi}{2}) + N(\pi,\frac{\pi}{2},\frac{\pi}{2}) + N(0,\frac{\pi}{2},\frac{3\pi}{2}) + N(\pi,\frac{\pi}{2},\frac{3\pi}{2}) \\ &+ N(0,\frac{3\pi}{2},\frac{\pi}{2}) - N(\pi,\frac{3\pi}{2},\frac{\pi}{2}) + N(0,\frac{3\pi}{2},\frac{3\pi}{2}) + N(\pi,\frac{3\pi}{2},\frac{3\pi}{2})\Big), \end{aligned} \tag{20}$$

where for example $N(0,\pi/2,\pi/2)$ is the count rate for $\chi = 0$, $\gamma = \pi/2$ and $\alpha = \pi/2$.

2.2.1. Measurement procedure

As seen from Δ_{tot} in Eq.(11) in principal each of the three degrees of freedom can be manipulated independently and the associated observables are separately measurable. A schematic illustration of the setup, together with a Bloch sphere description of the associated measurement directions os path, spin and total energy are depicted in Fig. 1. The projective measurements of each degree of freedom are carried out exploiting to phase manipulations between the eigenstates of the individual sub systems, resulting in the desired azimuthal angle on the Bloch sphere.

Path phase: The phase manipulation of the path subspace is accomplished with the auxiliary phase shifter made of a parallel-sided Si plate. Typical oscillations when rotating the phase shifter slab are depicted in Fig. 3.

Spin phase: In this experiment the tunable accelerator coil was used to tune the spin phase α. In particular the spin phase is accumulated by Larmor precession within the static magnetic field of the accelerator coil ($B_{\rm ACC} \cdot \hat{z}$) with $\alpha = \omega_L \tau_{\rm ACC}$, where ω_L is the Larmor frequency depending on $B_{\rm ACC}$ and $\tau_{\rm ACC}$ is the propagation time through the accelerator coil.

Energy phase: As explained above the manipulation of the energy phase is achieved by zero-field precession between the two RF-flippers. The first RF spin-flipper induces the energy difference $\hbar\omega$, which is balanced by the second RF flipper by choosing a frequency of $\omega/2$, resulting in the zero field phase difference $\gamma = \omega T$. Here T can by varied, since the second RF coil is mounted on a translation stage. This displacement of the second RF-flipper is the crucial point in this experiment, since by increasing the distance between the RF-flippers not only the zero field phase $\gamma = \omega T$ is changed, but also the Larmor precession angle within the static guide field B_0 which induces an additional undesired spin phase contribution $\alpha' = \omega_{\rm L(B_0)}$. However the aim is to address the zero field precession independently from Larmor precession. This is achieved by an auxiliary DC-flipper, which is mounted on the same translation stage subsequently to the second RF-flipper. Thus no additional phase shift, induced by Larmor precession, resulting from the change of Δd is observed. Phase contributions with the same sign occur in the regions between first and second RF-flippers ($d + \Delta d$) and between DC-flipper and the DC $\pi/2$ spin-rotator (denoted $d' - \Delta d$), compensating each other. Thus the total Larmor precession angle remains constant ($(d + d')\omega_{\rm L(B_0)}$), thought changing the position (Δd) of the second RF-flipper. The displacement only affects the zero field precession angle $\gamma = \omega T \propto \omega(d + \Delta d)$. To demonstrate this individual tuning of Larmor and zero field precession an independent polarimetric experiment was carried out [49].

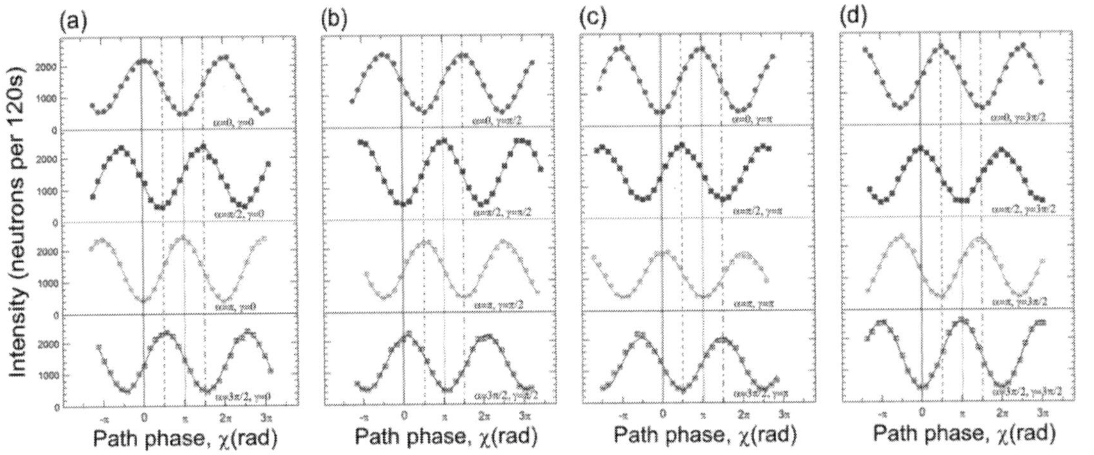

Figure 3. Typical interference patterns of the O-detector obtained by varying the path phase χ. The phases α and γ, for the spin and the energy, respectively, are tuned at 0, $\pi/2$, π and $3\pi/2$ in order to accomplish project measurements associated with $\hat{P}^j(0)$, $\hat{P}^j(\pi/2)$, $\hat{P}^j(\pi)$ and $\hat{P}^j(3\pi/2)$, with $j =$ spin, path and energy.

Table 1. The four experimentally determined expectation values and the final M value.

Observable	Settings χ	Settings γ	Settings α	Determined Values
$\sigma_x^{(P)}\sigma_x^{(E)}\sigma_x^{(S)}$	$(0;\pi)$	$(0;\pi)$	$(0;\pi)$	0.659(2)
$\sigma_x^{(P)}\sigma_y^{(E)}\sigma_y^{(S)}$	$(0;\pi)$	$(\pi/2;3\pi/2)$	$(\pi/2;3\pi/2)$	-0.603(2)
$\sigma_y^{(P)}\sigma_x^{(E)}\sigma_y^{(S)}$	$(\pi/2;3\pi/2)$	$(0;\pi)$	$(\pi/2;3\pi/2)$	-0.664(2)
$\sigma_y^{(P)}\sigma_y^{(E)}\sigma_x^{(S)}$	$(\pi/2;3\pi/2)$	$(\pi/2;3\pi/2)$	$(0;\pi)$	-0.632(2)

$M = 2.558(4)$

2.2.2. Experimental results

For the required projective measurements spin phase α and energy phase γ each was tuned at 0, $\pi/2$ and $3\pi/2$, while path phase scans χ (i.e., oscillation measurements) were performed. The resulting sixteen oscillations are depicted in Fig. 3, the dashed lines denote the values $\chi = 0$, $\pi/2$ and $3\pi/2$, which are required for the determination of M, as defined in Eq.(15). The average contrast of the oscillations were just below 70 per cent, which is clearly above the threshold visibility of 50 per cent, required for a violation of the Mermin like inequality.

Measured intensity oscillations were fitted to sinusoidal curves by applying a least squares method. The four expectation values, as defined in Eq.(15), were extracted from the fit curves. Statistical errors were estimated to be about ± 0.001, taking all fit errors from single measurement curves into account. One set of measurements consists of thirty-two oscillation measurements, since intensities with and without spin flipper (reference measurement) were recorded. This was done in order to allow corrections of the path phase χ instability, afterward. Four sets of thirty-two oscillations to reduce statistical errors. A final result of $M = 2.558 \pm 0.004$, exhibiting a clear violation of the classical limit $M = 2$ of the noncontextual border. The individual results of the four expectation values, together with settings of variables and the final value of M, are listed in Tab. 1. A more detailed discussion of this measurement is given in [50].

3. Polarimetric Experiment

Neutron polarimetry has several advantages compared to perfect crystal interferometry. It is insensitive to ambient mechanical and thermal disturbances and therefore provides better phase stability. Efficiencies of the manipulations, including state splitting and recombination, are considerably high (typically $> 98\,\%$) resulting in a better contrast compared to interferometry. In addition, while single-crystal interferometers accept neutrons only within an angular range of a few arc seconds, which leads to a significant decrease in intensity, polarimeters accept beams with a broad momentum distribution. Neutron polarimetry has been used to demonstrate fundamental quantum-mechanical properties. Just to mention the noncommutation of the Pauli spin operator [51] and a number of geometric phase measurements [52, 53, 54]. In recent experiments a test of an alternative model for non-local correlations [55], as well as demonstration of a universally valid uncertainty relation [56] has been performed successfully.

3.1. GHZ-States in Neutron Polarimetry

In this experiment we report on a correlation measurement in the regime of neutron polarimetry, where three independent degrees of freedom, namely total-energy, momentum and spin, are utilized and manipulated separately. The motion of a free neutron propagating in $+y$-direction, with polarization in arbitrary direction, in a static magnetic field with a field gradient parallel to its propagation direction (with $\vec{B}(y) = 0, \forall\, y > 0$ and $\vec{B}(y) = B_0 \cdot \vec{e}_z = B_z(y), \forall\, y > 0$) is described by a nonrelativistic Schrödinger equation, also referred to as Pauli equation given by

$$\Big(-\frac{\hbar^2}{2m}\vec{\nabla}^2 - \mu\sigma_z B_z(y)\Big)\Psi(y,t) = i\hbar\frac{\partial}{\partial t}\Psi(y,t). \tag{21}$$

Zeeman splitting within $\vec{B}(y)$ leads to the following solutions

$$\Psi(y,t) = \phi_+(y,t)\cos\frac{\vartheta}{2}|\Uparrow\rangle + \phi_-(y,t)e^{i\alpha}\sin\frac{\vartheta}{2}|\Downarrow\rangle. \tag{22}$$

$|\Uparrow\rangle$ and $|\Downarrow\rangle$ denote the spin-up and spin-down eigenstates referring to a quantization axis along the static magnetic field (i.e. $\vec{e}_z$), and $\phi(y,t)_\pm$ describe the wave packets of the up and down spin components

$$\phi(y,t)_\pm = \int a_\pm(k\mp\Delta k)e^{i\big((k\mp\Delta k)\cdot y-\omega(k)t\big)}, \tag{23}$$

where k is the momentum of the free particle and

$$k_\pm \simeq k \mp \frac{m|\mu|B_z(y)}{(\hbar^2 k)} = k\mp\Delta k. \tag{24}$$

The polarization vector is given by $\vec{P} = \langle\Psi|\vec{\sigma}|\Psi\rangle$, where the azimuthal angle α, being the relative phase between $|\Uparrow\rangle$ and $|\Downarrow\rangle$, equals $2y\Delta k$. The contrast of the measured oscillations, due to Larmor precession of the polarization, vanishes for large Δk, when the spatial separation of the wave packets exceeds the coherence length ($l_c \sim \lambda^2/\Delta\lambda$). This is referred to as longitudinal Stern-Gerlach effect. From this effect one can infer that another two-level system consisting of accelerated and decelerated part shows up. This system can be included in our representation by

$$|\Psi_{\mathrm{Bell}'}\rangle = \frac{1}{\sqrt{2}}\Big(|k_-\rangle\otimes|\Uparrow\rangle + |k_+\rangle\otimes|\Downarrow\rangle\Big), \tag{25}$$

where $|k_\pm\rangle$ is the eigenvector of the momentum operator $\hat{P}_\pm = \hbar\hat{k}_\pm$ with

$$\hat{k}_\pm|k_\pm\rangle = k_\pm|k_\pm\rangle. \tag{26}$$

3.2. Experimental setup and strategy

The experiment was carried out at the tangential beam port of the 250 kW research reactor facility TRIGA Mark II of the Vienna University of Technology (TU Vienna). A neutron beam, incident from a pyrolytic graphite crystal, of mean wavelength $\lambda = 1.99$ Å and spectral width of $\Delta\lambda/\lambda = 0.02$, is polarized to $r \sim 99\,\%$ along the $+z$-direction, by reflection from a bent Co-Ti supermirror array. The polarizer is adjusted to higher incident angles so that the second order harmonics in the incident beam are suppressed. A ^{3}He monitor detector is used for normalization in order to correct statistical fluctuations of the reactor power. A BF_3 detector with high efficiency. (more than 99 %) is used for count rate detection. A final maximum intensity of about 500 counts/s at a cross section of 10 x 10 mm^2 was recorded. To avoid unwanted depolarization a uniform guide field in Helmholtz configuration, pointing in $+z$-direction with a strength of about 13 Gauss is applied over the entire setup. A schematic illustration of the experimental arrangement is depicted in Fig.4 (a).

In the present experiment the spin-energy entanglement is achieved by a single RF-$\pi/2$ spin-rotator. The RF-flipper is operating at a frequency of $\omega_{\mathrm{rf}}/2\pi = 40$ kHz. The effective field, perpendicular to the initial polarization, is adjusted such that it induces a spin flip with a probability of 1/2 ($B_1^{(\omega)} = \pi\hbar/(2\tau|\mu|)$, with μ as magnetic moment and τ as propagation

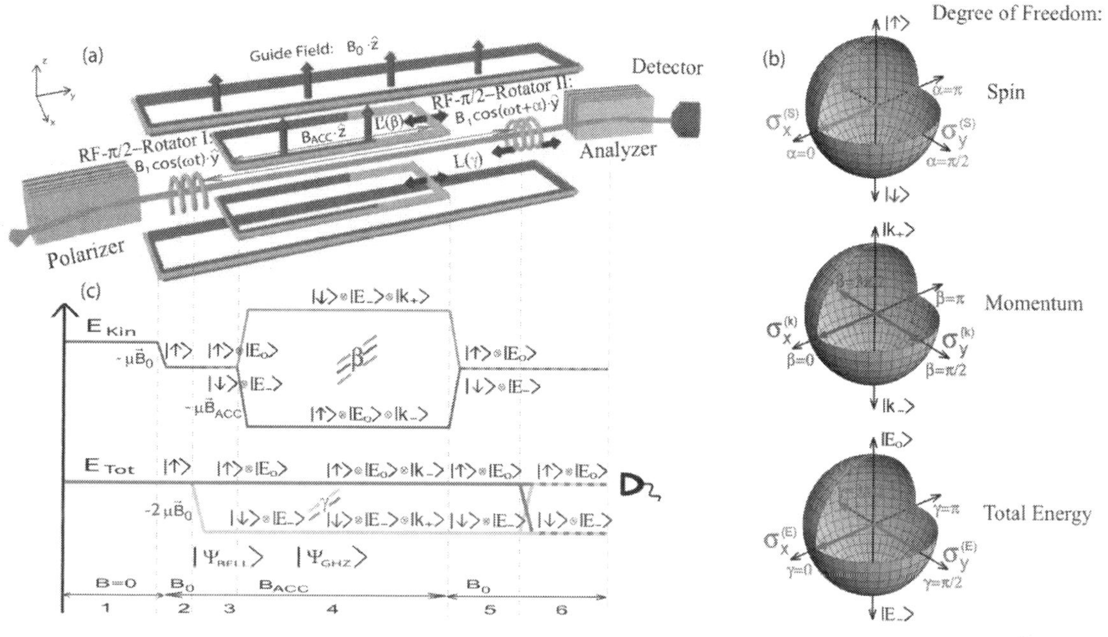

Figure 4. Experimental apparatus for observation of stronger-than-classic correlation between the spinor, momentum and energy degree of freedom. (b) Bloch-sphere description includes measurement settings of α, β and γ, determining the projection operators, used for joint measurement. The incident neutron beam is polarized by a supermirror polarizer. (c) Energy level diagram for momentum and total energy, including corresponding energy eigenstates $|E_0\rangle$, $|E_-\rangle := |E_0 - \hbar\omega\rangle$ and spin states $|\uparrow\rangle$. A $\pi/2$-RF spin-turner creates the spin energy entanglement followed by accelerator coil in Helmholtz configuration, preparing the GHZ-like state that exhibits an entanglement between spin, energy and momentum degree of freedom.

time in RF-field region). Therefore only the flipped spin component is affected by the energy manipulation yielding an entangled state vector, which can be represented as a Bell state

$$|\Psi_{\mathrm{Bell}}\rangle = \frac{1}{\sqrt{2}}\Big(|E_0\rangle \otimes |\Uparrow\rangle + |E_0 - \hbar\omega\rangle \otimes |\Downarrow\rangle\Big). \tag{27}$$

Together with the time dependent interaction within the first RF-coil and the static magnetic pointing in $+z$-direction, denoted as B_{acc}, the total system can be described by a GHZ-like state given by

$$|\Psi_{\mathrm{GHZ}}\rangle = \frac{1}{\sqrt{2}}\Big(|\Uparrow\rangle \otimes |k_-\rangle \otimes |E_0\rangle + |\Downarrow\rangle \otimes |k_+\rangle \otimes |E_0 - \hbar\omega\rangle\Big). \tag{28}$$

Each of the three degrees of freedom can be manipulated independently and the associated observables are separately measurable, which is essential for determining a sum of four expectation values defined in Eq.(15) where $\sigma^{(i)}_{x,y}$ are the Pauli operators in spin, total energy and momentum subspace. The projection operators, required for the decomposition of the Pauli operators (see Eq.(16)) are defined as

$$\hat{P}^{(\mathrm{S})}(\alpha) \;=\; \frac{1}{\sqrt{2}}(|\Uparrow\rangle + e^{-i\alpha}|\Downarrow\rangle)(\langle\Uparrow| + e^{i\alpha}\langle\Downarrow|)$$

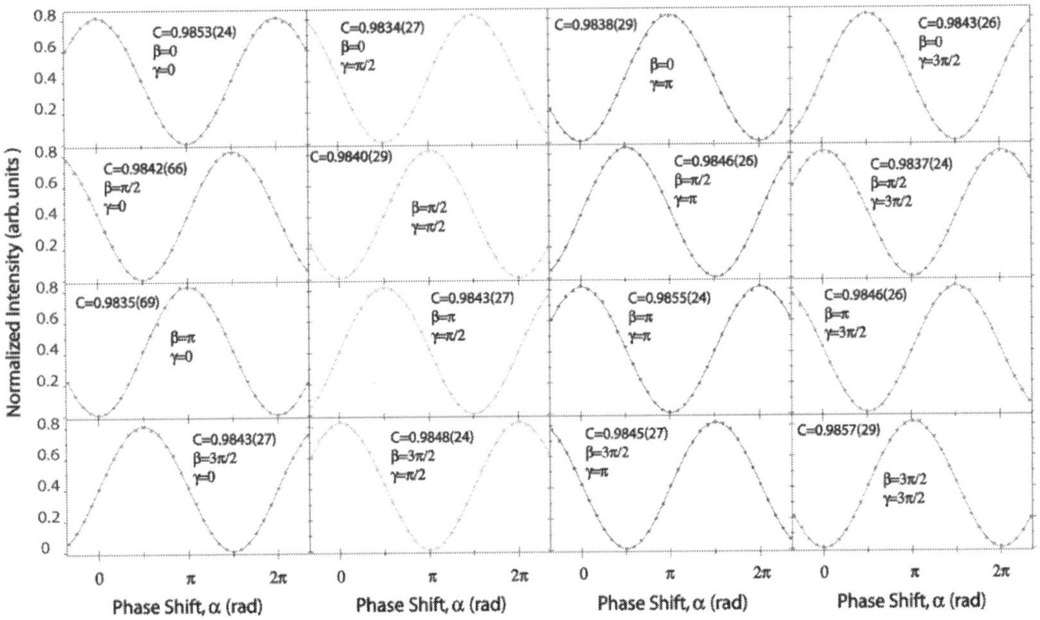

Figure 5. Typical intensity oscillation when varying the spin phase α for different setting of the momentum phase β and energy phase γ.

$$\begin{aligned}\hat{P}^{(\mathrm{k})}(\beta) &= \frac{1}{\sqrt{2}}(|k_+\rangle + e^{-i\beta}|k_-\rangle)(\langle k_+| + e^{i\beta}\langle k_-|)\\ \hat{P}^{(\mathrm{E})}(\gamma) &= \frac{1}{\sqrt{2}}(|E_0\rangle + e^{-i\gamma}|E_0-\hbar\omega\rangle)(\langle E_0| + e^{i\gamma}\langle E_0-\hbar\omega|),\end{aligned} \tag{29}$$

where α, β and γ are the azimuthal angles on the Bloch spheres, having only the values 0 and π or $\pi/2$ and $3\pi/2$, sine only σ_x and σ_y (see Fig. 4 (b)). Each expectation value $E(\sigma^{(\mathrm{S})}_{x,y}\sigma^{(\mathrm{k})}_{x,y}\sigma^{(\mathrm{E})}_{x,y})$ is experimentally determined by a combination of normalized count rates, using appropriate setting of α, β and γ. The projective measurements of each degree of freedom are carried out due to phase manipulations between the eigenstates of the individual sub systems, resulting in the desired azimuthal angle on the Bloch sphere.

The *spin-phase* measurement direction α is tuned by the phase of the oscillating field within the second RF spin-rotator denoted as $B_1^{(\omega)}\cos(\omega t+\alpha)\cdot\hat{\mathbf{x}}$, corresponding to the physical action of the projection operator Eq.(29). As all spin states to be analyzed lie in the $\hat{\mathbf{x}}\hat{\mathbf{y}}$-plane again a $\pi/2$ spinor-rotation is performed. Typical intensity modulations when scanning α are depicted in Fig. 5, where an average contrast of 0.9844(9) was achieved.

The measurement direction of the *momentum phase* is tuned by the propagation time within the accelerator coil. The acquired phase in momentum space is given by $\beta = \int B_{\mathrm{acc}}ds$. However in practice the strength of the magnetic field was varied, instead of the length, due to experimental convenience. The accelerator coil, just like the uniform guide field, is realized in Helmholtz configuration, for higher field homogeneity and accurate alignment of accelerator and guide field.

The measurement direction of the *energy phase* is tuned by the position of the second RF-$\pi/2$ spin-rotator, mounted on a motorized translation stage. Adjustment of the position of RF2 results in precise tuning of the relative phase γ between the two energy eigenstates. However, a change of the position of RF2 by ΔL, induces an undesired additional relative phase between the two spin eigenstates, due to Larmor precession within the guide field, denoted as $\Delta\delta = \omega_{\mathrm{L}}\Delta L/v$.

Table 2. Experimentally determined expectation values and the resulting M value.

Observable	Settings α	β	γ	Determined Values
$\sigma_x^{(S)}\sigma_x^{(k)}\sigma_x^{(E)}$	$(0;\pi)$	$(0;\pi)$	$(0;\pi)$	0.9843(12)
$\sigma_x^{(S)}\sigma_y^{(k)}\sigma_y^{(E)}$	$(0;\pi)$	$(\pi/2;3\pi/2)$	$(\pi/2;3\pi/2)$	-0.9839(10)
$\sigma_y^{(S)}\sigma_x^{(k)}\sigma_y^{(E)}$	$(\pi/2;3\pi/2)$	$(0;\pi)$	$(\pi/2;3\pi/2)$	-0.9840(10)
$\sigma_y^{(S)}\sigma_y^{(k)}\sigma_x^{(E)}$	$(\pi/2;3\pi/2)$	$(\pi/2;3\pi/2)$	$(0;\pi)$	-0.9837(11)
		M=3.936(2)		

Here ω_L is the Larmor frequency and v the velocity of the neutrons ($\sim$2000m/s). To retrieve a pure tuning of the energy phase γ this additional Larmor phase has to be compensated as in the interferometer experiment. This is done by appropriately tuning the phase of the oscillating magnetic field in RF1, yielding a reversed spin phase shift of $-\Delta\delta$. As this compensation depends on the relative position (ΔL) of RF2, the associated Larmor precession angle at each position has to be determined in an individual measurement using two DC- instead of RF-$\pi/2$ spin rotators. By shifting the position of DC2 (ΔL) pure Larmor precession is observed, from which the Larmor precession angle, in terms of ΔL, is determined.

If the resonance condition were exactly $\omega = \omega_L$, Larmor- and energy-phase, for spin- and energy-subspace respectively, would make phase contributions for the state, of the same amount but opposite directions. Thus, when varying L there should be no intensity modulation. In practice this does not hold, due to the so called Bloch-Siegert shift [47].

The second RF-$\pi/2$ spin-turner converts the $\hat{\mathbf{x}}$-component of the polarization back to the $\hat{\mathbf{z}}$-direction to be analyzed by the second supermirror. This is expressed by applying a projection operator for the spin $\hat{P}^{(S)} = |\Uparrow\rangle\langle\Uparrow|$. Finally the stationary intensity oscillations are given by

$$N(\alpha,\beta,\gamma) = \frac{1}{2}\Big(1 + C\cos(\alpha+\beta+\gamma)\Big), \tag{30}$$

with C being the contrast, experimentally determined as 0.9844(9). By tuning the momentum phase β and the energy phase γ at 0, $\pi/2$, π and $3\pi/2$, sixteen spin phase α scans (which are plotted in Fig. 5) were carried out for a determination of M, yielding a final value $M = 3.936(2)$. The individual results for each expectation value are summarized in Tab. 2.

In conclusion, the obtained violation of a Mermin-like inequality for a triply entangled GHZ-like state in a single-neutron system clearly confirm the prediction of quantum mechanics in explicit terms, seen in the final value $M = 3.936(2) \gg 2$. The deviation of less than 2 per cent from the theoretical value $M_{\text{theo}} = 4$ is worth noting here. This work has been partly supported by the Austrian Science Foundation, FWF (P17803-N02).

4. Conlusion and Outlook

Multi-entanglement has been studied in perfect Si-crystal interferometer experiments as well as polarimetric measurements. Entanglement of various degrees of freedom has been prepared and analyzed in detail. In Section 2.1 a technique of coherent energy manipulation in neutron interferometry has been presented. By utilizing the interferometer in combination with two RF fields time-independent interference patterns have been observed. After successful preparation and manipulation of a triply entangled GHZ-like state (i.e. entanglement of spin, path and energy degree of freedom), presented in Section 2.2, further investigation in the field of multi-entanglement in neutron interferometry are anticipated. In a next step preparation of a W-state, which requires three RF-flippers inside a neutron interferometer operating at different frequencies, is expected. In addition, generation of these typical triply entangled (pure) states will allow further mixture of these states. For instance measurements of gradual translation from the GHZ-state to the W-state are of special interest. At the second stage, neutron interferometer measurements will be carried out in order to determine and distinguish a variety of entangled states, by means of state tomography or entanglement witness.

Concerning neutron polarimetry, in addition to the experiments with spin-energy entangled states, induced by RF-flippers, multi-energy splitting, using Ramsey's resonance method of separated oscillating fields [57], will be topic of a forthcoming experiment.

References

[1] Einstein A, Podolsky B and Rosen N 1935 *Phys. Rev.* **47** 777
[2] Bohm D 1951 *Quantum Theory* (Prentice-Hall, Englewood Cliffs, NJ)
[3] Bell J S 1964 *Physics (Long Island City, N.Y.)* **1** 195
[4] Clauser J F, Horne M A, Shimony A and Holt R A 1969 *Phys. Rev. Lett.* **23** 880
[5] Bertlmann R A and Zeilinger A 2002 *Quantum [Un]speakables, from Bell to Quantum Information* Springer Verlag, Heidelberg
[6] Freedman S J and Clauser J F 1972 *Phys. Rev. Lett.* **28** 938
[7] Aspect A, Grangier P and Roger G 1981 *Phys. Rev. Lett.* **47** 460
[8] Kwiat P G, Mattle K, Weinfurter H, Zeilinger A, Sergienko A V and Shih Y 1995 *Phys. Rev. Lett.* **75** 4337
[9] Weihs G, Jennewein T, Simon C, Weinfurter H and Zeilinger A 1998 *Phys. Rev. Lett.* **81** 5039
[10] Tittel W, Brendel J, Zbinden H and Gisin N 1998 *Phys. Rev. Lett.* **81** 3563
[11] Rowe M A, Kielpinski D, Meyer V, Sackett C A, Itano W, Monroe C and Wineland D J 2001 *Nature (London)* **409** 791
[12] Moehring D L, Madsen M J, Blinov B B and Monroe C 2004 *Phys. Rev. Lett.* **93** 090410
[13] Sakai H, Saito T, Ikeda T, Itoh K, Kawabata T, Kuboki H, Maeda Y, Matsui N, Rangacharyulu C, Sasano M, Satou Y, Sekiguchi K, Suda K, Tamii A, Uesaka T and Yako K 2006 *Phys. Rev. Lett.* **97** 150405
[14] Matsukevich D N, Maunz P, Moehring D L, Olmschenk S and Monroe C 2008 *Phys. Rev. Lett.* **100** 150404
[15] Hasegawa Y, Loidl R, Badurek G, Baron M and Rauch H 2003 *Nature (London)* **425** 45
[16] Greenberger D M, Horne M A and Zeilinger A 1989 *Bell's Theorem, Quantum Theory, and Concepts of the Universe* (Kluwer Academics, Dordrecht, The Netherlands)
[17] Greenberger D M, Shimony A, Horne M A and Zeilinger A 1990 *Am. J. Phys.* **58** 1131
[18] Bouwmeester D, Pan J W, Daniell M, Weinfurter H and Zeilinger A 1999 *Phys. Rev. Lett.* **82** 1345
[19] Zhao Z, Chen Y, Zhang A, Yang T, Briegel H J and Pan J W 2004 *Nature (London)* **430** 54
[20] Walther P, Resch K, Rudolph T, Schenck E, Weinfurter H, Vedral V, Aspelmeyer M and Zeilinger A 2005 *Nature (London)* **434** 169
[21] Lu C Y, Zhou X Q, Guhne O, Gao W B, Zhang J, Yuan Z S, Goebel A, Yang T and Pan J W 2007 *Nature Physics* **3** 91
[22] Leibfried D, Knill E, Seidelin1 S, Britton J, Blakestad R B, Chiaverini J, Hume D B, Itano W M, Jost J D, Langer C, Ozeri R, Reichle R and Wineland D J 2005 *Nature (London)* **438** 639
[23] Häffner H, Hänsel W, Roos C F, Benhelm J, al kar D C, Chwalla M, Körber T, Rapol U D, Riebe M, Schmidt P O, Becher C, Gühne O, Dür W and Blatt R 2005 *Nature (London)* **438** 643
[24] Mermin N D 1990 *Phys. Rev. Lett.* **65** 1838
[25] Bell J S 1966 *Rev. Mod. Phys.* **38** 447
[26] Mermin N D 1993 *Rev. Mod. Phys.* **65** 803
[27] Simon C, Żukowski M, Weinfurter H and Zeilinger A 2000 *Phys. Rev. Lett.* **85** 1783–1786
[28] Kochen S and Specker E P 1967 *J. Math. Mech.* **17** 59
[29] Cabello A 2008 *Phys. Rev. Lett.* **101** 210401
[30] Cabello A, Filipp S, Rauch H and Hasegawa Y 2008 *Phys. Rev. Lett.* **100** 130404
[31] Kirchmair G, Zahringer F, Gerritsma R, Kleinmann M, Guhne O, Cabello A, Blatt R and Roos C F 2009 *Nature (London)* **460** 130404
[32] Amselem E, Radmark M, Bourennane M, Cabello A 2009 *Phys. Rev. Lett.* **103** 160405
[33] Hasegawa Y, Loidl R, Badurek G, Baron M and Rauch H 2006 *Phys. Rev. Lett.* **97** 230401
[34] Bartosik H, Klepp J, Schmitzer C, Sponar S, Cabello A, Rauch H and Hasegawa Y 2009 *Phys. Rev. Lett.* **103** 040403
[35] Rauch H, Treimer W and Bonse U 1974 *Phys. Lett. A* **47** 369
[36] Rauch H and Werner S A 2000 *Neutron Interferometry* (Clarendon Press, Oxford)
[37] Hasegawa Y, Loidl R, Badurek G, Filipp S, Klepp J and Rauch H 2007 *Phys. Rev. A* **76** 052108

[38] Badurek G, Rauch H and Summhammer J 1983 *Phys. Rev. Lett.* **51** 1015
[39] Jaynes E T and Cummings F W 1963 *Proc. IEEE* **51** 89
[40] Muskat E, Dubbers D and Schärpf O 1987 *Phys. Rev. Lett.* **58** 2047
[41] Golub R, Gähler R and Keller T 1994 *Am. J. Phys.* **62** 779
[42] Gähler R and Golub R 1987 *Phys. Lett. A* **123** 43
[43] Grigoriev S V, Kraan W H and Rekveldt M T 2004 *Phys. Rev. A* **69** 043615
[44] Alefeld B, Badurek G and Rauch H 1981 *Z. Phys. B* **41** 231
[45] Summhammer J, Hamacher K A, Kaiser H, Weinfurter H, Jacobson D L and Werner S A 1995 *Phys. Rev. Lett.* **75** 3206
[46] Summhammer J 1993 *Phys. Rev. A* **47** 556
[47] Bloch F and Siegert A 1940 *Phys. Rev.* **57** 522
[48] Sponar S, Klepp J, Loidl R, Filipp S, Badurek G, Hasegawa Y and Rauch H 2008 *Phys. Rev. A* **78** 061604(R)
[49] Sponar S, Klepp J, Badurek G and Hasegawa Y 2008 *Phys. Lett. A* **372** 3153
[50] Hasegawa Y, Loidl R, Badurek G, Durstberger-Rennhofer K, Sponar S and Rauch H 2010 *Phys. Rev. A* **81** 032121
[51] Hasegawa Y and Badurek G 1999 *Phys. Rev. A* **59** 4614
[52] Wagh A G, Badurek G, Rakhecha Buchelt R J and Schricker A 2000 *Phys. Lett. A* **268** 209
[53] Klepp J, Sponar S, Hasegawa Y, Jericha E and Badurek G 2005 *Phys. Lett. A* **342** 48
[54] Klepp J, Sponar S, Filipp S, Lettner M, Badurek G and Hasegawa Y 2008 *Phys. Rev. Lett.* **101** 150404
[55] Hasegawa Y, Schmitzer C, Bartosik H, Klepp J, Sponar S, Durstberger-Rennhofer K and Badurek G 2010 *submitted to Nature Physics*
[56] Erhart J, Sponar S, Sulyok G, Badurek G, Ozawa M and Hasegawa Y 2010 *submitted to Nature Physics*
[57] Grigoriev S, Chetverikov Y, Kraan W and Rekveldt M 2003 *Physica B* **335** 243

Methods and applications of gravity resonance spectroscopy within the qBounce experiment

G Cronenberg[1,2], K Durstberger-Rennhofer[1], P Geltenbort[2], T Jenke[1], H Lemmel[1,2] and H Abele[1]

1 Atominstitut, Technische Universität Wien, Stadionallee 2, 1020 Vienna, Austria
2 Institut Laue-Langevin, 6, Rue Jules Horowitz, 38042 Grenoble Cedex 9, France

E-mail: abele@ati.ac.at

Abstract. We give a short overview of the gravity resonance spectroscopy methods used recently by the *qBounce* experiment. It has been demonstrated that mechanically induced transitions between bound neutron states in the gravitational field can be observed. Applications of this new method test Newton's gravity law, search for extra dimensions of space-time, and study the neutrality of the neutron.

1. Introduction

A new method allows to study resonant transitions between discrete energy-states of ultra-cold neutrons (UCNs) bound in the gravity potential of the Earth [1]. It has the potential to shed new light on a broad range of questions. For example, the method provides a new way of looking on the limits of Newton's gravitational law and is sensitive to hypothetical fifth forces in the micrometer regime. New mechanisms like axion-like couplings within this length regime can be detected. Furthermore, it allows to probe neutrons neutrality.

Ultra-cold neutrons with mass m_n are placed on top of a flat neutron mirror. This works when their energy E_n is less than the Fermi potential V_{Fermi} of the mirror. The neutron, subject to the gravity field, forms hence bound states with discrete eigen energies E_n with $n > 0$ which are in the pico-eV regime. The first observation of bound neutrons in the gravity field were published in 2002 [2]. A review about gravity tests with neutrons can be found in [3, 4]. A speciality of the solution of the Schrödinger equation is that the energy levels are non-equidistant. Each pair of states has thus a unique energy difference $\Delta E_{pq} = \hbar(\omega_q - \omega_p)$. Quantum mechanical transitions with a characteristic energy exchange between an externally driven modulator and these energy levels are observed on resonance. The technique presents a novelty in the sense that the quantum mechanically transition is driven by an oscillating mirror and is not a consequence of a direct coupling of an electromagnetic charge or electromagnetic moment to an external field. This is an additional approach to the high sensitivity method used in atomic clocks [5], nuclear magnetic resonance [6] and atom interferometry [7].

2. Rabi's method

Transitions between quantum states can be realized by a coupling to a modulator using Rabi's method [8]. Several ways are possible: by applying varying magnetic fields and by mechanical

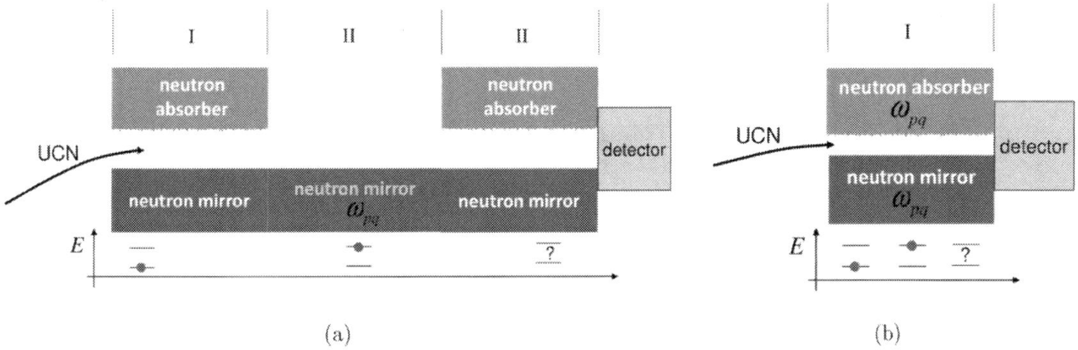

Figure 1. (a) A three-part Rabi-like setup is shown for illustration. Neutrons fly through the setup from left to right. The first region projects the state onto the ground state, the transition takes place in the second region. The projection measurement is realized in the third region together with the detector. (b) The experimental realization used for the *qBounce* experiment in 2009 and 2010.

vibrations, which modify the border condition periodically. The latter has been recently realized [1]. A different proposal can be found in [9, 10].

A typical Rabi-like setup can be seen in Figure 1(a). Neutrons fly through the three-part setup with the steps: preparation, excitation and analysis. The first part projects the state onto the ground state as higher states are suppressed by being removed from the neutron scatterer on top. The second region induces the transition by vibrating the mirror with a tuneable frequency and vibration amplitude. The third region, identical to the first, projects the state again onto the ground state. A detector at the end captures the neutron flux. This flux will be reduced if in region two a transition from the ground state to an excited state takes place.

Transitions in region two between two states $|p\rangle$ and $|q\rangle$ can be studied by treating the states as a spin-1/2 system. This is valid if the frequency of the other transitions is sufficiently different from ω_{pq}.

The time dependence of the states takes the following form:

$$|\Psi(t)\rangle = C_p(t)\, e^{-i\omega_p t}|\psi_p\rangle + C_q(t)\, e^{-i\omega_q t}|\psi_q\rangle, \tag{1}$$

with time dependent coefficients $C_{p,q}$ which can be obtained by changing into the rotating frame of reference:

$$C_p(t) = \tilde{C}_p(t)\cdot e^{-\frac{i}{2}\delta\omega t}, C_q(t) = \tilde{C}_q(t)\cdot e^{+\frac{i}{2}\delta\omega t}, \tag{2}$$

where $\delta\omega = \omega - \omega_{pq}$ is the detuning from the resonance frequency ω_{pq}. This results in a coupled differential equation for the time dependent coefficients $\tilde{C}_n$:

$$\frac{d}{dt}\begin{pmatrix}\tilde{C}_p(t)\\ \tilde{C}_q(t)\end{pmatrix} = \frac{i}{2}\begin{pmatrix}\delta\omega & \Omega_R\\ \Omega_R^* & -\delta\omega\end{pmatrix}\begin{pmatrix}\tilde{C}_p(t)\\ \tilde{C}_q(t)\end{pmatrix}, \tag{3}$$

with Ω_R being the Rabi frequency, a measure for the coupling to the vibrating mirror.

When initially in the state p, the probability of finding the neutron in state q after a time t is given by:

$$|\Psi(t)|^2 = \left(\frac{\Omega_R}{\Omega_R'}\right)^2 \sin^2\frac{\Omega_R' t}{2}, \tag{4}$$

where $\Omega_R' = \sqrt{\Omega_R^2 + \delta\omega^2}$ incorporates the detuning.

Let us now consider the experimentally realized setup shown in Figure 1(b). In comparison to the three-part setup, one region performs all three tasks at the same time. For that purpose, a second mirror with a rough surface is placed on top of the first mirror which modifies the problem. The orthonormal eigen states, previously Airy Ai functions, are now a superposition of Airy Ai and Bi functions and the eigen energies are increased. The rough mirror removes higher, unwanted states as described by [11]. Such an additional loss mechanism modifies the system and is different for each state as it is proportional to the overlap of the wave function with the additional absorbing layer. Each state has thus an additional damping term γ_n which enters the equation:

$$\frac{d}{dt}\begin{pmatrix}\tilde{C}_p(t)\\ \tilde{C}_q(t)\end{pmatrix} = \frac{i}{2}\begin{pmatrix}(\delta\omega + i\gamma_p) & \Omega_R \\ \Omega_R^* & (-\delta\omega + i\gamma_q)\end{pmatrix}\begin{pmatrix}\tilde{C}_p(t)\\ \tilde{C}_q(t)\end{pmatrix}. \tag{5}$$

The solution of this differential equation is:

$$\begin{pmatrix}\tilde{C}_p(t)\\ \tilde{C}_q(t)\end{pmatrix} =$$

$$e^{-(\gamma_p+\gamma_q)t/4}\begin{pmatrix}\cos\frac{\Omega_R' t}{2} + \frac{(\gamma_q-\gamma_p)/2+i\delta\omega}{\Omega_R'}\sin\frac{\Omega_R' t}{2} & i\frac{\Omega_R}{\Omega_R'}\sin\frac{\Omega_R' t}{2} \\ i\frac{\Omega_R^*}{\Omega_R'}\sin\frac{\Omega_R' t}{2} & \cos\frac{\Omega_R' t}{2} - \frac{(\gamma_q-\gamma_p)/2+i\delta\omega}{\Omega_R'}\sin\frac{\Omega_R' t}{2}\end{pmatrix}\begin{pmatrix}\tilde{C}_p(0)\\ \tilde{C}_q(0)\end{pmatrix}, \tag{6}$$

with $\Omega_R' = \sqrt{-\left((\gamma_q - \gamma_p)/2 + i\delta\omega\right)^2 + \Omega_R^2}$ now including the damping. Even though the structure of the solution is more complex the same behavior as in the three-part setup is recovered.

Without vibration the solution simplifies to

$$C_n(t) = e^{-\gamma_n t/2} C_n(0). \tag{7}$$

The recorded flux at the detector is proportional to the norm of the state:

$$|\Psi|^2 = |\tilde{C}_p|^2 + |\tilde{C}_q|^2. \tag{8}$$

2.1. *Experimental Realization*

The experimentally feasible version, shown in Figure 1(b), combines the three regions into one and has been used within the *qBounce* experiment performed at the fundamental physics instrument PF2 at the Institut Laue-Langevin in 2009 [1] and 2010. The setup consists of a neutron mirror on bottom and a neutron scatterer on top. They are vibrating with a tuneable frequency and desirable amplitude. The vibration of the setup is induced by piezo actuators, which allow vibrations in the range 100 – 800 Hz and a vibration amplitude of up to 5 mm/s. The neutron mirrors were made of BK7 glass with a Fermi potential of about $V_{\mathrm{Fermi}} \approx 100$ neV, which is several orders higher than the eigen energy of the neutron. The bottom mirror is polished and has a roughness less than 2 nm, whereas the top mirror has a surface roughness within the micrometer regime to provide the loss mechanism described in previous section. The velocity selection of the UCNs was performed by a collimation system in front of the setup. Only neutrons following a certain classical flight path in the gravity field are accepted into the setup resulting in an allowed velocity spectrum in the direction along the mirrors. This ensured that neutrons with a horizontal velocity $5.7 \leq v_h \leq 9.5$ m/s enters the setup resulting in a mean interaction time of $\tau \approx 20$ ms. To avoid a shift of states energy due to the magnetic moment of the neutron, magnetic shielding was provided by one layer of μ-metal around the vacuum chamber. This was sufficient to reduce the ambient magnetic field inside our setup by more than one order.

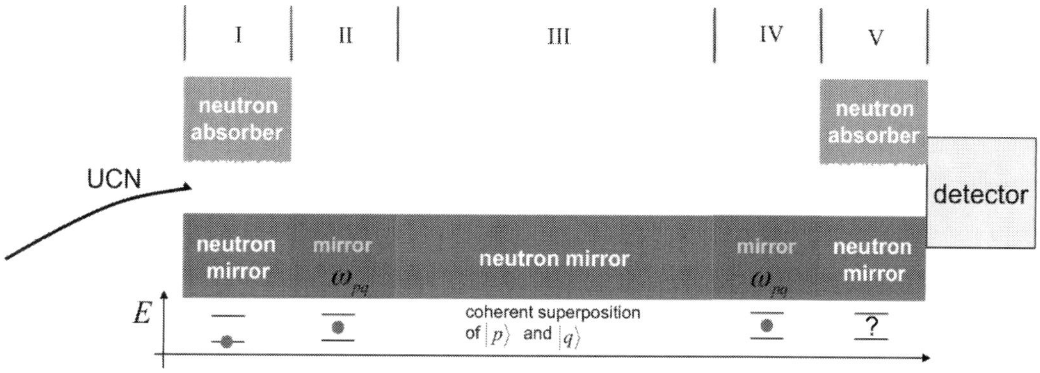

Figure 2. A Ramsey-like setup consisting of five regions. The first and last region suppress excited states allowing mainly transmission of the ground state. The vibrating mirrors in region two and four have a defined phase relation. The long region in the middle allows a free propagation of the coherent superposition of the states $|p\rangle$ and $|q\rangle$.

3. Ramsey's method of separated oscillating fields

The sensitivity of a Rabi-setup is inverse to the interaction time which is fixed by the length of the region two in Figure 1(a). The experimental limit lies in the scalability of this vibrating mirror setup. With the same length of the vibrating regions the interaction time can be increased by applying Ramsey's method of separated oscillating fields [12]. Another advantage of the Ramsey-like setup is the self focusing. This means that a velocity distribution of the neutrons will not deteriorate the effect as strongly as for a Rabi-like setup. Less demanding requirements allow a greater neutron flux.

In the Ramsey-like setup, shown in Figure 2, interaction takes place in two smaller regions instead of in one longer. They are separated by a distance, where the wave function can propagate freely. If a $\pi/2$-pulse is applied in the second region, it creates a coherent superposition between the states $|p\rangle$ and $|q\rangle$. Region four vibrates in phase with region two. The reversal of the transition in region four depends on the relative phase ϕ_{pq} between the states $|p\rangle$ and $|q\rangle$ obtained in the intermediate region. The neutron flux is hence dependent on this relative phase. For further details see [13].

4. Applications

The presented methods allow for a search for deviations from Newton's gravity law at short distances [14, 15]. A hypothetical energy shift is detectable via a change of the resonance frequency ω_{pq}. A generic parametrisation of such a deviation between two masses m_i and m_j is accounted for by adding a Yukawa-like term [16]

$$V(r) = -G\frac{m_i m_j}{r}\left(1 - \alpha e^{-r/\lambda}\right), \tag{9}$$

where α characterizes the strength of the deviation compared to Newton's gravity law and λ defines an interaction length. There are many theoretical predictions giving rise to such a deviation [16]. Large extra dimensions of space-time introduced by string theory which are compactified at length scales λ would modify the exponent of the Newton's inverse squared law [17]. The observed smallness of the cosmological constant and dark energy could be explained by Supersymmetric large extra dimensions [18, 19] or by a scalar chameleon field [20].

Possible axion-like couplings between matter and spin would also affect neutrons. Depending on their spin orientation a signal resolvable by such an experiment would be obtained.

Comparing the resonance frequency ω_{pq} for different spin orientations allows to detect an additional potential of the form:

$$V(\vec{r}) = \hbar g_s g_p \frac{\vec{\sigma} \cdot \vec{n}}{8\pi mc} \left(\frac{1}{\lambda r} + \frac{1}{r^2}\right) e^{-r\lambda}, \tag{10}$$

where $g_s g_p$ denotes the pseudo-scalar coupling constants. The geometry of the studied matter, ie. the neutron mirrors is noted by $\vec{n}$ and $\vec{\sigma}$ indicates the neutrons spin. Axion-like particles are possible candidates for dark matter [17] in the so-called axion window.

As has been shown in [21] such a setup can be used to provide new limits for the neutrons charge. Applying an electric field of 6×10^6 V/m in the intermediate region modifies the energy of the states and leads to a change in the relative phase ϕ_{pq} of the states $|p\rangle$ and $|q\rangle$. Current limits on the neutron charge are based on the deflection of neutrons within a present electric field [22].

To visualize the methods presented here fly-through setups have been used. However they can been realized as well in storage mode. All the applied steps are then not different parts in space but sequences performed at different times. This removes the constraint to choose neutrons with a narrow velocity distribution for Rabi like setups. It would allow a significant increase of the interaction time [23].

Gravity experiments with neutrons are motivated mainly due to the fact that in contrast to atoms the electrical polarizability of neutrons is extremely low [25]. Neutrons are not disturbed by short range electric forces such as van der Waals or Casimir forces. Together with its neutrality one holds the key for a sensitivity many orders of magnitude below the strength of electromagnetism.

Acknowledgments
We gratefully acknowledge support from the Austrian Science Fund (FWF) under Contract No. I529-N20 and No. I531-N20 and the German Research Foundation (DFG) within the Priority Programme (SPP) 1491 Precision experiments in particle and astrophysics with cold and ultracold neutrons, and DFG support under Contract No. Ab128/2-1.

References
[1] Jenke T, Geltenbort P, Lemmel H and Abele H 2011 *Nature Physics* **7** 468
[2] Nesvizhevsky V V et al. 2002 *Nature* **415** 299
[3] Abele H 2008 *Prog. Part. Nucl. Phys.* **60** 1
[4] Dubbers D, Schmidt M G 2011 *(Preprint hep-ph/1105.3694)*
[5] Ramsey N F 1990 *Rev. Mod. Phys.* **62** 541552
[6] Vandersypen L M K and Chuang I L 2005 *Rev. Mod. Phys.* **76**, 10371069
[7] Cronin A, Schmiedmayer J and Pritchard D 2009 *Rev. Mod. Phys.* **81** 10511129
[8] Rabi I I, Millman S, Kusch P, Zacharias J R 1939 *Phys. Rev.* **55** 526
[9] Pignol G et al. 2007 *Preprint quant-ph/0708.2541*
[10] Kreuz M et al. 2009 *Preprint physics/0902.0156*
[11] Westphal A et al. 2007 Eur. Phys. J. C **51** 367
[12] Ramsey N F 1950 *Physical Review* **78(6)** 695
[13] Abele H et al. 2010 *Phys. Rev.* D **81** 065019
[14] Abele H, Baeßler S and Westphal A 2003 *Lect. Notes Phys.* **631** 355
[15] Nesvizhevsky V and Protasov K V 2004 *Classical Quantum Gravity* **21** 4557
[16] Fischbach E & Talmadge C L 1999 Springer-Verlag New York
[17] Adelberger E G, Heckel B R & Nelson A E 2003 *Ann. Rev. Part. Sci.* **53** 77
[18] Sundrum R 1999 *J. High Energy Phys.* **07** 001
[19] Callin P and Burgess C P 2006 *Nucl.Phys.* B **752** 60-79
[20] Brax P and Pignol G 2011 *Phys. Rev. Lett.* **107**, 111301
[21] Durstberger-Rennhofer K et al. 2011 *Phys. Rev.* D **84** 036004
[22] Baumann J, Gähler R, Kalus J and Mampe W 1988 *Zh. Tekh. Fiz.* **58** 951

[23] Jenke T et al. 2010 *Phys. Rev.* D **81**, 065019
[24] Jenke T 2011 Gravitationstests mit Quantenobjekten *Dissertation*
[25] Schmiedmayer J et al. 1991 *Phys. Rev. Lett.* **66** 10151018

Structural characterisation of subcritical carbon dioxide confined in nanoporous carbon by *in situ* neutron diffraction

F K Katsaros[1*], Th A Steriotis[1], K L Stefanopoulos[1], N K Kanellopoulos[1], A C Hannon[2] and J D F Ramsay[3]

[1] Institute of Physical Chemistry, NCSR "Demokritos", Ag. Paraskevi Attikis, 15310, Greece
[2] ISIS Facility, Rutherford Appleton Laboratory, Chilton, Didcot, Oxon OX11 0QX, UK
[3] Laboratoire des Matériaux et Procédés Membranaires, UMR CNRS 5635, Université Montpellier II, Montpellier, France

*fkats@chem.demokritos.gr

Abstract. *In situ* neutron diffraction measurements of adsorbed carbon dioxide in an ordered mesoporous carbon (CMK-1) have been carried out along an isotherm at 253 K and at a range of pressures up to 18 bar. The experiment has been performed with the aid of a novel high-pressure adsorption apparatus and the GEM diffractometer (ISIS, Rutherford Appleton Laboratory, UK). Diffraction measurements of bulk liquid carbon dioxide have also been carried out. The structure factors and the total differential correlation functions of the adsorbed carbon dioxide suggest that the confined fluid has *liquid-like* properties at all thermodynamic states studied; however, some subtle differences were observed pointing to enhanced adsorption because of the presence of micropores.

1. Introduction

Sorption of fluids on nanoporous solids is very important in a series of applications, such as catalysis, H_2 and natural gas upgrade and storage (chromatographic pressure swing adsorption and membrane) separations, biological and geological processes etc. [1-3]. These processes are extremely complex from a fundamental point of view as the properties of sorbed fluids are in many cases different from the bulk due to confinement. For instance, confined geometries play significant role on first-order phase transitions and also in the glass formation [4]. For the above reasons there is an increasing interest in the experimental study of fluids confined in nanoporous materials [5], including phase transitions [6] and dynamical properties of liquids in nanometer-sized porous materials [7]. On the same ground, powerful theoretical and computer simulation techniques have been used for the study of adsorption in model nanopores [8-11].

On the other hand, adsorption measurements [12] are abundantly used for the study of pore-confined fluids, however such methods can only reveal the statistical "ensemble" macroscopic properties and do not provide information on their molecular structure. On the other hand techniques such as small-angle scattering (SAS) of x-rays or neutrons (SAXS, SANS) [13-17], contrast variation or contrast matching SANS [18-27] are essential tools for elucidating the microstructure of nanoporous materials and for resolving structural characteristic of each phase in multiphase materials.

Further, adsorption in conjunction with *in situ* SAS or diffraction can provide in depth information about both the adsorption mechanism and the structural organisation of the confined phase [28-42].

The discovery of ordered mesoporous materials opened new opportunities in the synthesis of structured carbons using templating techniques [43]. In particular, CMK-1 is an ordered mesoporous carbon molecular sieve [44] prepared by using ordered mesoporous silica MCM-48 as a template after filling its pores with a carbon source (e.g. sucrose), which is pyrolysed resulting in the formation of rod-type carbon inside the silica channels. Upon silica removal, the remaining pure carbonaceous material has a porous texture consisting of a three dimensional regular array of uniform mesopores, also containing micropores in the mesopore walls [45]. In a recent study [46], the results of *in situ* neutron diffraction measurements of adsorbed CO_2 in a purely siliceous MCM-41 revealed that the confined phase has *liquid-like* properties. In the present study *in situ* neutron diffraction measurements of CO_2 confined in a purely carbonaceous CMK-1 have been carried out along along the same isotherm at 253 K. Furthermore, neutron diffraction measurements of bulk liquid CO_2 (T=253 K, P=19.7 bar) have also been performed.

2. Experimental

The material CMK-1 was prepared using MCM-48 template by a procedure involving impregnation with sucrose in the presence of sulfuric acid, followed by carbonisation and removal of the silica template [47]. More specific, 2 g of MCM-48 were added to a solution obtained by dissolving 2.5 g of sucrose and 0.28 g of H_2SO_4 in 12 g of H_2O. The mixture was placed in a drying oven for 6 h at 373 K and subsequently the oven temperature was increased to 433 K and maintained there for 6 h. The sample turned dark brown or black during the treatment in the oven. The silica sample, containing partially polymerised and carbonised sucrose at the present step, was treated again at 373 and 433 K using the same drying oven after the addition of 1.5 g of sucrose, 0.16 g of H_2SO_4 and 12 g of H_2O. The carbonisation was completed by pyrolysis with heating to typically 1173 K under vacuum. The carbon -silica composite obtained after pyrolysis was washed with 48 wt% hydrofluoric acid at room temperature, to remove the silica template. The template-free carbon product thus obtained was washed with H_2O and ethanol, redissolved with EtOH after centrifuging, poured on to a petri dish and dried at 393 K.

The N_2 adsorption isotherms have been performed volumetrically at 77 K on an Autosorb-1 gas analyzer, Micropore version (Quantachrome instruments). The CO_2 adsorption isotherm was measured at 253 K on an Intelligent Gravimetric Analyser (IGA). For adsorption measurements, the sample was initially outgassed at 300 °C for 12 h under high vacuum (<10^{-6} mbar). For the gravimetric measurements, at 253 K an ice/NaCl bath was used. N_2 (99.999%) and CO_2 (99.995%) purchased from Messer-Griesheim were used directly from pressure reducer equipped gas cylinders.

The adsorption/neutron diffraction experiment was carried out at General Materials diffractometer (GEM), at ISIS pulsed neutron and muon source (UK Rutherford). GEM is one of the most advanced materials diffractometer in the world. For a detailed description of the instrument, see ref 48.

A novel high-pressure adsorption device (high vacuum up to 150 bar) has been constructed for performing *in situ* diffraction measurements and a thin wall cylindrical titanium zirconium (Ti/Zr) null scattering alloy was used as sample container (see refs 46 and 49 for a more extended discussion). After cooling down to room temperature the whole stick was inserted in the CCR, sample temperature was set to 253 K, and doses of CO_2 (99.9996%) were admitted, while equilibration was monitored by pressure readings. After measuring the diffraction spectrum at 18 bar, the sample was removed from the cell and measurements of empty cell as well as bulk liquid CO_2 were performed (T=253 K, P=19.7 bar).

The scattered neutrons were counted as a function of neutron time-of-flight for eight detector banks. The raw data were corrected for background, attenuation and multiple scattering and were normalised to a standard vanadium scatterer using the gudrun program [50-51]. Gudrun is preferred for this purpose because it takes into account the azimuthal detector angle, which is an important consideration for GEM. The Placzek inelasticity correction and the merging of data from different detector banks, were performed using the ATLAS suite of programs [52]. Finally, the calculated structure factors were Fourier transformed up to Q_{max}=15 Å^{-1} for the derivation of the differential correlation functions, using the Lorch modification function for reducing the termination ripples [53]; However, this gain results in some loss of resolution with Δr_{Lorch}= 0.36 Å.

3. Results and discussion

3.1. Adsorption isotherms

Figure 1 shows the adsorption isotherms of carbon sample prepared, as well as that of MCM-48, used as template. The N_2 adsorption isotherm for MCM-48 is of type IV, according to the IUPAC classification [54]. At low pressures, a steep increase is observed due to the formation of a monolayer on the pore walls, while at higher pressures an extended multilayer region and a sharp pore condensation step can be observed. On the other hand, the isotherm of CMK-1 showed an initial sharp uptake followed by a steadily increase up to about p/p_0=0.3-0.4. The distinct adsorption step observed in MCM-48, in the relative pressure range of 0.2–0.3, disappeared and the nitrogen adsorption isotherms changed from type IV (MCM-48) to type I/II for CMK-1. According to the IUPAC classification, the type I/II isotherms may be often observed on microporous solids. The CMK-1 carbon only showed a small increase in the amount of adsorption at relative pressures higher than 0.30, mainly, due to adsorption on the wider pores and the external surface. In addition to the mesopores with diameter of 22 Å, the structure of CMK-1 contains abundant micropores, which were formed in the amorphous carbon framework prior to the removal of silica wall, while the mesopores were formed upon the removal of the silica wall. One should note that the isotherms of all samples exhibited an upturn at pressures close to the vapour pressure which can be attributed to large (presumably macro-) pores formed between the particles. Again, the absence of hysteresis loop in all isotherms is due to their pore size, lying between the margins of micro- and mesopore regions [55].

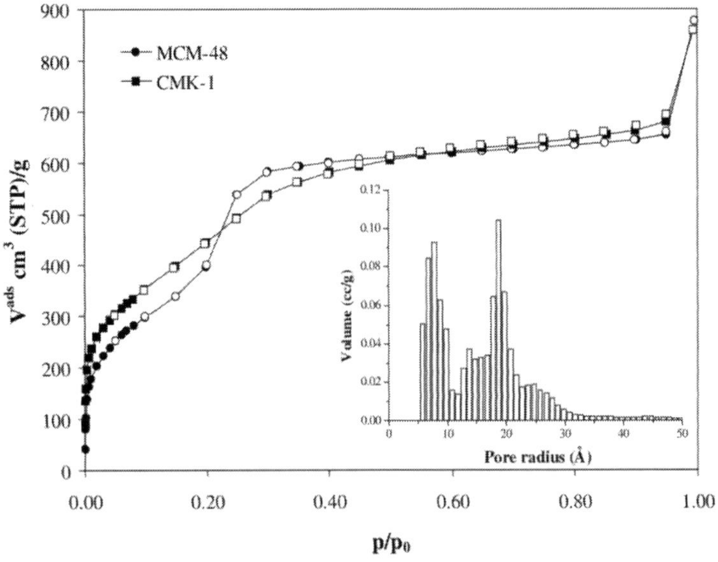

Figure 1. N_2 adsorption isotherms at 77 K on CMK-1 and MCM-48 precursor respectively. Insert: Pore size distribution of CMK-1 using NLDFT method.

The structural characteristics of CMK-1 including the specific surface area, the pore volume and the mean pore radius were determined from N_2 adsorption measurements at 77 K. A specific surface area, *S*, of 1760 m^2/g was determined by the BET (Brunauer-Emmett-Teller) method (surface area of N_2 molecules = 16.2 $Å^2$), while the total pore volume, V_p, equal to 1.02 cm^3/g, was derived from the amount adsorbed at p/p_0=0.975, assuming that N_2 has completely filled the pores in its normal liquid state (density =0.807 g/cm^3). Finally, pore size distribution was deduced by non-local density functional theory (NLDFT) [56] methods, revealed a bimodal pore size distribution (PSD), which range from 0.7 to 3 nm. Therefore, CMK-1 seems to have a PSD from micropores to small mesopores. Both large micropores (average size ~10-12 Å) and small mesopores (about 36 Å) are predominant in CMK-1. It must be noted that for the latter class of pores, the corresponding value obtained by BJH (Barrett–Joyner–Halenda) is about 22-24 Å. The average pore size obtained by BJH method is quite similar to that for MCM-48 using the same method, although it is quite different from the pore wall thickness of MCM-48 (ca. 10 Å) [57]. The observed changes can be attributed to the fact that the carbon precursor shrinks during carbonisation, giving rise to the greater pore width than the pore wall thickness of MCM-48 [57].

The CO_2 adsorption isotherm of CMK-1 at 253 K is shown in figure 2. The initial adsorption is due to the presence of micropores, while a gradual increase is obvious at higher pressures. The absence of a steep condensation step suggests a wider pore size distribution. Similar to N_2 isotherms no hysteresis loop was observed. Finally, the pore volumes calculated from N_2 and CO_2 isotherms are almost identical (Gurvitz rule is valid) indicating full accessibility of both molecules to the pore network.

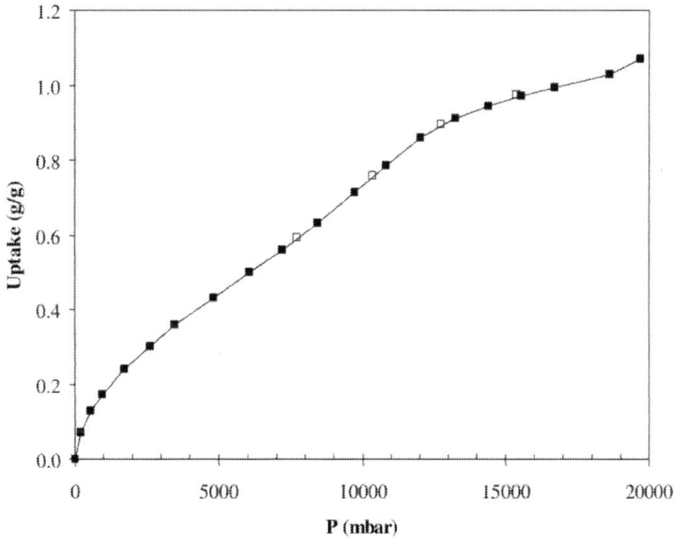

Figure 2. CO_2 adsorption isotherm at 253 K on CMK-1.

3.2. Structural characteristics of confined carbon dioxide

The structure factor of bulk liquid CO_2 is shown as the top curve of figure 3. The intermolecular structure peak is located at about 1.79 $Å^{-1}$, while a long range oscillation corresponding to intramolecular correlations is observed at larger *Q* values. The main diffraction peak has a minimum at Q~3 $Å^{-1}$ followed by a very tiny bump near 4 $Å^{-1}$. The main peak possesses a degree of asymmetry, which can be attributed to intermolecular orientational correlations between neighbouring CO_2 molecules arising mainly from electrical quadrupolar interactions [58]. The results are in excellent agreement with previous similar neutron diffraction measurements of bulk CO_2 [59-60]. It is worth mentioning that in some cases (e.g. 220 K, 58 bar) [60] a shoulder has been observed on the high Q side of the main diffraction peak (Q~2.2 Å^{-1}). Based on theoretical calculations, Gubbins *et al* [61] attributed the origin of this feature to the quadrupole - quadrupole contribution of the anisotropic part of the structure factor. Adya and Wormand [60] claimed that this hump is both temperature and density dependent. Chiappini *et al* [58] argued that this feature was only density dependent based on neutron diffraction experiments of bulk carbon dioxide in dense supercritical states. In our case, such a shoulder is not expected due to the lower density of the bulk phase.

Further, the structure factors of the confined carbon dioxide in CMK-1 at 10, 14 and 18 bar respectively are also presented at figure 3. The patterns exhibit clearly a peak corresponding to the most probable distance between nearest neighbour molecules, while the long range oscillations at larger Q values observed for the bulk liquid are also present although to a less extent. The peak position shifts slightly to the higher Q region with increasing pressure and reaches that of the bulk liquid at P=18 bar. As in the case of the bulk liquid, the main peaks of the confined CO_2 have a minimum at about Q=3 Å^{-1}. One may note that the main peak of the confined phase possesses a lower degree of asymmetry compared to that of the bulk liquid. In addition, during all adsorption stages, the first intramolecular oscillation of the confined CO_2 has been slightly moved towards the low-Q regime. The most striking difference, however, is that the bump near 4 Å^{-1} has become a prominent feature. This maximum has also been observed as a tiny bump for dense CO_2 thermodynamic states, such as bulk liquid close to the triple point [59-60] or supercritical fluid under very high pressures [58]. The observed densification can be explained by considering that within micropores, which are no more than a few molecular diameters wide, the potential fields from neighbouring walls overlap and produce a very deep potential well. As a result, the interaction energy between the gas molecules and the adsorbent matrix is increased and adsorption is enhanced while the adsorbed molecules within the confined pore space are arranged in very dense states to compensate the lack of free volume [33].

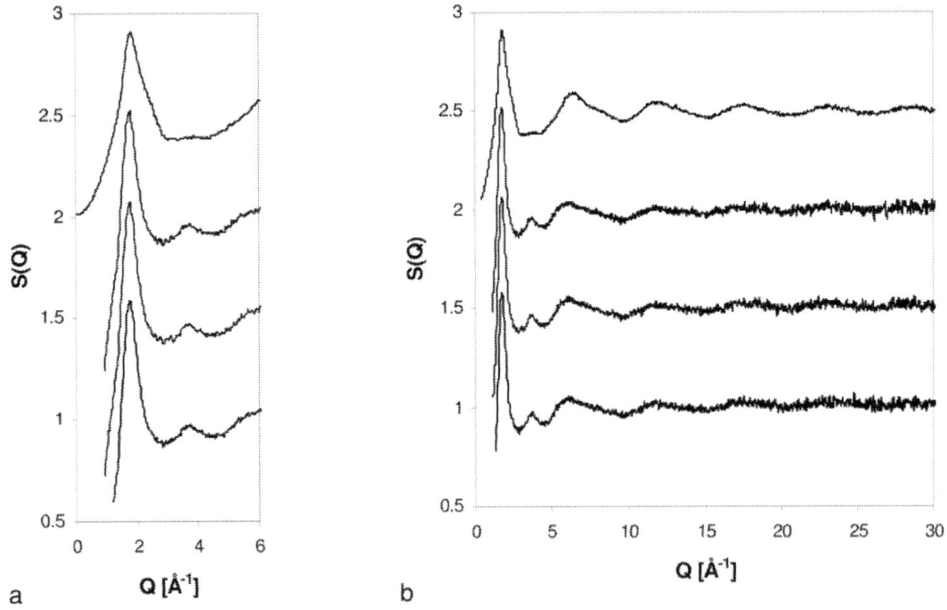

Figure 3. The total-scattering structure factors at 253 K for confined CO_2 at 10, 14 and 18 bar, respectively, and the total-scattering structure factor for the bulk liquid at 19.7 bar; the order is from the bottom to the top. The structure factors have shifted by 0.5 for clarity. (a) the low Q region and (b) the whole Q range.

Figure 4 illustrates the total differential correlation function, $D(r)$, for the confined CO_2 at pressures 10, 14 and 18 bar. The correlation function for the bulk liquid is also shown as the top curve. In all cases one clearly observes two intramolecular peaks (completely separated from the intermolecular interactions) occurring at 1.16 Å and 2.32 Å which correspond to the C-O (r_{CO}) and O-O (r_{OO}) distances, with $r_{OO} = 2r_{CO}$. Further, the intermolecular part of the correlation functions of the bulk liquid presents clearly two broad features centered at ~4 Å and ~8 Å, arising respectively from the first and second neighbour interactions. The first neighbour peak reveal three structures located at about r_1=3.3 Å, r_2=4 Å and r_3=5.2 Å respectively, in agreement with literature data on bulk liquid CO_2 [60]. Molecular dynamics simulations have revealed that the structure at r_1 is attributed mainly to O-O as well as to C-O pair correlations. Moreover, both C-C and C-O correlations give a positive contribution at r_2, whereas the structure at r_3 arises from bumps present in the partial C-O and O-O atom-atom pair correlation functions, when quadrupole forces are included in the simulations [58, 62].

On the other hand the correlation functions of CO_2 confined in CMK-1 do not exhibit these three structures located at the first neighbour peak, although there is a tendency for the structure r_3 to appear at about 5 Å. Based on simulation results, this feature is proved to be strongly potential dependent, becoming evident only when electric quadrupolar interactions are taken into account and can be ascribed to a sharpening of the O-O correlations, arising from increased orientational ordering of the fluid [58]. This seems to be a reasonable explanation, as the micropore filling is subject to enhanced Lennard-Jones interactions with the surface atoms. In addition, the main peak has been slightly shifted to larger distances (~4.2 Å) and does not seem to vary notably with increasing pressure. Further details about the molecular organisation of the confined phase can be deduced by advanced simulations in combination with our results obtained by the *in situ* diffraction measurements.

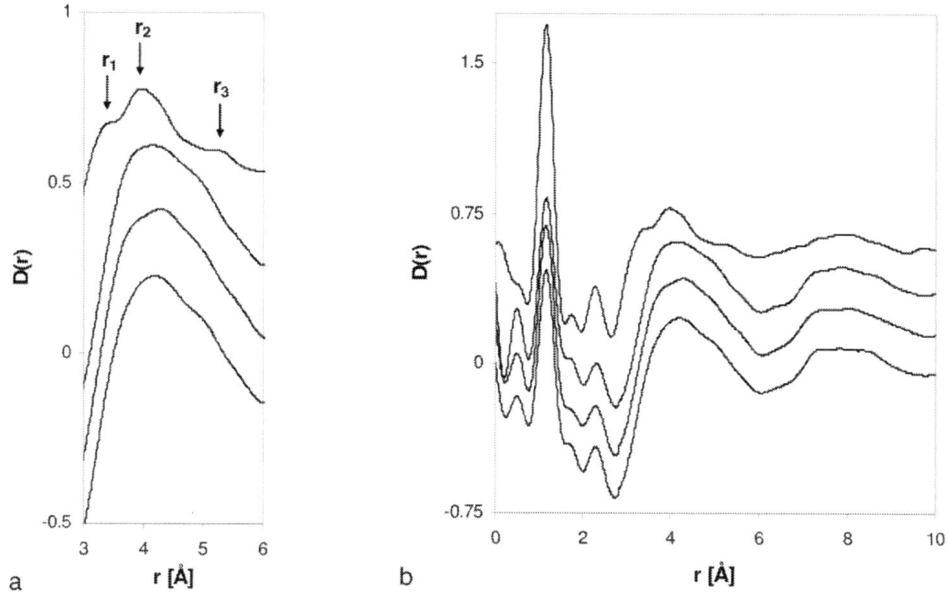

Figure 4. The differential correlation functions for confined CO_2 at 10, 14 and 18 bar, respectively, and for the bulk liquid; the order is from the bottom to the top. The correlation functions have shifted by 0.2 for clarity. (a) details of the first neighbour region and (b) the whole real space distance.

4. Conclusions

In conclusion, the structural characteristics of CO_2 confined in an ordered carbonaceous material have been investigated by *in situ* neutron diffraction measurements. Both the structure factors of confined CO_2 together with the total differential correlation functions have been derived and have been compared to those of the bulk CO_2 in liquid state. The results suggest that the confined fluid has *liquid-like* properties at all thermodynamic states examined. However, some certain differences observed can be attributed to enhanced CO_2 adsorption because of the strong interaction potential arising from the pore walls.

Acknowledgements

We are grateful to the ISIS technical staff for their great assistance provided for the construction of the high-pressure adsorption device. This research project has been supported by the European Commission under the 7th Framework Programme and it is part of the Research Infrastructures action of the Capacities Programme (NMI3), Contract No: 226507.

References

[1] Ruthven D M, Shamasuzzaman F and Knaebel K S 1994 *Pressure Swing Adsorption* (New York: VCH)
[2] Noh J S, Argaval R K and Schwarz J A 1987 *Int. J. Hydrogen Energy* **12** 693
[3] Matranga K R, Myers A L and Glandt E D 1982 *Chem. Eng. Sci.* **47** 1569
[4] Gelb L D, Gubbins K E, Radhakrishnan R and Sliwinska-Bartkowiak M 1999 *Rep. Prog. Phys.* **62** 1573
[5] Frick B, Zorn R and Büttner H (ed) 2000 *Proceedings of the International Workshop on Dynamics in Confinement, J. Phys* IV **10** pp 3-343
[6] Christenson H K 2001 *J. Phys.: Condens. Matter* **13** R95
[7] Frick B, Koza M and Zorn R (ed) 2003 *Proceedings of the Second International Workshop on Dynamics in Confinement, Eur. Phys J.* E **12** pp 5-194
[8] Tan Z and Gubbins K *J. Phys. Chem.* 1990 **94** 6061
[9] Benard P and Chahine R 1997 *Langmuir* **13** 808
[10] Egorov S A 2001 *J. Phys. Chem.* B **105** 6583
[11] Fu K, Robinson R L and Slattery J C 2004 *Chem. Eng. Sci.* **59** 801
[12] Cregg S J and Sing K S W 1982 *Adsorption, Surface Area and Porosity* (New York: Academic Press)
[13] Mitropoulos A C, Kanellopoulos N K, Stefanopoulos K L and Heenan, R K 1998 *J. Coll. Inter. Sci.* **203** 229
[14] Ramsay J D F 1998 *Adv. Coll. Inter. Sci.* **76-77** 13
[15] Makri P K, Stefanopoulos K L, Mitropoulos A C, Kanellopoulos N K and Treimer W 2000 *Physica* B **276-278** 477
[16] Hoinkis E, Lima E B and Schubert-Bischoff P 2004 *Langmuir* **20** 8823
[17] Kikkinides E S, Stefanopoulos K L, Steriotis T A, Mitropoulos A C, Kanellopoulos N K and Treimer W 2002 *Appl. Phys.* A **74** S954
[18] Stuhrmann H B and Miller A. 1978 *J. Appl. Cryst.* **11** 325
[19] Ramsay J D F and Wing G 1991 *J. Coll. Interf. Sci* **141** 475
[20] Charalambopoulou G C, Steriotis T A, Mitropoulos A C, Stefanopoulos K L, Kanellopoulos N K and Ioffe A 1998 *J. Invest. Derm.* **110** 988
[21] Charalambopoulou G C, Steriotis T A, Hauss T, Stefanopoulos K L and Stubos A K 2002 *Appl. Phys.* A **74** S1245
[22] Steriotis T A, Stefanopoulos K L, Keiderling U, De Stefanis A and Tomlinson A A G 2002 *Chem. Commun.* **20** 2396
[23] Calo J M and Hall P J 2004 *Carbon* **42** 1299
[24] Lelong G, Price D L, Brady J W and Saboungi M L 2007 *J Chem. Phys.* **127** 065102

[25] Shin T, Findenegg G H and Brandt A 2006 *Prog. Coll. Polym. Sci.* **133** 116
[26] Mergia K, Stefanopoulos K L, Ordás N and García-Rosales C *2010 Microporous Mesoporous Mater.* **134** 141
[27] Müter D, Shin T, Demé B, Fratzl P, Paris O and Findenegg G H 2010 *J Phys. Chem. Lett.* **1** 1442
[28] Mitropoulos A C, Stefanopoulos K L and Kanellopoulos N K 1998 *Microporous Mesoporous Mater.* **24** 29
[29] Katsaros F, Makri P, Mitropoulos A, Kanellopoulos N, Keiderling U and Wiedenmann A 1997 *Physica* B **234-236** 402
[30] Katsaros F K, Steriotis T A, Stefanopoulos K L, Kanellopoulos N K, Mitropoulos A C, Meissner M and Hoser A 2000 *Physica* B **276-278** 901
[31] Stefanopoulos K L, Romanos G E, Mitropoulos A C, Kanellopoulos N K and Heenan R K 1999 *J. Membr. Sci.* **153** 1
[32] Steriotis T A, Stefanopoulos K L, Mitropoulos A C, Kanellopoulos N K, Hoser A and Hofmann M 2002 *Appl. Phys.* A **74** S1133
[33] Steriotis T A, Stefanopoulos K L, Kanellopoulos N K, Mitropoulos A C and Hoser A 2004 *Colloids Surf.* A **1-3** 239
[34] Zickler G A, Jähnert S, Wagermaier W, Funari S S, Findenegg G H and Paris O 2006 *Phys. Rev.* B **73** 184109
[35] Roussel T, Pellenq R J-M, Bienfait M, Vix-Guterl C, Gadiou R, Béguin F and Johnson M 2006 *Langmuir* **22** 4614
[36] Hoinkis E and Röhl-Kuhn B 2006 *J. Coll. Interf. Sci.* **296** 256
[37] Eschricht N, Hoinkis E and Mädler F 2007 *Langmuir* **23** 2145
[38] Sel O, Brandt A, Wallacher D, Thommes M and Smarsly B 2007 *Langmuir* **23** 4724
[39] Muroyama N, Yoshimura A, Kubota Y, Miyasaka K, Ohsuna T, Ryoo R, Ravikovitch P I, Neimark A V, Takata M and Terasaki O 2008 *J. Phys. Chem.* C **112** 10803
[40] Jähnert S, Müter D, Prass J, Zickler G A, Paris O and Findenegg G H 2009 *J. Phys. Chem.* C **113** 15201
[41] Mascotto S, Wallacher D, Kuschel A, Polarz S, Zickler G A, Timmann A and Smarsly B M 2010 *Langmuir* **26** 6583
[42] Erko, M Wallacher D, Brandt A and Paris O 2010 *J. Appl. Cryst.* **43** 1
[43] Kruk M, Jaroniec M, Ryoo R and Joo S H 2000 *J. Phys. Chem.* B **104** 7960
[44] Ryoo R, Joo S H, Kruk M and Jaroniec M 2001 *Adv. Mater.* **13** 677
[45] Joo S H, Jun S and Ryoo R 2001 *Microporous Mesoporous Mater.* 2001 **44-45** 153
[46] Steriotis T A, Stefanopoulos K L, Katsaros F K, Gläser R, Hannon A C and Ramsay J D F 2008 *Phys. Rev.* B **78** 115424
[47] Ryoo R, Joo S H and Jun S 1999 *J. Phys. Chem.* B **103** 7743
[48] Hannon A C 2005 Nucl. *Instr. Meth. Phys. Res.* A **551** 88
[49] Stefanopoulos K L, Katsaros F K, Steriotis T A, Gläser R, Hannon A C and Ramsay J D F *J. Phys.: Conf. Ser.,* in press
[50] Day P, Enderby J E, Williams W G, Chapon L C, Hannon A C, Radaelli P G and Soper A K 2004 *Neutron News* **15** 19
[51] Radaelli P G, Hannon A C and Chapon L C 2003 *Notiz. Neut. Luce. Sinc.* **8** 19
[52] Hannon A C, Howells W S and Soper A C 1990 Neutron Scattering Data Analysis *IOP Conference Proceedings* No 107, ed M W Johnson , (Bristol, IOP) pp.193-211
[53] Lorch E 1969 *J. Phys.* C **2** 229
[54] Sing K S W, Everett D H, Haul R A W, Moscou L, Pierotti R A, Rouquérol J and Siemieniewska T 1998 *Pure Appl. Chem.* **57** 603
[55] Inoue Y, Hanzawa Y and Kaneko K 1998 *Langmuir* **14** 3079
[56] Ravikovitch P I, Vishnyakov A and Neimark A V 2000 *Langmuir* **16** 2311
[57] Ohkubo T, Miyawaki J, Kaneko K, Ryoo R and Seaton N A *J. Phys. Chem.* B **106** 6523

[58] Chiappini S, Nardone M, Ricci F P and Bellissent-Funel M C 1996 *Mol. Phys.* **89** 975
[59] van Tricht J B, Fredrikze H and van der Laan J 1984 *Mol. Phys.* **52** 115
[60] Adya A K and Wormald C J 1991 *Mol. Phys.* **74** 735
[61] Gubbins K E, Gray C G and Egelstaff P A 1973 *Mol. Phys.* **25** 1353
[62] De Santis A, Frattini R, Gazzillo D and Sampoli M 1987 *Mol. Phys.* **60** 21

Journal of Physics: Conference Series **340** (2012) 012047 doi:10.1088/1742-6596/340/1/012047

Study of Neutron-Neutron Scattering in *nd* Breakup Reaction

E Konobeevski[1]
Institute for Nuclear Research, Russian Academy of Sciences, Moscow, 117312 Russia

E-mail: konobeev@inr.ru

Abstract. To study the $n+d \rightarrow p+n+n$ breakup reaction the experimental setup allowing registration of all secondary particles in different kinematical arrangements was installed at the neutron channel of the Moscow Meson Factory of the Institute for Nuclear Research. The experiment is performed in broad energy range of neutrons (20-100 MeV) incident on deuterium target. The first preliminary data on neutron-neutron scattering length a_{nn} obtained in the final state interaction (FSI) geometry are presented. For E_n = 40 MeV and $\Delta\Theta$ = 6° the value a_{nn} = -17.9 ± 1.0 fm is obtained. Test measurements in the np quasi-free scattering (QFS) arrangement showed that we are able to obtain data on *nn* QFS in a broad range of neutron energy.

1. Introduction

In 1932, Heisenberg established the so–called charge independence, which implies that if electromagnetic effects are eliminated the scattering of protons on protons, protons on neutrons and neutrons on neutrons yields nearly the same results. A weaker condition is the so–called charge symmetry, which states that the *pp* and *nn* force should be equal. Now it is known that the charge independence is broken, for example by the small difference between the proton and neutron mass. The exact form of the *NN* potential can be obtained most strictly by studying the simplest systems, for example, the deuteron and nucleon-nucleon scattering experiments. But the absence of free neutron target prevents study of direct neutron-neutron scattering and direct comparison of *nn* and *pp* interaction. Thus investigation of *nn* scattering is performed only by studying the final state interaction (FSI) of two neutrons or quasi-free neutron-neutron scattering in reactions with few-body systems.

Among the few-body reactions the *nd* breakup is the most powerful tool for studying *nn* interaction. In view of absence of a free neutron target the use of deuterium target and neutron beam is as a matter of fact a unique and effective way of studying *nn* interaction. The rather small number of nucleons in the system allows the accurate strict solution of the three-body problem. Despite the simplicity of final *pnn* system the experiments can be performed in different kinematical arrangements of the outgoing three nucleons and their results can be compared with rigorous theoretical predictions.

2. Discrepancies in *nn* quasi-free scattering and *nd* Space Star configuration

Important arguments for continuation both experimental and theoretical works in this area are the clear discrepancies between the theory and existing data. The strongest discrepancies occur in the nn quasi-free scattering (QFS). In this geometry the two neutrons are detected at angles close to those of *nn* elastic scattering while the outgoing proton is at rest in the laboratory system. In *nd* breakup also the *np* QFS is possible.

The experimental *nn* QFS cross sections data were obtained at $E_n = 26$ MeV [1] and $E_n = 25$ MeV [2]. These data overestimate the *nd* theory by ~18%. It is contrasted with the fact that the theory described nicely the *np* QFS cross section data obtained in nd breakup reaction at $E_n = 26$ MeV [1]. Theoretical analysis performed by H.Witala and W.Glökle [3] shows that the theoretical results are quite stable under exchange of the standard nuclear potentials. Also the modern 3*N* forces have negligible effect on the QFS configurations. Authors proposed that the disagreement is connected with the unsettled properties of the *nn* force [3]. Thus, by simply multiplying the *nn* $^1S_0$ force matrix element by a factor $\lambda = 1.08$ one can perfectly well reconcile theory and data. Thereby it turned out that the *nn* QFS peak height is very sensitive to effective range parameter r_{eff} and hardly sensitive to a_{nn}. The outcome for an agreement with the data is the requirement that r_{eff} decreases from the value $r_{\mathrm{eff}} =$ 2.75 fm to a significantly smaller one: $r_{\mathrm{eff}} = 2.41$ fm. That strongly breaks charge symmetry and charge independence and is not supported by the present day chiral potential theory.

In the space star (SST) configuration, three nucleons are flying away in the plane perpendicular to the incoming beam in the center of mass system with momenta of equal magnitude. The low-energy *nd* SST cross sections are clearly underestimated by rigorous *nd* theoretical predictions. The theoretical cross sections practically do not depend on the *NN* potential used in the calculations and do not change significantly if any of the present day 3*NF* models is included [4]. Even such a large increase of the *nn* $^1S_0$ force strength by a factor $\lambda \approx 1.08$ [3] will not remove the discrepancy for the *nd* SST configuration.

3. Discrepancies in *nn* scattering length data

The final state interaction (FSI) geometry, where two neutrons fly together with small relative energy, is widely used for determination of singlet *nn* scattering length characterizing the *nn* scattering at zero energy. Replacing one of the neutrons by a proton we may determine the neutron-proton scattering length in the same *nd* breakup reaction.

Due to existence of the virtual singlet state of two nucleons with close to zero energy, the corresponding scattering lengths a_{nn} and a_{pp} have large negative values and are rather sensitive to small differences in *nn* and *pp*-potentials (a 1% change in the potential strength results in 20-30% shift in the scattering length). Thus, precise determination of singlet scattering lengths and their difference $a_{nn}-a_{pp}$ from experimental data is a convenient way for determining the measure of charge symmetry breaking (CSB) of nuclear forces.

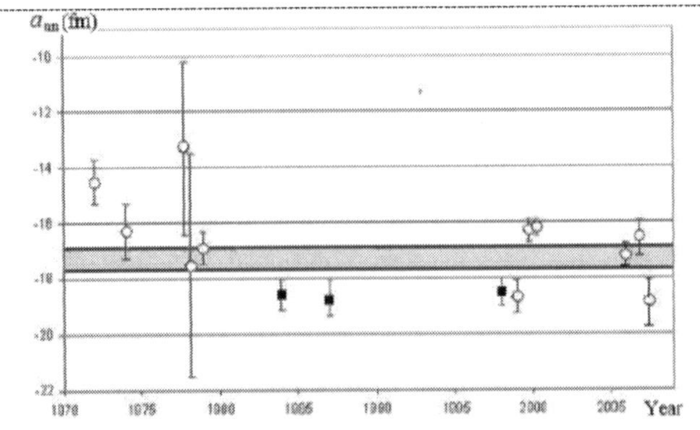

Figure 1. The recent experimental data on neutron-neutron scattering length. The gray region corresponds to a_{pp} values. Circles – data from $\pi^-+d\rightarrow\gamma+n+n$ reaction, squares – from $n+d\rightarrow p+n+n$ reaction.

In Figure 1 the recent results of determination of *nn* scattering length are shown. As on can see, the results for neutron-neutron scattering length obtained by now testify significant uncertainty of a_{nn} values which are clustered near (−16.3±0.4) fm [5,6] and (−18.5±0.4) fm [7,8]. The value a_{pp} = (−17.3±0.3) fm is determined in experiments on free *pp* scattering, and its uncertainty is related mainly to the model dependent procedure of exclusion of electromagnetic component of the *pp* interaction. So there is even uncertainty about the sign of the difference $a_{nn}-a_{pp}$, which in turn is a measure of charge symmetry breaking (CSB). In order to remove the existing uncertainty in the value of *nn* scattering length the new precise experiments are needed.

4. The experimental setup for studying nd breakup reaction

To study the *nd* breakup reaction the experimental setup allowing registration of all secondary particles was installed at the neutron channel of the Moscow Meson Factory of the Institute for Nuclear Research (INR) [9]. The setup allows us to obtain data in different kinematical arrangements in a broad energy region of neutrons (20–100 MeV).

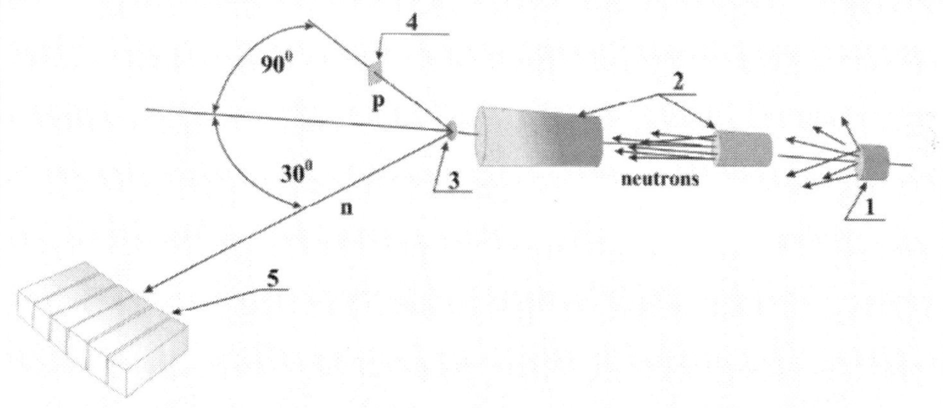

Figure 2. Schematic of the experimental setup for determining the a_{nn} scattering length in the nd breakup reaction: *1* – neutron-producing target; *2* - neutron beam collimators; *3* - deuterium (CD_2) target; *4* - proton detector; *5* - neutron hodoscope. For details see [10].

A schematic of the facility is shown in Figure 2. The beam stop of 200 MeV protons of the INR linear accelerator is used as a neutron source. The neutrons produced in the tungsten target (60 mm W) are collimated at zero angle at a length of 12 m in order to form a beam with a diameter of about 60 mm at the reaction CD_2 target. We plan to exchange the passive deuterium target by the active scintillation one. In this case the proton detection will occur in the target and accordingly the proton signal will serve as a starting signal for the time of flight neutron spectrometer.

In the case of passive target, protons are detected by a ΔE-E telescope consisting of two fast plastic scintillation detectors. Thus the proton telescope determines the proton energy and gives the start signal for TOF spectrometer. The angle of proton detector (placed on the right from the incident neutron beam axis) and its distance from the CD_2 target may be changed depending on the geometry used (FSI, QFS, SST). The neutron time-of-flight hodoscope consists of several (5-6 at the moment) detectors, located at different angles with respect to the primary neutron direction at a flight distance of 4-5 m from the CD_2 target, on the left from the incident neutron beam axis. The time resolution of all detectors was of about 0.6 ns. The angular resolution was defined by the detector cross section size and the time-of-flight distance; it was about ±0.5°.

The neutrons incident on the deuterium target have a continuous energy spectrum, with an endpoint energy equal to the proton beam energy. However, detection of three particles in coincidence (proton and two neutrons) makes it possible to reconstruct the primary neutron energy in the $n+d \rightarrow p+n+n$ reaction and obtain data on the reaction yield in a wide range of neutron energies, determined only by the statistics of the experiment.

5. Preliminary data on a_{nn}

The first experiments at our setup have been performed in the FSI geometry with the goal obtaining data on neutron-neutron scattering length. In our study we can obtain data on nd breakup reaction yield for various neutron opening angles and various energies of incident neutrons. In an experiment like this, the neutron–neutron FSI manifests itself as a peak in the reaction yield distribution versus the relative energy of two neutrons $\varepsilon = \frac{1}{2}(E_1 + E_2 - 2\sqrt{E_1 E_2}\cos\Delta\Theta)$, whose shape is very sensitive to a_{nn}.

To obtain this distribution it is necessary to detect in coincidence two neutrons flying in a narrow cone of angles with respect to the direction of motion of their center of mass, then measure the energies of both neutrons E_1 and E_2 and the angle between them, and finally analyze the shape of the dependence obtained. To extract the value of the neutron-neutron scattering length, the experimental dependence of the reaction yield on ε has to be compared with the simulation results based on the Watson-Migdal approximation [11].

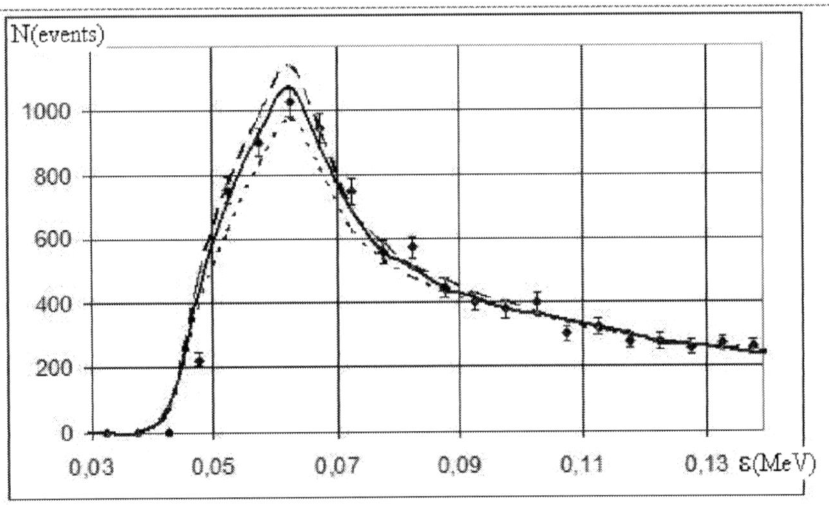

Figure 3. Comparison of the experimental dependence of reaction yield (E_n=40 MeV, $\Delta\Theta$=6°) with simulation results for different values a_{nn}. Curves with the maximum (dotted line) and minimum yield (dashed line) correspond to a_{nn} = -21.5 fm and -15.5 fm, respectively, central curve (solid line) corresponds to the value of the best fitting: –17.9 fm.

The best statistics in the experiment was obtained for an incident energy of 40 MeV and an opening angle of secondary neutrons of 6°. In Figure 3 these experimental data are compared with the simulation results for three *nn* scattering lengths (–15.5, –17.9, and –21.5 fm). During fitting, we calculated the χ^2 value for experimental and theoretical (simulated for different scattering lengths) points. To determine the scattering length and its statistical uncertainty, the values of $\chi^2(a_{nn})$ were approximated by a quadratic polynomial (see Figure 4). In this case, the minimum value χ^2_{min} determines the scattering length, and the statistical error in determining a_{nn} is given by $\Delta a_{nn} = \left| a_{nn}(\chi^2_{min}) - a_{nn}(\chi^2_{min}+1) \right|$, i.e., is defined by values of a_{nn} at χ^2_{min} and $\chi^2_{min}+1$. So the value of neutron-neutron scattering length and its statistical error determined from the data shown in Figure 4 are $a_{nn} = (-17.9 \pm 0.9)$ fm.

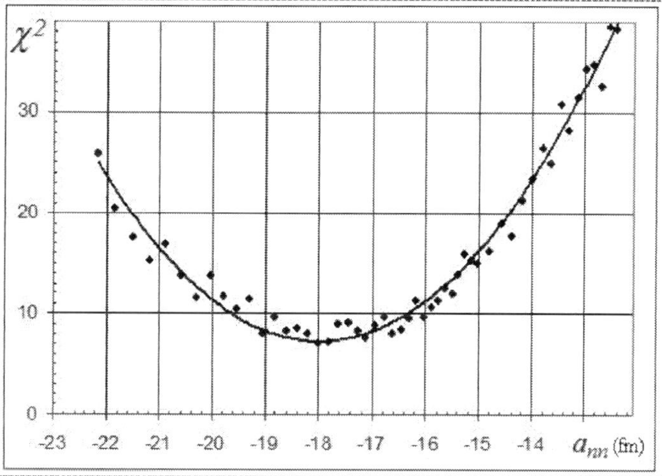

Figure 4. Dependence of χ^2 on a_{nn} for $E_n = 40 \pm 5$ MeV, $\Delta\Theta = 6^o$. The curve is an approximation of the χ^2-dependence by a quadratic polynomial. The minimum of the curve corresponds to the best fitting - $a_{nn} = -17.9$ fm.

6. Conclusion

The *nd* breakup reaction is a powerful tool to study *nn* interaction. Despite the simplicity of final *pnn* system the experiments can be performed in different kinematical arrangements of the three outgoing nucleons and their results can be compared with rigorous theoretical predictions. Important arguments for continuation both experimental and theoretical works in this area are the clear discrepancies between the theory and existing data. The strongest discrepancies occur in the *nn* quasi-free scattering (QFS) and in the *nd* space star geometries. Also the results for neutron-neutron scattering length obtained by now testify significant uncertainty of a_{nn} values.

The *nd* breakup reaction has been investigated at the neutron channel of the Moscow Meson Factory of the Institute for Nuclear Research. We presented the preliminary data on neutron-neutron scattering length, obtained in FSI geometry. Also, the test experiment in the np QFS geometry was performed. The results of the study show that we are able to perform measurements of $n+p \rightarrow p+n+n$ reaction in various kinematical geometries – FSI, QFS and SST.. The data obtained will give new valuable information on the neutron-neutron interaction in a broad range of neutron energies.

This work was supported by the Russian Foundation for Basic Research (grant 10-02-00603).

References

[1] Siepe A et al 2002 *Phys. Rev.* C **65** 034010
[2] Ruan X C et al 2007 *Phys. Rev.* C **75** 057001
[3] Witala H and Glöckle W *The nn quasi-free nd breakup cross section: discrepancies to theory and implications on the $^1S_0$ nn force* arXiv:1011.4714v1 [nucl-th]
[4] Witala H and Glöckle W 2010 *J. Phys. G: Nucl. Part. Phys.* **37** 064003
[5] Huhn V, Wätzold L, Weber C, et al. 2000 *Phys. Rev.* C **63** 014003
[6] von Witsch W, Ruan X, Witala H 2006 *Phys. Rev.* C **74** 014001
[7] Howell C R et al. 1998 *Phys. Lett.* B **444** 252
[8] Gonzalez Trotter D E et al. 2006 *Phys. Rev.* C **73** 034001
[9] Zuyev S V, Konobeevski E S, Mordovskoy M V 2008 *Instrum. Exp. Tech.* **51** 327

[10] Burmistrov Yu M, Zuyev S V, Konobeevski E S et al. 2009 *Instrum. Exp. Tech.* **52** 769
[11] Watson K M 1952 Phys. Rev. **88** 1163;
Migdal A B 1955 JETP (Sov.Phys.) **1** 2.

Neutron Decay with PERC: a Progress Report

G Konrad[1,2], H Abele[1], M Beck[2], C Drescher[3], D Dubbers[3], J Erhart[1], H Fillunger[1], C Gösselsberger[1], W Heil[2], M Horvath[1], E Jericha[1], C Klauser[1,4,5], J Klenke[6], B Märkisch[3], R K Maix[1], H Mest[3], S Nowak[1], N Rebrova[3], C Roick[3], C Sauerzopf[1], U Schmidt[3], T Soldner[4], X Wang[1] and O Zimmer[4]
(The PERC collaboration)

[1]Atominstitut, Technische Universität Wien, Stadionallee 2, A-1020 Wien, Austria
[2]Institut für Physik, Universität Mainz, D-55099 Mainz, Germany
[3]Physikalisches Institut, Universität Heidelberg, D-69120 Heidelberg, Germany
[4]Institut Laue-Langevin, F-38042 Grenoble, France
[5]Excellence Cluster Universe, Technische Universität München, D-85748 Garching, Germany
[6]Forschungs-Neutronenquelle Heinz Maier-Leibnitz, D-85747 Garching, Germany

E-mail: gkonrad@ati.ac.at

Abstract.
The PERC collaboration will perform high-precision measurements of angular correlations in neutron beta decay at the beam facility MEPHISTO of the Forschungs-Neutronenquelle Heinz Maier-Leibnitz in Munich, Germany. The new beam station PERC, a clean, bright, and versatile source of neutron decay products, is designed to improve the sensitivity of neutron decay studies by one order of magnitude. The charged decay products are collected by a strong longitudinal magnetic field directly from inside a neutron guide. This combination provides the highest phase space density of decay products. A magnetic mirror serves to perform precise cuts in phase space, reducing related systematic errors. The new instrument PERC is under development by an international collaboration. The physics motivation, sensitivity, and applications of PERC as well as the status of the design and preliminary results on uncertainties in proton spectroscopy are presented in this paper.

1. Introduction

High-precision measurements of observables in neutron beta decay address a number of questions which are at the forefront of particle physics [1–3], and are generally complementary to direct searches in high-energy physics. Main emphasis lies on the search for new physics beyond the Standard Model (SM) of elementary particles and fields. Possible extensions require new symmetry concepts like left-right symmetry, fundamental fermion compositeness, new particles, leptoquarks, supersymmetry, supergravity, or many more [4, 5]. Free neutron decay is a very active field, with a dozen new instruments planned or under construction worldwide. For recent reviews see Refs. [2, 3, 6, 7]. With the new Proton and Electron Radiation Channel (PERC) [8] several symmetry tests based on neutron beta decay data become competitive [9].

In the modern form of the SM, the differential decay rate of neutrons can be written as [10]:

$$d^3\Gamma = \frac{1}{(2\pi)^5}\frac{G_F^2|V_{ud}|^2}{2}p_e E_e (E_0 - E_e)^2 \, dE_e d\Omega_e d\Omega_\nu \times$$

$$\xi\left[1+a\frac{\mathbf{p}_\mathrm{e}\cdot\mathbf{p}_\nu}{E_\mathrm{e}E_\nu}+b\frac{m_\mathrm{e}}{E_\mathrm{e}}+\frac{\langle\mathbf{s}_\mathrm{n}\rangle}{s_\mathrm{n}}\left(A\frac{\mathbf{p}_\mathrm{e}}{E_\mathrm{e}}+B\frac{\mathbf{p}_\nu}{E_\nu}+D\frac{\mathbf{p}_\mathrm{e}\times\mathbf{p}_\nu}{E_\mathrm{e}E_\nu}\right)\right]. \tag{1}$$

Here, G_F is the Fermi weak coupling constant, V_ud is the upper left element of the Cabibbo–Kobayashi–Maskawa (CKM) matrix [11, 12], $\mathbf{p}_\mathrm{e}$, $\mathbf{p}_\nu$, E_e, and E_ν are the electron (neutrino) momenta and total energies, respectively, E_0 is the electron spectrum endpoint total energy, m_e is the electron mass, $\mathbf{s}_\mathrm{n}$ is the neutron spin, and the Ω_i denote solid angles. Quantity ξ is a factor inversely proportional to the neutron decay rate, a, A, B, and D are the angular correlation coefficients, while b is the Fierz interference term. The neutrino electron correlation coefficient a and the Fierz term b are measurable in decays of unpolarized neutrons, while the beta asymmetry parameter A, the neutrino asymmetry parameter B, and the triple correlation coefficient D require polarized neutrons. We mention that in the presence of left-handed (LH) scalar (S) and tensor (T) couplings B depends on the electron energy: $B = B_0 + b_\nu\frac{m_\mathrm{e}}{E_\mathrm{e}}$, where b_ν is another Fierz-like parameter similar to b [10, 13]. We note that a, A, and B_0 are sensitive to non-SM couplings only in second order, while b and b_ν depend in first order on LH S and T couplings. A non-zero Fierz term b would indicate the existence of LH S and T interactions. A non-zero triple correlation D would violate time reversal invariance. The most sensitive measurement of D in nuclear beta decay has been conducted in the beta decay of polarized neutrons [14] (see also [15]).

If the electron spin $\mathbf{s}_\mathrm{e}$ is observed more correlation coefficients called G, N, Q, and R appear in the differential decay rate of neutrons [10]:

$$\begin{aligned}\mathrm{d}^2\Gamma &= \frac{1}{(2\pi)^4}\frac{G_\mathrm{F}^2|V_\mathrm{ud}|^2}{2}p_\mathrm{e}E_\mathrm{e}\left(E_0-E_\mathrm{e}\right)^2\mathrm{d}E_\mathrm{e}\mathrm{d}\Omega_\mathrm{e}\times\xi\left[1+b\frac{m_\mathrm{e}}{E_\mathrm{e}}+A\frac{\langle\mathbf{s}_\mathrm{n}\rangle}{s_\mathrm{n}}\frac{\mathbf{p}_\mathrm{e}}{E_\mathrm{e}}\right.\\ &\left.+\,\mathbf{s}_\mathrm{e}\left(G\frac{\mathbf{p}_\mathrm{e}}{E_\mathrm{e}}+N\frac{\langle\mathbf{s}_\mathrm{n}\rangle}{s_\mathrm{n}}+Q\frac{\mathbf{p}_\mathrm{e}}{E_\mathrm{e}+m_\mathrm{e}}\left(\frac{\langle\mathbf{s}_\mathrm{n}\rangle}{s_\mathrm{n}}\cdot\frac{\mathbf{p}_\mathrm{e}}{E_\mathrm{e}}\right)+R\frac{\langle\mathbf{s}_\mathrm{n}\rangle}{s_\mathrm{n}}\times\frac{\mathbf{p}_\mathrm{e}}{E_\mathrm{e}}\right)\right].\end{aligned} \tag{2}$$

We note that N and R depend linearly on S and T couplings. First experimental limits on N and R have been presented in [16, 17].

Another observable is C, the proton asymmetry relative to the neutron spin. Neglecting recoil-order effects and radiative corrections, the proton asymmetry C is expressed by [13, 18]:

$$C = -x_C(A + B_0), \tag{3}$$

where $x_C = 0.27484$ is a kinematical factor[1].

The present status on the correlation coefficients a, A, B, C, D, N, and R, the Fierz terms b and b_ν, and the neutron lifetime τ_n is summarized in Ref. [3]. According to the Particle Data Group's (PDG) 2011 review [19], the relative errors on a, A, B, and C are 4 %, 0.9 %, 0.3 %, and 1 %, respectively. Recently, three beta asymmetry experiments have completed their analyses, namely UCNA [20], PERKEO II [21][2], and PERKEO III [23]. The PERKEO III collaboration improved the uncertainty on A by about a factor of 5 compared to the PDG 2011 average (preliminary). For D, R, b, and b_ν only upper limits have been derived.

Within the framework of the SM, neutron beta decay is described as a purely left-handed, V–A interaction. Then, $b = 0$ and $b_\nu = 0$, and also $D = 0$ if time reversal invariance is assumed, and the correlation coefficients a, A, B, and C depend only on the ratio $\lambda = g_\mathrm{A}/g_\mathrm{V}$ of the weak axial-vector to the vector coupling constant:

$$a = \frac{1-|\lambda|^2}{1+3|\lambda|^2},\quad A = -2\frac{|\lambda|^2+\lambda}{1+3|\lambda|^2},\quad B = 2\frac{|\lambda|^2-\lambda}{1+3|\lambda|^2},\quad \text{and}\quad C = x_C\frac{4\lambda}{1+3|\lambda|^2}. \tag{4}$$

[1] Note that we define the proton asymmetry parameter C with opposite sign compared to [13]. This retains the convention that a positive asymmetry indicates more particles to be emitted in the spin direction.
[2] Publication of the result of the last PERKEO II run [22] is underway.

Near the value $\lambda = -1.27$, the sensitivities of a, A, B, and C to λ are:

$$\frac{\mathrm{d}a}{\mathrm{d}\lambda} = 0.298, \qquad \frac{\mathrm{d}A}{\mathrm{d}\lambda} = 0.374, \qquad \frac{\mathrm{d}B}{\mathrm{d}\lambda} = 0.076, \qquad \text{and} \qquad \frac{\mathrm{d}C}{\mathrm{d}\lambda} = -0.124. \tag{5}$$

The size of the weak coupling constants is important for applications in cosmology (e.g., primordial nucleosynthesis), astronomy (e.g., solar physics), and particle physics (e.g., neutrino detectors) [2, 3, 6]. The value of λ can be determined from several independent neutron decay observables, each with different sensitivity to non-SM physics. Comparing the various values of λ therefore provides an important test of the validity of the SM (see, e.g., [9, 24]).

The neutron lifetime τ_{n} is inversely proportional to $|V_{\mathrm{ud}}|^2(1+3|\lambda|^2)$ [25]. Hence, independent measurements of τ_{n} and of an observable sensitive to λ allow the determination of V_{ud}. Along with V_{us} and V_{ub} from K-meson and B-meson decays, respectively, the unitarity of the CKM matrix is tested [19, 26].

2. Facility PERC

PERC is a new type of beam station for the measurement of angular correlations in the beta decay of free neutrons. In contrast to existing neutron decay spectrometers, PERC is a user instrument which delivers at its exit not neutrons but an intense beam of decay electrons and protons, under well defined and precisely variable conditions. Depending on the observable, different secondary spectrometers can be used, cf. Secs. 2.3 and 2.4. Thus, with PERC, we can measure the shapes and magnitudes of electron or proton energy spectra from polarized or unpolarized neutron decay, in pulsed or continuous neutron beam mode, with or without electron spin analysis. Many quantities can be derived from such spectra, some for the first time. From single particle spectra we can obtain:

- the correlation coefficients a, A, and C,
- the Fierz interference term b, yet unmeasured in neutron beta decay, and
- the electron helicity H_{e}.

We note that the achievable accuracies of a and C depend heavily on the systematic uncertainties in the spectroscopy of decay protons, which have not yet been completely analyzed, cf. Sec. 2.3. From the measured values of a, b, A, C, and the precise energy or momentum spectra, we derive:

- the ratio λ of the weak axial-vector to the vector coupling constant,
- the element V_{ud} of the CKM quark-mixing matrix,
- the neutrino asymmetry parameter B and the Fierz-like parameter b_ν,
- scalar g_{S} and tensor g_{T} admixtures,
- mass m_2 and mixing angle ζ of a W_{R} boson, mediating right-handed interactions,
- the weak magnetism f_2 and second class g_2 form factors, etc.

PERC will be built by an international collaboration with the Universities of Heidelberg and Mainz, the Technische Universität München, the Institut Laue-Langevin in Grenoble, and the Vienna University of Technology. The instrument will be set up at a new position of the beam facility MEPHISTO of the Forschungs-Neutronenquelle Heinz Maier-Leibnitz (FRM II) in Munich, Germany.

The final design of PERC has to be fixed in close cooperation with the manufacturers. In the following, we discuss only one possible layout.

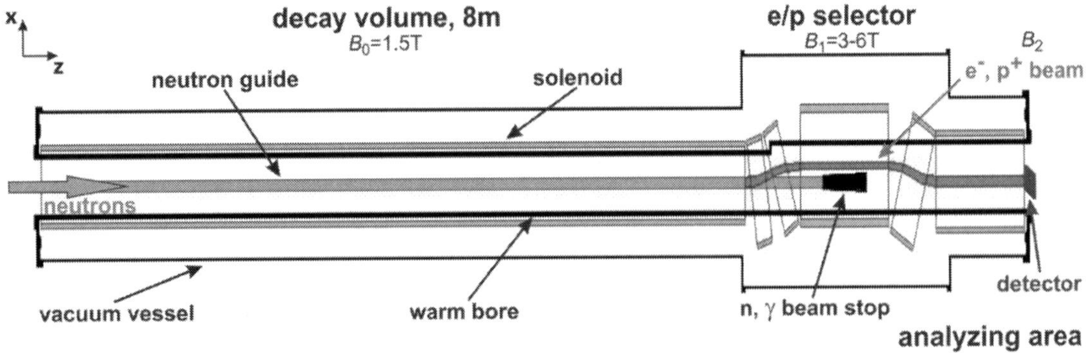

Figure 1. Scheme of the facility PERC: Cold neutrons (green) pass through the decay volume where only a small fraction decays. The decay products (red) are guided by the strong magnetic field towards the detector (blue). The superconducting coils are drawn in gray. We note that the tilted coils are also solenoids; their view is due to different longitudinal and transverse scales of the drawing. The equipment for neutron beam preparation, like velocity selector, polarizer, spin flipper, or chopper, is located in front of the instrument (to the left of the scheme) and not shown here. For details see [8].

2.1. Measurement principles and instrument

PERC is a beam station, which delivers at its exit neutron decay products. The set-up is schematically shown in Fig. 1 and in Ref. [8]; its design principles are thoroughly discussed in [8]. Here, we will only summarize the essential parts: Cold neutrons pass through an 8 meters long neutron guide, with a cross section of $6 \times 6\,\mathrm{cm}^2$, where about 10^6 neutrons decay per second and per meter of guide. The neutron guide is surrounded by a superconducting solenoid ($B_0 = 1.5\,\mathrm{T}$) of equal length. Decay electrons and protons are guided by the strong longitudinal magnetic field towards the electron/proton (e/p) detection system, i.e., detectors specialized for certain tasks (see also Secs. 2.3 and 2.4). This combination provides the highest phase space density of decay products. The count rates of neutron decay products will be increased by a factor of 100 compared to competing experiments, with the exception of PERKEO III[3]. At the end of the neutron guide, the decay products can be separated from the neutron beam by means of bending coils. In the e/p selector, the decay products pass a region of strongly enhanced magnetic field $B_1 > B_0$ ($B_1 = 3 - 6\,\mathrm{T}$) before they reach the detector (at, e.g., $B_2 = 0.5\,\mathrm{T}$), as can be seen from Fig. 2. The field B_1 acts as a magnetic mirror and transmits only the fraction $(1 - \cos\theta_C)/2$ ($\approx B_0/4B_1$, for $B_0 < B_1$) of all decay products, namely those emitted upstream under angles $\theta_0 \le \theta_C = \arcsin\sqrt{B_0/B_1}$ to the $z-$axis.

Field variations $B(z)$ must be slow enough such that the decay electrons and protons are transported adiabatically. Then, the magnetic transport no longer depends on the energy of the decay products. The condition for adiabatic transport can be formulated as the requirement that the quantity γ, defined in [28] as:

$$\gamma = \frac{2\pi p_e \cos\theta_0}{eB^2}\left|\frac{\partial B}{\partial z}\right|, \tag{6}$$

is small ($\gamma \ll 1$). Our magnetic field calculations show that adiabatic transport is guaranteed along the particle trajectories, for all design concepts under consideration.

[3] In PERKEO III, the count rates have already increased by about an order of magnitude [27].

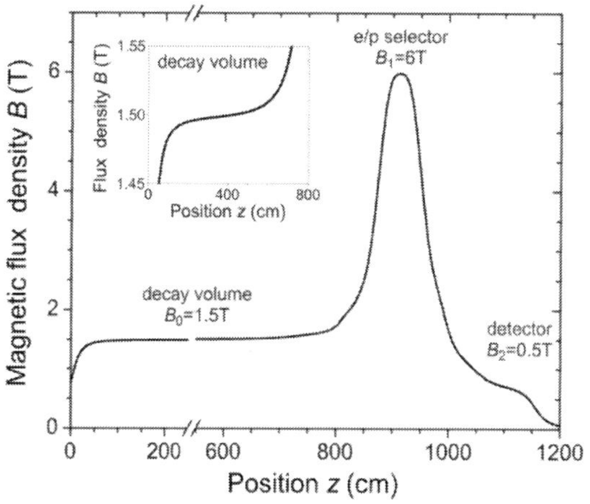

Figure 2. Magnetic flux density $B(z)$ along the central field line (preliminary), for $B_1 = 6\,\mathrm{T}$. The magnetic mirror field ratio B_1/B_0 determines the divergence of the emerging beam of decay electrons and protons by the cut-off angle $\theta_C = \arcsin\sqrt{B_0/B_1}$.

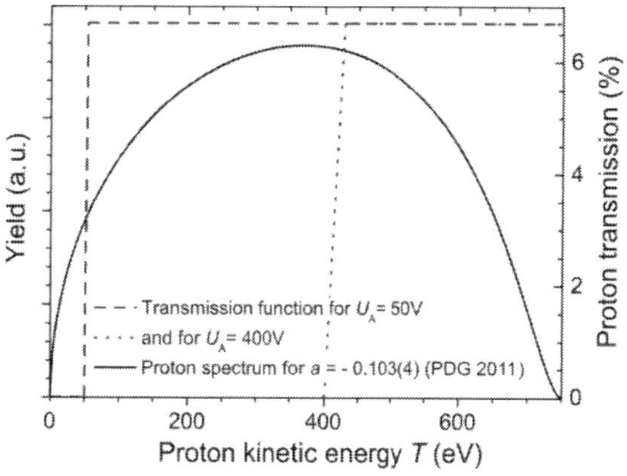

Figure 3. The dashed and dotted line indicate the transmission function for protons in unpolarized neutron decay (preliminary), for $B_1 = 6\,\mathrm{T}$ and aSPECT as detection system, with $B_2 = 0.4\,\mathrm{T}$ and $U_A = 50\,\mathrm{V}$ respectively $U_A = 400\,\mathrm{V}$. For elucidation, the solid line shows the proton energy spectrum.

2.2. *Measurement uncertainties and systematics*

The magnetic mirror serves to limit the phase space precisely, reducing related systematic errors. Systematic errors related to electron spectroscopy have been shown to be on the level of 10^{-4}, more than 10 times better than that achieved today [8]. There is one exception related to the knowledge of the neutron beam polarization, where presently the error is on the 10^{-3} level [29]. Techniques for the polarization of a roughly monochromatic ($\Delta\lambda/\lambda \approx 10\,\%$) cold neutron beam will be improved towards the 10^{-4} level, cf. Sec. 2.4.

Further details on the sensitivity and the applications of PERC may be found in [8]. For details on the neutron beam preparation and polarization analysis we refer to [30–32].

2.3. *Dominant uncertainties in the analysis of decay protons*

Depending on the decay parameters studied with PERC, the analysis of the decay electrons and protons will be performed with specialized detectors. As far as electrons are concerned, this can be done with an energy sensitive detector. For protons, PERC will feed a charged particle spectrometer, for instance an adapted spectrometer which can partially be based on the aSPECT [33–36] detection system. aSPECT is a retardation spectrometer which measures the proton recoil spectrum by counting all decay protons that overcome an electrostatic barrier U_A. Such a measurement allows the determination of the correlation coefficients a and C. The most important associated systematic uncertainties for aSPECT as detection system for PERC are:

- *Homogeneity of the magnetic field*: In the adiabatic approximation, the transmission function as shown in Fig. 3 can be calculated analytically, cf. Refs. [33, 34]. Our initial calculations indicate that the magnetic fields B_0, B_1, and B_2 must be controlled at the level of $\Delta B_0/B_0 < 1\,\%$, $\Delta B_1/B_1 = 1 \times 10^{-4}$, and $\Delta B_2/B_2 = 1 \times 10^{-4}$, respectively, in order to keep $\Delta a/a < 0.1\,\%$. The condition $\Delta B_0/B_0 < 1\,\%$ can be fullfilled in pulsed neutron beam mode. Then, the decay products will be counted in the detector only while the neutron pulse is fully contained within the central, homogeneous part of the magnetic field, as can be seen from Fig. 2 (zoom to the decay volume). Our magnetic field calculations show that the condition $\Delta B_1/B_1 = 1 \times 10^{-4}$ is fulfilled within the flux of neutron decay products. For aSPECT, it has been shown that $\Delta B_2/B_2$ is on the level of 10^{-4} [37].

- *Homogeneity of the electric field*: Our first estimates demonstrate that the electric potential between decay volume, e/p selector, and electrostatic barrier will have to be known with an accuracy of better than 10 mV, comparable to aSPECT and Nab [38].
- *Doppler effect due to neutron motion*: Unlike in aSPECT or Nab, the Doppler effect is not negligible as the neutron beam is collinear to the detection system. For a Gaussian neutron spectrum, preliminary estimates show that the mean neutron energy has to be known with a precision of better than 10^{-2}, whereas the uncertainty in width of the spectrum is negligible. An approximately Gaussian neutron spectrum can be obtained, e.g., by a neutron velocity selector.
- *Adiabaticity of proton motion*: For the determination of a, adiabatic transport must be guaranteed between decay volume and electrostatic barrier. Equation 6 is a first estimate for the adiabaticity of the proton motion.
- *Residual gas*: Simulations for aSPECT and Nab [39] have demonstrated that changes in the extracted value of a are tolerable if the residual gas pressure can be reduced to 10^{-8} mbar independent of the type of gas, and negligible if the pressure can be reduced to 10^{-9} mbar.
- *Particle trapping*: Surface potential variations in the order of 100 mV have been found in electrodes prepared within the aSPECT project [35]. Such variations can lead to local field extrema within the decay volume, the e/p selector, or the electrostatic barrier, and can therefore give rise to potential penning traps. For the same reason, local magnetic field minima must be avoided.

All these systematic effects are on the simulation agenda, and will be analyzed in due course. Measures to reduce and to precisely control systematic errors, like an online NMR system for monitoring of the magnetic field with 10^{-4} accuracy [37] or investigations of the patch effect with surface potential variations < 10 mV [24] (see also [40]), have already been initiated.

2.4. Sub-Projects

In spring 2010, the PERC instrument was approved by the German Research Foundation (DFG). Within the Priority Programme (SPP) 1491 'Precision experiments in particle and astrophysics with cold and ultracold neutrons' the DFG established seven projects which comprise, among others, the design and the construction of

- the superconducting magnet system,
- a non-depolarizing neutron guide,
- a pulsed spatial neutron magnetic spin resonator [30, 31],
- detectors for electron energy spectroscopy and proton spectroscopy,

the design of

- a spectrometer for simultaneous electron and proton momentum spectroscopy,
- a spectrometer for proton spectroscopy, and
- the neutron beam line (neutron polarization and polarization analysis with 10^{-4} accuracy; first results of the development phase see [32]).

The neutron beam line will be provided by the FRM II.

3. Summary

The new beam station PERC is under development by an international collaboration. The collaboration plans high-precision measurements of correlation coefficients in neutron beta decay, some of them for the first time. With PERC, the sensitivity of neutron decay studies will be improved by one order of magnitude. Thus, precise symmetry tests of various kinds are coming within reach with the new facility PERC.

The instrument PERC received approval by the DFG in spring 2010. It will be set up at the beam facility MEPHISTO of the FRM II in Munich, Germany. The first milestone, the design of the magnet system together with magnetic field calculations for strong magnetic fields, has been achieved. We are now ready for the call for tenders for the superconducting magnet system.

Acknowledgments

We would like to thank F. Glück for useful discussions. We are grateful to A. Senft who carried out first calculations of the Doppler effect, cf. Sec. 2.3. This work is supported by the German Research Foundation as part of the Priority Programme 1491 under contracts No. HE 2308/9-1, MA 4944/1-1, SCH 2708/1-1, SO 1058/1-1, and ZI 816/1-1, the Austrian Science Fund under contracts No. I 528-N20 and I 534-N20, the Vienna University of Technology, and the Forschungs-Neutronenquelle Heinz Maier-Leibnitz.

References

[1] Severijns N, Beck M and Naviliat-Čunčić O 2006 *Rev. Mod. Phys.* **78** 991

[2] Abele H 2008 *Prog. Part. Nucl. Phys.* **60** 1

[3] Dubbers D and Schmidt M G 2011 *Rev. Mod. Phys.* , in print URL `arXiv:1105.3694v1 [hep-ph]`

[4] Herczeg P 2001 *Prog. Part. Nucl. Phys.* **46** 413

[5] Ramsey-Musolf M J and Su S 2008 *Phys. Rev.* **456** 1

[6] Nico J S 2009 *J. Phys. G: Nucl. Part. Phys.* **36** 104001

[7] Soldner T *et al.* (eds) 2009 *Proceedings of the International Workshop on Particle Physics with Slow Neutrons, Grenoble, France, 2008* (Nucl. Instr. and Meth. A **611**)

[8] Dubbers D *et al.* 2008 *Nucl. Instr. and Meth. A* **596** 238 ,
for an extended version, see URL `arXiv:0709.4440v1 [nucl-ex]`

[9] Konrad G *et al.* 2011 *Proceedings of the 5th International Conference Beyond 2010, Cape Town, South Africa* ed Klapdor-Kleingrothaus H V, Krivosheina I V and Viollier R (World Scientific, Singapore) p 660 ISBN 978-981-4340-85-4 URL `arXiv:1007.3027v2 [nucl-ex]`

[10] Jackson J D, Treiman S B and Wyld H W 1957 *Phys. Rev.* **106** 517

[11] Cabibbo N 1963 *Phys. Rev. Lett.* **10** 531

[12] Kobayashi M and Maskawa T 1973 *Prog. Theor. Phys.* **49** 652

[13] Glück F, Joó J and Last J 1995 *Nucl. Phys. A* **593** 125

[14] Mumm H P *et al.* 2011 URL `arXiv:1104.2778v2 [nucl-ex]`

[15] Soldner T *et al.* 2004 *Phys. Lett. B* **581** 49

[16] Kozela A *et al.* 2009 *Phys. Rev. Lett.* **102** 172301

[17] Kozela A *et al.* 2011 *Acta Physica Polonica B* **42** 789

[18] Glück F 1996 *Phys. Lett. B* **376** 25

[19] Nakamura K *et al* (Particle Data Group) 2010 *J. Phys. G: Nucl. Part. Phys.* **37** 075021 and 2011 partial update for the 2012 edition URL `http://pdg.lbl.go`

[20] Liu J *et al.* 2010 *Phys. Rev. Lett.* **105** 181803

[21] Abele H 2009 *Nucl. Instr. and Meth. A* **611** 193

[22] Mund D 2006 Ph.D. thesis Ruprecht-Karls-Universität Heidelberg

[23] Mest H 2011 Ph.D. thesis Ruprecht-Karls-Universität Heidelberg

[24] Konrad G 2011 Ph.D. thesis Johannes Gutenberg-Universität Mainz , in progress

[25] Marciano W J and Sirlin A 2006 *Phys. Rev. Lett.* **96** 032002

[26] Czarnecki A, Marciano W J and Sirlin A 2004 *Phys. Rev. D* **70** 093006

[27] Märkisch B *et al.* 2009 *Nucl. Instr. and Meth. A* **611** 216

[28] Kruit P and Read F H 1983 *J. Phys. E* **16** 313

[29] Kreuz M *et al.* 2005 *Nucl. Instr. and Meth. A* **547** 583

[30] Gösselsberger C *et al.* 2011 *Physics Procedia* , accepted,
see also: Badurek G, Gösselsberger C and Jericha E 2011 *Physica B* **406** 2458

[31] Gösselsberger C *et al.* 2011 *J. Phys.: Conf. Series* , these proceedings

[32] Klauser C *et al.* 2011 *J. Phys.: Conf. Series* , these proceedings

[33] Zimmer O *et al.* 2000 *Nucl. Instr. and Meth. A* **440** 548

[34] Glück F *et al.* 2005 *Eur. Phys. J. A* **23** 135

[35] Baeßler S *et al.* 2008 *Eur. Phys. J. A* **38** 17

[36] Konrad G *et al.* 2009 *Nucl. Phys. A* **827** 529c

[37] Ayala Guardia F 2011 Ph.D. thesis Johannes Gutenberg-Universität Mainz , in progress

[38] Počanić D *et al.* 2009 *Nucl. Instr. and Meth. A* **611** 211

[39] Alarcon R *et al.* 2010 Nab proposal update and funding request
`http://nab.phys.virginia.edu/nab_doe_fund_prop.pdf`

[40] McGovern S 2010 Master's thesis University of Virginia

Journal of Physics: Conference Series **340** (2012) 012049 doi:10.1088/1742-6596/340/1/012049

Structural study of supercritical carbon dioxide confined in nanoporous silica by *in situ* neutron diffraction

K L Stefanopoulos[1*], Th A Steriotis[1], F K Katsaros[1], N K Kanellopoulos[1], A C Hannon[2] and J D F Ramsay[3]

[1] Institute of Physical Chemistry, NCSR "Demokritos", Ag. Paraskevi Attikis, 15310, Greece

[2] ISIS Facility, Rutherford Appleton Laboratory, Chilton, Didcot, Oxon OX11 0QX, UK

[3] Laboratoire des Matériaux et Procédés Membranaires, UMR CNRS 5635, Université Montpellier II, Montpellier, France

*stefan@chem.demokritos.gr

Abstract. *In situ* neutron diffraction measurements of adsorbed carbon dioxide in an ordered mesoporous silica (MCM-41) have been carried out along an isotherm slightly above the critical point (308 K) and at a range of pressures below and above the critical one (30-125 bar). The experiment has been performed with the aid of a novel high-pressure adsorption apparatus and the GEM diffractometer (ISIS, Rutherford Appleton Laboratory, UK). Diffraction measurements of both bulk supercritical and bulk liquid carbon dioxide have also been carried out. The structure factors and the total differential correlation functions of the adsorbed carbon dioxide suggest that the confined carbon dioxide has *liquid-like* properties. However, some differences are observed when the confined phase is compared to those of bulk liquid and bulk supercritical fluid respectively, mainly arising from orientational correlations between adsorbed molecules. These differences could be attributed either to strong interactions between the silica walls and CO_2 molecules or to the confinement of the fluid combined with the relatively large quadrupole moment of carbon dioxide.

1. Introduction

Supercritical fluids are gaseous compounds, which are compressed to a pressure higher than the critical pressure (P_c) above their critical temperature (T_c) and have nowadays been attracting scientific as well as industrial interest because of their unusual physical properties, since their density can vary significantly even with a slight change of the temperature and pressure. Based on these properties, supercritical fluids, supercritical fluids are widely used in industry as solvents for extraction and for chemical and biochemical reactions. [1-3]. In many cases carbon dioxide, the simplest liquid composed of linear molecules, is the solvent of choice because it is environmentally benign, abundant, inexpensive, non-flammable, non-toxic and its supercritical state is easily accessible with a T_c=304.1 K and P_c=73.8 bar. On the other hand, sorption of fluids on nanoporous solids is very important in a series of applications such as catalysis, H_2 and natural gas upgrade and storage, (chromatographic pressure swing adsorption and membrane) separations, biological and geological processes etc. [4-6]. These processes are extremely complex from a fundamental point of view as the properties of sorbed fluids are different from the bulk due to confinement. As a result, there is an increasing interest in the experimental study of fluids confined in nanoporous materials [7-10].

Adsorption measurements [11] are abundantly used for the study of pore-confined fluids, however such methods can only reveal the statistical "ensemble" macroscopic properties and do not provide information on their molecular structure. On the other hand techniques such as small-angle scattering (SAS) of x-rays or neutrons (SAXS, SANS) [12-16], contrast-variation or contrast-matching SANS [17-26] are essential tools for elucidating the microstructure of nanoporous materials and for resolving structural characteristic of each phase in multiphase materials. Further, adsorption in conjunction with *in situ* SAS or diffraction can provide in depth information about both the adsorption mechanism and the structural organisation of the confined phase [27-41].

In the present study *in situ* neutron diffraction measurements of CO_2 confined in a purely siliceous MCM-41 have been carried out along an isotherm slightly above the critical point (T=308 K) and at a range of pressures below and above the critical one (30, 50 and 125 bar respectively). Furthermore, neutron diffraction measurements of both bulk supercritical and bulk liquid CO_2 thermodynamic states have also been performed. For this reason, a high-pressure adsorption apparatus has been constructed. MCM-41 is an ideal candidate for adsorption studies because it consists of a two-dimensional hexagonal arrangement of cylindrical mesopores with uniform size.

2. Experimental

The adsorption/neutron diffraction experiment was carried out at General Materials diffractometer (GEM), at ISIS pulsed neutron and muon source (UK Rutherford). GEM is one of the most advanced materials diffractometer in the world. The most important reason in achieving this has been the detector array, which has a very large area of 7,270 m^2 and a very wide range in scattering angles correaponding to a Q range varying between 0.02 $Å^{-1}$ and 40 $Å^{-1}$. For a full description of the diffractometer, see ref 42.

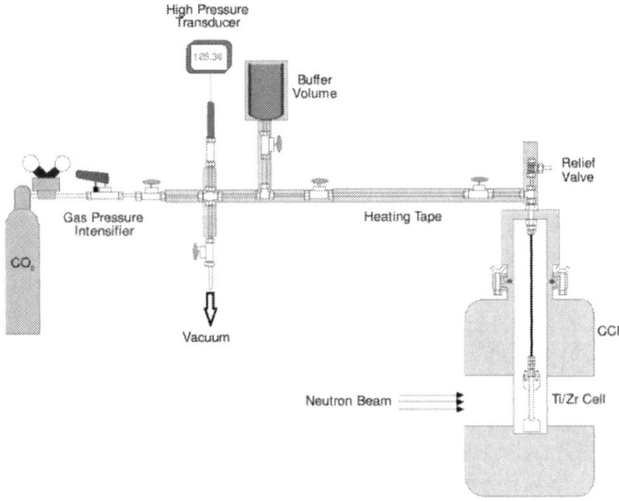

Figure 1. High-pressure adsorption device for carrying out *in situ* neutron diffraction measurements.

A high-pressure adsorption device (high vacuum up to 150 bar) has been constructed for performing *in situ* diffraction measurements (figure 1). A thin wall cylindrical titanium zirconium (Ti/Zr) null scattering alloy, having an overall nominal coherent scattering length of zero, was used as sample container. The cell was connected via a copper flange and a Swagelok® union to a 1/8" (OD) 316 SS capillary. The set-up is inserted in a CCR (close-cycle-refrigerator) stick and connected to a manifold equipped with a relief valve (250 bar), a pressure transducer (0-150 bar) as well as gas and vacuum connections. The gas line was equipped with a pressure intensifier and connected with a high purity CO_2 cylinder (99.9996%) system with the aid of a reducer. The sample was weighted and then was inserted in the cell and after sealing (with copper flange) and leak testing (helium), it was outgassed outside the CCR (500 K, 12 h 10^{-6} mbar). After cooling down to room temperature the whole stick was inserted in the CCR, sample temperature was set to 308 K, and doses of CO_2 were admitted, while equilibration was monitored by pressure readings. After measuring the diffraction spectra at 30 50 and 125 bar, the sample was removed from the cell and measurements of empty cell as well as bulk supercritical (T=308 K, P=125 bar) and bulk liquid (T=253 K, P=19.7 bar) CO_2 were performed.

The scattered neutrons were counted as a function of neutron time-of-flight for eight detector banks. The raw data were corrected for background, attenuation and multiple scattering and were normalised to a standard vanadium scatterer using the *gudrun* program [43-44]. The Placzek inelasticity correction and the merging of data from different detector banks, were performed using the ATLAS suite of programs [45]. Finally, the data were Fourier transformed up to Q_{max}=15 Å^{-1} using the Lorch modification function for reducing the termination ripples [46];

3. Results and discussion

The synthesis and the structural characteristics of MCM-41 based on powder x-ray diffraction, N_2 and CO_2 adsorption isotherms can be found elsewhere [47-48]. In brief, the specific surface area and the total pore volume have values 1204 m^2/g and 0.87 cm^3/g respectively. The unit cell parameter is approximately 40 Å^{-1} whereas the mean pore radius was estimated to be 12 Å according to BJH (Barrett–Joyner–Halenda) method and 17.5 Å according to non-local density functional theory (NLDFT) method.

The structure factor of bulk liquid CO_2 is shown in figure 2 (top curve). The intermolecular structure peak is located at about 1.79 Å^{-1}, whereas a long range oscillation corresponding to intramolecular correlations is observed at larger Q values. The main diffraction peak has a minimum at Q~3 Å^{-1} followed by a tiny bump near 4 Å^{-1}. The results are in excellent agreement with previous similar neutron diffraction measurements of bulk CO_2 [49-50]. It is worth mentioning that in some cases (e.g. 220 K, 58 bar) a shoulder has been observed on the high Q side, (Q~2.2 Å^{-1}), of the main diffraction peak [50]. Based on theoretical calculations, Gubbins *et al* [51] attributed the origin of this feature to the quadrupole-quadrupole contribution of the anisotropic part of the structure factor. Adya and Wormand [50] claimed that this hump is both temperature and density dependent. On the other hand, Chiappini *et al* [52] argued that this feature was only density dependent based on neutron diffraction experiments of bulk carbon dioxide in dense superctitical states. In our case, such a shoulder is not expected due to the lower density of the bulk phase. The main peak possesses a degree of asymmetry, which can be attributed to intermolecular orientational correlations between neighbouring CO_2 molecules arising mainly from electrical quadrupolar interactions. In particular, Chiappini *et al* [52] performed diffraction experiments on bulk supercritical CO_2 under high pressures and they also observed a strongly asymmetric main peak. By comparing their data with a calculated intermolecular structure factor for an orientationally uncorrelated model CO_2, they concluded that the asymmetry of the peak is due to the existence of the orientational correlations. In addition, the structure factor of the bulk supercritical fluid has many similarities to that of the bulk liquid. The main peak asymmetry, the minimum at about 3 Å^{-1}, the tiny bump at Q~3.7 Å^{-1} (to a less extent) and the long range intramolecular peaks are also present. One should note, however, that the position of the main peak has shifted slightly to lower angles (Q~1.74 Å^{-1}) due to the fact that the supercritical fluid has lower density than that of the liquid.

The structure factors, for the confined fluid at 30, 50 and 125 bar are also presented in figure 2. As in the case of bulk thermodynamic states, the patterns exhibit clearly a peak corresponding to the most probable distance between nearest neighbour molecules, while the long range oscillations at larger Q values are also present. The peak position shifts slightly to the higher Q region with increasing pressure and reaches that of the bulk supercritical fluid. As in the case of the bulk states, the main peaks of the confined CO_2 have a minimum at about Q=3 $Å^{-1}$. The structure factors of adsorbed CO_2, however, exhibit additionally a shoulder, q_1, on the high Q side of the peak (Q~2.7 $Å^{-1}$) followed by a more enhanced bump, q_2, at lower scattering vectors (Q~3.5 $Å^{-1}$) compared to the one observed at the bulk liquid. One should also mention that both features are mostly pronounced at 30 bar and to a less extent at 50 bar (figure 2b). A possible explanation is that at lower pressures the scattering is dominated by the adsorbed monolayer. This argument is further enhanced by the fact that these features have also been observed for carbon dioxide adsorbed in the same material at a subcritical thermodynamic state (T=273 K, P=5 bar) corresponding to monolayer coverage [48]. The confined phase seems to be in a quasi-liquid state where orientational correlations are somehow enhanced. This might be attributed to the effect of Van der Waals and Coulombic surface landscape in combination with interactions between neighbouring CO_2 molecules directly attached to the surface.

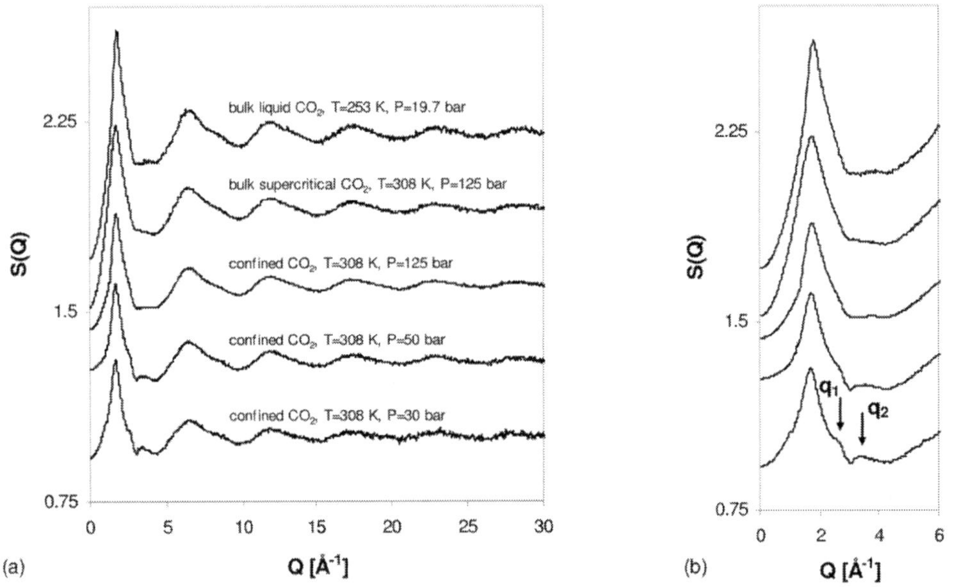

Figure 2. The total-scattering structure factor for confined CO_2 (308 K) at 30 bar, 50 bar, 125 bar, for bulk supercritical fluid at 125 bar (308 K) and for bulk liquid at 19.7 bar (253 K); the order is from the bottom to the top. The structure factors have shifted by 0.3 for clarity. (a) the whole Q range and (b) the low Q region.

Figure 3 illustrates the total differential correlation function, $D(r)$, for all thermodynamic states (bulk and confined respectively). In all cases, one clearly observes two intramolecular peaks (completely separated from the intermolecular interactions) occurring at 1.16 Å and 2.32 Å which correspond to the C-O (r_{CO}) and O-O (r_{OO}) distances, with $r_{OO}=2r_{CO}$. Furthermore, the intermolecular part of all the correlation functions shows two broad features centered at ~4 Å and ~8 Å, arising respectively from the first and second neighbour interactions. In the case of bulk liquid CO_2, however, the first neighbour peak reveals three structures located at about r_1=3.3 Å r_2=4 Å, and r_3=5.2 Å respectively (figure 3a), in agreement with literature data on bulk liquid CO_2 [50]. Molecular dynamics simulations have shown that the structure at r_1 is attributed mainly to O-O as well as to C-O pair correlations. Moreover, both C-C and C-O correlations give a positive contribution at r_2, while the structure at r_3 arises from bumps present in the partial C-O and O-O atom-atom pair correlation functions, when quadropole forces are included in the simulations [52, 53]. On the other hand, in the case of bulk supercritical fluid the structure r_1 is not present and there is also a slight shift of the position of the first neighbour peak at ~4.2 Å due to its lower density.

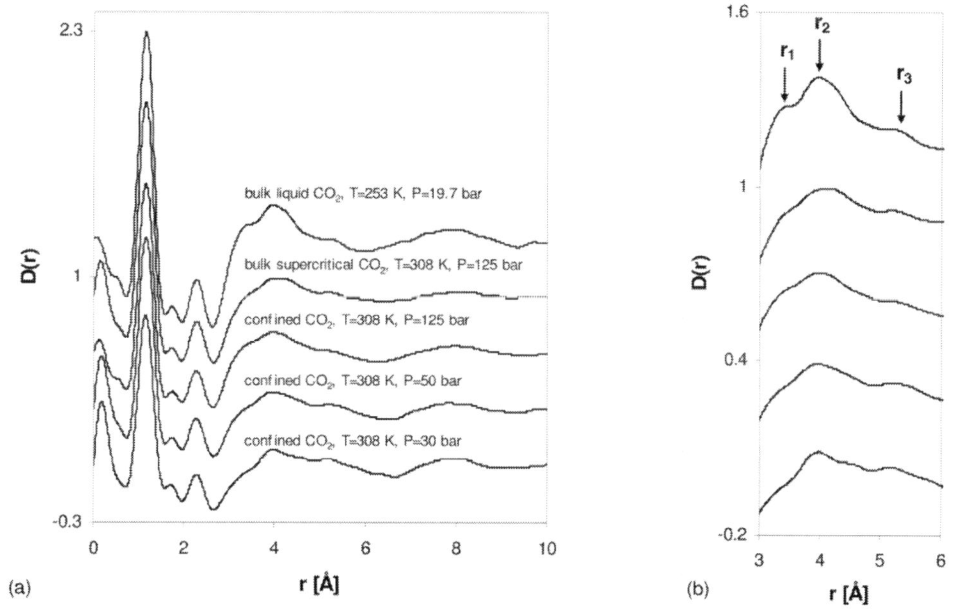

Figure 3. The differential correlation function for confined CO_2 (308 K) at 30 bar, 50 bar, 125 bar, for bulk supercritical fluid at 125 bar (308 K) and for bulk liquid at 19.7 bar (253 K); the order is from the bottom to the top. The correlation functions have shifted by 0.3 for clarity. (a) the whole real space distance and (b) details of the first neighbour region.

The molecular arrangement in confined states seems to be closer to that of the bulk supercritical fluid, however, certain differences are observed. First of all, the structure r_2 is narrower and a new feature is observed at ~4.6 Å for the confined carbon dioxide state at 30 bar. Further, the structure r_3 is more enhanced, especially for the confined fluid at 30 and 50 bar respectively (figure 3b). Again, r_3, was a prominent feature in the case of adsorbed CO_2 under subcritical conditions corresponding to the monolayer coverage [48]. Based on simulation results, this structure was proved to be strongly potential dependent, becoming evident only when electric quadrupolar interactions are taken into account and can be ascribed to a sharpening of the O-O correlations, arising from increased orientational ordering of the fluid [52]. This seems to be a plausible explanation, as at lower pressures the adsorbed monolayer is subject to enhanced Lennard-Jones and Coulombic interactions with the surface atoms. Based on this approach the silica surface seems to somehow organise the monolayer in a more "oriented" manner in order to counterbalance the strong adsorption potential with the finite size of the surface.

4. Conclusions

The structural characteristics of supercritical carbon dioxide molecules confined in a mesoporous MCM-41 material have been studied by sorption with *in situ* neutron diffraction measurements. For this purpose, a novel high-pressure adsorption device has been constructed and a diffractometer with a wide accessible Q range was selected. The structure factors of CO_2 and the total differential correlation functions suggest that the confined fluid has at all stages studied *liquid-like* properties, however, some subtle differences pointing to stronger orientational correlations inside the pores were observed. These differences were attributed to either pore wall-fluid interactions or the confinement of the fluid, combined with the relatively large quadrupole moment of carbon dioxide. Sorption with *in situ* neutron diffraction provides an excellent tool to study confined phases within pores In addition, a future combination of molecular simulation approaches coupled with our results would throw more light in the molecular arrangement under confinement.

Acknowledgements
We are grateful to the ISIS technical staff for their great assistance provided for the construction of the high-pressure adsorption apparatus. This research project has been supported by the European Commission under the 7th Framework Programme and it is part of the Research Infrastructures action of the Capacities Programme (NMI3), Contract No: 226507.

References

[1] Clifford T 1999 *Fundamentals of Supercritical Fluids* (London: Oxford University Press)
[2] Kiran E, Debenedetti P G and Peters C J 2000 *Supercritical Fluids Fundamentals and Applications* (Dordrecht: Kluwer)
[3] Knox D E, 2005 *Pure Appl. Chem.* **77** 513
[4] Ruthven D M, Shamasuzzaman F and Knaebel K S 1994 *Pressure Swing Adsorption* (New York: VCH)
[5] Noh J S, Argaval R K and Schwarz J A 1987 *Int. J. Hydrogen Energy* **12** 693
[6] Matranga K R, Myers A L and Glandt E D 1982 *Chem. Eng. Sci.* **47** 1569
[7] Gelb L D, Gubbins K E, Radhakrishnan R and Sliwinska-Bartkowiak M 1999 *Rep. Prog. Phys.* **62** 1573
[8] Frick B, Zorn R and Büttner H (ed) 2000 *Proceedings of the International Workshop on Dynamics in Confinement, J. Phys* IV **10** pp 3-343
[9] Christenson H K 2001 *J. Phys.: Condens. Matter* **13** R95
[10] Frick B, Koza M and Zorn R (ed) 2003 *Proceedings of the Second International Workshop on Dynamics in Confinement, Eur. Phys J.* E **12** pp 5-194
[11] Cregg S J and Sing K S W 1982 *Adsorption, Surface Area and Porosity* (New York: Academic Press)
[12] Mitropoulos A C, Kanellopoulos N K, Stefanopoulos K L and Heenan, R K 1998 *J. Coll. Inter. Sci.* **203** 229
[13] Ramsay J D F 1998 *Adv. Coll. Inter. Sci.* **76-77** 13
[14] Makri P K, Stefanopoulos K L, Mitropoulos A C, Kanellopoulos N K and Treimer W 2000 *Physica B* **276-278** 477
[15] Hoinkis E, Lima E B and Schubert-Bischoff P 2004 *Langmuir* **20** 8823
[16] Kikkinides E S, Stefanopoulos K L, Steriotis T A, Mitropoulos A C, Kanellopoulos N K and Treimer W 2002 *Appl. Phys.* A **74** S954
[17] Stuhrmann H B and Miller A. 1978 *J. Appl. Cryst.* **11** 325
[18] Ramsay J D F and Wing G 1991 *J. Coll. Interf. Sci* **141** 475
[19] Charalambopoulou G C, Steriotis T A, Mitropoulos A C, Stefanopoulos K L, Kanellopoulos N K and Ioffe A 1998 *J. Invest. Derm.* **110** 988
[20] Charalambopoulou G C, Steriotis T A, Hauss T, Stefanopoulos K L and Stubos A K 2002 *Appl. Phys.* A **74** S1245
[21] Steriotis T A, Stefanopoulos K L, Keiderling U, De Stefanis A and Tomlinson A A G 2002 *Chem. Commun.* **20** 2396
[22] Calo J M and Hall P J 2004 *Carbon* **42** 1299
[23] Lelong G, Price D L, Brady J W and Saboungi M L 2007 *J Chem. Phys.* **127** 065102
[24] Shin T, Findenegg G H and Brandt A 2006 *Prog. Coll. Polym. Sci.* **133** 116

[25] Mergia K, Stefanopoulos K L, Ordás N and García-Rosales C *2010 Microporous Mesoporous Mater.* **134** 141
[26] Müter D, Shin T, Demé B, Fratzl P, Paris O and Findenegg G H 2010 *J Phys. Chem. Lett.* **1** 1442
[27] Mitropoulos A C, Stefanopoulos K L and Kanellopoulos N K 1998 *Microporous Mesoporous Mater.* **24** 29
[28] Ramsay J D F and Hoinkis E 1998 *Physica* B **248** 322
[29] Katsaros F K, Steriotis T A, Stefanopoulos K L, Kanellopoulos N K, Mitropoulos A C, Meissner M and Hoser A 2000 *Physica* B **276-278** 901
[30] Stefanopoulos K L, Romanos G E, Mitropoulos A C, Kanellopoulos N K and Heenan R K 1999 *J. Membr. Sci.* **153** 1
[31] Steriotis T A, Stefanopoulos K L, Mitropoulos A C, Kanellopoulos N K, Hoser A and Hofmann M 2002 *Appl. Phys.* A **74** S1133
[32] Steriotis T A, Stefanopoulos K L, Kanellopoulos N K, Mitropoulos A C and Hoser A 2004 *Colloids Surf.* A **1-3** 239
[33] Zickler G A, Jähnert S, Wagermaier W, Funari S S, Findenegg G H and Paris O 2006 *Phys. Rev.* B **73** 184109
[34] Roussel T, Pellenq R J-M, Bienfait M, Vix-Guterl C, Gadiou R, Béguin F and Johnson M 2006 *Langmuir* **22** 4614
[35] Hoinkis E and Röhl-Kuhn B 2006 *J. Coll. Interf. Sci.* **296** 256
[36] Eschricht N, Hoinkis E and Mädler F 2007 *Langmuir* **23** 2145
[37] Sel O, Brandt A, Wallacher D, Thommes M and Smarsly B 2007 *Langmuir* **23** 4724
[38] Muroyama N, Yoshimura A, Kubota Y, Miyasaka K, Ohsuna T, Ryoo R, Ravikovitch P I, Neimark A V, Takata M and Terasaki O 2008 *J. Phys. Chem.* C **112** 10803
[39] Jähnert S, Müter D, Prass J, Zickler G A, Paris O and Findenegg G H 2009 *J. Phys. Chem.* C **113** 15201
[40] Mascotto S, Wallacher D, Kuschel A, Polarz S, Zickler G A, Timmann A and Smarsly B M 2010 *Langmuir* **26** 6583
[41] Erko, M Wallacher D, Brandt A and Paris O 2010 *J. Appl. Cryst.* **43** 1
[42] Hannon A C 2005 *Nucl. Instr. Meth. Phys. Res.* A **551** 88
[43] Day P, Enderby J E, Williams W G, Chapon L C, Hannon A C, Radaelli P G and Soper A K 2004 *Neutron News* **15** 19
[44] Radaelli P G, Hannon A C and Chapon L C 2003 *Notiz. Neut. Luce. Sinc.* **8** 19
[45] Hannon A C, Howells W S and Soper A C 1990 Neutron Scattering Data Analysis *IOP Conference Proceedings* No 107, ed M W Johnson , (Bristol, IOP) pp.193-211
[46] Lorch E 1969 *J. Phys.* C **2** 229
[47] Boger S T, Roesky R, Gläser R, Ernst S, Eigenberger G and Weitkamp J 1997 *Microporous Mater.* **8** 79
[48] Steriotis T A, Stefanopoulos K L, Katsaros F K, Gläser R, Hannon A C and Ramsay J D F 2008 *Phys. Rev.* B **78** 115424
[49] van Tricht J B, Fredrikze H and van der Laan J 1984 *Mol. Phys.* **52** 115
[50] Adya A K and Wormald C J 1991 *Mol. Phys.* **74** 735
[51] Gubbins K E, Gray C G and Egelstaff P A 1973 *Mol. Phys.* **25** 1353
[52] Chiappini S, Nardone M, Ricci F P and Bellissent-Funel M C 1996 *Mol. Phys.* **89** 975
[53] De Santis A, Frattini R, Gazzillo D and Sampoli M 1987 *Mol. Phys.* **60** 21

Effect of non-Newtonian gravity on the amplitude of neutron reflection from bulk materials

Kaoru Taketani
High Energy Accelerator Research Organization,
1-1 Oho, Tsukuba, Ibaraki 305-0801, Japan

E-mail: kaoru.taketani@kek.jp

Abstract. We have analysed the effect of non-Newtonian gravity on the amplitude of neutron reflection from bulk materials. We found that the first-order contribution of non-Newtonian gravity is significant when the neutron wavelength is nearly equal to the radius of curvature of the neutron refractive index due to non-Newtonian gravity and that it becomes negligible when the neutron wavelength is much smaller than the radius of curvature.

1. Introduction

Several speculative theories have being proposed with the aim of unifying quantum theory and general relativity. Some of these theories predict that the law of gravity will be modified at low accelerations. Several experiments have been performed in attempts to confirm the existence of this modification. However, no deviations have been observed from the predictions of Newtonian gravity [1], although the experiments do not provide strong constraints for length scales smaller than a millimetre. Attempts at verification in this range are attracting interest because several theories involving extra dimensions [2–4] or theories that postulate scalar particles originating from supersymmetry breaking [5] predict a non-Newtonian gravity-like force in this range. In experimental attempts to verify these predictions, measurements have been performed by varying the positions of several materials while measuring the forces between them [6]. When the distance between the materials is less than 100 nm, the effect of the Casimir force [7] cannot be neglected because the materials have finite dielectric constants. Therefore, to obtain experimental constraints on non-Newtonian gravity requires precisely evaluating Casimir forces.

On the other hand, it is not necessary to consider Casimir forces when performing experiments using neutrons because neutrons have a very small dielectric constant. Consequently, experiments employing neutrons can provide the strongest constraints for length scales smaller than 100 nm [8]. The neutron scattering amplitudes of materials are used when determining the constraints [9–12]. Because the neutron–nucleon scattering length, the neutron–electron scattering length, and the non-Newtonian gravity all contribute to the scattering amplitude and they have different momentum transfer dependences, three amplitudes with different momentum transfers have to be measured to estimate the non-Newtonian contribution. Further improvements could be achieved if experiments could be performed that are not affected by the neutron–electron scattering amplitude. Because the neutron charge radius is less than 10^{-11} m, the neutron-electron scattering contribution to the total scattering amplitude can be ignored when very cold neutrons are used. Nesvizhevsky and coworkers have proposed an experiment that involves measuring two neutron scattering amplitudes from rare noble gases by using very cold neutrons [13].

In this paper, we investigate the effects of non-Newtonian gravity on the reflection amplitude of very cold neutrons from bulk materials. In section 2, we introduce the notation for the Schrödinger equations inside and outside the bulk material. As the contribution of non-Newtonian gravity to the neutron potential is limited to the region near the bulk surface, we can select the two independent solutions of the Schrödinger equations that approach plane waves far from the surface. Taking the product of the reflection and transmission coefficients of these solutions, we obtain the wave functions for this setup. Using the continuities of the wave function and its logarithmic derivative on the surface of the material, we obtain the reflection amplitude from the material in section 3. In section 4, we analyse the reflection amplitude analytically by assuming that the non-Newtonian gravity potential is much smaller than the kinetic energy. In section 5, we numerically evaluate the contribution of non-Newtonian gravity to the reflection amplitude. Finally, we summarise the results obtained in this paper in section 6.

2. Schrödinger equations inside and outside a bulk material

We first parameterise the correction to Newton gravity as follows:

$$V(r) = -\frac{GMm}{r}\left(1 + \alpha_g e^{-\frac{r}{\lambda_g}}\right). \tag{1}$$

Here, G is the gravitational constant, M is the mass of the bulk material's element, m is the neutron mass, r is the distance from the bulk material's element to the neutron, α_g is a dimensionless parameter that indicates the strength of the deviation from the Newtonian potential, and λ_g is the effective range of the anomalous potential. As we are interested in the case for which α_g is much larger than unity, we ignore the Newtonian part of the gravitational potential. We define the z-axis as the normal to the bulk surface and take the origin of the z-axis to be on the surface. Moreover, we assume the bulk extends infinitely in the xy-plane and that the material exists in the positive z range. In this case, we obtain that the potential for a neutron along the z-axis by integrating (1) over the volume of the bulk material, which gives:

$$V(z) = \begin{cases} V_g e^{+2k_g z} & z < 0 \\ V_f + V_g\left(2 - e^{-2k_g z}\right) & z > 0, \end{cases} \tag{2}$$

where

$$k_g = \frac{1}{\lambda_g}, V_g = \frac{\alpha}{2k_g^2}, \text{and } \alpha = -Gm\pi\rho\alpha_g. \tag{3}$$

Here, V_f and ρ are the Fermi potential and the weight density of the bulk material, respectively.

In this study, we consider the case when the kinetic energy is larger than potential energy. In this case, the Schrödinger equations along the z-axis outside and inside the material can be written as:

$$\begin{cases} \Psi''(z) + \left(k_0^2 - \eta e^{+2k_g z}\right)\Psi(z) = 0 & z < 0 \\ \Psi''(z) + \left(k_1^2 + \eta e^{-2k_g z}\right)\Psi(z) = 0 & z > 0, \end{cases} \tag{4}$$

where

$$k_0 = k, k_1 = \sqrt{k^2 - \frac{2m}{\hbar^2}\left(V_f + 2V_g\right)}, \text{and } \eta = \frac{2m}{\hbar^2} V_g. \tag{5}$$

Here, $\hbar$ is Planck's constant and k is the z-component of the neutron wavenumber.

3. Exact value of the reflection amplitude

The analytical solution of (4) can be obtained by using Bessel functions. Outside the material, the two independent solutions that approach $e^{\pm ik_0 z}$ far from the boundary can be written as:

$$\phi_{o,\pm}(z) = \Gamma(1 \pm i\kappa_o)\left(\frac{\sqrt{\eta}}{2k_g}\right)^{\mp i\kappa_o} I_{\pm i\kappa_o}\left(e^{+k_g z}\frac{\sqrt{\eta}}{k_g}\right). \tag{6}$$

Here, κ_o is the ratio of the neutron wavenumber to the inverse of the effective range of non-Newtonian gravity, k_o/k_g, $\Gamma(\mu)$ is the gamma function, and $I_\nu(\zeta)$ is the modified Bessel function of the first kind. Far from the boundary, the parameter of the Bessel function, ζ, becomes small, which allows us to employ the following approximation:

$$I_\nu(\zeta) \sim \left(\frac{\zeta}{2}\right)^\nu \frac{1}{\Gamma(1+\nu)}. \tag{7}$$

Using this approximation, we can verify that (6) approaches $e^{\pm ik_o z}$. Inside the material, the two independent solutions that approach $e^{\pm ik_i z}$ far from the boundary are given by:

$$\phi_{i,\pm}(z) = \Gamma(1 \mp i\kappa_i)\left(\frac{\sqrt{\eta}}{2k_g}\right)^{\pm i\kappa_i} J_{\pm i\kappa_i}\left(e^{-k_g z}\frac{\sqrt{\eta}}{k_g}\right). \tag{8}$$

Here, κ_i is the ratio of neutron wavenumber inside the material to the inverse of the effective range of non-Newtonian gravity, k_i/k_g, and $J_\nu(\zeta)$ is the Bessel function of the first kind. When z is far from the boundary, the parameter of the Bessel function, ζ, becomes small. In this case, we can use the following approximation:

$$J_\nu(\zeta) \sim \left(\frac{\zeta}{2}\right)^\nu \frac{1}{\Gamma(1+\nu)}. \tag{9}$$

Applying this approximation to (8) confirms that it approaches $e^{\pm ik_i z}$.

If we normalise the absolute value of the incident wave function outside the material, $\phi_{o,+}(z)$ to unity, the wave functions outside and inside the material can be written as:

$$\phi_{o,+}(z) + R\phi_{o,-}(z), \text{and } T\,\phi_{i,+}(z). \tag{10}$$

Here, R and T are determined by the continuities of wave function and its logarithmic derivative at the material surface. The result can be written as:

$$R = -\frac{\phi_{o,+}(0)}{\phi_{o,-}(0)}\left.\frac{\frac{d}{dz}\ln\phi_{i,+}(z) - \frac{d}{dz}\ln\phi_{o,+}(z)}{\frac{d}{dz}\ln\phi_{i,+}(z) - \frac{d}{dz}\ln\phi_{o,-}(z)}\right|_{z=0}. \tag{11}$$

4. Perturbation analysis of the reflection amplitude

We now analytically analyse the reflection amplitude by assuming that the potential due to non-Newtonian gravity is much smaller than the kinetic energy. As η is small in this case, the parameter of the Bessel functions, ζ, becomes small and the following approximation are valid:

$$\begin{pmatrix} J_\nu(z) \\ I_\nu(z) \end{pmatrix} \sim \frac{\left(\frac{\zeta}{2}\right)^\nu}{\Gamma(1+\nu)} \begin{pmatrix} 1 - \left(\frac{\zeta}{2}\right)^2 \frac{1}{1+\nu} \\ 1 + \left(\frac{\zeta}{2}\right)^2 \frac{1}{1+\nu} \end{pmatrix}. \tag{12}$$

Using these approximations, we obtain the following approximations

$$\begin{pmatrix} \phi_{o,\pm}(0) \\ \left.\frac{d}{dz}\ln\phi_{o,\pm}(z)\right|_{z=0} \\ \left.\frac{d}{dz}\ln\phi_{i,\pm}(z)\right|_{z=0} \end{pmatrix} \sim \begin{pmatrix} 1 - x_o \pm i\,y_{o,a} \\ \pm i\,k_o\left(1 + 2x_o \mp i\left(y_{o,a} - y_{o,b}\right)\right) \\ \pm i\,k_i\left(1 + 2x_i \pm i\left(y_{i,a} - y_{i,b}\right)\right) \end{pmatrix}, \tag{13}$$

where

$$x_o = \frac{\kappa_o}{1+\kappa_o^2}\theta_o, y_{o,a} = \frac{\kappa_o^2}{1+\kappa_o^2}\theta_o, y_{o,b} = \frac{1}{1+\kappa_o^2}\theta_o, \tag{14}$$

$$x_1 = \frac{\kappa_1}{1+\kappa_1^2}\theta_1, y_{1,a} = \frac{\kappa_1^2}{1+\kappa_1^2}\theta_1, \text{and } y_{1,b} = \frac{1}{1+\kappa_1^2}\theta_1.$$

Here, θ_o is defined as

$$\theta_o = \int_{-\infty}^{0}\left(\sqrt{k_o^2 - \eta e^{+2k_g z}} - k_o\right)dz \sim -\frac{\eta}{2k_o}\int_{-\infty}^{0} e^{+2k_g z}dz = -\frac{mV_g}{2k_o k_g \hbar^2}, \tag{15}$$

and θ_i is defined as

$$\theta_i = \int_{0}^{\infty}\left(\sqrt{k_i^2 - \eta e^{-2k_g z}} - k_i\right)dz \sim -\frac{\eta}{2k_i}\int_{0}^{\infty} e^{-2k_g z}dz = \frac{mV_g}{2k_i k_g \hbar^2}. \tag{16}$$

θ_o and θ_i are equal to the phase shifts due to the potential with finite curvature outside and inside the material, respectively. Figure 1 shows the effect of κ_o on the ratio of the phase shift to the correction terms, x_o/θ_o, $y_{o,a}/\theta_o$, and $y_{o,b}/\theta_o$. When k_o is much larger than k_g, the anomalous potential only changes the phase of the wave functions at the boundary, $\phi_{o,\pm}(0)$. On the other hand, when $k_o = k_g$, the anomalous potential changes not only the phase but also the amplitude of the wave function at the boundary, $\phi_{o,\pm}(0)$. When k_o is much smaller than k_g, the effective range of the anomalous potential becomes smaller than the interatomic distance of the bulk material. In this case, the potential shown in equation (2) becomes inappropriate. Therefore, we ignore this case in this study.

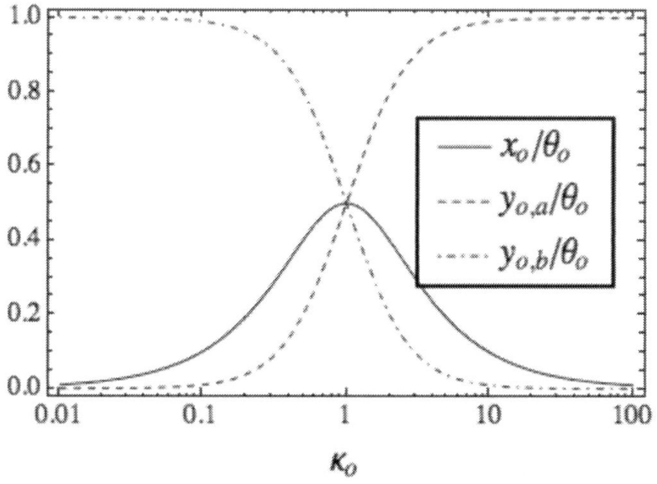

Figure 1. Effect of κ_o on the dependence of the ratio of the phase shift to the correction terms, x_o/θ_o, $y_{o,a}/\theta_o$, and $y_{o,a}/\theta_o$

Next, we approximate the amplitude of neutron reflection from bulk material using the above approximations for the wave functions. In this study, we restrict our attention to situations in which the kinetic energy far exceeds the Fermi potential and the difference of the refractive index inside the material from unity, dn, is small. In this case, both θ_o and dn are small. Therefore, we approximate the reflection amplitude by considering only first-order terms of both θ_o and dn. The resulting approximated reflection amplitude is $\theta_o - dn/2$ when $k_o = k_g$ and $-dn/2$ when $k_o >> k_g$. When $k_o = k_g$, the sign of the reflection amplitude, $\theta_o - dn/2$, can be reversed if θ_o is sufficiently large or dn is sufficiently small. In these cases, the argument of the reflection amplitude changes π. On the other hand, this phenomenon does not happen when $k_o >> k_g$.

5. Constraints on non-Newtonian gravity

We now numerically evaluate the obtained reflection amplitudes. As we are interested in measuring the effect of non-Newtonian gravity, we have to identify the contribution from the Fermi potential. Because the absolute uncertainty is expected to be proportional to the absolute value of the Fermi potential, it seems best to measure the reflection amplitude from materials with small Fermi potentials.

On the other hand, the contribution of non-Newtonian gravity is proportional to the weight density of the material. Therefore, we numerically evaluate the reflection amplitude from materials that have small Fermi potentials and are solids at normal temperature.

As an example, we consider the precision required to achieve the strongest constraints when $\lambda_g =$ 10 nm by measuring the reflection amplitude. In this case, the strongest constraint on α_g is that α_g is smaller than about 10^{20} [8,13]. Therefore, we numerically evaluated dn and θ_0, when $k = k_g = 1/10$ nm and $\alpha_g = 10^{20}$. Table 1 shows the results and parameters used for two materials that have small Fermi potentials and are solid at normal temperature. As the reflectivity is the square of the reflection amplitude, the result shows that we can obtain comparable constraints if we can measure the reflectivity due to the Fermi potential with a relative uncertainty of about 1%.

Table 1. Numerical values of dn and θ_0 when $k = k_g = 1/10$ nm and $\alpha_g = 10^{20}$ for two materials.

Material	ρ (kg/m^3)	V_f (neV)	dn	θ_0
Vanadium	6110	−7.19	-1.73×10^{-2}	-6.46×10^{-4}
Samarium	7353	6.13	$+1.48 \times 10^{-2}$	-7.78×10^{-4}

6. Conclusions

We have analysed the effect of non-Newtonian gravity on the reflection amplitude from bulk materials and found that the first-order contribution of non-Newtonian gravity is significant when the neutron wavelength is nearly equal to the radius of curvature of the neutron refractive index due to non-Newtonian gravity and the first-order contribution is negligible when the neutron wavelength is much smaller than the radius of curvature.

Acknowledgments

The author would like to thank Hirohiko Shimizu for his comments on this work. This work was supported by JSPS Kakenhi (19GS0120).

References

[1] Fischbach E and Talmadge C L 1999 *The Search for Non-Newtonian Gravity* (New York: Springer)
[2] Arkani-Hamed N, Dimopoulos S and Dvali G 1998 *Phys. Lett.* B **429** 263
[3] Arkani-Hamed N, Dimopoulos S and Dvali G 1999 *Phys. Rev.* D **59** 086004
[4] Antoniadis I, Arkani-Hamed N, Dimopoulos S and Dvali G 1998 *Phys. Lett.* B **436** 257
[5] Dimopoulos S and Giudice G F 1996 *Phys. Lett.* B **379** 105
[6] Adelberger E G, Heckel B R and Nelson A E 2003 *Annu. Rev. Nucl. Part. Sci.* **53** 77
[7] Bordag M, Mohideen U and Mostepanenko V M 2001 *Phys. Rep.* **353** 1
[8] Newman R D, Berg E C and Boynton P E 2009 *Space Sci. Rev.* **148** 175
[9] Leeb H and Schmiedmayer J 1992 *Phys. Rev. Lett.* **68** 1472
[10] Frank A, Isacker P V and Gómez-Camacho J 2004 *Phys. Lett.* B **582** 15
[11] Zimmer O and Kaiser N 2006 *Classical Quantum Gravity* **23** 6077
[12] Greene G L and Gudkov V 2007 *Phys. Rev.* C **75** 015501
[13] Nesvizhevsky V V, Pignol G and Protasov K V 2008 *Phys. Rev.* D **77** 034020

Phonon density of states in different clathrate hydrates measured by inelastic neutron scattering

M Celli[1], D Colognesi[1], L Ulivi[1], M Zoppi[1], A J Ramirez-Cuesta[2]

[1] Consiglio Nazionale delle Ricerche – Istituto dei Sistemi Complessi, via Madonna del Piano 10, Sesto Fiorentino (FI), Italy,

[2] ISIS Facility, Rutherford Appleton Laboratory, Chilton, Didcot, OX11 0QX, United Kingdom

E-mail: milva.celli@isc.cnr.it

Abstract. Clathrate hydrates, depending on the guest molecule type, generally exhibit one out of three different crystal structures: two cubic structures, *sI* and *sII* and one hexagonal structure, *sH*. In the past, our inelastic neutron scattering measurements on hydrogen clathrates have provided information on the quantum dynamics of the guest molecules in the water cages. Besides the guest dynamics, the dynamics of the water lattice itself has a large interest, due to the analogy with ice (e.g. proton disorder), and to the existence of various possible structures. Additionally, in these inclusion compounds, a coupling between the host and the guest motions is generally observed, and is considered to be relevant to explain the anomalous features of some macroscopic properties, such as thermal conductivity. Here, we present a systematic study of the H-projected phonon density of states (H-PDoS) of the lattice modes in clathrate hydrates. We have experimentally investigated the three existing structures (i.e. *sI*, *sII*, and *sH*) through inelastic neutron scattering measurements, and we have extracted the acoustic-optic and the librational H-PDoS's. By using proper isotopic substitutions, we have been able to tune the host scattering intensity with respect to the guest one. The studied samples consisted in three clathrates made of light water (namely, simple *sI* structure with Xe, simple *sII* structure with fully deuterated THF, and binary *sH* structure with MTBE and D_2), and two made of heavy water (namely, simple *sII* structure with Ne, and simple *sII* structure with fully deuterated THF). The experimental results have been compared with lattice dynamics simulations performed by us.

1. Introduction

Clathrate hydrates are crystalline inclusion compounds in which guest atoms or molecules are trapped inside cages (host) formed by a network of hydrogen-bonded water molecules [1]. As it is well known, depending on the guest molecule types and the thermodynamic conditions, there exist three different crystal structures of clathrates, named *sI*, *sII*, and *sH*. The cubic structure *sI* has 46 water molecules in the primitive unit cell, giving rise to two (small) dodecahedral cages and six (large) tetrakaidecahedral cages. The face-centered cubic structure *sII* has 136 H_2O molecules in the unit cell, giving rise to sixteen (small) dodecahedral cages and eight (large) hexakaidecahedral cages. Finally, the hexagonal structure *sH* is characterized by 34 water molecules in the unit cell, giving rise to three (small) dodecahedral, two (medium) irregular dodecahedra, and one (large) icosahedral cages.

In these last decades, a renewed interest in these compounds grew up for various reasons. From the applied point of view, methane hydrates can represent a possible source of energy in the future, since these clathrates are accumulated in the permafrost regions and in sediments on the deep ocean floor. Hydrogen clathrates represent promising systems as hydrogen storage materials, and carbon dioxide clathrates could be a possible method to sequestrate greenhouse gases. From the fundamental point of view, these systems are very interesting for analogies and differences with the physical properties of ice. Therefore understanding the dynamical properties of clathrate hydrates is very important in condensed matter physics, chemistry, geology as well as planetary astronomy [2], both from the theoretical and the practical points of view.

In past, we have performed inelastic neutron scattering experiments on clathrates made of heavy water and hydrogen guest molecules, both with and without fully deuterated tetrahydrofuran in stoichiometric proportion. From these measurements we have obtained information on the quantum dynamics of the guest molecule inside the cage [3, 4, 5]. Besides the guest motion, the lattice dynamics itself has a large interest, due to the analogies with ice [6, 7], and to the existence of various structures. Additionally, in these inclusion compounds, a coupling of host and guest motions is generally observed and considered to be relevant to explain the thermal conductivity anomaly of clathrates [8, 9]. A complete knowledge of the phonon density of states (PDoS) both in ice and in clathrate hydrates can provide help to understand the anomalous features of some macroscopic properties of clathrates, such as the aforementioned thermal conductivity, but also yield strength [10] and thermal expansivity [11, 12, 13].

Incoherent inelastic neutron scattering (IINS) can be used to study the phonon spectrum (i.e. the PDoS) of the clathrates hydrates themselves [14, 15, 16, 17, 18]. To our knowledge, most of the experiments were devoted to the frequency range of the acoustic mode (<15 meV) [14, 15, 16]. In general, the choice of the guests is not optimized to the lattice PDoS extraction. Finally, no experimental PDoS for the sH structure was found. Here, we present a systematic study of the H-projected density of states (H-PDoS) of the lattice and librational modes of various clathrate hydrates. We have experimentally investigated all the three more common structures (*sI*, *sII*, and *sH*) by means of inelastic neutron scattering measurements on the TOSCA spectrometer at ISIS (UK), and we have extracted the acousto-optical translational and the librational H-PDoS. By using proper isotopic substitutions, we have been able to tune the host scattering intensity with respect to the guest one. The samples studied were three clathrates made of light water (i.e. simple *sI* structure with Xe, simple *sII* structure with fully deuterated tetrahydrofuran (TDF), and binary *sH* structure with methyl tert-butyl ether (MTBE) and D_2), and two clathrates made of heavy water (i.e. simple *sII* structure with Ne and simple *sII* structure with TDF). The experimental results are compared with original lattice dynamics calculations performed by the authors. In literature simulation calculations, performed with similar potentials, are present [17, 19, 20, 21, 22, 23, 24]. These theoretical results are quite similar to ours, but generally exhibit a slightly lower accuracy level.

In section 2 we will give a brief description of the experimental procedure and the data analysis. Section 3 will be devoted to the simulation calculations. In section 4 we will discuss the results. Finally, conclusions will be presented in section 5.

2. Experiment and data analysis

All the samples (see Table 1) were prepared in the laboratories of ISC-CNR (Florence, Italy) using deionized light water and commercial heavy water, TDF and MTBE, plus research-purity gasses (Xe, Ne, and D_2). Every sample needed a specific synthesis procedure. The simple *sII* structure with Ne was prepared by adding pressurized Ne gas ($p \approx 2.5$ kbar) to ground D_2O ice. Binary *sH* clathrate was synthesized by mixing a calculated amount of liquid MTBE to ground H_2O ice at $T \approx -4$ °C, and then adding D_2 gas at about 1 kbar. For the H_2O-TDF and D_2O-TDF clathrates, a liquid mixture of water and tetrahydrofuran, in stoichiometric proportion (i.e. 17:1 mol) was frozen at $T \approx +2$ °C. Finally, the simple *sI* structure with Xe was produced by adding low pressure Xe gas ($p \approx 7$ bar) to liquid water at

T=-2°C, and stirring it during the clathrate formation until the sample froze up. The clathrate formation time depends on the synthesis procedure.

Table 1. Details of the measured clathrate hydrates samples, including their structures, chemical compositions, masses (m), measurement temperatures (T), and integrated proton current values (I.P.C.).

Sample no.	Structure	Composition	m (g)	T (K)	I.P.C. (μAh)
1	*SI*	H_2O + Xe	3.34	20.0±0.2	941.5
2	*sII*	H_2O + TDF	1.74	12.2±0.3	1033.7
3	*sH*	H_2O + MTBE +D_2	1.37	20.1±0.6	2008.3
4	*sII*	D_2O + Ne	4.26	20.2±0.6	1758.5
5	*sII*	D_2O + TDF	5.08	15.0±0.1	3965.6

Once the clathrate is formed, this is quenched by immersing the vessel into liquid nitrogen. Then the pressure can be released and the solid sample is recovered to fill the experimental cell. The transfer process has to be accomplished at low temperature (≈77 K), in a dry-nitrogen atmosphere, taking particular care to avoid sample heating and, especially for deuterated samples, air contamination.

Samples were shipped to the ISIS pulsed neutron source (Rutherford Appleton Laboratory, UK) in a cryogenic container kept at liquid nitrogen temperature. All the inelastic neutron scattering measurements described here were performed on the high-resolution, inverse-geometry TOSCA spectrometer [25]. IINS is a powerful technique for the study of hydrogenous materials due to the large incoherent scattering cross section of the proton, which is almost two orders of magnitude greater than the average value of the other nuclei. Thanks to this property, the IINS experiments allow for a quite simple access to the microscopic self-dynamics of hydrogen.

In figure 1 we report the neutron spectra of three clathrates, representing all the three possible crystal structures: *sI*, *sII* and *sH*. Due to the neutron scattering cross section value of the selected guest molecules, in these clathrates we mainly see the scattering from the protons of the host. By means of the data analysis described below, we can finally extract the H-PDoS.

The experimental time-of-flight data were transformed into energy-transfer (E) spectra, detector by detector, making use of standard routines available on the spectrometer including the removal of the $(E_1/E_0)^{1/2}$ kinematic factor (with E_0 and E_1 being the initial and final neutron energy, respectively) and correction for the incoming neutron flux. At this stage data were added together producing two distinct data blocks: one containing the backscattering detectors and the other the forward-scattering ones. This procedure was justified by the narrow angular-range spanned by each set of detectors [25]. Then the empty-can contribution was properly subtracted, taking into account the E_0-dependent sample transmission. At this level the important corrections for self-absorption attenuation and multiple scattering contamination were performed through the analytical approach suggested by Agrawal and Sears in the case of a flat slab-like sample [26, 27]. Both procedures were carried out in the framework of the incoherent approximation, totally justified for samples 1-3 by the preponderance of scattering from the H nuclei and by the polycrystalline nature of the samples. As for the fully deuterated samples 4 and 5, the rationale of this approximation is based on previous satisfactory results obtained for various forms of deuterated ice [6]. Making use of the method explained in ref. [28], the inputs needed for these two corrections were reduced to the knowledge of the atom-projected density of phonon states, which were approximately set up using the procedure reported in details in reference [29]. Multiple scattering contributions containing two or more inelastic scattering events were found to be modest (but not negligible) in the energy transfer interval of interest (i.e. 3 meV < E < 140 meV, the range of the fundamental bands). Finally these estimates were removed from the processed neutron spectra, which are reported in figure 1 in the form of the well-known self-inelastic structure factors,

$S_s(Q,E)$, essentially related to the H or D dynamics, the contribution from heavier atoms (i.e. O, C, Ne, and Xe) being fully negligible.

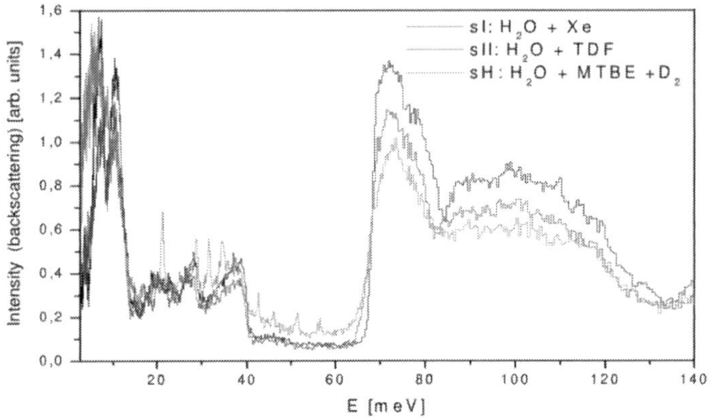

Figure 1. Neutron scattering spectra measured on TOSCA in backscattering configuration. samples no. 1 (black line), no. 2 (blue line), and no. 3 (red line), as in Table 1. The small narrow peaks present in the low-frequency part of the *sH* clathrate spectrum (red line) are due to methyl tert-butyl ether contributions and have been properly subtracted in the data analysis.

Finally, in order to determine and remove the multiphonon contributions in the TOSCA neutron spectra, we have used a well-known self-consistent procedure [28, 29] already positively tested for a number of systems (i.e. ice, metal hydrides etc.). Despite the approximations involved (i.e. a purely harmonic, isotropic, and single-site treatment of the multiphonon terms), the results reveal a satisfactory convergence of the method, allowing for a reliable extraction of the one-phonon components of the processed neutron spectra, $S_{s,+1}(Q,E)$, and finally for the evaluation of the hydrogen (or deuterium) density of phonon states, $G_{H(D)}(E)$:

$$G_{H(D)}(E) \cong \frac{4M_{H(D)}E}{\hbar^2 Q^2}\left[\coth\left(\frac{E}{2k_B T}\right)+1\right]^{-1} \exp\left(Q^2\left\langle \mathbf{u}^2_{H(D)}\right\rangle/3\right) S_{s,+1}(Q,E) \qquad (1)$$

where $\langle \mathbf{u}^2_{H(D)} \rangle$ is the proton (deuteron) mean square displacement, also worked out by the mentioned self-consistent procedure. The experimental estimates of the H(D)-projected density of phonon states are reported in figures 2 and 3 for all the samples.

3. Lattice dynamics simulations

Original lattice dynamics simulations have been performed on empty H_2O and D_2O clathrate hydrates exhibiting various structures, making use of the semi-empirical pair potential model named *extended simple point charge* (SPC/E), proposed by Berendsen *et al.* [30] for solid and liquid water. This widely used approach was meant to improve, through a minor reparametrization, the older *simple point charge* model with a modified value for the effective oxygen charge, q_O. Each water molecule is modeled as a rigid isosceles triangle, having point charges situated on each of the three atoms, with the effective hydrogen charge $q_H = -q_O/2$. In addition to the Coulombic terms, the water molecules interact also via Lennard-Jones sites situated on the oxygen atoms only. The SPC/E inter-atomic potentials have been plugged into the general utility lattice program GULP [31], together with initial clathrate super-cell configurations, created assuming the experimental crystallographic details (i.e. lattice parameters, internal coordinates and symmetry) and imposing the compliance of the well-known Bernal-Fowler rule for protons or deuterons [32]. GULP, after performing the Gibbs free energy minimization at T=15 K (results for the various cell volumes are reported in Table 2), evaluated $G_{H(D)}(E)$ for all the four simulated samples (see figures 4 and 5) using the Monkhorst-Pack scheme. Additional constant-volume ("conv") simulations, freezing the experimental values of the lattice parameters and optimizing the internal coordinates only, have been also performed and reported in the same figures. No symmetry constraints have been used in both sets of simulations. Details about the simulation set-up and results can be found in table 2.

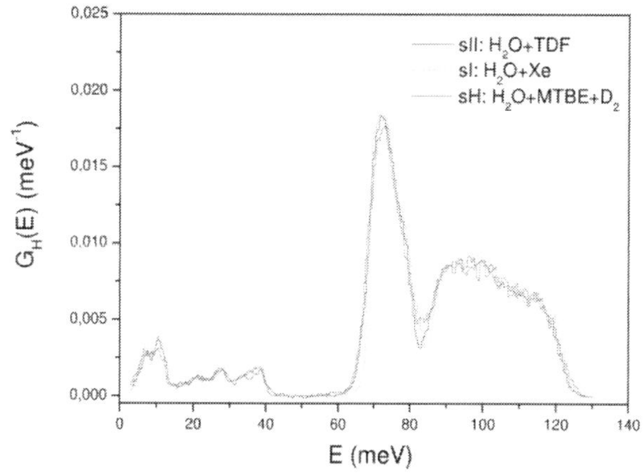

Figure 2. Experimental estimates of the H-projected density of phonon states for samples no. 1 (green line), no. 2 (blue line), and no. 3 (red line), as in table 1.

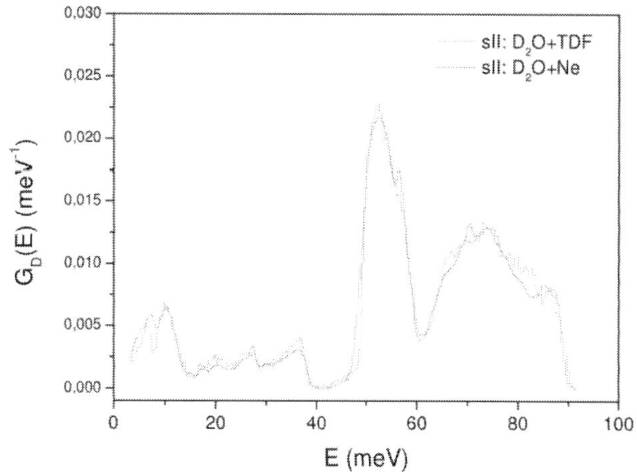

Figure 3. Experimental estimates of the D-projected density of phonon states for samples no. 4 (red line) and no. 5 (blue line), as in table 1.

Table 2. Details of the lattice dynamics simulations performed on empty clathrate hydrates, including: type of simulation (i.e. either Gibbs free energy minimization, "G min.", or constant volume calculation, "conv"), clathrate structure, number of molecules contained in the super-cell, super-cell size, initial cell volume (experimental), and fully-relaxed cell volume.

Simulation	Structure	Molecules	Super-cell	V_0 (Å^3)	V (Å^3)
G min.	*sI*	46 H_2O	1×1×1	1728.00	1632.68
conv	*sI*	368 H_2O	2×2×2	1728.00	-
G min.	*sII*	136 H_2O	1×1×1	5002.84	4894.93
conv	*sII*	136 H_2O	1×1×1	5002.84	-
G min.	*sH*	272 H_2O	2×2×2	1276.05	1254.80
conv	*sH*	272 H_2O	2×2×2	1276.05	-
G min.	*sII*	136 D_2O	1×1×1	5002.84	4841.46
conv	*sII*	136 D_2O	1×1×1	5002.84	-

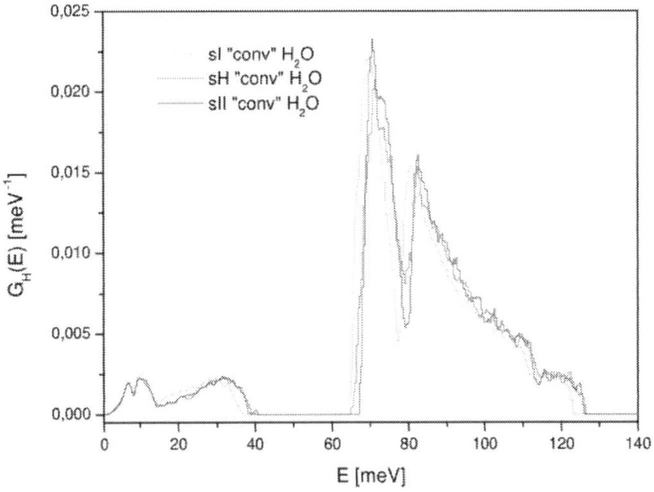

Figure 4. Simulated H-projected density of phonon states, at constant volume ("conv"), for empty *sI* (green line), *sH* (red line), and *sII* (blue line) H_2O clathrate.

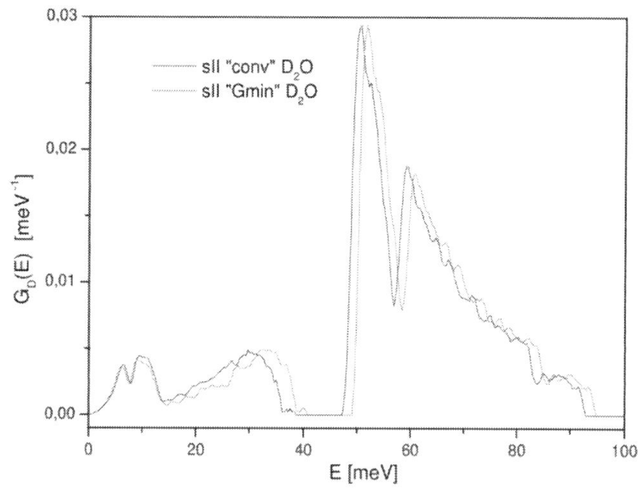

Figure 5. Simulated D-projected density of phonon states for empty *sII* D_2O clathrate obtained with two different simulation methods: at constant volume (black line) and minimizing the Gibbs free energy (red line).

4. Discussion

By comparing the experimental H-projected density of phonon states of the three clathrate structures (see figure 2), one observes a strong overall similarity among them, much more pronounced than that found by comparing these three phonon spectra with the H-projected density of phonon states of ice *Ih*. [33]. From a qualitative point of view, this general agreement is relevant. As a matter of fact, the lattices of all the clathrate structures are made of cages, formed by water rings (square, pentagonal, hexagonal etc.) linked together. Thus clathrate lattices turn out to be completely different from the normal ice one, which exhibits a well-known channel-like structure. Going back in more detail to figure 2, one can start focussing on the low energy part (i.e. E<40 meV) of $G_H(E)$, which is related to the lattice phonons, both acoustic and optic. The strength of this band is less intense than the other spectral parts because here one observes only the proton contribution to the low energy part of the total density of phonon states, $G(E)$, weighted by the corresponding eigenvector square moduli. In the librational region (60 meV <E<130 meV), one notes stronger differences in the H-projected densities of phonon states of the three structures. In particular, the first steep rise, located at about 60 meV, appears at a slightly higher frequency in the *sI* clathrate, while the dip around 85 meV is less pronounced in the *sH* clathrate than in the other two structures.

In figure 3, D-projected densities of phonon states are reported. In this case the hypotheses used to extract $G_D(E)$ from experimental neutron data are less obviously satisfied than in the case of hydrogen containing samples. As a matter of fact, in hydrogen containing samples, the coherent scattering contributions, neglected in the data analysis, are much less important than the incoherent ones. As a consequence, the results for $G_D(E)$ reported in figure 3 are intrinsically less accurate than those for $G_H(E)$ plotted earlier. Keeping in mind this general observation, by comparing the protonated (figure 2) and deuterated (figure 3) samples, one immediately notices the expected different frequency changes for the lattice-phonon band and for the librational band. The lattice-phonon frequency range roughly scales with the inverse square root of the molecular water mass. So it changes very little in moving from protonated to deuterated water. Differently, in the librational band, the frequencies approximately scale with the inverse square root of the H(D) atomic mass, for light and heavy water respectively, as already observed in various ice types [6].

Another point, concerning all the measured samples, is the fact that we have fully neglected the scattering contributions from the guest molecules trapped inside the clathrate cages. Obviously, among all the possible guest molecules, the most suitable ones have been selected, i.e. those with the lowest neutron scattering cross sections. However, when necessary (e.g. using protonated MTBE), we have removed the spurious peaks due to the guest by subtracting the scattering contributions from appropriate additional clathrate measurements including a fully deuterated host plus a hydrogenated guest. On the other hand, the guest inside the cages interacts with the host in a slightly different way in the two cases, due to the differences in the host phonon distribution; so, at least in principle, the removal procedure is not fully rigorous, although is surely sensible.

In figure 4 we have reported a constant-volume simulation of $G_H(E)$ for all the three structures, assuming empty cages (i.e. no guest at all). The lattice-phonon bands of the three structures appear very similar to one another, whereas the librational bands show some differences, and are qualitatively in agreement with the experimental results (see figure 2). More in detail, in the *sH* sample the minimum around 85 meV is less deep than in the other two structures, exactly like in the experimental case; while in the *sI* sample the simulated librational band is placed in a lower frequency range if compared to the same quantity in the other two samples. We recall that for the experimental *sI* sample, only the onset of the librational band is at a higher frequency value. At the moment the origin of such a peculiar behaviour is unknown and a simple effect of the cell volume variation can be easily ruled out just by checking table 2.

For completeness sake, in figure 5 we have reported a comparison between the D-projected densities of phonon states obtained through two different calculation methods, described in section 3, and named "conv" (constant volume calculation) and "Gmin" (Gibbs free energy minimization calculation). We observe that the general behaviour is the same, but the two spectra are slightly shifted one with respect to the other as far as the phonon frequencies are concerned. Contrary to the simulation *sI* results, in this case, such a shift is clearly due to the small differences in the cell volumes, which affect the intermolecular force constants. The same trend (i.e.: identical shape plus a slight frequency shift) has been observed by comparing "conv" and "Gmin" calculations for all the other clathrate structures.

5. Conclusions

We have presented a systematic study of the hydrogen-projected density of phonon states in the region of the lattice and the librational modes for various clathrate hydrates. We have experimentally investigated the three possible structures, namely *sI*, *sII*, and *sH*, by means of inelastic neutron scattering measurements performed on TOSCA (ISIS, UK), and we have extracted the acousto-optical translational and the librational phonon spectra. By using proper isotopic substitutions, we have been able to tune the host scattering intensity with respect to the guest one. The samples studied consisted in three clathrates made of light water (i.e. simple *sI* with Xe, simple *sII* with fully deuterated tetrahydrofuran, and binary *sH* with methyl tert-butyl ether and D_2), and two clathrates made of heavy water (i.e. simple *sII* with Ne, and simple *sII* with fully deuterated tetrahydrofuran). The similarities and the differences among our experimental hydrogen (deuterium)-projected density of phonon states are discussed. The authors have also performed original lattice dynamics calculations in order to evaluate the hydrogen (deuterium)-projected density of phonon states for all the measured structures, employing empty cages clathrates arranged in fixed-symmetry supercells. The experimental results have been compared with these simulations showing an overall good agreement, but still at a qualitative level. This is not surprising, since the same level of agreement between neutron scattering phonon data and lattice dynamics calculations with semi-empirical potentials is also found in much simpler ice structures, and has to be ascribed to the intrinsic difficulty of high accuracy H-bond modelling. Finally, one has also to conclude that, despite incoherent inelastic neutron scattering is a very sensitive characterization technique able to accurately single out several ice forms, in the case of clathrate hydrates, it turns out to be less powerful, given the strict similarities among the spectra of the water cages forming the three distinct clathrate structures.

References

[1] Sloan E D and Koh C A 2008 *Clathrate Hydrates of Natural Gases* (Boca Raton: CRC Press)
[2] Dartois E, Deboffle D and Bouzit M 2010 Methane clathrate hydrate infrared spectrum. II. Near-infrared overtones, combination modes and cages assignments *A&A* **514** A49
[3] Ulivi: L, Celli M, Giannasi A, Ramirez-Cuesta A J, Bull D J and Zoppi M 2007 Quantum rattling of molecular hydrogen in clathrate hydrate nanocavities *Phys. Rev.* B **76** 161401
[4] Ulivi L, Celli M, Giannasi A, Ramirez-Cuesta A J and Zoppi M 2008 Inelastic neutron scattering from hydrogen clathrate hydrates *J. Phys.:Condens. Matter* **20** 104242
[5] Celli M, Colognesi D, Giannasi A, Ulivi L, Zoppi M, Garcia Sakai V and Ramirez-Cuesta A J 2010 Simple and binary hydrogen clathrate hydrates: synthesis and microscopic characterization through neutron and Raman scattering *Adv. Sci. Tech.* **72** 196
[6] Li J 1996 Inelastic neutron scattering studies of hydrogen bonding in ices *J. Chem. Phys.* **105** 6733
[7] Petrenko V F and Whitworth R W 2006 *Physics of ice* (Oxford: University Press)
[8] Ross R G, Anderson P and Backstrom G 1981 Unusual *PT* dependence of thermal conductivity for a clathrate hydrate *Nature* **290** 322
[9] Tse J S, Shpakov V P, Belosludov V R, Trouw F, Handa Y P and Press W 2001 Coupling of localized guest vibrations with the lattice modes in clathrate hydrates *Europhys. Lett.* **54** 354
[10] Durham W B, Kirby S H, Stern L A and Zhang W 2003 The strength and rheology of methane clathrate hydrate *J. Geophys. Res.* **108** (**B4**) 2182
[11] Tse J S, McKinnon W R and Marchi M 1987 Thermal expansion of structure I ethylene oxide hydrate *J. Phys. Chem.* **91** 4188
[12] Gutt C, Asmussen B, Press W, Tse J S, Handa Y P and Johnson M R 2000 The structure of deuterated methane–hydrate *J. Chem. Phys.* **113** 4713
[13] Tanaka H, Tamai Y and Koga K 1997 large thermal expansivity of clathrate hydrates *J. Phys. Chem.* B **101** 6560
[14] Tse J S, Ratcliffe C I, Powell B M, Sears V F and Handa Y P 1997 Rotational and translational motions of trapped methane. Incoherent inelastic neutron scattering of methane hydrate *J. Phys. Chem.* **101** 4491
[15] Tse J S, Powell B M, Sears V F and Handa Y P 1993 The lattice dynamics of clathrate hydrates. An incoherent inelastic neutron scattering study *Chem. Phys. Lett.* **215** 383
[16] Gutt C, Baumert J, Press W, Tse J S and Janssen S 2002 The vibrational properties of xenon hydrate: An inelastic incoherent neutron scattering study *J. Chem. Phys.* **116** 3795
[17] Wang L and Dong S L 2010 Lattice dynamical simulation of methane hydrate *J. Nat. Gas Chem.* **19** 43
[18] Struzkin V V, Militzer B, Mao, W L, Mao H K, Hemley R J 2007 Hydrogen storage in molecular clathrates *Chem. Rev.* **107** 4133 figure 13
[19] Tse J S, Klein M L and McDonald I R 1983 Dynamical properties of the structure I clathrate hydrate of xenon *J. Chem. Phys.* **78**, 2096
[20] Tanaka H and Kiyohara K 1993 On the thermodynamic stability of clathrate hydrate. I *J. Chem. Phys.* **98** 4098
[21] Subbotin O S and Belosludov V R 2002 Relationship between the dynamic properties of Ih ice and xenon hydrate and the interplay of intramolecular and intermolecular vibratons *J. Struct. Chem.* **43** 545
[22] Ikeda-Fukazawa T, Yamaguchi Y, Nagashima K and Kawamura K 2008 Structure and dynamics of empty cages in xenon clathrate hydrate *J. Chem. Phys.* **129** 224506
[23] Nemov N A, Subbotin O S and Belosludov V R 2010 Modeling of phase transformation sII-sH

in Ar hydrates at high pressure *Comput. Mater. Sci.* **49** S256
[24] Jiang H and Jordan K D 2010 Comparison of the properties of Xenon, methane, and carbon dioxide hydrates from equilibrium and nonequilibrium molecular dynamics simulations *J. Phys. Chem.* C **114** 5555
[25] Colognesi D, Celli M, Cilloco F, Newport R J, Parker S F, Rossi-Albertini V, Sacchetti F, Tomkinson J and Zoppi M 2002 TOSCA neutron spectrometer: the final configuration *Appl. Phys.* A **74** [**Suppl. 1**] 64
[26] Agrawal A K 1971 Multiple scattering of neutrons in gaseous and liquid methane *Phys. Rev.* A **4** 1560
[27] Sears V F 1975 Slow-neutron multiple scattering *Adv. Phys.* **24** 1
[28] Dawidowski J, Bermejo F J and Granada J R 1998 Efficient procedure for the evaluation of multiple scattering and multiphonon corrections in inelastic neutron-scattering experiments *Phys. Rev.* B **58** 706
[29] Colognesi D, Andreani C and Degiorgi E 2003 Density of phonon states from a crystal-analyzer inverse-geometry spectrometer: a study on ordered solid hydrogen sulfide and hydrogen *Journal of Neutron Research* **11** 123
[30] Berendsen H J C, Grigera J R and Straatsma T P 1987 The Missing Term in effective pair potentials *J. Phys. Chem.* **91** 6269
[31] Gale J 1997 A computer program for the symmetry-adapted simulation of solids *J. Chem. Soc. Faraday Trans.* **1** 629
[32] Bernal J D and Fowler R H 1933 A theory of water and ionic solution, with particular reference to hydrogen and hydroxyl ions *J. Chem. Phys.* **1** 515
[33] Li J C, Ross D K, Howe L, Hall P G, and Tomkinson J 1989 Inelastic incoherent neutron scattering spectra of single crystalline and polycrystalline ice Ih *Physica B* **156-157** 376

Journal of Physics: Conference Series **340** (2012) 012052 doi:10.1088/1742-6596/340/1/012052

Stability of phases at high temperatures in CoRe based alloys being developed for ultra-high temperature applications

R Gilles [a], P Strunz [b], D Mukherji [c], M Hofmann [a], M Hoelzel [a] and J Roesler [c]

[a] *Technische Universität München, Forschungsneutronenquelle Heinz Maier-Leibnitz (FRM II) ZWE, Lichtenbergstr. 1, 85747 Garching, Germany*
[b] *Nuclear Physics Institute ASCR, CZ-25068 Řež near Prague, Czech Republic*
[c] *Technische Universität Braunschweig, Institut für Werkstoffe, Langer Kamp 8, 38106 Braunschweig, Germany*

E-mail: ralph.gilles@frm2.tum.de

Abstract

In the development of new high-temperature alloys for gas turbine applications various candidates are under consideration. This contribution deals with a CoRe based alloy strengthened by $Cr_{23}C_6$ type carbide and Cr_2Re_3 type σ phase precipitations (here designated as CoRe-1 alloy). High-temperature cycling experiments show how the influence of heating, cooling and the hcp<->fcc phase transformation of the Co-matrix on the stability of these phases. Neutron diffraction experiments with high-temperature vacuum furnace show that $Cr_{23}C_6$ carbides starts to dissolve around 1100°C and above 1250°C are almost completely dissolved. On the other hand σ phase is still present at 1300°C. This contribution describes the evolution of the different phases during the heating and cooling cycles which are repeated two times. Further, the influence of boron addition to CoRe-1 alloy was studied for samples in the first heating/cooling cycle. A newly developed tensile rig was also tested up to 980°C to combine in situ loading and heating for the neutron diffraction measurements.

1. Introduction

The Co-Re based alloys, a new class of high temperature alloy with relatively high melting point, were introduced by the Technische Universität Braunschweig, Germany in 2007 [1] to supplement Ni-based superalloys in future gas turbines for application at and above 1200°C. This project is implemented in "Beyond Ni-Base Superalloys" DFG Forschergruppe 727, a German research project. Co-based alloys are already used in gas turbines and achieved sustained success. Their disadvantage is the limited strength and the low service temperature of around 1000°C. To increase application temperatures the melting point of Co-based alloys has to be enhanced. This is achieved by addition of Re (which has the third highest melting point in the periodic table T_M = 3182°C) to the Co-alloy. In this way, an improvement of the melting point of the alloy by approximately 200°C - in comparison with Ni-based superalloys - is obtained [1].
The microstructure of Co-Re alloys is complex and at RT it has a ε Co (hcp) matrix. CoRe-1 alloy has other phases, particularly $Cr_{23}C_6$ type carbides and Cr_2Re_3 type σ phase. Massive and blocky Cr-carbides in the form of $Cr_{23}C_6$ are present at the grain boundaries and within the grains and very coarse particles of Cr_2Re_3 (σ phase) are also present. However, from the strengthening aspect, the most important phase is the fine lamellar form of the $Cr_{23}C_6$ carbides which are homogenously distributed in the grains [2]. Long term stability of the fine microstructure, particularly at high temperatures in the range of 1000°C to 1300°C, is very critical for the application of the Co-Re alloys.

Previous in-situ investigation at high temperatures by neutron diffraction detected a hcp (ε) to fcc (γ) phase transformation of the Co-matrix in the experimental alloys CoRe-1 (Co-17Re-23Cr-2.6C)[1] and CoRe-2 (Co-17Re-23Cr-1.2Ta-2.6C) [3, 4]. In these alloys the Co-matrix undergoes a phase transformation from hcp (ε) to fcc (γ) phase on heating between 1200°C and 1300°C. Moreover, the phase transformation although reversible exhibits a large hysteresis (around 100°C) in both of the alloys. It was concluded that the hysteresis is a result of the composition interplay between matrix and the minority phases, namely the $Cr_{23}C_6$ and σ phase [3, 4].
As the high temperature microstructure and its stability is of great importance for the application the simpler composition of the CoRe-1 alloy was further examined in more details. As previous neutron diffraction experiments [3, 4] have shown that the amount of σ phase volume increased after the thermal cycle, we monitored the structural evolution very carefully by repeating the thermal cycling to see if the irreversible changes in σ phase after the first cycle affect the transformation in the second cycle. Further, the CoRe alloy development has shown that small addition of boron (which is known to segregate to the grain boundaries) is needed, in order to strengthen the weak grain boundaries in polycrystalline alloys and to mitigate environmental embrittlement [5]. Therefore, an alloy with the same composition like the CoRe-1 but with a further addition of 500 wt. ppm boron (here designated as alloy CoRe-1B) was also studied. Even such small amount of boron has a significant effect on the initial microstructure of the fine Cr-carbide [5].
Moreover, the evolution of the microstructure at high temperatures up to 980°C under an applied uniaxial load was studied by means of a deformation rig on the Stress Spec instrument at FRM II. The results from the first test of the rig are presented.

2. Experimental

A sample of alloy CoRe-1 which already underwent a thermal cycling as in reference [4] was subjected to a second identical thermal cycling. It was intended to study the structural stability and equilibration during and after the second cycle. In addition a boron-containing alloy CoRe-1B (Co-17Re-23Cr-2.6C + 500 wt. ppm B) was subjected to a single thermal cycle for determination of the influence of boron on high temperature microstructure of CoRe-1 alloy.
The alloys CoRe-1 and CoRe-1B were melted in a vacuum arc furnace and cast into square rods in a solid Cu mould. After a heat treatment of the bars between 1350°C and 1450°C for a total time of 15 hours to homogenize them, the samples were water quenched. Cylinders of around 6 mm diameter and 30 mm length were cut for neutron diffraction measurements.
A Zeiss scanning electron microscope (SEM) Leo 1550 (FEG) was used to study the microstructure of the alloys. Microstructural images were recorded with standard and in-lens secondary electron (SE) detectors as well as with back scatter electron (BSE) detector at high resolution and fine particles in the sizes of 20 to 50 nm could be easily resolved. While BSE images were recorded using an accelerating voltage of 15kV, SE and in-lens SE images were taken with only 5 kV. The Z (atomic number) contrast from heavy elements like Re in the BSE images allows to distinguish the phases from their grey level.
In-situ neutron diffraction measurements at RT and in the high temperature region between 1100°C and 1300°C were performed at FRM II, Garching, Germany to study the microstructural stability of experimental Co-Re alloys. With a high temperature furnace in situ heating/cooling in 5°C steps (reached in 5 seconds) were effected to carry out measurements during a 10 minute hold at a given temperature. Using an area detector, with a sufficient angular region where all strong scattering Bragg reflections are present, measurements were performed at the Stress-Spec [6] instrument with a wavelength of 0.16225 nm. Further, high resolution measurements at RT were done at SPODI instrument [7, 8] using a wide angular region with 80 He^3 tube detector bank and a wavelength of 0.15482 nm.
At Stress-Spec, we adjust the same temperature program for the first and the second thermal cycle of CoRe-1 alloy. The sample with boron (CoRe-1B) was examined at the high temperature diffractometer SPODI. We followed basically the same thermal cycle as previously with the CoRe-1 sample without

[1] All compositions in this paper are expressed in atomic percentage (at. %) except for the boron additive which is always given as wt. ppm

boron [5]. The evolution of main phases (hcp, fcc) as well as of the strengthening phases ($Cr_{23}C_6$ and σ phase) during the thermal cycle RT – 1300°C – RT was studied in this way and compared to the previously measured phase evolution in CoRe-1 alloy.

As a first test of a new tensile rig to combine loading and heating in situ during the neutron diffraction measurement, tensile samples of CoRe-1 with 40 mm gauge length and diameter of 6 mm were prepared. These cylinders were connected with a M8 thread to the grips of the tensile rig. Slow increase of the load (1N/sec) allowed easy acquisition of diffraction patterns in the 2Θ range of 42° up to 56° with a 2 dimensional area detector at Stress-Spec instrument. The experiments were carried out at three different temperatures, namely T = 538°C, 800°C and 980°C.

3. Results and Discussion

The microstructure of the CoRe-1 and CoRe-1B alloys are shown in Figure 1. The low magnification of SEM images (Fig. 1a and c) shows the coarse microstructures in both alloys. The CoRe-1 alloy has Cr-carbides at the grain boundaries and also blocky Cr-carbides within the grains. The CoRe-1B alloy shows instead the presence of coarse (> 20 μm) σ phase particles within the grains (Fig. 1a and c). In addition the fine structure of lamellar $Cr_{23}C_6$ carbide is seen in both alloys in the higher magnification SEM images (Fig. 1b and d).

The neutron diffractograms were recorded during temperature increase (RT-1300°C) and temperature decrease (1300°C-RT) with various temperature steps (denser around the transformation temperatures). Usually, a selected angular region in 2Θ (43°-70°) where the large peaks of important phases are present was chosen at Stress-Spec, whereas the full angular region (38°-104°) was measured only at defined temperatures.

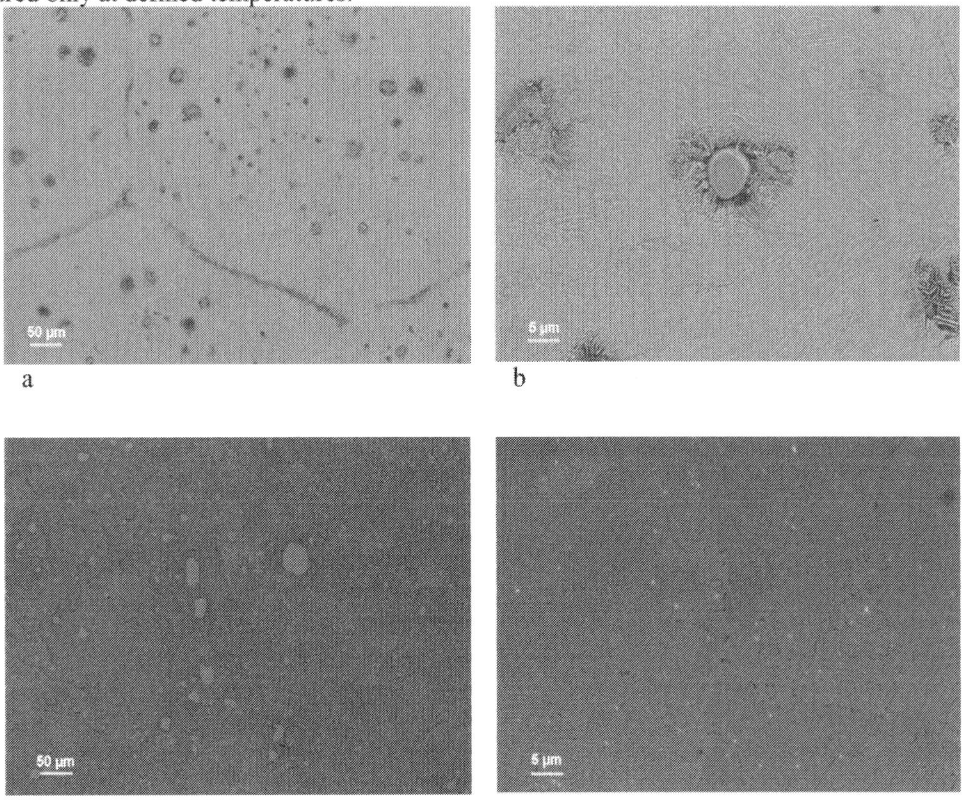

Fig. 1: SEM micrographs of CoRe-1 alloy (a and b) and CoRe-1B (c and d) in standard heat treatment condition. The CoRe-1 alloy shows presence of grain boundary and blocky Cr-carbides while CoRe-1B alloy shows presence of σ phase particles. Both alloys have a fine lamellar structure of $Cr_{23}C_6$ carbide (b and d).

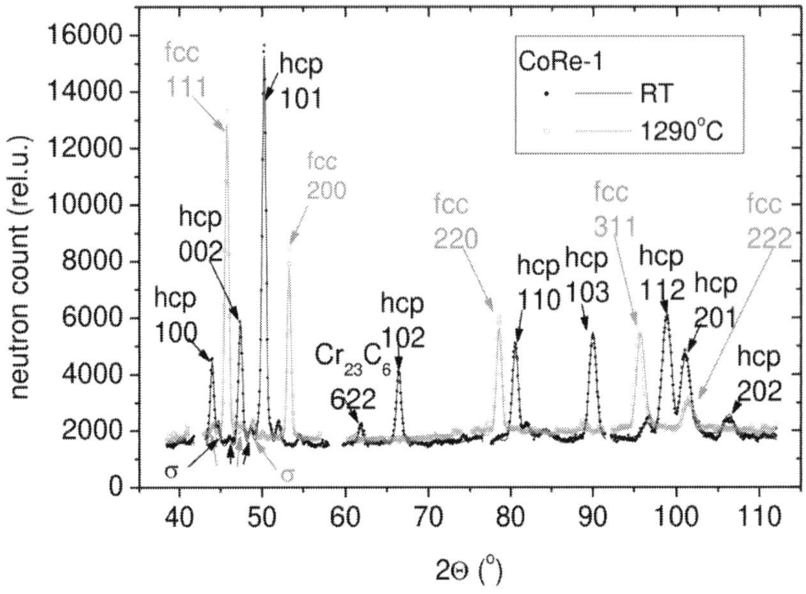

a

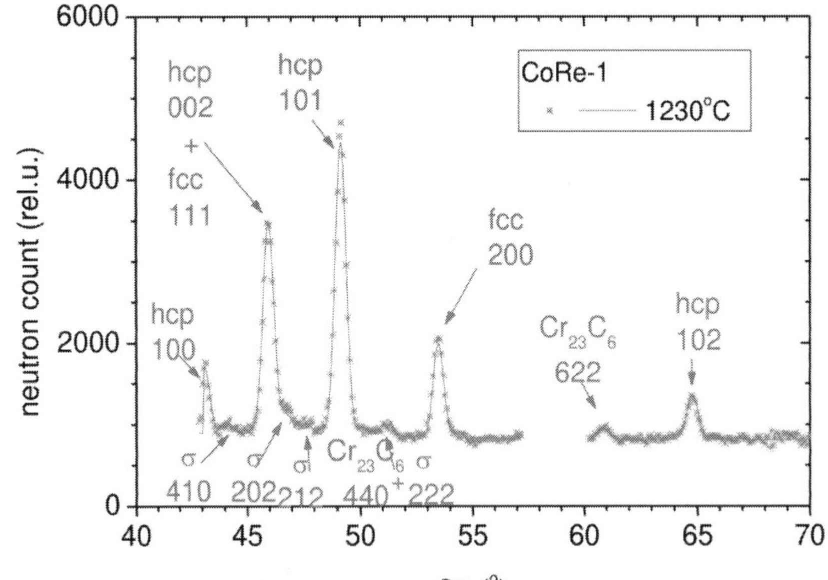

b

Fig. 2: Full diffractograms of CoRe-1 at RT between the first and the second cycle (hcp phase: before phase transformation) and at 1290°C (fcc phase: after phase transformation) measured at Stress-Spec (a). Limited 2Θ scattering range of CoRe-1 during phase transformation at 1230°C (hcp and fcc phase) measured at Stress Spec (b).

As an example of the measured neutron diffraction data, Figure 2a shows full diffractograms of CoRe-1 alloy taken at RT and at 1290°C. It contains the peaks from the matrix and the minority phases present in the alloy. The main phases are the Co solid solution with Re as the matrix, the Cr-carbide phase $Cr_{23}C_6$ and the σ phase Cr_2Re_3. Due to the ε to γ phase transformation the hcp peaks from the Co-matrix are present at the RT diffractogram, while the fcc peaks from the Co-matrix are to be found at the 1290°C. The peaks of the $Cr_{23}C_6$ and the σ phase differ in intensities in the two diffractograms. Figure 2b displays a typical confined angular region used at Stress-Spec during the temperature loop. From the intensities of the peaks of the individual phases, the temperature dependence of their relative volume fraction was deduced. All the diffractograms were analyzed by FullProf program [9]. Resulting phase fractions can be estimated from the peak integral intensities because the scattering contribution in the 2Θ range (42°-68°) covers all strong peaks of the relevant phases. It should be mentioned that due to the small amounts of σ phase and $Cr_{23}C_6$ phase in the sample the low measured values of integral intensity are difficult to determine and have a certain inaccuracy.

The volume fractions of the different phases in CoRe-1 alloy during the two cycles (called first and second cycle) are shown in Figures 3a-d. The second cycle on CoRe-1 sample does not change the temperature of the hcp<->fcc transformation of the Co matrix compared to the first cycle [4], neither during the temperature increase (centre of transformation at 1235°C), nor during the temperature decrease (centre of transformation at 1140°C) (Fig. 3a, b). The only difference is the lower volume fraction part of hcp phase at RT before and after the second cycle with respect to RT before the first cycle.

In CoRe-1 alloy the hysteresis exhibited by the $Cr_{23}C_6$ carbides (Fig. 3c) in the second cycle is almost identical to the one observed in the first cycle and also very much follows the hysteresis of Co-hcp phase. On heating the Cr-carbide starts to dissolve at around 1120°C and it is completely dissolved above 1250°C (Fig. 3c). During cooling the precipitation of carbides do not start before the temperature reaches 1150°C, but then the volume fraction reaches the original value within a short time around this temperature only. In the second cycle (measured at Stress-Spec) the hysteresis is better defined as the number of measured data points is much larger than in the first cycle (Fig. 3c). After cooling in both of the cycles, the $Cr_{23}C_6$ phase reaches the same volume fraction value which is a little bit above the starting value before the first cycle.

On the other hand, the σ phase evolution is significantly different in the first and in the second cycle (Fig. 3d). Its volume decreases on temperature increase above 1120°C, i.e. earlier than the decrease in the hysteresis loop of hcp and $Cr_{23}C_6$ phases. The σ phase amount increases after the first cycle but after the second cycle return to this same value. The σ phase evolution during heating and cooling follows a slightly different path in the first and the second cycle. Much less change in volume fraction is observed in the second cycle for this phase. The volume fraction of σ phase is similar after the first and the second cycle and the second cycle starts at RT with higher volume fraction than the first cycle. The change of σ content at RT before and after cycling is connected with the change of hcp matrix content. Similarly, the change of sigma content at HT between the first and the second cycle is connected with the change of fcc matrix content.

The alloy CoRe-1B was also subjected to similar heating cycles like the CoRe-1 alloy but only one thermal cycle was done in this case. A comparison of the hysteresis in CoRe-1 sample and CoRe-1B sample during the first cycle was performed. The corresponding evolution of the peak intensities of the individual phases are displayed in Figure 4. In the CoRe-1B sample, the transformation hcp<->fcc (Fig. 4a, b) on temperature increase occurs around 1225°C (centre of transformation) and on temperature decrease around 1135°C (centre of transformation). It differs slightly from the transformation temperature of CoRe-1 sample without boron addition (centre of transform was at 1235°C and 1140°C in CoRe-1 on temperature increase and on temperature decrease, respectively). The shift in alloys with and without boron is only less than 10°C and part of the error is likely due to the uncertainty of temperature synchronization in the respective in-situ experiments. The $Cr_{23}C_6$ phase shows no qualitative difference in CoRe-1 and CoRe-1B alloys. Again, only small shift of the hysteresis loop in the temperature scale towards lower temperatures can be observed (see Fig. 4c).

As in the case of CoRe-1, the alloy CoRe-1B stabilizes a higher amount of σ at the end of the thermal cycle compared to the initial amount before the thermal cycle (Fig. 4d). The difference in volume fraction is at the expense of hcp matrix phase as before (Fig. 4a) and [4]. It should be stressed, that the σ phase content in CoRe-1B follows the same evolution as in the alloy CoRe-1, so apparently B addition has not a strong influence on the hysteresis.

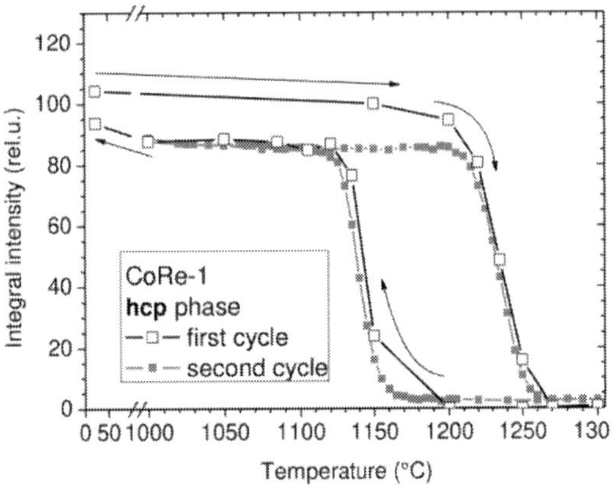

a

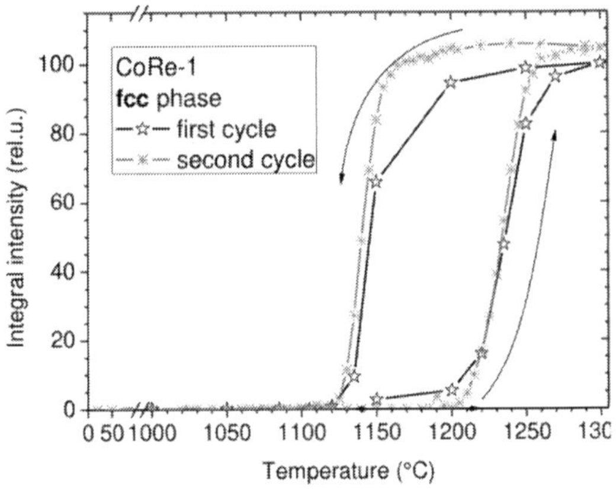

b

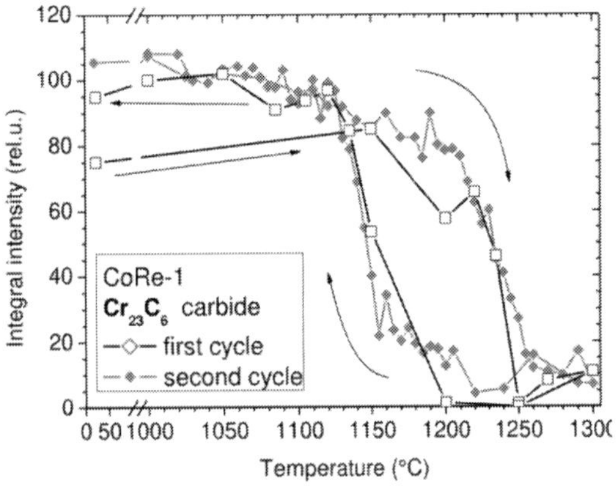

c

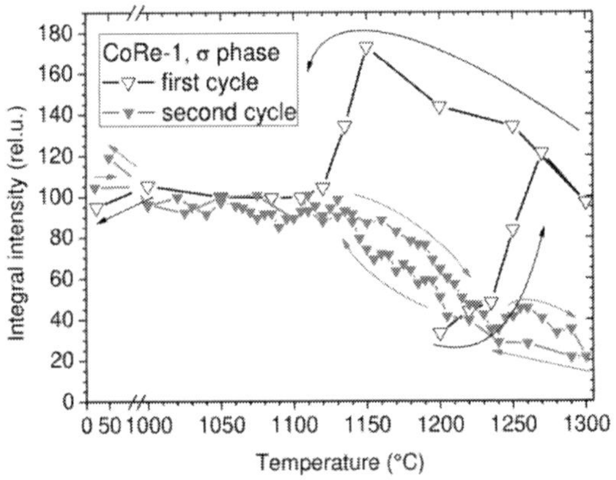

d

Fig. 3: Evolution of hcp (a), fcc (b), $Cr_{23}C_6$ (c), and σ (d) phase in CoRe-1 sample between RT and 1300°C during the first and second thermal cycle. For a better overview error bars are not implemented. The error bars are +/- 1 for fcc and hcp phase, +/- 5 for $Cr_{23}C_6$ and +/- 6 for σ phase.

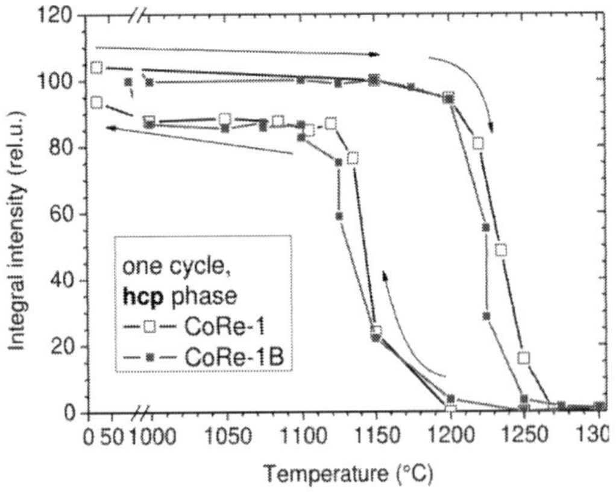

a

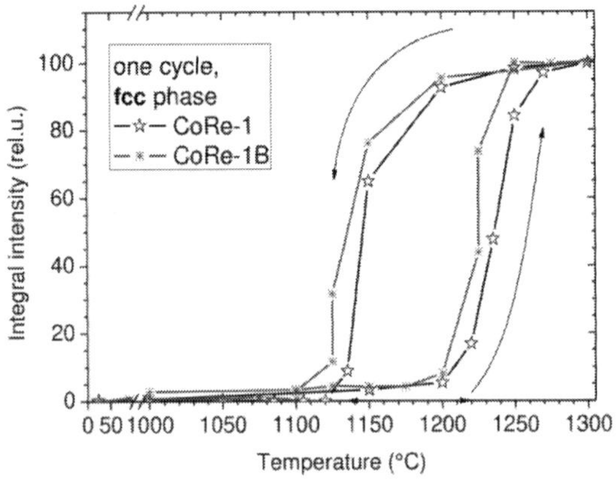

b

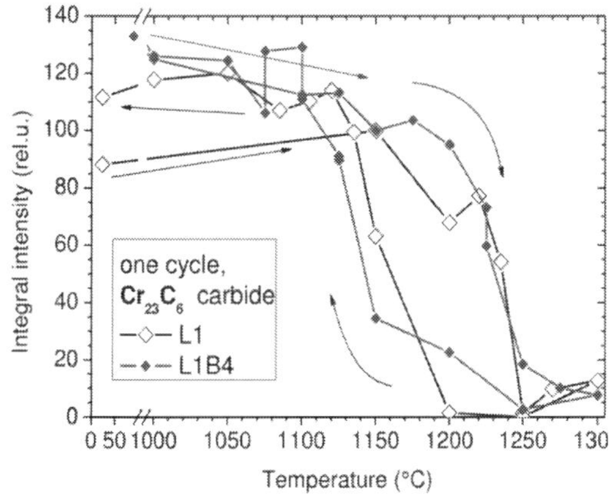

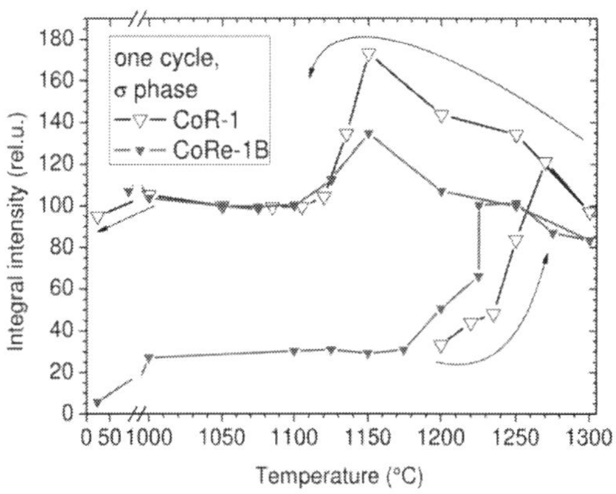

c d

Fig. 4: Evolution of hcp (a), fcc (b), $Cr_{23}C_6$ (c), and σ (d) phase in CoRe-1B alloy at high temperatures at the first cycle together with a comparison of CoRe-1 sample measured at the first cycle. For a better overview error bars are not implemented. The error bars are +/- 1 for fcc and hcp phase, +/- 7 for $Cr_{23}C_6$ and +/- 3 for σ phase.

First tensile-load tests at 538°C, 800°C and 980°C were successfully conducted at Stress-Spec using a newly developed tensile rig equipped with a light furnace (Figure 5). The stability of the setup, which is to be used in future experiments with Co-Re alloys, was tested during tensile loading at the mentioned temperatures. In all tests the alloy failed already before plastic yielding.

Fig. 5: The installed tensile rig with the high-temperature equipment including four lamps at the neutron diffractometer Stress-Spec at FRM II. The heating shield is dismounted to provide an insight into the setup.

The results of the tensile test in the elastic region with heating at 980°C are shown for the applied stress versus lattice strain in Figure 6. The linear increase of lattice strain versus applied stress for (10-11) reflection of CoRe alloy results in a modulus of elasticity value of about 230 GPa for this lattice plane. Further these tests resulted in the optimization of the in situ heating system. During the initial test grips were partially heated together with the sample (mainly due to the limited length of 40 mm sample gauge length). Under such conditions 400°C was the maximum achievable temperature. After fabrication of the new set of samples with a longer gauge length and the fabrication of new grips made from Ni-base superalloy MarM247 and with additional cooling of grips, significantly higher temperatures were achieved in the follow-up experiments. Also, polishing of the mirrors, new high-power light source and adding simple heat shields using aluminum foil made the heating setup more efficient so that maximum temperature of T = 980°C could be reached.

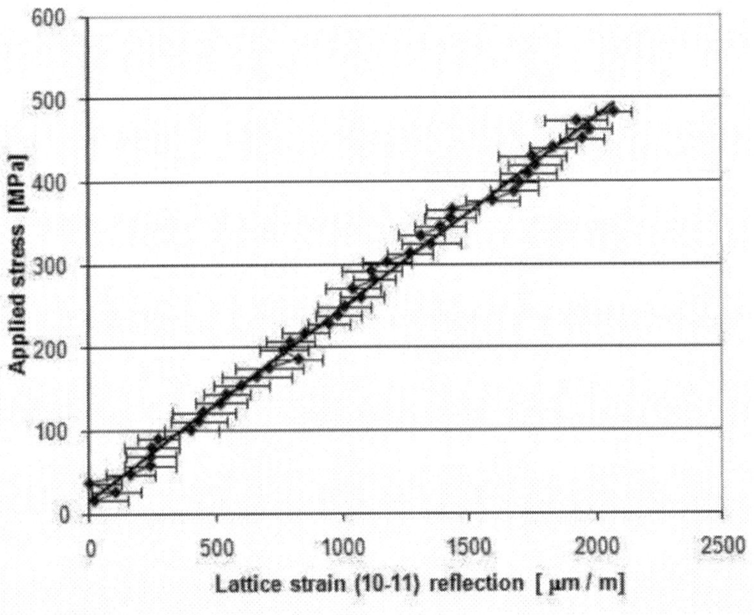

Fig. 6: Result of first tensile test at high temperature (980°C) with linear fit of applied stress versus lattice strain of (10-11) hcp reflection from Co-matrix.

4. Conclusion

The neutron diffraction method provided a detailed characterization of phases at high temperatures in experimental Co-Re based alloys. In particular the evolution of the various strengthening phases could be observed in-situ and quantitatively determined. The information gained through these in-situ neutron diffraction measurements is not easy to obtain through conventional microstructural studies involving ex-situ heating/cooling cycles. The second thermal cycle performed on CoRe-1 alloys brought first information on the high-temperature structure and thus on the basic stability during thermal cycling. Neutron diffraction experiments show dissolution of $Cr_{23}C_6$ above 1250°C, and still presence of Cr_2Re_3 (σ phase) at 1300°C. Performed cycling experiments yield in an increase of σ phase volume by heating in the second cycle and again increase by cooling in the second cycle. The diffraction data taken from CoRe-1B sample containing boron addition showed that the basic phase volume fraction evolution is the same as for the alloy CoRe-1 without boron. However, it is worth to note that it cannot be concluded on the morphology of the individual phases. Such a study can be done using small-angle neutron scattering and will be reported elsewhere.

The present experiment successfully tested the capabilities of the new sample environment. For the first time tensile tests close to 1000°C (i.e. at 980°C) could be performed at Stress-Spec instrument and we are confident that - with further minor improvements (e.g. an improved heat shielding) - a temperature capability of 1050°C or 1100°C will be achievable.

5. Acknowledgement

The authors would like to thank Alexander Schmidt and Manfred Teufelhart for support on the high-temperature furnace, Jens Krueger for support on the temperature control and Christian Randau on the evaluation process [10]. The travel support to P. Strunz for the experiment in the frame of the programme 'Research Infrastructures' - contracts No. CP-CSA_INFRA-2008-1.1.1-226507 (NMI3) - is acknowledged as well. P. Strunz also acknowledges the support by MPO project FR-TI1/378. We gratefully acknowledge financial support of the research unit 727 'Beyond Ni-base Superalloys' by the Deutsche Forschungsgemeinschaft.

6. References

[1] J Rösler, D Mukherji and T Baranski, Adv. Eng. Mater. 9, 2007 876-881.
[2] D Mukherji, M Klauke, P Strunz, I Zizak, G Schumacher, A Wiedenmann and J Rösler, Int. J. Mat. Res., 2010, vol. 101, 340-48.
[3] D Mukherji, P Strunz, R Gilles, M Hofmann, F Schmitz, J Rösler: Materials Letters 64 (2010), 2608-2611.
[4] D Mukherji, P Strunz, S Piegert, R Gilles, M Hofmann, M Hoelzel and J Roesler, Met. Trans. A, submitted.
[5] D Mukherji and J Rösler: Journal of Physics: Conference Series 240 (2010) 012066 (4 pp).
[6] M Hofmann, R Schneider, GA Seidl, J Kornmeier, R Wimpory, U Garbe and HG Brokmeier, Physica B, 2006, vol. 385 – 368, 1035-37.
[7] R Gilles, M Hoelzel, M Schlapp, F Elf, B Krimmer, H Boysen and H Fuess, Z. Kristallogr. Suppl., 2006, vol. 23, 183-188.
[8] M Hoelzel, A Synyshyn, R Gilles, H Boysen, H Fuess Neutron News (2007), 18, Number 4, 23-26.
[9] T Roisnel, J Rodriguez-Carvajal WinPLOTR: a Windows tool for powder diffraction patterns analysis Materials Science Forum, Proceedings of the Seventh European Powder Diffraction Conference (EPDIC 7), 2000, 118-123, Ed. R. Delhez and E.J. Mittenmeijer
[10] C Randau, U Garbe & HG Brokmeier, Journal of Applied Crystallography, 44(3), 2011, 641-646. doi: 10.1107/S0021889811012064.

Heterogeneous Catalysis under pressure - *In-situ* neutron diffraction under industrial conditions

Timur Kandemir[1], Frank Girgsdies[1], Igor Kasatkin[1], Edward Kunkes[1], Klaus-Dieter Liss[2], Vanessa K. Peterson[2], Robert Schlögl[1], Malte Behrens[1]

[1]Department of Inorganic Chemistry, Fritz-Haber-Institute of the MPG, Berlin, 14195, Germany

[2]The Bragg Institute, ANSTO, Lucas Heights, NSW 2232, Australia

E-mail: `behrens@fhi-berlin.mpg.de`

Abstract. The present work describes the application of a tubular reactor that allows in-situ neutron diffraction on working catalysts at high pressures. The designed reactor enables the application to a sample of industrially-relevant reaction conditions, *i.e.*, in a temperature range up to *330°C* and *60 bar* pressure, coupled with online gas-analysis. Application of the cell is demonstrated by ammonia synthesis over a commercial catalyst with diffraction data obtained from the high-resolution powder diffractometer, Echidna, at the Australian Nuclear Science and Technology Organisation, ANSTO.

1. Introduction

Neutron diffraction is a unique tool to gain information on crystalline phases, the crystal structure, structural disorder, crystalline domain size and shape, lattice strain and defect structure, of materials including catalysts [1]. Nowadays, catalyst are considered as dynamic materials whose active centres are formed or transformed under reaction conditions. *In-situ* characterization methods are needed to obtain information about such structural dynamics, which are typically applied in the pressure range from ultra-high vacuum to ambient. The structural properties of the catalyst may change with the partial pressure of the reactants or products, and investigations of catalysts at high pressures and near industrially-relevant working conditions using *in-situ* neutron diffraction can provide valuable information on the nature of the catalytically active sites. Due to the high penetration depth of neutrons, they can scatter from samples deep within conventional metallic reactors without the use of specialized neutron-beam transparent windows, enabling the application of demanding sample environments that are not easily used with conventional X-ray diffraction. Additionally, the relatively large Q ($=2\pi/d$) coverage and non-decreasing scattering factors at high Q make neutron diffraction a suitable tool for *in-situ* studies of working catalysts, enabling crystallographic information at reasonable counting times to be obtained. This contribution presents the application of a custom-made high-pressure cell to the ammonia-synthesis reaction.

2. Experimental

For the purpose of *in-situ* neutron diffraction a high-pressure continuous flow reactor was designed. The flow cell, made out of an $AlMg_3$ alloy, is suitable for reaction temperatures from ambient to 330°C and a maximum pressure of 60 bar.[2]. Mass spectrometry was used during

Figure 1. Flow reactor mounted on the Echidna instrument. The incident neutron-beam window is shown in the background.

the *in-situ* neutron diffraction experiment for online gas analysis. Previous experiments in the laboratory have demonstrated the reactor functionality for the pretreatment of the catalysts, and continuously-running, stable catalytic conversions have been achieved. Relatively fast changes of the gas atmosphere are possible to introduce various probe gases (*e.g.*, N_2O reactive-frontal chromatography for copper-based catalysis [3]). The example described here was carried out on the high-resolution powder diffractometer, Echidna, at the OPAL reactor facility using a Ge 335 monochromator to provide a beam at a wavelength of λ=1.622Å. A large array of 128 position sensitive ^{3}He detectors enabled an angular range of 4°-164° in 2Θ to be covered, corresponding to a Q-range of 0.28Å^{-1} to 7.7Å^{-1} [4, 5].

3. Application example: Ammonia synthesis

Ammonia synthesis is not only one of the largest-volume synthetic chemical-processes applied in chemical industries, but also serves as a prototype reaction for heterogeneous catalysis research. Significant efforts to investigate the surface chemistry of model catalysts have been made in the last few decades using this reaction, as well as practical tests of high-performance catalysts [6]. Ammonia synthesis from nitrogen and hydrogen/deuterium (*see [7] for detailed mechanism*) is a structure-sensitive reaction, which does not occur on every form of metallic iron:

$$N_2 + 3D_2 \rightleftharpoons 2ND_3$$

Commercial catalysts are so called "bulk-catalyst" with varying amounts of promoter species. The catalytically-inactive precursor material typically consists of fused iron oxides, e.g., approximately 80 vol.% magnetite, Fe_3O_4, and 20 vol.% wuestite, FeO. The specific BET-surface

area of the catalyst is around 10 m^2g^{-1}, while the mean particle size of the wuestite/magnetite-precursor is roughly 10 μm. The catalytically-active form is called "ammonia iron" [8], which is produced from the fused iron-oxides by a complex thermal treatment in a reducing reactant gas-feed. Studies have shown [9] that the precursor material exhibits a disk-like internal microstructure, leading to long-term stable aggregates of α-iron-platelets separated by oxide spacers, which are favorable for the conversion of di-nitrogen. There is evidence that the active surface undergoes reversible reconstruction [10] under working conditions leading, for instance, to the formation of nitrides. In addition to pure iron [11], a metastable non-iron para-crystalline metal structure [12], a skin of Fe_4N or its under-stoichiometric variants [13] are discussed as catalytically-active phases.
The pre-activated α-iron catalyst was transferred under inert gas into the flow cell. After inserting the cell into the reactor the catalyst bed was carefully heated at 1 K min^{-1} under the reaction feed-gas to the reaction temperature of 330°C, a guard-reactor with the same catalyst was operated at 250°C. After reaching the reaction temperature, the cell was pressurized to 60 bar. When the detected ammonia concentration at the exhaust was stable, diffraction patterns with an exposure time of 2 h were acquired. For a stoichiometric gas feed of 100 Nml min^{-1}, a coversion of 3.63% was achieved while the productivity was 7.59 $mg_{ammonia}\ g_{catalyst}{}^{-1}h^{-1}$. For accurate determination of the instrumental peak profile a LaB_6-standard was measured. Line broadening analysis using a pseudo-voigt-type peak-function with an axial-divergence term allowed to deconvolute angular dependencies on the integral breaths of the Lorentzian and Gaussian components.

4. Results

Fig. 2 shows the diffraction pattern of a commercial iron-catalyst in operation at 330°C and 60 bar, wherein diffraction peaks arising from the cell were removed. Strong diffraction peaks of α-Fe (- - - -) can be clearly distinguished and indexed. After deconvolution of the instrumental broadening from the diffracted intensities, the complex microstructure of the catalytically active α-Fe phase of the catalyst is indicated by the asymmetric peak broadening of the reflections as exemplified for the 211 profile shown in Fig. 2 (inset). In addition to metallic iron (a = 2.87840 (±0.00013) Å), there is also a noticeable contribution of a spinel phase (a = 8.38415 (±0.00022) Å), which can be refined using magnetite (Fe_3O_4) as a structural model (——). The refined lattice parameter is significantly shorter than expected for pure magnetite at 330 °C (8.4179 and 8.4245 Å at 300°C and 350°C, respectively [14]). On the other hand, it is much larger than expected for $FeAl_2O_4$ (Hercynite, 8.1695 Å at 400 °C [15]), which has been reported to form in ammonia-synthesis catalyst from FeO and the γ-Al_2O_3 promoter during reduction [16]. Our results suggest the presence of a non-stoichiometric and defective $(FeO)_x{\cdot}(Al_2O_4)_\gamma$ spinel-phase, in which the iron-cations are not reduced under *in-situ* conditions probably due to the interaction with Al in the spinel lattice. Transmission electron microscopy (TEM) studies (Fig. 3) revealed a complex composite microstructure of the catalyst with metallic iron particles and an embedding oxide matrix, wherein the promoter species (Ca, Al, K, Si, Ti, V) are enriched. This suggests that additional faint reflections arising in the level of background noise are attributable to small calcium ferrite crystallites. A careful correction for background artefacts will be mandatory for further evaluation.

In-situ neutron diffraction experiments as shown here are useful as a starting point for the further investigation of the structural changes occurring to a working catalyst during variation of the reaction conditions. Importantly, this information can be correlated to changes to catalytic performance. More detailed results of the ammonia-synthesis catalysts will be published elsewhere.

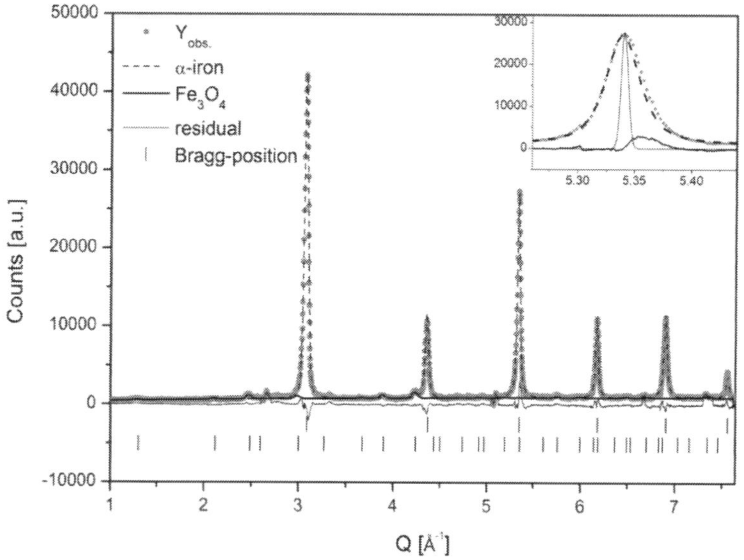

Figure 2. Neutron powder diffraction pattern of a working commercial ammonia catalyst at 330 °C and under 60 bar of syngas. A Rietveld-refinement based phase deconvolution was applied to the data. Bragg-positions of α-iron (top) and a spinel phase (structural model: Fe_3O_4, bottom) are marked, the difference between the model and the data is shown by the solid line above. The inset compares the asymmetric peak broadening of the 211 reflex of the α-iron-phase to the instrumental resolution function (narrow peak).

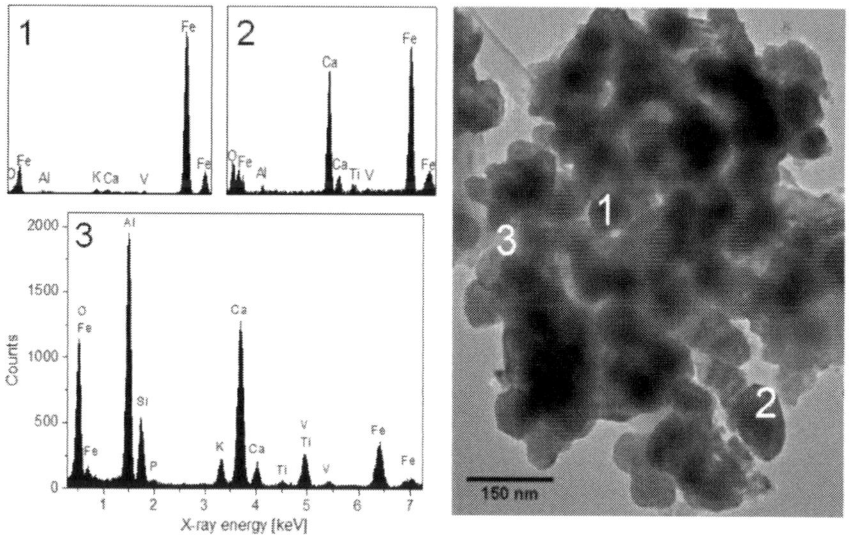

Figure 3. TEM micrograph and energy dispersive X-ray (EDX) spectra of the sample used during in-situ neutron diffraction. The point analysis spectra reveal that in addition to the α-iron particles (1) areas which are enriched in oxygen and in the promoter species are also present (2,3). These areas are found near the edges of the iron particles (2) and as an embedding matrix consisting of partially crystalline calcium ferrite/iron aluminum spinel (3).

5. Conclusions

The constructed high-pressure continuous-flow reactor was successfully used in *in-situ* neutron diffraction on an ammonia-synthesis catalyst, working at 330°C under a pressure of 60 bar. The preliminary data analysis confirms that all microstructural deviations of "ammonia-iron" that were derived at ambient pressure qualitatively also prevail at the technical harsh operation conditions. In particular, the absence of a stoichiometric iron nitride phase and the non-equilibrium structure of working metallic iron were confirmed to be no artefacts of low-pressure experimentation.

Acknowledgements

The authors would like to thank Scott Olsen, Friedl Bartsch, Mai Anh Burke for technical support and the DFG (German research foundation) for financial support (BE 4767/1-1). Friedrich Karl Seitz and Frank Rosowski (BASF) are acknowledged for providing the catalyst and fruitful discussions.

References

[1] Snyder R, Fiala J and Bunge H J (eds) 1999 *Defect and Microstructure Analysis by Diffraction* (New York: Oxford Science Publications)
[2] Kandemir T, Wallacher D, Hansen T, Liss K D, Schlögl R and Behrens M 2011 *preprint*
[3] Chinchen G C, Hay C M, Vandervell H D and Waugh KC 1987*J. Catal.* **51** 326
[4] Avdeev M, Hester J R, Peterson V K and Studer A J 2009 *Neutron News* **20** 22-33
[5] Liss K D, Hunter B A, Hagen M E, Noakes T J and Kennedy S J 2006 *Physica B* **385-386** 1013-1015
[6] Schlögl R 2008 *Ammonia Synthesis* 2nd ed (*Handbook of Heterogeneous Catalysis* vol 5) ed Ertl G, Knözinger H, Schüth F and Weitkamp J (Weinheim: Wiley-VCH)
[7] Ertl G 1991 *Elementary steps in ammonia synthesis: the surface science approach* (*Catalytic Ammonia Synthesis, Fundamentals and Practise*) ed Jennings R J (New York: Plenum)
[8] Mittasch A 1950 *Adv. Catal.*D **2** 81-104
[9] Egeberg R C 2001 *Surf. Sci.* **491** 183-194
[10] Emmet P H and Love K S 1993 *J. A. Chem. Soc.* **5** 4043-4050
[11] Schlögl R 1991 *Preparation and activation of the technical ammonia synthesis catalyst* (*Catalytic Ammonia Synthesis, Fundamentals and Practise*) ed Jennings R J (New York: Plenum)
[12] Ludwiczek H, Preisinger A, Fischer A, Hosemann R and Schönfeld A and Vogel W 1978 *J. Catal.* **51** 326
[13] Ertl G, Huber M and Thiele N 1979 *Z. Naturforsch.* **34a** 30
[14] Okudera H, Kihara K, Matsumoto T 1996 *Acta Cryst.* **52** 450-457
[15] Harrison R J, Redfern S A T, O'Neill H S C O 1998 *American Mineralogist* **83** 1092-1099
[16] Brill R 1932 *Ztschr. Elektrochem.* **8b** 669-673

Journal of Physics: Conference Series **340** (2012) 012054 doi:10.1088/1742-6596/340/1/012054

Lattice dynamics of $PbTiO_3$

Izumi Tomeno[1], Jaime A Fernandez-Baca[2], Karol J Marty[2], Yorihiko Tsunoda[3] and Kunihiko Oka[4]

[1] Faculty of Education and Human Studies, Akita University, Akita 010-8502, Japan
[2] Neutron Scattering Science Division, Oak Ridge National Laboratory, Oak Ridge, Tennessee 37831-6393, USA
[3] Department of Applied Physics, School of Science and Engineering, Waseda University, Shinjukuku, Tokyo 169-8555, Japan
[4] Nanoelectronics Research Institute, National Institute of Advanced Industrial Science and Technology, Tsukuba, Ibaraki 305-8568, Japan

E-mail: tomeno@gipc.akita-u.ac.jp

Abstract. Inelastic neutron scattering experiments were performed to investigate the phonon dispersion relations along the [1,0,0], [1,1,0] and [1,1,1] directions in cubic $PbTiO_3$. All of the transverse optic (TO) branches soften significantly toward the zone center around T_c. The zone-center TO mode stiffens considerably at 1173 K. The energies of the $[\xi, 0, 0]$ TO branch for $PbTiO_3$ measured at 793 K are considerably higher than those for $Pb(Zn_{1/3}Nb_{2/3})O_3$ at 423 K in the entire zone. The $[\xi, 0, 0]$ transverse acoustic (TA) branch is found to soften slightly at 793 K. The $[\xi, \xi, \xi]$ TA branch exhibits two minima at the middle point (1/4,1/4,1/4) and the zone boundary (1/2,1/2,1/2). These two modes are weakly temperature dependent up to 1173 K.

1. Introduction

Lead titanate $PbTiO_3$ undergoes only a single phase transition at $T_c = 763$ K from a cubic paraelectric to tetragonal ferroelectric phase. The covalent nature of Pb $6s$ and O $2p$ states plays an important role in the stability of tetragonal $PbTiO_3$ [1]. In the 1970s, Shirane *et al.* [2] investigated the lattice dynamics of $PbTiO_3$ by inelastic neutron scattering. The $[\xi, 0, 0]$ TO branch softens significantly around the zone center as the temperature approaches T_c. Recently, Kempa *et al.* [3] determined the phonon dispersion curves along the [1,0,0] and [1,1,0] directions in cubic $PbTiO_3$. However, the results for the $[\xi, 0, 0]$ TO branch are in disagreement with those reported previously [2]. Tomeno *et al.* [4] studied systematically the phonon dispersion relations in tetragonal $PbTiO_3$. The experimental results for tetragonal $PbTiO_3$ are generally in agreement with the first-principles calculations for the zone-center and zone boundary phonon energies [5]. Ghosez *et al.* [6] computed the full phonon dispersion curves in cubic $PbTiO_3$ from first-principles calculations. Unfortunately, until recently, the lack of the experimental phonon data above T_c could neither confirm nor disprove the predicted phonon behavior.

Lead titanate $PbTiO_3$ is also the end member of relaxor ferroelectrics $(1\text{-}x)Pb(Zn_{1/3}Nb_{2/3})O_3\text{-}xPbTiO_3$ (PZN-PT) and $(1\text{-}x)Pb(Mg_{1/3}Nb_{2/3})O_3\text{-}xPbTiO_3$ (PMN-PT). The donation of approximately 10% Ti atoms to PZN is crucial to piezoelectric constant enhancement [7]. The low-lying TO phonon branch in these relaxors drops sharply near the zone center [8,9]. The

Journal of Physics: Conference Series **340** (2012) 012054 doi:10.1088/1742-6596/340/1/012054

soft TO mode in Pb-based relaxors has been discussed in connection with the polar nanoregions [8,10], but the origin of the sharp drop remains controversial [11].

2. Experiment

The inelastic neutron scattering experiments were performed on the triple-axis spectrometers TAS-1 and T1-1 at JRR-3M in the Japan Atomic Energy Agency, and HB-3 at the High Flux Isotope Reactor, Oak Ridge National Laboratory. The TAS-1 and T1-1 spectrometers were operated with fixed incident energy, $E_i = 13.7$ or 14.7 meV. Collimation was 40'-40'-40'-40' for TAS-1, and guide-40'-40'-40' for T1-1. The HB-3 spectrometer was operated with fixed final energy, $E_f = 14.7$ meV. Collimation was either 48'-20'-20'-70' or 48'-40'-40'-70'. The energy resolution for TAS-1 and T1-1 was 0.66 meV in full width at half maximum (FWHM) at zero energy transfer, whereas that for HB-3 was 0.91 or 1.07 meV. The single crystals used in this work were grown using the top-seeded growth technique described in detail previously [12]. Most of the measurements were performed at 793 K, but selected phonons were also determined up to 1173 K.

3. Phonon dispersion relations

Figure 1 shows the phonon dispersion relations for cubic $PbTiO_3$ measured at 793 K. The TO Δ_5, Σ_3, Σ_4 and Λ_3 branches soften significantly toward the zone center. A set of TO branches shows the isotropic nature of the soft modes. The softening of these TO modes corresponds to the ferroelectric instability of the Γ_{15} TO mode. It is difficult to determine accurately the energies of the Γ_{15} TO mode near T_c. However, a series of constant-E scans indicates that the energy of the zone center TO mode is lower than 3 meV at 793 K. The present value for the soft mode energy is lower than the value reported by Shirane *et at* [2] or by Kempa *et at* [3]. The discrepancy in the soft-mode energy suggests that the phase-transition properties of $PbTiO_3$ are influenced by the crystal growth conditions. Figure 2 shows typical constant-E scans in search of the TO Δ_5 branches. Here the data are taken using the HB-3 spectrometer. The Γ_{15} TO mode stiffens considerably with increasing temperature. The energies of the Γ_{15} TO mode are approximately 6 meV at 1073 K and 8 meV at 1173 K.

In figure 3, the Δ_5 branches in cubic $PbTiO_3$ are compared with the corresponding Δ_2 branches in tetragonal $PbTiO_3$ measured at 295 K [4]. The significant softening of the TO Δ_5 branch is confined to the region $q \leq 0.2$.

The energies of the TA Δ_2 branch at 295 K are higher than those of the TA Δ_5 branch at 793 K. In the cubic phase, the long-wavelength TA Δ_5 and Λ_3 modes soften gradually with decreasing temperature. Constant-Q scans in figure 4 demonstrate that the TA Δ_5 mode softens simultaneously in cubic $PbTiO_3$. A similar tendency has been observed for the TA Δ_5 branch in $KTaO_3$ [13]. The significant softening of the zone-center TO mode in ABO_3 appears to lower the relevant TA phonon branch.

Moreover, the TA Λ_3 branch exhibits two minima around the middle point (1/4,1/4,1/4) and the zone boundary (1/2,1/2,1/2). Constant-E and constant-Q scans in figure 5 indicate the existence of the energy minimum around (1/4,1/4,1/4). The broad minimum of the TA Λ_3 mode around (1/4,1/4,1/4) suggests the tendency toward forming a fourfold periodicity along the [1,1,1] direction. Note in figure 1 that the other TA branches show the normal phonon dispersion. A possible interpretation is that the octahedron rotation of successive oxygen layers follows the sequence of same, same, opposite, and opposite senses. To our knowledge, our findings on the middle-point softening are the first report on the phonon dispersion in perovskite oxides ABO_3. One possibility is that the covalent nature of Pb and O atoms gives rise to the middle-point softening. The present results for the TA Λ_3 branch support the predictions of the unstable mode at (1/4,1/4,1/4) and the flat $[\xi,\xi,\xi]$ TA dispersion from first-principles calculations [6,14]. Figure 1 clearly shows the coexistence of the Γ_{15} and R_{25} mode softening. The lowering of

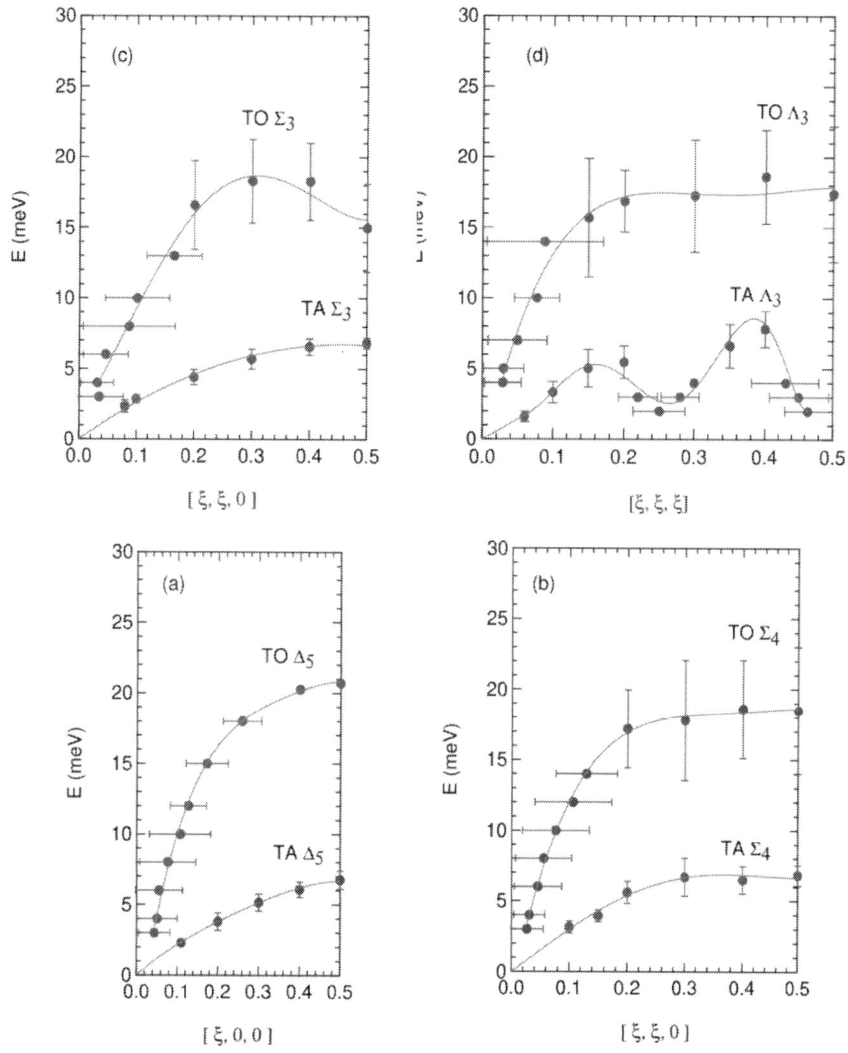

Figure 1. Phonon-dispersion curves for $PbTiO_3$ measured at 793 K. Each solid circle and the attached bar refer to the phonon peak and its FWHM, respectively. Solid lines are guides for the eye.

the R_{25} phonon energy represents the softening of the oxygen octahedron rotation. The zone-boundary instability in $PbTiO_3$ has been predicted from first-principles calculations [15]. In ABO_3, the tendency toward the R_{25} mode softening generally depends on the tolerance factor, $t = (r_O + r_A)/\sqrt{2}(r_O + r_B)$, where r_O, r_A, r_B are the ionic radii of the O, A, and B ions, respectively. The R_{25} mode softening has been observed for $SrTiO_3$ [16], and $NaNbO_3$ [17]. Lead titanate $PbTiO_3$ has the tolerance factor $t = 1.03$, slightly larger than 1.01 for $SrTiO_3$ and 0.97 for $NaNbO_3$. Theoretically $PbTiO_3$ shows a weak instability at the (1/2,1/2,1/2) point [6,14,15]. In this experiment, we found that the TA phonon energies around (1/4,1/4,1/4) and (1/2,1/2,1/2) are weakly temperature dependent up to 1173 K. Consequently, the softening of the zone-center TO mode dominates the single phase transition in $PbTiO_3$.

In figure 3(b), the TA Δ_5 and TO Δ_5 branches for $PbTiO_3$ at 793 K are compared with those

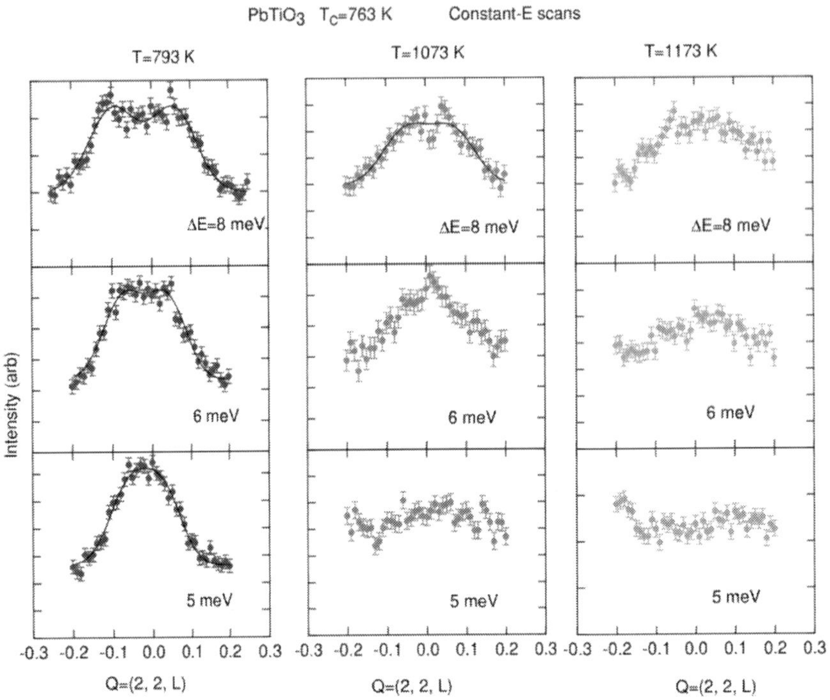

Figure 2. Constant-E scans in $PbTiO_3$ for the Δ_5 mode measured at $T = 793, 1073$ and 1173 K. Solid lines indicate the best fits to a double Gaussian.

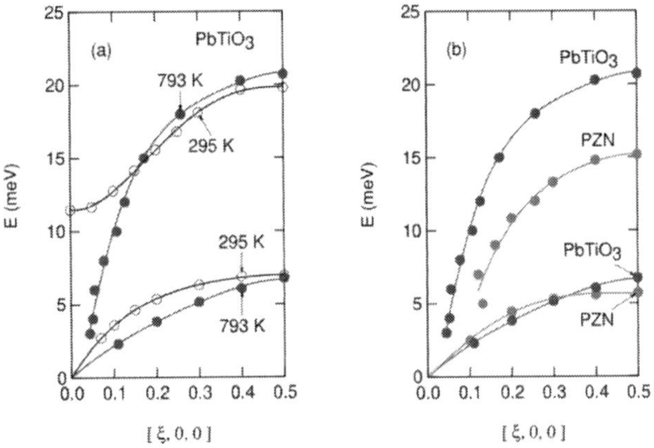

Figure 3. Transverse phonon dispersion curves along [1,0,0] in cubic $PbTiO_3$, tetragonal $PbTiO_3$, and PZN. The Δ_5 modes in cubic $PbTiO_3$ at 793 K are compared with the Δ_2 modes in tetragonal $PbTiO_3$ at 295 K [4], and the Δ_5 modes PZN at 423 K [9].

for PZN at 423 K [9]. The energies of the TO Δ_5 branch for $PbTiO_3$ are considerably higher than those for PZN in the entire range. On the other hand, there is a slight difference in the TA Δ_5 branch between $PbTiO_3$ and PZN. The replacement of the B site atom has a strong influence on the TO phonon dispersion. The average mass of the B atom in PZN is 1.75 times that of Ti in $PbTiO_3$. Therefore, the reduction of the X point TO energy is roughly explained by the square root of the B atom mass. The force constants for $PbTiO_3$ appear to be comparable to those for the Pb-based relaxors. The randomness at the B site in PZN broadens the long-wavelength TO phonon spectra. The important role of Pb atom can account for the similarity in the TA phonon dispersion between $PbTiO_3$ and PZN.

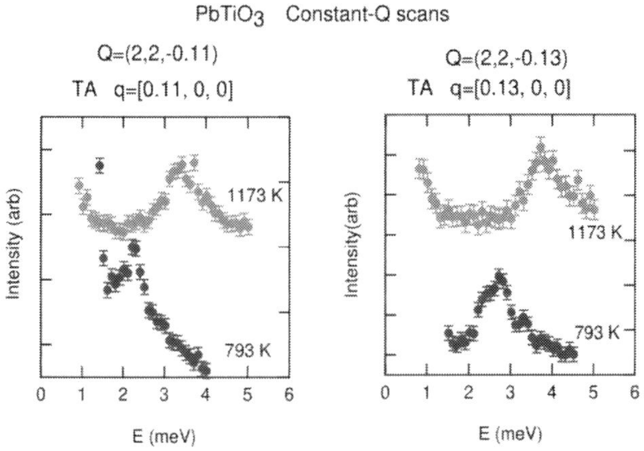

Figure 4. Constant-Q scans in $PbTiO_3$ for the TA Δ_5 mode measured at $T = 793$ and 1173 K.

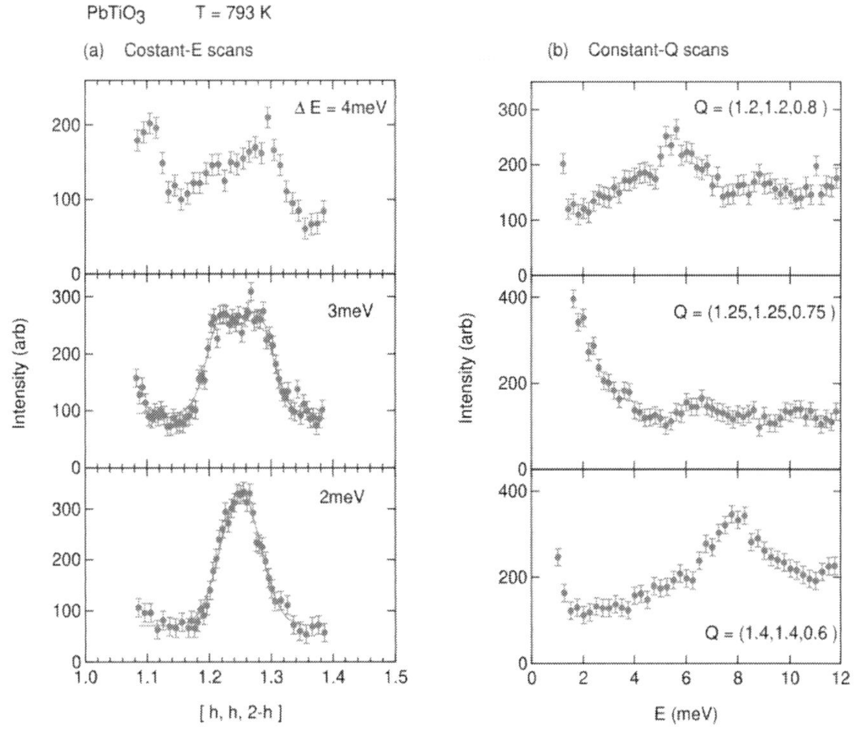

Figure 5. (a) Constant-E scans in $PbTiO_3$ for the Λ_3 mode through (1.25,1.25,0.75) measured at 793 K. (b) Constant-Q scans at Q=(h,h,2-h).

Acknowledgments
We thank K. Kohn, H. Unoki, Y. Ishii, and M. Nishi for helpful discussions. This study was supported in part by the U.S.-Japan Cooperative Program on Neutron Scattering. The work at Oak Ridge National Laboratory was supported by the U.S. Department of Energy under Contract No. DE-AC05-00R22725 with UT-Battelle, LLC.

References

[1] Cohen R E 1992 *Nature* **358** 136
[2] Shirane G, Axe J D, Harada J and Remeika J P 1970 *Phys. Rev.* B **2** 155
[3] Kempa M, Hlinka J, Kulda J, Bourges P, Kania A and Petzelt J 2006 *Phase Transit.* **79** 351
[4] Tomeno I, Ishii Y, Tsunoda Y and Oka K 2006 *Phys. Rev.* B **73** 064116
[5] García A and Vanderbilt D 1996 *Phys. Rev.* B **54** 3817
[6] Ghosez Ph, Cockayne E, Waghmare U V and Rabe K M 1999 *Phys. Rev.* B **60** 836
[7] Kuwata J, Uchino K and Nomura S 1982 *Jpn. J. Appl. Phys.* **21** 1298
[8] Gehring P M, Park S -E and Shirane G 2000 *Phys. Rev. Lett.* **84** 5216
[9] Tomeno I, Shimanuki S, Ishii Y and Tsunoda 2001 *J. Phys. Soc. Jpn.* **70** 1444
[10] Gehring P M, Wakimoto S, Ye Z -G and Shirane G 2001 *Phys. Rev. Lett.* **87** 277601
[11] Hlinka J, Kamba S, Petzelt J, Kulda J, Randall C A and Zhang S J 2003 *Phys. Rev. Lett.* **91** 1076021
[12] Oka K, Unoki H, Yamaguchi H and Takahashi H 1996 *J. Cryst. Growth* **166** 380
[13] Axe J D, Harada J and Shirane G 1970 *Phys. Rev.* B **1** 1227
[14] Waghmare U V and Rabe K M 1997 *Phys. Rev.* B **55** 6161
[15] Zhong W and Vanderbilt D 1995 *Phys. Rev. Lett.* **74** 2587
[16] Shirane G and Yamada Y 1969 *Phys. Rev.* **177** 858
[17] Tomeno I, Tsunoda Y, Oka K, Matsuura M and Nishi M 2009 *Phys. Rev.* B **80** 104101

Journal of Physics: Conference Series **340** (2012) 012055 doi:10.1088/1742-6596/340/1/012055

Structural modulation in $Bi_2Sr_2Ca_{0.4}Y_{0.6}Cu_2O_{8+\delta}$

A I Beskrovnyy[1,3] and Z Jirák[2]

[1]Joint Institute for Nuclear Research, 141980 Dubna, Moscow Region, Russia
[2]Institute of Physics of ASCR, Cukrovarnická 10, 16200 Prague 6, Czech Republic
[3]National Research Centre "Kurchatov Institute", 123182 Moscow, Russia

E-mail: jirak@fzu.cz

Abstract. The neutron diffraction study has been performed on a single crystal specimen of yttrium doped compound $Bi_2Sr_2Ca_{0.4}Y_{0.6}Cu_2O_{8+\delta}$, which is a structural analog of the Bi-2212 superconducting cuprates. The average structure is orthorhombic of *Fmmm* symmetry (a=5.448 Å, b=5.439 Å and c=30.416 Å). Two kinds of distortion are observed. The first one, manifested by satellite reflections of wavevector [τ,0,1] with τ=0.220, is associated with an incorporation of extra oxygen (δ~0.44) into BiO bilayer. The respective modulation consists of a flexing of the bilayer and corrugation of other structural layers with period 4.5a. The second distortion is associated with displacements in the BiO planes due to Bi^{3+} bonding and lone pair ordering, which results in common cases in a formation of infinite ribbons of Bi-O-Bi- chains and decrease of average symmetry to *Amaa*. In the present system, the reflections characteristic for *Amaa* are absent and, instead, a diffuse scattering along the [001] direction is observed, pointing to a presence of 2-dimensional superstructure with in-plane period 9.0a – twice the one of the regular modulation. Based on these data, we propose a structural model in which bonding chains in the BiO planes of $Bi_2Sr_2Ca_{0.4}Y_{0.6}Cu_2O_{8.44}$ are segmented and rearranged in a tweed pattern.

1. Introduction

The transition metal oxides with layered structures offer many interesting properties with prospects for applications. Among them, the systems based on BiO bilayers are particularly attractive. Two structural classes can be distinguished. The first one is represented by superconducting cuprates of general formula $Bi_2(Sr,Ca)_{n+1}Cu_nO_{2n+2+\delta}$ (n=1,2,3). These phases consist of $[Bi_2Sr_2O_{4+\delta}]$ rock-salt blocks stacked with oxygen deficient perovskite-like layers $[Ca_{n-1}Cu_nO_{2n}]$. Conducting parts are the single or multiple CuO_2 planar sheets, in which copper ions are arranged in a 2D square lattice. Analogous structures are reported also for other transition metal systems, in particular the n=1 phase $Bi_2Sr_2CoO_{6+\delta}$ or $Bi_2Sr_2MnO_{6+\delta}$ [1-3]. The second structural class is represented by so-called misfit cobaltites, where analogous rock-salt blocks alternate with CoO_2 layers of hexagonal CdI_2 type. In that case the cobalt ions form 2D trigonal lattice sheet and are octahedrally coordinated with oxygen above and below the sheet. Because of lack of an ideal stacking with the rock-salt blocks, the resulting phase is a kind of composite structure. An example is the important thermoelectric system $[Bi_{1.72}Sr_2O_4]$ $[CoO_2]_{1.82}$ [4, 5].

Structural complexity is encountered also in the systems of the first structural class, namely in Bi-2212 cuprates (the n=2 phase) schematically depicted in figure 1. The lattice mismatch between the perovskite and rock-salt blocks and character of the Bi-O bond result in a structural modulation – the corrugation of all atomic planes (see figure 2) and large in-plane displacements within the BiO and adjacent SrO planes. The modulation depends on the actual composition of the compound and its period varies between commensurate values 4*a* and 5*a* (*a*~5.40 Å). Important point is the insertion of extra oxygen into the BiO plane after each modulation period, which makes preferable to describe the structure in a multiple cell, sometimes very large, rather than to use the four-dimensional formalism. In such pseudo-commensurate approach we have solved previously the structures of the single crystals of supercoducting $Bi_2Sr_{3-x}Ca_xCu_2O_{8.2}$ and non-superconducting $Bi_2Sr_{3-x}Y_xCu_2O_{8.5}$ with modulation periods of 4.75*a* and 4.25*a*, respectively [6, 7]. Both systems are characterized with a formation of ribbons – the Bi-O double chains in the bilayers (see figure 3b, figure 3c). The present study is done on crystals of composition $Bi_2Sr_2Ca_{0.4}Y_{0.6}Cu_2O_{8+\delta}$ where ribbons are apparently absent and a more complex Bi-O bonding is formed.

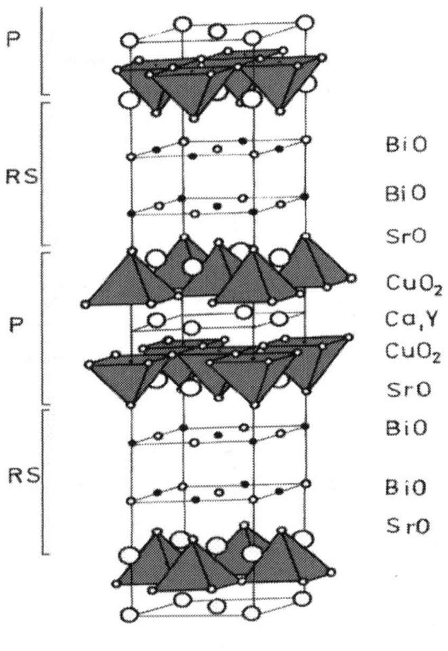

Figure 1. A simplified model of the $Bi_2Sr_2Ca_{1-x}Y_xCu_2O_{8+\delta}$ structure (space group *F4/mmm*; typical cell dimensions *a*=*b*~5.40 Å, *c*~30.5Å).

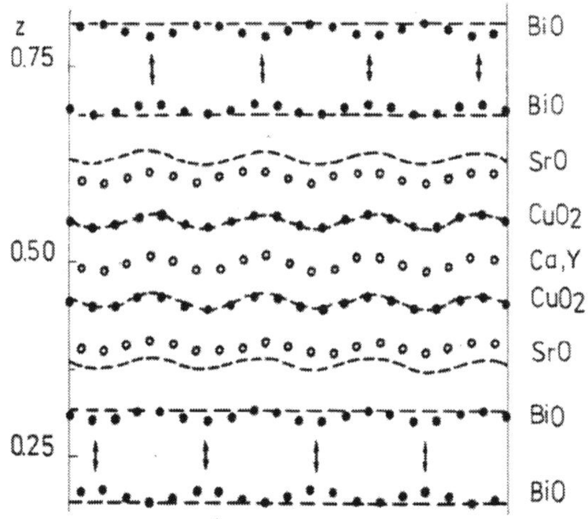

Figure 2. Schematic picture of the (x, z) plane of $Bi_2Sr_2Ca_{1-x}Y_xCu_2O_{8+\delta}$ (cations shown by dots and circles, oxygen atoms by the dashed lines). The modulation of atomic planes along x is due to insertion of extra oxygen atoms into the BiO bilayers with period ~4-5*a* (locations shown by the arrows).

2. Experimental

The $Bi_2Sr_2Ca_{0.4}Y_{0.6}Cu_2O_{8+\delta}$ crystals were grown by the travelling solvent floating zone technique as described elsewhere [8]. For determination of the structural modulation, a small specimen with dimensions 2 x 2 x 0.2 mm^3 consisting of a single domain was selected. The neutron diffraction investigation was performed on the DN-2 time-of-flight neutron diffractometer at the Joint Institute of Nuclear Research in Dubna (Russia) [9]. The use of a two-dimensional position sensitive detector allowed us to record the fundamental and satellite reflections practically in the whole octant of the reciprocal space.

3. Results and discussion

The fundamental reflections (*hkl*) measured by single crystal neutron diffractometry show systematic extinctions characteristic for the face centered orthorhombic cell of average *Fmmm* symmetry (a=5.448 Å, b=5.439 Å and c=30.416 Å). In addition, there are sharp satellites ($h\pm\tau$ k $l\pm1$) with τ=0.220(1) that evidence a pseudo-commensurate modulation of 4.5a period. Some satellites of the second order ($h\pm2\tau$ k $l\pm2$) are also observable. In spite of closeness of a and b parameters, no twinning is detected. This may suggest that the structural modulation, in particular the incorporation of extra oxygen, is inherent to the phase formation (see also the high-temperature study of the Bi-2212 cuprate in [10]).

In contrast to previously studied systems of the average *Amaa* symmetry, the diffraction peaks (*h*0*l*) with h,l =2n+1, which occur due to Bi-O ribbons, are completely missing in the present sample $Bi_2Sr_2Ca_{0.4}Y_{0.6}Cu_2O_{8+\delta}$. Instead, new diffuse satellites corresponding to the doubled modulation period 9.0a are detected in other parts of reciprocal space. These satellites form streaks (diffraction ridges) along the [001] directions. Similar behavior was observed earlier on $Bi_2Sr_2Ca_{1-x}Y_xCu_2O_{8+\delta}$ by electron microscopy [11]. There are, however, some important differences. Both the 4.5a and 9.0a periodicities occur for the present sample within one phase and no reflections violating the *Fmmm* symmetry are observed.

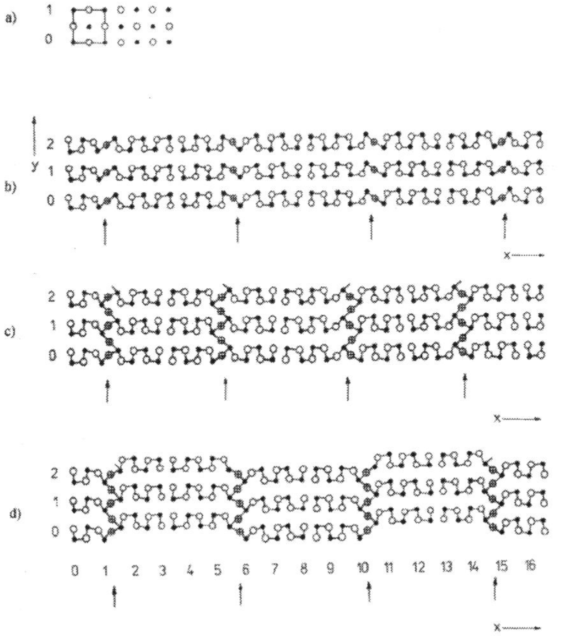

Figure 3. Scheme of the BiO plane in the simplified model (a), $Bi_2Sr_{3-x}Ca_xCu_2O_{8.2}$ (b), $Bi_2Sr_{3-x}Y_xCu_2O_{8.5}$ (c) and the arrangement found for the present compound $Bi_2Sr_2Ca_{0.4}Y_{0.6}Cu_2O_{8.44}$ (d). The Bi and O positions are shown are by the dots and circles, respectively; extra oxygen atoms are marked by crosses, the arrows show the modulation periods.

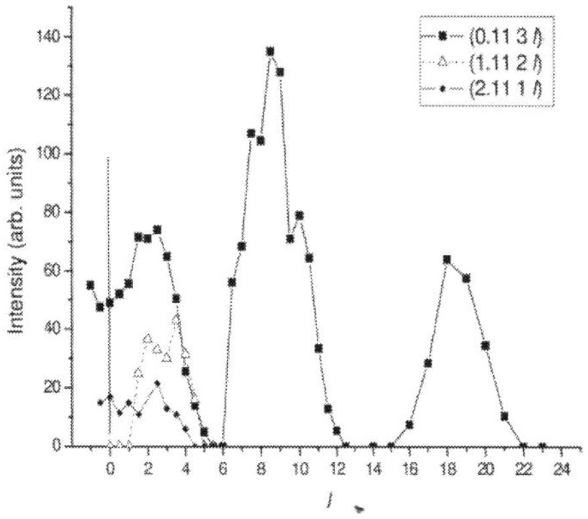

Figure 4. The intensity variation along the diffraction ridge (0.11 3 *l*). Data for (1.11 2 *l*) and (2.11 1 *l*), obtained in a limited *l* range, are also included.

The intensities of sharp reflections (*hkl*), ($h\pm\tau$ k $l\pm1$) and ($h\pm2\tau$ k $l\pm2$) observed on $Bi_2Sr_2Ca_{0.4}Y_{0.6}Cu_2O_{8+\delta}$ were analysed within a structural model with supercell nine-times enlarged in the *a*-direction. The refinement based on 105 fundamental reflections and 135 satellites showed that actual oxygen amount in $Bi_2Sr_2Ca_{1-x}Y_xCu_2O_{8+\delta}$ is close or equal to the ideal value δ~0.44 (experimental uncertainty ±0.10), which corresponds to two extra oxygen atoms incorporated into BiO layers after each 4.5*a* period. The insertion of two oxygen atoms instead of one is a situation similar to what we observed earlier for the related compound $Bi_2Sr_{3-x}Y_xCu_2O_{8.5}$ [7]. Also the extent of the corrugation of atomic planes (the *c*-axis modulation sketched in figure 2) is comparable.

The main issue is the absence of ribbons and origin of 9.0*a* modulation in the present sample. Relevant information can be deduced from diffuse satellites forming streaks of slowly variable intensity along the [001] direction (see figure 4). This is indicative for a 2-dimensional superstructure, most likely associated with ordering in the BiO bilayer and adjacent SrO planes. The streaks are narrow (resolution limited) in the [010] direction but broadened in the [100] direction, which means that the planar ordering is of a short range in the *a*-direction (the estimated coherence length is below 1000 Å). In order to substantiate the 2-dimensional superstructure, let us inspect the diffraction intensities in figure 4 in more detail. The most significant data are obtained along the ($\tau/2$ 3 l) diffraction ridge. The scan shown in figure 4 exhibits a very broad maximum around l=0 and other two maxima spreading over l= 7 – 11 and l= 16 – 21. As the first maximum is concerned, the diffraction intensity at the middle is rather low while it is stronger on shoulders at l=3 (and –3). Similar strong diffraction at l=3 is evidenced also in the (1+$\tau/2$ 2 l) and (2+$\tau/2$ 1 l) scans (their l range was unfortunately limited due to experimental reasons – see figure 4). We presume that the shoulders might occur due to *a*-axis shifts in the SrO planes in which the strontium and oxygen atoms adjust to the inserted extra oxygen in adjacent BiO plane. On the other hand, the ($\tau/2$ 3 l) diffraction centred at l=0, 9 and 18 is associated most likely with the *b*-axis displacements of bismuth and oxygen atoms in the bilayer and the maxima can be understood as an interference effect of BiO planes, located at z~0.25±0.06 levels (see figure 2). Based on an analogy with displacements that form ribbons in the *Amaa* systems, we propose for $Bi_2Sr_2Ca_{0.4}Y_{0.6}Cu_2O_{8+\delta}$ a model with segmented Bi-O double chains, schematically depicted in figure 3d.

The complete structure determination is out of the scope of present study. It might be noted that the actual system could be described as a composite structure (see e. g. [12, 13]).

4. Conclusions

Structural modulation is a general property of Bi-2212 cuprates. The reason is in incompatibility of the ideal tetragonal structure of the *F4/mmm* symmetry shown in figure 1 with the bonding character of the Bi^{3+} cation and its lone electron pair. In particular, the atoms in the ideal BiO plane are regularly spaced with a distance *a*/2~2.7 Å (see figure 3a), which is outside the range of possible Bi^{3+}-O bonds (2.0-2.5 Å). It appears that an appropriate coordination is achieved by two mechanisms - by the ordering of short (bonding) and long (non-bonding) Bi-O distances along the *b*-axis and by a periodic insertion of extra oxygen (typically with 4-5*a* spacings) and significant rearrangement of atoms in the BiO and adjacent SrO planes along the *a*-axis. The former mechanism results in an orthorhombic distortion, most commonly of the *Amaa* symmetry. In these cases, characteristic ribbons in the BiO plane as seen in figure 3b or figure 3c are formed. The insertion of extra oxygen defines the superstructure and is responsible for flexing of BiO bilayers and subsequent corrugation of other atomic planes.

The present $Bi_2Sr_2Ca_{0.4}Y_{0.6}Cu_2O_{8+\delta}$ system is characterized by the *Fmmm* symmetry. Extra oxygen atoms (δ~0.44) are inserted into BiO bilayers with period 4.5*a*. Though the flexing of BiO bilayers and corrugation of other atomic planes is similar to the *Amaa* systems, the planar pattern of bismuth and oxygen atoms in bilayers is significantly modified. Instead of the infinite strips of the Bi-O-Bi-bonds running along the *a*-directions, a tweed arrangement of Bi_9O_{11} segments is realized (figure 3d). This gives rise to a novel kind of 2-dimensional superstructure with in-plane period 9*a*.

In the conclusion, it is worth mentioning that the present sample is not superconducting. This is most likely related to an underdoping of conducting CuO_2 layers since the formal valence calculated for composition $Bi_2Sr_2Ca_{0.4}Y_{0.6}Cu_2O_{8.44}$ is $Cu^{2.14+}$, inferior to $Cu^{2.20+}$ in the $Bi_2Sr_{3-x}Ca_xCu_2O_{8.2}$ superconductors.

References

[1] Tarascon J. M. et al. 1989 *Phys. Rev.* B **39** 11587
[2] Jakubowicz N, Grebille D, Leligny H and Evain M 1999 *J. Phys.: Condens. Matter* **11** 3997
[3] Jirák Z, Pollert E, Sedmidubský D, Dlouhá M and Vratislav S 1992 *Physica* C **196** 68
[4] Hervieu M, Maignan A, Michel C, Hardy V, Creon N and Raveau B 2003 *Phys. Rev.* B **67** 045112
[5] Leligny H, Grebille D, Perez O, Masset A-Ch, Hervieu M, Michel C and Raveau B 1999 *C. R. Acad. Sci. Paris IIc* **2** 409
[6] Beskrovnyi A I, Dlouhá M, Jirák Z, Vratislav S and Pollert E 1990 *Physica* C **166** 79
[7] Beskrovnyi A I, Jirák Z, Nevřiva M and Shelkova I G 1993 *Physica* C **206** 27
[8] Nevřiva M, Jirák Z, Šíma V¸ Beskrovnyi A I, Martin C and Raveau B 1997 Book Series: *Electrochemical Society Series* **97-39** pp 480-486
[9] http://flnp.jinr.ru/137/
[10] Beskrovnyi A I, Dlouhá M, Jirák Z and Vratislav S 1990 *Physica* C **171** 19
[11] Zhang X F, Van Tandeloo G, Emmen J H P M and Brabers V A M 1993 *Physica* C **215** 39
[12] Perez-Mato J M, Etrillard J, Kiat J M, Liang B and Lin C T 2003 *Phys. Rev.* B **67** 024504
[13] Etrillard J, Bourges P and Lin C T 2000 *Phys. Rev.* B **62** 150

Journal of Physics: Conference Series **340** (2012) 012056 doi:10.1088/1742-6596/340/1/012056

Local disorder in yttrium doped ceria ($Ce_{1-x}Y_xO_{2-x/2}$) probed by joint X-ray and Neutron Powder Diffraction

M Coduri[1], M Scavini[1,2,3], M Allieta[1], M Brunelli[4], C Ferrero[5]

[1)] Dipartimento di Chimica Fisica ed Elettrochimica, Università di Milano, Via C. Golgi 19, I-20133 Milano, Italy

[2)] CNR, Ist Sci & Tecnol Mol, I-20133 Milano, Italy

[3)] Consorzio Interuniv Nazl Sci & Tecnol Mat INSTM, Unita Ric Milano, I-20133 Milano, Italy

[4)] Institut Laue-Langevin, BP 156, 38042 Grenoble Cedex 9, France

[5)] European Synchrotron Radiation Facility, 6 av. J. Horowitz, BP 220, 38043 Grenoble Cedex, France

E-mail: mauro.coduri@unimi.it

Abstract. Yttrium doped ceria materials ($Ce_{1-x}Y_xO_{2-x/2}$) are widely studied for their application in Solid Oxide Fuel Cells devices. An anomalous decrease in the isothermal ionic conductivity at increasing Y^{3+} concentration above a critical value has been observed and attributed to the formation of defect clusters / domains at the nanometric scale, the crystallographic structure of which is still under debate. In this context we present a combined Synchrotron Radiation and Neutron Powder Diffraction study. In particular, neutrons allow to determine accurately oxygen related parameters, the contribution of which in terms of X-ray scattering power is almost negligible when compared to that of cations. The effect of doping on the average structure is investigated using conventional Rietveld analysis, while the Pair Distribution Function (PDF) technique is used to explore structural distortions and the spatial extent of disorder as well. The local structure observed in the real space is not consistent with the mean crystallographic one and is better modeled considering a biphasic model.

1. Introduction

In recent years, doped ceria materials ($Ce_{1-x}Y_xO_{2-x/2}$, hereafter CYO) have attracted increasing attention as electrolytes for Solid Oxide Fuel Cells (SOFCs) because of their high ionic conductivity [1]. In a SOFC, oxygen gas on the cathode side of the cell reduces to O^{2-}, diffuses through the electrolyte layer and combines with hydrogen (or other fuels) producing water and electrons. The most common electrolyte used in SOFC is Yttria-Stabilised-Zirconia (YSZ), which is very stable in both reducing and oxidizing atmosphere, but requires high operating temperature (~1000°C) in order to exhibit high ionic conductivity [2]. Such high temperature requires an all-ceramic construction which raises the cost of the fuel cell. Overall manufacturing costs can be considerably reduced by operating at lower temperature. This is the case of doped ceria electrolytes, which maintain a high ionic conductivity even at intermediate temperatures (e.g. 500-700°C).

Although pure ceria (CeO_2) does not show a significant ionic conductivity, this can be raised by doping its fluorite structure with a lower valent oxide, such as Y_2O_3. For each trivalent ion substituting a Ce(IV) ion 0.5 oxygen vacancies appear to maintain electroneutrality. Ionic conductivity occurs by oxygen diffusion via the vacancy mechanism:

$$Y_2O_3 \xrightarrow{CeO_2} 2Y'_{Ce} + 3O_O^x + V_O^{\cdot\cdot} \qquad (1)$$

As a consequence the isothermal ionic conductivity σ should increase monotonically at increasing doping concentration. Conversely a maximum of σ is generally observed at a critical Y^{3+} concentration x, above which σ decreases (see e.g. [3,4]).

Local probes like EXAFS and NMR have been adopted in order to investigate the structural origin of the anomalous behavior of the conductivity. In particular EXAFS measurements [5,6] suggested the appearance of defect clusters formed by substitutional cations and oxygen vacancies. Kim *et Al.* [7] drew the same conclusion using solid state NMR experiments. These results are also supported by first principle atomistic calculations which suggest that: (*i*) it is energetically favorable for a vacancy to be in a first neighbor position with respect to the Y^{3+} substitutional ion [8]; *ii*) the reduced concentration of effective free mobile vacancies (i.e. the conductivity drops) is due to the formation of defect associates of the oxygen vacancy with the solute cation [9].

It should be noted that the EXAFS technique can be successfully employed to explore only the local structure of Ce^{4+} and Y^{3+} ions and cannot provide further information in case of more extended defects (e.g. on the nanometric scale). The same applies for the solid state NMR measurements. Anyway, HRTEM [10], SAED and EELS [11] measurements suggested that at higher doping concentrations complex defect clusters (a few nanometers large) form and act as deeper traps for oxygen vacancies. However, it is well known that the main limitation of electron beam-based techniques is that the investigated region could be not representative of the whole sample, particularly in highly defective oxides.

For this reason a bulk technique, able to investigate both the very local structure (first coordination shells) and the mesoscopic scale (i.e. a few nanometers) would be required to shed light on such a complex system. The Pair Distribution Function (PDF, i.e. the real space analysis of diffraction data) can be considered to be the right technique to solve this problem: Since both Bragg and diffuse scattering contributions are taken into account, it is a unique tool to determine the local and medium range deviations with respect to an ideally periodic structure [12, 13].

The Pair Distribution Function $G(r)$ is experimentally determined via sine Fourier transform of the corrected total scattering function $S(Q)$:

$$G(r) = \frac{2}{\pi}\int_0^\infty Q[S(Q)-1]\sin(Qr)\,dQ \qquad (2)$$

Where r is the radial distance. In order to compare or to fit the experimental $G(r)$ against a structural model, the $G(r)_{calc}$ can be calculated from a set of structural parameters according to:

$$G(r)_{calc} = \frac{1}{r}\sum_{\upsilon}\sum_{\mu}\left[\frac{b_\nu b_\mu}{\bar{b}^2}\delta(r - r_{\upsilon\mu})\right] - 4\pi r\rho_0 \qquad (3)$$

where ν and μ are the ions involved, b is the coherence scattering length and ρ_0 is the number density. The function G(r) indicates the probability of finding a couple of atoms at a certain r-distance with an integrated intensity dependent on the coherence scattering length b of the elements involved (see equation 3). The effect of the reciprocal space resolution on real space data is reported elsewhere [14].

Aim of this work is to provide a crystallographic description of the local disorder induced by yttrium doping in CYO materials. For that reason we performed X-Ray (XRPD) and Neutron Powder Diffraction (NPD) experiments since both techniques can be seen as complementary probes for the CYO system: X-ray scattering lengths b are proportional to the atomic number, thus the main contributions are due to the cations, especially cerium. On the other hand, neutron scattering lengths depend on the single isotopes. In particular, considering a natural isotopic compositions, the neutron b-values for Ce, Y and O are 4.84, 7.75 and 5.803 Å, respectively [15]. For that reason, neutrons are far more sensitive to oxygen related parameters than X-rays. Moreover, Ce, Y and O have an almost negligible neutron absorption cross section, which makes them very suitable for a NPD investigation.

Experimental data have been analyzed using both reciprocal space Rietveld analysis and atomic Pair Distribution Function.

2. Experimental

Micro-crystalline CYO samples with Y concentrations x = 0.125, 0.250 were prepared with the *Pechini* sol-gel method and fired at 900°C for 72 hours. XRPD patterns were collected at the ID31 beamline of the ESRF, Grenoble [16] in the 2θ range $0<2\theta<120°$ up to Q_{max}=29.4 $Å^{-1}$, where $Q=4\pi \sin\theta/\lambda$ is the transferred momentum. We collected experimental data (plus empty capillary and air background) at incident X-ray wavelength λ=0.35413(3) Å. Additional data were collected also on CeO_2 (Aldrich ≥99.0%) and Y_2O_3 (Aldrich 99.9%). In order to attain XRPD data quality for experimental PDF all patterns were obtained summing intensity counts over several scans performed at fixed temperature (T = 90 K).

NPD patterns for reciprocal space analysis were collected at the D20 instrument at ILL, Grenoble [17], warming the samples at the rate of 0.75 K/min from 4 K to room temperature at incident neutron wavelength λ=1.3595(3) Å in the 2θ range $13<2\theta<144°$. About 10 minutes acquisition time diffraction patterns were collected at 4 K and room temperature, whilst shorter acquisitions (2 minutes) were collected upon heating.

NPD patterns for real space analysis were collected at the D4c instrument at ILL [18] at incident neutron wavelength λ=0.4975(3) Å, covering a Q-range between 0.35 and 23.6 $Å^{-1}$ at fixed temperature (T = 90 K) for all the samples. Empty cryostat, vanadium rod, empty cans and boron powder patterns were also measured in order to properly subtract the background and for normalization. The backgrounds were measured periodically during the experiment in order to check their stability. PDF quality patterns were collected for each sample within 3-4 hours.

3. Results and discussion

We performed the data analysis using a multi-step approach. *First* the average crystallographic structure (i.e. cell parameter, atom positions and thermal parameters) was obtained for each sample by means of the reciprocal space Rietveld method. *Then* the *average* structural models were applied to the same data in real space, using the PDF approach to detect local deviations from the mean structure. *Finally* a structural model is proposed to interpret the local structure of CYO samples.

3.1. Reciprocal space analysis

The XRPD and NPD patterns were analyzed via the Rietveld method [20] as implemented in the GSAS software suite of programs [21] which features the graphical interface EXPGUI [22]. The background was fitted by Chebyshev polynomials. For XRPD data, absorption correction was performed through the Lobanov empirical formula [23] and line profiles were fitted using a modified pseudo-Voigt function [24] accounting for asymmetry correction [25]. In the final refinement cycles, cell parameter, positional coordinates (in Y_2O_3) and isotropic thermal parameters were allowed to vary as well as background and line profile parameters. For NPD data, absorption correction was considered as negligible. Line profiles were fitted only in the longer acquisition patterns. Since line profile parameters do not vary passing from low to high temperature, they were kept fixed in the refinements upon warming, where only the cell parameter and the isotropic thermal parameters were allowed to vary.

Cerium oxide exhibits fluoritic structure (space group $Fm\bar{3}m$). Cerium lies in the (0, 0, 0) site and O in (¼, ¼, ¼). The stable form of the Y_2O_3 sesquioxyde in the investigated temperature range is the so called C-type structure ($Ia\bar{3}$) [19]. The cations lie in two non equivalent positions: M1 (¼, ¼, ¼) and M2 (x, 0, ¼), while oxygen occupies a (x, y, z) site.

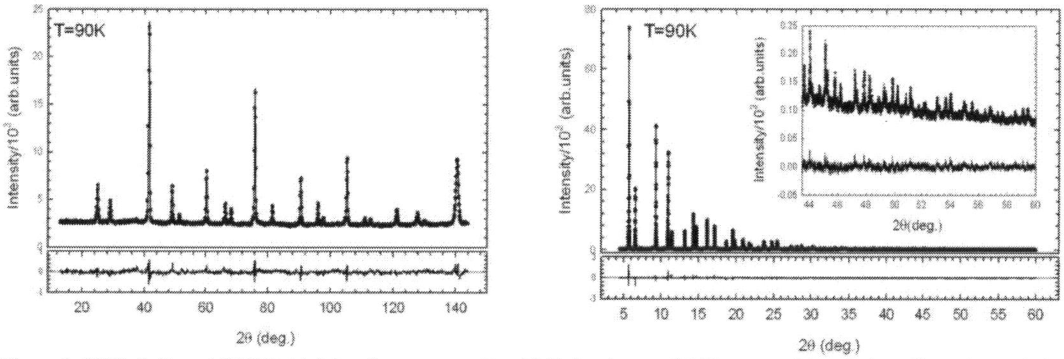

Figure1. NPD (left) and XRPD (right) refinements at T = 90 K for the x = 0.250 composition. Points: Experimental data, solid lines: Calculated. Residuals are shown below. For XRPD data the inset highlights the high angle region.

The C-type structure is very close to the CeO_2-fluorite: It can be obtained from a 2x2x2 fluorite supercell by removing two oxygen ions in each of the single fluorite cells. In this way the space group symmetry is lowered to $Ia\overline{3}$ and the cations gain a 6-fold coordination.

Both x=0.125 and x=0.250 samples both exhibit a CeO_2-like fluoritic structure. In figure 1 the NPD and XRPD patterns collected at T = 90 K for the x = 0.250 composition are shown as an example. All the peaks are well fitted by the fluoritic model and no noteworthy residual is observed. The same applies for the whole temperature range investigated. Since no positional degree of freedom is allowed in such a structure owing to symmetry constraints, the only structural parameters varied in the Rietveld refinements were the lattice parameter and the isotropic mean square displacements (*msd*) parameters. The Ce/Y site occupations were fixed based on the nominal compositions.

The cell parameter is found to decrease upon Y doping. In particular, from the XRPD analysis at 90 K, a=5.407228(3), 5.405631(5) and 5.403207(9) Å for CeO_2, x=0.125 and x=0.250, respectively. In figure 2 the isotropic mean square displacement parameters of metal (left) and oxygen (right) ions are shown.

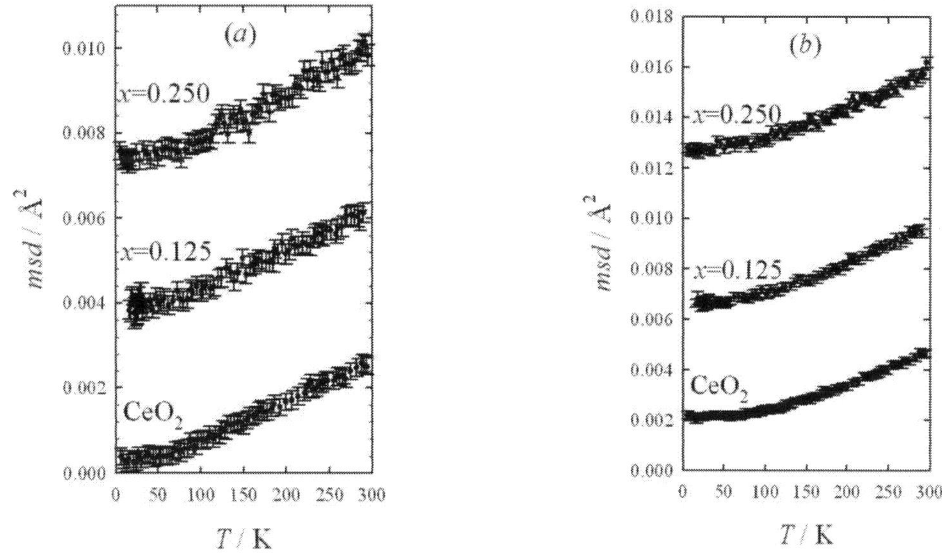

Figure2. Mean square displacements for metal (a) and oxygen (b) ions obtained by NPD patterns.

A rigid shift of the *msds* of doped samples towards higher values can be clearly observed. In absence of disorder, *msds* are related only to the square of the amplitude of atomic vibration around their equilibrium position. When disorder is present, ions occupying the same site in different cells can occupy equilibrium positions which differ from the average one, adding a contribution to the thermal parameter often defined as "static contribution". The observed shift can be then regarded as a clear fingerprint of doping induced disorder [26]. The *msd* parameters obtained at $T = 90$ K through XRPD and NPD refinements are in fairly good agreement and are shown in table 1 for a sake of comparison.

Table 1. Mean square displacements *msd* (in Å^2) of cations (M) and anions (O) obtained by XRPD and NPD refinements at 90 K. The samples exhibit fluoritic structure (space group $Fm\bar{3}m$): Cations lie in (0, 0, 0) and anions in (¼, ¼, ¼).

	X-Rays		Neutrons	
	msd (M)	***msd*** (O)	***msd*** (M)	***msd*** (O)
CeO_2	0.00091(1)	0.00302(6)	0.0007(2)	0.0023(1)
***x*=0.125**	0.00338(1)	0.00633(7)	0.0043(2)	0.0069(2)
***x*=0.250**	0.00813(2)	0.0127(2)	0.0077(2)	0.0132(2)

The analysis of *msds* clearly indicates the increasing of structural disorder upon Y doping.

3.2. Real space analysis

In order to gain insight into the local deviations that occur in such a disordered material, a PDF study was performed. As stated above, data were collected for both X-ray and neutron at $T = 90$ K in order to minimize thermal motion. X-ray real space data were processed using the PDFGetX2 software [27]: Data were corrected for background, sample self-absorption, multiple scattering and Compton scattering. Neutron real space data were processed by regrouping repeated scans of the D4c detectors over the available range of scattering angles in the reciprocal space. For each sample, empty container and sample environment contributions were subtracted from the raw data, taking into account absorption effects [28] as well as the incoherent-scattering contribution. A standard multiple-scattering [29] and Placzeck [30] corrections were applied. The PDF analysis was performed by fitting the experimental G(*r*) using the software PDFGui [31]. The agreement between model and experimental data is given by the R_w factor:

$$R_W = \left(\frac{\sum_{i=1}^{n} \omega(r_i)[G_{obs}(r_i) - G_{calc}(r_i)]^2}{\sum_{i=1}^{n} \omega(r_i) G_{obs}^2(r_i)}\right)^{\frac{1}{2}} \tag{4}$$

where $\omega(r_i) = 1/\sigma^2(r_i)$ and $\sigma(r_i)$ is the standard deviation at a distance r_i.

Due to its well ordered structure, CeO_2 can be considered a reference material for a PDF study. Moreover, since only special positions are occupied, it is possible to know a priori the positions of the peaks and which couples of atoms are involved, by simple geometric considerations. In table 2 the list of the first distances is reported considering the cell parameter a=5.407Å, as measured at 90 K.

Table 2. List of distances of the first coordination shells in a CeO_2-like structure. Values are referred to the case of CeO_2 measured at 90 K.

$d(Ce-O)NN = a\frac{\sqrt{3}}{4} \sim 2.34$ Å	$d(Ce-O)NNN = a\sqrt{\frac{11}{16}} \sim 4.48$ Å
$d(O-O)NN = \frac{a}{2} \sim 2.70$ Å	$d(O-O)NNN = a\frac{\sqrt{3}}{2} \sim 4.68$ Å
$d(Ce-Ce)NN \equiv d(O-O)NN = a\frac{\sqrt{2}}{2} \sim 3.82$ Å	$d(Ce-Ce)NNN \equiv d(O-O)NNN = a$

In figure 3*a* the G(*r*) curves obtained for CeO_2 using NPD (red) and XRPD (blue) are shown. As stated above, one advantage of neutrons is that they allow to study O-O distances which are undetectable through X-rays. On the other hand, high energy and high reciprocal-space resolution X-ray data from ID31 allow to reach higher Q_{max} values (i.e. to increase the real space resolution) and to explore a wider *r* range, due to the very high angular resolution and very narrow instrumental resolution function [16]. In figure 3*b* the experimental G(*r*) curves for the Y_2O_3 sample are also reported.

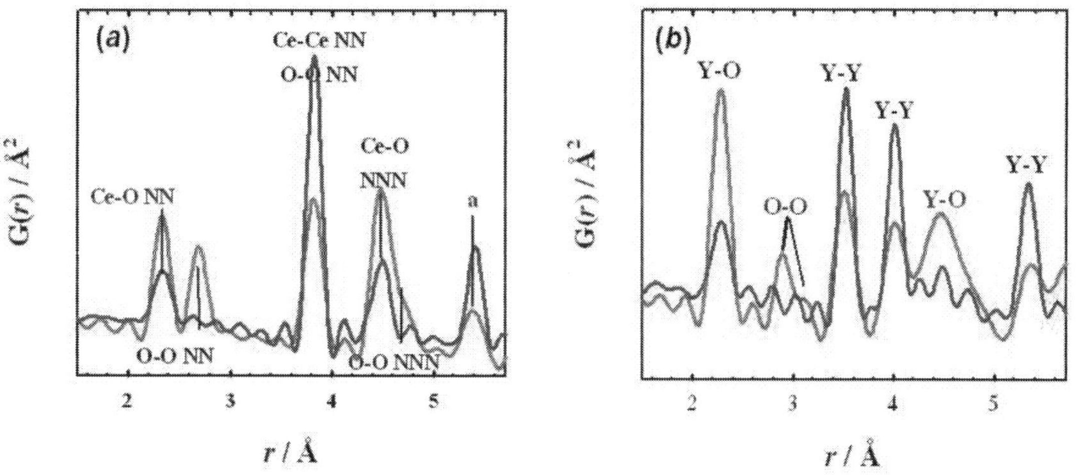

Figure3. Experimental G(*r*) curves for CeO_2 (*a*) and Y_2O_3 (*b*) obtained by NPD (red) and XRPD (blue). Labels indicate to which couple of atoms every peak is referred.

The NPD G(*r*) curves are shown in figure 4. The arrows are guides to the eye to indicate the amplitude and position shifts of the different peaks at increasing Y concentration. According to a considerable local structural disorder, the G(*r*) peaks related to doped sample look broader than in pure ceria and in most cases Y doping causes the peak features to resemble those of Y_2O_3.

Moreover, since the cell parameter decreases by enhancing the doping concentration, all peaks are expected to shift toward lower distances. However, this is true only for the NN (Nearest Neighbors) M-O and the NN M-M distances, while the NN O-O distances increase as well as the third M-M and O-O neighbors (which correspond to the cell parameter). All these findings indicate that disorder affects not only the first coordination shells: Shifts of peak positions are observed even at higher distances (see figure 4). It's worthwhile noticing that such a widespread structural disorder cannot be detected with other techniques such as EXAFS, which is sensitive only to the first coordination shells. Besides, interatomic distances involving only light atoms, such as O-O, are precluded.

The extent of the disorder induced by doping was determined by fitting the XRPD G(*r*) curves monitoring the residual R_w as a function of *r*. A fluoritic model has been applied: The cell parameter is fixed to the average structure value and the presence of two metals ions is taken into account by randomly populating the cation positions with Ce^{4+} and Y^{3+} ions in due proportions. Thermal parameters and scale factor only have been refined.

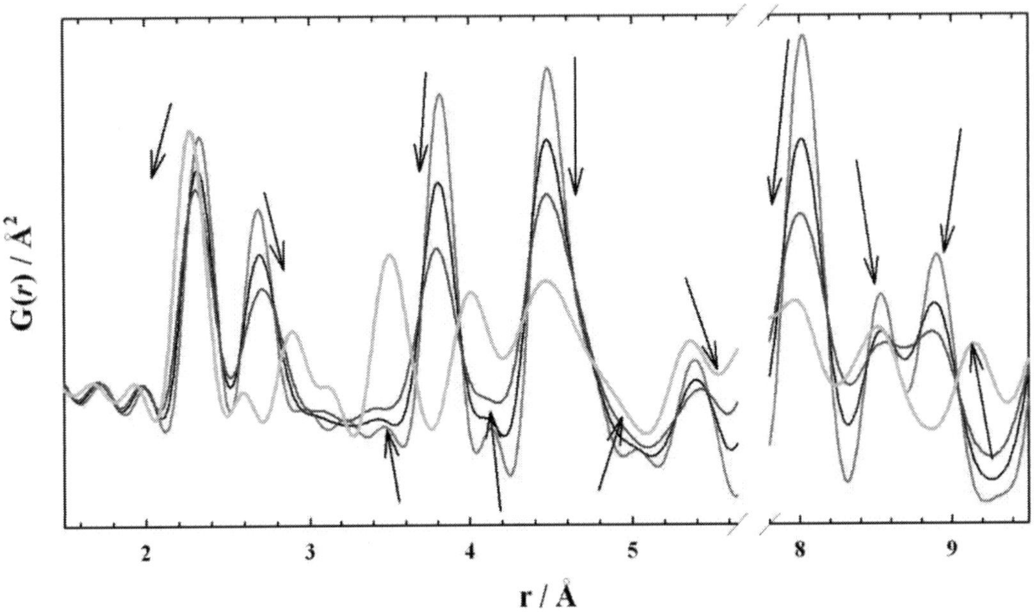

Figure4. NPD experimental $G(r)$ curves for CeO_2 (red), x=0.125 (black), x=0.250 (blue) and Y_2O_3 (grey). Arrows are guides for the eye to highlight the effect of doping on peak positions and width.

X-Ray diffraction patterns from ID31 were chosen for this purpose since they don't suffer from limited reciprocal space resolution effects. Conversely, oscillations in neutrons $G(r)$ curves are subjected to damping at increasing r, due to the lower resolution of the D4c instrument.

Whilst for CeO_2 the fluoritic model fits very well even the shortest distances, this is not true for the doped samples. At the subnanometric scale, (i.e. for $r < 10$ Å) R_w values increase dramatically by enhancing the doping rate and, most importantly, the peak positions of the calculated $G(r)$ curves are not consistent with the experimental ones (see figure 5*a* for x=0.250). In particular, the fluoritic model cannot account for the peak at ~4.1 Å which roughly appears at the same r-distance of the long M-M distance in the C-type structure (see figure 3*b*). The refinement against NPD data yielded similar outcomes (see figure 5*b*). In particular, in the $1 < r < 6$ Å range the following R_w values were obtained: 0.080 for CeO_2, 0.127 for x=0.125 and 0.197 for x=0.250. This indicates that the structure of CYO is distorted only locally to accommodate for the presence of two distinct metal ions, but on average it still exhibits a fluoritic symmetry.

A possible explanation could be the tendency of the two cations to retain their local environment within some coordination shells. This can be tentatively accounted for by a model composed of the two end-member phases, CeO_2 and Y_2O_3.

In figure 5*d* the refinements performed using the biphasic model are shown for the NPD case. Scale factors, atoms positions and two thermal parameters, one for cations and the other for oxygen ions, were refined. A similar model has been successfully employed to describe the local structure in the CeO_2-ZrO_2 solid solution [32]. The biphasic model (69.4% of CeO_2 and 30.6% of Y_2O_3) yields a reduction of the residual down to R_w=0.141, from 0.197 using the single fluorite phase.

Furthermore, when the results of the neutron PDF biphasic model are applied to the X-ray $G(r)$ (see figure 5*c*), the peak at ~4.1 Å is well fitted by refining only the overall scale factor and the thermal factors, whereas the single fluoritic model fails (see red circle in figure 5*a*). At the same time a noticeable decrease of R_W values is observed.

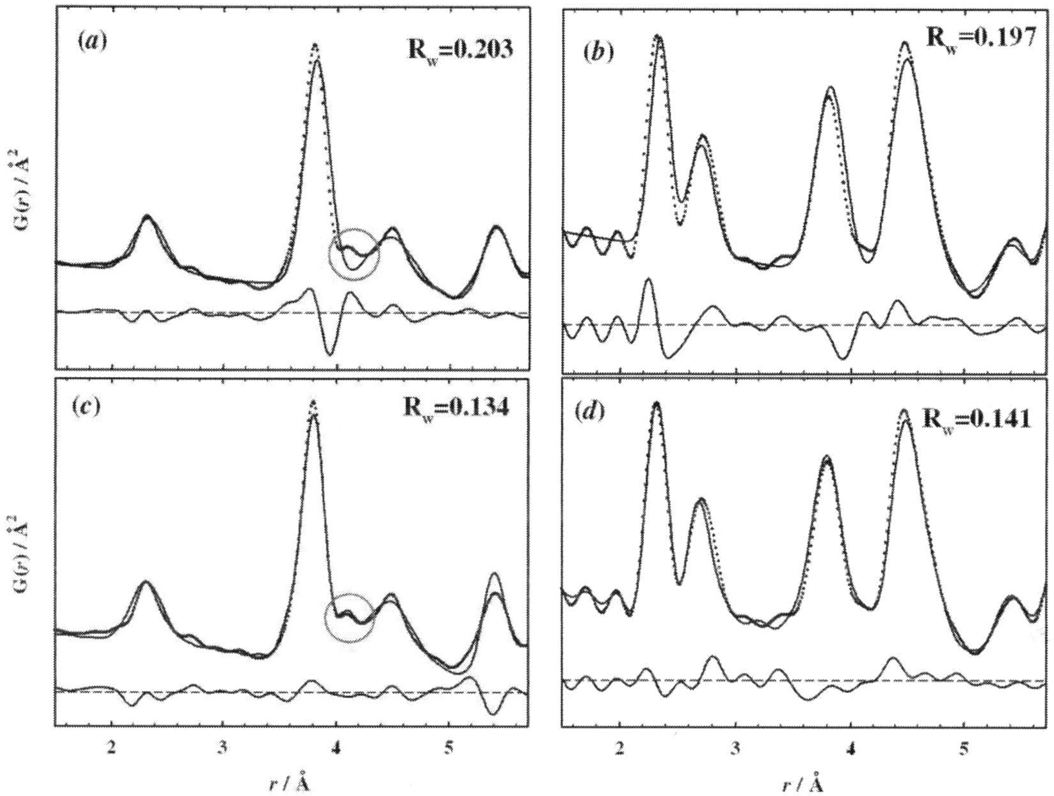

Figure5. XRPD and NPD G(r) curves for the x=0.250 composition. (a): fluoritic model against XRPD data, (b): fluoritic model against NPD data; (c): biphasic model against XRPD data; (d): biphasic model against NPD data. Points: experimental data; solid lines: calculated (up: fluoritic model; down: biphasic model). Residuals are also shown. The red circles in (a) and (c) highlight the improvement obtained with the biphasic model in fitting the 4.1 Å peak from the XRPD data.

The coordination of Ce^{4+} and Y^{3+} cations in fluorite and C-type structure, respectively, are shown in figure 6. The distance values are obtained by applying the biphasic model and refer to cation-cation pairs. The fluorite CeO_2 structure exhibits only one Ce-Ce distance (see figure 6a) and all ions are arranged in a perfect cubic environment. In the C-type structure (see figure 6b) two non-equivalent cation sites are present, labeled in black (Y1) and in blue (Y2), respectively: They differ in their oxygen vacancies distributions and lie in a slightly distorted cubic environment, thus gaining a 6-fold coordination. Two cation-cation distances are hence present.

It should be noted that the fit yields cell parameters which are quite different from the ones of the average structure: In the biphasic model the CeO_2 cell parameter changes from 5.4072 Å (pure ceria) to 5.3849 Å, while for the x=0.25 composition the measured cell parameter is 5.4032 Å; the biphasic model cell parameter for Y_2O_3 is 10.879 Å, which is larger than that of pure Y_2O_3, which is 10.602 Å.

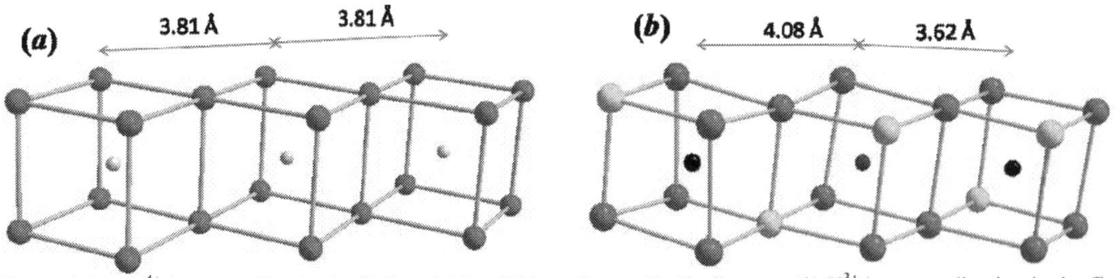

Figure6. (*a*) Ce^{4+} ions coordination in CeO_2, which exhibits only one Ce-Ce distance. (*b*) Y^{3+} ions coordination in the C-type Y_2O_3 structure. Two different Y-Y distances are present, depending on the oxygen vacancies distribution. White: Ce; blue and black: Y1 and Y2 sites, respectively; red: O; green: Oxygen vacancies. Arrows indicate cation-cation distances, the values of which are referred to the biphasic model.

However, rather than actual cell parameters of the CeO_2 and Y_2O_3 pure phases, the refined cell parameter values obtained through the biphasic model were considered here just variable parameters. Indeed, since disorder is localized on a sub-nanometric scale, M-M and M-O distances are not supposed to relax to the CeO_2 and Y_2O_3 values as expected for a true phase separation. However, the fit improvement after the application of this simple biphasic model suggests that complex defects as well as compositional fluctuations occur at the sub-nanometric scale.

4. Conclusion

Although there is growing evidence that short- and medium-range structural distortions deeply affect the physical properties of materials, a comprehensive description of the local structure in doped ceria materials is still lacking. To this purpose, an experimental approach involving XRPD, NPD and the related Pair Distribution Function analysis was performed in order to determine the average crystallographic structure and the local distortions arising from disorder. The study of the thermal parameters in the reciprocal space clearly indicates that the doping introduces disorder in the ceria structure. X-Ray real space refinements show that the coherence length of disorder is about 8-10 Å.

The first coordination shells are not consistent with the average fluoritic structure and are better described by a biphasic model, i.e. describing the chemical environment of the cations in their pure oxides. Although the model proposed in this work is very simple, it allows to improve considerably the fit of the first coordination shells. However, deviations from the mean structure are not limited to the first coordination shells and complex defect clusters may need to be considered to provide a comprehensive description of the local structure in the CYO solid solution.

Acknowledgments

The authors acknowledge the European Synchrotron Radiation Facility and the Institut Laue-Langevin for provision of beam time; they also wish to thank Dr. A. N. Fitch for assistance in using the ID31 beamline and Dr. H.E. Fischer for assistance in using the D4c instrument.

References

[1] Goodenough J B 2003 *Annu. Rev. Mater. Res.* 33 91–128
[2] Carrette L, Friedrich K A and Stimming U, *Fuel Cells* 1, (2001), 5
[3] Zhang T S, Ma J, Kong L B, Chan S H and Kilner J A 2004 *Solid State Ionics* 170 209–217
[4] Wang D Y, Park D S, Griffith J and Nowick A S 1981 *Solid State Ionics* **2** 95-105
[5] Deguchi H, Yoshida H, Inagaki T and Horiuchi M 2005 *Solid State Ionics* 176 1817 – 1825
[6] Wang Y, Kageyama H, Mori T, Yoshikawa H, Drennan J 2006 *Solid State Ionics* 177 1681–1685
[7] Kim N and Stebbins J F 2007 *Chem. Mater* 19 5742-5747
[8] Minervini L, Zacate M O and Grimes R W 1999 *Solid State Ionics* 116 339–349
[9] Ye F, Mori T, Ou D R and Cormack A N 2009 *Solid State Ionics* 180 1127–1132
[10] Ou D R, Mori T, Ye F, Takahashi M, Zou J and Drennan J 2006 *Acta Materialia* 54 3737–3746

[11] Ou D R, Mori T, Ye F and Takahashi M 2006 *Appl. Phys. Lett.* 89 171911.
[12] Billinge S J L and Kanatzidis M G 2004 *Chem. Commun.* 749–760.
[13] Egami T 1995 *J.Phys. Solids* **56** 1407–1413
[14] Qiu X, Bozin E S, Juhas P, Proffen T and Billinge S J L 2004 *J. Appl. Cryst.*. 37 110-116
[15] 1992 *Neutron News* 3, 29-37
[16] Fitch A N 2004 *J. Res. Natl. Inst. Stand. Technol.* 109 133–142.
[17] Hansen T C, Henry P F, Fischer H E, Torregrossa J and Convert P 2008 *Meas. Sci. Technol.* 19 034001
[18] Fischer H E, Cuello G J, Palleau P, Feltin D, Barnes A C, Badyal Y S and Simonson J M 2002 *Appl. Phys.* A74 S160-S162
[19] Adachi G and Imanaka N, 1998 *Chem. Rev.* 98 1479-1514
[20] Rietveld H M 1969 *J. Appl. Crystallogr.* 2 65-71
[21] Larson A C and Von Dreele R B 2004 *General Structural Analysis System (GSAS)*; Los Alamos National Laboratory Report LAUR 86-748.
[22] Toby B H 2001 *J. Appl. Crystallogr. 34* 210–214.
[23] Lobanov N N and Alte da Veiga, L. *6th European Powder Diffraction Conference*, Abstract P12-16, 1998.
[24] Thompson, P, Cox D E and Hastings J B 1987 20 *J. Appl. Cryst.* 79–83.
[25] Finger, L W, Cox D E and Jephcoat A P 1994 *J. Appl. Cryst.* 27 892–900.
[26] Argyriou D N 1994 *J. Appl. Cryst.* 27 155-158
[27] Qiu X, Thompson J W and Billinge S J L 2004 *J. Appl. Cryst.*, 37 678–678.
[28] Paalman H H and Pings C J 1962 *J. Appl. Phys.* 33 2635
[29] Blech I A and Averbach B L 1965 *Phys. Rev.* A137 1113.
[30] Yarnell J L, Katz M J, Wenzel R G and Koenig S H 1973 *Phys. Rev.* A7, 2130.
[31] Farrow C L, Juhàs P, Liu J W, Bryndin D, Bozin E S, Bloch J, Proffen T and Billinge S J L 2007 *J. Phys.: Condens. Matter* 19 335219.
[32] Gateshki M, Niederberger M , Deshpande A S, Ren Y and Petkov V 2007 *J. Phys.: Condens. Matter* 19 156205

Effect of high intensity ultrasound on the mesostructure of hydrated zirconia

G.P. Kopitsa[1*], A.E. Baranchikov[2], O.S. Ivanova[2], A.D. Yapryntsev[3], S.V. Grigoriev[1], P. Klaus Pranzas[4], V.K. Ivanov[2]

[1]Petersburg Nuclear Physics Institute, Gatchina 188300, Russia
[2]Institute of General and Inorganic Chemistry, Moscow 119991, Russia
[3]Moscow State University, Department of Materials Science, Moscow 119991, Russia
[4]GKSS Research Centre, Geesthacht, 21502, Germany
E-mail: kopitsa@lns.pnpi.spb.ru

Abstract. We report structural changes in amorphous hydrated zirconia caused by high intensity ultrasonic treatment studied by means of small-angle neutron scattering (SANS) and X-ray diffraction (XRD). It was established that sonication affects the mesostructure of $ZrO_2{\times}xH_2O$ gels (i.e. decreases their homogeneity, increases surface fractal dimension and the size of monomer particles). Ultrasound induced structural changes in hydrated zirconia governs its thermal behaviour, namely decreases the rate of tetragonal to monoclinic zirconia phase transition.

1. Introduction

Recently considerable progress was achieved in the field of application of high intensity ultrasonic treatment to nanomaterials synthesis. In the last few years a number of papers and reviews devoted to the development of advanced sonochemical methods of nanopowders and nanocomposites preparation were published [1].

Specific physical and chemical effects taking place during sonication of liquids arise mostly from acoustic cavitation, i.e. formation of vapour-gas microbubbles and their further oscillations and collapse. The mechanism of sonochemical processes in homogeneous liquids is well studied and thus obtained data are usually treated in the frames of two major theories, namely, hot spot theory [2] and the theory of local electrization [3]. The former supposes that high local temperatures (up to several thousand degrees) and pressures (several kilobars) arise during the collapse of cavitation bubbles. The latter explains the mechanism of sonochemical reactions through high local electric potentials formed on the surface of microcavities leading to intensive electrical discharges. In both cases extreme conditions in the cavities lead to decomposition of volatile compounds dissolved in the liquid (for instance, organometallic compounds, metal complexes etc.) and formation of nanostructured products.

On the contrary, information on the specific action of high intensity ultrasound on heterogeneous liquid-solid systems is still rather poor. It is generally accepted that the collapse of cavitation bubbles in the vicinity of the phase boundary (e.g. solid-liquid interface) occurs asymmetrically and subject it to shockwaves and local microjets of liquid. Thus, acoustic treatment of suspensions consisting of relatively large crystalline particles, which size is comparable to the size of a collapsing cavitation

* Author to whom correspondence should be addressed

bubble (d > 0.5–1 μm) leads to several specific effects including de-agglomeration, decrease of mean particle size, increase of the surface area due to erosion, etc. [1]. In turn, investigations concerning the influence of sonication on heterogeneous systems (for instance, sols and gels) consisting of substantially smaller particles were performed on a quite limited number of systems, mostly silica xerogels [4]. It was established that ultrasonication leads to the increase of the rate of gel formation and changes the mesostructure (e.g. density, size of monomers and size of aggregates) of gels. Nevertheless the effects of ultrasonication on such systems are to be further ascertained. For instance, recently we have established that ultrasonication affects the fractal structure of amorphous iron(III) oxide and governs its crystallization behaviour under hydrothermal conditions [5].

In view of this circumstance the present work aims to analyze the influence of high intensity ultrasound on the mesostructure and fractal structure of amorphous hydrated zirconia xerogels. Our investigation was primarily based on the use of small-angle neutron scattering method (SANS) which is widely used in the study of nuclear and magnetic inhomogeneities in various amorphous materials [6]. Structural changes in ZrO_2 caused by ultrasonication were also corroborated by means of X-ray diffraction (XRD) and low-temperature nitrogen adsorption. This paper continues the series of our research reports connected with studies of mesostructure and fractal properties of amorphous hydrated oxides. Experimental results reported in this paper not only comply with our previous findings on the ultrasonic-assisted synthesis of oxide materials, but also significantly expand our understanding of the principles of evolution of amorphous hydrated oxides in course of crystallization.

2. Experimental

2.1. Synthesis of samples

Hydrated zirconia gels were precipitated from zirconium oxynitrate (99%, Aldrich) solutions as follows: 50 ml of zirconium oxynitrate aqueous solution (C_0 = 0.10, 0.25 or 1.00 mole·L^{-1}) were added dropwise to 50 ml of ammonia aqueous solution (7.5 mole·L^{-1}) under vigorous stirring. The total duration of the precipitation procedure was 15 min. Gels of hydrated zirconia were also precipitated under the same conditions but under ultrasonic treatment performed using Bandelin Sonopuls HD 3200 generator operated at 20 kHz and equipped with titanium horn (TT-13 titanium tip). The output power measured using standard calorimetric technique was 59 W·cm^{-2}. During precipitation procedure ultrasonic horn was immersed 10 mm below the surface of the treated media. To prevent uncontrollable rise of the temperature due to dissipation of acoustic energy the precipitation process was conducted in the thermostated cell (25°C).

Upon completion of the sedimentation procedure the suspensions were additionally stirred for 30 min with (for sonicated samples) or without (for control ones) ultrasonic treatment. As synthesized hydrated zirconia gels were centrifuged (8000 min^{-1}) and washed several times with distilled water until the conductivity of mother liquor as measured using Hanna HI 98312 conductometer was less than 0.3 mS. Gels were further dried in air at 60°C overnight to produce hydrated zirconia xerogels. Throughout the paper hydrated zirconia xerogels synthesized from 0.10, 0.25 and 1.00 mole·L^{-1} solutions under ultrasonic treatment (US) and without sonication (C) are named Z-1US, Z-2US, Z-3US, Z-1C, Z-2C and Z-3C, respectively.

Thermal decomposition of the xerogels was conducted in a muffle furnace in air at 500 and 600 °C for 5 h. Heating rate was 10 °C/min. After the annealing samples were cooled down to ambient temperature within the furnace. Resulting zirconia powders are named hereafter X–T (here, X is a designation of the initial xerogel, and T is the temperature of treatment – 500 or 600 °C).

2.2. Methods of analysis

SANS measurements were performed on SANS-2 setup (FRG-1 neutron reactor, GKSS Research Centre, Geesthacht, Germany). The experiments were performed at neutron wavelength λ = 0.58 nm with $\Delta\lambda/\lambda$ = 10% and for four sample-detector distances SD = 1, 3, 7 and 20.7 m, which allowed to perform the measurements of the neutron scattering intensity for momentum transfers in the range $2.6 \cdot 10^{-2} < q < 2.6$ nm^{-1}. The scattered neutrons were detected by a two-dimensional position-sensitive ^3He detector (256×256 cells, *x* slope/ *y* slope – 2.2mm/ 2.2mm).

Samples of the xerogels were placed in a 1 mm thick quartz cells. Apparent density ρ_H of each sample was calculated as a weight of a powder divided by its volume. The initial spectra for each q range were corrected using the standard procedure [7] taking into account the scattering from the setup equipment and cell, as well as background from incoherent scattering of neutrons on hydrogen atoms which are present in samples in the form of physically and chemically bounded water. Resulting 2D isotropic spectra were averaged azimuthally and their absolute values were determined by normalizing to the incoherent scattering cross section from vanadium with inclusion of the detector efficiency and apparent density (ρ_H) for each sample. All measurements were done at room temperature.

The SANS intensity analyzed hereafter was defined as

$$I_s(q) = I(q) - T \cdot I_0(q) \tag{1}$$

where $I(q)$ and $I_0(q)$ are the momentum-transfer distributions of scattered neutrons behind the sample and beam without the sample, respectively; and T is the reduction of the transmitted beam, *i.e.*, the transmission coefficient.

The measured value of the SANS intensity, $I_s(q)$, is related to the scattering law $S(q)$ through the following expression:

$$I_s(q) = I \cdot L \cdot \int F(q - q_1) S(q) dq_1 \tag{2}$$

where L is the sample thickness and $F(q)$ – setup resolution function, which was approximated by a Gaussian and was calculated separately for each *SD* distance with the use of the standard procedure [8].

In addition to SANS, information on the structure of gels was obtained from the value of transmission coefficient $T = I(0)/I_0$. It is known [9] that intensities $I(0)$ and I_0 are related as

$$I(0) = I_0 \cdot \exp(-\Sigma_{tot} \cdot L) \tag{3}$$

where $\Sigma_{tot} = \sigma_s + \sigma_a$ is the integral scattering cross section which includes nuclear scattering σ_s and absorption σ_a.

Thermogravimetric and differential thermal analyses (TGA/DTA) of the xerogels were performed in air using Pyris Diamond thermoanalyzer (Perkin-Elmer) in the temperature range 20–800 °C (heating rate 10 °C/min).

X-ray powder diffraction (XRD) patterns were obtained using Rigaku D/MAX 2500 diffractometer (CuK_α radiation) over a 2θ range of 10–85 ° with an increment 0.02 °/step at the rate 2 °/min. Particle size was estimated using Scherrer equation

$$D = \frac{K\lambda}{[\beta(2\theta) - s]\cos\theta_0} \tag{4}$$

Here, θ_0 is the peak position, λ is the radiation wavelength, $\beta(2\theta)$ – FWHM of a corresponding peak, s – instrumental broadening (is equal to 0.1 ± 0.01 °2θ). The value of K (shape factor) was set equal to 1. The $\beta(2\theta)$ values were determined by subtraction of background followed by fitting profiles of (111) and $(1\bar{1}1)$ reflections of monoclinic zirconia and (111) reflection of tetragonal zirconia to the Voigt pseudo-functions.

Volume fractions of monoclinic (m-ZrO_2) and tetragonal (t-ZrO_2) phases in zirconia samples were estimated using equations proposed by Toraya et al. [10].

Specific surface area (S_{BET}) of samples was measured by low temperature nitrogen adsorption conducted using ATX-6 analyzer (Katakon, Russia). Before measurements the samples were outgased at 200 °C for 30 min under dry helium flow. Calculation of specific surface area was carried out within 8-point Brunauer-Emmett-Teller (BET) method.

3. Results and discussion

X-ray diffraction analysis has confirmed that all the initial xerogels are of amorphous nature giving two extremely broad diffraction peaks at ~30 and ~50–60 °2θ (see figure 1). Such X-ray patterns are quite common for as precipitated hydrous zirconia and according to [11] arise from short-range ordering in zirconium hydroxide polymers.

Thermal analysis of the xerogels has shown that their decomposition occurs in an identical way. Namely, overall weight loss is equal to 77–80 mass.% regardless of application of ultrasound during precipitation process and regardless of concentration of zirconium oxynitrate in the starting solutions. Typical thermal analysis data are presented in figure 1.

Here, weight loss at temperatures up to approximately 400 °C corresponds to elimination of physisorbed and chemically bounded water, which is accompanied by broad endothermic effect. Sharp exothermic peak at ~450 °C is obviously due to formation of nanocrystalline zirconia. These data are in line with our previous results. It has been shown [12] that hydrated zirconia synthesized in acidic media (when pH is lower than the point of zero charge) contains high amount of adsorbed NO_3^- ions. Thermal treatment of resulting xerogels is accompanied not only by elimination of physically and chemically bounded water, but also by a significant weight loss at $t > 400$ °C caused by decomposition of nitrates. On the other hand, hydrated zirconia precipitated under alkaline conditions, as it was done in this work, contains only trace amount of such impurities.

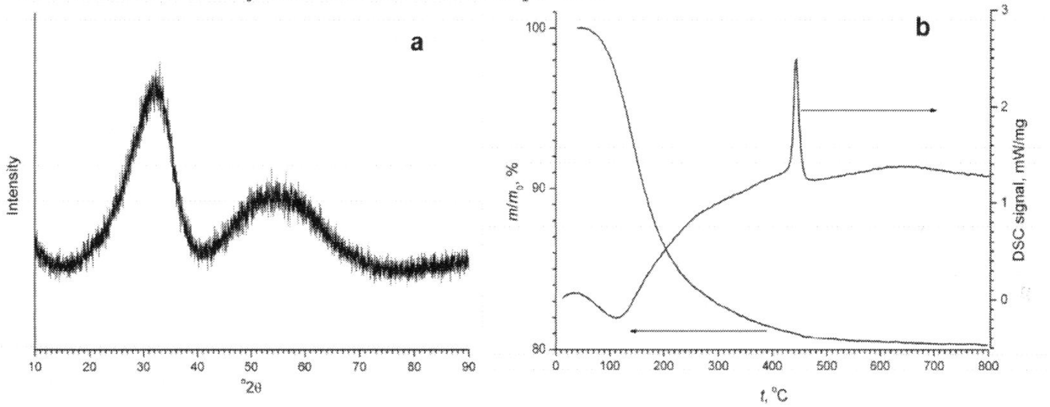

Figure 1. XRD (a) and TGA and DTA (b) data for Z-2US hydrated zirconia xerogel.

Thus basing on TGA data we can conclude that chemical composition of all the xerogels can be approximately ascribed as $ZrO_2 \cdot 1.7H_2O$ and it is almost identical regardless of synthetic procedure. This fact gives us an additional basis for the validity of SANS data comparison.

Figure 2 shows the experimental curves of differential neutron cross section $d\Sigma(q)/d\Omega$ *vs.* momentum transfer q for amorphous zirconia xerogels precipitated under various conditions. Comparison of these spectra gives an evidence that in all the cases increase in the concentration of starting zirconium oxynitrate solution results in an increase in the SANS intensity, indicating a decrease in the nuclear density homogeneity of corresponding xerogels in the range of 1–100 nm. One can also see that in the case of control samples dependence of SANS intensity upon the concentration of starting solution is much more pronounced than in case of ultrasonicated gels.

This effect is even more pronounced when the integral SANS cross-sections are compared (figure 3). Integral cross sections for the samples prepared from concentrated solutions (0.25 and 1.00 mole·L^{-1}) with or without ultrasonication are nearly identical indicating that the difference in their nuclear homogeneity is negligible. On the other hand, hydrated zirconia precipitated from diluted oxynitrate solution (0.10 mole·L^{-1}) under ultrasonic action possesses lower nuclear homogeneity in comparison with the corresponding control sample.

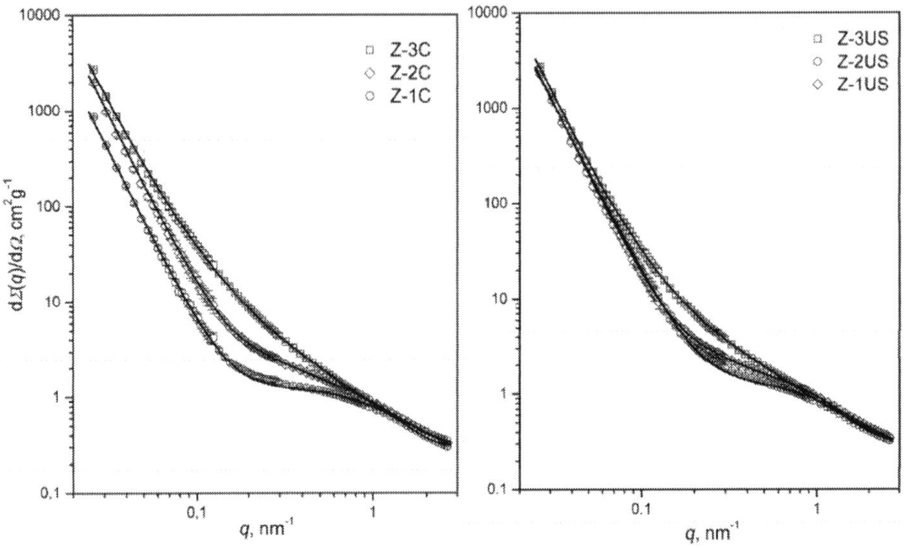

Figure 2. SANS differential cross-sections $d\Sigma(q)/d\Omega$ for amorphous $ZrO_2 \cdot xH_2O$ xerogels *vs.* momentum transfer q. Designations of corresponding samples are given in the pictures. Fitting of experimental data with formula (6 and 9) is given with firm lines.

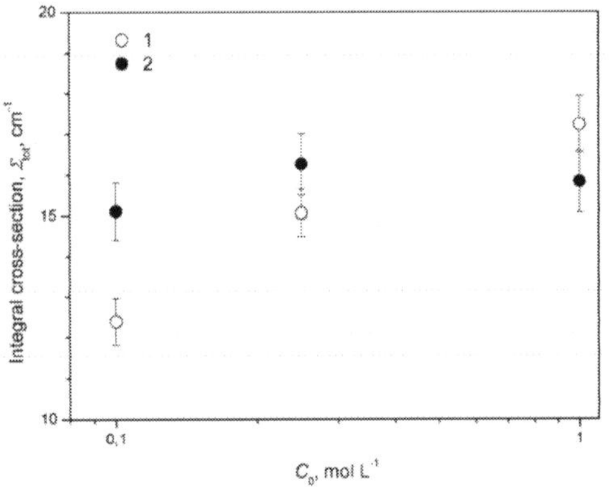

Figure 3. Integral SANS cross-sections Σ_{tot} for amorphous $ZrO_2 \cdot xH_2O$ xerogels *vs.* concentration of starting zirconium oxynitrate solution (1 – control samples; 2 – sonicated ones).

Figure 3 also demonstrates that in the control series of the gels increase in the concentration of starting solutions results in notable decrease in nuclear homogeneity. Application of ultrasound diminishes this effect so that nuclear homogeneity of samples synthesized from different starting solutions is practically identical.

In SANS experiments, the power-law momentum-transfer dependence of the scattering intensity is usually observed in the form $I(q) \sim q^{-n}$ ($n \leq 6$) in a certain momentum-transfer region $q > 1/R$, where R is the characteristic scale of the scattering system. The fractal dimension of the system and the correlation function of scattering inhomogeneities are determined by the n value or, more precisely, by the deviation from the Porod asymptotic ($n = 4$). For volume and mass fractals, n coincides with the fractal dimension D_V, $1 \leq D_V \leq 3$. For the scattering from three-dimensional objects with fractal surfaces $3 < n = 6 - D_S \leq 4$, where D_S is the surface fractal dimension, $2 \leq D_S \leq 3$.

It should be emphasized that a common property for all the samples under study is that corresponding spectra can be divided into two various q ranges, where the behaviour of the SANS cross section $d\Sigma(q)/d\Omega$ is significantly different (see figure 2). In particular, for low q values ($q < q_c$, where $q_c \approx 0.15$ nm^{-1} is the transition point between one scattering regime to another) the scattering cross section for all the samples satisfies the power law q^{-n}. As known, such a power-law dependence is observed for wide size distribution of scattering inhomogeneities with the relation $R_{max} >> R_{min}$ if the condition

$$R_{max}^{-1} << q << R_{min}^{-1}. \tag{5}$$

is satisfied. In addition, the power scattering law means that inhomogeneities making the dominant contribution to scattering are sufficiently large, so that $q_{min}R >> 1$. Bale and Schmidt [13] proposed a more exact applied criterion for a certain characteristic size of inhomogeneities, $q_{min}R \approx 3.5$. In our case, $q_{min} = 2.6 \cdot 10^{-2}$ nm^{-1} and the characteristic size of inhomogeneities is equal to $R \approx 135$ nm.

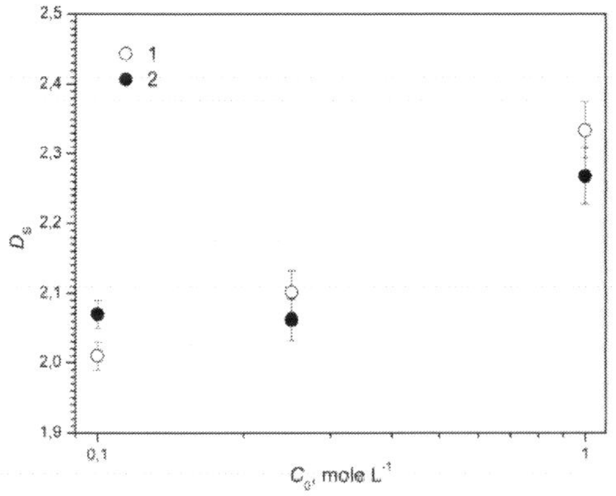

Figure 4. Surface fractal dimension (D_S) of amorphous zirconia xerogels versus concentration of initial solution (1 – control samples; 2 – sonicated ones).

The exponent n values found from the slope of the straight-line parts of the curves plotted in log-log scale (see figure 2) lie in the range from 3.67 to 3.99 (see table 1). This corresponds to the scattering from the fractal surfaces with the dimension $2.01 < D_S = 6 - n < 2.33$ (figure 4).

Thus, the further analysis of scattering in the region $q < q_c$ has been performed using the two-phase model (solid phase–pore) of the porous structure with the fractal surface of the phase interface [14]. According to this model, an object consists of inhomogeneities possessing strongly developed surface, so that, if the total area of the inhomogeneity surface measured in the scale of the inhomogeneity surface size R is proportional to R^2, the area of the surface of the scale $r << R$ is equal to $R^2(R/r)^{\Delta}$, where $0 < \Delta < 1$ and $n = 4 - \Delta$. In this case, the fractal dimension of the surface, $D_S = 2 + \Delta$, is larger than two.

Table 1. Parameters of mesostructure of amorphous hydrated zirconia samples according to SANS measurements.

Sample designation	Z-1C	Z-1US	Z-2C	Z-2US	Z-3C	Z-3US
ρ_H, g·cm^{-3}	1.485	1.68	1.62	1.794	1.58	1.78
T	0.73 ± 0.01	0.66 ± 0.01	0.67 ± 0.01	0.62 ± 0.01	0.64 ± 0.01	0.63 ± 0.01
Σ_{tot}, cm^{-1}	12.4 ± 0.6	15.1 ± 0.7	15.1 ± 0.6	16.3 ± 0.8	17.2 ± 0.6	15.8 ± 0.6
$A_1 \times 10^3$, cm^{-2}·g^{-1}	1.22 ± 0.07	5.77 ± 0.21	2.42 ± 0.20	5.53 ± 0.33	4.83 ± 0.57	3.57 ± 0.34
$D_S = 6 - n$	2.01 ± 0.02	2.07 ± 0.02	2.10 ± 0.03	2.06 ± 0.03	2.33 ± 0.04	2.27 ± 0.04
A_2, cm^2·g^{-1}	0.95 ± 0.01	1.02 ± 0.01	3.20 ± 0.08	1.91 ± 0.11	–	–
r_g, nm	1.41 ± 0.03	1.42 ± 0.03	4.62 ± 0.05	4.68 ± 0.04	–	–
$D_V = m$	–	–	–	–	1.80 ± 0.05	1.45 ± 0.05

Deviation from the power law q^{-n} is observed for all the samples (sonicated and non-sonicated) at higher values of momentum transfer. For the samples prepared from 0.10 and 0.25 mole·L^{-1} zirconium oxynitrate solutions (with or without ultrasonication), in the region $q > q_c$, the scattering curves exhibit a so-called "shoulder" indicating the presence of small inhomogeneities with characteristic size r_c. In this region the function $d\Sigma(q)/d\Omega$ exhibits Guinier behaviour. Thus, the observed scattering pattern indicates that the Z-1US, Z-2US, Z-1C and Z-2C xerogels contain two types of scattering inhomogeneities with strongly different characteristic scales which correspond to relatively large aggregates with the strongly developed surface that are formed from the initial small monomer particles. Analysis of the curve's slopes in Guinier region ($\ln(d\Sigma(q)/d\Omega)$ versus q^2) allows one to evaluate the values of the gyration radii r_g of small inhomogeneities as well as their characteristic size r_c. For instance, characteristic size of spheres can be calculated as $r_c = \sqrt{5/3}\, r_g$ [15]

The power-law dependence of the scattering cross section $d\Sigma(q)/d\Omega$ is usually observed for large q values, whereas the Guinier regime from which an upper estimate can be obtained for the scale of scattering inhomogeneities is observed for small q values. The absence of the Guinier region on the scattering curves for low q values means that the gyration radius R_g or, for the fractal systems, the upper self-similarity boundary is larger than the maximum size R_{max} of the inhomogeneities scattering from which can be detected in the experiment with a given resolution. In turn, the lower self-similarity boundary is likely determined by the size r_c of the small inhomogeneities.

In turn, SANS cross section for Z-3US and Z-3C samples prepared from concentrated solutions with or without ultrasonication, in the region $q > q_c$ obeys q^{-m} law, where the exponent m values are 1.45 and 1.8, respectively. This corresponds to the scattering from the volume fractals with the dimension $D_V = m = 1.45$ and 1.8, respectively [14]. Results obtained indicate that synthesis of xerogels from concentrated solutions lead to formation of volume fractal clusters consisting of primary particles. These clusters, in turn, form large-scale aggregates possessing well-developed fractal surface.

In general, we can conclude that scattering from hydrous zirconia samples under this study is in a good accordance with our previously reported data [12] indicating that scattering on amorphous zirconia xerogels precipitated at pH > 6 can be described within the model of porous structure with fractal surface.

In view of this circumstance, we used the following expression to analyze scattering from Z-1US, Z-2US, Z-1C and Z-2C xerogels over the entire q range [12]:

$$\frac{d\Sigma(q)}{d\Omega} = \frac{A_1(D_S)}{q^n} + A_2 \cdot \exp\left(-\frac{q^2 r_g^2}{3}\right) + I_{inc}. \qquad (6)$$

Here, $A_1(D_S)$ and A_2 are the free parameters first of which depends on the system fractal dimension [13] and the second is proportional to the product of the number of monodisperse inhomogeneities in the scattering volume and the neutron scattering amplitude density ρ [15]. The parameter I_{inc} is a constant that is independent of q and is associated with incoherent scattering on hydrogen atoms in xerogels.

It is well known that for porous materials consisting of two homogeneous phases A_1 is related to the phase-interface surface as [13]:

$$A_1(D_S) = \pi\rho^2\Gamma(5 - D_S)\sin[(D_S - 1)(\pi/2)]N_0, \qquad (7)$$

where N_0 is the characteristic of the fractal boundary, Γ is the gamma function, ρ_h is the solid-phase density, and ρ for a molecule containing several elements is defined as:

$$\rho = \sum_i b_i N_i \frac{\rho_h N_A}{M}. \qquad (8)$$

Here, N_A is the Avogadro constant, M is the molar mass, b_i is the scattering length for the i-th element in the molecule, and N_i is the number of atoms of this element. The constant N_0 in Eq. (7) is related to the specific surface (S_0) of the surface fractal as $S_0 = N_0 r^{2-Ds}$, where r^{2-Ds} is determined by the length of the yardstick. For smooth surfaces, $D_S = 2$ and $N_0 = S_0$.

Taking into account that for the Z-3C and Z-3US xerogels the crossover from one power-law dependence to another occurs at $q_c \approx 0.15$ nm^{-1}, we used the following expression to analyze scattering from these samples:

$$\frac{d\Sigma(q)}{d\Omega} = \frac{A_1(D_S)}{q^n} + \frac{A_3(D_V)}{q^m} + I_{inc}. \quad (9)$$

To obtain the final results, expressions (Eq. (6) and (9)) were convolved with the setup resolution function. The experimental curves of the differential cross section $d\Sigma(q)/d\Omega$ versus q were processed by the least mean squares method over the entire q range under investigation. The results of this analysis are given in figures 2 and 4 and in Table 1.

To analyze quantitatively the influence of ultrasonic treatment on the structure of amorphous xerogels we have compared the scattering from sonicated and non-sonicated samples. The results of such a comparison are presented in figure 5 as a difference in scattering as follows:

$$\frac{d\Sigma_{Dif}(q)}{d\Omega} = \frac{d\Sigma_{US}(q)}{d\Omega} - \frac{d\Sigma_C(q)}{d\Omega}. \quad (10)$$

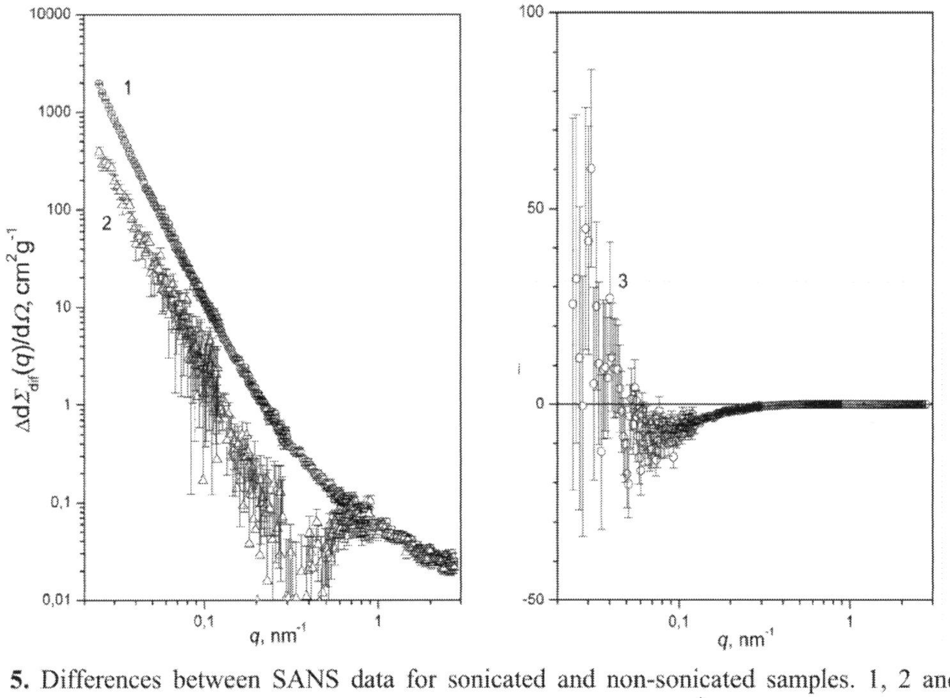

Figure 5. Differences between SANS data for sonicated and non-sonicated samples. 1, 2 and 3 – differential curves for samples obtained from 0.10, 0.25 and 1.00 mole·L^{-1} solutions, respectively.

This difference is positive over the entire q range for the xerogels prepared from 0.10 and 0.25 mole·L^{-1} solutions. This indicates that sonication results in formation of less homogeneous xerogels. For the xerogel prepared from concentrated solution (1.00 mole·L^{-1}) the opposite pattern is observed. In this case, the difference in scattering between sonicated and non-sonicated samples is positive at low and high q range only. These results are in a good agreement with the values of the integral cross-sections Σ_{tot} presented in figure 3.

Data presented in figure 4 demonstrate that surface fractal dimension of hydrated zirconia xerogels also depends upon both concentration of starting zirconium oxynitrate solution and conditions of synthesis (application of ultrasound). One can see that in general increase in concentration results in increase of surface fractal dimension. The same effect has been also observed previously for amorphous iron(III) hydroxide [5]. Probably xerogels synthesized from concentrated solutions possess higher surface fractal dimension due to the higher frequency of colloid particles collisions during precipitation that leads to formation of more ramified structures.

Effect of ultrasonication on the surface fractal dimension of xerogels synthesized from 0.25 and 1.00 mole·L^{-1} solutions is negligible. On the other hand, application of ultrasound results in notable growth of D_S in xerogels obtained from 0.10 mole·L^{-1} solution. We believe that such an effect also originates from higher frequency of colloid particles collisions in ultrasonic field caused by microjets of liquid being formed during the collapse of cavitation bubbles.

According to XRD data thermolysis of hydrated zirconia affects phase composition and even particle size of crystalline ZrO_2 obtained through thermal treatment. Thermolysis of initial xerogels at 500 °C results in formation of tetragonal zirconia with no traces of amorphous phase or thermodynamically stable monoclinic ZrO_2. Upon the rise of the temperature up to 600 °C phase transition partially occurs and monoclinic ZrO_2 is formed. Figure 6 presents examples of X-ray diffraction patterns and the dependence of the volume fraction of tetragonal zirconia from the concentration of initial zirconium oxynitrate solution.

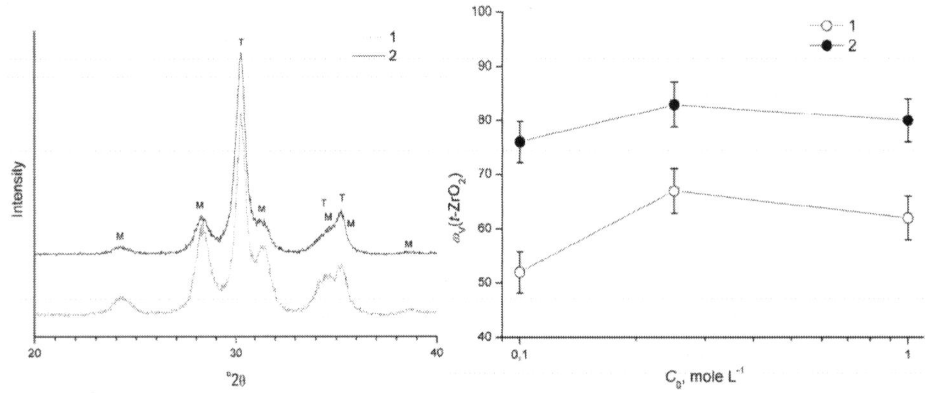

Figure 6. X-ray diffraction patterns of zirconia samples Z-1C-600 and Z-1US-600 (left) and volume fraction of tetragonal zirconia *vs*. concentration of initial solution (right). 1 – control samples; 2 – sonicated ones.

One can see that the application of ultrasound during synthesis of hydrated zirconia results in notable changes in phase composition of resulting nanocrystalline ZrO_2. Volume fraction of metastable tetragonal zirconia is higher for the samples obtained under sonication over all the concentration range.

X-ray diffraction data also demonstrate that ultrasound does not significantly affect particle size of tetragonal zirconia (see figure 7): in all the cases corresponding differences lay within the experimental error. On the other hand, increase in the concentration of starting solutions leads to a certain growth of t-ZrO_2 particle size. This effect is probably due to higher density of hydrated zirconia formed from concentrated solutions in comparison with diluted ones. Sonication of the initial gels results in slight growth of monoclinic zirconia particle size while increase in concentration of starting solution does not affect it.

Specific surface area data (see figure 7) is in a good accordance with t-ZrO_2 particle size determined from X-ray diffraction. Estimation of particle size from low-temperature nitrogen adsorption data for samples obtained from 0.10, 0.25 and 1.00 mole·L^{-1} solutions and heated at 500 °C results in the values of 15, 17 and 19–23 nm, respectively. On the other hand, one can see that thermal treatment at higher temperature (600 °C) results in notable decrease of specific surface area (S_{BET}) while particle size of t-ZrO_2 does not change significantly. This effect can not be referred to m-ZrO_2 as its particle size as estimated by Scherrer equation lies within the range of 13–17 nm for all the samples. Thus we can conclude that thermal treatment at 600 °C results in agglomeration of individual particles while this temperature is not high enough for coarsening and growth of nanocrystalline zirconia. On the other hand these data do not confirm the supposition described elsewhere [16] dealing with stabilization of tetragonal zirconia for sub-30 nm particles of ZrO_2.

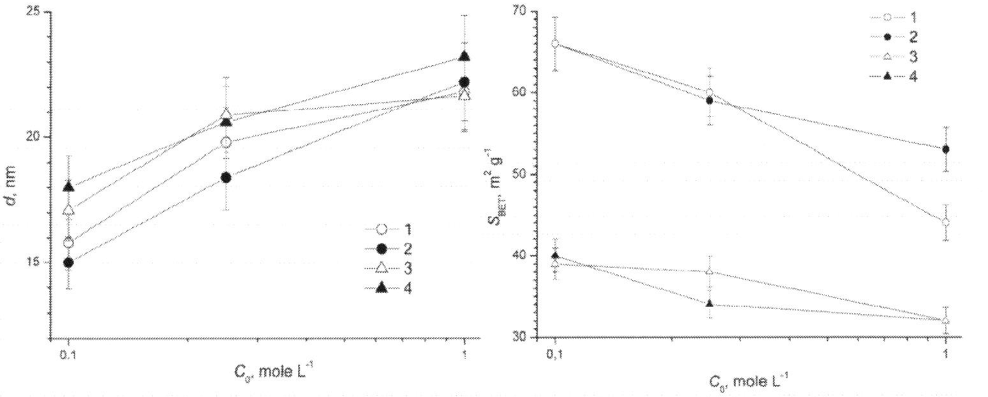

Figure 7. Particle size (left) and specific surface area (right) of tetragonal zirconia in samples heated at 500 °C (1, 2) and 600 °C (3, 4). 1,3 – control samples; 2, 4 – sonicated ones.

This work was supported by Russian Ministry of Science and Education (project 14.740.11.0281) and Russian Foundation for Basic Research (11-03-00981). G.K., A.B., S.G. and V.I. would like to thank GKSS Forschungszentrum for hearty welcome.

References

[1] Gedanken A 2004 *Ultrasonics Sonochemistry* **11** 47
Suslick K S 1999 *Ann. Rev. Matls. Sci.* **29** 295
Baranchikov A Ye, Ivanov V K and Tretyakov Yu D 2007 *Russ. Chem. Rev.* **76** 133
Meskin P E, Ivanov V K, Barantchikov A E, Churagulov B R and Tretyakov Yu D 2006 *Ultrasonics Sonochemistry* **13** 47
[2] Neppiras E A 1980 *Physics Report* **61** 159
[3] Margulis M A 2000 *Physics-Uspekhi* **43** 259
[4] Barrera-Solano M C, de la Rosa-Fox N and Esquivias L 1992 *J. Non-Cryst. Solids* **147–148** 194
Vollet D R, de Castro W C, Donatti D A and Ibañez Ruiz A 2005 *Phys. Stat. Sol.* **202** 411
Donatti D A, Vollet D R, Ibañez Ruiz A, Mesquita A and Silva T F 2005 *Phys. Rev.* B **71** 014203
[5] V.K. Ivanov, G.P. Kopitsa, F.Yu. Sharikov, A.Ye. Baranchikov, A.S. Shaporev, S.V. Grigoriev and P. Klaus Pranzas 2010 *Phys. Rev.* B **81** 174201
[6] Kopitsa G P, Ivanov V K, Grigoriev S V, Meskin P E, Polezhaeva O S and Garamus V M 2007 *JETP Letters* **85** 122
[7] Wignall G D and Bates F S 1986 *J. Appl. Cryst.* **20** 28
[8] Schmatz W, Springer T, Schelten J and Ibel K 1974 *J. Appl. Cryst.* **7** 96
[9] Gurevich I I and Tarasov T V 1965 *Physics of Low Energy Neutrons* (Moscow: Nauka) p 608
[10] Toraya H, Yoshimura M and Somiya S 1984 *J. Am. Ceram. Soc.* **67** C–119
[11] Southon P D, Bartlett J R, Woolfrey J L and Ben-Nissan B 2002 *Chem. Mater.* **14** 4313
Zyuzin D A, Moroz E M, Ivanova A S and Shmakov A N 2003 *Crystallography Reports* **48** 413
[12] Ivanov V K, Kopitsa G P, Baranchikov A E, Sharp M, Pranzas K and Grigoriev S V 2009 *Russ. J. Inorg. Chem.* **54** 2091
[13] Bale H D and Schmidt P W 1984 *Phys. Rev. Lett.* **53** 596
[14] Teixera J 1986 *On Growth and Form-Fractal and Non-Fractal Pattern in Physics* ed H E Stanley and N Ostrovsky (Boston: Martinus Nijloff) p 145
[15] Schmidt P W 1995 *Modern Aspects of Small-Angle Scattering* ed H. Brumberger (Dordrecht: Kluwer Academic Publishers) p 1

[16] Garvie R C 1965 *J. Phys. Chem.* **69** 1238
Djurado E, Bouvier P, Lucazeau G 2000 *J. Solid State Chem.* **149** 399
Shukla S, Seal S 2003 *Rev. Adv. Mater. Sci.* **5** 117

Reverse Monte Carlo modeling of Li_2S-P_2S_5 superionic conductors

Y Onodera[1], K Mori[1], T Otomo[2], A C Hannon[3], M Sugiyama[1] and T Fukunaga[1]
[1]Research Reactor Institute, Kyoto University, 2-1010 Asashiro-Nishi, Kumatori, Sennan, Osaka 590-0494, Japan
[2]Institute of Materials Structure Science, High Energy Accelerator Research Organization, 1-1 Oho, Tsukuba, Ibaraki 305-0801, Japan
[3]ISIS Facility, Rutherford Appleton Laboratory, Chilton, Didcot, Oxon, OX11 0QX, UK

E-mail: dera@post3.rri.kyoto-u.ac.jp

Abstract. Synchrotron X-ray and neutron diffraction measurements were carried out for $(^7Li_2S)_x(P_2S_5)_{100-x}$ glasses and a $^7Li_7P_3S_{11}$ metastable crystal. The Reverse Monte Carlo (RMC) modeling provided us the three-dimensional structures of $(^7Li_2S)_x(P_2S_5)_{100-x}$ and $^7Li_7P_3S_{11}$, particularly the distribution of Li^+ cations.

1. Introduction

$(Li_2S)_x(P_2S_5)_{100-x}$ glasses prepared by mechanical alloying (MA) are an excellent lithium ion conductor, exhibited a high ionic conductivity of ~10^{-4} S/cm at room temperature (RT) [1]. In addition, a $Li_7P_3S_{11}$ metastable crystal, known as glass-ceramic, can be obtained by aging the $(Li_2S)_{70}(P_2S_5)_{30}$ glass at 513 K, and its ionic conductivity is enhanced by more than one order of magnitude [2]. Therefore, the Li_2S-P_2S_5 system has attracted much attention as a solid electrolyte of lithium ion batteries.

So far, we synthesized a series of $(^7Li_2S)_x(P_2S_5)_{100-x}$ glasses in a wide x range by the MA, and reported that the electrical conductivity of $(^7Li_2S)_x(P_2S_5)_{100-x}$ glasses at RT increases exponentially from ~10^{-10} to ~10^{-4} S/cm with doping $^7Li_2S$ [3], where ^7Li is the isotope of lithium. Moreover, we confirmed that the electrical conductivity of $^7Li_7P_3S_{11}$ metastable crystal is close to 10^{-3} S/cm. In order to understand such electrochemical behaviors of the Li_2S-P_2S_5 system, it is very important to clarify the detailed structures, however they are still unclear.

In this work, synchrotron X-ray diffraction (SXRD) and neutron diffraction (ND) experiments were conducted with the $(^7Li_2S)_x(P_2S_5)_{100-x}$ glasses, together with the $^7Li_7P_3S_{11}$ metastable crystal. It is noteworthy that the use of ^7Li is to reduce the effect of absorption from the sample. The aim of this study is to clarify the three-dimensional structures of $(^7Li_2S)_x(P_2S_5)_{100-x}$ glasses and $^7Li_7P_3S_{11}$ metastable crystal, particularly the distribution of Li^+ cations, by Reverse Monte Carlo (RMC) modeling based on the ND and SXRD data.

2. Experimental Procedure

Prior to the synthesis of the $(^7Li_2S)_x(P_2S_5)_{100-x}$ glasses (x = 20, 40, 60, and 70), P_2S_5 was prepared by the MA from P (Mitsuwa Chemical Industry, 99.999%) and S (Koujundo Chemical Lab., 99.99%) powders. They were put into a zirconia pot (80 cc) with 20 zirconia balls (10 mm in diameter) in high purity argon gas, and then mixed mechanically at 300 rpm for 150 h, using a planetary ball mill apparatus (Fritsch, P-5). After that, we added $^7Li_2S$ powder (Koujundo Chemical Lab., 99.9%, istropic enrichment of ^{7}Li $\leq$ 99.94%) to P_2S_5 in the molar proportion of $(^7Li_2S)_x(P_2S_5)_{100-x}$, and carried out the additional MA at 300 rpm for 50 h. Moreover, the $^7Li_7P_3S_{11}$ metastable crystal was synthesized; the $(^7Li_2S)_{70}(P_2S_5)_{30}$ glass was heated at 523 K for 2 h in a vacuum-packed borosilicate glass tube under argon, followed by quick cooling to RT.

The SXRD experiments were carried out with a horizontal two-axis diffractometer, installed at the BL04B2 beam line in SPring-8 synchrotron radiation facility [4]. The photon energy is 61.60 keV. The powder samples of $(^7Li_2S)_x(P_2S_5)_{100-x}$ glasses (x = 20, 40, 60, and 70) and $^7Li_7P_3S_{11}$ metastable crystal were put into a stainless seal cell with capton windows in argon gas. The SXRD data were collected in the 2θ range 0.3 ° to 58.0 °. The scattering intensity was converted to the structure factor, $S(Q)$, by the polarization, absorption, and incoherent scattering corrections [5,6]. We also measured ND data with the General Material Diffractometer (GEM), installed at the ISIS facility in the UK Rutherford Appleton Laboratory [7]. The powder samples of $(^7Li_2S)_x(P_2S_5)_{100-x}$ glasses (x = 40 and 70) and $^7Li_7P_3S_{11}$ metastable crystal were put into a cylindrical vanadium can (8 mm in diameter) in argon gas. Similarly, the $S(Q)$ was obtained from the ND data by the absorption, multiple scattering, and background corrections [8,9].

The RMC modeling was carried out for the $S(Q)$ of $(^7Li_2S)_x(P_2S_5)_{100-x}$ glasses and $g(r)$ of $^7Li_7P_3S_{11}$, metastable crystal using RMC++ and RMCA [10,11], where $g(r)$ is the atomic pair distribution function. The $g(r)$ can be given by the Fourier transformation of $S(Q)$, as follows,

$$g(r)=1+\frac{1}{2\pi^2 r\rho}\int_{Q_{\min}}^{Q_{\max}} Q\{S(Q)-1\}\sin(Qr)dQ\,,$$

where ρ is the atomic number density.

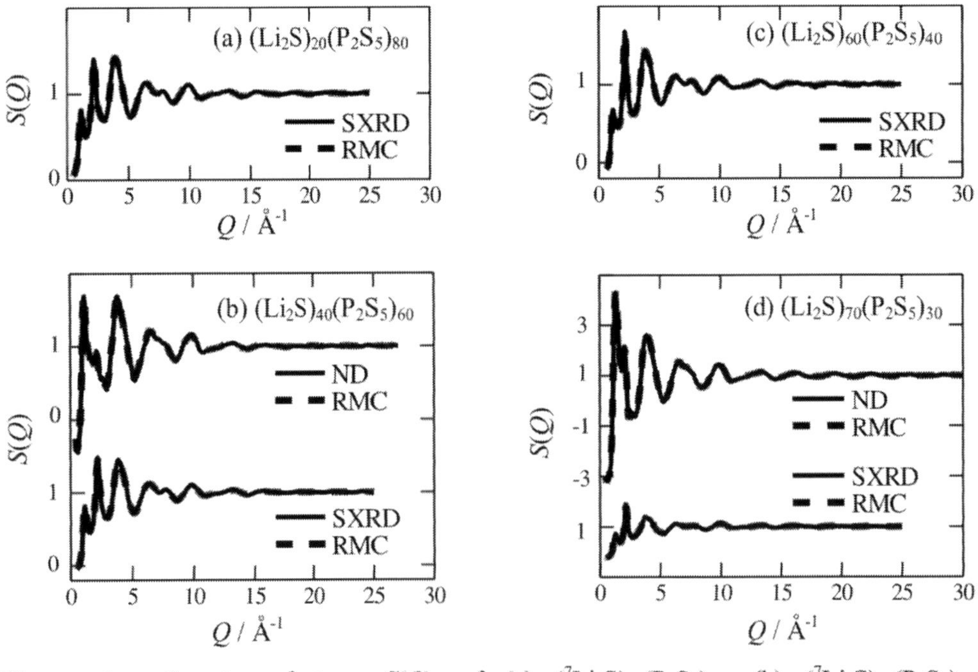

Figure 1. Structure factors, $S(Q)$, of (a) $(^7Li_2S)_{20}(P_2S_5)_{80}$, (b) $(^7Li_2S)_{40}(P_2S_5)_{60}$, (c) $(^7Li_2S)_{60}(P_2S_5)_{40}$, (d) $(^7Li_2S)_{70}(P_2S_5)_{30}$, glasses obtained by SXRD and ND experiments. The broken lines are $S(Q)$s obtained by the RMC modeling.

Table 1 Achievement rate [%] of coordination constraints for RMC modeling

	$N_{P\text{-}NBS} + N_{P\text{-}BS} = 4$	$N_{NBS\text{-}P} = 1$	$N_{BS\text{-}P} = 2$
$(^7Li_2S)_{20}(P_2S_5)_{80}$	94.7	93.2	91.7
$(^7Li_2S)_{40}(P_2S_5)_{60}$	96.9	95.2	96.6
$(^7Li_2S)_{60}(P_2S_5)_{40}$	97.4	93.5	94.2
$(^7Li_2S)_{70}(P_2S_5)_{30}$	98.7	97.8	97.1

(a)$(Li_2S)_{20}(P_2S_5)_{80}$ (b)$(Li_2S)_{40}(P_2S_5)_{60}$ (c)$(Li_2S)_{60}(P_2S_5)_{40}$ (d)$(Li_2S)_{70}(P_2S_5)_{30}$

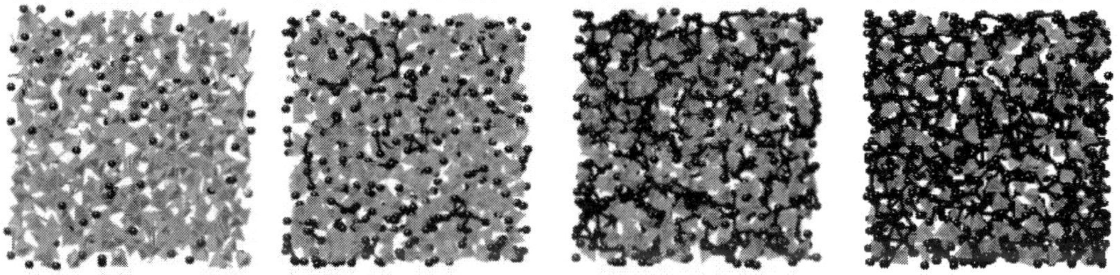

Figure 2. Three dimensional structures of (a) $(Li_2S)_{20}(P_2S_5)_{80}$, (b) $(Li_2S)_{40}(P_2S_5)_{60}$, (c) $(Li_2S)_{60}(P_2S_5)_{40}$, and (d) $(Li_2S)_{70}(P_2S_5)_{30}$. Grey tetrahedra represent the PS_4 tetrahedra. Black spheres and lines represent the Li^+ cations and Li-Li chains within 4.0 Å, respectively.

3. Results and discussion

From the first peak analysis of radial distribution functions (RDF) for $(^7Li_2S)_x(P_2S_5)_{100-x}$ glasses, the coordination number of S around P was found to be four independent of x. The S atoms in a PS_4 tetrahedron are classified into two: bridging S (BS; –P–S–P–) and non-bridging S (NBS; –P–S), respectively. From this standpoint, we carried out the RMC modeling of $(^7Li_2S)_x(P_2S_5)_{100-x}$ glasses under the following constraints; (1) the PS_4 tetrahedron is composed of BS and NBS (i.e., $N_{P\text{-}BS} + N_{P\text{-}NBS} = 4$); (2) the coordination number of P around BS must be 2 (i.e., $N_{BS\text{-}P} = 2$); (3) the coordination number of P around NBS must be 1 (i.e., $N_{NBS\text{-}P} = 1$). As an initial configuration, all of atoms (^{7}Li, P, BS, and NBS) were randomly put into a cubic box; 4991 atoms (Li: 322, P: 1288, BS: 1771, NBS: 1610) in the cubic box with 50.1 Å on a side for $(^7Li_2S)_{20}(P_2S_5)_{80}$; 4860 atoms (Li: 720, P: 1080, BS: 1260, NBS: 1800) 49.1 Å for $(^7Li_2S)_{40}(P_2S_5)_{60}$; 4991 atoms (Li: 1302, P: 868, BS: 651, NBS: 2170) 48.5 Å for $(^7Li_2S)_{60}(P_2S_5)_{40}$, and 4998 atoms (Li: 1666, P: 714, BS: 238, NBS: 2380) 47.0 Å for $(^7Li_2S)_{70}(P_2S_5)_{30}$. In the RMC modeling, excellent fits between the observed and calculated $S(Q)$'s were obtained, as shown in Fig. 1. Moreover, the coordination constraints were satisfied to better than 91%, as shown in table1. The three-dimensional structures of $(^7Li_2S)_x(P_2S_5)_{100-x}$ glasses are illustrated in Fig. 2, which indicates that $(^7Li_2S)_x(P_2S_5)_{100-x}$ glasses consist of a network of cornersharing PS_4 tetrahedra (as a grey tetrahedron), together with Li^+ cations (as a black sphere). Simultaneously, the RMC modeling was carried out for the $^7Li_7P_3S_{11}$ metastable crystal. As the initial configuration, we employed 4×8×4 unit cells of $^7Li_7P_3S_{11}$ metastable crystal refined by the Rietveld method [12]. That is, the triclinic box (a = 49.9 Å, b = 48.2 Å, c = 50.0Å, α = 102.906°, β = 113.304°, γ = 74.495°) containing 5376 atoms (Li: 1792, P: 768, S: 2816) was used in the RMC modeling. Figures 3 and 4 show the aspect of the RMC modeling on $g(r)$ and the three-dimensional structure of $^7Li_7P_3S_{11}$ metastable crystal, respectively.

The electrochemical behavior of $(^7Li_2S)_x(P_2S_5)_{100-x}$ glasses can be qualitatively understood by considering the distribution of Li^+ cations. For the lower x ($x \leq 40$) the Li^+ cations are located individually in the PS_4 network, meaning that the ionic conductivity is quite low. In contrast, for the higher x ($x \geq 60$) Li-Li correlations within 4.0 Å (as a solid line) increase drastically, which brings about the higher ionic conductivity. Furthermore, the Li-Li correlations are put in order by the transformation from the $(^7Li_2S)_{70}(P_2S_5)_{30}$ glass to the $^7Li_7P_3S_{11}$ metastable crystal. This is part of the reason why $^{7}Li_7P_3S_{11}$ metastable crystal exhibits the excellent ionic conductivity in the Li_2S-P_2S_5 system.

4. Conclusions

Synchrotron radiation X-ray and neutron diffraction measurements were carried out for the $(^{7}Li_2S)_x(P_2S_5)_{100-x}$ glasses and $^{7}Li_7P_3S_{11}$ metastable crystal. The three-dimensional structures of $(^{7}Li_2S)_x(P_2S_5)_{100-x}$ glasses and $^{7}Li_7P_3S_{11}$ metastable crystal were visualized by the Reverse Monte Carlo modeling. For the lower x ($x \leq 40$) Li^+ cations were located individually in the PS_4 network, whereas for the higher x ($x \geq 60$) Li-Li correlations within 4.0 Å increased drastically.

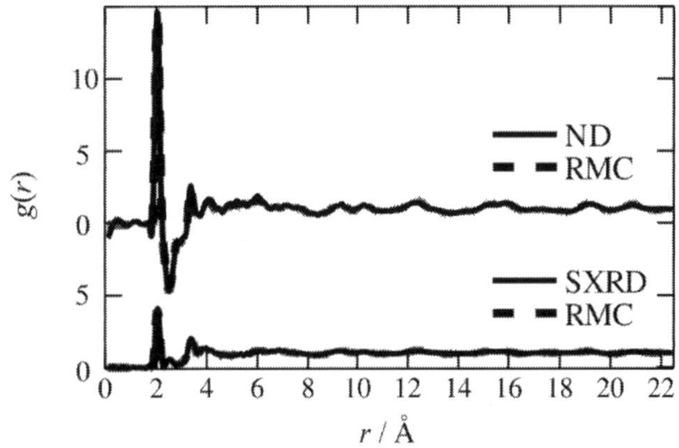

Figure 3. Pair distribution functions, $g(r)$, of $^{7}Li_7P_3S_{11}$ metastable crystal obtained by SXRD and ND experiments. The broken lines are the calculated $g(r)$s from the RMC modeling.

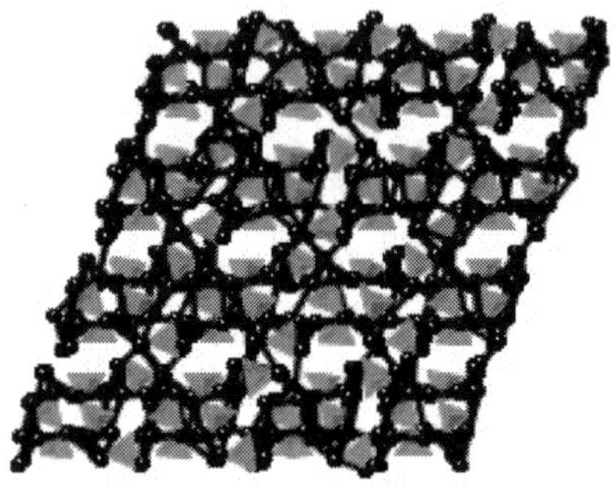

Figure 4. Three dimensional structure of $Li_7P_3S_{11}$ metastable crystal. Grey tetrahedra represent the PS_4 tetrahedra. Black spheres and lines represent the Li^+ cations and Li-Li chains within 4.0 Å, respectively.

Acknowledgements
We wish to acknowledge Dr. S. Kohara of SPring-8 for the help in the synchrotron radiation X-ray diffraction experiments. The synchrotron radiation X-ray diffraction experiments were performed at the BL04B2 in the SPring-8 with approval of the Japan Synchrotron Radiation Research Institute (JASRI) (Proposal No. 2009A1644).

References

[1] Tatsumisago M, Hama S, Hayashi A, Morimoto H, Minami T 2002 *Solid State Ionics* **154-155** 635.
[2] Mizuno F, Hayashi A, Tadanaga K, Tatsumisago M 2005 *Solid State Ionics* **17** 918.
[3] Onodera Y, Mori K, Otomo T, Hannon A C, Sugiyama M, Fukunaga T 2011 *IOP Conference Series: Material Science and Engineering* (in press).
[4] Kohara S, Suzuya K, Kashihara Y, Matsumoto N, Umesaki N, Sakai I 2001 *Nucl. Instrum. and Meth.* **A 467** 1030.
[5] Sakai S, 1991 *KEK report* **No. 90-16**.
[6] Hubbell J H, Veigele W J, Briggs E A, Brown R T, Cromer D T, Howerton R J, 1993 *J. Phys. Chem. Ref. Data* **4** 61.
[7] Williams W G, Ibberson R M, Day P, Enderby J E 1998 *Physica B* **241-243** 234.
[8] Paalman H H, Pings C J 1962 *J. Appl. Phys.* **33** 2635.
[9] Blech I A, Averbach B L 1965 *Phys. Rev.* **137** 1113.
[10] Evard E, Pusztai L 2005 *J. Phys.: Cond. Matter.* **17** S1.
[11] McGreevy R L, Pusztai L 1988 *Mol. Simul.* **1** 359.
[12] Onodera Y, Mori K, Otomo T, Hannon A C, Kohara S, Itoh K, Sugiyama M, Fukunaga T 2010 *J. Phys. Soc. Jpn. Suppl. A* **79** 87.

Neutron powder diffraction investigation in ammonium iron(III) bis (hydrogenphosphate)

B. F. Alfonso[1], C. Piqué[1], C. Trobajo[2], J. R. García[2], E. Kampert[3], I. Mirebeau[4], N. Rey[4], J. Rodríguez Fernández[5], M. T. Fernández-Díaz[6], J. A. Blanco[1]

[1]Departamento de Física, Facultad de Ciencias, Universidad de Oviedo, E–33007 Oviedo, Spain
[2]Departamento de Química Orgánica e Inorgánica, Facultad de Química, Universidad de Oviedo, E–33006 Oviedo, Spain
[3] Helmholtz-Zentrum Dresden-Rossendorf, Dresden High Magnetic Field Laboratory, D-01314 Dresden, Germany
[4]Laboratoire Léon Brillouin. (CEA-CNRS). CEN-Saclay, F-91191. Gif-sur-Yvette. Cedex, France
[5]CITIMAC, Facultad de Ciencias, Universidad de Cantabria, E–39005 Santander, Spain
[6]Institute Laue-Langevin, B. P. 156X, F–38042, Grenoble, France

E-mail: `jabr@uniovi.es`

Abstract. The deuterated form of ammonium iron(III) bis (hydrogenphosphate), $ND_4Fe(DPO_4)_2$, was investigated in detail from neutron powder diffraction data with a wavelength $\lambda = 4.724$ Å. The material undergoes two successive magnetic phase transitions which are associated with the Fe^{3+} magnetic moments. One at $T_C = 17.82 \pm 0.05$ K is attributed to the ferrimagnetic order with the magnetic moments, $\mu_{FI} = 4.19 \pm 0.02\ \mu_B$ at 4 K, lying on the crystallographic plane *ac*. The other transition is found to be at $T_t = 3.52 \pm 0.05$ K due to an antiferromagnetic arrangement, with an equal moment antiphase structure that is characterized by a long–period propagation vector close to $\vec{k}_{AF} \approx (1/16, 0, 1/16)$ and a magnetic moment for the Fe^{3+} ions of $\mu_{AF} = 4.41 \pm 0.03\ \mu_B$ at 1.5 K. The low symmetry of its triclinic crystal structure and the complex pattern of competing superexchange pathways seem to be responsible for the existence of this double magnetic phase transition.

1. Introduction

During the last century a considerable attention has been paid to porous crystalline materials, known as open metal framework systems, which are currently a hot topic of research. Porous materials such as zeolites have an important industrial and environmental applications ranging to catalysis, ion-exchange, sensing, separation and removal of contaminants and more [1, 2, 3]. In particular, iron phosphates are attractive because of their striking magnetic properties [4, 5, 6]. Recently we have focused our interest on $ND_4Fe(DPO_4)_2$ because it appears to behave magnetically in a complex manner. In previous works of $ND_4Fe(DPO_4)_2$ we have investigated the crystal and magnetic structures from neutron powder diffraction [7, 8] together with the physical properties [9]. The magnetic and specific heat measurements revealed the existence of two magnetic phases at low temperatures. The magnetic moments of the Fe^{3+} ions order ferrimagnetically below $T_C = 17.82 \pm 0.05$ K and they undergo a magnetic phase transition below

$T_t = 3.52 \pm 0.05$ K for adopting an antiferromagnetic arrangement, as was clearly evidenced from the temperature dependence of the magnetization under low applied magnetic field.

The intermediate and low-temperature magnetic structures were investigated from neutron powder diffraction on D2B (λ = 1.594 Å) and D1B (λ = 2.524 Å). The patterns collected between T_t and T_C indicate the existence of a ferrimagnetic (FI) arrangement involving the three Fe magnetic moments within the crystallographic cell. While the powder patterns below T_t show only some few new additional magnetic reflections, which could be indexed with a propagation vector close to a long-period commensurate vector. The best refinement (R_{mag} = 10.9 %) corresponds to an antiphase magnetic structure which is formed from the same magnetic unit of the FI ordering observed at higher temperatures. The determination of this magnetic structure is not straightforward. On the one hand, the number of the magnetic Bragg reflections is not large and they have moderate intensity, overlapping in some cases with the nuclear peaks. And, on the other, the absence of the Bragg reflections corresponding to the principal and the high-order harmonics (perhaps because they could be found at very low Bragg angles inside the central neutron beam on D1B and D2B diffractometers) makes very difficult the determination of the propagation vector. In order to overcome these problems, we have performed further neutron experiments with a neutron-wavelength larger than λ = 2.524 Å (λ = 4.724 Å), with the aim to obtain information on large magnetic interplanar distances, which could allow us to detect the existence of the principal and the higher-order harmonics of the propagation vector.

2. Experimental details

The neutron powder diffraction experiments were performed on the G61 (λ = 4.724 and 4.755 Å) diffractometer at the Laboratoire Léon Brillouin (LLB), Paris (France). Data were collected at several selected temperatures: 30, 4 and 1.5 K. Rietveld refinements of the crystal and magnetic structures were carried out using the FULLPROF program [10].

3. Results

The nuclear pattern collected at 30 K on G61 was fitted using a triclinic $P\bar{1}$ space group (see Fig. 1). The atomic model reported from our neutron powder experiments on D2B diffractometer [7, 8] was used as starting atomic model for the G61 experiment. Within the experimental uncertainties the results are the same: the primitive unit cell of $ND_4Fe(DPO_4)_2$ contains three Fe^{3+} ions on two different crystallographic sites with different local symmetry, one Fe1 located at the spatial inversion center and two Fe2 ions, Fe2(1) and Fe2(2), located on a crystalline site with only the identity symmetry.

As expected, the pattern collected at 4 K on G61 does not show any extra peaks different from those found at 30 K. All the magnetics peaks can be indexed with a propagation vector $\vec{k} = (0, 0, 0)$ referring to the 30 K crystallographic unit cell, indicating that the magnetic and nuclear unit cells are the same (Fig. 1). The directions of the magnetic moments of the two Fe1 and Fe2 sublattices are antiparallel lying on the ac-plane of the crystal structure. Therefore, the magnetic moments have a FI ordering with an arrangement ↑↓↑ (see Fig. 2 (a)), the Fe2 magnetic moments being ↑, while the Fe1 one is ↓. Table 1 shows the final refinement of components for the magnetic moments, in the ferrimagnetic phase, along the crystallographic directions of the primitive unit cell (a, b, c) obtained from G61 and those reported on D1B. The agreement between the values of the magnetic moment obtained in the two experiments is quite good, it allows us to confirm the FI character of the intermediate magnetic structure obtained from the experiment carried out on D2B and D1B.

The neutron powder diffraction experiment at 1.5 K (Fig. 1) was used to determine the low-temperature magnetic structure. The magnetic pattern shows additional satellite reflections, characteristic of an antiferromagnetic ordering. Their position appears to be incommensurate with regard to the nuclear lines. The determination of the propagation vector has been done

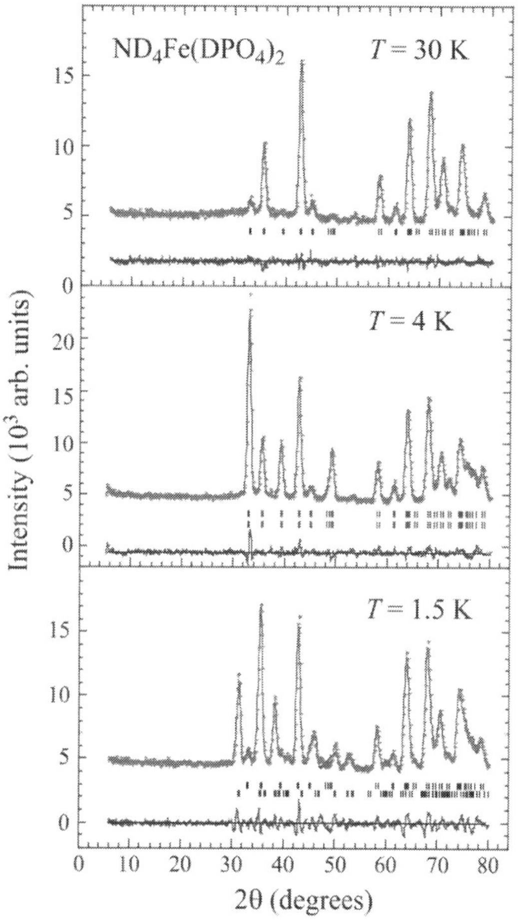

Figure 1. (Color online) Neutron powder diffraction refinement for $ND_4Fe(DPO_4)_2$. Observed (points) and calculated (solid line) collected on G61 at 30, 4 and 1.5 K. Bragg reflection positions are represented by vertical bars. The first (second) raw of Bragg diffraction reflections corresponds to the nuclear (magnetic) peaks. The observed-calculated difference is depicted at the bottom of each pattern.

graphically by following the method described by Chevalier et al. [11]. First, the basis vectors $\vec{a}^*$, $\vec{b}^*$, $\vec{c}^*$ of the reciprocal space are calculated from the cell parameters of the direct space. The reciprocal space of the crystal structure of $ND_4Fe(DPO_4)_2$ and the first Brillouin zone associated

Table 1. Components of the iron magnetic moments in $ND_4Fe(DPO_4)_2$ at 4 K (a) on G61 and (b) on D1B, with the following constraints: $\vec{\mu}_{Fe1} = -\vec{\mu}_{Fe2}$ and $\vec{\mu}_{Fe2(1)} = \vec{\mu}_{Fe2(2)}$.

	$\mu_{Fe_{1a}}$ (μ_B)	$\mu_{Fe_{1b}}$ (μ_B)	$\mu_{Fe_{1c}}$ (μ_B)	$\lvert\vec{\mu}\rvert$ (μ_B)	R_{mag} (%)	Ref.
(a)	-3.56(9)	0	-5.81(7)	4.19(2)	3.8	This work
(b)	-3.77(15)	0	-5.84(13)	4.20(12)	5.3	[8]

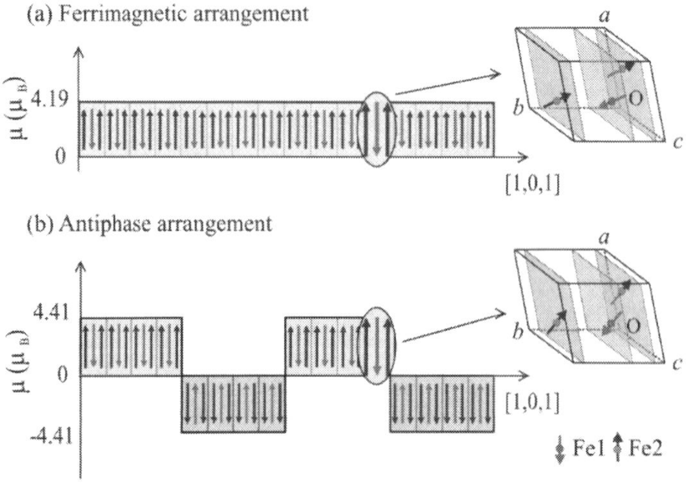

Figure 2. (Color online) Schematic representation of (a) the ferrimagnetic structure with a $\mu = 4.19(2)\ \mu_B$ at 4 K, and (b) the antiphase magnetic structure at 1.5 K for $ND_4Fe(DPO_4)_2$ along the direction of propagation vector $\vec{k}_{AF} = (\tau, 0, \tau)$ with $\tau \equiv 1/16$ and $\mu = 4.41(3)\ \mu_B$. In all the cases the magnetic moments are lying in the ac–plane (see text).

with the Bravais lattice are drawn in Fig. 3. Using the Bragg relation, Eq. (1), we calculate the modulus of the scattering vector $\vec{Q}$ for the Bragg's angles corresponding to the magnetic reflections (found at low angles, see Fig. 1). We plot spheres of radius $|\vec{Q}|$ centered at the origin (000) of the reciprocal space.

$$| \vec{Q} |= 4\pi sin\theta/\lambda \tag{1}$$

$$| \vec{Q} |=| \vec{K} \pm \vec{k} | \tag{2}$$

Due to the Bragg condition expressed by the Eq. (2), where $\vec{K}$ is a reciprocal lattice vector and $\vec{k}$ is the magnetic propagation vector, it is necessary to find the intersection of these spheres of radius $|\vec{Q}|$ with the small sphere of radius $|\vec{k}|$, centered at each point in the reciprocal space, in such a way that the ends of the diameter at the point of intersection cut the spheres of radius $|\vec{Q}|$ in the same direction for all the points of the reciprocal space. Then, the radius of the small sphere give us the magnetic propagation vector: $\vec{k} = (0.06, 0, 0.06)$ rlus (reciprocal lattice units). This propagation vector has been successfully used for indexing the magnetic neutron powder diffraction pattern collected at 1.5 K. The nearest long-period commensurate propagation vector is close to $\vec{k}_{AF} = (\tau, 0, \tau)$ with $\tau \equiv 1/16$ (the same propagation vector as that determined on D1B). The magnetic periodicity is repeated after 16 crystallographic unit cells along the a and c directions. The best fit to the experimental data corresponds to a longitudinal–amplitude modulated structure (sine–wave type) with the magnetic moments lying in the ac–plane, the maximum amplitude of the sine–wave modulated moment being 5.61(3) μ_B. The reliability factor for the Rietveld refinement obtained from experiments on G61 is $R_{mag} = 4.4$ %, which is better than that from D1B, $R_{mag} = 10.9$ %. Table 2 shows the position and intensity of the calculated and observed magnetic Bragg reflections that appear for angles 2θ less than 53°. Due to the fact that Fe^{3+} ion is a Kramers ion, this means that the magnetic moment over each site can not be zero at zero temperature, as it would be expected for a sine-wave magnetic structure. This circumstance implies that the magnetic arrangement found at 1.5 K for $ND_4Fe(DPO_4)_2$

Table 2. Positions and intensity of the magnetic Bragg reflections in $ND_4Fe(DPO4)_2$ at 1.5 K (see Fig. 1) [calc: calculated, obs: observed, and $(hkl)^{\pm\tau} \equiv (hkl)^{\pm} = (hkl) \pm (\tau, 0, \tau)$ with $\tau \equiv 1/16$ in rlus]. Both intensities are corrected by a Lorentz factor of $\sin 2\theta \sin\theta$. The intensities are calculated with a Fe sine–wave amplitude modulated moment having a maximum amplitude of 5.61(3) μ_B (see text).

2θ (°)	(hkl)	I_{calc} (arb. units)	I_{obs} (arb. units)
31.240	$(010)^-$	542.3	563.5
35.373	$(010)^+$	668.9	669.0
35.471	$(\bar{1}01)^-$	144.8	143.9
38.200	$(\bar{1}\bar{1}1)^+$	538.1	546.7
40.865	$(\bar{1}\bar{1}1)^-$	126.0	116.6
43.398	$(001)^-$	150.4	143.1
46.278	$(1\bar{1}0)^-$	233.6	239.8
52.340	$(1\bar{1}0)^+$	275.9	296.0

must evolve at low temperatures towards an equal–moment magnetic structure or antiphase ordering, where the magnetic moments over an Fe site are described by the squaring–up [12]:

$$\vec{\mu}_{lj} = \frac{4}{\pi} \times \sum_{p=0}^{\infty} \frac{\mu_j}{2p+1} sin((2p+1)\vec{k}\vec{R}_{lj}) \tag{3}$$

where $\vec{\mu}_{lj}$ is the magnetic moment of the Fe^{3+} ions located at the vector position $\vec{R}_{lj}$ (with the index j runs for Fe1 and Fe2, l is a composite index for a lattice translation, $\vec{R} = l_1\vec{a} + l_2\vec{b} + l_3\vec{c}$, where l_i are all integers), $2p+1$ (with $p = 0, 1, 2, ...$) are the 1st, 3rd, 5th, ... order harmonics and $\vec{\mu}_j$ is the equal–magnetic moment value of the iron ions. Similarly to the ferrimagnetic magnetic structure observed above $T_t = 3.52 \pm 0.05$ K, we have found that the best solution for the antiphase ordering corresponds to the case with the same constraints of the magnetic moments, *ie*, $\vec{\mu}_{Fe1} = -\vec{\mu}_{Fe2}$ and $\vec{\mu}_{Fe2(1)} = \vec{\mu}_{Fe2(2)}$.

It is worth noting that the principal harmonic $(000)^{\pm\tau}$ and the higher-order harmonics $(000)^{\pm n\tau}$ were not observed. To start with the first harmonics, as can see in Fig. 3, the sphere whose radius is the modulus of the propagation vector is inside the sphere corresponding to the lower angle measured on the three diffraction experiments: D2B ($\lambda = 1.594$ Å, $2\theta_{min} = 8.5^o$), D1B ($\lambda = 2.524$ Å, $2\theta_{min} = 6.2^o$) and G61 ($\lambda = 4.724$ Å, $2\theta_{min} = 5.5^o$). Then, in our case the principal reflection $(000)^{\pm\tau}$ is masked by the central beam. Other high–order harmonics such as $(000)^{\pm 3\tau}$ and $(000)^{\pm 5\tau}$ are expected to be found in the Bragg–angular range 2θ $(14°, 26°)$, but they are not observed, probably because these high–order harmonics are difficult to be

Table 3. Components of the iron magnetic moment in $ND_4Fe(DPO_4)_2$ (a) on G61 at 1.5 K and (b) on D1B at 1.89 K, with the following constraints: $\vec{\mu}_{Fe1} = -\vec{\mu}_{Fe2}$ and $\vec{\mu}_{Fe2(1)} = \vec{\mu}_{Fe2(2)}$.

	μ_{Fe1_a} (μ_B)	μ_{Fe1_b} (μ_B)	μ_{Fe1_c} (μ_B)	$\lvert\vec{\mu}\rvert$ (μ_B)	R_{mag} (%)	Ref.
(a)	-5.81(12)	0	-6.15(9)	4.41(3)	4.4	This work
(b)	-6.13(30)	0	-6.98(30)	4.78(8)	10.9	[8]

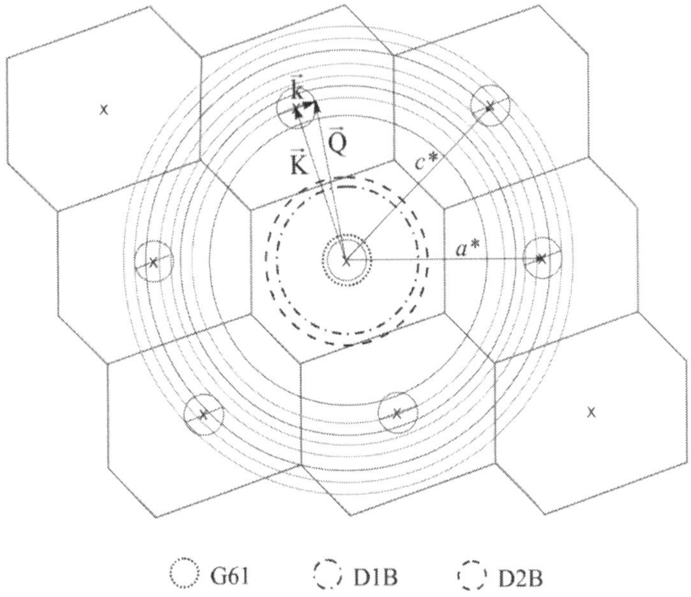

Figure 3. (Color online) Graphical determination of the propagation vector associated with the magnetic structure for $ND_4Fe(DPO_4)_2$ at 1.5 K (see text). The spheres corresponding to the lower angle measured on the three diffractometers: D2B (dashed lines), D1B (dot-dashed lines) and G61 (dots).

seen from a powder experiment, and they are often developed at very low temperatures with regard the magnetic transition temperature [13]. Apart from that, the direction of the magnetic moment in the ac–plane is quite close to direction of the propagation vector. In order to elucidate these different possibilities, we need to perform further experiments with a wavelength larger than $\lambda = 4.724$ Å. Taking into account all these points, the equal–moment magnetic structure has then a magnetic moment of $|\vec{\mu}_j| = 4.41(3)\ \mu_B$, as a result from Eq. (3), where the maximum amplitude associated with the sinus–wave, $5.61(3)\ \mu_B$, can be equalized to the expression $\frac{4}{\pi}\times\mu_j$. Table 3 shows the final refinement of the components for the magnetic moment along the crystallographic directions of the primitive unit cell (a, b, c) in the antiphase structure, obtained from G61 at 1.5 K and those reported on D1B at 1.89 K. The magnetic model is based on the high-temperature FI arrangement $\uparrow\downarrow\uparrow$ adopting an antiphase ordering following the sequence +-+-, where + (-) means $\uparrow\downarrow\uparrow$ $\uparrow\downarrow\uparrow$ $\uparrow\downarrow\uparrow$ $\uparrow\downarrow\uparrow$ ($\downarrow\uparrow\downarrow$ $\downarrow\uparrow\downarrow$ $\downarrow\uparrow\downarrow$ $\downarrow\uparrow\downarrow$) (see Fig. 2 (b)). This complex magnetic behavior where the system undergoes two magnetic phase transitions must be interpreted mainly as a consequence of the particular competition of superexchange interactions in $ND_4Fe(DPO_4)_2$ through Fe-O-P-O-Fe. This issue will be investigated in future works.

4. Conclusions

In summary, the $ND_4Fe(DPO_4)_2$ shows two magnetic phase transitions. Below $T_C = 17.82$ K the magnetic moments of the two sublattices of Fe^{3+} ions are coupled antiparallel on the ac-plane of the crystal structure having a FI ordering. When the temperature decreases below $T_t = 3.52$ K, the magnetic arrangement evolves to an antiphase one maintaining the same FI magnetic unit.

Acknowledgments
Financial support has been received from Spanish MEC and FEDER Grant No. MAT2008-06542-C04.

References
[1] Maspoch D, Ruiz-Molina D and Veciana J 2007 *Chem. Soc. Rev.* **36** 770
[2] Murugavel R, Choudhury A, Walawalkar M G, Pothiraja R and Rao C N R 2008 *Chem. Rev.* **108** 3549
[3] Mahata P, Sarma D, and Natarajan S 2010 *J. Chem. Sci.* **122** 19
[4] Riou-Cavellec M, Riou D and Férey G 1999 *Inorg. Chim. Acta* **291** 317
[5] Natarajan S and Mandal S 2008 *Angew. Chem. Int.* **47** 4798
[6] Alfonso B F, Trobajo C, Piqué C, Fernández-Díaz M T, Rodríguez Fernández J, Salvadó M A, Pertierra P, García-Granda S, García J R and Blanco J A 2010 *Acta. Materialia* **58** 1741
[7] Alfonso B F, Blanco J A, Fernández-Díaz M T, Trobajo C, Khainakov S A, and García J R 2010 *Dalton Trans.* **39** 1791
[8] Alfonso B F, Piqué C, Trobajo C, García J R, Kampert E, Zeitler U, Rodríguez Fernández J, Fernández-Díaz M T and Blanco J A 2010 *Phys. Rev. B* **82** 144431
[9] Alfonso B F, Piqué C, Trobajo C, García J R, Rodríguez Fernández J, Fernández-Díaz M T and Blanco J A 2011 Acepted for publication *J. Phys. Conference Series*
[10] Rodríguez–Carvajal J 1993 *Physica B* **192** 55
[11] Chevalier B, Etourneau J, Hagenmuller P, Quezel S and Rossat-Mignod J. 1985 *J. Less Common Metals* **111** 161
[12] Arfken G, 1970 *Mathematical Methods for Physicists* (Academic Press, New York)
[13] Blanco J A, Fåk, B, Ressouche E, Grenier B, Rotter M, Schmitt D, Rodríguez-Velámazan J A, Campo J and Lejay P 2010 *Phys. Rev. B* **82** 054414

Evidence of Magnetic Correlations Among the Charge Stripe Electrons of Charge-Stripe Ordered $La_2NiO_{4.11}$.

P. G. Freeman[1,2], S. M. Hayden[3], R. A. Mole[4,5], M. Enderle[2], D. Prabhakaran[6] and A. T. Boothroyd[6].

[1]Helmholtz-Zentrun Berlin, Hahn-Meitner-Platz 1, D-14109 Berlin, Germany
[2]Institut Laue-Langevin, BP 156, 38042 Grenoble Cedex 9, France
[3]Department of Physics, University of Bristol, Bristol, BS8 1TL, United Kingdom
[4]Bragg Institute, ANSTO, New Illawarra Road, Lucas Heights, NSW, Australia
[5]FRM II, Lichtenbergstraße, 1 85747 Garching, Germany
[6]Department of Physics, Oxford University, Oxford, OX1 3PU, United Kingdom

E-mail: paul.freeman@helmholtz-berlin.de

Abstract.
The magnetic excitations of charge-stripe ordered $La_2NiO_{4.11}$ where investigated using polarized- and unpolarized-neutron scattering to determine the magnetic excitations of the charge stripe electrons. We observed a magnetic excitation mode consistent with the gapped quasi-one-dimensional antiferromagnetic correlations of the charge stripe electrons previously observed in $La_{2-x}Sr_xNiO_4$ $x = 1/3$ and $x = 0.275$.

1. Introduction

Recent observations of a hourglass shaped excitation spectrum in cuprate superconductors point towards an underlying universal magnetic excitation spectrum for all of the cuprates[1]. In insulating charge stripe ordered $La_{5/3}Sr_{1/3}CoO_4$ an hourglass magnetic excitation spectrum has been observed, and explained within a charge stripe model with a weak inter-stripe spin interaction, relative to the intra-stripe spin interaction[2]. The magnetic excitations of $La_{5/3}Sr_{1/3}CoO_4$ indicate that the underlying magnetism of the cuprates can originate from charge stripe fluctuations. Furthermore, the lack of an hour glass excitation spectrum in charge stripe ordered $La_{2-x}Sr_xNiO_4$ is explained by the strong inter-stripe spin interaction of these materials.

The dispersion of the magnetic excitations of charge stripe ordered $La_2NiO_{4.11}$ deviates from the theory that describes the magnetic excitations in $La_{5/3}Sr_{1/3}CoO_4$ and $La_{2-x}Sr_xNiO_4$ $x \sim 1/3$ [3]. In a limited energy range, ~10-30 meV, the centre of the magnetic excitations shifts inwards towards the two dimensional antiferromagnetic wavevector $\mathbf{Q}_{\mathbf{AFM}} = (0.5, 0.5)$. If the inter-stripe spin interaction was weak, the excitations would be expected to disperse towards $\mathbf{Q}_{\mathbf{AFM}}$, but the inter-stripe spin interaction of $La_2NiO_{4.11}$ is strong[3, 4]. As the inter-stripe spin interaction is mediated by the charge stripe electrons, it is important to ascertain what are the magnetic properties of the charge stripe electrons of $La_2NiO_{4.11}$. In $La_{2-x}Sr_xNiO_4$ $x \sim 1/3$ we have observed the charge stripe electrons have gapped quasi-one-dimensional antiferromagnetic spin correlations (q-1D) along the charge stripes[5, 6].

2. Experimental details

A single crystals of $La_2NiO_{4.11}$ ($\delta = 0.11$) was grown by the floating-zone technique[7]. The sample used here was also used in our previous neutron studies of $\delta = 0.11$ [3], and is a rod of 8 mm diameter and 40 mm length. Thermo-Gravimetric Analysis (TGA) of an as-grown crystal determined the oxygen excess to be $\delta = 0.11 \pm 0.01$. Data on the bulk magnetization of an as-grown crystal are published elsewhere[8].

Neutron scattering measurements were performed on the triple-axis spectrometers IN20 at the ILL, and on PUMA at FRM-II. The energies of the incident and scattered neutrons were selected by Bragg reflection from crystal Heusler arrays (IN20), or arrays of pyrolytic graphite crystals (PUMA). The monochromators were horizontally focused on both instruments, and vertically focused on PUMA, to maximise neutron flux on the sample position. The analyzers were horizontally focused on both instruments and vertically focused on PUMA. Data were collected with a fixed final neutron wavevector of 2.662 Å^{-1}. A pyrolytic graphite filter was placed between the sample and analyzer to suppress higher-order harmonic scattering. On IN20 polarized neutrons were employed, and the neutron spin polarization $\mathbf{P}$ was maintained in a specified orientation with respect to the neutron scattering vector $\mathbf{Q}$ by an adjustable guide field of a few mT at the sample position. A monitor is placed between the monochromator and sample position to determine the number of neutrons incident on the sample position. We aligned the crystal so that the horizontal scattering plane was (h, h, l) (we refer here to the tetragonal unit cell of the space group $I4/mmm$ with unit cell parameters $a = 3.8$ Å and $c = 12.7$ Å). On IN20 the sample was mounted in a standard orange cryostat, and on PUMA the sample was mounted in a closed cycle refrigerator.

3. Results

We performed longitudinal polarized neutron scattering measurements on IN20. For these measurements we constrained the neutron polarization $\mathbf{P}$ to be parallel to the scattering vector $\mathbf{Q}$. In this configuration scattering from electronic magnetic moments causes the neutron spin to flip, whereas scattering via non-magnetic processes does not.

In figure 1 we show an elastic scan of the $\delta = 0.11$ parallel to $(hh0)$ through the the magnetic Bragg reflections at $l = 1$. In the Spin Flip (SF) data two sharp Bragg reflections are observed at $\xi = 0.363 \pm 0.001$ r.l.u. and $\xi = 0.637 \pm 0.001$ r.l.u. from the magnetic order. In the Non-Spin Flip channel (NSF) weaker Bragg reflections are observed due to the imperfect polarization of the neutron beam. The ratio of the intensities of the NSF and SF Bragg reflections gives a flipping ratio of 19.

In figure 2(a) we show scans along $(hh0)$ at $E = 2.5$ meV of $\delta = 0.11$, for both the SF and NSF channels. The data has been corrected for the imperfact neutron polariztion[9]. In the NSF channel no excitation peaks are observed, whereas in the SF channel a broad magnetic excitation peak centred on ~ 0.68 r.l.u. is observed. As the dispersion gradient of excitations from the ordered Ni^{2+} $S = 1$ ordered spins is effectively infinite at 2.5 meV, they only account for the left hand side of the SF peak. The right hand side of the excitation centred on $\xi \sim 0.75$ is an unaccounted for antiferromagnetic excitation, that has no elastic component.

At 2.5 meV we are below the out of plane anisotropy gap in $\delta = 0.11$, and the spins fluctuate around the ordered moment in the Ni-O plane[3]. For a wavevector along $(hh0)$ at 2.5 meV only the component of the spin fluctutaions perpendicular to this direction are probed. If the spin excitations are observed at a wavector of the same length that is appproximately along $(00l)$, all the in plane spin fluctuations are probed and the intensity of the observed excitation will increase. The q-1d in the $x = 1/3$ was determined to preferentially fluctuate out of the Ni-O

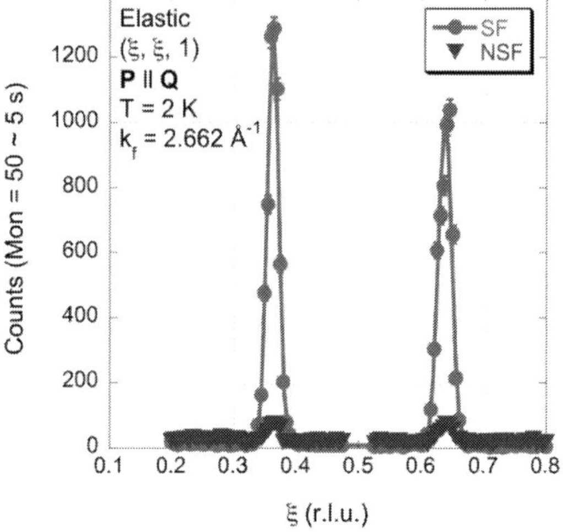

Figure 1. Spin-flip (SF) and non-spin-flip (NSF) diffraction from $La_2NiO_{4.11}$ at T = 2 K. The scan is parallel to $(hh0)$ with $l = 1$. No correction has been made for the imperfect polarization of the neutron beam. The solid line is the result of a fit to the SF data with 2 Gaussian lineshapes on a flat background.

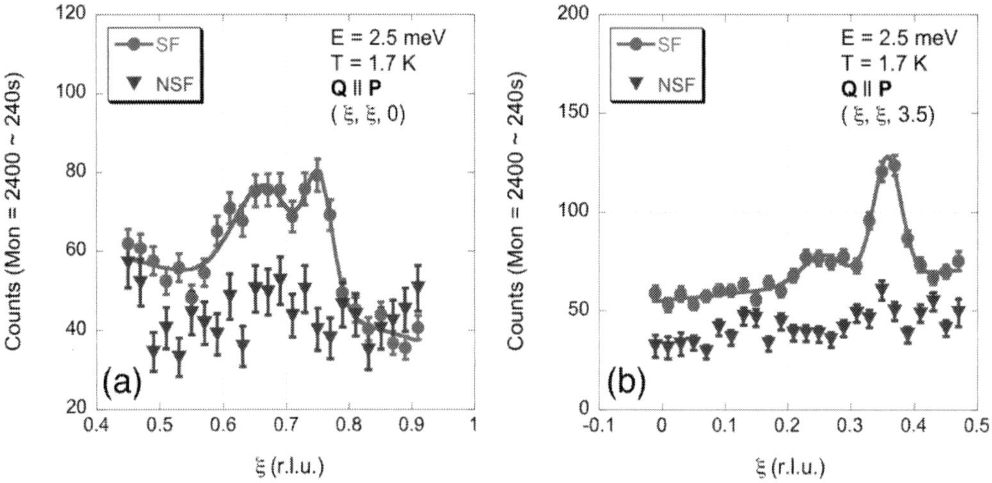

Figure 2. Spin-flip (SF) and non-spin-flip (NSF) scattering of $\delta = 0.11$ at 2.5 meV along directions parallel to $(h, h, 0)$ for (a) $l = 0$, and (b) $l = 3.5$. The solid line is a fit to the SF data of 2 Gaussian lineshapes on a sloping background. The sense of the two peaks is reversed in the two scans, i.e. the ordered Ni^{2+} spins result in the left excitation peak in (a), and the right peak in (b).

plane[5]. If the q-1D is measured at a wavector appproximately along $(00l)$ the intensity will decrease compared to a wavevector along $(h, h, 0)$. Figure 2 (b) shows a polarized neutron scan of the magnetic excitations along $(\xi,\ \xi,\ 3.5)$ which shows two peaks in the SF channel. In Fig.2

(a) and (b) the symmetry of scans reverses, so that the right hand side peak in (b) is from the ordered Ni^{2+} spins, which has increased in intensity as expected. If the SF peak on the left of Fig.2 (b) was from q-1D excitations that preferential fluctuate out of the Ni-O plane, the intensity of these excitations will be weaker in Fig.2 (b) than Fig.2 (a), as observed.

The polarized neutron scattering measurements of $\delta = 0.11$ at 2.5 meV show two magnetic excitations. We have shown that one of these excitation modes is gapped with no elastic component, and has the same wavevector dependence for intensity as the q-1D in $x = 1/3$ and $x = 0.275$. After confirming the existence of a second magnetic excitation in $\delta = 0.11$, we studied the energy and temperature dependence of this magnetic excitation using unpolarized neutron scattering.

Figure 3 (a) we show three scans parallel to $(h, h, 0)$ at $l = 3$ of the $\delta = 0.11$ at 4.2 K. As we were able to collect significant neutron counts in the data, we clearly observe the Lorentzian lineshape of the excitations, consistent with our previous results[3]. At 2 meV and $\xi = 0.270 \pm 0.006$ a small side peak is clearly resolved from the excitations from the order moments. The centring of this excitation agrees with the flexing of the q-1D observed in $x \sim 1/3$[5, 6]. At 4 meV the excitations at $\xi = 0.270$ and $\xi = 0.73$ are observed to have reduced counts compared to 2 meV, and by 6 meV the excitations have broadened to become an extended tail to the excitation peaks from the ordered moments. When the q-1d of $x = 1/3$ is scanned in an equivalent way to Fig. 3 (a), the q-1d is qualitatively observed to follow the same trend. Unfortunately the fitting uncertainties for the data fits of the 4 meV and 6 meV data in Fig. 3 (a) do not allow for a meaningful quantative comparison.

In figure 3 (b) we show a scan parallel to $(h, h, 0)$ at $l = 3$ for 4 meV of the $\delta = 0.11$, taken at 20 K. The scan at 20 K is well described by a fit of two Lorentzians on a sloping background, as indicated by the solid line of Fig. 3 (b). In the $x = 1/3$, on increasing temperature the

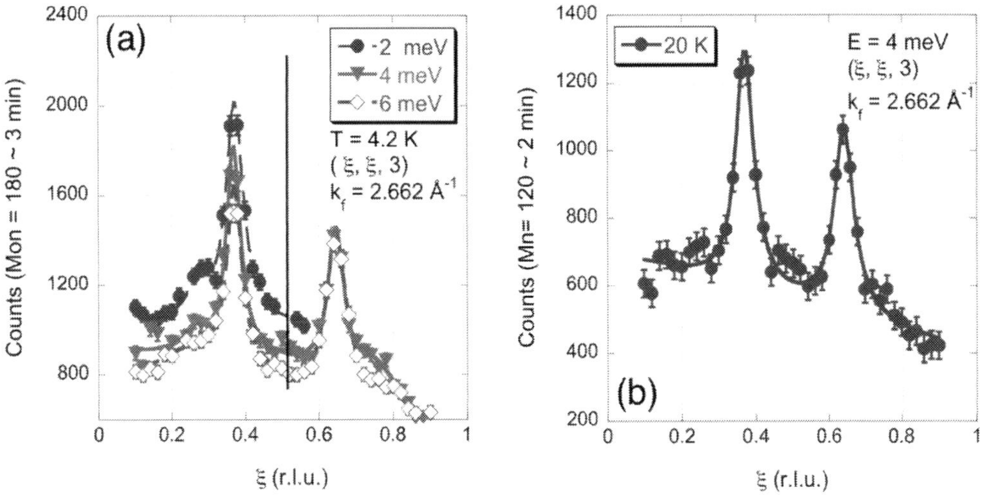

Figure 3. (a) Scans parallel to $(h, h, 0)$ of the magnetic excitations of $\delta = 0.11$ at T= 4.2 K. A solid vertical line indicates the left and right hand sides of the scans, that were fitted independently. The other solid and dashed lines indicate fits to the data of 2 Lorentzian lineshapes on a sloping background. (b) A scan parallel to $(h, h, 0)$ of the magnetic excitations of $\delta = 0.11$ for 4.0 meV taken at 20 K. The solid line is a fit to the data with 2 Lorentzian lineshapes on a sloping background.

Journal of Physics: Conference Series **340** (2012) 012060 doi:10.1088/1742-6596/340/1/012060

q=1D remains observable to temperatures well in excess of 20 K[10], unlike the scan at 20 K of the $\delta = 0.11$. The lack of observable q-1D at 20 K for $\delta = 0.11$ may be a result of the fit, with the sloping background and the Lorentzian lineshape incorrectly accounting for the q-1D. Additional measurements in the $(hk0)$ scattering plane away from the excitations from the order Ni^{2+} spins, should resolve this issue.

4. Discussion and Conclusions

We have presented experimental evidence of a second magnetic excitation mode in charge stripe ordered $La_2NiO_{4.11}$. A preferential polarization of the fluctuations out of the Ni-O plane, and the energy dependence of the magnetic excitations is consistent with the gapped quasi-one-dimensional antiferromagnetic excitations of the charge stripe electrons (q-1D) in $La_{2-x}Sr_xNiO_4$ $x \sim 1/3$. The main components of the ground state magnetic excitations in $La_2NiO_{4.11}$, the out of plane anistropy gap, the inter- and intra-stripe interactions, and the q-1D observed in this work, are unchanged compared to $La_{2-x}Sr_xNiO_4$ $x \sim 1/3$[3, 5, 6].

Despite the lack of differences in the magnetic interactions of $La_2NiO_{4.11}$ and $La_{2-x}Sr_xNiO_4$ $x \sim 1/3$, the ground state magnetic excitations from the order Ni^{2+} spins in the energy range 10-30 meV have remarkable differences. In $La_2NiO_{4.11}$ there is an inward dispersion of the magnetic excitations towards $\mathbf{Q_{AFM}} = (0.5, 0.5)$[3], whereas in $La_{2-x}Sr_xNiO_4$ $x \sim 1/3$ there is a partial gapping of the spin excitations accompanied by a resonance like feature at the magnetic zone centre[10, 11, 12, 13]. As the main magnetic interactions in these two materials are unaltered, we require further interactions to fully explain the magnetic excitation spectrums in the charge stripe ordered nickelates. As our previous studies for $x \sim 1/3$ have indicated, we need to research beyond the magnetic interactions to understand these striking phenomena [10, 11, 12].

5. Acknowledgements

This research project has been supported by the European Commission under the 7th Framework Programme through the 'Research Infrastructures' action of the 'Capacities' Programme, Contract No: CP-CSA_INFRA-2008-1.1.1 Number 226507-NMI3. Crystal growth was supported by the Engineering and Physical Sciences Research Council of Great Britain.

References

[1] Hayden S. M., Mook H. A., Dai Pengcheng, Perring T. G. and Doan F., 2004 Nature (London) **429**, 531; Tranquada J. M., Woo H., Perring T. G., Goka H., Gu G. D., Xu G., Fujita M. and Yamada K., 2004 Nature (London) **429**, 534.
[2] Boothroyd A. T., Babkevich P., Prabhakaran D., and Freeman P. G., 2011 Nature (London) **375**, 561.
[3] Freeman P. G., Enderle M., Hayden S. M., Frost C. D., Yao D. X., Carlson E. W., Prabhakaran D. and Boothroyd A. T., 2009 Phys. Rev. B **80**, 144523.
[4] Yao D. X., Carlson E. W., Campbell D. K., 2006 Phys. Rev. Lett. **97**, 017003.
[5] Boothroyd A T, Freeman P G, Prabhakaran D, Hiess A., Enderle M., Kulda J., and Altorfer F., 2003 Phys. Rev. Lett. **91**, 257201.
[6] Freeman P.G., Prabhakaran D., Nakajima K., Stunault A., Enderle M., Niedermayer C., Frost C. D., Yamada K., and Boothroyd A.T., 2011 Phys. Rev. B **83**, 094414.
[7] Prabhakaran D., Isla P., and Boothroyd A. T., 2002 J. Cryst. Growth **237**, 815.
[8] Freeman P. G. , Boothroyd A. T. , Prabhakaran D. and Lorenzana J., 2006 Phys. Rev. B **73**, 014434.
[9] A flipping ratio of f = 15 was used to correct the data, realistically reduced from the value of 19 observed in diffraction. Correcting the data using $f = 10 - 19$ results in insignificant differences to the data of Fig. 2.
[10] Freeman P. G., Boothroyd A. T., Ewings R. A., Hücker M., Prabhakaran D. and Tranquda J. M., 2008 J. Phys.: Condens. Matter **20**, 104229.
[11] Boothroyd A. T., Prabhakaran D., Freeman P. G., Lister S. J. S., Enderle M., Hiess A., and Kulda J., 2003 Phys. Rev. B **67**, 100407(R).
[12] Boothroyd A. T., Freeman P. G. , Prabhakaran D., Enderle M., and Kulda J., 2004 Physica B **345**, 1-5.
[13] Woo H., Boothroyd A. T., Nakajima K., Perring T. G., Frost C. D., Freeman P. G., Prabhakaran D., Yamada K. and Tranquada J. M., 2005 Phys. Rev. B **72**, 064437.

Journal of Physics: Conference Series **340** (2012) 012061 doi:10.1088/1742-6596/340/1/012061

Magnetic impurity effect on charge and magnetic order in doped $La_{1.5}Ca_{0.5}CoO_4$

K Horigane[1], H Hiraka[2], K Tomiyasu[3], K Ohoyama[2], D Louca[4] and K Yamada[1]
[1]WPI-AIMR, Tohoku University, Sendai 980-8577, Japan
[2]Institute for Materials Research, Tohoku University, Sendai 980-8577, Japan
[3]Department of Physics, Tohoku University, Sendai 980-8578, Japan
[4]Department of Physics, University of Virginia, Charlottesville 22904-4714, U.S.A.

E-mail adress: kh2dg@virginia.edu

Abstract. Neutron scattering experiments were performed on single crystals of magnetic impurity doped cobalt oxides $La_{1.5}Ca_{0.5}CoO_4$ to characterize the charge and spin orders. We newly found contrasting impurity effects. Two types of magnetic peaks are observed at q=(0.5,0,L) with L=half-integer and integer in $La_{1.5}Ca_{0.5}CoO_4$, while magnetic peak at L=half-integer (integer) was only observed in Mn (Fe)-substituted sample. Although Mn and Fe impurities degrade charge and magnetic order, Cr impurity stabilizes the ordering at x=0.5. Based on the crystal structural analysis of Cr doped sample, we found that the excess oxygen and change of octahedron around Co^{3+} were realized in Cr doped sample.

1. Introduction

Layered transition metal oxides have attracted much attention due to their wide variety of magnetic, electrical and structural properties. In some of the doped transition metal compounds, there is a real space ordering due to the charge carriers in a certain carrier concentration, resulting in an orbital ordering (OO) and sometimes a charge ordering (CO). In a neutron scattering study, Zaliznyak et al. investigated a checkerboard charge order of Co^{2+}/Co^{3+} and Co^{2+} magnetic order in the half-doped cobaltate $La_{1.5}Sr_{0.5}CoO_4$ [1]. From the analysis of the magnetic and charge order scattering, they concluded a checkerboard arrangement of Co^{2+} and non-magnetic Co^{3+} ions in the CoO_2 plane. However, Y. Moritomo et al. reported the drastic change of effective moment μ_{eff} in $La_{2-x}Sr_xCoO_4$($0.4<x<1.0$), suggesting a spin-state transition of Co^{3+} ions from the high-spin to intermediate spin state [2]. Thus, spin state of Co^{3+} is still controversial.

Recently, Fujita et al. found that Fe substitution significantly enhanced the charge and spin density wave in $La_{2-x}Sr_xCuO_4$(x~1/8) [3]. This result indicates that dynamical charge stripes can be stabilized into static ones. Although spin state of Co^{3+} ions was determined as low-spin state by previous neutron scattering measurements, we considered that Co^{3+} ions were not non-magnetic but paramagnetic state because of magnetic susceptibility measurements. Based on this report, we expected that paramagnetic Co^{3+} spin can be stabilized by tiny magnetic impurities. In this paper, we report the magnetic impurity effect of spin and charge ordering in $La_{1.5}Ca_{0.5}CoO_4$.

2. Experimental procedure

Polycrystalline samples were prepared by solid-state reaction. First, stoichiometric mixture of La_2O_3, $CaCO_3$, CoO, Cr_2O_3, Mn_3O_4 and Fe_2O_3 powders were ground and calcined at 1200°C for 24h in N_2

atmosphere. After repeating this grinding and sintering process two times, the powder sample was pressed into rod with a size of 8mmφ×100mm and sintered at 1200°C for 24h. Single crystals of nominal composition of $La_{1.5}Ca_{0.5}Co_{0.97}M_{0.03}O_4$ (M=Cr, Mn, Fe) were grown by floating-zone method at feeding speed of 3mm/h in air. The ceramic compounds were confirmed to be a single phase using x-ray powder diffraction. The cylindrical crystals studied here for neutron scattering were about 6mm in diameter and 25mm in length. The crystal has a checkerboard charge order on Co^{2+}/Co^{3+}. In this paper we use a unit cell with dimensions $a_{co}\times a_{co}\times c$ (eg. a_{co}~5.436Å, c~12.373Å for Cr doped sample).

Neutron scattering experiments were carried out on triple-axis spectrometer AKANE and powder diffractometer HERMES in Japan Atomic Energy Agency, Tokai. A combination of Ge(3,1,1) monochromator and PG(0,0,2) analyzer was employed for AKANE and no filters were attached owing to forbidden Ge (6,2,2) reflection. The incident energy of AKANE was fixed to be E_i =19.4meV, typical with a sequence of sequence of collimations of guide(20')-open-S-60'-blank(180'). The sample was mounted in a ^{4}He-cycle–type refrigerator, allowing wave vector transfer in the (H, 0, L) scattering plane. The wavelength of the neutron beam of HERMES was 1.8484Å, monochromatized by Ge(331). The fine powder samples were sealed in a vanadium cylinder (Φ10mm) with helium gas, and mounted on the cold head of a ^{4}He-cycle–type refrigerator.

3. Result and Discussion

Figure 1 shows (h, 0, 1) scans in nominal composition of $La_{1.5}Ca_{0.5}Co_{0.97}M_{0.03}O_4$ (M=Cr, Fe, Mn). Peaks due to checkerboard charge order of Co^{2+}/Co^{3+} were observed in all samples. Indeed this order persists up to room temperature (not shown). The normalized intensity of the Charge ordered peak was much stronger in Cr doped sample, however clear enhancements were not detected in the case of Fe and Mn doping. This result suggests that Cr doping effectively stabilized the checkerboard charge order. Figure 2 shows Magnetic scattering at (0.5, 0, L) in (a) pristine $La_{1.5}Ca_{0.5}CoO_4$ [4] and (b) Cr (c) Fe (d) Mn doped $La_{1.5}Ca_{0.5}CoO_4$. In pristine sample, two types of magnetic peaks are observed at q=(0.5,0,L) with L=half-integer and integer in $La_{1.5}Ca_{0.5}CoO_4$. On the other hand, magnetic impurity doping process drastically changes the magnetic diffraction pattern. The magnetic peaks with L=half-integer appear in Cr and Mn doping, whereas it switched to the reflection with L=integer in Fe doping. In order to determine correlation length of in plane and out of plane, the profiles of the charge order peaks are fitted to a Lorentzian form. We summarize the Neel order temperature and the correlation length of spin and charge ordering in table I. Although Mn and Fe impurities degrade charge and magnetic order, Cr impurity stabilizes the ordering at x=0.5. For example, correlation lengths of charge and magnetic order along c-axis of Cr doped system (ξ_{mag}=139(7)Å, ξ_{co}=129(5)Å) are much larger than those of LCCO system (ξ_{mag}=22(1)Å, ξ_{co}=59(2)Å). This result suggests that that Co^{3+} site is selectively replaced by Cr^{3+}.

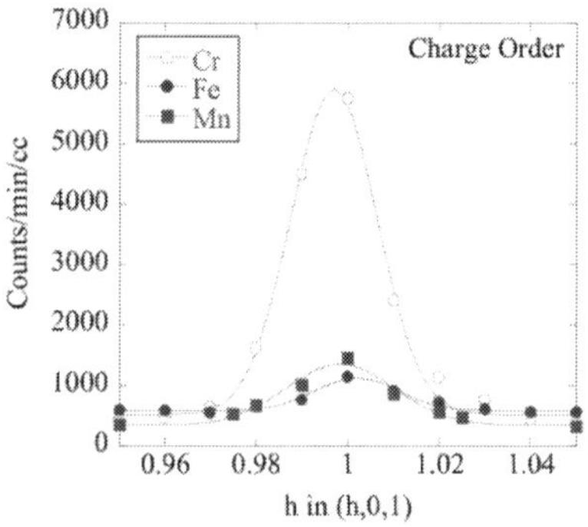

Fig.1 Checkerboard charge order peak in impurity doped $La_{1.5}Ca_{0.5}Co_{0.97}M_{0.03}O_4$ with M=Cr, Fe and Mn (nominal composition).

Table I Charge and magnetic correlation lengths in pristine and magnetic impurity doped sample.

	T_N (K)	$\xi_{co(ab)}$ (Å)	$\xi_{co(c)}$ (Å)	L = half-integer $\xi_{so(ab)}$ (Å)	L = half-integer $\xi_{so(c)}$ (Å)	L = integer $\xi_{so(ab)}$ (Å)	L = integer $\xi_{so(c)}$ (Å)
LCCO	52	115(12)	59(2)	195(4)	22(1)	98(8)	22(1)
Cr doping	52	124(10)	129(5)	219(6)	139(7)		
Fe doping	45	76(4)	19(1)			56(2)	8(1)
Mn doping	50	74(4)	24(1)	86(2)	35(2)		

To understand this mechanism from a view of crystal structure, we carried out neutron powder diffraction measurements on Cr doped sample. The diffraction data was analyzed by using the Rietveld analysis program RIETAN2000 [5]. The crystal structure parameters of Cr doped sample are listed in Table II. The diffraction pattern of Cr doped sample indicates an orthorhombic structure with the space group *A2mm* at 10K. It should be notice that excess oxygen is incorporated at an interstitial site between LaO layers. By Cr doping, *c*-axis changes from 12.313(6)Å to 12.373(6)Å. This result

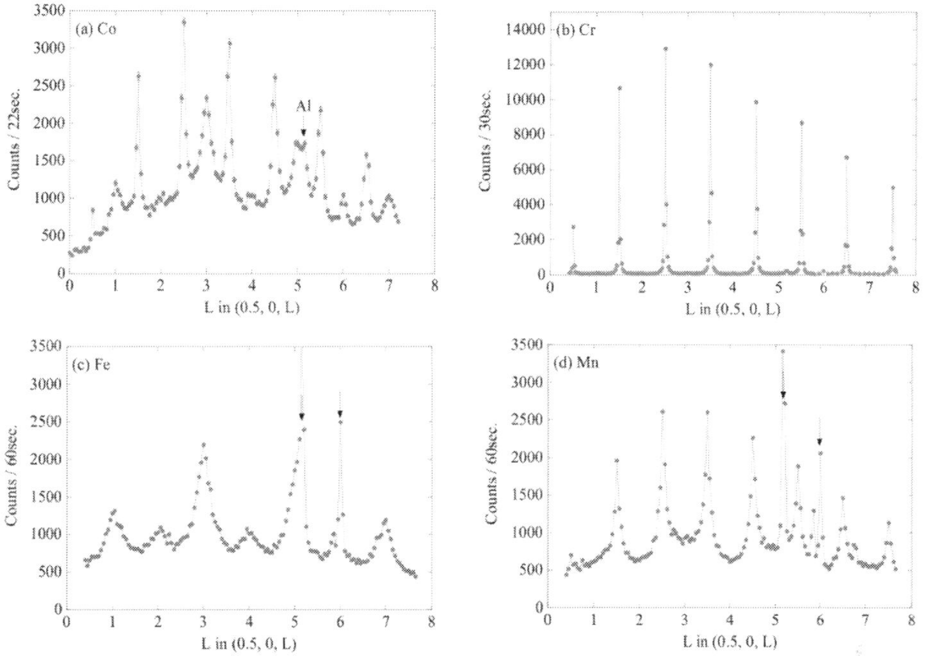

Fig.2 Magnetic scattering at (0.5, 0, L) in (a) pristine $La_{1.5}Ca_{0.5}CoO_4$ [4] and (b)Cr (c)Fe (d)Mn doped $La_{1.5}Ca_{0.5}CoO_4$. The arrows indicate aluminium scattering.

supports the incorporation of oxygen into interstitial site. The *c*-axis of Mn and Fe doping sample that is determined by triple-axis spectrometer is 12.366 Å and 12.336 Å, respectively. The enhancement of *c*-axis can be explained by incorporation of excess oxygen. From above result, excess oxygen is important factor to understand the drastic change of magnetic structure. As shown in table II, two Co sites are separated to two independent sites and are arranging a checkerboard charge ordering pattern. The average Co(1)-O bond length of in-plane and out of plane are 1.91(2)Å and 2.035(6)Å, respectively. On the other hand, those of Co(2)-O bond length are 1.94(2)Å and 2.149(7)Å, respectively. These result indicate that a volume of $Co(1)O_6$ octahedron is smaller than that of $Co(2)O_6$ octahedron, suggesting the Co(1) should be Co^{3+} and Co(2) be Co^{2+}. In the

Table II The crystal structure parameters of $La_{1.5}Ca_{0.5}Co_{0.97}Cr_{0.03}O_4$ (nominal composition) at 10K refined by Rietveld analysis. Cr are selectively occupied Co(1) site in this model.

	g	*x*	*y*	*z*	*B*
Co(1)	1	0	0	0	0.85(14)
Co(2)	1	0.5	0.5	0	0.85(14)
O(1)	1	0.245(5)	0.247(4)	0	0.78(6)
O(2)	1	0.751(5)	0.747(4)	0	0.78(6)
O(3)	1	-0.033(4)	0	0.1638(5)	0.78(6)
O(4)	1	0.488(4)	0.5	0.1736(6)	0.78(6)
O(5)	0.060(4)	0.25	0.25	0.25	0.78(6)
O(6)	0.060(4)	0.75	0.75	0.75	0.78(6)
La,Ca(1)	1	0.004(4)	0	0.3634(7)	0.38(6)
La,Ca(2)	1	0.505(4)	0.5	0.3602(7)	0.38(6)

A2mm, R_{wp}=6.56, *S*= 1.78
a=5.4357(26) Å, *b*=5.4304(25) Å, *c*=12.3731(59) Å

previous crystal structure study of $La_{1.5}Ca_{0.5}CoO_4$ [6], we also observed similar tendency of bond length change and determined Co(1) site as Co^{3+} by using resonant x-ray diffraction. Therefore, we concluded that Co(1) is Co^{3+} in Cr doped sample. In order to confirm that Cr is selectively occupied at Co(1) site, we compared the structural parameters. The structural parameters correspond to Co^{2+} do not change within error, while O(3)-Co(1)-O(3) angle clearly change from 174.8(4)° to 169.9(4)°. This result supports that Cr atoms selectively occupies the Co(1) site first.

In order to understand the spin fluctuation Cr doped system, we performed on the inelastic scattering in this system. Figure 3 shows inelastic neutron-scattering spectra for various energies at T=15K in Cr doping sample. At 4meV magnetic excitation appears at commensurate position. It stars to split around 8meV and spin wave is seen up to 16meV. Spin dispersion of this system is almost as same as that of $La_{1.5}Ca_{0.5}CoO_4$ [7]. Figure 4 shows the energy-Q slice showing around magnetic zone center. It reveals an energy gap of ~5meV in the magnetic spectrum. In $La_{1.5}Sr_{0.5}CoO_4$ this gap feature also appear at 3meV and the magnetic dynamics is completely uncorrelated along the c-axis above 4meV [8]. This implies that the spectrum can be considered as three dimensional below this gap. Comparing the gap energy of these system, we concluded that the charge and spin orderings can be stabilized by tiny Cr doping.

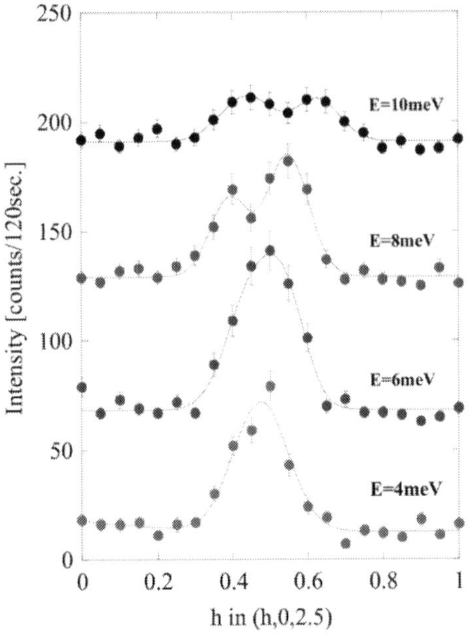

Fig.3 Inelastic neutron-scattering spectra for various energies at T=15K in $La_{1.5}Ca_{0.5}Co_{0.97}Cr_{0.03}O_4$ (nominal composition).

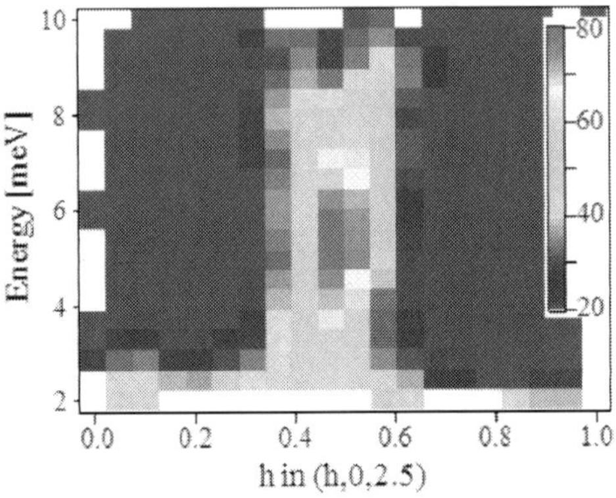

Fig.4 Energy-Q slice showing around magnetic zone center in $La_{1.5}Ca_{0.5}Co_{0.97}Cr_{0.03}O_4$ (nominal composition).

First, we discuss about drastic enhancement of the out of plane correlation length. In $La_{1.5}Ca_{0.5}CoO_4$, two types of magnetic structure that correspond to the difference of spin stacking along c direction are realized. On the other hand, two magnetic domains change to single domain by magnetic impurity doping. It's expected to enhance the correlation length of the out of plane because of suppression of stacking fault along c direction. However, checkerboard charge ordering is degraded in Fe and Mn cases. As a result, spin correlation is also degraded in these cases. As for the Cr doped case, charge ordering is not degraded because of selective replacement of Cr atoms. Therefore, drastic enhancement of correlation length along c-axis is realized.

Next, we consider the drastic change of magnetic structure by tiny magnetic impurity doping. One possible scenario for explaining the impurity effect is hole doping effect by excess oxygen. In this case, it expects that all impurity doped samples show magnetic reflections at L=integer position because magnetic peaks at L=integer were observed above x=0.5 in Ca doped sample. However, magnetic reflections at L=half-integer were observed in Cr and Mn doped sample. It means hole doping is not major effect for changing magnetic structure. Another parameter to explain this effect is c-axis. Ca doping changes not only hole concentration but also lattice constant. If we focus on the c-axis, magnetic structure changes from L=integer to L=half-integer by increasing c-axis. In Mn and Cr doping case, it should be observed at L=half-integer because c-axis increases by doping. However, enhancement of c-axis was only about 0.5%. Therefore, it is more appropriate to consider another parameter casing the drastic change of magnetic structure in this system. We propose that Co^{3+} spin is one plausible parameter to understand this impurity effect because Cr doping introduces not only excess oxygen but also distortion of octahedron around Co^{3+}. Indeed, magnetic peaks and excitation can be explained by Co^{2+} spins. However, the strongest exchange interaction J occurs through superexchange along straight bonds Co^{2+}-O-Co^{3+}-O-Co^{2+}, suggesting degrees of Co^{3+} is also important for magnetic structure. To clarify the clues, magnetic impurity effect of local structure and inelastic neutron experiments at high energy region are now in progress by our group.

4. References

[1] Zaliznyak I A, Hill J P, Tranquada J M, Erwin R and Moritomo Y 2000 *Phys. Rev. Lett.* **85** 4353
[2] Moritomo Y, Higashi K, Matsuda K and Nakamura A 1997 *Phys. Rev. B* **55** 14725
[3] Fujita M, Enoki M, Iikubo S, Kudo K, Koyayshi N and Yamada K arXiv:0903.5391
[4] Horigane K, Hiraka H, Uchida T, Yamada K and J. Akimitsu 2007 *J. Phys. Soc. Jpn.* **76** 114715
[5] Izumi F and Ikeda T 2000 *Master. Sci. Forum* **321-324** 198
[6] Horigane K, Hiraka H, Kousaka Y, Murata T, Noda Y, Murakami Y and J. Akimitsu 2008 *J. Phys. Soc. Jpn.* **77** 044601
[7] Horigane K, Yamada K, Hiraka H and Akimitsu J 2007 *J. Magn. Magn. Mater.* **310** 774
[8] Helme L M, Boothroyd A T, Coldea R, Prabhakaran D, Frost C D, Keen D A, Regnault L P, Freeman P G, Enderle M and Kulda J 2009 *Phys. Rev. B* **80** 134414

Uniaxial-stress enhancement of spin-driven ferroelectric polarization in a multiferroic $CuFe_{1-x}Ga_xO_2$

S Mitsuda[1], K Yoshitomi[1], T Nakajima[1], C Kaneko[1], H Yamazaki[1], M Kosaka[2], N Aso[3], Y Uwatoko[4], Y Noda[5], M Matsuura[6], N Terada[7], S Wakimoto[8], M Takeda[8], K Kakurai[8]

[1] Department of Physics, Tokyo University of Science, Tokyo 162-8601, Japan
[2] Graduate School of Science and Engineering, Saitama University, Saitama 338-8570, Japan
[3] Faculty of Science, University of the Ryukyus, Nishihara, Okinawa 903-0213, Japan
[4] Institute for Solid State Physics, University of Tokyo, Chiba 277-8581, Japan
[5] Institute of Multidisciplinary Research for Advanced Materials, Tohoku University, Sendai 980-8577, Japan
[6] Institute for Materials Research, Tohoku University, Sendai 980-8577, Japan
[7] National Institute for Materials Science, Tsukuba, Ibaraki 305-0044, Japan
[8] Quantum Beam Science Directorate, Japan Atomic Energy Agency, Tokai, Ibaraki 319-1195, Japan

mitsuda@nsmsmac4.ph.kagu.tus.ac.jp

Abstract. Delafossite compound $CuFeO_2$ is a spin-driven magneto-electric multiferroic, in which magnetic field-induced or nonmagnetic impurity-induced proper helical magnetic ordering generates a spontaneous electric polarization through the spin-orbit-interaction mediated modulation of Fe 3d-O 2p hybridization. As is expected from spin-lattice tightly coupled character of this system, we found an enhancement of spin-driven ferroelectric polarization under $[1\bar{1}0]$ uniaxial stress as spin-mediated piezoelectric effect. Preparing almost mono-domain state for multiferroic $CuFe_{1-x}Ga_xO_2$ sample with x=0.035, we have performed polarized neutron diffraction and in-situ ferroelectric polarization measurements to investigate how the helical magnetic structure is modified and how much ferroelectric polarization increases under application of $[1\bar{1}0]$ uniaxial stress up to 100 MPa. The uniaxial-stress variation of magnetic structural parameters characterizing helical magnetic ordering suggests decreasing of the ferroelectric polarization according to the calculation on the basis of d-p hybridization mechanism. Therefore, it is suggested that the enhancement of ferroelectric polarization originates not in variation of the magnetic structure but in the enhancement of the d-p hybridization coupling constants.

1. Introduction

Spin-driven magnetoelectric (ME) multiferroics, in which magnetic inversion symmetry breaking triggers off ferroelectricity, have been intensively investigated in recent cross-correlated materials research, since giant ME effects were discovered in $TbMnO_3$ [1]. The magnetic oxide $CuFeO_2$ to be

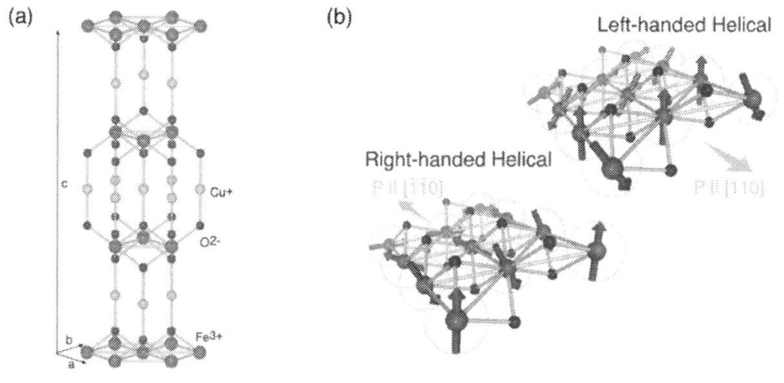

Figure 1. (a) Crystal structure of $CuFeO_2$ with the hexagonal basis. (b) The relationship between the spin helicity and the polarity of the electric polarization.

investigated here is a triangular lattice antiferromagnet with delafossite crystal structure, in which each Fe^{3+} triangular layer is well separated by one Cu^{1+} and two O^{2-} layers, and stacks along the hexagonal c axis, as is illustrated in figure 1(a). This frustrated magnet $CuFeO_2$ is a rare spin-driven magneto-electric multiferroic, in which magnetic field-induced [2] or nonmagnetic impurity-induced [3,4,5] proper helical magnetic ordering generates a spontaneous electric polarization *parallel* to the helical axis through the spin-orbit-interaction mediated modulation of Fe 3d-O 2p hybridization [6]. This d-p hybridization mechanism is distinct from the spin-current (inverse Dzyaloshinskii–Moriya) mechanism [7,8] accounting for the ferroelectricity generated by cycloidal magnetic orderings [9,10], although, similarly based on spin–orbit interactions. As was revealed in recent Xray diffraction measurement [11,12], $CuFeO_2$ is also a spin-lattice coupled system; to partially lift the degeneracy in the frustrated exchange interactions, the triangular lattice spontaneously distorts so that the triangular lattice elongates in the direction of the c-plane projection of magnetic propagation vector and contracts in the perpendicular direction. Therefore, as a cross-correlated phenomena, it can be expected that uniaxial stress applied perpendicular to the c-axis, which is conjugate to the anisotropic lattice distortion, modifies the helical magnetic structure and consequently controls the emerged ferroelectric polarization, namely, spin-mediated piezoelectric effect. In this study, to demonstrate uniaxial stress control of the ferroelectric polarization via modification of the magnetic structure, we investigated, by polarized neutron diffraction and in-situ ferroelectric polarization measurements, how the helical magnetic structure is modified and how much ferroelectric polarization changes under application of $[1\bar{1}0]$ uniaxial stress p up to 100 MPa, using a $CuFe_{1-x}Ga_xO_2$ (x=0.035) sample which exhibits the ferroelectric phase below T_C~8 K in zero magnetic field.

2. Preliminary Details

A single crystal of $CuFe_{1-x}Ga_xO_2$ (x=0.035) of nominal composition was prepared by floating zone method and was cut to dimensions of ~ $1.7 \times 6 \times 1.5$ mm^3 with the $[1\bar{1}0]$ direction parallel to the shortest dimension.

The magnetic structure of a ferroelectric phase of $CuFe_{1-x}Ga_xO_2$ (x=0.035) has been studied by experimentally and theoretically. Experimentally [13], complementary use of spherical neutron polarimetry and a four-circle neutron-diffraction measurement revealed that the proper-helical-type magnetic structure in the ferroelectric phase has a finite ellipticity of ~0.9, spin phase shift of ~73 ° and the magnetic propagation wave vector $\vec{Q}=(q,q,\frac{3}{2})$ with q~0.205. The helical axis of the magnetic structure is parallel to the c-plane projection of the magnetic propagation vector. Theoretically [14], on energy-considerations using the exchange parameters obtained by inelastic neutron measurements, a complex but similar distorted proper-helical-type spin configuration with a distribution of turn angles was proposed. In this study we employ the experimentally obtained magnetic structure [13] to characterize the spin order.

Because of the trigonal symmetry of the crystal structure, this system has three magnetic domains whose wave vectors of $(q,q,\frac{3}{2})$, $(q,-2q,\frac{3}{2})$ and $(-2q,q,\frac{3}{2})$ are crystallographically equivalent to each other. We refer to the three domains as (110)-, $(1\bar{2}0)$- and $(\bar{2}10)$-domains, along with our previous works [13]. In each of the domains, the ferroelectric polarization emerges along the helical axes, and as shown in figure 1(b), there is a one-to-one correspondence between the direction of the ferroelectric polarization and the spin helicity of the helical magnetic structure: left-handed (LH) or right-handed (RH) helical arrangements of the spins. Therefore, the multiferroic (ferroelectric and magnetic) domains are classified into six types with volume fraction of $V^{LH}_{(110)}$, $V^{RH}_{(110)}$, $V^{LH}_{(1\bar{2}0)}$, $V^{RH}_{(1\bar{2}0)}$, $V^{LH}_{(\bar{2}10)}$ and $V^{RH}_{(\bar{2}10)}$, as shown in figure 2(c). Taking the multiferroic domain structure into account, the macroscopic electric polarization P along [110] axis under applied uniaxial stress p in the $[1\bar{1}0]$ direction, which is experimentally obtained by time-integration of the polarization current flowing between two Ag-electrodes on [110] surfaces as shown in figure 2(c), is determined by sum of the three contribution from each multiferroic domain and can be described as

$$P = P_0^{(110)}(p)\,D_{(110)}(p)\,V_{(110)}(p) + \frac{1}{2}\left(P_0^{(1\bar{2}0)}(p)\,D_{(1\bar{2}0)}(p)\,V_{(1\bar{2}0)}(p) + P_0^{(\bar{2}10)}(p)\,D_{(\bar{2}10)}(p)\,V_{(\bar{2}10)}(p)\right) . \quad (1)$$

Here, the asymmetry of the volume fractions having RH- and LH- helical magnetic ordering in the (110)-domain, $D_{(110)}$, is defined by $(V^{LH}_{(110)} - V^{RH}_{(110)})/(V^{LH}_{(110)} + V^{RH}_{(110)})$, and total volume fraction of the (110) -domain, $V_{(110)}$, is defined by $(V^{LH}_{(110)} + V^{RH}_{(110)})$. $P_0^{(110)}$ is the value of P for mono-domain state, that is, $V_{(110)} : V_{(1\bar{2}0)} : V_{(\bar{2}10)} = 1:0:0$ and $D_{(110)} = 1$. The symbols for other trigonal domains are also defined similarly. The factor of 1/2 in front of the second term in equation (1) denotes that the [110] projection of the ferroelectric polarization is half the magnitude of the ferroelectric polarization emerged in the domain.

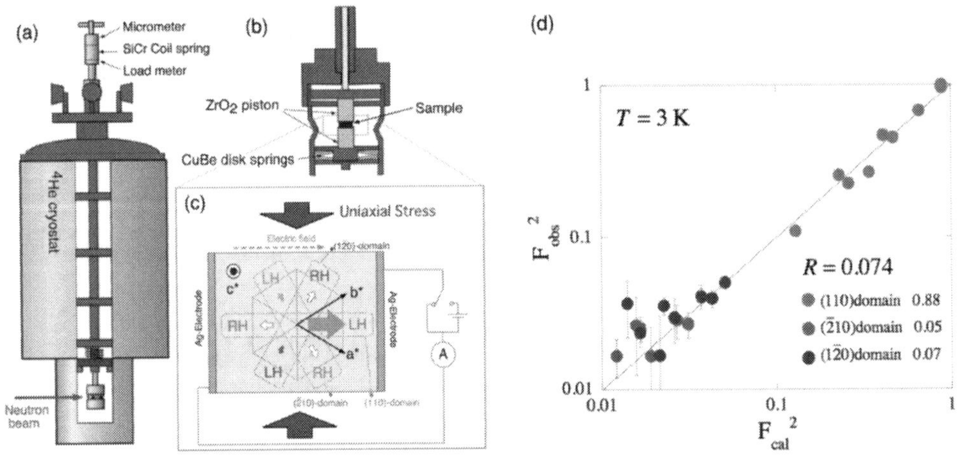

Figure 2. (a) Schematic drawing of the uniaxial stress device inserted in the cryostat. (b) Magnification of the sample space of the uniaxial stress device. CuBe disk springs supply uniaxial bias-stress of 5 MPa. (c) Relationships among the six possible directions of the ferroelectric polarization, applied electric field and uniaxial stress. Open and filled arrows denote the directions of the ferroelectric polarization induced by the RH- and the LH-helical magnetic domains, respectively. The sizes of the arrows qualitatively show the volume fractions of the six domains. (d) Comparison between the observed and the calculated magnetic structure factors. Red, green and blue circles denote data for the (110)-, $(1\bar{2}0)$- and $(\bar{2}10)$-domains, respectively.

All quantities appearing in equation (1) are generally uniaxial-stress-p-dependent and have different p-dependence for each of three trigonal domains. What we call spin-mediated piezoelectric effect here can be seen in p-dependence of the quantities of $P_0^{(110)}(p)$, $P_0^{(1\bar{2}0)}(p)$ and $P_0^{(\bar{2}10)}(p)$. However, the $[1\bar{1}0]$ uniaxial stress p enhances the anisotropic spontaneous lattice distortion for (110)-domain, whereas that may rather suppress the anisotropic spontaneous lattice distortion for $(1\bar{2}0)$- and $(\bar{2}10)$-domains. Thus, the p-dependence of $P_0^{(110)}(p)$ can be different from that of $P_0^{(1\bar{2}0)}(p)$ and $P_0^{(\bar{2}10)}(p)$, and therefore different spin-mediated piezoelectric effect must be mixed up in the macroscopic electric polarization P along [110] axis under the $[1\bar{1}0]$ uniaxial stress p. Moreover, applying the $[1\bar{1}0]$ uniaxial stress p causes the repopulation of three trigonal domains from $(1\bar{2}0)$- and $(\bar{2}10)$-domains to (110)-domain so as to give rise to distinct increasing of the macroscopic electric polarization P along [110] axis, as was quite recently demonstrated in our work [15]. Therefore, to properly study the spin-mediated piezoelectric effect seen in $P_0^{(110)}(p)$, almost mono-domain state, which satisfies the conditions of $V_{(110)} >> V_{(1\bar{2}0)} + V_{(\bar{2}10)}$ and $D_{(110)} \sim 1$, are indispensable. Fortunately, in our inelastic neutron scattering experiment [16] to obtain the spin wave dispersion in 4-sublattice (4SL) ground state of $CuFeO_2$, we have demonstrated that applying only 10 MPa of the $[1\bar{1}0]$ uniaxial stress *at room temperature* results in almost mono-domain state with $V_{(110)}$ ~0.97 in the 4SL phase at low temperature.

Taking these things into consideration, in the present study of the spin-mediated piezoelectric effect in the $CuFe_{1-x}Ga_xO_2$ (x=0.035) sample, we applied the $[1\bar{1}0]$ uniaxial stress using two different springs: CuBe disk springs and SiCr Coil spring. As shown in figure 2 (b), CuBe disk springs were set in the stress cell to keep, throughout the entire experiments, the uniaxial stress of 5 MPa applied *at room temperature* as a bias-stress to realize almost mono-domain state. On the other hand, as shown in figure 2 (a), the variable-uniaxial stress up to 100 MPa is additionally applied at low temperature of 20 K above T_c~8 K by a SiCr coil spring and a micrometer attached on top of the cryostat, and is monitored by a load meter. In present study, we have successively performed three different neutron experiments of (i) four-circle neutron diffraction measurement, (ii) spherical neutron polarimetry and (iii) polarized neutron diffraction with in-situ pyroelectric measurement. Note that the uniaxial bias-stress of 5 MPa has been applied throughout the entire neutron experiments, including off-bench pyroelectric measurement in between successive neutron experiments.

3. Results and Discussion

3.1 Four-circle neutron diffraction measurement

In order to evaluate the volume fractions of the three magnetic domains under the uniaxial bias-stress of 5 MPa, we performed four-circle neutron diffraction measurement using FONDER installed at JRR-3 in Japan Atomic Energy Agency (JAEA). The incident neutron beam with wavelength 1.24 Å was obtained by a Ge(311) monochromator. The sample with the stress cell was mounted on a closed-cycle He-gas refrigerator, and was cooled down to 3 K. We measured intensities of 23 magnetic Bragg reflections, of which 9, 8, and 6 magnetic reflections belong to the (110)-, $(1\bar{2}0)$- and $(\bar{2}10)$-domains, respectively. The effect of neutron absorption was corrected by the DABEX software. We have performed least-square fitting analysis, in which we used the established magnetic structural parameters of the elliptic helical magnetic structure presented in ref. 13 and refined the volume fractions, $V_{(110)}$, $V_{(1\bar{2}0)}$ and $V_{(\bar{2}10)}$. The comparison between the observed magnetic structure factor F_{obs} and the calculated values F_{cal} is shown in Fig. 2(d). The reliability factor was obtained to be 7.4%, and the volume fractions of the three magnetic domains were determined to be $V_{(110)}:V_{(1\bar{2}0)}:V_{(\bar{2}10)}$ = 0.88 : 0.07 : 0.05. This shows that the uniaxial bias-stress of 5 MPa applied onto the $[1\bar{1}0]$ surface successfully produces almost mono-domain state.

3.2 Spherical neutron polarimetry

Our tasks to study the spin-mediated piezoelectric effect are (i) obtaining the uniaxial-stress p variation of magnetic structural parameters characterizing elliptic helical magnetic ordering in the (110)-domain such as the direction of helical axis, the ellipticity, propagation wave number q and the spin phase shift δ (see figure 4 (b)), and (ii) obtaining the p-dependence of $P_0^{(110)}(p)$ which can be deduced through the relation of

$$P_0^{(110)}(p) \sim P(p)/(D_{(110)}(p)\,V_{(110)}(p)) \quad . \tag{2}$$

This relation would be applicable under almost mono-domain state in which the (100)-domain dominates over the $(1\bar{2}0)$- and $(\bar{2}10)$-domains. To determine the magnetic structural parameters of the direction of helical axis and the ellipticity, spherical neutron polarimetry was carried out at the triple-axis spectrometer TAS-1 with the CRYOPAD option installed at JRR-3 in JAEA. The sample was mounted in the uniaxial stress device inserted in the pumped ^{4}He cryostat, as shown in figure 2 (a)~(c). For p=5(bias) and 80 MPa, we measured polarization matrix terms at seven different magnetic Bragg reflections belonging to the (110)-domain at T = 2 K. As a result, it was turned out that the magnetic structural parameters are consistent with our previous results [13] and dose not show any clear uniaxial stress dependences within the experimental accuracy.

3.3 Polarized neutron diffraction with in-situ pyroelectric measurement

In order to obtain the uniaxial stress p dependence of the asymmetry $D_{(110)}(p)$, the volume fraction $V_{(110)}(p)$, the propagation wave number $q(p)$, the spin phase shift $\delta(p)$ and the macroscopic ferroelectric polarization $P(p)$ along the [110] direction, polarized neutron-diffraction and in-situ pyroelectric measurements were carried out at the triple-axis neutron spectrometer PONTA installed at JRR-3 in the JAEA. The sample was mounted in the uniaxial stress device inserted in the pumped ^{4}He cryostat, as shown in figure 2 (a)~(c).

In the in-situ pyroelectric measurements, as shown in the figure 3 (a), the temperature variation of the ferroelectric polarization P along the [110] direction, which were obtained by time-integration of the observed pyroelectric current, shows pronounced uniaxial stress p dependences in ferroelectric phase transition temperature T_C as well as in the value of P at low temperature. As is clearly seen in the inset of figure 3 (a), uniaxial-stress enhancement of the ferroelectric polarization P at T=2 K was observed. It should be noted that this enhancement of P was reversible in the p-range of 5 MPa (bias) ~ 100 MPa and also reproducible over off-bench pyroelectric measurements in between successive neutron experiments.

In the present polarized neutron-diffraction experiment, we measured only two magnetic Bragg reflections of $(q,q,\frac{3}{2})$ and $(\frac{1}{2}-q,\frac{1}{2}-q,\frac{3}{2})$. As shown in Fig.3 (b), these reflections can be assigned as $\vec{\tau}+\vec{Q}$ and $\vec{\tau}-\vec{Q}$ reflections, where $\vec{\tau}$ is a reciprocal lattice vector and $\vec{Q}=(q,q,\frac{3}{2})$ is the magnetic propagation vector. The scattering cross-section for these reflections is given by

$$\left(\frac{d\sigma}{d\Omega}\right)_{\vec{\kappa}=\vec{\tau}\pm\vec{Q}} \propto S(\vec{\kappa})\{(\mu_{x\perp}^2+\mu_{z\perp}^2)[V_{(110)}^{RH}+V_{(110)}^{LH}] \mp 2\mu_{x\perp}\mu_{z\perp}(\vec{p}_N \bullet \hat{\kappa})[V_{(110)}^{RH}-V_{(110)}^{LH}]\}\,. \tag{3}$$

Here $S(\vec{\kappa})$ is a factor dependent on the magnetic form factor $f(\vec{\kappa})$ and the spin phase shift δ, and is expressed by

$$S(\vec{\kappa}) \propto f(\vec{\kappa})^2\left|1 \pm e^{i\delta}\right|^2, \tag{4}$$

where + and − signs correspond to $(q,q,\frac{3}{2})$ and $(\frac{1}{2}-q,\frac{1}{2}-q,\frac{3}{2})$ reflections, respectively. $\mu_{x\perp}$ and $\mu_{z\perp}$ are the length of the spin components projected onto the plane normal to the scattering vector. $\vec{p}_N$ and $\hat{\kappa}$ are unit vectors of neutron polarization vector and the scattering vector, respectively. From equation (3), $D_{(110)}$ is approximately given by $(\frac{I_{ON}-I_{OFF}}{I_{ON}+I_{OFF}})/p_0$ for the two magnetic Bragg reflections [13], where p_0~0.975 is the instrumental beam polarization of the incident neutrons. I_{ON} and I_{OFF} are the intensities of a magnetic Bragg reflection measured when the spin flipper is ON ($\vec{p}_N \bullet \hat{\kappa} = 1$) and OFF ($\vec{p}_N \bullet \hat{\kappa} = -1$), respectively. As is seen in large difference in between flipper On- and flipper Off- intensities shown in figure 3 (c), the deduced asymmetry $D_{(110)}(p)$ was close to 1, suggesting that the poling-electric-field E_P=250 KV/m is strong enough to achieve a almost spin-helicity-mono-domain state. The asymmetry $D_{(110)}(p)$ is almost the uniaxial stress p-independent, as is seen in figure 3 (d).

As shown in figure 3 (c), with increasing the stress, the intensity at $(q,q,\frac{3}{2})$ Bragg reflection increases, whereas the intensity at $(\frac{1}{2}-q,\frac{1}{2}-q,\frac{3}{2})$ Bragg reflection decreases. Taking accounts of equation (4), these clearly indicate that the spin phase shift δ becomes small as the stress p increasing. In deducing the p-variation of volume fraction $V_{(110)}(p)$ and spin phase shift $\delta(p)$, we analyzed integrated intensity-data set, paying attention to that integrated intensity-data set at p=5 MPa(bias) contain the information of $\delta(p=5\text{MPa})$~73° [13] and $V_{(110)}(p=5\text{MPa})$~0.88 (see section 3.1). While the volume fraction $V_{(110)}(p)$ is almost the uniaxial stress p-independent as shown in figure 3 (d), the spin phase shift $\delta(p)$ shows pronounced p-variation as shown in figure 4 (a).

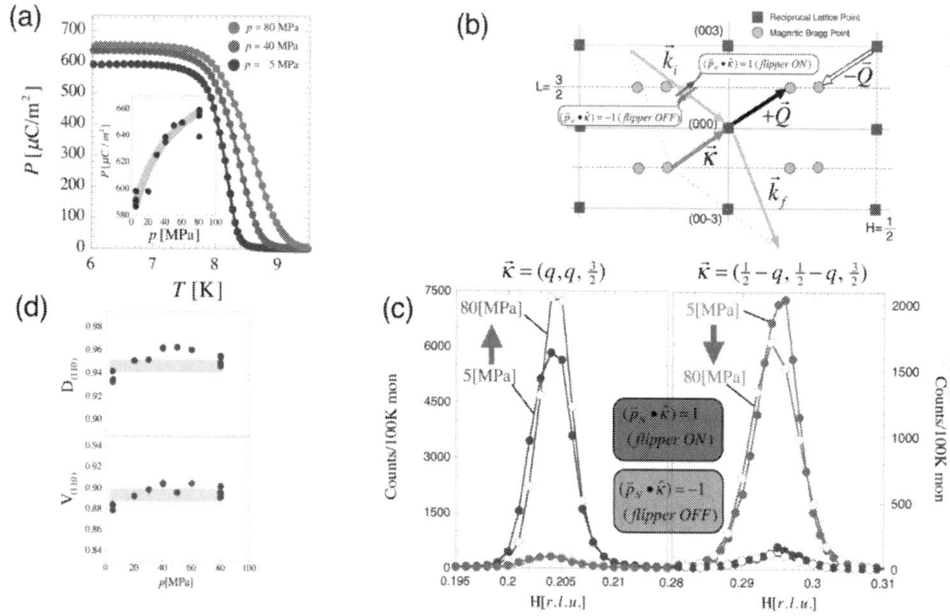

Figure 3. (a) Temperature variations of P measured on heating after cooling process with poling electric field E_P=250 KV/m at p =5, 40 and 80 MPa. The inset shows uniaxial stress dependence of P at T=2.0 K. (b) The location of the magnetic reflections surveyed in the present polarized neutron measurement using PONTA spectrometer in the (HHL) scattering zone. (c) The diffraction profiles of (H,H,$\frac{3}{2}$) reciprocal lattice scans for the $(q,q,\frac{3}{2})$ and $(\frac{1}{2}-q,\frac{1}{2}-q,\frac{3}{2})$ magnetic Bragg reflections at T=2K. (d) Uniaxial stress p dependence of $D_{(110)}$ and $V_{(110)}$. The solid lines are guides to the eyes.

3.4 Discussion

To summarize the results up to here, as for the task (i) obtaining the uniaxial-stress p variation of magnetic structural parameters characterizing elliptic helical magnetic ordering in the (110)-domain, it was turned out that only the spin phase shift δ among magnetic structural parameters shows pronounced p-variation as shown in figure 4 (a). And, as for the task (ii) obtaining the p-dependence of $P_0^{(110)}$ by equation (2), as is clearly seen in figure 4 (c-1), uniaxial-stress enhancement of the ferroelectric polarization $P_0^{(110)}$ at T=2 K was found as the spin-mediated piezoelectric effect.

According to the calculation on the basis of d-p hybridization mechanism proposed by Arima [6], the ferroelectric polarization along [110] direction $P_0^{(110)}$ has magnetic structural parameters dependences as is described by

$$P_0^{(110)} = C\, Sin(4\pi q)\{Sin(2(\theta_1 - \theta_2 - \delta)) + Sin(2(\theta_1 - \theta_2 + \delta))\} \quad , \qquad (5)$$

where finite ellipticity μ_x / μ_z~0.9 [13] was neglected and the original expression(the equation 6 in ref. 6) is extending to the case that Fe_1 site and Fe_2 site are not equivalent, as shown in figure 4 (b). Putting the almost p-independent propagation wave number $q(p)$ shown in figure 4 (a), the value of $\theta_1 - \theta_2$ ~100° [6] and p-dependent phase shift $\delta(p)$, we obtained the relative variation R of calculated $P_0^{(110)}$ to that at p =5MPa, as shown in figure4 (c-1). Thus, the uniaxial-stress variation of magnetic structural parameters characterizing helical magnetic ordering suggests decreasing of the ferroelectric polarization according to the calculation on the basis of d-p hybridization mechanism. Therefore, it is suggested that the enhancement of ferroelectric polarization seen in the figure 4 (c-1) originates not in variation of the magnetic structure but in the enhancement of the microscopic d-p hybridization coupling constants [6] which are contained in the coefficient C in equation (5).

As already mentioned, ferroelectric (magnetic) phase transition temperature T_c significantly increases under application of uniaxial stress. This is the evidence that uniaxial stress slightly modifies the competing exchange interactions in this frustrated magnetic system, which can be also seen in pronounced uniaxial stress variation of the spin phase shift $\delta(p)$. Therefore, the enhancement of the microscopic d-p hybridization coupling constants contained in the coefficient C would be similarly expected.

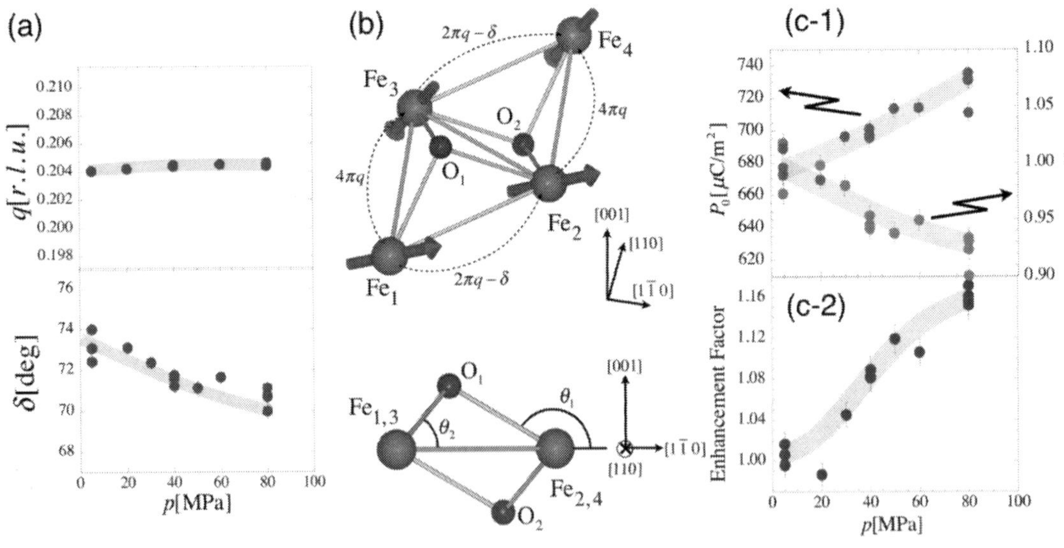

Figure 4. (a) Uniaxial stress p dependence of the magnetic propagation wave number q and the spin phase shift δ. (b) A Fe_4O_2 cluster used for the calculating the local electric dipole moment for the proper helical magnetic structure with finite phase shift δ. As the spin phase difference is depicted between Fe sites, Fe_1 site and Fe_2 site are not equivalent, and there is the spin phase shift δ in between. (c-1) Uniaxial stress p dependence of P_0 and the relative variation R of calculated ferroelectric polarization based on d-p hybridization model to that at p =5MPa. (c-2) Uniaxial stress p dependence of enhancement factor defined by $P_0(p)/P_0(p = 5\text{MPa})/R$. The solid lines are guides to the eyes.

Finally, let us mention the smallness of the uniaxial stress of ~100 MPa providing the spin-mediated piezoelectric effect. This is in contrast to the 100 time larger isotropic high pressure ~10 GPa which is necessary to suppress the long-range ordering accompanied by spontaneous lattice distortion [17]. Key point would be that $[1\bar{1}0]$-uniaxial stress is anisotropic and conjugate to the anisotropic spontaneous lattice distortion. To conclude, using spin-driven magneto-electric multiferroic $CuFeO_2$, we have demonstrated uniaxial stress control of the ferroelectric polarization via the modification of the magnetic structure as spin-mediated piezoelectric effect, in which the microscopic d-p hybridization coupling constants are suggested to be enhanced.

Acknowledgments

This work was supported by a Grant-in-Aid for Scientific Research (C) (Grant No. 23540424), from JSPS, Japan. The neutron diffraction measurements at JRR-3 were carried out as part of proposals 10783 and 10573B. Spherical neutron polarimetry measurement at TAS-1 was performed under the Common-Use Facility Program of JAEA.

References

[1] Kimura T, Goto T, Shintani H, Ishizaka K, Arima T and Tokura Y 2003 *Nature (London)* **426** 55
[2] Kimura T, Lashley J C, Ramirez A P 2006 *Phys. Rev.* B **73** 220401 R
[3] Kanetsuki S, Mitsuda S, Nakajima T, Anazawa D, Katori H A and Prokes K 2007 *J. Phys. Condens. Matter* **19** 145244
[4] Seki S, Yamasaki Y, Shiomi Y, Iguchi S, Onose Y and Tokura Y 2007 *Phys. Rev.* B **75** 100403 R
[5] Terada N, Nakajima T, Mitsuda S, Kitazawa H, Kaneko K and Metoki N 2008 *Phys. Rev.* B **78** 014101
[6] Arima T 2007 *J. Phys. Soc. Jpn.* **73** 094434
[7] Katsura H, Nagaosa N and Balatsky A V 2005 *Phys. Rev. Lett.* **95** 057205
[8] Sergienko I A and Dagotto E 2006 *Phys. Rev.* B **83** 020403(R)
[9] Yamasaki Y, Sagayama H, Goto T, Matsuura M, Hirota H, Arima T and Tokura Y 2005 *Phys. Rev. Lett.* **98** 147204
[10] Cabrera I, Kenzelmann M, Lawes G, Chen Y, Chen W C, Erwin R, Gentile T R, Leao J B, Lynn J W, Rogado N, Cava R J and Broholm C 2009 *Phys. Rev. Lett.* **103** 087201
[11] Terada N, Narumi Y, Sawai Y, Katsumata K, Staub U, Tanaka Y, Kikkawa A, Fukui T, Kindo K, Yamamoto T, Kanmuri R, Hagiwara M, Toyokawa H, Ishikawa T and Kitamura H 2007 *Phys. Rev.* B **75** 224411
[12] Ye F, Ren Y, Huang Q, Fernandez-Baca J A, Dai P, Lynn J W and Kimura T 2007 *Phys. Rev.* B **73** 220404 R
[13] Nakajima T, Mitsuda S, Takahashi K, Yamano Y, Masuda K , Yamazaki H, Prokes K, Kiefer K, Gerischer S, Terada N, Kitazawa H, Matsuda M, Kakurai K, Kimura H, Noda Y, Soda M, Matsuura M and Hirota K 2009 *Phys. Rev.* B **79**, 214423 and refferences therein
[14] Haraldsen J T, Ye F, Fishman R S, Fernandez-Baca J A, Yamaguchi Y, Kimura K and Kimura T 2010 *Phys. Rev.* B **82**, 020404 R
[15] Nakajima T, Mitsuda S, Nakamura T, Ishii H, Haku T, Honma Y, Kosaka M, Aso M and Uwatoko Y 2011 *Phys. Rev.* B **83**, 220101 R
[16] Nakajima T, Mitsuda S, Haku T, Shibata K, Yoshitomi K, Noda Y, Aso N, Uwatoko Y and Terada N *J. Phys. Soc. Jpn.* **80** (2011) 014714
[17] Terada N, Osakabe T, and Kitazawa H 2011 *Phys. Rev.* B **83** 020403(R)

Giant magnetoelastic interaction in UFe_2

L. Paolasini[1], F. Formisano[2], R. Caciuffo[3], G. H. Lander[3], G. Lapertot[4].

[1] European Synchrotron Radiation Facility. BP 220, 38043 Grenoble cedex, France.
[2] Consiglio Nazionale delle Ricerche, Istituto per l'Officina dei Materiali, c/o ILL, BP 156, F-38042, Grenoble Cedex 9, France.
[3] European Commission, JRC-ITU, Postfach 2340, D-76125 Karlsruhe, Germany.
[4] CEA-Grenoble, INAC-SPSMS, 17 rue des Martyrs, F-38054 Grenoble Cedex 9, France.

E-mail: paolasini@esrf.eu

Abstract. Using the unique coverage of (Q, ω) space available on the BRISP spectrometer, we have investigated the low Q dynamical response of a single crystal of UFe_2. A strong inelastic signal was found with a dispersion corresponding to the transverse acoustic phonon branch, forbidden in the first Brillouin zone. A detailed analysis of the neutron intensities as a function of temperature reveals a possible magnetic contribution superposed to this signal, which could originate from the magnon-phonon interaction, explaining the anomalous magnetoelastic behavior in this compound.

1. Introduction

The cubic Laves phase Rare-Earth iron compounds $REFe_2$ (RE=Ho, Dy, Tb, Er) form a wide class of ferromagnetic materials with important technological applications in the domain of magnetostriction. In these systems the spontaneous transition to the magnetically ordered state produces a large crystal structure distortion, and the microscopic magnetic behavior is explained well by a model assuming localized trivalent RE ions that provide large single-ion anisotropy arising from the unfilled $4f$ shells, combined with the iron magnetism, which provides high Curie temperature by the strong Fe-Fe exchange between the $3d$ electrons. The RE-Fe interaction is mediated by $5d$ electrons of RE (RKKY interaction), which are spin-polarized by the local exchange interaction with the $4f$ electrons [1]. $CeFe_2$ and UFe_2 are recognized as itinerant system and so this simple localized picture is not applicable. In these systems, the $4f$ and the $5f$ electrons, respectively, take part directly in the metallic bonding by the hybridization with the Fe $3d$ electrons. In particular, in the itinerant magnet UFe_2 the magnetic anisotropy is small and due mainly to the Fe magnetic moments (μ_{Fe}= 0.6 μ_B), because the U $5f$ electrons are almost completely delocalized and the spin and orbital magnetic moment ($\mu_U^S \sim -\mu_U^L \approx 0.23\ \mu_B$) contributions cancel each other, producing a nearly zero total magnetic moment [2]. However, below the Curie temperature T_c=165K, a large rhombohedral distortion is found in UFe_2, with a magnetostriction coefficient similar to that in $TbFe_2$, indicating an unusual situation in which a large magnetoelastic distortion of the crystal lattice is present despite the very small magnetic anisotropy and moments [3]. A magnon-phonon interaction mechanism was suggested many years ago to explain its peculiar magneto-elastic anomalies in the Mössbauer free recoil energy [4]. As a result, the excitations in this material cannot be regarded as non-interacting pure spin (magnons) and pure elastic (phonons) waves, and the phonon spectrum is strongly affected by magneto-vibrational contributions. Inelastic-neutron scattering experiments performed with $REFe_2$ crystals have provided a clear picture of the magnetic exchange interactions in these compounds, by fitting the magnon dispersion curves with a linear spin-wave theory and by using the Holstein-Primakov transformations [5, 6] in order to extract the relevant exchange constant parameters. In the localised rare-earth Laves phases, e.g. $ErFe_2$, the magnetic dynamics consists of three isotropic magnetic modes, as shown in Fig.1-A: (a) a dispersionless crystal-field-like mode arising from the precession of RE magnetic moments in the molecular field of Fe; (b) a steeply dispersive mode corresponding to the Fe spin waves; (c) an acoustic magnon consisting of all moments moving in phase. Fig.1-A shows also the dispersion curves of the longitudinal (LA) and transverse (TA) acoustic phonon branches of $ErFe_2$ at room temperature.

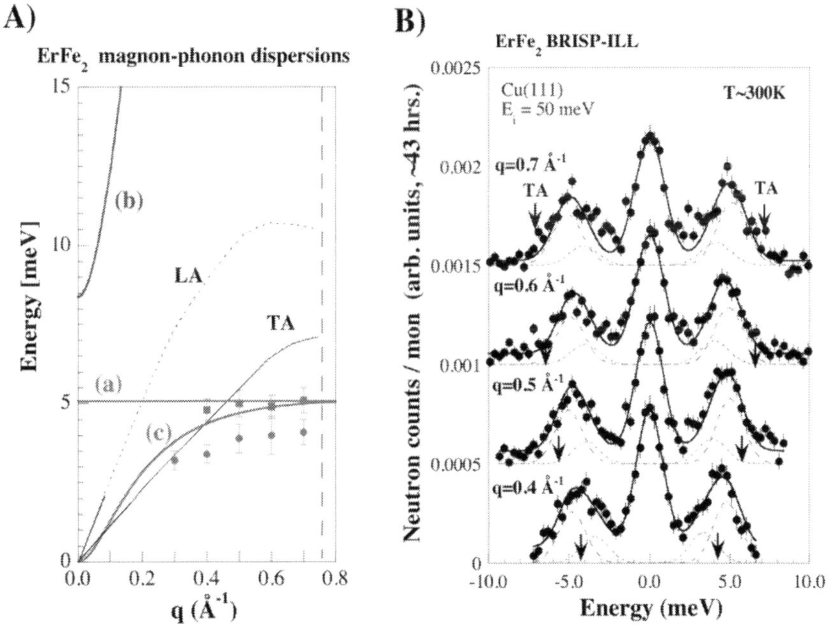

Figure 1. **A)**: Room temperature low energy spin wave modes in $ErFe_2$ and acoustic phonon dispersion curves: (a) dispersionless Er mode (green line); (b) steeply dispersive mode corresponding to the in-phase precession of Fe spins (blue line) (c) acoustic magnon consisting of in-phase precession of Fe and RE spins (red line) [5, 6]. Black lines represent the dispersion of transverse (TA) and longitudinal (LA) acoustic phonons determined at large Q wavevectors. Symbols are the fits of the experimental data. **B)**: Integrated intensities as a function of energy, taken at room temperature on $ErFe_2$ for different values of $||\mathbf{q}||$ wavevector. The arrows indicate the expected positions of TA phonons. Note that the spectrometer averages over all directions of $\mathbf{q}$.

Considerable experimental efforts were made to understand the nature of the magnetic excitations in UFe_2, and both the phonon and the magnon dispersion spectra were determined by inelastic neutron scattering on a large and well-characterized single crystal, the same used in this work [8, 9, 10, 11]. All neutron experiments to date on UFe_2 have succeeded in identifying only one mode corresponding to the sharply dispersive Fe spin waves (i.e. b-mode in Fig.1-A), with the stiffness constant substantially hardened at low temperature when compared to that of pure iron and isostructural $REFe_2$ compounds. The dispersionless mode (a-mode) is not expected to be present in UFe_2 as the $5f$ states are itinerant, but we would expect to see the overall acoustic magnon (c-mode). The linear spin wave theory applied to UFe_2 predict an acoustic magnetic mode with a dispersion close to the TA phonon branch [8], but this has not been positively identified. By working at small $\|\mathbf{q}\|$ within the first Brillouin zone we minimise the transverse phonon cross section (it should be zero, as seen in the work in Fig.1-B on $ErFe_2$) allowing a measurement of the magnetic contribution only. Theoretical considerations insist that this mode should be present, but cannot easily predict its intensity. Indeed, despite the fact that the U $5f$ electrons are almost delocalized and that the total magnetic moment of U is close to zero due to a nearly exact cancellation of the spin and orbital contributions, the spin dynamics in a Heisenberg itinerant ferromagnets are governed by the Fe and U spin moments, which are finite on both atoms [2].

2. Experimental results

Inelastic low $\|\mathbf{q}\|$ spectra of UFe_2 and $ErFe_2$ have been collected at the small-angle thermal neutrons time-of-flight (ToF) BRIllouin SPectrometer BRISP [12] at ILL. BRISP is a spectrometer devoted to the study of the low-q dynamics in disordered and magnetic systems and has three main features: i) a crystal monochomator, ii) a Fermi chopper producing the short (monochromatic) neutron pulses needed to combine the ToF technique with good energy resolution, and iii) a large ($\sim$ 2 m^2) bidimensional position sensitive detector composed of 160 3He filled tubes, with high spatial resolution (the dimension of the single detection pixel is 1.27 $\times$0.85 cm^2). This experiment was performed using monochromatic neutrons diffracted from the (111) reflection of a Cu crystal monochromator (incident energy E_0=49.9 meV), while the detector was moved to a distance of 4 m from the sample. Such an experimental setup allowed measuring the dynamical response of our sample at exchanged momentum q ranging from 0.2 to 1.2 $Å^{-1}$. The energy resolution was determined using a vanadium standard, is Gaussian in shape with a FWHM=1.9 meV independent of $\mathbf{q}$. The data are collected on the detector as neutron counts as a function of pixel (i.e. scattering angle) and ToF, and then transformed into ($\|\mathbf{q}\|$, E) by using programs specifically developed for BRISP and integrated in the ILL standard large array manipulation program (LAMP) [13].

UFe_2 crystallizes in the Laves Phase C15 structure, with a_0=7.057 Å at 300 K. Below the Curie temperature (T_c=165 K) it becomes ferrimagnetic, with the magnetic moments aligned along $\langle 111 \rangle$. The measurements have been performed on a large single-crystal specimen of UFe_2 of about 2.7 cm^3 in volume, with a cylindrical shape and oriented with the [110] axis very close to the cylinder axis. The crystal was oriented with the [111] axis parallel to the incident beam direction and the [110] axis along the vertical direction of the spectrometer, and cooled below the Curie temperature inside a helium evaporation cryostat. A test experiment was also performed at room temperature on a single crystal of $ErFe_2$ with the same size (2.1 cm^3) and shape as UFe_2 (a_0=7.276 Å, T_c=596 K, μ_{Fe}= 1.6 μ_B, μ_{Er}= 9 μ_B). The data are shown in Fig.1-B for different $\|\mathbf{q}\|$ values and where I($\|\mathbf{q}\|$, E) is integrated over all the scattering angle directions. The magnetic response is dominated by the non-dispersive excitation (a-mode) at 5 meV and superposed to a dispersive acoustic magnon mode (c-mode). According to Refs.[5, 6], the linear spin wave theory provides a generally good representation of the $ErFe_2$ data at 300 K and above. In this condition, the non-dispersive mode is related to a precession motion of $ErFe_2$ magnetic moments rather than a magnetic exciton, that is a collective mode arising from crystal field (CF) excitations coupled by exchange interactions. Random phase approximation theory includes CF effects, which dominate the low temperature regime in $ErFe_2$ [7]. Despite the difficulty to disentangle the magnon groups, being at the limit of the energy resolution, it is important to notice that any forbidden transverse acoustic (TA) phonon contribution is superposed to these inelastic magnetic excitations, as expected from the phonon polarization dependence of coherent nuclear inelastic cross section in the first Brilloiun zone (FBZ).

In the case of UFe_2, a strong acoustic mode was observed both at neutron energy gain and neutron energy loss, near the zone center, as shown in Fig.2-B. The dispersion curve of this mode is linear and corresponds exactly to the TA phonon branch dispersion previously determined at high Q scattering vectors [10]. The intensity decreases rapidly toward the Brillouin zone boundary (ZB), and a softening arises below the Curie temperature as shown in Fig.2-A, in which both the magnon and phonon branches are reported.

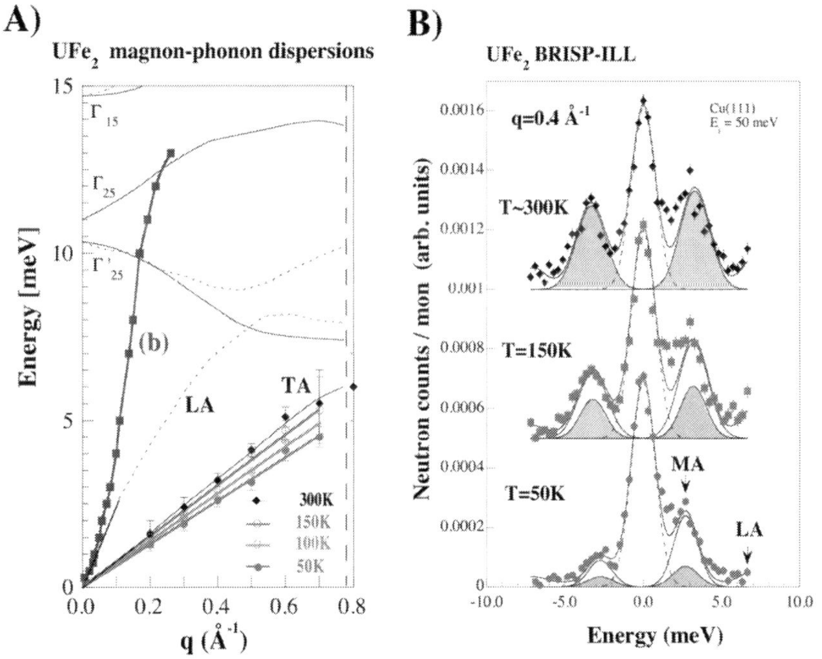

Figure 2. A): UFe_2 phonon dispersion curves (black thin lines, from Ref.[9], Fe in-phase precession mode (blue square symbols, from Ref.[10] and temperature dependence of magneto-acoustic mode. The vertical dashed line indicates the Brillouin zone boundary. **B)**: Temperature dependence of the neutron scattering intensities integrated over all the directions of scattering wavevector $\|\mathbf{q}\| = 0.4$ Å^{-1}. The inelastic spectra show the shift of the magneto-acoustic (MA) mode. The grey-shaded area represents the estimated phonon contribution at low temperature, calculated by applying the Bose factor to the signal determined at 300K.

At room temperature this mode is still present, albeit a broadening occurs close to the ZB. In order to estimate a possible magnetic contribution to this signal, we have calculated the phonon contribution at low temperature by applying the Bose factor to the signal determined at 300K. Fig.2-B shows the temperature dependence of this magneto-acoustic mode (MA) in which the shaded grey area indicates this phonon estimation at low temperature. Clearly, at 50K not all the signal can be attributed to a TA phonon contribution and we believe that a possible magnetic contribution exists. It is worth noticing that a strong paramagnetic scattering is expected in the itinerant systems also at room temperature, as already observed in UFe_2 for the b-mode. In conclusion, our results on UFe_2 strongly suggest the existence of an acoustic branch in the FBZ, with the same linear dispersion as a TA phonon, and which does not exist in the isostructural $ErFe_2$ by using the same experimental conditions. It is possible that the origin of this excitation is a coupled magneto-acoustic (MA) elastic wave arising from the interaction between the magnon acoustic mode (c-mode in Fig.1-A) and the TA phonon in presence of a strong magnetoelastic interaction. As explained in Ref.[4], when a spin wave and a phonon have the same frequency and the same symmetries, both the branches are renormalized and the repulsion affects the dispersions over the entire Brillouin zone. In this case, due to the dispersion crossing, the phononic or the magnonic pure character is lost and a mixed character (magnon-like or phonon-like) appears. Note that a magnon-phonon interaction was also observed in the optic phonon Γ_{15} crossing with the Fe b-mode around 15 meV [11]. The observation in the FBZ of this MA branch, forbidden for a pure TA phonon, arises from the magnetic part of the magnon-phonon interaction, explaining the anomalous magneto-elastic behavior in this compound with respect to the isostructural $REFe_2$ compounds, and resolving a long-standing problem about the missing acoustic magnon c-mode in UFe_2. Further experiments with polarized neutrons in the small $||q||$ region could shed light on the interplay between spin-lattice interactions in this itinerant system, opening new perspectives in the studies of magnetically induced lattice distortion in actinide intermetallic compounds.

3. Acknowledgments

We acknowledge the precious assistance of Alessio de Francesco, Alessio Laloni, and Lorenzo Sani during the BRISP experiments.

4. References

[1] Brooks M S S *et al* 1991 *J. Appl. Phys.* **70** 5972.
[2] Wulff M *et al* 1989 *Phys. Rev.* B **39** 4719.
[3] Popov Yu F *et al* 1980 *Sov. Phys. JETP* **51** 1223.
[4] Eremenko V V *et al* 1988 *Sov. Phys. JETP* **67** 2093.
[5] Rhyne J *et al* 1977 *Physica* B **86-88** 149.
[6] Clausen K *et al* 1982 *J. Phys. C: Solid State Phys.* **15** 3587.
[7] Koon N C and Rhyne J 1980 *JMMM* **15-18** (1980) 349.
[8] Paolasini L *et al* 1996 *Phys. Rev.* B **54** 7222.
[9] Paolasini L and Lander G H 2000 *J. Alloys and Compounds* **303-304** 232.
[10] Paolasini L *et al* 1998 *Phys. Rev.* B **58** 12125.
[11] Paolasini L *et al* 1999 *Phys. Rev.* B **59** 6867.
[12] Aisa D *et al* 2005 *Nucl. Instrum. Meth. Phys. Res.* A **544** 620
[13] http://www.ill.eu/instruments-support/computing-for-science/cs-software/all-software/lamp.

Journal of Physics: Conference Series **340** (2012) 012064 doi:10.1088/1742-6596/340/1/012064

Neutron diffraction studies on an exotic magnetic system, Nd_7Rh_3

S Rayaprol[1,*], V Siruguri[1], A Hoser[2], P Henry[2,$] and E V Sampathkumaran[3]

[1]UGC-DAE Consortium for Scientific Research, Mumbai Centre, R-5 Shed, B A R C Campus, Trombay, Mumbai – 400085, India

[2]Helmholtz-Zentrum Berlin, Hahn-Meitner Platz 1, 14-109 Berlin, Germany

[3]Tata Institute of Fundamental Research, Homi Bhabha Road, Colaba, Mumbai – 400005, India

[$]*Presently at* ESS AB, Box. No. 176, 22100 Lund, Sweden

[*]sudhindra@csr.res.in

Abstract. Binary intermetallic compound Nd_7Rh_3 crystallizes in a Th_7Fe_3 type hexagonal structure in space group $P6_3mc$ and has been reported to show two antiferromagnetic (AFM) phase transitions at 32 K and 10 K from magnetic susceptibility, and a field induced first-order magnetic transition at a field strength of 1 Tesla at 2 K from magnetization measurements. These magnetic properties are different from the isostructural counterparts like Tb_7Rh_3, Ho_7Rh_3, and Nd_7Ni_3. In order to understand the differences in the magnetic behaviour between Nd_7Rh_3 and other isostructural compounds, we have carried out neutron diffraction (ND) studies on polycrystalline Nd_7Rh_3 at various temperatures between 2 and 45 K. ND patterns were also recorded at T = 2K in the presence of applied magnetic fields from 0 to 1.5 Tesla. ND spectra on Nd_7Ni_3 and Ho_7Rh_3, for example, at 2 K in zero magnetic field exhibit a strong AFM peak at a Q (= $4\pi \mathrm{Sin}\theta/\lambda$) value of 0.32 and 0.38 $Å^{-1}$, respectively. However, ND spectra on Nd_7Rh_3, carried out at Focusing Crystal Diffractometer (FCD-Dhruva-Mumbai), diffractometers E6 and E9 (HZB-Berlin) using wavelengths λ = 1.48Å, 2.45Å and 2.8Å respectively, at T = 2K in zero field do not exhibit any AFM peaks in the entire Q-range studied. Instead, long-range ferromagnetic order is established as evidenced by the intensity enhancement of the nuclear Bragg peaks. The lack of extra peaks in the ND patterns would imply that the magnetic order seen in magnetization could be of a $k = 0$ type which changes from one type to another on the application of magnetic field. The results of nuclear and magnetic structure refinement are discussed to explain the magnetic behaviour of Nd_7Rh_3 at low temperatures.

1. Introduction

The compound Nd_7Rh_3 crystallizes in Th_7Fe_3 type hexagonal structure in the space group $P6_3mc$ [1-2]. In this structure, Nd occupies three non-equivalent sites, *6c*, *12d* and *2b* sites, whereas Rh occupies the *12d* site, respectively. Magnetization studies on Nd_7Rh_3 exhibit two magnetic phase transitions, one at 32 K and the other at 10 K. At 32 K, Nd_7Rh_3 undergoes a transition from paramagnetic to an antiferromagnetic phase and that below 10 K; it undergoes another antiferromagnetic to ferromagnetic transition [2]. In addition to anomalous magnetic behavior, it has been reported by Tsutaoka *et al* [2] that Nd_7Rh_3 exhibits drastic changes in thermoelectric power below 50 K owing to changes in the electronic structure, which might influence the magnetic structure at low temperatures. Nd_7Rh_3 exhibits a field-induced first-order magnetic phase transition at 1.8 K [3]. On the basis of isothermal magnetization and magnetoresistance measurements, it was shown that Nd_7Rh_3 exhibits magnetic phase co-existence phenomenon in the super-cooled state following this transition [3]. The anomalous transition at 10 K was also probed in detail using magnetic susceptibility, specific heat and magnetoresistance measurements, and it was observed that this transition is first-order like [4]. Recent studies on single-crystal of Nd_7Rh_3 have also conclusively established the anomalous magnetic behavior of Nd_7Rh_3 [5]. All these detailed studies revealed Nd_7Rh_3 is an interesting magnetic material which makes it different from isostructural compounds such as Ho_7Rh_3, Tb_7Rh_3, and Nd_7Ni_3 etc. Therefore, it is expected that the magnetic structure at low temperatures, for Nd_7Rh_3 will also be different from its isostructural counterparts. This is the primary motivation of this investigation.

2. Experimental Details

The intermetallic Nd_7Rh_3 was prepared in polycrystalline form by repeated arc-melting of stoichiometric ratio of Nd and Rh. The ingot thus obtained was annealed in a sealed quartz ampoule for over a few weeks to ensure homogeneity of the sample. The phase formation was checked by powder X-ray diffraction as well as by powder neutron diffraction at room temperature.

Initial characterization of the sample was done by measuring magnetization as a function of temperature and magnetic field using a commercial vibrating sample magnetometer. Powder neutron diffraction studies were carried out at room temperature at FCD (Dhruva, India) and at low temperatures on diffractometers E6 and E9 at BENSC (HZB, Germany). The neutron diffraction (ND) data has been refined by Rietveld analysis method using the FULLPROF software package.

To overcome the problem of preferred orientation during ND measurements under applied magnetic field, the powder sample of Nd_7Rh_3 was mixed with a 1:1 solution of deuterated methanol and deuterated ethanol. Such a mixture helps in holding the grains of the polycrystalline sample intact under the application of magnetic field. The signal from this organic mixture is very minimal and is observed as a hump in the background. The sample and the mixture were contained in an aluminum can and loaded onto a liquid helium cryo-magnet for low temperature and in-field ND experiments.

3. Results and Discussions

3.1. Magnetization

In figure 1, magnetization data measured on the sample used for neutron diffraction has been shown. The two magnetic transitions expected at 32 and 10K and the irreversibility of the M(H) curve is clearly seen in the figure 1. The ZFC-FC bifurcation is also clearly seen in the inset figure. The observed behavior of M(H) and dc $\chi(T)$ are qualitatively same as reported in Refs. 3 and 4. It must be noted here that the magnetization measurements were carried out on a chunk of a polycrystalline sample.

3.2. Neutron Diffraction

3.2.1. *Room temperature neutron diffraction:* Neutron diffraction measurements at room temperature were carried out with a view to refine the crystal structure in the paramagnetic state of Nd_7Rh_3. The experiment was carried out in air on the Focussing crystal diffractometer (FCD) at Dhruva reactor at a wavelength of 1.48 Å. The Rietveld refined ND pattern is shown in figure 2. The refined structural parameters are listed in Table 1.

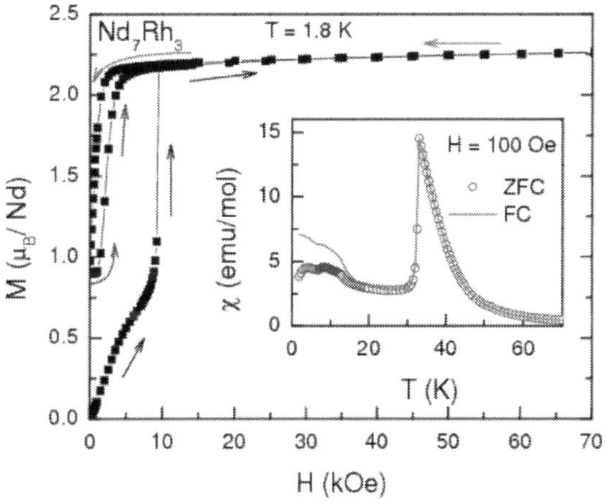

Figure 1 Magnetization (M) for Nd_7Rh_3 at T = 1.8 K is plotted against the ramping field (H). The black, green and blue arrows are indicative of the up down and up cycle of ramping field. The magnetic susceptibility measured in a field of 100 Oe is shown as an inset. The data has been measured in zero field cooled (ZFC) and field cooled (FC) state of the samples.

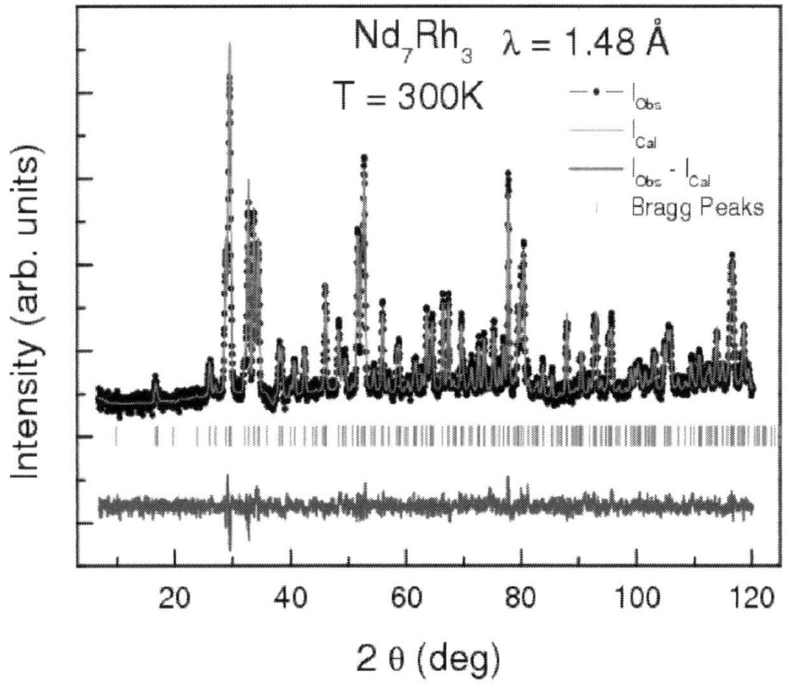

Figure 2 Rietveld refinement of the ND data of Nd_7Rh_3

Table 1 Refined structural parameters of Nd_7Rh_3 at 300 K.

Space group: $P6_3mc$			a =10.0030(3) Å		c = 6.3184(3) Å	
Atom	Site	Symmetry	*x*	*y*	*z*	*B* (Å^2)
Nd1	6*c*	.m.	0.5394	0.4605	0.0075	0.0114
Nd2	12*d*	1	0.8774	0.1259	0.3060	0.0100
Nd3	2*b*	3m.	1/3	2/3	0.0220	0.0124
Rh	12*d*	1	0.1773	0.7998	0.2535	0.0127

3.2.2 Low temperature neutron diffraction: ND studies on isostructural compounds such as Nd_7Ni_3 [2, 6], Ho_7Rh_3 [7] exhibit a strong magnetic peak around Q = 0.32 – 0.38 $Å^{-1}$. Q is the reciprocal space scattering vector, and is defined by $4\pi Sin\theta/\lambda$. If the magnetic structure of Nd_7Rh_3 is similar to that of Nd_7Ni_3 or Ho_7Rh_3 etc, it could also exhibit a similar strong magnetic peak around similar Q values. Therefore, we carried out further ND experiments on Nd_7Rh_3 at various temperatures between 2 and 45 K, and the patterns are shown in figure 3. The figure also shows the Rietveld refinement of T = 2 K data. The first and most significant difference in the ND pattern of Nd_7Rh_3 and its isostructural compounds is the absence of the strong magnetic peak at lower angles (in terms of Q also there is no anomaly observed in the entire Q-range studied). ND data measured on both E6 and E9 diffractometers at T = 2K, is shown as a function of Q in the inset of figure 3. In the entire Q-range, no trace of the strong magnetic peak is found at lower angles.

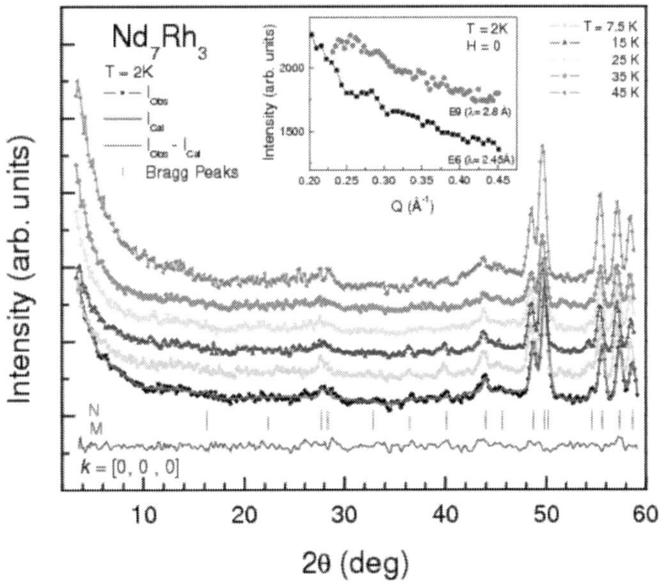

Figure 3 ND patterns recorded at various temperatures on E6. The Rietveld refinement of T = 2K data is also shown in the figure. The inset of the figure exhibits the T = 2K data measured on both E6 and E9, plotted on Q-scale to show the absence of the expected strong magnetic peak in Nd_7Rh_3.

Xu *et al* could refine the magnetic structure of Nd_7Ni_3 assuming a helical magnetic structure and using propagation vector, k = (0, 0, 1/3) [6]. Owing to the difference in the magnetic behaviour when compared to Nd_7Ni_3 or Ho_7Rh_3, and the absence of a strong magnetic peak at lower angles in Nd_7Rh_3, it can be proposed that the magnetic structure of Nd_7Rh_3 might be different from the one reported for Nd_7Ni_3. Since there are no additional peaks attributable to magnetic ordering alone, the magnetic structure could be assumed to be having similar dimensions as that of the crystal structure and a propagation vector, k = (0 0 0). All possible orientations for the three Nd sites were tried and it was found that the magnetic lines in the ND pattern of Nd_7Rh_3 could be fitted fairly well assuming a collinear magnetic ordering along *c*-axis. With increasing temperature there is a slight decrease in the intensity of some nuclear peaks, affecting the magnetic moment on three non-equivalent Nd sites. No moment was observed on Rh even at lowest temperature. The magnetic moment on Nd calculated from the refinement is listed in Table 2.

In order to ascertain the effect of magnetic field on the magnetic structure, we have also measured ND patterns at T = 2 K under various applied fields. In figure 4, the ND patterns recorded at H = 0, 1.5 and 5 Tesla are shown. The refined ND pattern for H = 0 is shown again for direct comparison. As can be clearly seen from the ND patterns, the effect of magnetic field is very minimal on the diffraction peaks, even for a strong field of 5 Tesla. The magnetic moment on Nd obtained from the Rietveld analysis of all the in-field diffraction patterns is also recorded in the Table 2. It is seen that the moments at the three Nd sites respond in an unequal manner along *c*-axis while in reality, it is likely that the magnetic ordering in this compound is quite complex whose net effect is to suppress any strong long-range antiferromagnetic interactions.

The moments on Nd sites are plotted as a function of temperature of measurement in figure 5. The moment values increase with decreasing temperature and reach maxima at around 15 K and fall below 10 K, thus signifying the changes in the magnetic structure corresponding to the anomalies observed in magnetic measurements. The values of magnetic moment on Nd atom observed at T = 2K from ND experiments are close to the saturation magnetization values observed at T = 1.8K. We see a finite moment even at 45 K, which can imply the existence of short range magnetic correlations well above 30 K.

Table 2 Magnetic moments (M) on Nd atom obtained from the Rietveld analysis of ND data at various temperatures and also at various fields at T = 2K.

$M(\mu_B)$ \ T	2 K			7.5 K	15 K	25 K	35 K	45 K
	H = 0	1.5 T	5 T					
μ_{Nd1}	1.938	2.064	0.599	2.297	2.878	2.171	1.930	0.886
μ_{Nd2}	2.426	1.886	3.273	2.418	2.725	1.428	2.195	1.102
μ_{Nd3}	1.675	1.203	1.212	1.270	2.128	1.718	1.888	0.612

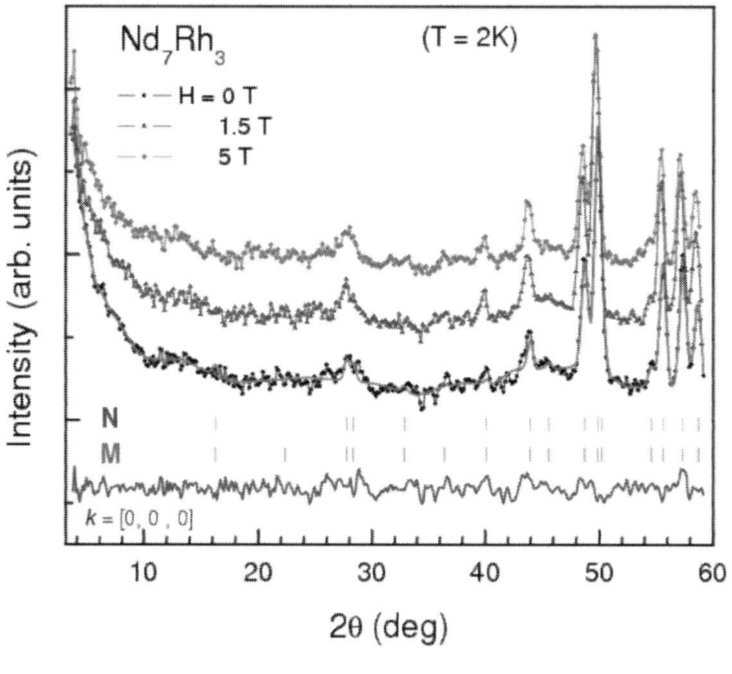

Figure 4 ND patterns recorded at T = 2K under fields of 0, 1.5 and 5 T on E6. The Rietveld refinement of H = 0T data is also shown in the figure.

4. Conclusions

Neutron diffraction experiments were carried out on polycrystalline sample of Nd_7Rh_3 to determine the magnetic structure at low temperatures. A magnetic structure with unequal magnetic moments on three non-equivalent Nd sites has been observed. ND studies show that Nd_7Rh_3 exhibits a kind of magnetic structure which changes from one type to another in the presence of magnetic field, as observed by the changes in the magnetic moments on Nd. Another observation that makes Nd_7Rh_3 an interesting compound is that the diffraction patterns look similar before and after the first order transition. Naturally, the magnetization of this compound is quite different from its isostructural Nd_7Ni_3. More detailed studies are required to understand the exact nature of magnetic structure and neutron diffraction experiments on single crystal of Nd_7Rh_3 can give more insight on the magnetic structure and order of this magnetically exotic compound.

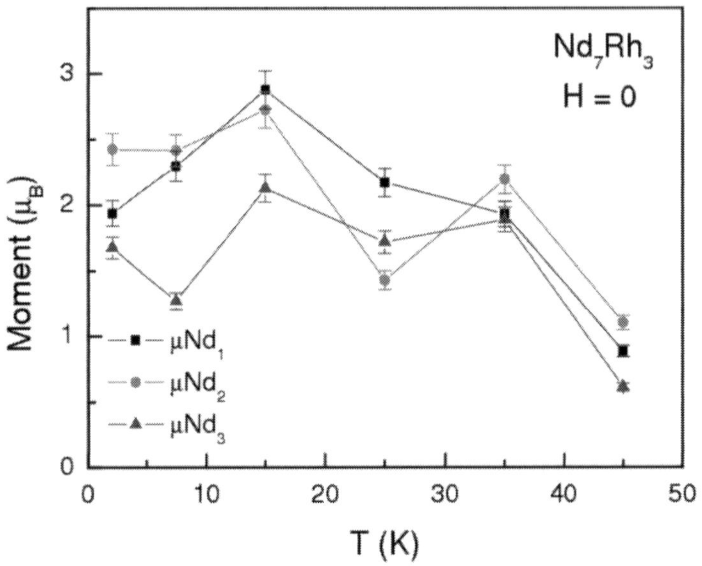

Figure 5 The variation of the magnetic moments on three non-equivalent Nd sites is plotted against the temperatures at which ND patterns were recorded. The moments were obtained from the Rietveld analysis.

Acknowledgements

Authors thank Kartik K. Iyer for sample preparation and magnetic measurements. SR and VS thank Department of Science and Technology (DST, India) for providing financial support to carry out ND experiments at HZB-Berlin.

References

[1] Olcese G. L, 1973 *J. Less-Common Met.* **33** 71
[2] Tsutaoka T, Nakamori Y, Tokunaga T, Kadomatsu H, Itoh Y, 1998 *J. Alloys. Compund.* **270** 53
[3] Sengupta K, Sampathkumaran E V, 2006 *Phys. Rev. B* **73** 020406R
[4] Sengupta K, Sampathkumaran E V, 2006 *J. Phys.:Condens. Mater* **18** L401
[5] Tanaka A, Tsutaoka T, Shigeoka T, 2008 *Physica B* **403** 3248
[6] Xu X, Kawano S, Tustaoka T, Fukuda H, Tokunaga T, Andoh Y, Kurisu M and Nakamoto G, 1998 *Physica B* **241 – 243** 742
[7] Tsutaoka T, Nishiume Y, Tokunaga T, Nakamori Y, Andoh Y, Kawano S, Nakamoto G, and Kurisu M, 2003 *Physica B* **327** 352

Magnetic structure of iron borate $DyFe_3(BO_3)_4$: A neutron diffraction study

C Ritter[1], A Pankrats[2], I Gudim[2] and A Vorotynov[2]

[1] Institute Laue-Langevin, BP 156, F-38042 Grenoble, France

[2] L V Kirenskii Institute of Physics, Siberian Branch of RAS, Krasnoyarsk 660036, Russia

E-mail: ritter@ill.fr

Abstract. Neutron diffraction measurements were performed on the iron borate $DyFe_3(BO_3)_4$ to investigate details of the crystallographic structure, the low temperature magnetic structure and its magnetic properties. $DyFe_3(BO_3)_4$ adopts at room temperature the $P3_121$ symmetry and becomes antiferromagnetic below T_N = 39 K. Both, the rare earth and the iron sublattice, follow the same magnetic propagation vector $\tau = [0, 0, ½]$ which leads to a doubling of the crystallographic unit cell in the *c*-direction. The easy axis anisotropy of the rare earth determines the moment orientation to be mainly along *c*. No spin reorientation is found between T_N and 1.5 K, however, a small anomaly in the thermal dependence of the unit cell *a*-parameter is found at about 27 K which could be connected to repopulation of low lying Kramers doublets of Dy^{3+}. The magnetic moment value of the Fe-moment is at 1.5 K with $\mu_{Fe} = 4.5\ \mu_B$ only slightly smaller than expected for an S = 5/2 ion while the Dy moment is strongly reduced and amounts only to $\mu_{Dy} = 6.4\ \mu_B$.

1. Introduction

The family of rare earth ferroborates of general formula $RFe_3(BO_3)_4$ where R is a rare earth or Y has recently attracted considerable attention due to their interesting magnetic, magnetoelectric and multiferroic properties [1-5]. The presence of two magnetic sublattices, a structure which sees helicoidal Fe-Fe-chains with a strong one-dimensional character and the absence of direct R-O-R-interactions [6] opens the way for a variety of magnetic ordering phenomena where the temperature dependent competition between the anisotropy of the rare earth sublattice and that of the iron sublattice leads to different magnetic structures and may induce spin reorientations and metamagnetic behavior. Previous neutron diffraction studies have established an easy-axis orientation of the magnetic moments for R = Tb, Pr [7, 8] over the whole temperature range below T_N and for R = Ho [9] below the spin reorientation temperature while for R = Nd, Er, Ho (above 5 K) and Y an easy-plane orientation has been found [10, 8, 9]. Magnetic resonance data indicate for R = Gd a spin reorientation from easy-plane to easy-axis on lowering the temperature [11]. Depending on the ionic size of the rare earth a structural transition from a high temperature *R*32 phase to a low temperature $P3_121$ phase can take place in the $RFe_3(BO_3)_4$ compounds where the presence of two different iron sites can lead in the low symmetry $P3_121$ phase to slightly different Fe-Fe distances along the helicoidal chains and even to slightly different orientations of the two magnetic Fe sublattices [12, 13, 9]. The presence of short-range magnetic order within the helicoidal Fe-spin chains was determined to persist in $YFe_3(BO_3)_4$ up to temperatures several times T_N adding a further piece to the highly interesting puzzle of the magnetic behavior of this family of compounds [14]. The second order antiferromagnetic transition leading to the long range ordered magnetic state seems to be driven by Fe-O-B-O-Fe superexchange interactions while the superexchange Fe-O-R-O-Fe pathway leads in all up to now known $RFe_3(BO_3)_4$ compounds with magnetic R to $T_{NR} = T_{NFe}$.

Magnetization and specific heat data by Popova et al. [15] determined $T_N = 38$ K for $DyFe_3(BO_3)_4$ and qualified it as an easy-axis antiferromagnet. The authors of [15] described the magnetic behavior of the iron and the rare-earth sublattices as qualitatively different with the Dy-sublattice apparently staying paramagnetic down to lowest temperatures. Gudim et al [16] came from their magnetization data to the same conclusions as [15] concerning T_N and the easy-axis antiferromagnetic state of $DyFe_3(BO_3)_4$. Several groups [15-17] found a maximum in the magnetic susceptibility at about 25 K; this was explained [15, 17] to be caused by a repopulation of the two lowest Dy^{3+} energy levels. As part of a systematic study on the influence of the different rare earths on the magnetic behavior of $RFe_3(BO_3)_4$ compounds we performed temperature dependent high resolution and high intensity neutron diffraction studies on $DyFe_3(BO_3)_4$.

2. Experimental details

2.1. Sample preparation

Single crystals of $DyFe_3(^{11}BO_3)_4$ were prepared at the Institute of Physics at Krasnoyarsk following the preparation procedure described in detail in [16]. Because of the strong neutron absorption by natural boron 99% with ^{11}B enriched B_2O_3 was used. Powder samples were then made from the grown crystals.

2.2. Neutron diffraction measurements

Neutron diffraction data were taken at the Institut Laue-Langevin in Grenoble, France, using the high resolution powder diffractometer D2B ($\lambda = 1.594$ Å) and the high flux powder diffractometer D20 ($\lambda = 2.418$ Å). The sample was placed inside a thin vanadium container of 3 mm diameter in order to reduce the absorption from dysprosium. The temperature dependence was measured on D20 between 1.5 K and 50 K with a temperature resolution of about 0.6 K taking a spectrum every 10 min. High resolution data were taken on D2B at 1.5 K, 50 K and at room temperature. All data were refined by the Rietveld method using the FULLPROF [18] program.

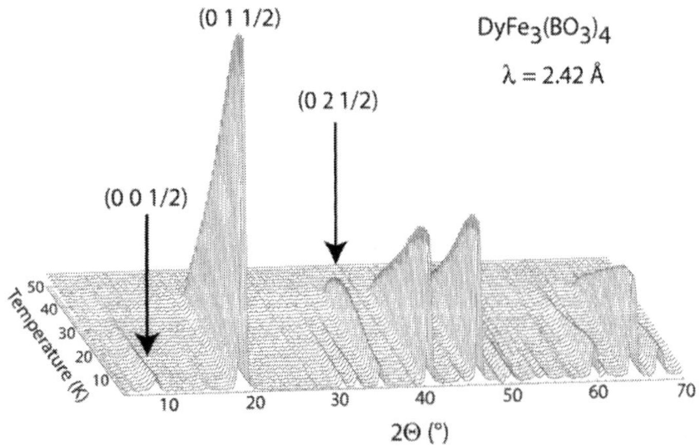

Figure 1. Thermal dependence of the magnetic scattering in $DyFe_3(BO_3)_4$.

3. Results and discussion

The high resolution neutron diffraction data taken at 300 K and 50 K (figure 2a) on D2B were used to refine the room temperature crystal structure and to verify the purity of the sample. No impurity phases are present and the fit proceeded smoothly in the spacegroup $P3_121$. Results of this refinement are shown in table 1 together with some selected interatomic distances. Error bars are slightly higher than usual due to the closeness of the structure to the higher symmetric phase *R*32 which should be adopted by $DyFe_3(BO_3)_4$ at about 340 K [12]. A fit of the 300 K data against the *R*32 model resulted in significantly higher R-factors.

Table 1. Structural data at 1.5 K, 50 K and 300 K of $DyFe_3(BO_3)_4$ refined in $P3_121$ together with the most important interatomic distances.

$P3_121$	1.5 K	50 K	300 K
a (Å)	9.5365(1)	9.5379(1)	9.5439(1)
c (Å)	7.5621(1)	7.5609(1)	7.5676(1)
Dy (3*a*) *x*	0.3358(5)	0.3344(4)	0.3329(8)
Fel (3*a*) *x*	0.8825(7)	0.8829(6)	0.8850(10)
Fe2 (6*c*) *x*	0.7841(6)	0.7837(6)	0.7857(7)
y	0.4543(5)	0.4539(5)	0.4522(8)
z	0.3414(5)	0.3412(5)	0.3294(11)
O1 (3*b*) *x*	0.9238(10)	0.9239(10)	0.9289(19)
O2 (6*c*) *x*	0.4180(8)	0.4174(7)	0.4146(11)
y	0.7244(7)	0.7246(7)	0.7375(11)
z	0.1273(6)	0.1281(6)	0.1507(7)
O3 (6*c*) *x*	0.8779(10)	0.8786(10)	0.8832(14)
y	0.6951(10)	0.6941(10)	0.6947(14)
z	0.8153(11)	0.8139(10)	0.8197(14)
O4 (6*c*) *x*	0.8495(10)	0.8494(10)	0.8562(15)
y	0.6357(10)	0.6355(9)	0.6469(14)
z	0.1804(10)	0.1810(10)	0.1771(11)
O5 (6*c*) *x*	0.4760(9)	0.4753(11)	0.4832(12)
y	0.1464(8)	0.1470(8)	0.1423(12)
z	0.8453(10)	0.8444(10)	0.8299(18)
O6 (3*b*) *x*	0.1863(13)	0.1856(12)	0.1882(22)
O7 (6*c*) *x*	0.4717(7)	0.4726(7)	0.4842(13)
y	0.4652(9)	0.4654(8)	0.4572(13)
z	0.8181(10)	0.8134(10)	0.8123(11)
Bl (3*b*) *x*	0.3352(16)	0.3373(15)	0.3256(12)
B2 (6*c*) *x*	0.5536(11)	0.5559(10)	0.5517(9)
y	0.8806(10)	0.8818(9)	0.8824(13)
z	0.1498(8)	0.1525(9)	0.1685(15)
B3 (3*b*) *x*	0.7828(16)	0.7847(15)	0.7801(17)
Fel-Fel (Å)	3.181(5)	3.177(4)	3.159(7)
Fe2-Fe2 (Å)	3.199(6)	3.195(6)	3.198(11)
2 x Dy-O3 (Å)	2.360(10	2.372(10)	2.375(15)
2 x Dy-O4 (Å)	2.350(11)	2.352(10)	2.332(16)
2 x Dy-O7 (Å)	2.356(7)	2.347(6)	2.362(10)
R_{Bragg}	3.4	3.5	3.3
R_{Mag}	-	-	3.5

The high intensity temperature dependent data taken on D20 revealed the onset of additional magnetic scattering in the neutron diffraction patterns below about 40 K. Figure 1 shows for $2\Theta < 70°$ the thermal changes in the spectra in form of a difference thermodiffractogam where the data taken at 50 K were subtracted from all spectra. This eliminates the nuclear scattering and leaves only the purely magnetic scattering. All magnetic peaks appear at positions not corresponding to nuclear Bragg peaks and can be indexed using the magnetic propagation vector τ = [0, 0, ½]. The same propagation vector has been already found in all other $RFe_3(BO_3)_4$ compounds studied with neutrons [7-10]. It leads to a doubling of the magnetic unit cell in *c*-direction and is determined by the antiferromagnetic coupling between neighboring Fe spins along the helicoidal chain. The thermal dependence of the intensity of the different magnetic peaks seen in figure 1 is not identical (e.g. the (0 2 ½) reflection decreases on lowering the temperature) and indicates the presence of more than one magnetic sublattices having different temperature dependences. Magnetic symmetry analysis as implemented in the program BASIREPS which is part of the FULLPROF suite of programs [18, 19] was used to determine for τ = [0, 0, ½] the allowed irreducible representations and their basis vectors for the two possible magnetic sites (Fe and Dy) in $P3_121$. There are three allowed irreducible representations which were tested by refinement against the high resolution data taken at D2B at 3 K (figure 2b). It was found that only the basis vectors of one irreducible representation (IR) are describing the magnetic structure correctly. Table 2 lists the basis vectors of this IR and the refined values of their coefficients, the crystallographic details of the refinement are included in table 1.

Table 2. Basis vectors (BV) of the irreducible representation 3 for τ = [0, 0, ½] in $P3_121$. Refined values of their coefficients determined from the refinement of the magnetic structure of $DyFe_3(BO_3)_4$ at 3 K. The coefficients describing the magnetic moment of Fel (BV2) and Fe2 (BV3) in *z*-direction were coupled.

	BV1	BV2	BV3
Fel (3*b*)			
x, *y*, *z*	1 -1 0	0 0 1	
-*y*, *x*-*y*, *z*+1/3	-1 -2 0	0 0 -1	
-*x*+*y*, -*x*, *z*+2/3	-2 -1 0	0 0 1	
Coefficient	0.09(14)	4.45(5)	
Fe2 (6*c*)			
x, *y*, *z*	1 0 0	0 1 0	0 0 1
-*y*, *x*-*y*, *z*+1/3	0 -1 0	1 1 0	0 0 -1
-*x*+*y*, -*x*, *z*+2/3	1 1 0	-1 0 0	0 0 -1
y, *x*, -*z*	0 1 0	1 0 0	0 0 -1
x-*y*, -*y*, -*z*+2/3	-1 0 0	1 1 0	0 0 1
-*x*, -*x*+*y*, -*z*+1/3	1 1 0	0 -1 0	0 0 1
Coefficient	0.00(1)	0.27(12)	-4.45(5)
Dy (3*b*)			
x, *y*, *z*	1 -1 0	0 0 1	
-*y*, *x*-*y*, *z*+1/3	-1 -2 0	0 0 -1	
-*x*+*y*, -*x*, *z*+2/3	-2 -1 0	0 0 1	
Coefficient	1.28(8)	-6.27(7)	

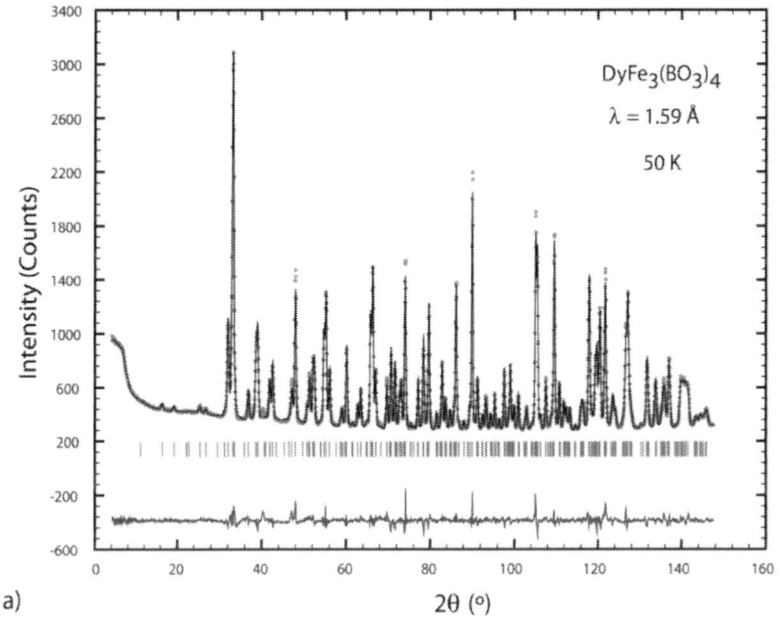

a)

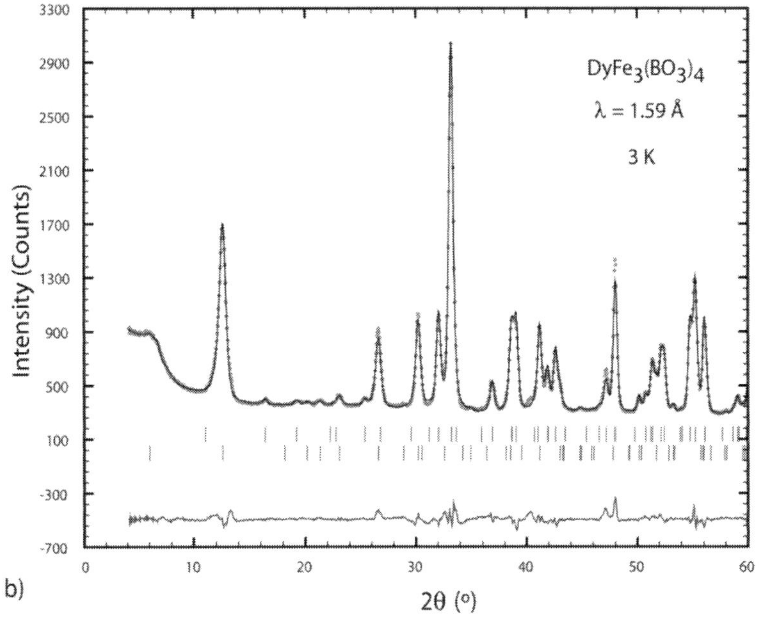

b)

Figure 2. Observed (dots, red), calculated (line, black) and difference pattern of $DyFe_3(BO_3)_4$ at a) 50 K in the paramagnetic state refined in $P3_121$ and b) at 3 K in the magnetic state. The tick marks indicate the calculated position of the nuclear (upper row) and (for b)) magnetic (lower row) Bragg peaks.

The magnetic structure found (figure 3) resembles strongly those found for $TbFe_3(BO_3)_4$ [7] and $PrFe_3(BO_3)_4$ [8] with a dominant easy axis orientation of the magnetic moments. The coefficients of the basis vectors describing an easy plane orientation of the two iron sites are zero or very close to zero (Table 2). Only the rare earth site sees a non negligible component within the hexagonal basis plane. As in the case of the easy axis compounds with R = Tb [7] and Ho (below 5 K) [9] the spins of the iron and the rare earth sublattice are aligned antiparallel within the hexagonal layers while the alignment had been shown to be parallel in the case of R = Pr. The different relative orientations of the two sublattices have been discussed in [8] and connected to the spin-orbit coupling of the rare earth. The magnetic moment value of iron amounts to about $\mu_{Fe} = 4.5\ \mu_B$ and corresponds to the values found in the other $RFe_3(BO_3)_4$ compounds adopting the easy axis orientation approaching the value expected for a S = 5/2 ion. The total magnetic moment of the Dy^{3+} ion is with $\mu_{Dy} = 6.4\ \mu_B$ largely inferior to the expected free ion value of 10 μ_B. The same situation had been found before in compounds with R = Ho ($\mu_{Ho} = 5.1\ \mu_B$) [9] and Er ($\mu_{Er} = 5.1\ \mu_B$) [8] but not in $TbFe_3(BO_3)_4$ [7] where the ordered magnetic moment corresponds with $\mu_{Tb} = 8.6\ \mu_B$ nearly to the expected free ion value. The ordering of the rare earth sublattice is known to be induced first by the polarizing influence of the ordered Fe-sublattice before at lower temperatures indirect R-O-Fe-O-R or R-BO_3-R interactions start to play a role. One has to recall here that in the corresponding $RA1_3(BO_3)_4$ compounds no long rang ordered magnetism exists [20]. Popova et al. [15] suggested from the interpretation of their magnetization data that the dysprosium subsystem in $DyFe_3(BO_3)_4$ would stay essentially paramagnetic down to lowest temperatures. This has to be put into the context of the above discussed strong polarizing influence of the iron sublattice and our own neutron results: The ordering of the Dy-sublattice is not resulting from direct R-O-R exchange interactions but depending on the ordering of the Fe-sublattice, the value of the ordered magnetic moment $\mu_{Dy} = 5.1\ u_B$ is strongly reduced compared to the free ion value of Dy^{3+}, however, the neutron data clearly evidence the long range antiferromagnetically ordered nature of the Dy-sublattice. We would like to mention here the highly interesting results of Chaudhury et al. [21] who found in $TmAl_3(BO_3)_4$ the existence of a large magnetoelectric effect generated in absence of any d-electron spins solely by the rare earth moment.

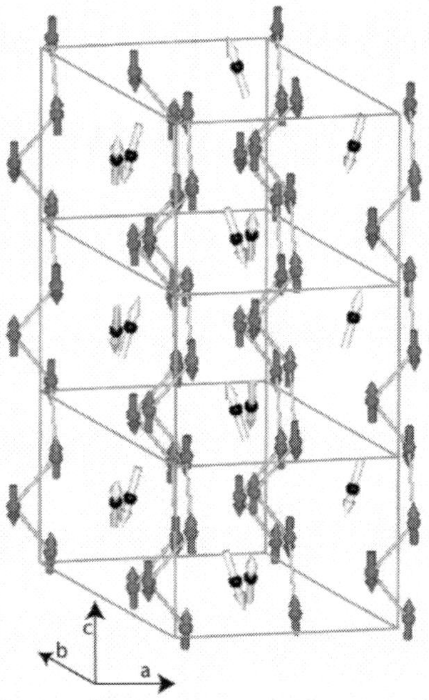

Figure 3. Magnetic structure of $DyFe_3(BO_3)_4$. Fe spins in red (black), Dy moments in yellow (light grey), the direct Fe-Fe exchange along the helicoidal chains is indicated by light blue (grey) lines.

In order to check for the presence of a possible spin reorientation (as found in $HoFe_3(BO_3)_4$) and in order to verify whether both sublattices possess the same Néel temperature the temperature dependent high flux data of D20 were sequentially refined. For this refinement the atom coordinates were kept fixed at values determined from a fit of the high resolution data taken just above T_N at 50 K on D2B which presents purely nuclear scattering. Results of this refinement at 50 K are included in table 1 and show that the crystallographic structure does not undergo any significant changes between 50 K and 3 K validating the approach of fixing the atom coordinates for the sequential refinement. The refinement proceeded smoothly over the whole temperature range using the magnetic structure model found at 3 K. Figure 4 displays the temperature dependence of the total Dy moment, of the Fe moment and of the Dy basal plane component. It becomes clear that both sublattices start to get long range ordered at the same temperature, $T_{NDy} = T_{NFe} = 38$ K. No sign of a clear spin reorientation is found, one sees, however, that the basal plane component of the Dy sublattice is only present between 20 K and lowest temperatures. This coupling leads to a 120° arrangement of the Dy spins occupying the threefold Wykoff site *3b* within a nuclear unit and has been already found as minority spin structure in $ErFe_3(BO_3)_4$ [8]. It does not correspond to the simple collinear easy plane configuration as found in $HoFe_3(BO_3)_4$ above 5 K [9] or in $ErFe_3(BO_3)_4$ above 10 K [8]. Analyzing in more detail the temperature dependence of the total Dy moment increase (figure 5) one can notice that on lowering the temperature the moment increase is following the increase of the polarizing Fe sublattice magnetization. At about 25 K, however, one sees a reinforced increase and in parallel at about 20 K the arising of the basal plane component of the Dy moment. The basal plane component cannot be induced by the anisotropy of the Fe-sublattice as it does not - as explained before - correspond to the collinear easy plane arrangement, but has to be related to the symmetry of indirect R-R exchange interactions. This indicates, together with the accelerated total Dy-moment increase below 25 K, that the influence of the indirect R-R interactions becomes important at this relatively high temperature.

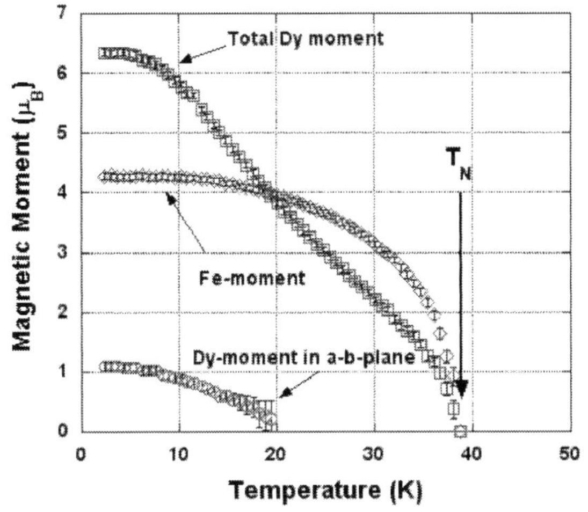

Figure 4. Temperature dependence of the magnetic moment values of the Fe-sublattice (diamonds), of the Dy-sublattice (squares) and of the basal plane component of the Dy-moment.

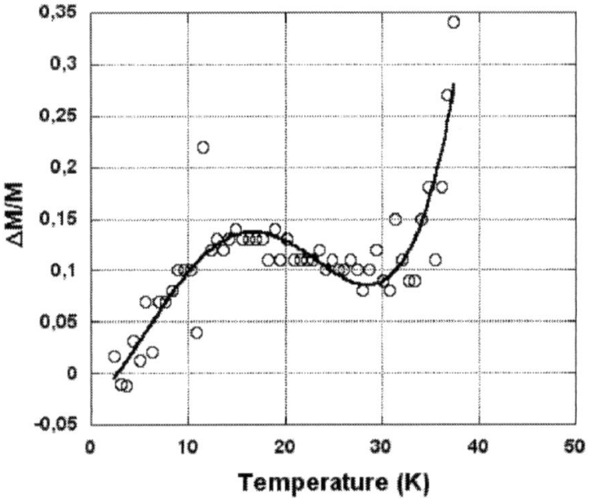

Figure 5. Temperature dependence of the Dy-moment increase.

Figure 6 displays the temperature dependence of the unit cell *a*-parameter where a small anomaly can be detected at about 27 K. This is the first evidence of such an thermal expansion anomaly in the family of $RFe_3(BO_3)_4$ compounds. Its interpretation as a magnetostriction effect is not supported by our analysis of the temperature dependence of the sublattice magnetizations (figure 5). It can, however, be connected to the Schottky type anomalies seen in the susceptibility and specific heat data at about 23 K presented by Popova et al. [15] which were related to a repopulation of two low lying Kramers doublets. Changes in the anisotropy of the Dy-sublattice induced by this repopulation could eventually be as well responsible for the appearance at about 20 K of the basal plane magnetic moment component.

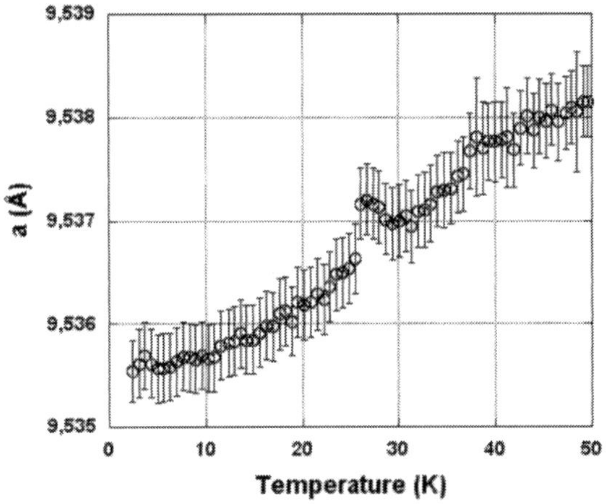

Figure 6. Temperature dependence of the unit cell *a*-parameter of $DyFe_3(BO_3)_4$.

4. Summary

Temperature dependent neutron diffraction studies on $DyFe_3(BO_3)_4$ show that this compounds magnetically orders at about $T_N = 38$ K into a magnetic structure of predominantly easy axis type with a magnetic propagation vector $\tau = [0\ 0\ \frac{1}{2}]$. Both magnetic sublattices order at the same temperature, magnetic moment values amount to $\mu_{Fe} = 4.5\ \mu_B$ and $\mu_{Dy} = 6.4\ \mu_B$. A small magnetic component forming a 120° type arrangement of the Dy-spins appears below about 20 K. The appearance of this component and a small anomaly visible in the thermal expansion of the unit cell *a*-parameter at about 27 K might be connected to a repopulation of the lowest Dy^{3+} energy levels.

Acknowledgement

This work was supported by RFBR, grant no. 10-02-00765.

References

[1] Jaque D 2001 *J. Alloys Compounds* **323/324** 204
[2] Chen X, Luo Z, Romero J J, Sole J G, Huang Y, Jiang A and Tu Ch 2001 *J. Phys.: Condens. Matter 13* 1171
[3] Kalashnicova A M, Pavlov V V, Pisarev R V, Bezmaternykh L N, Bayer M and Rasing Th 2004 *JETP Lett. 80* 293
[4] Zvezdin A K, Krotov S S, Kadomtseva A M, Vorob'ev G P, Popov Yu F, Pyatov A P, Bezmaternykh L N, Kuvardin A V and Popova E N 2005 *JETP Lett.* **81** 272
[5] Zvezdin A K, Vorob'ev G P, Kadomtseva A M, Popov Yu F, Pyatakov A P, Bezmaternykh L N, Kuvardin A V and Popova E N 2006 *JETP Lett.* **83** 509
[6] Campá J A, Cascales C, Gutiérrez-Puebla E, Monge M A, Rasines I and Ruiz-Valero C 1997 *Chem. Mater.* **9** 237
[7] Ritter C, Balaev V, Vorotynov A, Petrakovskiĭ G, Velikanov D, Temerov V and Gudim I 2007 *J. Phys.: Condens. Matter* **19** 196227
[8] Ritter C, Vorotynov A, Pankrats A, Petrakovskiĭ G, Temerov V, Gudim I and Szymczak R 2010 *J. Phys.: Condens. Matter* **22** 206002
[9] Ritter C, Vorotynov A, Pankrats A, Petrakovskiĭ G, Temerov V, Gudim I and Szymczak R 2008 *J. Phys.: Condens. Matter* **20** 365209
[10] Fischer P, Pomjakushin V, Sheptyakov D, Keller L, Janoschek M, Roessli B, Schefer J, Petrakovskiĭ G, Bezmaternykh L N, Temerov V and Velikanov D 2006 *J. Phys.: Condens. Matter* **18** 7975
[11] Pankrats A, Petrakovskiĭ G, Bezmaternykh L N and Bayukov O A 2004 *JETP* **99** 766
[12] Hinatsu Y, Doi Y, Ito K, Wakeshima M and Alemi A 2003 *J. Solid State Chem.* **172** 438
[13] Klimin S A, Fausti D, Meetsma A, Bezmaternykh L N, van Loosdrecht P H M and Palstra T T M 2005 *Acta Cryst.* B**61** 481
[14] Rebbouh L, Desautels R D, Ritter C, Cadogan J M, Temerov V, Pankrats A and van Lierop J 2011 *Phys. Rev.* B **83** 140406(R)
[15] Popova E A, Tristan N, Vasiliev A N, Temerov V L, Bezmaternykh L N, Leps N, Büchner B and Klingeler R 2008 *Eur. Phys. J.* B **62** 123
[16] Gudim I A, Pankrats A I, Durnaĭkin E I, Petrakovskiĭ G A, Bezmaternykh L N, Szymczak R and Baran M 2008 *Cryst. Rep.* **53** 1140
[17] Volkov D V, Demidov A A and Kolmakova N P 2008 *JETP* **106** 723
[18] Rodríguez-Carvajal J 1993 *Physica* B **192** 55 http://www.ill.eu/sites/fullprof/php/downloads.html
[19] Ritter C 2011 *Sol. St. Phenomena* **170** 263
[20] Neogy D, Chattopadhay K N, Chakrabarti P K, Sen H and Wanklyn B M 1996 *JMMM* **154** 127 and 1997 *J. Phys. Chem. Solids* **59** 783
[21] Chaudhury R P, Lorenz B, Sun Y Y, Bezmaternykh L N, Temerov V L and Chu C W 2010 *Phys. Rev.* B **81** 220402(R)

Journal of Physics: Conference Series **340** (2012) 012066 doi:10.1088/1742-6596/340/1/012066

Neutron scattering studies of the spin-5/2 antiferromagnetic linear trimer substance $SrMn_3P_4O_{14}$

Masashi Hase[1], Andreas Dönni[1], Kiyoshi Ozawa[1], Hideaki Kitazawa[1], Osamu Sakai[1], Vladimir Yu. Pomjakushin[2], Lukas Keller[2], Koji Kaneko[3], Naoto Metoki[3], Kazuhisa Kakurai[3], Masaaki Matsuda[4], Tao Yang[5], Rihong Cong[5] and Jianhua Lin[5]

[1]National Institute for Materials Science (NIMS), 1-2-1 Sengen, Tsukuba, Ibaraki 305-0047, Japan

[2]Laboratory for Neutron Scattering, Paul Scherrer Institut (PSI), CH-5232 Villigen PSI, Switzerland

[3]Japan Atomic Energy Agency (JAEA), 2-4 Shirakata Shirane, Tokai, Naka, Ibaraki 319-1195, Japan

[4]Oak Ridge National Laboratory (ORNL), Oak Ridge, TN 37831, USA

[5]College of Chemistry and Molecular Engineering, Peking University, Beijing 100871, People's Republic of China

E-mail: HASE.Masashi@nims.go.jp

Abstract.
We studied the spin-5/2 antiferromagnetic (AF) linear trimer substance $SrMn_3P_4O_{14}$ using neutron powder diffraction and inelastic neutron scattering (INS) techniques. We determined the magnetic structure below $T_{N1} = 2.2(1)$ K. Ordered magnetic moments have x and z components. The angle between two neighboring moments in the trimer is 164 or 166° at 1.5 K, indicating that the intra-trimer (J_1) interaction is AF as expected. The propagation vector is expressed as $\mathbf{k} = [0, k_y, 0]$. The value of k_y is 0.317 at 1.5 K, indicating an incommensurate magnetic structure. The value of k_y changes abruptly between 1.7 and 1.8 K and reaches values close to 0.33 above 1.8 K, indicating another phase transition at $T_{N2} = 1.75(5)$ K. We observed plural magnetic excitations. The peak positions are 0.46, 0.68 and 1.02 meV. Our INS results are consistent with those expected in the trimer model. We evaluated the J_1 value as 0.29 meV (3.4 K).

1. Introduction

This paper describes an investigation of the spin-5/2 substance $SrMn_3P_4O_{14}$ [1]. The temperature T dependence of the magnetic susceptibility indicates occurrence of a magnetic long-range order (LRO) below about 2.6 K. A 1/3 magnetization plateau was observed at both 1.3 K and 4.2 K (below and above the transition temperature) [1, 2]. The plateau was observed in a powder sample. Therefore, the plateau can appear irrespective of the applied magnetic field direction. As a result, the plateau is a quantum-mechanical magnetization plateau generated by an energy gap in the magnetic excitation spectrum. Although the spin value is large (5/2), the quantum-mechanical nature remains even in the ordered state.

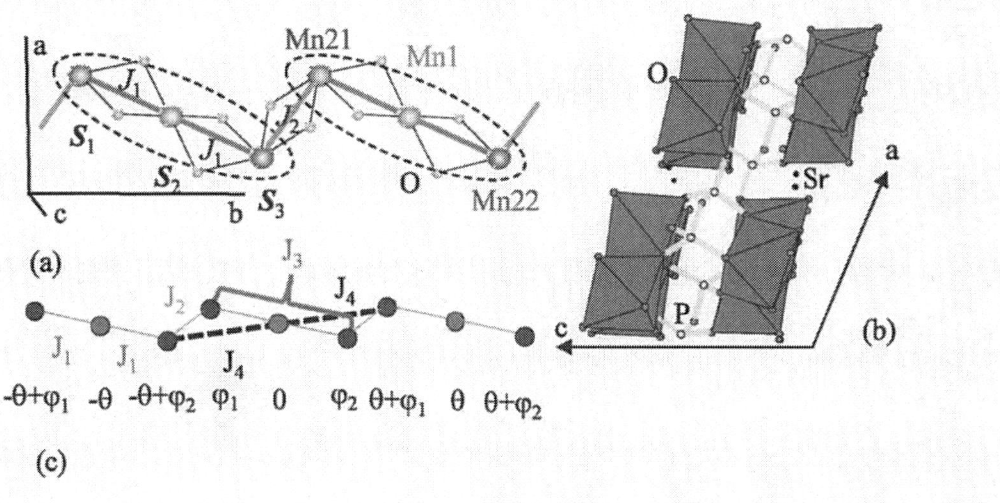

Figure 1. (a) A schematic drawing of positions of Mn^{2+} ions ($3d^5$) having localized spin 5/2 in $SrMn_3P_4O_{14}$ [1]. Two crystallographically independent Mn^{2+} sites (Mn1 and Mn2) exist. The Mn2 sites are divided into two sites (Mn21 and Mn22) for the magnetic structure analysis. Two kinds of short Mn-Mn bonds exist and have Mn-O-Mn paths. The Mn-Mn distances are 3.26 and 3.31 Å at 4.0 K. The exchange interaction parameters are respectively defined as J_1 and J_2. The dominant AF J_1 interactions form the spin trimers indicated by the ellipses. The Hamiltonian is expressed as $\mathcal{H} = J_1(S_1S_2 + S_2S_3)$. The J_1 and J_2 interactions form a trimerized chain that is almost parallel to the b axis. The Mn-Mn distances in the other bonds are more than 4.89 Å. These bonds have no Mn-O-Mn paths. (b) Projection of the structure along the b axis. Trimerized chains in octahedra might be mutually coupled magnetically in the a and c directions through Mn-O-P-O-Mn paths. (c) The spin system in which next-nearest-neighbor exchange interactions (J_3 and J_4) in a chain are taken into account. The parameters φ_1, φ_2, θ and their linear combinations indicate angles between the center Mn1 moment and each moment.

We could explain the magnetic-field H and T dependence of the magnetization using the spin-5/2 linear trimer formed by the antiferromagnetic (AF) J_1 interaction (J_1 = 4.0 K) depicted as ellipses in Fig. 1(a) [2]. We infer that weak inter-trimer interactions stabilize the magnetic LRO. In order to confirm our model, we performed neutron powder diffraction and inelastic neutron scattering experiments. We deduce the signs of exchange interactions from the magnetic structure determined from neutron powder diffraction results. We expect to observe magnetic excitations of the trimer.

2. Methods of experiments

We synthesized $SrMn_3P_4O_{14}$ under hydrothermal conditions [1]. We carried out neutron powder diffraction experiments at the SINQ spallation source of Paul Scherrer Institute (Switzerland). We used the high-resolution powder diffractometer for thermal neutrons HRPT [3] (wavelength λ = 1.886 Å, high intensity mode $\Delta d/d \geq 1.8 \times 10^{-3}$) and the high-intensity cold neutron powder diffractometer (DMC) with λ = 2.458 Å. We performed Rietveld refinements of crystal and magnetic structures using the FULLPROF [4] program with the use of its internal tables for scattering lengths and magnetic form factors.

We carried out inelastic neutron scattering (INS) measurements using the cold neutron triple-axis spectrometer LTAS installed at JRR-3M in JAEA. The final neutron energy was fixed at 2.6 meV. Higher-order beam contamination was effectively eliminated using a cooled Be filter before the sample. The horizontal collimator sequence was guide-80'-Be-sample-120'-open. This setup yields an energy resolution of 0.1 meV (full width at half maximum, FWHM) at an energy transfer $\omega = 0$ meV. The resolution was determined from incoherent scattering of the sample.

3. Neutron powder diffraction results

The crystal structure modeled in space group $P2_1/c$ (no. 14) proposed in [1] fits our data well. The structure parameters were refined from a HRPT diffraction pattern collected at 4.0 K (a paramagnetic state) and were kept fixed in the subsequent refinements of the magnetic structure. The lattice constants at 4.0 K amounted to $a = 7.661(1)$ Å, $b = 7.784(1)$ Å, $c = 9.638(1)$ Å and $\beta = 111.70(2)°$.

Many magnetic reflections were observed below $T_{\rm N1} = 2.2(1)$ K and could be indexed with a propagation vector $\mathbf{k} = [0,\ k_y,\ 0]$ with $k_y \sim 0.3$. Using the determined propagation vector we performed the symmetry analysis according to Izyumov *et al.* [5] to derive possible magnetic configurations for Mn1 (2d) and Mn2 (4e) sites. For the calculation of the basis functions we used program BASIREP [4]. The little group of a propagation vector for the space group $P2_1/c$ contains only two symmetry operators: 1 and $\{2_y|0\frac{1}{2}\frac{1}{2}\}$ resulting in the splitting of Mn2 (4e) into two independent orbits Mn21 and Mn22 sites. The little group has two one-dimensional 1D irreducible representations (IRs) τ_1 and τ_2 with the characters for 2_y: $\pm e^{-i\pi k_y}$, respectively. Both IRs enter three times the magnetic representation for all three Mn-sites. The basis functions for τ_2 dictate the absence of the $(0, 1-k, 0)$ magnetic reflection, which indeed has zero intensity. Therefore, we can immediately choose τ_2. Nevertheless, we have performed a simulated annealing search of the magnetic structures for both IRs. The τ_2 produced excellent results, although the τ_1 failed. We found that the y-components of magnetic moments can be ignored and performed final refinements under the following conditions. Only the x and z components of magnetic moments are considered. The Mn21 and Mn22 sites are crystallographically identical. Therefore, the magnitudes of Mn21 and Mn22 moments are constrained as identical. In general, the magnetic structure consists of ellipsoidal spirals running on all three Mn-sites with independent phases. The spin 5/2 of Mn^{2+} ions is almost isotropic [2]. Therefore, we assumed a circular spiral magnetic structure.

We determined the magnetic structure using DMC diffraction patterns (Fig. 2). Magnetic moments have x and z components. The angle between Mn1 and Mn21 (Mn22) moments in the trimer is 164° (166°) at 1.5 K. The angle between Mn21 and Mn22 moments in the trimer is 2°. The angle between two Mn2 moments in the J_2 bond is 111°. These angles indicate that both the J_1 and J_2 interactions are AF. Figure 3 shows the T dependence of k_y and magnitudes of the Mn1 and Mn2 moments ($M1$ and $M2$). As was described, the magnitudes of Mn21 and Mn22 moments were constrained as identical. The value of k_y is 0.317 at 1.5 K, indicating an incommensurate magnetic LRO. The value of k_y changes abruptly between 1.7 and 1.8 K and reaches values close to 0.33 above 1.8 K, indicating another phase transition at $T_{\rm N2} = 1.75(5)$ K. We cannot determine whether the phase between $T_{\rm N2}$ and $T_{\rm N1}$ is commensurate or incommensurate. $M1$ and $M2$ increase with decreasing T and change rapidly around $T_{\rm N2}$.

We consider the origin of the long-range period of the magnetic structure. Competition among plural exchange interactions must be necessary. Therefore, we take next-nearest-neighbor exchange interactions (J_3 and J_4) in a chain into account [Fig. 1(c)]. According to the experimental results, we consider the coplanar magnetic structure. The relative angle is defined by one parameter for each magnetic moment. The classical magnetic energy per trimer is expressed as follows.

$$\begin{aligned} E &= J_1 M_1 M_2 (cos\varphi_1 + cos\varphi_2) + J_2 M_2 M_2 cos(\theta + \varphi_1 - \varphi_2) \\ &\quad + J_3 M_2 M_2 cos(\varphi_1 - \varphi_2) + J_4 M_1 M_2 [cos(\theta + \varphi_1) + cos(\theta - \varphi_2)]. \end{aligned} \tag{1}$$

We examine whether we can reproduce the experimental value of θ (119°) at 1.8 K under the

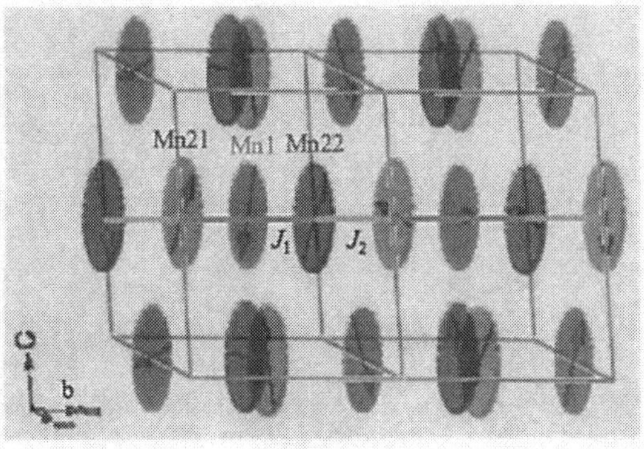

Figure 2. The incommensurate magnetic structure of $SrMn_3P_4O_{14}$ at 1.5 K.

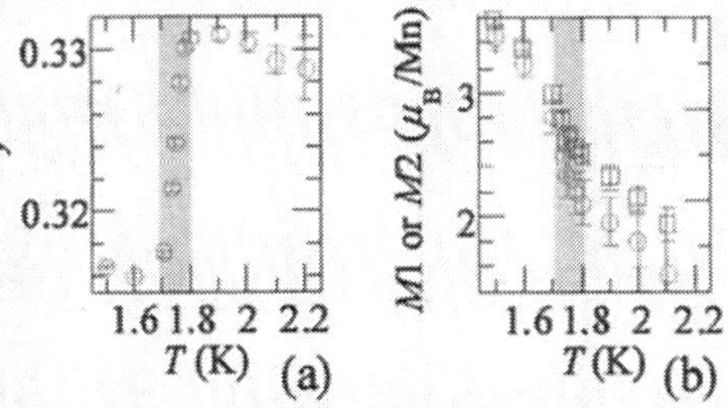

Figure 3. Temperature dependence of several parameters of $SrMn_3P_4O_{14}$. The hatched area shows the transition region. (a) The y component of the propagation vector. (b) The magnitudes of the magnetic moments of Mn1 (circles) or Mn2 (squares). The magnitudes of Mn21 and Mn22 moments were constrained as identical.

condition that $\varphi_1 = 170°$, $\varphi_2 = 160°$ and $M_1/M_2 \equiv \alpha = 0.84$ are given. The θ dependent part of the energy is expressed as follows.

$$E'(\theta)/(J_2M_2^2) = cos(\theta + \varphi_1 - \varphi_2) + j_4\alpha[cos(\theta + \varphi_1) + cos(\theta - \varphi_2)]. \tag{2}$$

$E'(\theta)/(J_2M_2^2)$ shows a minimum at $\theta = 119°$ when $j_4 = 0.58$. Consequently, the long-range period of the magnetic structure may be explained by the competition between nearest-neighbor and next-nearest-neighbor exchange interactions in a chain. The following problems should be solved in further investigations. E is unchanged when we exchange φ_1 and $-\varphi_2$. E can be a minimum when $\varphi_1 = -\varphi_2$. Therefore, this model cannot explain the experimental values of φ_1 and φ_2. In addition, this model cannot explain the reason why the coplanar magnetic structure appears. We should consider anisotropy. The origin of the phase transition at $T_{N2} = 1.75(5)$ K is not determined. Dipole-dipole interactions may be relevant at low T.

4. Inelastic neutron scattering (INS) results

We performed all the INS measurements above $T_{N1} = 2.2(1)$ K. Circles in Fig. 4(a) show the ω dependence of INS intensity (constant-Q scan spectrum) at $Q = 1.0$ Å^{-1} and 5.0 K. The value of Q is the magnitude of the scattering vector. Excitations are apparent between 0.5 and 1.5 meV. Two kinds of excitations with different peak positions seem to overlap each other. The spectrum is almost independent of Q except for difference in intensities (not shown). Probably, dispersion of the excitations is weak if it exists. The weak Q dependence indicates that the excitations are intrinsically transitions between discrete energy levels. We consider that the intensities below 0.5 meV cannot be explained only by incoherent scattering. Probably, low-energy excitations exist. Circles in Fig. 4(b) show the constant-Q scan spectrum at $Q = 1.0$ Å^{-1} and 11.2 K. The excitations between 0.5 and 1.5 meV are again seen. Intensities around 0.5 meV are larger at 11.2 K than at 5.0 K. We will explain curves in Fig. 4 later.

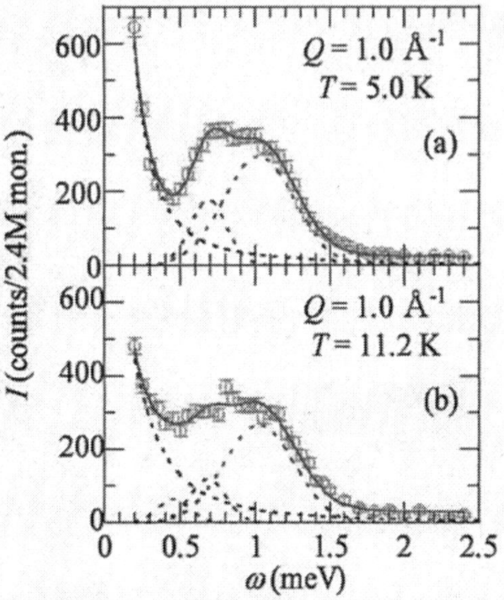

Figure 4. The inelastic neutron scattering intensity *vs.* energy transfer (ω) [constant-Q (magnitude of the scattering vector) scan spectra] of $SrMn_3P_4O_{14}$ (circles) at $Q = 1.0$ Å^{-1} at 5.0 K (a) and 11.2 K (b). The solid line represents the sum of two or three Gaussians and one Lorentzian (plus constant backgrounds). The dashed line indicates each Gaussian or Lorentzian.

We examined whether the spin-5/2 AF trimer model with $J_1 = 4.0$ K can account for the observed excitations or not. Figures 5(a) and 5(b) depict respectively a schematic drawing of low-lying energy levels and the T dependence of calculated occupation ratio of five energy levels. The INS intensity is proportional to the occupation ratio except for a coefficient determined from corresponding eigen states. Excitations from the ground state (GS) are dominant at 5.0 K. We consider that the two excitations between 0.5 and 1.5 meV correspond to transitions from GS to the first-excited and second-excited states (1ES and 2ES). The respective energy differences are $2.5J_1$ and $3.5J_1$. We compared the spectrum above 0.2 meV in Fig. 4(a) with a sum of two Gaussians and one Lorentzian (plus constant backgrounds).

$$I(\omega) = \frac{I_0 a_0}{\pi} \frac{1}{\omega^2 + a_0^2} + \Sigma_i \frac{I_i}{\sqrt{\pi} a_i} \exp[-\frac{(\omega - \omega_i)^2}{a_i^2}] + I_{\rm BG}. \quad (3)$$

Here the sum is from $i = 1$ to 2. The peak position is assumed finite and 0 for the Gaussians and Lorentzian, respectively. The sum of two Gaussians and one Lorentzian (solid line) reproduces well the spectrum in Fig. 4(a). The peak position is 0.68 meV or 1.02 meV. The value of J_1 is evaluated as 0.27 meV (3.2 K) or 0.29 meV (3.4 K). These values are slightly smaller than the value determined in the magnetization results ($J_1 = 4.0$ K). The peak width (FWHM) is 0.28 meV and 0.55 meV for the 0.68 meV and 1.02 meV excitation, respectively. These widths are larger than the energy resolution of 0.1 meV at $\omega = 0$ meV, indicating existence of weak dispersion.

We compare the constant-Q scan spectrum at 11.2 K in Fig. 4(b) with a calculated curve. From the calculated results in Fig. 5(b), transitions from 1ES or 2ES are also expected at 11.2 K. The transition from 2ES to 4ES may increase the intensities around 0.5 meV. The energy difference is $1.5J_1$. We compared the spectrum above 0.2 meV at 11.2 K with a sum of three Gaussians and one Lorentzian (plus constant backgrounds) given in Eq. (3) with $i = 1$ to 3. To reduce variable parameters, we assumed that the peak position (0 meV) of the Lorentzian, and the peak positions (0.68 and 1.02 meV) and widths (0.28 and 0.55 meV) of the two Gaussians were constant. We used the values obtained at 5.0 K. This assumption is reasonable for transitions between discrete energy levels. The sum of the three Gaussians and

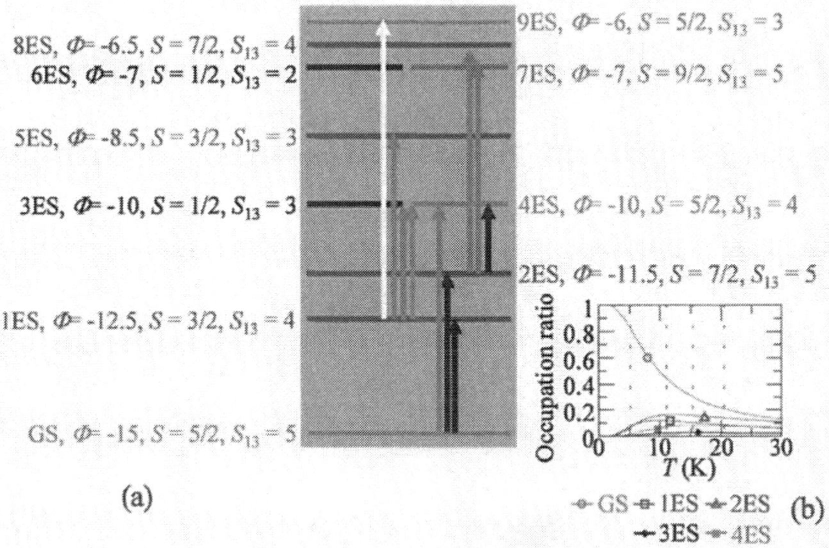

Figure 5. (a) A schematic drawing of low-lying energy levels in the spin-5/2 AF linear trimer [ground state (GS) and excited states (ESs)]. The parameters $\epsilon \equiv E/J_1$ and S indicates the eigen energy and the total spin, respectively. S_{13} is defined in the text. To distinguish two degenerate eigen states with $\epsilon = -10$, we name the two states 3ES and 4ES. To distinguish two degenerate eigen states with $\epsilon = -7$, we name the two states 6ES and 7ES. The arrows indicate allowed transitions from GS, 1ES or 2ES. We observed the three black-arrow transitions. The gray-arrow transitions may exist. We could not detect the white-arrow transition. (b) The temperature dependence of calculated occupation ratio of the low-lying energy levels when $J_1 = 4.0$ K.

one Lorentzian (solid line) reproduces well the spectrum at 11.2 K. The peak position of the third Gaussian is 0.46 meV. The value of J_1 is evaluated as 0.30 meV (3.5 K) and is close to the values evaluated from the other two transitions.

The following selection rules are derived theoretically.[6]

$$\Delta S = 0, \pm 1, \quad \Delta M = 0, \pm 1 \quad \text{and} \quad \Delta S_{13} = 0, \pm 1. \tag{4}$$

$S_i (i = 1, 2, 3)$ is the spin operator in the trimer. S and S_{13} are defined as $S_1 + S_2 + S_3$ and $S_1 + S_3$, respectively. M is the z component of S. Probably, the transitions from GS, 1ES or 2ES are dominant below 11.2 K. We can observe transitions with ω up to $8J_1$ in the present experiments. The arrows in Fig. 5(a) indicate allowed and observable transitions from GS, 1ES or 2ES. We observed the three black-arrow transitions. The gray-arrow transitions may exist. However, we could not prove the existence of the gray-arrow transitions. Energy differences of some gray-arrow transitions ($2.5J_1$ and $4J_1$) are the same as or close to the energy differences of black-arrow transitions ($2.5J_1$ and $3.5J_1$). Therefore, we could not extract contribution of the gray-arrow transitions from the experimental results. Energy differences of the other gray-arrow transitions are $4.5J_1 = 1.3$ meV and $5J_1 = 1.4$ meV when $J_1 = 0.29$ meV. Small INS intensities are seen around these energies in Fig. 4. In our analyses, the small intensities correspond to the tail of the 1.02 meV excitation. However, the peak width is larger in the 1.02 meV excitation (FWHM = 0.55 meV) than in the 0.46 meV excitation (FWHM = 0.25 meV) or the 0.68 meV excitation (FWHM = 0.28 meV). The gray-arrow transitions may exist in the tail. We could not detect the white-arrow transition. We do not have calculated intensities of transitions. Therefore, we could not determine the reason why we could not detect the white-arrow transition. The INS intensity of the white-arrow transition may be very small.

5. Summary

We performed neutron powder diffraction and inelastic neutron scattering (INS) experiments of powders of the spin-5/2 antiferromagnetic (AF) linear trimer substance $SrMn_3P_4O_{14}$. Magnetic reflections appear below $T_{N1} = 2.2(1)$ K. We determined the magnetic structure. Ordered magnetic moments have x and z components. The angle between two neighboring moments in trimers is 164 or 166° at 1.5 K, indicating that the J_1 interaction is AF as expected. The propagation vector is expressed as $\mathbf{k} = [0, k_y, 0]$. The value of k_y is 0.317 at 1.5 K, indicating an incommensurate magnetic structure. The value of k_y is abruptly changed between 1.7 and 1.8 K and reaches values close to 0.33 above 1.8 K, indicating another phase transition at $T_{N2} = 1.75(5)$ K. We observed plural magnetic excitations in INS experiments. The peak positions are 0.46, 0.68 and 1.02 meV. These experimental results are consistent with those expected in the trimer model with the J_1 value of 0.29 meV (3.4 K).

References

[1] Yang T, Zhang Y, Yang S, Li G, Xiong M, Liao F and Lin J 2008 *Inorg. Chem.* **47** 2562
[2] Hase M, Yang T, Cong R, Lin J, Matsuo A, Kindo K, Ozawa K and Kitazawa H 2009 *Phys. Rev.* B **80** 054402
[3] Fischer P, Frey G, Koch M, Konnecke M, Pomjakushin V, Schefer J, Thut R, Schlumpf N, Burge R, Greuter U, Bondt S and Berruyer E 2000 *Physica* B **276-278** 146; http://sinq.web.psi.ch/hrpt/
[4] Rodriguez-Carvajal J 1993 *Physica* B **192** 55; http://www.ill.eu/sites/fullprof/
[5] Izyumov Y A, Naish V E and Ozerov R P 1991 *Neutron Diffraction of Magnetic Materials* (New York: Consultants Bureau)
[6] Furrer A and Güdel H U 1979 *J. Magn. Magn. Mater.* **14** 256

Effect of uniaxial pressure on helimagnetic structure in Lu_2Fe_{17}

J Kamarád[1], M Míšek[1], K Prokeš [2], S Maťaš [2] and Z Arnold[1]

[1] Institute of Physics, ASCR, v.v.i., Na Slovance 2, Prague 8, Czech Republic
[2] Helmholtz Zentrum Berlin, BENSC, Hahn-Meitner-Platz 1, Berlin, Germany

E-mail: kamarad@fzu.cz

Abstract. The incommensurate helimagnetic structure (IHS) in the Lu_2Fe_{17} intermetallics has been induced at 5 K by hydrostatic pressure above 0.3 GPa and characterized by a propagation vector ($00t_z$). The observed dependence of t_z on external variables is discussed within a simple model based on a complex dependence of exchange interactions $J_{n,m}$ (n, m = f, j, k and g) on distances between Fe-atoms in inequivalent positions 4f, 12j, 12k and 6g of the hexagonal $P6_3/mmc$ crystal structure. To verify this model, we have studied magnetization and representative neutron reflections of Lu_2Fe_{17} single crystals under high hydrostatic pressure and uniaxial compression with stress σ || a-axis and σ || c-axis. Both, the metamagnetic transitions and magnetic satellites of the ($00t_z$) type that verify the presence of IHS in Lu_2Fe_{17} at 5 K were observed under hydrostatic pressure and under uniaxial stress σ || c-axis only. We show that the dominant role in an appearance of IHS in the Lu_2Fe_{17} under compression is played by the J_{ff} and J_{fj} exchange interactions.

1. Introduction

The family of the R_2Fe_{17} (R = rare earth and Y) intermetalics is studied for decades, due to very diverse ferro- and ferrimagnetic properties that exhibit almost all members of this family [1,2]. The origin of a variety of magnetic structures, magnetoelastic effects and magnetocrystalline anisotropy is mostly ascribed to the rare earth element characteristics and their interaction with the ferromagnetic Fe-sublattice [2]. From this point of view, the R_2Fe_{17} intermentallics with nonmagnetic R-elements (R = Lu and Y) offer a possibility to study magnetic behavior of the only Fe-sublattice in dependence on different external (T, P, H) variables.

The ground state of the Lu_2Fe_{17} intermentallic compound is ferromagnetic, but, an incommensurate helimagnetic structure (IHS) was detected in a very narrow temperature range just below the Néel temperature T_N [3]. A possible existence of non-collinear arrangement of magnetic Fe-moments in Fe-rich metallic systems (including Invar alloys) is discussed by theoreticians and searched for by experimentalists for almost a century. A similarity of magnetic behavior of Lu_2Fe_{17} and Y_2Fe_{17} intermetalics and Invar alloys is documented by their negative linear thermal expansions at temperatures below T_N or the Curie temperature T_C, respectively. Both, the thermal expansion and the linear compressibility are strongly anisotropic with a pronounced decrease of the crystal c-parameter with increasing both, temperature and pressure [4]. Moreover, despite of a complex structure of these intermetallics, the ab-initio calculated local electronic structure of inequivalent Fe-sites exhibit identical features as the electronic structure of Invar alloys [5]. These exciting problems were the reason for intensive studies of both, the Lu_2Fe_{17} and the Y_2Fe_{17} intermentallics under pressure during the last decade.

The ferromagnetic ground state is totally suppressed in Lu_2Fe_{17} and Y_2Fe_{17} intermentallics by hydrostatic pressures above 0.3 GPa and 0.8 GPa, respectively [6-8]. The pressure induced IHS that is described by a propagation vector $(00t_z)$ [3,7], is however, closely related to the original ferromagnetic structure and underlining hexagonal crystal structure of these intermetallics that is shown in figure 1. The ferromagnetic arrangement remains preserved in the basal *ab*-planes (easy magnetization plane), but, magnetization of the planes rotates as one moves along the crystal *c*-axis (that is the hard magnetization direction). Under the hydrostatic pressure, the ferromagnetic state in the intermetallics can be restored by external magnetic field only. The strong pressure dependence of the critical field H_C applied in the basal plane of Lu_2Fe_{17} along *a*-axis is shown in inset of figure 2, where magnetization isotherms measured at 5 K under different hydrostatic pressures are presented. Magnetic phase diagram of the Lu_2Fe_{17} [9], constructed from magnetization measurements under hydrostatic pressure, is shown in figure 2.

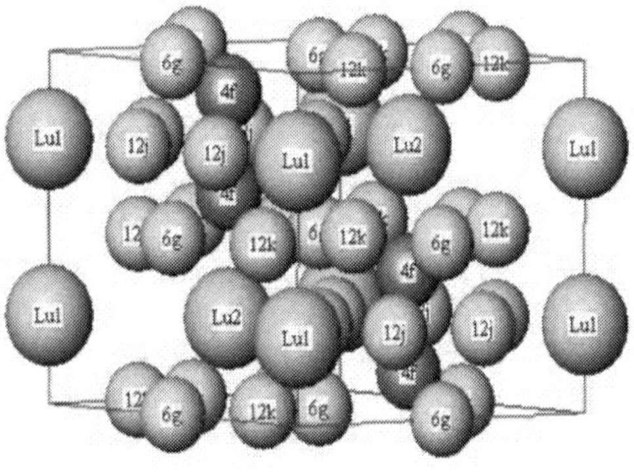

Figure1. The hexagonal $P6_3/mmc$ crystal structure of Lu_2Fe_{17}.

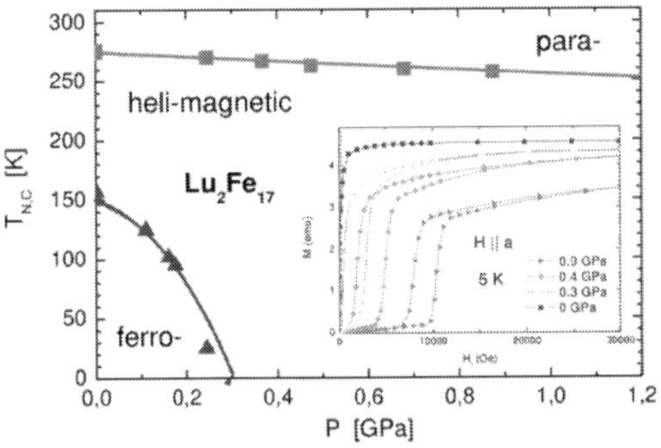

Figure 2. The magnetic phase diagram of Lu_2Fe_{17}.

A use of uniaxial stresses to a study of the Lu_2Fe_{17} intermentallics in this work has been motivated by a need to understand on a more detailed level the role of direct exchange interactions in the occurrence of the non-collinear Fe-moments arrangement in the these intermetallics. Based on the recent experimental data [10,11], we present a simple phenomenological model for the description of interplay of the Fe-Fe exchange interactions in Lu_2Fe_{17} . The observed effects of uniaxial stresses and hydrostatic pressures on the magnetic transitions in the Lu_2Fe_{17} intermentallics are presented and discussed.

2. Experimental

The hexagonal Lu_2Fe_{17} single crystal ($P6_3/mmc$ space group, a = 8.395 Å, c = 8.293 Å) has been grown by the Czochralski method and oriented using the Laue X-ray diffraction method [12]. Miniature pressure cells (up to 1.2 GPa) were designed and fabricated with respect to the experimental arrangement of the E4 neutron diffractometer (λ = 2.42 Å) in connection with vertical and horizontal superconducting magnets installed at the Helmholtz Zentrum Berlin. A detail description of the hydrostatic CuBe cells for both, the microscopic (neutron scattering) and the macroscopic (magnetization) studies of single crystals under pressure can be found elsewhere [13]. In contrast to the isotropic effect of hydrostatic pressure, the uniaxial compression of a single crystal induces a decrease of one crystal parameter (along the stress) and an increase of other (perpendicular) crystal parameters. The uniaxial cell presented in figure 3 allows to compress the Lu_2Fe_{17} single crystal along the *c*-axis inside the E4 diffractometer (or a SQUID magnetometer) while keeping its easy magnetization direction (EMD) *a*-axis || H || axis of rotation of diffractometer (or magnetometer). To induce similar decrease of the crystal a-parameter, a single crystal can be compressed along the *a*-axis with the EMD *b*-axis || H || axis of rotation of diffractometer (or magnetometer). Both experimental arrangements enable the study of the ($00t_z$) propagation vector by reciprocal space scans due to the permanent orientation of the *c*-axis within the diffraction plane. The uniaxial pressure cell for neutron diffraction experiments is made of non-magnetic $TiAl_6V_4$ alloy and sliding parts of the prism (part 4 in figure 3) are made of a zirkon ceramics. The prism ensures compression of a sample in direction perpendicular to the axial force induced by a barrel spring (part 3 in figure 3). In a working position, the prism is fixed inside a tube (part 6 in figure 3) and all parts are inserted into the cell. The axial force is increased by a compression of the barrel spring by supporting screws (parts 2 in figure 3). The uniaxial pressure can be increased up to the yield stress of a studied material.

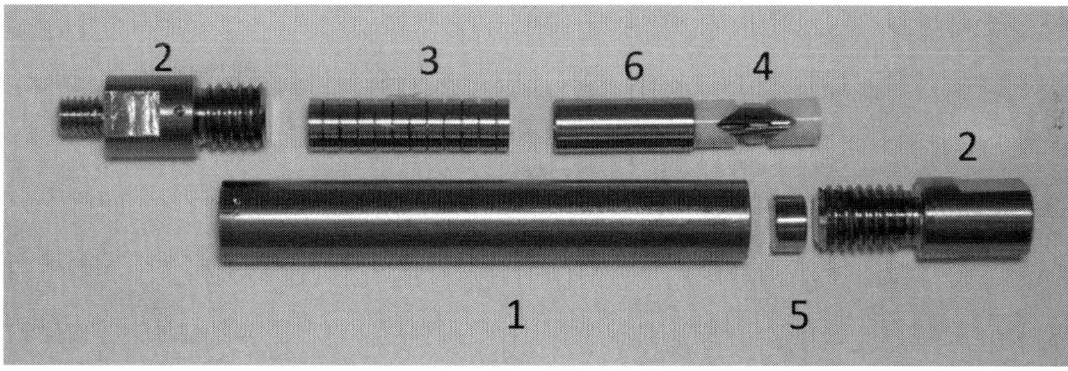

Figure 3. Uniaxial pressure cell for neutron diffraction studies: 1 - cell, 2 - supporting screws, 3 - barrel spring, 4 - prism, 5 - washer, 6 - fixing tube.

3. **Results**

3.1. Experimental results

The magnetization measurements were accompanied by representative neutron diffraction studies under identical external (T, P, H) conditions. Figure 4 shows reciprocal space scans along [00*l*] direction in the vicinity of the nuclear (002) reflection at ambient pressure. Magnetic satellites ($002\pm t_z$) were observed at temperatures below the Néel temperature, T_N = 275 K, down to a critical temperature for the ferromagnetic state, T_C = 150 K. Similar magnetic satellites were also found around (101), (300) and (004) nuclear reflections. Satellites of a (t_x t_y 0) type have never been detected. The propagation vector t_z decreases with decreasing temperature and intensity of the nuclear reflections increases after transition of Lu_2Fe_{17} into ferromagnetic state (see figure 4). IHS induced by hydrostatic pressure or by uniaxial compression of crystal along c-axis exhibits magnetic satellites of the same type. Figure 5 shows magnetic satellites ($002\pm t_z$) with a much larger value of t_z. These reflections are induced at 5 K by hydrostatic pressure of 0.9 GPa and are suppressed by external magnetic field of 3 T without a change of intensity of the nuclear reflections.

The temperature dependence of the propagation vector t_z of the pressure induced IHS is presented in figure 6. Generally, the value of t_z increases with increasing pressure or uniaxial compression.

However, a linear decrease of t_z with decreasing temperature at ambient pressure changes its character under hydrostatic pressure. In the case of uniaxial compression of crystal along c-axis, t_z decreases with increasing temperature up to T_C and then increases along the ambient $t_z(T)$ curve. To some extent, a more complex character of the $t_z(T)$ curve measured under hydrostatic pressure can be ascribed to a decrease of hydrostatic pressure inside the Cube cell with decreasing temperature. This effect is caused by a difference in thermal expansions of the CuBe cell and the pressure transmitting medium.

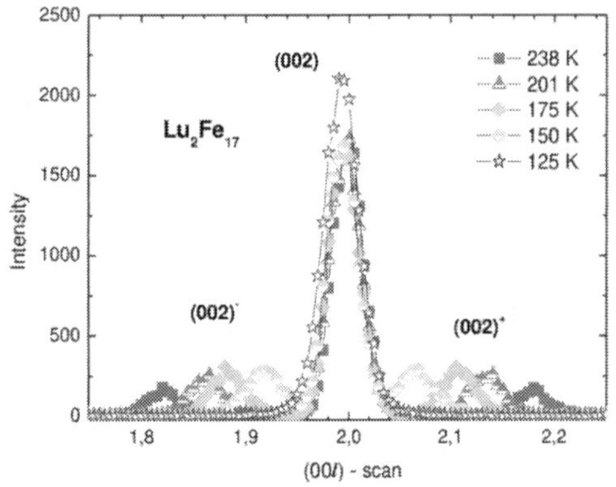

Figure 4. Temperature dependence of magnetic ($002\pm t_z$) satellites of nuclear (002) reflection at ambient pressure.

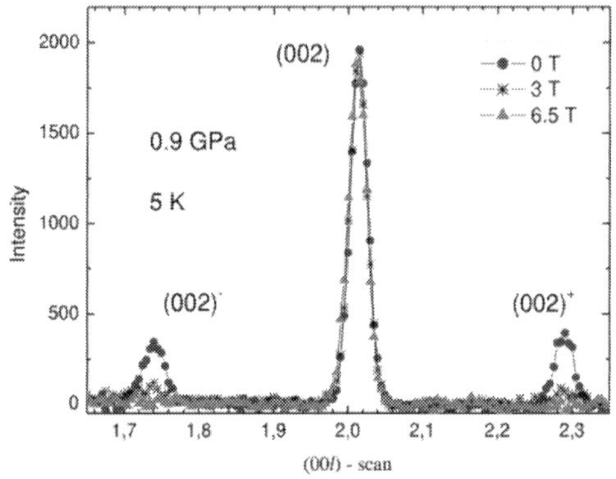

Figure 5. Magnetic satellites of the nuclear (002) reflection induced by pressure 0.9 GPa at 5 K at different magnetic fields.

The magnetization isotherms that were measured at 5 K under different pressures are shown in figure 7. Identical magnetization isotherms have been measured at ambient pressure and under uniaxial compression of crystal along a-axis up to the yield stress of Lu_2Fe_{17}. Hence, we were not able to induce IHS in this case by the uniaxial compression and these two magnetization isotherms at 5 K reflect the ferromagnetic ground state at ambient pressure and under uniaxial $\sigma \parallel$ a-axis. The metamagnetic transitions that verify the presence of IHS in Lu_2Fe_{17} at 5 K were observed under hydrostatic pressure and under uniaxial stress $\sigma \parallel$ c-axis only, as shown in figure 7.

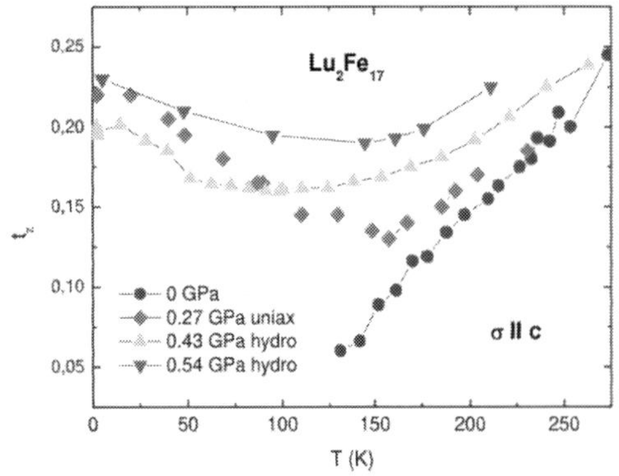

Figure 6. Temperature and pressure dependence of the propagation vector t_z.

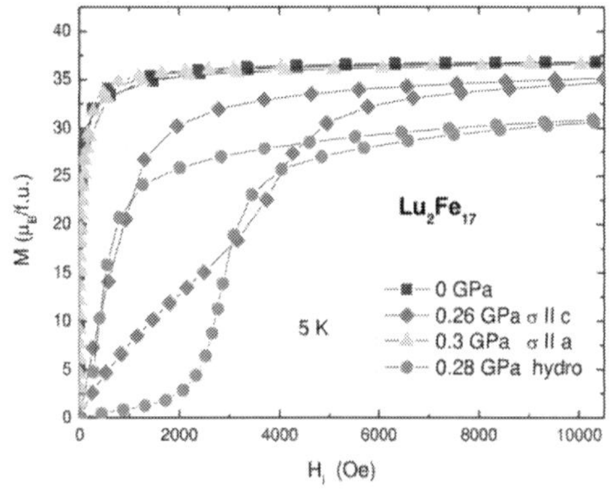

Figure 7. Magnetization isotherms of Lu_2Fe_{17} at 5K under different pressures.

3.2. Phenomenological model of the incommensurate helimagnetic structure

A simple phenomenological model of IHS in the Lu_2Fe_{17} intermetallic compound is based on a complex dependence of exchange interactions $J_{n,m}$ (n,m = f, j, k, g) on distance d_{Fe-Fe} between Fe atoms placed in the inequivalent 4f, 12j, 12k and 6g crystal positions, see figure 1. Values of $J_{n,m}$ were determined from experimental Mössbauer data recently [10,11]. These dependencies nicely follow the known Bethe-Slater curve with extremely sharp slope of negative exchange interaction below d_{Fe-Fe} = 2.45 Å, see figure 8.

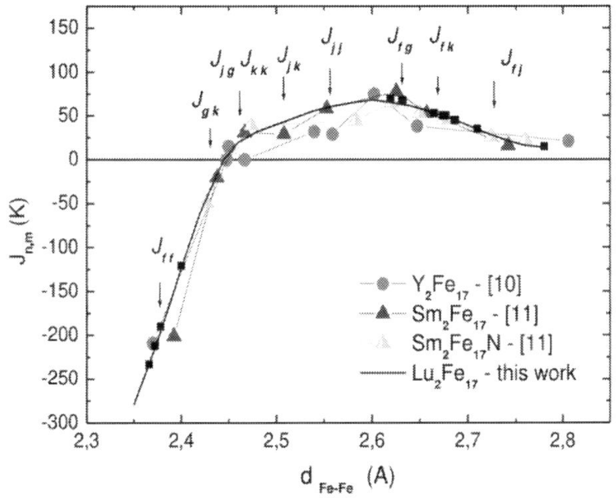

Figure 8. Dependence of exchange interactions $J_{n,m}$ on interatomic distances between Fe_n atoms placed in the inequivalent crystal lattice positions n= f, j, k, g

Table 1. Values of relative Fe-moment μ_n and direct exchange interactions $J_{n,m}$

n	μ_n	$J_{f,n}$(0 GPa)		$J_{f,n}$(1 GPa)	$J_{f,n}$(uniax)
		300K	5K	5K	
f	1.198	-212 K	-121 K	-233 K	-190 K
g	1.039	45 K	45 K	50 K	45 K
k	0.931	68 K	68 K	70 K	68 K
j	0.965	10 K	50 K	53 K	35 K
$-I_1/4I_2$		0.41	1.07	0.59	0.61
t_z		0.18	--	0.15	0.14

Considering the fact that the R-Fe interactions are much weaker than the direct exchange interaction $J_{n,m}$ between Fe atoms, we can assume that the experimentally determined dependence of $J_{n,m}$ on interatomic Fe-Fe distances presented in figure 8 is valid for all the R_2Fe_{17} intermetallics. Then, the values of $J_{n,m}$ in Lu_2Fe_{17} can be easily determined using the curve in figure 8 and known distances $d_{Fe\text{-}Fe}$ between Fe_n atoms placed in the inequivalent positions, n = f, g, k and j. The positions of Lu- and Fe-atoms in the studied Lu_2Fe_{17} intermetallics were precisely determined by Rietveld method from neutron diffraction experiments at room temperature [14]. The received experimental values agree well with results in literature [3,15]. The refinement also gave relative values of Fe-moments in these inequivalent positions, μ_n, with respect to μ_{Fe} = 2.15μ_B/at.Fe [14,15]. Data are presented in table 1. Thermal expansion and compressibility of Lu_2Fe_{17} were measured by both, the micro-strain gauges under hydrostatic pressures up to 1 GPa and by the neutron diffraction method [7,14]. The simple model of IHS is based on the layered crystal structure of Lu_2Fe_{17}, where the Lu+Fe_j plane and the Fe_k+Fe_g plane are the nearest neighbor planes of each the Fe_f plane. The next nearest plane of the Fe_f plane is again the Fe_f plane, see figure 1. Due to a significant decrease of exchange interaction with increasing $d_{Fe\text{-}Fe}$ above 2.7 Å, the total exchange interaction between the planes is proportional to a sum of the exchange interactions between Fe_f atom and its nearest neighbors, i.e. Fe_j, Fe_k and Fe_g atoms only. Hence, the exchange interactions J_{fj}, J_{fk}, J_{fg} and J_{ff} must be taken into account (black squares on the curve in figure 8). The positive values of the J_{jj}, J_{kk}, J_{kg} interactions ensure a collinear (ferromagnetic) arrangement of Fe-moments inside the Lu+Fe_j and the Fe_k+Fe_g planes. IHS appears when a positive interaction between the Fe_f plane and the nearest neighbor Lu+Fe_j and Fe_k+Fe_g planes is overcome by a negative interaction between the next nearest neighbor Fe_f planes. Then, Fe-moments in the Lu+Fe_j and Fe_k+Fe_g planes are frustrated and magnetization of these planes rotates with respect to Fe-moments in the Fe_f plane by an angle α.

The exchange energy is proportional to the following sum:

$$E_{ex} \approx -\mu_{Fe}^2 \{ I_1.\cos(\alpha) + I_2.\cos(2\alpha) \} \text{ with } I_1 = \mu_f N_f\{\mu_j N_j J_{fj} + \mu_g N_g J_{fg} + \mu_k N_k J_{fk}\} ; \; I_2 = \mu_f^2 N_f^2 J_{ff} ,$$

where N_n is number of the relevant nearest neighbors and α is an angle between the orientation of Fe-moments in the adjacent Fe_n planes. The minimum energy E_{ex} with respect to the angle α is given by a general relation $\partial E_{ex}/\partial\alpha = 0$ that leads to a relation, $\cos(\alpha) = -I_1/4I_2$. Hence, IHS appears when the absolute value of the negative I_2 interaction is larger than one quarter of the positive interaction I_1. The results describing the appearance of IHS at temperatures 300 K and 5 K at ambient pressure and at temperature 5 K under hydrostatic pressure 1 GPa and uniaxial compression along c-axis (σ = 0.3 GPa) are collected in table 1. The distances $d_{n,m}$ were calculated from the refined crystal lattice parameters of Lu_2Fe_{17}, a = 8.383 Å, c = 8.335 Å at temperature 5 K and a = 8.395 Å, c = 8.293 Å at temperature 300 K. The used values of a linear compressibility, $\kappa_a = 3\text{x}10^{-3}$ GPa^{-1}, $\kappa_c = 3.5\text{x}10^{-3}$ GPa^{-1} at 300K and $\kappa_a = 4.1\text{x}10^{-3}$ GPa^{-1}, $\kappa_c = 7.5\text{x}10^{-3}$ GPa^{-1} at 5K, well agree with values of κ_a and κ_c at temperature 300 K presented in [15].

4. Discussion of the effects of hydrostatic and uniaxial pressures

The negative thermal expansion plays a very important role in the appearance of IHS in Lu_2Fe_{17} at ambient pressure. Due to the larger d_{ff} distance at temperature 5K than at 300 K, the negative J_{ff} interaction at 5 K is not able to compete with the relatively strong positive interaction J_{fj} and the ferromagnetic state is stable at 5 K at ambient pressure. Vice versa, a decrease of the lattice c-parameter with increasing temperature leads to a strengthening of the negative J_{ff} interaction and a weakening of the positive J_{fj} interaction, see table 1. As a result, IHS appears at temperatures above 150 K, up to T_N at ambient pressure. In the case of Y_2Fe_{17}, the lattice parameters, a = 8.466 Å, c = 8.315 Å at room temperature, do not allow to reach a balance between I_2 and $I_1/4$. This intermetallic compound is ferromagnetic up to the temperature of its transition into paramagnetic state.

Under hydrostatic pressure, the high linear compressibility κ_c at 5 K induces a very strong increase of the negative J_{ff} interaction, see figure 8 that overcomes even an increase of the positive J_{fj} interaction, see table 1. Hydrostatic pressure increases with increasing temperature in the clamped CuBe cell. This effect probably compensates a decrease of compressibility κ_c with increasing temperature. So, the lattice parameters and the exchange interactions remain almost constant during heating of sample under hydrostatic pressure and values of the propagation vector t_z in dependence on temperature exhibit a flat minimum only. The lower pressure, the more pronounced minimum has been observed as it was expected from the temperature dependence of t_z at ambient pressure, see figure 6.

The effect of uniaxial compression of Lu_2Fe_{17} single crystal along its c-axis on the negative J_{ff} interaction at temperature 5 K is well comparable with the effect of hydrostatic pressure. However, a small increase of the crystal a-parameter induced by uniaxial compression leads to a smaller increase of the positive J_{fj} interaction than in the case of hydrostatic pressure. The uniaxial compression seems to be more effective at 5 K than the compression by hydrostatic pressure. On the other hand, the uniaxial stress is not able to compensate the mentioned decrease of compressibility κ_c with increasing temperature because the uniaxial stress σ is not affected by increasing or decreasing temperature (there is no pressure transmitting medium). Hence, the effect of uniaxial compression weakens with increasing temperature up to a temperature where the effect of negative thermal expansion becomes dominant, see figure 6. Moreover, the maximum of uniaxial stress σ is restricted by the yield stress of the studied material.

The direct exchange interaction exhibits the maximum in vicinity of d_{fk} and d_{fg} distances. Variations of these distances induced by pressure or temperature influence the relevant exchange interactions very slightly and so, J_{fk} and J_{fg} interactions do not play an important role in the studied effects. The same conclusion is also valid in the case of the positive J_{jj} and J_{kk} exchange interaction, see figure 8, and so, pressure does not destroy the ferromagnetic arrangement of Fe-moment inside the crystal Lu+Fe_j and Fe_k+Fe_g planes.

5. Conclusions

The ferromagnetic ground state of the Lu_2Fe_{17} intermetallic compound has been suppressed by hydrostatic pressure and by uniaxial compression of the compound along its crystal c-axis. However, the ground state of Lu_2Fe_{17} is not affected by a uniaxial compression along its crystal a-axis. The propagation vector (00t_z) of the pressure induced incommensurate helimagnetic structure exhibits the minimum as a function of temperature. This is a consequence of anisotropy of magneto-elastic properties of Lu_2Fe_{17} that becomes evident in the pronounced temperature dependence of the linear compressibility κ_c and in the negative thermal expansion along the crystal c-axis. The presented simple model of IHS elucidates almost all basic features of the effect of pressure on magnetic structures of the Lu_2Fe_{17} intermetallics taking into account the direct exchange interaction between Fe atoms only. The model accentuates namely the dominant role of both, the negative exchange interaction J_{ff} between Fe_f atoms and the positive J_{fj} exchange interaction in the appearance of IHS in the Lu_2Fe_{17} intermetallic compound.

Acknowledgement

The authors acknowledge the financial support of this work by the project No. 202/09/1027 of the Grant Agency of the Czech Republic. This research project has been also supported by the European Commission under the 7th FP through the Key Action: Strengthening the European Research Area, Research Infrastructures. Contract n°: 226507 (NMI3).

References

[1] Buschow K H J 1980 *Ferromagnetic Materials* vol 1, ed E P Wohlfarth (Amsterdam: North-Holland) p 334
[2] Cambell I A 1972 *J. Phys.* F **2** L47
[3] Givord D and Lemaire R 1974 *IEEE Trans. Mag.* MAG **10** 109
[4] Kamarad J, Skorokhod Y, Machatova Z, Arnold Z and Andreev A V 2004 *J. Magn. Magn. Mats.* **272-276** 2098
[5] Sabiryanov R F and Jaswal S S 1998 *Phys. Rev.* B **57** 7767
[6] Kamarad J, Prokhnenko O, Prokes K, Arnold Z and Andreev A V 2007 *J. Magn. Magn. Mats.* **310** 1801
[7] Prokhnenko O, Ritter C, Medvedeva I, Arnold Z, Kamarad J and Kuchin A 2003 *J. Magn. Magn. Mats.* **258-259** 564
[8] Prokhnenko O, Kamarad J, Prokes K, Arnold Z and Andreev A V 2005 *Phys. Rev. Lett.* **94** 107201
[9] Kamarad J, Arnold Z, Medvedeva I V and Kuchin A G 2002 *J. Magn. Magn, Mats.* **242-245** 876
[10] Anagnostou M S, Panagiotopoulos I, Kostikos A, Niarchos D and Zouganelis G 1994 *J. Magn. Magn. Mats.* **130** 57
[11] Li Z W and Morrish A H 1997 *Phys. Rev.* B **55** 3670
[12] Andreev A V, Rafaja D, Kamarad J, Arnold Z, Homma Y and Shiokawa Y 2003 *J. Alloys and Compound* **361** 48
[13] Kamarad J, Machatova Z and Arnold Z 2004 *Rev. Sci. Instrum.* **75** 5022
[14] Prokhnenko O 2003 *Magnetic Structures and Magnetovolume Anomalies in R_2Fe_{17} Intermetallic Compounds* (Prague: Institute of Physics ASCR, Doctoral thesis) p 131
[15] Kozlenko D P, Glazkov V P, Savenko B N and Voronin V I 2004 *Eur. Phys. J.* B **41** 445

Symmetry analysis of the magnetic structures in TbIG and Tb:YIG at low temperature

M Lahoubi

Department of Physics, Badji Mokhtar-Annaba University, PO Box-12, 23000 Annaba, Algeria

E-mail: mlahoubi@hotmail.fr

Abstract. Magnetic structures in TbIG and in terbium-yttrium ferrite garnet $Tb_{0.37}Y_{2.63}IG$ are revisited using Bertaut' representation analysis for the highest crystallographic subgroups of the space group $Ia\bar{3}d$ with the wave vector $\boldsymbol{k} = 0$. Powder neutron diffraction experiments performed only on TbIG are combined with single crystal magnetization measurements. Three field-induced phase transitions are observed at 4.2 K in high magnetic fields up to 150 kOe along the <001> direction while <111> remain the easy axis of magnetization. New parameters of the previous Tb^{3+} double umbrella magnetic structure are found at 5 K in a modified model of the rhombohedral subgroup $R\bar{3}c$. From magnetizations on $Tb_{0.37}Y_{2.63}IG$ single crystal, non collinear magnetic structure is evidenced at 4.2 K along the easy axis <001>. The parameters of a 'square pyramid' umbrella are predicted using both the basis vectors of the tetragonal subgroup $I4_1/acd$ and a model based on the effective spin Hamiltonian with anisotropic exchange $\tilde{\boldsymbol{G}}$ and magnetic $\tilde{\boldsymbol{g}}$ tensors with the requirement $g_yG_y < g_zG_z << g_xG_x$. Anisotropy of magnetization is observed by comparison with the hard axis <111> below $Ta = 80$ K where the earlier spin reorientation phase transitions <001>↔<uuw>↔<111> occur between $T_1 = 40$ K and $T_2 = 140$ K. The basis vectors of the monoclinic subgroup C2/c are tentatively proposed for the low symmetry angular phases <uuw>. All results are compared to previous works.

1. Introduction

The rare earth iron garnets $\{RE^{3+}{}_3\}[Fe^{3+}{}_2](Fe^{3+}{}_3)O_{12}$ or REIG are ferrimagnetic materials with a crystallographic structure of the cubic space group $Ia\bar{3}d$-(O_h^{10}) No. 230, where the rare earth ions RE^{3+} or the Yttrium Y^{3+} are in the dodecahedral {24c}(222) site and the two iron ions Fe^{3+} in the octahedral [16a]($\bar{3}$) and the tetrahedral (24d)($\bar{4}$) sites [1 and Refs. herein]. They are now the subject of a great interest after the famous paper of Hur *et al.* [2] which have revealed that TbIG exhibits a magnetodielectric (MD) response at low temperature as well when a very small external magnetic field ($H_{ex} < 0.2$ T) is applied. This large MD effect appears to be correlated with the previous magnetostriction and high anisotropic behavior of this compound [3, 4]. Recently, Kang *et al.* [5] have discovered in a high external field up to 100 kOe a possible coupling between the magnetic exchange and ligand-field excitations which occurs at a specific temperature range situated between 60 and 80 K with a distinct behavior above and below another characteristic temperature (140 K). In the same time, Louca *et al.* [6], using neutron scattering on TbIG single crystal have found a surprisingly result that the previously suggested reduced symmetry from the cubic to a rhombohedral structure in cooling below T_N (568 K) cannot adequately describe the data in any range while a tetragonal phase is more appropriate. Zhang *et al.* [7], by introducing the spin fluctuations in the improved exchange field, the magnetic and magneto-optical properties of the Tb^{3+} ions in TbIG and Tb:YIG have been calculated without taking into account the differences between the two non collinear magnetic structures which appear at low temperature. Due to these recent controversies, it is imperative to revisit the magnetic ordering of the Tb^{3+} ions in TbIG and in a mixed terbium-yttrium ferrite garnet compound of the $Tb_xY_{3-x}Fe_5O_{12}$ system (hereafter Tb_xY_{3-x}IG) with precise experiments combined by symmetry analysis.

2. Experimental

The neutron diffraction experiments on TbIG powder sample were performed at ILL Grenoble, using the diffractometer D1B with the value of the wavelength λ equal to 2.52 Å. All patterns are recorded in the 5–300 K temperature range. The resolution of the multidetector is equal to 0.2°. We present only the results of the treatment of the pattern at $T = 5$ K. The other patterns used to reexamine the temperature evolution of the double umbrella magnetic structure will be published elsewhere [8]. For the magnetic study the samples are two spherical single crystals. These two samples have been grown from a $PbO/PbF_2/B_2O_3$ solvent using a spontaneous nucleation method with the mixture of pure oxides Fe_2O_3, Y_2O_3 and Tb_2O_3. The electron microprobe analysis performed directly on the sphere gives the terbium concentrations ($x = 3$) for the pure TbIG and ($x = 0.37$) for the mixed compound $Tb_{0.37}Y_{2.63}$IG with respectively the associated weights and diameters: (0.4313 g; 5.5 mm) and (0.0413 g; 3.1 mm). The two samples was oriented along the [111] and [001] crystallographic directions by the X-ray Laüe technique with an error less than 1°. The isothermal magnetizations $M_T(H)$ in function of the internal magnetic field H (the applied field H_{ex} minus the demagnetizing field of the sphere) are performed at the LNCMI Grenoble. For $x = 3$, the high dc applied field produced by a Bitter magnet is varied by steps of 2 kOe up to 150 kOe at 4.2 K. For $x = 0.37$, a superconducting magnet with H_{ex} up to 80 kOe is used at 4.2 and 40 K and also around 80 K where the magnetization anisotropy is assumed to appear. Some of our unpublished $M_T(H)$ data [9] are also chosen in order to complete the study. All results are reported in Bohr magneton (μ_B) for two formula unit's 2Tb_xY_{3-x}IG.

3. Results for TbIG

3.1. *High field magnetization measurements at 4.2 K*

The $M_T(H)$ curves at 4.2 K are plotted in the figure 1. The <111> direction is confirmed as the easy axis of magnetization. Along the <100> direction, when the field is increasing three critical fields H_{c0}, H_{c1} and H_{c2} appear at respectively ~ 20, 68 and 104 kOe. These results are slightly different to our previous results [3, 4] where only two field-induced phase transitions have been reported at H_{c1} (66.66 kOe) and H_{c2} (82.49 kOe). These transitions are characterized by a first order nature with the onset for the first time in dc fields of a hysteresis around H_{c0} and also H_{c2}. The technical saturation magnetization $M_S[001] = 31.60$ μ_B/mol is found for $H >> H_{c2}$ by extrapolating the $M_T(H)$ curve to zero internal field H. The resulting terbium magnetization $M_{Tb3+} = 41.60$ μ_B/mol leads to 6.93 μ_B/Tb^{3+} related to a reduction of 23 % by comparison with the free ion value (9 μ_B).

3.2. *Powder neutron diffraction pattern at 5 K*

The pattern recorded at 5 K is presented in the figure 2. The observed reflections (hkl) are indexed with the general extinction rule (h + k + l = 2n) of $Ia\bar{3}d$ with the wave vector $\boldsymbol{k} = 0$. Four superstructure lines $(110)^*$, $(310)^*$, $(411,330)^*$ and $(433,530)^*$ appear without ambiguity. Contrary to previous findings at 1.5 K [10, 11] and 4.2 K [3] we have also evidenced additionally the previous observed superstructure lines $(200)^*$ and $(600,442)^*$ with a sensibility (0.5 and 1% respectively to $(110)^*$) in the same order of magnitude [12]. These two lines are not resolved at 13 K (not show) [8].

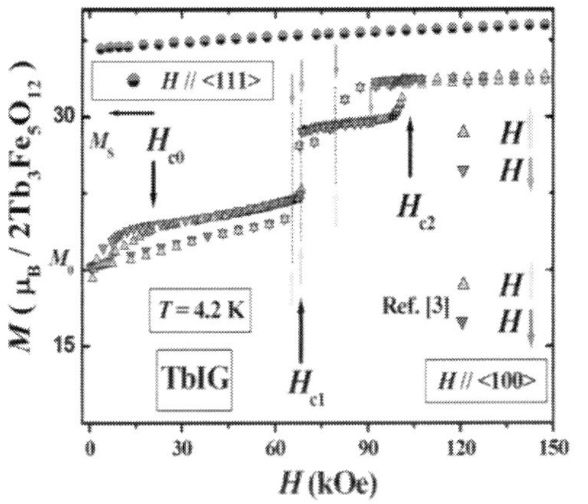

Figure 1. TbIG: $M_T(H)$ curves at 4.2 K.

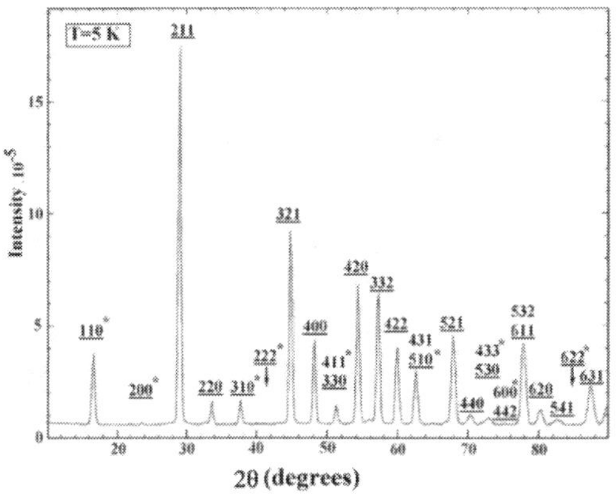

Figure 2. TbIG: D1B pattern at 5 K.

3.3. *Representation analysis*

We have reported on table 1, the notations C_j, C'_j (j = 1-3) and {i = 1-6} [13, 14], the numbers and the crystallographic positions of the RE^{3+} ions in the Wyckoff site {24c}. The local axis (***U***,***V***,***W***) in the $D_2(222)$ symmetry are identified to the parameters g_α ($\alpha = x,y,z$) of the magnetic tensor $\tilde{g}$. The Bertaut' representation analysis of the cubic space group $Ia\bar{3}d$-(O_h^{10}) [15] was applied in the past by Tchéou *et al.* [10] and Bertaut *et al.* [11] in order to determine the magnetic modes of the Tb^{3+} ions in TbIG. However, in their original works, two three-dimensional irreductible representations were used: $\Gamma_{4g} = T_{1g}$ and $\Gamma_{5g} = T_{2g}$. The representation analysis of $Ia\bar{3}d$ has been developed later completely by us taking into account the coupling between the Tb^{3+} ions with the two iron ions Fe^{3+} in sites [16a] and (24d) [8, 9]. In the mixed $Tb_xY_{3-x}IG$ system ($0 < x \leq 3$), $\Gamma_{4g}(T_{1g})$ can describe in a first approximation the magnetic structures below T_N. Using the earlier linear combinations of the spins [15]: $\boldsymbol{F}_j = +\boldsymbol{S}_j + \boldsymbol{S}_{j+3} + \boldsymbol{S}_{j+6} + \boldsymbol{S}_{j+9}$; $\boldsymbol{G}_j = +\boldsymbol{S}_j - \boldsymbol{S}_{j+3} + \boldsymbol{S}_{j+6} - \boldsymbol{S}_{j+9}$; $\boldsymbol{C}_j = +\boldsymbol{S}_j + \boldsymbol{S}_{j+3} - \boldsymbol{S}_{j+6} - \boldsymbol{S}_{j+9}$ and $\boldsymbol{A}_j = +\boldsymbol{S}_j - \boldsymbol{S}_{j+3} - \boldsymbol{S}_{j+6} + \boldsymbol{S}_{j+9}$ with (j = 1-3), the basis vectors and the multiplicity (m) for the RE^{3+} inequivalent sites are listed on table 2. The chemical cell is equal to the magnetic cell, thus the primitive translation noted (1| 1/2,1/2,1/2) of the crystallographic lattice (I) is a primitive translation of the magnetic lattice (I). The spin vectors $\boldsymbol{S}_j$ and $\boldsymbol{S}_{j+12}$ (j = 1-3) are coupled ferromagnetically. The forbidden superstructure lines $(222)^*$ and $(622)^*$ are not observed in the figure 2. Then, the symmetry operation ($\bar{1}$|0,0,0) is an inversion center and the magnetic modes $\boldsymbol{C}_j$ and $\boldsymbol{A}_j$ are equals to zero. The spins vectors $\boldsymbol{S}_j$ and $\boldsymbol{S}_{j+6}$ (j = 1-3) are coupled ferromagnetically in the magnetic modes $\boldsymbol{F}_j$ and $\boldsymbol{G}_j$. Consequently, the site {24c} splits into six magnetically inequivalent sublattices noted C_j and C'_j (j = 1-3) (see table 1) so that in any of a given point the crystalline field effect being exactly the same: C_2 and C_3 are related to C_1 by a rotation of 120 and 240° around the 3-fold symmetry [111] axis (and respectively for C'_1, C'_2 and C'_3).

3.4. *Investigation of the novel double umbrella for TbIG*

The appearance of the three critical fields H_{c0}, H_{c1} and H_{c2} at 4.2 K along the <100> direction (see figure 1), confirms that the true subgroup of $\mathrm{Ia}\bar{3}\mathrm{d}$ is $\mathrm{R}\bar{3}\mathrm{c}$-($D_{3d}^{6}$) No. 167, in agreement with the previous observed rhombohedral distortions [16]: the irreductible representation ($\Gamma_{4g} = T_{1g}$) is reduced to ($\Gamma_{2g} = A_{2g}$) + E_g, where the E_g modes are not considered. The preceding magnetic inequivalency becomes a crystallographic inequivalency: the sublattices (C_j) corresponding to the site 6e(2) and C'_j to the site 6e′(2) of $\mathrm{R}\bar{3}\mathrm{c}$ in positions $\pm(X,1/2-X,1/4)$ (and cyclic permutation) with $X = 3/8$ and $X = 7/8$ respectively. The associated basis vectors described by the ferromagnetic mode $\boldsymbol{f}_j = \boldsymbol{S}_j + \boldsymbol{S}_{j+6}$ (j = 1-3) in the rhombohedral axis *X*, *Y*, *Z* are gathered on table 2.

Table 1. Notations of the sublattices, numbers, positions and local axis (***U***,***V***,***W***).

Notations of the sublattices C_j; C'_j (j = 1-3) and {i=1-6} from refs. [13], [14]	Numbers and positions of the RE^{3+} ions in the Wyckoff site {24c}	Local axis in D2(222) symmetry ***U*** g_z ***V*** g_x (or g_y) ***W*** g_y (or g_x)
C_1 1 3	**(1)** 1/8,0,1/4 **(7)** 7/8,0,3/4 **(13)** 5/8,1/2,3/4 **(19)** 3/8,1/2,1/4	100 $01\bar{1}$ 011
C'_1 2 4	**(4)** 3/8,0,3/4 **(10)** 5/8,0,1/4 **(16)** 7/8,1/2,1/4 **(22)** 1/8,1/2,3/4	100 011 $0\bar{1}1$
C_2 3 5	**(2)** 1/4,1/8,0 **(8)** 3/4,7/8,0 **(14)** 3/4,5/8,1/2 **(20)** 1/4,3/8,1/2	010 $\bar{1}01$ 101
C'_2 4 6	**(5)** 3/4,3/8,0 **(11)** 1/4,5/8,0 **(17)** 1/4,7/8,1/2 **(23)** 3/4,1/8,1/2	010 101 $10\bar{1}$
C_3 5 1	**(3)** 0,1/4,1/8 **(9)** 0,3/4,7/8 **(15)** 1/2,3/4,5/8 **(21)** 1/2,1/4,3/8	001 $1\bar{1}0$ 110
C'_3 6 2	**(6)** 0,3/4,3/8 **(12)** 0,1/4,5/8 **(18)** 1/2,1/4,7/8 **(24)** 1/2,3/4,1/8	001 110 $\bar{1}10$

We suggest a new combination of these magnetic basis vectors and we propose four modified double umbrella models for the six sublattices C_j and C'_j (table 3a). Around the 3-fold symmetry [111] axis, the sublattices C_j and C'_j are situated in the three glide planes c, c.3 and c.3^2 of the space group $\mathrm{R}\bar{3}\mathrm{c}$.

Table 2. Splittings of Wyckoff positions, symmetry, multiplicity m and basis vectors.

Space groups	Sites	Symmetry	m	Basis vectors of RE^{3+} ions with $\boldsymbol{k} = 0$
$Ia\bar{3}d$ ($\Gamma_{4g} = T_{1g}$)	24c	(222)	6	$\Psi_{11} = F_{1x}$ $\Psi_{12} = F_{2x} + F_{3x}$ $\Psi_{13} = G_{2z} + G_{3y}$ $\Psi_{21} = F_{2y}$ $\Psi_{22} = F_{1y} + F_{3y}$ $\Psi_{23} = G_{3x} + G_{1z}$ $\Psi_{31} = F_{3z}$ $\Psi_{32} = F_{1z} + F_{2z}$ $\Psi_{33} = G_{1y} + G_{2x}$
$R\bar{3}c$ ($\Gamma_{2g} = A_{2g}$)	6e	(2)	3	$\Psi_{11}(I) = f_{1X} + f_{2Y} + f_{3Z}$ $\Psi_{11}(II) = f_{1Y} + f_{1Z} + f_{2Z} + f_{2X} + f_{3X} + f_{3Y}$
	6e′	(2)	3	$\Psi_{11}(IV) = f_{4X} + f_{5Y} + f_{6Z}$ $\Psi_{11}(V) = f_{4Y} + f_{4Z} + f_{5Z} + f_{5X} + f_{6X} + f_{6Y}$
C2/c ($\Gamma_{1g} = A_g$)	8f	(1)	2	$\Psi_{11} = G_{1x}; \Psi_{11} = G_{1y}; \Psi_{11} = F_{1z}$ $\Psi_{11} = f_{1x} - f_{2y}; \Psi_{11} = f_{1y} - f_{2x}; \Psi_{11} = f_{1z} - f_{2z}$
	8f′	(1)	2	$\Psi_{11} = G_{2x}; \Psi_{11} = G_{2y}; \Psi_{11} = F_{2z}$ $\Psi_{11} = f_{4x} - f_{5y}; \Psi_{11} = f_{4y} - f_{5x}; \Psi_{11} = f_{4z} - f_{5z}$
	4e	(2)	1	$\Psi_{11} = f_{3z}$ $\Psi_{11} = f_{3x} - f_{3y}$
	4e′	(2)	1	$\Psi_{11} = f_{6z}$ $\Psi_{11} = f_{6x} - f_{6y}$ First choice (I112/b) Second choice (R12/c)
$I4_1/acd$ ($\Gamma_{2g} = A_{2g}$)	16e	(2)	4	$\Psi_{11}(I) = G_{1y} + G_{2x}$ $\Psi_{11}(II) = F_{1z} + F_{2z}$
	8b	(222)	2	$\Psi_{11}(III) = F_{3z}$

Table 3. TbIG: validation of the novel double umbrella model at 5 K.

(a) Four modified models for the RE^{3+} ions in A_{2g}	(b) The exclusive model at 5 K $I_M\{200\}^*_{cal}$ and $I_M\{600,442\}^*_{cal}$	(c) Parameters at 5 K for C_1 and C'_1 Factor of reliability $R = 6.7$ %
f ≠ f ' and a ≠ a' (6e): $C_j : \boldsymbol{S}_j = \boldsymbol{S}_{j+6} = f\boldsymbol{n} \pm a\boldsymbol{g}_j$	f ≠ f ' and a' ≈ a − ε $C_1 : \boldsymbol{m}_1 = \boldsymbol{S}_1 = \boldsymbol{S}_7 = (f - a/3)\boldsymbol{n} - (2\sqrt{2}/3)a\boldsymbol{p}_1$	Spherical coordinates (m_1, θ_1, ϕ_1) for C_1 $m_1 = 8.07\ \mu_B$; $\theta_1 = 32°$; $\phi_1 = 180°$ $(m'_1, \theta'_1, \phi'_1)$ for C'_1
(6e′): $C'_j : \boldsymbol{S}'_j = \boldsymbol{S}'_{j+6} = f'\boldsymbol{n} \pm a'\boldsymbol{g}_j$	$C'_1 : \boldsymbol{m}'_1 = \boldsymbol{S}_4 = \boldsymbol{S}_{10} = (f' + a'/3)\boldsymbol{n} + (2\sqrt{2}/3)a'\boldsymbol{p}_1$	$m'_1 = 8.99\ \mu_B$; $\theta'_1 = 27°$; $\phi'_1 = 0°$ ϕ_1 and ϕ'_1 are the angles from the unit vector $\boldsymbol{p}_1$ in the (111) plane
	$I_M\{200\}^*_{cal} = A(a - a')^2$ $= A\varepsilon^2$ with the coefficient A $= 256\{0.27f(Tb^{3+})\}^2$ $I_M\{600,442\}^*_{cal} = B(a - a')^2$ $= B\varepsilon^2$ with the coefficient B $= 798.10\{0.27f(Tb^{3+})\}^2$	$\varepsilon = 0.20\ \mu_B$; ($\varepsilon \to 0$ at $T > 5$ K) $M_S^{cal}<111> = 34.56\ \mu_B$/mol $M_S^{mes}<111> = 34.53\ \mu_B$/mol $\delta(M_S)<111> = 0.03\ \mu_B$/mol

Each glide plane containing [111] (unit vector $\boldsymbol{n}$) and one of the rhombohedral axis $\{[\bar{1}11], [1\bar{1}1], [11\bar{1}]\}$ represented by the unit vectors $\boldsymbol{g}_j$, (j = 1-3). In this condition, we have: c = $(\boldsymbol{n}, \boldsymbol{g}_3)$, c.3 = $(\boldsymbol{n}, \boldsymbol{g}_1)$ and c.3^2 = $(\boldsymbol{n}, \boldsymbol{g}_2)$. The moments C_1 and C'_1 (figure 3) are drawed in the (π) plane associated to the glide plane c.3 which remains a principal plane of the magnetic tensor $\tilde{g}$. In contrast, the cubic axis [100] and [011] cease to be principal axes. Furthermore, the perpendicular projection of $[\bar{1}11]$ (unit vector

$\boldsymbol{g}_1$) on the plane (111) is along the low symmetry axis $[\bar{2}11]$ (unit vector $\boldsymbol{p}_1$) and equivalent directions. We choose the mixed system of crystalline axis between Ia$\bar{3}$d and R$\bar{3}$c: [111] (unit vector $\boldsymbol{n}$); the low symmetry axis $[\bar{2}11]$ (unit vector $\boldsymbol{p}_1$) and the perpendicular to the (π) plane of the local axis $[0\bar{1}1]$ ($\boldsymbol{W}$) (not drawed in the figure 3). If one takes into account the smallness superstructure lines (200)* and (600,442)* at 5 K (table 3b), new parameters are found for C_1 (m_1, θ_1, ϕ_1) and C'_1 (m'_1, θ'_1, ϕ'_1) sites with a good reliability factor R (table 3c) for a refinement based on all the reflections. The results obtained in the subgroup R$\bar{3}$ of Ia$\bar{3}$d [12] lead to values of the components m_{1x} and m'_{1z} above the free Tb^{3+} ion value (9 μ_B). Previous calculated temperature evolution [17] is not consistent with the onset of all the superstructure lines [8].

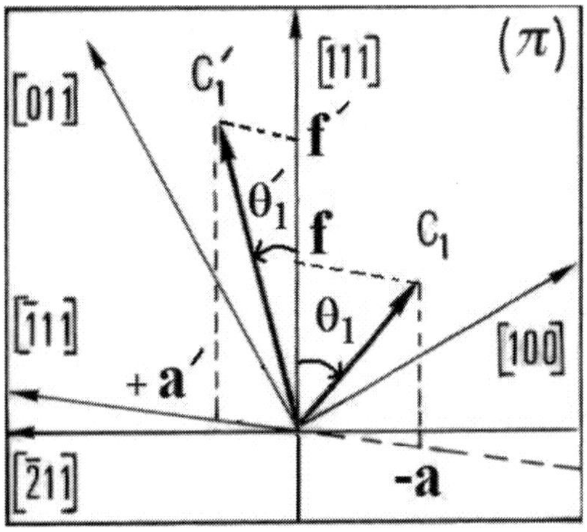

Figure 3. TbIG: the novel model at 5 K.

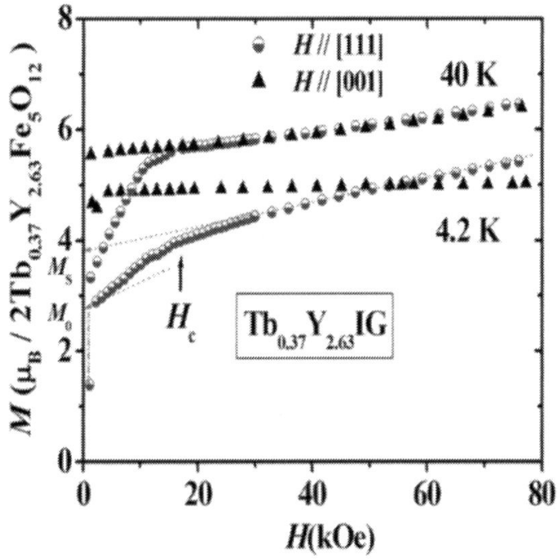

Figure 4. $Tb_{0.37}Y_{2.63}IG$: $M_T(H)$ at 4.2 and 40 K.

4. Results for $Tb_{0.37}Y_{2.63}IG$

4.1. *High field magnetization measurements*

For $Tb_{0.37}Y_{2.63}IG$, two typical isothermal magnetization curves $M_T(H)$ at 4.2 and 40 K along [111] and [001] directions are presented in the figure 4. For H parallel to [111], we observe an increasing of the magnetization corresponding to the rotation of the magnetization from the spontaneous direction. So, field-induced phase transitions at a critical field H_c are observed at 4.2 (~ 17 kOe) and 40 K (~ 11 kOe) in increasing fields contrary to our previous magnetization measurements at 1.5 K on the same sample [18]. We can define by extrapolating the $M_T(H)$ curves to zero internal field H, firstly the spontaneous magnetization M_0[111] = 2.86 μ_B/mol for $H < H_c$ and secondly the technical saturation magnetization M_S[111] = 3.86 μ_B/mol for $H >> H_c$. The rapid approach to saturation observed along [001] shows that the easy axis is in this direction and remains until 40 K as evidenced by comparison of the curves and consequently, [111] is the hard axis. At 4.2 K, the value of M_S[001] = 4.88 μ_B/mol gives the resulting M_{Tb3+} = 5.12 μ_B/mol with $\langle m\rangle_{mes}$ = 6.92 μ_B/Tb^{3+} which is in agreement with the above pure TbIG value contrary to that found previously [19] (6.52 μ_B/Tb^{3+}) for a similar composition (x ~ 0.37). An average reduction of 23.1 % is obtained by comparison with the free ion. The terbium moments have a non collinear ordering at 4.2 K. The thermal variation of the magnetization difference $\Delta M_1(T) = \{M_S[001] - M_0[111]\}(T)$ is plotted in the figure 5. We can observe a rapid decreasing in the curve after a maximum at $T_1 = 40$ K. A particular inflexion point is observed around 80 K and $\Delta M_1 = 0$ in the region above $T_2 = 140$ K when the [111] direction becomes the easy axis until room temperature.

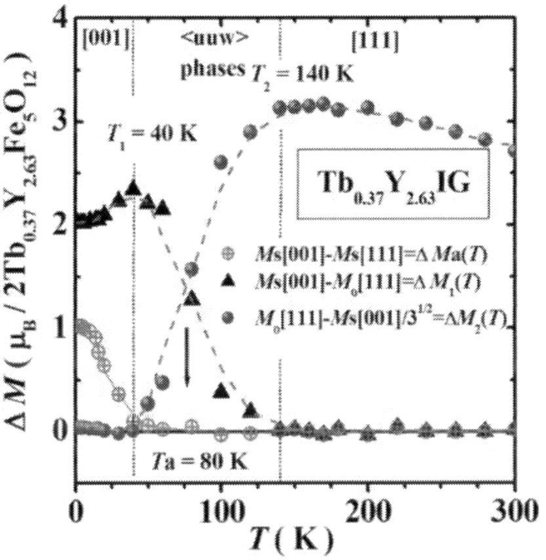

Figure 5. $Tb_{0.37}Y_{2.63}IG$: the anisotropic properties

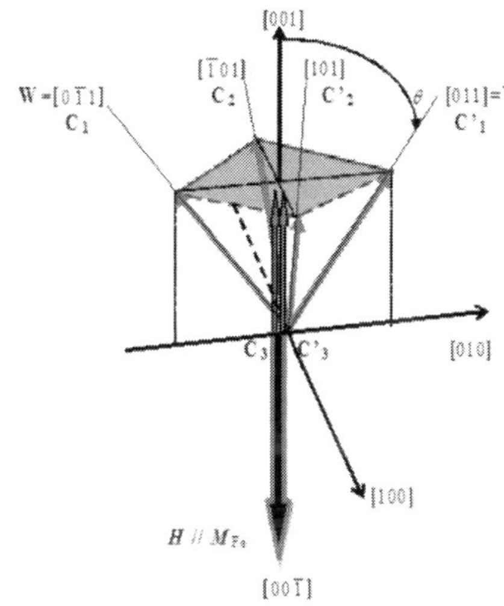

Figure 6. $Tb_{0.37}Y_{2.63}IG$: the 'square pyramid' model.

This inflexion point is always present in the thermal variation of the second magnetization difference curve $\Delta M_2(T) = \{M_0[111] - M_S[001](1/3)^{1/2}\}(T)$ (figure 5). A magnetization anisotropy is evidenced in the plot of the third magnetization difference curve $\Delta Ma(T) = \{M_S[001] - M_S[111]\}(T)$ which reaches a maximum value of 1.02 μ_B/mol at 4.2 K. This new result gives the first confirmation that non collinear magnetic structures take place in the so-called angular phases <uuw> below the typical temperature of anisotropy $Ta = 80$ K. The spin reorientation phase transitions <001>↔<111> introduce another complexity due to the low symmetry of the <uuw> phases [18, 20] where the iron sublattice magnetization $\boldsymbol{M}_{Fe}$ lies in an arbitrary direction in one of the cubic $\{0\bar{1}1\}$ type planes.

4.2. Representation analysis

Rhombohedral distortions have been observed until 50 K for $x = 1.17$ in the previous X-ray analysis of magnetoelastic distortions of the system $Tb_xY_{3-x}IG$ for single crystals in the 5–300 K temperature range [21]. For $x \leq 0.7$ no tetragonal distortions which are about one order smaller than rhombohedral distortions have been revealed (also for monoclinic distortions). The high <001>, <111> and low <uuw> symmetry phases observed by magnetic measurements must be connected with the allowing maximal subgroups of $Ia\bar{3}d$. They are obtained by the 'symmetry lowering device' [22, 23]: the tetragonal $I4_1/acd$ ($\Gamma_{2g} = A_{2g}$) below T_1 and the rhombohedral $R\bar{3}c$ ($\Gamma_{2g} = A_{2g}$) above T_2 (table 2) are concerned. When a spin reorientation phase transition temperature region exists, *the space group must be a subgroup of the two groups at the ends* [24] with the requirement that the two ends of the spin reorientation phase transitions are characterized by second order nature in the T-range (40–140 K) [18]. Despite the previous theoretical predictions [20] that the nature of the upper transition at the end of the spin reorientation phase transition <uuw>↔<111> might be first order like the transition observed at T_{SR} = 14.5 K in DyIG [25, 26], good agreement is found with another second order transition at T_2 = 135 K on $Tb_{0.26}Y_{2.74}IG$ single crystal [27]. The monoclinic space group (C2/c) (Γ_{1g} = A_g) [23] (table 2) is a subgroup of the two space groups $I4_1/acd$ and $R\bar{3}c$: it may be used in the intermediate *T*-range.

4.2.1. Tetragonal basis vectors of $I4_1/acd$ ($\Gamma_{2g} = A_{2g}$).
Using the above linear combinations $\boldsymbol{F}_j$ and $\boldsymbol{G}_j$, the tetragonal basis vectors for the two RE^{3+} inequivalent sites (8b) and (16e) are given on table 2. A 'square pyramid' umbrella structure is found in the subgroup $I4_1/acd$ ($\Gamma_{2g} = A_{2g}$): site (8b) C_3,C'_3 with $\boldsymbol{S}_3 = +\boldsymbol{S}_6 = +\boldsymbol{S}_9 = +\boldsymbol{S}_{12}$ = (0,0,F) along [001]; site (16e): C_1,C'_1 with $\boldsymbol{S}_1 = +\boldsymbol{S}_7$ = (0,a,b); $\boldsymbol{S}_4 = +\boldsymbol{S}_{10}$ = (0,–a,b) in the (100) plane; C_2,C'_2 with $\boldsymbol{S}_2 = +\boldsymbol{S}_8$ = (a,0,b); $\boldsymbol{S}_5 = +\boldsymbol{S}_{11}$ = (–a,0,b) in the (010) plane with the general condition: $F \neq (a^2 + b^2)^{1/2}$.

4.2.2. Monoclinic basis vectors of C2/c ($\Gamma_{1g} = A_g$).
The magnetic structures which appear in the intermediate *T*-range are complicated. A T_1 = 40 K, the further splitting will depend on the choice of the monoclinic axis of the subgroup C2/c of $I4_1/acd$. We choose at first the case of 'unique axis c, cell choice 3' of the subgroup I112/b [23] with the origin at $\bar{1}$, a cubic binary axis 2 = (2_z| 0,1/2,0) in 0,0,*z* and a glide plane b = ($\bar{1}.2_z$| 0,1/2,0) in *x*,*y*,0. We obtain the corresponding sites: (8f) C_1, C'_1; (8f') C_2, C'_2; (4e) C_3 and (4e') C'_3. At T_2 = 140 K, we choose in this second case the following generators for the non-isomorphic subgroup R12/c(C2/c) of $R\bar{3}c$ [23]: the identity (1| 0,0,0), the inversion ($\bar{1}$| 0,0,0), a diagonal binary axis 2_d = ($2_{x\bar{y}}$| 1/2,1/2,1/2) in $x,\bar{x}$+1/2,1/4 and a glide plane ($\bar{1}.2_d$) = ($\bar{1}.2_{x\bar{y}}$| 1/2,1/2,1/2) in *x*,*y*,*z*. We obtain the following additional splitting of the sites C_1, C_2, C_3 (6e) and C'_1, C'_2, C'_3 (6e'): sites (8f) C_1,C_2; (8f') C'_1, C'_2; (4e) C_3 and (4e') C'_3. The corresponding basis vectors are given on table 2. This preliminary result of the Bertaut' representation analysis of C2/c provides a first but an incomplete answer in our research of the temperature evolution of magnetic structures in $Tb_{0.37}Y_{2.63}IG$.

4.3. Investigation of the 'square pyramid' umbrella for $Tb_{0.37}Y_{2.63}IG$.

4.3.1. The effective spin Hamiltonian model.
We suppose that the quasi doublet model with a pseudo spin S = 1/2 seems to apply to the Tb^{3+} ion, suggesting that the next energy level is lying at least 80 cm^{-1} above [18]. The contribution of the RE^{3+} ions of the site {i} in the {i = 1-6} notation [13, 14] to the Zeeman and exchange energies is represented by the spin Hamiltonian: $\hat{H} = +\mu_B \boldsymbol{H}\tilde{g}_i \boldsymbol{S}_i + (-M_{Fd}/M_{Fe})\tilde{G}_i \boldsymbol{S}_i$. In the system of the local axis (*U*,*V*,*W*) the anisotropic exchange $\tilde{G}$ tensor has the same axis as the magnetic tensor $\tilde{g}$. If the magnetic field vector $\boldsymbol{H}$ is equal to zero, the components of the magnetic moment $\boldsymbol{m}$ are reduced to: $m_\alpha = (\mu_B/2)\times n_\alpha g_\alpha G_\alpha/(\sum n_\alpha^2 G_\alpha^2)^{1/2}$ (by omitting the indice i). The anisotropy of these two tensors is assumed to be similar and diagonal and we shall consider in our investigation only six parameters g_α and G_α (α= *x*,*y*,*z*). The similar reduction (~ 23 %) obtained for our two terbium concentrations (*x* = 3) and (*x* = 0.37) imply that the new parameters of TbIG at 5 K along the [111] axis confirmed by the recent method of validation of magnetic structures [28] can be used. For the sublattice C_1, in the notation C_j, C'_j (j = 1-3) the moment vector $\boldsymbol{m}_1$ with the value (8.07 μ_B) is in the plane (*U*,*W*) = ([100],[011]) and originate an angle of 30.79 and 23.95° with [111] and [100] respectively; in this case, we have: $g_yG_y/g_zG_z = tg(23.95°)/\sqrt{2} = 0.3134$. In the same manner for the sublattice C'_1, the moment vector $\boldsymbol{m}'_1$ (8.99 μ_B) is in the plane (*U*,*V*) = ([100],[011]) and originate with [111] an angle of 28.07° and with

[011] and [100] the angles of 7.19 and 82.81° respectively. We have then: g_xG_x/g_zG_z = tg(82.81°)/$\sqrt{2}$ = 5.605. We obtain a *particular requirement* from the following inequalities: $g_yG_y < g_zG_z << g_xG_x$ with 1 < 3 << 18. For the sublattices C'$_1$ (or C$_1$), the moment vector ***m'***$_1$ (or ***m***$_1$) which is in the same equivalent plane (***U***,***W***) ≡ (***U***,***V***) = ([100],[011]), the ratio of the components (***W***/***V***) is proportional to (g_yG_y/g_xG_x) = 0.056 with δ = arctg(0.056) = 3.2°, the angle with the local axis ***V*** = [011]. In the tetragonal model, if we take into account the angle of 45°–3.2° = 41.8° with [001] we can found the value 8.49 μ_B for the moments of the sublattices C$_1$,C'$_1$ and C$_2$,C'$_2$. The moment vectors ***m***$_3$ and ***m'***$_3$ of the sublattices C$_3$ and C'$_3$ are parallel to the [001] axis with the new *specific condition* for the low moment value: F < $(a^2 + b^2)^{1/2}$ < 9 μ_B (free ion value). We can assume that if g_y = 0, we obtain θ= +45° and the moment vector ***m'***$_1$ is *aligned directly* along the direction of the local axis ***V*** = [011]. Finally at 4.2 K, the predicted parameters of our single 'square pyramid' umbrella (figure 6) with $\boldsymbol{M}_{Fe}$ along [00$\bar{1}$] are defined by: site (8b), C$_3$,C'$_3$: $m_3 = m'_3$ = 8.1±0.02 μ_B and θ= 0°; site (16e), C$_1$,C'$_1$ and C$_2$,C'$_2$: $m_1 = m'_1 = m_2 = m'_2$ = 8.97±0.02 μ_B and θ= ∓45° respectively in the (100) and (010) planes.

4.3.2. Discussion.

Previous parameters obtained with crystal field calculations on $Tb_{0.26}Y_{2.74}IG$ single crystal [29] give a calculated average moment $<m>_{cal}$ = 7.66 μ_B/Tb^{3+} close in agreement with their measured magnetization value $<m>_{mes}$ = 7.69 μ_B/Tb^{3+}. But, two different *conical* arrangements are found: the sublattices C$_1$,C'$_1$ and C$_2$,C'$_2$ are decoupled and situated in two planes with respectively (3, 5) and (4,6) in the {i = 1-6} notation of ref. [14]. Their model is also based only the two anisotropy constants K_1 and K_2 where only first order transitions exist. At the opposite, our results are in good agreement with the magnetization measurements and the required conditions on the three anisotropy constants K_1, K_2 and K_3 > 0 in their relative form $K'_1 = K_1/K_3$ and $K'_2 = K_2/K_3$ found previously [18, 20]. NMR investigations are made also on the same compound $Tb_{0.26}Y_{2.74}IG$ [30]. The value of $<m>_{cal}$ = 6.62 μ_B/Tb^{3+} leads to a poor agreement with $<m>_{mes}$ and with the constraint of our model in which the four large moments build the 'square pyramid' umbrella while the two low moments are along [001].

5. Conclusion

The magnetic structures in TbIG and $Tb_{0.37}Y_{2.63}IG$ have been revisited using the Bertaut's representation analysis. Taking into account the observed smallness superstructure lines (200)* and (600,442)*, the refinement of the TbIG powder neutron diffraction data at 5 K give rises new parameters of the double umbrella in the subgroup R$\bar{3}$c along the easy axis [111]. A good agreement is found with the high field magnetization at 4.2 K where a third critical field H_{c0} is observed and added to previous H_{c1} and H_{c2} along the hard axis [001]. For $Tb_{0.37}Y_{2.63}IG$, a critical field H_c is observed at 4.2 and 40 K along the hard axis [111]. Parameters of a 'square pyramid' umbrella structure are predicted at 4.2 K in the tetragonal subgroup I4$_1$/acd where the easy axis belongs to the high symmetry phase (<001>) until T_1 (40 K). For $T > T_2$ (140 K), the easy axis is in the high symmetry phase (<111>) and the magnetic structures can be described in the subgroup R$\bar{3}$c. Between the two preceding second order spin reorientation phase transition temperatures where the angular and low symmetry phases <uuw> exist, an anisotropy of magnetization is observed below a specific temperature of anisotropy *Ta* (80 K). This result seems to be associated to a change of the spontaneous and the field-induced non collinear magnetic structures. Basis vectors of the subgroup C2/c are proposed for the intermediate *T*-range. Such results look reasonable and are indeed a first approximation without neutron diffraction study on $Tb_{0.37}Y_{2.63}IG$.

Acknowledgments This work is dedicated to the father of symmetry analysis of magnetic structures by group theory, the Academician Dr. E F Lewy Bertaut who died in 2003. We thank Dr. B Ouladdiaf (ILL Grenoble) for his assistance during the D1B neutron diffraction experiments.

References

[1] Guillot M 1994 'Magnetic Properties of Ferrites' in R. W. Cahn, P. Haasen and E. J. Kramer Editors, *Materials Science and Technology-A Comprehensive Treatment* **Vol. 3B**: Electronic and Magnetic Properties of Metals and Ceramics, **Part II** Chapter **8** 7–92 K. H. J. Buschow (Vol. Ed.) VCH Verlagsgesellschaft mbH/VCH Publishers Inc. Weinheim/New York
[2] Hur N *et al.* 2005 *Appl. Phys. Lett.* **87** 042901
[3] Lahoubi M *et al.*1984 *I.E.E.E. Transactions on Magn.* MAG **20** 1518
[4] Lahoubi M *et al.*1985 Proc. of the Fourth Inter. Conf. on Ferrites (ICF4) San Francisco Oct.31-Nov.2 1984 *Advances in Ceramics* **Vol. 15** Part I ed. Wang F F Y 275–82
[5] Kang T D *et al.* 2010 *Phys. Rev.* B **82** 014414
[6] Louca D *et al.* 2009 *Phys. Rev.* B **80** 214406
[7] Zhang G y *et al.* 2009 *J. Magn. Magn. Mater.* 3**21** 3077
[8] Lahoubi M 2011 Chapter in 'neutron diffraction' book project accepted *in Intech open access*
[9] Lahoubi M 1986 Thesis, Doctorat of State in Physics, University of Grenoble France 1–45
[10] Tchéou F *et al.* 1970 Colloque International C.N.R.S. 'Les Terre Rares' **180 Vol. II** 313
[11] Bertaut E F *et al.*1970 *Solid State Communications* **8** 239
[12] Hock R *et al.*1990 *J. of Solid State Chemistry* **84** 39
[13] Wolf W P *et al.* 1962 *J. Phys. Soc. Japan* **17 Suppl. B-I** 443
[14] Pearson R F 1965 *Proc. Phys. Soc.* **86** 1055
[15] Bertaut E F 1968 *Acta Cryst.* A **24** 217
[16] Levitin R Z *et al.*1984 *Sov. Phys. Solid State* **26** (5) 841
[17] Druzhinina R F and Shkarubskii V V 1988 *Sov. Phys. Solid State* **30** (2) 342
[18] Lahoubi M *et al.* 2000 *Physica B* **284-288** 1503
[19] Guillot M *et al.* 1989 *Z. Phys. Condensed Matter* **74** 335
[20] Borodin V A *et al.* 1983 *Phys. Met. Metall.* **56** 7
[21] Levitin R Z *et al.*1983 *Sov. Phys. Solid State* **25** (6) 1074
[22] Bertaut E F 1981 *J. Magn. Magn. Mater.* **24** 267
[23] Hahn Th 2002 *International Tables for Crystallography* **Vol. A** Direct Space
[24] Geller S and Balestrino G 1980 *Solid State Communications* **33** 315
[25] Lahoubi M *et al.* 2009 *Journal of Physics: Conference Series* **150** 042108
[26] Lahoubi M *et al.* 2010 *Journal of Physics: Conference Series* **200** 082018
[27] Tsitskishvili K F *et al.*1982 *Sov. Phys. Solid State* **24** (11) 1964
[28] Wills A S 2007 *Zeitschrift für Kristallographie* **suppl. 26** 56
[29] Druzhinina R F and Lagutin A S 1989 *Sov. Phys. Solid State* **31** (6) 1091
[30] Doroshev V D and Savosta M M 1992 *Sov. Phys. Solid State* **34** (5) 831

Journal of Physics: Conference Series **340** (2012) 012069 doi:10.1088/1742-6596/340/1/012069

Structure of Correlations in $Fe_{90}Zr_{10}$ Spin Glass

K Mergia and S Messoloras
Institute of Nuclear Technology and Radiation Protection, National Centre for Scientific Research "Demokritos", GR-15310 Aghia Paraskevi Athens, Greece

E-mail: kmergia@ipta.demokritos.gr

Abstract. There are various representations for the long time evolution and the dynamics of spin glasses, but a coherent, overall accepted real space description remains lacking. Small-Angle Neutron Scattering measurements reveal the long range spin-spin correlations from which the correlation length as a function of the temperature, time or magnetic field can be deduced. The scattering of the amorphous $Fe_{90}Zr_{10}$ alloy during zero magnetic field cooling arises from long range correlations described by a Lorentzian function and scattering from spin clusters. From the temperature dependence of the correlation length two transitions are observed without, however, the correlation length to diverge. The application of magnetic field suppresses the transitions. The experimental findings are discussed in conjunction with different theoretical models.

1. Introduction

The study of amorphous FeZr alloys has attracted a lot of interest, mainly, because different magnetic behaviours can be realized by varying the iron content [1-10]. Three different types of magnetic behaviour have been observed in $Fe_{100-x}Zr_x$ alloys, specifically i) ferromagnetic alloys for x=12 which are characterized by a Curie temperature T_c, ii) re-entrant spin glasses (RSG) for x=9-10 that are characterised by two transition temperatures, T_c which refers to the transition from the paramagnetic (PM) to the ferromagnetic (FM) state and T_f which refers to the transition from the FM state to the spin glass (SG) state and iii) spin glasses for $x \leq 8$, in which one transition takes place from the PM to SG state at the temperature T_g [1-6]. The temperature T_c varies strongly with the Zr concentration and decreases as the Zr content increases [1, 7], whereas T_f varies in the opposite way and coincides with T_g for x=7 [1].

TEM and Lorentz microscopy studies on $Fe_{90}Zr_{10}$ amorphous films showed the existence of magnetic domains with a configuration that remains the same upon cooling through the re-entrant spin-glass phase [8, 9]. Senousi et al [9] concluded that the main reason for the existence of the re-entrant behaviour is the anisotropy of the magnetic domains and that the long range ferromagnetic order is little perturbed by the spin-glass re-entrant transition.

The focus of the paper is a neutron scattering study of the magnetic correlations, the nature of the phase transitions and the response to applied magnetic field of the amorphous $Fe_{90}Zr_{10}$.

2. Small Angle Neutron Scattering from Spin Glasses

The scattering from a magnetic system, in the static approximation, is described by its spin-spin correlation function $g(r)$ (isotropic system) which can be separated into two parts, the one describing the long range correlation $g_{lr}(r)$ and the other the short range one $g_{sr}(r)$. The neutron scattering is then given by [11]

$$S(Q)=F\{g_{lr}(r)\}+F\{g_{sr}(r)-\langle g(r)\rangle\} \quad (1)$$

where $F\{\ \}$ denotes the sine Fourier transform i.e. for a system with spherical symmetry

$$F\{g(r)\}=\frac{1}{A}\int g(r)r^2\frac{\sin Qr}{Qr}dr,\ A=\int g(r)r^2\,dr \quad (2)$$

$\langle g(r)\rangle$ is the averaging over the whole sample volume and $|\mathbf{Q}|=4\pi\sin\theta/\lambda$ (2θ is the scattering angle and λ is the neutron wavelength). The first term in equation (1) gives the scattering at high Q values (Bragg scattering in a polycrystalline sample), whereas the second term corresponds to SANS scattering. As the scattering is conserved, loss of scattering from the SANS region implies that the short range order has been transformed to a long range one (within the term long range is included the scattering from constant magnetization). Notwithstanding that in the low Q region the length scale observed is of the order $Qd\sim 2\pi$, abrupt changes in the observed scattering indicate the transition to long order or absence of them indicate that the system has not undergone a long range order transitions (e.g. from paramagnetic to FM).

In a ferromagnetic system the neutron scattering in the static approximation for temperatures $T>T_c$ is described by the a Lorentzian [6]

$$I(Q)=\frac{A\kappa^2}{\kappa^2+Q^2}\quad \text{(Lor)} \quad (3)$$

where κ is the inverse spin correlation length, ξ, given by $\xi=1/\kappa$. This form of scattering is derived from the spin correlation function of Ornstein-Zernicke type, i.e.

$$\langle \mathbf{S}_i(0)\circ\mathbf{S}_j(\mathbf{r})\rangle_T\sim\frac{1}{r^{(d-1)/2}}\exp(-\kappa r) \quad (4)$$

for dimensionality d equal to 3.

In a system in which random fields exist the correlation function can be described by [13-14]

$$\langle \mathbf{S}_i(0)\circ\mathbf{S}_j(\mathbf{r})\rangle_T\sim\frac{1}{r^{(d-3)/2}}\exp(-\kappa r) \quad (5)$$

This correlation function for $d=3$ gives the following form for the neutron scattering

$$I(Q)=\frac{B\kappa^4}{\left(\kappa^2+Q^2\right)^2}\quad (\text{Lor}^2) \quad (6)$$

If d_l the dimensionality of the system below which the system cannot preserve long range order, then, for lower dimensionalities the neutron scattering is described by the sum of a Lor and a Lor^2 function. The correlation length, ξ, might be different for the Lor or Lor^2 functions [15]. The form of the correlation function recently has attracted new attention [16]. Both Lor and Lor^2 curves can be easily distinguished from the usual Guinier approximation at low Q values.

3. Experimental

Initially the polycrystalline alloy $Fe_{90}Zr_{10}$ was fabricated from pure (99.99%) constituent metals by arc melting under argon atmosphere. Weight losses were small and less than 0.1%. Amorphous ribbons were produced by fast cooling of the melt alloy on a rotating wheel giving a cooling rate of 10^6 K/s and in an argon atmosphere. No weight loss was detected during the fabrication of the amorphous ribbons from the melt alloy. The thickness of the produced ribbons was around 10 μm and their width around 3 mm, exhibiting high elasticity consistent with their amorphous state. In addition XRD measurements showed no crystallinity. The SANS measurements were carried out at the LOQ instrument at ISIS neutron spallation source facility, in U.K. in the temperature range 10 to 300 K. The magnetic field was applied in the horizontal direction

In this work we shall discuss two types of experimental procedures, a) zero field cooling (ZFC) and b) field cooling (FC).

4. Results

4.1 Zero field cooling

SANS measurements in zero magnetic field at low temperatures show the spontaneous changes of the magnetic state of the sample as structural changes can not be induced. However, the reversibility of the magnetic state prepared during a cooling process can not be taken for granted [17]. For example, the melting temperature of spin clusters produced at low temperatures will depend on the heating rate. Thermodynamic reversibility, even if it exists, will be observed for a process as slow as restricted by the dynamics of the system. It is, therefore, important to determine the reversibility of either a ΔT change or of a full cycle. The holding time for a SANS measurement is about 30 min whereas 10 K temperature change has a duration of about 5 min. So reversibility or irreversibility is discussed in this time context.

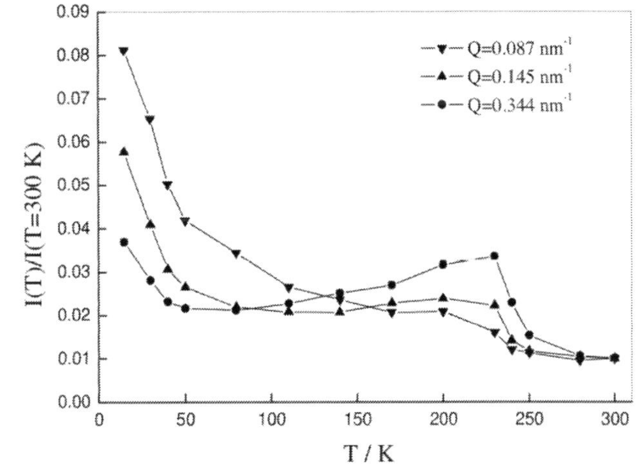

Figure 1. SANS for ZFC versus temperature for different values of Q.

The SANS increases during ZFC (from 300 down to 15 K) and it has been verified that the scattering is fully reversible, both step-wise or as a cycle, for ZF warming up. From 300 down to 250 K the scattering remains almost constant. The ratio of the scattering to that at T=300 K versus temperature and for three different Q values is presented in figure 1. As discussed above all changes in the SANS arise from changes in the spin-spin correlation function and not from any structural changes. Therefore, the increase in SANS at 230 K reflects that the uncorrelated spins of the paramagnetic state of room temperature develop correlations. Magnetization measurements show that there is a transition at T_c=240 K which has been attributed to a PM to FM transition [18]. However, these correlations can not be of long range as the scattering at, for example, $Q=0.145$ nm^{-1} reflects structure in the range of 45 nm. Around 130 K a cross over for the different Q values is observed and as the temperature decreases the scattering increases with an apparent maximum at zero temperature. The faster increase of the smaller Q values shows that now spin correlated sizes larger than those of the transition at 230 K are being grown. The strong development of the second transition could be assigned at around 40 K whereas the transformation continues down to 15 K.

As discussed in section 2 the SANS data can be fitted by a Lor or Lor+Lor^2 functions (equations 3 and 6) as it has been for the amorphous $Fe_{91}Zr_9$ alloy [19]. However, such functional forms do not fit the SANS curves in any temperature range. That Lor or Lor+Lor2 functions do not fit the SANS data have been also been found in the SANS measurements in a series of FeZr alloys [20]. They have chosen to fit the SANS data using a Lor function plus a complicated form presenting spin clusters.

Also in our case, we find that the SANS curves show that two sources of scattering exist. The first can be represented by a Lor function, whereas the second has a Q^{-4} dependence which is the Porod approximation [21]. The presence of the Porod law implies that two magnetic phases exist which have different magnetizations (SANS arises form a different magnetic contrast) and that they are separated by sharp interfaces. This is in agreement with Lorentz microscopy studies on $Fe_{90}Zr_{10}$ films which show that magnetic regions exist in all temperatures down to 6 K and that the structure of these regions does not change during the cooling below T_c [9]. All the SANS measurements have been fitted by the functional form

$$I(Q)=\frac{A\kappa^2}{\kappa^2+Q^2}+\frac{C}{Q^4}=\frac{A\kappa^2}{\kappa^2+Q^2}+\frac{\alpha S\left|(\mathbf{M}_p-\mathbf{M}_m)\times\mathbf{n}_Q\right|^2}{Q^4} \quad (7)$$

where S the total surface of the scattering particles, $\mathbf{M}_p$ the mean magnetization of one particle, $\mathbf{M}_m$ the mean magnetization of the matrix and $\mathbf{n}_Q$ the unit vector along the direction of the scattering vector $\mathbf{Q}$ and α a constant that accounts for the correct units.

Figure 2 shows the scattered intensity as a function of Q at several temperatures during ZFC and the continuous lines are least square fits of equation 7 to the experimental data. From the least square fits the values for ξ, A, and C have been determined and are depicted in figures 4 and 5 as a function of temperature.

The ξ value at room temperature (RT) is large (22.4 nm) and decreases as the temperature decreases to 250 K, followed by an increase as the temperature approaches $T_c \approx 230$ K. Further it presents a slow decrease towards 60 K, whereas further decrease of the temperature results in its abrupt increase to around 10 nm. This behaviour of the correlation length is completely different to that observed in a $Fe_{91}Zr_9$ alloy [19]. Also, later measurements on the same concentration alloy [20] show completely different behavior. Sample differences should be excluded as the same group carried out both measurements. However, it should be noted that between these two measurements different not overlapping Q-ranges and different fitting procedures were used. Our Q-range, due to the wide wavelength range available at the LOQ instrument, covers both Q-ranges used. Therefore, whether a function fits the data is more stringently tested and both long and short wavelength correlations are measured. It should also be added that it is not apparent if the Lor function fitted to the previous SANS data corresponds to our equation 3 which is the Fourier transform of Ornstein-Zernicke correlation function i.e. our constant A is correlation length independent.

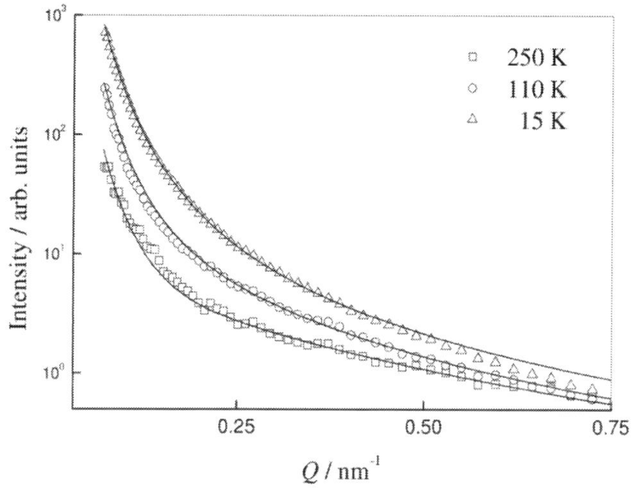

Figure 2. SANS versus Q at different temperatures during ZFC. The continuous lines are least square fits to the data according to equation (7).

This temperature variation of both correlation length, ξ, and constant, A, indicates the presence of the two transitions as described before. The constant C presents a continuous increase as the temperature decreases, which becomes abrupt below 50 K. This variation of C can be explained either with the size increase of the spin clusters or with the increase of the magnetic contrast between the clusters and the matrix (equation 7). This will be discussed later on, after the description of the sample behaviour under the other processes.

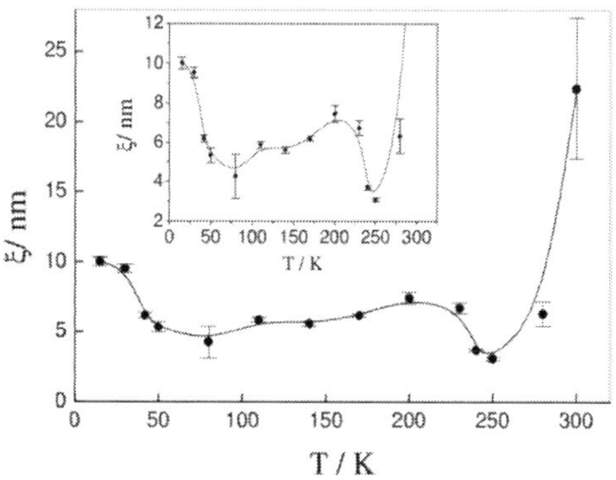

Figure 3. Spin correlation length as a function of temperature during ZFC. The inset is a zoom. The continuous lines are guide to the eye.

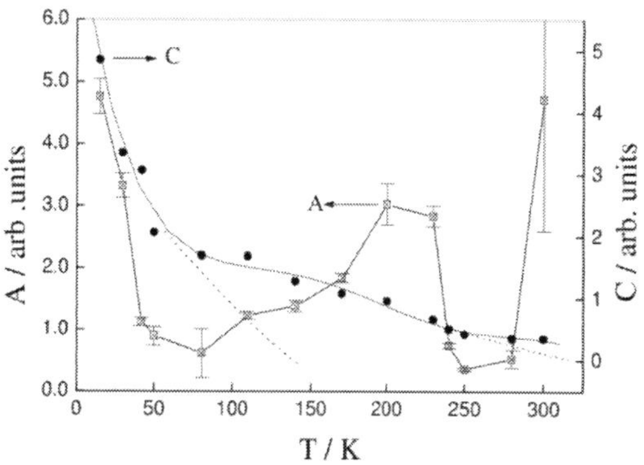

Figure 4. The constants A and C of equation 10 as a function of temperature during ZFC. The continuous lines are guide to the eye.

4.2 Field cooling

The application of a magnetic field at 300 K reduces the scattered intensity. This reduction is initially proportional to the magnetic field magnitude and for fields larger than 0.5 T no further change is observed. The scattering at RT with field off is described by the equation 7. With the application of a magnetic field (>0.16 T) the second term is not required to describe the scattering and $I(Q)$ presents a $1/Q^2$ behaviour. This is proved by the linear dependence of $\ln(I(Q))$ versus $\ln(Q)$ having a slope equal to 2 for all field values larger than 0.16 T (figure 5).

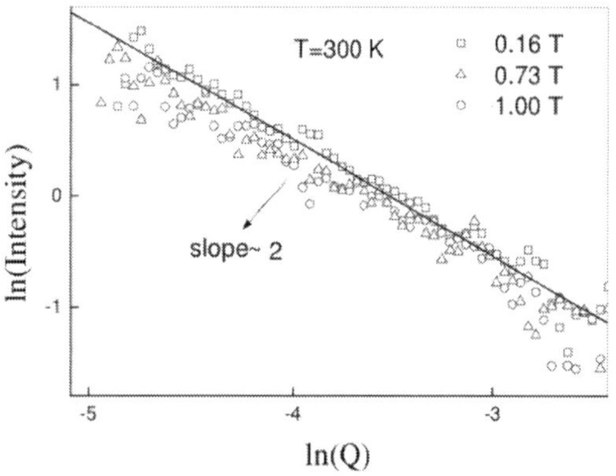

Figure 5. Variation of the logarithm of the scattered intensity versus the logarithm of Q at RT for various field values.

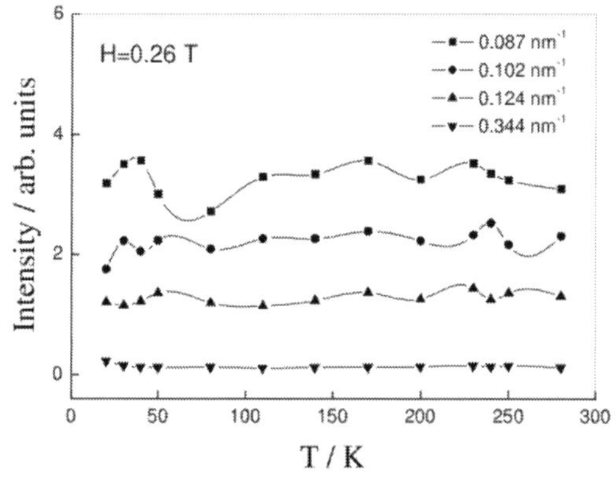

Figure 6. Variation of the scattered intensity versus temperature for various Q values, during FC in a field of 0.26 T.

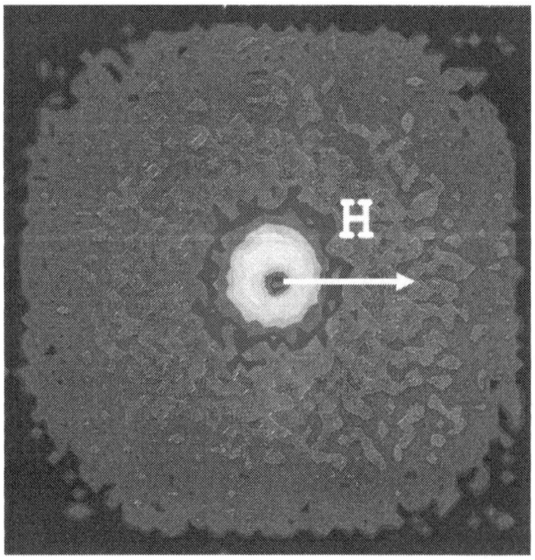

Figure 7. Constant scattered intensity contour from the area detector for H=0.26 T at 30 K during FC.

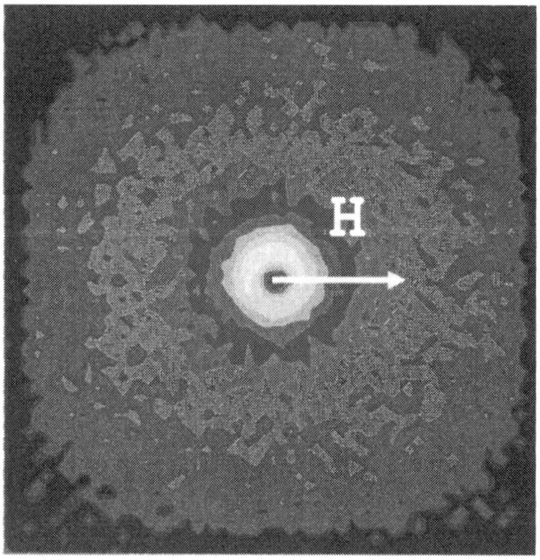

Figure 8. Constant scattered intensity contour from the area detector at 20 K after removal of the field.

During FC from 280 to 20 K in a magnetic field of 0.26 T no variation in the scattered intensity versus temperature is observed, as it is depicted in figure 6, and none of the phase transitions observed in the ZFC occurs. The scattering in this temperature range can be described by the equation

$$I(Q)=\frac{B}{Q^2}+\frac{C}{Q^4} \tag{8}$$

During FC the value for constant C is at least one order of magnitude lower than that determined during ZFC.

In all temperatures during FC the scattered intensity on the 2D detector presents elliptical distortion with the maximum scattering occurring perpendicular to the applied field direction (figure 7). Such prolate distortion is expected from an ensemble of ferromagnetic clusters. After the removal of the field at 20 K the scattering becomes oblate (figure 8).

5. Discussion and Conclusions

The purpose of this work is to understand the nature of the two transitions of this amorphous alloy at the so called T_c and T_f temperatures. Initially, we have to understand the magnetic state of the sample at room temperature. The SANS from the amorphous $Fe_{90}Zr_{10}$ alloy at RT is described by the equation 7 which involves two terms. This fact indicates that there are two scattering sources. One comes from a spin correlation function (equation 4, $d=3$) that gives rise to the Lor type scattering and the other comes from a magnetization contrast that gives rise to a Porod like scattering. The presence of Porod scattering suggests the existence of magnetic clusters with different magnetization than that of the matrix. It should also be noted that the Porod law dependence implies sharp interfaces between the two magnetic phases. So, two magnetic phases co-exist at room temperature. The local extent of either phase has to be above 120 nm in diameter, otherwise the Porod law would not have been observed (the Porod law is an asymptotic form of scattering which applies for $QR>4$). The word phase here has the meaning of a magnetic moment spatial configuration. From the second term in equation (7) it is apparent that configurationally different phases could arise from magnetic moment differences either in orientation or in magnitude. The first possibility would reflect magnetic domain formation whereas the second should be the result of structural differences either in the Fe concentration or in sample crystallinity. However, it should be remembered that these regions are large and separated with sharp interfaces. The Lor term indicates that within one phase or both or at the interface of these two phases spin-spin correlations of about 24 nm exist.

In the previous paragraph it has been established that at RT two regions with different magnetizations co-exist and these two regions are separated with sharp interfaces. For the description of the system the two phase magnetic model is most suitable [23]. Within this picture the spin misalignment scattering [24] is thought insignificant and it is ignored. This conclusion is based on that the explicit angular dependence of the spin misalignment contribution is not observed before or after the application of the magnetic field (figure 7). Therefore the model we are thinking is that of a magnetic inhomogeneity embedded in a matrix. We shall call the magnetic inhomogeneity spin cluster implying simply that its spins behave the same way. The magnetic moments of spin clusters and matrix at RT are different both in magnitude and possibly in direction and one of them might be zero (paramagnetic). Also, if they undergo a phase transition it would be expected to occur at different temperatures. In order to understand the magnetism of these two regions, we shall next discuss the dependence of the SANS on magnetic field at RT and the temperature change of the constant C (figure 4 and equation (7)).

The SANS at 300 K in a magnetic field above 0.6 T shows a $1/Q^2$ dependence. For lower fields the SANS is described by equation (7). This implies that the application of the magnetic field has two effects, a) the correlation length arising for the Lor function becomes very large and b) the magnetic contrast difference between the two phases becomes zero (Porod scattering, second term of equation

7). The Porod scattering without magnetic field is isotropic, with the application of a small field it becomes anisotropic and as the field increases, it decreases to become zero for fields equal and above 0.6 T. The anisotropic SANS implies magnetic inhomogeneities and that the magnetic moment of the inhomogeneities (or of the surrounding matrix) has a definite angle with the magnetic field. Since the magnetic contrast decreases as the field increases, we have to assume that a) the spins in the matrix are aligned with the application of a small magnetic field and b) as the field increases the magnetic moments of the cluster rotate to be finally aligned with the field. As, finally, the magnetic contrast becomes zero we have to accept that the magnitude of the magnetic moments in both the spin cluster and the matrix are almost equal. The fact that a small magnetic field destroys both the short (Lor function, equation (7)) and long wavelength (Porod scattering) correlations indicates that both the spin clusters and the matrix are very close to a ferromagnetic transition. The FM transition here has the meaning that the spins on average are aligned to the applied magnetic field. Therefore, from the application of the magnetic field at RT the picture which emerges is as follows: a) a small magnetic field aligns the spins of the matrix but its average magnetic moment is different than that of the spin clusters (this explains that the isotropic without magnetic field SANS becomes anisotropic after the application of a small magnetic field), b) as the field increases the mean magnetic moment of the spin clusters increases as it is demonstrated by the decrease of the constant C and c) at field of 0.6 T both the magnetic moments of the spin clusters and the matrix are equal giving $C = 0$. In this picture it has been implied that as the magnetic field increases the size of the spin clusters remains constant but their average magnetic moment increases. However, another possibility consistent with the experimental results exists. The spins at the surface of the cluster are partially influenced by the molecular field arising from the magnetization of the matrix. As the matrix spins are aligned in the applied magnetic field their magnetization increases. Now the spins on the cluster experience a stronger molecular field and the external field. The combination of these two fields becomes strong enough to align to the external magnetic field a shell of the spins of the cluster, thus reducing its size. In this picture the reduction of the constant C is attributed to the reduction of the size of the spin cluster and not predominantly to the alignment of its magnetic moments to the external magnetic field. This second interpretation of the experimental results is more appealing as it includes a plausible physical mechanism for the homogenization of the two magnetic regions.

During ZFC a "cusp" is observed at around 230 K (figure 1) the presence of which indicates a phase transition in agreement with ac susceptibility measurements on the same sample. Furthermore, at the same temperature a maximum in the values of ξ and A is observed, whereas this does not apply for the constant C (figures 3 and 4). In the previous paragraphs it has been established that there are two different in magnetization regions which are described with the constant C (see equation (7)). In figure 4 the temperature change of the constant C can be interpreted as the superposition of two curves one which would indicate a phase transition above 300 K and the other at around 150 K (see extrapolated lines in figure 4). Keeping in line with the discussion of the previous paragraph in which the response of the system to the application of a magnetic field was discussed, the high temperature transition is attributed to the matrix whereas the lower temperature transition to the spin clusters.

Summarizing the above we may conclude that there is strong experimental evidence for the correctness of the model for re-entrant spin glasses that visualize such systems, in the ferromagnetic state, to be composed of an infinite three-dimensional (3D) FM cluster (matrix) and finite spin clusters [1, 25-27]. However, if for T_c we take the temperature of 240 K we do not observe a long range ferromagnetic order (figure 4, constant C) as magnetization measurements have been interpreted [28] but the onset of a FM type transition of the matrix. The SANS results demonstrate that even small magnetic fields destroy the differentiation between matrix and clusters (figure 5), thus measurements even in a small magnetic field do not reveal the ground state of the system. SANS corresponds to the ground state of the system unperturbed by the application of a magnetic field thus, these type of measurements are invaluable in elucidating the re-entrant behaviour [29].

Having outlined above the physical processes reflected in the second term of equation 7, we now focus our discussion to the results arising from the Lor part of equation (7) which corresponds to the correlation function given by equation 4. The physical parameters derived form the SANS data fitting are the correlation length, ξ, and the constant, A, which is proportional to the strength of the magnetization i.e. $A = \langle \mathbf{S}^2 \rangle_T$. Both constants show a transition at 230 and 40 K (see figures 3 and 4). In the magnetization measurements a bifurcation of the heating and cooling curves is observed at a temperature so called freezing temperature T_f [30]. The extrapolated freezing temperature to zero field derived from magnetization measurements is 40 K [31] which corresponds to the fast increase of the SANS observed in figure 1. The lower temperature transition has been attributed to a transition from the ferromagnetic state to a mictomagnetic (cluster spin-glass state) state. It should be noted that the magnetization fails to saturate at 4.2 K and in fields of 15 T [32]. The continuous increase of SANS (figure 1) as the temperature decreases is consistent with the lack of magnetic moment saturation. These two magnetic phase transitions are also observed in resistivity measurements [33]. Initially the room temperature specific resistivity decreases as the temperature decreases which is expected if the main contribution arises from electron-phonon scattering. At around 240K there is a plateau and below this temperature an almost linear increase in the specific resistivity is observed which apparently could be attributed to an increase of the electron-spin scattering. Further, at around 30 K an abrupt change in the slope is observed. The magnetoresistance at a magnetic field of 8 T is positive at all temperatures and shows a maximum at around T_c [34]. From both high field magnetization and Mössbauer measurements at 4.2 K the magnetic moment is 1.75 μ_B/Fe atom [35]. We conclude this paragraph by restating that the two magnetic transitions demonstrated in figures 1, 2 and 3 have been observed at around the same temperatures by different experimental techniques.

The two transitions of the correlation length and constant A at 230 and 40 K are obviously connected with the transition of the matrix (240 K) and the spin cluster (40 K). That the phase transitions denoted by the parameters of the scattering described by the Lor function (equation 7) are connected with the phase transition of the matrix and the spin clusters points out that this scattering most probably arises from the interface of the two regions. It is, thus, proposed that at the spin cluster-matrix interface the competing interactions give rise to Ornstein-Zernicke type spin correlations i.e. the interface is always just before a FM transition. This proposition is corroborated with the fact the application of a magnetic filed (ZFC) suppresses both transitions at 230 and 40 K.

Concluding we may suggest that this work offers two contributions in the understanding of the re-entrant spin glasses. It establishes that there are two magnetically different regions which have different transitions temperature to FM order. Small magnetic fields can switch both these regions to a common FM state. Further the competing interaction at the interface of these regions results to short wavelength (10 nm) spin correlations which also follow the phase transitions of the two regions. A small magnetic field suppresses these transitions as it is stronger than the internal fields generated at the interface of the two magnetic regions. The findings of this work are in broad agreement with a recent theoretical description of the re-entrant spin glass system [36].

Acknowledgments

We would like to thank the British Council and the Greek General Secretariat of Research and Technology for financial support. The experiments at the ISIS Pulsed Neutron and Muon Source were supported by a beamtime allocation from the Science and Technology Facilities Council. The contribution of Dr. S. King during the experiments is appreciated.

References

[1] Kiss L F, T. Kemény T, Vincze I and Gránásy L 1994 *J. Magn. Magn. Mater.* **135** 161
[2] Hiroyoshi H and Fukamichi K 1982 *Physics Letters* **85A** 242
[3] Ryan D H and Coey J M D 1987 *Phys. Rev. B* **35** 242

[4] Kaul S N 1983 *Phys. Rev. B* **27** 6923
[5] Read D A, Moyo T and Hallam G C 1984 *J. Magn. Magn. Mater.* **44** 279
[6] Beck W and Kronmueller H 1985 *Phys. Stat.. Solidi (b)* **132** 449
[7] Ryan D H, Coey J M D, Batalla E, Altounian Z and Strom-Olsen J O 1987 *Phys. Rev. B* **35** 8630
[8] Wronski Z, Janicki A J and Matyja H 1983 *J. Mater. Sci. Lett.* **2** 5
[9] Senoussi S, Hadjoudj S, Jouret P, Bilotte J and Fourmeaux R 1988 J. Appl. Phys. **63** 4086
[10] Mergia K, Messoloras S, Nicolaides G, Niarchos D and Stewart R J 1996 J. Appl. Phys. **76** 6380
[11] Mergia K and Messoloras S 2008 J. Phys.: Condens. Matter **20** 104219
[12] Marschall W and Lovesey S W 1971 *Theory of Thermal Neutron Scattering*
[13] Aharony A, Imry Y and Ma S -K 1976 *Phys. Rev. Lett.* **37** 1367
[14] Grinstein G 1976 *Phys. Rev. Lett.* **37** 944
[15] Aharony A and Pytte E 1983 *Phys. Rev. B* **27** 5872
[16] Dominicis C de, Giardina I, Marinari E, Martin O C and Zuliani F 2005 *Phys. Rev. B* **72** 014443
[17] Cugliandolo L F and Kurchan J 1999 *Phys. Rev. B* **60** 922
[18] Buschow K H J and Smit P H 1981 *J. Magn. Magn. Mater.* **23** 85
[19] Rhyne J J and Fish G E 1985 *J. Appl. Phys.* **57** 3407
[20] Rhyne J J, Erwin R W, Fernandez- Baca J A and Fish G E 1988 *J. Appl. Phys.* **63** 4080
[21] Porod G 1951 *Kolloid-Z* **124** 83; Porod G 1952a *Kolloid-Z* **125** 51; Porod G 1952b *Kolloid-Z* **125** 108
[22] Burke S K, Cywinski R and Rainford B D 1978 *J. Appl.. Cryst..* **11** 644
[23] Löffler J F, Braun H B, Wagner W, Kostorz G and Wiedenman A 2005 *Phys. Rev. B* **71**, 134410
[24] Michels A and Weissmüller J 2008 *Rep. Prog. Phys.* **71** 066501
[25] Coles B R, Sarkissian B V B and Taylor R H 1978 *Phil. Mag.* B **37** 489
[26] Kaul S N 1980 *Solid State Commun.* **36** 279
[27] Kaul S N 1985 *J. Magn. Magn. Mater.* **53** 5
[28] Kaul S N 1988 *J. Phys. F: Met. Phys.* **18** 2089
[29] Srinath S and Kaul S N 2000 *Europhys. Lett.*, 51 441
[30] Kaul S N and Srinath S 1998 *J. Phys.: Condens. Matter* **10** 11067
[31] Kaul S N 1983 *Phys. Rev. B* **27** 6923
[32] Krishnan R, Rao K V and Liebermann H H 1984 *J. Appl. Phys.* **55** 1823
[33] Obi Y, Wang L C, Motsay R and Onn D G 1982 *J. Appl. Phys.* **53** 2304
[34] Dahlberg D, Rao K V and Fukamichi K 1984 *J. Appl. Phys.* **55** 1942
[35] Unruh K M and Chien C L 1983 *Phys. Rev. B* **30** 4968
[36] Niidera S and Matsubara F 2007 *Phys. Rev. B* **75** 14413

Magnetic and structural properties of $Fe_{3-x}Cr_xAl_{0.5}Si_{0.5}$

K Rećko[1], D Satuła[1], L Dobrzyński[1,2], M Biernacka[1], J Waliszewski[1], A Go[1], J J Milczarek[2]

[1]Faculty of Physics, University of Białystok, Lipowa 41, 15-424 Białystok, Poland

[2]The Soltan Institute for Nuclear Studies, 05-400 Otwock-Świerk, Poland

E-mail: karo@alpha.uwb.edu.pl

Abstract. Results of X-ray, neutron, magnetization and Mössbauer measurements on polycrystalline samples of $Fe_{3-x}Cr_xAl_{0.5}Si_{0.5}$ (x = 0, 0.125, 0.250, 0.375 and 0.5) alloys are presented. The alloys crystallize in the structure of DO_3 type, and their unit cell volumes are practically independent of the chromium concentration. X-ray and neutron diffraction confirmed the phase homogeneity of all the samples. Neutron and Mössbauer measurements disclosed that Cr atoms occupy preferentially B-sites, while D sites are almost entirely occupied by Al and Si. The total magnetisation as well as the site and individual magnetic moments $\mu_{Fe(A,C)}$, $\mu_{Fe(B)}$ and $\mu_{Cr(B,D)}$ have been found to be linearly dependent on the chromium concentration.

1. Introduction

Fe_3Al and Fe_3Si alloys crystallize in DO_3-type of structure, see Fig.1. Its unit cell within space group F-43m (no. 216) can be described in terms of four interpenetrating fcc lattices: (A,C) - sites originating from (0,0,0) and (½, ½, ½), respectively, B-site originating from (¼, ¼, ¼), and D-site originating from (¾, ¾, ¾). A, B and C-sites in fully ordered structure are occupied by iron atoms, whereas the D-site is occupied by aluminium or silicon.

When chromium is substituted for iron a number of interesting properties occur. The lattice parameter of alloys based on Fe_3Si increases linearly with chromium content, while it decreases linearly with same (absolute) slope in $Fe_{3-x}Cr_xAl$ alloys [1, 2]. Thus the composition of $Fe_{3-x}Cr_x(Al_{0.5}Si_{0.5})$ should exhibit lattice parameter independent of the concentration of chromium. The locations of Cr impurities, as well as their magnetic moments are also different [1, 2]. From the crystal structure point of view, one should note that the nearest-neighbour configurations [1] make the B-sites inequivalent with respect to the symmetry-equivalent A and C ones. This results in the diffraction pattern characterized by three types of structure factors:

$$F_{hkl}^{(1)} \sim \sqrt{(f_A - f_C)^2 + (f_B - f_D)^2} \quad (1)$$

corresponding to the superstructure lines with h,k,l - all odd and

$$F_{hkl}^{(2)} \sim (f_A + f_C) - (f_B + f_D) \quad (2)$$

corresponding to the $h+k+l=4n+2$, where n is integer, and

$$F_{hkl}^{(3)} \sim f_A + f_C + f_B + f_D \quad (3)$$

corresponding to the fundamental reflections with $h+k+l=4n$, where the f_i is the effective scattering amplitude of the *i-th* site. The symmetry non-equivalence of B and (A,C) iron sites is reflected in different on-site magnetic moments. The experiments [2-5] and calculations [6, 7] show that the 3d impurities tend to occupy preferentially one or another of iron sites, (A,C) or B. The element placed to the left of Fe in the Periodic Table of Elements should locate preferentially in B sites, while the elements to the right of Fe enter preferentially the (A,C) sites [7-9]. Accepting such rules, and assuming the same form-factors $f_A = f_C = f_{A,C}$, the appropriate formulas for magnetic structure factors can be written in the simple form:

$$F_{hkl}^{(1)} \sim i\left(f_B - f_D\right) \tag{4}$$

$$F_{hkl}^{(2)} \sim 2f_{A,C} - f_B - f_D \tag{5}$$

$$F_{hkl}^{(3)} \sim 2f_{A,C} + f_B + f_D \tag{6}$$

The studies of $Fe_3(Al_xSi_{1-x})$ alloys [8, 9] showed that the magnetic moments depend on the relative concentration of Al and Si. In contrast to the weak moments found at D-sites, the values of magnetic moments at (A,C) and B positions turned out not to vary systematically with Cr content [8]. The analysis of the spin-density distributions, carried out by the Maximum Entropy Method analysis [10] showed that magnetic moments of $\mu_{Fe(B)}$ vary linearly with silicon content.

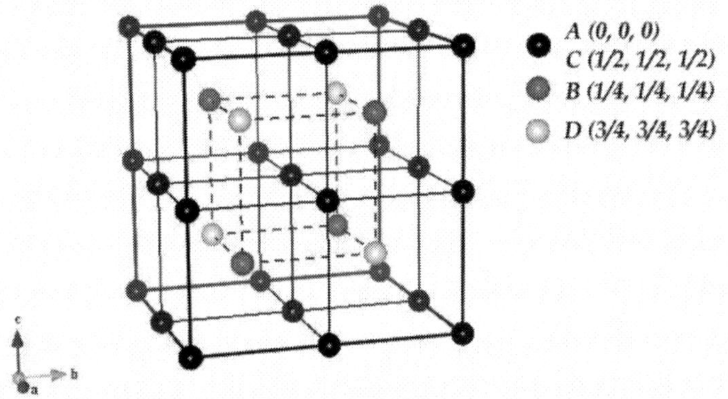

Figure 1. The unit cell of the DO_3-type structure.

Ab-inito total energy calculations of $Fe_{3-x}Cr_xAl$ [11] strongly suggested the B-site preference of Cr atoms. Similarly, the self- consistent band structure calculations [12] as well as recently published self-consistent TB-LMTO calculations disclosed the existence of a strong preference of the B-site occupation by chromium [13]. On the other hand, CPA calculation of $Fe_{3-x}Cr_xAl$ [14] did not show definite preference of Cr to the B-sites, especially at larger Cr content. According to ab-initio calculations, chromium in Fe_3Si perturbs the neighbouring iron atoms and this perturbation seems to be local and most efficient when Cr atoms are n.n. to iron atoms at both (A,C) - and B-positions [15]. In light of the experimental observation, chromium in $Fe_{3-x}Cr_xSi$ alloys enters both (A,C) and B sites [1, 2, 16-18] while certain preference to the occupation of B-site in Fe_3Al was observed [1, 2, 14]. One should note that in the case of Fe_3Al it is very difficult to achieve full DO_3-type ordering, and B–D disorder is usually observed. This little B–D disorder, included in the calculations [19], appeared to be not responsible for the lack of preferential occupation of sites in the case of chromium in Fe_3Si. Our recent measurements were carried out on $Fe_{3-x}Cr_xAl_{0.5}Si_{0.5}$ [20] with the aim to have same volume per atom, which should permit to study in better way the effects of local surroundings.

2. Sample preparation

The sample preparation has been described in [20]. The sample annealing procedure, however, was slightly modified. Because the losses of mass came from the evaporation of chromium and aluminium due to their high vapour pressure, the composition of the alloys was checked by X-ray and Mössbauer measurements after each stage of thermal history. The experience of previous experiments [1, 2, 16-18] taught that special care should be dedicated to the sample annealing procedures.

3. X-ray diffraction

X-ray diffraction measurements were performed using conventional Bragg-Brentano powder diffractometer (HZG-4, University of Bialystok, Poland) with $Co_{K\alpha}$ radiation, λ=1.78893 Å. These measurements have revealed the homogeneity of all Cr-doped samples at room temperature. With an increase of the chromium content, the intensities of the superstructure reflections decrease. It is well-known that certain B-D disorder appears systematically in Cr-doped Fe_3Al [21]. The superstructure lines with h,k,l - all odd are directly sensitive to the difference of amplitudes at B and D positions (see eq. 4). Our Mössbauer spectroscopy results exclude the presence of iron in these positions, especially for alloys with higher concentration of chromium. Thus, this type of disorder may concern the partial occupation of D-sites by chromium only. Unfortunately, the intensities of (200) lines are not sensitive to this kind of disorder (see eq. 5). This is due to the fact that the atomic scattering factors of iron and chromium as well as aluminum and silicon are hardly distinguishable in X-ray diffraction experiment.

According to the results presented in Table 1, in $Fe_{3-x}Cr_xAl_{0.5}Si_{0.5}$ alloys the effective scattering amplitude of the (A,C) - sites were found to be close to iron atomic scattering factors, which means that Al and Si atoms do not enter (A,C) positions. However, from the occupancies of D – sites it follows that the samples are not exactly stoichiometric and contain both Al and Si in the ratio slightly deviating from the nominal 1:1 proportion. Such situation was discussed in detail in ref. [20]. Alternative explanation for the increase of the scattering amplitude at D sublattice seems to be B-D disorder, caused by Cr atoms occupying partly D site. Therefore, the model with the B-D disorder and untouched (A,C) sites had also to be considered. The latter conclusions are supported by Mössbauer experiments (see section below).

In the light of previous results [20] where the $Ag_{K\alpha}$ radiation, λ = 0. 55926 Å, was used, the obtained isotropic Debye-Waller factors versus increasing chromium concentration were as follows: 0.376(12) - for undoped alloy and 0.353(19), 0.357(14), 0.380(14) and 0.344(18) $[Å]^2$ for the consecutive chromium-containing alloys. Therefore in order to reduce the free parameters during the refinements presented below the weighed average value B = 0.366 $[Å]^2$ was fixed for all alloys. Moreover, according to results reported in [20] it was assumed that (A,C) positions are fully occupied by iron atoms and the chromium atoms enter predominantly (in the range of error bars) B sublattice. The diffraction patterns were analysed by Rietveld – type profile refinement method using FullProf program [22]. For better readability the goodness-of-fit parameters defined in ref. [22] are also quoted.

Table 1. Structural parameters for $Fe_{3-x}Cr_xAl_{0.5}Si_{0.5}$ alloys, i.e. lattice parameter – a, site occupancies - and agreements factors [22] obtained from X-ray measurements at room temperature.

Measured composition	a(Å)	Fe(B)	Si(D)	R_F	R_{Bragg}	χ^2
$Fe_{2.909}Cr_{0.107}Al_{0.492}Si_{0.492}$	5.7199(6)	0.993(6)	0.976(11)	0.12	0.06	6.18
$Fe_{2.779}Cr_{0.231}Al_{0.495}Si_{0.495}$	5.7200(7)	0.983(8)	0.973(15)	0.21	0.19	7.93
$Fe_{2.662}Cr_{0.384}Al_{0.477}Si_{0.477}$	5.7199(6)	0.974(7)	0.997(17)	0.20	0.11	7.41
$Fe_{2.571}Cr_{0.481}Al_{0.474}Si_{0.474}$	5.7202(8)	0.967(6)	1.003(11)	0.11	0.06	5.85

The analysis of the occupations of the B and D – sites allow to compare the estimated compositions of the powder samples with the nominal ones according to the equations:

$$\zeta_{Fe(B)} \cdot f_{Fe} = \left(1 - x_{Cr(B)}\right) f_{Fe} + x_{Cr(B)} \cdot f_{Cr} \tag{7}$$

and

$$\zeta_{Si(D)} \cdot f_{Si} = \left(1 - y_{Cr(D)}\right) \cdot \left(\frac{f_{Si} + f_{Al}}{2}\right) + y_{Cr(D)} \cdot f_{Cr} \tag{8}$$

where ζ - site occupancy. In the later equation the ratio aluminium to silicon equal to 1:1 was fixed. Results obtained by X-ray diffraction powder technique show that the samples are single DO_3 phase. This conclusion means that for the heat treatment applied here the estimated sample compositions do not differ from the nominal ones.

Another important finding is observed independence of the lattice parameter (see 2nd row of Table 1) of the Cr concentration. Lets recall that the experiments previously carried out on $Fe_{3-x}Cr_xSi$ [1, 16, 17] showed a contraction of the unit cell volume against chromium concentration in the range up to x = 0.4, while the opposite trend was observed in $Fe_{3-x}Cr_xAl$ system [18] up to x = 0.6. For better insight into the nature of magnetic moments it is important to be sure that the parameters of the unit cell as well as interatomic distances remain independent of chromium content.

4. Neutron scattering

In order to determine magnetic structure, the neutron powder data were collected at the reactor MARIA (IEA, Swierk) at 10 K (see Fig. 2) and at room temperature using 2-axis diffractometer and an unpolarized incoming-neutron wavelength of $\lambda = 0.985$ Å. The measurements were carried out in the range of $10^0 - 110^0$ scattering angles 2θ with the 0.1^0 step size.

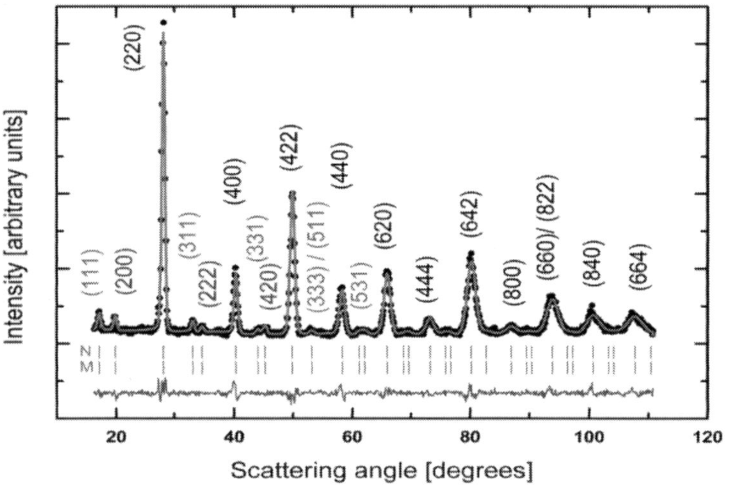

Figure 2. RT neutron powder diagrams experimental, calculated and the residual ones obtained for nominal $Fe_{2.875}Cr_{0.125}Al_{0.5}Si_{0.5}$ alloy, with marked N-nuclear and M- magnetic Bragg peaks. All peaks can be indexed within DO_3-type of structure. However, for better readability of the figure, some of the peaks shown have been left without indices.

Analysis of the neutron intensities cannot distinguish aluminium and chromium due to similar values of their neutron scattering lengths. Fortunately, in the light of X-ray results, rather small B-D disorder is present in measured samples. The neutron data confirm again the expectation that the lattice parameter of the alloys should be practically independent of the chromium concentration [20]. They show unambiguously that the chromium atoms locate predominantly at B-sites. As a result, (A,C) are treated as containing iron atoms only, while B – sites are occupied by iron and chromium ions. In order to reduce the number of free parameters, the chemical compositions at D positions were set to the ones given by the X-ray data (see Table 2). The distribution of atoms among the (A,C) and B crystallographic sites obtained by the use of X-ray and neutron scattering as well as the on-site occupancies remain in very good agreement. Moreover, judging from the temperature dependence of the magnetic moments measured for $Fe_{2.875}Cr_{0.125}Al_{0.5}Si_{0.5}$ at 10 K and at RT (the only sample for which the measurements have been carried out at RT-see Fig. 2), Curie temperature must be well above RT. The magnetic structure is commensurate with the nuclear one, thus all observed intensities disclose both nuclear and magnetic contributions of scattering amplitudes, i.e.:

$$I_{hkl} \propto |F_{hkl}|^2 = |F_N|^2 + \frac{2}{3}|F_M|^2 \tag{9}$$

where the appropriate structure factor components are equal to

$$F_{hkl}^N = \sum_j b_j e^{-i\vec{G}\cdot\vec{r}_\alpha} = \sum_j b_j \exp\left[-i\cdot\left(\frac{2\pi}{a}h,\frac{2\pi}{a}k,\frac{2\pi}{a}l\right)\cdot\left(x_j,y_j,z_j\right)\right] \tag{10}$$

and

$$\vec{F}_{hkl}^M = \vec{\chi}\sum_j \vec{p}\left(\vec{K}\right)e^{i\vec{K}\cdot\vec{r}_j} \tag{11}$$

with

$$\vec{p}\left(\vec{K}\right) = 0.2695\cdot\mu_j\cdot f_j\left(\vec{K}\right) \tag{12}$$

$$\vec{\chi} = \left[\hat{\vec{m}} - \left(\hat{\vec{K}}\cdot\hat{\vec{m}}\right)\cdot\hat{\vec{K}}\right] \text{ - the Halpern's vector} \tag{13}$$

where b_j - is the neutron scattering amplitude of the *j-th* atom, f_j - magnetic form factor of the *j-th* atom and μ_j - magnetic moment of the *j-th* atom.

Table 2. Structural and magnetic parameters for $Fe_{3-x}Cr_xAl_{0.5}Si_{0.5}$ alloys as well as GOF (goodness of fit) parameters obtained from neutron measurements collected at 10 K and at room temperature.

Chromium nominal concentration - x		0.125_{RT}	0.125_{10K}	0.25_{10K}	0.375_{10K}	0.5_{10K}
Site occupancy	Fe(A,C)	Fixed to 1	Fixed to 1	Fixed to 1	Fixed to 1	Fixed to 1
	Fe(B)	0.91(4)	0.91(3)	0.78(5)	0.66(6)	0.57(4)
	Cr(B)	0.09(3)	0.09(3)	0.22(5)	0.34(6)	0.43(4)
	Al/Si(D)[1]	0.984	0.984	0.989	0.956	0.949
	Cr(D)	0.016	0.016	0.011	0.044	0.051
Individual magnetic moments [μ_B/atom]	$\mu_{Fe(A,C)}$	1.49(13)	1.51(12)	1.45(18)	1.36(15)	1.29(17)
	$\mu_{Fe(B)}$	2.38(17)	2.44(14)	2.34(15)	2.20(17)	2.06(19)
	$\mu_{Cr(B,D)}$	1.44(14)	1.44(11)	1.52(17)	1.62(20)	1.69(21)
On-site magnetic moments [μ_B]	$\mu_{(A,C)}$	1.49(13)	1.54(12)	1.45(18)	1.36(15)	1.29(17)
	$\mu_{(B)}$	2.04(32)	2.09(25)	1.49(35)	0.90(41)	0.45(35)
	$\mu_{(D)}$	-0.023	-0.023	-0.017	-0.071	-0.086

[1] The ratio aluminum to silicon was fixed to 1:1 during whole of the series refinement.

Total magnetic moment [μ_B/f.u.]	5.0(6)	5.1(5)	4.4(7)	3.5(7)	2.9(7)
R_F	0.08	0.06	0.11	0.09	0.08
R_{Bragg}	0.12	0.08	0.15	0.07	0.11
R_M	0.11	0.10	0.16	0.10	0.14
χ^2	4.1	3.6	4.5	3.8	4.8

The best description of the magnetic part of the neutron powder diagram is obtained for antiferromagnetically coupled Cr and Fe. All magnetic moments of iron are oriented parallel to each other. According to [20] the concentrations of iron at (A,C) sites were equal to: 0.98(1) for x = 0.125 at RT, and 0.99(3), 0.94(6), 0.97(5), 0.95(3) at 10K for x = 0.125, 0.25, 0.375 and 0.5, respectively. Therefore in order to reduce the number of free parameters during data refinements for all measured alloys the Fe(A,C) occupancy was assumed to be equal to one. The most likely distribution of atoms in the samples as well as their magnetic moments arrangement are presented in Table 2. The individual magnetic moments and the magnetisation per formula unit, μ_{tot} versus chromium concentration diminish linearly wit Cr concentration, see Table 2. The behaviour of the moments at A and C sites can be easily understood in terms of Cr atoms appearing at neighbouring B-sites. The decrease of the moment at B-site is clearly connected with negative contribution coming from chromium moments. Magnetic moments per formula units are in good agreement with the ones measured in magnetisation studies at 10 K, see Table 3. Such measurements were carried out using vibrating sample magnetometer (VSM Lake Shore model 7300) under magnetic fields up to 1.2 T in the temperature 10 K and 300 K.

Table 3. Spontaneous magnetisation of $Fe_{3-x}Cr_xAl_{0.5}Si_{0.5}$ samples M_s [μ_B/f.u.] versus chromium concentration obtained for 10K and 300K.

Cr content	x = 0	x = 0.125	x = 0.25	x = 0.375	x = 0.5
10K	5.5(3)	4.8(4)	4.2(3)	3.6(2)	3.1(2)
300K	4.7(3)	4.2(3)	3.6(2)	3.1(2)	2.7(3)

6. Mössbauer spectroscopy

The ^{57}Fe Mössbauer measurements have been carried out in the constant acceleration mode at room temperature using ^{57}Co source in Cr matrix. The spectrometer was calibrated before and after measurements. The measured spectra were fitted using superposition of elementary sextets broadened by Gaussian distribution of hyperfine magnetic field. The values of hyperfine fields obtained from the measurements are presented in Table 4. The values with stars in this Table denote those ones, which were fixed ones during fitting procedure.

For perfect ordering of the alloy the spectrum should contain only two sextets connected with iron atoms Fe(A,C) surrounded by 4Fe(B), 4Si/Al(D) (Fe(4)) and Fe(B) surrounded by 8Fe(A,C) (Fe(8)). The intensity ratios of those sextets are equal to 2. For better description of the spectrum of Fe_3(Si/Al) one needs to use an additional broad sextet Fe(7) with the hyperfine magnetic field between values obtained for Fe(8) and Fe(4). It means that there exists a small disorder in the studied alloy. The value of isomer shift of Fe(7) is equal within the error bars to the value observed for Fe(8). It means that observed disordering are probably between (A,C) and (D) sites. Small part of Fe(A,C) atoms occupies the Al/Si(D) sites.

Table 4. The hyperfine fields parameters obtained from the best fits of the Mössbauer spectra of the $Fe_{3-x}Cr_xAl_{0.5}Si_{0.5}$ alloys. B denotes the magnetic hyperfine field, σ_B - standard deviation of Gaussian distribution, IS - the isomer shift relative to α-Fe and A means the intensity of the local components.

Chromium content -x	sextet	B[T]	σ_B [T]	IS [mm/s]	A [%]
	S(8)	29.87 ± 0.04	0.41 ± 0.11	0.06 ± 0.01	31.5 ± 2.9
0	S(7)	25.74 ± 0.79	2.2 ± 1.2	0.05 ± 0.03	10.8 ± 4.4
	S(4)	20.73 ± 0.03	1.01 ± 0.06	0.21 ± 0.01	58.1 ± 2.9
	S(8)	29.29 ± 0.09	0.91 ± 0.07	0.061 ± 0.005	24.4 ± 1.3
0.125	S(7)	25.2±0.6	2.11 ± 0.36	0.048 ± 0.011	16.4 ± 2.3
	S(4)	20.32 ± 0.04	1.26 ± 0.05	0.205 ± 0.005	43.7 ± 1.3
	S(3)	14.95 ± 0.09	1.2 (*)	0.178 ± 0.009	15.5 ± 0.5
	S(8)	28.06 ± 0.15	1.47 ± 0.13	0.053 ± 0.007	21.4 ± 1.9
	S(7)	24.35 ± 0.28	1 (*)	0.041 (*)	14.1 ± 1.5
0.25	S(4)	20.19 ± 0.11	1 (*)	0.201 ± 0.006	28.7 ± 1.6
	S(3)	14.87 ± 0.15	2.26 ± 0.19	0.17 (*)	30.3 ± 1.9
	S(2)	8.31 ± 0.23	1 (*)	0.151 ± 0.024	5.4 ± 0.7
	S(8)	26.96 ± 0.17	1.5 (*)	0.053 ± 0.011	16.1 ± 1.5
	S(7)	23.13 ± 0.37	1.8 (*)	0.023 ± 0.018	15.5 ± 1.7
0.375	S(4)	19.69 ± 0.17	1.24 ± 0.25	0.199 ± 0.014	18.7 ± 4.5
	S(3)	14.48 ± 0.33	2.42 ± 0.51	0.16 (*)	39.8 ± 7.4
	S(2)	7.9 ± 1.1	3.0 ± 0.49	0.14 (*)	9.9 ± 3.8
	S(8)	25.31 ± 0.48	1.79 ± 0.33	0.046 ± 0.013	14.8 ± 3.7
	S(7)	21.62 ± 0.51	1.8 (*)	0.026 ± 0.031	13.3 ± 4.2
0.5	S(4)	18.94 ± 0.28	1.67 ± 0.34	0.18 (*)	18.6 ± 4.7
	S(3)	14.05 ± 0.25	1.94 ± 0.21	0.14 (*)	28.6 ± 3.2
	S(2)	7.5 (*)	2.6 (*)	0.13 (*)	24.5 ± 0.5

For quantitative description of the site preference of Cr atoms in the studied alloys let us assume that the relative intensities of given sextets are proportional to the probability of finding of the iron atom with given chemical environments. Assuming random distribution of iron atoms at given sublattices one can calculate probabilities of finding iron atoms surrounded by n other Fe atoms in the n.n. shell for given x.

Full analysis of these data and their comparison with ab-initio calculations will be presented in a separate paper.

7. Conclusions

The studies have confirmed that the lattice parameter of the alloys is independent of chromium content, so the results obtained for the magnetic moments of iron and chromium reflect mainly an influence of local surroundings. The linear variation of magnetic moments with chromium makes it tempting to assume an influence of the nearest-neighbour shells only. This, however, needs to be confirmed by calculations that are underway. One should point out that the data obtained may be qualified as high-quality ones as rather negligible disorder could be claimed. Apparently, the chromium magnetic moment orients antiparallel to iron, and it is relatively large, about 1.5 Bohr magneton.

Acknowledgements

We are grateful to K. Szymański for fruitful and stimulating discussions. This work was partly sponsored by the State Committee of Scientific Research NN 202172335.

References

[1] Waliszewski J, Dobrzyński L, Malinowski A, Satuła D, Szymański K, Prandl W, Brückel Th, Schärpf O 1994 *J. Magn. Magn. Mater.* **132** 349
[2] Satuła D, Dobrzyński L, Waliszewski J, Szymański K, Rećko K, Malinowski A, Brückel Th, Schärpf O, Blinowski K 1997 *J. Magn. Magn. Mater.* **169** 240
[3] Niculescu V, Raj K, Burch T J, Budnick J I 1976 *Phys. Rev. B* **13** 3167
[4] Stearns M B 1964 *J. Appl. Phys.* **53** 1095
[5] Niculescu V, Burch T J and Budnick J I 1983 *J. Magn. Magn. Mater.* **39** 223
[6] Swintendick A C 1976 *Solid State Comm.* **19** 511
[7] Garba E J D and Jacobs R L 1986 *J. Phys. F. Metal Phys.* **16** 1485
[8] Dobrzyński L, Petrillo C, Sacchetti F 1990 *Phys. Rev. B* **42** 1142
[9] Moss J and Brown P J 1972 *J. Phys. F* **2** 358
[10] Dobrzyński L 1995 *J. Phys.: Cond. Matter* 7 1373
[11] Go A, Pugaczowa-Michalska M, Dobrzyński L 2004 in Proceedings of the XVI International School on Physics and Chemistry of Solids Bialowieża, Poland
[12] Pugaczowa-Michalska M, Go A, Dobrzyński L, Lipiński S 2003 *J. Magn. Magn. Mater.* 256 46
[13] Go A, Pugaczowa-Michalska M, Dobrzyński L 2006 *Phys. stat. sol. (b)* **243** 124
[14] Fuks D, Dorfman S, Liubich V and Kutsenko 2002 *Lnt. J. Quant. Chem.* **90** 1478
[15] Go A, Pugaczowa-Michalska M, Dobrzyński L 2004 J. *Magn. Magn. Mater.* **272-276** e217
[16] Satuła D, Szymański K, Dobrzyński L, Waliszewski J 1993 *J. Magn. Magn. Mater.* **119** 309
[17] Szymański K, Dobrzyński L, Satuła D 1995 Proceedings of the International Conferrence on Physics, Leuven, Hiperfine Interavctions (c) **1** 76
[18] Satuła D, Szymański K, Dobrzyński L, Waliszewski J and Malinowski A 1995 *J. Magn. Magn. Mater.* **140-144** 61
[19] Go A, Pugaczowa-Michalska M, Dobrzyński L 2009 *Acta Phys. Pol. A* **115** 191
[20] Go A, Rećko K, Dobrzyński L, Milczarek J J 2011 *Site preference and magnetism of* $Fe_{3-x}Cr_xAl_{0.5}Si_{0.5}$ *JPCM* in press
[21] Satuła D, Dobrzyński L, Malinowski A, Szymański K, Waliszewski J 1995 *J. Magn. Magn. Mater.* **151** 211
[22] Rodriguez – Carvajal J 1993 *Physica B* **192** 55

Journal of Physics: Conference Series **340** (2012) 012071 doi:10.1088/1742-6596/340/1/012071

Magnetic order and spin-reorientation in HoGa

R A Susilo[1], S Muñoz Pérez[1], R Cobas[1], J M Cadogan[1] and M Avdeev[2]

[1] Department of Physics and Astronomy, University of Manitoba, Winnipeg, Manitoba, R3T 2N2, Canada

[2] Bragg Institute, ANSTO, PMB 1, Menai, NSW 2234, Australia

E-mail: cadogan@physics.umanitoba.ca

Abstract. We have determined the magnetic structure of the intermetallic compound HoGa by high-resolution neutron powder diffraction. This compound crystallizes in the orthorhombic (*Cmcm*) CrB-type structure and the magnetic structure comprises ferromagnetic order of the Ho sublattice along the *c*-axis. The Curie temperature is 66(3) K. Upon cooling below 20 K, the Ho magnetic moments cant away from the *c*-axis towards the *ab*-plane. At 3 K, the Ho moment is 8.8(2) μ_B and the Ho magnetic moments point in the direction $\theta = 30(2)^\circ$ and $\phi = 49(4)^\circ$ with respect to the crystallographic *c*-axis. The observation of an *ab*-plane component at around 50° from the *a*-axis is in contrast with the suggested magnetic structure of *ac* order ($\theta = 32^\circ$ and $\phi = 0^\circ$) reported by Delyagin *et al.* [1] on the basis of a ^{119}Sn Mössbauer spectroscopy study of a Sn-doped HoGa sample. However, we find that these two sets of orientations are in fact indistinguishable by Mössbauer spectroscopy.

1. Introduction

The orthorhombic RGa (R = rare earth) compounds crystallize in the CrB-type *Cmcm* (#63) structure. There is one R site, the 4*c*, and one Ga site, also a 4*c*. The RGa compounds were first prepared in the early 1960's [2, 3, 4, 5]. They order ferromagnetically with a Curie temperature that ranges from a high of 183 K in GdGa [6, 7, 8] to a low of ~15 K in TmGa [9]. Neutron diffraction work on TbGa was reported in 1964 by Cable *et al.* [10] who showed that the Tb moments order along the *c*-axis. In 1971, Barbara *et al.* used neutron diffraction to show that ErGa orders along the *b*-axis [11]. In 1992, Nesterov *et al.* [12] used ^{119}Sn Mössbauer spectroscopy on Sn-doped samples to show that NdGa, HoGa and ErGa undergo spin-reorientations upon cooling. This Mössbauer work was recently extended by Delyagin *et al.* [1]. Within the last year or two, the RGa compounds have attracted some interest due to their potential for use as magnetocaloric-effect-based low-temperature refrigeration applications [13, 14, 15].

In this paper we present the results of high-resolution neutron powder diffraction carried out on HoGa to examine the spin-reorientation reported to occur at ~ 20 K. HoGa is a ferromagnet with a Curie temperature of 66(3) K [1, 6, 12, 15]. Its initial easy direction of order is the *c*-axis and the reorientation was suggested by Delyagin *et al.* [1] to involve a canting towards the *a*-axis, within the *ac*-plane. We show that the canting actually tips the Ho moments towards an intermediate planar arrangement, which differs from the structure suggested on the basis of ^{119}Sn Mössbauer spectroscopy. However, we find that the two magnetic structures are, in fact, indistinguishable by ^{119}Sn Mössbauer spectroscopy.

2. Experimental Methods

The HoGa sample was prepared in an argon-arc furnace. Stoichiometric amounts of the pure elements (Ho (99.9%) and Ga (99.999%)) were melted several times under pure (less than 1 ppm impurity) argon to ensure homogeneity. X-ray powder diffraction measurements were made at room temperature using Cu–$K\alpha$ radiation on a PANalytical X'Pert Pro diffractometer. Neutron diffraction experiments were carried out on the *Echidna* high-resolution powder diffractometer at the OPAL reactor in Sydney, Australia [16]. The neutron wavelength was 1.6220(5) Å. The neutron diffraction data were corrected for absorption effects and all patterns were refined using the Rietveld method and the FullProf/WinPlotr suite [17, 18].

3. Results

Refinement of the x-ray powder diffraction pattern obtained at 295 K confirmed the formation of the CrB-type *Cmcm* orthorhombic phase. Any impurity phases present amounted to less than 1 wt%, undetectable in the powder diffraction pattern. The refined lattice parameters are a = 4.2794(5) Å, b = 10.7869(11) Å, c = 4.0445(5) Å.

3.1. Neutron Diffraction

In figure 1 we show the neutron powder diffraction patterns of HoGa, obtained at 80 K, 45 K, 22 K and 3 K. The Curie temperature of HoGa is 66(3) K, as mentioned earlier, so the 80 K pattern represents the nuclear scattering from the orthorhombic *Cmcm* cell. In Table 1 we show the atomic position parameters for HoGa derived from a refinement to the 80 K nuclear diffraction pattern. The conventional R-factors for this refinement are R_B = 5.65% and R_F = 4.06%.

Table 1. Crystallographic data for HoGa in the orthorhombic *Cmcm* cell, refined from the neutron diffraction patterns. The 4c site occupied by both the Ho and Ga atoms is defined by the atomic positions (0 y $\frac{1}{4}$).

T(K)	a(Å)	b(Å)	c(Å)	Atom	y
80	4.2779(5)	10.7595(10)	4.0353(4)	Ho	0.3592(3)
				Ga	0.0750(5)
45	4.2762(5)	10.7571(11)	4.0344(5)	Ho	0.3595(4)
				Ga	0.0752(6)
22	4.2760(5)	10.7589(11)	4.0337(5)	Ho	0.3596(3)
				Ga	0.0752(6)
3	4.2758(5)	10.7582(11)	4.0337(5)	Ho	0.3594(3)
				Ga	0.0751(6)

The diffraction patterns obtained at 45 K, 22 K and 3 K show considerable magnetic contributions from the Ho sublattice. There are no additional peaks that would signal antiferromagnetic order and all magnetic contributions simply add to the existing Bragg nuclear peaks i.e. the propagation vector k = 0. The dominant magnetic contributions occur at the (020), (110) and (021) positions, with scattering angles $2\theta \sim 17.3°$, 23.6° and 29.1°, respectively.

In order to consider all possible magnetic structures allowed for HoGa, we carried out Representational Analysis for the Ho site using the *BASIREPS* program, part of the

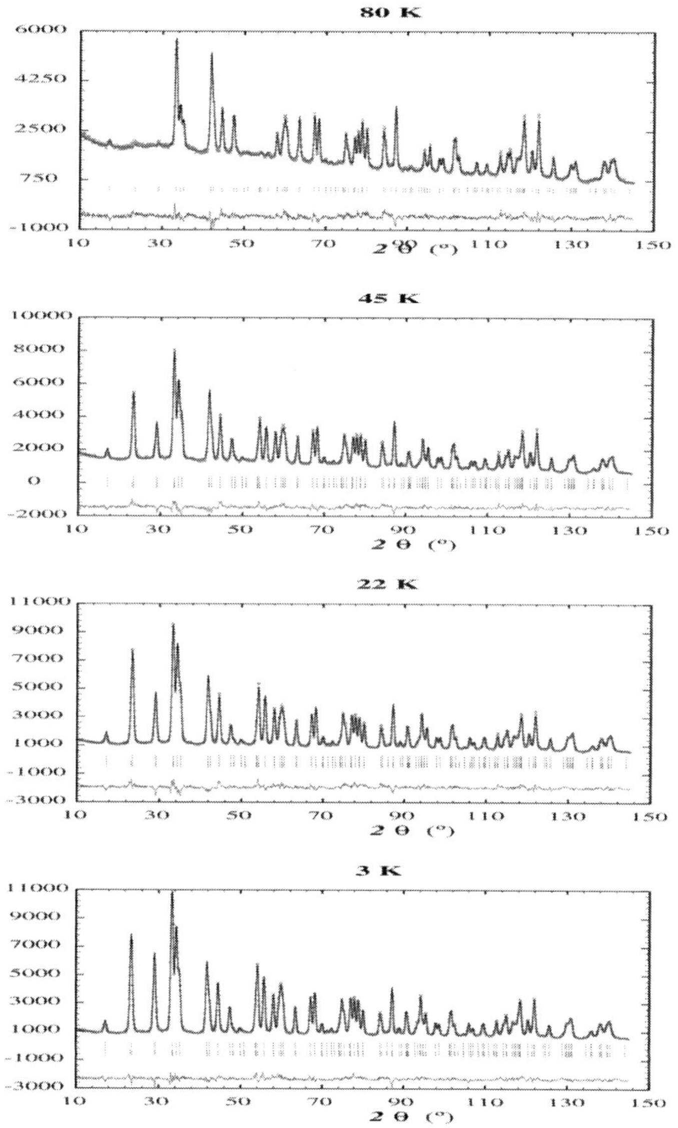

Figure 1. Neutron powder diffraction patterns of HoGa obtained at 80 K, 45 K, 22 K and 3 K, with a neutron wavelength of 1.6220(5) Å. The two sets of Bragg markers represent the nuclear and magnetic contributions to the patterns.

FullProf/WinPlotr suite [17, 18]. The decomposition of the magnetic representation comprises six one-dimensional representations, each appearing once:

$$\Gamma^{4c}_{Mag} = 1\Gamma_2 + 1\Gamma_3 + 1\Gamma_4 + 1\Gamma_5 + 1\Gamma_7 + 1\Gamma_8 \quad (1)$$

and the basis vectors of these irreducible representations are given in table 2.

We can rule out the antiferromagnetic modes Γ_2, Γ_4 and Γ_8 on the strength of the magnetometry and susceptibility work which clearly indicates ferromagnetic order. Thus, we are left with ferromagnetic order along one of the orthorhombic crystal axes. The observation of

Table 2. Representational Analysis for the Ho(4c) site in HoGa with a propagation vector k = 0. The respective atomic positions are $(x,\ y,\ z)$, $(-x,\ -y,\ -z)$, $(\frac{1}{2}+x,\ \frac{1}{2}+y,\ z)$, and $(\frac{1}{2}-x,\ \frac{1}{2}-y,\ -z)$.

Representation	Ordering Mode	Moment directions
Γ_2	G_Y	$+-+-$
Γ_3	F_Z	$++++$
Γ_4	G_X	$+-+-$
Γ_5	F_Y	$++++$
Γ_7	F_X	$++++$
Γ_8	G_Z	$+-+-$

substantial magnetic contributions to the (020) and (200) peaks, combined with the absence of any magnetic intensity at the (002) peak, clearly indicates ferromagnetic order along the *c*-axis, corresponding to the Γ_3 mode. At 45 K we find a refined Ho magnetic moment of 6.2(2) μ_B, which increases to 8.2(2) μ_B by 22 K. The conventional R-factors for the 45 K refinement are R_B = 6.44%, R_F = 4.32% and R_{mag} = 7.27%. The R-factors for the 22 K refinement are R_B = 7.30%, R_F = 5.29% and R_{mag} = 6.93%.

Upon further cooling below 22 K we observe an increase in the magnetic intensity at the (021) peak at $2\theta = 29.1°$, relative to the magnetic intensity at the (110) peak at $2\theta = 23.6°$. Similar changes occur at other peaks, indicating that the magnetic order tips away from the crystal *c*-axis. This spin-reorientation has been observed in ac-susceptibility measurements [15]. At 3 K, the Ho magnetic moments are canted away from the *c*-axis by 30(2)°. The moments amount to 8.8(2) μ_B at 3 K, i.e. the 'free-ion' value (9 μ_B). The R-factors for the 3 K refinement are R_B = 5.27%, R_F = 4.04% and R_{mag} = 4.38%. Our refinement of the 3 K pattern also shows that the canting Ho moments tip towards an intermediate planar arrangement between the crystal *a*- and *b*-axes. We find polar angles of $\theta = 30(2)°$ and $\phi = 49(4)°$. This reorientation is driven by competition between the different crystal-field orders acting on the Ho^{3+} ions, with the higher-order (4th and 6th) terms becoming significant as the temperature is lowered. The observation of *c*-axis order just below T_c is consistent with the signs and relative magnitudes of the diagonal and off-diagonal second-order crystal-field terms derived from single-crystal susceptibility measurements by Shohata *et al.* [7, 8].

3.2. ^{119}*Sn Mössbauer spectroscopy*

Delyagin *et al.* [1] used ^{119}Sn doping in the RGa compounds to study the magnetic order of the R sublattice by ^{119}Sn Mössbauer spectroscopy. The Sn dopant is non-magnetic so any hyperfine field at this dopant site is transferred from the neighbouring magnetic order of the R sublattice. Information about the ordering directions can be deduced by determining the orientation of the hyperfine field (assumed collinear with the R magnetic order) within the principal axis frame of the Electric Field Gradient (EFG) at the ^{119}Sn site. Delyagin *et al.* showed that the EFG axes are aligned along the orthorhombic axes, as expected from the $m2m$ point symmetry of the Ga 4c sites, the location of the ^{119}Sn dopant. The particular identification of the EFG axes is (xyz)=(bca). Within this EFG frame, the temperature dependence of the quadrupole splitting showed that above about 20 K the hyperfine field makes an angle of 90° with the z(EFG) axis, along the y(EFG) axis. In other words, Ho magnetic ordering is along the *c*-axis, as we found by neutron diffraction. Below 20 K, this angle decreases to about 58° at 4 K, with the hyperfine field lying in the yz(EFG) plane, i.e. Ho magnetic order in the crystal *ac*-plane. In figure 2 we show the orientational relationship between the Mössbauer and neutron diffraction measurements, in terms of the crystal (abc) and EFG (xyz) axes.

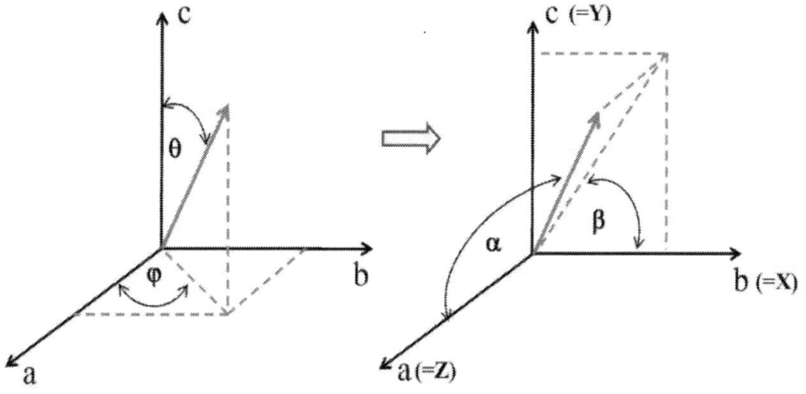

Figure 2. Orientation of the Ho magnetic moment (and hence hyperfine field at ^{119}Sn sites) relative to the crystal (abc) and EFG(xyz) axes, used in the interpretation of the neutron powder diffraction and Mössbauer data.

The quadrupole shift in a magnetically split Mössbauer spectrum, as defined by Delyagin *et al.*, is

$$QSH = \frac{eQV_{zz}}{8} \times [3cos^2\alpha - 1 + \eta(sin^2\alpha)cos(2\beta)] \quad (2)$$

using standard notation. Here, α and β are the polar angles of the hyperfine magnetic field in the principal axis frame of the EFG at the ^{119}Sn nuclei. The asymmetry of the EFG tensor is described by the η parameter. Using the accepted conventions for describing the components of the EFG tensor, Delyagin *et al.* deduced $eQV_{zz} = 1.10(8)$ mm/s and $\eta = 0.65(9)$. Thus, taking $\theta = 58°$ and $\phi = 90°$, one obtains the experimentally observed value of QSH i.e. $-0.087(4)$ mm/s.

As shown earlier, the principal z-axis of the EFG frame lies along the crystal's orthorhombic a-axis. Our neutron diffraction results are determined relative to the crystal c-axis. It is straightforward to show that our angles $\theta = 30(2)°$ and $\phi = 49(4)°$ correspond to $\alpha = 71(3)°$ and $\beta = 66(3)°$ in the EFG frame employed in the Mössbauer work and using equation (2), our magnetic structure yields a QSH value of $-0.15(4)$ mm/s, fully consistent with the Mössbauer result.

4. Conclusions

We have used high-resolution neutron powder diffraction to show that the magnetic order of the Ho sublattice in HoGa is ferromagnetic along the orthorhombic c-axis below its Curie temperature of 66(3) K. Below ~20 K, the Ho moments cant away from the c-axis towards the ab-plane and at 3 K the Ho magnetic order is defined by the polar angles of $\theta = 30(2)°$ and $\phi = 49(4)°$, relative to the orthorhombic abc axes.

Acknowledgments

We are grateful for financial support from the Natural Sciences and Engineering Research Council of Canada and the Canada Foundation for Innovation. JMC acknowledges support from the Canada Research Chairs programme.

References

[1] Delyagin N N, Krylov V I and Rozantsev I N 2007 *J. Magn. Magn. Mater.* **308** 74–9.

[2] Iandelli A 1960 in *The Physical Chemistry of Metallic Solutions and Intermetallic Compounds* **1** 376. New York: Chemical Publishing Co., Inc.
[3] Baenziger N C and Moriarty Jr. J L 1961 *Acta Cryst.* **14** 946–7.
[4] Schob O and Parthé E 1965 *Acta Cryst.* **19** 214–24.
[5] Dwight A E, Downey J W and Conner Jr. R A 1967 *Acta Cryst.* **23** 860–2.
[6] Fujii H, Shohata N, Okamoto T and Tatsumoto E 1971 *J. Phys. Soc. Japan* **31** 1592.
[7] Shohata N, Fujii H and Okamoto T 1974 *J. Phys. Soc. Japan* **37** 567.
[8] Shohata N 1977 *J. Phys. Soc. Japan* **42** 1873–80.
[9] Barbara B., Nguyen V N and Siaud E 1972 *C.R. Acad. Sci.* **B274** 1053.
[10] Cable J W, Koehler W C and Wollan E O 1964 *Phys. Rev.* **136** A240–2.
[11] Barbara B, Bècle C, Nguyen N N and Siaud E 1971 *Conf. Digest No. 3: Rare-Earths and Actinides*, Institute of Physics, Durham, 219–221.
[12] Nesterov V I, Reiman S I and Rozantsev I N 1992 *Sov. Phys. Solid State* **34** 671–3.
[13] Zhang J Y, Luo J, Li J B, Liang J K, Wang Y C, Ji L N, Liu Y H and Rao G H 2009 *J. Alloys and Compounds* **469** 15–9.
[14] Chen J, Shen B G, Dong Q Y, Hu F X and Sun J R 2009 *Appl. Phys. Lett.* **95** 132504 (3pp).
[15] Chen J, Shen B G, Dong Q Y and Sun J R 2010 *Solid State Commun.* **150** 157–9.
[16] Liss K-D, Hunter B A, Hagen M E, Noakes T J and Kennedy S J 2006 *Physica B* **385-6** 1010-2.
[17] Rodríguez-Carvajal J 1993 *Physica* B **192** 55–69.
[18] Roisnel T and Rodríguez-Carvajal J 2001 *Mater. Sci. Forum* **378–81** 118–23.

Spin Wave Dispersion Relation and Damping in Sendust Alloy in 8 to 295 K Temperature Range

K Świderska[1], J Jankowska Kisielińska[1], L Dobrzyński[2,3]

[1] Institute of Atomic Energy POLATOM, Świerk, Poland

[2] The Soltan Institute for Nuclear Studies, Świerk, Poland

[3] Faculty of Physics, University of Bialystok, Białystok, Poland

E-mail: k.swiderska@cyf.gov.pl

Abstract. Previous neutron scattering experiments on single crystal of Sendust alloy (73.5 at. % Fe, 17 at. % Al, and 9.5 at. % of Si) revealed extremely high spin wave damping at room temperature. The aim of present studies was to test if the spin wave damping originates from lattice disorder or comes from mutual spin waves interactions or other excitations that would be temperature dependent. The spin wave dispersion relation and damping were studied in temperature range from 8 to 295 K. No regular changes of both spin wave damping and stiffness constant (which characterises dispersion relation) with temperature were found. Thus, within this temperature range, spin waves interactions with dynamical excitations of all kinds must be negligible, and their damping is most likely produced by lattice disorder alone. It is of interest to note that the inelastic background intensity below the spin wave peaks increases with temperature and for energy transfers higher than 30 meV this increase is larger than that of spin-wave peak intensity.

1. Introduction

Sendust is a ternary alloy of Fe, Si, and Al (85 wt. % Fe, 5 wt. % Al, and 10 wt. % of Si). It has no crystalline anisotropy and its magnetostriction is negligible. Sendust structure is of the DO_3 type, same as for Fe_3Si and Fe_3Al alloys, and constitutes of four sublattices with basis atomic sites A(0,0,0), B(1/4,1/4,1/4), C(1/2,1/2,1/2), and D(3/4,3/4,3/4) [1, 2]. In the stoichiometric alloy (like Fe_3Si or Fe_3Al) A, B and C positions are occupied mainly by Fe atoms, while Si and Al atoms occupy D positions. In Fe-deficient alloys appropriate parts of Si and Al atoms occupy iron positions [3]. Atom in position B is surrounded by 8 equivalent nearest neighbours (NN) of the A and C type. The surrounding of symmetry equivalent A and C sites consists of 4 NN of the B type and 4 NN of the D type [1].

Previous neutron scattering measurements on the same sample of Sendust alloy at room temperature revealed extraordinarily high spin wave damping [1, 2]. To find the origin of the damping, the temperature dependence of spin wave dispersion relation and damping at 8 to 295 K were studied by means of inelastic neutron scattering. The results do not show the changes of stiffness constant and damping expected for cubic ferromagnetic metals on the assumption that spin waves interact with each other and with electrons only [4]. The measurements reveal surprisingly high increase of q-independent neutron scattering with temperature.

2. The sample and experimental technique

The sample used in the experiment was a single crystal of Sendust (73.5 at. % Fe, 17 at. % Al, and 9.5 at. % Si), produced by Bridgman method, which was already investigated by other authors [1, 2]. Its composition is nearly stoichiometric, thus mainly Fe atoms occupy A, B, and C positions. Si or Al atoms probably occupy D positions randomly. It follows from the stoichiometry that 6 at. % of Al/Si atoms must enter iron positions. Sample dimensions are 20 × 15 × 27 mm^3, the latter being parallel to [110] crystallographic direction. Curie temperature of the sample is 794±1 K [2].

The experiments were carried out with the neutron triple axis spectrometer H6 in the Institute of Atomic Energy, Świerk, Poland. The monochromator and the analyzer were made of pyrolytic graphite. Inelastic neutron scattering (INS) was investigated for three crystallographic directions: [100], [110] and [111], close to (111) and (022) reciprocal lattice points. The spin waves studies were carried out in the temperature range from 8 to 295 K (well below the Curie temperature), and energy transfers ranged from 8 to 40 meV. To obtain better resolution the energies of scattered neutrons were fixed to 14.8 meV and 20.5 meV ($\lambda = 2.35$ Å and $\lambda = 2$ Å) for energy transfers below and above 20 meV, respectively. The range of investigated energy transfers did not contain values for which $k_i \approx 2k_f$ so no filter was used. The number of neutrons emerging from reactor core was monitored and determined the duration of the intensity measurement at every point of measured distribution.

3. Results and data analysis

Scattered neutron intensities were measured for fixed neutron energy transfers and varying scattering vectors. Figure 1 shows examples of the intensity distributions for chosen temperature values.

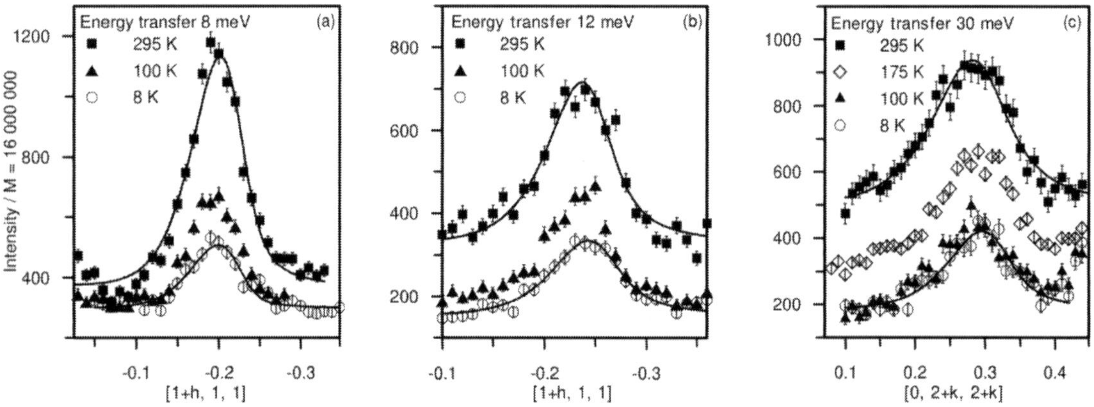

Figure 1. Examples of the neutron scattering intensity distributions at several temperatures in the vicinity of (111) reciprocal lattice point in [100] direction and 8 meV (a) and 12 meV (b) energy transfer, and in the vicinity of (022) reciprocal lattice point in [022] direction and 30 meV energy transfer (c). Solid lines are fits for 8 K and 295 K.

Apparently, not only the intensity of spin wave peak increases with temperature but one also detects an increase of the q-independent intensity, B (see equation (1)), especially for energy transfers above 30 meV. For the 8 meV energy transfer in the range from 8 to 295 K this increase is three-fold slower than that for neutron scattering on spin waves at the distribution maximum. For 12 meV energy the increase of the background and that of spin scattering are roughly proportional to each other. For 30 meV it is ~1.5-fold faster than that of spin wave scattering. The increase of B with temperature indicates that it likely comes from the incoherent scattering.

Every distribution was fitted with the convolution of spectrometer resolution function $R(Q,\omega,Q',\omega')$ and neutron scattering cross-section, with a simultaneous background (B) adjustment.

$$I = I_0 \int dQ' d\omega' R(Q,\omega,Q',\omega') \frac{d^2\sigma(Q',\omega')}{d\Omega' d\omega'} + B \tag{1}$$

where I is scattered neutrons intensity and I_0 is the intensity of incoming neutrons. Neutron scattering cross-section on damped spin waves was assumed in the form [4]

$$\frac{d^2\sigma}{d\Omega d\omega} \propto \frac{k_f}{k_i} f^2(Q)\left(n_B(\omega)+1\right)\left[\frac{\hbar\Gamma_q}{\left(\hbar\omega-\hbar\omega_q\right)^2+\left(\hbar\Gamma_q\right)^2}-\frac{\hbar\Gamma_q}{\left(\hbar\omega+\hbar\omega_q\right)^2+\left(\hbar\Gamma_q\right)^2}\right] \tag{2}$$

where $\hbar\Gamma_q$ is the energy broadening of a spin wave of energy $\hbar\omega_q$.

Since Sendust is a ferromagnet with negligible anisotropy, if any, the dispersion relation was assumed in the form

$$\hbar\omega_q = Dq^2 \tag{3}$$

The general q dependence characterizing energy broadening $\hbar\Gamma_q$ was fitted with [4, 5]

$$\hbar\Gamma_q = Aq^n \tag{4}$$

According to previous results [2]: $n_1 = 2.83$ (for the [111] direction) and $n_2 = 1.95$ (for the [110] direction). In our fittings n was taken to be equal to 2 or 3.

Parameters D, A_2 and A_3 determined for several distributions are listed in table 1 (a) - (f).

Table 1. Fitted stiffness constants D, damping parameters A_2 and A_3, and energy broadenings for neutron inelastic scattering intensity distributions at chosen temperatures. The displayed results concern scans in the vicinity of (111) reciprocal lattice point in [100] direction for energy transfers 8 meV (a, b) and 12 meV (c, d) energy transfer, and in the vicinity of (022) reciprocal lattice point in [022] direction for the energy transfer of 30 meV (e, f).

(a) Spin wave energy 8 meV (111) [100]

T [K]	χ^2	D [meV·Å^2]	A_2 [meV·Å^2]	$\hbar\Gamma$ [meV]
8	1.6	158±4	15±10	0.77±0.48
100	1.4	163±2	13±5	0.65±0.23
295	4.6	160±2	22±5	1.08±0.26

(b) Spin wave energy 8 meV (111) [100]

T [K]	χ^2	D [meV·Å^2]	A_3 [meV·Å^3]	$\hbar\Gamma$ [meV]
8	1.6	161±4	80±41	0.88±0.45
100	1.3	165±2	62±21	0.67±0.22
295	4.0	163±2	103±23	1.12±0.25

(c) Spin wave energy 12 meV (111) [100]

T [K]	χ^2	D [meV·Å^2]	A_2 [meV·Å^2]	$\hbar\Gamma$ [meV]
8	1.0	159±2	29±6	2.22±0.44
100	0.9	169±2	16±4	1.11±0.25
295	3.0	170±3	28±7	1.95±0.52

(d) Spin wave energy 12 meV (111) [100]

T [K]	χ^2	D [meV·Å^2]	A_3 [meV·Å^3]	$\hbar\Gamma$ [meV]
8	0.9	161±2	112±22	2.29±0.44
100	1.0	171±2	56±14	1.05±0.26
295	3.4	172±3	100±30	1.85±0.56

(e) Spin wave energy 30 meV (022) [022]

T [K]	χ^2	D [meV·Å^2]	A_2 [meV·Å^2]	$\hbar\Gamma$ [meV]
8	1.4	137±3	30±7	6.5±1.7
100	1.5	127±4	66±12	15.6±2.9
125	1.1	138±3	42±7	9.2±1.4
175	1.4	138±3	38±6	8.3±1.3
295	1.0	147±3	49±8	9.9±1.7

(f) Spin wave energy 30 meV (022) [022]

T [K]	χ^2	D [meV·Å^2]	A_3 [meV·Å^3]	$\hbar\Gamma$ [meV]
8	1.3	139±3	67±17	6.7±1.8
100	1.4	125±12	146±31	17.2±4.4
125	1.1	139±4	99±17	10.0±2
175	1.3	139±3	88±15	8.8±1.5
295	1.1	147±6	127±24	11.8±2. 4

The values of D and $\hbar\Gamma_q$ obtained for both $n = 2$ and $n = 3$ assumptions do not differ within standard deviations for individual distributions. Both exponents, n, turned out to describe the spin wave linewidths equally for $|q| < 0.45$. For $|q| > 0.45$, however, the q dependence is apparently stronger.

The spin wave dispersion relation and energy broadening were calculated for fitted D and A_2 parameters. The results are illustrated in figure 2.

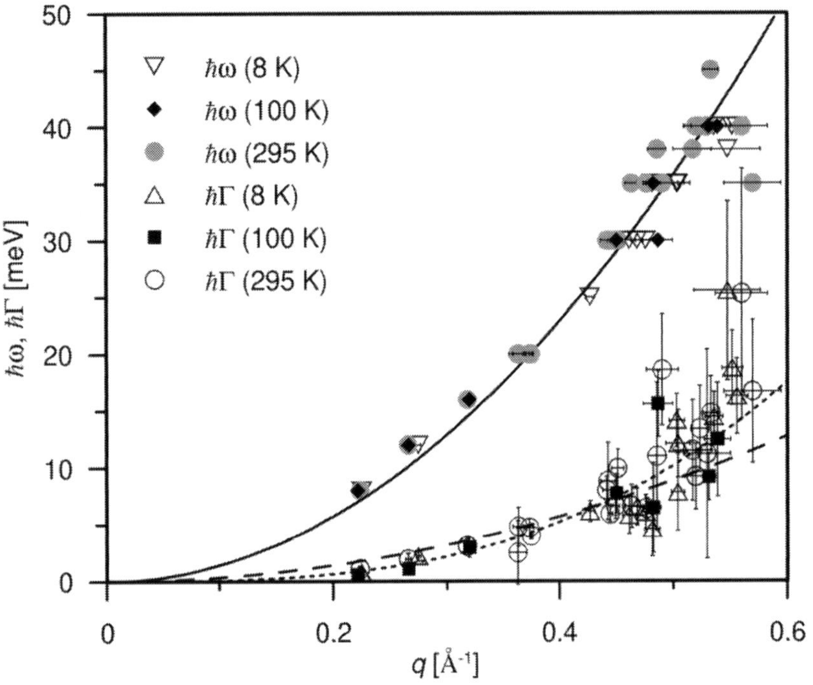

Figure 2. The spin wave energy and broadening as a function of wave vector at three temperatures. The solid line represents the dispersion relation of the form $\hbar\omega_q = Dq^2$, dashed and dotted lines represent the energy broadening fit $\hbar\Gamma_q = A_2q^2$ and $\hbar\Gamma_q = A_3q^3$ respectively, fitted to the spectra measured at 8 K. We do not mark the results measured along separate directions of measurements as no anisotropy was observed.

4. Discussion and conclusions

The number of spin waves propagating in the lattice ruled by the Bose statistics increases when the temperature is increasing. The greater number of spin waves should result in greater number of their interactions with each other, with other types of excitations in the lattice, and interactions with electrons. This is why the damping should increase with temperature. The various types of interactions of spin waves may affect not only the spin wave damping but also dispersion relation - stiffness constant is expected to decrease with temperature [4, 5].

To see the reason for previously observed high spin-wave damping, the temperature dependences of dispersion relation and spin wave damping in single crystal of Sendust were measured. The measurements were performed in temperature range 8 to 295 K for various neutron energy transfers.

The determined spin wave stiffness constant and damping parameter do not show any regular changes with temperature. The lack of temperature dependence of damping as represented by the spin wave peak widths indicates that spin wave energy broadening probably is produced neither by their mutual interactions nor by their interactions with other types of thermal excitations. The source of extraordinarily large energy broadening must be tracked to other features of the material. The broadening may be induced by some disorder within the crystal arising e.g. from the randomness of the occupation of D positions by Si and Al atoms. Also the slight deviation from stoichiometry of the alloy (73.5 at. % Fe instead of 75%) may be an additional source of the lattice disorder. Apparently different metallic radii of aluminum and silicon can result in local stretching and contraction of the lattice. Different local Fe surroundings and distances may induce local changes in exchange integrals producing broadening of dispersion relation or affect spin wave propagation in the lattice, producing their damping.

We have observed that inelastic background intensity below the spin wave peaks increases with temperature and for energy transfers higher than 30 meV this increase is larger than that of spin-wave peak intensity. The elucidation of this effect as well as that of the rapid increase of damping at larger quasi-momentum needs further work.

References

[1] Parzych G, Jankowska-Kisielińska J and Dobrzyński L 2010 *Acta Phys. Polonica A* **117** 578
[2] Dobrzyński L, Wiśniewski A, Uemura Y J, Shapiro S M and Wicksted J P 1988 *Phys. Rev. B* **37** 7175
[3] Suwalski S, Schneeweiss O, Zemčík T and Tucholski Z 1979 *Physica Status Solidi (a)* **53** K195
[4] Marshall W and Lovesey S W 1971 *Theory of Thermal Neutron Scattering. The Use of Neutrons for the Investigation of Condensed Matter* (Oxford, Oxford University Press) pp 261-290
[5] Singh A and Tešanović Z 1989 *Phys. Rev. B* **39** 7284

Journal of Physics: Conference Series **340** (2012) 012073 doi:10.1088/1742-6596/340/1/012073

Effects of impurities in $CeCoIn_5$ using inelastic neutron scattering

J Panarin[1], S Raymond[1], G Lapertot[1] and J-M Mignot[2]
[1]CEA-Grenoble, INAC-SPSMS-MDN, 38054 Grenoble Cedex 9, France
[2]Laboratoire Léon Brillouin, CEA/Saclay, 91191 Gif sur Yvette, France

E-mail: justin.panarin@gmail.com

Abstract. The influence of Nd magnetic impurities on the spin dynamics of $CeCoIn_5$ was studied by inelastic neutron scattering. In the Nd-substituted compound $Ce_{0.95}Nd_{0.05}CoIn_5$, the spin resonance peak (observed at $\Omega_{res} = 0.55$ meV in the pure system) is shifted to lower energies but the ratio Ω_{res}/k_BT_c remains almost unchanged. These observations are compared with non-magnetic La-substitution.

1. Introduction

Superconductivity is a macroscopic quantum state resulting from the condensation of electrons in Cooper pairs. In the case of conventional superconductivity the pairing mechanism is the weak electron-phonon interaction. However in strongly correlated systems exhibiting a superconducting (SC) state the pairing mechanism is thought to be of a different nature. Examples of such unconventional superconductors are high-temperature superconducting cuprates (HTSC), heavy-fermion compounds (HF) and the iron-based superconductors. In these compounds, the origin of the pairing is strongly suspected to be the magnetic interaction, and a seminal study of the magnetic excitation spectra by inelastic neutron scattering (INS) in the cuprate $YBa_2Cu_3O_{6+x}$ showed the appearance of a sharp excitation called a magnetic spin resonance [1] in the SC state. This was later generalized to other cuprates. Such a feedback of superconductivity on the magnetic excitation spectra was backed up by theories of a pairing mechanism of magnetic origin. The recent discovery of similar excitations in HF superconductors UPd_2Al_3[2] $CeCoIn_5$[3] and $CeCu_2Si_2$[4], as well as in the iron superconductors [5] suggests that the magnetic resonance could be a universal feature of unconventional superconductors.

Among HF superconductors the compound $CeCoIn_5$ has the highest critical temperature, $T_c = 2.3$ K [6]. It crystallizes in the tetragonal space group $P4/mmm$ and consists of alternating $CeIn_3$ and $CoIn_2$ layers. A quasi-2D nature is supported by de Haas van Alphen measurements, which established a Fermi surface composed of nearly cylindrical sheets [7]. As concerns the low-energy magnetic excitations measured by INS, a quasi-elastic signal is measured above T_c with a linewidth of 0.3 meV. Below T_c, the spectral shape changes from a quasi-elastic to a sharp inelastic peak, which appears at an energy of $\Omega_{res} \approx 0.55$ meV ($\approx 2.7k_BT_c$) at the antiferromagnetic position $\mathbf{Q} = (\frac{1}{2}, \frac{1}{2}, \frac{1}{2})$ [3].

The introduction of impurities is a useful probe to investigate the microscopic nature of the SC state [8]. For a d-wave superconductor like $CeCoIn_5$, impurities are expected to have a Cooper pair breaking effect [9]. In $CeCoIn_5$ the substitution of other lanthanid on the Ce site make it possible to probe the effect of non-magnetic (La)[10][11] and magnetic impurities (Nd)[12] on the spin resonance. The effect of these impurities on the SC state differs from those observed in conventional superconductors for which non-magnetic impurities (at low concentration) do no affect the superconductivity. Indeed La ($4f^0$) reduces the critical temperature T_c by $(-0.056T_c)/(1$ % of La substitution)[10], the Nd($4f^3$) reduces T_c by $(-0.042T_c)/(1$% of Nd substitution). Moreover Nd induces a magnetic order within the SC state at very low temperature. The ordering temperature T_m increases with Nd substitution [12].

2. Experimental details

In this article, we report INS experiments performed on single crystal samples of $Ce_{0.95}Nd_{0.05}CoIn_5$ and compared the results with those obtained in Ref.[13] for $Ce_{0.98}La_{0.02}CoIn_5$. We find that the ratio between the resonance energy Ω_{res} and T_c remains almost constant for Nd impurities as found previously from La impurities but the excitation broadening is less pronounced with Nd substitution than with La substitution.

Single-crystal samples of $Ce_{0.95}Nd_{0.05}CoIn_5$ were grown by the self-flux method [14] and characterized by specific heat measurements performed using a commercial Quantum Design *Physical Properties Measurement System* (PPMS) down to T = 400 mK. From these measurements we deduce the T_c which is equal to 1.77 K. For this composition, the magnetic ordering temperature T_m is around 0.8 K.

The INS experiment was performed on the cold neutron triple-axis spectrometer 4F1 at the Laboratoire Léon Brillouin, Saclay. In this experiment, the incident beam was provided by a double pyrolytic graphite (PG) monochromator. A liquid-nitrogen-cooled Be filter was placed before the sample in order to avoid any higher order contaminations. Measurements were performed with a fixed final wavevector $k_f = 1.3$ Å^{-1}. The collimations were 60'-open-open. The energy resolution determined from the full width at half-maximum (FWHM) of the incoherent signal was 0.16 meV. The sample consisted of an assembly of about 20 single crystals of $Ce_{0.95}Nd_{0.05}CoIn_5$ co-aligned and glued with Fomblin oil on two thin aluminum plates. The mosaic spread of the assembly, as derived from a rocking curve through the (1,1,1) Bragg reflection, was around 1 degree. The sample was put in a ^{3}He-^{4}He dilution insert with [1 1 0] and [0 0 1]axes defining the scattering plane.

In a triple-axis experiment performed at constant k_f, the background corrected neutron intensity is proportional to the scattering function $S(\mathbf{Q}, E)$, which is itself proportional to the imaginary part of the dynamic susceptibility $\chi''(\mathbf{Q}, E)$,

$$S(\mathbf{Q}, E) = n(E, T)\ \chi''(\mathbf{Q}, E) \tag{1}$$

χ" was analysed using an inelastic Lorentzian spectral function where $n(E,T) = 1/(1 - e^{-E/k_BT})$ is the detailed balance factor,

$$\chi''(\mathbf{Q}, E) = \frac{1}{2}\left[\frac{\chi_Q \Gamma_Q E}{(E - \Omega_{res})^2 + \Gamma_Q^2} + \frac{\chi_Q \Gamma_Q E}{(E + \Omega_{res})^2 + \Gamma_Q^2}\right] \tag{2}$$

Here Γ_Q is the relaxation rate, Ω_{res} the resonance energy and χ_Q the static susceptibility at the wave-vector $\mathbf{Q}$.

3. Results and Discussion

The magnetic excitation spectrum measured at $\mathbf{Q} = (\frac{1}{2}, \frac{1}{2}, \frac{1}{2})$ is shown in Fig.1 in the upper panel. The dashed line corresponds to the background signal obtained by performing an energy scan at a different $\mathbf{Q}$ vector, $\mathbf{Q}=(\frac{1}{2}, \frac{1}{2}, 0.2)$. Without substitution, the resonance peak was reported [15] to occur at 0.55 meV,with a relaxation rate of 0.07 meV. In the 5 % Nd-substituted

Journal of Physics: Conference Series **340** (2012) 012073 doi:10.1088/1742-6596/340/1/012073

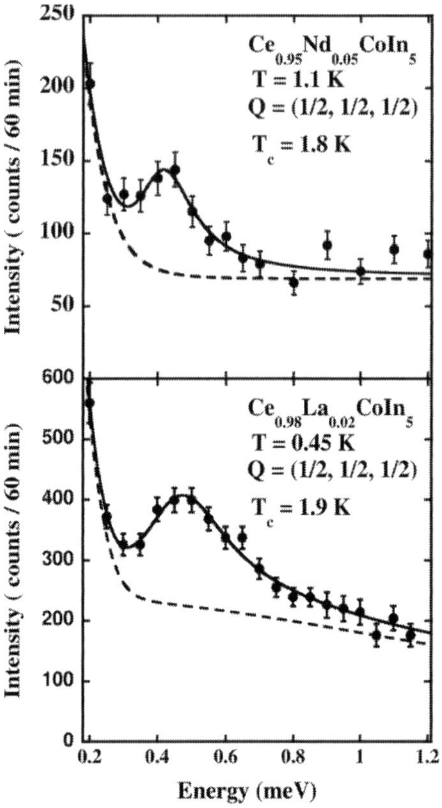

Figure 1. Excitation spectra measured at $\mathbf{Q} = (\frac{1}{2}, \frac{1}{2}, \frac{1}{2})$ for $Ce_{0.95}Nd_{0.05}CoIn_5$ and $Ce_{0.98}La_{0.02}CoIn_5$. The solid line is an inelastic Lorentzian fit and the dashed line indicates the background as described in the text.

system the resonance peak shifts to $\Omega_{res} = 0.41$ meV and undergoes a weak broadening, $\Gamma_Q = 0.10 \pm 0.012$ meV. In consequence the ratio $\Omega_{res}/k_B T_c$ is about 2.7 as in the pure compound . For comparison, the lower panel of Fig.1 shows the spectrum obtained on a 2% La-substituted compound reported in [13] with $T_c = 1.9$ K close to the T_c of $Ce_{0.95}Nd_{0.05}CoIn_5$ ($T_c = 1.8$ K) . For that La-concentration, Ω_{res} is equal to 0.45 meV but the linewidth is substantially increased, with $\Gamma_Q = 0.15 \pm 0.01$ meV. For a 3.5% La-substituted compound (not shown here) with $T_c = 1.7$ K, Ω_{res} is equal to 0.35 meV and the linewidth is also $\Gamma_Q = 0.15 \pm 0.02$ meV.
A constant-energy scan was performed along the c-axis around $(\frac{1}{2}, \frac{1}{2}, \frac{1}{2})$ in order to measure the correlation length of the resonance upon Nd-substitution. Fig.2 shows the spectrum measured at E = 0.45 meV . The scan is analyzed using a Gaussian lineshape. The background is determined by measurements at high temperature where the magnetic spectrum is no longer peaked in **Q**-space. The signal still show a maximum at $(\frac{1}{2}, \frac{1}{2}, \frac{1}{2})$, as for the pure compound. The correlation length obtained from the inverse of the Gaussian half-width at half maximum is $\xi_c = 6.5 \pm 0.1$ Åexactly as in the pure compound [15]. In the 2% La-substituted sample ξ_c shows a small decrease with a value of $\xi_c = 5.1 \pm 0.1$ Å. A constant energy scan performed along the a-axis cannot be performed for the 5%-substituted compound due to geometrical constraints of the spectrometer limiting the range in **Q**-space.

The comparison between Nd and La substitutions can be summarized as follow. The evolution

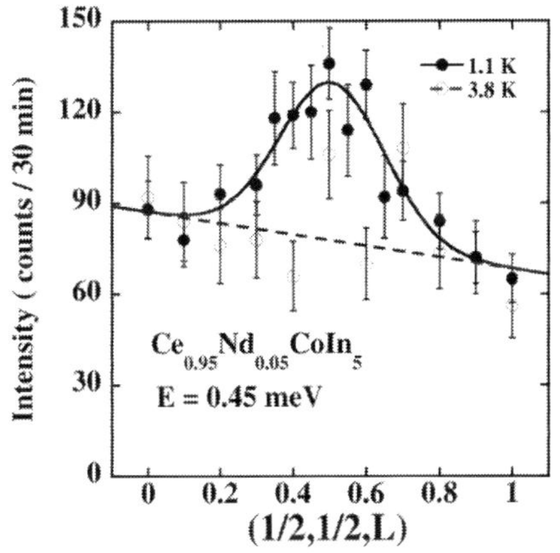

Figure 2. Excitation spectrum measured at E =0.45 meV in the direction $[\frac{1}{2}, \frac{1}{2}, L]$ for $Ce_{0.95}Nd_{0.05}CoIn_5$. The solid line is a Gaussian fit and the dashed line indicates the background as described in the text.

of the linewidth is different in the two cases. Indeed whereas the La-substitution induces a substantial broadening, Nd-substitution weakly increases the width of the excitation. Such effect means that non-magnetic impurities have a stronger effect on the life-time of the spin resonance. However both substituents decrease the resonance energy while conserving the ratio Ω_{res}/k_BT_c. This decrease could be explained by an impurity effect on the SC gap [9] leading to a conservation of this ratio as it was proposed in Ref.[13] for La-substitution. To our knowledge the only INS experiment showing the effect of magnetic and non-magnetic impurities on the spin resonance [16] compared the effects of non-magnetic Zn and magnetic Ni substitution in $YBa_2Cu_3O_7$. A detailed comparison between that experiment and La-substitution was presented in Ref.[13]. There we could complete this parallel for magnetic impurities in $CeCoIn_5$. As in the cuprate, the magnetic substitution decreases T_c less than its non-magnetic equivalent. Moreover, both impurities have no effect on the ratio Ω_{res}/k_BT_c and no influence on the correlation lengths. Whereas there is stark contrast between La and Zn substitution [13], the effect of magnetic impurities in $CeCoIn_5$ and $YBa_2Cu_3O_7$ seem very similar reinforcing the idea that magnetic impurities do not disrupt spin correlations over large scales [16].

4. Conclusion

Our INS experiments showed that the substitution of magnetic Nd impurity in $CeCoIn_5$ has one pmaineffect on the spin resonance: a decrease the resonance energy which is interpreted by a SC gap diminution. Contrary to the non-magnetic case, the resonance shows a very weak broadening indicating that the spin resonance is weakly affected by magnetic impurities. The next step of this work will be to performed INS experiments at low temperatures to observe the evolution of the spin resonance in the magnetically ordered phase and to extend the measurements to other Nd concentration to confirm the present results.

References

[1] Rossat-Mignod J, Regnault L P, Vettier C, Bourges P, Burlet P,Bossy J, Henry J Y and Lapertot G 1991 *Physica* C **185** 82-86

[2] Metoki N, Haga Y, Koike Y, Aso N and Onuki Y 1997 *J. Phys. Soc. Jpn.* **66** 2560
[3] Stock C, Broholm C, Hudis J, Kang H J and Petrovic C 2008 *Phys. Rev. Lett.* **100** 087001
[4] Stockert O, Arndt J, Schneidewind A, Schneider H, Jeevan H S, Geibel C, Steglich F and Loewenhaupt M 2008 *Physica* C **403** 973
[5] Lumsden M D and Christianson A D 2010 *J. Phys: Condens. Matter* **22**, 203203 and references therein.
[6] Petrovic C, Pagliuso P G, Hundley M F, Movshovich R, Sarrao J L, Thompson J D, Fisk Z and Monthoux P 2001 *J. Phys: Condens. Matter* **13** L337
[7] Shishido H, Settai R, Aoki D, Ikeda S, Nakawaki H, Nakamura N, Iizuka T, Inada Y, Sugiyama K, Takeuchi T, Kindo K, Kobayashi T C, Haga Y, Harima H, Aoki Y, Namiki T, Sato H and Onuki Y 2002 *J. Phys. Soc. Jpn.* **71** 162
[8] Balatsky A V, Vekhter I and Zhu J X 2006 *Rev. Mod. Phys.***78** 373
[9] Sun Y and Maki K 1995 *Phys. Rev.* B **51** 6059
[10] Petrovic C, Pagliuso P G, Hundley M F, Movshovich R, Sarrao J L, Thompson J D, Fisk Z and Monthoux P 2002 *Phys. Rev.* B **66** 054534
[11] Tanatar M A, Paglione J, Nakatsuji S, Hawthorn D G, Boaknin E, Hill R W, Ronning F, Sutherland M, Taillefer L, Petrovic C, Canfield P C and Fisk Z 2005 *Phys. Rev. Lett.* **95** 067002
[12] Hu R, Lee Y, Hudis J, Mitrovic V F and Petrovic C 2008 *Phys. Rev.* B **77** 165129
[13] Panarin J, Raymond S, Lapertot G, Mignot J M and Flouquet J 2011 *Phys. Rev.* B **84** 052505
[14] Canfield P C and Fisk Z 1992 *Phil. Mag.* B **65** 1117
[15] Panarin J, Raymond S, Lapertot G and Flouquet J 2009 *J. Phys. Soc. Jpn.* **78** 113706
[16] Sidis Y, Bourges P, Fong H F, Keimer B, Regnault L P, Bossy J, Ivanov A, Hennion B, Gautier-Picard P, Collin G, Millius D L and Aksay I A 2000 *Phys. Rev. Lett.* **84** 5900

Journal of Physics: Conference Series **340** (2012) 012074 doi:10.1088/1742-6596/340/1/012074

Phonon density of states and the search for a resonance mode in $LaFeAsO_{0.85}F_{0.15}$ (T_c=26 K)

Z Yamani[1], D H Ryan[2], J M Cadogan[3], F Canepa[4], A Palenzona[5] and Andrea Orecchini[6,7]

[1] NRC Canadian Neutron Beam Centre, Chalk River, ON, K0J 1J0, Canada
[2] Physics Department, McGill University, Montreal, QC, H3A 2T8, Canada
[3] Department of Physics and Astronomy, University of Manitoba, Winnipeg, MB, R3T 2N2, Canada
[4] Dipartimento di Chimica e Chimica Industriale, Università di Genova, and CNRIMEM, Genova, Italy
[5] Dipartimento di Chimica e Chimica Industriale, Università di Genova, and CNRSPIN, Genova, Italy
[6] Institut Laue-Langevin, 156X, 38042 Grenoble Cédex, France
[7] Dipartimento di Fisica and CNR-IOM, Universit di Perugia, 06123 Perugia, Italy

E-mail: dhryan@physics.mcgill.ca

Abstract. While the high transition temperatures suggest that the conventional BCS phononmediated mechanism may not provide the main pairing mechanism in the recently discovered $RFeAsO_{1-x}F_x$ (1111-type) superconductors, there is, as yet, no consensus, despite extensive experimental and theoretical study. We report here the results of an inelastic neutron scattering investigation of an overdoped polycrystalline sample of $LaFeAsO_{1-x}F_x$ with x=0.15 (T_c=26 K). Four excitation peaks were observed at 13.6±1.5, 24.2±0.8, 32.2±0.5, and 41.4±1.0 meV. They were identified as phonon modes based on their wavevector and temperature dependence. The peak positions agree well with first-principles calculations of phonon density of states as well as experimental data on both the insulating parent and optimally doped $LaFeAsO_{1-x}F_x$ compounds. No evidence for the presence of a resonance mode was found. We found that the phonon density of states of the x=0.15 sample remains unchanged below T_c and is similar to samples with other fluorine concentrations. This suggests that a standard electron-phonon pairing mechanism cannot explain the high transition temperatures observed in these materials.

1. Introduction

The recent discovery [1] of high transition temperature superconductivity in $R\mathrm{FeAsO}_{1-x}\mathrm{F}_x$ (1111-type), where R is La or a rare earth element, has once again generated intense interest in the field of superconductivity. Since then, four other types of iron-based compounds have been found to also superconduct with doping or under high pressure. They are: BFe_2As_2 (122-type with B=Ba, Sr, Ca) [2], AFeAs (111-type with A=alkali metal) [3], $\alpha-$FeSe(Te) (11-type) [4], and $K_{0.8}Fe_2As_2$ (isostructural to $BaFe_2As_2$) [5]. These new Fe-based superconductors are the only alternative to the cuprate superconductors for exploring high temperature superconductivity. Despite many experimental and theoretical studies, there is still no consensus on the pairing mechanism that leads to high superconducting transition temperatures in these Fe-based materials. The high transition temperatures might suggest that a phonon-mediated mechanism cannot be the main pairing mechanism, and firstprinciples calculations based on density functional theory [6] also appear to support the notion that an exotic mechanism may play a role. However there is some evidence, including the strong relationship found between the magnetic state of iron sublattice [7] and T_c [8] to the FeAs configuration that indicates the lattice (i.e. phonons) could also be involved.

The magnetic and superconducting phase diagrams of the cuprate and the Fe-based superconducting materials have some striking similarities. In both systems, the parent material exhibits antiferromagnetic order and superconductivity appears when the antiferromagnetically ordered phase has been suppressed by chemical doping, although the long range AF ordered phase in the Fe-based parent materials is metallic rather than Mott insulating as it is with cuprates. One prominent neutron scattering feature of cuprate superconductors is the presence of a resonance excitation mode [9]. The resonance mode is localized in both wavevector and energy. It is believed that this resonance mode is intimately related to superconductivity in cuprates as its intensity increases below T_c and its energy scales with T_c as a function of doping [10]. Resonance modes have also been observed [11] in Co, Ni, and K-doped $BaFe_2As_2$ as well as in the FeSe(Te) superconductors, indicating that magnetic interactions play a similar role in the superconductivity of the Fe-based materials as they do in the cuprates. Recent experiments have also found [12] a resonance mode in an optimally doped $LaFeAsO_{0.92}F_{0.08}$ polycrystalline sample with T_c=29 K. In order to investigate whether the resonance energy in Fe-based superconductors scales with T_c as it does in the cuprates, further experiments on Fe-based superconductors with different doping levels across the phase diagram are required.

Among all families of the Fe-based superconductors, the $R\mathrm{FeAsO}_{1-x}\mathrm{F}_x$ superconductors exhibit the highest superconducting transition temperatures (55 K for R=Sm [13]). Therefore a complete investigation of the $R\mathrm{FeAsO}_{1-x}\mathrm{F}_x$ properties, including the phonon and magnetic excitations, is required to gain a comprehensive understanding of high superconducting transition temperatures found in these materials. For most of the rare earths, the excitation spectra of the $R\mathrm{FeAsO}_{1-x}\mathrm{F}_x$ materials is complicated due to the presence of the crystal field excitations of the rare earth 4f electrons. $R\mathrm{FeAsO}_{1-x}\mathrm{F}_x$ compounds with R=La (no 4f electrons) provide an alternative to study the phonon and magnetic excitation spectra in these Fe-based superconductors without the complication of the crystal field excitations. We report here the wavevector and energy dependence of the excitations in an overdoped $LaFeAsO_{0.85}F_{0.15}$ polycrystalline sample with T_c=26 K at temperatures below and above T_c.

2. Experiments

The experiments were performed on a polycrystalline sample of $LaFeAsO_{0.85}F_{0.15}$ prepared by solid state reaction at ambient pressure from La, As, Fe, Fe_2O_3 and FeF_2. LaAs was first synthesized starting from pure elements in an evacuated sealed glass tube at a maximum temperature of 550°C. The final sample was synthesized by mixing LaAs, Fe, Fe_2O_3 and FeF_2 powders in stoichiometric proportions, using uniaxial pressing to make powders into a pellet. To avoid loss of fluorine, the heat treatment of the pellet was performed in a tantalum crucible welded under argon, at 1250°C for 72 hours, followed by cooling. Resistivity measurements indicated a superconducting transition temperature of 26 K. The sample was analyzed by powder x-ray diffraction in a Guinier camera, with Si as internal standard. The powder pattern showed the sample to be almost pure, with the space group P4/nmm (a = 4.028(1) Å and c = 8.720(1) Å). However, two faint lines were detected at low angles and attributed to LaOF.

The inelastic neutron scattering spectra were measured using the thermal time-of-flight neutron spectrometer IN4 at the Institut Laue Langevin (ILL), France. A crystal monochromator is used to set the incident energy and a time-of-flight technique is employed to analyse the scattered neutron energies, through proper phasing of a Fermi chopper with the acquisition electronics of the detectors. The spectrometer is equipped with a wide angular range detector assembly designed for powder work. The incident neutron wavelength of 1.1 Å (E_i=67.6 meV) was chosen to allow measurements with an energy transfer of up to 60 meV (neutron energy loss).

About 2 g of the polycrystalline $LaFeAsO_{0.85}F_{0.15}$ sample was placed inside a thin aluminium platelike mount. The inelastic spectra were collected in the temperature range 2 K to 50 K. The scattering from the empty aluminium mount was also measured to be used as background.

3. Results and discussion

The inelastic spectrum of the $LaFeAsO_{0.85}F_{0.15}$ sample was measured at several temperatures below and above T_c (=26 K). The observed scattering from the empty mount was subtracted from the signal. Figure 1 shows the background subtracted inelastic intensity as a function of wavevector and energy transfer at 2 K with an incident energy of E_i=67.6 meV. There are four excitation peaks that appear as bands of scattering at constant energy transfers: 13, 24, 32, and 41 meV. As the intensity of all four excitations increases with Q, we have identified them as phonon modes.

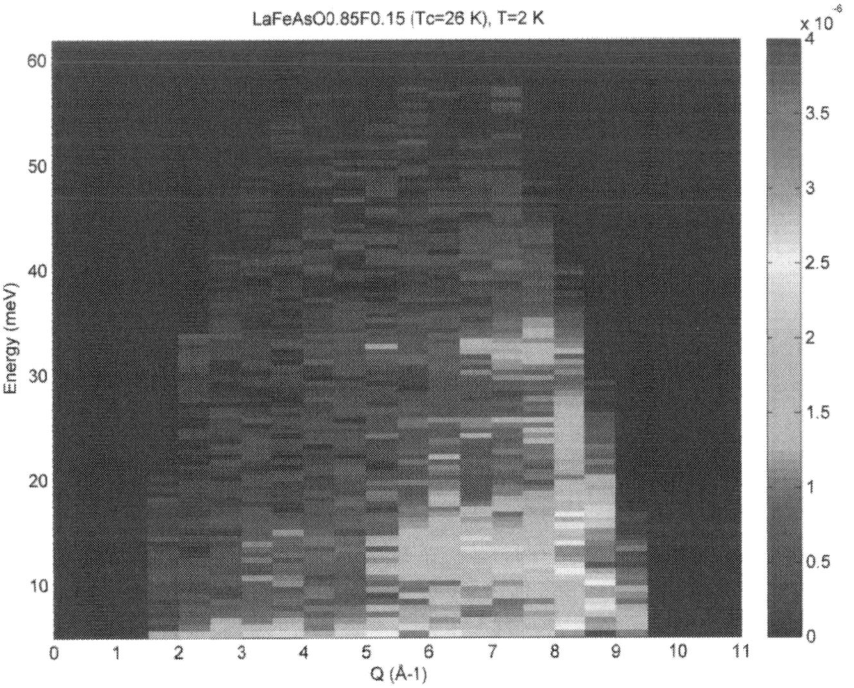

Figure 1. Observed scattering as a function of wavevector and energy transfer at 2 K. The data from the empty sample mount has been used to subtract the background. Four bands of excitations are observed at constant energy transfers: 13, 24, 32, and 41 meV. The intensity of all four excitations increases with Q indicating they are phonon modes. Similar data have been observed at temperatures higher than T_c. We find no evidence for presence of resonance characterized by an increase in intensity below T_c at any Q within our reach.

Within the limit of the single-phonon scattering and large wavevector transfer, the intensity from neutron down-scattering by phonons is given by [14]:

$$I(\mathbf{Q},\omega) = \frac{k_f}{k_i}\left(N\sum_{i=1}^{n} b_i^2 \frac{\hbar Q^2}{2M_i} e^{-2W_i(Q)} Z_i(\omega)\right)\frac{[n(\omega)+1]}{\hbar\omega} \quad (1)$$

Where k_f and k_i are final and incident neutron wavevectors, respectively; N is the number of unit cells; the sum runs over all atoms in the unit cell; b_i, M_i, $W_i(Q)$, are the scattering length, mass, and the Debye-Waller factor of the atom at site i in the unit cell, respectively; $Z_i(\omega)$ is the contribution to the phonon density of states from the atom at site i. The generalized phonon density of states defined by

$$g(\omega) = \sum_{i=1}^{n} \frac{b_i^2}{2M_i} e^{-2W_i(Q)} Z_i(\omega) / \sum_{i=1}^{n} \frac{b_i^2}{2M_i} e^{-2W_i(Q)} \quad (2)$$

can be extracted from the observed inelastic intensity integrated over all measured Q's:

$$I(\omega) = \int dQ\, I(\mathbf{Q},\omega) = Ag(\omega)\frac{[n(\omega)+1]}{\hbar\omega} \quad (3)$$

The observed integrated intensities $I(\omega)$ at 2 K (below T_c) and 50 K (above T_c) are shown in figure 2(a). In figure 2(b), we show $g(\omega)$ calculated from equation (3) at two temperatures. It is clear that no noticeable change is observed below T_c. We were able to fit four Gaussian peaks to the data centered at 13.6±1.5, 24.2±0.8, 32.2±0.5, and 41.4±1.0 meV. The peak positions for all four modes agree reasonably well with the first principles calculations of the phonon density of states for LaFeAs(O,F) [6,15] as well as with the previously reported phonon modes in the insulating parent LaFeAsO and optimally doped $LaFeAsO_{0.9}F_{0.1}$ [16]. The consistency between our results and those of first principles calculations that predict weak electron–phonon (e–p) coupling suggests that the e–p coupling is indeed weak and therefore that the standard e–p mediated coupling cannot explain the high transition temperatures observed in $LaFeAsO_{1-x}F_x$ superconductors. This view is further supported by the observation of similar phonon density of states for insulating and superconducting materials from optimally doped to overdoped with different T_c. If e–p coupling were indeed responsible for the superconductivity in this system, one would expect it to change as superconductivity appeared and the transition temperature evolved, and thus lead to a composition dependence in the phonon density of states.

We also searched for a neutron resonance peak characterized by an increase in intensity below T_c. A comparison of the data at 2 K (figure 1) and 50 K indicated no increase in intensity below T_c at any wavevector and energy transfer within our reach. The resonance peak observed [11] in Co, Ni, and K doped $BaFe_2As_2$ as well as in FeSe(Te) superconductors by neutron scattering, scales with T_c: $E_{res} = 4.9\,k_B T_c$. Recent experiments have also found [12] the resonance mode at 13 meV in an optimally doped $LaFeAsO_{0.92}F_{0.08}$ polycrystalline sample with T_c=29 K indicating a scaling ratio of 5.2. For our sample with a T_c of 26 K, the resonance peak would be expected within energy range of 10-12 meV. However, as indicated we did not detect any increase in intensity in this region below T_c. This could be due to the low sensitivity (small sample volume) of our experiment or because of the fact that the measurements were performed with a large E_i and hence the low-Q region (<1.5 Å^{-1}) in the first Brillouin zone was not accessible in this energy range. However, the second Brillouin zone was accessible, but here the failure to observe a resonance could be due to reduced sensitivity as a result of the form factor [12]. We note that no resonance peak was observed in another overdoped $LaFeAsO_{0.87}F_{0.13}$ [17]. Further experiments are required to establish whether the resonance mode is absent in the overdoped region of Fe-based superconductors.

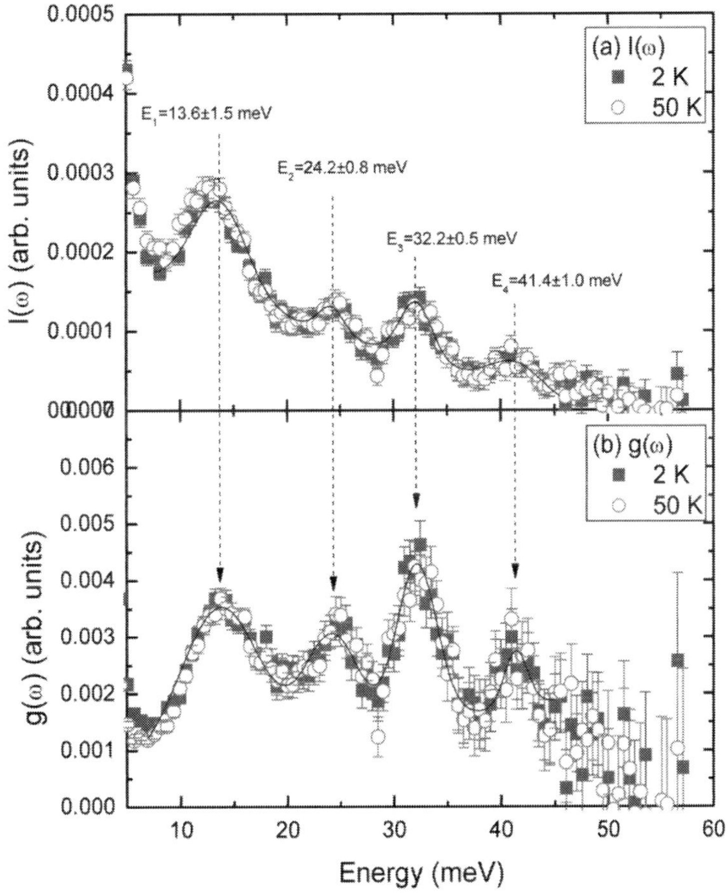

Figure 2. The observed intensity integrated over all wavevectors, $I(\omega) = \int dQI(\mathbf{Q}, \omega)$, at 2 and 50 K. (b) The generalized phonon density of states at 2 and 50 K, obtained by multiplying the data in (a) by $\frac{\hbar\omega}{[n(\omega)+1]}$. See text for details. No noticeable change is observed below T_c.

4. Conclusions

We have measured the inelastic spectrum of an overdoped $LaFeAsO_{0.85}F_{0.15}$ with T_c=26 K. We found four excitation peaks at 13.6±1.5, 24.2±0.8, 32.2±0.5, and 41.4±1.0 meV. Based on their wavevector and temperature dependence we have identified all four peaks as phonon modes. The peak positions agree well with the first principle theoretical calculations of phonon density of states as well as experimental data on both the insulating parent and optimally doped $LaFeAsO_{1-x}F_x$ compounds. This suggests a standard electron-phonon pairing mechanism cannot explain the high transition temperatures observed in these materials. We have found no evidence for the presence of a resonance peak.

Acknowledgements

J.M.C. acknowledges support from the Canada Research Chairs program. Financial support for various stages of this work was provided by the Natural Sciences and Engineering Research

Council of Canada and Fonds Québécois de la Recherche sur la Nature et les Technologies.

References

[1] Kamihara Y, Watanabe T, Hirano M, and Hosono H 2008 *J. Am. Chem. Soc.* **130** 3296
[2] Rotter M, Tegel M, and Johrendt D 2008 *Phys. Rev. Lett.* **101** 107006
[3] Wang X C, Liu Q Q, Lv Y X, Gao W B, Yang L X, Yu R C, Li F Y and Jin C Q 2008 *Solid State Commun.* **148** 538
[4] Hsu Fong-Chi, Luo Jiu-Yong, Yeh Kuo-Wei, Chen Ta-Kun, Huang Tzu-Wen, Wu Phillip M, Lee Yong-Chi, Huang Yi-Lin, Chu Yan-Yi, Yan Der-Chung and Wu Maw-Kuen 2008 *Proc. Natl. Acad. Sci.* 105 **14262**
[5] Guo Jiangang, Jin Shifeng, Wang Gang, Wang Shunchong, Zhu Kaixing, Zhou Tingting, He Meng and Chen Xiaolong 2010 *Phys. Rev. B* **82** 180520(R)
[6] Boeri L, Dolgov O V and Golubov A A2008 *Phys. Rev. Lett.* **101** 026403
Wang C, Li L, Chi S, Zhu Z, Ren Z, Li Y, Wang Y, Lin Z, Luo Y, Jiang S, Xu X, Cao G and Zhu'an X 2008 *Eur. Rev. Lett.* **83** 67006
Mazin I I, Singh D J, Johannes M D and Du M H 2008 *Phys. Rev. Lett.* **101** 057003
[7] Yin Z P, Lebègue S, Han M J, Neal B P, Savrasov S Y and Pickett W E 2008 *Phys. Rev. Lett.* **101** 047001
Yildirim T 2009 *Physica (Utrecht)* **469C** 425
[8] Mukuda Hidekazu, Terasaki Nobuyuki, Kinouchi Hiroaki, Yashima Mitsuharu, Kitaoka Yoshio, Suzuki Shinnosuke, Miyasaka Shigeki, Tajima Setsuko, Miyazawa Kiichi, Shirage Parasharam, Kito Hijiri, Eisaki Hiroshi and Iyo Akira 2008 *J. Phys. Soc. Japan* **77** 093704
[9] Dai Pengcheng, Mook H A, Aeppli G, Hayden S M and Dogan F 2000 *Nature* **406** 965
[10] Bourges P, Keimer B, Pailhes S, Regnault L P, Sidis Y, Ulrich C 2005 *Physica C* **424** 45
For general references also see: Tranquada J M in *Handbook of High-Temperature Superconductivity: Theory and Experiment*, edited by Schrieffer J R and Brooks J S 2007 Springer, New York, p. 257; also available at: arXiv:cond-mat/0512115.
[11] For a review and general references see: Lynn J W and Dai P 2009 *Physica C* **469** 469
Lumsden M D and Christianson A D 2010 *J. Phys.: Condens. Matter* **22** 203203
[12] Shamoto Shin-ichi, Ishikado Motoyuki, Christianson Andrew D, Lumsden Mark D, Wakimoto Shuichi, Kodama Katsuaki, Iyo Akira and Arai Masatoshi 2010 *Phys. Rev. B* **82** 172508
[13] Ren Z A, Lu W, ang J, Yi W, Shen X L, Li Z C, Che G C, Don X L, Sun L L, Zhou F and Zhao Z X2008 *Chinese Phys. Lett.* **25** 2215
[14] Ernst G, Broholm C, Kowach G R and Ramirez A P 1998 *Nature* **396** 147
[15] Singh D J and Du M H 2008 *Phys. Rev. Lett.* **100** 237003
[16] Higashitaniguchi Satoshi, Seto Makoto, Kitao Shinji, Kobayashi Yasuhiro, Saito Makina, Masuda Ryo, Mitsui Takaya, Yoda Yoshitaka, Kamihara Yoichi, Hirano Masahiro and Hosono Hideo 2008 *Phys. Rev. B* **78** 174507
Christianson A D, Lumsden M D, Delaire O, Stone M B, Abernathy D L, McGuire M A, Sefat A S, Jin R, Sales B C, Mandrus D, Mun E D, Canfield P C, Lin J Y, Lucas M, Kresch M, Keith J B, Fultz B, Goremychkin E A and McQueeney R J 2008 *Phys. Rev. Lett.* **101** 157004
[17] Qiu Y, Kofu M, Bao Wei, Lee S H, Huang Q, Yildirim T, Copley J R D, Lynn J W, Wu T, Wu G and Chen X H 2008 *Phys. Rev. B* **78** 052508

Dynamical Spin Susceptibility Studied by Inelastic Neutron Scattering on $LaFeAsO_{1-x}F_x$

S. Shamoto[1,2,*], M. Ishikado[1,2], S. Wakimoto[1,2], K. Kodama[1,2], R. Kajimoto[2,3], M. Arai[2,4]

[1]Quantum Beam Science Directorate, Japan Atomic Energy Agency, Tokai, Ibaraki 319-1195, Japan

[2]JST-TRIP, Tokyo 102-0075, Japan

[3]Research Center for Neutron Science and Technology, Comprehensive Research Organization for Science and Society (CROSS), Tokai, Ibaraki 319-1106, Japan

[4]J-PARC Center, Japan Atomic Energy Agency, Tokai, Ibaraki 319-1195, Japan

E-mail: shamoto.shinichi@jaea.go.jp

Abstract. Low-energy spin excitations on polycrystalline $LaFeAsO_{1-x}F_x$ samples have been studied by inelastic neutron scattering. The Q-integrated dynamical spin susceptibility $\chi''(\omega)$ at 11 meV decreases with increasing fluorine content x across the phase transition from antiferromagnetic to superconducting phases. For superconducting samples, a peak in $\chi''(\omega)$ develops at $\omega=E_{res}$ below T_c, accompanied by its reduction below E_{res}. Fe 3d orbital contribution on the dynamical spin susceptibility and the electron scattering between Γ and M Fermi surfaces are discussed.

1. Introduction

Since the discovery of high-T_c superconductivity (SC) in iron-based materials [1], a number of iron-based superconductors have been discovered. These superconductors have similar cylindrical Fermi surfaces at Γ and M points as schematically shown in Fig. 1 [2]. Because of the strong nesting condition from hole Fermi surfaces at Γ point to electron Fermi surfaces at M point, the Lindhard response function calculated from the band structure has a sharp peak at antiferromagnetic (AF) wave vector Q_{AF}=(1/2, 1/2, L) [3]. The non-interacting magnetic susceptibility is renormalized by the superexchange interaction J. The imaginary part of renormalized dynamical spin susceptibility $\chi''(Q,\omega)$ in itinerant magnetic materials can be probed by inelastic neutron scattering. The dynamical spin susceptibility in one of the optimally doped iron-based superconductors, $BaFe_{1.85}Co_{0.15}As_2$, has been described as antiferromagnetic metal within an itinerant electron picture [4]. Here, in addition to the dynamical spin susceptibility of $LaFeAsO_{1-x}F_x$ measured by inelastic neutron scattering [5,6,7,8], Fe 3d orbital contribution on the dynamical spin susceptibility and the electron scattering between Γ and M Fermi surfaces are discussed.

2. Experimental methods

High quality polycrystalline $LaFeAsO_{1-x}F_x$ samples with x =0, 0.057, 0.082, and 0.158 have been synthesized by solid state reaction. The fluorine concentration, x, in the synthesized sample was determined by secondary ion microprobe mass spectrometry measurement. Based on the Meissner signal measured with a superconducting quantum interference device magnetometer, the superconducting transition temperatures (T_c) in $LaFeAsO_{1-x}F_x$ with x =0, 0.057, 0.082, and 0.158 were determined to be 0, 25, 29, and 7 K, respectively, as shown in Fig. 2. The unit cell volume decreased monotonically with increasing fluorine content x. Details of syntheses and characterizations of all samples were reported elsewhere [5]. Inelastic neutron scattering measurements were performed at the following spectrometers, the triple-axis spectrometers TAS-1 at JRR-3 (JAEA), HB-3 at HFIR (ORNL), the chopper spectrometers 4SEASONS at J-PARC, and MERLIN at ISIS (RAL), using large amounts of polycrystalline samples with masses of 25-36 g. Observed scattering intensities were normalized among samples by using (0 0 2) nuclear Bragg peak intensity.

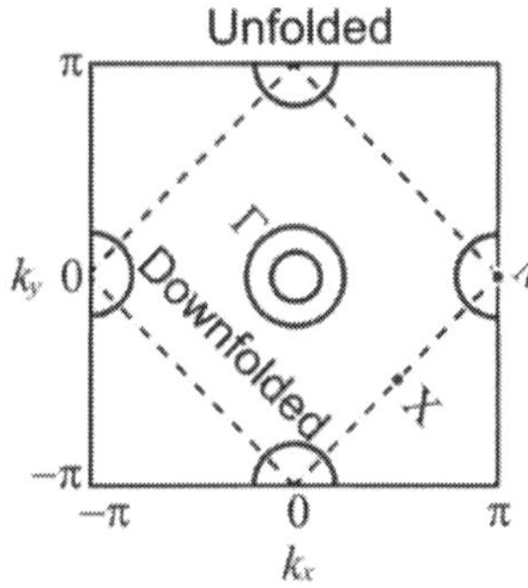

Figure 1. Schematic Fermi surfaces of LaFeAsO. AF wave vector corresponds to the scattering vector from Γ to M points.

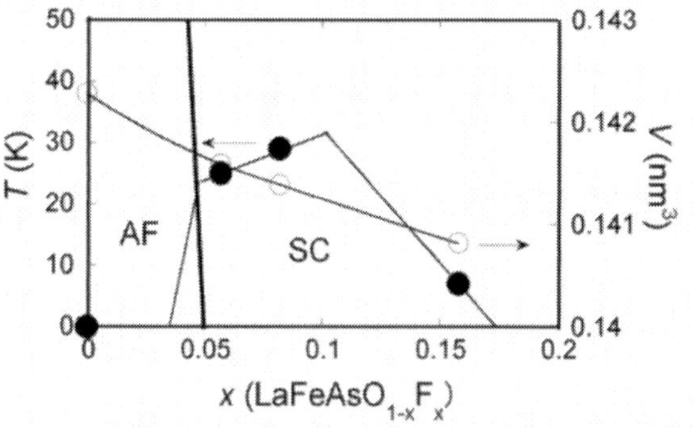

Figure 2. Phase diagram of $LaFeAsO_{1-x}F_x$ system. Closed circles show T_c, while open circles are their unit cell volumes.

3. Results and Discussion

3.1. Dynamical spin susceptibility and spin resonance mode of $LaFeAsO_{1-x}F_x$

In the parent LaFeAsO compound, the spin excitation at Q_{AF}~1.2 Å^{-1} has a long tail to the higher momentum transfers. This is due to the powder averaging of a magnetic rod in reciprocal lattice space [5,9]. In the superconducting compounds, the low-energy spin excitations remain with the similar intensity. However, the peak width in Q becomes broad in the superconducting compounds, due to the shorter AF spin correlation length. The broad peak is approximately fitted by a Gaussian peak.

The non-interacting susceptibility, $\chi_0(q,\omega)$, is proportional to the number of electron transitions from occupied state to unoccupied state, as follows.

$$\chi_0(q,\omega)=\sum_k \frac{f(\varepsilon_k)-f(\varepsilon_{k+q})}{\varepsilon_k-\varepsilon_{k+q}-\omega-i\delta} \tag{1}$$

The non-interacting susceptibility is renormalized by the superexchange interaction J, leading to the dynamical spin susceptibility as follows.

$$\chi(q,\omega)=\frac{\chi_0(q,\omega)}{1-J\chi_0(q,\omega)} \tag{2}$$

The q-integrated imaginary part of dynamical spin susceptibility, $\chi''(\omega)$, of $LaFeAsO_{1-x}F_x$ at 11 meV around Q_{AF} in the normal state is shown in Fig. 3. The value at x=0.082 was scaled from that below T_c in ref. 6 based on the spin resonance ratio in ref. 7. Volume shrinkage in Fig. 2 affects the $\chi''(\omega)$ value. Except for the effect, the decrease of dynamical spin susceptibility with increase of x can be attributed to two possible reasons. One is the decrease of the number of electron transitions in equation (1). The other is the decrease of the superexchange interaction J in equation (2). In the case of $LaFeAsO_{1-x}F_x$, the electron number increases with increase of doping. Consequently, the hole Fermi surface may disappear, as observed in overdoped $Ba(Fe,Co)_2As_2$ [10]. Recent NQR measurement of the similarly overdoped $LaFeAsO_{1-x}F_x$ superconductor exhibits a Hebel-Slichter peak in $1/T_1$ below T_c [11]. This result suggests that the symmetry of superconducting order parameter changes from $s_\pm$ to normal s wave with electron doping. These changes correspond to the disappearance of the hole Fermi surface. It should be noted that the normal s wave SC in this iron-based compound results only in low T_c. Recent inelastic neutron scattering of overdoped $Ba(Fe,Co)_2As_2$ also shows the disappearance of dynamical spin susceptibility at nearly the same electron concentration (x=0.15) [12] where angle resolved photoemission spectroscopy (ARPES) shows the disappearance of hole Fermi surface [10]. Therefore, the decrease of dynamical spin susceptibility is consistent with the decrease of density of states at the hole Fermi surface, leading to no electron transition at low energies from occupied state to unoccupied state at the momentum transfer Q_{AF} in equation (1).

In optimally doped $LaFeAsO_{0.92}F_{0.08}$, an enhancement in $\chi''(Q_{AF},\omega)$ has been observed below T_c [6,7]. The observed peak energy of $E_{res}\sim$ 13 meV corresponds to 5.2 k_BT_c. This intensity enhancement is attributed to spin resonance mode [3,13]. The peak appears as a result of coherence in superconducting state. Because of the time reversal symmetry of spin scattering between two Fermi surfaces at Γ and M, dynamical spin susceptibility exhibits a peak when superconducting order parameters in two Fermi surfaces has a sign reversal as in the $s_\pm$ pairing symmetry [14,15]. This is because the dynamical spin susceptibility is multiplied by a coherence factor in the superconducting state described as follows.

$$\frac{1}{2}\left(1-\frac{\varepsilon_k\varepsilon_{k+Q}+\Delta_k\Delta_{k+q}}{E_kE_{k+q}}\right)\approx\frac{1}{2}\left(1-\frac{\Delta_k\Delta_{k+q}}{|\Delta_k||\Delta_{k+q}|}\right) \tag{3}$$

where $E_k=\sqrt{\varepsilon_k^2+\Delta_k^2}$ is quasi-particle dispersion relation and ε_k is band energy measured relative to Fermi energy. When the signs of order parameter on different FSs are inverse, this coherence factor remains to be one, leading to the resonance mode. In contrast, it becomes zero in the case of same signs, resulting in the suppression of dynamical spin susceptibility. Meanwhile, the superconducting gaps may not have a simple structure. One side of Fermi surface may have different gap structure from the other. The SC gap may not be uniform along k_z. Therefore, such a superconducting gap structure can be studied from inelastic neutron scattering spectrum through the coherence factor. Low energy gap of 7 meV observed in the measurement suggests that optimally doped $LaFeAsO_{0.92}F_{0.08}$ exhibits fairly uniform full gap structure along k_z on all the Fermi surfaces. In $BaFe_{1.84}Co_{0.16}As_2$, the resonance mode has been observed independent on L [16]. On the other hand, it becomes L dependent in $BaFe_{1.92}Co_{0.08}As$ [17] and $BaFe_{1.9}Ni_{0.1}As_2$ [18]. Recent work about thermal conductivity along c-axis of the $Ba(Fe,Co)_2As_2$ system shows nodal behavior in the wide range of Co doping level except for the optimally doped compound [19]. This proximity to the nodal gap structure in the optimally doped compound, $BaFe_{1.85}Co_{0.15}As_2$, may correspond to the low energy gap of about 2.5 meV observed in the dynamical spin susceptibility at 4 K [4]. These results suggest that the superconducting gap may have a complicated structure corresponding to 3-dimensional band dispersion in iron-based superconductors, possibly resulting in common broad spin resonance peaks. In any case, the optimally doped high-T_c superconductor shows about twice enhancement in the dynamical spin susceptibility at the peak energy, E_{res}. In addition, the details of electronic band structure of an iron-based superconductor are sensitive to structural parameters, such as the pnictogen height [20]. According to a band structure calculation [21], z^2–orbital-type Fermi surface in LaFePO with low pnictogen height appears instead of *xy*-orbital-type Fermi surface. Note that the orbital-type is the main orbital contribution to the Fermi surface at Γ point. For this case, the nesting condition between two Fermi surfaces at Γ and M points may disappear. In other words, the superexchange interaction J between different orbitals in the iron square lattice may become zero depending on the hopping integral between those orbitals. According to the orbital dependence of J, $\chi''(\omega)$ obtained by inelastic neutron scattering mainly shows identical orbital contribution in two Fermi surfaces separated by Q_{AF}. Theoretically, the importance of the intra-orbital pair scattering for spin fluctuation mediated SC has been proposed [3]. Based on the spin fluctuation mechanism, observed $\chi''(\omega)$ reflects important part associated with SC gaps.

Note that a recent theoretical calculation can also explain the peak structure by s_{++} wave with strong dumping effect, which is consistent with impurity effects such as Co doping on the SC [22]. The SC gap opening effect seems similar to that of $s_{\pm}$ wave. One of the important differences in spin excitation between $s_{\pm}$ and s_{++} waves is the peak energy of $\chi''(\omega)$ relative to the summation of two superconducting gap energies at hole and electron Fermi surfaces.

Anyway, these results may suggest that an intimate relationship between electron scattering between Γ and M Fermi surfaces and high-T_c SC in iron-based superconductors.

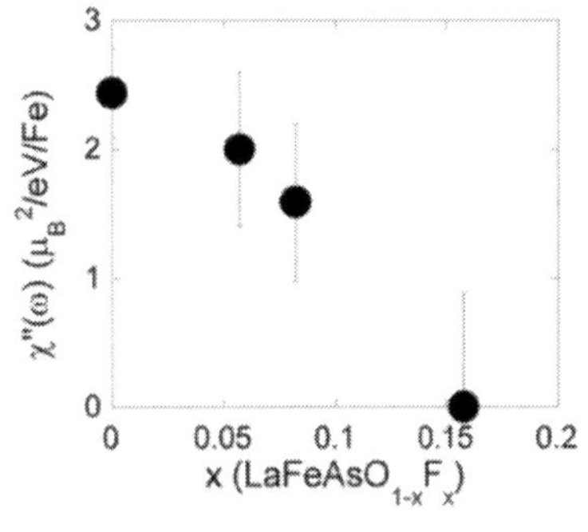

Figure 3. Q-integrated dynamical spin susceptibility $\chi''(\omega)$ at 11 meV in the normal state as a function of F concentration x in $LaFeAsO_{1-x}F_x$.

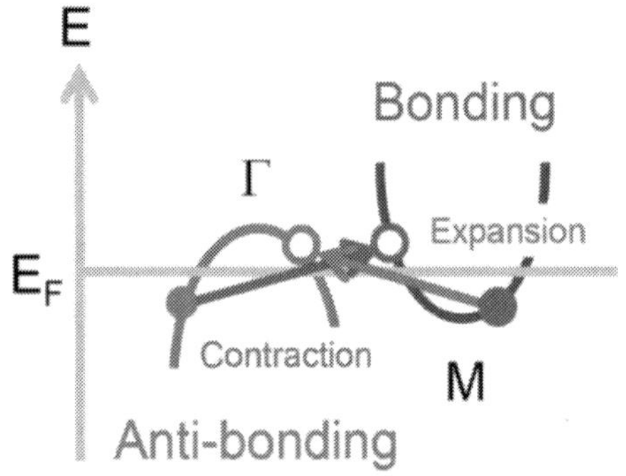

Figure 4. Schematic illustration of electron scattering between Γ and M Fermi surfaces. Γ and M Fermi surfaces have anti-bonding and bonding characters, respectively. Their electron scattering between them corresponds to lattice contraction or expansion, resulting in the lattice breathing.

3.2. Lattice breathing by electron round scattering between Γ and M Fermi surfaces

Γ and M Fermi surfaces in iron-based superconductors have hole and electron characters, respectively. In a band, a bottom part has bonding character whereas the top part has anti-bonding character, because of the energy dependence on the node number. Carriers in the bottom and top parts are usually called as hole and electron, respectively. In the sense, a lattice expands in the case of electron loss or hole loss, while a lattice contracts in the opposite case, *i.e.*, electron gain or hole gain. When an electron carrier is scattered from Γ to M Fermi surfaces, Γ Fermi surface gains a hole and M Fermi surface gains an electron, resulting in lattice contraction. Alternatively, when an electron carrier is scattered from M to Γ Fermi surfaces, Γ Fermi surface loses a hole and M Fermi surface loses an electron, resulting in lattice expansion. Therefore, electron round scattering between Γ and M Fermi surfaces is lattice breathing. As observed by inelastic neutron scattering, strongly enhanced dynamical spin susceptibility at Q_{AF} due to the good nesting condition between Γ and M Fermi surfaces suggests the frequent electron hopping between them. It may result in active lattice breathing. The importance of breathing mode has also been discussed in superconductors, $BaPb_xBi_{1-x}O_3$ and $Ba_{1-x}K_xBiO_3$, in relation to charge disproportionation on bismuth atoms [23]. In this sense, there might be possible similarity between iron-based superconductors and these high-T_c oxide superconductors.

4. Summary

The dynamical spin susceptibility, $\chi''(\omega)$, observed in $LaFeAsO_{1-x}F_x$ samples by inelastic neutron scattering is discussed as the result of good nesting condition between Γ and M Fermi surfaces separated by Q_{AF}. It consists mainly of the same orbital components in two Fermi surfaces. The decrease of $\chi''(\omega)$ as a function of fluorine content x in $LaFeAsO_{1-x}F_x$ samples is mainly attributed to the disappearance of hole Fermi surface except for the volume shrinking effect. The electron round scattering between hole and electron Fermi surfaces is introduced in possible relation to lattice breathing.

Acknowledgements
This work was supported by JST, Transformative Research-Project on Iron Pnictides (TRIP). Inelastic neutron scattering experiments were partially conducted under UK-Japan and US-Japan collaborations. Work at ORNL was supported by the Scientific User Facilities Division, Office of Basic Energy Sciences, US Department of Energy.

References

[1] Kamihara Y, Watanabe T, Hirano M, Hosono H 2008 *J. Am. Chem. Soc*. **130** 3296
[2] For example, Singh D J and Du M H 2008 *Phys. Rev. Lett*. **100** 237003
[3] Mazin I I, Singh D J, Johannes M D, Du M H 2008 *Phys. Rev. Lett*. **101** 057003
[4] Inosov D S, Park J T, Bourges P, Sun D L, Sidis Y, Schneidewind A, Hradil K, Haug D, Lin C T, Keimer B, Hinkov V 2010 *Nature Phys*. **6** 178
[5] Ishikado M, Kajimoto R, Shamoto S, Arai M, Iyo A, Miyazawa K, Shirage P M, Kito H, Eisaki H, Kim S-W, Hosono H, Guidi T, Bewley R, Bennington S M 2009 *J. Phys. Soc. Jpn*. **78** 043705
[6] Wakimoto S, Kodama K, Ishikado M, Matsuda M, Kajimoto R, Arai M, Kakurai K, Esaka F, Iyo A, Kito H, Eisaki H, Shamoto S 2010 *J. Phys. Soc. Jpn*. **79** 074715
[7] Shamoto S, Ishikado M, Christianson A D, Lumsden M D, Wakimoto S, Kodama K, Iyo A, Arai M 2010 *Phys. Rev*. B **82** 172508
[8] Shamoto S, Ishikado M, Wakimoto S, Kodama K, Kajimoto R, Arai M, Fukuda T, Nakamura H, Machida M, Eisaki H 2010 *Physica* C **470** S284
[9] Warren B E 1941 *Phys. Rev*. **59** 693
[10] Sekiba Y, Sato T, Nakayama K, Terashima K, Richard P, Bowen J H, Ding H, Xu Y-M, Li L J, Cao G H, Xu Z-A, Takahashi T 2009 *New J. Phys*. **11** 025020
[11] Mukuda H, Nitta M, Yashima M, Kitaoka Y, Shirage P M, Eisaki H, Iyo A 2010 *J. Phys. Soc. Jpn*. **79** 113701
[12] Matan K, Ibuka S, Morinaga R, Chi Songxue, Lynn J W, Christianson A D, Lumsden M D, Sato T J 2010 *Phys. Rev*. B **82** 054515
[13] Kuroki K, Onari S, Arita R, Usui H, Tanaka Y, Kontani H, Aoki H 2008 *Phys. Rev. Lett*. **101** 087004
[14] Maier T A, Scalapino D J 2008 *Phys. Rev*. B **78,** 020514R
[15] Korshunov M M, Eremin I 2008 *Phys. Rev*. B **78** 140509R
[16] Lumsden M D, Christianson A D, Parshall D, Stone M B, Nagler S E, MacDougall G J, Mook H A, Lokshin K, Egami T, Abernathy D L, Goremychkin E A, Osborn R, McGuire M A, Sefat A S, Jin R, Sales B C, Mandrus D 2009 *Phys. Rev. Lett*. **102** 107005
[17] Christianson A D, Lumsden M D, Nagler S E, MacDougall G J, McGuire M A, Sefat A S, Jin R, Sales B C, Mandrus D 2009 *Phys. Rev. Lett*. **103** 087002
[18] Chi S, Schneidewind A, Zhao J, Harriger L W, Li L, Luo Y, Cao G, Xu Z, Loewenhaupt M, Hu J, Dai P 2009 *Phys. Rev. Lett*. **102** 107006

[19] Reid J-Ph, Tanatar M A, Luo X G, Shakeripour H, Doiron-Leyraud N, Ni N, Bud'ko S L, Canfield P C, Prozorov R, Taillefer L 2010 *Phys. Rev.* B **82** 064501
[20] Kuroki K, Usui H, Onari S, Arita R, Aoki H 2009 *Phys. Rev.* B **79** 224511
[21] Vildosola V, Pourovskii L, Arita R, Biermann S, Georges A 2008 *Phys. Rev.* B **78** 064518
[22] Onari S, Kontani H, Sato M, 2010 *Phys. Rev.* B **81** 060504R
[23] For example, Pei S, Jorgensen J D, Dabrowski B, Hinks D G, Richards D R, Mitchell A W, Newsam J M, Sinha S K, Vaknin D and Jacobson A J 1990 *Phys. Rev.* B **41** 4126

Journal of Physics: Conference Series **340** (2012) 012076 doi:10.1088/1742-6596/340/1/012076

Neutron study of non-Gaussian self dynamics in liquid parahydrogen

Ubaldo Bafile[1], Milva Celli[1], Daniele Colognesi[1], Marco Zoppi[1], Eleonora Guarini[2], Alessio De Francesco[3], Ferdinando Formisano[3], Martin Neumann[4]

[1] Consiglio Nazionale delle Ricerche, Istituto dei Sistemi Complessi, via Madonna del Piano 10, I-50019 Sesto Fiorentino, Italy

[2] Dipartimento di Fisica e Astronomia, Università di Firenze, via G. Sansone 1, I-50019 Sesto Fiorentino, Italy

[3] Consiglio Nazionale delle Ricerche, Istituto Officina dei Materiali, Operative Group in Grenoble (OGG), c/o Institut Laue Langevin, 6 rue Jules Horowitz, BP 156 F-38042 Grenoble Cedex, France

[4] Fakultät für Physik der Universität Wien, Strudlhofgasse 4, A-1090 Wien, Austria

E-mail: `ubaldo.bafile@isc.cnr.it`

Abstract. A time-honoured approach to single-molecule, or self, dynamics of liquids is based on the so-called Gaussian approximation (GA), where it is assumed that, in the whole dynamical range between hydrodynamic diffusion and free-particle streaming, the motion of a particle is fully determined by a unique function of time directly related to the velocity autocorrelation function. An evident support to the GA is offered by the fact that the approximation becomes exact in both above limit conditions. Yet, experimental inquiries into the presence of non-Gaussian dynamics are very scarce, particularly in liquid parahydrogen in spite of its importance as the prototype of a "quantum Boltzmann liquid" which has also served as a benchmark for the development of quantum dynamics simulation algorithms. Though experimental evidence of the breakdown of the GA was obtained by some of the authors a few years ago, the localization in Q space of non-Gaussian behaviour was still undetermined, and no quantitative assessment of the effect was ever obtained. These issues have been tackled and solved by a new neutron investigation, which provides the first determination of non-Gaussian behaviour in the framework of the well-known theoretical approach by Rahman, Singwi and Sjölander.

Single-particle microscopic dynamics has always been a major research subject in the physics of fluid matter [1], and a huge amount of work has been devoted to its understanding both theoretically and experimentally, while an ever-growing role has been played by computer simulations. In the standard formulation of statistical mechanics, where the dynamical properties of a system composed of N identical particles are expressed through time-dependent correlation functions, the relevant quantity is the *self* intermediate scattering function which for an isotropic fluid is written as [2]

$$F_{\mathrm{s}}(Q,t) = \frac{1}{N}\sum_{j=1}^{N}\left\langle \exp\left[-\mathrm{i}\mathbf{Q}\cdot\mathbf{r}_j(0)\right]\exp\left[\mathrm{i}\mathbf{Q}\cdot\mathbf{r}_j(t)\right]\right\rangle = \left\langle \exp\left[-\mathrm{i}\mathbf{Q}\cdot\mathbf{r}(0)\right]\exp\left[\mathrm{i}\mathbf{Q}\cdot\mathbf{r}(t)\right]\right\rangle \quad (1)$$

and represents the time autocorrelation of the spatial Fourier components, with wave vector $\mathbf{Q}$, of microscopic density fluctuations. Here $\langle ... \rangle$ indicates a quantum statistical average, $Q = |\mathbf{Q}|$, and $\mathbf{r}_j(t)$ is the Heisenberg position operator for the j-th particle at time t, while the absence of a subscript indicates that one tagged, but arbitrary, particle is referred to.

One of the most important approaches to self dynamics is surely the one based on the so-called Gaussian approximation (GA) [3], expressed by

$$F_s(Q,t) = \exp[-Q^2\gamma(t)], \tag{2}$$

which consists in the assumption that, in the whole dynamical range between hydrodynamic diffusion ($Q \to 0$ and long times) and free-particle motion ($Q \to \infty$ and short times), the motion of a particle is completely determined by a function of time only, $\gamma(t)$, while the exponential dependence on Q^2 is what justifies the name of GA.

Equation (2) is valid in both the above limit conditions. If the simpler case of classical fluids is considered, one has [4] $\gamma(t) = t^2/(2M\beta)$ for a free gas and $\gamma(t) = Dt$ for a system where self diffusion dominates, where M is the mass of the particles, D is the diffusion coefficient, $\beta = 1/(k_B T)$, k_B is the Boltzmann constant, and T is the temperature. Moreover, equation (2) also applies to the case of a perfect solid with harmonic vibrations. Therefore, the GA is a good starting point to approach the problem of self dynamics in a liquid because, even if shown to disagree in detail with real dynamical data, still the discrepancy is probably not too large in any specific range of Q and t, including those most interesting for the study of molecular motions, at the nanometer-picosecond time and length scales.

Turning to the more general, but relevant here, case of quantum systems, where $F_s(Q,t)$ becomes a complex quantity, it was shown by Rahman *et al.* [5] that in an isotropic system one can rigorously write

$$F_s(Q,t) = \exp\left[\sum_{p=1}^{\infty} (\mathrm{i}Q)^{2p}\gamma_p(t)\right], \tag{3}$$

where the γ_p's are functions related to increasing-order correlation functions of different-time velocities of one particle. The $p = 1$ term, which also includes the recoil effect, is given by

$$\gamma_1(t) = -\mathrm{i}\frac{\hbar t}{2M} + \frac{1}{3}\int_0^t dt_1 (t - t_1) u(t_1) \tag{4}$$

where $\hbar$ is the Planck constant and $u(t) = \langle \mathbf{v}(0) \cdot \mathbf{v}(t) \rangle$ is the velocity auto-correlation function (VACF). If only the first term of (3) is retained, one immediately recovers the GA with $\gamma(t) = \gamma_1(t)$. The following γ_p's describe then the non-Gaussian behaviour, and most important of them, $\gamma_2(t)$, is related to a multiple time integral of a four-velocity correlation function of the type $\langle v_Q(t_1)v_Q(t_2)v_Q(t_3)v_Q(t_4)\rangle$, with $v_Q = \mathbf{v} \cdot \hat{\mathbf{Q}}$ denoting the projection of the tagged-particle velocity along the wave vector direction.

If the GA is assumed to be valid, $F_s(Q,t)$ can be directly computed from the knowledge of $u(t)$ alone without any further approximation. We recall here only a few important formulas and definitions, referring the interested reader to the more detailed treatments in Refs. [5, 6].

The VACF obeys the relationships $u(-t) = u^*(t) = u(t + \mathrm{i}\beta\hbar)$ with the asterisk denoting complex conjugation. As a consequence, its power spectrum

$$p(\omega) = \frac{1}{2\pi}\int_{-\infty}^{\infty} dt\, e^{-\mathrm{i}\omega t} u(t) \tag{5}$$

is a real but in general not an even function of ω that can be split into the sum of its symmetric part $p_S(\omega) = [p(\omega) + p(-\omega)]/2$ and its asymmetric part $p_A(\omega) = [p(\omega) - p(-\omega)]/2$, related to each other by virtue of the fluctuation-dissipation theorem [7]. Following Ref. [5] we define the spectral function

$$f(\omega) = \frac{4M}{3\hbar}\frac{p_{\mathrm{A}}(\omega)}{\omega}, \tag{6}$$

which is symmetric and obeys $\int_0^\infty d\omega\, f(\omega) = 1$ and $f(0) = 2MD\beta/\pi$. $f(\omega)$ is the quantity of interest, since it can be shown [5] that

$$\gamma_1(t) = \frac{\hbar}{2M}\int_0^\infty d\omega\, \frac{f(\omega)}{\omega}\left[(1-\cos\omega t)\coth\left(\frac{\beta\hbar\omega}{2}\right) - \mathrm{i}\sin\omega t\right]. \tag{7}$$

Within the GA, the self intermediate scattering function is then immediately calculated as

$$F_{\mathrm{s}}(Q,t) = \exp[-Q^2\gamma_1(t)] \tag{8}$$

and, from this, a Fourier transformation leads to the self dynamic structure factor

$$S_{\mathrm{s}}(Q,\omega) = \frac{1}{2\pi}\int_{-\infty}^{\infty} dt\, e^{-\mathrm{i}\omega t}F_{\mathrm{s}}(Q,t), \tag{9}$$

which is the suitable quantity to make contact with the results of spectroscopic experiments.

Despite its fundamental significance, the VACF itself is not directly accessible experimentally, and is a typical dynamical property to be studied through molecular dynamics (MD) simulation. However, while MD has proven effective and reliable in determining the microscopic dynamics of classical fluids [8], the situation is far less settled, and thus far more interesting, for systems belonging to what has often been termed the class of quantum Boltzmann liquids (QBLs), where the denomination summarizes a peculiar situation in which the fluid can still be thought of as made of distinguishable particles obeying Boltzmann statistics, with negligible quantum exchange effects, while quantum delocalization may be relevant, as evidenced by the structural properties [9].

Intense research activity has been carried out in the last two decades in order to come to a satisfactory understanding of QBL dynamics, driven by both a *per se* theoretical interest [10, 11, 12, 13] and, even more, by the need of developing still missing algorithms for a quantum dynamics simulation, for which various approaches have been devised [10, 11, 14, 15, 16].

Molecular hydrogen in its liquid state is the system most often studied, for a number of reasons. An obvious interest exists in extending to the dynamical side the investigation of the microscopic properties of such a fundamental substance, for which only recently a full account of the static structure was obtained by the joint use of Path Integral Monte Carlo computations and neutron diffraction experiments [17]. It has to be remarked that the molecular nature of hydrogen does not constitute a problem, as it is known that, even in the condensed phases, the rotovibrational degrees of freedom of H_2 molecules can be considered as decoupled from the translational motion [18], and the same concepts and formalism used for a monatomic system apply to hydrogen as well provided that the dynamical variables such as position and velocity are referred to the molecular centre of mass and M is taken to be the total mass of the molecule. Hydrogen is also considered as a prototype QBL featuring evident quantum effects while still remaining composed of distinguishable particles. Moreover, from the experimental point of view, H_2 is by far the best sample for experimental studies of the self dynamics due to its very large cross section for incoherent neutron scattering, which is the most important probe of single-particle microscopic dynamics. The possibility of experimental checks, together with the other reasons mentioned above, has therefore made liquid hydrogen a default benchmark for quantum simulation techniques [13, 15, 16, 19].

In talking of H_2 one should be more specific by mentioning that, at the liquid temperature (typically in the range between 15 and 20 K), the equilibrium species is practically 100% parahydrogen, i.e. the one having zero total nuclear spin and only the even-J rotational states populated. Actually, the temperature is such that all molecules find themselves in the $J = 0$ level.

A critical assessment of GA is still lacking, mainly because experimental studies have been rare. As far as classical fluids are concerned, only very few results were obtained a long time ago through neutron scattering [20] and MD simulation [21] on liquid argon, which showed the existence of regions of the (Q, t) space where the GA does not hold precisely. Non-Gaussian dynamics was also found in the simulation of a much more complex system such as a glass-forming polymer [22]. In contrast, no advance was to be recorded until recently on the question whether the GA can be more than a working hypothesis in the field of QBL in general, and of H_2 in particular, where it still kept the status of a conjecture to be either validated or disproved. In any case, most of the experimental studies of QBLs have so far concerned liquid parahydrogen [6, 23].

A first important result came a few years ago from a neutron scattering experiment [24] carried out on the TOSCA spectrometer at the pulsed neutron source ISIS. The validity of GA was questioned without resorting to calculations based on a quantum-simulated VACF. Rather, a consistency check was performed with the following procedure. The centre-of-mass $S_s(Q, \omega)$ was computed within GA with an adjustable model of $f(\omega)$, written in the form suggested by Levesque and Verlet [25] with free parameters that could be fitted so as to give good agreement with the neutron data in a Q range around 20 nm^{-1}. This $f(\omega)$ was then employed in the computation, again within GA, of the spectrum in a higher-Q range ($\sim$ 40 nm^{-1}), where a clear discrepancy appeared with the measured one. This result evidenced a breakdown of GA in liquid para-H_2, since it was not possible to find a unique $f(\omega)$ reproducing the experimental data in both investigated regions of Q. In other words, one was forced to conclude that, besides the explicit quadratic dependence on Q^2, a residual Q dependence was to be allowed for in the exponential function of equation (2).

A different question, however, is whether the actual $S_s(Q, \omega)$ at a given Q can be obtained on the basis of the GA, i.e. from (7), (8), and (9), with the $f(\omega)$ corresponding to an independently determined VACF. The experiment reported on in Refs. [24] could not provide an answer. Actually, the above described analysis might induce one to think that the GA is a less exact representation of the real spectrum at the higher Q's, at least in the explored range, while, as indeed shown below, the opposite conclusion is true.

The above situation motivated a further effort towards the experimental assessment of this still unsolved issue, by combining a new neutron study carried out on a wider (Q, ω) region in constant-Q conditions with a GA analysis of data computed using a reliable simulated VACF. For the latter, the Path Integral Centroid Molecular Dynamics (PICMD) method was applied [10, 14, 15]. In this paper we report on the results of this investigation as far as the validity of GA is concerned.

The inelastic scattering of neutrons from a liquid para-H_2 sample, at temperature $T = 15.7$ K and a molecular number density of 22.53 nm^{-3} [26], was recorded with two thermal-neutron time-of-flight spectrometers at the Institut Laue-Langevin (ILL) in Grenoble in order to cover different Q ranges. The small-angle BRISP spectrometer was used for Q values not exceeding 10 nm^{-1}, while for $20 \leq Q/\text{nm}^{-1} \leq 45$ the experiment was carried out on IN4C. The incident neutron energies were $E_0 = 50$ and 65.2 meV, respectively.

The already mentioned separability of the internal motions from the translation of the molecule as a whole is effectively exploited in the Young-Koppel modelling of neutron spectra from fluid hydrogen [27] as those of a system of harmonically vibrating free rotors. Taking into account that only the ground rotational and vibrational levels are thermally populated at the temperature of the liquid, and that no vibrational excitations are possible with the available neutron energy, the double-differential cross section for neutron scattering consists of replicas of $S_s(Q,\omega)$ shifted by the rotational energies of the $J \rightarrow J'$ transitions with odd J' [28]. Due to the small moment of inertia of the molecule, these lines are well separated and only the most intense of them, the $J = 0 \rightarrow 1$ transition corresponding to a shift $\hbar\omega_{0\rightarrow 1} = 14.69$ meV [29], is probed in the energy window of the experiment. The very small contribution of coherent scattering is, instead, located around the quasielastic line and plays no role at all. Then, to an extremely good approximation, the studied sample scatters neutrons according to

$$\frac{d^2\sigma}{d\Omega d\omega} = \frac{k'}{k} a(Q)_{0\rightarrow 1} S_s(Q, \omega - \omega_{0\rightarrow 1}) \tag{10}$$

where k and k' are the neutron wave vectors before and after scattering. Here the factor $a(Q)$ is a known, transition-specific, form factor [28]. Equation (10) allows for the extraction of $S_s(Q,\omega)$ from an experimental set of constant-Q spectral intensity data, eliminating the rotational frequency shift, although the result remains broadened by the instrumental energy resolution function $R(\omega)$. This was measured from the scattering of a vanadium sample to be a Gaussian line with a half width at half maximum (HWHM) of 0.62 and 1.39 meV for BRISP and IN4C, respectively. An absolute normalization of the dynamic structure factor was not attempted at this stage.

As far as the simulation is concerned, we note that the PICMD method is well suited to the computation of the so-called *canonical* (or Kubo-transformed [30]) VACF

$$u_c(t) = \frac{1}{\beta} \int_0^\beta d\lambda \left\langle e^{\lambda \mathcal{H}} \mathbf{v}(0) \cdot e^{-\lambda \mathcal{H}} \mathbf{v}(t) \right\rangle \tag{11}$$

where $\mathcal{H}$ is the Hamiltonian operator of the system. $u_c(t)$ is a real, even function of t, whose power spectrum

$$p_c(\omega) = \frac{1}{2\pi} \int_{-\infty}^{\infty} dt\, e^{-\mathrm{i}\omega t} u_c(t) \tag{12}$$

is related to that of the complex $u(t)$ through

$$p(\omega) = p_c(\omega) \frac{\beta\hbar\omega}{1 - \exp(-\beta\hbar\omega)} \tag{13}$$

and is directly proportional to $f(\omega)$, since

$$p_c(\omega) = \frac{3}{2M\beta} f(\omega). \tag{14}$$

We have applied the PICMD algorithm to a system of 256 molecules interacting with the Silvera-Goldman potential [31] in a cubic box. The Trotter number (i.e. the number of beads on the classical ring polymers replacing the quantum mechanical particles) was $P = 64$. Our implementation of PICMD consisted in a combination of alternating MD and Monte Carlo phases. At each time step of the centroid MD, i.e. while keeping the centroid positions fixed, a short PIMC simulation of 160 passes was performed in the space of bead coordinates relative to the centroid, and then the centroid positions and velocities were moved forward in time with the leapfrog integrator in the isokinetic ensemble. The important point is that, in contrast to the usual implementation, where the sampling of "intramolecular" bead coordinates is performed on the fly by molecular dynamics, we have done it by Monte Carlo, thus avoiding the ergodicity problems [32] affecting the intramolecular modes in MD. Due to this splitting between MD and

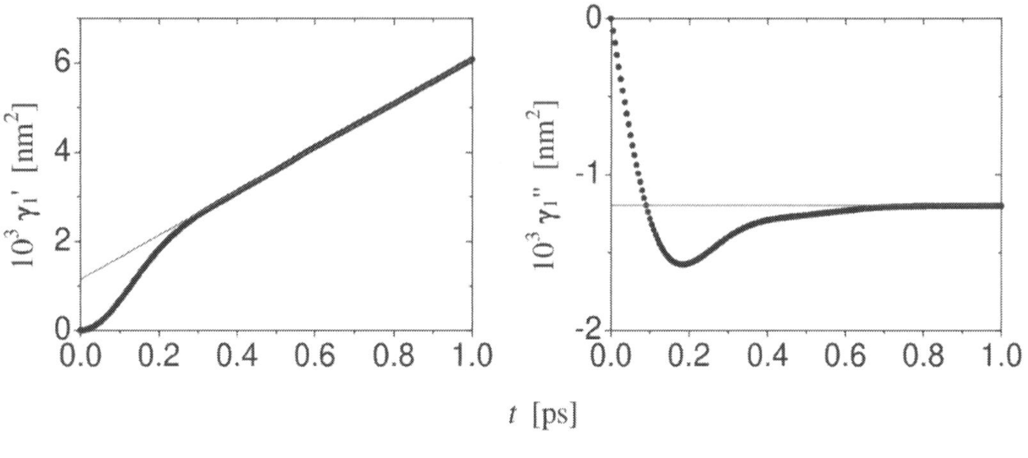

Figure 1. Short-time behavior of the real (left) and imaginary (right) parts of $\gamma_1(t)$ as calculated by PICMD (blue dots). The red straight line in the left frame shows the large-time linear behaviour fitted to the data in the time interval between 0.6 and 1.0 ps. In the right frame the large-time limit is calculated from $\gamma_1''(t) \approx -(\hbar\beta/2)D_1$ where $D_1 = 4.94 \cdot 10^{-5}$ cm^{2}s^{-1} is the fitted real-part asymptotic slope, in excellent agreement with the diffusion coefficient $D = 4.95 \cdot 10^{-5}$ cm^{2}s^{-1} [33].

PIMC a large time step of $5 \cdot 10^{-3}$ ps could be used, and the results obtained after $2 \cdot 10^5$ steps (i.e. 1 ns) are expected to be statistically accurate.

Once $u_c(t)$ is determined from the PICMD simulations, the above equations together with (7) can be used to evaluate $\gamma_1(t)$, which is shown in figure 1. Employing prime and double prime notation for the real and imaginary parts of complex quantities, the theory of Ref. [5] predicts large-time asymptotic linear behaviours of $\gamma_p(t)$ given by $\gamma_p'(t) \approx D_p t - C_p$ and $\gamma_p''(t) \approx -(\hbar\beta/2)D_p$ for all values of p, with, in particular, $D_1 = D$. The PICMD calculations agree with this theoretical prescription.

At all Q values of the experiments, the spectra resulting from the use of (8) and (9), convoluted with $R(\omega)$, can be compared with the measured ones, once the latter have been normalized by a scale factor determined at the highest Q values, where they have the same shape as the computed ones (see below).

We first consider the case $Q = 10$ nm^{-1} using BRISP data. As shown in the first panel of figure 2, the agreement is good, implying the validity of the GA. This statement has to be taken with caution, since the resolution broadening strongly affects the shape of the spectrum which consists of a narrow line. It cannot be rigorously excluded that the calculated $S_s(Q,\omega)$ actually differs from the true one. If so, however, one should admit that the resolution broadening is such as to exactly mask the non-Gaussian contributions, a result which seems rather unlikely. A different result is found, instead, at the Q values explored in the IN4C experiment, as displayed in the other panels of figure 2. A substantial agreement observed at the highest Q changes into an emerging discrepancy with decreasing Q.

Thus, while confirming the already found overall breakdown of the GA, the present study provides interesting new results. First, non-Gaussian effects in the self dynamics are confined in the Q range between 10 and 40 nm^{-1}, approximately. Second, the maximum deviation from GA is located somewhere between 10 and 20 nm^{-1}, where its detection could probably be made difficult by resolution limitations of the available spectrometers.

The above results suggest to try to get insight into the time dependence of non-Gaussian

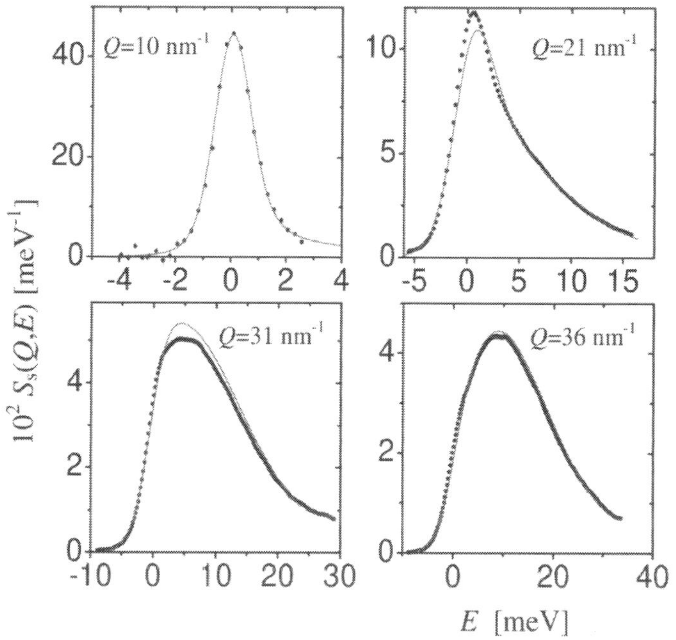

Figure 2. Self dynamic structure factor obtained from neutron data (blue dots) and calculated within the GA using the quantum simulated VACF (red line) at various Q values indicated in each frame. The top left frame refers to data from the BRISP experiment, all other frames show results of the IN4C experiment. All spectra are broadened by the respective instrumental resolution. The right shift of the peak position with increasing Q is due to the molecular recoil.

contributions to $F_s(Q,t)$. In this respect the present BRISP data do not provide useful information, covering a Q range where self dynamics shows an essentially Gaussian behaviour. Using the convolution theorem to express the resolution broadening in t space as the product with the Fourier transform of $R(\omega)$, the IN4C experimental spectra are seen to correspond to the function

$$\tilde{F}(Q,t) = R(t)F_s(Q,t) = R(t)\ \exp[-Q^2\gamma(Q,t)]$$

where equation (2) is now replaced by a more general expression allowing for non-Gaussian dynamical behaviour through an explicitly indicated Q dependence of γ. Making the assumption that most of the non-gaussianity is contained in the $p = 2$ term of equation (3), we set up a simplified model of $\gamma(Q,t)$, written as

$$\gamma(Q,t) = \gamma_1(t) - Q^2\gamma_2(t) + Q^4x(t), \tag{15}$$

where $\gamma_2(t)$ is retained, but all possible contributions beyond the Q^2 term are, somewhat roughly, included in a remainder denoted $x(t)$, whose characterization is likely beyond feasibility and is not attempted, since the object of our present analysis was the determination of $\gamma_2(t)$ at least at a semi-quantitative level.

To this aim, we Fourier-transformed the experimental spectra into the time domain to obtain $\tilde{F}(Q,t)$ and, from this, $\gamma(Q,t)$. Then, $\gamma'(Q,t)$ and $\gamma''(Q,t)$ have been separately analysed by performing at each value of t parabolic fits of the real and imaginary part of equation (15) as functions of Q^2. In the fits, the PICMD simulated $\gamma_1'(t)$ and $\gamma_1''(t)$ have been used, while $\gamma_2'(t)$

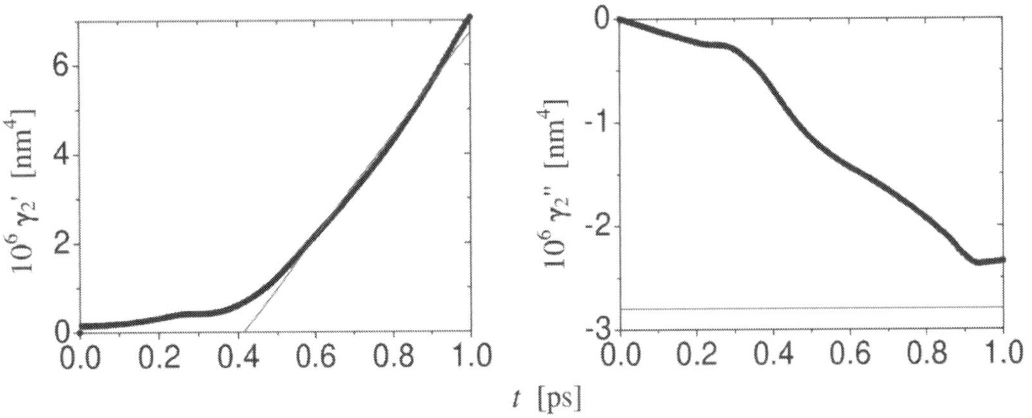

Figure 3. Short-time behavior of the real (left) and imaginary (right) parts of $\gamma_2(t)$ as derived by means of the procedure described in the text (blue dots). The red straight line in the left frame shows the large-time linear behavior fitted to the data in the time interval between 0.6 and 1.0 ps. In the right frame the large-time limit is calculated from $\gamma_2''(t) \approx -(\hbar\beta/2)D_2$ where D_2 is the fitted real-part asymptotic slope.

and $\gamma_2''(t)$ are obtained as fitted free parameters, displayed in figure 3. It can be observed that $\gamma_2(t)$ is nearly three orders of magnitude smaller than $\gamma_1(t)$, while the remainder $x(t)$ turns out to be of the order of 10^6 times smaller, supporting the assumption that its contribution is fully negligible.

With the present data, $\gamma_2(t)$ can only be studied at short times. In fact, the extraction of $\gamma(Q,t)$ requires to divide $\tilde{F}(Q,t)$ by $R(t)$, which is a Gaussian function with a HWHM of ~ 0.6 ps. The fast decaying tails of $R(t)$ produce unrealiable data at times larger than about 1 ps, since the main effect of the division is to magnify the unavoidable noise. This limitation is, therefore, related to the fact that the $J = 0 \rightarrow 1$ rotational line, located at $\sim$ 15 meV, requires for its excitation a neutron energy of tens of meV, which, with the currently available spectrometers, does not allow for a higher resolution power.

Nevertheless, the time range $t \lesssim 1$ ps is shown in figure 1 to be sufficient for $\gamma_1(t)$ to reach its asymptotic behaviour, and the same seems likely to be case for $\gamma_2(t)$. It has to be noted that, in addition to the limitation mentioned above, the extraction of $\gamma_2''(t)$ as the argument of an oscillatory function (the complex exponential of equation (15)) is a further difficulty, so that we have to consider our determination of $\gamma_2''(t)$ as probably not much more than a semi-quantitative estimate. However, it shows the right order of magnitude and most of its non-trivial time dependence seems confined to the explorable time range.

In summary, a careful neutron study of the self dynamics of liquid parahydrogen, besides confirming the presence of non-Gaussian dynamics, has provided as a new result the so-far missing knowledge about the Q ranges that can or cannot be properly accounted for by GA calculations. The comparison of neutron data with GA spectra computed from the PICMD-simulated VACF reveals non-Gaussian behaviour below $Q \sim 40$ nm^{-1} and a smooth approach to the expected high-Q recovery of Gaussian dynamics. On the other hand, a detailed study of the onset of non-gaussianity at low Q, where parahydrogen spectra consist of an intrinsically narrow line, still remains an experimental technical challenge given the resolution performances of present neutron spectrometers. The data collected in the experiments here reported have also provided the first determination of non-Gaussian effects in the $\gamma_p(t)$ expansion of equation (3) for a quantum liquid, at least in the very important subpicosecond regime.

In perspective, the efforts currently being made in the field of quantum dynamics simulation will hopefully lead to effective and reliable algorithms for the calculations not only of the VACF, but also, directly, of the dynamic structure factor, a goal which is presently within reach only for classical fluids. Once the simulation results can be matched directly with the experimental data, independently of any underlying approximation, studies on the GA will be possible with a higher degree of confidence.

Acknowledgments

The Scientific Membership Agreement between Consiglio Nazionale delle Ricerche and ILL and the skillful technical assistance of ILL staff are gratefully acknowledged.

References

[1] Boon J P and Yip S 1980 *Molecular Hydrodynamics* (New York: McGraw-Hill)
Balucani U and Zoppi M 1994 *Dynamics of the Liquid State* (Oxford: Clarendon)
[2] Van Hove L 1954 *Phys. Rev.* **95** 249
[3] Vineyard G H 1958 *Phys. Rev.* **110** 999
[4] Squires G L 1996 *Introduction to the Theory of Thermal Neutron Scattering* (New York: Dover)
[5] Rahman A, Singwi K S and Sjölander A 1962 *Phys. Rev.* **126** 986
[6] Celli M, Colognesi D and Zoppi M 2002 *Phys. Rev.* E **66** 021202
[7] Schofield P 1960 *Phys. Rev. Lett.* **4** 239
[8] Allen M P and Tildesley D J 1987 *Computer Simulation of Liquids* (Oxford: University Press)
[9] Zoppi M 2003 *J. Phys.: Condens. Matter* **15** R1
[10] Cao J and Voth G A 1994 *J. Chem. Phys.* **100** 5106
[11] Martyna G J 1996 *J. Chem. Phys.* **104** 2018
Cao J and Martyna G J 1996 *J. Chem. Phys.* **104** 2028
[12] Rabani E and Reichman D R 2002 *J. Chem. Phys.* **116** 6271
Makri N 1999 *Annu. Rev. Phys. Chem.* **50** 167
[13] Reichman D R and Rabani E 2002 *J. Chem. Phys.* **116** 6279
[14] Cao J and Voth G A 1994 *J. Chem. Phys.* **101** 6168
Hone T D and Voth G A 2004 *J. Chem. Phys.* **121** 6412
[15] Calhoun A, Pavese M and Voth G A 1996 *Chem. Phys. Lett.* **262** 415
Pavese M and Voth G A 1996 *Chem. Phys. Lett.* **249** 231
Kinugawa K 1998 *Chem. Phys. Lett.* **292** 454
Yonetani Y and Kinugawa K 2003 *J. Chem. Phys.* **119** 9651
[16] Georgescu I, Deckman J, Fredrickson L J and Mandelshtam V A 2011 *J. Chem. Phys.* **134** 174109
[17] Zoppi M, Celli M and Soper A K 1998 *Phys. Rev.* B **58** 11905
Bermejo F J, Kinugawa K, Cabrillo C, Bennington S M, Fåk B, Fernández-Díaz M T, Verkerk P, Dawidowski J and Fernández-Perea R 2000 *Phys. Rev. Lett.* **84** 5359
Zoppi M, Neumann M and Celli M 2002 *Phys. Rev.* B **65** 092204
Cunsolo A, Pratesi G, Colognesi D, Verbeni R, Sampoli M, Sette F, Ruocco G, Senesi R, Krisch M H and Nardone M 2002 *J. Low Temp. Phys.* **129** 117
Dawidowski J, Bermejo F J, Ristig M L, Fåk B, Cabrillo C, Fernández-Perea R, Kinugawa K and Campo J 2004 *Phys. Rev.* B **69** 14207
Celli M, Bafile U, Cuello G J, Formisano F, Guarini E, Magli R, Neumann M and Zoppi M 2005 *Phys. Rev.* B **71** 014205
[18] Van Kranendonk J 1983 *Solid Hydrogen* (New York: Plenum)
[19] Craig I R and Manolopoulos D 2004 *J. Chem. Phys.* **121** 3368
Liu J and Miller W H 2008 *J. Chem. Phys.* **128** 144511
[20] Sköld K, Rowe J M, Ostrowski G and Randolph P D 1972 *Phys. Rev.* A **6** 1107
[21] Desai R C 1978 *J. Chem. Phys.* **44** 77
Tsang T 1978 *Phys. Rev.* A **17** 393
[22] Colmenero J, Alvarez F and Arbe A 2002 *Phys. Rev.* E **65** 041804
[23] Zoppi M, Colognesi D and Celli M 2001 *Europhys. Lett.* **53** 34
Celli M, Colognesi D and Zoppi M 2002 *J. Low Temp. Phys.* **126** 585
[24] Colognesi D, Celli M, Neumann M and Zoppi M 2004 *Phys. Rev.* E **70** 061202
Celli M, Colognesi D, Ramirez-Cuesta A J and Zoppi M 2004 *Physica* B **350** e1083
[25] Levesque D and Verlet L 1970 *Phys. Rev.* A **2** 2514

[26] Roder H M, Childs G E, McCarty R D and Angerhofer P E 1973 *National Bureau of Standards Technical Note* n. 641
[27] Young J A and Koppel J U 1964 *Phys. Rev.* A **135** 603
[28] Guarini E 2003 *J. Phys.: Condens. Matter* **15** R775
[29] Huber K P and Herzberg G 1979 *Constants of Diatomic Molecules* (New York: Van Nostrand)
[30] Kubo R 1966 *Rep. Prog. Phys.* **29** 255
[31] Silvera I F and Goldman V V 1978 *J. Chem. Phys.* **69** 4209
[32] Pérez A, Tuckerman M E and Müser M H 2009 *J. Chem. Phys.* **130** 184105
[33] O'Reilly D E and Peterson E M 1977 *J. Chem. Phys.* **66** 934

Self diffusion in liquid aluminium

F Kargl[1], H Weis[1], T Unruh[2,3] and A Meyer[1]

[1] Institut für Materialphysik im Weltraum, Deutsches Zentrum für Luft- und Raumfahrt (DLR), 51170 Köln, Germany
[2] Forschungsneutronenquelle Heinz Maier-Leibnitz (FRM II), Technische Universität München, 85747 Garching, Germany
[3] Present address: Lehrstuhl für Kristallographie und Strukturphysik, Friedrich-Alexander-Universität Erlangen-Nürnberg, Staudtstr. 3, 91058 Erlangen, Germany

E-mail: florian.kargl@dlr.de

Abstract. Here we report temperature dependent self-diffusion coefficients of liquid aluminium measured on absolute scale by using incoherent quasielastic neutron scattering at temperatures of 980 K, 1020 K, and 1060 K. Aluminium self-diffusion coefficients follow an Arrhenius law with an activation energy of 280±70 meV. The Sutherland-Einstein equation relating viscosity to the diffusion coefficient well captures the temperature dependence and absolute values of the here reported aluminium self-diffusion coefficients using the covalent radius of aluminium. A comparison to published molecular dynamics simulation data helps to further narrow down the choice of the potential.

1. Introduction

Important parameters which can not only be used to calibrate atomistic simulations [1] but also serve as input parameters for modeling of solidification [2, 3] are diffusion coefficients of the equilibrium and undercooled melt. Self-diffusion coefficients in pure liquid metals or simple binary alloys are a first important parameter in this respect. Whereas numerous data exist for solids [4] the database for liquids is relatively scarce and subject to larger error. This is due to difficulties in measuring liquid diffusion coefficients with classical capillary experiments under gravity conditions. Therefore, new methods had to be established to determine diffusion coefficients with high accuracy. With novel X-ray [5] and neutron [6] radiography techniques high-accuracy interdiffusion data can now also be determined in capillary experiments provided a stable density layering can be achieved . However, for measurements of self-diffusion coefficients quasielastic neutron scattering (QNS) delivers highly accurate data over a large temperature range as shown recently for liquid nickel (Ni) [7], titanium (Ti) [1], and copper (Cu) [8]. Further, since physical properties as e. g. shear-viscosity are often experimentally more easily to determine it is important to establish a relation that allows to predict one from the other. Often the relation between diffusion and shear-viscosity is discussed using the Stokes-Einstein relation and variations thereof [9]. However, their general validity or a generic relation are still to be established for liquid metals.

Lightweight aluminium(Al)-based alloys are vastly used for applications in the aerospace and automotive industries as well as the energy sector. However, to date no measured diffusion coefficients exist for liquid Al. Whereas for capillary experiments suitable tracers are lacking, QNS is challenged by the orders of magnitude lower incoherent cross section of only $\sigma_i = 0.008$ barn compared with e. g. 0.55 barn for liquid Cu. Nonetheless, we here report temperature dependent self-diffusion coefficients of liquid Al measured by QNS. The data are finally compared with viscosity data [10] using the Sutherland-Einstein relation.

2. Experiment and Data Analysis

High purity aluminium (99.9999%, Hydro-Aluminium) was used for the experiments. A solid Al-cylinder of 21 mm diameter and 40 mm height was contained and subsequently melted in a thin walled Al_2O_3 hollow cylinder of 23 mm outer diameter at 0.5 mm wall thickness. Using this type of cylinder, scattering contributions of the container in the energy region of interest at and around the elastic line are significantly reduced.

The QNS experiments were performed on the neutron time-of-flight spectrometer TOFTOF at FRM-II [11] using incoming neutron wavelengths of 7.0 Å and 10 Å, respectively. Results obtained at these two wavelengths compare well with each other. The chopper settings of the instrument were chosen such that a virtually equal elastic energy resolution of about $\sim 80\,\mu$eV (full width at half maximum) was achieved for both incoming wavelengths. A standard high temperature vacuum furnace having a niobium heating element and heat shields was used as sample environment. Experiments were carried out at 980 K, 1020 K, and 1060 K in the liquid.

Data analysis was performed using the program FRIDA-1 [12]. Normalization of the measured intensity to a vanadium standard, correction for background contributions, correction for self-absorption, and interpolation to constant q resulted in the scattering law $S(q,\omega)$. The measured $S(q,\omega)$ can be well described by

$$S(q,\omega) = b_q + R(q,\omega) \otimes \left\{ A \cdot \frac{\Gamma}{\omega^2 - \Gamma^2} \right\}, \tag{1}$$

with b_q denoting an energy transfer independent but q dependent background, and the instrumental energy resolution function, denoted by $R(q,\omega)$, convolved with a single Lorentzian line with Γ denoting its half width at half maximum. In the hydrodynamic limit (low q, long timescales) and for an incoherent scatterer, Γ linearly depends on q^2 as

$$\Gamma = \hbar D_s q^2, \tag{2}$$

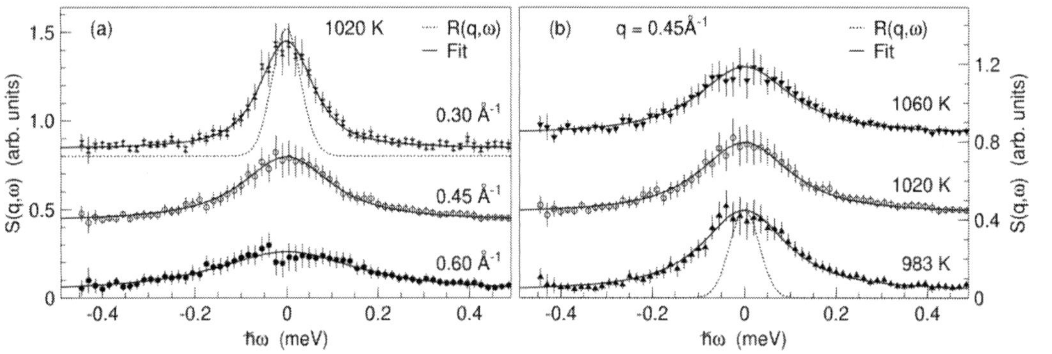

Figure 1. Scattering law $S(q,\omega)$ of liquid Al: (a) At 1020 K for three different wavenumbers. (b) At 0.45 Å^{-1} for three different temperatures. Data for different q values and different temperatures have been offset for clarity reasons by 0.4 and 0.8, respectively. The measured $S(q,\omega)$ were rescaled on an equidistant $\hbar\omega$ grid. The instrumental energy resolution function is represented by the dashed line. It was rescaled by multiplication with a constant to match the measured data for clarity reasons. Solid lines represent fits to the data using equation (1).

with the self-diffusion coefficient represented by D_s.

For Al coherent scattering contributions at low q are in principle not negligible due to its incoherent cross section of only 0.008 barn. Based on a calculated value for the coherent contribution to the total structure factor at the melting point [13] the ratio of coherent to incoherent scattering can be calculated as two third to one third at low q. However, reported thermodiffusion coefficients are two orders of magnitude larger than D_s. This means that the Rayleigh line for coherent scattering is extremely broad, hence, resulting in a virtually constant background [14, 15]. Moreover, Brillouin lines are due to the high speed of sound expected to be well separated from the quasielastic line in the here accessed q and $\hbar\omega$ range [9, 15, 16, 17].

With the large sample diameter chosen some concerns may be raised about multiple scattering affecting the measured signal. Multiple scattering can only be properly accounted for in Monte-Carlo simulations by either having a valid model of $S(q,\omega)$ at hand or by using an iterative approach on the measured data. With the chosen wavelengths the kinematically allowed region in (q,ω)-space is rather restricted. Hence, we resort to a study using several generic models and a transmission of 69% [18] in order to estimate the impact of multiple scattering on the here presented data. In this study it is shown that multiple scattering impacts mostly on the amplitude but has only a mild effect on the measured relaxation times. Most noteworthy is a more pronounced stretching and a decrease in relaxation times by about 10%. In the here presented case the transmission is 97% by direct comparison of normalized measured intensities to a thinner and mostly incoherently scattering sample. Hence, in terms of transmission one is quite far off the mentioned study. In the here presented case stretching of $S(q,\omega)$, which would be indicative of a second apparent relaxation process caused by multiple scattering, can not be seen. Multiple scattering depends on the accessible (q,ω)-space and therefore should depend on the incoming neutron wavelength. No differences around the quasielastic line for 7 Å and 10 Å incoming neutron wavelength can be observed. However, as expected some minor changes in the inelastic region where multiple scattering is known to have some more pronounced impact on the data are visible. Hence, it can be concluded that multiple scattering does not greatly impact on the quasielastic data at wavenumbers q relevant for determining D_s.

3. Results and Discussion

Measured and fitted $S(q,\omega)$ obtained at 10 Å incoming wavelength are shown in figure 1 (a) for a single selected temperature and three selected q values. Further, figure 1 (b) shows $S(q,\omega)$ for the three measured temperatures at a single selected q value of 0.45 Å^{-1}. In both figures data are shown on a linear scale. The increasing broadening of the quasielastic signal with increasing temperature at constant q (b) as well as with increasing q at constant temperature (a) can be well observed. The data are well described by equation (1).

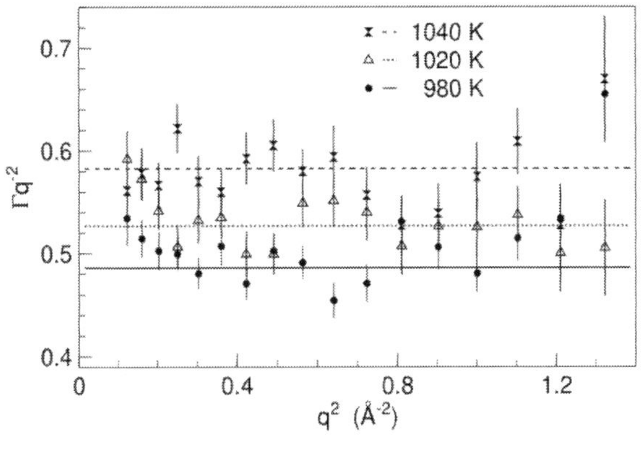

Figure 2. Half width at half maximum Γ divided by q^2 shown as a function of q^2 for three different temperatures. The data are well described by horizontal lines as expected in the hydrodynamic limit. Multiplying the y-axis by $1.51 \cdot 10^{-8}$ results in values equal to the self-diffusion coefficient D_s in units of $m^2\,s^{-1}$ detailed in table 1.

Table 1. Al self-diffusion coefficients measured by quasielastic neutron scattering.

T (K)	D_s (10^{-9} m^2 s^{-1})
980 ± 2	7.2 ± 0.6
1020 ± 2	7.9 ± 0.7
1060 ± 2	8.8 ± 0.7

The obtained Γ values divided by q^2 are shown in figure 2 as a function of q^2 for the three different temperatures. The data are well described by a horizontal line for q^2 greater than about 0.2 Å^{-2}. Hence, equation 2 can be applied to the data yielding self-diffusion coefficients for liquid Al. The resulting temperature dependent self-diffusion coefficients are presented in table 1. The fit of an Arrhenius equation to the diffusion data results in an activation energy of 280±70 meV.

In figure 3 the self-diffusion coefficients measured for Al are finally compared with self-diffusion coefficients calculated using the Sutherland-Einstein (SE) relation

$$D = k_B T / 4\pi\eta r \,, \tag{3}$$

with shear-viscosity η, diffusivity D, Boltzmann constant k_B, and a radius r of the moving particle. This equation is valid for slip-boundary conditions. The calculated diffusivities are represented by a solid line in figure 3. The covalent radius of Al equaling to 1.21 Å was used in the calculation. Calculated and measured diffusivities agree surprisingly well. Al viscosities are hereby subject to a 14% error [10]. For Al the SE equation well captures the temperature dependence of the measured diffusion coefficients around the melting point. On the other hand using the covalent radius assuming non-slip boundary conditions whereby 4 (Sutherland-Einstein) is replaced by 6 (Stokes-Einstein) in the denominator worked quite well for the glass forming PdNiCuP system [19] but not in general [9]. Hence, predicting diffusion coefficients from viscosity data still remains ambiguous for liquid metals. However, both flavours of this relation at least are able to predict diffusion coefficients for liquid metals within a factor of two.

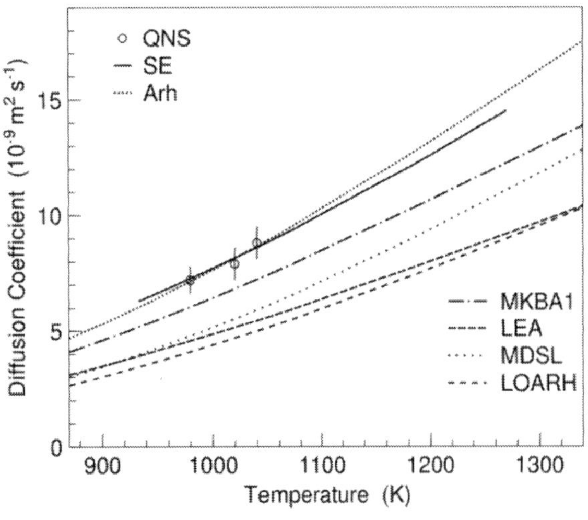

Figure 3. Temperature dependent self-diffusion coefficient of liquid Al (symbols). Data are compared with diffusion coefficients derived from viscosity data for liquid Al [10] using equation (3) (solid line). The reported fit curves to D_s obtained from molecular dynamics simulations [20] using different potentials (— · —MKBA1, - - - -LEA, ······MDSL, – – –LOARH) are shown and compared with the best Arrhenius fit to our data (finely dotted line).

Finally, the data can be compared with self-diffusion coefficients reported from an extensive set of molecular dynamics (MD) simulations [20] as shown in figure 3. A comprehensive study has been undertaken by those authors using different potentials and comparing their results with experimental data provided it was available. The best agreement with experimental structure factors was reported for the MDSL and the MKBA1 potential. Taking not only structural data but also the here reported and previously lacking diffusion coefficients into account the authors of reference [20] could now conclude that the MDSL potential appears to capture the overall temperature dependence well with the activation energy of 310±20 meV close to the here reported value of 280±70 meV. However, the absolute values of the diffusion coefficient at and around the melting point are underestimated by between a factor of 1.5 to 1.7 by this potential. The MKBA1 potential on the other hand delivers D_s values much closer to the measured ones with the activation energy slightly underestimating the measured one but still being well within error bars. Hence, it can be concluded that the MKBA1 potential appears to work overall best in describing the structural and mass transport properties of liquid Al.

4. Conclusion

In conclusion temperature dependent diffusion coefficients for Al measured by QNS have been presented. The diffusion coefficients compare well with diffusion coefficients calculated using the Sutherland-Einstein equation for slip boundary conditions, published viscosity data for Al and the covalent radius of Al atoms. A comparison with molecular dynamics simulations on liquid Al enabled to identify the overall best potential. The experimental data hereby provides a benchmark against which simulation data can be further calibrated. The here presented Al diffusion coefficients are also important for understanding metal impurity diffusion in liquid Al.

Acknowledgments

We thank the FRM-II for providing us with beamtime and Philipp Kuhn who critically read our manuscript.

References

[1] Horbach J, Rozas R E, Unruh T and Meyer A 2009 *Phys. Rev.* B **80** 212203
[2] Kasperovich G, Meyer A and Ratke L 2010 *Int. Found. Res.* **4** 2
[3] Zhang L J, Du Y, Steinbach I, Chen Q and Huang B Y 2010 *Acta Mat.* **58** 3664-75
[4] Mehrer H 2007 *Diffusion in Solids* (Springer, Heidelberg)
[5] Zhang B, Griesche A and Meyer A 2010 *Phys. Rev. Lett.* **104** 035902
[6] Kargl F, Engelhardt M, Yang F, Weis H, Schmakat P, Schillinger B, Griesche A and Meyer A 2011 *J. Phys.: Condens. Matter* **23** 254201
[7] Meyer A, Stüber S, Holland-Moritz D, Heinen O and Unruh T 2008 *Phys. Rev.* B **77** 092201
[8] Meyer A 2010 *Phys. Rev.* B **81** 012102
[9] Iida T and Guthrie R 1993 *The Physical Properties of Liquid Metals* (Oxford Science Publications, Oxford).
[10] Assael M J *et al*, 2006 *J. Phys. Chem. Ref. Data* **35** 285-300
[11] Unruh T, Neuhaus J, Petry W 2007 *Nucl. Instr. Methods* A **580** 1414-22
[12] http://www.messen-und-deuten.de/frida/frida.html
[13] Blairs S and Joasoo U 1980 *J. Inorg. Nucl. Chem.* **42** 1555-58
[14] Touioukiam Y and Ho C 1973 *Thermal Diffusivity* (Vol. 10 of Thermophysical Properties of Matter) (IFI-Plenum, New York)
[15] Scopigno T, Balucani U, Ruocco G and Sette F 2002 *J. Non-Cryst. Sol.* **312-314** 121-27
[16] Sjögren L 1978 *J. Phys. C: Sol. State Phys.* **11** 1493-1503
[17] Ebbsjö I, Kinell T and Waller I 1978 *J. Phys. C: Sol. State Phys.* **11** L501-04
[18] Wuttke J 2000 *Phys. Rev. E* **62** 6531-9
[19] Meyer A 2002 *Phys. Rev.* B **66** 134205
[20] Becker C A and Kramer M J 2010 *Modelling Simul. Mater. Sci. Eng.* **18** 074001

Journal of Physics: Conference Series **340** (2012) 012078 doi:10.1088/1742-6596/340/1/012078

Two-dimensional Fermi liquids sustain surprising roton-like plasmons beyond the particle-hole band

A Sultan[1], H Godfrin[1], M Meschke[2], H-J Lauter[3,4], H Schober[3], H Böhm[5], R Holler[5], E Krotscheck[5] and M Panholzer[5]

[1] Institut Néel, CNRS et Université J. Fourier, BP 166, F-38042 Grenoble Cedex 9, France
[2] Low Temperature Laboratory, Aalto University, PO. Box 3500, 02015 TKK, Finland
[3] Institut Laue-Langevin, BP 156, 38042 Grenoble Cedex 9, France
[4] Oak Ridge National Laboratory, PO BOX 2008 MS6475, Oak Ridge TN 37831-6475, USA
[5] Institute for Theoretical Physics, Johannes Kepler University, A-4040 Linz, Austria

E-mail: henri.godfrin@grenoble.cnrs.fr

Abstract. Using inelastic neutron scattering, we have investigated the elementary excitations of an isotropic two-dimensional Fermi liquid, ^{3}He adsorbed on graphite. We provide in this article a detailed account of the principles and methods which allowed measuring for the first time inelastic spectra on a liquid monolayer of ^{3}He, a strong neutron absorber. We also summarise the results presented at this Conference, and review our recent experimental and theoretical work on this this interacting many-body system. At low wave-vectors, near the edge of the particle-hole band, a mode identified as the zero-sound excitation by comparison to our theoretical calculations, is found as predicted at energies much lower than in bulk ^{3}He. The mode enters the particle-hole band, where it undergoes Landau damping. Surprisingly, however, intensity is observed in the neutron spectra at wave-vectors larger than twice the Fermi wave-vector. This new branch is interpreted as the high wave-vector continuation of the zero-sound mode, in agreement with the theory. The results open new perspectives in the understanding of the dynamics of correlated fermions.

1. Introduction

The concept of elementary excitations was introduced by Landau to describe the low temperature behavior of quantum fluids: bosonic liquid ^{4}He and fermionic liquid ^{3}He. His elegant theory turned out to be the seed of the theoretical framework used presently to describe not only quantum fluids, but more generally condensed matter, nuclear and particle physics: i.e., all systems where strongly correlated matter plays an essential role. Quantum fluids ^{4}He and ^{3}He are still the unavoidable reference and benchmark: these simple monoatomic and isotropic systems display macroscopic quantum phenomena: Bose-Einstein condensation in liquid ^{4}He and a pure Fermi Liquid state with a spherical Fermi surface, Cooper pairing of fermions in liquid ^{3}He... Thanks to these extraordinary properties, theoretical predictions on topological excitations, particle physics and cosmologic effects can be tested in helium at very low temperatures, as we would do in a test tube! For this reason, the helium isotopes are named "model systems" and their properties are intensively investigated in low temperature laboratories.

In addition to thermodynamic studies, many efforts have been devoted by the neutron community to understand the dynamics of the strongly interacting quantum fluids. A milestone for understanding correlated bosons was the observation of the "phonon-roton" collective mode of liquid ^{4}He, predicted by Landau [1] to explain the system's thermodynamics. For fermions, the situation is more complicated, since the spectrum hosts two types of modes: collective ("zero-sound" in ^{3}He, or "plasmon" in charged systems), and incoherent particle-hole (PH) excitations. Both are described by Landau's theory of Fermi liquids [2, 3] in the small wave-vector/ low energy region of the spectrum. At higher wave-vectors/energies, the collective mode enters the PH band, where it is strongly damped [2, 3]. It was thus believed that the dynamics at high wave-vectors is essentially incoherent.

We have recently reported preliminary inelastic neutron scattering data on a monolayer of liquid ^{3}He at very low temperatures [4, 5]. The most striking features were the observation of a zero-sound branch very close to the PH band and the presence of a roton-like excitation at high wave-vectors. We shall discuss in this Conference the neutron data showing the remarkable behaviour of this two-dimensional Fermi fluid, as well as the experimental techniques. Moreover, a new theoretical framework is proposed [6, 7] where we introduce intermediate states that are not describable by the quantum numbers of a single (quasi-)particle, leading to an accurate interpretation of the spectra measured for ^{3}He films. By exploring the dynamics of Fermi many-body systems in the region outside the scope of Landau's theory, we open new perspectives in the understanding of highly correlated fermions.

2. Fermi liquids

Fermi liquids constitute one of the most active research areas in condensed matter, and the reader is familiar with the fundamental publications: Landau's original papers, Pines and Nozières [2], Wilks [8], Fetter and Walecka [9], Dobbs [10] and other books. Liquid ^{3}He has been investigated in detail, owing to the simplicity of its (spherical) Fermi surface, and the possibility to tune the interactions by simple changing the pressure from essentially zero (saturated vapour presure) to 3.4 Mpa. Thermodynamic results have completely confirmed Landau's phenomenological theory, but a microscopic theory is still lacking even at zero temperature. Several open questions are described in a remarkable article published by David Pines in Physics Today [11]. In the present article we are more particularly concerned by the dynamics of Fermi liquids. This information is contained in the dynamic structure factor, which has also been investigated in bulk liquid ^{3}He. This function of energy and wave-vector has been measured by neutron scattering; the main results are the direct observation of the particle-hole excitations as well as the zero-sound and the paramagnon collective mode, as predicted by Landau's theory of Fermi liquids.

2.1. Excitations in Fermi liquids

The fundamental excitation of a Fermi liquid at zero temperature is the creation of a particle-hole pair [2, 3]. The existence of the Fermi surface, and this is a fundamental topological effect, causes the excitations of the system to be confined to a region of the energy-momentum space, the particle-hole band (PHB). The boundaries of the PHB for a non-interacting system (Fermi gas) are given by the expression $E/E_F=(k/k_F)^2 \pm 2(k/k_F)$ where E_F is the Fermi energy $E_F=\hbar^2 k^2/2m$, k_F is the Fermi wave-vector and m the mass of the ^{3}He atom. The Fermi wave-vector is simply calculated knowing the density and the dimensionality of the system.

Landau theory of Fermi liquids postulates the existence of "quasi-particles" having an effective mass m^* which is a measure for the density of states at the Fermi level. In liquid ^{3}He, m^*/m, as seen by heat capacity measurements, is substantially larger than one, and increases with pressure [3, 8]. One could expect the PHB to shift to lower energies by a substantial amount, due to the increased effective mass. However, in the energy investigated here this does not happen: the Landau picture is only valid for very small energies and wave-vectors.

According to Landau's theory, two collective modes are present in a Fermi liquid [2, 3]: a collective oscillation of the Fermi sphere (zero-sound collective density wave), and a spin wave

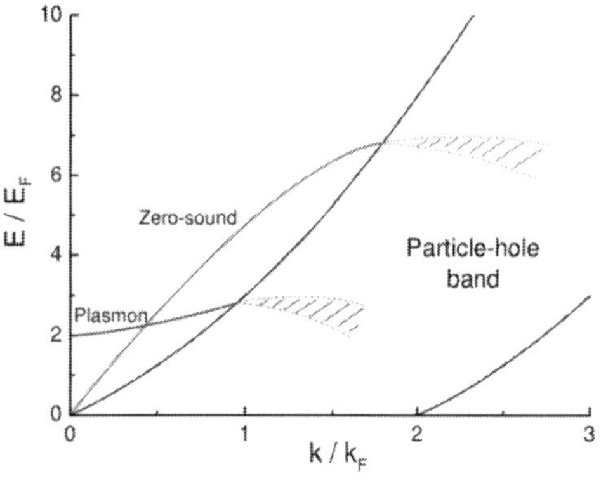

Figure 1. Schematic representation of the elementary excitations of a Fermi liquid. The spectrum displays an incoherent particle-hole band, and collective modes (zero-sound in a neutral Fermi liquid, like ^{3}He, or plasmons in a charged system, like the conduction electrons of a metal).

(paramagnon mode). Zero-sound has been detected first at low wave-vectors and very low temperatures by ultrasonic techniques. More recently, several neutron scattering investigations [3, 12, 13, 14] have been performed at wave-vectors on the order of k_F. At relatively low wave-vectors, zero-sound is observed as a well-defined mode with a linear dispersion relation, located above the PHB. It displays then a negative curvature, finally entering the PHB. There, the zero-sound branch becomes very broad, due to Landau damping: it decays into incoherent PH excitations [2, 3]. In electronic systems, the mode also exists, although its dispersion relation is modified due to the charged nature of the fermions. The collective electronic density mode is named "plasmon" [2]. A gap is present at small wave-vectors, but the high wave-vector behaviour is silmilar to that of liquid ^{3}He zero sound, the dispersion relation apparently finishes within the particle-hole band. We shall see that this is not necessarily true for two-dimensional systems.

3. Two-dimensional Fermi liquids

Two-dimensional liquid ^{3}He has been investigated using thermodynamic techniques by many research groups. Pionnering heat capacity investigations on liquid ^{3}He adsorbed on graphite were realised in Seattle [15], Grenoble (neutrons and NMR) [16, 17, 18], Bell Laboratories (heat capacity) [19], and London (heat capacity, NMR) [20, 21], among others. The results display all the characteristics expected from Fermi liquids: the heat capacity is linear in temperature, and the magnetization does not depend on temperature near T=0. Since there is no critical point in 2D liquid ^{3}He, one can investigate systems whose densities range essentially from zero (Fermi gas) to very high densities, until the system solidifies, forming a "commensurate solid" with respect to the graphite, or to the underlying adsorbed layer. The effect of the interactions appears, following Landau's theory, through an enhancement of the thermodynamic properties with respect to those of the Fermi gas. According to heat capacity and magnetization measurements, the effective mass increases as a function of density from $m^* = m$ to very large values, substantially larger than in bulk liquid ^{3}He.

Before our experiments, nothing was known experimentally about the dynamics of two-dimensional liquid ^{3}He. The measurements brought several surprises. The first one was the observation of a substantial change of the zero-sound mode, found to be at much lower energies that in bulk liquid ^{3}He. The main challenge, however, was the observation of substantial intensity in the inelastic structure factor at wave-vectors on the order of $2k_F$, indicating the emergence of a new excitation branch beyond the particle hole continuum. We present in the following sections the principles and methods of the experiment, we show typical results of the neutron scattering measurements, and we discuss finally the associated theoretical developments.

4. Experimental details

The *principle* of the experiment is simple. The excitations created in a liquid ^{3}He film are measured by inelastic neutron scattering. The liquid ^{3}He film of atomic thickness (submonolayer characterised by a carefully determined areal density) is adsorbed on an exfoliated graphite substrate of relatively large surface area. The substrate is enclosed in an aluminium experimental cell. The system is cooled by a dilution refrigerator to temperatures well below 100 mK, where k_BT is much lower than characteristic excitation energies. At such low temperatures, the pressure of the ^{3}He gas in equilibrium with the adsorbed ^{3}He film is essentially zero, and therefore all the ^{3}He present in the cell is adsorbed on the substrate.

The recorded neutron spectra correspond to neutrons reaching a set of detectors placed in a horizontal plane (scattering angle) at different times (time-of-flight technique to determine their energy). The neutron spectra reveal the creation of excitations in the ^{3}He film. This experiment requires a detailed data analysis, which involves a careful subtraction of the background signal due to the substrate, the experimental cell and the instrument. We describe in the following subsections the experimental set-up, the measuring techniques, as well as the details of the data analysis.

4.1. Neutron scattering on ^{3}He samples

The neutron scattering inelastic measurements were performed at the Institut Laue-Langevin (ILL - Grenoble) on the time-of-flight spectrometer IN6, situated on the cold neutron guide of the high flux reactor. IN6 is a time-focussing time-of-flight spectrometer designed for quasi-elastic and inelastic scattering for incident wave-lengths in the range of 0.4 to 0.6 nm. Experiments were performed at two wave-lengths: at 0.51 nm giving acess to the main part of the interesting energy-momentum range, and 0.41 nm in order to extend the range to higher energies and momenta. The energy resolution for a wave-length of 0.51 nm is on the order of 0.08 meV, increasing with the wave-vector up to 0.12 meV. At the smaller incident wave-length (0.41 nm), the resolution is about twice these values. A large number (337) of elliptical ^{3}He detectors cover the scattered neutrons angular range from 11.9 to 113 degrees. This corresponds, for elastic scattering, to wave-vectors in the range up to 20.5 nm^{-1} at an incident wave-length of 0.51 nm, and 24 nm^{-1} at an incident wave-length of 0.41 nm). The useful energy range is 0.1 to 2 meV (0.2 to 4 mev) repectively, for the two different wave-lengths.

The energy-momentum range of IN6 is particularly well adapted to the investigation of quantum fluids dynamics [22, 23, 24]. In particular, this instrument was used by Scherm and coworkers for bulk liquid ^{3}He studies [13, 14]. These experimental data have been very useful as a reference for the analysis of the results obtained on the two-dimensional samples.

One of the major difficulties encountered in measuments on ^{3}He is the very large absorption cross section of this isotope [3, 12, 14]. At the long wavelenghts needed for the inelastic measurements, the neutron penetration is on the order of a tenth of a millimetre in bulk liquid ^{3}He. For this reason, a high flux (the flux at the sample is on the order of 10^5 cm^{-2} sec^{-1}) and very long measuring times (about 20 hours) are necessary in order to collect a few neutron counts on each detector. The sample geometry and environment must therefore be carefully studied to reduce the background.

The experiments presented here on liquid ^{3}He films share with bulk liquid ^{3}He measurements the difficulties discussed above: high absorption and small signal. They present additional experimental complications, the most obvious one being the necessity of a solid substrate onto which the liquid ^{3}He monoatomic film is adsorbed. We use exfoliated graphite, as decribed below. The substrate introduces intense elastic scattering, in particular at Q=0, and at the graphite (0002) reflection; a phonon branch emerging from this Bragg peak also contaminates the signal. The graphite sample is oriented with the (0002) direction normal to the scattering plane, and therefore, the Bragg reflection should not occur in the scattering plane. However, the mosaic spread of the substrate leads to a Bragg peak in the scattering plane at 18.767nm^{-1}. In addition, instrumental background as well as neutrons incoherently scattered on impurities resulting from the graphite exfoliation procedure lead to quasielastic scattering. These spurious signals are not negligible, compared to the small number of counts originating from the ^{3}He inelastic scattering we want to measure. For this reason, the spectra of two-dimensional ^{3}He cannot be exploited at small energies.

Since the adsorbed layer consists of ^{3}He, the neutron absorption of this isotope reduces the intensity of the signal. The absorption measured at the graphite Bragg peak is of about 20% in our experiments. Subtracting the background is thus a delicate task: one cannot simply use the measurement with the bare graphite substrate as background. A further complication is the modification of the graphite peak position, height and width due to the adsorption of one or more helium layers. This effect, due to the modification of the intereference pattern of a finite number of planes (additional planes are introduced by the helium adsorption), is well known in neutron surface studies (see references in [16, 24]). It is taken into account in the analysis by adequate fits of the peak shape and position; the procedure, performed for each Q value, is delicate and time consuming. A consistency check is made by comparing the peak evolution for samples of different areal density.

4.2. The exfoliated graphite substrate

Exfoliated graphite is an excellent substrate for the investigation of two-dimensional adsorbed systems [16, 17, 24]. It presents many advantages: homogeneous adsorption due to the presence of relatively large atomically flat platelets corresponding to the (0002) graphite planes; it can be cleaned by simple procedures (baking at a few hundred degrees), and pumping on the system is sufficient to remove most of the contamination of a sample exposed to air. For this reason, exfoliated graphite samples are used in neutron scattering, NMR and heat capacity investigations of two-dimensional ^{3}He, where the ultrahigh vacuum techniques used in many other surface studies cannot be applied.

Exfoliated graphite consists of natural graphite flakes, exfoliated using an intercalation and a rapid heating procedure. The exfoliated material is then recompressed to form a self-sustaining material. Diverse levels of exfoliation are found in commercially available materials (Grafoil, Papyex...). They are characterised by their specific surface area (20 m^2/gram for Grafoil), by the size of the atomically flat regions (the coherence length is typically 20 nm), by the mosaic spread of the crystallites orientation (typically 30 degrees), and by their thermal properties (in particular the thermal conductivity at low temperatures) [16].

The choice of the substrate is dictated by several considerations. A large specific surface area is in principle convenient, since the signal is proportional to the amount of adsorbed gas. However, the strong neutron absorption of ^{3}He introduces a severe limitation. A Grafoil sample covered with ^{3}He would absorb all the beam within a thickness of a few millimetres. For this reason, we have chosen a ZYX-UCAR exfoliated graphite sample of smaller specific area (2 m^2/gram), but of mucher higher quality in terms of coherence lenth (190 nm), mosaic spread (only 10 degrees) and thermal conductivity (the large graphite crystals ensure a more efficient heat transport). This sample has therefore the advantage to allow forming relatively large two-dimensional liquid ^{3}He regions, relatively well oriented within the neutron scattering plane, with a moderate total neutron absorption using a sample of dimensions comparable to the neutron beam size (25 mm×50 mm) [16]. In our experiments, a stack of ZYX graphite sheets of dimensions 25×25×5 mm^3, with a total height of 60 mm, was used.

4.3. Experimental cell and cryogenic set-up

The ZYX sample is placed in a copper holder consisting of two plates, at the top and bottom, connected by a copper rod for thermal purposes. The ZYX sheets are thermally connected to the copper plates by means of copper foils. These are located at the back side of the stack with respect to the neutron detectors. Copper to graphite bonding is achieved by thermal diffusion, at a temperature on the order of 800 C for several hours, under secondary vacuum. This procedure also ensures the main cleaning (outgasing) of the graphite sample. Later on in the experimental procedure, pumping at room temperature will be sufficient to ensure that the full surface area of the sample is available for the sample absorption. The ZYX and its copper holder are placed in a thin walled aluminium cell. The top, made out of copper, is dismountable; an indium o-ring ensures that the cell is leak-tight. The copper top part is then bolted to the mixing chamber of a dilution refrigerator, providing a base temperature lower than 30 mK, determined by means of a calibrated carbon resistor thermometer. In the absence of the neutron beam, the sample temperature is essentially that of the mixing chamber. Opening the beam gives rise to a small temperature increase. This is not due to the absortion of neutrons by the ^{3}He film, since the energy released in the capture process is only 764 keV: heating originates in the absorption of gamma rays generated at the level of the instrument monochromator. Most of the power deposited in the experimental cell corresponds to absorption by the aluminium walls. The sample itself remains below 100 mK, which is sufficient for the purposes of this experiment. A vanadium sample is attached to the bottom of the cell. Measurements with this incoherent scatterer are used for the calibration of the relative efficiency of the 337 detectors. The dilution unit equipped with the sample cell are placed in the vacuum chamber of the instrument IN6. This allowed us to remove the room temperature vessel of the dilution unit, thus reducing the background. IN6 is also equipped of a helium filled box between the sample and the detectors, again reducing the background.

4.4. Adsorption isotherms system

In this experiment we handle small amounts of ^{3}He and ^{4}He gas, which must be accurately measured, both during the adsorption isotherm measurements and for the introduction of the helium sample. Typical quantities are on the order of 10 cm^3 STP of gas. For this purpose we use a stainless steel manyfold (typical diameter 6 mm) including small dead-volume valves, 2 calibrated stainless steel volumes (a volume of about 1000 cm^3 for the initial introduction of gas, and one of about 20 cm^3 for small increments of the amount of gas). The pressure in the system is measured with a precision Baratron MKS pressure gauge (100 Torr range). The temperature of the room is carefully recorded, in order to convert the amounts of gas measured in the calibrated volumes to amounts of gas in STP conditions.

4.5. Calibration of the substrate surface area

The ZYX sample is characterized by ^{3}He and ^{4}He adsoption isotherms at 4.2 K [16]. This procedure, well known in surface science, yieds the "surface area" of the sample. The resulting figure, however, has to be used with caution. It is very accurate in the sense that the measurement can be repeated, for a given sample, obtaining the same result. Also, for samples of the same type, the ratio of the measured areas can be determined quite precisely. This allows a direct scaling of experimental results obtained in different laboratories. Comparing the adsorption isotherms, we can determine the amount of gas corresponding, for our sample, to a given density.

The phase diagram of ^{3}He adsorbed on graphite has been determined by heat capacity and NMR techniques as a function of the ^{3}He coverage and temperature [15, 16, 17, 18, 19]. These measurements yield in particular the amount of gas needed to fill the $\sqrt{3}$ phase, the density of which is determined by the graphite in-plane lattice parameter as 6.366 atoms/nm^2. The "Commensurate surface area" A_c can therefore be determined from the ratio of the measured adsorption isotherm and that of another experiment where the commensurate phase coverage could be determined.

In addition, extensive neutron diffraction studies have been performed on the solid phases observed in the helium/graphite system: commensurate and incommensurate phases with respect to the graphite substrate have been observed [16, 17, 25]. The determination of the lattice parameter of these triangular phases gives immediately the specific area (nm^2/atom) for a given coverage. In previous experiments we have determined the amount of gas corresponding to the perfect filling of the commensurate phase using neutron diffraction on this sample to be 14.351 cm^3 of gas in standard conditions of temperature and pressure (STP: T=273.15 K and P=100 kPa). For comparison, a measurement performed using D_2, again on the same sample, gave the value 14.336 cm^3 STP for the $\sqrt{3}$ phase. One can then calculate the sample "commensurate area" A_c=59.73 m^2. This is the area that one should use for areal densities close to that of the $\sqrt{3}$ phase, at submonolayer coverages. At high coverages, however, adsorption of helium does not only take place at the atomically flat parts of the graphite platelets, but also at some places of weaker adsoption potential (defects, edges). The effective surface area near monolayer coverage is therefore somewhat larger than the "commensurate area" (which measures essentially the adsorption area of the homogeneous sites) [16, 17]. More specifically, we have determined by neutron diffraction on the same sample the commensurate area A_c, and also the effective area A_I which is available for adsorption at high coverages, on the dense incommensurate solid phase, in the vicinity of monolayer completion. The corresponding ratio of areas is 1.093, thus yielding an effective area for adsorption at high coverages A_I=65.3 m^2.

4.6. Preplating with ^{4}He

The results presented here correspond, as will be seen later, to a film of relatively high areal density: 4.7 atoms/nm^2. Such a density cannot be achieved at submonolayer coverages, since the fluids enters into a coexistence with the $\sqrt{3}$ phase at relatively low densities [16, 17]. In order to study a system where the interactions are strong, we have chosen to work with a fluid in the second layer, where the phase diagram offers a wider density range; although a commensurate phase also occupies a substantial part of the phase diagram, this effect only happens for densities of about 5.5 atoms/nm^2.

On the other hand, a dense ^{3}He solid first layer would have led to substantial neutron absorption. For this reason, we have used a ^{4}He complete monolayer. This isotope, due to its smaller zero point energy in the well formed by the adsoption potential and the interaction with neighbouring atoms, adsorbs preferentially with respect to ^{3}He, the lighter isotope. One can therefore engineer a bilayer system where the first layer consists of solid ^{4}He at high density, and the second layer of pure liquid ^{3}He. Preplating with ^{4}He also improves substantially the quality of the substrate, by smoothing out the imperfections of the bare graphite, in particular by filling the strong binding sites ("heterogeneities") which represent a few percent of the adsorbed first layer atoms. The high density first layer ^{4}He solid acts as a high quality, atomically smooth, substrate, onto which the ^{3}He layer can be deposited. The binding energy is much weaker than on the bare graphite: about 30 K compared to 140 K. The adsortion potential is nevertheless sufficiently strong, for experiments well below 1 Kelvin, to ensure that the motion of the adatoms is confined in the plane. The second layer ^{3}He fluid is therefore a practically ideal two-dimensional system, and confinement effects within the planes are very small, due to the large coherence length, increased with respect to the bare ZYX to values larger than 20 nm.

The amount of ^{4}He to be introduced to cover exactly the graphite by one atomic layer of ^{4}He can be determined using our neutron diffraction measurements on ^{4}He adsorbed on this

ZYX sample. Near monolayer completion the solid density is about 11.1 atoms/nm^2, and this number increases up to 11.63 atoms/nm^2 at very high coverages (maximum layer density) under the pressure of the additionally adsorbed atoms [16, 17]. The areal density of the complete ^{4}He first layer needed for the present experiment is found to be 11.3 atoms/nm^2, taking into account the compression due to the liquid ^{3}He second layer. The amount of ^{4}He gas introduced is 28.38 cm^3 of gas STP (standard temperature and pressure conditions, defined as T=273.15 K and P=100 kPa)

4.7. Two-dimensional ^{3}He sample

The amount of ^{3}He gas introduced in the cell is 11.03 cm^3 of gas STP. With the surface area of the substrate determined in the previous section, this corresponds to an areal density ρ_c=4.9 atoms/nm^2 in the commensurate coverage scale, and ρ_i=4.5 atoms/nm^2 in the incommensurate coverage scale. The first value is appropriate if the ^{3}He atoms of the liquid phase in the second layer explore only the homogeneous regions of the substrate, and the second value should be used if the fluid covers the whole area of the substrate, an effect favoured by the ^{4}He preplating. These values provide an upper and lower bound the actual density of the sample, which has therefore the value ρ=4.7±0.2 atoms/nm^2.

5. Measured neutron spectra

We show in this section results obtained on a two-dimensional liquid ^{3}He system of areal density ρ=4.7 atoms/nm^2 (see previous section). The Fermi wave-vector is obtained directly from the areal density; in two dimensions $k_F=\sqrt{2\pi\rho}$. The Fermi energy is $E_F=\hbar^2 k_F{}^2/2m$, and hence $E_F=\hbar^2\pi\rho/m$. The Fermi temperature is defined as $T_F=E_F/k_B$ (note than in 3D T_F is often defined including a 2/3 factor). With this notation, in the 2D system T_F [K]=0.5053 ρ [atoms/nm^2] and E_F [meV]=0.04355 ρ [atoms/nm^2]. For ρ=4.7±0.2 atoms/nm^2, k_F=5.43±0.12 nm^{-1}, E_F=0.204±0.009 meV

We present results for different scattering angles. Grouping of detectors provides a wave-vector resolution of 0.1 nm^{-1}. The subtraction of the large elastic signal of the cell (background) does not allow a precise determination of the spectrum for low energies, typically less than about 0.3 meV. We first discuss the "low wave-vector" region. We show in Figure 2 the spectrum obtained in our 2D system for a momentum transfer Q=5.5 nm^{-1}, comparable to the Fermi wave-vector. The most striking result is the absence of the collective mode observed in the bulk liquid at different pressures for energies of about 1 meV [3, 12, 13, 14]. Instead, we observe a mode whose energy is just above the PHB [4, 5]: the two-dimensional zero-sound collective mode, identified by the evolution of this signal in our data at different coverages (not shown here). Also, the mode corresponds well with theoretical results, which cover broad regions of the energy-momentum plane, including those where the experimental results are spoiled by the graphite background [26, 6, 7].

In Figure 3 we show the spectrum obtained for a momentum transfer Q= 11.5 nm^{-1}. It corresponds to the region dominated by the damping of excitations within the PHB [4, 5]. As expected, the signal is weak. However, one can clearly observe a broad mode, which corresponds to the damping of zero-sound within the PHB expected from theoretical calculations.

The behavior at higher wave-vectors is presented in Figure 4, where the momentum transfer is Q=16.5 nm^{-1}. The striking features are now the appearence of a large intensity peak (log scale!) at low energies, clearly *below* the PHB [4, 5]. Using data in the whole energy-momentum plane accessible with IN6 we have shown [5] that this intensity corresponds to the collective zero-sound mode reappearing beyond the PH continuum, at elevated wave-vectors.

The neutron spectra reflect the pronounced damping obtained within a rather well defined region of the energy-wave-vector plane, the particle-hole band. At this point, it is interesting

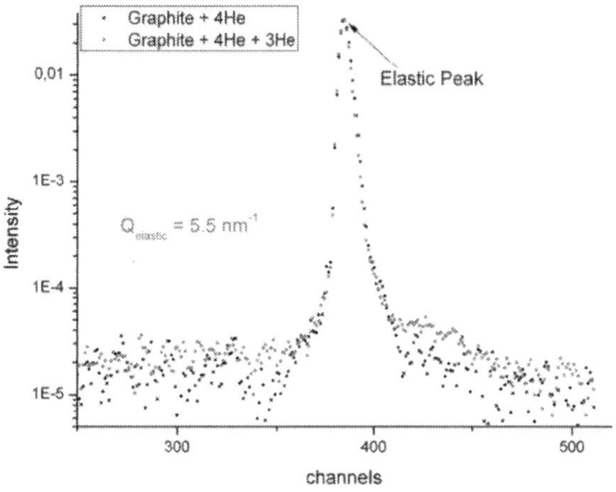

Figure 2. Time-of-flight inelastic neutron scattering spectrum measured at an angle of 25.5 degrees, corresponding to a wave-vector of 5.5 nm^{-1}. Note the logarithmic vertical (counts normalised to monitor) scale. A time channel corresponds to 9.625μsec. Squares are measurements of the background performed on the substrate preplated by a monolayer of ^{4}He. Circles correspond to the signal measured with the additional ^{3}He layer, after absorption corrections. A small signal is visible after the elastic peak, it corresponds to the creation of the zero-sound excitation at the upper edge of the particle-hole band.

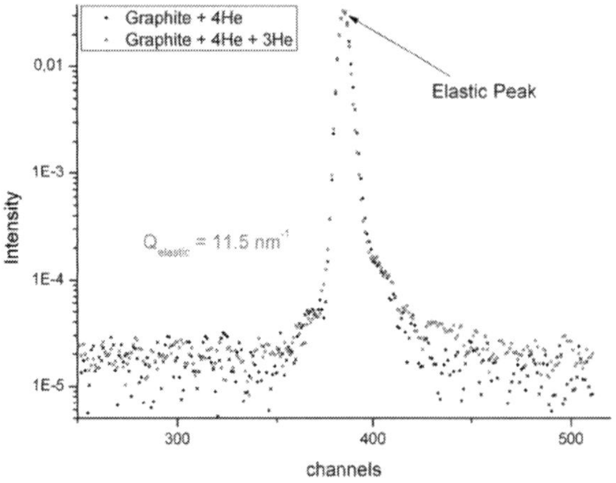

Figure 3. Time-of-flight inelastic neutron scattering spectrum measured at an angle of 55.3 degrees, corresponding to a wave-vector of 11.5 nm^{-1}. The weak signal after the elastic peak corresponds to the creation of highly damped zero-sound excitations inside the particle-hole band.

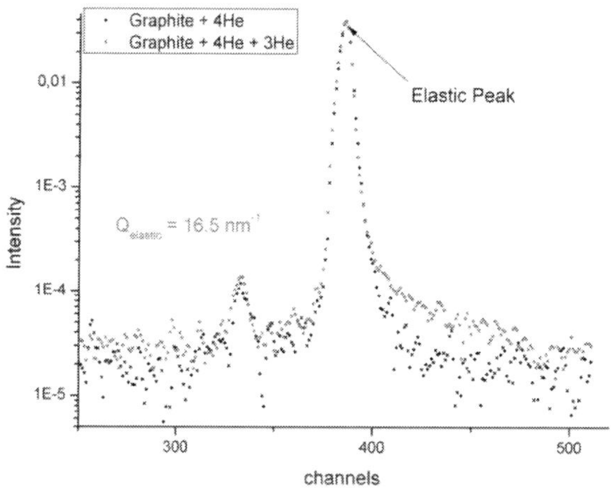

Figure 4. Time-of-flight inelastic neutron scattering spectrum measured at an angle of 84.1 degrees, corresponding to a wave-vector of 16.5 nm^{-1}. The rather intense signal observed after the elastic peak at very low excitation energies (short TOF times) corresponds to the creation of unexpected zero-sound excitations below the particle-hole band.

to reconsider the problem of the effective mass. The effective mass for a second-layer two-dimensional liquid ^{3}He film can be determined using the heat capacity data of Greywall [19] with the appropriate corrections for the coverage scale [18]. We obtain $m^* \approx 3.8$; for comparison, this is the effective mass observed in bulk liquid ^{3}He at a pressure of about 0.75 MPa. The PHB calculated using this value for the effective mass would be depressed to four time smaller energies, which is in contradiction with our data.

Including a large number of spectra like those displayed in the figures above, we obtained after a preliminary analysis [4], spectra which could be directly compared to the theory [4]. The complete set of data leads to an impressive agreement between the experiments and the dynamic many-body theory [5], presented during this Conference.

6. Conclusions

The inelastic neutron scattering, surface physics and low temperature techniques described here have allowed us to investigate directly the elementary excitations of a model 2D Fermi liquid, ^{3}He adsorbed on graphite. In spite of the technical difficulties of such an experiment, we were able to observe the zero-sound mode of a this interacting 2D Fermi Liquid in a large wave-vector range. At low wave-vectors the zero sound collective mode was found very close to the particle-hole band upper limit, substantially lower in energy than in the bulk, in agreement with theoretical calculations. The presence of the mode at high wave-vectors, damped in the region of coexistence with the incoherent particle-hole excitations, but well-defined for wave-vectors on the order of $2k_F$, is an important and surprising finding. The dynamic many-body theory developed by Krotscheck and coworkers [7] provides calculated spectra for our experimental conditions in excellent agreement with the experimental data. This effect is not specific to liquid ^{3}He, it should be observable in other two-dimensional non-localised many-body Fermi systems. This led us to propose a novel superconductivity mechanism mediated by high wave-vector density or spin-density fluctuations [5].

Acknowledgments
We acknowledge the financial support from the EU FRP7 low temperature infrastructure grant 228464 "Microkelvin", the Austrian-French grant FWF-ANR "High-Q Fermions" ANR-2010-INTB-403-01 and the CNRS-Air Liquide grant "Contrat de collaboration Recherche-Entreprise" 72B204/00.

References

[1] Landau L D 1947 *J. of Phys. Moscow* **11** 91
[2] Pines D and Nozières P 1966 *The Theory of Quantum Liquids* (New York : Benjamin)
[3] Glyde H R 1966 *Excitations in Liquid and Solid Helium* (Oxford : Clarendon Press)
[4] Godfrin H, Meschke M, Lauter H-J, Böhm H M, Krotscheck E and Panholzer M 2010 *J. of Low Temp. Phys.* **158** 147
[5] Godfrin H, Meschke M, Lauter H-J, Sultan A, Böhm H M, Krotscheck E and Panholzer M *to be published*
[6] Böhm H M, Krotscheck E, Panholzer M, Godfrin H, Lauter H-J and Meschke M 2010 *J. of Low Temp. Phys.* **158** 194
[7] Böhm H M , Holler R, Krotscheck E, Panholzer M 2010 *Phys. Rev. B* **82** 224505
[8] Wilks J 1967 *Liquid and Solid Helium* (Oxford : Clarendon Press)
[9] Fetter A L and Walecka J D 1971 *Quantum Theory of Many Particle Systems* (New York : McGraw-Hill)
[10] Dobbs E R 2001 *Helium Three* Oxford University Press
[11] Pines D 1981 *Physics Today* **34** 106
[12] Sköld K, Pelizzari C A, Kleb R and Ostrowski GE 1976 *Phys. Rev. Lett.* **37** 842
[13] Scherm R, Glucksberger K, Fak B, Sköld K, Dianoux A J, Godfrin H and Stirling W G 1987 *Phys. Rev. Lett.* **59** 217
[14] Glyde H R, Fak B, Dijk N H, Godfrin H, Glucksberger K and Scherm R 2000 *Phys. Rev. B* **61** 1421
[15] van Sciver S W and Vilches O E 1978 *Phys. Rev. B* **18** 285
[16] Godfrin H and Lauter H-J 1995 *Progress in Low Temp. Physics* Vol.XIV, Chapter 4, p.213-320, ed. W.P. Halperin, (Amsterdam : Elsevier Science B.V.)
[17] Godfrin H and Rapp RE 1995 *Advances in Physics* **44** 113
[18] Morhard K D, Bäuerle C, Bossy J, Bunkov Y M, Fisher S N and Godfrin H 1996 *Phys. Rev. B* **53** 2658
[19] Greywall D S 1990 *Phys. Rev. B* **41** 1842
[20] Casey A, Patel H, Nyéki J, Cowan B P and Saunders J 2003 *Phys. Rev. Lett.* **90** 115301
[21] Neumann M, Nyéki J, Cowan B P and Saunders J 2007 *Science* **317** 1356
[22] Stunault A, Andersen K H, Blanc Y, Fak B, Godfrin H, Guckelsberger K and Scherm R 1992 *Physica B* **180-181** 926-928
[23] Andersen K H, Stirling W G, Scherm R, Stunault A, Fak B, Godfrin H and Dianoux A J 1992 *Physica B* **180-181** 851-853
[24] Lauter H-J, Godfrin H, Frank V L P and Leiderer P 1992 *Phys. Rev. Lett.* **68** 2484
[25] Lauter H J, Godfrin H, Frank V L P and Schildberg H P 1990 *Physica B* **165-166** 597-598
[26] Hernandez E S and Calbi M M 2000 *J. Low Temp. Phys.* **120** 1

What is the structure of liquid Bismuth?

El'ad N Caspi[1], Yaron Greenberg[1], Eyal Yahel[1], Brigitte Beuneu[2], Guy Makov[3]

[1]Physics Department, Nuclear Research Centre-Negev, 84190 Beer-Sheva, Israel

[2]Laboratoire Léon Brillouin (CEA-CNRS), CEA/Saclay – 91191 Gif-sur-Yvette cedex, France, EU

[3]Materials Engineering Department, Ben-Gurion University of the Negev, 84105 Beer-Sheva, Israel

caspie@nrcn.org.il

Abstract. The structure of liquid Bismuth is probably the best-studied among elemental liquid metals because of a combination of the interesting physical properties of Bismuth and its excellent neutron scattering properties. Over the last six decades there have been more than 10 independent studies of the structure of liquid Bi, near the melting temperature. This remarkable number of measurements provides an opportunity to compare these results and to analyze the different sources of error contributing to the calculated pair distribution function. In the present contribution we analyze possible sources of error by varying the analysis procedure for a given measurement. By repeating a previous measurement in a new experimental configuration we demonstrate that an invariant (i.e. up to an absolute error) structure factor can be obtained. Transforming the structure factor into the radial distribution introduces new sources of error which causes the scatter to be greater than that required to resolve issues such as the existence of liquid-liquid phase transitions in Bi and to obtain correlation between thermo-physical properties and structure. We consider the contribution of different parameters when transforming the structure factor to the radial distribution function.

1. Introduction

In the last decade the structure of liquids has received renewed attention in the context of liquid-liquid phase transitions (LLPT) and anomalous thermo-physical properties [cf. 1-5]. These phenomena are probably associated with small changes in the three dimensional liquid structure. The most common method to describe the structure of the liquid is through the radial distribution function (RDF), which is a limited one dimensional representation of the full three dimensional structure. Identification of small changes in the three-dimensional liquid structure by their effect on the RDF requires its knowledge to high accuracies.

This challenge is even greater when we consider that the RDF cannot be measured directly in real space. Instead, diffraction patterns from the liquid are collected and the liquid structure factor is then obtained by a series of approximations to account for incoherent scattering, multiple scattering, inelastic scattering (in the case of neutron diffraction), Compton scattering (in the case of X-ray diffraction), absorption, and scattering by the experimental setup. All these issues have been known for a long time and a standard set of procedures has been developed to analyze the experimental data and extract the structure factor. Following the application of these correction procedures, the accumulating systematical error in the case of neutron scattering from heavy elements is of the order of 4-5 % [6-9].

To determine the RDF from the structure factor an additional error is introduced by the use of finite range in the momentum transfer, q, space. This finite cutoff introduces errors into the RDF, most notably short wavelength oscillations, which may be considered as numerical noise. These oscillations may affect the interpretation of the RDF, and in particular prevent identification of small changes therein. Several attempts have been made in the literature to propose procedures which reduce this source of error in the RDF, but it has not been possible to eliminate it [10,11].

Bismuth is an ideal element for neutron scattering measurements due to its high coherent scattering cross-section (9.1477 barns), its negligible incoherent scattering cross-section (0.0084 barns), and a very small absorption cross-section (0.0338 barns (at λ=1.8 Å). Adding these unique properties to the bismuth relatively low melting temperature (~550 K) made this element the focus of a relatively large number of neutron scattering studies in its liquid phase [cf. 5,9,12-17]. Moreover, the bismuth's rather complex pressure-temperature phases diagram both in the liquid, and in the solid states [1] enhances researchers interests in its physical properties in elevated temperature and pressure.

In the present work, we report on two separate studies of the structure of liquid Bi at ambient pressure that we have performed at the 7C2 diffractometer at Saclay, using different experimental setups, separated in time by 5 years, and show that the structure factors obtained agree to better than 1%, thus establishing the consistency of our analysis procedures. Using this structure factor as a reference, we compare a large number of experimental studies of liquid Bi at ambient pressure in order to investigate the experimental scatter in the structure factor and find out its significance. We then explore how the errors in the experimental determination of the structure factor affect the calculated RDF, focusing on two main issues: sensitivity to convergence in the high q range and choice of q_{max}.

2. Experimental setup

We have measured the structure factor of liquid bismuth near the melting point on two different occasions, separated over 5 years, at the 7C2 diffractometer at Saclay. In both experiments the neutron wavelength was ~0.7 Å, and the detector setup was similar. However, the experimental setup in the vicinity of the sample was significantly different between the two experiments.

In the first experiment, dated to 2006 [5], the sample was placed inside an 8 mm inner-diameter quartz tube, with walls thickness of 1 mm. The sample-holder assembly was placed inside a cylindrical double walled vanadium furnace, the inner diameter of each wall being 30 mm, and 90 mm. The dimensions of the neutron collimator close to the sample (in the upstream direction) were 10 cm × 5 cm × 1.2 cm (L×H×W).

In the second experiment, dated to 2011, the sample was placed inside a 10 mm inner-diameter quartz tube, with walls thickness of 1 mm. The sample-holder assembly was placed inside a cylindrical single walled vanadium furnace, the inner diameter the wall being 30 mm. The dimensions of the neutron collimator close to the sample (in the upstream direction) were 10 cm × 5 cm × 1.6 cm (L×H×W).

It is worth noting that between the years 2006, and 2011 the 7C2 diffractometer and its monochromator were twice removed for maintenance, and then were reassembled and recalibrated.

Even though the above mentioned differences are significant in an experiment aiming at the absolute measurement of the structure factor, the form of the corrections and the analysis procedures applied in both measurements were identical with the exception of the different experimental parameters used. Thus, in both experiments we performed conventional data reductions for background, cell, and furnace contributions, incoherent, multiple, and inelastic scattering [6-8].

3. Results

3.1. Structure factor – S(q)

The validity of the procedure described above can be determined by requiring that different experimental measurements of I(q) produce the same structure factor. This comparison is difficult to achieve, since the corrections to the measured signal applied by different researchers are not identical, nor are the details of the analysis procedures.

In this work, we have the opportunity to compare our two independent measurements separated over 5 years using different setups, on the same instrument. This comparison can be undertaken to a high accuracy, while removing extraneous, non-physical sources of experimental scatter. Before comparing the two measurements we first determine their statistical convergence.

In Figure 1 we present the structure factor as determined in a series of measurements taken over different collection times. We see, trivially, that the experimental signal to noise ratio decreases as the collection time is increased. More significantly, we see that the signal to noise ratio decreases as the transferred momentum, q, increases, reflecting the strong momentum dependence of the atomic scattering function. Specifically, in the high q region the signal has very weak q dependence so we can estimate the signal to noise ratio quantitatively. As an example, the signal in our S(q) data, collected over 6 hours (Figure 1) at $q=12.4$ Å^{-1}, equals $(S(q)-1)\approx 0.01$. The statistical noise in that region can be estimated from the scatter of the data points, and is roughly 0.003 for one standard deviation (σ). As we will see later this high q signal to noise ratio could be used as an indicator for the statistical quality of the measured S(q).

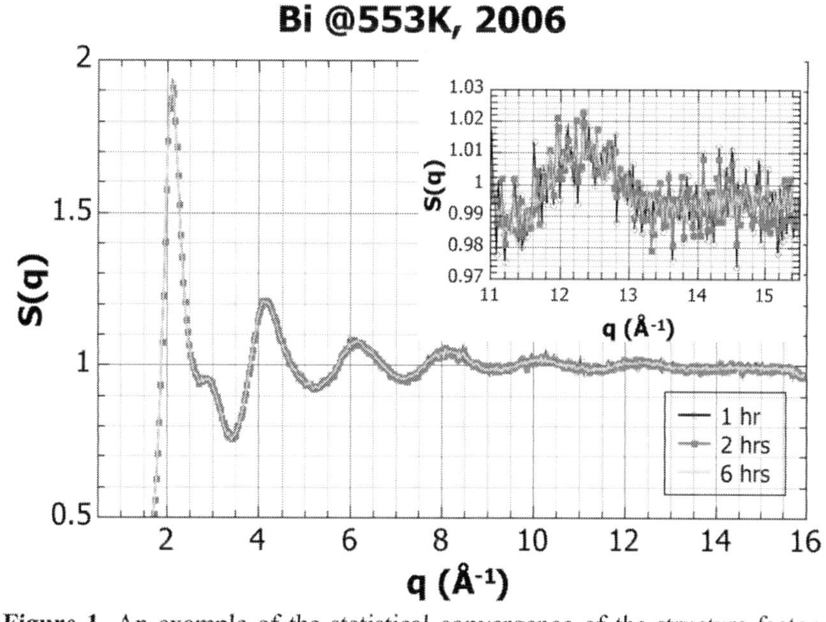

Figure 1. An example of the statistical convergence of the structure factor, S(q), of liquid bismuth measured on the 7C2 neutron diffractometer at 2006. Results from 1 hour, 2 hours, and 6 hours measurement of data collection of the same experiment are depicted.

Having established the statistical convergence, we now proceed to compare the two experimental determinations (from 2006, and 2011) of the liquid bismuth structure factor S(q) presented in Figure 2. We see that the results are in excellent agreement to better than 1% i.e the differences are of the order of the statistical noise. Thus, we have an indication that the structure factor obtained by our data analysis procedures in these two different experiments is indeed the physical structure factor. Nevertheless, since the analysis methods were the same in both cases a systematic error due to uncertainties in the overall 7C2 diffractometer setup, the knowledge of the exact dimensions of sample holder, furnace, and sample chamber within the beam path used for the absorption corrections [6], and the approximation made in the inelastic scattering corrections [8], cannot be ruled out.

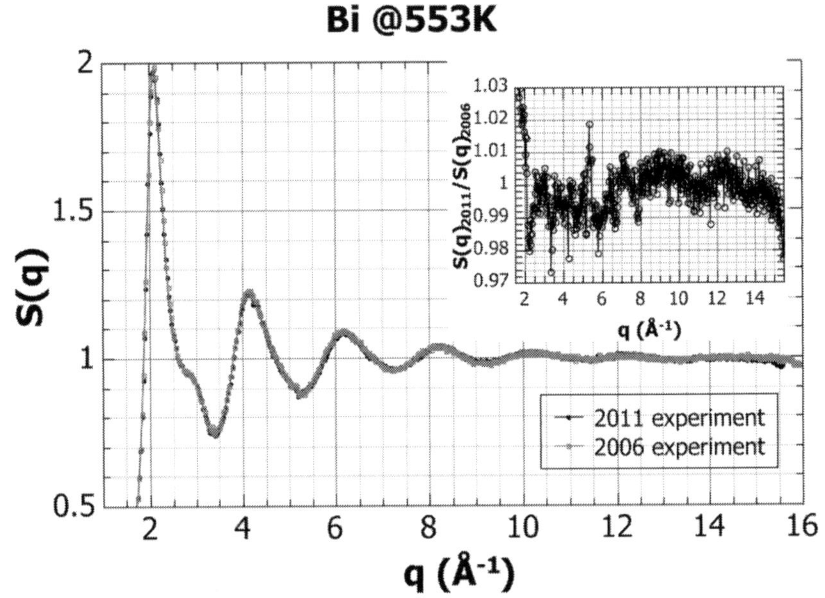

Figure 2. Comparison of two measurements of liquid bismuth structure factor, S(q), undertaken on the 7C2 neutron diffractometer at Saclay. The ratio of the S(q) signals is presented in the inset and shows that the deviation between the two measurements is typically less than 1%.

Having established that we have an estimate of the true physical structure factor, we compare our results with previous studies over the last 60 years, shown in Figure 3a.

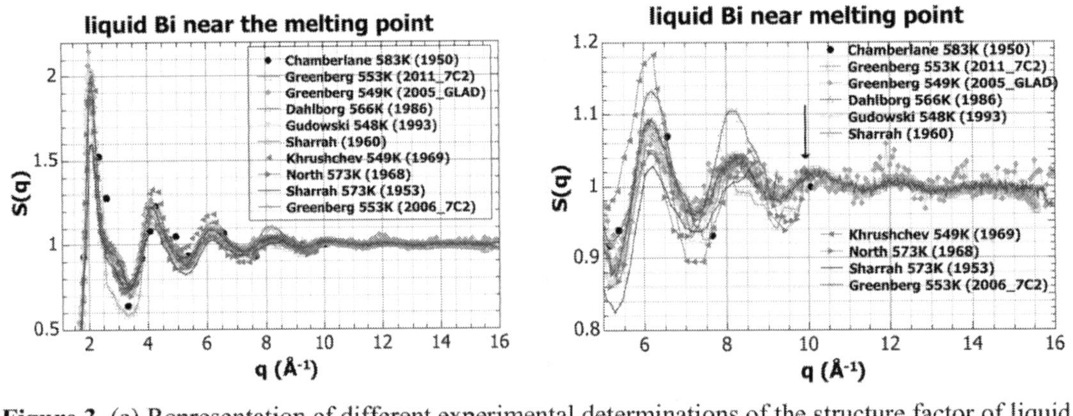

Figure 3. (a) Representation of different experimental determinations of the structure factor of liquid bismuth near the melting point over the last 60 years. (b) The high q range of the structure factor. The arrow indicates the maximal momentum transfer, q, measured in all but our measurements.

We see that the positions of the maxima and minima are essentially the same over 60 years of measurements, as expected from the relatively exact determination of the momentum transfer in diffraction measurements. In contrast, the amplitude of S(q) varies considerably between the measurements across the years. The magnitude of the variation is approximately 30% in (S(q)-1) even for the first peak in S(q). Moreover, relative scatter between measurements increases with increasing q. Finally, all measurements reduce to noise with the typical q_{max} of 10 $Å^{-1}$, with the exception of our results at 7C2 which extend to approximately 14 $Å^{-1}$ keeping a reasonable signal to noise ratio (see Figure 3b).

3.2. Radial distribution function – G(r)

The RDF, G(r), is the manifestation in direct space of the structure factor and as such is easier to interpret as representing a one dimensional average liquid structure. However, due to the experimental constraints, S(q) can be determined up to a certain q_{max} value. This may prompt the question of how to obtain the "best" representation of G(r) from the limited data. A naïve approach might be to use the maximum amount of q range available in the data, and, trivially, to aim for the longest measurement time.

The effect of measurement statistical noise in determining the structure factor on the RDF is negligible as can be seen from Figure 4 where we compare the RDFs obtained from structure factors measured by us with twice and six times the number of neutron counts. A small effect of the increased measurement time is a reduction in the amplitude of the oscillations in G(r), especially pronounced in the shoulder region (4-5Å).

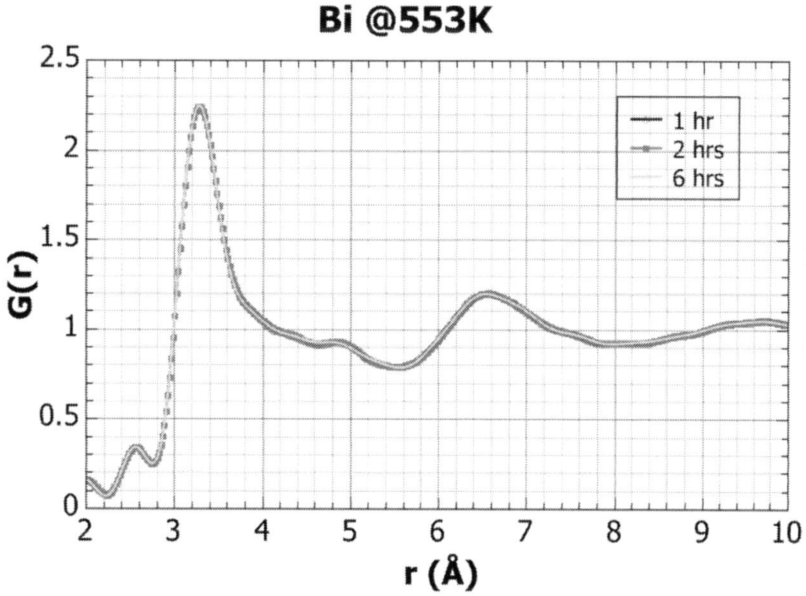

Figure 4. Calculation of the RDF, G(r), using 1, 2, and 6 hour S(q) measuredt in the 2006 neutron diffraction experiment on the 7C2 diffractometer at Saclay.

In Figure 5, we see that increasing the cut-off wavevector, q_{max}, used to obtain the RDF, does not lead to a monotonous decrease of the amplitude of the data oscillations. Instead, we find that the oscillations in the RDF associated with the finite q-space measurements may increase with increasing q_{max} (Figure 5). We note that the main impact of the oscillations is in the region between the first and second hard sphere peaks in the RDF. This is to be expected as the high q range in the structure factor affects mainly the finer details of the liquid structure, such as the existence of shoulder, or hump between the 1st and 2nd hard sphere peaks in G(r) in anomalous liquid metals [5,20].

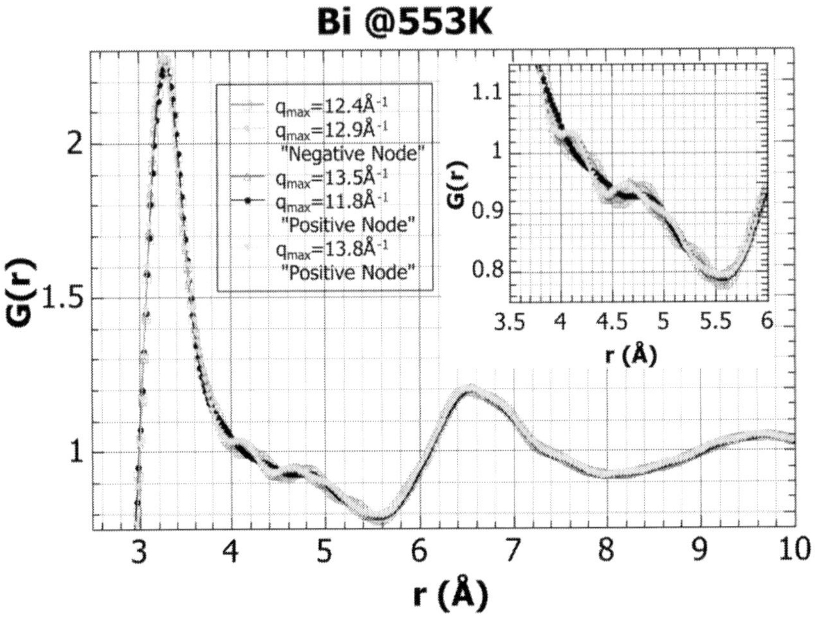

Figure 5. Liquid bismuth RDF as determined from the S(q) measured at 2006 on the 7C2 diffractometer at Saclay. Different curves represent different q_{max} cut-offs in the S(q) data. A positive node is when $S(q_{max})=1$, and $dS(q)/dq>0$ (at $q=q_{max}$). A negative mode is when $S(q_{max})=1$, and $dS(q)/dq<0$ (at $q=q_{max}$).

Further progress may be obtained if we consider the effect on the calculation of G(r) of using different cut-off values [q_{max} at the S(q) measured data] at selected positions from a positive to negative node in S(q) (cf. Figure 3). We see that choosing the cut-off at a node, i.e. at $S(q_{max})=1$, leads to a reduction in the oscillations (Figure 5) with better results at a positive node. However, choosing the cut-off at the highest available positive node in the data requires some caution if the chosen q_{max} is in the region where the statistical noise is too high. In our case, choosing the cut-off at the positive node q_{max}=13.8 Å^{-1} (Figure 1) leads to high numerical oscillations (Figure 5).

The sensitivity of G(r) to q_{max} is examined by considering alternative choices determined by the positive nodes of S(q) (Figure 6). We see that as expected from the analysis above, as q_{max} is increased the results for G(r) converge and the oscillations are reduced. From these results we see that measurements limited to 10 Å^{-1} still introduce a spurious contribution to the RDF, which is significantly reduced, but not eliminated at 11.8 Å^{-1} (Figure 6).

4. Discussion

We have shown that the structure factor can be determined up to an accuracy determined by the statistical noise and should not be limited by inconsistencies in the data analysis procedures. This does not mean that this is the exact physical structure factor as there can be systematic methodological errors which may only be uncovered by detailed comparison with determinations of the structure factor by other independent means, such as X-ray diffraction. Such a comparison is beyond the scope of the present work.

The RDF is well-known to be sensitive to the S(q) data cut-off limit, i.e. q_{max} up to which the S(q) data is chosen for the calculation of G(r). We see that extending the measurement range in q space, even significantly, does not ensure convergence of the RDF. We also note that statistical convergence does not in itself ensures the convergence of the RDF, which is also affected by the choice of q_{max}.

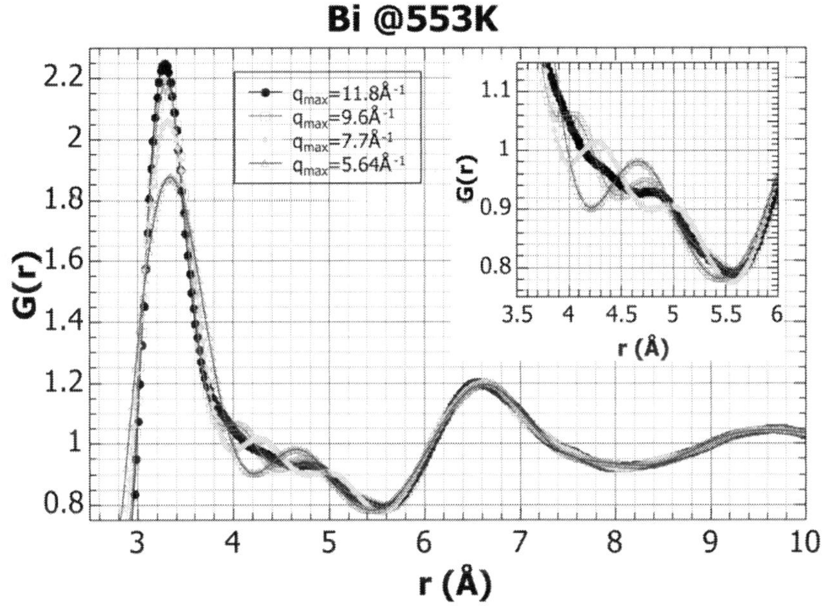

Figure 6. Liquid bismuth RDF as determined from the S(q) measured at 2006 on the 7C2 diffractometer at Saclay. Different curves represent different q_{max} cut-offs in the S(q) data. In this case all cut-offs were chosen at a positive node, i.e. $S(q_{max})=1$, and $dS(q)/dq>0$ (at $q=q_{max}$).

A more detailed analysis of the choice of q_{max} shows that if it is located at a node defined by $S(q_{max})=1$, the spurious wavelengths introduced by the Fourier transform are reduced. We start from the well known definition of the exact RDF [cf. 16],

$$G(r)=1+\frac{1}{(2\pi^2 r\rho)}\int_0^\infty dqq(S(q)-1)\sin(qr) \tag{1}$$

where ρ is the atomic density. We now define $G(r,q_{max})$, as the experimental RDF obtained from Fourier transforming the measured S(q) data up to q_{max}. The error in $G(r,q_{max})$, ε, is defined by $\varepsilon(r,q_{max})=G(r,q_{max})-G(r)$, and it is a function describing the oscillations about the exact G(r). These oscillations vary in wavelength and amplitude with q_{max}. As q_{max} is increased the oscillations vary. This variation may be estimated by considering the locations of the extrema of the oscillations obtained from (after leaving out the constants multiplying the integral),

$$\frac{\partial\varepsilon}{\partial r}=\frac{\partial G(r)}{\partial r}-\frac{\partial G(r,q_{max})}{\partial r}=\int_{q_{max}}^{\infty} dqq(S(q)-1)\frac{qr\cos qr-\sin qr}{(qr)^2}=0 \tag{2}$$

to be insensitive to choice of q_{max}, i.e. requiring that,

$$\frac{\partial}{\partial q_{\max}}\left(\frac{\partial \varepsilon}{\partial r}\right)=0=q_{\max}^{3}\left(S\left(q_{\max}\right)-1\right)\frac{q_{\max} r \cos q_{\max} r-\sin q_{\max} r}{\left(q_{\max} r\right)^{2}} \tag{3}$$

This requirement is met if we choose q_{max} at a node, i.e. at $S(q_{max})=1$. If we introduce the further requirement that the magnitude of variations in $\partial\varepsilon/\partial r$ is minimal, i.e.,

$$\frac{\partial^{2}}{\partial q_{\max}{}^{2}}\left(\frac{\partial \varepsilon}{\partial r}\right)=\frac{\partial}{\partial q_{\max}}\left(q_{\max}^{3}\left(S\left(q_{\max}\right)-1\right)\frac{q_{\max} r \cos q_{\max} r-\sin q_{\max} r}{\left(q_{\max} r\right)^{2}}\right)>0 \tag{4}$$

this condition is met at a node, if $S'(q_{max})>0$, i.e. the node is positive. Thus, we come to the conclusion that the optimal choice of q_{max} is at the largest value of a positive node of S(q) at which the signal to noise ratio is acceptable.

This conclusion also suggests a criterion for determining the signal to noise required at the experiment. The noise should be considerably less than the signal abs(S(q)-1) determined at the minimum before q_{max}, e.g., in our case if $q_{max}=11.8$ Å^{-1} then the noise should be considerably less than the signal of ~0.01. Returning to our experimental data as shown in Figure 1 we see that in our 6 hours run the scatter of data points in the vicinity of $q=11.8$ Å^{-1} ±0.003 Å^{-1} for one standard deviation, three times less than the value of the signal.

5. Conclusions

Due to its excellent neutron scattering properties, its rather low melting temperature, and its complex solid and liquid phase diagrams, the structure factor, S(q), of liquid bismuth was studied extensively by means of neutron scattering over the last 6 decades. No less than 10 different measurements were undertaken at ambient pressure and near the melting point temperature, with additional studies conducted over the years as a function of pressure, temperature, and alloying [cf. 18,19].

Therefore, liquid bismuth presents a good study ground for the re-examination of error propagation and the main uncertainties (statistical, and systematical) in the process of obtaining S(q), and from it the RDF by neutron scattering. In this context, we focus here on the comparison of the determination of liquid bismuth S(q) over the years, and on the numerical error propagating into the calculation of the RDF from it.

We suggest the requirements on the accuracy needed in measuring S(q) for adequate G(r) calculations: choosing the highest value of "positive-node" q_{max} in the S(q) data, while keeping significant (factor of 3) signal to noise ratio in the dip (peak) just before (after) this q_{max} value. Moreover, we demonstrate here how the measurement of S(q) up to a q_{max} value, which is too low, hinders the ability to study the fine details in the calculated RDF, even when the statistics of the measured S(q) is adequate. We would like to stress, that it is only through careful study of liquid structure factors up to large q_{max} that it will be possible to identify the small structural changes occurring in liquids.

References

[1] Umnov A G, Brazhkin V V, Popova S V and Voloshin R N, 1992 *J. Phys. Cond. Mater.* **4** 1427
[2] Zu F -Q, Zhu Z -G, Guo L -J, Qin X -B, Yang H and Shan W -J 2002 *Phys. Rev. Lett.* **89** 125505
[3] Katayama Y, Mizutani T, Utsumi W, Shimomunra O, Yamakata M and Funakoshi K -I 2000 *Nature* **403** 170
[4] Greaves G N, Wilding M C, Fearn S, Langstaff D, Kargl F, Cox S, Vu Van Q, Majérus O, Benmore C J, Weber R, Martin C M and Hennet L 2008 *Science* **322** 566
[5] Greenberg Y, Yahel E, Caspi E N, Benmore C J, Beuneu B, Dariel M P and Makov G 2009 *Europhys. Lett.* **86** 36004

[6] Paalman H H and Pings C J 1962 *J. Appl. Phys.* **33** 2635
[7] Blech I A and Averbach B L 1965 *Phys. Rev.* **137** A1113
[8] Placzek G. 1952 *Phys. Rev.* **86** 377
[9] North D M, Enderby J E and Egelstaff P A 1968 *J. Phys. C* **1** 1075
[10] Waser J and Schomaker V 1953 *Rev. Mod. Phys.* **25** 671.
[11] Kaplow R, Strong S L and Averbach B L 1965 *Phys. Rev.* **138** A1336.
[12] Chamberlain O. 1950 *Phys. Rev.* **77** 305
[13] Sharrah P C and Smith G P 1953 *J. Chem. Phys.* **21** 228
[14] Sharrah P C Petz J I and Kruh R F 1960 *J. Chem. Phys.* **32** 241
[15] Khrushchev B I and Bogomolov A M 1969 *Fiz. Metal. Metalloved.* **27** 1011
[16] Dahlborg U and Davidovic M 1986 *Phys. Chem. Liquids* **15** 243
[17] Gudowski W, Mellegård A, Dzugutov M, Howells W S and Zetterström P 1993 *J. Non-Cryst. Solid.* **156-158** 130
[18] Yaoita K, Tsuji K, Katayama Y, Imai M and Chen J –Q 1992 *J. Non-Cryst. Solid.* **150** 25
[19] Isherwood S P and Orton B R 1969 *J. Appl. Cryst.* **2** 219
[20] Greenberg Y, Yahel E, Caspi E N, Beuneu B, Dariel M P and Makov G 2010 *J. Chem. Phys.* **133** 094506

Quasi-elastic neutron scattering of dense molecular liquid selenium bromide

H Shimakura[1,2], Y Kawakita[2], Y Inamura[2], T Kikuchi[2], H Ueno[1,2], M Nakamura[2], K Nakajima[2], S Ohira-Kawamura[2], K Shibata[3], M Arai[2], S Takeda[4]

[1]Graduate School of Sciences, Kyushu University, 6-10-1 Hakozaki, Higashi-ku, Fukuoka 812-8581, Japan
[2] J-PARC center, Japan Atomic Energy Agency, 2-4 Shirakata Shirane, Tokai, Naka, Ibaraki 319-1195, Japan
[3] Tokai Center, Comprehensive Research Organization for Science and Society (CROSS), 162 Shirakata, Tokai, Naka, Ibaraki 319-1106, Japan
[4] Faculty of Sciences, Kyushu University, 6-10-1 Hakozaki, Higashi-ku, Fukuoka 812-8581, Japan

shimakura.hironori@jaea.go.jp

Abstract. Quasi-elastic neutron scattering spectra of liquid Se_2Br_2 at 298 K, 373 K and 473 K were measured by utilizing the cold-neutron disk-chopper spectrometer AMATERAS in J-PARC. The obtained dynamic structure factors are reproduced well by a two-Lorentzian model consisting of slow and fast modes. With increasing temperature, the maxima in the Q-dependence of the magnitude for both modes shift to a low Q direction, while the temperature dependence of the magnitude is opposite between slow and fast modes. The temperature dependence of the magnitude and the analysis of the energy width for these modes reveal that the fast mode is originated from dynamics in strongly interacted neighbouring molecules, while the slow mode is derived from inter-molecular dynamics without the strong interaction.

1. Introduction

Chalcogen-halogen equi-atomic liquids consist of chain molecule with X-A-A-X (A: S, Se, X: Cl, Br) form, where the bond angle and the dihedral angle between X-A-A and A-A-X planes are about 100° and 85°, respectively [1,2]. For Se_2Cl_2 and Se_2Br_2, the static structure factors obtained from neutron diffraction measurements were analyzed by Misawa's method [3] where total structure factor, $S(Q)$, as a function of momentum transfer Q were dissolved into intra-molecular correlation with molecular form factor, inter-molecular correlation by hard-sphere (HS) packing of the molecules and orientational correlation among neighbouring molecules. It is found in this paper that $S(Q)$ can be reproduced by intra-molecular correlation in the Q-region beyond 5 Å^{-1}, while $S(Q)$ in relatively low Q region is strongly affected by inter-molecular correlations. The HS packing of molecules creates a relatively sharp peak in $S(Q)$ at a specific low Q corresponding to an averaged number density of molecules. The orientational correlation introduced among neighbouring molecules is dominant in a middle Q range beyond the sharp peak by HS packing and below a prominent peak of the intramolecular correlation (molecular form factor) at around 4 $Å^{-1}$, so that the adjacent molecules lie with the Se-Se bond parallel with each other ([1] and Fig.5 in [4]).

Recently ab initio molecular-dynamics (MD) simulations of liquid Se_2Cl_2 [5] suggested that the interaction energy between Se_2Cl_2 molecules has local minimum at around 3.2 Å, when the adjacent molecules arrange with the Se-Se bond parallel to each other and that an attractive interaction of this molecular arrangement is derived from p-p hybridization between the occupied lone-pair orbital of Se atom and the unoccupied anti-bonding σ* orbital of Cl-Se bond on the neighbouring molecule and also p-d hybridization.

Quasi-elastic neutron scattering (QENS) measurements of liquid A_2X_2 performed at room temperature by Yao et al. [4] revealed that there are two kinds of dynamical modes with fast and slow relaxation times and that the former depends on the kind of halogen species while the latter depends on the molecular mass. However, detailed Q-dependence of QENS was not discussed in this study because of shortage of scattering angles where good statistics was achieved. The temperature dependence of QENS of liquid Se_2Br_2, was studied [6], suggesting that there are two terms of non-correlating inter-molecular contribution and the orientational correlation between strongly coupled neighbouring molecules. The precise profile analysis including fitting with model functions was not performed because of limited instrumental energy resolution.

To obtain dynamic structure factors, $S(Q,E)$, over a wide Q-E ranges with sufficient energy resolution at various temperatures and to discuss inter-molecular dynamics associating with static structure information, new QENS measurements of liquid Se_2Br_2 has been performed.

2. Experimental procedure

The QENS experiments of liquid Se_2Br_2 were performed with time-of-flight technique using the cold-neutron disk-chopper spectrometer AMATERAS installed in Materials and Life Science Experimental Facility (MLF), Japan Proton Accelerator Research Complex (J-PARC), Tokai, Japan. AMATERAS covers the scattering angles of 4.7° ~ 111.7° in horizontal and -16° ~ 22° in vertical [7]. To access dynamics in different time scales easily, multiple-E_i technique [8] was applied, where the monochromating chopper operating at 150 Hz, 6 times faster than neutron pulse emerging at 25Hz, and the band chopper at 25Hz yielded incident energies (E_i's) of 41.8, 7.7 and 3.1 meV simultaneously with energy resolutions of 4.0, 0.2 and 0.04 meV for elastic scattering, respectively. The accessible Q- and E-ranges were $0.37 < Q < 7.4$ $Å^{-1}$ and $-20.0 < E < 20.0$ meV for $E_i = 41.8$ meV, $0.18 < Q < 3.1$ $Å^{-1}$ and $-6.0 < E < 6.0$ meV for $E_i = 7.7$ meV, and $0.13 < Q < 2.0$ $Å^{-1}$ and $-2.0 < E < 2.0$ meV for $E_i = 3.1$ meV, respectively.

The liquid sample of Se_2Br_2 was sealed in a cylindrical silica cell with an inner diameter of 10mm and a wall thickness of 0.3 mm. The sample cell was put into an aluminium container to obtain good thermal contact and to gain temperature uniformity. The measurements were carried out at 298, 373 and 473 K by utilizing a cryo-furnace. The Q-E maps of the scattering intensity were calculated by the software suite Utsusemi [9] from the scattering event data of the time-of-flight. To deduce $S(Q,E)$ from the obtained TOF spectra, subtraction and absorption corrections were made by applying the following formula,

$$S(Q,E) = \frac{1}{C}\frac{\left(I_{S+C}(Q,E) - I_B(Q,E)\right) - \frac{A_C^{S+C}}{A_C^C} \times \left(I_C(Q,E) - I_B(Q,E)\right)}{A_S^{S+C}},$$

where $I_{S+C}(Q,E)$, $I_C(Q,E)$ and $I_B(Q,E)$ are the measured scattering intensity for the sample containing in the cell, that for the empty cell and that for the instrumental back ground with crio-furnace, respectively. A_S^{S+C} presents a coefficient for absorption of neutrons with a scattering event at the sample position (the subscription of S) due to the sample and the cell (the superscription of S+C), and A_C^{S+C} and A_C^C are absorption coefficients of neutrons with a scattering event at the cell position. Those coefficients are calculated from the integration over scattering positions with the geometry of the sample and the cell at each scattering angle and energy transfer. Therefore, those have also Q- and E-

dependences. C is a normalized factor depending on number of atoms, scattering cross sections and total number of incident neutrons. We evaluated C from comparison between the energy integration of QENS at several Q values and the known structure factor, $S(Q)$ from the earlier neutron diffraction experiment [1], since Se_2Br_2 is almost coherent scatterer of neutrons.

3. Results of QENS measurements

Figure 1 shows the intensity map of $I_{S+C}(Q,E)$ for liquid Se_2Br_2 in the silica tube and the Al container at 298 K with $E_i = 41.8$ meV. The overall profile of $I_{S+C}(Q,E)$ consists of three hallow components at around $Q = 2$, 3.5 and 5.5 Å^{-1} contributed by the scattering from the liquid sample in the silica glass tube and Bragg peaks from the aluminium container observed at $Q = 2.7$, 3.1, 4.4, 5.2, 5.4, 6.2, 6.8, 6.9 Å^{-1}. Quasi-elastic scattering is clearly observed over an energy transfer range from -10 to 10 meV. These hallow components accompanied by QENS are consistent with the former experiments [6]. Since we are interested in inter-molecular dynamics of Se_2Br_2 molecules dominant in a relatively low Q region, the result of $E_i = 7.7$ meV which focuses on Q-range from 0.3 to 3.3 Å^{-1} is mainly discussed in this paper.

Figure 2 shows QENS spectra of $E_i = 7.7$ meV accumulated over a Q- range of $1.9 < Q < 2.1$ Å^{-1} for the measurement of the liquid sample in the silica cell and the aluminium container at 473 K, the measurement of the empty cell, the instrumental back ground and the deduced $S(Q,E)$ of liquid Se_2Br_2 at 473 K by subtraction of the cell contribution with taking into account the absorption correction. The overall $S(Q,E)$ at 298K corrected by detailed valance factor is shown in Figure 3. Each spectrum was obtained by accumulating scattering events over a Q width of ±0.1 Å^{-1} at around a representative Q value. The elastic envelope exhibits a maximum at around 1.7 Å^{-1} and the energy width is broader with increasing Q value.

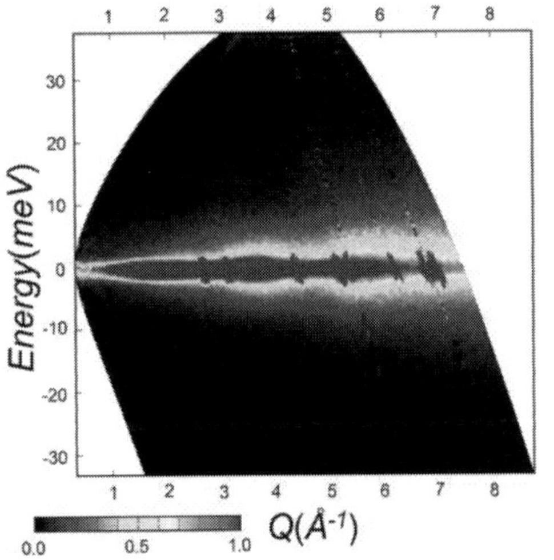

Figure 1. Scattering intensity map of liquid Se_2Br_2 containing a silica tube and an aluminium container at 298 K measured by the cold-neutron disk-chopper spectrometer AMATERAS with multi E_i technique. The spectrum is for $E_i = 41.8$ meV.

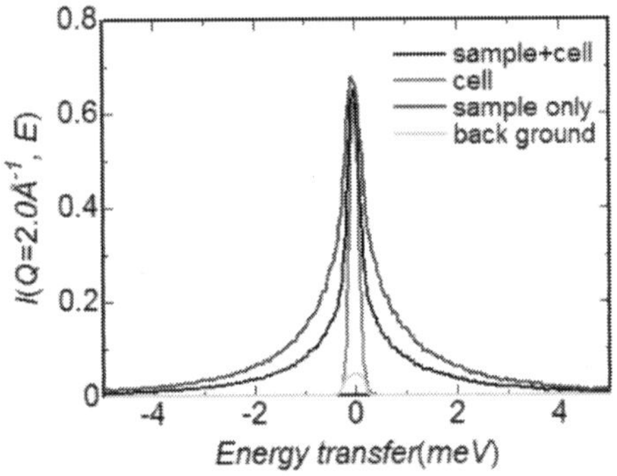

Figure 2. Scattering intensity $I(Q,E)$ for liquid Se_2Br_2 at 473 K with $E_i = 7.7$ meV accumulated over $1.9 < Q < 2.1$ Å^{-1}. Black, red and green lines are that for the measurement of the liquid sample in the silica cell and the aluminium container, that for empty cell, and that for instrumental back ground with a crio-furnace. Blue line is $S(Q,E)$ for liquid Se_2Br_2 at 473 K obtained by subtraction and absorption corrections.

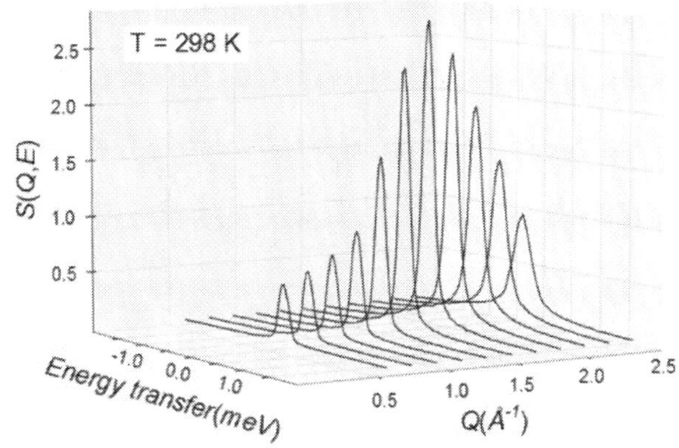

Figure 3. $S(Q, E)$ of liquid Se_2Br_2 at 298 K obtained from the spectrum by $E_i = 7.7$ meV.

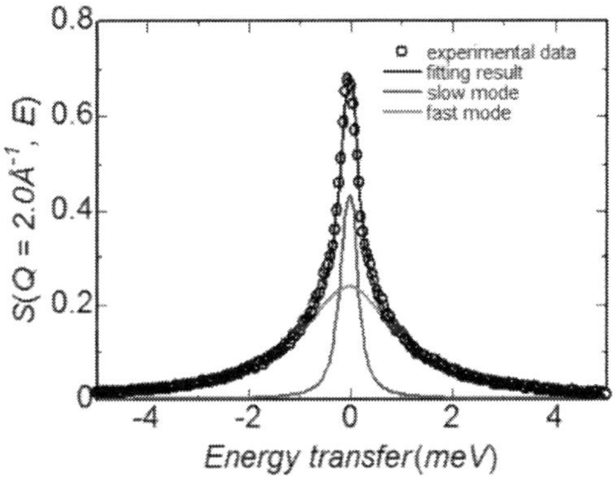

Figure 4. Shows the fitting result of $S(Q = 2.0 Å^{-1}, E)$ at 473 K , which integrated over $1.9 < Q < 2.1$. Blue line and red line shows narrow component and broad component of Lorenz function, respectively. Black line shows sum of two components and blank dot shows experimental data.

4. profile analysis of QENS and discussion

The profile of the resultant $S(Q,E)$ appears a relatively-sharp peak at $E = 0$ meV and wide tails toward large energy gain and loss, which can be reproduced well by a sum of two Lorentz functions as earlier study demonstrated [4]. Figure 4 shows a fitting result of $S(Q = 2.0 Å^{-1},E)$ at 473 K. The model function is a sum of two Lorentz functions, $L(Q,E)$, convoluted by the instrumental resolution function, $R(Q,E)$ as follows; $S_{Model}(Q)=\left(\sum_{i=1}^{n} L_i(Q,E)\right)\otimes R(Q,E)$,

where $n = 2$ and the convolution is performed on E axis. The experimental $S(Q,E)$ is quite well reproduced by the two-Lorentzian model so that liquid Se_2Br_2 involves two quasi-elastic modes with slow and fast relaxation times corresponding to narrow and broad Lorentzians, respectively. These results are consistent with the earlier work [4]. Hereafter we denote these slow and fast modes as $L_s(Q,E)$ and $L_f(Q,E)$.

Figure 5 shows Q-dependences of the magnitudes of both (a) $L_f(Q,E)$ and (b) $L_s(Q,E)$, which describe as $A_f(Q)$ and $A_s(Q)$, respectively. With increasing temperature, the maxima in the Q-dependence of the magnitude for both components shift to a low Q direction, while the temperature dependence of the magnitude is opposite between narrow and broad components. Taking into account that energy integral of $S(Q,E)$ gives $S(Q)$ for coherent scatter, we confirm that the sum of $A_f(Q)$ and $A_s(Q)$ at 298 K is consistent with the profile of the total $S(Q)$ obtained by neutron diffraction [1]. Moreover we observe reasonable temperature dependence of $S(Q)$ by x-ray diffraction [10] which is consistent with the temperature dependence. Figure 6 shows the half widths at the half maxima (HWHM) for $L_f(Q,E)$ and $L_s(Q,E)$ which describe as $\Gamma_f(Q)$ and $\Gamma_s(Q)$, respectively. $\Gamma_f(Q)$ has a minimum at the Q value where the maximum of $A_f(Q)$ locates, while $\Gamma_s(Q)$ does not show clearly a minimum associated with the peak of $A_s(Q)$. The former is so called de Gennes narrowing. $S(Q,E)$ shows narrowing in energy transfer at the Q value where $S(Q)$ has a peak by strong correlation.

The Q positions at the maximum intensity of A_s recall us that the HS packing of molecules creates a sharp peak at a low Q region of the first peak in S(Q), corresponding to an averaged number density. t ([1],). The temperature dependence of A_s can be reasonably interpreted as thermal expansion with

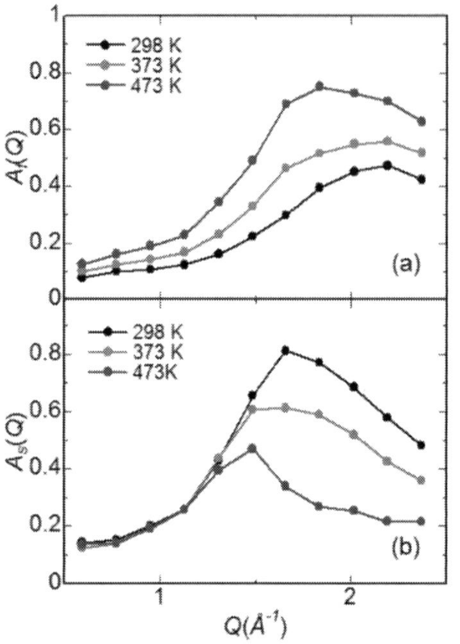

Figure 5. Black, red and blue points depict the magnitudes of Lorentz functions in the model fitting for QENS of liquid Se_2Br_2 at 298, 373 and 473 K, respectively. (a) broad and (b) narrow components . Lines are guides for eye.

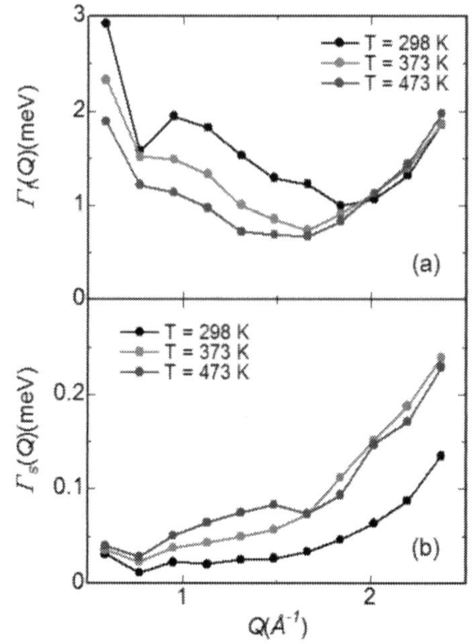

Figure 6. Black, red and blue line shows the half width of half maximum of broad components (a) and narrow components (b) for liquid Se_2Br_2 at 298, 373 and 473 K, respectively.

increasing temperature, because the Q position where the maximum in A_s locates shifts to a lower Q direction and the intensity of A_s become weaker. Moreover, it is natural that the relaxation time of the slow mode becomes shorter at a higher temperature due to thermal agitation, which explains the temperature dependence of Γ_s. This interpretation on the slow mode is consistent with the suggestion by Yao et al. that the slow mode depends on molecular weight.

The fast mode may be assigned as inter-molecular correlation with strong coupling. Since the Q position where the maximum in A_f locates also shifts to lower Q direction with increasing temperature, the fast mode may depend on density. Yao et al. observed the dependency on halogen species for the fast mode [4]. Although they interpreted from this dependency the fast mode as rotational motion of single molecule, we interpret it as the rotational motion of an orientationally-ordered pair of molecules, i.e., a dimmer. The interesting behavior of the Q dependence of Γ_f that, with increasing temperature, it decreases in a relatively low Q region below 1.8 $Å^{-1}$ while it does not show temperature dependence in a higher Q region is comprehensible. Dimmer has a relatively short intermolecular distance which makes an effect on structure factor in a relatively high Q region beyond 1.8 $Å^{-1}$. When the coupling force is stronger than thermal agitation at the measuring temperatures, the relaxation time may exhibit no temperature dependence. The increase in A_f with increasing temperature is explained as increase of number of isolated dimmers by thermal expansion. In contrast, when coupled molecules start dissolving, the intermolecular correlation length increases, which relates to dynamic structure factor in a lower Q region below 1.8 $Å^{-1}$. The dimmers during dissolving process may have a short relaxation time, which makes Γ_f increase up toward low Q direction in a low Q region. The dimmer isolation at higher temperatures may cause slowing down in relaxation, and decreasing Γ_b.

5. Summary

The results of QENS experiments for liquid Se_2Br_2 have been reported. The dynamic structure factor in Q region, where the first peak of $S(Q)$ lies, can separate into the fast mode and the slow mode. The temperature dependence of the maximum intensity of A_s and $\Gamma_s(Q)$ reveals that the slow mode depends on inter-molecular dynamics without orientational correlation. On the other hand, the fast mode depends on rotation of the dimmer because Γ_f has two aspects, which the relaxation time at high-Q shows similar value and the relaxation time at low-Q becomes longer with increasing temperature. These interpretations are consistent with the earlier works [1, 4, 5, 6].

6. Acknowledgement

QENS experiments were performed with approval of MLF(proposal number 2010A0075). The authors are grateful to Prof. Masayuki Tokita, Prof. Kenji Maruyama and Dr. Kazuya Aizawa for valuable discussions.

References

[1] Maruyama K, Kawakita Y,Yao M, Endo H, and Misawa M 1991 *J.Phys. Soc. Jpn.* **60** 3032.
[2] Prascher B P. and Wilson A K 2007 *J. Mol. Struct.: THEOCHEM* **814** 1.
[3] Misawa M 1989 *J. Chem. Phys.* **90** 6563.
[4] Yao M, Kawakita Y, Shiga T, Yamamoto I and Endo H, 1997 *J. Phys. Soc. Jpn.* **66** 3115.
[5] Koga A, Ohmura S and Shimojo F, 2009 *J. Phys. Soc. Jpn.* **78** 074601.
[6] Kawakita Y, Chiba A, Yokota Y, Yao M, Shibata K, Ohmasa Y and Takeda S, 2006 *Phisica B* **256** 385
[7] Nakajima K et al. 2007 *J. Neutron Res.* **15** 13.
[8] Nakajima K et al. submitted to *J. Phys. Soc. Jpn.*
[9] Y. Inamura et al. 2010 *Proc. 19th Meet, Int. Collaboration on Advanced Neutron Sources, PSI-Proc.* 10-01
[10] Shimakura H, Ohara K, Kawakita Y, Wakisaka Y, Kohara S and Takeda S, in prepartion.

Journal of Physics: Conference Series **340** (2012) 012081 doi:10.1088/1742-6596/340/1/012081

Fluctuations in medium-range structure of Bi-based metallic liquid alloys

H Ueno[1,2], Y Kawakita[2], K Ohara[3], S Tahara[4], S Kohara[3], M Itou[3], S Takeda[1]

[1] Department of Physics, Faculty of Sciences, Kyushu University 6-10-1 Hakozaki,Higashi-ku, Fukuoka 812-8581, Japan
[2] J-PARC Center, Japan Atomic Energy Agency, 2-4 Shirakata Shirane, Tokai-mura, Naka-gun, Ibaraki 319-1195, Japan
[3] Research and Utilization Division, Japan Synchrotron Radiation Research Institute (JASRI, SPring-8), 1-1-1 Kouto, Sayo-cho, Sayo-gun, Hyogo 679-5198, Japan
[4] Department of Physics and Earth Sciences, Faculty of Science, University of the Ryukyus, 1 Senbaru, Nishihara-cho, Nakagami-gun, Okinawa 903-0213, Japan

ueno@phys.kyushu-u.ac.jp

Abstract. Liquid structure of $Bi_{50}Zn_{50}$, which is situated at around the Bi-rich end of miscibility gap in Bi-Zn system, has been investigated by neutron and x-ray diffraction experiments and following analysis using reverse Monte Carlo (RMC) structural modelling. Among the partial correlations calculated from the structural model obtained by RMC, the Zn-Zn partial has a large temperature variation. It is found that there are medium-range fluctuations in Zn distribution which have a scale of 10 Å.

1. Introduction

Structural fluctuations in liquid metallic alloys consisting of polyvalent metals have been drawn attentions in liquid metal physics. Especially Bi-based liquid alloys still stimulate researchers on temperature variation of physical properties such as viscosity, sound velocity, thermodynamics and so on because some of them observed peculiar behaviour of the properties but the others claim their ways to treat liquid sample during measurements. These controversial situations may be related to large structural fluctuations existing even in miscible region [1]. We aim at revealing such structural fluctuations by diffraction measurements and following reverse Monte Carlo (RMC) structural modelling. Bi-Zn system has a miscibility gap with a critical composition of $Bi_{18}Zn_{82}$ ranging from 0 to 60 at. % Zn [2]. Earlier structural study on liquid Bi-Zn system by neutron diffraction measurements just above the coexistence curve of miscibility gap show that there are large differences in the profile of the total structure factor, $S(Q)$, between the Zn and Bi sides from the critical composition [3]. To know from microscopic viewpoint what fluctuations evolve by approaching to the two-phase region with decreasing temperature is of fundamental importance to understand a leading factor which induces phase separation in metallic liquid alloys. In this study as a first stage of our research to understand structural fluctuations in Bi-based liquid alloys, we focus on liquid $Bi_{50}Zn_{50}$ which is situated near at the Bi-rich end of miscibility gap, because reduction of partial structures is easy and under competition in extending local ordering between the constituents predominance of one type of fluctuations can be expected.

2. Experiments

2.1. Neutron diffraction

Neutron diffraction measurements for $Bi_{50}Zn_{50}$ were carried out using the Kinken powder diffractometer for high efficiency and high resolution measurements, HERMES, installed at the JRR-3M reactor in Japan Atomic Energy Agency, Tokai, Japan [4]. HERMES has a movable multi-detector system with 150× ^{3}He detectors in each 1° interval in scattering angle. In our experiment, two angle-sets were applied to cover the scattering angle from 7.0 to 156.5° with 0.5° interval. The sample of $Bi_{50}Zn_{50}$ alloy was obtained by mixing pure elements of Bi and Zn in an evacuated silica tube at 800 °C and then quenched to prevent macroscopic phase separation. The sample sealed under vacuum in a cylindrical fused silica cell with dimension of 8 mm in an inner diameter and 0.5 mm in a wall thickness was located in a furnace chamber filled by He gas. The incident neutron beam was 1.125 Å in wave length which was provided by a Ge(5 5 1) monochromator. After subtraction of the contributions of the cell and the furnace background from the scattering intensity taking into account of neutron absorption by the sample and the cell, structure factors, $S^N(Q)$, of neutron diffraction were obtained over a momentum-transfer range from $Q = 0.65$ to 10.85 Å^{-1} at 520, 750 and 850 °C.

2.2. X-ray diffraction

X-ray diffraction measurements for $Bi_{50}Zn_{50}$ were performed with a two axis diffractometer installed at BL08W in SPring-8, Sayo-gun, Hyogo, Japan [5]. Incident X-ray was 182.6 keV in energy and 0.0679 Å in wavelength. The $Bi_{50}Zn_{50}$ sample was shaped into a disk with 0.7 mm thickness to optimize x-ray transmission and located into a gap of quartz plates which were made from a couple of glass tube with one-end closed and polished down to 0.3mm thickness. The sample in the cell was set in a furnace chamber filled up by He gas. High temperatures from 520 to 850 °C were achieved by a nickel chrome resistance wire set in the chamber. Structure factors, $S^X(Q)$, of x-ray diffraction were obtained from $Q = 0.70$ to 24.00 Å^{-1} in a momentum transfer space at 520, 600, 750 and 850 °C. For the sake of comparison, pure liquid Bi was also measured at 500 °C by using BL08W and pure liquid Zn was done at 470 °C using BL04B2 [6].

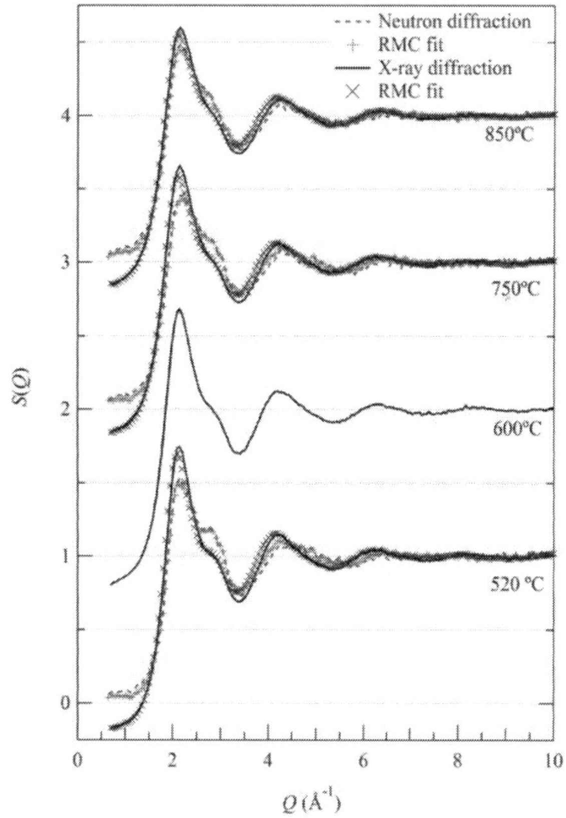

Figure 1. Total structure factor of liquid $Bi_{50}Zn_{50}$ alloy obtained by neutron (dotted blue lines) and x-ray (solid black lines) diffraction measurements. Fitting results of reverse Monte Carlo are drawn as black diagonal crosses (×) and blue plus marks (+) for neutrons and x-rays respectively.

3. Results

The total structure factors of liquid $Bi_{50}Zn_{50}$ obtained by neutron and x-ray diffraction measurements are compared in figure 1. In $S^N(Q)$ at 520 °C, the first peak is located at 2.0 Å^{-1} accompanied by a shoulder at a higher Q side. This characteristic shoulder structure was observed in the earlier work [3]. In $S^X(Q)$, the ratio of the intensity of the shoulder to the first peak is smaller than that in $S^N(Q)$. It is interesting to notice that the oscillations beyond 3.5 Å^{-1} are not coincide well between $S^N(Q)$ and $S^X(Q)$. With increasing temperature, the first peak remains at 2.0 Å^{-1} and the shoulder structure becomes smaller and even broader. At 850 °C, the differences between $S^N(Q)$ and $S^X(Q)$ are small compared with those at a lower temperature. The differences between $S^N(Q)$ and $S^X(Q)$ originate from the difference in the scattering abilities of x-ray and neutron beams between Bi and Zn atoms. For neutron diffraction, the coherent scattering length of Bi is 1.5 times larger than that of Zn (Bi: 8.532 fm, Zn: 5.680 fm). On the other hand, the number of electrons in the Bi atom is more than twice of Zn atoms (Bi: 83, Zn: 30). As a result, Bi-Bi correlations are emphasized in x-ray diffraction while Zn-Zn correlations are more detectable in neutron diffraction, as tabulated in table 1.

Table 1. Weighting factors of partial structures of Bi-Bi, Bi-Zn and Zn-Zn for x-rays and neutron diffraction spectra of $Bi_{50}Zn_{50}$.

	Bi-Bi	Bi-Zn	Zn-Zn
Neutron	0.360	0.480	0.16
X-ray(Q = 0)	0.539	0.390	0.07

Total pair distribution functions, $g^N(r)$ and $g^X(r)$, by neutron and x-ray diffractions were derived by a conventional Fourier transformation from $S^N(Q)$ and $S^X(Q)$, respectively as shown in figure 2. There are significant differences in the first peak position and profile between $g^N(r)$ and $g^X(r)$. The first peak in $g^X(r)$ at 520 °C is located at 3.3Å and slightly shifts to a shorter r direction with increasing temperature. Although that in $g^N(r)$ is settled at 3.1 Å at every temperature, slightly shorter than that in $g^X(r)$, the profile changes significantly from an asymmetric shape with a shoulder at a shorter r side at the lowest temperature to a rather symmetric shape at the highest temperature. Taking into account the weighting factors of the partial structure shown in table 1, the shoulder structure locate at the larger r side of the first peak in $g^N(r)$ is reasonably attributed to Zn-related partials while the peak at 3.3 Å in $g^X(r)$ is assigned as Bi-Bi partial. The second peaks observed in a r region between 6 to 8 Å shifts the peak position to a shorter r direction with increasing temperature, regardless of thermal expansion [2].

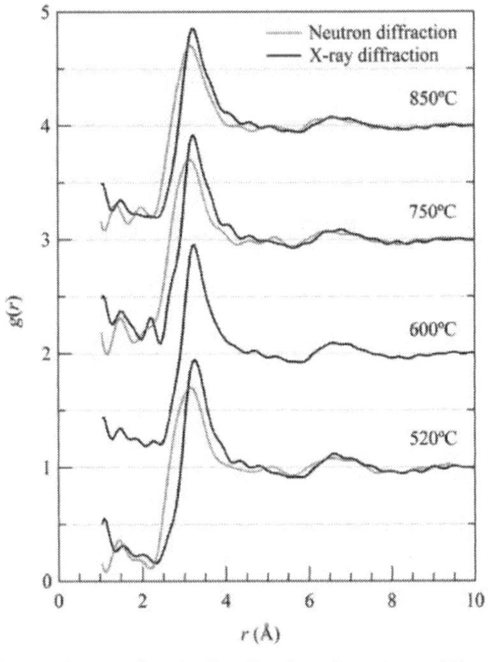

Figure 2. Total pair distribution function of liquid $Bi_{50}Zn_{50}$ alloy obtained by neutron (dotted blue lines) and X-ray (solid black lines) diffraction measurements.

4. Reverse Monte Carlo structural modelling

To approach more detailed partial structures and medium-range fluctuations in the atomic distribution of liquid $Bi_{50}Zn_{50}$, the reverse Monte Carlo (RMC) structural modelling [7, 8] was applied to $S^N(Q)$ and $S^X(Q)$ simultaneously. The RMC program creates an atomic configuration model in three-dimensional space reproducing the experimental structure factor and the pair correlation function. In our modeling procedure, 5000 particles (Bi: 2500, Zn: 2500) were distributed at random in a cubic box as the initial configuration. The cube sizes were selected to give the proper number density [2] at each temperature and periodic boundary condition was applied to the simulation box. In addition, the particles are subject to a physical restriction of the cutoff distance of 2.0 Å which represents minimum distance between particles. The results of the RMC applied to x-ray and neutron diffraction data of liquid $Bi_{50}Zn_{50}$ are shown also in figure 1 as diagonal crosses(×) for x-ray and plus marks(+) for neutron. The neutron- and x-ray-weighted total structure factors calculated by the structural model obtained by the RMC procedure reproduce well simultaneously the experimental data of $S^N(Q)$ and $S^X(Q)$.

Partial pair distribution functions, $g_{\alpha\beta}(r)$ (α, β = Bi, Zn), calculated from the atomic configuration of the RMC model are shown in figure 3. As expected from the difference between $S^N(Q)$ and $S^X(Q)$, Zn-related partials are distributed in a shorter distance than Bi-Bi correlation. The large temperature dependence of the asymmetrical shape of the first peak in $g^N(r)$ is mainly attributed to the Zn-Zn partial correlation since $g_{Zn\text{-}Zn}(r)$ exhibits a large temperature dependence in the height of the first peak located at 2.7 Å from 4.0 at 520 °C to 2.4 at 850 °C. The Bi-Bi partial correlation shows an interesting distribution that the first peak located at 3.3 Å has a tail at a longer side decreasing gradually up to about 6 Å. The first peak of $g_{Bi\text{-}Zn}(r)$ is situated at an intermediate atomic distance between $g_{Zn\text{-}Zn}(r)$ and $g_{Bi\text{-}Bi}(r)$. The second peaks are diverged among three partials, i.e., 4 - 6 Å for $g_{Zn\text{-}Zn}(r)$, 5 - 7 Å for $g_{Bi\text{-}Zn}(r)$, and 6 - 8 Å for $g_{Bi\text{-}Bi}(r)$. The spikes seen at 2 Å in all partials are spurious by the cut off length set at 2 Å for all partial pairs.

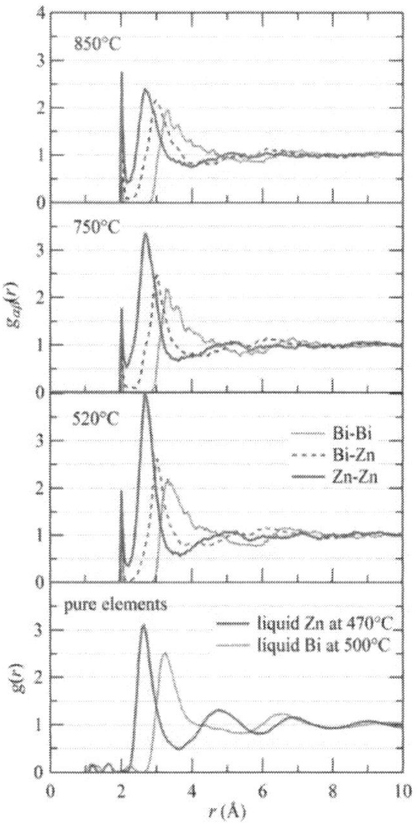

Figure 3. Partial pair distribution functions of liquid $Bi_{50}Zn_{50}$ obtained by reverse Monte Carlo structural modelling
Dotted blue line: Bi-Bi correlation, broken green lines: Bi-Zn correlation and solid red line: Zn-Zn correlation.
As references, the pair distribution functions of pure liquid Bi and Zn are shown in the lower.

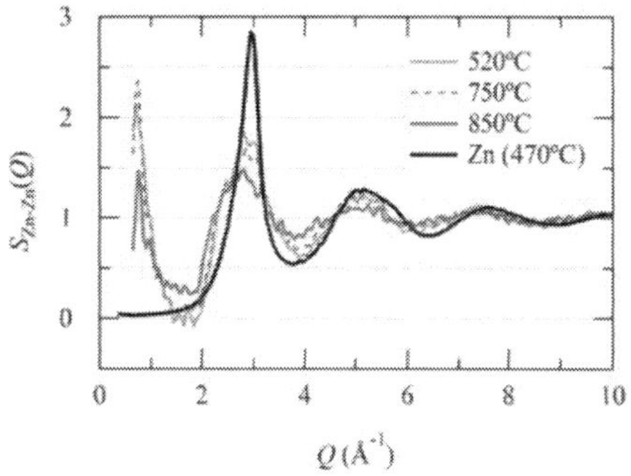

Figure 4. Temperature dependence of Zn-Zn partial structure factors, $S_{Zn\text{-}Zn}(Q)$, of liquid $Bi_{50}Zn_{50}$ obtained by reverse Monte Carlo. Solid red line: 850 °C, broken green lines: 750 °C and dotted blue lines: 520 °C. For the sake of comparison, the structure factor of pure liquid Zn on 470 °C is depicted as black full thick line.

5. Discussion

In general, liquid alloys with miscibility gap exhibit concentration fluctuations even in much higher temperatures far from two-phase region and then increases them by approaching the coexistence curve with decreasing temperature. The present composition of $Bi_{50}Zn_{50}$ is far from the critical composition for the miscibility gap of around $Bi_{18}Zn_{82}$ in liquid Bi-Zn system, but still situated in the two-phase region. Among three partials in our RMC results, only $g_{Zn\text{-}Zn}(r)$ shows a large temperature variation. The position of the first peak in $g_{Zn\text{-}Zn}(r)$ is almost coincident with the first neighbour distribution of pure liquid Zn. However, the oscillations beyond the first neighbour in $g_{Zn\text{-}Zn}(r)$ are suppressed compared with those of pure liquid Zn where the second, the third and the fourth peaks clearly distinguished at around 4.7, 7.8 and 9.0 Å.

The corresponding partial structure factors, $S_{Zn\text{-}Zn}(Q)$, of liquid $Bi_{50}Zn_{50}$ are shown in figure 4 together with the structure factor of pure liquid Zn, $S^{Zn}(Q)$, at 470°C obtained by x-ray scattering. $S^{Zn}(Q)$ has a prominent but asymmetrical first peak at 2.95 Å. First peaks of $S_{Zn\text{-}Zn}(Q)$ of liquid $Bi_{50}Zn_{50}$ are located at a slightly lower Q than that of $S^{Zn}(Q)$, accompanied by a hump at a lower Q side which is remarkable at the highest temperature. There is a significant prepeak at 0.75 Å^{-1} in $S_{Zn\text{-}Zn}(Q)$ of liquid $Bi_{50}Zn_{50}$, which originates from medium-range structural fluctuations. Zn distribution is extracted from the three-dimensional structural model of liquid $Bi_{50}Zn_{50}$ created by RMC as shown in figure 5. To visualize fluctuations in Zn distribution, sticks are attached to atomic pairs locating within 3.6 Å which corresponds to the first neighbouring shell. The medium-range structural fluctuations of the scale of about 10 Å can be seen in the Zn distribution.

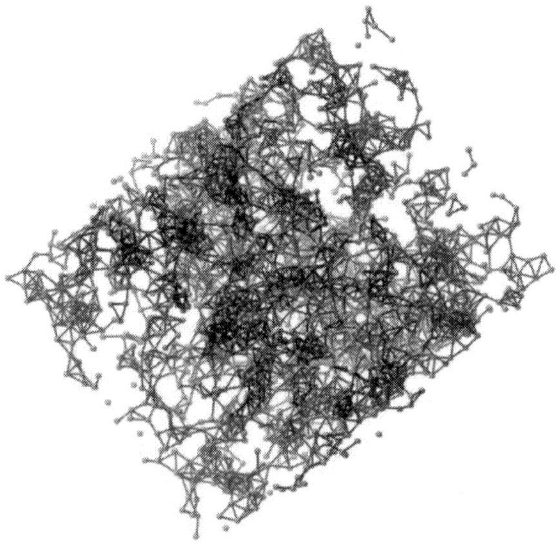

Figure 5. Zn distribution of the structural model of liquid $Bi_{50}Zn_{50}$ at 520 °C obtained by reverse Monte Carlo. The bonds are attached to atom pairs locating within 3.6 Å corresponding to the first neighbouring shell to visualize fluctuations in Zn distribution.

6. Summery

Neutron and x-ray diffraction measurements of liquid $Bi_{50}Zn_{50}$ were performed at 520, 600, 750 and 850 °C which are all above the coexistence curve of the miscibility gap (only x-ray for 600°C). The results of RMC structural modelling simultaneously applied to the neutron and x-ray experimental data reveal that, among three partials of Zn-Zn, Bi-Zn and Zn-Zn which all contribute to the first neighbouring shell, Zn-Zn partial correlation has shortest atomic distance and exhibits a large temperature variation. The medium-range structural fluctuations of the scale of about 10 Å in Zn distribution cause the prepeak at 0.7 Å^{-1} in $S_{Zn\text{-}Zn}(Q)$.

Acknowledgements

The authors express thanks to Dr. László Pusztai for helpful discussions and his encouragements to this study. They also appreciate to Dr. László Temleitner for his helps on x-ray experiments and to Prof. Kenji Ohoyama and Mr. Manabu Ohkawara for their helps on neutron experiments. The neutron experiment was performed with approval of the Institute of Solid State Physics, the University of

Tokyo (Proposal Number 10760). The x-ray diffraction measurements were performed at SPring-8 with approval of JASRI (Proposal Numbers 2009A1411 and 2010A1480).

References

[1] Mudry S, Plevachuk Yu, Sklyarchuk V, Yakymovych A 2008 *J. Non-Cryst. Solids* **354** 4415
[2] Kanai S and Tsuchiya Y 1993 *J. Phys. Soc. Jpn.* **62** 2388
[3] Tamaki S, Takeda S, Harada S, Waseda Y, Matsubara E 1986 *J. Phys. Soc. Jpn.* **55** 4296
[4] Ohoyama K, Kanouchi T, Nemoto K, Ohashi M, Kajitani T, Yamaguchi Y 1998 *Jpn. J. Appl. Phys.* 37 3319-26
[5] Kohara S, Itou M, Suzuya K, Inamura Y, Sakurai Y, Ohishi Y and Takata M 2007 *J. Phys.: Condens. Matter* **19** 506101
[6] Isshiki M, Ohishi Y, Goto S, Takeshita K and Ishikawa T 2001 Nucl. Instrum. Meth. Phys. Res. A **467-468** 663
[7] McGreevy R L and Pusztai L 1988 *Mol. Simul.* **1** 359
[8] Gereben O, J´ov´ari P, Temleitner L and Pusztai L 2007 *J. Optelectron. Adv. Mater.* **9** 3021

Vibrational density of states measurements in disordered systems

N Violini [1,2], E Cornicchi[2,3], A Orecchini[2,3], A Paciaroni[2,3], C. Petrillo[2,3] and F Sacchetti[2,3]

[1] Julich Center of Neutron Scattering, Forschungszentrum Julich, D-52425 Jülich, Germany
[2] Dipartimento di Fisica, Università di Perugia, Via A. Pascoli, I-06123, Perugia, Italy
[3] Istituto Officina dei Materiali del Consiglio Nazionale delle Ricerche, Unità di Perugia, c/o Dipartimento di Fisica, Università di Perugia, Via A. Pascoli, I-06123, Perugia, Italy

E-mail: n.violini@fz-juelich.de

Abstract. We present a data treatment procedure, based on an iterative technique, properly developed to subtract the multi-phonon contributions from the dynamic structure factor in a self-consistent way. With this technique, we derive the one-phonon vibrational density of states from the dynamic structure factor of different disordered systems, in the framework of the incoherent scattering approximation. We present results on glassy glucose ($C_6H_{12}O_6$), a nearly perfect incoherent scatterer, due to high hydrogen content. The data treatment procedure has been found to work well also for the more complex case of dry and hydrated DNA.

1. Introduction

Neutron Scattering is a useful technique to access the atomic dynamics of solids. The density of states (v-DOS) is an important quantity obtainable by inelastic neutron scattering data, because it provides the weight of each frequency, or energy, of the vibrational modes. Nevertheless working out the v-DOS from the measured double differential cross section can be a rather difficult task. Indeed a wide range of custom techniques is found in literature, depending on the system under investigation and on the particular characteristics of the experiment. When the system can be reasonably approximated as an incoherent scatterer, the incoherent one-phonon approximation can be safely used to work out the v-DOS from the dynamic structure factor, in the limit where the multi-phonon contributions are negligible [1]. Unfortunately the most common systems have both coherent and incoherent cross sections. In principle the incoherent approximation is valid in coherent systems when the correlations between atomic motions can be ignored, that is at high enough wave vector transfer [2, 3], however in such a limit the multiphonon contribution cannot be neglected [1, 4, 5]. Usually experiments are performed at smaller wave vector transfer for reasons of intensity, available dynamic range or other questions concerning the physical problem under investigation. This makes the data analysis difficult and often requires a comparative investigation with molecular dynamic simulations [6]. Of course the double differential cross section includes several contributions from modes of different microscopic origin, especially when treating data from experiments on multi-component and macromolecular systems. In these cases, librational and vibrational modes belonging to specific and bound atomic groups contribute to the v-DOS, because their eigenfrequencies can be comparable to that of the translational ones [7, 8]. From a more general point of view, it is interesting to stress that the true v-DOS is in general different from that obtained from a straightforward analysis of the inelastic neutron scattering experiments [9], which often does not consider the limits of the one-phonon incoherent approximation. We discuss here the details of an iterative technique to derive the v-DOS from neutron inelastic data and present the results of its application to two different topologically disordered systems: glassy glucose and DNA from salmon testes. Details of the performed experiments will be presented in separated papers, but a description can also be found in ref.[10].

2. Methods

Firstly we introduce the incoherent intermediate scattering function:

$$F_{inc}(Q,t) = \int_{-\infty}^{+\infty} \cos(\omega t) S_{inc}(Q,\omega) d\omega \tag{1}$$

defined as the time Fourier transform of the incoherent dynamic structure factor:

$$S_{inc}(Q,\omega) = \frac{1}{\pi}\int_{0}^{+\infty} \cos(\omega t) F_{inc}(Q,t) dt. \tag{2}$$

In the incoherent approximation, the dynamic structure factor (Eq.(1) and Eq.(2)) accounts for single-particle motions under the assumption that each vibrational unit of mass M has a Gaussian-distributed time-dependent displacement from its equilibrium position, $\gamma(t)$. In the case that only harmonic vibrational motions are present, the relation between $\gamma(t)$ and the v-DOS $g(\omega)$ can be derived from the Bloch identity [11] as

$$\gamma(t) = \frac{k_B T}{M}\int_{0}^{+\omega_M} (1-\cos(\omega t)) n(\omega) \frac{g(\omega)}{\omega^2} d\omega \tag{3}$$

where ω_M is the upper experimental available frequency and $n(\omega)$ is the Bose occupation number. The intermediate scattering function is then obtained as:

$$F_{inc}(Q,t) = \exp(-Q^2\gamma(t)). \tag{4}$$

To apply the approximation to measured data, the first step is the calculation of a guess $g(\omega)$ from the low angle data. Indeed at the lower scattering angles which are available in the actual experiments, contaminant multi-phonon processes are minimized and the one-phonon can therefore be roughly approximated by the total measured one. This allows to deduce the $g(\omega)$ directly from the experimental dynamic structure factor, using the rigorous result for a cubic crystal and for fully isotropic systems:

$$S_{inc}^{+1}(Q,\omega) = Q^2 e^{(-2W)} \frac{k_B T}{2M} n(\omega) \frac{g(\omega)}{\omega}, \tag{5}$$

where e^{-2W} is the Debye-Waller factor. The guess $g(\omega)$ permits to calculate back the theoretical one-phonon dynamic structure factor by means of the Eq.(5) and the total one through the sequential use of Eq.(3), Eq.(4) and Eq.(2), at all desired wavevector transfers and scattering angles. Thus an estimation of the multiphonon contributions can be obtained by trivially calculating the difference between them. Indeed the peculiar aspect of the technique is the subtraction of multiphonon terms from the total dynamic structure factor. This is achieved through a specifically developed iterative procedure which minimizes the difference between the total measured dynamic structure factor $S^{exp}(Q,\omega)$, and the calculated one. That means simply to require, omitting the $Q-$ and $\omega-$dependences for clarity:

$$S^{exp} = S_{inc} = S_{inc}^{+1} + (S_{inc} - S_{inc}^{+1}) \tag{6}$$

which implies that $S_{inc}^{+1}(Q,\omega)$ must be

$$S_{inc}^{+1} = S^{exp} - (S_{inc} - S_{inc}^{+1}). \tag{7}$$

Thus a new $S_{inc}^{+1}(Q,\omega)$ is obtained by means of Eq.(7) and the latter gives a new $g(\omega)$ through Eq.(5). The $g(\omega)$ thus derived can be used to calculate again $S_{inc}^{+1}(Q,\omega)$ and $S_{inc}(Q,\omega)$. When we applied this method to real experiments we found that the procedure rapidly converges so that a small number of iterations is necessary. Thus the calculated v-DOS has been derived self-consistently with the experimental dynamic structure factor, in such a way that the so-derived $g(\omega)$ can be used to describe the experimental dynamic structure factor in all the dynamic range. In case a good description of the experimental data is obtained, this also yields further support to the incoherent approximation, the validity of which is assumed as starting point of the adopted procedure.

3. Application to prototype disordered systems

We applied the iterative technique to work out the v-DOS from neutron data on two different disordered systems: glassy glucose and DNA. The use of the incoherent approximation is justified by the high incoherent cross-section of hydrogen atoms, largely and uniformly present in both systems. The mass term in Eq.(3) and Eq.(5) was treated as a fitting parameter during the application of the iterative technique (see below for details). As stated before the present investigations involved two different inelastic neutron scattering (INS) experiments on glucose and DNA. Both experiments were performed using the INS Time-Of-Flight (TOF) spectrometer IN4 at the Institute Laue Langevin (Grenoble, France). In the case of glucose, an incident energy of 16.88 meV was used. The resulting resolution was a gaussian-shaped function with full-width at half-maximum (FWHM) of 1.2 meV. The sample holder was an aluminium hollow cylinder of dimensions: 29 mm height, 14 mm inner diameter, and 16 mm external diameter. In the case of DNA data were acquired at two different incident neutron energies of about 67.5 and 4.8 meV, on both dry and hydrated DNA. The two instrumental setups have an energy resolution of 3.5 and 0.8 meV respectively. Only data at 67.5 meV incident energy are considered in this paper. The INS data treatment included the standard procedure: subtraction of empty container, correction for the absorption coefficient and for energy dependence of detectors efficiency, normalization to vanadium used as an elastic incoherent scatterer. Measurements were performed at 300 K for glucose and 100 K for dry and hydrated DNA, in order to work in a region where the system is far away from the melting point. Figure 1 shows the v-DOS obtained for the glucose sample, with the method described in the previous section. The comparison with the spectrum measured by Raman spectroscopy [12] displays that only a minor contribution from molecular internal motions is included in the measured v-DOS. The reduced v-DOS $\frac{g(\omega)}{\omega^2}$ as a function of ω is also displayed in figure 1. The plot shows clearly the presence of a broad excess of modes with respect to the Debye v-DOS calculated as $g_D(\omega) = \frac{3\omega^2}{\omega_D^3}$, where ω_D was obtained like in ref. [13], by means of the longitudinal and transverse velocities measured by ultrasound techniques ($v_L = 3200m/s$ and $v_T = 2200m/s$). Absolute normalization is obtained by the use of the hydrogen mass in Eq.(3) and Eq.(5). This value has been obtained as a fitting parameter resulting from the iterative routine and corresponds to the expected value of 1 atomic units. Indeed the present v-DOS is projected onto the hydrogen sites.

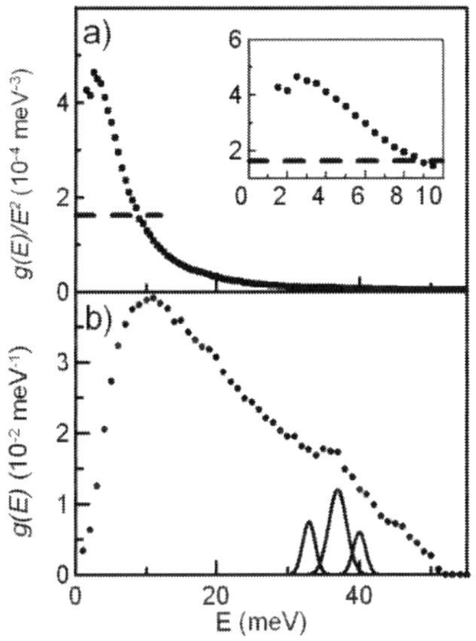

Figure 1. a) Reduced v-DOS $g(E)/E^2$ as a function of energy transfer E. (— — —) represents glucose v-DOS calculated in Debye approximation, as explained in the text. b) Generalized vibrational density of states measured with IN4 spectrometer as a function of energy transfer, together with a part of the Raman Spectrum (——) from ref.[12].

Figure 2 shows the v-DOS of both dry and hydrated DNA obtained by the one-phonon approximation applied to the experimental dynamic structure factor in wide momentum-energy range available in this experiment. The v-DOS obtained by integration of the measured dynamic structure factor over all wavevector transfer is shown for comparison. In this case the absolute normalization cannot be trivially obtained. We assumed an effective cross section deduced as the mean cross section per nucleus. The resulting fitting mass value was the mean mass per proton, inside the molecule, taking into account for the chemical composition of the DNA, as from ref [14], which is about 30 atomic units.

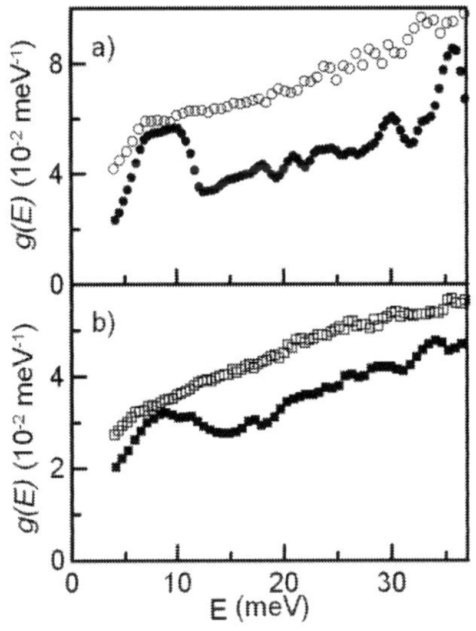

Figure 2. a) Density of states of hydrated (●) DNA obtained by the one phonon approximation as stated in the text, compared to the integral of the measured dynamic structure factor on the wave vector transfer which are available in this experiment (○). b) Density of states of dry (■) DNA is compared to the integral of the measured dynamic structure factor on the wave vector transfer which are available in this experiment (□).

Figure 3 shows the v-DOS of the DNA hydration water, obtained by subtraction of the dry DNA spectrum from the hydrated one. It is compared to the $g(\omega)$ of pure bulk water taken from [15]. The comparison evidences the differences between the spectra due to the different temperatures of the two measurements (100 K in the present experiment, 300 K in ref. [15] in the liquid phase). Indeed, the low temperature hydration water shows many sharp structures as it can be expected in this temperature range. The v-DOS of hydration water contains a considerable contribution from the phonon v-DOS of ice[16], which is also shown in Fig.3. This suggests a partial crystallization of the hydration water.

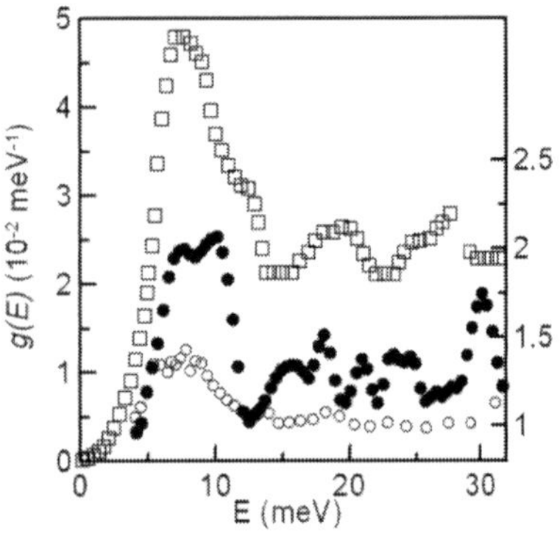

Figure 3. Density of states of three different systems: hydration water of DNA (●) as derived from the present data (see text), bulk liquid water (○) and solid ice (□) respectively from ref. [15] and [16]. Data are rescaled to be on the same scale. Data of bulk liquid water are shifted and referred to the scale on the right side.

4. Conclusions

The data treatment technique we presented was specifically developed to work out the vibrational density of states from neutron data. The advantage of this technique is the subtraction of multi-phonon contributions from the experimental dynamic structure factor by means of a self-consistent iterative procedure. Through this method we derived the one-phonon vibrational density of states of different disordered systems, by applying the incoherent approximation for isotropic systems to glassy glucose, and dry and hydrated DNA. The presented results are also discussed in comparison with available data for bulk water.

References

[1] Colognesi D, Celli M and Zoppi M 2004 *J. Chem. Phys.* **120** 5657
[2] Suck JB, Rudin H, Güntherodt HJ and Beck H 1981 *J. of Phys.: Solid State Phys.* C **14** 2305
[3] Chaplot SL, Mukhopadhyay R, Vijayaraghavan PR, Deshpande AS and Rao KR 1989 *J. of Phys.* **33** 595
[4] Dawidowski H, Bermejo FJ and Granada JR 1998 *Phys. Rev.* B *58* 706
[5] Caciuffoli R, Francescangeli O, Melonet S, Stefanon M, Gering E, Sucks JB, Bonnetll M, Alliav P and Vinait F 1989 *J. of Phys.: Cond. Matt.* **1** 5621
[6] Fabiani E, Fontana A and Buchenau U 2008 *J. Chem. Phys.* **128** 24 244507
[7] Fontana A, Rocca F, Fontana MP, Rosi B, Dianoux AJ 1990 *Phys. Rev.* B **41** 3778
[8] Sauvajol HL, Djurado D, Dianoux AJ, Theophilou N and Fischer JE 1991 *Phys. Rev.* B **43** 14305
[9] Taraskin SN and Elliott SR 1997 *Phys. Rev.* B **55** 117
[10] Violini N 2011 PhD Thesis (Perugia University)
[11] Squires GL 1996 *Introduction to the Theory of Thermal Neutron Scattering* (New York: Mineola-Dover)
[12] Soderholm S, Roos YH, Meinander N and Hotokka M 1999 *J. Raman Spectroscopy* **30** 1009
[13] Schirmacher W, Ruocco G, and Scopigno T 2007 *Phys. Rev. Lett.* **98** 025501
[14] Gaudin D and Jankowskit K 1980 *Organic Mass Spectrometry* **15** 78
[15] Zanotti JM, Bellissent-Funel MC and Chen SH 1999 *Phys. Rev.* E **59** 3084
[16] Koza MM, Schober H, Parker SF and Peters J 2008 *Phys. Rev.* B **77** 104306

Reflectivity from floating bilayers: can we keep the structural asymmetry?

V Rondelli [1], G Fragneto[2], S Motta[1], E Del Favero[1], L Cantù[1]

[1]Dept. of Chemistry, Biochemistry and Biotechnologies for Medicine, University of Milan, Segrate, Italy
[2]Institut Laue-Langevin, Grenoble Cedex, France

E-mail: valeria.rondelli@unimi.it

Abstract. To assess the structure of complex biomembranes, the use of asymmetric model systems is rare, due to the difficulty of realizing artificial membranes with desired heterogeneous composition and applicable for single membrane structural investigation. We developed an experimental model with a single macroscopic bilayer floating on top of another adhering to a silicon flat surface, prepared by Langmuir-Blodgett Langmuir-Schaefer technique, then investigated by neutron reflectivity. On the way to more complex systems, containing lipids of different nature, we tested whether a simple imposed asymmetry is kept in time and whether it can stand some standard experimental protocols commonly employed in treating model membranes. We focused on cholesterol, a basic component with a transverse distribution that is not symmetric in biomembranes, and may assume specific location in functional domains. So we forced different asymmetries in the "adhering + floating" bilayers system composed of phospholipids and cholesterol in bio-similar mole ratios. The neutron reflection accessible length-scale and its sensitivity, enhanced by the possibility to play with deuteration, allowed assessing the cross profile of the membrane and revealing that lipid redistribution can occur.

1. Introduction

The structural complexity of biomembranes, based on their heterogeneity in composition, dramatically involves the asymmetric disposition of different components, from lipids to proteins, in the transverse or the longitudinal directions. Furthermore, inhomogeneities in the two leaflets of a membrane can couple, constituting the basis for the structural stabilization and modulation of functional domains involved in transmembrane signaling. In order to build suitable mimics of complex biomembranes, it is then necessary to realize asymmetric model systems applicable for structural investigation. The experimental study of asymmetric model membranes is rare, due to the difficulty of realizing artificial membranes with wanted and defined heterogeneous composition. However some attempts have been made, for example by preparing phospholipid unilamellar vesicles (LUVs) containing small amounts of glycosphingolipids only in their outer layer. The mechanical properties of the membrane were found to be strongly affected by the doping glycolipids, producing a softening that turns to hardening in the case of symmetric redistribution of molecules [1]. Fine assessment of the inner structure and inhomogeneities of the membrane requires overall aggregate monodispersity. In this framework we developed an experimental model with a single macroscopic bilayer floating at 1.5-2 nm on top of another adhering to a silicon flat surface, prepared, layer-by-layer, by combination of Langmuir-Blodgett and Langmuir-Schaefer techniques. The system so built has led in the past to stable and reproducible floating bilayers [2]. With this technique the composition of each layer can be chosen suitably. On the way to more complex systems, containing lipids of different nature, we tested whether a simple imposed asymmetry was kept in time and whether it could stand some standard experimental protocols commonly employed when dealing with model membranes (i.e. annealing and solvent exchange). We focused on cholesterol, a basic component with a transverse distribution that is not symmetric in biomembranes, and may assume specific location in functional asymmetric domains. It has been shown that floating bilayer systems containing cholesterol can be prepared, obtaining stable membranes for low cholesterol:phospholipid ratios, in the range 1-10 mol % [3-9].

In the present work, we forced asymmetry in an "adhering + floating" bilayers system composed of phospholipids and cholesterol in much higher amounts, in bio-similar mole ratios (11:2.5 mol:mol). We investigated the structure of the bilayers by neutron reflectivity, which couples the Angstrom scale of neutron accessibility to the possibility of changing the contrast profile of the system. In fact, selective deuteration of lipids or different H_2O/D_2O solvent composition can be exploited to allow or enhance the visibility of different components.

We present results on the feasibility of the desired asymmetric free floating membranes, and on the effects of common experimental protocols on the overall stability of the bilayer and on the lipid redistribution between different leaflets.

2. Materials and methods

Both hydrogenated and fully deuterated lipids were used. Cholesterol was purchased from Sigma-Aldrich Co, d_{85}-DSPC, d_{75}-DPPC were from Avanti Polar Lipids Co.. According to a well assessed standard protocol [10-11], lipids were dissolved in chloroform (99%) to a final concentration of 1mg/ml. Mixed lipid systems were obtained by mixing appropriate amounts of various chloroform lipid solutions. Lipids were then deposited on the surface of a Langmuir trough filled with pure water kept at T = 18°C. The various monolayers were compressed un to a surface pressure of 40 mN/m, similar to the one in real systems [2], while recording the corresponding (π-A) isotherms. All of the used monolayers are in the gel phase in these conditions. They were then layer-by-layer deposited on a silicon substrate. Asymmetric bilayers were realized by completely changing the monolayer in between different steps.

Pressure-area isotherms

Pressure-area (π-A) isotherms were recorded on a Nima Langmuir-Blodgett trough, using a Wilhelmy plate for pressure sensing. All (π-A) experiments were carried out at 18°C (±0.5), below the melting temperature of the lipids used. Water for the subphase was processed in a Milli-Q system (Millipore, Bedford, MA), to a resistivity of 18 MΩ.cm. Each lipid solution was spread over the water subphase and chloroform was let to evaporate completely, over 15min. All isotherms were recorded using a barrier speed of 25 cm^2/min.

Stability and reproducibility of Langmuir films were verified by performing various compression-expansion cycles on two different Nima Langmuir troughs.

Floating bilayer buildup

Substrates were single crystals of silicon (5 x 5 x 1.5 cm^3) polished on one large face (111) . The silicon blocks were cleaned before use in subsequent baths of chloroform, acetone, ethanol, water and treated with UV-Ozone for 30min [12]. The contrast solutions were H_2O (Milli-Q system), D_2O (99% pure, provided by ILL) and Silicon Matched Water, that is, a mixture of H_2O and D_2O with the same scattering length density of Silicon (SMW i.e. 0.62 H_2O and 0.38 D_2O volume fractions).

Double bilayer depositions were done in water coupling the Langmuir-Blodgett [13] and Langmuir-Schaefer Techniques [14], as follows. At the initial stage, the silicon block was immersed in water at 18°C in the Langmuir trough. A suitable lipid solution, the one chosen for the adhering bilayer was spread on the water surface and progressively compressed to 40 mN/m, the optimal for phosphocolines [2]. The silicon block was then almost completely risen and subsequently dipped again into water (speed 6 mm/min) across the monolayer, keeping the pressure constant. Two facing monolayers were adsorbed onto the block. Then, the water surface was completely cleaned and a different solution was spread, with the composition desired for the inner side of the floating membrane. The third layer was deposited by rising the block again, according to the same Langmuir technique. To prepare asymmetric floating bilayers, the surface of the trough was cleaned again and a solution with the composition desired for the outer side of the floating membrane was spread and compressed to 40 mN/m. The fourth closing monolayer was deposited by rotating the block by 90°, in the Langmuir Schaefer configuration, and lowering it carefully onto the surface.

The block was then closed in a teflon cell and fixed by an aluminum thermostated cage.

Neutron Reflectivity

Reflectivity measurements were performed on the D17 [15] reflectometer at ILL, Grenoble, France (TOF mode, resolution = 1-5%, λ range between 2 and 20 Å, with two incoming angles of 0.7 and 4°).The cell was oriented vertically and kept in position while changing solvents and temperature. Measurements were performed at the silicon-water interface, the beam coming from the side of the silicon block.

In a neutron reflectivity measurement R, the ratio between the intensities of the reflected and incoming beams, is collected, as a function of q, the momentum transfer perpendicular to the interface [16].

Reflectivity is related to the scattering length density across the interface by the approximate relation:

$$R(q_z) \approx \frac{(16\pi^2)}{q_z^2} |\rho(q_z)|^2$$

which is the reflectivity in the Born approximation [17]. $\rho(q_z)$ is the Fourier transform of the scattering length profile $\rho(z)$ along the normal to the interface, giving information about the composition of each layer and about its local structure. The scattering length density is given by:

$$\rho(z) = \sum_j b_j n_j = \sum_j b_j n_j$$

where n_j is the number of nuclei per unit volume and b_j is the scattering length of nucleus j.

The method of analysis often used for specular reflection data involves the construction of a model of the interface that may be represented by a series of parallel layers of homogeneous material. Each layer is characterised by a scattering length density (SLD) and a thickness, which are used to calculate a model reflectivity profile by means of the optical matrix method (Born and Wolfe, 1989). The interfacial roughness between any two consecutive layers may also be included in the model by the Abeles method. The calculated profile is compared to the measured profile and the quality of the fit is assessed by using χ^2 in the least-squares method.

Data were analyzed using the software Motofit [18], allowing simultaneous fit of data sets referred to the same sample in different contrast conditions, using the SLDs reported in table 1.

Table 1. Properties of materials used

Material	SLD (10^{-6}Å^{-2})[a]
Si	2.07
SiO_2	3.41
H_2O	-0.56
D_2O	6.36
cholesterol $C_{27}H_{46}O$	0.22
deuterated lipid heads $C_{10}D_{13}H_5O_8PN$	5.70
deuterated lipid chains $C_{30}D_{62}$ gel phase	7.66
deuterated lipid chains $C_{30}D_{62}$ fluid phase	6.13

[a] SLD values are from ref [19] and [20]

3. Results and discussion

3.1 Langmuir monolayers

Since the final goal of our study is to assess the structure of complex biomembranes, with this work we want to build up a basic model membrane with defined heterogeneous and asymmetric composition.

The effects generated by cholesterol addition to floating membranes have been studied [3-4] by means of neutron reflectivity in the case of low total amount of cholesterol, with particular attention to the phase behavior.

We now want to build up stable asymmetric floating cholesterol-containing bilayers with bio-similar composition. The aim of this work was to study the cholesterol transverse distribution inside the membrane and to test whether the forced asymmetry is kept.

First, we verified the effect brought about by addition of cholesterol to DPPC, by recording the (π-A) curves of the mixed Langmuir monolayers at the air-water interface as a function of different molar ratios between d_{75}-DPPC and cholesterol, from the gas to gel phase and, particularly, to the 40 mN/m pressure required for deposition. The corresponding curves are reported in fig 2 and display the expected features, following the existing literature. This guarantees the original integrity and compactness of the bilayer-forming monolayers. Moreover, the monolayers are stable in time and the pressure-area results are repeatable both after subsequent compression-expansion cycles and after monolayer re-spreading.

Progressive addition of cholesterol to d_{75}-DPPC results in progressively decreasing the occupied area per average molecule. For all the systems, at a given pressure, the measured area per molecule is lower than expected for ideal mixing. In fact, the effect of cholesterol is to rigidify and order the lipid chains, with a consequent reduction of the occupied area per average molecule at the air-water interface.

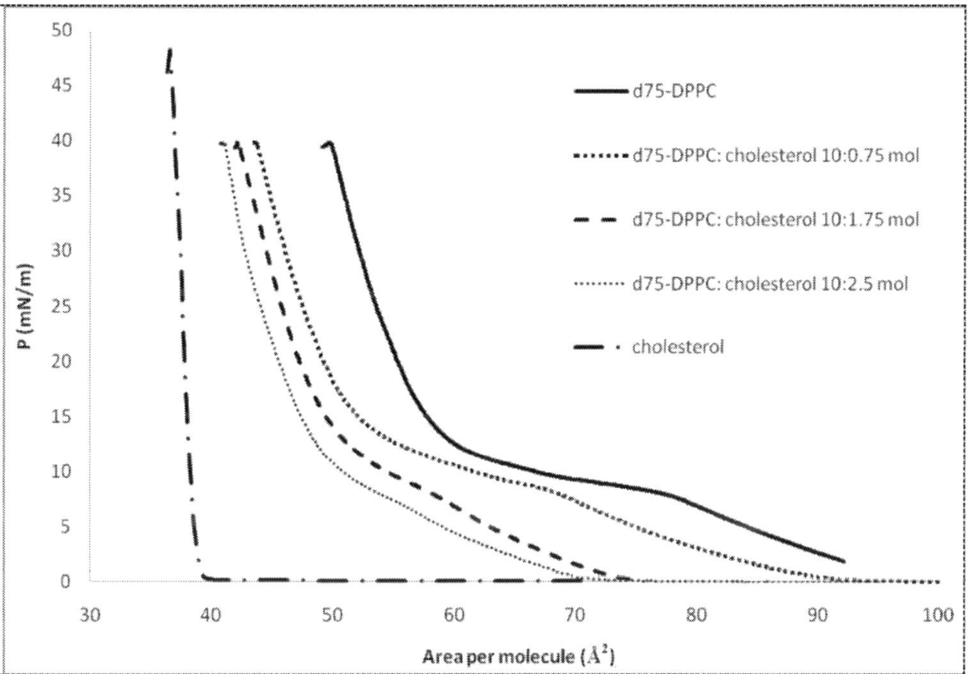

Figure 1. Pressure-area isotherms of various mixtures of d_{75}-DPPC and cholesterol at 18°C. The subphase is water.

To study the cholesterol transverse distribution inside the membrane we compared two floating model membranes: one was a 'blank', with no cholesterol, the other was a cholesterol-containing bilayer (see fig 2). For the bilayer adhering to the silicon support, we selected the long chain phospholipid DSPC, being in gel phase all over the investigated temperature range, from 22°C to 51°C. This guarantees the compactness and stability of the supporting bilayer. For the floating bilayer, DPPC was used as lipid matrix, being the lipid most likely to be found in membrane domains enriched in cholesterol and sphingolipids.

In biological membrane domains, the three components phospholipid : sphingolipid : cholesterol are present in a 10 : 1 : 2.5 molar ratio. A good approximation to simulate the composition of a biomembrane microdomain is obtained by inserting the 70% of the total amount of cholesterol in the inner layer of the membrane, and 30% in the outer. Accordingly, as reported in fig 1, for the cholesterol-containing system, the third layer was made of DPPC : cholesterol in molar ratio 11 : 1.75 and the fourth layer was made of DPPC : cholesterol in molar ratio 11 : 0.75.

We also applied selective deuteration to enhance or minimize the visibility of different components of the membranes under study. In particular, deuterated d_{75}-DPPC was used in order to highlight cholesterol with respect to the lipid matrix.

Journal of Physics: Conference Series **340** (2012) 012083 doi:10.1088/1742-6596/340/1/012083

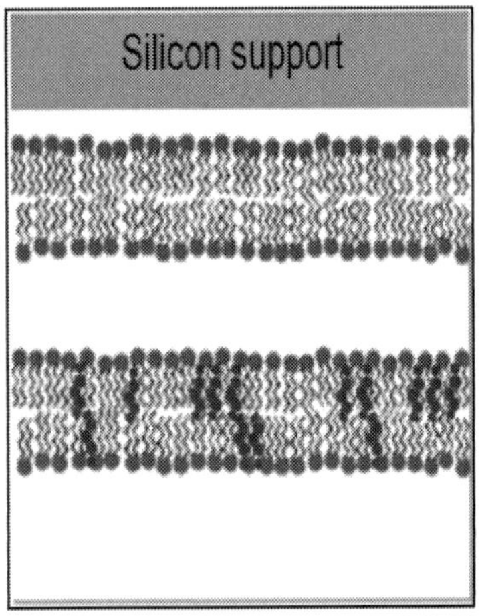

		Sample 1	Sample 2
Layer 1	ADHERENT BILAYER	DSPC	d_{85}-DSPC
Layer 2		DSPC	d_{85}-DSPC
Layer 3	FLOATING BILAYER	d_{75}-DPPC	d_{75}-DPPC:CHOL 11:1.75
Layer 4		d_{75}-DPPC	d_{75}-DPPC:CHOL 11:0.75

Figure 2. Scheme of the samples construction

3.2 Reflectivity measurements

Measurements on a bare silicon block were performed, in different solvents (H_2O, D_2O, SMW), to measure the characteristics of the silicon oxide layer formed at the silicon-air interface.

Then, both the bare matrix and the cholesterol containing samples were placed under the neutron beam as they came from the deposition procedure, without any annealing. They were thermostated at 22°C and four subsequent reflectivity measurements were taken, of one hour each, to evaluate any molecular exchange eventually occurring among the four lipid layers. The four spectra overlap, showing that the system is stable in these conditions over these times.

As a further step, the cholesterol containing system was submitted to a temperature increase to 51°C (the procedure currently meant by annealing) and, after a reflectivity measurement, it was lowered back to 22°C. Reflectivity was measured again at this point. Results are reported in fig 3.

Experiments at 22°C, after the annealing, have been performed in three contrast solutions: water, heavy water and silicon matched water. The adhering + floating system has shown to be stable against solvent exchange, consisting of progressive substitution of the aqueous solvent under continuous flow. This opens the possibility to use a biological relevant solvent in a true biomimetic perspective.

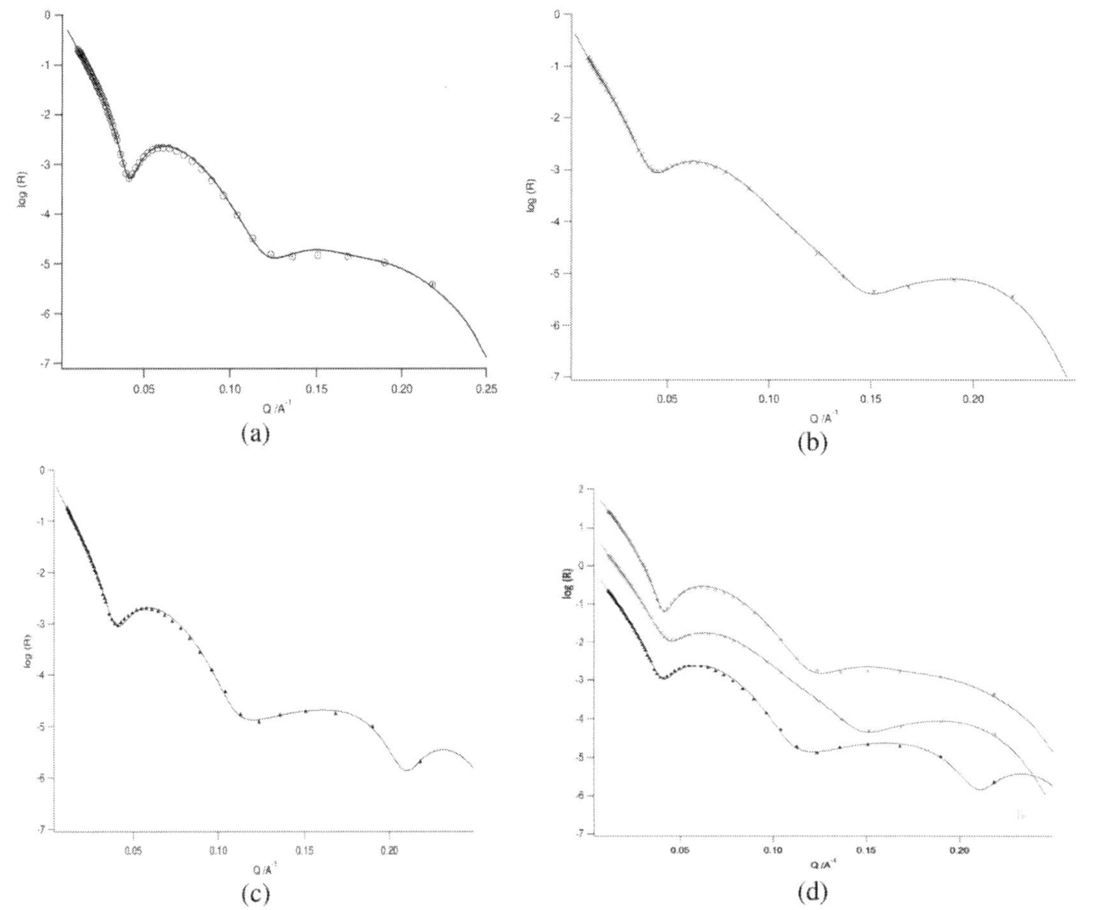

Figure 3. Reflectivity spectra of the cholesterol-containing sample in H_2O (a) at 22°C before annealing, (b) at 51°C, (c) at 22°C after annealing.

In insert (d) the curves have been arbitrarily shifted by 5 (51°C) and 10 (22 °C) for better clarity in comparing them: upper curve: 22°C before annealing. Middle curve: 51°C. Lower curve: 22°C after annealing.

The fit of the experimental data was performed with Motofit using a 11-layers model. This was composed by the silicon oxide, a water layer, the first bilayer (composed of 4 layers: heads, chains, chains, heads), a water layer and the floating bilayer (composed of 4 layers: heads, chains, chains, heads). We performed a simultaneous fit of the three solvents, gaining a better resolution for the complex structure. The obtained parameters are reported in tables 2 and 3. The errors are estimated from the maximum variation in the acceptable fit subject to the constraints of space filling and stoichiometry.

First, from table 2, we notice that the addition of cholesterol leads to a thickening of the floating membrane, despite a reduction of water content. This indicates that the bilayer is more compact, in agreement with the (π-A) curves. Also, it increases the roughness among the layers, suggesting that some cholesterol molecules can sit across the layers, without participating exclusively to either of them. This effect has been observed also in DPPC- cholesterol supported bilayers [21]. The cholesterol presence in the two layers seems not uniform, although not respecting the forced distribution. No significant differences between the two adherent DSPC bilayers are detected. No water was detected between the silicon oxide and the first bilayer headgroup, whereas between the two bilayers a 20±1 Å thick water layer was found.

Table 2. Parameters obtained fitting the reflectivity profiles of two samples studied. Heads and chains 3 refer to the inner layer of the floating bilayer, 4 refer to the outer layer.

	d_{75}-DPPC			d_{75}-DPPC + chol		
	t; r [a]	w [b]	chol [c]	t; r [a]	w [b]	chol [c]
Heads 3	8; 4	20	0	7; 1	15	0
Chains 3	16; 5	12	0	22; 8	6	43
Chains 4	18; 1	19	0	18; 8	11	57
Heads 4	9; 5	25	0	9; 4	16	0

[a] t=layer thickness (±1, Å); r= layer roughness (±2 Å).
[b] w= layer water content (±5 % volume).
[c] chol= layer cholesterol relative content (±2% over the total amount of cholesterol i.e. 11:2.5 molar ratio lipid:cholesterol).

Table 3. Parameters obtained fitting the reflectivity profiles at different temperatures of the asymmetric cholesterol containing sample. The numeration of layers goes from 1 to 4, starting from the layer towards the Silicon to the outer layer of the floating membrane.

	22°C			51°C			22°C back		
	t; r [a]	w [b]	chol [c]	t; r [a]	w [b]	chol [c]	t; r [a]	w [b]	chol [c]
Heads 1	10; 8	24	0	9; 8	22	**0**	6; 8	15	0
Chains 1	19; 1	17	0	19; 1	15	0	22; 6	7	0
Chains 2	19; 1	19	0	23; 8	10	0	23; 5	21	0
Heads 2	8; 8	25	0	7; 8	19	0	8; 5	25	0
water	19; 8		0	22; 3		0	20; 1		0
Heads 3	7; 1	15	0	6; 8	45	0	10; 8	45	0
Chains 3	22; 8	6	43	16; 8	38	58	16; 9	40	50
Chains 4	18; 8	11	57	16; 1	28	42	16; 7	40	50
Heads 4	9; 4	16	0	6; 1	45	0	10; 9	45	0

[a] t=layer thickness (±1, Å); r= layer roughness (±2 Å).
[b] w= layer water content (±5 % volume).
[c] chol= layer cholesterol relative content (±2% over the total amount of cholesterol i.e. 11:2.5 molar ratio lipid:cholesterol).

A different situation is found upon annealing, as shown in table 3.

No cholesterol was found in the d_{85}-DSPC adhering bilayer before, during or after the annealing. However the cholesterol was found to assume a symmetric distribution in the hydrophobic region of the floating d_{75}-DPPC bilayer. This indicates that there is no structural stringent constraint for cholesterol as such to stay in an asymmetric distribution. Kinetic or trapping effects could be involved in the initial asymmetry. Also, at 51°C, over the floating bilayer melting temperature, the water layer between the two bilayers thickens from 20 to 22 Å.

Annealing the sample is commonly used to reduce local packing defects originated by the preparation procedures of lipid systems, and then it is applied also to deposited bilayers. Data of Table

3 show that the first bilayer is presumably extensively affected by single-lipid protrusion. In fact, the strictly-hydrophobic region is shallower than expected for a DSPC bilayer, while the shells are thicker, rougher and more hydrated. After annealing, the structural parameters of the first bilayer become closer to expected, and it becomes more compact. On the other hand, after annealing, the boundaries of the different regions of the floating bilayer become more uniform and defined. But we show that annealing is likely to induce also a more extensive redistribution of lipids, impeding the realization of biomimetic systems of desired asymmetry. It can be seen that after annealing the distribution of cholesterol has been flattened, within the floating bilayer. In these conditions as reasonable no cholesterol migration between bilayers has occurred. We also observe that after annealing the hydration of the outer three layers increases. The floating bilayer chains contain 40% in volume of water and its thickness decreases.

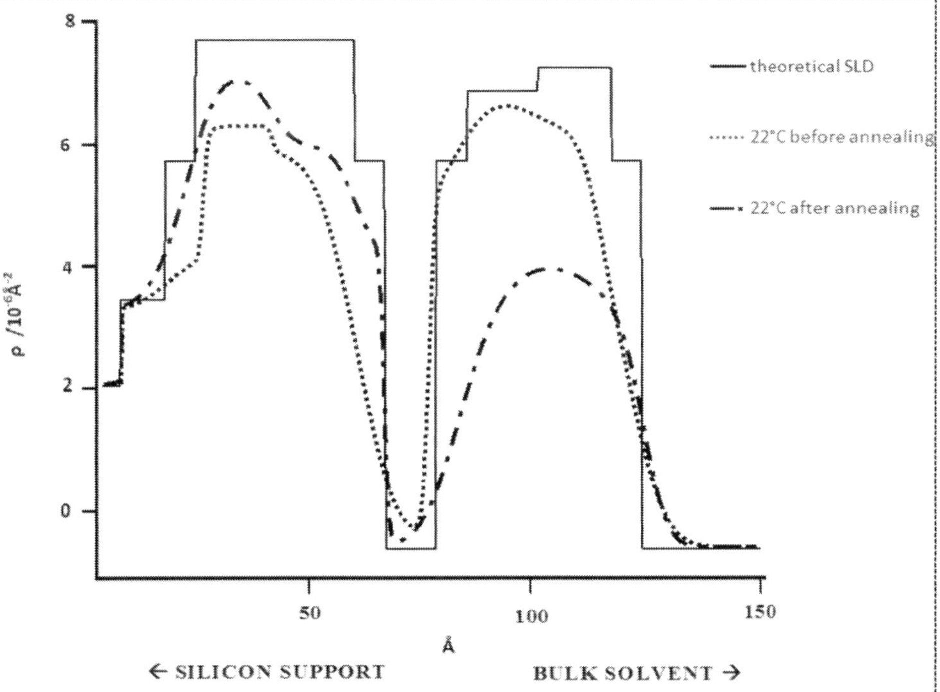

Figure 4. Theoretical and experimental (before and after the annealing) contrast profiles of the cholesterol-containing system at 22°C in water. The theoretical one refers to the system as prepared (asymmetric disposition of cholesterol in the floating bilayer), with no solvent penetration and no roughness.

The present experiment suggests that annealing should eventually be excluded from the protocol to be used with asymmetric samples, both for lipid redistribution and for macroscopic bilayer integrity. In fact, the estimated water penetration after annealing could correspond to bilayer punctuation or fragmentation occurring at high temperature.

On the other hand, some positive indication can be drawn. The role of coupling of molecules, shapes and morphology, in the build-up and in the non-covalent stabilization of uneven distribution of components is a debated topic. This is likely to show up even in the simplest cases, as the one presented here. In real membranes segregation, preferential distribution and domain formation are existing phenomena, also strongly involving cholesterol. Alternatively, this experiment shows that cholesterol is easily evenly redistributed in the forcedly flat, bicomponent membrane considered, despite the initial asymmetry. This seems to redirect to the concept of coupling as determinant for membrane structure: coupling between membrane curvature and molecular sorting, as well as coupling of pair molecules matched in the packing shape. Cholesterol uneven distribution could be preserved in a non-flat geometry or by a suitable shape-coupling with other lipids within a complex membrane (like glycosphingolipids) also without invoking a prominent role of proteins. In addition, the characteristic umbrella effect, typical of cholesterol and evident in the π-A behaviour, could be modulated by 3D morphological constraints.

Acknowledgements
The authors wish to thank the ILL for beam-time and use of the PSCM facilities.

References

[1] Brocca P, Cantù L, Corti M, Del Favero E, Motta S 2004 Shape Fluctuations of Large Unilamellar Vesicles Observed by Laser Light Scattering: Influence of the Small-Scale Structure *Langmuir* **20**(6) 2141-2148
[2] Charitat T, Bellet-Amalric E, Fragneto G, Graner F 1999 Absorbed and free bilayers at the solid liquid interface *Eur. Phys J B* **8**, 583-593
[3] Stidder B, Fragneto G, Roser S J 2005 Effect of Low Amounts of Cholesterol on the Swelling Behavior of Floating Bilayers *Langmuir* **21**, 9187
[4] Stidder B, Fragneto G, Cubitt R, Hughes A V, Roser S J 2005 Cholesterol Induced Suppression of Large Swelling of Water Layer in Phosphocholine Floating Bilayers *Langmuir* **21**, 8703-8710
[5] Smondyrew A M, Berkowitz M L 1999 Structure of dipalmitoylphosphatidylcholine/cholesterol bilayer at low and high cholesterol concentrations: Molecular dynamics simulation *Biophys. J.* **77**, 2075-2089
[6] Pasenkiewicz-Gierula M, Rog T, Kitamura K, Kusumi A 2000 Cholesterol effects on the phosphatidylcholine bilayer polar region: A molecular simulation study *Biophys. J.*, **78**, 1376-1389.
[7] Chiu S W, Jakobsson E, Mashl R J, Scott H L 2002 Cholesterol induced modifications in lipid bilayers: A simulation study *Biophys. J.*, **83**, 1842-1853.
[8] Hofsass C, Lindahl E, Edholm O 2003 Molecular dynamics simulations of phospholipid bilayers with cholesterol. *Biophys. J.*, **84**, 2192-2206.
[9] Tu K C, Klein M L, Tobias D J 1998 Constant-pressure molecular dynamics investigation of cholesterol effects in a dipalmitoylphosphatidylcholine bilayer *Biophys. J.*, **75**, 2147-2156.
[10] Blodgett K B, Langmuir I 1937 Built-Up Films of Barium Stearate and Their Optical Properties *Phys. Rev* **51**, 964-982.
[11] Fragneto G, Charitat T, Graner F, Mecke K, Perino-Gallice , Bellet-Amalric 2001 A fluid floating bilayer *Europhys Lett* **53** (1) 100–106
[12] Vid J R 1985 *Vac. Sci. Tecnol. A* **3**, 3, 1027
[13] Roberts G 1990 Langmuir Blodgett films *Plenum press* New York
[14] Tamm L K, McConnell H M 1985 *Biophys J* **47**, 105 1985
[15] Cubitt R , Fragneto G 2002 D17: The new reflectometer at the ILL *Applied Physics A* **74**, S329-S331
[16] Penfold J, Thomas R K 1990 *JPCM* **2**, 1369
[17] Born M, Wolf E 1959 Principles of Optics *Pergamon Press Ltd, London.*
[18] Nelson A 2006 Co-refinement of multiple contrast neutron/X-ray reflectivity data using MOTOFIT *Journal of Applied Crystallography* **39**, 273-276
[19] Fragneto G, Graner F, Charitat T, Dubos P, Bellet-Amalric E 2000 *Langmuir*, **16**, 4581.
[20] Deme B, Lay-Theng L, 1997 *J. Phys. Chem. B*, **101**, 8250.
[21] Vacklin H, Tiberg F, Fragneto G, Thomas R 2005 Composition of supported model membranes determined by neutron reflection *Langmuir* **21**, 2827-2837

Neutron magnetic resonance and non-specular reflection from a magnetic film placed in an oscillating magnetic field

S V Kozhevnikov[1,5], V K Ignatovich[1], Yu V Nikitenko[1], F Ott[2,3], F Radu[4], A Rühm[5] and J Major[5]

[1]Frank Laboratory of Neutron Physics, JINR, 141980 Dubna, Moscow Region, Russian Federation

[2]CEA, IRAMIS, Laboratoire Léon Brillouin, F-91191 Gif sur Yvette, France

[3]CNRS, IRAMIS, Laboratoire Léon Brillouin, F-91191 Gif sur Yvette, France

[4]Helmholtz-Zentrum Berlin für Materialien und Energie, Albert Einstein Strasse 15, D-12489 Berlin, Germany

[5]Max-Planck-Institut für Intelligente Systeme (formerly Max-Planck-Institut für Metallforschung), Heisenbergstr. 3, D-70569, Stuttgart, Germany

E-mail: kozhevn@nf.jinr.ru

Abstract. Experimental results on neutron reflection from a magnetic film placed in an oscillating magnetic field are reported. We found evidence for neutron spin resonance in the film and observed a spatial beam-splitting. The beam-splitting finds its origin in the exchange of an energy quantum $\hbar\omega$ between the oscillating field and the neutron.

1. Introduction

Acceleration and deceleration of neutrons in a magnetic field were experimentally demonstrated in the case of a constant [1] and an oscillating [2] magnetic field. In 1978, V.K. Ignatovich predicted [3] the effect of spatial beam-splitting upon reflection and refraction from media in which the magnetic induction is non-collinear with respect to the applied magnetic field. This effect was observed for reflected [4,5] and refracted [6] beams. The magnitude of the beam-splitting and its relation to the sign of the spin-flip transitions corresponded to the gain or loss of Zeeman energy by the neutron during a spin-flip event in a rather large applied magnetic field (~10 kOe). In references [7-9], beam-splitting upon reflection and refraction in a low magnetic field (about 100 Oe) applied parallel to the film surface was observed. In this case the magnitude of the beam-splitting and the types of neutron spin transitions involved found their origin in the presence of magnetic domains and domain walls with non-collinear magnetization.

Recently, spatial beam-splitting upon reflection and refraction was predicted [10-14] in the case of a magnetic film placed in an oscillating magnetic field. We report here on the experimental observation of this effect.

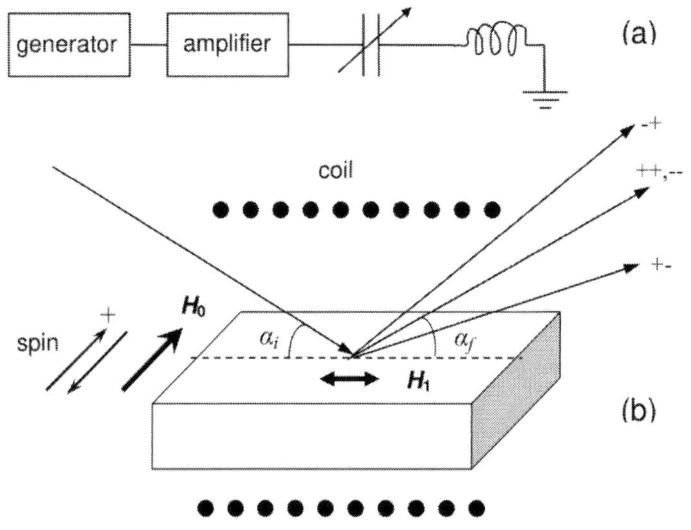

Figure 1. Experimental setup. (a) Schematic of the electric circuit generating the oscillating field. (b) Geometry of the experiment: H_0 is the constant magnetic field and H_1 is the oscillating magnetic field produced by the rectangular coil.

2. Experimental setup

The experimental setup is presented in figure 1. A radio-frequency generator (Rohde&Schwarz SMC100A) and a power amplifier (100 W CW, American Microwave Technology M3200/M3000) were used to create an oscillating field in a resonant LC circuit. The adjustable capacity could be varied in the range 5 – 100 pF. The magnetic coil had a length of 50 mm, a cross-section of 5×30 mm^2, and consisted of 50 turns of 1 mm diameter Cu wire.

The experiment was conducted at the polarized neutron reflectometer N-REX$^+$ (Forschungs-neutronenquelle Heinz Maier-Leibnitz, FRM II, Garching, near Munich, Germany) with the sample surface oriented horizontally. The neutron wavelength λ was 4.26 Å (FWHM = 1 %), the angular resolution of the incident beam was 0.006°. A supermirror in transmission mode was used as polarizer. The polarization of the incident beam was 97 %. The polarization of the reflected neutrons was not analyzed. A Mezei type spin-flipper of 100 % efficiency was used to flip the beam polarization. The reflectivity was measured using a two-dimensional ^{3}He position-sensitive detector (PSD) with a spatial resolution of 3 mm. The distance between the collimating diaphragm and the sample was 2200 mm, the distance between the sample and the detector was 2500 mm. The sample was a permalloy (Py) film of 5000 Å thickness deposited on a Si substrate with a size of 25×25×1 mm^3. The nominal composit-ion of the permalloy was Fe(20.6 % at.)Ni(79.4 % at.).

The geometry of the experiment is presented in figure 1b. The constant magnetic field H_0 = 20 Oe was produced by an electromagnet and applied perpendicular to the reflection plane. The polarization of the incident neutron beam was directed parallel ('+') or antiparallel ('–') with respect to this quantization field H_0. The sample was mounted on a glass holder inside the magnetic coil. The oscillating magnetic field $H_1(t) = H_1 \cos \omega t$ was applied in the film plane, perpendicular to the constant magnetic field H_0 and the neutron spin. The amplitude of the oscillating magnetic field was H_1 =10 Oe. The frequency of the magnetic field was varied in the interval f= 25…35 MHz. The grazing incidence angle of the neutron beam was fixed at $\alpha_i = 0.4°$. The reflected neutrons were measured as a function of the exit angle using the position-sensitive detector. The magnetic state of the sample was prepared by saturating it in an H_0 field of 1000 Oe and subsequently releasing the field to 20 Oe, where it was held throughout the measurements.

3. Sample characterization

The sample was characterized by polarized neutron reflectometry (PNR) and vibrating sample magnetometry (VSM) in CEA Saclay, France. In figure 2, the hysteresis loop of the film is presented. The film was close to magnetic saturation above 500 Oe and reached the full saturation equal to 9.5 kG for an applied field of 4 kOe. The coercive field of the film is of the order of about ± 60 Oe. The film is partially demagnetized for low fields and the average magnetization is 4 kG for an applied field of 20 Oe. VSM measurements show that in this state there is a small magnetization component perpendicular to the film plane (data not shown here). The fact that a small perpendicular anisotropy appears in thick permalloy films is documented in literature [15].

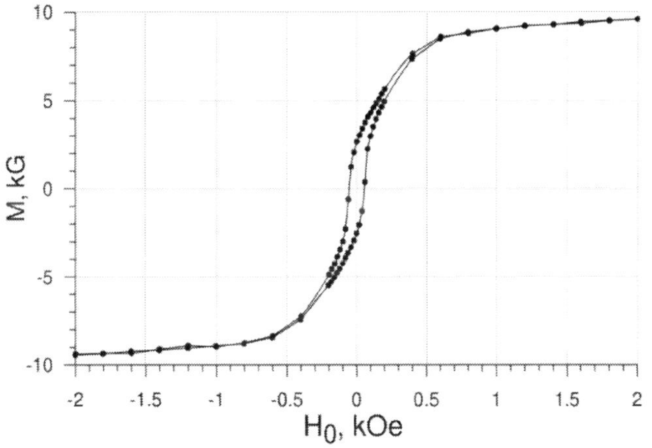

Figure 2. Hysteresis loop measured by VSM.

The characterization by neutron reflectometry was carried out at the polarized neutron reflectometer PRISM [16] installed at the reactor ORPHEE (LLB, Saclay, France). The neutron wavelength was set to 4.0 Å (FWHM = 0.37 Å), the angular resolution was 0.04°. In figure 3a, the spin-resolved experimental reflectivities R^{++} (solid symbols) and R^{--} (open symbols) are shown as a function of the vertical component of the wave-vector of the incident beam, $q = 2\pi \sin\alpha_i / \lambda$, for an applied field of 20 Oe (after saturation in a field of 4 kOe). The dashed line indicates the value of q corresponding to λ = 4.26 Å and $\alpha_i = 0.4°$ as used during the resonance experiment performed at N-REX$^+$. The solid lines correspond to fits obtained with the program *SimulReflec* [17]. The fit result U = 211.2 neV for the nuclear potential corresponds to 89 % of the theoretical value U = 237.3 neV for the nominal sample stoichiometry Fe(20.6 % at.)Ni(79.4 % at.). The layer density is reduced during the sample preparation which is the most likely explanation since reduced densities are usually observed for sputtered films. The fit result for the magnetization is M = 4.0 kG. This value is in agreement with the VSM data.

The reflectivities for a higher external magnetic field of 4.4 kOe are shown in figure 3b. The fit result for the nuclear potential U was the same as for the low field. The fit result for the magnetization is M = 9.0 kG. This is again very close to the VSM data. One can conclude that during the resonance experiments performed at H_0 =20 Oe the sample was partially demagnetized, and that it was initially saturated to a magnetization of about 9.0 kG by applying a high magnetic field of 1000 Oe.

4. Results

The first proof-of-principle experiment on beam-splitting at reflection from a permalloy film (2500 Å) placed in an oscillating magnetic field was carried out in 2006 at the time-of-flight polarized neutron reflectometer REMUR at the reactor IBR-2 (FLNP, JINR, Dubna, Russia). At a fixed frequency of

33.5 MHz, the neutron spin resonance was examined by scanning the constant external magnetic field in the range 100 – 6000 Oe. These results are not discussed here and will be published elsewhere.

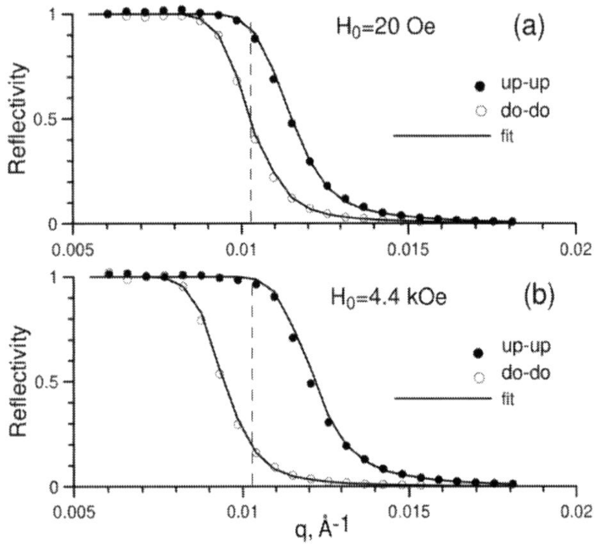

Figure 3. Experimental neutron reflectivities in the up-up (solid symbols) and down-down (open symbols) channel for a neutron wavelength of 4.0 Å. Solid lines are fit curves obtained with the program *SimulReflec* [17].
(a) demagnetized state in a field of 20 Oe; (b) saturated state in a field of 4.4 kOe. The dashed line corresponds to an incidence angle of 0.4° and a neutron wavelength of 4.26 Å as used during the resonance experiments at N-REX[+].

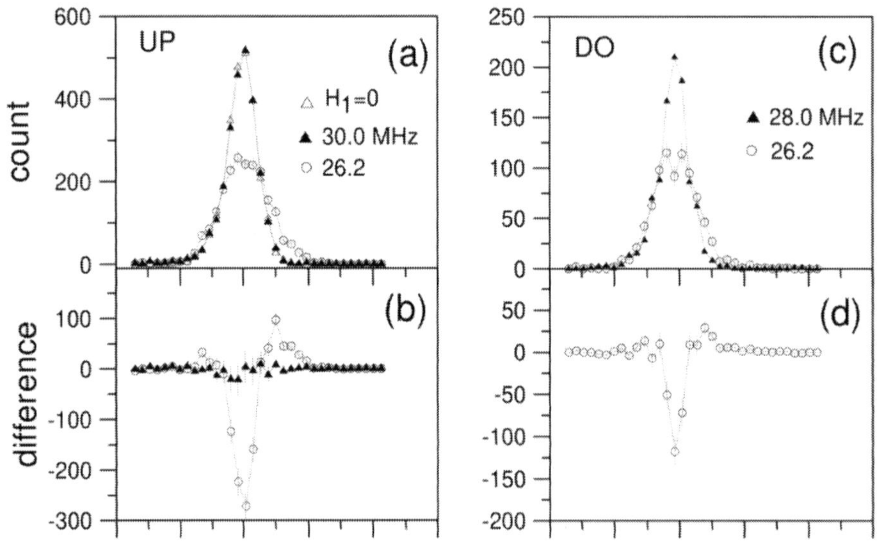

Figure 4. (Top row) Neutron counts versus reflection angle α_f without oscillating field (H_1=0 Oe) and with oscillating field (H_1=10 Oe) for the frequencies 26.2, 28.0 and 30.0 MHz. (Bottom row) Difference datasets obtained from the data plotted in the upper panels by subtraction of reference exit angle profiles as follows: In the spin up case (left column) the measurement for H_1=0 Oe was subtracted, and in the spin down case (right column) the measurement for 28 MHz was subtracted. Spin up and spin down refer to the polarization of the incident beam. The measuring time was 20 s per point.

In the experiment reported here, we scanned the neutron resonance by variation of the frequency of the oscillating magnetic field. In figure 4, the reflected intensity is plotted as a function of the reflection angle α_f for different oscillating magnetic fields. The measuring time was 20 s per point. Left and right panels correspond to up and down polarization of the incident beam, respectively. Figures 4a and 4c show raw neutron counts as a function of the reflection angle α_f for different frequencies (26.2, 28.0 and 30.0 MHz) of the oscillating field (H_1 = 10 Oe) and without an oscillating field (H_1 = 0 Oe). Figure 4b shows the difference of the signals obtained with oscillating field (26.2 and 30.0 MHz) and without oscillating field (H_1 = 0 Oe) for spin up polarization of the incident beam. Figure 4d shows the difference of the signals obtained at 26.2 MHz and at 28.0 MHz for the spin down case. In figure 4 one can distinguish three exit angle regimes:

- $\alpha_f = 0.338\ldots0.455°$, specular region ($\alpha_f \cong \alpha_i$)
- $\alpha_f = 0.199\ldots0.338°$, off-specular region I ($\alpha_f < \alpha_i$)
- $\alpha_f = 0.455\ldots0.641°$, off-specular region II ($\alpha_f > \alpha_i$)

One can see that the spin up signal at the frequency 30.0 MHz coincides with the signal without the oscillating field (H_1 = 0). At the frequency 26.2 MHz the signal decreases in the specular region and increases in the off-specular regions, in particular in region II. This observation holds for both initial beam polarizations.

The reflected signals shown in figure 4, namely the integral number of counts and the peak intensity in the specular region, are plotted as a function of the frequency of the oscillating field in figure 5. Left and right panels correspond to up and down polarization of the incident beam, respectively. One can identify a resonance feature at the frequency f_{res} = 26.2 MHz at which the reflected signal is minimized. The full width at half maximum (FWHM) of the resonance is Δf_{res} = 1.2 MHz. The relative signal reduction at the resonance is similar for up and down polarization of the incident beam: 25 % in case of the integrated counts and about 60 % in case of the peak maximum.

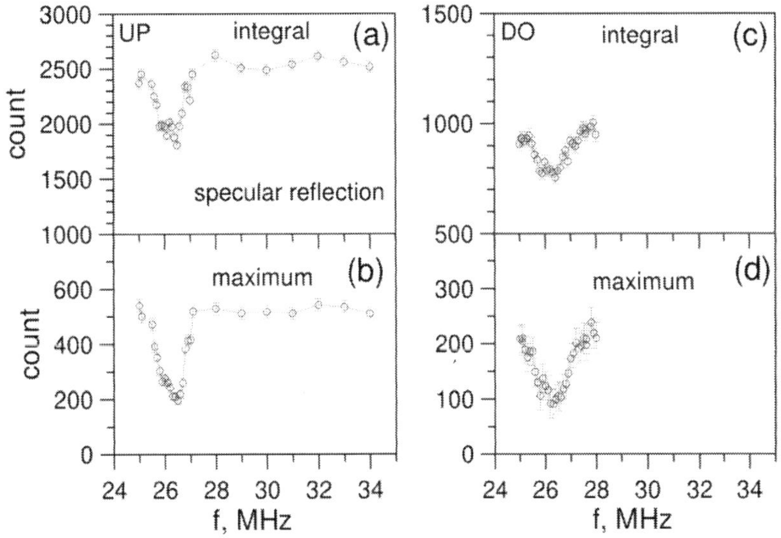

Figure 5. Reflected number of neutrons as a function of the frequency of the oscillating magnetic field. (top) integrated reflected signal, (bottom) maximum of the reflected peak, (left column) initial polarization spin up, (right column) initial polarization spin down.

The frequency dependence of the integrated signals as shown in figures 5a and 5c for the specular region is plotted again for the case of integration over the two off-specular regions I and II in figures 6a and 6b, for up and down polarization of the incident beam, respectively. Closed symbols correspond to region I ($\alpha_f < \alpha_i$), open symbols to region II ($\alpha_f > \alpha_i$). As opposed to figures 5a and 5c, however, reference datasets (obtained without oscillating field for spin up and for an oscillation frequency of 28.0 MHz for spin down, respectively) have been integrated over regions I and II and the results subtracted as a baseline before plotting the data in figure 6. Let us quantify our results by normalizing the integrated off-specular excess counts at the resonance as plotted in 6a and 6b by the integrated specular counts outside the resonance as plotted in figures 5a and 5c (i.e. 2500 counts for spin up, 1000 counts for spin down). In these terms, the spin-flip effect amounts to 3 % for region I ($\alpha_f < \alpha_i$) and 12 % for region II ($\alpha_f > \alpha_i$), both for up and down polarization of the incident beam. Similar numbers are obtained when comparing the neutron counts at the peak maxima of the specular and non-specular reflected beams accordingly. In this case, however, the statistical errors are larger since the contributing numbers of neutron counts are lower.

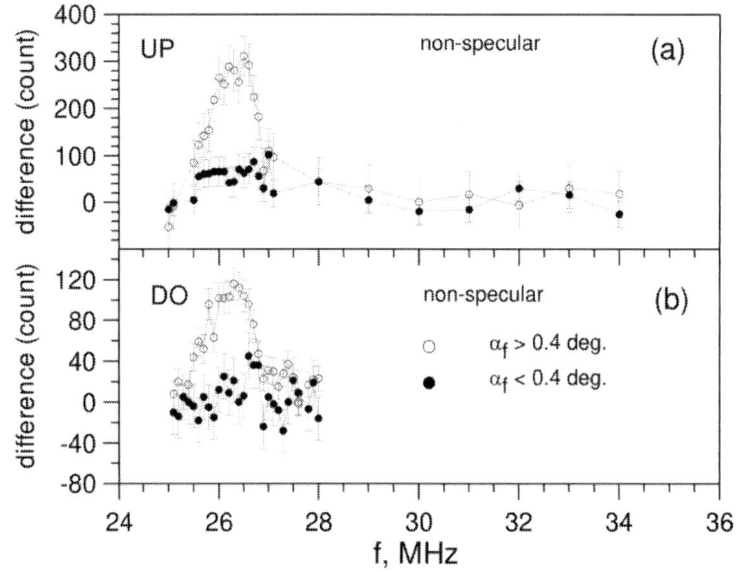

Figure 6. Excess counts of non-specularly reflected neutrons integrated over the exit angle intervals 0.455°-0.641° (open symbols) and 0.199°-0.338° (solid symbols) as a function of the frequency of the oscillating magnetic field. These excess counts were determined from the integrated raw data by subtraction of the corresponding integrated counts measured (a) without oscillating field for spin up and (b) at an oscillation frequency of 28.0 MHz for spin down polarization of the incident beam, respectively.

The observed fact is that the resonant off-specular intensities are proportional to the respective off-resonance specular reflectivities. A detailed quantitative explanation of the origin of this observation requires a more careful theoretical consideration. In this context it has to be investigated if the resonance phenomenon can be treated as a perturbation or if this observation is related to more general concepts such as particle flux conservation.

Having concluded that reflected neutrons are partially redirected into non-specular directions close to the resonance condition, we will now discuss the value of the measured resonance frequency f_{res}. It is clear that a neutron spin-flip should take place inside the film when the frequency of the applied oscillating field is close to the neutron Larmor precession frequency f_L inside the magnetic film. This Larmor frequency depends on the magnetic induction B inside the film as $f_L = \gamma B$, where $\gamma = 2\mu/\hbar = 2916\,\mathrm{G}^{-1}\mathrm{s}^{-1}$, μ is the magnetic moment of the neutron, and $\hbar = h/(2\pi)$ is the reduced

Planck constant. The resonance frequency $f_{res} = 26.2$ MHz found in our experiments corresponds to Larmor precession in a magnetic induction $B = 9.0$ kG. This value coincides with the saturation magnetization $M = 9.0$ kG obtained by VSM and by neutron reflectometry at PRISM under an external magnetic field $H_0 = 4.4$ kOe. The external magnetic field in the neutron spin resonance experiment at N-REX$^+$, however, was only $H_0 = 20$ Oe. In this low field the film was partially demagnetized and the average magnetization was $M = 4.0$ kG (as determined from VSM and reflectometry). However, at low external field the magnetic state of the permalloy film is characterized by large (several tens of μm) saturated domains in which the local magnetization is 9.0 kG, but whose orientations are distributed such that the average magnetization is reduced to 4.0 kG and oriented along the constant external field and the neutron beam polarization. This shows that the observed resonance phenomenon is caused by interaction of the neutrons with the local field in individual magnetic domains.

Let us finally discuss the magnitude of the observed beam-splitting. The Zeeman energy in the constant magnetic field H_0 is equal to $\pm\mu H_0$, where the signs '+' and '–' correspond to spin up and spin down neutrons, respectively. For $H_0 = 20$ Oe, the Zeeman energy is thus 0.12 neV which is totally negligible compared to the energy quantum $\hbar\omega_{res} = h f_{res} = 108.54$ neV of the oscillating magnetic field at the resonance frequency $f_{res} = 26.2$ MHz. The initial kinetic energy of the neutrons is $E = \frac{\hbar^2 k_i^2}{2m} = \frac{\hbar^2 k_{i\perp}^2}{2m} + \frac{\hbar^2 k_{i//}^2}{2m}$, where m is the neutron mass, and where $k_{i\perp} = k_i \sin\alpha_i$ and $k_{i//} = k_i \cos\alpha_i$ are the wave vector components perpendicular and parallel to the sample surface. During reflection and refraction from a homogeneous medium, the parallel wave vector component does not change: $k_{i//} = k_{f//}$. From energy conservation it thus follows for spin-flip reflection in the presence of an oscillating magnetic field of frequency f:

$$\frac{h^2 \sin^2\alpha_f}{2m\lambda^2} = \frac{h^2 \sin^2\alpha_i}{2m\lambda^2} \pm hf \ , \tag{1}$$

where the '+' sign corresponds to the spin-flip from the (Zeeman) low-energy '–' state (spin down) to the high-energy '+' state (spin up), and vice versa. Since incidence and exit angle are small $(\sin\alpha \approx \alpha)$, we can write:

$$\alpha_f = \left(\alpha_i^2 \pm f\frac{2m\lambda^2}{h}\right)^{\frac{1}{2}} \tag{2}$$

Note that this beam-splitting due to resonance with an oscillating field is opposite to the Zeeman beam-splitting. In the oscillating field we have $\alpha_f^{-+} < \alpha_i$ and $\alpha_f^{+-} > \alpha_i$, while in case of the Zeeman splitting one has $\alpha_f^{-+} > \alpha_i$ and $\alpha_f^{+-} < \alpha_i$ [6].

Figure 7 shows the beam-splitting as calculated from Eq. (2) for our experimental parameters, i.e. for a neutron wavelength of 4.26 Å and an external field oscillating at the resonance frequency $f_{res} = 26.2$ MHz. The two experimental data points obtained for $\alpha_i = 0.4°$ agree very well with the calculated exit angles for the spin-flipped neutrons. Note that the strong beam-splitting can clearly be attributed to the Zeeman energy difference $2\mu B$ related to the strong magnetic induction $B = 9.0$ kG in saturated domains, while the interaction energy $2\mu H_0$ with the constant external field would be way too small to explain the observed effect. On the other hand, the observed beam-splitting is of similar magnitude as the one reported in references [7-9] for reflection from a certain domain structure and domain wall morphology. In this case the external magnetic field was also low (20-100 Oe), but the local magnetic induction was reported to be high (20 kG), similar as in our case. Note, however, that this type of beam-splitting due to a certain static configuration of internal magnetic fields is fundamentally different from our observations, because in the study presented here the beam-splitting is absent outside the resonance condition. This proves that the strong beam splitting we observed is entirely due to local neutron spin resonance in saturated single domains.

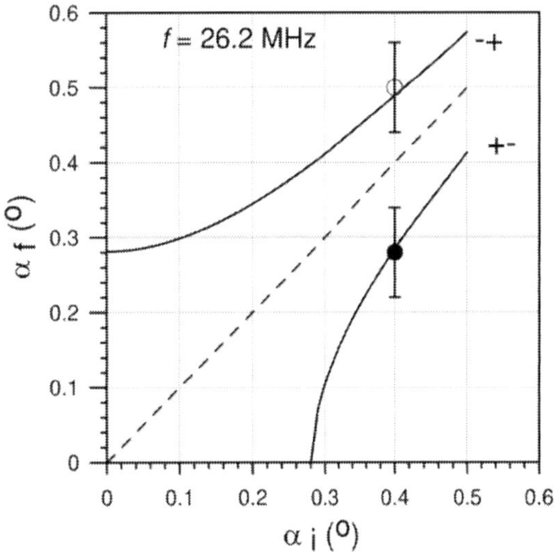

Figure 7. Calculated beam-splitting due to internal spin-flip induced by an external magnetic field oscillating at 26.2 MHz. The involved spin transitions, '−+' and '+−', are indicated. The symbols correspond to our experimental results. The error bars are defined by the spatial resolution of our position-sensitive detector.

5. Conclusions

We have obtained new experimental results from neutron reflection on a thin permalloy film placed in an oscillating magnetic field. The following key conclusions can be drawn.

- Neutron magnetic resonance takes place inside the film.
- The spatial beam-splitting is measurable only when the frequency of the oscillating external magnetic field is close to the Larmor precession frequency corresponding to the local magnetic induction in the sample.
- Non-specular reflected beams occur for both initial spin states of neutrons, but the intensities of the two non-specular beams are not identical.
- The resonant off-specular intensities are proportional to the respective off-resonance specular reflectivities.

The frequency of the oscillating magnetic field as applied in the experiments reported here can be used in the future as a new parameter for investigations of local magnetic properties in layered magnetic structures, and it may also be extended to the study of dynamic magnetic processes in such systems. This method can be called Neutron Resolved Magnetic Resonance.

Acknowledgements

The authors are thankful to Dr. T. Keller for discussions and J. Franke for the technical help during the experiment. S.K. is indebted to Doctor of Sciences A.I. Frank for the interest to this subject and fruitful discussions. The support of the management and staff at FRM II, Garching, and ORPHEE, Saclay, is gratefully acknowledged. This work has been supported by a focused Neutron Research Funding of the Max Planck Society, Munich. Also this work has been partially supported by the European Commission under the Sixth Framework Program through the Key Action: Strengthening the European Research Area, Research Infrastructures, contract No. RII3-CT-2003-505925, ECONET project 16273ZA and IMAMINE project.

References

[1] Alefeld B, Badurek G, and Rauch H, 1981 Phys. Lett. A **83** 32.

[2] Alefeld B, Badurek G, and Rauch H, 1981 Z. Phys. B **41** 231.
[3] Ignatovich V K, 1978 JETP Lett. Vol. **28**, No. 5, 286.
[4] Felcher G P, Adenwalla S, de Haan V O, and van Well A A, 1995 Nature **377** 409.
[5] Korneev D A, Bodnarchuk V I, and Ignatovich V K, 1996 JETP Lett. **63** 944.
[6] Aksenov V L, Nikitenko Yu V, and Kozhevnikov S V, 2001 Physica B **297** 94.
[7] Krist Th, Müller D J, and Mezei F, 1999 Physica **B 267-268** 194.
[8] Radu F, Leiner V, Westerholt K, Zabel H, McCord J, Vorobiev A, Major J, Jullien D, Humblot H, and Tasset F, 2005 J. Phys.: Condens. Matter **17** 1711.
[9] Aksenov V L, Kozhevnikov S V, Nikitenko Yu V, and Lauter H, 2000 Physica B **276-278** 179.
[10] Kozlov A V, and Frank A I, 2005 Phys. At. Nucl. **58** 1104.
[11] Frank A I, and Kozlov A V, 2009 Physica B **404** 2550.
[12] Ignatovich V K, Nikitenko Yu V, and Radu F, 2009 Nucl. Instr. and Meth. A **604** 653.
[13] Ignatovich V K, Nikitenko Yu V, and Radu F, 2010 Nucl. Instr. and Meth. A **620** 410.
[14] Nikitenko Yu V, Ignatovich V K, and Radu F, 2011 Physica B **406** 2473.
[15] Sugita Y, Fujiwara H, and Sato T, 1967 Applied Physics Letters **10** 229.
[16] Fermon C, Ott F, Legoff G, Glättli H, and Wintenberger V, 2000 Rev. Scient. Instrum. **71** 3797.
[17] SimulReflec. http://www-llb.cea.fr/prism/programs/simulreflec/simulreflec.html

Study of magnetization of a bilayer nanoststructure CoCu/Co (GF/F) by polarized neutron reflectometry

N.K. Pleshanov[1], V.L. Aksenov[2], A.P. Bulkin[1], A.A. Fraerman[3], V.A. Matveev[1], Yu.V. Nikitenko[2], V.G. Syromyatnikov[1], S.N. Vdovichev[3], V.M. Uzdin[4]

[1] Petersburg Nuclear Physics Institute, Orlova Roscha 1, Gatchina, Russia

[2] Joint Institute of Nuclear Research, Joliot-Curie str. 6, Dubna, Russia

[3] Institute of Microstructure Physics, GSP-105, N.Novgorod, Russia

[4] St.Petersburg State University, Ulyanovskaya str. 1, Petrodvorets, St.Petersburg, Russia

E-mail: matveev@lns.pnpi.spb.ru

Abstract. Magnetization reversal of the exchange-coupled granulated ferromagnetic (GF) CoCu and ferromagnetic (F) Co nanolayers is studied by polarized neutron reflectometry. The main parameters that specify the structure and saturation magnetization of the GF and F layers are determined for Si(substrate)/$Co_{0.5}Cu_{0.5}$(5nm)/Co(x)/Si(3nm) samples with a thickness x varying from 6 to 20 nm. Neutron data allow some suggestions to be made about the features of the GF/F bilayer magnetization reversal. For small x the main mechanism of the magnetization reversal is the domain wall motion hindered by the exchange interaction between GF and F. As a consequence, the magnetization of the Co layer contacting with numerous granules is reversed in fields exceeding its own coercivity by an order of magnitude. For large x the magnetization reversal undergoes several stages conditioned by interaction of GF and F nanolayers. These stages may be characterized by three fields. For an oppositely magnetized sample, when the field approaches H_1, there appear regions with flipped moments coupled to non-flipped moments in neighboring regions ("lateral magnetic springs") in the F layer. The field H_1 sets off the overturn of magnetic moments of the granules, which is accompanied by the pinning of the regions with reversed magnetization in the F layer. The fields H_2 and H_3 tag the completion of the overturn of the F layer magnetization and the granule moments, respectively. When the applied field reaches H_3, any further change of the magnetic state is due to a reversible rotation of the granule moments and the F layer magnetization. The reversibility of magnetic states persists with fields both increasing above H_3 and decreasing down to 0 and further to $-H_1$.

1. Introduction

Polarized neutron reflectometry is a technique for studying magnetic nanosystems, which yields a unique information needed for solution of fundamental and applied tasks motivated by development of nanotechnologies. Such tasks include, e.g., the study of dimensionality effects on nanosystems, the phase proximity effects, the role of phase interfaces in definition of magnetic properties of nano-objects, establishment of relations between morphology and magnetism of layered structures. Interest to magnetic nanolayers is conditioned by possibilities of their applications in electronic industry, and, in particular, by new perspectives and promises of spintronics [1]. The basic architecture of the simplest spintronic device is a magnetically soft layer, severed from a 'pinned' hard layer with a metal nonmagnetic interlayer or a sufficiently thin (for tunnelling) dielectric interlayer [1,2]. Structures consisting of exchange-coupled magnetically soft and hard ferromagnetic layers [3] may be of interest for using, e.g., as the magnetically hard layers in spin-tunnel structures. Adding a magnetically soft layer via a thin nonmagnetic interlayer, one can build high-resolution probes for the magnetic force microscope [4]. The interlayer exchange coupling oscillates as a function of the interlayer thickness. Optimizing the interlayer thickness, one can observe giant magnetoresistance in the external field. On the basis of such trilayer nanofilms, it is possible to make controlled sources of magnetic fields of submicron dimensions without using the shape anisotropy [5,6].

A layer of pure cobalt can be used as a soft ferromagnetics (F) and a solid solution of cobalt and copper can be used as a granulated ferromagnetics (GF) [7]. Samples of layered metal nanostructures were formed on standard silicon wafers by magnetron sputtering technique in atmosphere of Ar with a pressure 10^{-6} Torr for residual gases. It follows from magnetic measurements that the shape of hysteresis loops is close to rectangular for Co films of thickness from 10 to 20 nm, their coercivity steadily rising from 10 to 30 Oe, respectively. Also a series of samples with CoCu films was prepared. CoCu films were obtained by simultaneous sputtering from two (Co and Cu) targets. The vacuum annealing was then carried out at a temperature about 180°C for 1 hour for more efficient phase separation, as described earlier [7]. One could judge about the phase restructuring of annealed CoCu films from such indirect evidences as a decrease in resistivity and an increase in coercivity (from tens to hundreds oersted). The coercive field of isolated $Co_{0.5}Cu_{0.5}$ films with thicknesses 15 nm and 5 nm is, respectively, 180-200 and about 300 Oe. The next series of samples were CoCu/Co (GF/F) films prepared in a single technological cycle and annealed as a whole. In particular, the coercivity of bilayers with the same thicknesses (15 nm) of $Co_{0.5}Cu_{0.5}$ and Co films is 100-120 Oe. To avoid oxidation in air and under vacuum annealing, a layer of silicon of thickness 3 nm was deposited on top of the structures. In the same manner the Si(substrate)/CoCu(5nm)/Co(x)/Si(3nm) samples with CoCu films of the same thickness (5 nm) and Co films of different thickness x (10, 15, 17 and 20 nm) were prepared (two samples for each value of x). Dimensions of silicon substrates are 10x10x0.5 mm^3.

In the present paper the magnetization reversal of the exchange-coupled CoCu (GF) and Co (F) nanolayers is studied by polarized neutron reflectometry. Neutron reflectivities are sensitive to the depth profile of the magnetization, both its magnitude and direction (see, e.g., [8,9]). One sample for each thickness x of Co films (samples A, B, C and D, accordingly) were chosen for neutron measurements. All sample surfaces are homogeneous with a glitter of metal; except for one of the two samples with a Co film of nominal thickness 15 nm: its surface is slightly tinted with brown and bluish stains. The cause of this difference in samples was established after neutron measurements. All neutron measurements were carried out at the reflectometer NR-4M (beam 13 of the reactor WWR-M, PNPI, Gatchina). Theoretical reflectivities were calculated by the generalized matrix method [10] lately known as 'supermatrix method'.

2. Structural features of nanolayers

In order to obtain data on the parameters of structure and saturation magnetization, the samples were magnetized in a field 600 Oe, parallel to the sample surface, then the field was diminished to 470 Oe (to prevent overheating of the electromagnet) and time-of-flight measurements of spin-up (R_+) and spin-down (R_-) neutron reflectivities were carried out. These measurements with magnetically saturated samples were carried out without the analyzer, just by switching the flipper off and on to change the sign of the incident beam polarization. The results of fitting the reflectivities for samples A, B, C and D with the model in Fig. 1 are represented in Fig. 2. The thickness and magnetization of CoCu and Co layers are designated, accordingly, as d_1, M_1 and d_2, M_2. The fitting in the whole range of momentum transfers q is possible only on the assumption that the upper protective layer of Si is oxidized (d_3 is the thickness of unoxidized silicon, d_4 is the thickness of the surface oxide layer), and silicon substrate is also oxidized (d_0 is the thickness of its oxide layer). The roughness of all interfaces is assumed to be the same (σ).

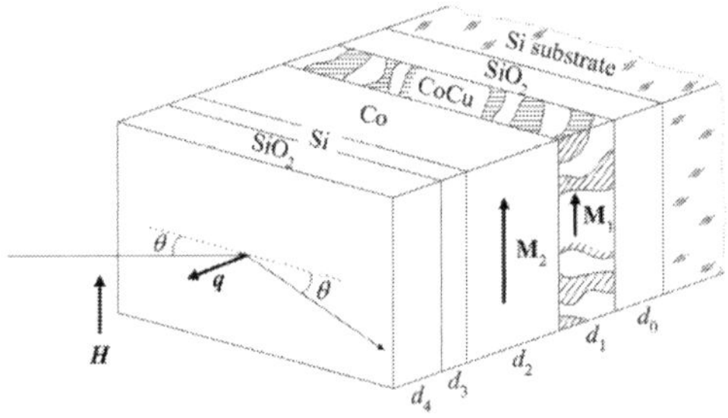

Figure 1. The layered structure model used to fit the neutron reflectivity data in Fig. 2. In calculations the interfaces are rough. The external field ***H*** is parallel to the sample surface. The wave vector transfer ***q*** for specular neutron reflection is perpendicular to the sample surface; θ is the glancing angle.

The model suggested allowed fitting the specular reflectivities of neutrons with the opposite spins (R_+ and R_-) for all samples with the exception of sample B. The structural parameters obtained by fitting and the respective errors are given in Table 1. The thicknesses of CoCu layers in samples well agree with the nominal value 5 nm. A large value of the magnetization of the CoCu layers (when averaged over all samples, $4\pi M_1$=6.5±0.7 kGs) indicates that sufficiently pure phases of cobalt and copper formed in solid solution of the metals. The thickness of the Co layer in the sample A (6.9±0.4 nm) turned out to be smaller than the nominal value 10 nm, whereas the thicknesses obtained for samples C (17.5±0.3 nm) and D (19.5±0.4 nm) agree, within the fitting error, with the nominal values, accordingly, 17 and 20 nm. The magnetization of the Co layer in sample C was found to be $4\pi M_2$=17.60±0.45 kGs, which is but slightly smaller than the saturation magnetization of the bulk cobalt ($4\pi M_S$=18.5 kGs). The magnetization of the Co layers in samples A and D is still smaller, respectively, 13.0±0.6 and 14.2±0.8 kGs.

The interfacial roughness in the samples seems to be determined by the roughness of substrates (1.5-2.0 nm are common values for silicon wafers with standard polishing). It is also to be noted that, as the volume expansion coefficient for silicon oxidation is known to be about 2, the thicknesses d_3 (unoxidized silicon) and d_4 (oxidized silicon) obtained for samples A,C,D correspond to the nominal thickness 3 nm of silicon layer before its oxidation in air (as-deposited).

It is to be noted that the thickness of unoxidized silicon in the protective layer of sample A is only 1.0±0.6 nm. For sample B the neutron data can be satisfactorily fitted only on the assumption that not only the protective silicon layer is completely oxidized, but also the Co layer is partially oxidized, with the sample surface becoming quite friable. It follows from the fitting that only 6±1 nm of the cobalt layer is not oxidized, its magnetization being $4\pi M_2$ = 15.0±1.2 kGs. From its remaining upper part and the layer of oxidized silicon there formed two nonmagnetic layers, partially oxidized Co layer of thickness about 8 nm with 10-20% content of oxygen and a surface oxide layer of thickness about 3-4 nm. In order to satisfactorily fit the date, one has also to admit that the roughness of both interfaces of the surface oxide layer is quite significant (near 15 nm), remarkably exceeding its thickness. It indicates that the oxide layer is very friable and that the model but roughly describes the depth profile of the potential of the surface region of the structure, so the respective parameters in Table 1 are omitted.

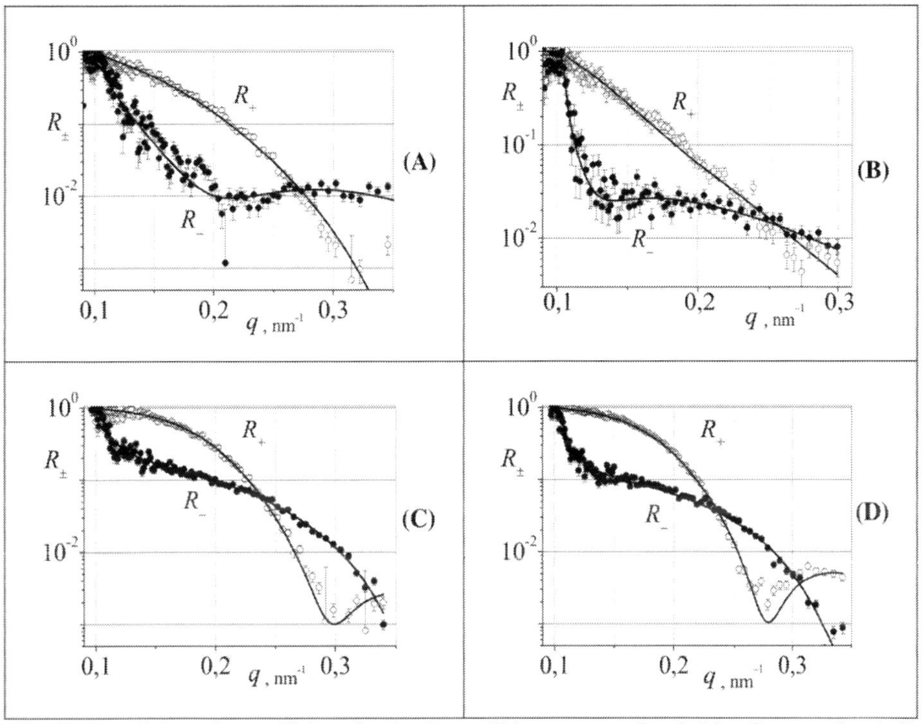

Figure 2. The fitting of the spin-up (R_+) and spin-down (R_-) neutron reflectivities for samples A-D. The structure parameters obtained are given in Table 1. The glancing angle is 3.6 mrad for all samples. The external field is 470 Oe.

Table 1. The structure parameters obtained from neutron data fitting.

sample	d_0, nm	d_1, nm (CoCu)	$4\pi M_1$, kGs	d_2, nm (Co)	$4\pi M_2$, kGs	d_3, nm	d_4, nm	σ, nm
A	7.7±0.7	5.3±0.7	7.0±1.2	6.9±0.4	13.0±0.6	1.0±0.6	3.5±0.5	2.0±0.7
B	4.5±1	5.6±0.8	5.6±1.2	6±1	15.0±1.2	-	-	-
C	5.5±0.6	4.9±0.7	6.7±1	17.5±0.3	17.60±0.45	2.1±0.6	2.0±0.4	1.5±0.4
D	3.5±1	4.9±0.7	6.7±1	19.5±0.4	14.2±1.2	2.1±0.6	2.0±0.6	1.5±0.5

The conclusion about a partial oxidation of cobalt layer in sample B is sustained by the above-mentioned fact that its surface is tinted with brown and bluish stains. Oxidation of ultrathin heterogeneous structures is not well studied, so the reason for a significant loosening of the surface region under oxidation of cobalt layer through the silicon oxide layer is not quite understood. It is not unlikely that the weakening of adhesion of layers during oxidation of cobalt and significant mechanical stresses in the silicon oxide layer may cause peeling-off of the surface layers of the structure.

3. Magnetization reversal of GF/F bilayer

In Fig. 3 are represented the field dependences of the relative difference of non-spin-flip reflected intensities (I_{++} and I_{--}),

$$P_{NSF} = (I_{++}-I_{--})/(I_{++}+I_{--}), \quad (1)$$

and the flipping efficiency of the coating,

$$F_{SF} = 2I_{SF}/(I_{++}+I_{--}+2I_{SF}) \quad (2)$$

(the ratio of the sum of two spin-flip intensities $I_{+-}=I_{-+}=I_{SF}$ to the total intensity of reflection of unpolarized beam, represented by the sum of oppositely polarized beams).

The intensities were obtained with a white beam of thermal neutrons in the standard measurements scheme "polarizer-flipper-sample-flipper-analyzer" [11,12] and in fields parallel to the sample surface. Since measurements without loss of the neutron beam polarization could be made only in the fields of one polarity (positive fields), the samples were preliminarily magnetized in the opposite field (-600 Oe), which was then dropped to 0. Further measurements were carried out in positive fields growing up to 600 Oe (the forward direction of the magnetization reversal) and then in decreasing fields (the inverse direction of the magnetization reversal). When the samples were rotated by 180° and 90°, the field dependences of the reflected intensities did not change, as it should be expected for magnetically isotropic CoCu and Co layers. Full hysteresis loops can be obtained by reflection of the curves P_{NSF} and F_{SF} obtained with respect to the origin of coordinates or coordinate axes, respectively. The latter means that the forward and inverse field dependences of F_{SF} intersect at the ordinate axis (Fig. 3).

Experimental neutron data allow some suggestions to be made about physical mechanisms of the magnetization reversal process in the system under consideration. Polarized neutron reflectometry is sensitive to the "local" magnetization, viz. to the magnetization laterally averaged by the neutron over coherently illuminated area of the order of 10^5 nm^2. When analyzing the results of measurements, one has to bear in mind that an increase (a decrease) in P_{NSF} and F_{SF} means, as a rule, that the local magnetization components, accordingly, parallel ($M_{\|}$) and perpendicular ($M_{\perp}$) to the applied field, are increased (decreased), the sign of P_{NSF} corresponding to that of $M_{\|}$. Therefore, neutron measurements are informative of the change of both $M_{\|}$ and $M_{\perp}$, though the neutron intensities in the measurement scheme used are not sensitive to the sign of $M_{\perp}$. The symmetry of $F_{SF}(H)$ with respect to the ordinate axis means that the components $M_{\perp}$ in the forward and inverse magnetization processes differ in zero field only in the sign.

When analyzing the neutron data, one has to take into consideration that the Co granules in ultrathin CoCu films may vary in shape from round ones, with moments easily aligned along the field, to elongated needle-like ones with moments directed along their largest size, as a consequence of the shape anisotropy. Since there always exist a certain number of needle-like granules oriented at a considerable angle or even perpendicular to the external field, their moments will not be parallel even to the largest field used in the experiment. It explains the fact that F_{SF} is not equal to 0 in the field 600 Oe (Fig. 3). Both external field and coupling with the Co layer and neighboring granules affect the magnetic granules. And vice versa, if the magnetic moment of a granule, due to its shape anisotropy, is inclined to the field, in the Co layer the moments contacting the granule will also be deviated to the field direction. This deviation of moments will be transferred into the depth of the Co layer. As a result, in general a multidomain state will be set up that corresponds to the minimum of the total energy including anisotropy energy, exchange and long-range dipole interaction terms, as well as the energy of coupling with the external field. As the granules are oriented randomly, the components $M_{\|}$ and $M_{\perp}$ will depend on the sample surface coordinates in a complicated manner. The perpendicular magnetization component averaged over the isotropic sample vanishes. However, neutrons average over a comparatively small, though containing many granules, area and are sensitive to $M_{\perp}$, not equal to 0. In a sufficiently large fields the magnetic state of the structure develops into a single-domain state and the local magnetization changes reversibly.

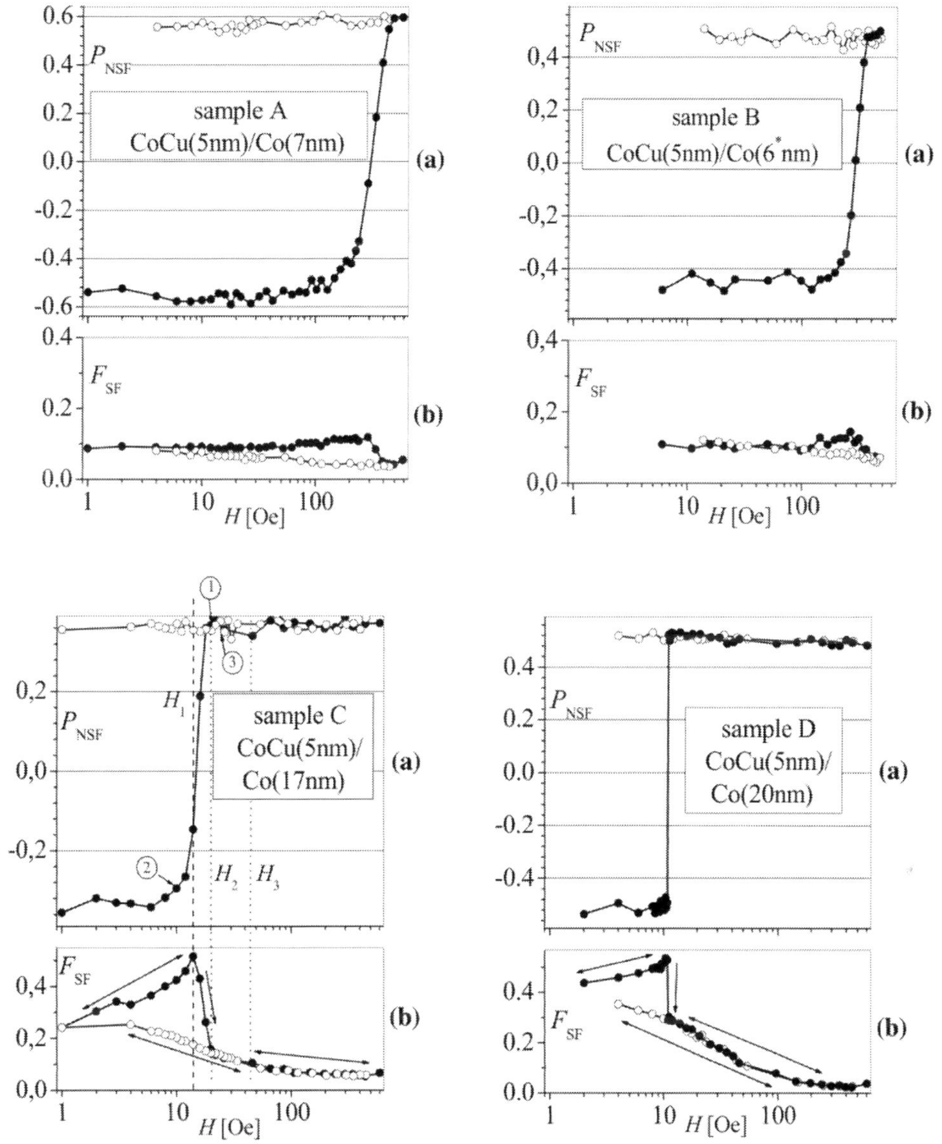

Figure 3. Quantities (a) P_{NSF} and (b) F_{SF} for samples A-D as functions of the field, growing (filled circles) and decreasing (blank circles). On magnetization curves of sample C: Double (single) arrows at the curves $F_{SF}(H)$ show field ranges with (ir)reversible magnetization. Time-of-flight measurements (Fig. 4) were carried out at points 1, 2 and 3. Characteristic fields (see the text): H_1 sets off the irreversible flip of magnetic moments of the granules, H_2 and H_3 tag the completion of flips of, respectively, the Co layer magnetization and the granule moments in the CoCu film.

It follows from neutron measurements (Fig. 3) that the coercive field of samples A and B is in tens of times larger than that of samples C and D. The structural data obtained in Sec. 2 explain why the magnetization reversal of sample B with a nominal thickness 15 nm of the cobalt film is similar to that of sample A. Because of the partial oxidation, the actual thickness of magnetic cobalt (6 nm) in sample B almost coincides with the thickness of cobalt in sample A (7 nm). It is evident that the magnetically hard CoCu (GF) layer with a thickness 5 nm has an essential effect on the magnetization reversal of magnetically soft Co layers of approximately the same thickness. Judging by the value of the coercive field of samples A and B, the granule moments in the CoCu layer control, through the exchange coupling, the Co layer magnetization. As a consequence, the magnetization of the Co layer contacting with numerous granules is reversed in a field by an order of magnitude greater than the isolated Co layer with the same thickness. The fact that the value F_{SF} is quite low in the whole range of the fields indicates that the main mechanism of the magnetization reversal of the Co layer is the domain wall motion hindered by the exchange coupling with GF. Reversed in a field -600 Oe, the granule moments do not flip even in positive fields (appr. up to 100 Oe) and hinder the motion of domain walls in the Co film. The domain walls start moving only when the granule moments flip in larger fields. It follows from neutron data that the coercive field of bilayers in samples A and B is near 300 Oe. In fields above 500 Oe the moments of all granules and the Co film magnetization overturn to the field direction, and the spin-flip reflection essentially reduces. Now the granule moments deviate from the field direction due to the shape anisotropy, but in the same fields this deviation is less, for the granule moments are coupled with the Co film, the moments of which are aligned in directions close to the external field. As a consequence, when the field reduces, the spin-flip reflection grows slowly enough, but in small fields the moments of the granules again deviate noticeably from the field direction, dragging the moments of the magnetically soft Co layer. In a sufficiently low fields the components $M_{\perp}$ are close in magnitude for the forward and inverse magnetization processes, so the two curves $F_{SF}(H)$ converge (Fig. 3-A,B).

The thicknesses of the Co films in samples C and D exceed in about three times that of the CoCu films. This is fundamental for the magnetization reversal scenario. Although the influence of CoCu is weakened, analysis of the whole set of measurements allows one to conclude that the CoCu layer also plays an important role in the magnetization reversal.

Sample C with a maximum magnetization of the Co layer was chosen for more detailed neutron measurements, incl. time-of-flight measurements, now with the analyzer (unlike the measurements represented in Fig. 2). The structural and magnetization parameters obtained above were used in fitting and analysis of spectral neutron reflectivities for different magnetic states of the sample. To take account of the coupling of the Co layer with numerous granules coupled also with each other is a rather complex problem. So the calculations of reflectivities as functions of the momentum transfer q are carried out with a simplified model, pertinent only for estimation of non-collinearity of the local magnetization to the external field. Namely, the angle χ between the local magnetization and the field is assumed to be the same for the entire Co film, though the sign χ may be different in different parts of the sample and the magnetization is equal to the value obtained by the fitting (Fig. 2-C). It is the magnetic state of the Co film that mainly determines the reflected neutron intensities, for its thickness in three times and its saturation magnetization in 2.5 times are larger than those of the CoCu film.

Double (single) arrows in Fig. 2-C,D show ranges with a (ir)reversible field dependence. One can see from the neutron data that the field dependences of F_{SF} are practically reversible everywhere except for a narrow range of fields in the forward magnetization curve for sample C from $H_1 \cong 14$ Oe to $H_2 \cong 20$ Oe; for sample D the width of this range (near $H_1 \cong 11$ Oe) is no more than 1 Oe.

In a field 600 Oe the moments of Co granules in the CoCu layer and the magnetization vector in the Co film are close to the guide field direction ($\chi \approx 4^{\circ}$ for sample C). When the field reduces, the directions of the granule moments are defined to a greater extent by the shape anisotropy and, while remaining in the film plane, the moments deviate more and more from the field direction, dragging with them the moments of the Co layer regions contacting these granules, the consequence being that the neutron SF reflection increases. Rotation of the moments is gradual and reversible. A satisfactory fitting of neutron reflectivities as functions of the momentum transfer q (Fig. 4-1) at $H = 19$ Oe is obtained on the assumption that the Co film magnetization inclination angle is $\chi = 28\pm3^{\circ}$. To determine the portions of regions with local magnetization inclinations χ and $-\chi$ did not seem to be possible without 3D neutron polarization analysis. On the basis of magnetic isotropy of the sample, one may assume that those portions are equal. The reflectivities are much less sensitive to magnetic parameters of the CoCu film, so that we could deduce only that the local magnetization in the CoCu film makes an acute angle with the field (28° in calculations in Fig. 4-1). It is evident that, when the field drops to 0, the moments retain an acute angle with respect to the positive field direction.

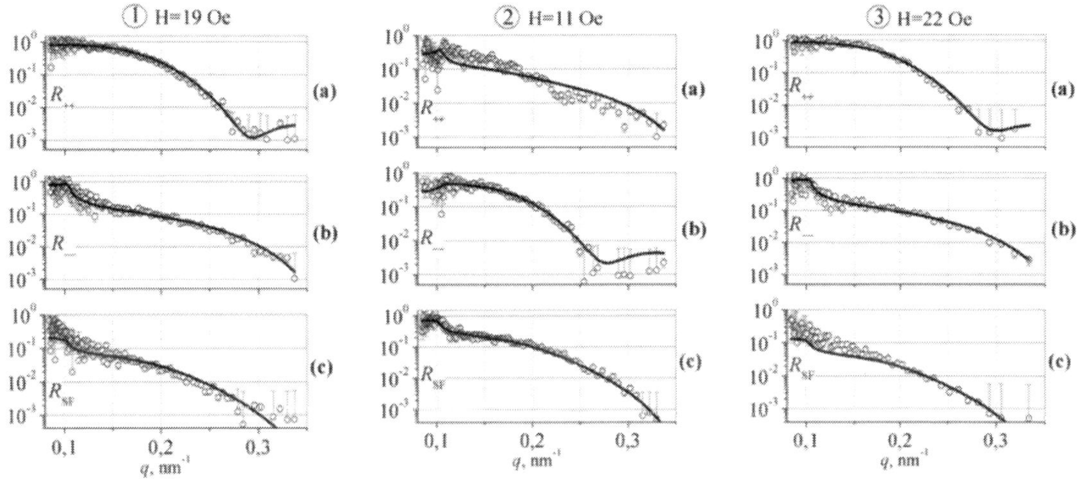

Figure 4. Time-of-flight reflectivities for neutrons reflected without spin-flip (a) R_{++} , (b) R_{--} and with spin-flip (c) $R_{+-} = R_{-+} = R_{SF}$ as functions of q for sample C at fields (1) 19 Oe, (2) 11 Oe and (3) 22 Oe (as marked in Fig. 3-C). Theoretical reflectivities (solid curves) are calculated with structural and magnetization parameters given in the caption of Fig. 2; the fitting parameter is the magnetization inclination angle χ: (1) 28±3°; (2) 120°; (3) 22±4°.

It follows from the magnetization reversal symmetry that, after the reversal of magnetization in a field -600 Oe, the local magnetization inclination angle χ (with respect to the positive field direction) in the Co film at the field -19 Oe is near 180°- 28°=152°. When the field drops to 0, it will further deviate from 180°.

With the growth of the field above 0 the angle χ still reduces (see F_{SF} in Fig. 3-C). Rotations of all moments are reversible in the fields up to H_1. The time-of-flight neutron reflectivities obtained for sample C in a field 11 Oe, close to H_1 =14 Oe, are represented in Fig. 4-2. In this case only R_{--} and R_{SF} could be fitted with χ = 120° in the simplified model. The fitting of R_{++} is just impossible, if all moments in the Co layer are assumed to make blunt angles with the field direction. A high level of the reflectivity R_{++} at small q can be explained only on the assumption that there exist regions in the Co layer with $M_{\|}$ >0. One may suggest that, due to the torque moment from an external field $H < H_1$, there appear regions in the Co layer with a local magnetization flipped ($M_{\|}$>0). Such regions should be coupled with regions with $M_{\|}$<0 so that the boundaries between them move reversibly, when the field changes. Such elastically coupled magnetic regions may be called "lateral magnetic springs". Lateral magnetic springs arising in fields approaching H_1 should reversibly disappear when the field falls. The Co granules in the CoCu film seem to play an important role in reversibility of the magnetic states. Their intrinsic coercivity is considerably larger than that of the Co film, so the granule moments under the Co layer regions with a flipped magnetization remain to be oriented in a direction opposite to the field. As a consequence, they impose a torque forcing the Co layer regions with a flipped magnetization to return to a state with $M_{\|}$<0. Such a coupling ensures elasticity of the lateral magnetic springs and reversibility of magnetic states. The observed (Fig. 3-C) enhanced growth in spin-flip reflection when the field approaches H_1 is indicative of the increasing growth rate for $|M_{\perp}|$. To satisfactorily fit all reflectivities in Fig. 4-2, one has to improve the model describing the moments distribution in the bilayer.

The field H_1 sets off flips of the granule moments in the CoCu layer. When a granule moment flips, the shape anisotropy energy does not change and the energy of the moment in the external field decreases. Such flips of the granule moments are most probable under the Co layer regions with a flipped magnetization, where the torque moment of the external field is enhanced by the ferromagnetic interaction. The reverse flipping of the granule moments is related to surmounting potential barriers, hence it is irreversible. Therefore, the granules with flipped moments pin the regions in the Co layer with the magnetization flipped and, enhancing the torque moment of the external field, diminish the perpendicular magnetization component in the Co layer. A fast decrease in F_{SF} in a short range of fields from H_1 to H_2 (Fig. 3-C) signifies that there is an intense growth of regions in the Co layer with a magnetization flipped and the pinning of such regions due to continuing flips of granule moments.

The overturn of the Co film magnetization is basically completed at the field H_2. This is confirmed by a satisfactory fitting of neutron reflectivities (Fig. 4-3) obtained in a field H = 22 Oe, slightly above H_2= 20 Oe, with χ = 22±4°. Some discrepancy of the theoretical and experimental spin-flip reflectivities R_{SF} at small q is likely to signify that the granules with moments not yet flipped increase noncollinearity of the magnetization in the Co layer (the portion of such granules is less than a third, as estimated with our simplified model). Note also that the field H_2 is less than the coercive field of the isolated Co film with the same thickness. A possible reason is a softening of the Co film in the sample subjected to annealing for efficient separation of phases in the CoCu film.

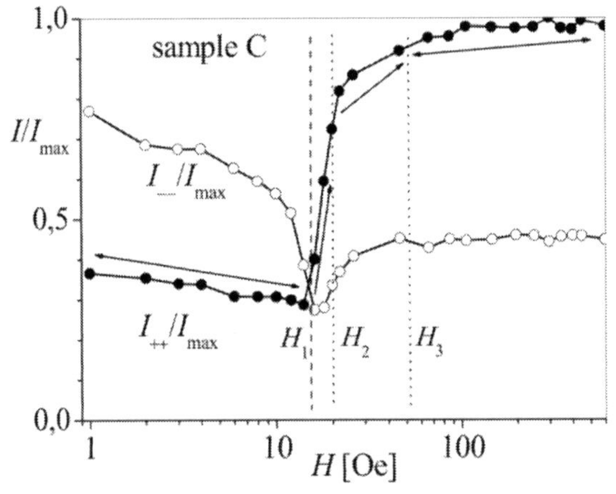

Figure 5. The integral intensities I_{++} (filled circles) and I_{--} (blank circles) for sample C as functions of the growing field (preliminarily the sample was oppositely magnetized in a field -600 Oe). Double (single) arrows show field ranges with (ir)reversible neutron intensities. The intensities are normalized with a maximum intensity I_{max} = max(I_{++}). Characteristic fields: H_1 sets off the irreversible flip of magnetic moments of the granules, H_2 and H_3 tag the completion of flips of, respectively, the Co layer magnetization and the granule moments in the CoCu film.

The irreversibility of the reflected intensities I_{++} and I_{--} (Fig. 5) in fields between H_2= 20 Oe and H_3 near 50 Oe indicates that in the field H_2 a considerable number of granule moments are still oriented against the field. This can be explained by the assumption that the coupling between magnetic granules and the Co layer depends on the contact area, on the presence of Cu atoms in the contacting region, on the mechanical tension distribution, etc. As a consequence, a part of granule moments do not flip even when the magnetization of the entire Co layer is overturned. The continuing flips of granule moments are responsible for irreversibility of I_{++} and I_{--} in fields between H_2 and H_3.

Irreversible flips of the granule moments are accompanied with a reversible decrease in the inclination angle of the moments already flipped. It seems that the latter determines the quasi-reversible behavior of F_{SF} between H_2 and H_3. The calculations for the given structure show that a decrease in χ for $\chi <$ 30° leads to a growth of not only I_{++}, but also I_{--}, so that P_{NSF} but slightly change (for this reason the two curves of P_{NSF} in Fig. 3-C merge between H_2 and H_3).

The fact that flips of the granule moments are completed in the field H_3, which is by an order of magnitude smaller than the coercive field of an isolated CoCu film (5 nm), means that the exchange coupling with a comparatively thick Co layer essentially contributes to the flips of the granule moments. The change in the magnetic structure in fields above H_3 is associated with a reversible rotation of moments in the bilayer. As it has been mentioned before, the configuration of magnetic moments depend only on the field magnitude, the magnetic states are reversible in fields above and below H_3 (down to $-H_1$).

4. Discussion

Neutron reflectometry yields an important information to elucidate the relation between structural features and magnetic properties of the layers. Neutron data allowed the main parameters to be determined and some suggestions to be made about the features of the magnetization reversal of $Co_{0.5}Cu_{0.5}$(5nm)/Co(x) (GF/F) bilayers with x varying from 6 to 20 nm. For small x the main mechanism of the magnetization reversal is the domain wall shifts essentially hindered by the exchange interaction between GF and F. For large x the magnetization reversal undergoes several stages conditioned by interaction of GF and F nanolayers. To comply with the neutron data, we suggested that, just below the field that sets off the overturn of magnetic moments of the granules, due to the torque from the field, there appear regions in the F layer with a reversed ($M_{\|}>0$) local magnetization, which are coupled to regions with the initial magnetization $M_{\|}<0$. Such "lateral magnetic springs" reversibly disappear when the field drops. The Co granules in the CoCu film seem to play an important role in this reversibility by imposing a torque forcing the Co layer regions with $M_{\|}>0$ to return to a state with $M_{\|}<0$ in the field decreased.

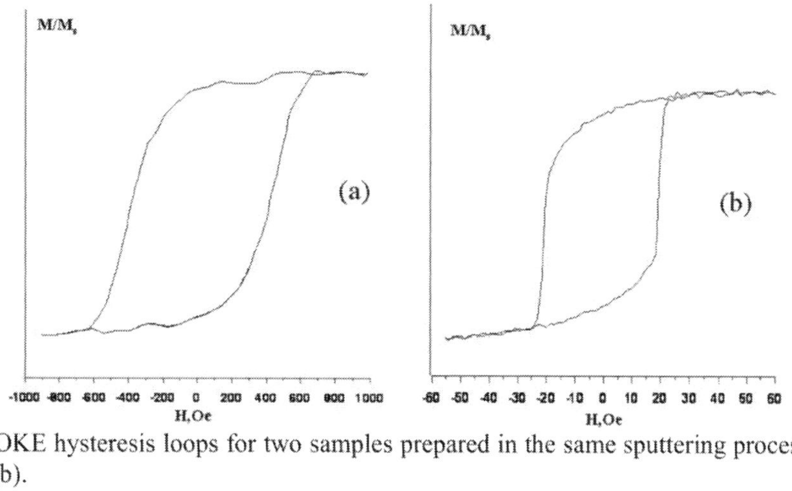

Figure 6. MOKE hysteresis loops for two samples prepared in the same sputtering process as samples A (a) and C (b).

It is to be noted that, as a rule, the field dependence of the polarizing efficiency $P=(R_+-R_-)/(R_++R_-)$ is used in the study of the magnetization reversal of layered structures by polarized neutron reflectometry. The hysteresis loops obtained in this manner are similar to those obtained by standard magnetometry (cf. $P_{NSF}(H)$ in Fig. 3 and MOKE hysteresis loops in Fig. 6). Using, in addition, the selectivity of neutron reflection to the thickness of layers, from the field dependence of the polarizing efficiency at different wavelengths, we inferred that thicker layers in CoFe/TiZr supermirrors are magnetically harder and the magnetization reversal starts from thinner layers [13]. In the present paper the intensities of neutrons reflected without spin-flip and with spin-flip were used, and it allowed us to independently analyze magnetization components parallel and perpendicular to the external field. In order to fully interpret the neutron data, a detailed theoretical analysis and an additional experimental work with neutron and other techniques are required.

Whenever a time-of-flight measurement was carried out at a certain glancing angle, the reflected beam profile was also scanned with a thin slit on the detector window. In all cases the off-specular scattering was found to be practically at the background level, in spite of the presence of such inhomogeneities as granules. It remained insignificant even for samples in small fields, when the cobalt layer broke up into magnetic domains giving rise to magnetic off-specular scattering. We attribute this to the facts that (a) the volume of granulated ferromagnetic is very small (a layer of thickness 5 nm); (b) granules with lateral dimensions 10-20 nm produce off-specular scattering at large angles with quite small cross sections; (c) lateral dimensions of magnetic domains in the cobalt layer are quite large. As a consequence, most neutron scattering is concentrated within the specular peak. Colder neutrons and greater luminosity are required to obtain additional information about the samples from off-specular scattering.

Many thanks to Prof. S.M. Dunaevskiy for his interest to this work. The authors acknowledge RFBR for support (grants RFBR 08-02-00467a and RFBR-NNIO 11-02-91337).

References

[1] Hirota E, Sakakima H and Inomata K 2001 *Springer Series in surface science* **40**
[2] Parkin S, Jiang X, Kaiser C, Panchula A, Roche K and Samant M 2003 *Proceedings of the IEEE* **91** 661
[3] Yan S, Barnard J A, Xu F,Weston J L and Zangari G 2001 *Phys.Rev.* B **64** 184403
[4] Yihong W , Shen Y , Liu Z , Li K and Qiu Nano J 2003 *Appl. Phys. Lett.* **82** 1748
[5] Cowburn R P 2003 *J.Appl.Phys.* **93** 1748
[6] Fraerman A A, Gusev S A, Nozdrin Yu N, Samokhvalov A V and Vdovichev S N 2006 *Phys. Rev.* B **73** 100503(R)
[7] Childress J R and Chien C L 1991 *J.Fppl.Phys.* **70** 5885
[8] Felcher G P 1993 *Physica* B **192** 137
[9] Zabel H 1994 *Physica* B **198** 156
[10] Pleshanov N K 1996 *Physica* B **100** 423
[11] Moon R M, Riste T and Koehler W 1969 *Phys. Rev.* **181** 920
[12] Majkrzak C F, Satija S, Neumann D A and Ankner J F 1990 *NIST reactor: Summary of activities, July 1989 - June 1990* **55**
[13] Pleshanov N K, Bodnarchuk V, Gähler R, Korneev D A, Menelle A, Metelev S V, Pusenkov V M, Schebetov A F and Ul'yanov V A 2001 *Physica* B **297** 126

The study of the oxidation of thin Ti films by neutron reflectometry

V A Matveev, N K Pleshanov, A P Bulkin, V G Syromyatnikov

Petersburg Nuclear Physics Institute, 188300, Gatchina, St.Petersburg, Russia

E-mail: matveev@lns.pnpi.spb.ru, pnk@pnpi.spb.ru, apbulkine@pnpi.spb.ru, svg@pnpi.spb.ru

Abstract. Neutron reflectometry is found to be quite sensitive to oxidation of very thin layers of Ti. Preliminary results of the study of oxidation of thin Ti films in air are represented. The thickness of the oxide layers formed in air at room temperature on Ti films of thickness 20, 10 and 5 nm (as sputtered) is found to be 2.9±0.3, 3.6±0.3 and 6.1±0.3 nm, respectively. The annealing at temperatures in the range 100-300°C leads to the growth in the oxide layer thickness and in roughness of both (air/oxide and oxide/metal) interfaces. The study of the oxidation of thin Ti films is of interest for improvement of polarizing neutron mirrors and supermirrors.

1. Introduction

Polarized neutrons are known to be a unique probe allowing important physical information to be obtained, inaccessible by other experimental techniques. This information had a great impact in many fields of science and technology. An increase in the polarizing efficiency of neutron polarizers and analyzers means more direct, detailed and reliable information in experiments with polarized neutrons. Quite often, to achieve a reasonable resolution, polarized neutron beams with relatively small divergence are required and the polarization of scattered neutrons is analyzed with numerous channels of small angular aperture. Then neutron optical polarizers [1,2] may turn out to be more efficient than ^{3}He spin filters [3,4] and their further development remains an important task.

To increase the angular acceptance and the transmittance of the neutron optical polarizers, supermirror coatings [5] are used. Within a supermirror, the layer thicknesses are gradually varied, with the effect that neutrons are reflected up to fairly large glancing angles, exceeding the total reflection angle of Ni, one of the best neutron reflectors, by a certain factor *m*. Polarizing supermirrors are made of alternating magnetic and non-magnetic layers with a low neutron optical contrast for one of the spin components. The substrate is coated with an absorbing antireflective underlayer [6]. The composition and the thickness of the underlayer can be designed to minimize reflection, in principle, to any reasonable level [7]. By using efficient underlayers and matching the potentials of non-magnetic and magnetic layers, the specular reflectivities for neutrons with the spin opposite to the magnetization of the magnetic layers (spin-down neutrons) are reduced to a level about 1%, e.g., for CoFe(V)/TiZr [2] and CoFeV/TiN [8] supermirrors.

Among the remaining factors that enhance the unwanted reflection of the spin-down neutrons are structural imperfections, including an oxide layer on the supermirror surface [9] and "magnetically dead" regions [10,11] in magnetic layers near their interfaces, that give rise to appearance of potential barriers for the spin-down neutrons. Thin titanium layers with a negative potential can be used to suppress reflection of neutrons with the undesired spin [12]. We do not consider transmission polarizers without underlayers; their polarizing efficiency cannot be noticeably improved by the methods suggested in Ref. [12]. The efficiency of Ti interlayers of thickness 1- 2 nm at the interfaces and a Ti/TiO_2 surface bilayer of thickness 5-10 nm for suppression of the reflection of neutrons with the undesired spin by 1-2 orders of magnitude has been substantiated. The TiO_2 oxide layer is formed by natural oxidation of Ti layers in contact with air. The oxidation of Ti nanolayers and the stability of Ti/TiO_2 bilayers are of great interest for applications in polarized neutron optics. Therefore, a systematic study of oxidation of very thin (5-40 nm) titanium films is required.

2. Oxidation of thin Ti films: experimental results

The samples were prepared by magnetron sputtering of Ti films of thickness 20 (sample a), 10 (sample b) and 5 nm (sample c) on glass substrates. The study of oxidation on these films was carried out at the reflectometer NR-4M (PNPI, Gatchina, Russia) in time-of-flight mode using neutrons of wavelengths 0.1-0.4 nm.

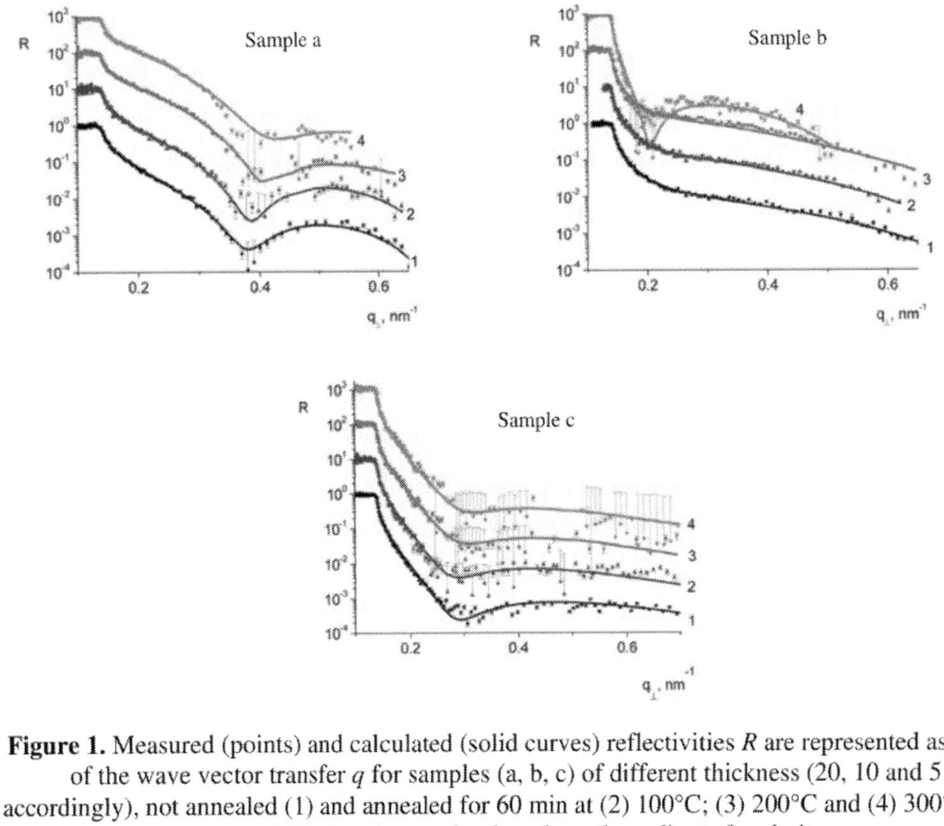

Figure 1. Measured (points) and calculated (solid curves) reflectivities R are represented as functions of the wave vector transfer q for samples (a, b, c) of different thickness (20, 10 and 5 nm, accordingly), not annealed (1) and annealed for 60 min at (2) 100°C; (3) 200°C and (4) 300°C. Curves are shifted against each other along the ordinate for clarity.

The layered structure of our samples can be described as follows: The glass substrate with roughness σ_S is coated with a titanium layer of thickness d_1, on top of which a titanium dioxide TiO_2 (rutile) layer of thickness d_2 is formed. The interface roughness between Ti and TiO_2 and the surface roughness of the sample are σ_1 and σ_2, respectively. In such a model of the film structure the specular reflection of neutrons can be described as reflection from a one-dimensional potential

$$V(z) = V_S \cdot erf(-z,\sigma_S) + V_1 \cdot [erf(d_1 - z,\sigma_1) - erf(-z,\sigma_S)] + \\ + V_2 \cdot [erf(d_1 + d_2 - z,\sigma_2) - erf(d_1 - z,\sigma_S)] \quad (1)$$

where z is the coordinate along the axis perpendicular to the sample surface, V_S, V_1 and V_2 are the neutron potentials of the substrate, titanium and its oxide, respectively. The error function

$$erf(z,\sigma) = \frac{1}{\sigma\sqrt{2\pi}} \int_{-\infty}^{z} \exp\left[-\frac{t^2}{2\sigma^2}\right] dt \quad (2)$$

describes the smoothing of the depth profile at an interface with the Gaussian roughness represented by the standard deviation σ of the interface heights from the mean surface. The potentials of the glass substrate, titanium and its oxide with bulk densities are 100, -50.8 and 68.3 neV, respectively. The numerical matrix method (see, e.g., Ref. [13]) was used in calculations of theoretical neutron reflectivities.

The neutron experimental and calculated reflectivities for Ti films before and after annealing are represented in Fig. 1. The best-fit results are represented in Table 1. One can see that the titanium films of smaller thickness are subject to deeper oxidation in air. The thicknesses of the oxide layers formed in air at room temperature are 2.9, 3.6 and 6.1 nm for Ti films with initial thicknesses d_{Ti} ("as sputtered"), of 20, 10 and 5 nm, respectively.

Table 1. The best-fit parameters used in the calculation of the fit curves displayed in Fig. 1.

Annealing temperature	T=20°C	T=100°C	T=200°C	T=300°C
Sample a: Ti 20 nm (nominal thickness)				
d_1, nm	17.4±0.4	16.8±0.5	14.3±0.4	13.7±0.5
σ_1, nm	0.8±0.6	1.4±0.8	2.4±0.5	2.6±0.6
d_2, nm	2.9±0.3	3.8±0.5	8.2±0.3	9.4±0.5
σ_2, nm	0.8±0.7	1.4±1.2	3.6±0.7	4.9±1.1
$d_1 + d_2$	20.3±0.7	20.6±1.0	22.5±0.7	23.1±1.0
d_{Ti}, nm	19.0±0.6	19.0±0.8	19.0±0.6	19.0±0.8
Sample b: Ti 10 nm (nominal thickness)				
d_1, nm	7.3±0.4	6.9±0.5	6.5±0.5	1.5±0.3
σ_1, nm	0.8±0.7	1.0±0.7	1.1±0.6	1.1±0.7
d_2, nm	3.6±0.3	4.2±0.4	5.4±0.4	13.7±0.4
σ_2, nm	0.9±0.6	2.2±0.5	3.4±0.8	3.4±0.6
$d_1 + d_2$	10.9±0.7	11.1±0.9	11.9±0.9	15.2±0.7
d_{Ti}, nm	9.3±0.6	9.3±0.8	9.3±0.8	9.3±0.6
Sample c: Ti 5 nm (nominal thickness)				
d_1, nm	2.0±0.4	1.9±0.4	1.6±0.4	0.7±0.6
σ_1, nm	1.1±0.5	1.7±0.5	2.2±0.6	3.1±1.0
d_2, nm	6.1±0.3	6.3±0.3	6.8±0.3	8.5±0.5
σ_2, nm	1.7±0.5	2.8±0.5	2.9±1.0	3.7±1.0
$d_1 + d_2$	8.1±0.7	8.2±0.7	8.4±0.7	9.2±1.1
d_{Ti}, nm	5.5±0.6	5.5±0.6	5.5±0.6	5.5±0.9

The thickness of titanium layers as sputtered were evaluated from the best-fit parameters as $d_{Ti} = d_1 + d_2/\alpha$, where α=1.76 is the expansion rate of titanium under its oxidation into rutile.

According to the neutron data fitting, native oxide layers formed on thinner Ti films are thicker (see Table 1). However, the native oxide layers on thinner Ti films seem to be more stable, since the oxide layer thickness d_2 for thicker Ti films grows faster in process of annealing. From oxidation of thicker Ti layers, one could expect that sample c with an as-sputtered Ti thickness 5.5±0.6 nm should be completely oxidized at temperature 200°C. Quite surprisingly, the growth of its oxide layer was not as significant as expected. For all samples the increase in the annealing temperature led not only to thicker oxide layers, but also to the growth of the surface roughness and the roughness of the oxide/metal interface. Generally, many questions still remain and we want to use not only neutron reflectometry, but also X-ray reflectometry in our further experiments to clarify them.

Acknowledgements.
The authors are grateful to V.Kharchenkov, N. Kolyvanova and V. Sukharev for preparation of the samples.

References
[1] Schärpf O 1989 *Physica* B **156–157** 639
[2] Schebetov A F, Pleshanov N K, Pusenkov V M, Peskov B G, Shmelev G E, Kraan W H, Por P T, Rekveldt M Th and Mikhailova V E 1994 *Nucl. Instr. and Meth.* B **94** 575
[3] Hussey D S, Rich D R, Belov A S, Tong X, Yang H, Bailey C, Keith C D, Hartfield J, Hall G D R, Black T C and Snow W M 2005 *Rev. Sci. Instrum.* **76** 053503
[4] Wolf M, Radu F, Petoukhov A, Humbliot H, Jullien D, Andresern K and Zabel H 2006 *Neutron News* **17** 26
[5] Mezei F 1976 *Commun. Phys.* **1** 81
[6] Drabkin G M, Okorokov A I, Schebetov A F, Borovikova N V, Gukasov A G, Egorov A I and Runov V V 1975 *J. Eksp. Teor. Fiz.* **69** 1916
[7] Pleshanov N K, Peskov B G, Pusenkov V M, Syromyatnikov V G and Schebetov A F 2006 *Nucl. Instr. and Meth.* A **560/2** 464
[8] Clemens D, Böni P, Friedli H P, Göttel R, Fermon C, Grimmer H, Van Swygenhoven H, Archer J, Klose F, Krist Th, Mezei F and Thomas P 1995 *Physica* B **213-214** 942
[9] Pusenkov V M, Metelev S V, Pleshanov N K, Syromyatnikov V G, Ul'yanov V A and Schebetov A F 2004 *Physica* B **348** 285
[10] Vergnat M, Dufour C, Bruson A, Houssani S, Marchal G, Mangin P H, Rhyne J J, Erwin R and Vettier C 1989 *J. Phys. Colloq.* **50** 207 C7
[11] Pleshanov N K, Pusenkov V M, Schebetov A F, Peskov B G, Shmelev G E, Siber E V and Soroko Z N 1994 *Physica* B **198** 27
[12] Pleshanov N K 2010 *Nucl. Instrum. Methods* A **613** 15
[13] Knittl Z 1976 *Optics of Thin Films* (New York: John Willie & Sons)

Investigation of Physically and Chemically Ionic Liquid Confinement in Nanoporous Materials by a Combination of SANS, Contrast-Matching SANS, XRD and Nitrogen Adsorption

G E Romanos[1*], K L Stefanopoulos[1], O C Vangeli[1], K Mergia[2], K G Beltsios[3], N K Kanellopoulos[1] and D Lairez[4]

[1] Institute of Physical Chemistry and [2] Institute of Nuclear Technology and Radiation Protection, NCSR "Demokritos", Ag. Paraskevi Attikis, 15310, Greece
[3] Department of Materials Science and Engineering, University of Ioannina, 45110, Greece
[4] Léon Brillouin, CEA/CNRS UMR 12, CEA-Saclay, 91191 Gif-sur-Yvette cedex, France

[*]groman@chem.demokritos.gr

Abstract. In the present study, [bmim][PF_6] ionic liquid (IL) was introduced into the pores of two ordered mesoporous silicas (MCM-41 and SBA-15) having different pore sizes by means of two different processes: a) with physical imbibition from a methanol solution under high vacuum and b) by chemically immobilising the IL with silanisation of the pore surface followed by reaction with butyl-methyl imidazolium chloride and anion exchange with PF_6, the process termed as the "grafting to" method. Both the extent of IL entrapment and the structural properties of the IL phase under confinement were investigated by SANS, contrast-matching SANS, XRD and nitrogen adsorption measurements. The results show that the pores of chemically prepared samples are not totally filled by IL and also suggest for ordering of the silylated IL phase. On the other hand, the physically prepared samples are almost or totally filled with IL whereas no evidence for ordering of the confined IL phase was observed.

1. Introduction

Room-temperature ionic liquids (ILs), an increasingly important set of electrolytes, have received much attention due to their unique characteristics: low vapour pressure, thermal stability, non-flammability, chemical tuneability and good electrolytic and solvation properties. Furthermore the incorporation of ILs into the pore structure of advanced porous materials (carbon nanotubes [1], ordered mesoporous silicas) for the formation of supported ionic liquid phase systems (SILPs), or into the nanopores of the separating layer of asymmetric membranes (carbon, ceramic, polymeric) is of key importance and consists the only way to overcome major drawbacks of ILs such as, their high viscosity, slow gas diffusivity and high price, making feasible their large scale application as solvents, catalyst dispersion agents and absorbents in the fields of organic synthesis, heterogeneous catalysis and gas separation. The development of well-stabilised thin IL films on the high pore surface area (1000 m^2/g) of advanced porous materials may considerably reduce the quantity of IL that is necessary for a certain scale of application and in parallel makes it possible to exploit all of the mass of the incorporated IL, since viscosity and diffusion effects may be diminished in such ultra thin layers. In this regard the interaction of ILs with solid surfaces [2-4] and, both their state and morphology after incorporation into the pore structure are issues that deserve further investigation.

For example if the IL phase, under extreme confinement into the nanopores, is crystallised then the benefits derived from the thin film formation and the high surface area, which is available for reaction or adsorption, are counterbalanced by the slower diffusion rates holding for the solid phase compared to the liquid one. Moreover, if the pore structure of the supporting material is blocked due to extended IL deposition onto the pore mouths or if the entire pore structure is completely filled with the liquid IL phase, then all the potential benefits of SILPs are taken out. Finally, it is often the case that high temperatures and differential pressures are required in some catalytic or gas separation processes, which despite the low vapour pressure of ILs, could lead to their significant evaporation or displacement out of the pore structure. In this regard it is preferable to further stabilise ILs through the formation of a chemical bond, rather than rely the stability solely on the physical and electrostatic interaction between the IL's ions and the pore walls. The chemical attachment of the cation or the anion on to the pore wall may, in addition, lead to a preferable orientation of the IL's layers towards the core of the pore, which accommodates the reactants or the adsorbates. This orientation may be favorable when selective penetration from a gas or liquid mixture is required i.e. exploiting the hydrophilic or hydrophobic character of the ions and can bring the advance of higher diffusion rates, especially for IL membranes, where the gas transport is described as a sequential process of complexation-decomplexation of the gas molecule with the IL's ions, usually the anions. For the aforementioned reasons, it is essential to develop and examine methods that are sensitive to the bulk of the pore structure and not to the pore opening as the widely used N_2 porosimetry and the advanced microscopy techniques.

Small-angle scattering of neutrons or x-rays (SANS or SAXS) [5-8], contrast-variation or contrast-matching SANS [9-13] and adsorption in conjunction with neutron (x-ray) diffraction or SANS (SAXS) [14-20] have been proved to be essential tools for elucidating the microstructure of nanoporous materials and the properties of confined phases within the pores. Moreover, combination of SANS and 3D stochastic reconstruction methods is also a powerful tool for visualising the evolution of the adsorption mechanism and show how sorption alters the pore space characteristics of the material [21-22]. SANS has the advantage over the conventional gas adsorption methods that it can provide information for the bulk of the structure, including pores inaccessible to gas molecules, such as closed pores. Previous SANS experiments on IL in cyclohexane oil emulsions demonstrated the formation of surfactant-stabilised dispersed nanodroplets with IL cores [23]. Recently, voltage-clamp measurements through single conical nanopore obtained by chemical etching of a single ion track in polyimide film were reported [24]. The results showed that the origin of the pink noise of the ionic current could be ascribed to a cooperative effect on ions motion in confined geometry. Furthermore, silica based ordered mesoporous materials like MCM-41 and SBA-15 consist of a two-dimensional hexagonal arrangement of cylindrical mesopores of uniform size (*P6mm* space group). They are materials exhibiting Bragg peaks in the small-angle regime due to the regular packing of the cylindrical pores and they are considered as model systems for both adsorption studies and mesopore size determination. They are synthesised by using micellar aggregates of surfactans (MCM-41) [25] or amphiphilic block copolymers (SBA-15) [26] as structure-directing template and their pore size can be tailored by the choice of the amphiphile.

We are now proposing the potentiality of SANS and contrast-matching SANS for studying both the extent of pore filling and the organisation of confined [bmim][PF_6] IL in the nanopores of two ordered nanoporous materials (MCM-41 and SBA-15) with different pore sizes. The IL was introduced a) by physical imbibition under vacuum and b) by the "grafting to" method involving chemical immobilisation by means of silanisation of the pore surface followed by reaction with butyl-methyl imidazolium chloride and anion exchange with PF_6. In addition, N_2 adsorption isotherms at 77 K and x-ray diffraction (XRD) measurements have been carried out.

2. Experimental

2.1. Materials and chemicals
The materials MCM-41 and SBA-15 (named as MCM and SBA) were provided by Sigma-Aldrich® and Claytec Inc. respectively. Solvents were used as purchased without 'further drying'. Diethyl ether (puriss, Reag ACS, Aldrich), chloroform (GC, Aldrich), acetonitrile (GC, Aldrich), methanol (99.8%, Aldrich). Reaction reagents were used as purchased without 'further purification'. Chlorobutane (99%, Fluka), 1-methylimidazole (99% redistilled, Aldrich), (chloropropyl)-trimethoxysilane (96%, Aldrich), sodium hexafluorophoshate (98%, Aldrich). The [bmim][PF_6] IL was prepared in a similar way previously reported [27-28] and its purity was confirmed by NMR.

2.2. Pore filling procedures
The pore filling of the samples with IL was achieved by physical imbibition of [bmim][PF_6] from its methanol solution under vacuum (prepared samples named as MCM_PHYS and SBA_PHYS) and by chemical grafting of the IL under reflux conditions (prepared samples named as MCM_SILP and SBA_SILP).

2.2.1. *Physical imbibition.* A vacuum assisted physical imbibition method was involved in order to facilitate the pore filling process. More specific crimp top-flat bottom headspace vials of 20 ml were appropriately modified to permit the application of high vacuum (10^{-3} mbar) into their internal space. The modification consisted on the welding of the common port of a high vacuum three-way glass valve (ACE 8145-06 three-way stopcock) on the sidewall of the vials close to their bevelled neck. About 200 mg of the SBA and MCM powders, having an absolute pore volume of 0.19 and 0.25 cm^3 respectively, were placed on the bottom of the vials together with a stirring magnet and the vials were tightly sealed by means of crimp caps and PTFE/black butyl molded septa. The three-way valve was then turned to the vacuum working port and the vials were in parallel heated to 200 ^oC to allow the removal of the adsorbed water phase from the solid surface. After 24 h, the temperature was decreased to 60 ^oC the valve was turned to the off position and a binary mixture of [bmim][PF_6] and methanol (v/v) 1:1 (5 ml solution) was injected on the powder by means of a glass syringe that perforated the septa. The powder suspension was left for another 24 h under stirring to assist the infiltration of the IL solution into the pores. In order to break the vacuum and avoid the direct contact with the air moisture the three-way valve was turned to the gas working port permitting the introduction of 1 bar of argon into the vial. After several hours without stirring the supernatant aliquot was suctioned by means of the glass syringe and replaced with pure methanol. The washing with methanol was repeated 5 times to remove most of the excess IL before proceeding with the solvent evaporation (rota-vapour) to yield the IL impregnated solid powder.

2.2.2. *Chemical-"grafting to" method.* A 200 mg portion of MCM and SBA was dried at 423 K under a vacuum for 1 h. To the solid support, a solution of (3-chloropropyl)-trimethoxysilane (2.5 mL, 13.5 mmol) in $CHCl_3$ (10 mL) was added (67.5 mmol silylant/g sample) and the mixture was refluxed under stirring and N_2 atmosphere for 48 h. After filtration, the hybrid material was washed consequently by refluxing in pentane (30 mL), CH_3CN (30 mL), and diethyl ether (30 mL) followed by drying for 1 h at 343 K. Following, the chloropropylated hybrid material was reacted with 1-methylimidazole (5 mL) under N_2 atmosphere at 358 K and remained under rotation for 48 h. The reaction mixture was filtered and the hybrid material was refluxed with diethyl ether (30 mL) for 30 min and then filtered and dried at 343 K for 1 h. To the synthesised [spmim][Cl] hybrid material a solution of $NaPF_6$ (68 mg, 0.4 mmol) in ethanol (6 mL) was added under N_2 atmosphere and the mixture remained under stirring for 5 d at room temperature. Reaction progress for the ion exchange step was monitored with the $AgNO_3$ test, which was positive from the first 24 h from reaction onset. The hybrid material was consequently filtered, washed by refluxing with methanol (30 mL) for 30 min, filtered, and dried in a vacuum for 1 h.

2.3. Instrumentation

The nitrogen adsorption isotherms have been performed volumetrically at 77 K on an Autosorb-1 MP gas analyzer, (Quantachrome instruments). The samples were initially outgassed at 160 °C for 12 h, under high vacuum (<10^{-6} mbar). Pore size distributions, according to BJH [29] and non-local density function theory (NLDFT) [30-31] methods, have been derived from the commercial software AS1win (provided by Quantachrome instruments).

The XRD measurements were performed on a D8 Advance Bruker diffractometer using Cu-Kα radiation and parallel beam stemming from a Göbel Mirror. The measurements were carried out in a 2θ region between 15 and 55^0 corresponding to a Q range varying from 10.7 to 37.7 nm^{-1} .

The SANS measurements were carried out at the PACE spectrometer (LLB, Saclay, France) in a Q-range varying from 0.035 to 3 nm^{-1} corresponding to two sample-detector distances (4.56 m and 1.06 m respectively). The samples were in powdered form and were mounted in a 1 mm thick sealed quartz container (provided by Hellma). For the contrast-matching SANS measurements, they were immersed in an appropriate H_2O/D_2O mixture. The "wet" samples, MCM_w, MCM_PHYS_w, MCM_SILP_w, SBA_w, SBA_PHYS_w and SBA_SILP_w were allowed for 4 h to come to equilibrium before measurement [32]. Absolute values for the intensity, $I(Q)$, were obtained by correcting the raw data for sample transmission, scattering of the empty cell and instrumental background; as a final step, the data were calibrated against a water standard [33]. All corrections were treated by using the software PASINET (for a detailed procedure of data treatment, see ref. 33).

3. Results and discussion

3.1. N_2 adsorption isotherms

The figures 1a and 1b illustrate the N_2 adsorption isotherms at 77 K of both set of samples (MCM, MCM_SILP, MCM_PHYS) and (SBA, SBA_SILP, SBA_PHYS) respectively. Further, the structural characteristics of the samples are reported at table 1. The specific surface area, S, was determined by the BET method [34], while the total pore volume, V_p, was derived from the amount adsorbed at relative pressure, $p/p_0 \approx 0.97$. Finally, pore size distributions were deduced based on both BJH (Table 1) and NLDFT methods (insets of figures 1a and 1b and table 1).

Table 1. Structural parameters of the samples obtained by nitrogen adsorption and SANS measurements*.

Sample	S (m^2/g)	V_p (cm^3/g)	R_p^{BJH} (nm)	R_p^{NLDFT} (nm)	a_0 (nm)
MCM	1070	1.27	1.5	2.1	4.4
MCM_SILP	320	0.42	1.9	2.5	4.4
MCM_PHYS	110	0.33			4.4
SBA	700	0.93	2.8	3.7	10.3
SBA_SILP	480	0.67	3.0	3.8	10.3
SBA_PHYS	17	0.02			10.3

*Abbreviations: S: Specific surface area, V_p: pore volume,: R_p^{BJH} pore radius evaluated by the BJH method, R_p^{NLDFT}: pore radius calculated by the NLDFT method, a_0: lattice parameter estimated from the interplanar d_{10} spacing of SANS data ($a_0 = 2d_{10}/\sqrt{3}$).

The isotherms of the pristine samples (MCM and SBA respectively, figures 1a and 1b) are of type IV, according to the IUPAC classification [35]. Both samples exhibit a hysteresis loop (type H1 by IUPAC classification) corresponding to capillary condensation and evaporation on open cylindrical pores at both ends. The step associated with capillary condensation in the cylindrical mesopores commences at $p/p_0 \approx 0.4$ for the MCM sample whereas for the SBA the correspondent relative pressure is about 0.6. The fact that capillary condensation occurs at a higher relative pressure in the case of SBA compared to the MCM is related to its larger pore size. The upturn of the isotherms from both pristine samples at pressures close to the vapour pressure is attributed to large mesopores, formed between the silica particles.

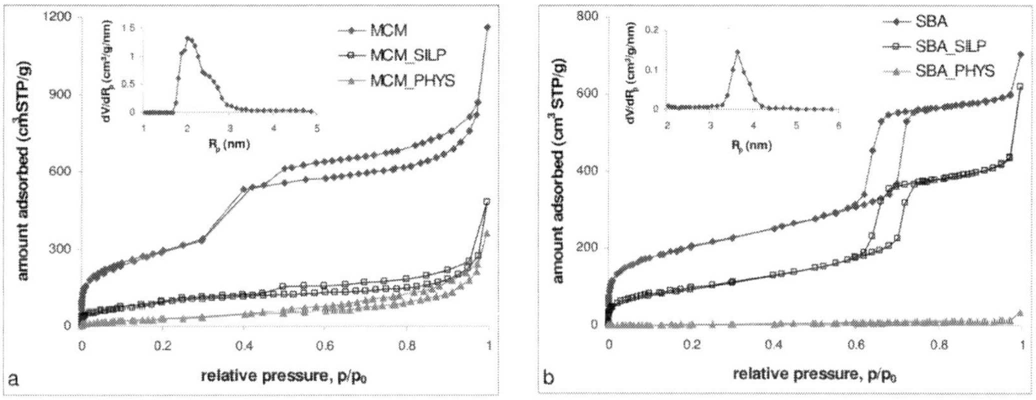

Figure 1. Nitrogen adsorption isotherms at 77 K of (a) MCM, MCM_SILP, MCM_PHYS and (b) SBA, SBA_SILP, SBA_PHYS samples. (Insets: pore size distributions of pristine (a) MCM and (b) SBA samples according to NLDFT method).

On the other hand, in the case of the physically-loaded sample (SBA_PHYS), the amount adsorbed is almost zero (figure 1b). Since N_2 is not dissolved in the particular IL phase the result indicates a total absence of accessibility of N_2 molecules to the pore network, suggesting that the pores are totally filled with IL or the pore mouths are completely blocked rendering the bulk of the pore inaccessible for adsorption. Obviously, pore size distribution is not possible to be extracted. Further, the adsorption isotherm of the MCM_PHYS sample (figure 1a), does not provide reliable information on the pore blocking extent. In particular, the isotherm nearly coincides with that of a non-porous reference silica sample [36], suggesting that the measured adsorbed volume can be attributed to the interparticle space. This argument is also supported by the fact that no reliable pore size distribution can be deduced by either BJH or NLDFT methods. However, as it will be further discussed, the SANS results imply that the pore "core" of the MCM_PHYS sample is empty despite the fact that the MCM pore size is smaller compared to the SBA one (table 1).

The N_2 isotherms of the SILP samples indicate that there is still significant pore space which remains open after grafting the IL molecule. Indeed, the maximum loading that can be achieved with the "grafting to" method is limited to the formation of a monolayer of the silanised IL on to the pore surface. Concerning the SBA_SILP sample, it is certain that the size of the silanised IL, silyl-propyl methyl-imidazolium hexafluorophosphate [spmim][PF_6], is not large enough to result in the effective blockage of the accessibility to the bulk of the pore (3.7 nm). Moreover, the "grafting to" method led solely to a significant reduction of the pore volume but not to alteration of the pore size (table 1). One may then conclude that the grafted IL molecules do not form a dense monolayer on the pore surface (see figure 2). This is something expected due to the sparse population of the surface silanol groups [27] that serve as the anchoring agents of the silyl part of the silanised IL. In this regard, there are still significant areas along the parallel cylindrical pores that are unaffected and where capillary condensation of N_2 occurs at relative pressures identical to these observed for the untreated samples.

Interesting to note is that the pore volume reduction is much more enhanced in the MCM_SILP compared to the SBA_SILP sample. In addition, there is a remarkable shift of the mean pore size to larger values (table 1). The difference in the pore volume reduction can be attributed to the different extent of IL loading, which in the case of the SILP samples is strongly related to the initial population of the silanol groups on the surface of the ordered mesostructured silicas. In this regard, we can state that the silanol population of MCM_SILP is much higher compared to this of the SBA_SILP.

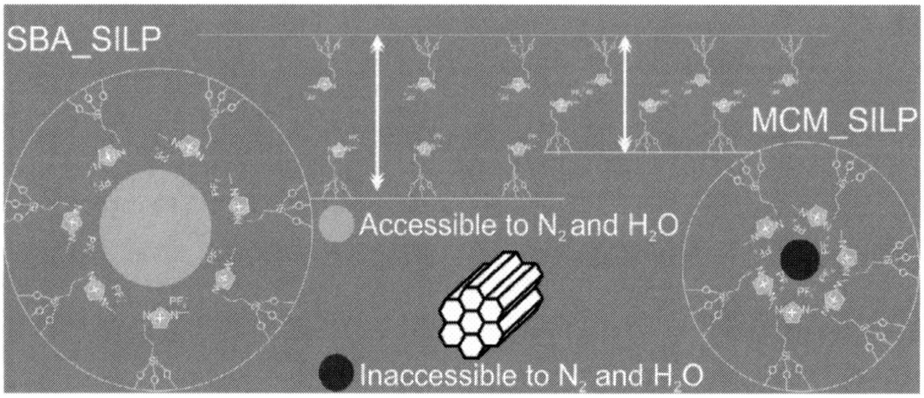

Figure 2. Schematic drawing of the pore loaded with IL in MCM_SILP and SBA_SILP samples according to the "grafting to" method (see text for details).

However, the enhanced pore loading itself, cannot reasonably explain the significant shift of the pore size. This effect can be rather attributed to the complete blocking of a small portion of the MCM_SILP's nanopores (the smaller ones) by the [spmim][PF_6] molecules protruding from the pore walls (figure 2). The size of the [bmim][PF_6] has been recently calculated at 0.9 nm [37]. As a result, this argument seems quite reasonable by taking into account the silane part of the silylated IL [spmim][PF_6] and the space occupied by the surface hydroxyls, a size corresponding to approximately 1.3-1.5 nm for the grafted IL. Thus the ratio between the MCM's mean pore radius (2.1 nm) and the silylated IL's molecular size (1.5 nm) is close to unity and in this respect, a small negative deviation of some of the MCM's pores from the average size can lead to their complete blockage by the grafted IL. The complete blockage of a fraction MCM's pores is further evidenced by the comparison between the incoherent backgrounds of the contrast-matching SANS measurements.

3.2. XRD measurements

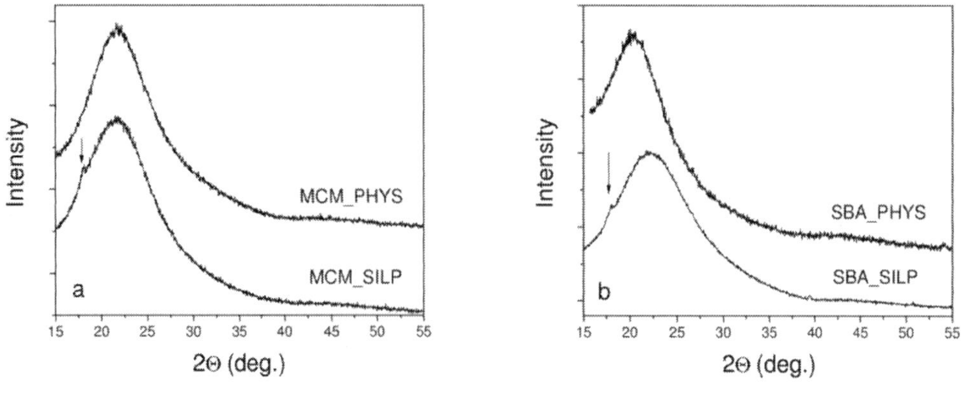

Figure 3. X-ray patterns of physically and chemically IL confinement on (a) MCM-41 and (b) SBA-15.

Figure 3 presents the diffraction spectra from all IL-loaded samples. In all cases, the broad reflection located about 22^0 is due to silica. Both SILP samples (MCM_SILP and SBA_SILP) show a diffraction peak appearing around 18^0 revealing ordering of the grafted silylated IL molecules. On the other hand, this Bragg peak is not observed for the samples loaded with IL via physical imbibition. A first remark is that the interaction of [bmim][PF_6] with the silica pore walls does not affect its ordering. This is in contrast with a case previously studied [38], concerning the interaction of the same IL with graphitic pore walls (carbon nanotubes). In a previous study [28], a degree of ordering was also observed on SILP samples developed on MCM-41 supports of various pore sizes. Based on current and previous XRD results [28] we can conclude that the extent of ordering cannot be directly related to the surface area, the pore volume and the pore size of the support but rather to the different degree of the IL loading because of the constriction effects.

3.3. SANS and contrast-matching SANS

Figure 4a (including the inset) shows the SANS curves from the dry pristine (MCM) and IL-loaded (MCM_SILP and MCM_PHYS) samples respectively. The hexagonal network of all samples produces one intense peak (10) and one weak (11). The high-order peaks are not observed because they are outside of the experimental Q window. In addition, the intensity of Bragg peaks for the samples treated with IL is decreased. In the case of the pristine SBA sample, except for the strong (10) reflection, the overlapping (11), (20) and the faint (21) Bragg peaks are visible.

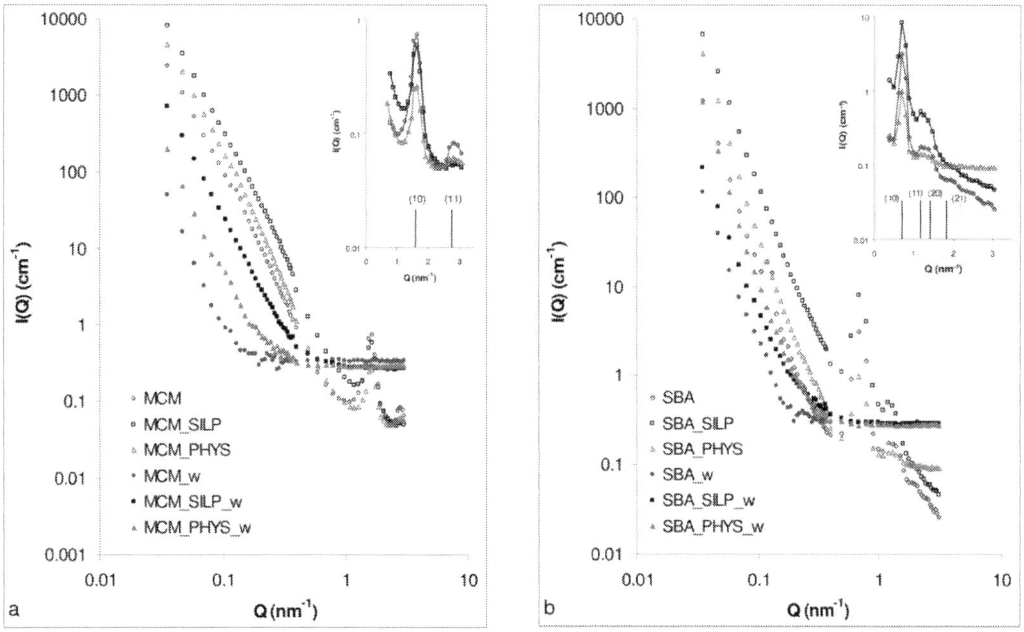

Figure 4. SANS and silica contrast-matching SANS curves from pristine and IL loaded (a) MCM and (b) SBA samples. (Insets: high-Q region spectral details of the pristine and IL loaded (a) MCM and (b) SBA samples; lines have been added as a guide to the eye for the peak positions; peak positions are also indicated by vertical lines).

On the other hand, the scattering from the SBA_PHYS sample becomes flat for Q>1.6 nm^{-1}, mainly due to the incoherent background arising from the hydrogen nuclei contained in the IL; this results in very weak (11) and (20) reflections and the absence of the (21) peak (figure 4b, including the inset). The presence of the incoherent background suggests that a larger amount of IL is confined on the pores via physical imbibition compared to the amount of IL confined by the "grafting to" method. This finding is in agreement with the N_2 adsorption measurements as discussed previously. Another observation is that the intensity of all Bragg peaks for the samples treated with IL, except for the case of SBA_SILP sample, is decreased. As it will be further discussed, this intensity increase implies a small IL loading into the pore space. The calculated values of the lattice parameter, a_0, are given at table 1. The fact that the lattice parameter for all samples (pristine and IL loaded respectively) remains unchanged implies a rather rigid solid matrix that does not alter upon filling with IL either physically or chemically.

In the low-Q region (0.03<Q (nm^{-1})<~0.3) the intensity follows a power law ($I(Q) \sim Q^{-a}$). According to the SANS theory, when a=4, the intensity decreases asymptotically as Q^{-4} following the so-called *Porod's law* [39]. In our case, the observed gradients, especially for the SBA samples, are close to Porod scattering, indicating a contribution from objects considerably larger than the cylindrical mesopores. In specific, it is normally observed in the tail of a SANS curve when Qd>4, where d is the size of the scattering object. Following this argument the size of such objects is >100 nm and can therefore be ascribed to intergrain interfacial scattering, as has also been observed in other studies [40-42]. The specific surface areas for pristine samples derived from the absolute intensities and the Porod function have values less than 2 m^2/g, also in agreement with previous studies [43]. These areas are far smaller than the correspondent BET surface areas (table 1) verifying that the scattering is indeed from the outer surfaces of the particles. Another interesting observation is an increase in intensity for all IL-loaded samples compared to the pristine ones. This behaviour can be attributed to the increase in the average contrast due to the fact that the intergrain space is filled with IL [44], also indicating that apart from the pore surface the external surface of the material is also covered by the IL phase.

SANS can be proved to be a powerful, if not a unique, technique for investigating whether the IL has been partially or completely filled within the pores of ordered mesoporous materials. This is one of the advantages of the SANS technique over the conventional gas adsorption methods because neutrons can also "see" pores which are inaccessible to gas molecules (such as closed pores). For instance, IL could have been dispersed around the pore surface preventing, thus, the access to N_2 molecules. According to the scattering theory, for a two-phase system (for instance, SiO_2-empty pore and SiO_2-IL), the intensities of the Bragg reflections are related to the square of the contrast, defined as the difference of the scattering length density (SLD) between the silica matrix and the pore content:

$$I(Q) = A(\rho_{SiO_2} - \rho_p)^2 \quad (1)$$

where A is a constant term,

$$\rho = \sum_i b_i \frac{dN_A}{M_w} \quad (2)$$

ρ is the SLD of a molecule consisting of i atoms (in an empty pore, ρ_p=0, while in a pore filled with IL, $\rho_p=\rho_{IL}$), b_i is the coherent neutron scattering length of the individual atoms in the molecule, d is the bulk density of the scattering object, M_w is its molecular weight and N_A is the Avogadro number. The method has previously been utilised for measuring the pore filling of SBA-15 by carbon for the synthesis of ordered nanoporous carbons [45]. Further, it has also been employed for evaluating the density, d, of fluids confined within the pores of ordered mesoporous materials under various pressures and temperatures [46-47] In this work, since the SANS experiment has been carried out under ambient conditions, one would assume that the density of the confined IL will be approximately the same with that of the bulk one. Indeed, XRD results confirmed this argument in the case of samples loaded with IL via physical imbibition, where no evidence for room temperature IL crystallisation under confinement was observed. Following eq 1, when the pores are completely filled with IL, the intensity ratio of the reflections of the pristine to the IL-loaded matrices will be equal to the squared SLD ratio:

$$\frac{\int_{Q_1}^{Q_2} I_{SiO_2-empty\ pore}(Q)Q^2\mathrm{d}Q}{\int_{Q_1}^{Q_2} I_{SiO_2-IL}(Q)Q^2\mathrm{d}Q} = \left(\frac{\rho_{SiO_2}}{\rho_{SiO_2}-\rho_{IL}}\right)^2 \quad (3)$$

where the first term is the calculated integrated intensity ratio, R_I, of the Bragg reflections; the limits of the integrals, Q_1 and Q_2 extend over the tails of the peaks. The diffuse scattering arising from pore correlations was treated as a continuous background [48]. In particular, for the fitting of each peak a linear background was used. Further, the incoherent background has been subtracted from the measured intensities in the case of IL treated samples. The second term, R_{SLD}, can be deduced by calculating the SLD of IL and silica respectively; in the case of IL, according to eq 2, ρ_{IL}=1.6·10^{10} cm^{-2} (d=1.38 g/cm^3 at 20 °C). We should note that this is the value for the IL confined via physical imbibition (MCM_PHYS and SBA_PHYS samples). On the other hand, the SLD of silanised-IL confined by the "grafting to" method (MCM_SILP and SBA_SILP) cannot be estimated because of the presence of the silane part. In this regard, the estimation of the IL skeleton density, d, is not possible.

Further, contrast-matching SANS can be employed for the experimental SLD calculation of MCM and SBA samples. The fact that hydrogen and deuterium have scattering lengths of opposite sign means that, neutrons not only can "see" hydrogen isotopes but they can also differentiate between them. It is then possible to match the SLD of the pristine materials (mostly consisted of silica) with an appropriate mixture of water/heavy water. As a result, contrast-matching between the water and the host material can be achieved resulting in the minimisation of the SANS signal ($\rho_p = \rho_{H_2O/D_2O}$ in eq 1). Based on previous contrast-matching SANS measurements performed on these materials [46, 49-54], we used a H_2O/D_2O mixture of 0.40 H_2O volume fraction for both pristine MCM and SBA samples which should "match" the SLD of the silica matrix ($\rho_{SiO_2} \approx 3.6\cdot10^{10}$ cm^{-2}). Indeed, when the samples were immersed in this particular mixture (MCM_w, SBA_w), the Bragg reflections were entirely suppressed and only a constant background remained, mostly due to the incoherent scattering from the protons of H_2O (figures 4a and 4b). In addition, the remaining SANS signal at lower Q values (Q<0.2 nm^{-1}), arising mainly from intergrain scattering, was still observed. However, it was much lower (1-2 orders of magnitude) compared to the dry samples. This suggests that an almost complete contrast-matching has been achieved.

In the case of SBA samples (SBA and SBA_PHYS), the value of the integrated intensity ratio of (10) Bragg peaks, R_I=3.6. In comparison with the value of the squared SLD ratio, R_{SLD}=3.35±0.25, one can claim safely a complete pore filling with IL. The result is also in agreement with the nitrogen adsorption measurements, previously discussed. For the MCM samples, however, the peak integrated intensity ratio, R_I =3.0, is lower than R_{SLD}, implying that the pores of MCM_PHYS sample have not been totally filled with IL. The argument is further enhanced by taking into consideration the small nitrogen uptake observed from the adsorption isotherm data. The result can be attributed to the enhanced attractive forces between the confined IL molecules and the pore wall. The degree of pore filling by IL in the MCM_PHYS sample has been further investigated by modeling the SANS data. In this case, eqs 1 and 3 are not longer valid because we are now dealing with a three-phase system consisting of the silica matrix, the IL layer and the empty pore "core". The modulation of the peak intensities before and after the adsorbed IL can be reproduced by calculating the form factor of pores and IL via appropriate density models [55]. In particular, the scattering amplitude of a uniform core-shell cylinder with three density levels i.e. a dense silica matrix, a confined IL layer and an empty pore "core" can be analytically expressed [56]. The application of this model, despite the lack of high-order reflections, fits reasonably with the experimental data for a pore of radius R_p=1.6 nm (in agreement with N_2 adsorption results, table 1) covered with an IL layer having thickness, t=1.4 nm (figure 5). As a result, the pores of MCM_PHYS sample are almost filled with IL and only the pore "core", having radius R_p-t= 0.2 nm, remains empty.

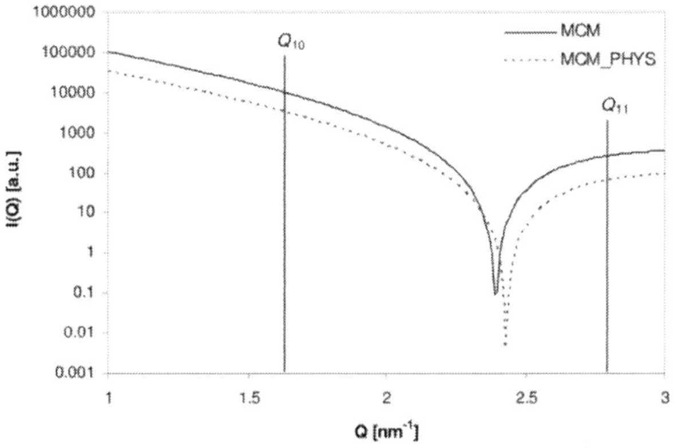

Figure 5. SANS fit for the intensities for the pristine and for loaded with IL via physical imbibition MCM samples for R_p=1.6 nm and t=1.4 nm. The vertical lines denote the positions of the (10) and (11) diffraction peaks.

As previously mentioned, no quantitatively analysis can be performed for the samples treated with the "grafting to" method. However the increased intensity of Bragg peaks for the SBA_SILP sample compared to the MCM_SILP one, suggests a smaller IL loading, in agreement with N_2 porosimetry results. More specific, this intensity increase, especially of the first-order (10) reflection, has also been observed by the performance of *in situ* SANS [40, 47] and SAXS [53, 57] measurements during the initial stages of fluid adsorption on ordered mesoporous MCM-41 and SBA-15 respectively.

As a further step, the pores of the samples treated with IL were filled with the same water mixture that "matches" the silica matrix (MCM_PHYS_w, MCM_SILP_w, SBA_PHYS_w, SBA_SILP_w). In comparison with the dry IL loaded samples, one would not expect significant changes to the SANS signal since their pores are almost (MCM_IL) or completely filled (SBA_IL) with IL. For instance, in a recent study, surfactant adsorption in SBA-15 (with R_p=4 nm) has been studied by SANS [56] The influence of the layer thickness of the adsorbed surfactant films on the scattering curve was investigated by contrast-matching the silica with the appropriate H_2O/D_2O mixture. The calculated thickness of the surfactant films varied between 1 and 3 nm. As the film thickness decreased, a systematic lowering (but not elimination) of the intensities of the Bragg reflections was observed. Our results, however, show clearly that in all (loaded with confined IL) wet samples the peaks are completely vanished, as similarly observed with the pristine contrast-matched samples (figures 4a and 4b). It should be stressed that this behaviour is observed even at physically-loaded samples where the pores are almost (MCM_PHYS) or completely filled (SBA_PHYS) with IL. This is an experimental confirmation of the enhanced solubility of the water in the IL.

Another interesting feature of the pristine and physically-loaded wet samples is that, in the high Q region, the dominated flat incoherent scattering is roughly the same for all samples, except for the pristine one (MCM_w) for which it is higher. This could be attributed to the larger pore volume of MCM compared to that of the SBA (table 1), resulting in a larger amount of adsorbed water and, thus, the presence of more hydrogen atoms. The solubility of the water mixture in IL would not differ between the silica samples (MCM_PHYS and SBA_PHYS respectively), and by taking into consideration that the incoherent background of MCM_PHYS_w is lower than that of the MCM_w and comparable to the correspondent SBA_PHYS_w sample one would possibly suggest that the pore "core" remains empty after the water immersion in the MCM_PHYS sample. As a result, the amount of adsorbed water in both contrast-matched samples loaded with IL via physical imbibition would be roughly the same. Although the size of the "core" diameter (0.4 nm) would allow diffusion of water molecules, the contrast-matching SANS measurements imply that this is not the case. A possible explanation could be that the IL molecules in the adsorbed layer are oriented with the negative ion towards the silica wall and the positive (hydrophobic) towards the pore "core". This is the reason why there is no water entrance into the empty pore "core". It is important to note that issues related to the orientation of ILs ions in accordance to the surface charge are very important for their application in electrified interfaces and electrochemical systems. For this reason, many studies both theoretical and experimental have been devoted to explain the distinct capacitance properties of ILs at the electrode surface, in which the cations and anions of the ILs can form an alternate layer [58-61]. However, there are major improvements still needed [62]. Another interesting observation concerning the wet samples prepared by the "grafting to" method is that the incoherent background for SBA_SILP_w is a bit higher than that of MCM_SILP_w. This result also implies that some of the pore "cores" of MCM_SILP would be inaccessible to water molecules resulting in a less amount of adsorbed water. One should mention that the complete blockage of a fraction of MCM's pores was also concluded from the N_2 adsorption measurements as previously discussed. In this work we have showed that SANS has the potentiality to indirectly predict the ion layers formation of [bmim][PF_6] when confined in mesostructured silicas. Further attempts will be made in the direction to correlate our results with the findings of other advanced electrochemical and AFM methods [63].

4. Conclusions

SANS, contrast-matching SANS, XRD and nitrogen adsorption measurements of IL confined on the pores of ordered mesoporous materials have been presented. IL was introduced by two different methods such as physical imbibition and chemical grafting. In the case of IL confined by physical imbibition both SANS and nitrogen adsorption measurements suggest a complete pore filling with IL for SBA-15, and reveal that only a small fraction of pore volume, the pore "core" is empty for MCM-41. The pore size and the thickness of the IL layer were evaluated by modeling the SANS data. In addition, according to XRD results, there was no evidence for ordering of the confined IL phase. On the other hand, SANS and adsorption measurements suggest a partial pore filling with IL for the chemically grafted SILP samples whereas XRD results indicate a degree of ordering of the grafted IL phase. Moreover, the contrast-matching SANS results from all IL-loaded samples confirm the enhanced solubility of water in IL. They also provide evidence that water does not enter the empty pore "core" of the physically-loaded MCM-41 possibly due to the orientation of IL molecules in the adsorbed layer. Finally, in the case of samples prepared by the "grafting to" method both the adsorption and contrast-matching SANS experiments suggest a complete blockage of a fraction of MCM's pores by the silylated IL resulting in the inaccessibility of both nitrogen and water molecules towards the pore "cores".

Acknowledgements

This research project has been supported by the European Commission under the 7th Framework Programme through the Key Action: Strengthening the European Research Area, Research Infrastructures. Contract n°: 226507 (NMI3).

References

[1] Chen, S, Kobayashi K, Miyata Y, Imazu N, Saito T, Kitaura R and Shinohara H. 2009 *J. Am. Chem. Soc.* **131** 14850
[2] Millefiorini S, Tkaczyk A H, Sedev R, Efthimiadis J and Ralston J 2006 *J Am. Chem. Soc.* **128** 3098

[3] Gao L and McCarthy T J 2007 *J. Am. Chem. Soc.* **129** 3804
[4] Wang J and Li Y J 2009 *J. Am. Chem. Soc.* **131** 5364
[5] Ramsay J D F 1998 *Adv. Coll. Inter. Sci.* **76-77** 13
[6] Mitropoulos A C, Kanellopoulos N K, Stefanopoulos K L and Heenan, R K 1998 *J. Coll. Inter. Sci.* **203** 229
[7] Makri P K, Stefanopoulos K L, Mitropoulos A C, Kanellopoulos N K and Treimer W 2000 *Physica* B **276-278** 477
[8] Hoinkis E, Lima E B and Schubert-Bischoff P 2004 *Langmuir* **20** 8823
[9] Stuhrmann H B and Miller A. 1978 *J. Appl. Cryst.* **11** 325
[10] Charalambopoulou G C, Steriotis T A, Hauss T, Stefanopoulos K L and Stubos A K 2002 *Appl. Phys.* A **74** S1245
[11] Steriotis T A, Stefanopoulos K L, Keiderling U, De Stefanis A and Tomlinson A A G 2002 *Chem. Commun.* **20** 2396
[12] Calo J M and Hall P J 2004 *Carbon* **42** 1299
[13] Lelong G, Price D L, Brady J W and Saboungi M L 2007 *J Chem. Phys.* **127** 065102
[14] Mitropoulos A C, Stefanopoulos K L and Kanellopoulos N K 1998 *Microporous Mesoporous Mater.* **24** 29
[15] Katsaros F K, Steriotis T A, Stefanopoulos K L, Kanellopoulos N K, Mitropoulos A C, Meissner M and Hoser A 2000 *Physica* B **276-278** 901
[16] Steriotis T A, Stefanopoulos K L, Mitropoulos A C, Kanellopoulos N K, Hoser A and Hofmann M 2002 *Appl. Phys.* A **74** S1133
[17] Steriotis T A, Stefanopoulos K L, Kanellopoulos N K, Mitropoulos A C and Hoser A 2004 *Colloids Surf.* A **1-3** 239
[18] Eschricht N, Hoinkis E and Mädler F 2007 *Langmuir* **23** 2145
[19] Sel O, Brandt A, Wallacher D, Thommes M and Smarsly B 2007 *Langmuir* **23** 4724
[20] Mascotto S, Wallacher D, Kuschel A, Polarz S, Zickler G A, Timmann A and Smarsly B M 2010 *Langmuir* **26** 6583
[21] Kikkinides E S, Stefanopoulos K L, Steriotis T A, Mitropoulos A C, Kanellopoulos N K and Treimer W 2002 *Appl. Phys.* A **74** S954
[22] Eschricht N, Hoinkis E, Mädler F, Schubert-Bischoff P and Röhl-Kuhn B 2005 *J. Coll. Inter. Sci.* **291** 201
[23] Eastoe J, Gold S, Rogers S E, Paul A, Welton T, Heenan R K and Grillo I. 2005 *J. Am. Chem. Soc.* **127** 7302
[24] Tasserit C, Koutsioubas A, Lairez D, Zalczer G and Clochard M-C 2010 *Phys. Rev. Lett.* **105** 260602
[25] Kresge C T, Leonowicz M E, Roth W J, Vartuli J C and Beck J S 1992 *Nature* **359** 710
[26] Zhao D, Feng J, Huo Q, Melosh N, Fredrickson G H, Chmelka B F and Stucky G D 1998 *Science* **279** 548
[27] Romanos G E, Vangeli O C, Stefanopoulos K L, Kouvelos E P, Papageorgiou S K, Favvas E P and Kanellopoulos N K 2009 *Microporous Mesoporous Mater.* **120** 53
[28] Vangeli O C, Romanos G E, Beltsios K G, Fokas D, Kouvelos E P, Stefanopoulos K L and Kanellopoulos N K 2010 *J. Phys. Chem.* B **114** 6480
[29] Barret E P, Joyner L G and Halenda P H 1951 *J. Am. Chem. Soc.* **73** 373
[30] Tarazona P 1985 *Phys. Rev.* A **31** 2672
[31] Neimark A V, Ravikovitch P I and Vishnyakov A 2003 *J. Phys: Conden. Matter* **15** 347
[32] Mergia K, Stefanopoulos K L, Ordás N and García-Rosales C *2010 Microporous Mesoporous Mater.* **134** 141
[33] Brûlet A, Lairez D, Lapp A and Cotton J-P 2007 *J. Appl. Cryst.* **40** 165
[34] Brunauer S, Emmet P H and Teller E J 1939 *J. Am. Chem. Soc.* **60** 309
[35] Sing K S W, Everett D H, Haul R A W, Moscou L, Pierotti R A, Rouquérol J and Siemieniewska T 1998 *Pure Appl. Chem.* **57** 603

[36] Cregg S J and Sing K S W 1982 *Adsorption, Surface Area and Porosity* (New York: Academic Press)
[37] Singh M P, Singh R K and Chandra S, 2011 *J. Phys. Chem.* B **115** 7505
[38] Chen S, Wu G, Sha M and Huang S 2007 *J Am. Chem. Soc.* **129** 2416
[39] Porod G 1951 *Kolloid Z.* **124** 83
[40] Ramsay J D F, Kallus S and Hoinkis E 2000 *Stud. Surf. Sci. Catal.* **128** 439
[41] Kallus S, Hahn A and Ramsay J D F 2003 *Eur. Phys. J.* E **12** S31
[42] Liu D, Zhang Y, Chen C-G, Mou C-Y, Poole P H and Chen S-H 2007 *Proc. Nat. Acad. Sci.* **104** 9570
[43] Edler K J, Reynolds P A and White J W 1998 *J. Phys. Chem.* **102** 3676
[44] Ramsay J D F and Hoinkis E 1998 *Physica* B **248** 322
[45] Ehrburger-Dolle F, Morfin I, Geissler E, Bley F, Livet F, Vix-Guterl C, Saadallah S, Parmentier J, Reda M, Patarin J, Iliescu M and Werckmann J 2003 *Langmuir* **19** 4303
[46] Morineau D, Guégan R, Xia Y and Alba-Simionesco C 2002 *J. Chem. Phys.* **117** 8966
[47] Steriotis T A, Stefanopoulos K L, Katsaros F K, Gläser R, Hannon A C and Ramsay J D F 2008 *Phys. Rev.* B **78** 115424
[48] Zickler G A, Jähnert S, Wagermaier W, Funari S S, Findenegg G H and Paris O 2006 *Phys. Rev.* B **73** 184109
[49] Glinka C J, Nicol J M, Stucky G D, Margolese D, Huo Q, Higgins J B and Leonowicz M E 1996 *J. Porous Mater.* **3** 93
[50] Tun Z, Mason P C, Mansour F K and Peemoeller H 2002 *Langmuir* **18** 975
[51] Erko M, Wallacher D, Brandt A and Paris O 2010 *J. Appl. Cryst.* **43** 1
[52] Müter D, Shin T, Demé B, Fratzl P, Paris O and Findenegg G H 2010 *J Phys. Chem. Lett.* **1** 1442
[53] Schreiber A, Ketelsen I, Findenegg G H and Hoinkis E 2007 *Stud. Surf. Sci. Catal.* **160** 17
[54] Linton P, Rennie A R and Alfredsson V 2011 *Solid State Sci.* **13** 793
[55] Impéror-Clerc M, Davidson P and Davidson A 2000 *J. Am. Chem. Soc.* **122** 11925
[56] Shin T, Findenegg G H and Brandt A 2006 *Prog. Colloid Polym. Sci.* **133** 116
[57] Jähnert S, Müter D, Prass J, Zickler G A, Paris O and Findenegg G H 2009 *J. Phys. Chem.* C **113** 15201
[58] Maolin S, Guozhong W, Qiang D, Zhongfeng T and Haiping F 2010 *Langmuir* **26** 12667
[59] Pinilla C, Del Pópolo M G, Kohanoff J and Lynden-Bell R M 2007 *J. Phys. Chem.* B **111** 4877
[60] Buzzeo M C, Evans R G and Compton R G 2004 *Chem. Phys. Chem.* **5** 1106
[61] Baldelli S. 2005 *J. Phys. Chem.* B **109** 13049
[62] Kornyshev A A 2007 *J. Phys. Chem.* B **111** 5545
[63] Atkin R and Warr G G 2007 *J. Phys. Chem.* C **111** 5162

Journal of Physics: Conference Series **340** (2012) 012088 doi:10.1088/1742-6596/340/1/012088

Near Surface Crystallization of Pluronic P123

[1]S. Gerth, [1]M. Klimczak, [2]A. Nelson and [1]A. Magerl

[1]Chair of Crystallography and Structural Physics, University of Erlangen-Nürnberg, D-91058 Erlangen, Germany

[2]The Australian Nuclear Science and Technology Organization, Australia

E-mail: gerth@krist.uni-erlangen.de

Abstract. On a hydrophilic substrate the growth of a Bragg peak was observed at Q≈0.05 $Å^{-1}$ which we attribute to a layering of micelles. This ordering is destroyed when the system is sheared. When shear is removed, it builds up and reaches equilibrium at ambient temperature after approximately 700s. However, there is a waiting time of 200s before any structure becomes visible again. When lower shear rates are applied the ordering recovers more quickly, but it does not build up to the same extent.

1. Introduction

Aqueous solutions of tri-block copolymers PEO-PPO-PEO are well-known for the reduction of surface tension and friction at an interface. For these reasons they are used in many washing agents or lubricants [1, 2, 3, 4, 5]. In medical applications tri-block copolymers are used as drug delivery agents, wound stabilizers or structural gels [6, 7, 8, 9]. At low temperatures PPO and PEO show good solubility in water. With increasing temperature, however, especially the PPO block undergoes a distinct dehydration, accompanied by an increase in hydrophobicity. Above the critical micellization temperature (CMT), micelles are formed with a hydrophobic core (PPO) surrounded by a hydrophilic shell (PEO). This makes the surfactants an excellent model system for the study of crystallization in soft matter [10].

Different chemical surface treatments influence the near-interface ordering. In particular crystal growth is preferred at an interface which is attractive for the micelles [11]. Then, coming from a liquid state, micellar layers may build up at the interface driven by an increased polymer concentration even if the bulk concentration is too low for bulk crystallization.

2. Sample and Methods

Our interest focuses on Pluronic P123 (purchased from Sigma Aldrich and used without further purification), which has a central part of 70 propylene oxide (PO) units and two end groups of 20 ethylene oxide (EO) units (EO_{20}–PO_{70}–EO_{20}). It readily forms stable micelles [1, 12, 13, 14]. In concentrated bulk solutions (above 27 weight percent (wt-%)) the micelles self-assemble into crystalline structures at intermediate temperatures [15, 16]. This study focuses on the near surface crystallization of diluted Pluronic P123 below the concentration regime for bulk micelle crystallization with particular emphasis on the effect of shear on agglomerated polymers near an interface. A cone-plate shear device suitable for in-situ neutron reflectometry was built with a cone angle of 4°. To assure a constant shear rate the servo[1] controls a motor[2] coupled with a gear transmission ratio of 1:25 to the shear cone. Thus, the shear rate is calculated to

$$\dot{\gamma} = \frac{2\pi \cdot n_{\mathrm{rpm}}}{25 \cdot 60 \tan 4^\circ}.$$

It is important to control both the sample and the shear cone temperature to minimize the heat transfer to the sample due to friction at the sealing between sample and cooling liquid (see figure 1). The neutron path goes through a silicon crystal plate above which the liquid of interest and the shear cone are located. More details will be given in a forthcoming publication.

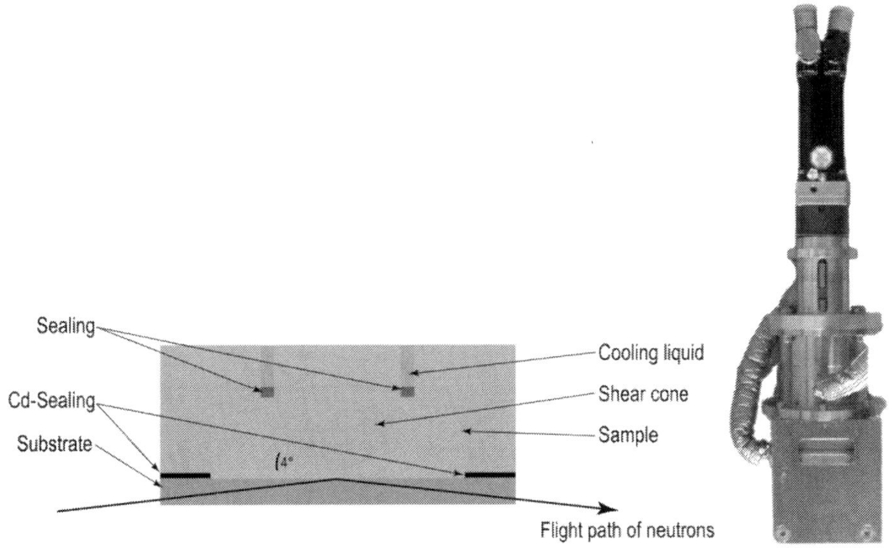

Figure 1. Schematic drawing of the shear cell (left), and a picture of the cell (right) with the motor on the top and the tubing system for the cooling water. The neutron entrance window is visible at the bottom of the shear cell.

The pluronic is dissolved in D_2O to enhance the neutron scattering contrast to the protonated polymer chains and to achieve total reflection at the solid-liquid interface. Taking the density differences for H_2O and D_2O in account, the P123 concentration (in wt-%) for a heavy water solutions is

$$c_{\mathrm{D_2O}} = \frac{1}{1 + \left(\frac{1}{c_{\mathrm{H_2O}}} - 1\right) \frac{\rho_{\mathrm{D_2O}}}{\rho_{\mathrm{H_2O}}}}.$$

Thus, the number density of polymers in the solvent is the same as in light water and a comparison with phase diagrams from literature is readily possible.

Reflectometry is the method of choice for the investigation of adsorbates at a molecular scale at a buried solid-liquid interface [17, 18, 19]. The measurements were done at the PLATYPUS reflectometer [20, 21] at ANSTO[3] in Australia. This is a vertical time of flight reflectometer with the capability of scattering downwards to $Q_{\max} = 0.4$ Å^{-1}. The time of flight option is valuable for analyzing the adsorption kinematics because all Q-points are measured simultaneously, in contrast to a monochromatic angular dispersive reflectometer.

[1] Schneider Electric Lexium 05
[2] Schneider Electric BSH0552T01A2A
[3] Australian Nuclear Science and Technology Organization

3. Results

The adsorption kinetics of 2 wt-% P123 dissolved in D_2O were investigated after the cessation of shear at a temperature of 68 °C on a hydrophilic silicon substrate. No bulk crystallization occurs in this temperature and concentration regime. Every single reflectivity profile has been accumulated for a period of 60 s. In figure 2 example datasets are shown after the cessation of a shear rate of 50 s^{-1} and 400 s^{-1}.

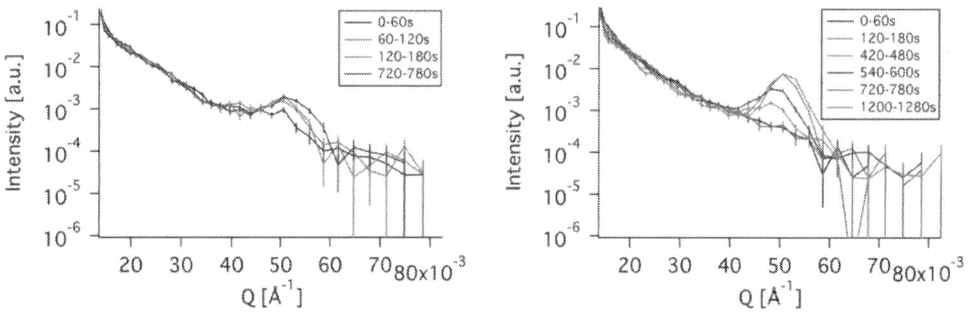

Figure 2. Comparison of successive reflectivity profiles measured for 60 seconds each after the cessation of a shear rate of 50 s^{-1} (left) and 400 s^{-1} (right).

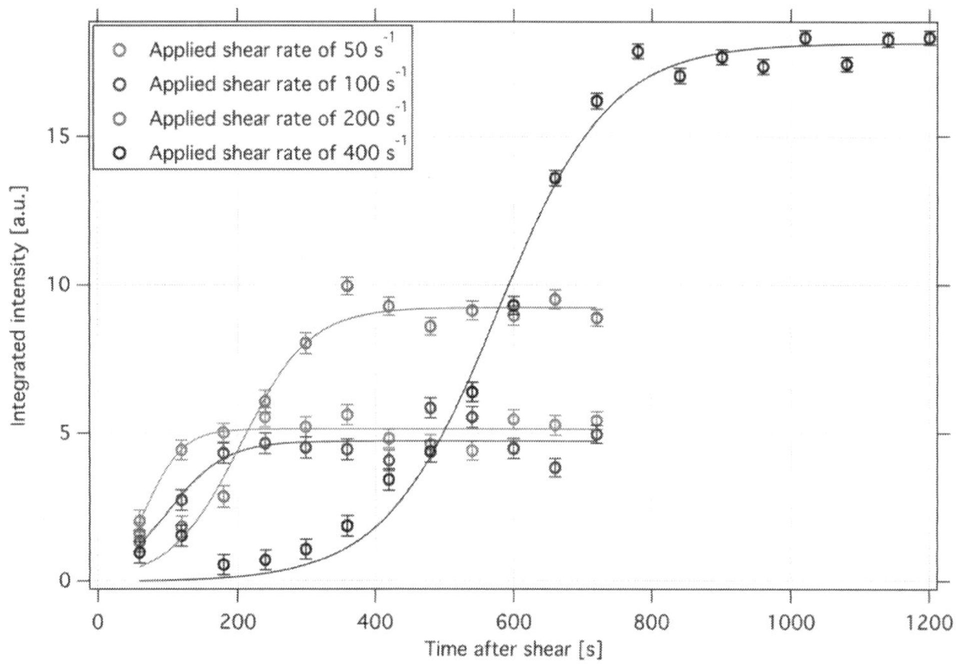

Figure 3. Integrated intensity of the Bragg peak after the cessation of different shear rates. The solid lines represent a fit for the model of a logistic function.

The reflectivity profiles were multiplied by Q^4 to compensate for the *Fresnel* reflectivity. A linear function was subtracted to account for background and the oscillation originating from the thin silicon dioxide. The intensity of the Bragg peak was integrated to characterize the recovery

Journal of Physics: Conference Series **340** (2012) 012088 doi:10.1088/1742-6596/340/1/012088

process. A logistic function was used for refinement of the inflection time t_0, the extent of the recovery of the Bragg peak $I_{\max}$ and the growth rate r defined by $I(t) = I_{\max}\left(\frac{1}{1+\exp(-r(t-t_0))}\right)$. This function models an exponential growth in the initial stage combined with a saturation when half of the surface is covered. The inflection time t_0 corresponds to the point of inflection of the growth process.

Fit parameters are shown in table 1 and plotted in figure 4. It is remarkable that after a shear rate of 50 s^{-1} and 100 s^{-1} the integrated intensity is about equal. After shearing with higher rates the intensity increases linearly. The inflection time seems to increase and the growth rate to decrease with the shear rate.

Table 1. Refinement results for the model of an exponential saturation. $I_{\max}$ is the integrated intensity of the Bragg peak, t_0 is the inflection time of the adsorption process and r the growth rate.

Applied shear rate [s^{-1}]	$I_{\max}$ [a.u.]	t_0 [s]	r [$10^{-3}s^{-1}$]
50	5.14 ± 0.11	70.97 ± 7.16	37.8 ± 10.4
100	4.74 ± 0.12	100.7 ± 9.68	24.6 ± 5.86
200	9.24 ± 0.13	204.4 ± 5.67	19.8 ± 1.94
400	18.1 ± 0.10	576.2 ± 3.20	12.4 ± 0.43

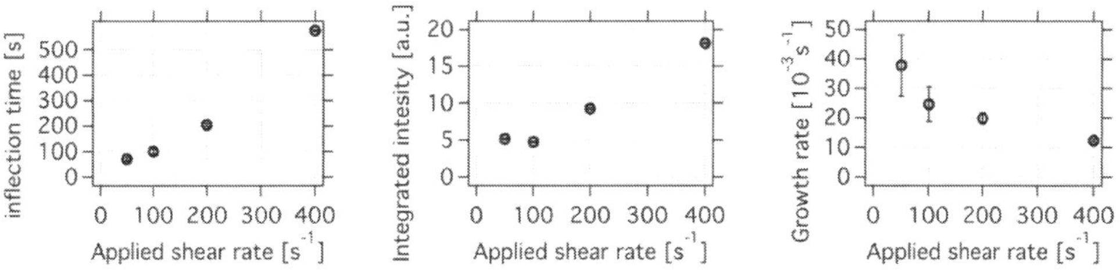

Figure 4. Inflection time t_0, integrated intensity of the Bragg peak $I_{\max}$ and growth rate r plotted as a function of the applied shear rate.

4. Conclusion and Outlook

We have constructed a compact cone-plate shear cell for inverse geometry with temperature stabilization and an encapsulated volume for long time measurements.

The recovery of an adsorbed micellar layer was observed in dependence of the previously applied shear rate. A 2 wt-% solution of P123 dissolved in D_2O was measured at 68 °C to emphasize the effect of the interface. It has to be noted, that in this concentration and temperature regime no bulk crystallization occurs as observed in small angle X-ray scattering measurements and rheometry data [16]. Therefore, the ordering (figure 2) is only induced from the hydrophilic interface. Using hydrophobic interfaces, no ordering is observed at all.

The integrated intensity is about the same after stopping the shear rate of 50 s^{-1} and 100 s^{-1}. The growth rate r decreases as a function of the previously applied shear rate. The integrated intensity is increased after the cessation of shear with rates above 100 s^{-1}. This could be due to a shear induced ordering within the liquid. Also the inflection time of the recovery changes quadratically with the applied rotation speed of the cone. Maybe this is due to the low damping of the rotation of the liquid after stopping the shear cone, a topic which needs further attention in the future. Alternatively, it could also be attributed to shear induced ordering of the polymer within the liquid. The polymer chains may be entangled on the surface and as soon as the shear rate reaches a critical value, they become free yielding an improved ordering. This would cause a depletion like zone at the interface as a function of the applied shear rate and thus a change in the inflection time and the growth rate. Below this critical shear rate the polymer chains stay entangled close to the interface and the ordering after stopping the shear remains unchanged.

References

[1] Loh W 2002 *Encyclopedia Surface Colloid Sci.* **8** 802–813
[2] Alexandridis P and Hatton T 1995 *Colloids and Surfaces A* **96** 1–46
[3] Förster S and Konrad M 2003 *J. Mater. Chem.* **33** 2671–2688
[4] Bahadur P and Riess G 1991 *Tenside Surf. det.* **28** 173
[5] Schmolka I 1977 *J. Am. Oil Chem. Soc.* **54** 110
[6] Lin S and Kawashima Y 1985 *Pharm Acta Helv.* **60** 339
[7] Yokoyama M 1992 *Crit. Rev. Therapeutic Drug Carrier Syst.* **9** 213
[8] Gunzman M, Garcia F, Molpeceres J and Aberturas M 1992 *Int J. Pharm.* **80** 119
[9] Kabanov A, Bartakova E, Melik-Nubarov N, Fedoseev N, Dorodnich T, Alakhov V, Chekhonin V, Nazalova I and Kabanov V 1992 *J Controlled Release* **22** 141
[10] Mortensen K 1996 *J. Phys.: Condens. Matter* **8** A103–A124
[11] Wolff M, Scholz U, Hock R, Magerl A, Leiner V and Zabel H 2004 *Phys. Rev. Lett* **92** 255501
[12] Sasanuma Y 1995 *Macromolecules* **28** 8629–8638
[13] Guo C, Wang J, Liu H and Chen J 1999 *Langmuir* **15** 2703–2708
[14] Guo C, Liu H and Chen J 1999 *Colloid Polym Sci* **277** 376–381
[15] Wanka G, Hoffmann H and Ulbricht W 1994 *Macromolecules* **27** 4145–4159
[16] Walz M, Wolff M, Voss N, Zabel H and Magerl A 2010 *Langmuir* **26** 14391–14394
[17] Tolan M 1999 *X-ray scattering from soft-matter thin films* (*Springer tracts in modern physics* vol 148) (Springer-Verlag Berlin)
[18] Als-Nielsen J and McMorrow D 2001 *Elements of Modern X-Ray Physics* (John Wiley & Sons, Ltd)
[19] Daillant J and Gibaud A 1999 *X-Ray and Neutron Reflectivity: Principles and Applications* (Springer)
[20] James M, Nelson A, Holt S, Saerbeck T, Hamilton W and Klose F 2011 *Nuclear Instruments and Methods in Physics Research A* **632** 112–123
[21] Nelson A 2010 *Journal of Physics: Conference series* **251** 012094

Structural characterization of semicrystalline polymer morphologies by imaging-SANS

A Radulescu[1], L J Fetters[2] and D Richter[3]

[1]Forschungszentrum Jülich GmbH, Jülich Centre for Neutron Science, Outstation FRMII, 85747 Garching, Germany

[2]School of Chemical and Biomolecular Engineering, Cornell University, Ithaca, NY 14853-5021, USA

[3]Forschungszentrum Jülich GmbH, Jülich Centre for Neutron Science and Institute for Complex Systems, 52425 Jülich, Germany

a.radulescu@fz-juelich.de

Abstract. Control and optimization of polymer properties require the global knowledge of the constitutive microstructures of polymer morphologies in various conditions. The microstructural features can be typically explored over a wide length scale by combining pinhole-, focusing- and ultra-small-angle neutron scattering (SANS) techniques. Though it proved to be a successful approach, this involves major efforts related to the use of various scattering instruments and large amount of samples and the need to ensure the same crystallization kinetics for the samples investigated at various facilities, in different sample cell geometries and at different time intervals. With the installation and commissioning of the MgF_2 neutron lenses at the KWS-2 SANS diffractometer installed at the Heinz Maier-Leibnitz neutron source (FRMII reactor) in Garching, a wide Q-range, between $10^{-4}Å^{-1}$ and $0.5Å^{-1}$, can be covered at a single instrument. This enables investigation of polymer microstructures over a length scale from 1nm up to 1μm, while the overall polymer morphology can be further examined up to 100μm by optical microscopy (including crossed polarizers). The study of different semi-crystalline polypropylene-based polymers in solution is discussed and the new imaging-SANS approach allowing for an unambiguous and complete structural characterization of polymer morphologies is presented.

1. Introduction

The macroscopic behavior of semicrystalline polymers depends on the constitutive microstructures consisting of molecules arranged in the unit cell (~Å), lamellar crystals (~100Å), fibrils or boards (~1000Å) and spherulitic macro-aggregats (>1μm). Control and optimization of polymer properties require the global knowledge of these microstructural features in various conditions. Such a wide length scale can be explored by combining conventional pinhole-, focusing- and ultra-SANS techniques [1] which enable to cover a Q-range typically between $10^{-5}Å^{-1}$ and $0.2\text{-}0.3Å^{-1}$. This approach, which permits the exploration of independent or interrelated structural levels emerging within a length scale from 10Å up to 10μm, was often applied for the quantitative characterization of the morphology of a wide range of materials, from hydrogels to geologic samples [2]. Optical and transmission electron microscopy observations have often been used complementary to the ultra-SANS investigations, in order to help for the assessment of the scattering data.

This combination of scattering methods proved to be successful in resolving the structure and morphology of complex systems showing a hierarchical multi-scale organization [3-8]. The investigation of crystallization behavior of polymers from melt or solution requires in this case a special attention. Occurrence and evolution of microstructures in a semicrystalline polymer depend strongly on the crystallization conditions which are defined and controlled by the experimental items, like the variation and control of temperature and the amount and geometry of sample. Thus, major efforts that must ensure the same crystallization kinetics for the sample studied at various experimental facilities (diffractometers, neutron sources, etc.), in different sample cell geometries and at different time intervals are involved. The access of a wide Q-range at the same diffractometer solves such problems and allows for a complete and unambiguous characterization of the structures and morphologies arising from crystallization processes in an easier way.

In this paper we report about the exploration of the multilevel structures that form upon cooling in solution of polypropylene-based semicrystalline polymers by using two imaging-SANS approaches based on the combination of small-angle neutron scattering and microscopy methods. The first approach implies the combination of three SANS techniques: conventional SANS at the KWS-2 classical pinhole SANS diffractometer [9,10], focusing-mirror SANS (f-SANS) at the KWS-3 very-SANS instrument [11,12] and ultra-SANS at the double-crystal diffractometer DKD [13]. The configuration of the instruments at the FRJ-2 reactor in Jülich will be briefly presented and the experimental results discussed. Optical microscopy with crossed polarizers was used complementary to the ultra- and f-SANS investigations. The second experimental approach it is since recently in use at the KWS-2 diffractometer, which was re-installed in 2007 at the FRMII reactor in Garching. This approach implies for the observation of a wide-Q range, between 10^{-4}Å^{-1} and 0.5Å^{-1}, the use of neutron lenses (MgF_2 optical elements [14-16]) besides the classical pinhole set-up. The use of optical microscopy for the investigation of the large-scale morphologies several micrometers in size it is a prerequisite condition for the complete characterization of the complex aggregates in this case.

2. Experimental section

2.1. Samples

A syndiotactic polypropylene homopolymer *s*PP and a multiblock atactic-isotactic alternating polypropylene (*a*PP-*i*PP) copolymer were studied by SANS, f-SANS and ultra-SANS in dilute solutions (polymer volume fraction Φ_{pol}=0.5%) in deuterated decane (d-22) or dodecane (d-26). The polymers were prepared via a living polymerization in toluene at 0°C using a bis(salicylaldiminato)titanium complex. Details about the synthesis of polymers are presented elsewhere [3,17]. The self-assembling structural features of two materials, a *s*PP polymer with the molecular weight M_W=61.4kg/mol (M_W/M_n=1.1) and a *a*PP-*i*PP alternating copolymer with the M_W=107kg/mol (M_W/M_n=2), were investigated at room temperature.

For testing the performance of the focusing lenses at the KWS-2 SANS instrument a poly(methymethacrylate) (PMMA) colloidal sample (R=7760Å) was investigated. The obtained results have been compared with those obtained at the focusing-mirror KWS-3 instrument on a similar sample [18]. In order to minimize the effects arising from multiple scattering, h-PMMA dissolved in h-cis-decalin (0.25% colloids volume fraction) was measured, which provided a reasonable contrast condition and an acceptable sample transmission [18].

Finally, a crystalline-amorphous *s*PP-P(E-*co*-P) diblock copolymer was investigated in a deuterated decane (d-22) solution (Φ_{pol}=1%) at KWS-2 diffractometer using the combined conventional pinhole and focusing-lenses modes.

2.2. SANS techniques

Three types of SANS techniques were used at the FRJ-2 reactor. The conventional SANS measurements were performed on the KWS-2 pinhole instrument using a neutron wavelength of λ=7.2Å ($\Delta\lambda/\lambda$=10%) and a sample-to-detector distance varying from 2 to 20m, covering a Q range between 0.002 and 0.14Å^{-1}, where Q is the modulus of the scattering vector, $4\pi(\sin\theta)/\lambda$, with θ being half the scattering angle. The raw data were corrected two dimensionally for the empty cell scattering and detector sensitivity and were calibrated in absolute units using a Plexiglas secondary standard [9]. The KWS-3 focusing-mirror SANS instrument [11,12] works on a principle based on a one-to-one image of an entrance aperture on a two-dimensional high-resolution position-sensitive detector (0.5mm spatial resolution)by neutron reflection from a double focusing toroidal mirror. A small entrance aperture (typically 2x2mm^2) and the neutron wavelength λ=12.7Å ($\Delta\lambda/\lambda$=9%) permitted SANS studies within a Q range 1x10^{-4} to 2x10^{-3}Å^{-1}. The ultra-SANS measurements were performed at the DKD double-crystal diffractometer, which, operating with a neutron wavelength λ=4.5Å, could cover a Q range between 3x10^{-5} and 2x10^{-4}Å^{-1}. The high-Q resolution existed in one dimension only, therefore, in order to present the measured data in continuation to those measured at KWS-2 and KWS-3, slit corrections had to be applied [13].

At KWS-3 and DKD instruments the multiple scattering effects were very important and dominated for sample thicknesses within the range 1-2mm, typical for conventional pinhole-SANS experiments [11]. To minimize the contribution of such effects and to have a reasonable sample transmission (higher than 70%) one had either to decrease the path length, dilute the sample (to nearly infinite dilution), or one had to decrease the contrast by using a certain fraction of protonated solvent.

Following the shutdown of the FRJ-2 reactor the KWS-2 and KWS-3 instruments have been moved to the Heinz Maier-Leibnitz neutron source (FRMII reactor) in Garching. Here, the much higher flux allows at KWS-2 for the use of longer wavelengths, up to 20Å, thus enabling much lower values in Q, up to 7x10^{-4}Å^{-1}, in the pinhole configuration. A further decrease in Q up to 10^{-4}Å^{-1} is possible by means of focusing elements – the MgF_2 lenses [14-16], which are used in the combination with a small entrance aperture, at the longest collimation length of 20m. The conceptual design and the preliminary tests using aspherical lenses are reported in detail elsewhere [15]. Here, we present some additional results of testing the performance of lenses and report about the complete characterization of the *s*PP-P(E-*co*-P) structures in solution by combining the pinhole and the focusing-lenses modes. This proved to be a suitable experimental approach for the exploration of structural features over a wide length scale on samples like semicrystalline polymers requiring a special care with respect to variation of the environmental conditions. In the present study seven lenses have been used with a high-resolution detector (space resolution of 0.5mm) placed at 8.9m distance from the lenses and a small collimation aperture (6x6mm^2) placed at 18.2m in front of the lenses. The neutron wavelength used was λ=17.6Å and the sample size (sample aperture) was 1.5cm x 2.5cm. A beam-stop of 4x4mm^2 was used in order to mask the direct beam focused on the detector.

Finally, optical microscopy using a Zeiss-Axioplan-2 microscope in transmission bright field direct geometry equipped with a Linkam THMS600 temperature stage was involved in the study. Polarized light with crossed polarizers was used to complete the characterization of some of the morphologies identified. The technique was always used as a complementary method to the ultra- and f-SANS investigations and served for the assessment of the scattering data. For the case of exploring a wide Q range only at the KWS-2 instrument, the optical microscopy becomes necessarily used for the determination of the morphology and size of the large scale macro-aggregates within the micrometer range.

3. Modelling and Data Interpretation

Self-assembly and crystallization from solution of semicrystalline polymers often result in complex morphologies displaying multiple structural levels on length scale ranging from nanometers to microns. The combination of scattering data with fractal geometry concepts recently commenced to become a general path of investigating such systems. Fractal models describe power law regimes often observed in measured scattering profiles [19-21] with the exponents depending on the geometric structure of the scattering objects:

$$\frac{d\Sigma}{d\Omega}(Q) \approx Q^{-p} \tag{1}$$

Hence, one-dimensional (p=1), platelet-like (p=2), mass fractals ($1 \leq p \leq 3$), or surface fractals ($3 \leq p \leq 4$) structures can be identified and characterized. Objects with smooth surface yield p=4 (Porod scattering), whereas diffuse interfaces lead to the occurrence of exponents higher than p=4 [22].

The analysis of the extensive scattering data in Q has been done in terms of the Beaucage model [20,21], which describes material microstructure in terms of level of structure where each level is characterized by a Guinier regime combined with a structurally limited power law regime. With two size limits ($QR_{gi} \approx 1$ and $QR_{g(i+1)} \approx 1$) for one structural level, the model describes an arbitrary number of interrelated structural features on different size scales by:

$$\frac{d\Sigma}{d\Omega}(Q) = \sum_{i=1}^{m} G_i \exp\left(\frac{-Q^2 R_{gi}^2}{3}\right) + B_i \exp\left(\frac{-Q^2 R_{g(i+1)}^2}{3}\right)\left[\frac{\left(erf\left(QR_{gi}/\sqrt{6}\right)\right)^3}{Q}\right]^{p_i} \tag{2}$$

where m is the number of structural levels observed in the scattering pattern extending over many decades in Q. The first term describes a large structure of average size R_{gi} composed of small-scale structures of average size $R_{g(i+1)}$. The second term allows for different morphology for the large structure according to (1). G is the classical Guinier prefactor ("forward scattering"), while B is a prefactor specific tot the type of power law regime [20,21].

Alternatively, in some cases, the scattering data from aggregates were evaluated using the traditional form factor approach [23].

4. Results and Discussion

Figure 1 presents the scattering profile at room temperature from the 0.5% solution of *s*PP polymer in d-22 measured over four decades in Q by combining the SANS, f-SANS and ultra-SANS methods. The sample was cooled in a controlled way (1°C/min) from the single coil regime, at 160°C. The power law exponent specific for the characteristic structures formed by the polymer is also indicated.

The scattering profile at intermediate and high Q follows a Q^{-2} behavior, which is indicative of polymer two-dimensional or platelet-like aggregates. The intensity drop at high Q relates to the lateral profile of the polymer platelets. The lateral extension of the platelets must exceed 3500Å, which corresponds to the lowest extension in Q of the platelets scattering profile. At lower Q the scattering intensity increases dramatically following a steep power law with an exponent higher than 4 and terminates in a Guinier-like regime at very low Q. This observation shows the formation of large-scale aggregates exhibiting diffuse interfaces and sizes within the micrometer range. For a more detailed characterization of these aggregates, optical microscopy observations have been done complementary to the ultra- and f-SANS investigations.

Figure 2 presents two micrographs collected from the polymer solution at room temperature. Open spherical aggregates with sizes around 10µm made by radially disposed smaller structures can be observed. The crossed polarizers observation of these aggregates revealed a texture which is typical for spherulitic morphology. A polymer spherulite represents a spheriform cluster of primary fiber-like or elongated lamellar crystallites that start growing laterally from the nucleation site [24]. The typical pattern yielded by a spherulitic texture when is observed with crossed polarizers is the "Maltese cross", which is observed in figure 2-right.

The scattering and microscopy data are completing each other. The scattering profile collected over a wide Q-range revealed formation of lamellar structures and large aggregates with diffuse interfaces several micrometers in size. The microscopy observations proved that these structures are interrelated and represent different structural levels of a hierarchical morphology: the polymer spherulite represents the large scale structural level which is composed of small-scale elongated lamellar substructures that basically exhibit two structural levels, their lateral extension and width. Due the crossover between the high-Q scattering feature of spherulites and the low-Q scattering feature of the lamellae, in the scattering pattern collected over an extended Q range only two of these structural levels can be observed, namely the size of the spherulites and the thickness of the lamellae.

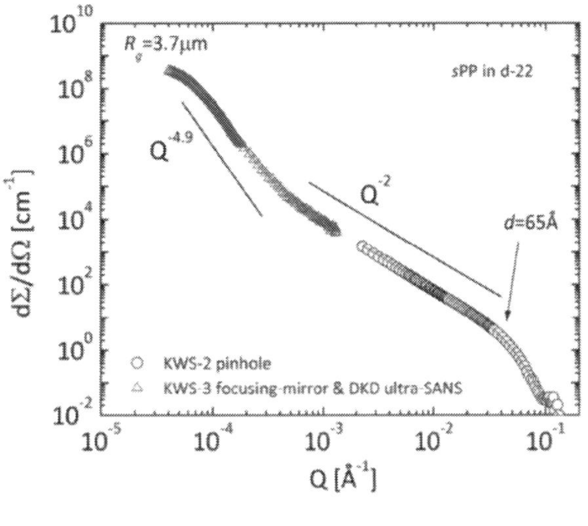

Figure 1. Wide-Q scattering pattern from solution of *s*PP homopolymer in d-22 at room temperature. The straight lines represent the power law behavior in different Q regions, whereas the solid red curve represents the model description of experimental data according to the multi-level Beaucage model (2). The main structural parameters delivered by the model analysis are also indicated.

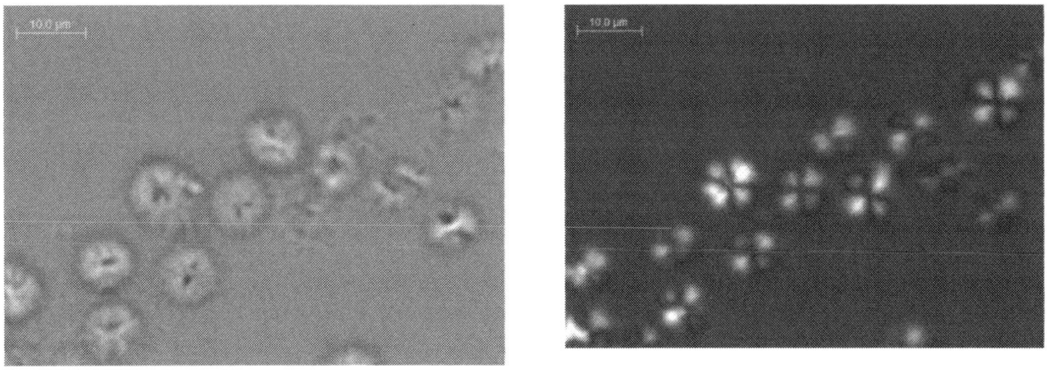

Figure 2. Morphologies of *s*PP homopolymer in d-22 observed under microscope using the bright field (left) and crossed polarizers (right); the scale bars represent 10μm.

A fit of the scattering data using the multilevel equation of Beaucage (2), which is also shown in figure 1, delivered the sizes (R_g of the spherulites and thickness d of the lamellae) and density parameters of the two structures. Using the obtained forward scattering and the lamellar thickness d and supposing compact lamellae, the total volume fraction of polymer inside the platelets coincides with the polymer volume fraction in the initial solution, which shows that all polymers are crystallizing at room temperature. With this information, using the "forward scattering" from the spherulites and their size, the volume fraction which is occupied by the polymer inside the spherulite is about 30%, which shows the open aspect of these macro-aggregates. It should also be noted that the polymer spherulites are aggregates with diffusive interfaces rather than showing a fractal character like it was reported [25]. This confirms the recent results [5] obtained on joint spherulitic aggregates of semi-crystalline multiblock PEB-*n* copolymers and paraffin molecules in common solution. We want to underline that for the complete characterization of such complex morphologies showing a hierarchical internal organization an exploration of a wide Q range is required in order to avoid misinterpretation of the scattering features.

Figure 3-left presents the scattering profile at room temperature from the 0.5% solution of *a*PP-*i*PP alternating copolymer in d-26 measured over a wide Q-range by combining the SANS, f-SANS and ultra-SANS methods. The sample was cooled in a controlled way (1°C/min) from the single coil regime, at 180°C.

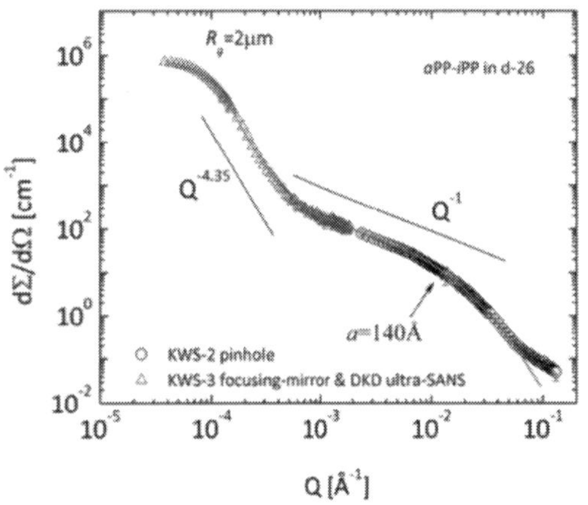

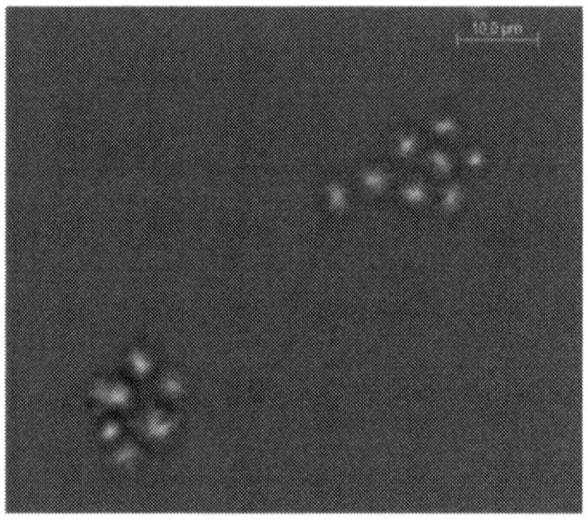

Figure 3. (left) Wide-Q scattering pattern from solution of *a*PP-*i*PP alternating copolymer in d-26 at room temperature; lines with the same meaning as in figure 1; (right) Morphology of *a*PP-*i*PP alternating copolymer in d-26 observed under microscope using the bright field; the scale bar represents 10μm.

The scattering pattern collected over four decades in Q reveals again formation of a complex morphology showing multiple structural levels and hierarchical internal organization. Large micron size aggregates with diffusive interfaces are composed of one-dimensional rod-like substructures, as revealed by the Q^{-1} behavior of the scattered intensity observed at the intermediate Q range. Two structural levels are revealed, the overall size of the macro-aggregates (R_g) at very low Q and the thickness a of the rods, towards high Q. A length of the rod L=0.8μm is deduced from the Beaucage fit in the crossover Q-region between the patterns from the macro-aggregates and the one-dimensional structures. The fact that the radius of the macro-aggregates is larger than the length of the rod-like substructures demonstrates that branching effects develop extensively in the *a*PP-*i*PP morphology. Optical micrographs (figure 3-right) show large scale aggregates displaying a spherical shape and irregular edges. With crossed polarizers no typical features of spherulitic texture were observed. This proves that the polymer aggregates are solvent-swollen dense objects showing a preponderant amorphous or diffuse behavior. The fit of the wide-Q scattering data using the multilevel equation of Beaucage (2) delivered the geometrical parameters (R_g of the macro-aggregates and the rod radius a, shown in figure 3-left) and the "forward scattering" from the two identified structures. The temperature evolution of the self-assembling of *a*PP-*i*PP alternating polymers [3] has shown that the scattered intensity level is still increasing in decreasing the temperature down to 0°C. Therefore, at room temperature there is a partition of the dissolved polymers between the self-assembled structures and the remaining chains in solution and it is not possible to obtain independent information on the polymer volume fraction inside the aggregates and the volume fraction of rod-like and large-scale aggregates in solution from the interpretation of the "forward scattering". Nevertheless, the scattered intensity level observed at the lowest Q is very low compared to that from the *s*PP solution. This shows that the polymer volume fraction inside the alternating copolymer macro-aggregates is much lower than that inside the sPP spherulites. Finally, the tendency of the *a*PP-*i*PP alternating copolymer to form one-dimensional open structures instead of lamellar morphologies may be explained taking into account the molecular architecture in a similar way with formation of rod-like aggregates in the case of the PEB-*n* random copolymers in solution [1]: the amorphous *a*PP blocks seem to form a loose corona around the crystalline nuclei formed by the crystallizing sections of the *i*PP blocks and largely screen the cocrystallization of these nuclei, promoting thus a one-dimensional growth.

Figure 4 shows the scattering from PMMA colloids corrected with background subtraction and in the radial averaged presentation as resulted from data measured at KWS-2 with seven MgF_2 lenses and a neutron wavelength λ=17.6Å. The result obtained on the same sample at KWS-3 focusing-mirror instrument with a neutron wavelength λ=12.7Å is shown in parallel. The minimum Q value reached at KWS-2 with neutron lenses goes down to 10^{-4}Å^{-1}, which shows that more than three decades in Q range, from 0.5Å^{-1} up to 10^{-4}Å^{-1}, can be covered at a single instrument. A fit with the solid sphere form factor including the polydispersity $\sigma(R)/R$ of the particle radii reported by static light scattering (σ_0=0.06 of the log-normal size distribution), which is depicted by the solid curve in figure 4, delivered a colloidal radius of about 8100Å, a value slightly higher than that obtained by static light scattering [18]. Because of finite resolution, the minima of the form factor are smeared out in the experimental data measured at both instruments.

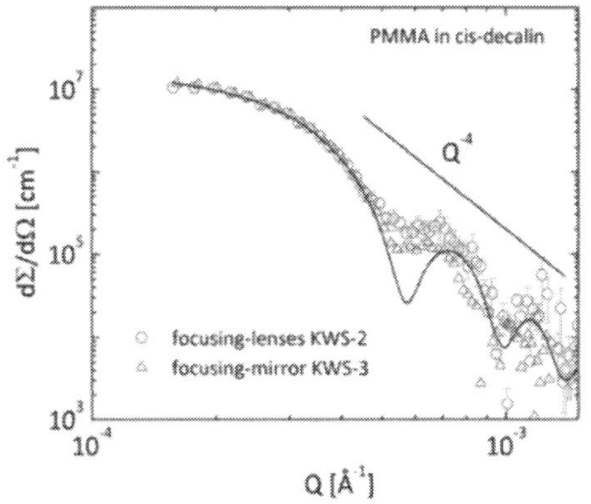

Figure 4. Intensity versus scattering vector of the 0.25% PMMA in cis-decalin solution obtained by focusing-mirror at KWS-3 and focusing-lenses at KWS-2. The solid curve represents the fit of the experimental data with the form factor of spheres (see text).

The scattering profile measured over a wide Q range from the 1% solution of the *s*PP-P(E-*co*-P) diblock copolymer in d-22 at 0°C is displayed in figure 5-left. The measurements have been performed by using the pinhole and focusing-lenses modes at KWS-2. The results obtained at KWS-3 are shown in parallel with the focusing-lenses data. The Q^{-1} power law behavior observed at low and intermediate Q-range indicates formation of long one-dimensional or rod-like aggregates. The length of these rods is not accessible since no Guinier-like plateau is observed towards small Q. Thus, the length of the aggregates must exceed 1.5μm. The scattering features observed towards high Q relate to internal structure of the rod-like aggregates. A detailed analysis of the pinhole data as a function of temperature and polymer volume fraction in solution was done elsewhere [3]. The weak increasing tendency of the scattered intensity which is observable at very low Q denotes the formation of larger scale aggregates several microns in size. Figure 5-right shows the morphology of these aggregates captured at 0°C by using a thermal stage: long semi-rigid needles with a thickness of about 3μm are observed.

The straightforward attempt to model the experimental data using the cylinder form-factor [23] is hindered by the presence of a weak shoulder burying the form-factor effect of the rod's lateral size at around $Q \cong 8 \times 10^{-3}$ Å^{-1}. This feature denotes a correlation effect arising from the association of polymer rods in bundles [3]. For a detailed model analysis of the scattering data the structure factor for aligned cylinders [26] must be taken into account. A lateral size of the polymer rods of about 450Å can be roughly estimated from the deviation of the scattered intensity from Q^{-1} behavior. Combining the scattering data and the microscopy results the hierarchical character of the diblock copolymer morphology in solution is again understood: thin rods of a thickness of about 450Å, revealed by scattering, associate in bundles which, at much larger scale, appear as thick one-dimensional semi-rigid cylinders, shown by the micrographs.

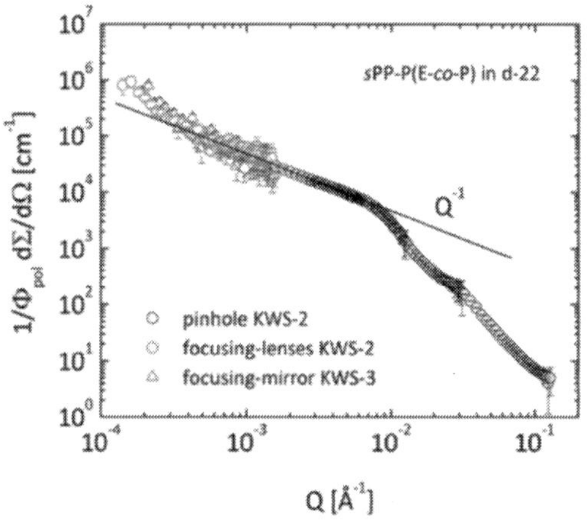

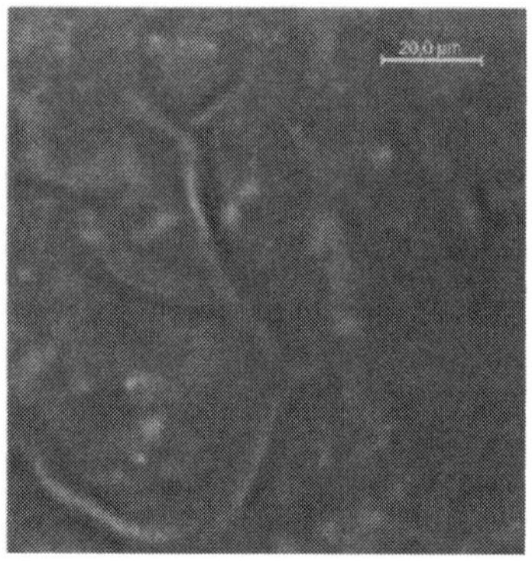

Figure 5. (left) Scattering pattern from the 1% solution of the *s*PP-P(E-*co*-P) diblock copolymer in d-22 at 0°C; (right) Morphology of *s*PP-P(E-co-P) diblock copolymer in d-22 observed under microscope; the scale bar represents 20μm.

5. Conclusion

Imaging-SANS approach combining wide-Q small-angle neutron scattering techniques and optical microscopy is a powerful method for the complete characterization of complex morphologies showing multiple structural levels and a hierarchical organization on a length scale between 10Å and a hundred of microns. The structural and texture features of the morphologies of interest are unambiguously elucidated, which would not be possible by using only one of the methods. A clear example is the understanding of the *s*PP spherulitic morphology which cannot be completely characterized using only microscopy observations or scattering data. Another essential message is that one needs to combine pinhole- and focusing-SANS in order to achieve a wide enough Q-range for avoiding misinterpretation of the scattering results. With the installation and commissioning of the MgF_2 neutron lenses at the

KWS-2 SANS diffractometer installed at the Heinz Maier-Leibnitz neutron source (FRMII reactor) in Garching, a wide Q-range, between 10^{-4}Å^{-1} and 0.5Å^{-1}, can be covered at a single instrument. This allows for the elimination of problems related to the sample geometry or the control of the in situ sample treatment, which are inherent when several scattering instruments are involved in the study at different time periods.

Acknowledgements
We thank Jörg Stellbrink (Forschungszentrum Jülich GmbH) for the kindness to provide us the f-SANS data on PMMA solution. Useful discussions with Dietmar Schwahn, Henrich Frielinghaus and Vitaliy Pipich (Forschungszentrum Jülich GmbH) about data interpretation are gratefully acknowledged.

References

[1] Radulescu A, Fetters L J and Richter D 2008 *Adv. Polym. Sci.* **210** 1
[2] Schaefer D W and Agamalian M M 2004 *Curr. Opin. Solid State Mater. Sci.* **8** 39
[3] Radulescu A, Mathers RT, Coates GW, Richter D and Fetters L J 2004 *Macromolecules* **37** 6962
[4] Radulescu A, Schwahn D, Stellbrink J, Kentzinger E, Heiderich M, Richter D and Fetters L J 2006 *Macromolecules* **39** 6142
[5] Radulescu A, Schwahn D, Stellbrink J, Monkenbusch M, Fetters L J and Richter S 2011 *J. Pol. Sci. Part B: Polym. Phys.* **49** 144
[6] Beaucage G, Ulibarri T A, Black E P and Schaefer D W 1995 Multiple Size Scale Structures in Silica-Siloxane Composites Studies by Small Angle Scattering *Hybrid Organic-Inorganic Composites* ed J E Mark, C I C Lee and P A Bianconi (ACS) chapter 9, pp 97-111
[7] Beaucage G, Aubert J H, Lagasse R R, Schaefer D W, Rieker T P, Erlich P, Stein R S, Kulkarni S and Whaley P D 1996 *J. Pol. Sci. Part B:Polym. Phys.* **34** 3063
[8] Schaefer D W, Rieker T, Agamalian M, Lin J S, Fischer D, Sukumaran S, Chen C, Beaucage G, Herd C and Ivie J 2000 *J. Appl. Cryst.* **33** 587
[9] Schwahn D, Meier G and Springer T 1991 *J. Appl. Cryst.* **24** 568
[10] Vad T, Sager W F C, Zhang J, Buitenhuis J and Radulescu A 2010 *J. Appl. Cryst.* **43** 686
[11] Radulescu A, Kentzinger E, Stellbrink J, Dohmen L, Alefeld B, Rücker U, Heiderich M, Schwahn D, Brückel T and Richter D 2005 *Neutron News* **16** 18
[12] Goerigk G and Varga Z 2011 *J. Appl. Cryst.* **44** 337
[13] Schwahn D and Yoo M H 1986 *Springer Proceedings in Physics* Springer Verlag: Berlin vol.2 pp 83-88
[14] Choi S M, Barker J G, Glinka C J, Cheng Y T and Gammel P L 2000 *J. Appl. Cryst.* **33** 793
[15] Koizumi S, Iwase H, Suzuki J, Oku T, Motokawa R, Sasao H, Tanaka H, Yamaguchi D, Shimizu H M and Hashimoto T 2007 *J. Appl. Cryst.* **40** s474
[16] Frielinghaus H, Pipich V, Radulescu A, Heiderich M, Hanslik R, Dahlhoff K, Iwase H, Koizumi S and Schwahn D 2009 *J. Appl. Cryst.* **42** 681
[17] Tian J, Hustad P D and Coates G W 2001 *J. Am. Chem. Soc.* **123** 5134
[18] Stellbrink J 2004 Multiple Scattering Effects in Concentrated Colloidal Solution *FRJ-2 Experimental Report, Forschungszentrum Jülich GmbH*
[19] Schaefer D W, Martin J E, Wiltzius P and Cannell D S 1984 *Phys. Rev. Lett.* **52** 2371
[20] Beaucage G 1996 *J. Appl. Cryst.* **29** 134
[21] Hammouda B 2010 *J. Appl. Cryst.* **43** 1474
[22] Ruland W 1987 *Macromolecules* **20** 87
[23] Pedersen J S 1997 *Adv. Colloid. Interface Sci.* **70** 171
[24] Granassy L, Pusztai T, Tegze G, Warren J A and Douglas J F 2005 *Phys. Rev.* E **72** 011605
[25] Owen A and Bergmann A 2004 *Polym. Int.* **53** 12
[26] Oster G and Riley D P 1952 *Acta Crystallogr.* **5** 272

Neutron diffraction on polymer nanocomposites - A tool for structural and orientation studies

A.A. Sapalidis[1], F.K. Katsaros[1], Th.A Steriotis[1], N.K. Kanellopoulos[1], S. Dante[2,3], and T. Hauss[2]

[1]N.C.S.R. "Demokritos" – Institute of Physical Chemistry, 15310, Athens, Greece

[2]Helmholtz-Zentrum Berlin für Materialien und Energie, Hahn-Meitner-Platz 1, D-14109 Berlin, Germany

[3]Facility of Nanobiotechnology, Department of Neuroscience and Brain Technology, Italian Institute of Technology, Via Morego 30, I-16163 Genova, Italy

asap@chem.demokritos.gr

Abstract. A series of Polyvinyl alcohol (Mowiol 5-88) – Bentonite nanocomposite films with predefined clay loading (up to 0-20%), were prepared via solvent casting technique. The developed films, due to the favourable polymer-particle interactions, revealed excellent dispersion of the clay particles in the polymer matrix and improved properties. Furthermore, the properties of PVA/clay nanocomposites as well as their structural changes as a function of the relative humidity were thoroughly investigated using neutron membrane diffraction experiments. The samples prior their measurement were equilibrated at different relative humidity levels (%RH) using either H_2O or D_2O. The application of contrast variation technique enabled us to investigate the contribution of both the polymer and the clay particles to the diffraction spectra. Thus, the use of H_2O enlightened the low Q region, providing information about the structure of the inorganic phase and specifically the stacking of the clay platelets. The diffraction patterns in this region obtained from perpendicular and in-plane sample positions revealed that there is a specific orientation of bentonite plates, parallel to the film surface. This conclusion is in agreement with the results obtained from XRD and gas permeability technique, in which the well organized and dispersed impermeable inorganic layers, increase the resistance in flow through the nanocomposites film, acting as gas barriers. On the other hand, diffraction experiments on pre-equilibrated with D_2O samples revealed the structural changes in polymeric matrix, due to hydration. The obtained peak revealed the presence of a new crystalline phase, presumably induced by the presence of the silicates, which is in agreement with DSC data reported in previous studies.

1. Introduction

Polymer-Layered Silicate Nanocomposites (PLSNs) are two-phase materials in which the polymers are reinforced by nanoscale fillers. In this type of materials, the nano-dimensioned inorganic platelets are finely dispersed in the polymer matrix offering remarkable improvement in performance properties of the polymer including high moduli [1,2,3] increased strength [4] and heat resistance [5], improved gas barriers properties [6,7], fire redundancy [8,9] etc. The control and the optimization of the abovementioned structural factors are expected to enable the development of nanocomposite materials with predefined superior properties.

Since the properties of the nanocomposites are defined by the dispersion of the inorganic nanoplatelets into the polymer, the degree of exfoliation /intercalation is of great importance. A number of methods have been reported in the literature for this purpose [10,11,12]. Generally, the state of dispersion and exfoliation of nanoparticles is typically studied by means of X-ray diffraction (XRD) analysis and transmission electron microscopy (TEM) observation. Due to its easiness and availability, XRD is most commonly used to probe the nanocomposites' structure [13,14,15]. The nanostructure, namely, intercalated or exfoliated, may be identified by monitoring the position, shape, and intensity of the basal reflections, from the distributed silicate layers. For example, in an exfoliated nanocomposite, the extensive layer separation associated with the delamination of the original silicate layers in the polymer matrix results in the eventual disappearance of any coherent X-ray diffraction from the distributed silicate layers. Alternatively, for intercalated nanocomposites, the finite layer expansion associated with the polymer intercalation results in the appearance of a new basal reflection corresponding to the larger gallery height. Additionally in some cases, XRD is employed to study the kinetics of the polymer melt intercalation [16].

Although XRD offers a convenient method to determine the interlayer spacing of the silicate layers in the original layered silicates and in the intercalated nanocomposites (within 1-4 nm), little can be said about the spatial distribution of the silicate layers or the structural homogeneity of the nanocomposites. In addition, some layered silicates do not exhibit well-defined basal reflections, rendering the determination of the intensity pattern and the shape of the relative peaks difficult [17]. Therefore, in many cases conclusions concerning the mechanism of nanocomposites formation and their structure based solely on XRD patterns are only tentative.

On the other hand, TEM allows a qualitative understanding of the internal structure, spatial distribution and dispersion of the nanoparticles within the polymer matrix, and views of the defect structure through direct visualization. However, special care must be exercised to guarantee a representative cross section of the sample.

To overcome the disadvantages of the conventional techniques, numerous advanced methods have been developed for the characterization of PLSNs. Recently, small angle scattering of both X-Rays and neutrons, has been employed to characterize the nanocomposite structure and provide a measure of nanoparticle orientation [18,19,20,21,22]. Furthermore, the effect of shear on the orientation of polymer–clay nanocomposites has been examined by several groups [23,24,25]. Additionally, contrast variation or contrast matching SANS has been considered as very powerful tool for investigation of multicomponent nanocomposite systems, e.g., clay-polymer solutions [26] silica-polymer suspension [27] etc.

The aim of this work is to employ neutron membrane diffraction to study the dispersion and orientation of the clay layers in the polymeric matrix, and mainly to monitor the structural changes (crystallinity, swelling, disordering and migration of clay layers) as a function of the relative humidity of the composites. To this end, PVA/Bentonite nanocomposites with different clay loading were prepared by effective dispersion of the inorganic platelets into the PVA matrix, via solvent casting technique. The developed nanocomposites were studied by a variety of conventional techniques. Due to the favourable polymer-particle interactions, the materials exhibited excellent dispersion of the clay particles in the polymer matrix with improved mechanical strength and advanced gas barrier properties, while they retained their optical transparency even at high clay loadings (20 wt %). Furthermore, the solvent equilibrated films were studied by neutron diffraction (V1 diffractometer, HZB, Germany) using H_2O and D_2O. The diffraction patterns obtained from different sample orientations relative to the neutron beam (in-plane and perpendicular) revealed that there is a specific orientation of bentonite plates, parallel to the films' surface. This conclusion is in agreement with the results obtained from XRD and gas permeability measurements.

2. Experimental

2.1. Materials and Methods

Low viscosity, partially hydrolyzed atactic Poly(vinyl alcohol) Mowiol® 5-88 (Average weight molecular weight: 37000 g/mol - Sigma) was used for the preparation of the nanocomposites. Bentonite (cationic exchange capacity of 80 meq/100g) was supplied by S&B Industrial Minerals S.A. The purity of raw clay used in this study was about 76%.

The dispersion of the clay platelets was investigated by classical XRD and TEM techniques. The XRD patterns were recorded on a Siemens XD-500 diffractometer using CuKα1 radiation source. Furthermore the Electron Transmission images were obtained using a Jeol JEM 2011 TEM.

The oxygen permeability experiments at 23 °C and variant %RH were performed on a PBI Dansensor OPT-5000 instrument, according to ASTM F2622-08 method. The samples, prior to their testing, were conditioned in oven at a temperature of 50 ± 2 °C for 48 h (ASTM D 618 – Procedure B).

The mechanical strength of the films was measured by means of a Thumler GmbH Tensile Tester Model (cell load 250N -PA 6110 Nordic Transducer), using specimens of 3 cm width and 6 cm length. Prior the testing, the samples were pre-equilibrated at constant relative humidity.

2.2. Preparation of PVA/Bentonite nanocomposites

A PVA solution in water (10 wt%) was produced (6 hours mixing at 90°C) and used as stock solution. The samples were prepared by mixing water suspensions of bentonite with the polymer solutions in quantities that gave 5, 10 and 20 wt% clay loading on the final films (PVAB05, PVAB10 and PVAB20, respectively). Initially, the mixture of polymer and bentonite (100ml) was stirred for half hour at 80°C and then sonicated for an extra half hour. The suspension (10 to 20 ml regarding the final film thickness) was poured in square (12x12 cm) polystyrene Petri dishes and left to dry slowly at 25°C for about 15 days. The average thickness of the samples used for permeability and mechanical tests was 0.1 ± 0.02 mm.

2.3 Neutron diffraction experiments

Neutron diffraction experiments were carried out on the V1 membrane diffractometer at the Berlin Neutron Scattering Centre of the Helmholtz Center Berlin for Materials and Energy (Berlin, Germany). V1 diffractometer is extremely well suited in this respect, not only due to its geometry but also because of the unique sample environment (in terms of controlled relative humidity) and pertinent know-how available [28,29,30,31].

Experiments were performed on pre-equilibrated with H_2O or D_2O at two different sample

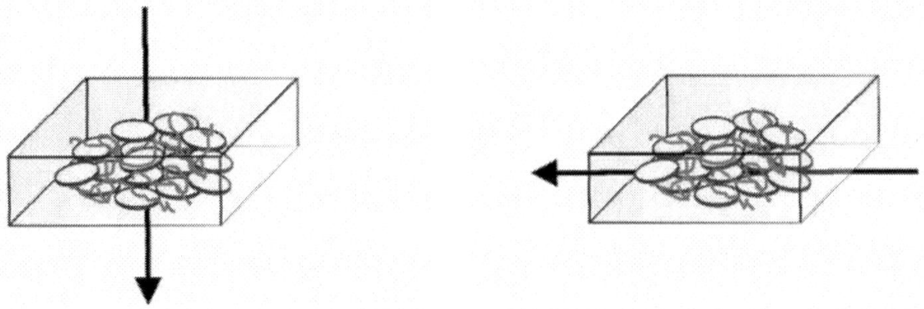

(a) (b)

Figure 1. Experimental setup for neutron diffraction measurements: (a) perpendicular and (b) in-plane geometry orientations, i.e. perpendicular and in-plane to neutron beam (Figure 1). To this end, film samples were placed between quartz microscope slides with the aid of a specially constructed holder. The holder was placed in a hermetically sealed aluminum can, equipped with a H_2O (or D_2O) reservoir in order to retain the equilibration conditions during diffraction measurements. The can was attached onto a goniometer head and was further connected to a circulating water bath for temperature control. Prior to the experiment the sample was aligned parallel to the beam with a laser pointer.

The diffraction patterns were obtained by rocking the sample at different fixed angles. The two-dimensional position sensitive 3He detector of V1 was set to move in steps of approximately 1-2° in 2θ, starting as close as possible to the main beam. At each detector angle (2θ), the sample was rocked though a range centered on half the detector angle (ω). The data were corrected for detector response and sample absorbance then integrated to give lamellar intensity as a function of q (Å^{-1}). The neutron wavelength and sample to-detector distance were 5.23Å and 102.38 cm, respectively.

3. Results and discussion

3.1. Characterisation of nanocomposites using XRD and TEM

Transmission Electron Microscopy (TEM) was used to for the preliminary characterization of the nanocomposites studied in this work. Emphasis was given on the investigation of the inorganic layers' dispersion into the polymer matrix. Typical TEM images of the 20 wt % clay loaded sample (PVAB20) are shown in Figure 2.

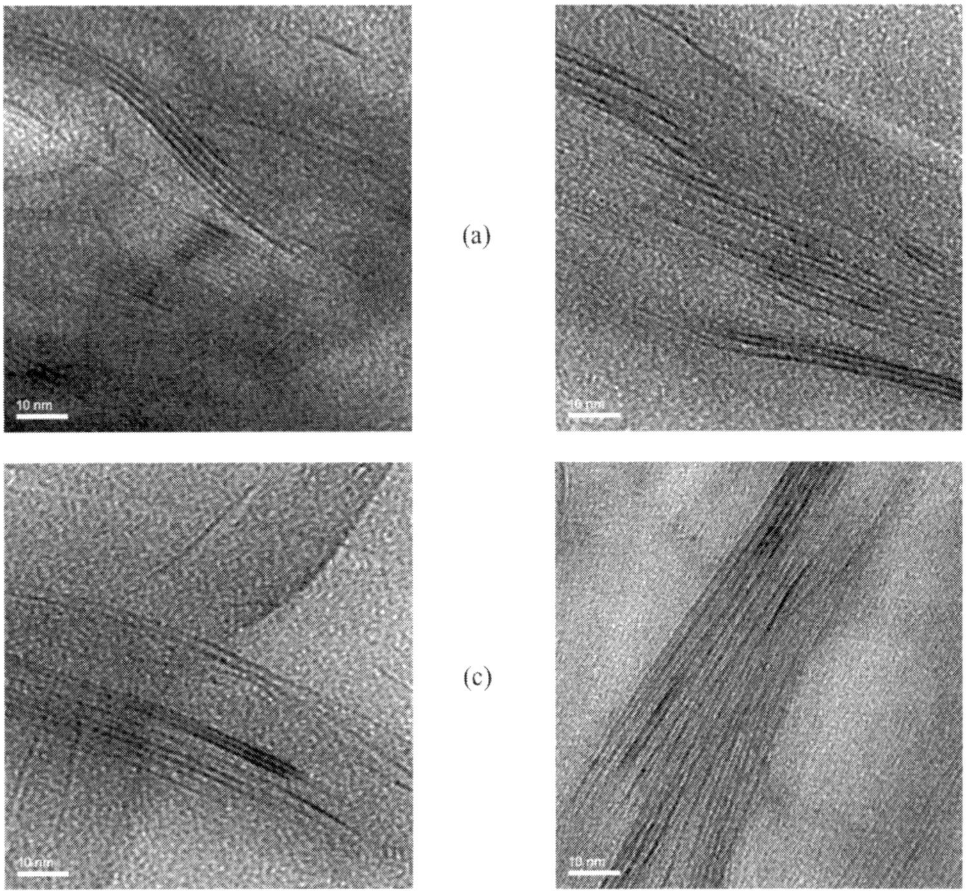

Figure 2 . TEM images of PVAB20 nanocomposites

TEM observations revealed the existence of silicate layers in the exfoliated state, while some larger intercalated tactoids could also be identified.

The XRD patterns of bentonite powder, pure PVA film and nanocomposite film (only the PVAB20 is presented for simplicity) studied in this work are shown in Figure 3. The peak at $2\theta = 19.4^{\circ}$, which corresponds to a dimension of 4.57 Å, can be attributed to crystal reflections ($10\bar{1}$ and 101) of the polymer [32]. In the case of clay sample, the peak at $2\theta = 7^{\circ}$ is referred to the basal spacing (~12.6 Å) of the bentonite platelets, while the other peak (around 29°) can be related to clay impurities. On the other hand, the observed increase in the basal spacing of PVAB20 sample (around 27Å) reveals the formation of intercalated structures, in which the polymeric chains were incorporated between the silicate layers resulting in a well ordered multilayer morphology. The increase in clay's basal spacing is due to favourable interactions between the clay's surface and the polymer groups. It must be noted that intercalated regions are obtained at lower clay concentration than previous studies have reported [33]. Furthermore, a small shoulder corresponding to the characteristic peak of the clay basal spacing ($2\theta = 7^{\circ}$) is also present in XRD pattern of PVAB20 sample, indicating that a small portion of the clay formed aggregates during the preparation of nanocomposite films.

3.2 XRD patterns in different forms

In order to investigate the potential orientation of the inorganic layers which lead to the enhanced properties of the produced nanocomposites, the XRD patterns of PVAB20 sample were measured in both film and powder forms. Initially the sample was placed on the top of the sample holder and the XRD spectra were obtained. Furthermore the sample was fragmented into small pieces and placed on the sample holder and measured again. The acquired patterns of PVAB20 sample (shown in Figure 4) reveal that the 001 peak of the clay is much more pronounced in the case of the film, giving evidence of well orientation of clay particles, parallel to the surface of the nanocomposite film.

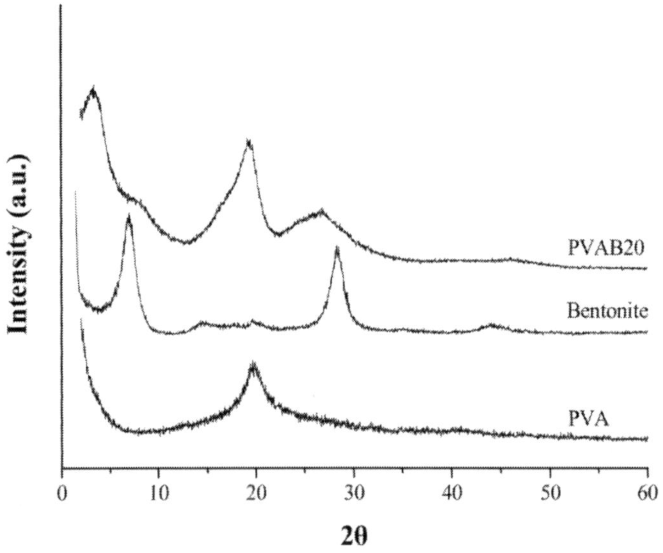

Figure 3. XRD patterns of pure clay (bentonite), PVA and PVAB20 film

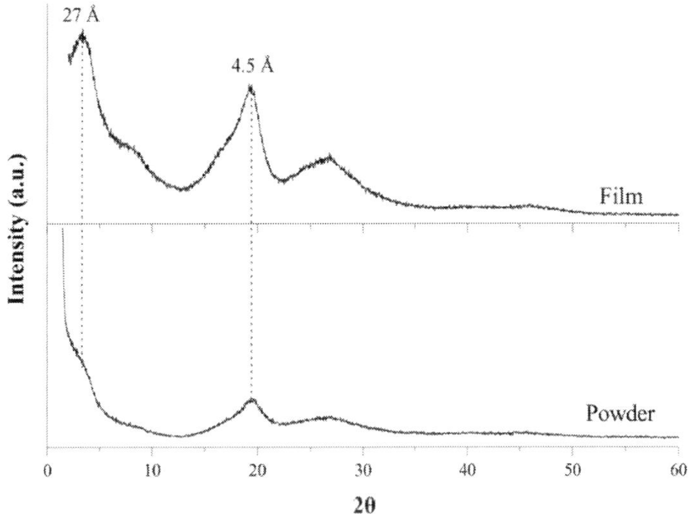

Figure 4. XRD patterns of PVAB20 sample in different forms

3.3 Properties of nanocomposites

In general, the final properties of a nanocomposite are determined by the competitive effects of polymer-polymer and polymer-clay interactions. In PVA/Bentonite nanocomposites, the polymer–clay mixing can be defined by the interactions between the functional groups of PVA and the negative charge on the clay surface. In the case of the produced films, the presence of the acetoxy groups in the selected partially hydrolised PVA together with the metal cations in the clay lattice, enabled the development of strong polymer – clay interactions. The abovementioned synergy in molecular level assisted the dispersion of the inorganic layers into the polymer matrix, leading to the formation of intercalated and exfoliated structures which enhances the overall material's performance. Therefore, the developed films, revealed improved mechanical strength, advanced gas barrier properties, increased heat resistance, while retained their optical transparency even at high clay loading (up to 20 wt %).

Figure 5 presents the oxygen permeability at 23°C at different relative humidity values in regards to the corresponding value of the neat polymer (P^{comp}/P, P^{comp}: permeability of composite, P: permeability of PVA). A significant improvement of the gas barrier properties of all samples studied is observed, due to the increased tortuosity of the permeation path which is caused by the presence of impermeable clay particles. The permeability of the PVAB5 reduced to about 50% while the corresponding value of PVAB20 (at 50% RH) decreases about 6 times in relation to that of pure PVA film. Nevertheless, this behavior is diminished at higher %RH, although the obtained permeability values of nanocomposite films are still significantly lower than of pure PVA, at the same relative humidity. This effect can be attributed to the fact that the water acts as plastisiser, facilitating the transport of oxygen molecules through the film.

In general, strong polymer–clay interactions can affect crystallinity, molecular orientation, and packing of the molecules in the vicinity of the nanoplatelets, leading to an enhancement in gas barrier properties, while poor adhesion between nanoplatelets and matrix usually results in opposite effect [34,35]. In addition, the developed films exhibited increased mechanical strength (Figure 6). At ambient conditions (45 %RH), the Young's modulus of the samples increased significantly with increasing clay loading (about 260% for 20% wt clay content). The great improvement in mechanical

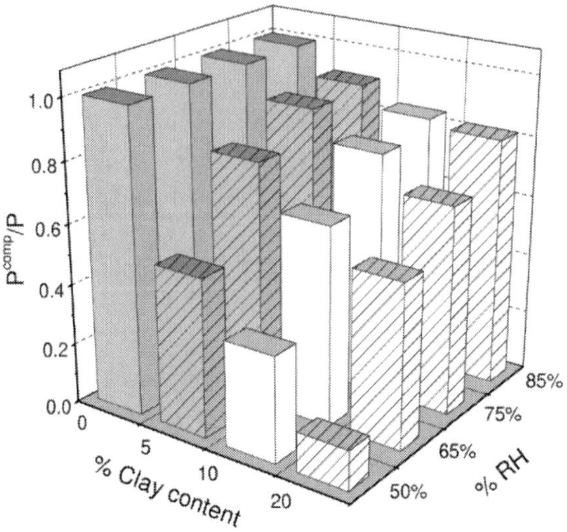

Figure 5. Oxygen permeability of PVA/Bentonite nanocomposites as well as pure PVA at 23°C

properties of the nanocomposites can be attributed to the strong interaction between matrix and silicate layers via formation of hydrogen bonds, due to the strong hydrophilicity of the clay edges [36]. Furthermore, it is clear that the effect of bentonite on the mechanical properties of the nanocomposites is more pronounced at higher %RH (increase up to 193 times for PVAB20 at 70% RH). This behavior can be attributed to the fact that, while PVA losses great percentage of its mechanical properties at

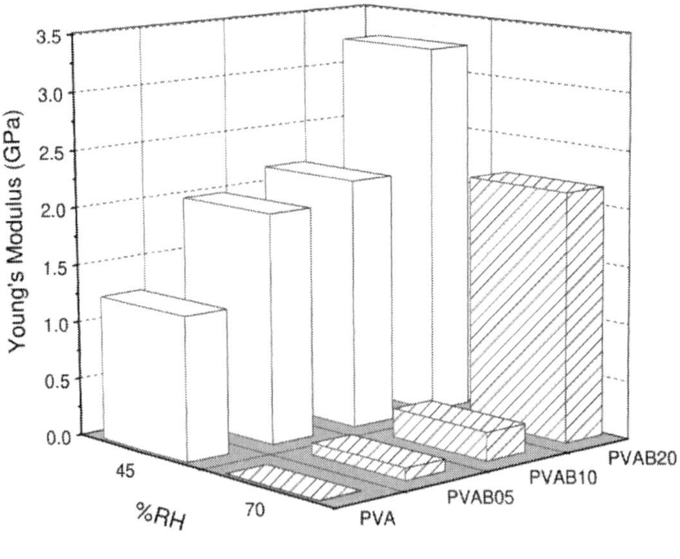

Figure 6. Mechanical properties (Young's modulus) of developed nanocomposites.

high relative humidity, clay's strength remains unaffected by water.

3.4 Neutron Diffraction

The diffraction patterns collected for PVAB20 sample equilibrated with H_2O and D_2O are illustrated in Figure 7 (a) and (b) respectively. In all samples a broad diffraction peak at around 1.39Å^{-1} is present, corresponding to a spacing of 4.5 Å. This peak can be attributed to the average value of the (101) and $(10\bar{1})$ reflections of the polymer crystals and it is typical for PVA [37,38]. On the other hand, the experimental data acquired by using the two sorbents, namely H_2O and D_2O, reveal pronounced differences especially in the range of 0.1-0.8 Å^{-1}.

Taking advantage of the different components studied, the contrast variation technique was employed to highlight scattering from the two different structural components that are present in the nanocomposite films, i.e polymer (PVA) and bentonite particles. Therefore, measurements on hydrated samples enabled us to highlight specific regions of the diffraction spectra. Due to the large differences between the scattering length density of H_2O and bentonite (Table 1), H_2O- hydrated samples were measured in order to monitor the structural changes as a function of the relative humidity of the composites at low Qregion (inorganic rich region). Thus, the shoulder at 0.16-0.17 Å^{-1} (Figure 7a) can be related to clay structural characteristics. Alternatively, based on the higher contrast of the system D_2O/PVA (Table 1), diffraction experiments on pre-equilibrated with D_2O samples revealed the structural changes in polymeric matrix. Therefore, the obtained peak at 0.68-0.70 Å^{-1} in the PVA/D_2O system (Figure b) can be related to polymer structure. The corresponding feature, which

Table 1. Scattering length densities of the materials used in this study

Sample	Scattering length density (Å^{-2})
Bentonite	3.67 x 10^{-6}
PVA	7.7 x 10^{-7}
H_2O	-5.54 x 10^{-7}
D_2O	6.34 x 10^{-6}

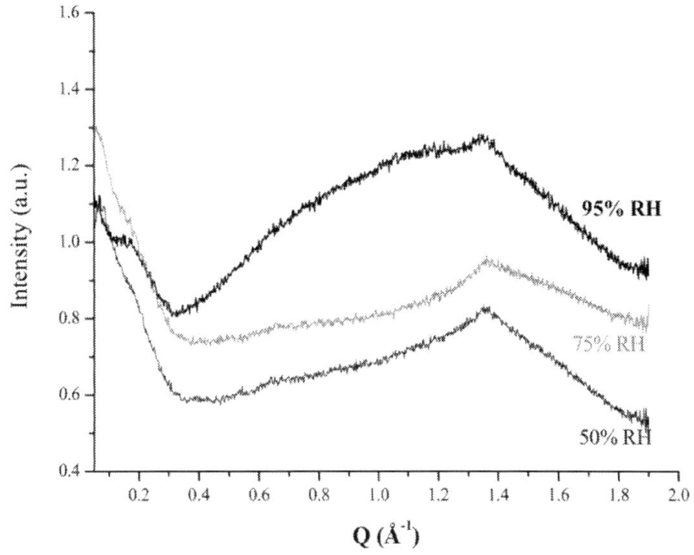

Figure 7a. Neutron diffraction patterns for PVAB20 sample pre-equilibrated with H_2O

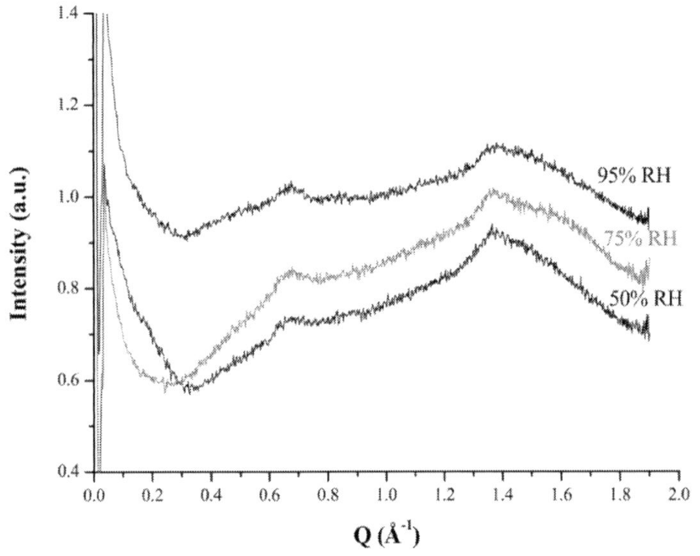

Figure 7b. Neutron diffraction patterns for PVAB20 sample pre-equilibrated with D_2O

is obvious in all %RH studied, may be attributed to the formation of a new polymeric crystalline phase, presumably induced by the silicates, during the development of the films. This assumption is in good agreement with calorimetric results (DSC) of previous studies for similar polymer-clay systems, in which the co-existence of two crystalline phases was reported [33].

Furthermore, the diffraction patterns obtained from lamellar beam incidence measurements are

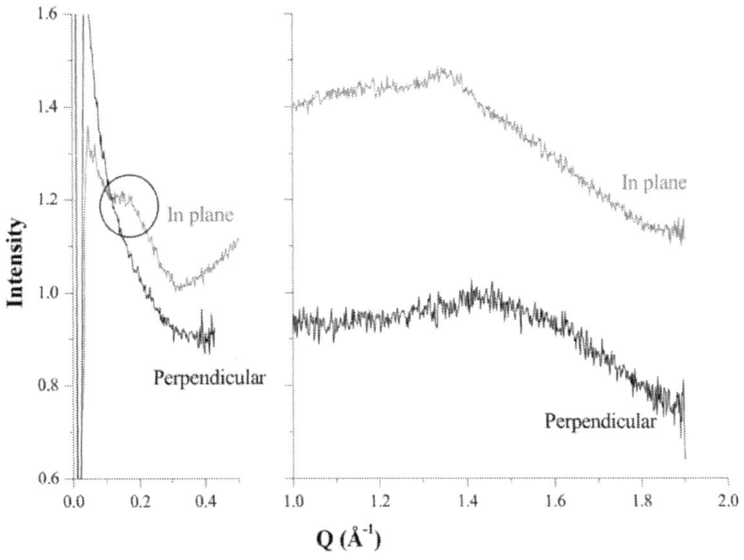

Figure 8. Perpendicular and in plane diffraction patterns for PVAB20 equilibrated at 95% H_2O.

quite different from those obtained from in-plane measurements, especially at low Q-region. In the case of lamellar measurements, a shoulder at about 0.16-0.17 $Å^{-1}$ is clearly observed. This feature can be attributed to the intercalated regions due to the effective incorporation of polymer chains between the silicate layers. Therefore, the PVAB20 sample exhibited well ordered multi-layer morphology. On the other hand this characteristic is not present in the other conformation, revealing that there is a specific orientation of the clay layers into the developed nanocomposite films, i.e. parallel to the film surface. This conclusion is in good agreement with the data obtained from XRD measurements (in powder and film form) as well as with the results derived by gas permeability technique and mechanical tests. In such a polymer-nanofiller system, oriented impermeable inorganic layers are expected to increase the resistance in flow through the nanocomposites film, acting as gas barriers.

Additionally, the excellent dispersion of the bentonite particles is anticipated to significantly boost the mechanical strength of the composites. It must be noted that the above mentioned pattern corresponds to a dimension of about 36-38 Å, which is notably larger than the basal spacing of ~26 Å, determined by XRD. The observed increase in clay's basal spacing can be ascribed to the considerable swelling of the nanocomposite films due to water sorption and it has been previous reported in literature [39,40].

4. Conclusions

The properties of well characterised PVA/clay nanocomposites and their structural changes as a function of relative humidity were thoroughly studied, using neutron membrane diffraction experiments (V1 diffractometer, HZB, Berlin).

The contrast variation technique was successfully employed to investigate the contribution of both the polymer and the clay particles to the diffraction spectra. Thus, due to scattering length densities of PVA and Bentonite, measurements on H_2O-hydrated samples provided information about the structure of the inorganic platelets, in low Q-region, while D_2O-hydrated samples enabled the investigation of the structural characteristics of polymeric matrix. As a result, the obtained peak at 0.68-0.70 $Å^{-1}$ (PVA/D_2O system) was ascribed to a new crystalline phase, which was induced by polymer –clay interactions.

Additionally, the diffraction patterns from perpendicular and in-plane geometry, in low Q region, revealed not only the dispersion of polymeric chains into the clay galleries, but also the orientation of inorganic platelets. The favourable arrangement of clay particles parallel to film surface leaded to nanocomposites with improved mechanical strength and advanced gas barrier properties. Therefore, neutron diffraction with contrast variation methodology can be considered as a very powerful tool for structural and orientation studies on polymer nanocomposites.

5. Acknowledgements

This research project has been supported by the European Commission under the 7th Framework Programme through "Research Infrastructures" action of the "Capacities" Programme, contract N° CP-CSA_INFRA-2008-1.1.1 (Number: 226507-NMI3).

References

[1] LeBaron P C, Wang Z and Pinnavaia TJ 1999 *Appl. Clay Sci.* **15** 11
[2] Giannelis E P, Krishnamoorti R and Manias E 1999 *Adv. Polym. Sci.* **138** 107
[3] Vaia R A, Price G, Ruth P N, Nguyen H T and Lichtenhan J 1999 *Appl. Clay Sci.* **15** 67
[4] Giannelis E P 1998 *Appl. Organomet. Chem.* **12** 675
[5] Gilman J W 1999 *Appl. Clay Sci.* **15** 31
[6] Xu R, Manias E, Snyder A J and Runt J 2001 *Macromolecules* **34** 337
[7] Sinha-Ray S, Yamada K, Okamoto M and Ueda K 2002 *Nano Lett.* **2** 1093
[8] Gilman J W, Jackson C L, Morgan A B, Harris R, Manias E, Giannelis E P, Wuthenow M, Hilton D and Phillips H 2000 *Chem. Mater.* **12** 1866
[9] Morgan A B 2006 Polym. Adv. Technol. **17** 206

[10] Krishnamoorti R, Vaia R A and Giannelis E P 1996 *Chem. Mater.* **8** 1728
[11] Morgan A B and Gilman J W 2003 *J. Appl. Polym. Sci.* **87** 1329
[12] Van der Hart D L, Asano A and Gilman J W 2001 *Chem. Mater.* **13**, 3796
[13] Giannelis E P, Krishnamoorti R and Manias E 1999 *Adv. Polym. Sci.* **138** 107
[14] Biswas M and Ray S S 2001 *Adv. Polym. Sci.* **155** 167
[15] Ray S S and Okamoto M 2003 *Prog. Polym. Sci.* **28** 1539
[16] Vaia R A, Jant K D, Kramer E J and Giannelis E P 1996 *Chem. Mater.* **8** 2628
[17] Zanetti M, Lomakin S and Camino G 2000 *Macromol. Mater.Eng.* **279** 1
[18] Bafna A, Beaucage G, Mirabella F and Mehta S. 2003 *Polymer* **44** 1103
[19] Nawani P, Burger C, Rong L, Chu B, Hsiao B S, Tsou A H, Weng W 2010 *Polymer* **51** 5255
[20] Bandyopadhyay J, Sinha Ray S 2010 *Polymer* **51** 1437
[21] Nusser K, Neueder S, Schneider G J, Meyer M, Pyckhout-Hintzen W, Willner L, Radulescu A and Richter D 2010 *Macromolecules* **43** 9837
[22] Chua Y-C, Chan A, Wong H-C, Higgins J S and Cabral J T 2010 *Macromolecules* **43** 9578
[23] Krishnamoorti R. and Yurekli K 2001 *Curr. Opin. Colloid. Interface Sci.* **6** 464
[24] Ogata N, Jimenez G, Kawai H and Ogihara T J 1997 *Polym. Sci. Part B: Polym. Phys.* **35** 389
[25] Fong H, Liu W D, Wang C S and Vaia R A 2002 *Polymer* 2002 **43** 775
[26] Schmidt G, Nakatani A I, Butler P D and Han C C 2002 *Macromolecules* **35** 4725
[27] Hecht A-M, Horkay F and Geissler E 2001 *Phys. Rev. E* **64** 041402
[28] Charalambopoulou G Ch. Steriotis Th A, Hauss Th, Stefanopoulos K L and Stubos A K 2002 *Appl. Phys. A* **74** [Suppl.]S1245
[29] Charalambopoulou G Ch, Steriotis Th A, Hauss Th, Stubos A K and Kanellopoulos N K 2004 *Physica B* **350** 603
[30] Milani S, Berti D, Dante S, Hauss T. and Baglioni P 2009 *Langmuir* **25** 4084
[31] Buchsteiner A, Hauss T, Dante S and Dencher N 2010 *Biochim. Biophys. Acta* **1798** 1969
[32] Assendert H E. and Windle A H 1998 *Polymer* **39** 4295
[33] Strawhecker K E and Manias E 2000 *Chem. Mater.* **12** 2943
[34] Chaiko D J and Leyva A A 2005 *Chem. Mater.* **17** 13
[35] Osman M A, Mittal V, Morbidelli M and Suter UW 2003 *Macromolecules* **36** 9851
[36] Sengwa R J, Choudhary S and Sankhla S 2009 *Colloid Surface A* **336** 79
[37] Assendert H E and Windle A H 1998 *Polymer* 39 4295
[38] Takahashi Y 1997 *J. Polym. Sci. Pol. Phys.* **35** 193
[39] Cebula D J, Thomas R K, Middleton S, Ottewill R H and White J W 1979 *Clay Clay Miner.* **27** 39
[40] Devineau K, Bihannic I, Michot L, Villiéras F, Masrouri F, Cuisinier O, Fragneto G and Michau N 2006 *Appl. Clay Sci.* **31** 76

Collective Dynamics of Intracellular Water in Living Cells

A. Orecchini[1,2], F. Sebastiani[1], M. Jasnin[3], A. Paciaroni[1], A. De Francesco[4], C. Petrillo[1], M. Moulin[5], M. Haertlein[5], G. Zaccai[2] and F. Sacchetti[1]

[1] Dipartimento di Fisica and CNR-IOM, Università di Perugia, I-06123 Perugia, Italy.
[2] Institut Laue-Langevin, 6 rue Horowitz, F-38042 Grenoble, France.
[3] Max-Planck-Institut für Biochemie, Am Klopferspitz 18, D-82152 Martinsried, Germany.
[4] CNR-IOM OGG c/o Institut Laue-Langevin, 6 rue Horowitz, F-38042 Grenoble, France.
[5] Deuteration Laboratory, Institut Laue-Langevin, 6 rue Horowitz, F-38042 Grenoble, France.

E-mail: `orecchini@ill.fr`

Abstract. Water dynamics plays a fundamental role for the fulfillment of biological functions in living organisms. Decades of hydrated protein powder studies have revealed the peculiar dynamical properties of hydration water with respect to pure water, due to close coupling interactions with the macromolecule. In such a framework, we have studied coherent collective dynamics in protein and DNA hydration water. State-of-the-art neutron instrumentation has allowed us to observe the propagation of coherent density fluctuations within the hydration shell of the biomolecules. The corresponding dispersion curves resulted to be only slightly affected by the coupling with the macromolecules. Nevertheless, the effects of the interaction appeared as a marked increase of the mode damping factors, which suggested a destructuring of the water hydrogen-bond network. Such results were interpreted as the signature of a "glassy" dynamical character of macromolecule hydration water, in agreement with indications from measurements of the density of vibrational states. Extending the investigations to living organisms at physiological conditions, we present here an *in-vivo* study of collective dynamics of intracellular water in *Escherichia coli* cells. The cells and water were fully deuterated to minimise the incoherent neutron scattering background. The water dynamics observed in the living cells is discussed in terms of the dynamics of pure bulk water and that of hydration water measured in powder samples.

1. Introduction

Water dynamics is well known to play a crucial role for the fulfillment of biological functions in living organisms. Indeed biological macromolecules, like proteins or nucleic acids, can accomplish their functional tasks only if allowed to explore the vast landscape of their conformational substates through thermal anharmonic atomic movements. Such macromolecular dynamics, mainly occurring on the picosecond time scale, are practically frozen out in dry samples and become active only in the presence of a sufficient quantity of hydration water, which provides the macromolecule with the necessary structural flexibility. Within this framework, the achievement of a better understanding of the dynamical behaviour of hydration water itself is obviously a challenge of major interest [1].

A vast literature exists on single-particle thermal motions of hydration water, which can involve diffusion phenomena of translational or rotational nature. The large majority of investigations by different methods have shown that water diffusion in a hydration shell extending over one or two molecular layers is significantly hindered by interactions with a biomolecular surface, resulting for example in diffusion coefficients a factor 2 to 10 smaller than in bulk water (see for instance Ref. [2]). Measurements of single-particle water motions in *Escherichia coli* (*E.coli*) [3, 4] and red blood cells [5] have revealed that a major fraction of approximately 80-90% of water in these cells is characterized by a translational diffusion coefficient similar to bulk water. A minor fraction of approximately 10% of cellular water, which exhibits reduced dynamics, was attributed to dynamically bound water on the surface of the macromolecules. On the other hand, collective density fluctuations in the hydration shell of biomolecules have started to draw some attention only recently, despite the propagation of such excitations might be involved in a number of biological actions requiring transmission of information, like for instance allosteric reactions. The study of collective modes on the picosecond time scale has been significantly boosted by the advent of Brillouin spectroscopies at THz frequencies, which can nowadays be achieved by inelastic neutron and X-ray scattering (INS and IXS respectively).

The state-of-art neutron Brillouin spectrometer BRISP [6] has allowed our group to undertake a systematic investigation of collective density fluctuations in various biological samples, with particular focus onto the dynamical behaviour of hydration water. The determination of THz dispersion curves for the hydration shell of a biomolecule was first achieved for the Ribonuclease A (RNase) protein [7], at hydration levels between 0.7 and 1.0 g of water per g of protein. In this system, two collective modes of different characteristic energies $\hbar\omega$ (or frequencies ω) were observed in the wavevector range $Q = 0.2 - 1.4$ Å^{-1}. As shown in the left panel of Fig. 1, the so-called low-frequency mode displays an optic-like character with a rather constant energy of about 6-7 meV. The so-called high-frequency mode is instead considered of acoustic-like nature and is characterised by linearly increasing energies with

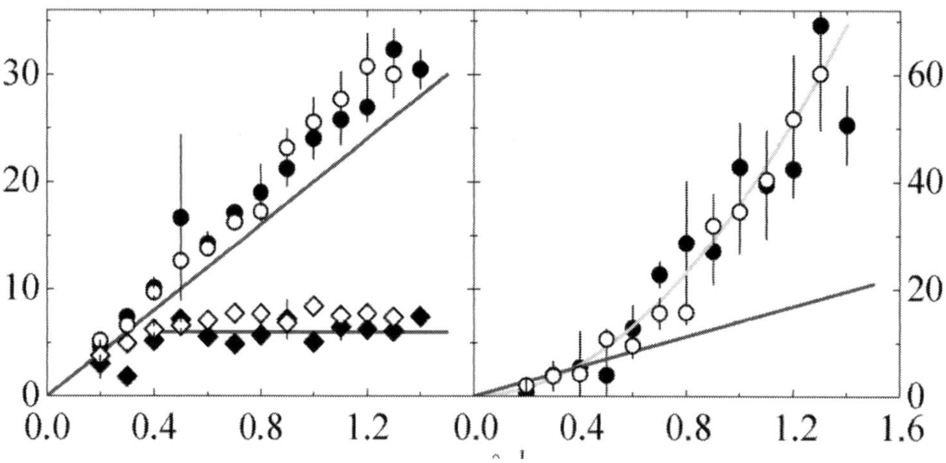

Figure 1. *Left:* Experimental dispersion curves of THz collective modes in RNase (full symbols) and DNA (empty symbols) hydration water. Circles refer to the high-frequency mode, lozenges to the low-frequency mode. The dispersion curves of bulk water (full lines) are reported for comparison. *Right:* Damping factors of the high-frequency mode for RNase (full circles) and DNA (empty circles) hydration water. The green solid line is a quadratic fit of the damping factors Q-dependence. The blue straight line represents the trend of the high-frequency damping factor in bulk water.

increasing wavevector: the low-Q slope of this mode provides a propagation speed of about 3400 m/s. Both excitations are strongly damped. The high-frequency mode, in particular, shows a Q^2 increase of the damping factor (right panel of Fig. 1), which crosses the corresponding dispersion curve at about 0.6 Å^{-1} (not shown). For higher Q-values, the density fluctuation becomes overdamped and consequently its propagation becomes more and more hindered and short-ranged. Very recently we have extended such studies to samples of hydrated DNA, at hydration levels between 1 and 15 g of water per g of DNA [8, 9]. Despite the different structural conformation and hydrophilic sites distribution of the DNA helix with respect to the globular protein RNase, the collective dynamics of hydration water behaves very similarly in the two systems. As shown in Fig. 1, the same two-mode dispersion curve, with similar characteristic energies, Q-dependences, propagation speeds and damping factors, is observed in the hydration water of both macromolecules.

Such a double-branched structure of the THz dispersion curve closely resembles what observed in pure water by both INS and IXS [10, 11, 12] (see Fig. 1). In the bulk liquid, the low-frequency mode is located at the same almost-constant energy of ~6 meV, while the high-frequency one propagates at the sligthly lower speed of 3040 ± 80 m/s. Nonetheless, the major difference between bulk and hydration water seems to reside in the damping factors amplitude and Q-dependence. Indeed, the damping factors of hydration water are larger than in the bulk and increase with a parabolic, rather than linear [10, 11], wavevector dependence. Such larger damping factors correspond to shorter propagation lifetimes and mean free paths of the collective modes, which is ascribed to a destructuring effect on the local tetrahedral order of water hydrogen bonds, induced by the interaction with the biomolecule surface [7]. Such a perturbation of the hydrogen-bond network confers on hydration water a "glassy" dynamical character, as also indicated by previous determinations of the density of vibrational states [13].

As a whole, the emerging picture of hydration water dynamics suggests that the interaction with the biomolecule produces, on one hand, a remarkable slowing down of diffusional single-particle processes and, on the other hand, a sizeable reduction of the propagation lengths of coherent collective modes, despite the speed and frequencies are only slightly affected.

In this framework, the characterisation of intracellular water collective dynamics in *E.coli* is complementary to the single-particle dynamics results obtained from quasi-elastic incoherent neutron scattering on both *E.coli* and red blood cells, which have shown that single-particle water motions beyond the macromolecule hydration shells were similar to bulk water [3, 5]. Consequently it appears interesting to consider whether a similar effect takes place for the collective dynamics as well. In this short paper, we present the preliminary results of a neutron Brillouin experiment performed on a culture of living *E.coli* bacteria, with the aim of observing whether the collective dynamics of intracellular water appears closer to that of the pure liquid or of the hydration layer of protein and DNA powders.

2. Experimental Methods

Due to the large incoherent scattering cross-section of hydrogen, with respect to the mainly coherent cross-section of deuterium, a neutron spectroscopy investigation on the coherent dynamics of cellular water requires the availability of a fully deuterated sample of living cells.

To this end, a fermenter culture of fully deuterated *E.coli* bacteria was grown at the Deuteration Laboratory of the Institut Laue-Langevin (ILL). From the fermenter culture, about 7 g of living *E.coli* were harvested by centrifugation, sealed in a slab-shaped aluminium mount (30mm × 50mm × 4mm) and placed in the neutron beam of the BRISP spectrometer at ILL. The centrifugation step was carefully controlled to get rid of the maximum amount of external water, yet without removing any intracellular water, which would destroy the cell envelope of *E.coli* and kill the cell. After the experiment, the dry weight of the sample was measured to be 20% of the total. Of the 80% water weight, 8% corresponded to extracellular and 72% to intracellular water [14]. Storage of the deuterated bacterial cell paste at room temperature for about 48 hours did not lead to any significant cell death. To ensure that the full set of measurements were performed on actually living cells, the sample in the neutron beam was replaced with a fresh one taken from the cell culture every 24 hours. The experiment lasted 7 days, 4 of which were devoted to the *E.coli* cells and the remaining 3 were employed for background, empty sample holder and vanadium standard measurements. All measurements were performed at room temperature. The BRISP spectrometer was used with an incident wavelength of 0.9885 Å and a sample-to-detector distance of 4.5 m.

The true scattering intensity $I(Q,\omega)$ of the *E.coli* cells was obtained from the raw data by following standard reduction procedures. In particular, the data were corrected for environmental and sample holder contributions, energy dependence of the detector efficiency, sample self-absorption and multiple scattering. The resulting spectra at selected Q-values are plotted in Fig. 2. Because of the fully deuterated nature of the samples, to a first approximation the residual incoherent contributions can be neglected. In addition, considering the very high water content in the *E.coli* cell, in the present very preliminary analysis the spectra in Fig. 2 will be ascribed to the dynamical behaviour of intracellular water only, while the contribution of the "dry" *E.coli* components will be neglected. More detailed and specific analyses are under development and will be published elsewhere.

3. Data Analysis and Discussion

For a quantitative analysis of the data, the spectra $I(Q,\omega)$ can be fitted to a model function based on a classical dynamic structure factor $S_{cl}(Q,\omega)$, properly multiplied by the Bose-Einstein occupation number $n(\omega,T)$ and convoluted with the energy resolution function $R(\omega)$:

$$I(Q,\omega) = [I_0(Q)R(\omega)] \otimes \left\{ \frac{k_f}{k_i} \frac{\hbar\omega}{k_B T} \left[n(\omega,T)+1\right] S_{cl}(Q,\omega) \right\} , \tag{1}$$

where k_f and k_i are the final and incoming neutron wavevector respectively, T is the absolute temperature, k_B is the Boltzmann constant and $I_0(Q)$ is an overall scaling factor. At the employed instrumental configuration, the BRISP resolution function $R(\omega)$ had a Gaussian shape with a full width at half maximum of 2.7 meV, as determined by the vanadium measurement. The choice of the most suited form for the dynamic structure factor $S_{cl}(Q,\omega)$ depends of course on physical considerations about the system under investigation.

From a first look at $I(Q,\omega)$ in Fig. 2, it emerges that clear inelastic features appear in the spectra well outside the experimental resolution function. Based on the approximations discussed in the previous section, we interpret such features as coherent collective modes propagating through *E.coli* intracellular water. As recalled in the Introduction, all previous determinations of collective density fluctuations in either pure [10, 12] or interacting water [7, 15] report the existence of two inelastic modes. In addition, to a closer inspection of the present spectra, the two modes are already visible in the raw data at intermediate Q-values (e.g. 0.6 Å^{-1}). On such grounds, it appears reasonable to chose a dynamic structure $S_{cl}(Q,\omega)$ containing two damped harmonic oscillators (DHOs):

$$\begin{aligned} S_{cl}(Q,\omega) &= A_{el}(Q)\delta(Q,\omega) + \\ &+ \frac{A_L(Q)}{\pi} \frac{\Omega_L^2(Q)\Gamma_L(Q)}{[\omega^2-\Omega_L^2(Q)]^2+\omega^2\Gamma_L^2(Q)} + \\ &+ \frac{A_H(Q)}{\pi} \frac{\Omega_H^2(Q)\Gamma_H(Q)}{[\omega^2-\Omega_H^2(Q)]^2+\omega^2\Gamma_H^2(Q)} , \end{aligned} \tag{2}$$

where the subscripts refer to the low (L) and high (H) frequency mode, while $\Omega(Q)$, $\Gamma(Q)$ and $A(Q)$ represent respectively the proper frequencies, damping factors and intensity factors of

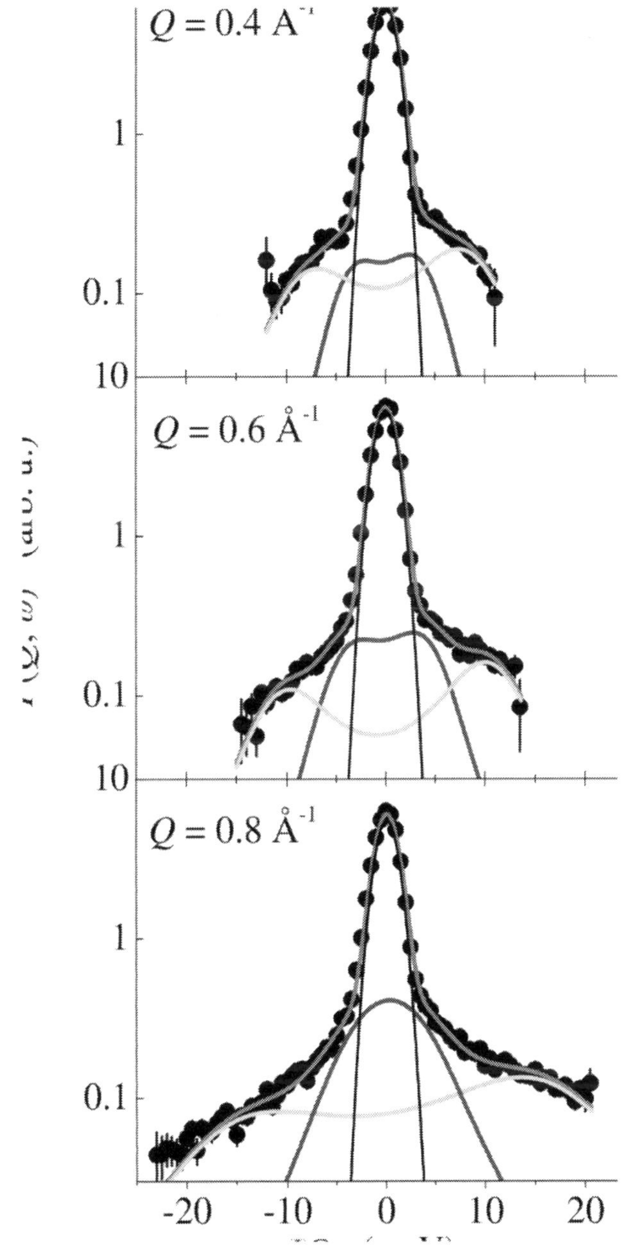

Figure 2. Coherent inelastic spectra of *E.coli* intracellular water at selected Q-values. The black line shows the experimental resolution function. Inelastic shoulders are well-visible outside the resolution function. The red line shows a fit of the data with two DHOs (see text). The blue and green lines represent respectively the resulting low- and high-frequency DHO.

each DHO. The first term containing a Dirac δ-function accounts for the quasielastic peak of the sample. Due to the relatively large energy resolution of the present experiment, a more specific definition of the quasielastic lineshape is not needed. The fits of the experimental $I(Q,\omega)$ with the model of Eqs. 1 and 2 were performed independently at each Q-value. The best fitting functions are shown in Fig. 2. It is readily seen that the fits are in excellent agreement with the measured spectra, thus supporting the existence of two well defined modes. The resulting oscillation frequencies $\hbar\Omega$ and damping factors $\hbar\Gamma$ are plotted in Fig. 3 as functions of Q.

The dispersion curves of intracellular water closely recall those of bulk water and display an even stronger resemblance to those of hydration water of DNA and RNase (cfr. Fig. 1). The low-frequency mode, with an almost constant characteristic energy, is located between 4 and 7 meV, slightly increasing with Q. The high-frequency mode has a fairly linear Q-dependence, with a steady slope corresponding to a propagation speed of 3460 ± 30 m/s. Similarly to what observed for protein and DNA hydration water, such results indicate that interactions with the biomolecular environment inside *E.coli* cells produce only minor changes in the THz dispersion curves of water.

A more interesting piece of information is brought about by the behaviour of the damping factors, in particular those concerning the high-frequency mode. The right panel of Fig. 3 shows that, although the values of $\hbar\Gamma_H$ increase steadily faster than the linear behaviour of bulk water, they remain well below the quadratic trend observed for RNase and DNA hydration water. Indeed, a fit of $\hbar\Gamma_H$ to a power law of the form aQ^x yields for the exponent the result $x = 1.5 \pm 0.1$, to be compared with $x = 2$ for hydration water and $x = 1$ for bulk water. As recalled in the Introduction, the quadratic damping factors of the solvent layer surrounding DNA and RNase are interpreted in terms of shorter propagation lengths due to a "glassy" dynamic character of hydration water. In view of such a picture, the observed trend of $\hbar\Gamma_H$ in *E.coli* suggests that intracellular water might be regarded as a mixture of both bulk and hydration water. Water molecules in close contact with proteins, nucleic acids, or other macromolecular

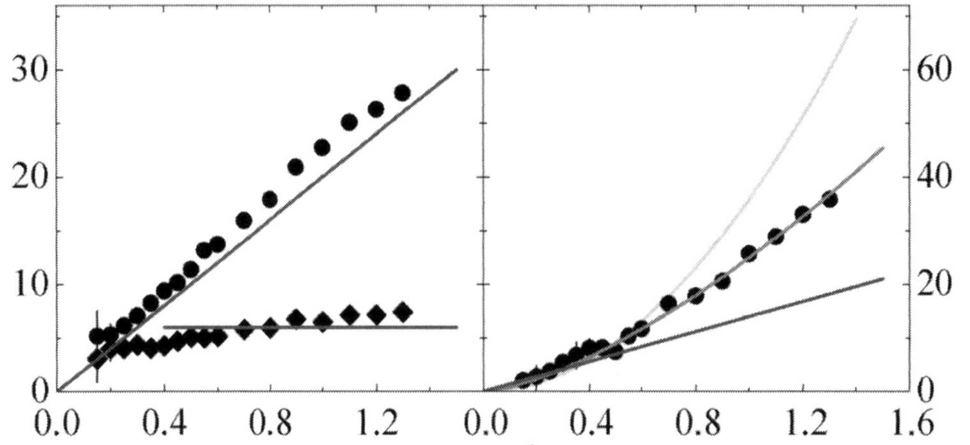

Figure 3. *Left:* Experimental dispersion curves of THz collective modes of *E.coli* intracellular water. Circles refer to the high-frequency mode, lozenges to the low-frequency mode. The dispersion curves of bulk water (full lines) are reported for comparison. *Right:* Damping factors of the high-frequency mode for *E.coli* intracellular water (circles). The red solid line is a fit to the high-frequency damping factor with a $Q^{1.5}$ power law. The green and blue solid lines report the behaviour oh hydration and bulk water respectively.

constituents of the *E.coli* interior would display the dynamical character of hydration water. The remainder of water would instead behave like the free normal liquid, in qualitative agreement with what found for the single-particle dynamics counterpart [3].

This hypothesis might be pushed further by considering the damping factors of *E.coli* water like a weighted average of the damping factors of bulk and hydration water. Such an average can be calculated by using the present damping factors data to work out an experimental estimate of the fraction of free water in *E.coli* cells. At the various Q values, it turns out that the fraction of normal water is about 0.6 with respect to the total amount of intracellular water. If one considers that the *E.coli* cell is composed by intracellular water for approximately 72% in weight [14], the above-estimated fraction of free liquid suggests that the remaining interacting water hydrates the cell inner components to a hydration level of about 0.9, which would approximately correspond to three layers of coordinated water around a protein or biomolecule in the cell. The fraction of intracellular water whose collective dynamics is affected by macromolecular interactions is higher than found previously for single-particle motions [3, 4]. Considering the approximations involved in the calculations, however, further experiments will be necessary before the significance of the difference is established.

4. Conclusion

The present experimental study reports about the observation by neutron Brillouin spectroscopy of THz collective modes propagating in *E.coli* intracellular water. Preliminary results indicate that, although the dispersion curves resemble those of protein and DNA hydration water, the corresponding damping factors suggest intracellular water to be composed by a considerable amount of free non-interacting water. This finding complements and corroborates the results recently obtained for the single-particle dynamics of *E.coli* water [3].

A more detailed and precise analysis should take into account the role of density fluctuations propagating throught the cell itself. This necessarily involves a careful separation of the experimental $I(Q,\omega)$ into partial dynamic structure factors, separately accounting for water-water, cell-cell and water-cell correlations. Although a reliable such procedure requires considerable efforts in terms of data reduction and analysis, work toward this direction is currently in progress.

References

[1] Ball P 2008 *Chem. Rev.* **74** 108 *and references therein.*
[2] Halle B 2004 *Philos. Trans. R. Soc. London B* **359** 1207.
[3] Jasnin M, Moulin M, Haertlein M, Zaccai G and Tehei M 2008 *EMBO Rep.* **9** 543.
[4] Persson E and Halle B 2008 *Proc. Natl. Acad. Sci. U.S.A.* **105** 6266.
[5] Stadler A M, Embs J P, Digel I, Artmann G M, Unruh T, Buldt G and Zaccai G 2008 *J. Am. Chem. Soc.* **130** 16852.
[6] Aisa D *et al* 2005 *Nucl. Instr. and Meth. A* **544** 620.
[7] Orecchini A, Paciaroni A, De Francesco A, Petrillo C and Sacchetti F 2009 *J. Am. Chem. Soc.* **131** 4664.
[8] Cornicchi E, Sebastiani F, De Francesco A, Orecchini A, Paciaroni A, Petrillo C and Sacchetti F 2011 *J. Chem. Phys.* **135** 025101.
[9] Orecchini A, Paciaroni A, Petrillo C, Sebastiani F, De Francesco A and Sacchetti F *J. Phys.: Cond. Matt.* accepted.
[10] Sacchetti F, Suck J-B, Petrillo C and Dorner B 2004 *Phys. Rev. E* **69** 061203 *and references therein.*
[11] Petrillo C, Sacchetti F, Dorner B and Suck J-B 2000 *Phys. Rev. E* **62** 3611.
[12] Pontecorvo E, Krisch M, Cunsolo A, Monaco G, Mermet A, Verbeni R, Sette F and Ruocco G 2005 *Phys. Rev. E* **71** 011501 *and references therein.*
[13] Paciaroni A, Orecchini A, Cornicchi E, Marconi M, Petrillo C, Haertlein M, Moulin M, Schober H, Tarek M and Sacchetti F 2008 *Phys. Rev. Lett.* **101** 148104.
[14] Record M T, Courtenay E S, Cayley D S and Guttman H J 1998 *Trends Biochem. Sci.* **23** 143.
[15] Sacchetti F, Orecchini A, Cunsolo A, Formisano F and Petrillo C 2009 *Phys. Rev. B* **80** 024306.

Changes in molecular dynamics of apomyoglobin during amyloid formation

Andreas Maximilian Stadler [1,*], **Elisa Fabiani** [2], **Giuseppe Zaccai** [3]

[1] Forschungszentrum Jülich GmbH, ICS-1 & JCNS-1, 52425 Jülich, Germany

[2] Polytech'Grenoble - Université Joseph Fourier, BP 53, 38041 Grenoble Cedex 9 & Institut de Biologie Structurale J.P. Ebel, 41, rue Jules Horowitz, 38027 Grenoble Cedex 1

[3] Institut Laue-Langevin, 6 rue Jules Horowitz, 38042 Grenoble, France

*a.stadler@fz-juelich.de

Abstract. Changes of molecular dynamics associated with amyloid fibril formation were explored using apomyoglobin as a model system. Protein dynamics was measured using elastic incoherent neutron scattering. The formation of amyloid fibres was evidenced by the presence of characteristic signatures in the X-ray diffraction pattern.

1. Introduction

Many neurodegenerative diseases such as Alzheimer's and Parkinson's are correlated with protein misfolding, aggregation and deposition of insoluble amyloid fibers in cells and tissues [1, 2]. It is of both fundamental and medical interest to characterize the molecular reasons and connections of protein misfolding, aggregation and amyloid formation. Different classes of proteins are able to form amyloid fibers. One hypothesis states that it is even a general property of the polypeptide chain itself under destabilizing conditions such as extremes of heat or pH [1-3]. Amyloid fibers are densely packed with a high degree of structural order. A common diffraction pattern of amyloid fibers with two peaks at $q = 0.62$ and 1.36 Å^{-1} termed the cross-β pattern is characteristic of amyloid fibers [4]. The diffraction signature corresponds to the stacking distance of $d = 2\pi/q = 10.1$ Å between different strands and to the characteristic repeat distance of $d = 4.6$ Å within β sheets.

Myoglobin (Mb) has served extensively as model system in molecular biophysics. Frauenfelder and co-workers called it once 'the hydrogen atom of biology' [5]. It is small enough to allow a fundamental understanding of the physical processes involved but sufficiently complex to represent typical protein behavior. The protein acts as a scaffold for a heme-group that can carry an oxygen molecule or other small ligands. The protein with heme-group is termed holo-Mb; without heme-group it is called apo-Mb. At temperatures above ~ 328 K (~ 55°C) and at basic solvent conditions of pH 9 apo-Mb is reported to form amyloid fibres [6, 7]. Disordered segments of apo-Mb that emerge at high temperature and conditions that are unfavorable for protein folding were found to be essential for amyloid fiber formation in apo-Mb [7].

We are using apo-Mb as a model system for the study of amyloid formation. Structural changes from the native-like α-helical state below 328 K to the β-sheet structure at temperatures above ~328 K have been monitored by small and wide angle X-ray scattering [8]. We have also studied the dynamics of apo- and holo-Mb during the α-to-β transition [9]. Using a simplified approach the results indicated a more resilient ("stiffer") β-sheet structure in apo-Mb at temperatures above ~328 K as compared to the native α-helical structure below ~ 328 K. However, we were not able to determine mean square displacements with sufficient accuracy to quantify changes in protein dynamics associated with amyloid formation and to deduce further information about protein resilience in absolute units.

In this article we report a study on changes in protein dynamics associated with amyloid formation. Elastic incoherent neutron scattering (EINS) was used to determine mean square displacements $\langle u^2 \rangle$ of protein thermal fluctuations with high accuracy. Thermal motions were investigated using two different neutron spectrometers with time resolutions of 0.1 and 1 ns, respectively. Effective force constants $\langle k' \rangle$ were calculated from the $\langle u^2 \rangle$ and are compared between the native state of apo-Mb during amyloid formation and of heat incubated apo-Mb which was shown to have characteristic signatures of amyloid fibers.

2. Material and Methods

2.1. Sample preparation

Horse heart Mb purchased from SIGMA-ALDRICH was dissolved in water. The pH of the Mb solution was then lowered to 1.5 with concentrated HCl on ice. The acidified solution was mixed with 4 volumes of 2-butanone. The upper organic layer was decanted, and the extraction was repeated twice. The aqueous layer was dialyzed against buffer solution (20 mM KH_2PO_4, pH 7) followed by pure water. After dialysis the protein solution was lyophilized to obtain apo-Mb powder.

For the neutron experiments, apo-Mb powder was hydrated by pipetting uniformly buffer solution (20 mM KH_2PO_4, pH 9) to the sample to a level of 0.73 g H_2O/g protein. This corresponds to approximately two hydration layers per protein and was intended to allow for some pH effects.

Around 120 mg hydrated protein powder was sealed in flat aluminum sample holders with internal spacing of 0.2 mm for the neutron scattering experiments. One sample was hermetically sealed in the sample holder and was incubated for 2 days at 338 K to promote the formation of amyloid fibers. The weight of the heat incubated sample was checked before and after the incubation time to ensure that no evaporation of hydration water occurred.

Small amounts of roughly 10 mg hydrated powder were measured in thin quartz tubes in the X-ray diffraction experiments. One hydrated powder sample was prepared at room temperature; the second sample was hermetically sealed and incubated at 338 K for 2 days to allow the formation of amyloid fibers, the third sample is a part of a sample which was measured on IN13 just before the X-ray experiments.

2.2. Elastic Incoherent Neutron Scattering

EINS is a well suited method for the study of the dynamics of complex biological macromolecules. Thermal and cold neutrons have wavelengths in the order of some Å and energies of some meV and are therefore sensitive probes to measure thermal molecular motions on an atomic length-scale and picosecond to nanosecond time-scale. In biological systems, the incoherent scattering is dominated by the signal of hydrogen atoms as their incoherent scattering cross section is much larger than that of the other elements that occur in biological matter [10]. In the time-scale of the neutron experiments, the H atoms reflect the motion of the chemical groups to which they are bound. As hydrogen atoms are uniformly distributed in proteins, neutron scattering detects average internal macromolecular motions. EINS experiments, so called 'elastic scans', provide the mean square displacements $\langle u^2 \rangle$ of the system as a function of temperature.

Mean square displacements $<u^2>$ were calculated within the Gaussian approximation according to

$$\langle u^2 \rangle = \frac{-6 \cdot \Delta(\ln I(q))}{\Delta q^2}, \quad (1)$$

where $I(q)$ represents the measured elastic intensity and q the modulus of the scattering vector [11]. We use in this work the definition of $<u^2>$ introduced by Smith that accounts for the full amplitude of motions [12]. The Gaussian approximation is strictly valid for localized motions at $q^2 \rightarrow 0$ but holds up to $\langle u^2 \rangle \cdot q^2 \sim 2$ for motions with the most compact geometry. The validity of the approximation is similar to that of the Guinier region of small angle scattering, and depends on the geometry of the motion or the shape. Kidera and Go used normal mode refinement of X-ray crystallography data and showed that protein dynamics are intrinsically anisotropic with ellipoidal shape and axes of 1:1:1.7 [13]. For thermal motions with ellipsoidal axes of 1:1:~2 the Gaussian approximation holds to significantly larger values of roughly $\langle u^2 \rangle \cdot q^2 \sim 8$ as reported by Tehei et al. [14].

The temperature variation of the $<u^2>$ was interpreted in terms of an empirical effective force constant $<k'>$ called resilience by Zaccai [15]

$$\langle k' \rangle = 0.00276 / \left(d\langle u^2 \rangle / dT \right). \quad (2)$$

The units are chosen such that $<k'>$ is in N/m when $<u^2>$ is in Å^2 and T is in K [15]. The force constants describe the rigidity, or resilience, of the protein. Neutron scattering experiments

2.3. Neutron scattering experiments

EINS was measured on the neutron backscattering spectrometers IN13 and IN10 at the Institut Laue-Langevin (Grenoble, France). IN13 is characterized by an energy resolution of ΔE = 8 μeV (full-width at half-maximum, FHWH) and a large scattering vector range of $0.2 < q < 4.9$ Å^{-1} [16] and uses thermal neutrons with a wavelength of λ=2.2 Å. The instrument therefore is sensitive to macromolecular motions in a space-time window of some Å in some 0.1 ns. IN10 has a very high energy resolution of ΔE = 1 μeV FWHM, a scattering vector range of $0.5 < q < 2.0$ Å^{-1} and uses cold neutrons with a wavelength of λ=6.3 Å [16]. The instrument parameters result in an observable space-time window of some Å in some ns.

The relation between the scattering angle 2θ and the modulus of the scattering vector is $q = \frac{4\pi}{\lambda}\sin(\theta)$ for elastically scattered neutrons. Measured data were analyzed in the q^2-range from 1.6 to 3.5 Å^{-2} for IN13 and 0.7 to 3.8 Å^{-2} for IN10. The samples were oriented with an angle of 135° with respect to the incident beam. Multiple scattering was neglected as the transmissions of all samples were larger or equal 0.9. The measured data were corrected for empty cell scattering and the neutron detectors were calibrated with a vanadium reference. The weight of the samples was measured before and after the measurements to ensure that no loss of material occurred.

2.4. X-ray diffraction

X-ray diffraction was measured on the macromolecular crystallography beamline ID14-1 at the ESRF (Grenoble, France). The beam line is optimized to measure X-ray diffraction of protein crystals, but is also excellently suited to record powder diffraction patterns. The beamline uses monochromatic hard X-rays with a wavelength of 0.934 Å.

3. Results and Discussion

Usually, heavy water solvent is used for incoherent neutron scattering measurements of protein dynamics [17-20] as the incoherent scattering cross section of D_2O is much smaller than that of H_2O. However, as water molecules move out of the observable space-time window - some Å in some 0.1 ns for IN13 and some Å in some ns for IN10 - measurements of protein dynamics in H_2O solvent using high resolution neutron backscattering spectroscopy are feasible above ~280 K. This was already demonstrated for protein solutions in H_2O [14] and also for H_2O hydrated protein powders [21]. As D_2O has a significant influence on protein thermal stability [22] it might modify the relation of protein unfolding and amyloid formation in apo-Mb. Therefore, we chose to perform the measurements using H_2O solvent.

Apo-Mb was investigated in the native state and in a heat incubated state. Heating of native apo-Mb above 328 K is reported to result in amyloid fiber formation [6], and the extended heat incubation phase of 2 days at 338 K should lead to a saturated formation of amyloid fibers in the sample. As a control, X-ray diffraction patterns were recorded of the different samples, see Figure 1: the native state of apo-Mb before neutron scattering measurements, apo-Mb after the experiment on IN13 and the heat incubated state of apo-Mb are compared directly. The broad peaks at 0.4 - 0.8 $Å^{-1}$ and 1.1 – 2.0 $Å^{-1}$ in the X-ray diffraction pattern are typical features of globular proteins [23]. Solvent scattering contributes to the shoulder at 2.0 $Å^{-1}$ but was not subtracted from the measured intensities. The small streak at 1.54 $Å^{-1}$ is not related to scattering from the protein and might be caused by absorption of the quartz windows of the sample holders. The shift of the peak from 0.59 $Å^{-1}$ in the native state to 0.63 $Å^{-1}$ in the heated samples and the emergence of the peak at 1.33 $Å^{-1}$ are characteristic signatures of amyloid formation in apo-Mb as reported by Onai et al. [8]. Obviously, there is already a small amount of amyloid fibers present in the native state of apo-Mb, which is visible by the presence of the shoulder at 1.33 $Å^{-1}$. Larger amounts of amyloid fibers formed both during the neutron scattering experiments and by heat incubation for 2 days at 338 K, which is clearly visible by the emergence of the strong peak at 1.33 $Å^{-1}$.

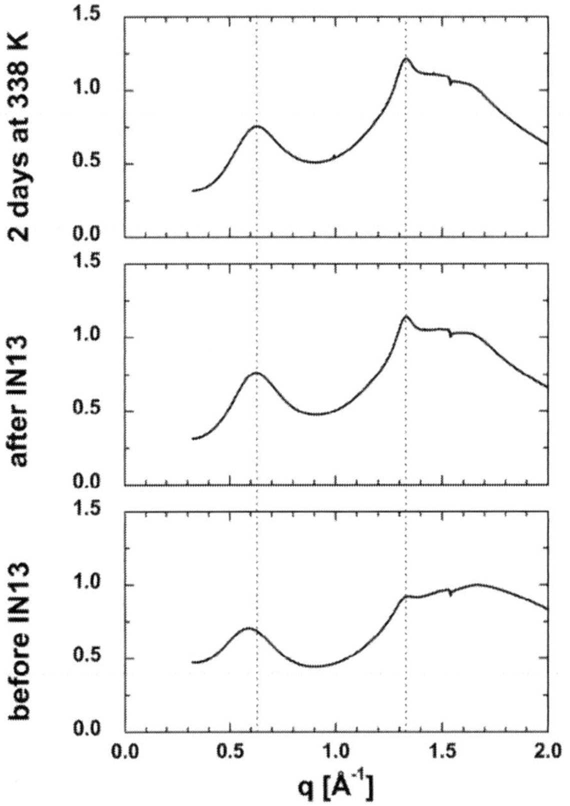

Figure 1: X-ray diffraction measured of the different samples. The shift of the broad peak from 0.59 to 0.63 Å^{-1} and the emergence of the narrow peak at 1.33 Å^{-1} are fingerprints of amyloid formation.

In the following we focus on the neutron scattering experiments. Measured elastic intensities using the neutron spectrometer IN10 are shown in Figure 2 (A) and (B), elastic data measured with the spectrometer IN13 are given in Figure 2 (C) and (D).

Mean square displacements $<u^2>$ were calculated from the slopes of the linear fits shown in Figure 2. The $<u^2>$ obtained with the IN13 spectrometer are given in Figure 3 (B), the $<u^2>$ determined with IN10 are reported in Figure 4 (B). In cases for which the data quality does not allow to derive sufficiently precise information about changes in the $<u^2>$, an analysis of the summed elastic intensity $\sum I(q)$ as a function of temperature in the same q range can give a good indication of the presence of a transition in dynamical behavior [9]. Normalized summed elastic intensities of native apo-Mb and heat incubated apo-Mb are shown in Figure 3 (A) measured with IN13 and in Figure 4 (A) measured with IN10. However, it should be noted that no information about the amplitudes of motions or about protein resilience in absolute units can be deduced from the summed elastic intensity analysis.

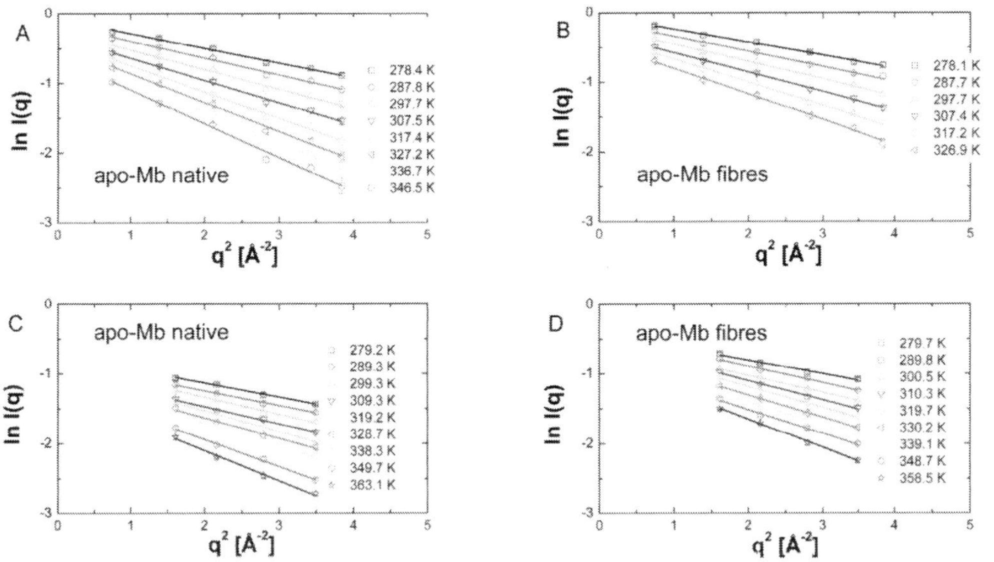

Figure 2: Measured EINS of the different samples. Data shown in A and B were measured on the instrument IN10, data in C and D were measured on IN13. Straight lines are linear fits to the data.

The transition in the dynamics of native apo-Mb at around 320 K is clearly visible in the summed elastic intensities on both instruments. The transition was observed before [9] and attributed to the change from a 'softer' α helical native structure of apo-Mb to a 'stiffer' β sheet structure in amyloid fibers. The break at 320 K is better visible in the IN10 than in the IN13 data. This seems to indicate that protein dynamics in the ns time scale depend more strongly on the secondary structure than protein dynamics in the 0.1 ns time scale. As expected, the heat incubated sample does not show any break at around 320 K, which indicates that amyloid formation has been completed during heat incubation for 2 days at 338 K.

On the other hand, the $\langle u^2 \rangle$ of the native state and of the heat incubated state measured with IN13 overlap within the error bars and no change in the dynamics of native apo-Mb at 320 K is visible. Mean force constants $\langle k' \rangle$ were derived from the $\langle u^2 \rangle$ measured with IN13 and are given in Table 1. Heat incubated apo-Mb was found to be only slightly stiffer than apo-Mb. Considering the error bars both native and heat incubated apo-Mb seem to have similar stiffness.

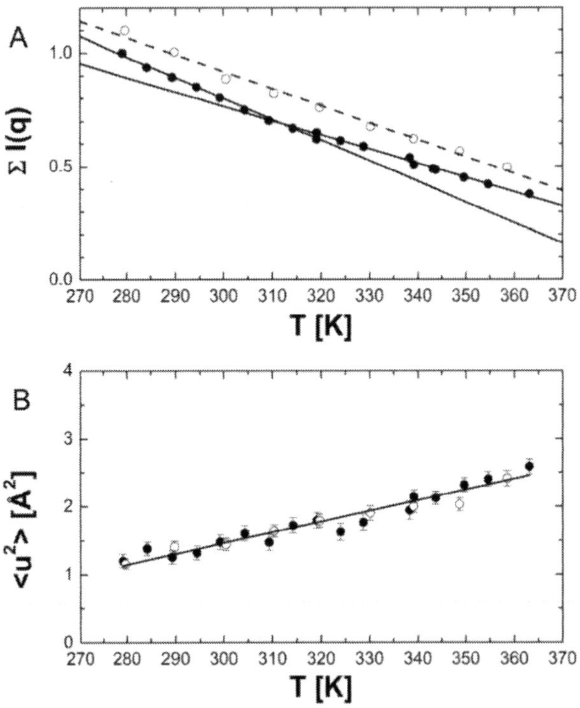

Figure 3: Native apo-Mb (filled symbols) and heat incubated apo-Mb (empty symbols) measured with IN13. (A) Summed elastic intensity $\sum I(q)$ and (B) mean square displacements $<u^2>$. The solid line in (B) corresponds to a linear fit to the $<u^2>$ of native apo-Mb.

Table 1: Effective force constants <*k*'> measured using the IN13 spectrometer.

Sample	<k'> [N/m]
Native apo-Mb	0.18 +/- 0.01
Heat incubated apo-Mb	0.20 +/- 0.01

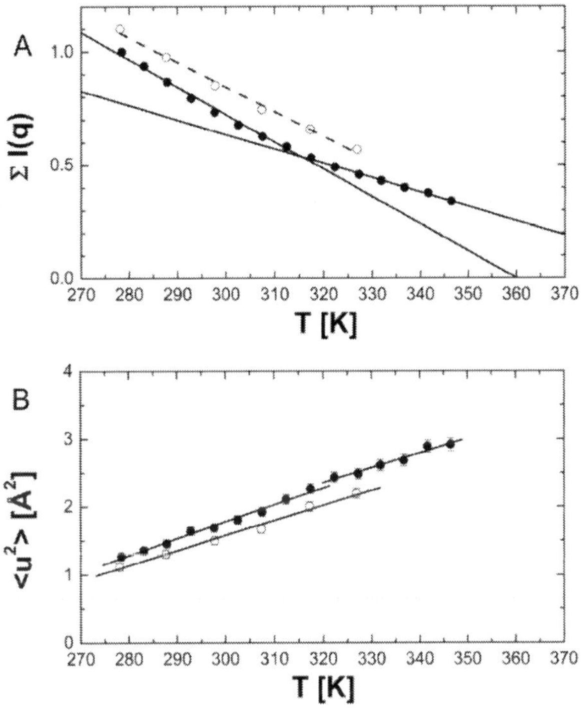

Figure 4: Native apo-Mb (filled symbols) and heat incubated apo-Mb (empty symbols measured with IN10. (A) Summed elastic intensity $\sum I(q)$ and (B) mean square displacements $<u^2>$.

Table 2: Effective force constants $<k'>$ measured using the IN10 spectrometer.

Sample	$<k'>$ [N/m]
Native apo-Mb $T < 320$ K	0.110 +/- 0.004
Native apo-Mb $T > 320$ K	0.126 +/- 0.012
Heat incubated apo-Mb	0.124 +/- 0.006

Comparing the $<u^2>$ measured with IN10, the amplitudes of motion of the heat incubated sample seem to be significantly smaller than the native state. This might indicate that thermal fluctuations in fully formed amyloid fibers are smaller than in the α helical native state. As the $<u^2>$ of the heat incubated apo-Mb are also smaller than the $<u^2>$ of native apo-Mb above 328 K, this would imply that amyloid formation in apo-Mb extends over a longer time and is not fully completed during the IN10 experiment. However, the reduced $<u^2>$ might also just be caused by a smaller hydration level in the heat incubated sample than in native apo-Mb. A clear break in the $<u^2>$ of apo-Mb at 320 K is absent. However, there might be a subtle change in the slope below and above 320 K. Effective force constants $<k'>$ were calculated from the mean square displacements and are given in Table 2. The changes are only small but it seems to be that the α helical native state of apo-Mb below 320 K is slightly softer than the β sheet state of apo-Mb above 320 K, which is in agreement with previous results using the summed elastic intensity analysis [9]. The heat incubated apo-Mb with fully developed amyloid fibers seems to have a similar resilience as the structure of apo-Mb above 320 K.

4. Summary

We present a study on protein dynamics during amyloid formation in apo-Mb using EINS. Changes in the dynamics during the α-to-β transition of apo-Mb at 320 K in the 0.1 ns and 1 ns time-scale are clearly visible in the summed elastic intensities on both instruments. Although we have collected data with high statistics, no changes in the mean square displacements in the 0.1 ns time scale are visible during the α-to-β transition. In the ns time-time scale there might be a small change of slope in the $\langle u^2 \rangle$ during the α-to-β transition in apo-Mb. The $\langle u^2 \rangle$ of heat incubated apo-Mb seem to be reduced in the ns but not in the 0.1 ns time-scale as compared to native apo-Mb. The effective force constants determined in the ns time-scale are in agreement with a α helical structure of apo-Mb below 320 K being softer than the β sheet structure above 320 K. Heat incubated apo-Mb seems to have a similar resilience as apo-Mb above 320 K. Furthermore, our study indicates that average protein dynamics in the 0.1 ns time scale doesn't seem to be correlated strongly to the secondary structure content of proteins. Thermal fluctuations in proteins in the ns time scale seem to depend more strongly on the secondary structure of proteins.

5. Acknowledgements

The author (A.M.S.) thanks Prof. Georg Büldt for continuous support. Trevor Forsyth and Estelle Moussou are gratefully acknowledged for their help with the X-ray measurements. We also thank Moeava Tehei, Francesca Natali, and Prof. Judith Peters for fruitful discussions and help with the experiments on IN13 and IN10. This work is based on experiments performed at the ILL and at the ESRF in Grenoble, France.

References

[1] Dobson, C.M. 1999 *Trends Biochem Sci* **24** 329-332
[2] Stefani, M. and C.M. Dobson 2003 *J Mol Med* **81** 678-99
[3] Dobson, C.M. 2003 *Nature* **426** 884-90
[4] Eanes, E.D. and G.G. Glenner 1968 *J Histochem Cytochem* **16** 673-7
[5] Frauenfelder, H., B.H. McMahon, and P.W. Fenimore 2003 *Proc Natl Acad Sci U S A* **100** 8615-8617
[6] Fändrich, M., M.A. Fletcher, and C.M. Dobson 2001 *Nature* **410** 165-166
[7] Fändrich, M., et al. 2003 *Proc Natl Acad Sci U S A* **100** 15463-15468
[8] Onai, T., et al. 2007 *J Appl Crystallogr* **40** S184-S189
[9] Fabiani, E., et al. 2009 *Eur Biophys J* **38** 237-244
[10] Sears, V.F. 1992 *Neutron News* **3** 26-37
[11] Gabel, F., et al. 2002 *Q Rev Biophys* **35** 327-67
[12] Smith, J.C. 1991 *Q Rev Biophys* **24** 227-291
[13] Kidera, A. and N. Go 1990 *Proc Natl Acad Sci U S A* **87** 3718-22
[14] Tehei, M., et al. 2004 *EMBO Rep* **5** 66-70
[15] Zaccai, G. 2000 *Science* **288** 1604-1607
[16] *http://www.ill.eu/instruments-support/instruments-groups/yellowbook/.* 2010.
[17] Doster, W., S. Cusack, and W. Petry 1989 *Nature* **337** 754-6
[18] Stadler, A.M., et al. 2009 *Biophys J* **96** 5073–5081
[19] Stadler, A.M., et al. 2008 *Biophys J* **95** 5449-5461
[20] Stadler, A.M., et al. 2011 *J R Soc Interface* **8** 590-600
[21] Wood, K., et al. 2007 *Proc Natl Acad Sci U S A* **104** 18049-54
[22] Bonnete, F., D. Madern, and G. Zaccai 1994 *J Mol Biol* **244** 436-47
[23] Hirai, M., et al. 2002 *J Synchr Rad* **9** 202-205

Journal of Physics: Conference Series **340** (2012) 012093 doi:10.1088/1742-6596/340/1/012093

Resolution Effects on the Mean Square Displacement as Obtained by the Self-Distribution-Function Procedure

A Benedetto[1], S Magazù[1,*], F Migliardo[1], C Mondelli[2] and M A Gonzalez[3]

[1]Department of Physics, University of Messina, Viale Ferdinando Stagno D'Alcontres n°31, P.O. BOX 55, 98166 S. Agata, Messina, Italy
[2]CNR-INFM-OGG and CRS Soft, Institut Max Von Laue Paul Langevin, 6 rue Jules Horowitz, BP 156, F-38042 Grenoble, France
[3]Institut Max Von Laue Paul Langevin, 6 rue Jules Horowitz, BP 156, F-38042 Grenoble Cedex 9, France

E-mail: smagazu@unime.it

Abstract. In the present contribution, a procedure for molecular motion characterization based on the evaluation of the Mean Square Displacement (MSD), through the Self-Distribution Function (SDF), is presented. It is shown how MSD, which represents an important observable for the characterization of dynamical properties, can be decomposed into different partial contributions associated to system dynamical processes within a specific spatial scale. It is also shown how the SDF procedure allows us to evaluate both total MSD and partial MSDs through total and partial SDFs. As a result, total MSD is the weighed sum of partial MSDs in which the weights are obtained by the fitting procedure of *measured* Elastic Incoherent Neutron Scattering (EINS) intensity. We apply SDF procedure to data collected,by IN13, IN10 and IN4 spectrometers (Institute Laue Langevin), on aqueous mixtures of two homologous disaccharides (sucrose and trehalose) and on dry and hydrated (H_2O and D_2O) lysozyme with and without disaccharides. It emerges that the hydrogen bond imposed network of the water-trehalose mixture appears to be stronger with respect to that of the water-sucrose mixture. This result can justify the higher bioprotectant effectiveness of trehalose. Furthermore, it emerges that partial MSDs of sucrose and trehalose are equivalent in the low Q domain (0÷1.7) $Å^{-1}$ whereas they are different in the high Q domain (1.7÷4) $Å^{-1}$. This suggests that the higher structure sensitivity of sucrose should be related to the small spatial observation windows. Moreover, the role of the instrumental resolution in EINS is considered. The nature of the *dynamical transition* is highlighted and it is shown that it occurs when the system relaxation time becomes shorter than the instrumental energy time. Finally, the bioprotectants effect on protein dynamics and the amplitude of vibrations in lysozyme are presented.

1. Introduction

During the last few years, considerable efforts have been addressed, through experimental, theoretical and computational studies, to clarifying the microscopic nature of the dynamics of biological macromolecules. One of the phenomena, which has so far been widely debated, although not fully clarified, is the so-called *dynamical transition* in protein systems which, in literature, is referred to as a sharp rise in the Mean Square Displacement (MSD) of hydrated proteins compared to the dry sample, usually registered in the temperature range T=200÷240 K [1-5]. The basic understanding of the mechanism underlying the observed *dynamical transition* remains controversial and various models have been proposed. Such a *transition* had been previously ascribed to a sudden change in an *effective elasticity* of the protein [3], to motions of specific side groups [6], to a specific fragile-to-strong crossover in dynamics of hydration water [7], to the microscopic manifestation of the glass transition in the hydration shell [8] and to a resolution effects due to a relaxation process that enters the experimentally accessible frequency window [9].

It is well known that neutron scattering allows us to characterize the structural and dynamical properties of a wide class of material systems, such as polymers and proteins. These properties can be described by the time-dependent spatial correlation function $G(\mathbf{r},t)$ introduced by Van Hove [10], whose space-time Fourier transform corresponds to the scattering function $S(\mathbf{Q},\omega)$. When the system scattering cross section is mainly incoherent, the relevant contribution is given by the self-distribution function (SDF) [10].

The experimentally obtained neutron scattering data are also connected with the employed spectrometer instrumental features. This implies that the system observables, e.g. the MSD, are influenced by instrumental effects: the energy window and the transferred wave vector values determine the time and space ranges of the observable motions. On this, several contributions are reported in literature[11-14]. In particular, Smith J C *et al.* have shown how the existence of a temperature-dependent relaxation frequency can lead to a transition in the measured MSD in the absence of any change in the Elastic Incoherent Neutron Scattering (EINS) function [13,14].

Recently, a procedure (the SDF procedure) for MSD evaluation from the EINS data has been formulated. The main aim of this work is to present the SDF procedure [15-19], to clarify how the measured intensity depends on the instrumental resolution [20-24] and to evaluate the effects of the instrumental energy resolution on the MSD extracted by EINS data [23,24].

2. Experimental Section

Experimental data were collected at the Institute Laue Langevin (Grenoble, France) using the IN4, IN13 and IN10 spectrometers. These spectrometers are characterized by a relatively high energy of incident neutrons (16 meV) and allow us to span quite a wide range of momentum transfer with three different energy resolutions.

More specifically, for the IN10 spectrometer, the incident wavelength was 6.27 Å, the Q-range was 0.30÷2.00 $Å^{-1}$ and the elastic energy resolution (FWHM) was 1 μeV, which corresponds to an elastic time resolution of 2192 ps; for the IN13 spectrometer, the incident wavelength was 2.23 Å, the Q-range was 0.28÷4.27 $Å^{-1}$ and the elastic energy resolution (FWHM) was 8 μeV, which corresponds to an elastic time resolution of 274 ps; for the IN4 spectrometer, the incident wavelength was 3.60 Å, the Q-range was 0.30÷4.5 $Å^{-1}$ and the elastic energy resolution (FWHM) was 200 μeV, which corresponds to an elastic time resolution of 11 ps.

Trehalose/19H_2O and sucrose/19H_2O mixtures and partially deuterated lysozyme in dry, D_2O, H_2O, with and without disaccharide environments at a hydration value of h=0.4 (h=g of water/g of protein) were used. The considered hydration value was chosen as the activity of proteins depends crucially on the presence of at least a minimum amount of solvent water [25,26].

Data were collected by the three spectrometers in the temperature range 20÷320 K. Empty cell contribution was subtracted and spectra were normalized to a vanadium standard. This data treatment was performed with the *Lamp* code relative to the three spectrometers used; other programs, i.e. *Mathematica*, and specific new codes were written and used for data analysis.

3. Instrumental Energy Resolution Effects on the Scattering Law

The scattering law and the intermediate scattering function are connected by a direct and an inverse time Fourier transform. The experimentally accessible quantity in the *ω-space*, due to the finite energy instrumental resolution $\Delta\omega$, is the convolution of the scattering law $S(Q,\omega)$ with the instrumental resolution function $R(\omega;\Delta\omega)$, i.e. the *measured* scattering law $S_R(Q,\omega;\Delta\omega)$.

The main aim of the present paragraph is the evaluation of the effects of the finite instrumental energy resolution. This study, which has been performed in the time domain through the time Fourier transform of the latter convolution product, yields [20-24]:

$$S_R(Q,\omega=0;\Delta\omega)=\int_{-\infty}^{\infty} I(Q,t)R(t)dt \quad (1)$$

In the ideal elastic case in which the resolution is a delta function in the *ω-space*, we obtain from Eq. (1) that $S_R(Q,\omega=0;\Delta\omega)$ coincides with $S(Q,\omega=0)$. In figure 1, the case of IN10 and IN13 spectrometers are considered. As can be seen, two general situations can occur. The first concerns the case in which the resolution time is longer than the system characteristic time, i.e. $\tau_{RES}>\tau$, as shown in figure 1, a and b; in this circumstance, the resolution effects are relatively small; the *measured* elastic scattering laws (shadowed areas) are very close to the areas of the intermediate scattering functions. The second case occurs when the resolution time is shorter than the system characteristic time, i.e. $\tau_{RES}<\tau$, as shown in figure 1c; in this case, the resolution effects become important: the *measured* elastic scattering law is close to the area subtended by the resolution function instead of to the area of the intermediate scattering function.

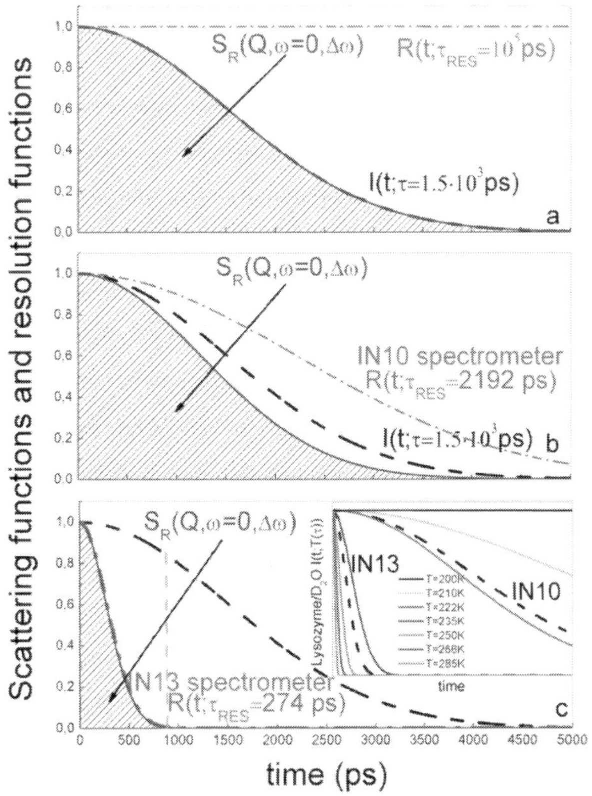

Figure 1. Resolution effects in elastic neutron scattering contribution. Comparison between the intermediate scattering function $I(t;\tau)$ at a fixed τ value (dashed curves), resolution functions $R(t;\tau_{RES})$ for different τ_{RES} values (dash-dotted curves) and the related *measured* elastic scattering laws $S_R(Q,\omega=0,\Delta\omega)$ (indicated areas). These latter, obtained considering Eq. (1), represent the *measured* elastic contribution of an elastic neutron scattering experiment on a system with a characteristic time τ in which the instrumental resolution time is τ_{RES}. (a) Case in which the resolution time is longer than the system characteristic time; no resolution effect is present: the *measured* elastic scattering law and the elastic scattering law are coincident. (b) Case in which the resolution time is slightly higher than τ:

in this case the *measured* elastic scattering law is slightly smaller than the elastic scattering law. (c) Case in which the resolution time becomes much smaller than the system characteristic time τ; the resolution effect becomes relevant: the *measured* elastic scattering law strongly differs from the elastic scattering law. (Insert) Intermediate scattering function of lysozyme/D_2O for different temperature values (data taken from Ref. [24]), and time resolution functions of IN13 and IN10 spectrometers. The intermediate scattering function crosses the resolution function at about T=220 K for IN10 and at about T=240 K for IN13.

To better understand what quantitatively happens, in figure 2 the *measured* elastic scattering law $S_R(\omega=0,\Delta\omega)$, as a function of the energy resolution, is shown. Starting from the same intermediate scattering function used for the case reported in Fig. 1 ($\tau=1.5*10^3$ ps), the *measured* elastic scattering law $S_R(\omega=0,\Delta\omega)$ has been evaluated as a function of the instrumental energy resolution according to Eq. (1); the areas of figure 1 have been calculated and their behaviour as a function of the instrumental energy resolution is shown. As can be seen, when $\tau_{RES}>\tau$ the resolution effects are negligible (i.e. in figure 2, the upper square, which represents the *measured* elastic scattering law corresponding to figure 1b, and the horizontal dashed-dotted line, which represents the scattering law, are close), whereas when $\tau_{RES}<\tau$ the resolution effects become important (i.e. in figure 2 the lower square, which represents the *measured* elastic scattering law corresponding to figure 1c, and the horizontal line, are far apart), and the *measured* elastic scattering law becomes equal to the area of the resolution function (dashed curve). It is important to highlight that, in the specific case reported in Fig. 1 and in Fig. 2 (i.e. Gaussian behaviour for both instrumental resolution and system scattering law functions), it can be shown that an inflexion point in the *measured* elastic scattering law occurs when the instrumental energy resolution of the used spectrometer intersects the inverse of the system relaxation time (as it can be seen in Fig. 2 considering the intersection between the vertical dashed arrow and the curve).

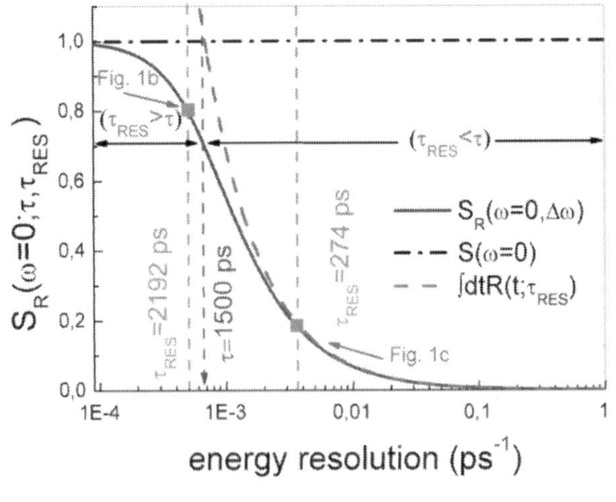

Figure 2. *Measured* elastic scattering law, $S_R(Q,\omega=0,\Delta\omega)$, as a function of energy resolution, $\Delta\omega=1/\tau_{RES}$, for system with fixed τ value. Starting from the same intermediate scattering function used for the cases reported in Fig. 1 (for which the system characteristic time is $\tau=1.5\cdot10^3$ ps), the $S_R(\omega=0,\Delta\omega)$ has been evaluated as a function of $\Delta\omega$ according to Eq. (1); in particular, the shadowed areas of Fig. 1 have been calculated as a function of the instrumental energy resolution and the result is here reported (full curve); the vertical dashed arrow indicates the system relaxation time. The two cases of Fig. 1 are considered (squares) and two regimes are found: when $\tau_{RES}>\tau$ the resolution effects are negligible (case of Fig. 1b); when $\tau_{RES}<\tau$, the resolution effects become important (case of Fig. 1c) and the distance between the *measured* elastic scattering law and the elastic scattering law (horizontal

dash-dotted line) increases. The change from one condition to the other one occurs when the system characteristic time intersects the resolution time; at the corresponding instrumental energy resolution value an inflection point occurs in the *measured* scattering law.

Furthermore, it is now important to take into account that the comparison between the *measured* elastic scattering laws for different instrumental energy resolutions (figure 1) is complementary to a case in which the same system is investigated at different temperature values with a fixed instrumental energy resolution (e.g. the system relaxation time obeys an Arrhenius or non-Arrhenius behaviour). To fix ideas, let us assume that by increasing temperature, the system characteristic time decreases, so giving rise to changes in the intermediate scattering function. To report a specific example, in the insert of figure 1, the behaviour of the main intermediate scattering function of hydrated lysozyme, taken at different temperature values, is reported together with the resolution function of the IN10 and IN13 spectrometers at ILL (black dashed lines). As can be seen, by increasing the temperature, the intermediate scattering function crosses the IN10 resolution function, at first, at about T=220 K and then the IN13 resolution function at about T=240 K.
Finally, in a complementary way, by applying the theorem of integral average in Eq. (1), the intermediate scattering function, evaluated at an equivalent time t^*, and $S_R(Q,\omega=0;\Delta\omega)$ are proportional [20-22]:

$$S_R(Q,\omega=0;\Delta\omega) \propto I(Q,t^*) \tag{2}$$

Now, evaluating the spatial Fourier transform of the latter equation one obtains:

$$F_r\left\{S_R(Q,\omega=0;\Delta\omega)\right\} \propto G(r,t^*) \tag{3}$$

The obtained relationship shows that, due to the fact that the SDF can be normalized, the normalized spatial Fourier Transform (F_r) of the *measured* EINS intensity profile corresponds to SDF evaluated at t^*, in which, following the standard convention for the distribution functions, r represents a displacement.

4. Self-Distribution-Function Procedure and Instrumental Energy Resolution on MSD from EINS experiments

In the following, the SDF procedure, a new procedure for MSD evaluation from EINS experiments, is presented. The SDF procedure is a recipe for MSD evaluation from EINS experiments and has been presented in previous works [15-19]. It is essentially based on the determination of the self-distribution function and on its use in the evaluation of the average statistical values of the physical quantity of interest $\langle A\rangle$, in agreement with the statistical mechanics definition:

$$\langle A\rangle = \int_{-\infty}^{\infty} A(\mathbf{r})G^{self}(\mathbf{r})d\mathbf{r} \tag{4}$$

in which the spatial self-distribution function, as a probability, may be normalized to unit:

$$\int_{-\infty}^{\infty} G^{self}(\mathbf{r})d\mathbf{r} = 1 \tag{5}$$

In the specific case of MSD evaluation, the dynamic observable A corresponds to the second power of the displacement, $\mathbf{r}^2$:

$$\langle \mathbf{r}^2\rangle = \int_{-\infty}^{\infty} \mathbf{r}^2 G^{self}(\mathbf{r})d\mathbf{r} \tag{6}$$

In the case in which the system can be considered isotropic, the volume integral becomes dependent only on the scalar r. In such a case the normalization condition and the MSD become:

$$\int_{-\infty}^{\infty} 4\pi r^2 G^{self}(r)dr = 1 \tag{7}$$

$$\langle \mathbf{r}^2\rangle = \int_{-\infty}^{\infty} r^2\left[4\pi r^2 G^{self}(r)\right]dr \tag{8}$$

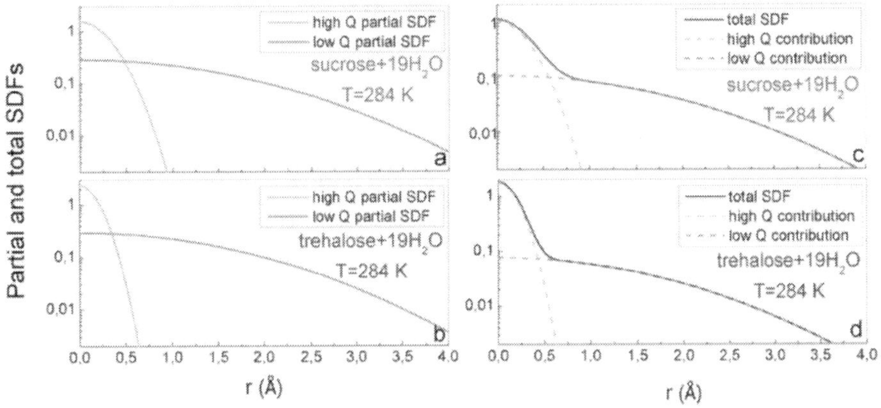

Figure 3. Partial SDFs as a function of r at T=284K for (a) sucrose/H_2O and (b) trehalose/H_2O are shown. As can be seen, the different kinds of motion are spatially well separated within the accessible Q range. Total normalized SDF as a function of r at T=284K for (a) sucrose/H_2O and (b) trehalose/H_2O together with their partial contributions are also shown. The SDF very closely follows the first partial contribution in the range (0÷0.5) Å and the second one in the range (0.5÷5.0) Å.

The SDF can be applied directly to the experimentally determined EINS profiles as well as to any function capable of reproducing EINS profile behaviour; it represents an integral procedure which takes into account the global Q behaviour and by so doing, it allows us to reduce the error on the $Q\to 0$ extrapolation. Considering that the self-distribution function can be written as a sum of Gaussian functions, as reported in Ref. [15-17,21], the normalization condition gives the following result:

$$G^{self}(r)=\sum_n A_n G_n^{self}(r)=\sum_n \frac{A_n}{16(\pi a_n)^{3/2}}\exp\left(-r^2/4a_n\right) \quad (9)$$

in which $\sum_n A_n = 1$; in addition the MSD is:

$$\langle \mathbf{r}^2\rangle = 6\sum_n A_n a_n = \sum_n A_n \langle \mathbf{r}^2\rangle_n \quad (10)$$

in which partial SDFs and partial MSDs are present. This formula highlights that the MSD corresponds to a weighted sum of the different displacement contributions present in the system. In figure 3a and 3b, the partial SDFs as a function of r at T=284 K for sucrose/H_2O and trehalose/H_2O, respectively, are also shown. It is interesting to observe that the high-Q SDFs are different from each other whereas the low-Q SDFs are equal. Figure 3c and 3d show the obtained normalized SDF and its partial contributions as a function of r at T=284K for sucrose/H_2O and trehalose/H_2O, respectively. These functions were obtained by the fit procedure of the measured scattering intensity with Eq. (10). As can be seen, different kinds of motion are spatially well separated within the accessible Q range; furthermore, the SDF very closely follows the first partial SDF in the range (0÷0.5) Å and the second one in the range (0.5÷5.0) Å.

$\langle \mathbf{r}^2\rangle$ represents the MSD in 3D space; if $\langle r^2\rangle$ represents the MSD in 1D space, for isotropic systems we have that:

$$\langle r^2\rangle = 1/3\langle \mathbf{r}^2\rangle \quad (11)$$

This implies that Eq. (10) in 1D yields:

$$\langle r^2\rangle = 2\sum_n A_n a_n = \sum_n A_n \langle r^2\rangle_n \quad (12)$$

Therefore, this procedure allows us to obtain the autocorrelation function $G_{self}(r,t^*)$ versus r, together with its different partial contributions, as well as to determine the partial MSDs, their weights and the total MSD.

Figure 4a shows the MSD for sucrose and trehalose. Figures 4b and 4c show the partial MSDs for sucrose and trehalose, evaluated by the SDF procedure, in the temperature range 20÷287K related to the *small-r* domain and the *high-r* domain, respectively. As can be seen, the partial MSD behaviours of sucrose and trehalose are equivalent in the *high-r* domain, whereas they are different in the *small-r* domain. This circumstance suggests that the higher structure sensitivity of sucrose compared to trehalose should be related to the small spatial observation window. This finding is in agreement with several findings on disaccharide systems [27-32], in particular with inelastic neutron scattering results on the relaxation versus vibration contribution analysis, i.e. the boson peak, indicating the presence of a higher suppression of the local fast dynamics in the trehalose/water system compared to the sucrose/water system. In particular, such dynamic suppression correlates with a weaker temperature dependence of viscosity namely with a lower fragility character, which justified the higher bio-protectant effectiveness of trehalose compare to sucrose [31].

Furthermore, this finding is in agreement with other experimental findings obtained by inelastic neutron scattering which reveal a higher downshift of the OH intramolecular stretching contribution for the trehalose/water system, indicating a stronger hydrogen bonded network in the trehalose/water system compared to the sucrose/water system [32].

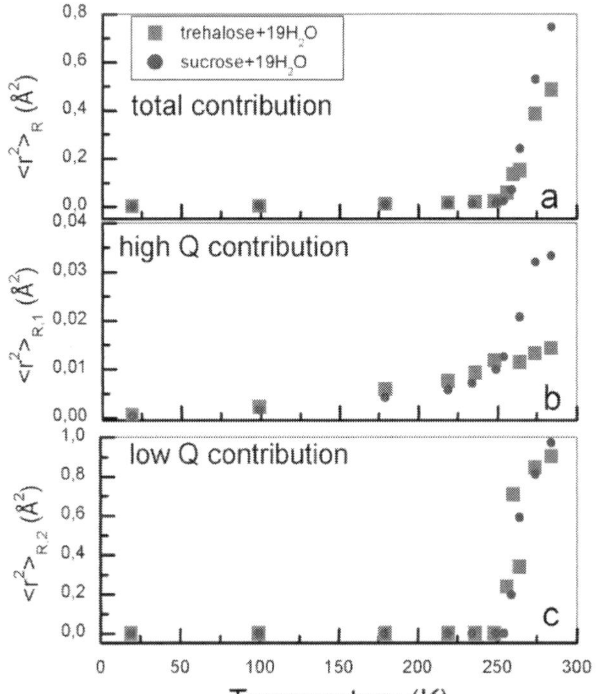

Figure 4. MSD, high Q partial MSD and low Q partial MSD temperature behaviour for trehalose/H_2O and sucrose/H_2O in the temperature range 20÷287K. As can be seen, the partial MSD behaviours of sucrose/H_2O and trehalose/H_2O are equivalent in the *high-r* domain, whereas they are different in the *small-r* domain.

To adapt the previous definition of the MSD to the case of the EINS experiments, it is necessary to consider that the instrumental energy resolution influences the physical observables as discussed in the previous paragraph; therefore Eq. (6) can be re-written as follows:

$$\left\langle r^2 \right\rangle_R = \int_{-\infty}^{\infty} r^2 G_R^{self}(r)\,dr \tag{13}$$

Starting from this relationship, in the following, we shall analyse the instrumental energy resolution effects on the MSD, in the framework of an EINS experiment. In this framework, as the output of an EINS experiment is the *measured* elastic scattering law, we have that:

$$\left\langle r^2 \right\rangle_R = \int_{-\infty}^{\infty} FT_r \left\{ S_R(Q, \omega = 0) \right\} r^2\,dr \tag{14}$$

where FT_r represents the spatial Fourier transform operator. For the evaluation of the MSD, let us consider Eq. (1); then Eq. (14) becomes:

$$\begin{aligned} \left\langle r^2 \right\rangle_R &= \int_{-\infty}^{\infty} FT_r \left\{ S_R(Q, \omega = 0) \right\} r^2\,dr \\ &= \int_{-\infty}^{\infty} FT_r \left\{ \int_{-\infty}^{\infty} I(Q,t) R(t)\,dt \right\} r^2\,dr = \int_{-\infty}^{\infty} \left\{ \int_{-\infty}^{\infty} G^{self}(r,t) R(t)\,dt \right\} r^2\,dr \\ &= \int_{-\infty}^{\infty} R(t) \left\{ \int_{-\infty}^{\infty} G^{self}(r,t) r^2\,dr \right\} dt = \int_{-\infty}^{\infty} \left\langle r^2 \right\rangle(t) R(t)\,dt \end{aligned} \tag{15}$$

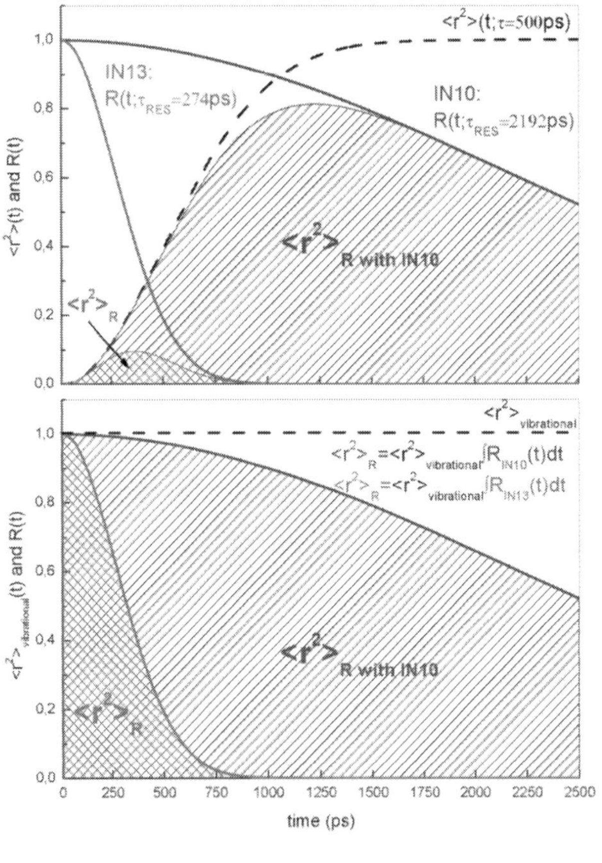

Figure 5. (a) Normalized time behaviour of the actual "system MSD" at a fixed τ value, $<r^2>(t;\tau)$ (dashed curve); resolution function for different τ_{RES} values (curves), $R(t;\tau_{IN13})$ and $R(t;\tau_{IN10})$; *measured* MSD $<r^2>_R$ evaluated by Eq. (15). (b) Effect on the *measured* MSD due to the different energy

resolution of the two spectrometers, IN13 and IN10, when only vibrational motions occur; in particular, as can be seen, different *measured* MSDs correspond to the same actual system MSD.

More specifically, $\langle r^2\rangle(t)$ is the actual "system MSD", while $\langle r^2\rangle_R$ is the "*measured* MSD". The first is a function of time, whereas the second is the MSD value as obtained by EINS; it is a pure number, not a function of time, and depends on the instrumental resolution used. Figure 5a shows the effect on the *measured* MSD due to the resolution function used: due to the different energy resolutions used different *measured* MSDs correspond to the same actual system MSD.
Finally, it is interesting to observe what occurs in the vibrational motions domain, e.g. at low temperature value. When only vibrational motions occur, the actual system MSD can be considered almost constant $\langle r^2\rangle(t)\rightarrow\langle r^2\rangle^{(V)}$ (such a condition is approximately satisfied in the low temperature range up to T=40 K); in such a case, the *measured* MSD from Eq. (15) is:

$$\langle r^2\rangle_R=\int_{-\infty}^{\infty}\langle r^2\rangle(t)R(t)\,dt=\int_{-\infty}^{\infty}\langle r^2\rangle^{(V)}R(t)\,dt=\langle r^2\rangle^{(V)}\int_{-\infty}^{\infty}R(t)\,dt \qquad (16)$$

Then, starting from Eq. (16), it is possible to determine the actual system MSD at the lowest temperature values. Figure 5b shows the effect on the *measured* MSD due to the use of a different energy resolution; as can be seen, different *measured* MSDs correspond to the same actual system MSD. A validity test of the proposed procedure is obtained by applying this procedure to the data collected on the same systems, i.e. dry and hydrated lysozyme (see the insert of figure 6), by the two spectrometers IN13 and IN10, working at a different energy resolution. As a result, at the lowest temperature values we obtain the same actual system mean displacement values, i.e. 0.016 Å [23].
It is also of interest to examine the normalization procedures most frequently applied both on the measured elastic scattered intensity, and on the extracted MSD.
It is well known that the *measured* scattering law, which is a function of Q, represents the number of elastically diffused neutrons within a given solid angle. Therefore, normalization consisting in a multiplicative factorization, i.e. $S_R\rightarrow nS_R$, is an allowed transformation able to rescale the data since it does not change the proportionality relationship between the scattering intensity at different Q values. This transformation may be useful, for example, to give the same intensity value (e.g. at the origin) to different spectra collected at different temperatures and does not produce any change in MSD evaluation.
In this regard, it should be taken into account that, on the contrary, the same kind of normalization applied to the logarithm of the *measured* scattering law is a misleading procedure since it leads to an incorrect evaluation of the MSD value. In fact, the normalization of the logarithm of the *measured* scattering law corresponds to a power elevation of the *measured* scattering law, i.e. $nLn(S_R)\rightarrow(S_R)^n$, which gives an incorrect MSD, i.e. $n\langle r^2\rangle_R$.
On the other hand, as far as the logarithm of the *measured* scattering law is concerned, it can be observed that the proper transformation is a translation, this latter corresponding to a normalization of the *measured* scattering law: $Ln(S_R)+Ln(n)\rightarrow nS_R$.
Finally, the normalization of the *measured* scattering law for the *measured* scattering law obtained at a given temperature (usually the lowest) is again wrong since it changes the relationship between the *measured* scattering law relative to different Q values; this procedure would bring a shift of the MSD, assigning it the value of zero at the lowest temperature: $S_R(T)/S_R(T_0)\rightarrow MSD=\langle r^2\rangle_R(T)-\langle r^2\rangle_R(T_0)$.
Regarding this, it is worth observing that the method previously described for obtaining the vibrational MSD value, which is based on the instrumental energy resolution effects evaluation, clarifies how a correct normalization of both the scattering functions and the MSD trends should be performed.

5. Protein *Dynamical Transition*

In the third paragraph, addressed to the description of the instrumental energy resolution effect in EINS, it is highlighted that an inflection point in the *measured* scattering law occurs when the system relaxation time intersects the instrumental energy resolution time. This finding also implies that sometimes, transitions in the *measured* elastic scattering laws may have mistakenly been attributed to real transitions in the dynamic properties of the systems, instead of the simple instrumental effect. More specifically, because a transition in the *measured* elastic scattering law as a function of temperature generates a transition in the extracted MSD temperature behaviour, also transitions in the MSDs may have been mistakenly attributed to real transitions in the dynamic properties of the systems instead of the simple instrumental resolution effect. As highlighted by Smith J C *et al.* [13,14] and by Sokolov A P *et al.* [9], this may be the case of the so-called *dynamical transition*, which represents an open question in the biophysical field. In particular, Sokolov A P *et al.* in Ref. [9] by combining neutron and dielectric spectroscopy data, have shown that there is no sudden change in the dynamics of the protein at temperatures around 200÷230 K, and, therefore, these authors have proposed that the *dynamical transition* is just a result of the protein's structural relaxation reaching the limit of the experimental frequency window.

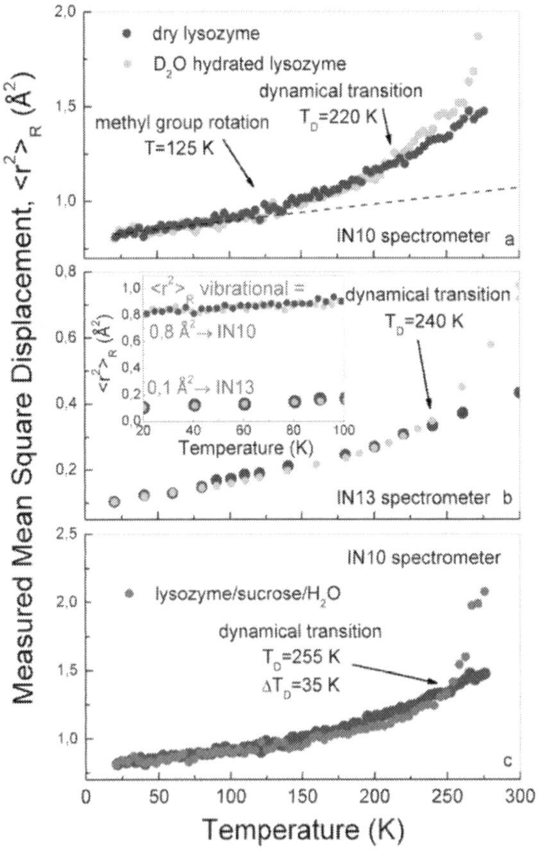

Figure 6. *Measured* MSDs of lysozyme obtained using two different spectrometers with two different instrumental energy resolutions. Comparison between the *measured* MSD temperature behaviour of dry and D_2O hydrated lysozyme, obtained from data collected by both (a) IN10 and (b) IN13 spectrometers. The *dynamical transition* temperature is T_D=220 K from IN10 data and T_D=240 K from IN13 data. (c) Comparison between the *measured* MSDs obtained for H_2O hydrated lysozyme with sucrose and dry lysozyme by the IN10 spectrometer. The presence of disaccharide shifts the *dynamical transition* to a higher temperature value of T=255 K. (Insert) Comparison between the dry

and D_2O hydrated lysozyme *measured* MSDs both from IN10 and IN13 spectrometers in the low temperature range when only vibrational motions occur. The *measured* MSD is about 0.8 $Å^2$ with IN10 and about 0.1 $Å^2$ with IN13. This discrepancy is connected to resolution effects and can be used to determine the system vibrational motion amplitudes.

More specifically, with regard to hydrated lysozyme samples, it has recently [24] been demonstrated that the so-called *dynamical transition*: (i) is a finite instrumental energy resolution effect, and, more specifically, appears when the characteristic system relaxation time intersects the resolution time, (ii) it does not imply any transition in the dynamical properties of systems, (iii) it is not due to the fragile-to-strong- dynamical crossover (FSC) in the temperature behaviour of the system relaxation time, differently to what Chen S H et al. proposed in Ref. [7].

In figure 6, the average *measured* MSDs as a function of temperature are shown for dry and hydrated lysozyme, with and without sucrose, evaluated by the SDF procedure in the range 20÷287 K. The MSDs are obtained using two different instrumental energy resolutions. The *dynamical transition* temperature is T_D=220 K from IN10 data, in which the resolution used is 1 μeV, and T_D=240 K from IN13 data, in which the resolution used is 8 μeV. These values are in agreement with the conclusion on the nature of the *dynamical transition* presented in Ref. [24]. In fact, as can be seen in the temperature behaviour of the system relaxation time [24], a relaxation time of τ=2192 ps corresponds to the temperature value of T=220 K, which is equal to the *dynamical transition* temperature value obtained with IN10 for which the instrumental resolution time is τ_{RES}=2191 ps; moreover, a relaxation time of τ=274 ps corresponds to the temperature value of T=240 K, which is equal to the *dynamical transition* temperature value obtained with IN13 for which the instrumental resolution time is τ_{RES}=274 ps.

Finally, a confirmation of this finding can be given considering that the presence of disaccharides shifts the system relaxation time towards higher values [33] and, therefore, the above *dynamical transition* must occur at the higher temperature when sucrose is added. As can be seen in Fig. 6c, the presence of disaccharide shifts the *dynamical transition* to a higher temperature value of T=255 K.

6. Conclusion

In this paper, a general presentation of the SDF procedure applied on two homologous disaccharides, i.e. sucrose and trehalose, was presented; the energy resolution effects on neutron scattering functions were also discussed.

The SDF procedure was presented for the first time in Ref. [15], then, in Ref. [16] it was theoretically improved and applied on polyethylene glycol with a mean molecular weight of 400 Dalton. An experimental check of the procedure was also performed in Ref. [16] by its application on hydrogenated and on two partially deuterated polyisoprene systems. In Ref. [17], the procedure was also improved and was applied to dry myoglobin in a trehalose matrix. Then, the effects of the instrumental energy resolution on the scattering functions was taken into account in Ref. [20].

The case of the two homologous disaccharides (trehalose and sucrose), which is partially discussed in the present paper, was treated in Ref. [19,21,22] within a comparison between the SDF procedure and the common Gaussian protocol for the MSD determination. For the first time, the higher mobility of sucrose/H_2O compared to trehalose/H_2O was related to a specific spatial domain. In particular, in Ref. [19] the effects of natural bioprotectants on the protein *dynamical transition* are shown. Then the effects of the instrumental energy resolution directly on the measured MSD were computed and presented in Ref. [23].

In the light of the obtained results, a complete analysis of the neutron scattering spectra has allowed us to solve the puzzling question of the so-called *dynamical transition*. In Ref. [23,24], the idea that the *dynamical transition* was due to the fact that the system characteristic time crosses the time corresponding to the finite energy resolution was presented. This idea was connected to the fact that the *dynamical transition* temperature obtained using different instrumental energy resolutions coincides case by case with the temperature for which the system relaxation time and the resolution time are equal (compare the *dynamical transition* temperature of figure 6 with the relaxation time reported in Ref. [7,9]). In the same paper [23], the determination of the amplitude of vibration motion in dry and hydrated (H_2O and D_2O) lysozyme was also shown.

7. References

[1] Doster W, Cusak S and Petry W 1989 *Nature* **337** 754
[2] Rasmussen B F, Stock A M, Ringe D and Petsko G A 1992 *Nature* **357** 423
[3] Zaccai G 2000 *Science* **288** 1604
[4] Doster W 2008 *Eur. Biophys. J.* **37** 591
[5] Sokolov A P, Roh J H, Mamontov E and Garcia Sakai V 2008 *Chem. Phys.* **345** 212
[6] Lee A L and Wand J 2001 *Nature* **411** 501
[7] Chen S H, Liu L, Fratini E, Baglioni P, Faraone A and Mamontov E 2006 *Proc. Natl. Acad. Sci. U.S.A.* **103** 9012
[8] Doster W, Busch S, Gasper A M, Appavou M S, Wuttke J and Scheer H 2010 *Phys. Rev. Lett.* **104** 098101
[9] Khodadadi S, Pawlus S, Roh J H, Garcia Sakai V, Mamontov E and Sokolov A P 2008 *J. Chem. Phys.* **128** 195106
[10] Van Hove L 1954 *Phys. Rev.* **95** 249
[11] Gabel F and Bellissent-Funel M C 2007 *Biophys. J.* **92** 4054
[12] Kneller G R and Calandrini V 2007 *J. Chem. Phys.* **126** 125107
[13] Becker T and Smith J C 2003 *Phys. Rev. E* **67** 021904
[14] Becker T, Hayward J A, Finney J L, Daniel R M and Smith J C 2004 *Biophys. J.* **87** 1436
[15] Magazù S, Maisano G, Migliardo F and Benedetto A 2008 *J. Molec. Struct.* **882** 140
[16] Magazù S, Maisano G, Migliardo F and Benedetto A 2008 *Phys. Rev. E* **77** 061802
[17] Magazù S, Maisano G, Migliardo F and Benedetto A 2008 *J. Phys. Chem. B* **112** 8936
[18] Magazù S, Migliardo F, Benedetto A, Gonzalez M and Mondelli C 2010 *Spectroscopy* **24** 387
[19] Magazù S, Migliardo F, Benedetto A, Mondelli C and Gonzalez M 2011 *J. Non Cryst. Sol.* **357** 664
[20] Magazù S, Maisano G, Migliardo F, Galli G, Benedetto A, Morineau D, Affouard F and Descamps M 2008 *J. Chem. Phys.* **129** 155103
[21] Magazù S, Maisano G, Migliardo F and Benedetto A 2009 *Phys. Rev. E* **79** 041915
[22] Magazù S, Maisano G, Migliardo F and Benedetto A 2010 *Biochim. et Biophys. Acta* **1804** 49
[23] Magazù S, Migliardo F and Benedetto A 2010 *J. Phys. Chem. B* **114** 9268
[24] Magazù S, Migliardo F and Benedetto A 2011 *J. Phys. Chem. B* **115** 7736
[25] Careri G 1998 *Prog. Biophys. Mol. Biol.* **70** 223
[26] Gregory R B 1995 *Protein-Solvent Interactions* Dekker, New York
[27] Ballone P, Marchi M, Branca C and Magazú S 2000 *J. Phys. Chem. B* **104** 9268
[28] Branca C, Magazù S, Maisano G and Migliardo F 2001 *Phys. Rev. B* **64** 224204
[29] Affouard F, Bordat P, Descamps M, Lerbret A, Magazù S, Migliardo F, Ramirez-Cuesta A J and Telling M F T 2005 *Chem. Phys.* **317** 258
[30] Magazù S, Maisano G, Migliardo P and Villari V 1999 *J. Chem. Phys.* **111** 9086
[31] Magazù S, Migliardo F, Affouard F, Descamps M and Telling T F 2010 *J. Chem. Phys.* **132** 184512
[32] Branca C, Magazù S, Maisano G, Bennington S M and Fak B 2003 *J. Phys. Chem. B* **107** 1444
[33] Lerbret A, Bordat P, Affouard F, Hedoux A, Guinet Y and Descamps M 2007 *J. Phys. Chem. B* **111** 9410

Acknowledgments
The authors wish to acknowledge the Institut Laue-Langevin (ILL), Grenoble, France for beam time on IN4, IN10 and IN13 spectrometers.

Journal of Physics: Conference Series **340** (2012) 012094 doi:10.1088/1742-6596/340/1/012094

Correlated motion of protein subdomains and large-scale conformational flexibility of RecA protein filament

Garmay Yu.[1], **Shvetsov A.**[1], **Karelov D.**[1], **Lebedev D.**[1], **Radulescu A.**[2], **Petukhov M.**[1,3], **Isaev-Ivanov V.**[1]

[1] Petersburg Nuclear Physics Institute, Russia

[2] Jülich Center for Neutron Science, Germany

[3] Research and Education Centre "Biophysics", St. Petersburg State Polytecnic University, St. Petersburg, Russia.

alexxy@omrb.pnpi.spb.ru

Abstract. Based on X-ray crystallographic data available at Protein Data Bank, we have built molecular dynamics (MD) models of homologous recombinases RecA from *E. coli* and *D. radiodurans*. Functional form of RecA enzyme, which is known to be a long helical filament, was approximated by a trimer, simulated in periodic water box. The MD trajectories were analyzed in terms of large-scale conformational motions that could be detectable by neutron and X-ray scattering techniques. The analysis revealed that large-scale RecA monomer dynamics can be described in terms of relative motions of 7 subdomains. Motion of C-terminal domain was the major contributor to the overall dynamics of protein. Principal component analysis (PCA) of the MD trajectories in the atom coordinate space showed that rotation of C-domain is correlated with the conformational changes in the central domain and N-terminal domain, that forms the monomer-monomer interface. Thus, even though C-terminal domain is relatively far from the interface, its orientation is correlated with large-scale filament conformation. PCA of the trajectories in the main chain dihedral angle coordinate space implicates a co-existence of a several different large-scale conformations of the modeled trimer. In order to clarify the relationship of independent domain orientation with large-scale filament conformation, we have performed analysis of independent domain motion and its implications on the filament geometry.

1. Introduction

There are wide range of supramolecular protein machines that are important for proper functioning of the living beings. Their function proceeds from their conformational flexibility, so that an understanding of the nature of the motion of a molecular machine is obviously important for understanding of its function. Recent advantages in modelling of the protein dynamics allows building realistic models of the protein mobility at the atomic level. Although the current state of molecular modelling and limited computational resources put limitations on the system size and modelling time, making it impossible to simulate real supramolecular systems on their characteristic time-scale, it seems however possible to simulate the motion of their essential building blocks and try to approximate the large-scale motion/conformation from the motion of the smaller parts of the system.

Homologous recombination protein RecA plays a key role in DNA repair following radiation damage in bacteria. A number of X-ray crystallographic structures of RecA proteins exist [1-4], including complexes of *E.coli* RecA pentamer with ss- and ds-DNA [5]. Although the structure and function of a RecA and its analogues (such as RadA in archae and Rad51 in higher eucariotes) are very well studied, the molecular mechanisms for the enzymatic activity of these types of proteins is yet to be elucidated in detail [6].

Protein mobility on the subdomain scale, such as C-terminal domain and L2 loop, are thought to play an important role in ATP hydrolysis and DNA binding, respectively [6]. In all RecA isoforms where this process has been described formation of a presynaptic complex upon binding of ATP and either single- or double-stranded DNA brings about a significant increase in the helix pitch and a moderate decrease in the cross-sectional gyration radius of the protein filament. This change in the filament geometry is required to accommodate the stretched form of the DNA [7], which seems to be an essential intermediate in the process of homologous pairing and strand exchange. Similar large-scale conformational flexibility of the filament can be observed under conditions of high ionic strength in the absence of DNA, thus being inherent property of the enzyme, and is accomplished largely by N-terminal domain motion [8]. There are indications, however, that the transition between the two conformational states may be more complex and involve the motion of the C-terminal domain and the unstructured regions of RecA monomer, particularly the unstructured C-terminal tail [8;9]. Whether these or other possible conformational changes also facilitate homological pairing of DNA are unclear.

Deinococcus radiodurans is the most radiation resistant species currently known (LD_{37} = 5.5 kGy, compared to 0.17 kGy for *E. coli*). Compared to RecA protein from *E. coli,* its analog from *D. radiodurans* has higher affinity toward double-stranded DNA [10] and seems to exhibit a higher large-scale flexibility [11]. These unusual properties, along with its role in DNA repair following damage by ionizing radiation makes it an interesting object to compare with *E. coli* protein.

In this work we use molecular dynamics to analyze the conformational motion of RecA proteins from *E. coli* and *D. radiodurans* to elucidate on the possible mechanisms of the filament stretching associated with DNA binding.

2. Materials and Methods

2.1. Molecular Dynamics

Regularized models of X-ray crystallographic structures of RecA proteins were prepared based on the crystal structure of RecA from *E.coli* (PDB: 2REB) [3] using standard protocols of ICM-Pro software package (Molsoft LLC, USA) [12]. Two loops absent in the crystal structure (residues number 156-165 and 194-210) were rebuild using Monte-Carlo energy minimization as implemented in loop modelling tools of ICM-Pro. During the energy minimization only the standard torsion angles (φ, ψ, ω and χ_i) of amino acids were allowed to vary. ICM's default set of energy parameters (ECEPP/3 potential) for van-der-Waals, electrostatic, torsion energy interactions and hydrogen bonding were used in the calculations.

Molecular dynamics simulations (MD) were done with GROMACS version 4.0, the widely used software complex for molecular modelling and dynamics of proteins [13] using multiprocessor cluster of NRC Kurchatov Institute (Moscow, Russia). MD modelling involved several standard steps including (a) creation of the protein topology file and preparation of input data for GROMACS using pdb2gmx tool, (b) construction of hydration model for the RecA proteins in a periodic water box with the minimal distance to water box border of 10Å, (c) thermodynamic equilibration of hydrated proteins and surrounding solvent, and (d) MD simulations at a constant temperature using the G53A6 parameter set and SPC water model [14]. The thermodynamically equilibrated system was used to simulate MD at 37°C with duration of 10 ns and MD step of 2 fs. Every 10 ps of MD time states of the model system were recorded for a future analysis. Neighbor searching was performed every 10 steps. The PME-Switch algorithm was used for electrostatic interactions with a cut-off of 1.5 nm as implemented in GROMACS. A single cut-off of 1.7 nm was used for Van der Waals interactions. Temperature coupling was done with the V-rescale algorithm [15]. Pressure coupling was done with the Parrinello-Rahman algorithm [16;17].

2.2. Rigid domain search

The task of determining rigid domains in a molecular dynamics trajectory analysis requires a measure of quality of a molecule segmentation. A molecular dynamics trajectory is an array of N frames, $T=\{T_i\}_{i=1}^{N}$. Each frame contains a set of n atomic coordinates $T_i=\{T_{ij}\}_{j=1}^{n}$. To determine whether a structure is "hard" or "soft", we can attempt to fit it to some reference structure *ref* with an LSQ fitting algorithm for each frame of the trajectory. Denoting the fitted structures with tilde, $T\rightarrow\tilde{T}$, we then calculate the RMSD $r(\tilde{T})=\frac{1}{Nn}\sum_{i,j}\|ref_j-\tilde{T}_{ij}\|^2$ that quantifies the changes of the structure over time. It should be noted that $r(T)$ depends on the choice of the reference structure. Obviously, for any set of structures T the reference that minimizes $r(T)$ is its average, consequently it is possible to use some kind of iterative algorithm to find the best reference structure, using the mean of the previous iteration as the reference for the current one. However, it seems reasonable to use some frame of the trajectory as the reference structure for fitting, and use the mean of the fitted structures for calculating RMSD. This measure does not provide the superposition with the minimal RMSD, but generally it is smaller for more similar structures, so that we used it for computation speed.
The principal points of the algorithm are:

1. define the number of domains D;
2. provide initial segmentation, $S=\{S_i\}_{i=1}^{n}$, where n is number of atoms in structure, and S_i is the index of the domain that i^{th} atom is reffered to, and
3. do global minimization of the function $F(S)=\sum_{k=1}^{D} r(\tilde{U}_k), U_k=\{U_{ki}\}_{i=1}^{N}, \{U_{kij}=T_{ij}\}_{j=1}^{n}|_{S_j=k}$, which is the sum of RMSD for all domains.

Global optimization was performed using simulated annealing algorithm realization provided by Gnu Scientific Library. Pairwise structure fitting was done using Kabsch algorithm [18].

2.3. Principal component analysis

PCA is the commonly used method of dimension reduction, that we use for describing and quantifying of the conformational changes of the modelled structures during MD simulation. The well-known method maps multidimensional input data into a lower-dimension space by a linear transformation, that diagonalizes the covariance matrix of the data. So it represents the data in the set of non-correlated variables that explains maximum possible (within linear transforms) part of the data variance.

We performed PCA using GROMACS molecular simulation package for analysis in atomic coordinate space, and Modular Data Processing library for analysis in the main chain dihedral angle coordinate space.

2.4. Filament construction

Filament was constructed using co-orientation of the central rigid domains of the neighboring monomers of the trimer. We evaluate the translation vector and the rotation matrix of the most identical domains of the neighbouring monomers using Kabsch algorithm and apply it to produce the coordinates of the monomers in the long filament.

2.5. SANS experiment and simulation
Samples of RecA protein from *D. radiodurans* in D_2O buffer containing 5 mM $MgCl_2$ and 20 mM TrisHCl (pH7.4) were prepared as described in [20]. SANS measurements were performed at 15°C on KWS-1 spectrometer (Juelich Centre for neutron science outstation at FRM-2 reactor, Muenich, Germany) using two detector positions in the range of momentum transfer 0.006 - 0.2 Å^{-1}. Simulated SANS spectra were calculated using Debye formula, $I(q)=\int_0^{r_{max}} p(r)\frac{\sin(rq)}{rq}dr$, where the pair distance distribution *p(r)* was calculated by Monte-carlo simulations. Instrument resolution function was assumed to be a Gaussian with 10% FWHM. Fitting the data to scattering by long uniform helix was performed as described in [20].

3. Results and discussion
SANS spectrum of RecA from *D. radiodurans* (figure 1) exhibits 1/Q relationship in the low-Q region and a diffraction maximum near the 0.1 Å^{-1}, characteristic for the helical filaments formed by RecA proteins [19;20]. While the experimental data show a reasonable agreement with scattering by a long uniform helix [20], there is a significant systematic deviation of the model curve near the diffraction maximum (figure 0, inset), evident from high χ^2 value for the fit. Such a discrepancy was previously indicated for this RecA protein [11], but not observed for RecA proteins from *E. coli* and *P. aeruginosa* [8;20]. This could indicate that compared to the latter enzymes, RecA from *D. radiodurans* either has a much lower conformational mobility, so that the monomer structure is not averaged out by the thermal motion, making the uniform body model inadequate, or has a substantial polydispersity in the filament structure.

The molecular dynamics simulations of 10 ns in length were performed for *E. coli* and *D. radiodurans* RecA trimers with and without ATP. To visualize the relative mobility of the protein parts, C_α atoms mobility maps of the center monomer of the trimers was obtained. Color maps, shown in figure 2, reflect the deviation of the C_α atoms of the monomers fitted to the initial structure relative to their initial position. As expected, both proteins exhibit high mobility in the regions of L1 and L2 loops and C-terminal domain. Mobility of C-terminal domain was significantly higher in RecA protein from *D. radiodurans* as compared with the *E. coli* protein, and was restricted by binding of ATP (figure 2).

Further analyzing the trimer motion of *D. radiodurans* RecA we attempted to pinpoint the correlated motions in the protein monomer. We used PCA in the coordinate space of the carbon backbone atoms of the structures obtained during MD simulation fitted to the initial structure. The first principle component of the central monomer of the trimer clearly reveals correlated motion in all three main domains of RecA protein, comprising C-terminal domain and N-terminal α-helix motion and β-sheet deformation in the central domain (figure 3).

Mobility of N-terminal α-helix that forms the intermonomer interface with the two helices of the central domain of the adjacent monomer is directly linked to the large-scale conformational changes of the RecA polymer, particularly to the increase in the helical pitch of the filament when the presynaptic complex is formed [5;8]. Results of our MD simulation imply that N-terminal domain mobility is a part of a global conformational motion in the protein monomer that also affects orientation of C-terminal domain through allosteric interaction. Motion of C-terminal domain of RecA protein upon presynaptic complex formation has been established experimentally [9], which could imply that the conformations seen in MD model of the protein trimer are stabilized by cofactors (ATP and DNA) binding.

For further characterization of the monomer dynamics, we have searched for the rigid parts of the monomer. We found that RecA monomer can be divided into seven regions which motion should be described independently: two parts of N-terminal domain, entire C-terminal domain, rigid part comprising most of the central domain, loop L1, loop L2, and the small region adjacent to it. Figure 4 shows the positions of C_α atoms of the monomer for all frames of the trajectory fitted by central domain. The fact that most of the central domain shows little flexibility allows us to use this part of the protein to extrapolate filament geometry from the trimer structure. The co-orientation (i.e. translation vector and rotation matrix) of the rigid parts of two central domains of the second and third monomers of the trimer was used for construction of the long filament. This extrapolation allowed for an estimate of the main parameters of the large-scale geometry of the filament: helical pitch and gyration radius. These two parameters are observable experimentally and could by measured by SANS. There appears to be a substantial variability in these parameters, particularly, even though in the absence of cofactors the protein is expected in the inactive form (helical pitch ~6.3 nm), a secondary maximum near 8 nm is observed, which is close to the active form geometry (figure 5).

To investigate the relationship between domain motion and large-scale filament conformation, we have performed analysis of the relative domain motion. We performed PCA of atomic coordinates of C_α atoms of each domain of the frames which were aligned by the positions of C_α atoms of the central domain. This procedure characterizes motion of each domain relative to the central domain by several principal components. We observe a strong correlation of the first principle motion of N-terminal α-helix with both the calculated cross-sectional gyration radius and the pitch of the protein filament extrapolated from the trimer (figure 6 a,b). We indeed would expect an effect of N-terminal domain motion on the filament geometry as it is a part of the intermonomer interface. At the same time, the first principle motion of C-terminal domain is also correlated with the filament pitch (figure 6 c). Projection of the C-terminus motion on the first principle axis shows two distinct clusters of values, which give rise to the two peaks in the helical pitch histogram. This may suggest that the existence of two distinct states of the filament, active and inactive, is determined by the allowed conformation of C-terminal domain, rather than by the character of motion of N-terminal α-helix.

To isolate quasi-stable states of the protein in the MD trajectory we performed PCA analysis in the space of polypeptide chain torsion angles. The vectors used for analysis were comprised of sine and cosine values of the main chain dihedral angles. For all three monomers of the trimer the projection of the trajectory on the first two principle axes had similar shape (figure 7 a,b,c). The corresponding population maps plotted in terms of free energy (figure 7 d,e,f) show three shallow energy minima, which suggests tree preferred states of the protein trimer. We went on to simulate neutron scattering experiment by calculating small-angle scattering spectra of the protein 50-mer, reconstructed from the trimer on the assumption of uniform geometry.

Spectra were calculated for each frame of the trajectory (10 ps step), and subsequently averaged over 1 ns time around each energy minimum shown in figure 7. Any single frame of the 10 ns trajectory provided a poorer fit as compared to the uniform helix (best-fit frame shown in figure 8, dashed line, χ^2~53). Thus we could not find any "frozen" conformation of this RecA protein that would explain the data better than the uniform helix model. On the other hand, comparison of the curves simulated based on the energy profile (figure 8, solid lines) with the experimental spectrum, obtained for RecA protein from *D. radiodurans* (figure 8, squares) also shows poor fit of the three-state model to the data.

Such a discrepancy likely points to a higher flexibility of the modelled trimer as compared to a long filament and could also indicate essential inhomogeneity of the real filaments. The latter is supported by the elevated scattering intensity near Q=0.07 $Å^{-1}$ that corresponds to the helical pitch values of ~80 Å, which, according of the MD simulation, is characteristic for one of the two clusters of conformational states of the trimer (figure 6c).

4. Conclusions

Molecular dynamics model of the trimer of RecA from *D. radiodurans* reveals pronounced motion of the C-terminal domain that is allosterically coupled with the motion of N-terminal domain and conformational changes in the β-sheet structure of the central domain.

The positional relationship of the rigid domains of the monomers in the trimer allows to use their co-orientation to extrapolate the large-scale geometry of the protein filament from the trimer structure. Analysis of the relative trajectories of individual domains suggests that motion of N-terminal α-helix is strongly correlated with the filament pitch and cross-sectional gyration radius. Motion of C-terminal domain is associated with the change in the filament pitch characteristic for the transition between active and inactive protein forms, which may point to the guiding role of C-terminal domain dynamics over large-scale filament conformation.

The extrapolated filament geometry shows high variability along the MD trajectory that can not be directly reconciled with SANS data, that implies higher flexibility of the modeled trimer as compared to a long filament and could also indicate essential inhomogeneity of the real filaments. This calls for simulation of larger RecA oligomers to inquire into long-range correlations of the inter-monomer co-orientation.

5. Acknowledgements

This work was supported by the Russian Academy of Science program 'Physics of Elementary Particles' (subprogram "Neutron physics", direction "Studies of structure, dynamics and non-ordinary properties of matter by neutron scattering"), Russian Foundation for Basic Research (grant № 07-04-01548-a and 11-04-01229-a) and by grants of the Ministry of Education and Science of the Russian Federation (RNP 2.2.1.1.4663 and state contract 02.740.11.5223).

References

[1] Datta S, Prabu MM, Vaze MB, Ganesh N, Chandra NR, Muniyappa K & Vijayan M 2000 *Nucleic Acids Res.* **28** pp. 4964-4973
[2] Rajan R & Bell CE 2004 *J. Mol. Biol.* **344** pp. 951-963
[3] Story RM, Weber IT & Steitz TA 1992 *Nature* **355** pp. 318-325
[4] Xing X & Bell CE 2004 *J. Mol. Biol.* **342** pp. 1471-1485
[5] Chen Z, Yang H & Pavletich NP 2008 *Nature* **453** pp. 489-484
[6] Bell CE 2005 *Mol. Microbiol.* **58** pp. 358-366
[7] Nishinaka T, Ito Y, Yokoyama S & Shibata T 1997 *Proc. Natl. Acad. Sci. U.S.A.* **94** pp. 6623-6628
[8] Petukhov M, Lebedev D, Shalguev V, Islamov A, Kuklin A, Lanzov V & Isaev-Ivanov V 2006 *Proteins* **65** pp. 296-304
[9] Yu X, Jacobs SA, West SC, Ogawa T & Egelman EH 2001 *Proc. Natl. Acad. Sci. U.S.A.* **98** pp. 8419-8424
[10] Battista JR 1997 *Annu. Rev. Microbiol.* **51** pp. 203-224
[11] Karelov DV, Lebedev DV, Suslov AV, Shalguev VI, Kuklin AI, Islamov AK, Lauter H, Lanzov VA & Isaev-Ivanov VV 2008 *Journal of Physics: Condensed Matter* **20** p. 104215
[12] Abagyan R & Totrov M 1994 *J. Mol. Biol.* **235** pp. 983-1002
[13] Van Der Spoel D, Lindahl E, Hess B, Groenhof G, Mark AE & Berendsen HJC 2005 *J Comput Chem* **26** pp. 1701-1718
[14] Oostenbrink C, Villa A, Mark AE & van Gunsteren WF 2004 *J Comput Chem* **25** pp. 1656-1676
[15] Bussi G, Donadio D & Parrinello M 2007 *J Chem Phys* **126** p. 014101
[16] Parrinello M & Rahman A 1981 *Journal of Applied Physics* **52** pp. 7182-7190
[17] Nose S 1984 *The Journal of Chemical Physics* **81** pp. 511-519
[18] Kabsch W 1976 *Acta Crystallographica Section A* **32** pp. 922-923
[19] DiCapua E, Schnarr M, Ruigrok RW, Lindner P, Timmins PA. 1990 *J Mol Biol* 214:557-570.
[20] Lebedev DV, Baitin DM, Islamov A, Kuklin AI, Shalguev V, Lanzov VA, Isaev-Ivanov VV.

2003 *FEBS Lett* 537:182-186.

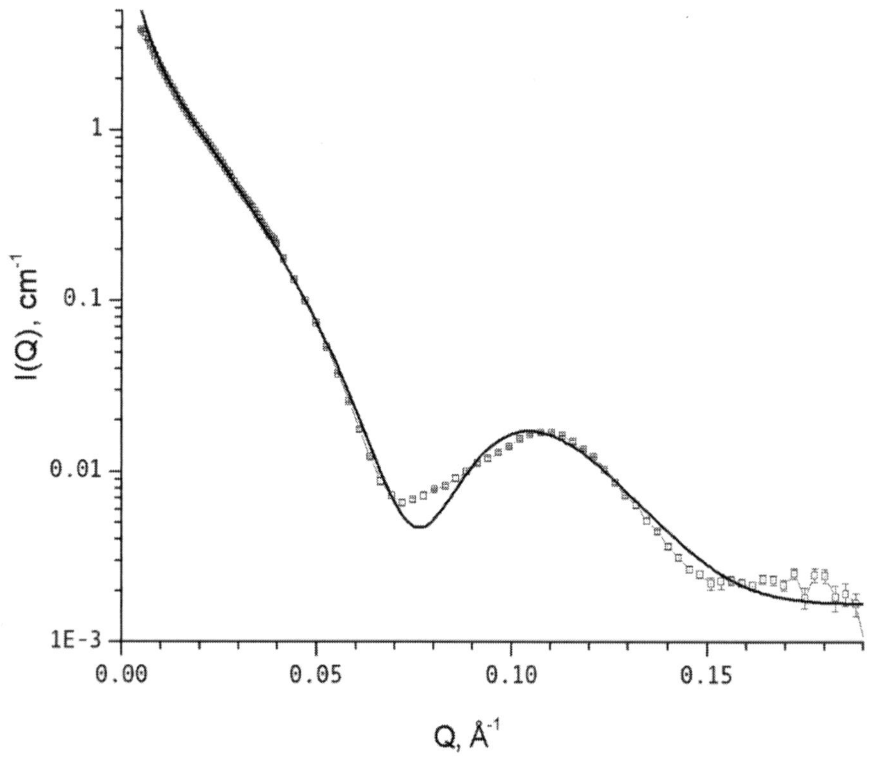

Figure 1: SANS spectrum of RecA self-polymer. Solid line shows the fit to scattering by a long uniform helix with the pitch of 67 Å and the cross-sectional gyration radius of 39.5 Å, χ^2=30.6.

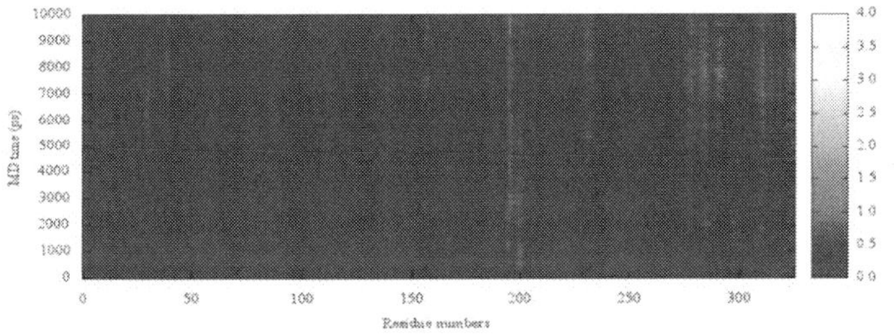

a)

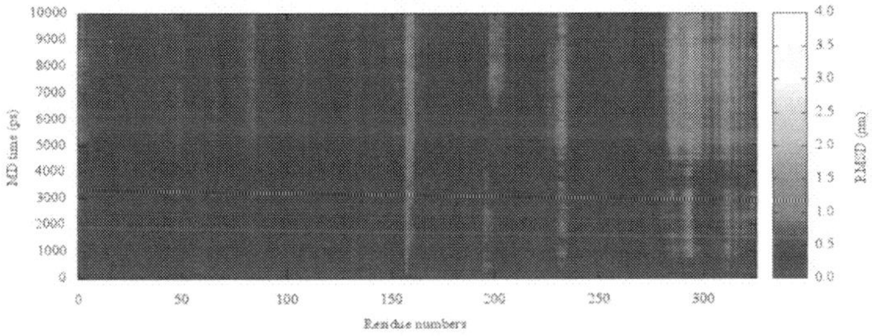

b)

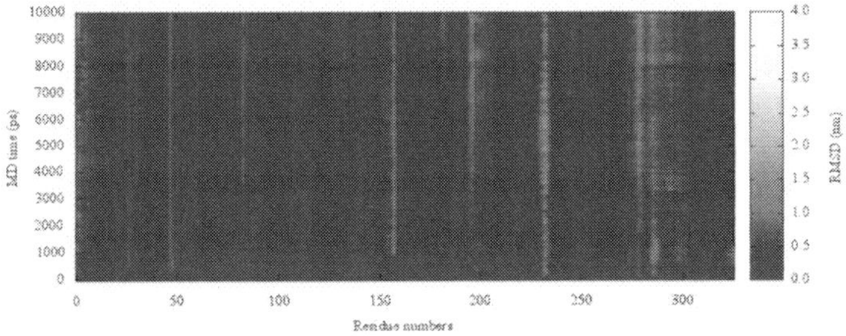

c)

Figure 2. Mobility maps from MD simulations for a) *E. coli* protein; b) *D. radiodurans* protein and c) *D. radiodurans* protein complex with ATP.

Figure 3. Extremal projections of trajectory of the middle monomer of a trimer of RecA from *D.radiodurans* on the first principal axis.

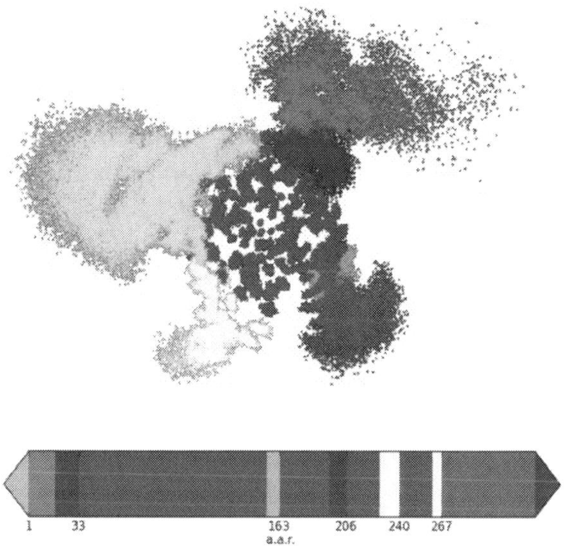

Figure 4. Relative domain motion in RecA protein from *D. radiodurans*. The bar below represents what parts of amino-acid sequence correspond to the respective domains.

Color	residue numbers
Red	1-17
Blue	18-32
Yellow	227-239, 261-266
Orange	155-162
Purple	195-205
Green	267-327
Brown	remaining residues

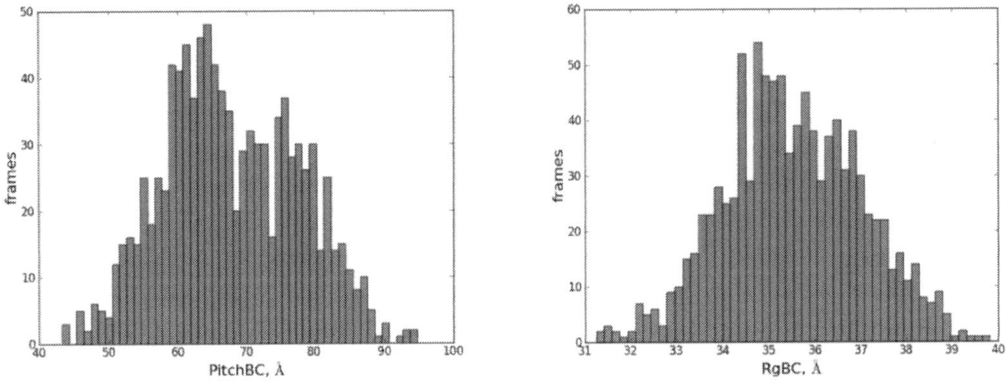

a. b.

Figure 5. Distribution of the geometric parameters of the filament, - helical pitch (a) and cross-sectional gyration radius (b) based on the MD trajectory.

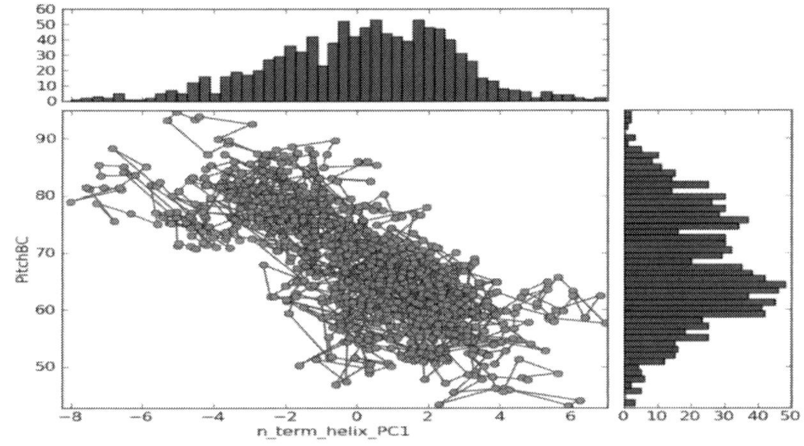

a.

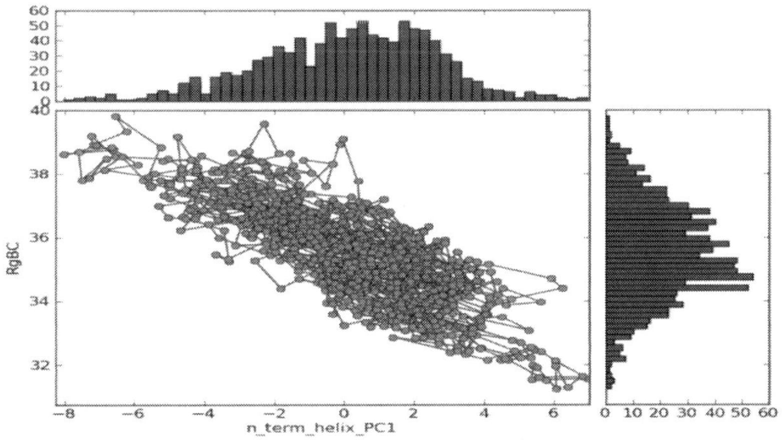

b.

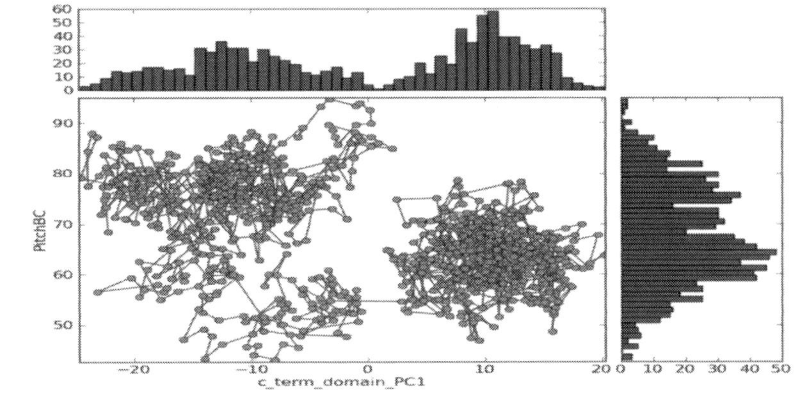

c.

Figure 6. Relationship between filament geometry and domain orientation:
a) N-terminal helix vs pitch (Å);
b) N-terminal helix vs gyration radius (Å) ; c) C-terminal domain vs pitch (Å)

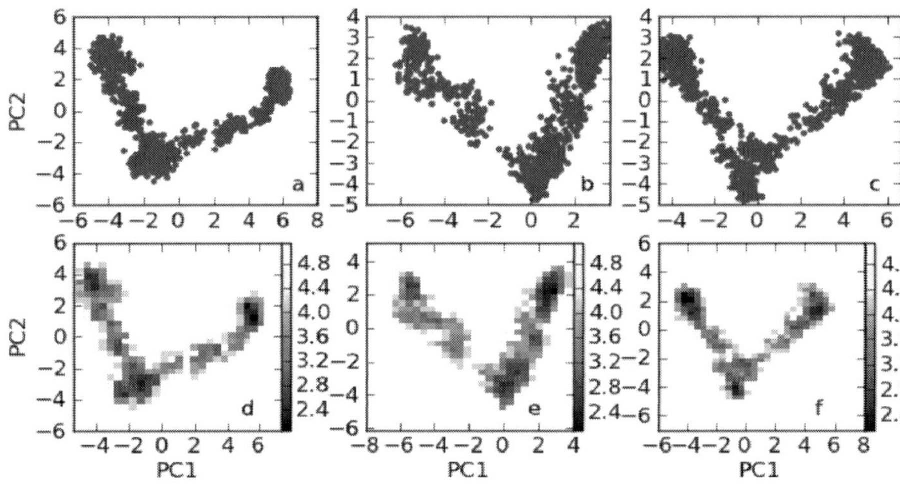

Figure 7. Projections of the monomers trajectory of *D. radiodurans* RecA monomers on the plane of the first two principal axes in dihidral angle space (a,b,c), and the corresponding estimated free energy maps (d,e,f). Color scale in kcal/mol.

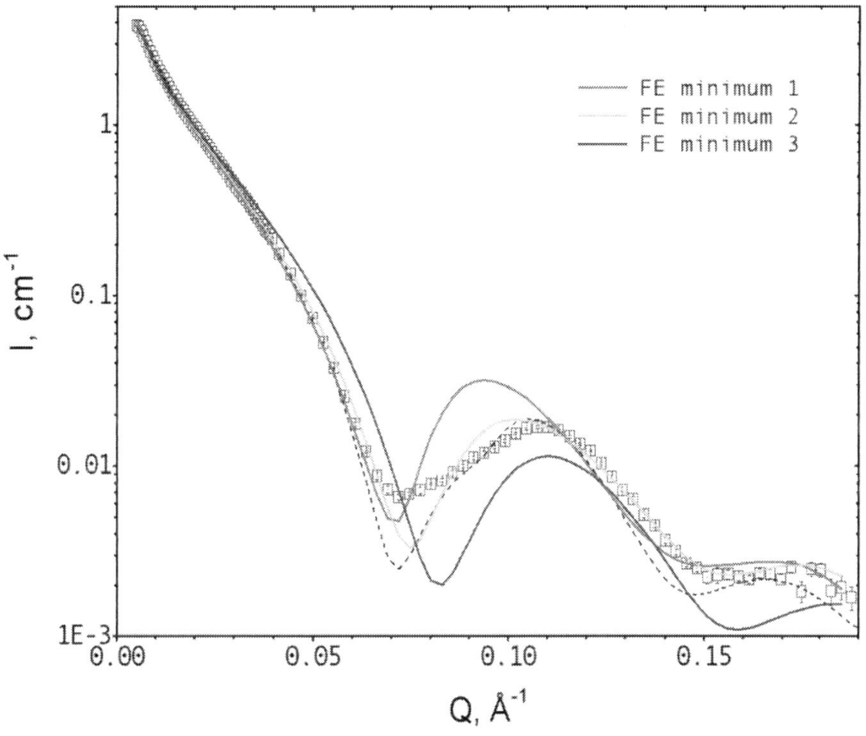

Figure 8. Simulated SANS spectra of RecA from *D. radiodurans*. Lines show the average of 100 spectra calculated for the structures yielded by MD around each of the three local minima of the free energy shown in figure 7. Open squares represent the experimental curve shown in figure 1. Dashed line is the best fit of the scattering spectrum to a single frame of the MD trajectory.

Neutronographic Texture Analysis of Zirconium Based Alloys

M Krůželová[1], S Vratislav[1], L Kalvoda[1], M Dlouhá[1]

[1] Department of Solid State Engineering, Faculty of Nuclear Sciences and Physical Engineering, Czech Technical University in Prague, Trojanova 13, 120 00 Prague 2, Czech Republic

E mail: monika.kruzelova@fjfi.cvut.cz

Abstract. Neutron diffraction is a very powerful tool in texture analysis of zirconium based alloys used in nuclear technique. Textures of five samples (two rolled sheets and three tubes) were investigated by using basal pole figures, inversion pole figures, and ODF distribution function. The texture measurement was performed at diffractometer KSN2 on the Laboratory of Neutron Diffraction, Department of Solid State Engineering, Faculty of Nuclear Sciences and Physical Engineering, CTU in Prague. Procedures for studying textures with thermal neutrons and procedures for obtaining texture parameters (direct and inverse pole figures, three dimensional orientation distribution function) are also described. Observed data were processed by software packages HEXAL and GSAS. Our results can be summarized as follows: i) All samples of zirconium alloys show the distribution of middle area into two maxima in basal pole figures. This is caused by alloying elements. A characteristic split of the basal pole maxima tilted from the normal direction toward the transverse direction can be observed for all samples. ii) Sheet samples prefer orientation of planes (100) and (110) perpendicular to rolling direction and orientation of planes (002) perpendicular to normal direction. iii) Basal planes of tubes are oriented parallel to tube axis, meanwhile (100) planes are oriented perpendicular to tube axis. Level of resulting texture and maxima position is different for tubes and for sheets. The obtained results are characteristic for zirconium based alloys.

1. Introduction

Zirconium is a metal with below-normal c/a ratio (c/a<1,633). The main characteristics of this element are described at [1]. Zirconium has very low absorption cross-section of thermal neutrons, high hardness, ductility, and corrosion resistance. Therefore, zirconium alloys are used in the nuclear industry (Zircaloy) as fuel rod cladding, especially in water reactors [2].

Most of the texture measurements are related to the basal pole distribution, which plays an important role. Less is known the distribution of prism or pyramidal plane poles. These are strong indicators for degree of annealing, and, moreover are important for quantitative determinations of textures, for example, by analysis of the orientation distribution function (ODF) [3]. The slip planes of hexagonal close packed metals should rotate into the plane of rolled sheet. Accordingly the texture in which the basal plane (which is the slip plane) lies in or near the rolling plane is predominant [4].

In our work the texture of two rolled sheets (Zrp, ZZ13) and three tubes (Zry2, ZiT, and ZrW) were investigated by direct pole figures, inversion pole figures, and orientation distribution function (ODF).

2. Experiment

The texture measurements were performed on the diffractometer KSN-2 at Laboratory of Neutron Diffraction, Department of Solid State Engineering, Faculty of Nuclear Sciences and Physical Engineering. Our laboratory developed, and tested experimental, and calculation techniques for quantitative texture analysis based on the ODF combined with the diffraction of thermal neutrons.

2.1. The KSN-2 diffractometer

The texture experiments were carried out on the KSN-2 diffractometer using the TG-1 texture goniometer with automatic data collection. The monochromatic wavelength was 0.1362 nm, and single-crystal Cu(200) was used like monochromator. The experimental data were measured in transmission or reflection arrangement. Depending on the requirements, the specimens consisted of one or several sheets of the examined material. Pole figures were measured for (100), (002), (101), (110), and (102) reflections of zirconium samples. The measured data were corrected for absorption, irradiated volume, and they were normalised. The experimental data processed in this manner were used to calculate the coefficients of expansion to express the measured direct pole figures using spherical functions.

The KSN-2 diffractometer was used simultaneously for diffraction patterns measurement of zirconium or zircaloy samples. By means of these experimental data the inversion pole figures were calculated, and then all these diagrams were used in the quantitative texture analysis by means of Rietveld procedures. The KSN-2 diffraction device offers good intensity and the best resolution value of $\delta d/d = 0.007$ in the region $d \sim 1.0 \div 0.1$ nm (d is interplanar spacing).

The high penetration of neutrons through the majority of materials is the main advantage for examination of the textures of coarse-grained materials like oriented zirconium sheets. Quantitative texture analysis (three-dimensional distribution function - ODF) in connection with neutron diffraction is extensively used in this research field. For example, texture influences both the behaviour of polycrystalline materials during thermal or mechanical treatment, and the anisotropy of various properties in technical materials. Consequently, there are several areas in physic and metallurgy where texture studies should be performed. Among these problems we can put nuclear technology (investigations of fuel elements and fuel tubes based on the zirconium alloys). Our work is focused on research of the textural properties of zirconia-based materials after thermal and mechanical processing using thermal neutron diffraction.

2.2. Samples

Rolled sheets

Sample Zrp was composed of 11 sheets of circular shape with a diameter of 50 mm. The thickness of one sheet was 0.25 mm. The sheets were cold – rolled with 92% reduction. Table 1 shows the composition of Zrp.

Table 1. Chemical composition of sample Zrp.

	Cu [at%]	Fe [at%]	Mo [at%]	W [at%]
Zrp	0.72	0.82	0.43	0.12

Sample ZZ13 was forged and hot rolled to the thickness 3.3 mm. Then the sample was annealed at 664°C. Table 2 shows the composition of ZZ13.

Table 2. Chemical composition of sample ZZ13.

	Hf [at%]	Ca [at%]	Mg [at%]	O [at%]	H [at%]
ZZ13	0.41	0.044	0.096	0.055	0.0011

Tubes

All tubes (Zry2, ZiT a ZrW) were extruded, and then cold-rolled with inter-annealing (700°C) to the final sizes. Table 3 shows the composition of tubes, and Table 4 shows parameters of tubes. There υ is tube length, φ_e is outside tube diameter, and φ_i is internal tube diameter.

Table 3. Chemical composition of tubes.

	Sn [at%]	Fe [at%]	Cr [at%]	O [at%]	H [at%]	N [at%]	C [at%]
Zry2	1.13	0.35	0.16	0.011	0.0010	0.0050	0.0012
ZiT	1.07	0.41	0.19	0.011	0.0020	0.0060	0.0012
ZrW	1.09	0.43	0.21	0.011	0.0015	0.0045	0.0012

Table 4. Parameters of tubes.

	υ [mm]	φ_i [mm]	φ_e [mm]
Zry2	36.9	11.2	12.1
ZiT	30	8	9.1
ZrW	30	8	9.2

3. Methods of texture analysis

The results of our texture measurement are represented by direct pole figures, inverse pole figures, and the orientation distribution function. Direct pole figures were measured for samples Zry2, and ZiT. Inverse pole figures were calculated for samples Zrp, ZZ13, Zry2, ZiT, and ZrW.

The direct pole figure is a stereographic projection of spatial distribution of poles of the particular plane to the project plane which is parallel to the significant sample plane. Two different methods were used for direct pole figures determinations.

First method is conventional, and the distribution pole density for selected reflection ($2\theta_{hkl}$ is fixed) is measured for the different direction in the sample coordination system defined by R, N, and T axis.

Secondly, the Rietveld refinement method was used for the pole figure determination. In this case the Rietveld refinement method was introduced by the harmonic analysis texture procedure to the Rietveld software code GSAS [5]. So it is taking into consideration the influence of preferred orientation on the level of intensity measured in the neutron diffraction patterns. In harmonic analysis procedure the crystal coordination is given by reflection index (hkl). Sample coordinates (α, β) are determined by the sample orientation on the diffractometer. For example, in case of rolling sheets (mmm sample symmetry), α is angle between RD (rolling direction), and TD (transverse direction), and β is angle between ND (normal direction), and SV (scattering vector). Before definition of the sample coordinations (α, β) we need to define instrument coordination system (I, J, K). In this case I is normal to the diffraction plane, and J is parallel with the direction of the incident neutron beam. Then we define a standard set of rotating goniometr system: Eulerian angles (φ_1, ϕ, φ_2). Experimental diffraction patterns were treated by the GSAS software procedure [5], and the value of selected pole figures was obtained. As a result of this refinement procedure is the matrix harmonic coefficients $C_l^{m,n}$. The harmonic coefficients are used to the calculations of the direct pole figure some selected (hkl) reflections.

The quantitative description of texture is given by ODF (Orientation Distribution Function) defined as

$$\frac{dV(g)}{V} = f(g)dg,$$

where f(g) is ODF, dV the volume of grains of orientation g within the element of orientation space dg, and V the volume from which data are collected.

The common representation of ODF is given by equation:

$$f(g) = \sum_{l,m,n} C_l^{m,n} T_l^{m,n}(g),$$

where $C_l^{m,n}$ are series coefficients, $T_l^{m,n}(g)$ are generalized spherical harmonic functions of degree l.

The obtained direct pole figures were then processed by HEXAL [6], and the sections of the Euler orientation space were calculated. The interpretation of ODF was realized by the ideal orientations [1].

In both cases, resulting pole figure are used like input data for ODF values determination by means software code HEXAL [6] or popLA [7]. HEXAL software package (the hexagonal symmetry of the crystals and orthorhombic symmetry of the specimen) was used, and the ODF values were obtained together with all texture characteristics (pole figures, inverse pole figures, ODF - f (g) values, fiber texture with <110> and <001> axis parallel to rolling direction, parameters of the ideal orientations (hkl)<uvw>, texture index J, volume fraction coefficient f).

Measured diffraction patterns were used at the same time for the inversion pole figure determinations. The inverse pole figure shows how the selected direction in the sample reference frame is distributed in the reference frame of the crystal. The pole density values were calculated by Mueller formula [6]. Figure 1 shows the distribution of poles in the inverse pole figure of zirconium.

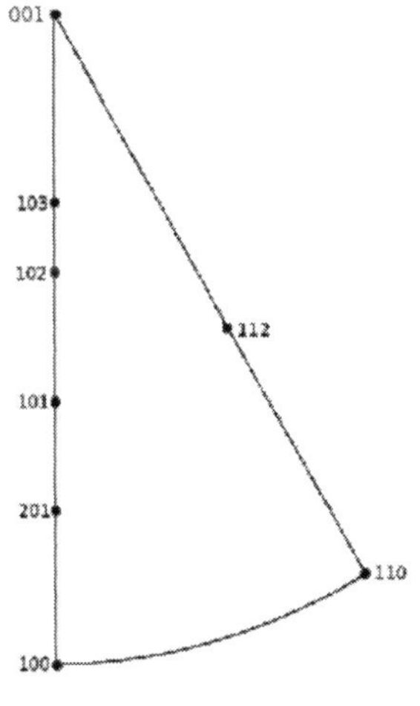

Figure 1. Distribution of poles in inverse pole figure.

4. Results

Direct pole figures obtained by using software package GSAS are show at Figure 2 (pole (002)), Figure 3 (pole (100)), and Figure 4 (pole (101)). Direct pole figures obtained by using software package HEXAL are show at Figure 5 (pole (002)), Figure 6 (pole (100)), and Figure 7 (pole (101)).

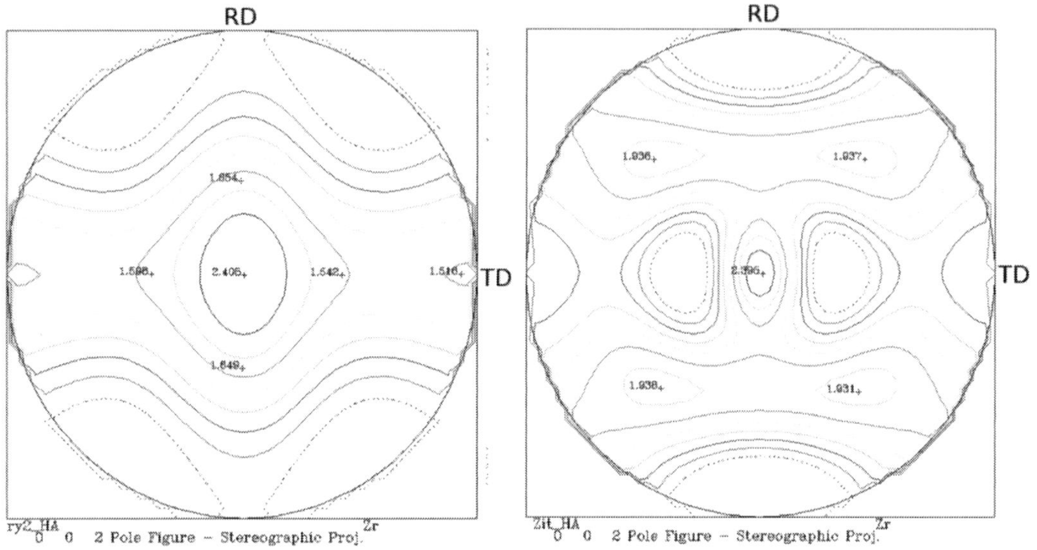

Figure 2. Direct pole figures (002) determined by means of Rietveld procedure (package GSAS) of sample Zry2 (left), and sample ZiT (right).

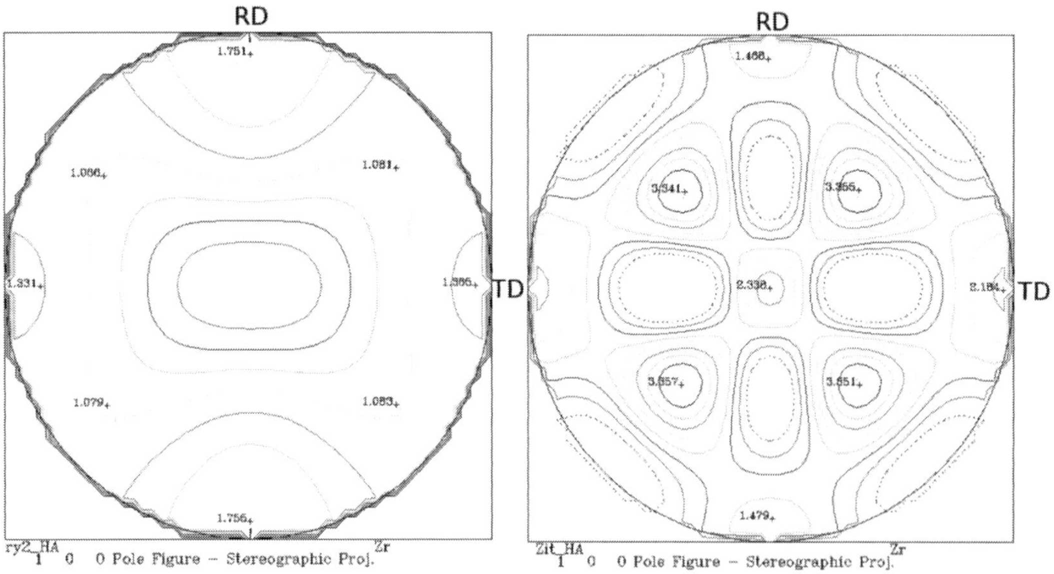

Figure 3. Direct pole figures (100) determined by means of Rietveld procedure (package GSAS) of sample Zry2 (left), and sample ZiT (right).

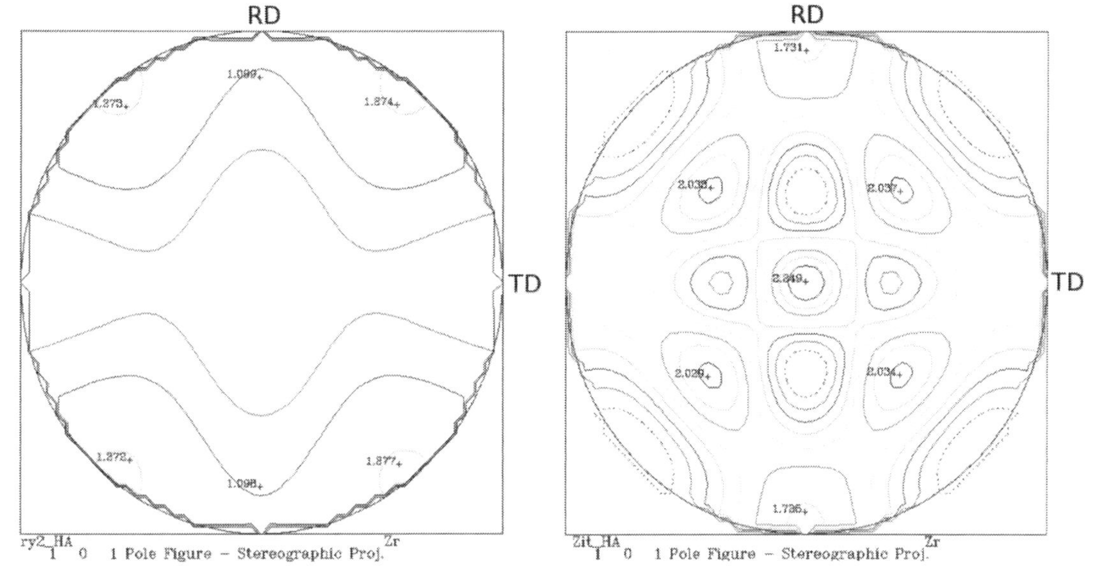

Figure 4 Direct pole figures (101) determined by means of Rietveld procedure (package GSAS) of sample Zry2 (left), and sample ZiT (right).

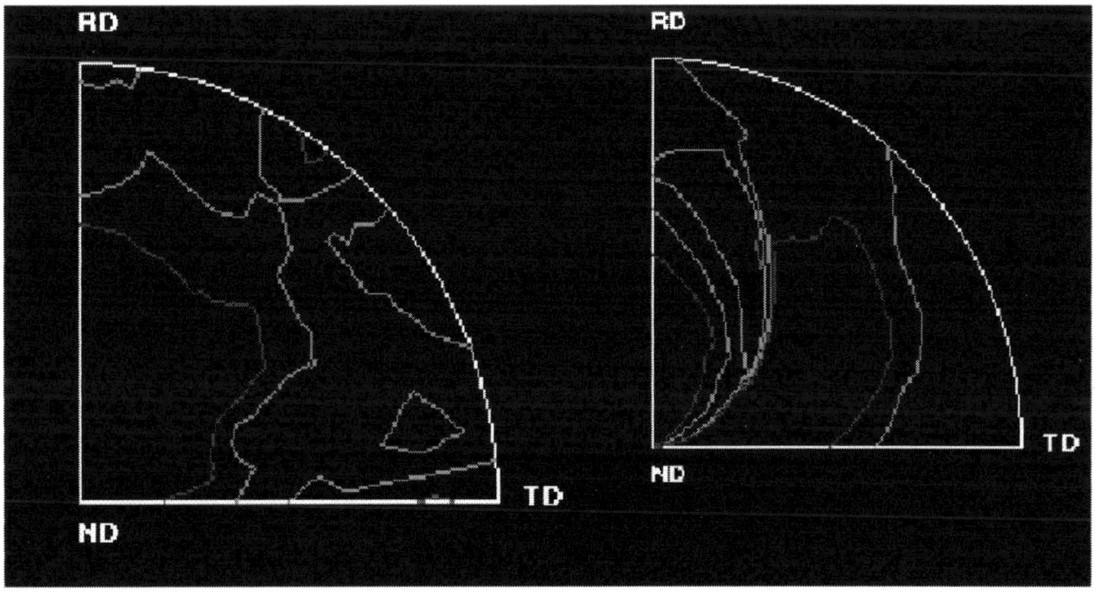

Figure 5. Direct pole figures (002) (obtained by using software package HEXAL) of sample Zry2 (left), and sample ZiT (right).

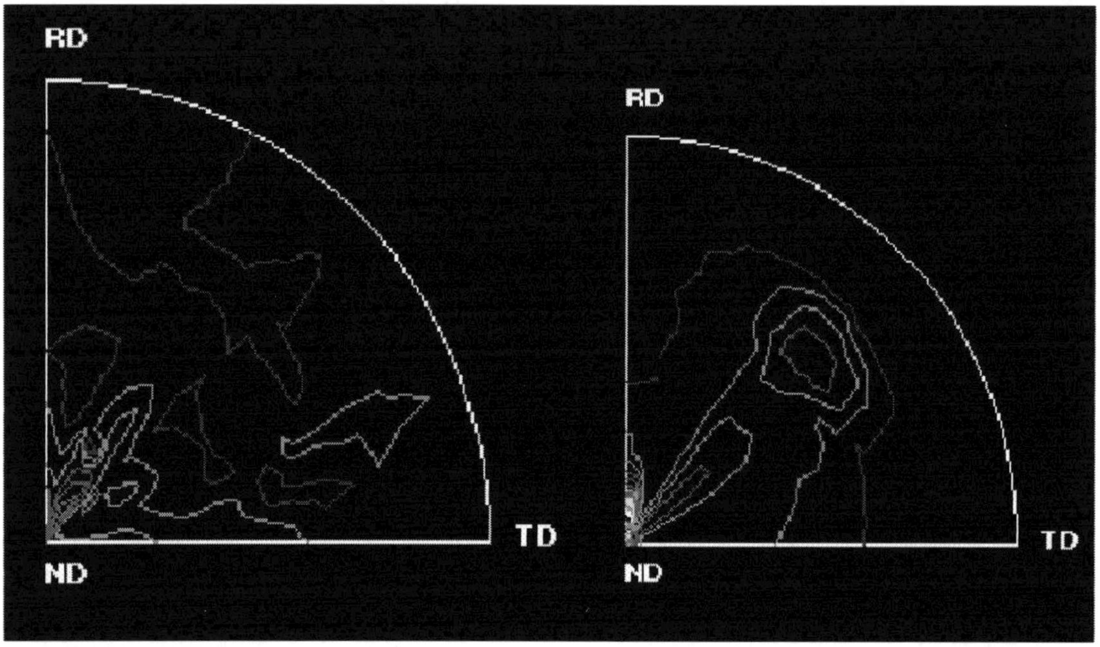

Figure 6. Direct pole figures (100) (obtained by using software package HEXAL) of sample Zry2 (left), and sample ZiT (right).

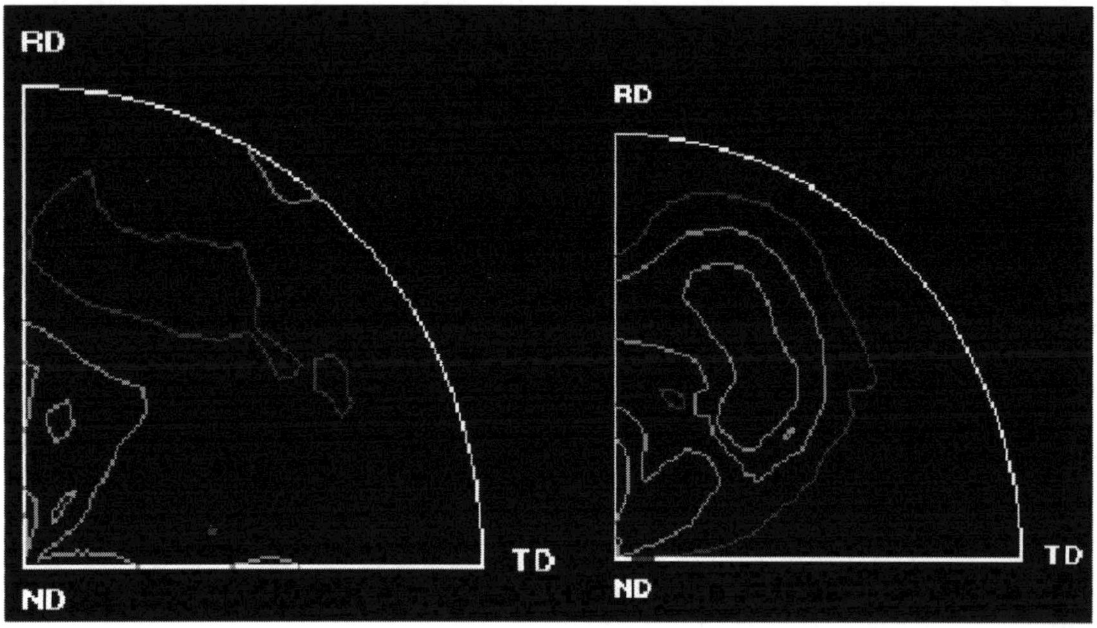

Figure 7. Direct pole figures (101) (obtained by using software package HEXAL) of sample Zry2 (left), and sample ZiT (right).

The ODF representations are shown at Figure 8 (sample Zry2), and Figure 9 (sample ZiT). Table 5 shows the ideal orientation of samples Zry2, and ZiT. ODF - f (g) values are also given at this Table.

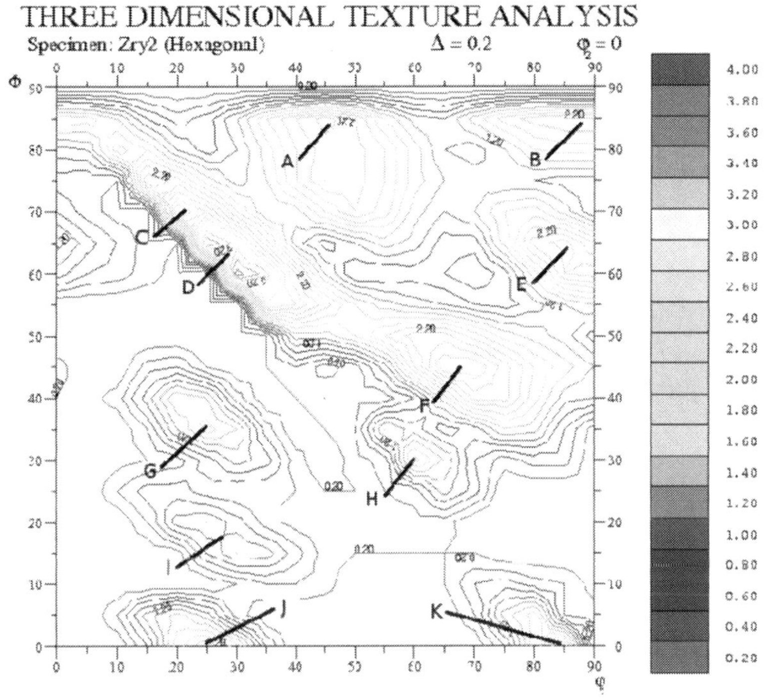

Figure 8. The orientation distribution (ODF) for $\varphi_2 = 0°$ calculated from measured pole figures (package HEXAL) of sample Zry2.

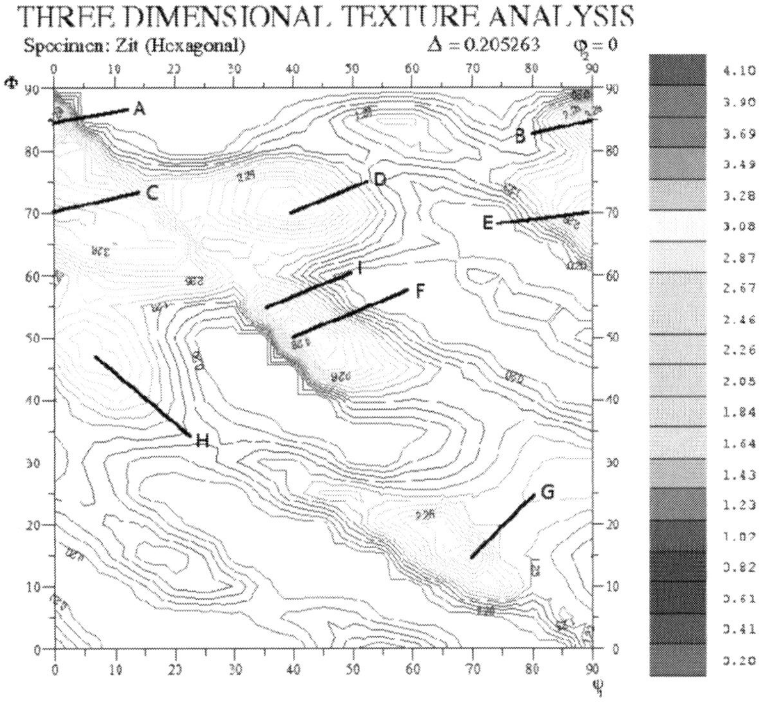

Figure 9. The orientation distribution (ODF) $\varphi_2 = 0°$ calculated from measured pole figures (package HEXAL) of sample ZiT.

Table 5. Ideal orientations, and ODF - f (g) values of samples Zry2, and ZiT.

Point	sample Zry2	ODF's values f(g)	sample ZiT	ODF's values f(g)
A	(110)[-555]	2.20	(110)[-550]	2.20
B	(110)[-115]	2.20	(110)[-115]	3.28
C	(111)[-321]	2.20	(111)[-550]	3.28
D	(112)[-311]	3.20	(111)[-523]	2.29
E	(112)[- 5-34]	2.20	(111)[-325]	2.26
F	(113)[-4-22]	2.20	(113)[-301]	3.28
G	(114)[-511]	1.20	(118)[-5-31]	2.26
H	(115)[-3-21]	1.20	(113)[-550]	2.26
I	(118)[-501]	1.20	(112)[-512]	2.26
J	(110)[-442]	1.20		
K	(110)[-114]	1.20		

The calculated inverse pole figures of zirconium samples Zrp, ZZ13, Zry2, ZrW, and ZiT are show at Table 6. There (0, 45) means direction between ND, and TD.

Table 6. Calculated inverse pole figures of samples Zrp, ZZ13, Zry2, ZrW, and ZiT.

	$p_{002, ND}$	$p_{002, (0,45)}$	$p_{100, RD}$	$p_{100, TD}$	$p_{110, RD}$
Zrp	3.01	N	2.74	1.73	1.68
ZZ13	2.78	N	2.60	0.98	3.12
Zry2	1.70	1.73	1.17	0.89	2.78
ZrW	1.16	1.81	3.77	0.75	0.72
ZiT	1.17	2.40	2.95	0.44	0.79

5. Discussion

Both methods used for direct pole figures calculations give comparable results (Figures 2 – 7).

The addition of alloying elements breaks the central area into two maxima (see basal pole figures - Figure 2, and Figure 5) [4].

A characteristic split of the basal pole maxima tilted from the radial toward the tangential direction for samples can be observed in basal pole figures for Zry2 (Figure 2), and ZiT (Figure 3). Tubes were mainly compressed in the radial direction, whereas the compressive forces in the tangential direction were comparatively small. Hence the resulting texture is consistent with the texture for sheets [3].

The middle orientation of cold-rolled zirconium is (002)[10-10] (Figure 2, and Figure 3). There is a tendency for [100] direction to align with the transverse direction.

The same results can be seen from inverse pole figures of tubes (Table 5).The (100) planes are oriented parallel to transverse direction. The (100) pole density maximum is tilted from the rolling direction towards the transverse direction.

The basal planes of tubes are oriented parallel to tube axis. The angle position of (002) pole maxima of sample Zry2 is tilted by 45° from the normal direction (Table 5). A similar situation occurs for sample ZiT, but the value of pole density is lower (Table 5).

The (110) planes are oriented perpendicular to rolling direction. Sample Zry2 shows the largest value of $p_{110, RD}$ ($p_{110, RD} = 2.78$, Table 5).

Sheet samples prefer orientation of planes (002) perpendicular to normal direction. The angle position of (002) pole maxima of sample ZrW is tilted by 45° from the normal direction (Table 5).

The (100) planes are oriented perpendicular to rolling direction. The (100) pole density maximum is tilted from the rolling direction towards the transverse direction.

An analogous situation can be observed for planes (110). These are also oriented perpendicular to rolling direction. Sample ZZ13 shows the largest value of $p_{110, RD}$ ($p_{110, RD} = 3.12$, Table 5).

From the ideal orientations, frequent occurrence of the (110)[-555], (110)[-115], and (118)[-5-31] is obvious (Table 4).

6. Conclusions

Our laboratory developed and tested experimental and computing techniques for quantitative texture analysis based on the ODF combined with the diffraction of thermal neutrons. Neutron diffraction is a very powerful and suitable tool for microstructure, and texture characterization of these zirconium based alloys, and generally grain-oriented materials.

Zirconium alloys have a very suitable physical, mechanics, corrosion or neutron absorption properties to be used in nuclear technique, and industry. From these reasons is very important to investigate the all scale properties of zirconium, and zirconium alloys.

Our results in quantitative texture analysis are in agreement with the results the other authors in this branch of research [2], [8], [9]. Level of resulting texture and maxima position is different for sheets, and for tubes. Substantial influence also has a final tube thickness. The results obtained are characteristic for zirconium based alloys.

References

[1] Kruzelova M 2010 *Study of Zirconium Based Alloys by Neutron Diffraction - Master's thesis* (Prague: Czech Technical University).
[2] Murty K L 1989 *Eighth International Symposium* **1023** 570 – 595.
[3] Tenckhoff E 2005 Journal of ASTM International **2**.
[4] Barrett C S 1952 *Structure of Metals* (McGraw-Hill 2nd ed.).
[5] Von Dreele R 1997 *J. Appl. Cryst* **30** 517 – 525.
[6] Dlouha M, Kalvoda L, Vratislav S, Cech B 1991 *Kov. Mater. - Metall. Mater.* **29** 289-299
[7] Kallend J S, Kocks U F, Rollett A D, Wenk H 1991 *Textures and Microstructures* **18** 1203-1208.
[8] Nikulina A V 2004 *Metal Science and Heat Treatment* **46** 458 – 462.
[9] Hsun H 1974 *Texture of metals - Technical report* (United States Steel Corporation).

Acknowledgments

This research has been supported by the Ministry of Education, Youth and Sports of the Czech Republic grant MSM6840770021, the Ministry of Industry and Trade of the Czech Republic grant MPOFRTI1-378.

Journal of Physics: Conference Series **340** (2012) 012096 doi:10.1088/1742-6596/340/1/012096

In-situ neutron diffraction and acoustic emission investigation of twinning activity in magnesium

K Máthis[1], P Beran[2], J Čapek[1] and P Lukáš[2]

[1]Department of Physics of Materials, Faculty of Mathematics and Physics, Charles University, Prague, 121 16, Czech Republic
[2]Nuclear Physics Institute, Řež, 250 68, Czech Republic

E-mail: mathis@met.mff.cuni.cz

Abstract. The neutron diffraction and acoustic emission (AE) techniques have been used for *in-situ* investigation of deformation twinning activity in cast polycrystalline magnesium at room temperature. The combination of these two techniques results in obtaining complementary information about the twinning mechanism during the straining. It is shown that loading mode significantly influences the twin nucleation and the twin size. In tension, the twin nucleation is observed during the entire test, while in compression only at the beginning of the plastic deformation. Nevertheless, the overall twinned volume does not differ. The optical micrographs support the above mentioned conclusions.

1. Introduction

The magnesium and its alloys are widely used in transportation industry due to their excellent strength to density ratio. However, the poor formability of magnesium alloys represents a substantial limitation for applications. Thus the developing of material with better strength-ductility ratio is of great importance.

The poor ductility of magnesium is most commonly rationalized in terms of dominancy of basal slip. In hexagonal materials the mechanisms of plastic deformation depend on the ratio c/a of crystallographic axes. In magnesium alloys, this value is close to the ideal, i.e. to $\sqrt{8/3}$. Therefore, in magnesium, the main slip system comprises the basal plane (0001) with the close packed directions of $1/3\langle 11\bar{2}0\rangle$ [1]. While these vectors are perpendicular to the c axis, the slip in the *a* direction cannot produce strain parallel to the c axis. According to von Mises criterion [2], the activity of five independent slip systems is required for the plastic deformation of polycrystals, but this condition in magnesium is not fulfilled [3]. The basal and prismatic slip planes together dispose with 4 independent slip systems. Thus either the second order pyramidal system or deformation twinning must be activated. At ambient temperatures, the critical resolved shear stress of the second order pyramidal system is much higher than that for twinning [4]. Consequently, in the common case, the twinning is the only active deformation mode that can provide straining along the *c*-axis at room temperature [5]. The $10\bar{1}2$ $\langle 10\bar{1}\bar{1}\rangle$ twinning mode has been found as the most significant in magnesium alloys [6].

The main advantage of the neutron diffraction technique consists in the deep penetration depth of thermal neutrons – the diffraction data usually characterize the entire volume of the specimen, especially its internal stress state [7, 8]. Recently, the *in-situ* neutron diffraction method has been utilized also for characterization of deformation mechanisms in different materials as e.g. shape memory alloys [9, 10] or hexagonal materials [11]. It has been shown, that the integrated intensity of particular reflections is sensitive on the structural change (e.g. martensitic transformation) or reorientation of the crystal lattice (e.g. twinning, deformation texture) during the loading.

The acoustic emission (AE) technique yields information on the dynamic processes involved in plastic deformation of alloys [12]. The AE originates from transient elastic waves generated within a material owing to sudden localized irreversible structure changes. It has been shown [13] that deformation twinning and dislocation glide are the major sources of AE in hexagonal closed packed materials.

The advantage of both methods consist in giving information from entire volume in real time, contrary to the e.g. microscopy or electron backscattered diffraction methods, where only a small volume of the specimen is investigated in *post-mortem* state. Both, the neutron diffraction and the acoustic emission have been found as excellent method for study deformation processes in magnesium alloys [14, 15], since they provide complementary information: the AE is sensitive only to twin nucleation, whereas neutron diffraction characterize their growth as well [15].

The present work focuses on the study of twinning activity in cast polycrystalline magnesium with a random texture. The influence of the loading mode on twinning is discussed in detail. The influence of the twinning on AE signal and neutron diffraction pattern is also presented.

2. Experimental

Cast polycrystalline magnesium with 1 wt.% Zr content was used for the experiment. The specimens possessed a random texture and had an average grain size of 110 μm. Tensile testing was carried out using cylindrical specimens with a diameter of 5 mm and gauge length of 25 mm. Compression testing was performed using samples with a rectangular cross-section of 15 mm x 15 mm and height of 30 mm. Both tensile and compression testing took place at room temperature, at a strain rate of $2x10^{-4}$ s^{-1} and were finished at 4% strain. The strain was controlled by using an extensometer. A standard neutron powder diffraction instrument MEREDIT (Medium Resolution Diffractometer) [16] equipped with a loading frame was used for *in-situ* neutron diffraction measurements. The loading frame was installed in a vertical position, this arrangement provide information on evolution of radial component of the lattice strain tensor. Mechanical loading of the sample was interrupted for ~6 hours in order to collect the neutron diffraction pattern from the strained part of the specimen at a constant level of applied strain. Diffraction patterns were collected between 4 and 144° 2θ with step of 0.08° and time per step of 210 sec. FullProf Suite Tool [17] was used for the refinement of the measured diffraction patterns.

AE was monitored using a computer controlled PCI-2 device (Physical Acoustic Corporation). The facility incorporated a piezoelectric transducer with a flat response between 50 and 650 kHz and a preamplifier giving a gain of ~40 dB. The threshold level of detection was set as 35 dB.

The specimens for light microscopy were mechanically polished and finally etched (10 s) in a solution of 10% nitric acid.

3. Results

Tensile and compression true stress-true strain curves of specimens deformed to fracture, are shown in Fig. 1. The loading mode influences the shape of the deformation properties. The work hardening rate in compression is significantly higher at the beginning of the plastic deformation than that in tension. Both the yield stress and elongation exhibit an asymmetry. This, so called tension-compression asymmetry is well known [18] and it is connected with twinning phenomenon [18].

The deformation twinning in magnesium is accompanied by intensity variations between the *hk.l* diffraction peaks corresponding to the parent and twin grain families, as it was shown by Gharghouri et al. [19]. The direction of the intensity change depends on the loading mode and the applied experimental diffraction geometry, i.e. on the mutual position of the diffraction and applied load vectors. In the given experimental set-up we used the radial mode, where all diffraction vectors are perpendicular to the direction of the applied load. The diffraction patterns of the initial state and after

4% compression strain are shown in Fig. 2 measured on compression sample. The intensities of reflections {10.0} and {00.2} change evidently with increasing strain.

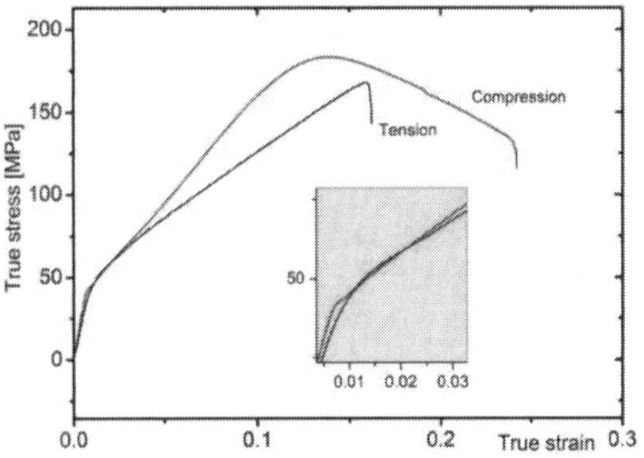

Figure 1. True stress-true strain curves deformed in tension and compression at room temperature.

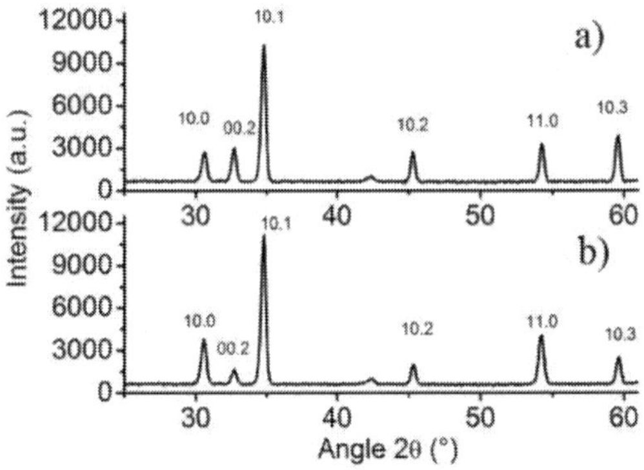

Figure 2. Diffraction patterns measured in compression a) initial state; b) after 4% strain.

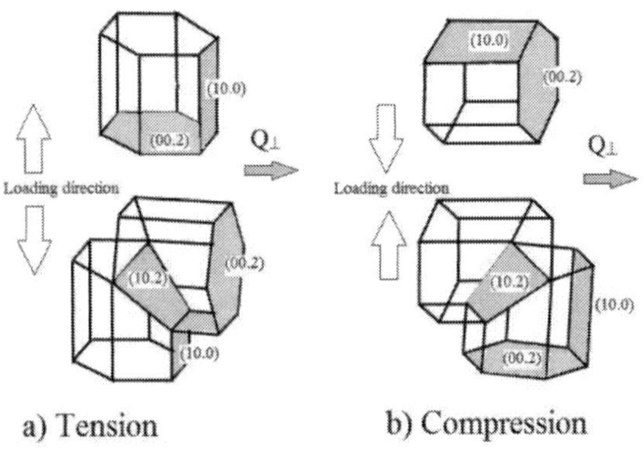

Figure 3. Schematic of the twinning mechanism with respect to the detector bank position a) in tension; b) in compression.

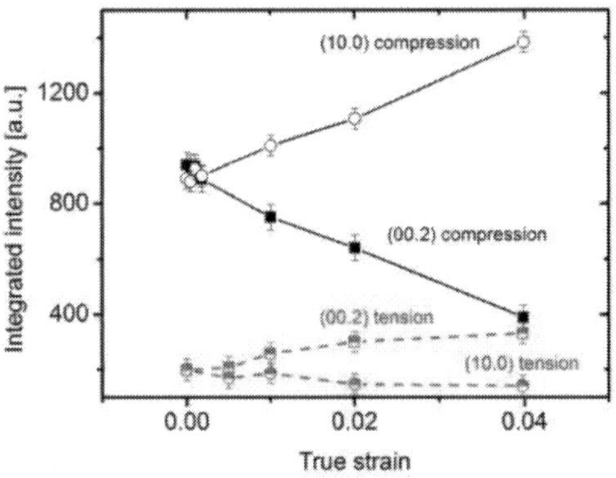

Figure 4. Evolution of the integrated intensity with strain for reflections (00.2) and (10.0) as a function of the loading mode.

In tension, the $10\bar{1}2$ twinning produces *c*-axis extension in grains, whose *c*-axis is parallel to the loading direction. Since this mechanism reorients basal planes by 86.3° (Figs. 3a and 4), an increase in the intensity of the {00.2} peak is expected. Consequently, the intensity of the parent {10.0} reflection decreases. On the other hand, during the compression test the prismatic {10.0} orientations have high Schmid factor for $10\bar{1}2$ twinning [20]. Therefore the intensities of the monitored peaks behave by the opposite way (Figs. 3b and 4). In order to compare the overall twinned volume during the tension and compression, respectively, we have normalized the measured integrated intensities. 0% is the integrated intensity of the particular peak measured in the initial state. Fig. 5 represents the absolute value (in percents) of the change of integrated intensity of the reflection {00.2} in comparison to the initial state as a function of the strain. It can be seen that the values are within the experimental error the same. The twinned volume in random textured cast magnesium alloys has been studied by several authors [21-24]. The numerical simulations indicate that in compression the twinned volume is larger (35% in compression vs. 10% in tension) [21, 24]. An opposite view was given by Xu et al. [23], who didn't find during their TEM investigation noticeable differences in the extent of twinning of the tensile and compressive AM50 samples. Our results support the latter opinion.

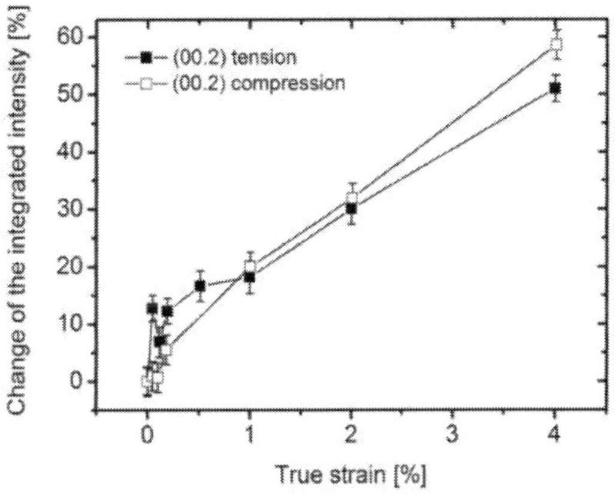

Figure 5. Change of integrated intensity of the reflection {00.2} in comparison to the initial state for both loading modes

The comparison of acoustic emission response for tension and compression depicted in Fig. 6 revealed a different evolution of twinning. The signal recorded in compression (Fig. 6a) reached its maximum at approx. 1% strain in the low stress plateau range. The jump in the intensity of the {00.2} reflection indicate that this maximum is connected with activation of the tensile $10\bar{1}2$ twinning system. Similar behavior has been observed in compressed cast magnesium by Klimanek and Pötzsch [25]. Using light microscopy they found a maximum in the apparent mean volume fraction of the twins in the range of 0.05 – 0.1ε. The rapid decrease of the AE signal after reaching of the maximum is caused by twin growth and change in the deformation mechanism. In compression, fast growth of $10\bar{1}2$ twins was observed [26]. The AE method is sensitive only to twin nucleation, not to twin growth [13], thus the AE signal decreased. The continuing of deformation is followed by activation of non-basal slip systems [27, 28]. This mechanism increases the dislocation density and consequently reduces the mean free path of moving dislocations. Since the latter parameter is directly proportional to the AE signal strength [12], the AE signal decreased under the detectable limit.

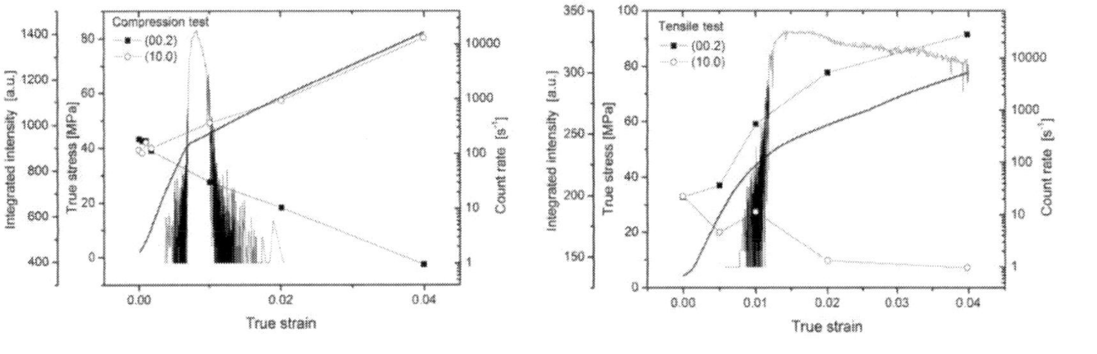

(a) (b)

Figure 6. Dependence of acoustic emission signal on loading mode (a) – compression, (b) – tension). The true stress-true strain curves are depicted in blue. The evolution of corresponding integrated intensities is represented by scatter charts.

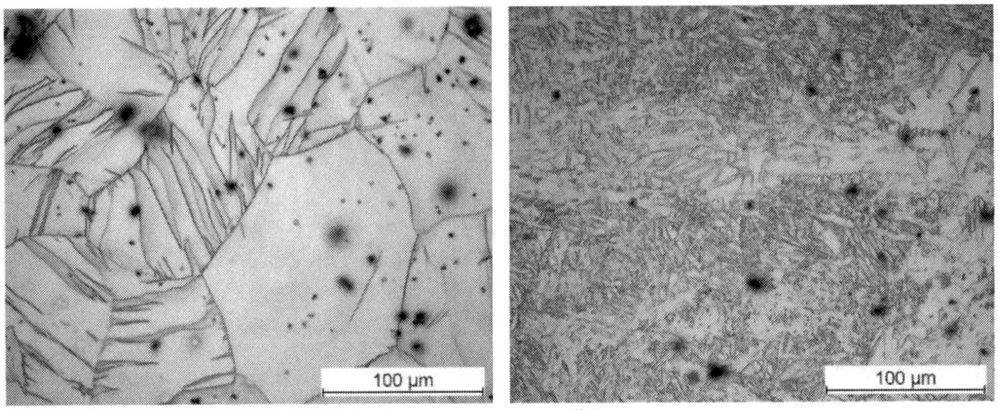

(a) (b)

Figure 7. Micrographs of samples deformed at room temperature in (a) compression and (b) tension. The loading direction is perpendicular to the image plane.

During the whole tensile test the burst AE signals were observed. After the reaching the maximum, the signal level decreases gradually (Fig. 6b). This behavior indicates that twin nucleation took place during the entire test. In tension, the nucleation of twins requires higher stresses and their growth is limited [29]. The variation of the integrated intensity corresponds to the AE findings. In the elastic region the intensity of the parent and twin plains is negligible. When the yield point is reached, a strong increase could be observed. The microstructures observed after both tension and compression tests support the above mentioned consideration. After the compression test large thick twins could be observed (Fig 7a). On the other hand, the tensile specimen exhibits a lot small twins (Fig 7b).

4. Conclusions

In the current study the deformation twinning in weak textured cast magnesium during both tensile and compression test was investigated. $10\bar{1}2$ tensile twinning was observed in both loading modes. The neutron diffraction measurements revealed that the overall twinned volume during the tensile and compression deformation does not differ within the experimental error. Nevertheless, differences in the twinning mechanism and twin size were found by means of acoustic emission measurement. In compression, the twin nucleation was observed entirely at the beginning of the plastic deformation. The subsequent straining caused twin growth. On the other hand in tension, significant twin nucleation was observed during the entire test. The light optical microscopy examination showed that the twin size in tension is significantly smaller.

Acknowledgement

This work is part of the research program MSM 0021620834 financed by the Ministry of Education of the Czech Republic. K.M. is grateful for the financial support of the Czech Grant Agency under the contract P108/11/1267.

References

[1] Roberts C S 1960 *Magnesium and Its Alloys* (New York John Wiley & Sons, Inc.)
[2] von Mises R 1928 *Z. Angew. Math. Mech.* **8** 161
[3] Lukac P 1985 *Czech. J. Phys.* **35** 275
[4] Chapuis A and Driver J H 2011 *Acta Mater* **59** 1986
[5] Yoo M H 1981 *Metallurgical Transactions a-Physical Metallurgy and Materials Science* **12** 409

[6] Barnett M R 2007 *Mat Sci Eng a-Struct* **464** 1
[7] Davydov V, Lukas P, Vrana M, Malard B, Pilch J, Maximov V and Sittner P 2010 *Mat Sci Eng a-Struct* **527** 3310
[8] Lukas P, Strunz P, Davydov V and Kuzel R 2009 *Powder Diffr* **24** S26
[9] Sittner P, Novak V, Landa M and Lukas P 2007 *Mat Sci Eng a-Struct* **462** 12
[10] Sittner P, Novak V, Lukas P and Landa M 2006 *J Eng Mater-T Asme* **128** 268
[11] Muransky O, Barnett M R, Carr D G, Vogel S C and Oliver E C 2010 *Acta Mater* **58** 1503
[12] Heiple C R, Carpenter S H and Carr M J 1981 *Metal Science* **15** 587
[13] Friesel M and Carpenter S H 1984 *Journal of Acoustic Emission* **6** 11
[14] Mathis K, Chmelik F, Janecek M, Hadzima B, Trojanova Z and Lukac P 2006 *Acta Mater* **54** 5361
[15] Muransky O, Carr D G, Sittner P and Oliver E C 2009 *Int. J. Plast.* **25** 1107
[16] *MEREDIT – Medium resolution neutron powder diffractometer at Nuclear Physics Institute in Řež* http://neutron.ujf.cas.cz/meredit
[17] Rodriguez-Carvajal J 1993 *Physica B* **192** 55
[18] Caceres C H, Lukac P and Blake A 2008 *Philosophical Magazine* **88** 991
[19] Gharghouri M A, Weatherly G C, Embury J D and Root J 1999 *Philos Mag A* **79** 1671
[20] Beyerlein I J, Capolungo L, Marshall P E, McCabe R J and Tome C N 2010 *Philosophical Magazine* **90** 2161
[21] Graff S, Steglich D and Brocks W 2007 *Adv. Eng. Mater.* **9** 803
[22] Jain J, Poole W J, Sinclair C W and Gharghouri M A 2010 *Scripta Mater* **62** 301
[23] Xu S, Gertsman V Y, Li J, Thomson J P and Sahoo M 2005 *Can. Metall. Q.* **44** 155
[24] Agnew S R, Yoo M H and Tome C N 2001 *Acta Mater* **49** 4277
[25] Klimanek P and Potzsch A 2002 *Mat Sci Eng a-Struct* **324** 145
[26] Knezevic M, Levinson A, Harris R, Mishra R K, Doherty R D and Kalidindi S R 2010 *Acta Mater* **58** 6230
[27] Agnew S R and Duygulu O 2005 *Int. J. Plast.* **21** 1161
[28] Clausen B, Tome C N, Brown D W and Agnew S R 2008 *Acta Mater* **56** 2456
[29] Koike J 2005 *Metall. Mater. Trans. A-Phys. Metall. Mater. Sci.* **36A** 1689

Journal of Physics: Conference Series **340** (2012) 012097 doi:10.1088/1742-6596/340/1/012097

Studying of PYSZ and FYSZ turbine blade coatings by small-angle neutron scattering

V Ryukhtin[1], B Saruhan[2], R Ochrombel[2], L Noirez[3] and A Wiedenmann[4]

[1] Helmholtz Zentrum Berlin for Materials and Energy, DE-14109 Berlin, Germany
[2] German Aerospace Centre, Linder Hoehe, DE-51147 Cologne, Germany
[3] Laboratoire Léon Brillouin, CEA, Saclay, 91191 Gif-Sur-Yvette Cedex, France
[4] Institut Laue-Langevin, 6 rue Jules Horowitz FR-38000 Grenoble, France

E-mail: vasyl.ryukhtin@helmholtz-berlin.de

Abstract. We present here results of small-angle neutron scattering (SANS) investigation of partially and fully yttria stabilized zirconia (PYSZ and FYSZ, respectively) turbine blade coatings (TBC). The samples have been prepared by electron beam physical vapour deposition method (EB-PVD) with various setup parameters. Structure parameters of porosity have been studied in-situ with high-temperature furnace. Temperature dependence of specific surface and anisotropy parameter of porosity has been obtained. Differences in character of microstructural morphology changing with in-situ thermal treatment as well as open and closed porosity ratio in PYSZ and FYSZ has been described and discussed.

1. Introduction

Turbine blade coatings (TBC) improve operational characteristics of advanced turbine blades. Engines protected with TBC blades exhibit increased their performance mainly due to the increase of gas-turbine inlet temperature [1]. Also coated blades have better corrosion resistivity and much longer life-time. TBC produced by electron beam physical vapour deposition (EB-PVD) technique has unique columnar microstructure. These columns are mainly perpendicular to surface and thus high-temperature stability is improved during operation temperature induced cyclic altering.
Conventional and most widely used material for TBC is Partially Yittria Stabilized Zironia - PYSZ (7 wt. % Y_2O_3 + ZrO_2). It is a relatively cheap composition and allows the so-called single-source-evaporation during EB-PVD processing relying on the very similar vapour pressure values of zirconia and yttria. However, the use of PYSZ becomes problematical as the inlet temperature in future engines reaches to temperatures as high as 1400 ^{0}C causing phase instability of the standard TBC-material [2, 3]. Use of Fully Yttria Stabilised Zirconia - FYSZ (14 wt.% Y_2O_3 + ZrO_2) allows the formation of more stable cubic phase and thus extending the application to higher operation temperatures. PYSZ exhibits better thermal shock and erosion resistances than FYSZ. Despite these facts, FYSZ seems to be a promising TBC-material due to its intrinsically lower thermal conductivity and phase stability for the envisaged higher operation temperatures. In general, the porous morphology and, thus the resulting thermal flux behaviour at EB-PVD TBCs can be altered by varying the deposition parameters such as substrate temperature, rotation speed of substrate and power used for evaporation of ingots, etc. [2].

Moreover, relying on above mentioned properties, it is expected that the coating morphology and thermal conductivity value can depend on the stabilisation level of yttria in zirconia.
SANS is widely used for investigation of zirconia-based ceramics [4-7]. This method provides information about bulk of studied material in contrast to *e.g.* electron microscopy (SEM, TEM). However, EB-PVD TBC deposits have a rather complex microstructure – wide range of sizes, high level of anisotropy and texturing of different kinds of pores, and as a consequence, the full analysis of SANS data is too complicated, it requires introduction of model with too many free parameters which are mutually independent. Nevertheless, neutron and X-ray small-angle scattering (SANS, SAXS) allows to characterize porosity in TBC materials through integral parameters of scattering function – apparent Porod constant and specific surface. One similar example of in-situ high temperature SANS measurements of plasma sprayed PYSZ TBC can be found in [9]. In present work, we show the influence of EB-PVD deposition parameters and the level of yttria stabilisation on morphology and porosity of the TBC coatings.

2. Experimental

"Von Ardenne" (Ardenne Anlagentechnik GmbH, Dresden/Germany) EB-PVD pilot equipment with max EB-power of 150 kW has been used for the preparation of the studied samples. Ingot powder has been evaporated and deposited at constant chamber pressure of about 0.8 Pa. The parameters during deposition (temperature in chamber and rotation speeds) are summarized in Table 1.

Table 1. Sample description.

Sample name	Stabilisation level Y_2O_3, wt %	Substrate rotation speed, rpm	Substrate temperature, °C	Thermal post treatment
PYSZ-03	7	3	1000	Virgin and aged at 100 hours at 1100 °C
PYSZ-12	7	12	950	Virgin and aged at 100 hours at 1100 °C
PYSZ-30	7	30	870	Virgin and aged at 100 hours at 1100 °C
FYSZ-03	14	3	950	Virgin and aged at 100 hours at 1100 °C
FYSZ-12	14	12	950	Virgin and aged at 100 hours at 1100 °C
FYSZ-20	14	20	975	Virgin and aged at 100 hours at 1100 °C

Materials have deposited on FeClAl-alloy substrate without having a bond coat. After deposition TBC specimens have been removed from substrate by means of etching in an acid solution. The thickness of final samples is about 0.4 mm. Such as-deposited TBC samples were used for in-situ SANS measurements at high temperature aging in vacuum. Scanning electron microscopy pictures are summarised in figure 1.
SANS measurements of studied FYSZ TBC samples in matching liquid have been conducted in HZB at V4, and partially at PAXY in LLB (Saclay) instruments. Mixture of D_2O (89 wt%) and H_2O (11 wt%) was used for "shadowing" of open pores since it has the same scattering length density (SLD=$5.55*10^{10}$ cm^{-2}) value. Measurements of PYSZ samples have been carried out at V4 instrument (HZB) with neutron wavelength of λ=6.05 Å and sample-detector (SD) distances of 1 m and 4 m. FYSZ TBC material was measured at PAXY (LLB) with neutron wavelength of λ=4.8 Å and sample-detector (SD) distances of 0.6 m and 5 m.

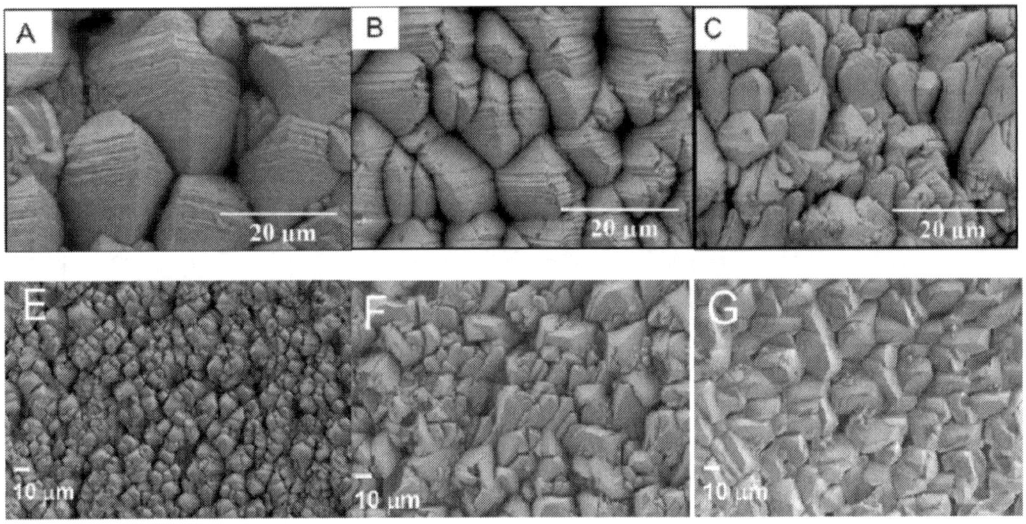

Figure 1. Scanning electron microscopy (SEM) images of PYSZ (A: PYSZ-03, B: PYSZ-12, C: PYSZ-30) and FYSZ (E: FYSZ-03, F: FYSZ-12, G: FYSZ-20) as-coated samples, growth direction is normal to image plane.

Experimental SANS data have been corrected for the instrument and sample holder and buffer background and calibrated by water scattering.

3. Results and discussion

3.1. In-situ high temperature SANS

Porosity structure of the studied coatings consists of three pore types – intercolumnar gaps sized about 0.5-2 µm, voids between feather-arms (100-500 nm) and intra-porous arrays (5-300 nm) [5]. Although SANS Q-range covers only nano-sized porosity *i.e.* about 1-50 nm for the both presented here SANS measurements, specific surface of whole porosity is reachable in measured high Q in so-called Porod regime [9]. Thus, a simple model of ellipsoids of rotation with preferred orientation has been chosen in order to follow change of anisotropy induced by in-situ temperature treatment. We have used SASProfit program for fitting of 2D scattering patterns [10]. This software allows carrying out fitting of 2D SANS data simultaneously for sample-to-detector distances 1 m and 4 m. Neutron beam has been monochromatized by means of velocity selector at wavelength 6.05 ′ with $\Delta\lambda/\lambda = 10$ %.

Irreversible decreasing of porosity anisotropy in given specimen during heat treatment can be observed in example of two-dimensional SANS diffraction pictures of in-situ measured in high temperature furnace (HTF-2) of as-coated PYSZ-12 sample (see in figure 2).

It should be mentioned that specimens were exposed by heat treatment in evacuated environment (at about 10^{-3} bar), however, usual heat aging procedure assume to be conducted at ambient air pressure. We expect that environmental pressure difference can just slightly influence sintering process and, thus, porosity structure of the end-point coatings.

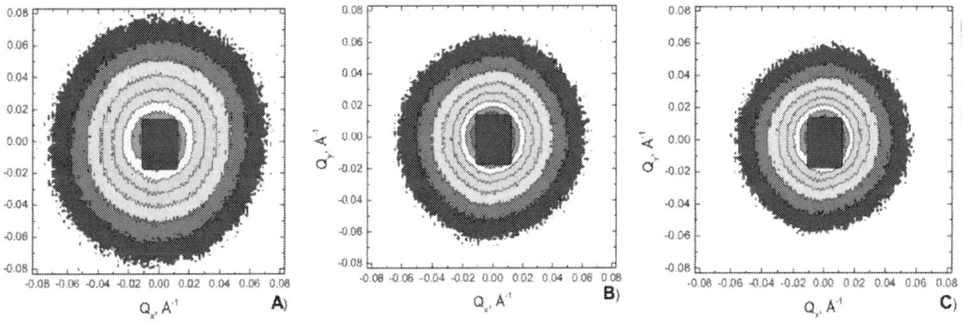

Figure 2. *In situ* SANS pattern of PYSZ-12 sample measured before high-T treatment at room temperature (A), during treatment at 1100 ^{0}C (B) and after treatment at room temperature (C).

Fitted aspect ratio (r/R_0) values for PYSZ and FYSZ samples are shown in figure 3. Obtained values of aspect ratio are highly dependent on sample orientation with respect to incident neutron beam because of high anisotropy of inter columnar pores. For both samples groups fitted r/R_0 values behave similarly with temperature. These anisotropy parameters keep constant quantity at the beginning of heating, then smoothly decrease and stay unchanged during cooling from maximum reached (1100 ^{0}C) down to room temperature. However, observed decrease of the r/R_0 ratio (see figure 3) starts at lower temperature in FYSZ (800 ^{0}C) than in PYSZ (900 ^{0}C) coatings. But, on the other hand, fitted specific surface values dependence on temperature is different for PYSZ and FYSZ materials. Porosity specific surface in partially yttria stabilised coatings slowly decreases or stays constant at the first stage of heating (up to T ~ 700 ^{0}C) and then decreases. In case of FYSZ, material specific surface significantly increases for temperatures up to T~600 ^{0}C and then decreases (700 ^{0}C < T ≤ 1100 ^{0}C). Finally, the measured specific surfaces in all samples slightly increase after cooling, it could be ascribed to thermal shrinkage and as consequence changing of scattering contrast, identical results have been described for $La_2Zr_2O_7$ and $SrZrO_3$ in '[11]'.

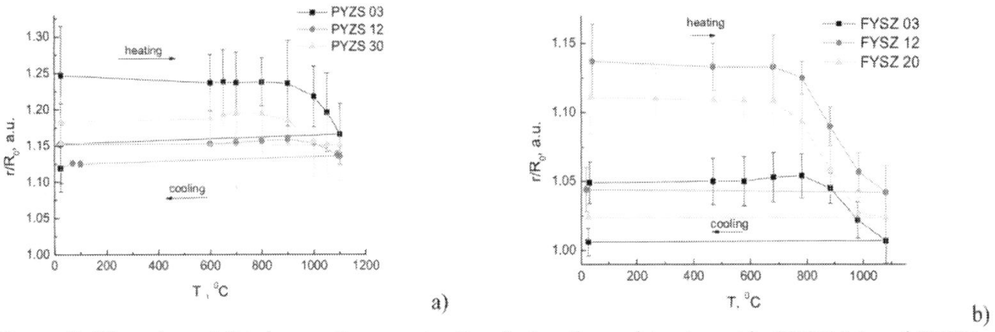

Figure 3. Changing of fitted porosity aspect ratios during thermal treatment in PYSZ (a) and FYSZ (b).

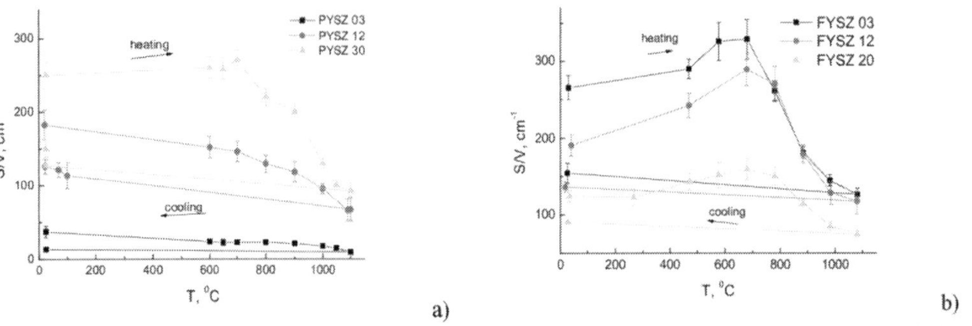

Figure 4. Fitted specific surface parameters from in-situ SANS during thermal treatment in partially-(a) and fully-(b) yittria stabilized zirconia coatings.

3.2. SANS with SLD matching liquid

As was mentioned above, the method of EB-PVD causes that most of pores in as-deposited material are highly anisotropic [12] thus, 3D analysis should be performed in order to obtain proper surface area values. Moreover, size range distributions of pores are rather wide, mutually overlapping and thus porosity model requires too many free parameters, which could give faulty results. Therefore, radially averaged apparent Porod constant [13] was used for qualitative characterization of total (*i.e.* open and closed) and closed porosity in PYSZ and FYZS samples. Apparent Porod constants were obtained from radial averaged SANS data measured for virgin (as-deposited) and thermally aged (100 hours at 1100 ^{0}C in air) PYSZ and FYSZ coatings. The rotation speed has strong influence on porous morphology in EB-PVD TBCs [3] and thus on different thermal resistivity properties. The open and closed porosity at the EB-PVD TBCs can be differentiated by employing comparable SANS analysis in scattering length density (SLD) matching liquid. Figure 6 shows obtained apparent Porod constants from the SANS data of heat-treated EB-PVD FYSZ coatings deposited with different rotation speeds (filled symbols in figure 6). Addition of matching liquid (D_2O/H_2O in 89/11 mass %) during SANS measurement results in SANS curves with lower intensity (open symbols in figure 6) due to "shadowing" of open porosity. Higher incoherent background level caused by hydrogen presented in light water.

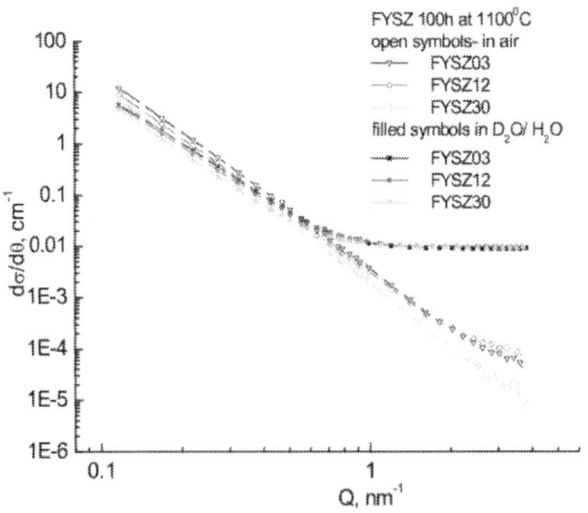

Figure 5. SANS scattering of FYSZ TBC samples measured in air (a) and in matching liquid (b).

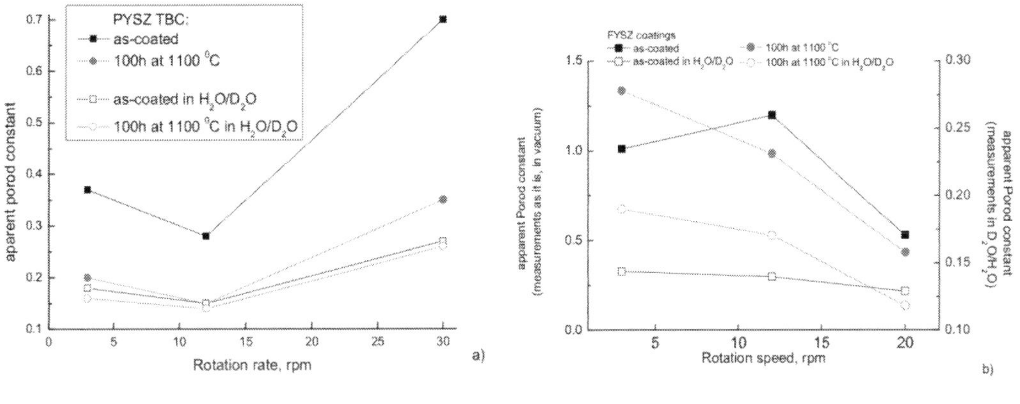

Figure 6. Apparent Porod constants of total and close porosity in PYSZ (a) and FYSZ (b) TBC samples.

Dependencies of these apparent Porod constants values on rotation speed during deposition for FYSZ are shown in figure 6. High-Q region of SANS data (Q>0.5 nm^{-1}) were used for fitting with the help of SASFit program [10]. Incoherent background was included in scattering procedure for data measured in D_2O/H_2O.

In all studied specimens significant reduction of specific surfaces was detected after sintering (see figure 6). It was obviously caused by smoothing of porosity surface due to thermal aging, as was observed by SEM earlier [5]. In PYSZ (so-called feather-arms) growth of secondary columns was found to relate with the vapour flux and the rotation axis in respect of the vapour incident angle (VIA). The rotation of the substrates produces a “sunrise-sunset” vapour incidence pattern, that determines the deposition rate over a differential area as a function of time. Thus, the interplay between neighbouring column tips and VIA of the incoming particles enhances the dominance of the shadowing effect over the surface diffusion. This shadowing phenomenon enhances the formation of secondary columns [14]. The data obtained on the EB-PVD PYSZ TBCs by employing the same experimental method display an increasing trend of the open porosity population with the rotation speed [12]. However, the highest rotation speed employed during the deposition of EB-PVD FYSZ results in the formation of the lowest total porosity population. This opposite tendency may be related to the difference in the preferred column growth direction in cubic and tetragonal phases.

Obviously, scattering from closed pores should be reduced more significantly with thermal aging then it was measured. This could be attributed with converting of open feather-arms to close porosity.

4. Conclusions

Deposition parameter induced changes of porous morphology in PYSZ and FYZS coatings has been measured and demonstrated by in-situ high temperature SANS experiments and SANS measurements with shadowing of open pores utilizing matching liquid immersion technique. FYSZ coating samples exhibited lower anisotropy of bulk porosity in comparison to PYSZ. Moreover, the reduction of average pore aspect ratios starts occurring at lower temperature in FYSZ (~800 ^{0}C) than in PYSZ (~900 ^{0}C). As the apparent specific surface area increases in FYSZ at the first stage of thermal treatment, does not change in PYSZ. Specific surface area value is highest in PYSZ coatings grown using the fastest substrate rotation speed, whereas, this is the lowest in FYSZ, which could be ascribed to multiple growth direction of material with tetragonal phase as often encountered in PYSZ and cubic – which is the overwhelming phase at FYSZ.

Acknowledgement

This work is supported by the DFG under the contract Nos. BA 1170/12-1, SA 1343/4 and WI 1151/5-1. The authors acknowledge with gratitude the contributions of Dr. A. Flores Renteria through his PhD study and of Dr. Dirk Wallacher for his extensive support with operation of high temperature equipment for SANS.

References

[1] DeMasi-Marcin J T and Gupta D K 1994 *Protective Coatings in the Gas Turbine Engine, Surface and Coatings Technology* **68/69** 1-9.
[2] Levi C G 2004 Current Opinion in Solid-State and Mater. Sci. **8** 77-91.
[3] Peters M, Schulz U, Saruhan-Brings B and Leyens C 2005 *Proc. 17th Int. Symp. On Air Breathing Engines (ISABE), The American Institute of Aeronautics and Astronautics Inc.* 1081.
[4] Renteria A F, Saruhan B, Schulz U, Raetzer-Scheibe H J, Haug J and Wiedenmann A 2006 *Surface and Coatings Technology* **201 (6)** 2611-2620.
[5] Renteria A F 2006 *Dissertation*, Fakultät für Georessourcen und Materialtechnik RWTH Aachen.
[6] Ryukhtin V, Saroun J, Harjo S, Motohashi Y, Wiedenmann A and Strunz P 2004 *Physica B: Condensed Matter* **350 (1-3)** E1019-E1022.
[7] Strunz P, Schumacher G, Vassen R and Wiedenmann A 2004 *Acta Materialia* **52 (11)** 3305-3312.
[8] Allen A J, Krueger S, Kerch H M, Hahn H and Skandan G 1996 *Nanostr. Mater.* **7** 113–126.
[9] Porod G 1982 in: Small Angle X-ray Scattering, *Academic Press* 17.
[10] Saroun J 2000 *J. Appl. Cryst* **33** 824-828.
[11] Strunz P, Schumacher G, Vaßen R, Wiedenmann A and Ryukhtin V 2006 *Scripta Materialia* **55 (6)** 545-548.
[12] Saruhan B, Ochrombel R, Ryukhtin V and Wiedenmann A 2009 *Advanced Engineering Materials* **11-6** 488-494.
[13] Ilavsky J, Long G G, Allen A J and Berndt C 1999 *Materials Science and Engineering A* **272** 215-221.
[14] Cho J, Terry S G, LeSar R and Levi C G, 2005 *Material Science and Engineering A* **391(1-2)** 390-401.

Investigation of metal-matrix composite containing liquid-phase dispersion

P Strunz[1], D Mukherji[2], R Gilles[3], T Geue[4] and J Rösler[2]

[1] Nuclear Physics Institute ASCR, CZ-25068 Řež near Prague, Czech Republic
[2] TU Braunschweig, Institut fuer Werkstoffe, Langer Kamp 8, 38106 Braunschweig, Germany
[3] TU München, Forschungsneutronenquelle Heinz Maier-Leibnitz (FRM II), Lichtenbergstrasse 1, 85747 Garching, Germany
[4] Laboratory for Neutron Scattering, Paul Scherrer Institute, 5232 Villigen PSI, Switzerland

E-mail: strunz@ujf.cas.cz

Abstract
Al-Pb binary system is a suitable model system for testing liquid phase dispersion strengthening in bulk materials for structural applications. Liquid Pb islands can be finely dispersed in still solid Al matrix due to the substantial difference of melting points. The Al-Pb system prepared by means of Equal Channel Angular Pressing (ECAP) process was investigated by Small-Angle Neutron Scattering technique (SANS) which enables in-situ measurement of size and morphology parameters of Pb inclusions at elevated temperatures. It was observed that the lead particles were elongated roughly in the direction of ECAP. During the subsequent in-situ thermal cycle RT-400°C-RT, the elongated Pb particles transformed to nearly spherical shape. The change of scattering contrast during melting of Pb mapped the transform of the confined lead particles to the liquid phase. The center of the transition region is around 342°C (615K) for both the as-cast and the ECAP deformed samples, which is a significant shift with respect to the free Pb melting point 327°C (600K). For the ECAP sample, the transition is not sharp, indicating a broad size distribution of lead particles.

1. Introduction

Over many decades, a large number of investigations were devoted to find ways to harden metallic materials by introducing a dispersion of second phase particles through precipitation or as dispersoids [1]. To obtain maximum strengthening at ambient temperatures, strong particles with large resistance against cutting by dislocations are required. However, at elevated temperatures, this aspect becomes less important as the dislocations are able to circumvent the particles by climb. The theory predicts that soft particles are particularly attractive because they facilitate relaxation of the dislocation stress field at the particle, thus ensuring a large attractive interaction between particle and dislocation [2-4]. In fact, this concept is utilized in potassium doped tungsten wire filaments, where evaporation of potassium leads to the formation of vapor-filled bubbles about 9 nm in diameter at service temperature [5]. Similarly, one would expect effective dispersion strengthening in bulk materials for structural applications if stable voids or a liquid phase could be finely distributed in the solid matrix.

It is, however, not trivial to test this hypothesis, as a fine and stable dispersion of liquid phase in a metal matrix is not easy to obtain. Al-Pb binary system can be a model system in this respect and was chosen to test this hypothesis. Pb (fcc, lattice constant a_{Pb} = 0.495 nm) has negligible solubility in Al (fcc, a_{Al} = 0.405 nm) in the solid state and no intermediate phases form in Al-Pb binary system [6]. Importantly, the melting points of Al (T_M=660.32°C) and Pb (T_M=327.46°C) are very different, therefore at high temperatures (>400°C) Pb can remain as a liquid dispersion in solid Al matrix. It has been reported that T_M of Pb increases in the confinement of the Al matrix and this increase is particle size dependent [7]. The melting of Pb nanoparticles was studied by means of dynamic mechanical analyzer / internal friction measurements [8] and transmission electron microscopy (TEM) [6].

Structural, thermodynamic and diffusion properties of crystalline particles completely embedded in a metallic matrix have been investigated in detail in a variety of homogenous systems of immiscible elements for the past two decades. The monotectic binary Al-Pb system, which has a large miscibility gap in the liquid phase and extremely limited mutual solubility in the solid phase, has been a key-type system for such investigations. Shapes and sizes of nanoscale Pb inclusions and their melting in Al have been studied by several authors, see e.g. [9, 10]. Nevertheless, neutron scattering carrying bulk information was not – to the autrhors' knowledge – yet employed.

However, Pb particles are very coarsely dispersed in Al matrix in the as cast Al-Pb system. For any interaction with the dislocations, a fine dispersion of the Pb particles within the grains is needed. Also for mechanical property tests, this dispersion should be available in bulk materials and not only on thin films. Equal Channel Angular Pressing (ECAP) was used to produce a uniform and fine dispersion of Pb in the present study.

As cast and ECAP deformed Al samples containing 4wt.% of Pb were investigated by small-angle neutron scattering (SANS) both ex situ and in situ at temperatures in the vicinity of the Pb melting point. The aim was (i) to test a possibility to investigate microstructure of such system in situ by SANS, (ii) to confirm T_M shift of the confined Pb particles and (iii) eventually conclude on details of microstructure during thermal cycle around solid – liquid transition temperature of lead particles.

2. Experimental

Al-4wt.%Pb system was melted in a vacuum arc furnace (VAF) with non-consumable tungsten electrode. High purity iron free aluminum (99.9995%) and lead (99.8%) was used for melting. It was seen that even a very small Fe impurity in Al resulted in a Fe-Pb eutectic formation in the cast system. The microstructure of the as cast Al-4%Pb is shown in Fig. 1 and it can be seen that the Pb particles are coarse (in μm scale) and dispersed mainly at the grain boundaries and at triple points (i.e. where three grains touch). In order to refine the Pb particle dispersion, cast bars of 10 mm x 10 mm square cross-section 50 mm long were processed by equal-channel angular pressing. ECAP was done in a die with 90° angle (Φ) following the so called route B (sample rotated 90° between each pressing [11]. Load of ~50 kN was needed for the pressing. The bars were smeared with Molykote BR2+ lubricant prior to pressing.

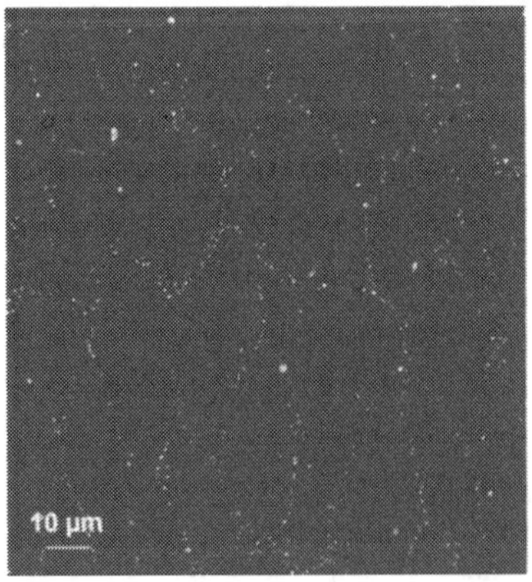

Figure 1. Dispersion of Pb in the as cast Al-4%Pb system.

Fig. 2 shows scanning electron microscopy (SEM) images of a Al-2.5%Pb system with increasing number of identical ECAP passes (up to 40). It can be seen that the Pb particles are elongated after ECAP. The particle sizes were measured from the micrographs assuming an ellipsoidal shape. The variation in particle sizes with increasing number of ECAP passes are shown in Fig. 3. It shows that ECAP considerably refines the as cast particle size and also particles are dispersed more homogenously. However, saturation is reached after ~20 passes (in the case of Al-2.5%Pb system) and further ECAP processing actually increases the particle size. In total, 8 ECAP passes were adopted in the Al-4%Pb system for optimal refinement. The microstructure after 8 passes is shown in Fig. 4.

ECAP deformed Al-4%Pb and pure Al samples were investigated by small-angle neutron scattering (SANS) both ex situ at RT and in situ at temperatures in the vicinity of Pb melting point. The measured samples are reported in Table 1.

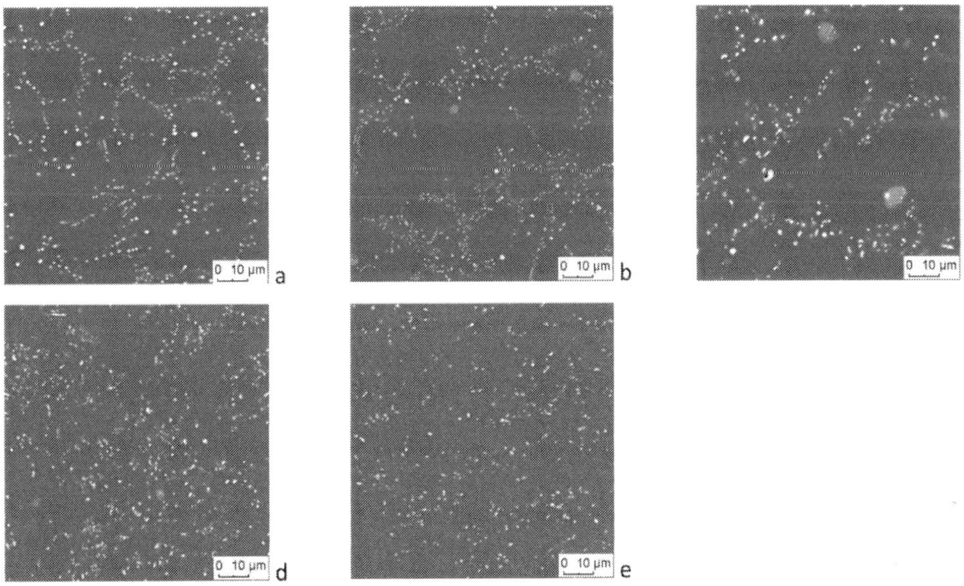

Figure 2. Pb dispersion in as cast sample and after different number of ECAP passes in Al-2.5%Pb system. The particle dispersion is shown with increasing ECAP passes. a) as cast, b) 4 pass, c) 8 pass, d) 16 pass and e) 40 pass.

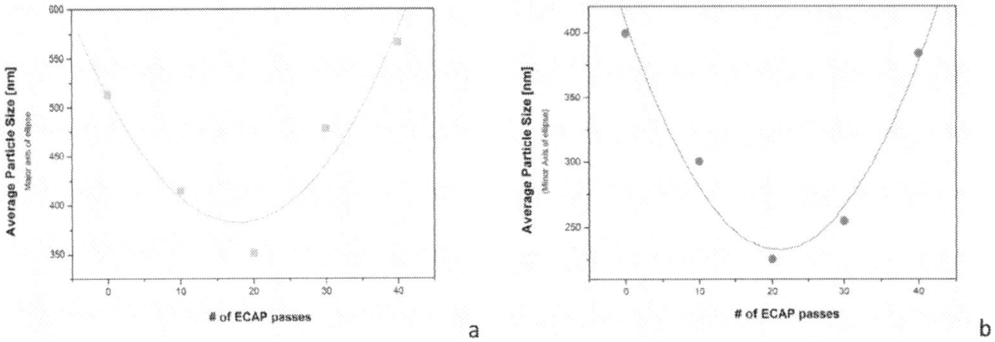

Figure 3. Evolution of Pb particle size versus increasing number of ECAP passes in Al-2.5%Pb system: (a) major axis of the ellipsoid, (b) minor axis of the ellipsoid

Figure 4. Pb dispersion in ECAP (after 8 pass) Al-4%Pb sample.

Table 1: Samples used for SANS measurements in different orientations

Sample	ECAP	Measurement mode (ex-, in- situ at HT)	thermal treatment:	orien-tation	Size (Å)	
					smaller particles	larger particles
as cast Al-4%Pb No. 2	No	ex situ	no thermal treatment	1, 2	1340	5670
ECAP Al-4%Pb, bar 3	Yes	ex situ	no thermal treatment	1,2,3	1490 (short axis), 1970 (long axis)	4790 8410
ECAP Al-4%Pb, bar 3, A	Yes	ex situ	RT-400°C, **1** cycle, prior to SANS	1, 2	1480 1480	5270 5870
ECAP Al-4%Pb, bar 3, B	Yes	ex situ	RT-400°C, **2** cycles, prior to SANS	1, 2	1480 1480	5270 5870
Pure Al with ECAP	Yes	ex situ	no thermal treatment	1, 2	1630 2200	4770 7370
As cast Al-4%Pb, bar 2,	No	in situ at HT, 1 cycle	RT-400°C, 1 cycle	1	1340	5670
ECAP Al-4%Pb, bar 3, A	Yes	in situ at HT, 1 cycle	RT-400°C, **1** cycle	1	1480 1480	5270 5870
ECAP Al-4%Pb, bar 3, B	Yes	in situ at HT, 2 cycles	RT-400°C, **2** cycles	1	1480 1480	5270 5870

Orientation 1: beam parallel to ECAP direction

Orientation 2: beam perpendicular to ECAP direction

Orientation 3: beam perpendicular to ECAP direction; turned by 90° (χ angle, i.e. around the ECAP direction) with respect to the orientation 2

Neutron scattering measurements were carried out at the SANS-II instrument of SINQ user laboratory of Paul Scherrer Institute (PSI), Villigen, Switzerland [12]. Samples in the form of cubes (approximately 10 x 10 x 10 mm^3) were used and a vacuum furnace was employed for the in-situ heating. The form of the sample enables its rotation by 90° around the sample axes and thus measurements at different orientations with respect to ECAP direction. The sample thickness of 10 mm is neither a problem from the transmission point of view (Al matrix practically does not absorb neutrons), nor from the multiple scattering point of view (scattering contrast is low and thus the multiple scattering does not play a role).

The scattering data were collected at several geometries; the sample-to-detector distance (SDD) varied from 1.2 m to 6 m and the neutron wavelength λ from 4.55 Å to 19.6 Å. The full covered range of the scattering vector magnitude Q = $|\mathbf{Q}|$ was approximately 2×10^{-3} Å^{-1} to 0.2 Å^{-1} (i.e. 2×10^{-2} nm^{-1} < Q < 2 nm^{-1}), where the magnitude $Q = |\mathbf{Q}| = |\mathbf{k}\ \mathbf{k}_0|$, $\mathbf{k}_0$ and $\mathbf{k}$ being the wave vectors of the incident and scattered neutrons, respectively, and $|\mathbf{k}| = |\mathbf{k}_0| = 2\pi/\lambda$ due to the elastic scattering. The measured raw data were corrected for background scattering and calibrated to absolute cross section.

Due to the relatively large size of Pb particles (hundreds of nanometers) - from the point of view of SANS - the important measurements were the ones at the lowest Q values, i.e. with SDD equal to 6 m and λ= 19.6 Å. This experimental geometry allowed – due to the rather low primary beam intensity – to remove the beamstop without a danger of overloading the detector. The used sample background subtraction ([13], see section 3.4 and scheme in Fig. 1 therein) enables evaluation of the scattered data even in the region of the primary beam. Scattering in or near this region provides information even on the large particles of size several hundreds of nanometers, although no details about size distributions can be extracted. It cannot be excluded, that there exist still larger particles (of micrometer size), scattering of them cannot be, however, distinguished (by scattering curve profile) from the primary beam.

The scattered intensity of the neutrons is proportional to the volume fraction as well as to the scattering contrast $(\Delta\rho_{SLD})^2$ (i.e. the square of the difference between scattering-length densities (SLDs) of the particles and the matrix, see e.g. [14]). The SLD of the particles and matrix depends on their composition, which is known precisely in this case, and on their densities (ρ_{Al} = 2.702 g/cm^3 at 293 K, ρ_{Pb} = 11.34 g/cm^3 at 293 K). At RT, the SLD of Pb is 31.02×10^9 cm^{-2} and SLD of Al is 20.77×10^9 cm^{-2}. Therefore, the scattering contrast at RT $(\Delta\rho_{SLD}(293K))^2$ is equal to 1.051×10^{20} cm^{-4}.

3. The cause of the possible SANS pattern change

There are expected changes in the mass density and shape of Pb particles during thermal cycle applied on Al-Pb composite (both as cast and ECAP processed). These changes can affect both the measured intensity and also the shape of 2D scattering curves.

The dominant cause for a change of scattered intensity is the change of the scattering contrast. The other possibilities (change of the specific surface, change of the size of particle due to the thermal expansion during heating or change of the size during phase transformation) are estimated to be small in comparison to this effect.

Change of the scattering contrast is connected to the density changes in dependence on temperature due to (i) thermal expansion coefficient difference (23.1 μm·m^{-1}K^{-1} for Al and 28.9 μm·m^{-1}K^{-1} for Pb at 25°C) and (ii) step-like change of average spacing of Pb atoms during phase transformation (density of liquid Pb at melting point is 10.66 g/cm^3).

The scattering contrast change due to the different thermal expansion coefficients of Al and Pb is estimated to be altogether not larger than 3% in the investigated high temperature range 300-400°C.

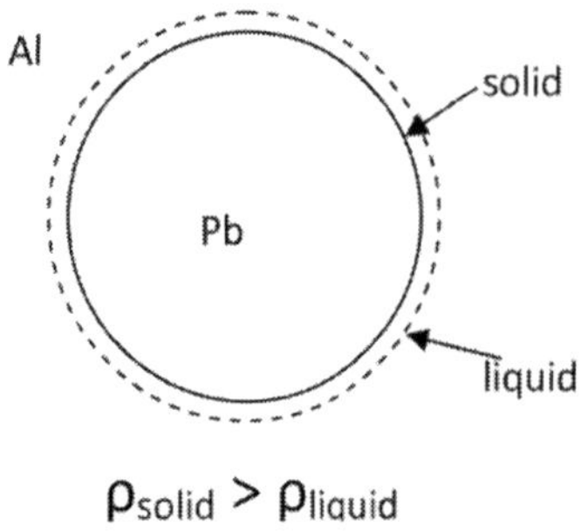

Figure 5. Scheme of the Pb particle expansion on solid-liquid transformation.

On the other hand, a stepwise expansion of Pb inclusion at its melting temperature (Fig. 5) causes large volume (and thus also density) change between the solid and liquid Pb. This volume change needs to be accommodated by plastic or elastic deformation (or both) of the surrounding Al matrix. It may eventually also need diffusional rearrangement of atoms in the Al matrix in case of particle shape changes. It is known [6] that thermal vacancies become mobile in Al at temperatures above 200°C, which enables such rearrangement.

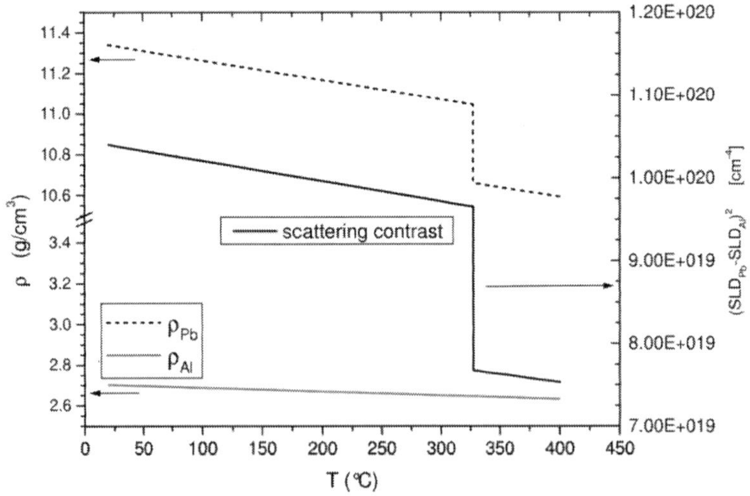

Figure 6. Theoretical temperature change of Al and Pb densities as well as of the neutron scattering contrast of Pb with respect to Al.

For pure Pb not confined into any matrix, the density change would be of the form shown in Fig. 6. The step in the dependency is at the melting point. The scattering contrast then changes (also shown in Fig. 6) by 28% in the measured temperature range RT-400°C. Out of this change, 19% is due to the step-like change at melting point and 9% due to thermal expansion coefficient difference. As lead particles are confined in the solid matrix, their expansion to the full extent on heating may not occur. Nevertheless, also partial thermal and transformational expansion leads to a certain lowering of Pb density with respect to the unmelted particle at RT. Therefore, although the change of scattering contrast of Pb confined in Al can be expected to be lower than for free Pb, there should be a change of the contrast and at least its stepwise part at the melting point should be visible in SANS intensities. The above mentioned magnitudes of the scattering contrast change can be taken as an upper limit of the expected scattering intensity change.

It should be also pointed out that different sizes of confined particles could possess slightly different melting points [6, 7] which would cause smearing of the step in the scattering contrast – and thus in the scattered intensity – dependency.

The quantity, which is reported later in the text and figures, is "volume fraction times the scattering contrast". Although scattering contrast is known at RT, its evolution with temperature T is unknown (a precise measurement of lattice parameter evolution with temperature using diffraction for both Al and confined Pb would be necessary [15] to determine this). Therefore, we can report only the compound parameter $c_{vf} \times (\Delta\rho_{SLD}(T))^2/(\Delta\rho_{SLD}(293K))^2$ where c_{vf} is the volume fraction of Pb particles seen by SANS. It is normalized in such a way that the parameter magnitude at RT is equal to the volume fraction. The further evolution of the compound parameter at higher temperature is, however, practically exclusively driven by the scattering contrast change with temperature.

On cooling back to RT, Pb particles should shrink on liquid-solid phase transformation. It could – in principle – produce voids between the Pb particle and the Al matrix in case the matrix deformation on heating was (at least partially) plastic.

The original shape of the solid lead inclusion is ellipsoid rather than sphere in the case of ECAP deformed material. Then, the shape of Pb particles can possibly change during the phase transformation (Fig. 7) as a consequence of Pb expansion described in the previous paragraphs. Due to this expansion, hydrostatic pressure on Pb particles increases and they have a tendency to change shape from ellipsoidal to spherical. This change in particle shape would influence the shape of 2D SANS profile.

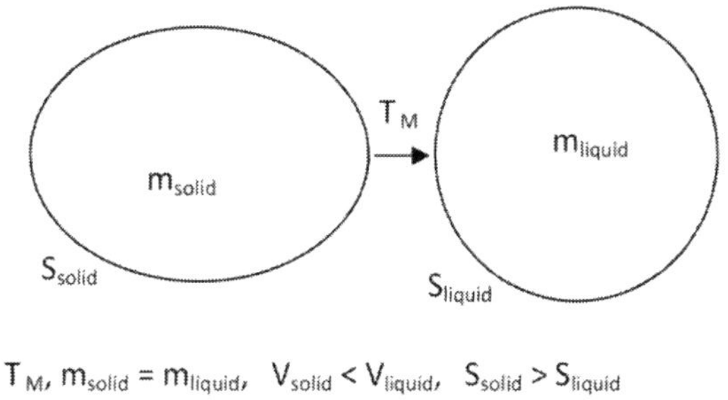

Figure 7. Scheme of the Pb particle shape change in ECAP deformed sample on solid-liquid transformation.

4. RT measurements at several orientations

4.1. The as cast sample

SANS measurement in two orientations on the as-cast sample exhibited no significant anisotropy. The scattering can be modeled well by two types of particles, which are most likely the parts of the same size distribution. The scattering near Q=0 is well described by spheres of mean diameter 5670 Å. In order to describe well the scattering also at larger Q-values, the model has additionally included spheres of mean diameter 1340 Å (fitted sizes are reported in Tab. 1). We decided to split the size distribution into these two parts in order to observe easily size changes during in-situ thermal cycle (see later in the text) by observing volume fraction evolutions of these two fixed particle sizes. The data does not allow detailed extraction of more parameters (e.g. size distribution width).

4.2. The ECAP samples

The ECAP samples were tested at different orientations both before and after in situ heat treatment. Before heat treatment, the ECAP sample shows significant anisotropy, mainly when the ECAP direction is perpendicular to the neutron beam. It means that the Pb particles are rather prolate ellipsoids than spheres in this case. The size of the individual axes found by SANS is written in Tab. 1.

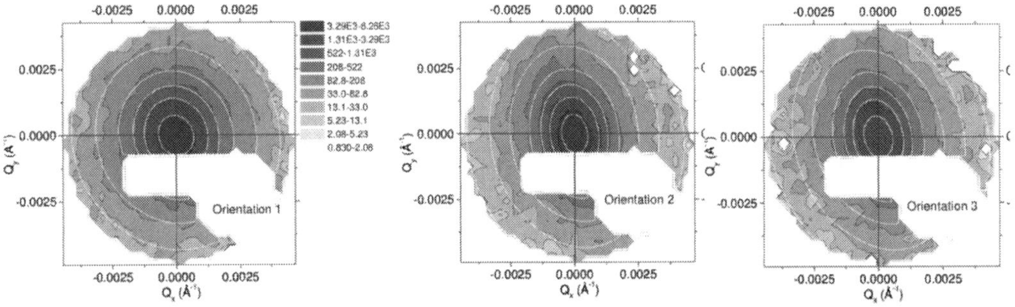

Figure 8. SANS data (differential cross section $d\Sigma/d\Omega(Q_x,Q_y)$ measured at RT for ECAP Al-4%Pb sample at low Q-values before the in-situ thermal cycle. The grey scale map (logarithmic scale) shows measured 2D data versus Q_x and Q_y after raw-data treatment and the white equi-intensity lines depict the fitted curves. The data for three orientations (definition see Tab. 1) of the cube-shaped sample are shown. The white parts of the PSD were masked as there were regions on the PSD with blind spots of the detector efficiency. Leaving the spots unmasked would significantly harm the evaluation.

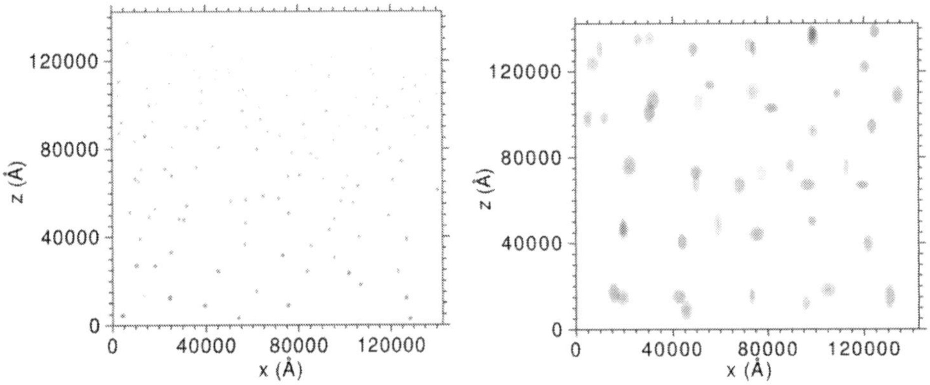

Figure 9. The depiction of the models resulting from the optimum fits for the ECAP Al-4%Pb sample. The microstructure was modeled in 3D. The used depiction procedure resumes imaging through transmission electron microscopy. The model is imaged using slices of the 3D model with the thickness equal to the mean precipitate size. The model of smaller (left) and larger (right) particles is projected to 2D and a partial transparency of the particles is enabled. As the precipitates are of various sizes and positions, this procedure results in the different grey color brightness of the imaged particles. Ellipsoidal shape of particles can be recognized.

The fitting of the 2D SANS patterns for three different orientations (Fig. 8) was performed at once (i.e. a 3D fit was in fact carried out). Therefore, it was found by SANS that the larger axis of the ellipsoid does not lie exactly in the direction of ECAP, but it is tilted in two directions always by approximately 15°. This fact explains orientation of anisotropy in the 2D SANS data measured at two different orientations of the sample with neutron beam perpendicular to ECAP direction (Fig. 8, orientations 2 and 3). The third orientation of the measured ECAP sample (beam parallel to ECAP) provides nearly isotropic scattering (Fig 8, orientation 1). The models of both types of particles are displayed in Fig 9

The measurement at two sample orientations on the ECAP sample after in situ thermal cycle (RT-400°C-RT) (Fig. 10) shows that the scattering is again nearly isotropic. However, the orientation 2 still exhibits small elongation of the pattern in the direction perpendicular to ECAP. This is possibly caused by small retained elongation of the large Pb particles along ECAP. Nevertheless, the change from the large anisotropy before thermal treatment to the small anisotropy after the treatment means, that the solid-liquid transition of Pb particles strongly influenced the shape of the Al cavity in which Pb is confined.

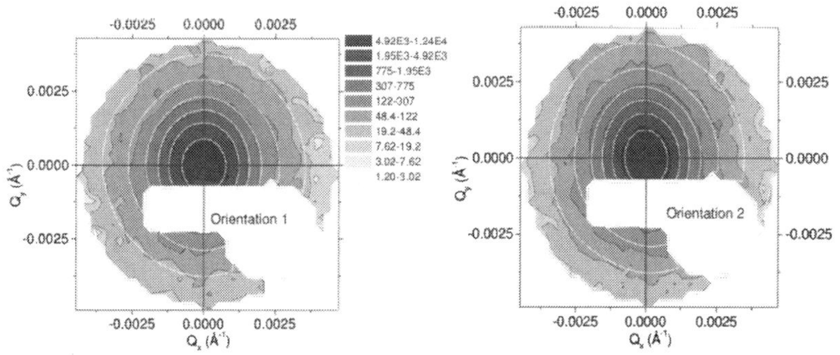

Figure 10. SANS data ($d\Sigma/d\Omega(Q_x,Q_y)$) measured at RT for ECAP Al-4%Pb sample at low Q-values after the in-situ thermal cycle.

4.3. The pure Al sample after ECAPs

The last sample which was tested ex situ was pure Al (i.e. without any Pb) after ECAP procedure. Surprisingly, it exhibits also rather strong scattering, moreover anisotropic (see Fig. 11). As the original material contained no impurities, the only explanation is that the elongated pores were created in the material during the ECAP procedure. Unfortunatelly, there was no virgin Al sample without the ECAP procedure measured by SANS, therefore this can be taken only as a speculation. Nevertheless, any other explanation for the scattering would be difficult to comprehend. If we assume scattering contrast of pores, then the measured volume fraction would be 0.06% for smaller (135 nm) and 0.11% for larger (570 nm) pores. At this point, it cannot be excluded that such pores are present also in the ECAP Al-4%Pb sample. However, it is highly probable that all such potential pores are filled by lead in the ECAP Al-4%Pb sample. Otherwise, the high temperature scattering (particularly around the Pb melting point) which will be described in the next section would be different.

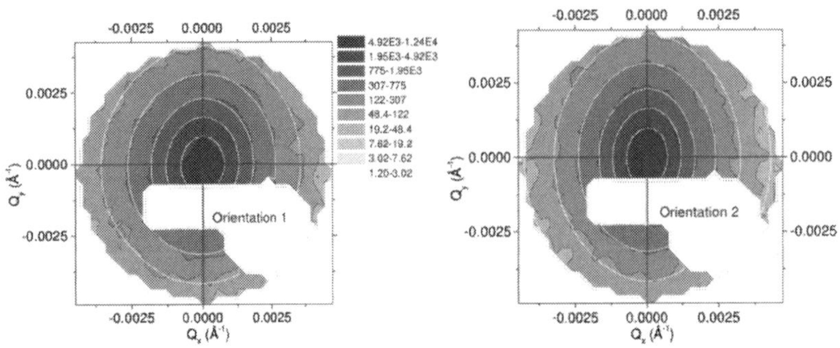

Figure 11. SANS data ($d\Sigma/d\Omega(Q_x,Q_y)$) measured at RT for pure Al after ECAP at low Q-values.

5. In situ measurements

All the in situ measurements were performed in orientation 1, i.e. beam parallel to the ECAP direction. The thermal cycle used for in situ SANS measurement was basically RT-300°C -400°C-RT, where the temperature increase was paused with 5 or 10K temperature steps between 300 and 400°C in order to collect the neutron scattering data.

5.1. As cast Al-4%Pb sample

From the evaluation of the in-situ measurement (RT-400°C-RT cycle) of the as cast Al-4%Pb sample, the temperature dependence of volume fraction times scattering contrast of Pb particles can be deduced (Fig. 12). The following conclusions can be drawn:

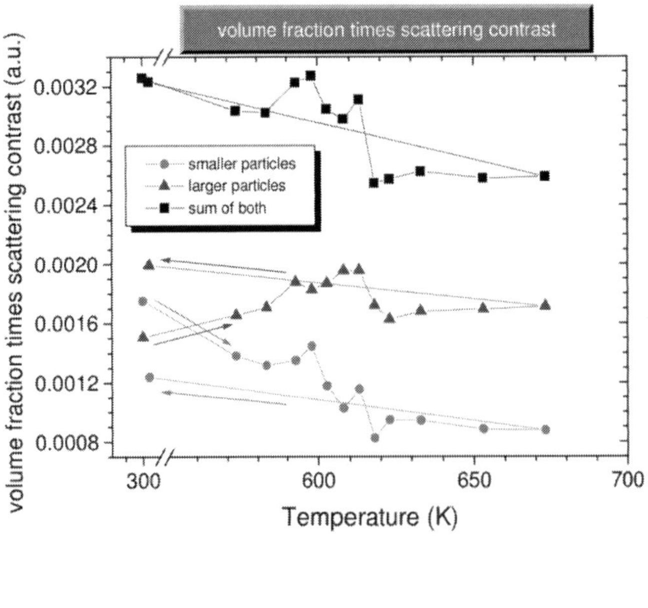

Figure 12. The temperature dependence of volume fraction times scattering contrast of the as cast Al-4%Pb sample during the thermal cycle.

1) The total amount of Pb particles is the same before and after in situ thermal cycle, which is in accordance with the expectations.

2) The total amount of Pb particles times scattering contrast is roughly constant during heating RT -> 613 K (340°C). A small decrease in this temperature range is possibly due to the small decrease of scattering contrast.

3) However, the proportion between smaller and larger particles changes: amount of smaller particles decreases while amount of larger Pb particles increases during heating to 613K (340°C). Small increase of size of Pb particles due to the thermal expansion is not the cause for this effect because then a corresponding decrease in size during cooling would also be present (i.e. such effect would be reversible). However, it is not the case: the amount of small particles is different at RT before and after the cycle. The same is valid for large particles. Further, when we exclude dissolution of smaller Pb particles (remembering Al and Pb are immiscible), then a possible explanation is joining of the very neighboring particles. Another, even more probable explanation, is smoothing of the surface of particles (wrinkles smoothing) which increases scattering to low angles on expense of large-angle scattering. It can be easily accommodated in the matrix as thermal vacancies are mobile in Al above 200°C [6].

4) Between 613K (340°C) and 618K (345°C), a step-like decrease of the scattered intensity occurs. It is due to the sudden scattering contrast change on the solid to liquid transition. Pb expands in a stepwise manner at this point and its scattering length density is lowered. Temperature T_M =615±3K (342±3°C) can be thus deduced as the melting point of Pb in confinement in Al. The observed melting is rather sharp: transformation occurs over a narrow temperature range (within 5K). The large majority of Pb particles melted within the given range 613 to 618K (340-345°C). The temperature of the transformation is 15K higher than for the free lead, and agrees with previous observations on confined Pb particles [6, 7].

5) The total volume fraction of Pb after the thermal cycle returns back to the original value observed before thermal cycle. There is also no indication of creation of additional scattering intensity at large Q-values which could indicate creation of small voids at Al-Pb interface. Therefore, the Al matrix deformation on heating was highly probably only elastic and no voids were thus produced on cooling.

5.2. Al-4%Pb ECAP sample, one thermal cycle

The same evaluation was performed for Al-4%Pb ECAP sample which underwent one thermal cycle (RT-400°C-RT). The resulting volume fractions are plotted in Fig. 13.

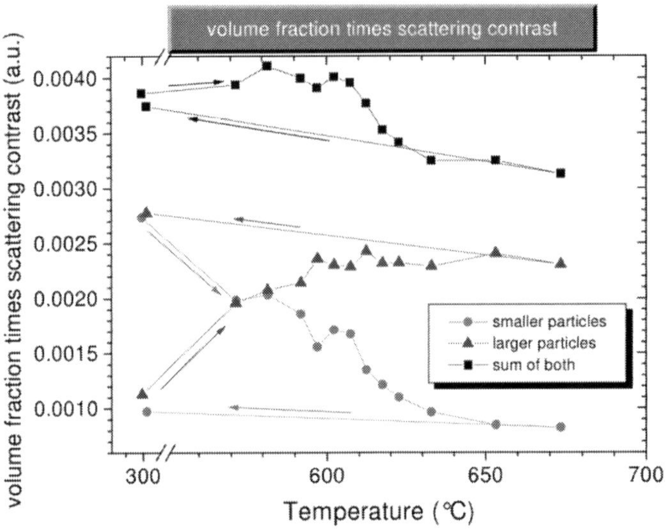

Figure 13. The temperature dependence of volume fraction times scattering contrast of the Al-4%Pb ECAP sample during the thermal cycle.

The amount of small particles at RT before the in-situ cycle is larger than for the as-cast sample. Nevertheless, the overall character of the volume fraction evolution with temperature is the same as for the as-cast sample. There is again a visible change of proportion between small and large particles up to the melting point very similar to the one observed by the as cast sample. The amount of small particles is also different at RT before and after the heating cycle. The same is observed for large particles as well.

However, there exist some differences in details between the ECAP and the as cast samples. Firstly, on temperature increase, there is no visible step down observed in the "volume fraction times scattering contrast" parameter of large particles in ECAP sample (blue line and triangles in Fig. 13) around the melting point. This parameter is nearly constant above 607K. It differs from the large-particle line for as cast sample in Fig. 12. It means that - for the ECAP sample - the increase in volume fraction of larger particles (or wrinkles smoothing) above 607K still acts and completely balances (in terms of the scattered intensity) the gradual decrease of the scattering contrast of large particles during melting.

Secondly, the change of scattering contrast in the vicinity of melting point is not as sharp as in the as cast sample. This can be seen clearly in Fig. 14, where both the as cast and the ECAP data are presented. In the ECAP sample, the transition region ranges from 607 to 633K (334-360°C), which is a 26 K broad range and is significantly broader (probably due to the broader Pb particle-size range) than the transition region of the as cast sample (5K). The center of the transition region is at 343°C (616K), i.e. practically at the same position as the center for the as cast sample (342°C or 615K) and thus 16K higher than melting point of free Pb.

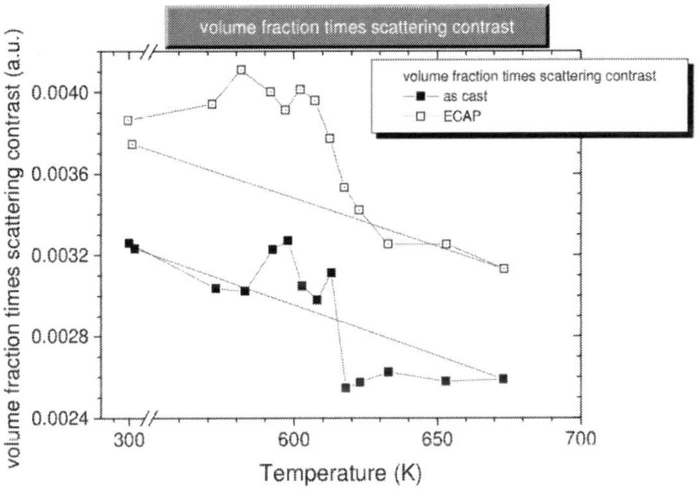

Figure 14. The temperature dependence of total volume fraction times scattering contrast of the as cast Al-4%Pb sample and the Al-4%Pb ECAP sample during the thermal cycle.

The stepwise change of the scattering contrast which can be deduced from Fig. 14 is around 17% for the as cast and around 22% for the ECAP deformed sample. Taking into account the experimental uncertainties, this is in good agreement with the theoretical value for scattering contrast change of a free lead particle (19%) shown in Fig. 6. This agreement means that – although confined in Al matrix - the Pb particle expands to its full extent (i.e. as a free particle) on the solid-liquid transition.

5.3. Al-4%Pb ECAP sample, two thermal cycles

In order to see evolution of the microstructure during the thermal cycling, two cycles RT-400°C-RT-400°C-RT were applied on another ECAP deformed Al-4%Pb sample. For this sample, however, less number of points on the temperature curve was measured due to the time restrictions. The evolution of the volume fraction of Pb particles during cycling is plotted in Fig. 15. It can be seen that the first cycle corresponds well to the previous Al-4%Pb sample (Fig. 13), also with respect to the proportion of smaller and larger particles. The second cycle, however, does not follow the same dependency. After the second cycle, the ratio between smaller and larger particles volume fractions returns to the same value as before the second cycle. That means that – unlike in the first cycle - there is no growth of particles (or smoothing of wrinkles) in the second cycle (within measurement errors) and that the microstructure is reversible. The largest influence on the morphology can be thus ascribed to the first cycle. The following thermal cycling would probably lead to much slower changes in the microstructure, as is usual in any cycling testing of metallic materials.

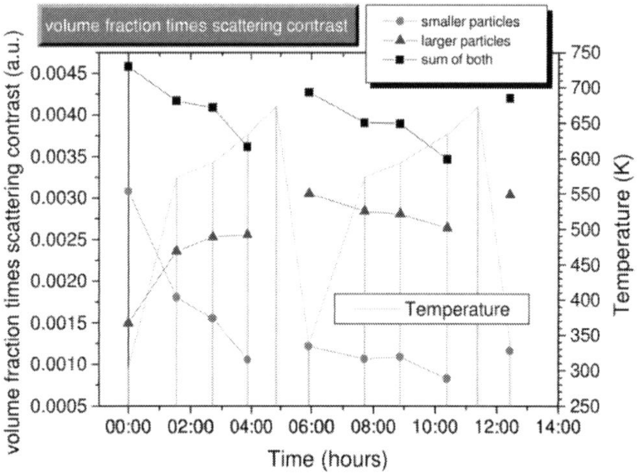

Figure 15. The time and temperature dependence of volume fraction times scattering contrast of the Al-4%Pb ECAP sample cycled two times between RT and 400°C.

6. Conclusions

2D SANS data for the as-cast Al-4%Pb sample shows no anisotropy. Anisotropy of Pb particles shape can be observed only after severe plastic deformation (ECAP). The lead particles were elongated roughly in the direction of ECAP.

Elongated pores in pure Al after ECAP processing were detected. They are of similar size and volume fraction like Pb particles in Al-4%Pb ECAP sample. It indicates that empty cavities are always created during ECAP in aluminium, but - when Pb is present - this element fills these empty spaces.

During the in-situ thermal cycle RT-400°C, morphological transformation of elongated Pb particles to nearly spherical shape during the solid-to-liquid transition of Pb for ECAP Al-4%Pb can be deduced.

On heating up to the melting point, the increase of volume of large particles at the expense of smaller inhomogeneities can be observed. It is caused either by joining of the next neighboring particles or by smoothing of the surface of particles (wrinkles smoothing).

The change of scattering contrast during melting of Pb maps the transform of confined Pb particles to the liquid phase. The center of the transition region is around 342°C (615K) for both as cast and ECAP deformed samples. This would mean a significant shift with respect to the free lead melting point (327°C or 600K). Nevertheless, it is not surprising as such effect was already observed for confined particles [7].

For the ECAP sample, the transition is not sharp: 26 K broad range indicates that there is a broad size distribution of Pb particles. The inclusions of different sizes exhibit different melting points, which then gives rise to the scattering contrast change within the 26 K broad range. Also this is a confirmation of the previous observations, i.e. the dependence of the melting point of confined Pb on the particle size [7]. On the other hand, the transformation solid-liquid seems to be sharp for the as cast sample, probably due to the narrower size distribution.

Although confinement has an effect on the temperature of phase transformation, it seems to have no effect on thermal expansion of the lead particle during solid-liquid transformation. The Pb particle expands to its full extent (i.e. as a free particle).

From SANS data, it can be also deduced that the Al matrix surrounding Pb particle undergoes no plastic deformation. However, for the ECAP sample, there is a change of Al cavity shape on the phase transform which remains unchanged on cooling. It would indicate that some diffusional process takes place at HT in Al surrounding the Pb particle which results in a permanent change of the cavity shape.

Acknowledgement
The authors thank SINQ (PSI Villigen, Switzerland) for providing beamtime and support for the SANS measurement. A travel support to P. Strunz for the experiment in the frame of the programme 'Research Infrastructures' - contract No. RII3-CT-2003-505925 (NMI3) - is acknowledged. P. Strunz also acknowledges the support by MPO project FR-TI1/378.

References

[1] Sims C T, Stoloff N S and Hagel W C 1987 *Superalloys II* (NY: John Wiley and Sons)
[2] Srolovitz D J, Petkovic-Luton R A and Luton M J 1988 *Acta Metall.* **31** 2151
[3] Arzt E and Wilkonson D S 1986 *Acta Metall.* **34** 1893
[4] Rösler J and Arzt E 1990 *Acta Metal.* **38** 671
[5] Wright P K 1978 *Metall. Trans.* **9A** 955
[6] Gabrisch H, Kjeldgaard L, Johnson E and Dahmen U 2001 *Acta Mater.* **49** 4259-4269
[7] Zhang D L and Cantor B 1991 *Acta Metal. et Mater.* **39** 1595
[8] Chen X M, Fei G T, Cui P, Zheng K, Zhang R L 2007 *Phys. Lettr.* **A 363** 150-153
[9] Johnson E, Johnson A, Dahmen U, Chen S and Fujii T 2001 *Mat Sci Eng.* **A 304-306** 187-193
[10] Shi Z, Wynblatt P, Srinivasan S G 2004 *Acta Mater.* **52** 2305-2316
[11] Iwahashi Y, Furukawa M, Horita Z, Nemoto M and Langdon T G 1998 *Metall Trans* **A 29A** 2245-2252
[12] Strunz P, Mortensen K and Janssen S 2004 *Physica* **B 350** e783–5
[13] Strunz P, Wiedenmann A, Gilles R, Mukherji D, Zrník J and Schumacher G 2000 *J. Appl. Cryst.* **33** 834-838
[14] Kostorz G 1979 *Neutron Scattering* (Treatise on materials science and technology) ed G Kostorz (New York: Academic Press) pp 227
[15] Strunz P, Gilles R, Mukherji D, Hofmann M, del Genovese D, Rösler J, Hoelzel M and Davydov V 2009 *J. Appl. Cryst.* **42** 981-989

Residual stresses in biaxially fatigued austenitic stainless steel sample of cruciform geometry

Yu V Taran[1,*], A M Balagurov[1], J Schreiber[2], A Evans[3] and A M Venter[4]

[1]Frank Laboratory of Neutron Physics, Joint Institute for Nuclear Research, Dubna, Russian Federation
[2]Fraunhofer Institute for Nondestructive Testing (Dresden branch), Dresden, Germany
[3]SINQ, Paul Scherrer Institute, Villigen, Switzerland
[4]Research and Development Division, NECSA Limited, Pretoria, South Africa

*E-mail: taran@nf.jinr.ru

Abstract. A specifically designed cruciform-shaped austenitic stainless steel AISI 321 sample was subjected to ex-situ biaxial tension-compression cycling to establish ferromagnetic martensitic phase conversion under the action of plastic deformation. The time-of-flight neutron diffraction technique was employed for in-plane residual stress determination in this sample for both the austenitic and martensitic phases. The 2D data enabled determination of macro-, micro-, hydro- and deviatoric contributions to the total phase stresses.

1. Introduction

Investigation of fatigue degradation and martensitic transformation of austenitic stainless steel AISI 321 (Cr18Ni10Ti) is of great practical relevance due to its wide application amongst others in the nuclear and chemical industries. The experimental simulation of these processes is usually performed by uniaxial mechanical tension-compression cyclic loading. However, machines and structures are generally subjected to more complex multiaxial and possibly nonproportional stresses rather than simple uniaxial stresses. The failure and fracture problem under multiaxial stresses is one of the most important topics to ensure the reliability of industrial components.

To investigate the effects of in-plane biaxial loading on stress-induced martensite phase transformation, the cruciform geometry samples of two kinds named Krest and Kreuz were specifically designed for nondestructive investigation using neutron diffraction techniques. The design comprises a central membrane 15 mm in diameter and 2 mm thick surrounded by thicker material 5 mm in thickness extending to the cruciform edges. Specific care was taken to minimize possible sharp surfaces at the membrane-surrounding material interface to eliminate stress concentrators. Samples of this geometry were subjected to ex-situ biaxial out-of-phase tension-compression cycling under load control at different amplitudes using an Instron 100 kN biaxial planar cruciform system. During plastic deformation a new ferromagnetic martensite phase is forming in the parent austenite matrix [1].

Sample Krest-2, cycled at 0.5 Hz frequency and 17 kN load amplitude for 407 cycles rendering about 40 % martensite volume fraction conversion, has been investigated earlier using time-of-flight (TOF) neutron diffraction techniques at the IBR-2 pulsed nuclear reactor at JINR [2, 3] along two perpendicular in-plane directions across the diameter of the membrane. A comprehensive 2D mapping could not be completed due to the IBR-2 facility being shut down for refurbishment. A 2D mapping of

the membrane section was pursued following the (220) austenite and (211) martensite diffraction reflections using a monochromatic neutron beam at the ILL high flux reactor, the SALSA instrument. The residual stress analysis from the experimental data was inconclusive due to excessive crystallographic texture in the intensity of the martensite reflection [4]. Thus, the investigation of the Krest-2 was not finished. As distinct from the study of the Krest-2 the investigation of the Kreuz-1 was carried out to its conclusion in the present work.
To overcome the problems associated with the crystallographic texture, it became essential that a number of diffraction peaks representative of each phase be investigated simultaneously to assist in the identification of those Bragg reflections for which the intensities were not influenced to adversely. This paper subsequently reports the results from a 2D diffraction analysis of the in-plane residual stresses in the membrane section of the Kreuz-1 sample, using TOF neutron diffraction at the SINQ spallation source (Paul Scherer Institute, Switzerland), the POLDI instrument. POLDI is a correlation type TOF diffractometer [5].

2. Experiment

The Kreuz-1 sample (Fig. 1) has been subjected to biaxial cycling at load amplitude of 15 kN and frequency of 0.5 Hz until rupture of the membrane. This rupture occurred during the tension cycle 1904 within two adjoining quarters of *A2-0-B1* and *A2-0-B2* at the interface of the membrane and the peripheral 5-mm-thick surrounding material as indicated in Fig. 1.

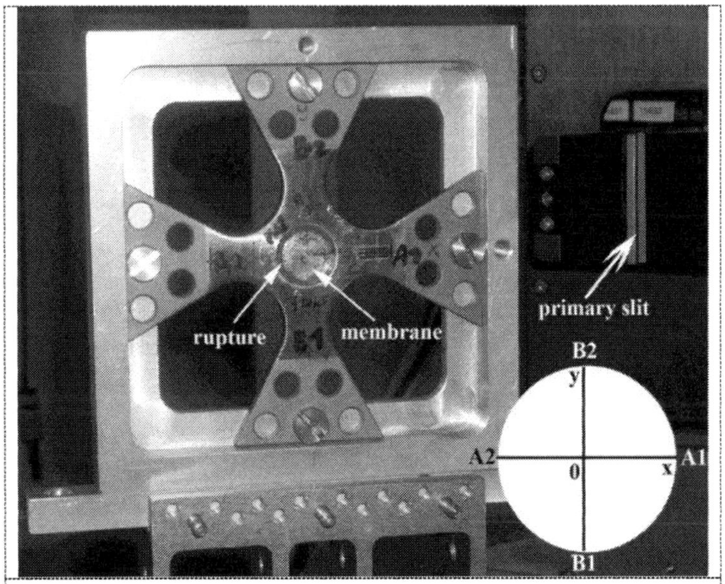

Figure 1. The Kreuz-1 sample mounted on the POLDI instrument for the in-plane x-component measurements. The right bottom insert is shown the membrane image with the sample coordinate system (x, y) and the loading directions *A1-A2* and *B1-B2*. The neutron scattering vector was along the loading direction *A1-A2*. The vertical slit is visible at the right top with the horizontal slit directly behind it. The y-component measurements were done by rotating the sample 90° clockwise to have this direction along the scattering vector.

The experimental set-up of the Kreuz-1 cruciform sample on the POLDI instrument is shown in Fig. 1. A neutron gauge volume fully illuminating the 2 mm membrane thickness, but limited to a size of 1.5 mm vertically and horizontally, was formed by using slits 1.5 mm in size for the primary neutron beam and a 4 mm radial collimator in the secondary neutron beam. Diffraction data were recorded by a one-dimensional ^{3}He position-sensitive detector covering an angular range of nearly 30° at a central scattering angle 2θ=90°. The 2D in-plane diffraction mapping of the membrane and the surrounding material was performed using a discrete 1×1 mm^2 rectangular measurement mesh. The sample coordinate system (x, y) corresponds to the loading directions *A1-A2* and *B1-B2*, respectively (both systems are shown in the insert to Fig. 1).
Strain scanning was performed on the two quarters of *A1-0-B2* (right) and *A2-0-B2* (left) assuming mirror symmetry relative to the *A1-0-A2* axis. Diffraction measurements were carried out for two perpendicular orientations of the sample with the neutron scattering vector being directed along either the x- or y-axis (x- or y-component hereinafter). Neutron diffraction patterns covering the d-spacing range 0.7-2.2 Å were measured at all sampling points. Selective measurements at points surrounding the membrane section were used as reference for the strain assessment in the austenite phase since no martensite formation was present here. Data processing was done on the (111) austenite and (110) martensite diffraction reflections using a Gaussian single-peak fit approach (POLDI in-house program "Poldiausfit").

3. Total austenite residual stress tensor

To calculate residual strains from the experimentally measured d-values of the (111) austenite and (110) martensite peaks respective stress-free reference d-spacing is needed. Unfortunately, owing to the complex nature of the austenite matrix phase transformation during fatigue cycling, it is impossible to obtain stress-free data for martensite inclusions. For the austenite phase, as a rough but plausible approach, the unstrained reference parameter $d_0(111)$=2.072418 Å was measured on the outer part of the sample surrounding the membrane where no plastic yielding and martensite phase formation had occurred. Using Hooke's law for a biaxial geometry, the austenite stress tensor in each measured point was calculated using Young's modulus E=170 GPa and Poisson ratio ν=0.3. In Fig. 2 both x- and y-components of the austenite total residual stresses are illustrated (two upper curves) as a function of radius r along the *A1-0-A2* diameter line (y=0). The qualitative estimation of the stress relief occurring in the austenite phase due to the rupture of the membrane was obtained by averaging the x- and y-components of the total stresses in both investigated quarters. The calculated values were within 40 MPa, implying that the stress relief in the austenite phase is marginal.

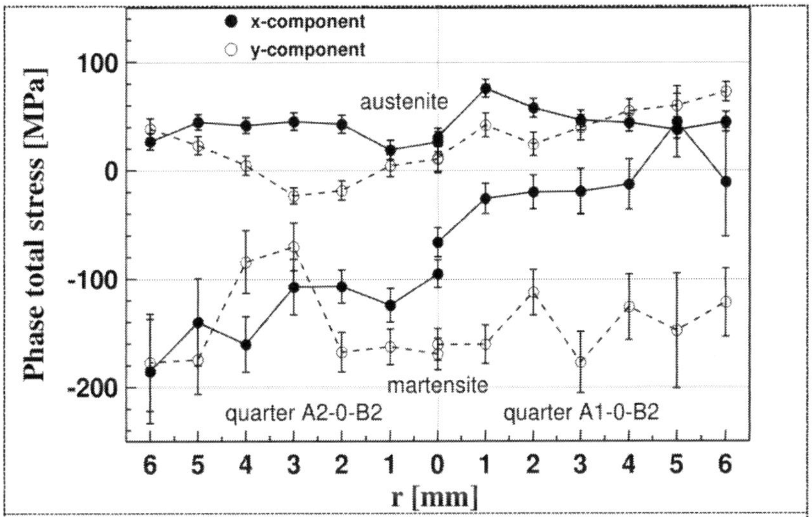

Figure 2. Radial distributions of x- and y-components of the phase total residual stresses in austenite (two upper curves) and martensite (two lower curves) as a function of radius r along the *A1-0-A2* diameter line with $y = 0$.

4. Solution of d_0-problem for martensite phase

For the determination of the martensite unstrained parameter, $d_0(110)$, three general conditions can be applied (e.g., see [6, 7]), namely, separation of the total stress, ${}^t\sigma$, into macrostress, ${}^M\sigma$ (the same in each phase) and phase microstress, ${}^\mu\sigma$, a phase mixture rule and a stress equilibrium approach, which successively are:

$$ {}^t\sigma_j^p = {}^M\sigma_j + {}^\mu\sigma_j^p \,, \tag{1} $$

$$ {}^\mu\sigma_j^a f^a + {}^\mu\sigma_j^m f^m = 0 , \tag{2} $$

$$ {}^M\bar{\sigma}_j = \frac{1}{V}\int_V {}^M\sigma_j dV = 0 \ , \tag{3} $$

where subscript j indicates a stress component at a sampled point, f^p represents a volume fraction of p-phase, p=a for austenite or m for martensite, respectively. Using Eqs. (1), (2), the macrostress can be expressed as:

$$ {}^M\sigma_j = {}^t\sigma_j^a f^a + {}^t\sigma_j^m f^m \,. \tag{4} $$

As the measurements were made in a discrete mesh, the integration over a free body volume can be replaced by the summation

$$ \sum_k [{}^t\sigma_j^a(\vec{r}_k) f^a(\vec{r}_k) + {}^t\sigma_j^m(\vec{r}_k) f^m(\vec{r}_k)] = 0 , \tag{5} $$

where the radius-vector, $\vec{r}_k$, corresponds to the k-th mesh cell.

If the phase volume fractions are known, the martensite unstrained parameter, $d_0(110)$, may be estimated from Eq. (5) using the planar presentation of Hooke's law for the total martensite stress ${}^t\sigma_j^m(\vec{r}_k)$ expressed through the martensite lattice spacing $d(110)$.

5. Determination of martensite phase volume fraction

Taking into account that $f^a + f^m = 1$, the martensite volume fraction in each sampled point was calculated from the ratio of the austenite and martensite diffraction peak areas:

$$A_a/A_m = (f^a/f^m)(F_a/F_m)^2 (j_a/j_m)(Q_a/Q_m), \tag{6}$$

where p-phase values F_p and j_p are the structure factor and reflection multiplicity. For the Q factor it can be shown that in the case of TOF diffraction the relation $Q_p \sim (1/d_p)^2$ is valid. The results of martensite fraction calculations for x- and y-components are shown in Fig. 3.

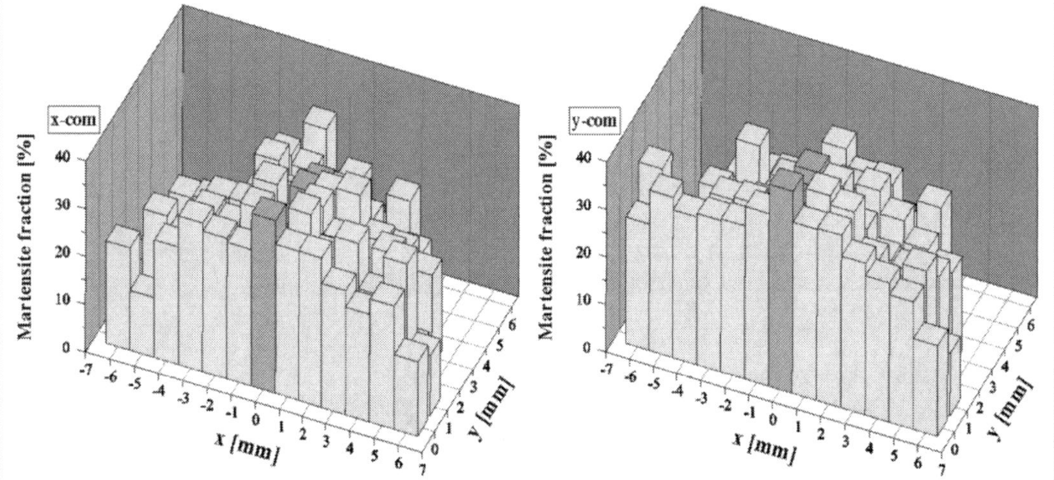

Figure 3. Two-dimensional distribution of the martensite volume fraction determined using Eq. (6) from respectively the x- (left plot) and y- (right plot) component investigations within the *A1-0-B2* ($x \geq 0$) and *A2-0-B2* ($x \leq 0$) membrane quarters.

As a first-order observation, some asymmetry seems to exist between the left and right membrane quarters. However, the mean values of martensite fraction within both quarters summarized in Table 1 do not show any marked asymmetry. Despite the rupture of the membrane in the left quarter, asymmetry is not expected as the martensite transformation is irreversible. The 4% difference between the mean values of martensite fraction for the x- and y-components can be ascribed to the presence of weak crystallographic texture in the (110) martensite intensity.

Table 1. Experimental data for martensite phase.

	x-component			y-component		
quarter	fraction [%]	d_0(110) [Å]	${}^t\sigma_x^m$ [MPa]	fraction [%]	d_0(110) [Å]	${}^t\sigma_y^m$ [MPa]
right	23.8	2.030077	-58.2(8.4)	27.5	2.029965	-120.9(5.8)
left	23.7	2.029929	-112.9(9.4)	26.7	2.029432	-114.7(7.6)

6. Total martensite residual stress tensor

Assuming no elastic mismatch between austenite and martensite phases and resolving Eq. (5) relative to $d_0(110)$, four d_0-values were respectively determined from the x- and y-component measurements in both quarters (Table 1). The mean value, 2.029851 Å, was subsequently used as the martensite unstrained reference parameter $d_0(110)$ for the strain calculations. The martensite stress tensor in each measured point was calculated taking the elastic constants as used for the austenite phase. In the case, we have neglected elastic anisotropy in each phase. In Fig. 2 (two lower curves) both *x*- and *y*-components of the martensite total residual stresses are illustrated as a function of radius r along the *A1-0-A2* diagonal line for *y*=0. The averaged martensite total residual stresses within each quarter are shown in Table 1 (${}^{t}\sigma_x^m$ and ${}^{t}\sigma_y^m$). A large difference exists in the values between the left and right membrane quarters for the *x*-component, while the *y*-component values are similar. This is indicative of potential stress relief in the martensite phase due to the rupture in the membrane.

7. Separation of macrostress from total phase residual stresses

By using Eq. (4), the residual macrostresses can be separated as ${}^{M}\sigma_x$ and ${}^{M}\sigma_y$ in a two-phase material. The 2D results of the *x*- and *y*-component macrostresses within the top half of the membrane are shown in Fig. 4 from diffraction measurements parallel to the *A1-0-A2* direction and for *y*-coordinate between 0 and 6 mm. The *x*-component (Fig. 4, left plot) demonstrates noticeable asymmetry between the right and left quarters. Its value, as such, in the quarter *A1-0-B2* varies from 60 to -40 MPa, while its dispersion in the quarter *A2-0-B2* is appreciably smaller varying between 35 to -20 MPa. Again this effect may be attributed to the stress relief of the *x*-macrostress component due to the membrane rupture. The *y*-component macrostress does not show obvious asymmetry between the quarters.

The residual macrostress calculations are based on the macrostress equilibrium condition (Eq. (3)), i.e., the mean values of the *x*- and *y*-components of macrostresses in both membrane quarters must equal zero. The results from the averaging macrostress procedure are presented in Table 2. The fact that the averaged macrostress components are not exactly equal to zero, but within 17 MPa, reflects the approximate nature of the assumptions used for the determination of phase strain-free parameters being satisfactory. This particularly concerns determination of the d_0-parameter for the martensite phase, potentially influenced by texture contributions on the martensite volume fraction parameter used in the calculations.

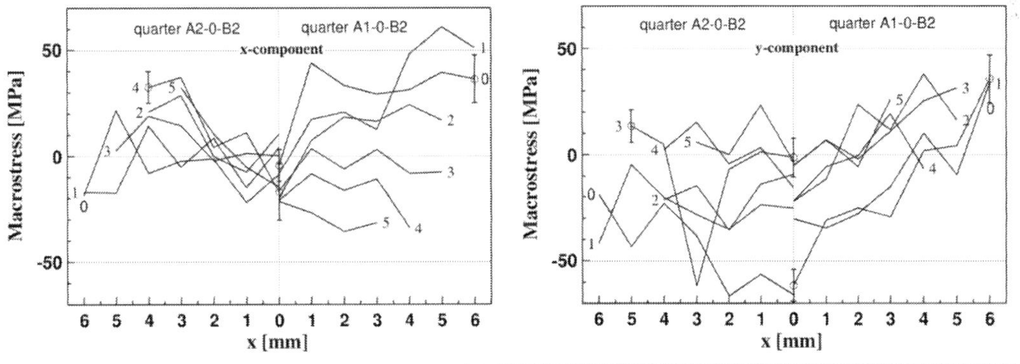

Figure 4. Dependences of the *x*- and *y*-component macrostresses (left and right plots, respectively) on x-coordinate in the top half section of the membrane at different *y*-coordinates shown as a number of a curve. A typical error is shown for some curves.

Table 2. Macro- and hydrostresses.

	averaged macrostress [MPa]		averaged hydrostress [MPa]	
quarter	*x*-comp.	*y*-comp.	austenite	martensite
right	6.8(4.5)	-1.4(3.9)	22.1(2.4)	-59.7(3.4)
left	2.5(2.8)	-17(4)	18.7(1.4)	-75.8(4.8)

8. Separation of microstress from total phase residual stresses

In addition to macrostresses, Eqs. (1), (2) allow the calculation of the planar components of the phase microstresses:

$$^{\mu}\sigma_j^a = f^m({}^t\sigma_j^a - {}^t\sigma_j^m), \tag{7}$$

$$^{\mu}\sigma_j^m = -f^a({}^t\sigma_j^a - {}^t\sigma_j^m). \tag{8}$$

As an illustration, the results of the microstress *x*- and *y*-component calculations for the austenite and martensite phases are shown in Fig. 5. This shows that the austenite phase is in a state of tension, whilst the martensite phase exhibits a balancing compressive stress with the larger value being in accordance with its smaller volume fraction. The asymmetry in the microstresses of the martensite phase between the left and right quarters of the membrane may be attributed to the crystallographic texture.

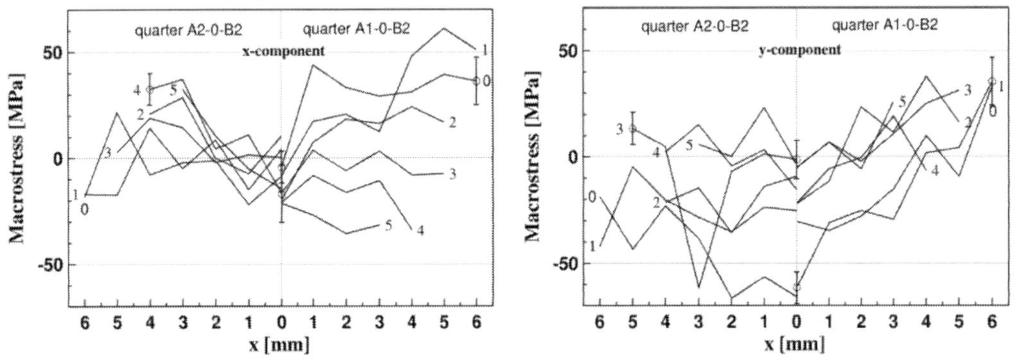

Figure 5. Dependences of the microstress *x*- and *y*-components (left and right graphs) in the austenite and martensite phases, respectively, as a function of radius *r*. Scan along y-axis was carried out twice for each sample orientation.

9. Separation of hydrostatic stress from total phase residual stresses

Since the plastic deformation does not occur under hydrostatic (volumetric) stress, the deviatoric stress tensor can adequately describe the residual stress state with respect to mechanical loading such as fatigue cycling. The total stress tensor of a *p*-phase may be separated into the hydrostatic τ_H^p and deviatoric ${}^t\tau_j^p$ components:

$${}^t\sigma_j^p = {}^t\tau_H^p + {}^t\tau_j^p, \tag{9}$$

where, by definition, $\tau_H^p = (1/3)\cdot Tr({}^t\sigma_i^p)$ and $Tr({}^t\tau_i^p) = 0$. The phase hydrostatic components τ_H^p calculated using the results of the planar total residual stresses of austenite and martensite phases are shown in Fig. 6. The results of the averaging hydrostress procedure presented in Table 2 point to practical absence of asymmetry between the right and left membrane quarters.

It is known that the martensite formation during plastic deformation of an austenite matrix is connected with volume dilatation. Since the specific volume of martensite is larger (about 2 %) than that of austenite, the martensite phase is generally expected to be in hydrostatic compression, whereas the austenite is in tension. Since the deviatoric stress tensor is a simple difference between the total and hydrostatic stress tensors, we do not present the result of its calculation.

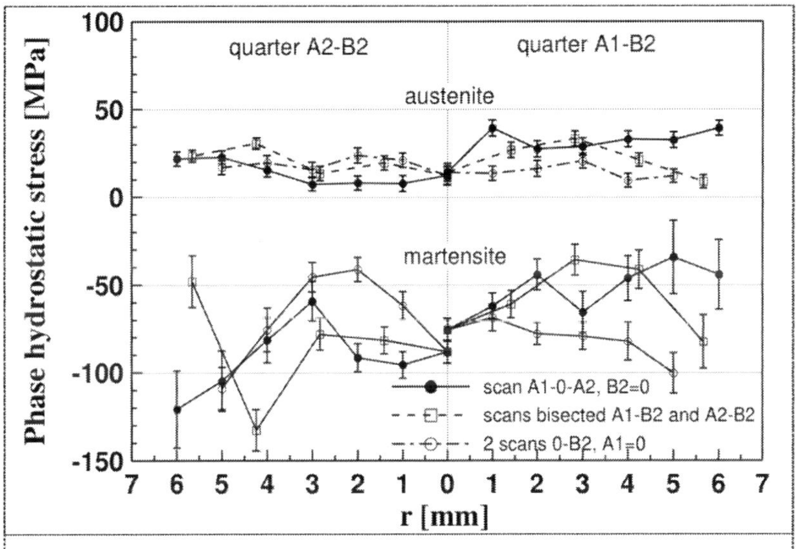

Figure 6. Radial distributions of the hydrostatic stresses in the austenite and martensite phases along the respective load directions, as well as along the measurement direction bisecting the primary load directions.

10. Conclusion

The main results of this work:

1) The biaxial fatigue cycling of austenitic stainless steel AISI 321 was performed in ex-situ on a sample of cruciform geometry with an Instron 100 kN biaxial planar cruciform system. The martensite transformation of austenite steel matrix was observed during plastic deformation.

2) Neutron diffraction 2D-strain scanning and residual stress determination for both the austenite and martensite phases were carried out with the time-of-flight high resolution stress-diffractometer POLDI.

3) Separation of the phase total residual stresses into macro-, micro-, hydro- and deviatoric residual stresses was possible using a phase mixture rule and a stress equilibrium approach.

The authors are grateful to the SINQ facility for the allocation of beam time at the POLDI instrument, as well as to Prof. R.A.Winholtz from the University of Missouri (USA) for very useful consultations. The financial support from the JINR-SA Bilateral Scientific Cooperation Programme is gratefully acknowledged.

The authors are grateful to the SINQ facility (PSI) for the allocation of beam time at the POLDI instrument. The financial support from the JINR-SA Bilateral Scientific Cooperation Programme is gratefully acknowledged.

References

[1] Taran Yu.V. et al. Investigation of In-Plane Biaxial Low Cycle Fatigued Austenitic Stainless Steel AISI 321. I. Mechanical Testing on the Planar Biaxial Load Machine. JINR Commun. E18-2007-110. Dubna, 2007. 24 p.

[2] Taran Yu.V. et al. Investigation of In-Plane Biaxial Low Cycle Fatigued Austenitic Stainless Steel AISI 321. II. Neutron Diffraction Stress Analysis at the IBR-2 Pulsed Nuclear Reactor. JINR Commun. E18-2007-111. Dubna, 2007. 18 p.

[3] Taran Yu.V. et al. Neutron Diffraction Analysis of Stresses in an In-Plane Biaxially-Fatigued Stainless Steel Sample of Cruciform Geometry. Materials Sci. Forum. 2008. V. 571-572. P. 131-136.

[4] Taran Yu.V. et al. Neutron Diffraction Research of In-Plane Biaxial Fatigued Stainless Steel Sample of Cruciform Geometry. The ILL Experimental Report No. 1-01-29. 2008; http://club.ill.fr/cv/.

[5] Stuhr U. et al. Time-of-Flight Diffraction with Multiple Frame Overlap. Part II: The Strain Scanner POLDI at PSI. Nucl. Instr. Meth. A. 2005. V. 545. P. 330-338.

[6] Winholtz R.A, Cohen J.B. Load Sharing of the Phases in 1080 Steel during Low-Cycle Fatigue. Metallurg. Transactions A. 1992. V. 23. P. 341-354.

[7] Winholtz R.A. Separation of Microstresses and Macrostresses. Measurements of Residual and Applied Stress Using Neutron Diffraction. Ed. by M.T.Hutchings and A.D.Krawitz. Kluwer Academic Publishers, Dordrecht, 1992. P. 131.

Journal of Physics: Conference Series **340** (2012) 012100 doi:10.1088/1742-6596/340/1/012100

Peak broadening and peak shift pole figures investigations by STRESS-SPEC diffractometer at FRM II

W M Gan[1], C Randau[1], M Hofmann[2], H G Brokmeier[1,3], M Mueller[1] and A Schreyer[1]

[1] Institute of Materials Research, Helmholtz-Zentrum-Geesthacht, D-21502 Geesthacht, Germany
[2] Research Neutron Reactor, Heinz-Maier-Leibnitz (FRM II), Technical University of Munich, D-85375 Garching, Germany
[3] Institute of Materials Science and Engineering, Clausthal University of Technology, D-38678 Clausthal-Zellerfeld, Germany

E-mail: weimin.gan@hzg.de

Abstract This paper studied for the first time peak intensity, peak position and FHWM pole figures with one time measurement at the neutron diffractometer STRESS-SPEC via in-situ tensile deformation on austenitic steel. Fibre distribution with its evolution from central tensile direction to normal direction of these three kinds of pole figures was obtained. Variation of peak position and FWHM can be correlated to the reorientation of the texture component.

1. Introduction

Diffraction (X-ray or neutron) analysis on polycrystalline metals is a non-destructive method which gives real information on materials, such as crystal size, their deformation, preferred orientation and structural defects [1]. Preferred orientation or crystallographic texture analysis through measuring the relative intensities of certain Bragg peaks by diffraction methods has been well established after the pioneered work of Bung [2]. With this method intensity variation corrected with background or absorption and defocusing (using X-ray) of the considered lattice plane is measured with respect to the sample orientation, intensity pole figure can therefore be built.

It is known that coherent X-ray or neutron diffractions of poly crystals result in plentiful peak profile information. The real crystals are not idea such as there always exits defects, micro strains, grain boundaries, stacking faults, etc. [3]. Generally, the diffraction peaks get broader when the grains in a material become smaller, especially for nano-crystals. It was also noted that contribution of dislocations to peak broadening was much more apparent than small grains when they were larger than 1 μm. In addition, strains that vary from grain to grain or within a grain (micro strains) due to local environment can also produce peak broadening. Macro strain due to an external load leads to the peak shift. X-ray diffraction peak profile for grain size and micro strain analysis based on Williamson-Hall and Rietveld methods has been widely investigated using powder diffraction data [4, 5]. However, researches on peak broadening and peak shift pole figures have been rarely done because of the requirement of high flux and high resolution instruments having a fast detector readout system. Especially, under loading condition the peak broadening pole figure caused by dislocations and peak shift pole figure caused by macro strains have not been performed so far. Peak shift pole figure can describe the deformation anisotropy. The relationship between micro strain and texture is still unclear.

Using neutron diffraction bulk samples rather than surfaces can be measured. Therefore, coarse-grained materials can be characterized easily; averaging over texture inhomogeneities of mechanical test samples can open the direct correlation between texture and material's properties. Moreover, environmental cells (heating, cooling and deformation, etc.) are available and the angular resolution in orientation space is better than for an X-ray goniometer because no defocusing occurs. The materials science neutron diffractometer STRESS-SPEC at FRM II was developed as a multi-purpose diffractometer for strain and crystallographic texture [6, 7]. This diffractometer is located at a thermal beam port and comprises a highly flexible monochromator arrangement, utilizing three different monochromators: Ge, bent Si and pyrolytic graphite (PG) which give continuous opening angle that allows a wavelength selection between 1 Å and 2.4 Å. In addition, STRESS-SPEC offers the users various sample environments such as tensile/compression, torsion, cyclic loading, temperature controlling, etc.

Current study was carried out to study the evolution of peak intensity, peak broadening and peak position pole figures with increasing loads using the neutron diffractometer STRESS-SPEC, which is the first time to perform such kind of measurements on this diffractometer. Microstructure and strain anisotropies, correlations among these three kinds of pole figures will be discussed.

2. Experimental

The in-situ tensile/compression device was designed under the cooperation with SPODI group at FRM II, as shown in Figure 1 (left). It achieves a chi 0~90° tilting and a phi 0~360° rotation which can cover a complete pole figure under loading by combining reflection and transmission methods. It can also be mounted and aligned easily on the STRESS-SPEC sample table.

Austenitic steel was used to prepare the tensile sample with its dimension shown in Figure 1 (right). Wavelength of 1. 73 Å produced by Ge monochromator was selected. Primary slit which controlled the coming beam size to Ø10 mm with the second slit open were utilized. An area detector of 30×30 cm^2 which can cover the (111) and (200) reflections simultaneously was installed. Pole figure measuring points were selected under the yield point at 0 MPa, 100 MPa, 200 MPa and 300 MPa, respectively. At each point the standard equal angular scan method for complete pole figure measurement was realized, of which χ tilts from 0° to 90° and φ rotates from 0° to 360° at each χ. Due to a large area detector and its distance to 850 mm only 6 times χ tilting can cover 90°. At STRESS-SPEC a continuous scanning routine of φ rotation with each 5° was implemented, i.e., the detector collects all the diffraction patterns when φ rotates 5° continuously in 1 min. Total time of 7.5h was needed at one load for a complete pole figure measurement.

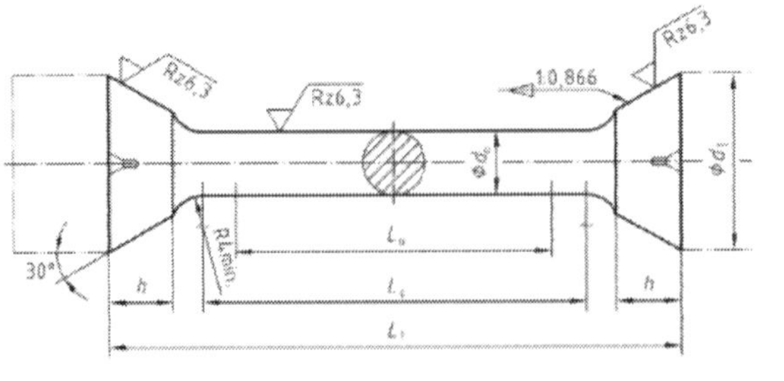

d_0	L_0	d_1	$h\approx$	L_C	L_1
6	30	20	6	68	90

Figure 1 View of the tensile machine (left); and right is the dimension of austenite sample (in mm).

A software package StressTexureCalculator (SteCa) developed mainly by C. Randau at STRESS-SPEC group was used for data treating of current measured data [8]. SteCa allows to extract pole figure data for intensity pole figures (crystallographic texture), for peak position pole figures (macro strain) or for peak broadening pole figures (micro strain) and to construct diffraction patterns (phase analysis) as well as strain profile patterns as function of x, y, z (strain mapping). Fundamentals and functions of SteCa have been described in detail from the reference paper [8]. Gaussian fitting was used to calculate the peak intensity, peak position and FWHM using linear background of an *hkl* reflection. The output ASC II data each containing intensity, peak position and FWHM variation with equal angular 5°×5° grid can be plotted using PfPlot.

3. Results and discussions

Figure 2 shows the variation of (200) peak position at start position of each different load. Clear is that the central peak position was shifted with increasing load. This is due to the lattice planes perpendicular to the loading direction are compressed.

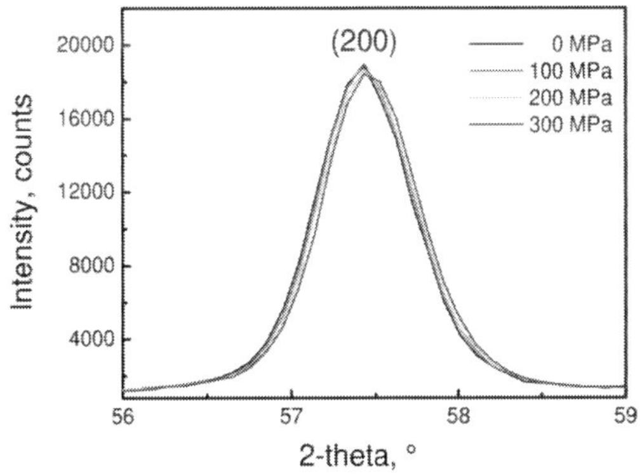

Figure 2 Variation of (200) peak with different loads.

Figure 3 shows the (111), (200) and calculated ODFs under load of 0 MPa and 300 MPa, respectively. The tensile direction is the pole figure centre. There exits a strong brass {110} <112>, a cube {001} <100> and a weak rotated goss {110} <110> components at the initial material. Looking on the pole figures at load=300 MPa they indicate that there exist a rotation from pole figure central tensile direction to the edge. But the maximum intensity was remained.

Figure 4 shows the peak position pole figures of (111) and (200) at two loads, respectively. At loads=0 MPa both (111) and (200) peaks show a position variation about 0.02°. When the load was increased to 300 MPa the (111) peak position distributes from χ=40° of 49.01° to χ=90° of 49.05°, and (200) peak from χ=60° of 57.43° to χ=90° of 57.49°. High load leads to the tensile effect on the lattice planes at central part and a compression effect on the planes at sample's edge. The 2θ angle was shifted about 0.06° for (200) peak under load 300 MPa which was still under yield point of this austenitic steel.

Looking on the (111) and (200) FHWM pole figures of load=0 MPa and load=300 MPa, as shown in Figure 5, both illustrate a similar as a fibre distribution along central tensile direction which demonstrate a precise sample alignment of the instrument setup. There exits a gradient decrease of FWHM in (200) from centre of 0.79° to edge of 0.73° at load= 0 MPa. This indicates that there could exit relatively high density of dislocations in the as received material; and with much higher density in the centre than at the edge. It has been reported that dislocation contributes mainly to the peak broadening when its size is not in the nano region [9]. FWHM at the central part was increased to 0.81° of (200) peak under load= 300 MPa, as shown in the figure, apparent is that much more peak broadening occurred from centre to edge of the pole figure than that at load=0 MPa.

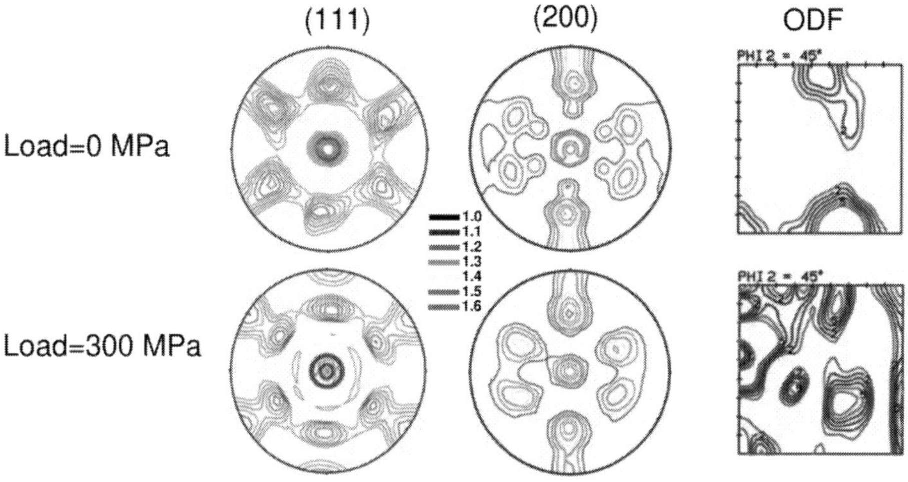

Figure 3 (111) and (200) peak intensity pole figures and calculated ODF at Phi2=45° for loads 0 MPa and 300 MPa, respectively.

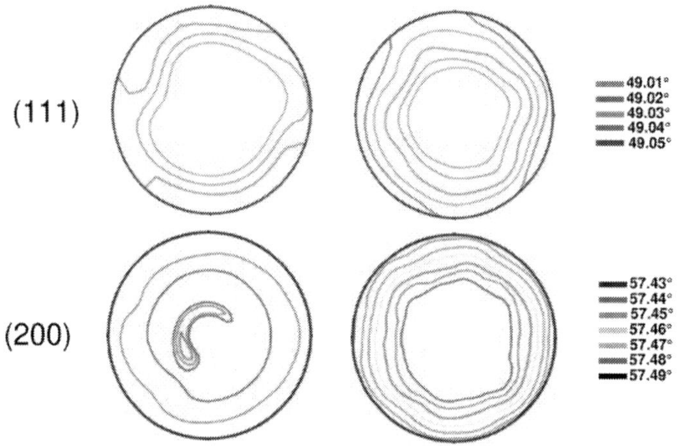

Figure 4 (111) and (200) peak position pole figures at 0 MPa (left) and 300 MPa (right).

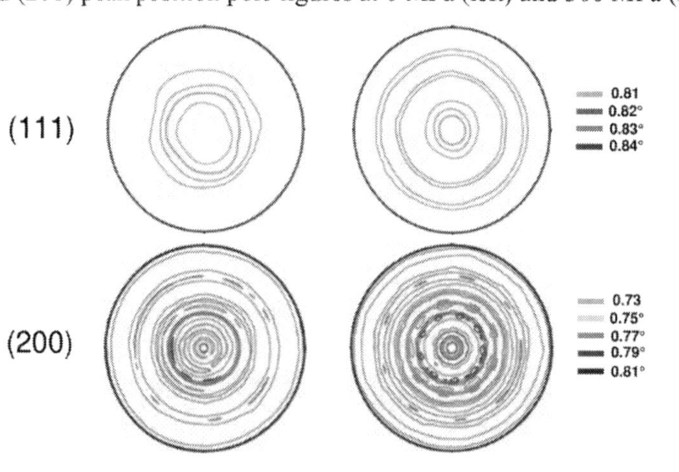

Figure 5 (111) and (200) FWHM pole figures at 0 MPa (left) and 300 MPa (right).

Two main points should be emphasized when comparing intensity, peak position and FWHM pole figures. Firstly, all the pole figures show a fibre distribution around the tensile direction. Reorientation of some grains at 300 MPa was noticed in intensity pole figure. The variations of peak position and FWHM after tensile are not so large since current load is still under yield strength. Different irradiated gauge volume at each χ tilt position might have some effects on the peak profile. However, tensile deformation of the rod sample, a relatively large coming beam and long counting time can decrease this effect.

Secondly, distribution of contour level in peak position and FWHM pole figures which are presented with α and β angles are similar as that of peak intensity in the (111) and (200) pole figures. In theory, there should be no plastic deformation under 300 MPa. This should be due to the modification of Young's modulus resulted resulted from the preferred orientation distributions.

4. Summary

From this investigation it indicated that the strain and texture distribution can be obtained simultaneously provides a good alignment of the neutron diffractometer. Gradient distribution of peak position and FWHM can be related to the preferred orientation distribution of crystals. This new development will greatly enlarge and improve the application of STRESS-SPEC in the filed of materials science.

Acknowledgements

We would like to thank G. Seidl and B. Schwebke for sample preparation and tensile machine test.

References

[1] Wilkens M 1975 *J. Appl. Cryst.* **8** 191
[2] Dahms M and Bunge H J 1989 *J. Appl. Cryst.* **22** 439
[3] Snyder R L, Fiala J and Bunge H J *Defect and Microstructure Analysis by Diffraction* (USA: Oxford University Press)
[4] Hall W H and Willianson G K 1951 *Proc. Phys. Soc.* **LXIV, II-B** (London) P937
[5] Rietveld H M 1969 *J. Appl. Cryst.* **2** 65
[6] Brokmeier H G, Gan W M, Randau C, Voeller M, Rebelo-Kornmeier J and Hofmann M 2011 *Nucl. Instr. & Meth. in Phy. Res.* **A642** 87
[7] Hofmann M, Schneider R, Seidl G, Rebelo-Kornmeier J, Wimpory R C, Garbe U and Brokmeier H G 2006 *Phy. B* **385-386** 1035
[8] Randau C, Garbe U and Brokmeier H G 2011 *J. Appl. Cryst.* **44** 641
[9] Ungár T, Ott S, Sanders P G, Borbély, A and Weertman J R 1998 *Acta mater.* **46** 3693

Neutron and X-ray diffraction study of residual and internal stress evolution in pearlitic steel during cold drawing

M Kriška[1], J Tacq[1], K Van Acker[1], M Seefeldt[1], S Van Petegem[2]

[1)] Catholic University of Leuven, Department of Metallurgy and Materials Engineering, B-3001 Heverlee (Leuven), Belgium

[2)] Paul Scherrer Institute, NUM/ASQ, CH-5232 Villigen PSI, Switzerland

Martin.Kriska@mtm.kuleuven.be

Abstract. Neutron and X-ray diffraction were used to study the residual and internal stress evolution during cold drawing in pearlitic steel wires. A selection of high strength filaments drawn to different reductions has been investigated. In order to compare the evolution of macro and micro residual phase stresses in ferrite, the lattice strain evolution has been studied in axial and transverse direction. In-situ neutron diffraction tests in "Poisson" geometry have been carried out at the TOF strain scanner POLDI at PSI, Switzerland. These tests revealed a significant scatter in mechanical response among differently oriented ferrite grains, including a peculiar response of the {200} reflection, cp. [1, 2].

1. Introduction

High-carbon pearlitic steel wires are used in many applications that require exceptional tensile strength combined with acceptable ductility, which hence offer potential for materials and energy savings in various industrial applications. In terms of quantity, the most important are steel cords in automotive tire reinforcement, production of cables for suspension bridges and others [3-5]. The microstructural evolution of heavily drawn pearlitic steel used for high strength applications has been extensively studied with different experimental techniques. As widely discussed in literature, the pearlitic microstructure consisting of ferrite and cementite is changed drastically with severe plastic deformation and cementite starts to dissolve [6-8]. This cementite decomposition strongly affects the mechanical properties of the cold drawn wires and therefore is a challenging phenomenon during wire drawing to be well managed. However, while extensive studies on plastically induced cementite decomposition have been done with local high resolution techniques [6-8], an integral characterization method giving results representative of the bulk mechanical behaviour is still missing. In general, diffraction techniques were found to be promising techniques for these observations. Especially, in situ neutron diffraction has been proved a powerful tool to study internal stresses and hence the mechanical behaviour of pearlitic steels [1, 9, and 10]. Moreover, static neutron diffraction can be used to reveal both the residual phase microstresses arising as a consequence of the drawing process due to the different elastic and plastic properties of particular phases, and - in combination with XRD - the macrostresses [3, 10 and 11]. The importance of these stresses must be taken into account when considering the mechanical behaviour of carbon steel wires that are mechanically inhomogeneous. In general, their presence may influence dimensional stability and static strength, onset of micro cracks and delamination and finally work-hardening and strengthening mechanisms in pearlitic steel.

Correspondingly, the residual stress evolutions during the cold drawing process and the internal stress among differently oriented crystallites have been investigated by complementary use of neutron and x-ray diffraction and the result are discussed in the present study.

2. Experimental

2.1. Material

A selection of hypereutectoid steel wires prepared by cold drawing has been used for the investigation. The studied material has been prepared with standard wire route geometry (standard alignment of the drawing machine). Specimens of different drawing strains (true strain (2ln (d_0/d_n) - ε_t - 0.0 - 3.9)) were used for static neutron/x-ray diffraction and in-situ neutron diffraction experiments.

2.2. Neutron and X-ray diffraction

The time-of-flight (TOF) neutron diffraction (ND) method has been used for lattice strain measurement. The present static and in situ experiments have been carried out at the Paul Scherrer Institute's spallation neutron source SINQ, on the instrument POLDI (Pulse-OverLap DIffractometer, designed mainly for strain-scanning experiments) [12]. The POLDI tensile rig with a load capacity of maximal 25 kN allows in situ measurements in horizontal and vertical position and thus measuring the lattice strain evolution both parallel and perpendicular to the tensile axis (axial and transverse strains). Correspondingly, static measurements on wires have been carried out in both (axial and transverse) directions. Further, in-situ neutron diffraction tests have been performed in transverse "Poisson" geometry. In general, the TOF-instrument has a definite advantage: it always yields peak shifts for more than a single reflection line [13]. Moreover, since neutron radiation can completely penetrate a bundle of wires and since the macrostress contribution in one phase has to balance out for equilibrium reasons, ND provide valuable information to check the presence of microstresses (stresses of the 2nd kind). In order to assess the residual macrostress evolution during cold drawing, X-ray diffraction measurements were done on the identical unloaded material on a Seifert MZ IV goniometer (by means of a "five point" measurement in the 0 - 0.5 $\sin^2\psi$ interval) at K.U. Leuven using a 40 kV Cr tube. Classical laboratory XRD with limited, shallow penetration offers in two phase material a possibility to determine the total phase stress. The measured total phase stress is then the sum of macro and microstresses.

Due to the low volume fraction and orthorhombic structure of cementite (spreading the scattering intensity into a large number of Bragg reflections), high transparency in comparison with ferrite [14] and the high level of deformation, only the ferrite phase was probed with both diffraction techniques. To convert the {110} lattice plane strains (measured by ND) into residual phase microstress in ferrite, orientation dependent Diffraction Elastic Constants (DECs E_{110} = 220 GPa and ν_{110} = 0.28) were used [15]. Assuming a principal (micro) strain and stress state, i.e. negligible shear stresses, the axial stress component was calculated from generalized Hooke's law as follows [16]:

$$\sigma_{Axial} = \frac{E_{110}}{(1+\upsilon_{110})(1-2\upsilon_{110})}[(1-\upsilon_{110}) * \varepsilon_{AD} + \upsilon_{110} * \varepsilon_{TD}] \quad (1)$$

where ε_{AD} and ε_{TD} are the axial and transverse lattice plane strains, respectively. The elastic strains were computed with a reference lattice spacing d_0 measured on the heat patented wire in "Poisson" geometry using a long counting time. Similarly hoop/radial stresses were obtained (by equation. (2)), taking into account a need of translation of the average lattice plane strains measured by ND in transverse direction which involves the radial and hoop contributions through $\overline{\varepsilon_{TD}} = \frac{\overline{\varepsilon_{rr}+\varepsilon_{\varphi\varphi}}}{2}$:

$$\sigma_{Hoop/Radial} = \frac{E_{110}}{(1+\upsilon_{110})(1-2\upsilon_{110})}[(1-\upsilon_{110})\frac{1}{2}\varepsilon_{TD} + \upsilon_{110}(\frac{1}{2}\varepsilon_{TD} + \varepsilon_{AD})] \quad (2)$$

3. Results and discussion

3.1. Ex-situ neutron and x-ray diffraction

The average lattice plane strains observed by ND for the individual reflections are shown for axial direction (AD) in figure 1 (a) and for transverse direction (TD) in figure 2 (a). Compressive strains were measured in AD for all depicted orientations. In TD, the lattice plane spacing patterns are more complex: tensile and compressive residual lattice plane strains were recorded at low ($\varepsilon_t = 0 - 0.6$) and at higher strain ($\varepsilon_t = 1.2$), respectively. At even larger strains, a strong tensile shift was observed for the {200} and {211} orientations, seen in figure 2 (a).

The evolution of the bulk average {110} lattice plane spacings in AD shows a monotonously decreasing tendency with increasing drawing strain. The same will thus apply to the average residual phase microstress in the {110} reflecting grains that are dominating in the arising {110} fibre texture. However, it turns out that at larger strains ($\varepsilon_t \geq 1.2$), the trend of the {211} lattice plane spacings evolves differently, namely shifting back towards less compression. The lattice plane strain evolutions of the individual reflections in TD (radial/hoop direction) are plotted in figure 2 (a). A good reproducibility of the measurements has been shown for the {110} and {211} oriented grains at a drawing strain of $\varepsilon_t = 0.6$ by successive measurements on three different samples of the same material (see the enclosed enlargement in figure 2 (a)). A similar lattice plane spacing evolution among differently oriented ferrite grains is found up to a true drawing strain of $\varepsilon_t = 1.2$, similar in terms of compressive and tensile strains in the range of $\varepsilon_t = 0 - 0.6$ and $\varepsilon_t = 0.6 - 1.2$, respectively. Beyond $\varepsilon_t \geq 1.2$, the lattice plane spacings start to diverge – comparable to what is observed for AD (see above) and during in-situ tensile tests on as-patented as well as deformed wire (see section 3.2).

The axial/hoop total stresses measured by XRD and an overview of the total, micro and calculated macrostress evolutions in the ferrite are shown in figure 1 (b) and (c) and 2 (b) and (c), respectively. Herein, the phase microstress has been calculated for the {110} reflecting grains, as described in section 2.2. In literature, the DECs E_{110} and E_{211} for ferritic and pearlitic steels have been reported to be similar for the {110} and {211} orientation families [14]. Own in-situ neutron diffraction tests on the present as-patented wires, however give significantly different DECs E_{110} and E_{211}, see section 3.2. The residual phase microstresses, as calculated above, will thus strongly vary among the different grains at large strains. Macro residual stresses then should not be calculated from the difference of XRD measured total lattice plane strains probing {211} oriented grains and ND measured micro phase lattice plane strains probing {110} oriented grains. The plotted total and macro stresses are thus not reliable beyond $\varepsilon_t = 1.2$. The corresponding strong intergranular stresses evolving at large strains might explain the remarkable total ferrite stress behaviour at large strains, see below.

In the total ferrite stress plotted in figure 1 (b), a low and a high deformation regime of residual stress evolution during cold drawing can be distinguished. They are characterized by a monotonous and a rapidly changing behaviour, respectively. The ferrite stress is compressive in the early passes of the low deformation regime, possibly due to the total stress being dominated by the microstress evolution as long as the cementite does not deform plastically. This interpretation of the initial dip in terms of the microstresses matches with the evolution of the full width at half maximum (FWHM) from the neutron and the X-ray diffraction peaks. It is well known that the broadening is primarily the result of two related phenomena, namely a reduction of the "crystallite size" (coherently diffracting domain size) and an increase in the range of microstrains, related, among others to the dislocation density [9, 17]. As one can see in figure 3 (b), it is after the first three passes that the FWHM takes a turn to a much lower slope - meaning that the accumulation of microstrain or the storage of excess dislocations is slowing down. Excess dislocations would pile up at the ferrite-cementite phase boundary as long as the cementite does not deform plastically. However, this trend of the total ferrite stress evolution is suddenly changed and gradually turns into tensile values. This matches with the traditional theory for residual stress evolution in drawn wires [18] where two distinct types of residual stress patterns (compressive or tensile) are distinguished for drawn wires, depending on the amount of reduction. For reductions per pass of less than about 1 percent (coupled with a deformation localized in a surface layer) the axial (macro) residual stresses are compressive at the surface (and then tensile in the centre). In case of larger reductions corresponding with industrial and also the present drawing treatment, the additional axial residual stresses introduced per pass are tensile at the surface of the wire. A microstress saturation at high deformation, as the evolution of the {110} reflections seems to suggest, would confirm that macro residual stresses are decisive for the total stress evolution at the surface at intermediate and large strains.

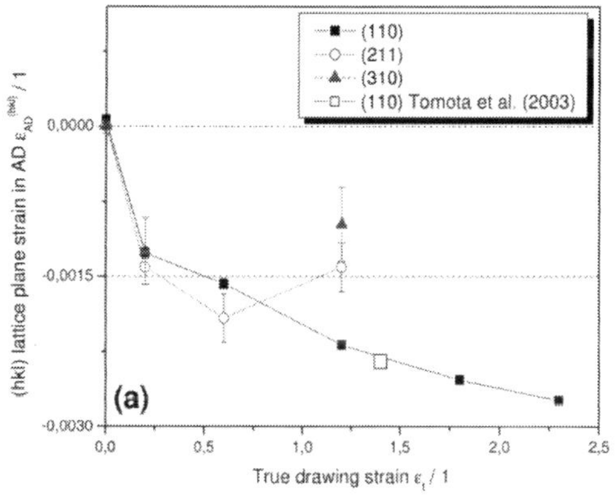

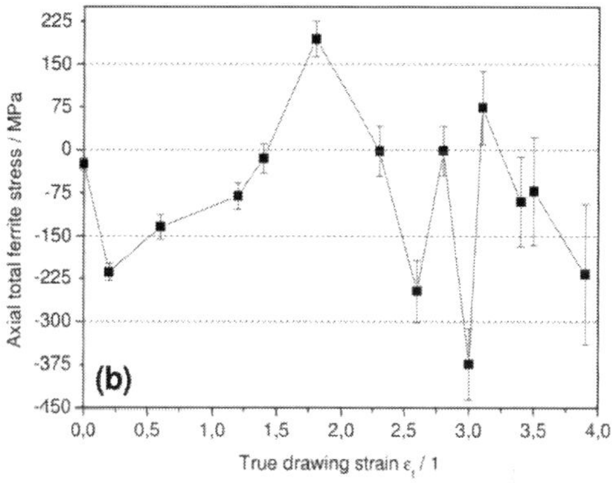

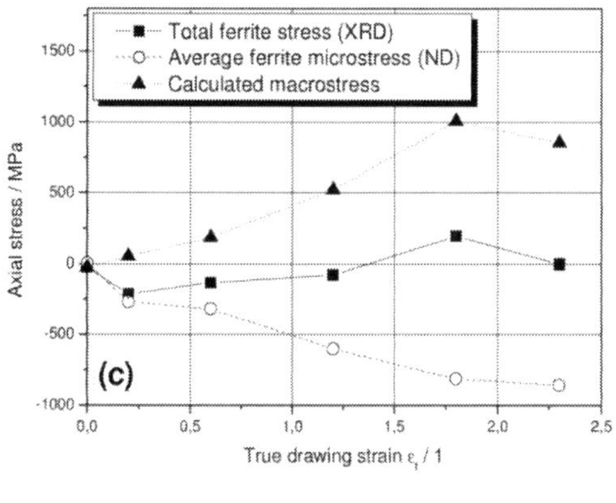

Figure 1. (a) Ferrite lattice plane spacings determined by ND in axial direction; (b) the axial total ferrite stress evolution by XRD and (c) overview of the total, micro and calculated macrostress evolutions as functions of drawing strain during cold drawing. The error bars reflect the uncertainties of the peak fitting procedure. In (a), displayed ferrite 110 lattice plane strain dependence indicating possible microstress saturation fits with value of accumulated residual ferrite strain after cold drawing to ε_t = 1.4 as reported by Tomota et al. [9].

If the macrostress indeed dominated the total phase stress evolution in the high deformation regime, one would also expect a corresponding monotonous evolution of the recorded total ferrite (tensile) stress. However, this is inconsistent with the observed rapid fluctuations in the present total stress evolution. The physical reason for these remarkable "oscillatory" behaviour at large strains is still unclear. In the static ND measurements, clearly for TD, and, as far as the signals are there, also for AD, two distinct regimes are observed as well, an early one where the different reflecting orientations are behaving similarly, and a late one where they are diverging. However, since all observed reflections still evolve monotonously, increasing intergranular stresses alone would not give rapid fluctuations in micro phase and total stresses. Therefore, sudden microstructural changes have to be taken into account as well, for instance loss of integrity, rupture or decomposition of the present cementite phase, as observed at large true strains (ε_t > 2) related to relaxation of interface energy and interface stresses [19, 21 and 22].

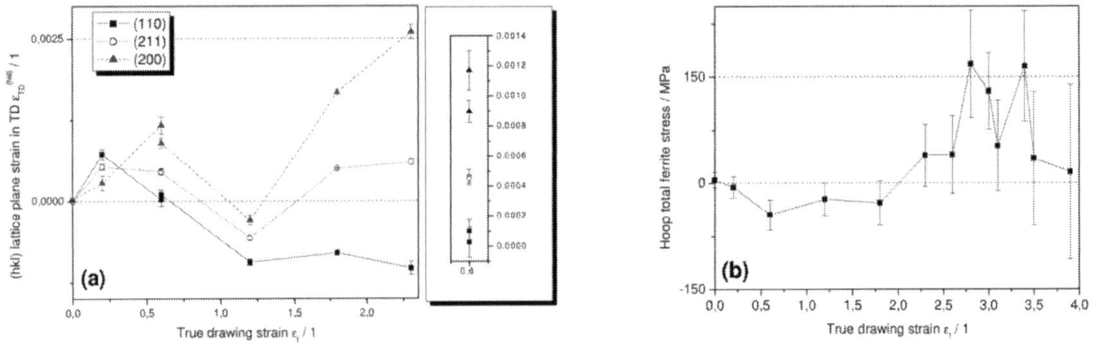

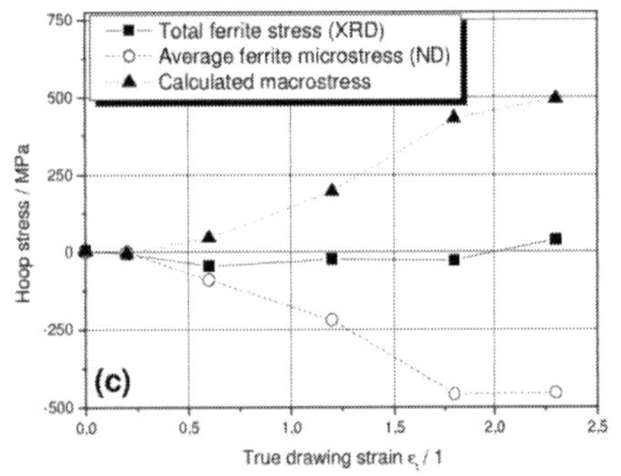

Figure 2. (a) Ferrite lattice plane spacings determined by ND in transverse direction, the enclosed enlargement shows repeated measurements for ε_t = 0.6; (b) the hoop total ferrite stress evolution by XRD and (c) overview of the total, micro and calculated macrostress evolutions as functions of drawing strain during cold drawing.

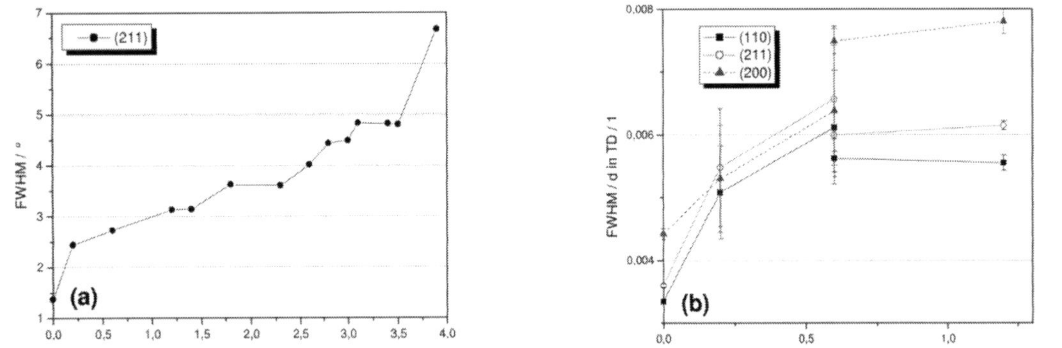

Figure 3. The FWHM determined by XRD (a) and ND (b). In (a), the recorded FWHM evolution corresponds to the experimentally observed significant increase of dislocation density (at large true strain, $\varepsilon_t > 3$) with further drawing [23].

3.2. In-situ neutron diffraction study

Figure 4 shows the lattice plane strain in transverse direction for different {hkl} ferrite grain families as a function of an external stress during in situ tensile loading. The measurement has been performed on as-patented wire (with a lamellar cementite phase). As one can see, a significant variation in the mechanical response is found among differently oriented ferrite grains, including a peculiar response of the {200} reflection, cp. [1, 2, 24]. The present results show a discrepancy with the behavior of high carbon steel studied by Oliver et al. [1] and Kanie et al. [25]. In the fully elastic regime, individual crystallites within the probed sample exhibit a stiffness corresponding to their inherent anisotropy [26] and yield in the following order: {200}, {310}, {211} and {110}. In contrast to [1] and [23], the compliant and (on average) soft {200} orientations here do yield much earlier than the others (regime A), and they undergo an even more dramatic tensile shift (regime B) that even changes the character of the internal stresses in these grains from compressive to tensile. In case of [1], this variation in the yielding might be due to the difference in the morphology of the hard phase, the cementite being globular there and lamellar here. Further, similar to behavior reported for dual phase steel [3], all probed orientations show a second turning point (regime C). As comes out from figure 4, all probed orientations are yielding clearly before these second turning points. Therefore, these turning points might indicate that a non-probed component of the microstructure, i.e. the cementite phase, is getting plastic.

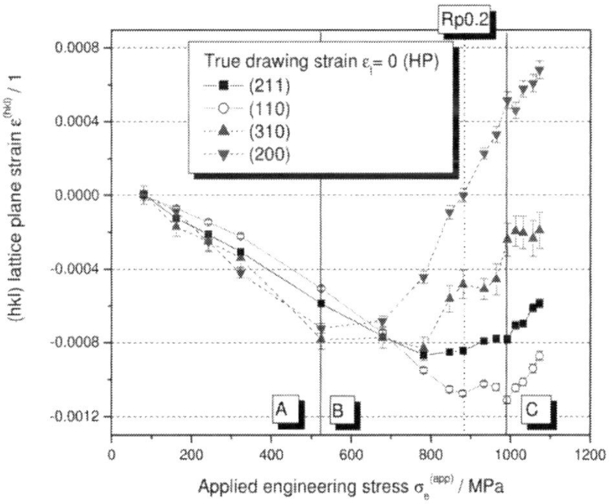

Figure 4. Ferrite lattice plane strain - stress responses of individual reflections for as-patented wire in transverse direction during an in situ tensile test in ND.

4. Conclusions

A complementary ex/in-situ neutron and X-ray diffraction study of residual and internal stress evolution in pearlitic steel during cold drawing was presented. The current XRD results indicate that in general, the total residual stress evolution during severe cold drawing can be described in terms of two distinct regimes. These are low and high deformation regime characterized by monotonous and rapidly fluctuating behavior, respectively. It was shown that the macrostress dominated the total residual stress evolution at low strain (ε_t = 0.6 - 1.8). Moreover, ex situ neutron diffraction has pointed to a strong orientation dependence of the lattice spacing evolution beyond a true strain of $\varepsilon_t \geq 1.2$. Finally, in situ neutron diffraction has – in contrast to some results reported earlier in literature – revealed strong differences in yielding and stress partitioning among particular {hkl} reflections.

Acknowledgements

The authors gratefully acknowledge support the IAP program of BELSPO, project no. P6/24, and from NV Bekaert SA.

References

[1] Oliver EC, Daymond MR and Withers PJ 2004 *Acta Materialia* **52** 1937
[2] Withers PJ 2007 *C.R. Physique* **8** 806
[3] Seefeldt M, Dillien S and Stuhr U 2011 *Materials Science Forum* **31**
[4] Elices M 2004 *J. Mater. Sci.* **39** 3889
[5] Li L and Virta J 2011 *Materials Science and Technology* **27** 845
[6] Gavrilyuk VG 2003 *Materials Science and Engineering A* **345** 81
[7] Sauvage X, Copreaux J, Danoix F and Blavette D 2000 *Philosophical Magazine A* **80** 781
[8] Borchers Ch, Al-Kassab T, Goto S and Kirchheim R 2009 *Materials Science and Engineering A* **502** 131
[9] Tomota Y, Lukáš P, Neov D, Harjo S and Abe YR 2003 *Acta Materialia* **51** 805
[10] Seefeldt M, Walentek A, Van Houtte P, Vrána M and Lukáš P 2006 *Materials Science Forum* **524-525** 375
[11] Van Acker K, Root J, Van Houtte P and Aernoudt E 1996 *Acta Materialia* **44** 4039
[12] Stuhr U, Egger J, Hofer A, Rasmussen P, Graf D, Bollhalder A, Schild M, Bauer G and Wagner W 2005 *Nuclear Instruments and Methods in Physics Research A* **545** 330
[13] Withers, PJ Johnson MW and Wright JS 2000 *Physica B* **292** 273
[14] Wilson D and Konnan Y 1964 *Acta Metallurgica* **12** 617
[15] Eigenmann B and Macherauch E 1996 *Mat.-wiss. U. Werkstofftech* **27** 426
[16] Ezeilo AN and Webster GA 1999 *Textures and Microstructures* **33** 151
[17] Hauk V *1997 Structural and Residual Stress Analysis by Nondestructive Methods* Amsterdam Elsevier Science B.V
[18] Dieter GE 1988 *Mechanical Metallurgy* London McGraw-Hill
[19] Zelin M 2002 *Acta Materialia* **50** 4431
[20] Zhang X, Godfrey A, Hansen N, Huang X and Qing Liu W 2010 *Materials Characterization* **61** 65
[21] Umemoto M, Todaka Y and Tsuchiya K 2003 *Materials Science Forum* **426-432** 859
[22] Zolotorevsky NY, Titovets YF and Vasiliev DM 2004 *Rev.Adv.Mater.Sci.* **7** 91
[23] Shiratori T, Shiota Y, Ryufuku S, Adachi Y, Suzuki Y and Tomota Y 2006 *The 3rd International conference on Advanced Structural Steels* Gyeongju Korea
[24] Sevillano JG, Alkorta J, González D, Van Petegem S, Stuhr U and Van Swygenhoven H 2008 *Advanced Engineering Materials* **10** 951
[25] Kanie A, Tomota Y, Torii S and Kamiyama T 2004 *ISIJ International* **44** 1952
[26] Dye D, Stone HJ and Reed RC 2001*Current Opinions in Solid State and Materials Science* **5** 31

Porosity determination in doped graphites using small-angle neutron scattering measurements

K Mergia[1], K L Stefanopoulos[2], M Martinez-Escandell[3] and Pavel Strunz[4]

[1] Institute of Nuclear Technology and Radiation Protection, National Centre for Scientific Research "Demokritos", GR-15310 Aghia Paraskevi Attikis, Greece
[2] Institute of Physical Chemistry, National Centre for Scientific Research "Demokritos", GR-15310 Aghia Paraskevi Attikis, Greece
[3] Department of Inorganic Chemistry, University of Alicante, E-03690 Alicante, Spain
[4] Nuclear Physics Institute, Neutron Physics Laboratory, CZ-25068 Řež near Prague, Czech Republic

E-mail: kmergia@ipta.demokritos.gr

Abstract. The porosity of undoped and Zr- and Ti-doped graphites has been determined using small-angle neutron scattering measurements. To differentiate between open and closed pores in the neutron scattering measurements the contrast matching technique was employed. In the pore volume distribution three distinct sizes of pores are present. Doping increases the volume fraction of pores, shifts the mean size to lower values and makes the size distributions wider. In addition it enhances drastically the closed porosity.

1. Introduction

Carbon and graphite are attractive materials for high-temperature applications due to their high thermal conductivity, low electrical resistivity, high strength, high modulus, excellent thermal shock resistance and light weight. They are widely used as engineering materials, such as heaters, electrical contacts, high-temperature heat exchangers, rocket nozzles, leading edges of aircraft wings and plasma facing materials in fusion devices etc. [1-3]. For Fusion application the main drawback of carbon-based materials is their chemical erosion under hydrogen bombardment which has to be avoided [4-5]. Doping of carbon with small amounts of certain elements is known to improve its properties. High thermal conductivity and high thermal shock resistance, [6-7] and elimination or reduction of their chemical erosion are achieved [8-9] through doping. Some dopants, such as TiC, present catalytic effects on the graphitization, leading to a higher degree of graphitization and thus enhancement of thermal conductivity. In addition, they may act as reinforcement improving the mechanical properties [10]. For an effective improvement of all the properties, a very fine and homogeneous dispersion of the dopants is required, since both the thermo-mechanical properties and the reduction of chemical erosion are particularly sensitive to the particle size of the dopants [8].

The production of doped-carbon materials via co-pyrolysis of a carbon precursor with a heteroatom source soluble in the carbon precursor is a good alternative to produce doped self-sintering carbon with a homogeneous particle distribution. This material can be used to produce graphites without the need of using a binder. The heteroatom precursor must be soluble in the carbon source (usually carbon or petroleum pitch) in order to achieve a good dispersion of the dopant.

A number of methods have been applied for the investigation of the porosity in carbon. Among these are gas adsorption [11], mercury intrusion porosimetry [12], TEM [13] and small-angle scattering techniques [14-17]. Gas adsorption and mercury porosimetry do not provide information about closed porosity and TEM only gives limited information concerning pore connectivity. Small-Angle Neutron Scattering (SANS) can provide information for both open and closed porosity.

In the current work we present a SANS study of the pore microstructure in samples of doped and undoped graphites with the aim to elucidate the effect of doping on the porosity. Three types of dopants were used, namely a) Ti in the form of TiC nanoparticles, b) Ti using as a precursor tetrabutyl-ortotitanate and c) Zr using as a precursor zirconium acetyl acetonate. Closed and open porosity have been determined using the contrast variation technique [18]. The volume distribution, average size and specific surface area (SSA, i.e. surface area per unit volume) of the pores in these materials have been determined.

2. Experimental

2.1 Sample fabrication

Titanium doped mesophase pitches were prepared by co-pyrolysis of an aromatic petroleum residue (ethylene tar) with two different titanium sources, tetrabutyl-ortotitanate and titanium carbide nano-powder with sizes around 100 nm. In the case of zirconium doped mesophase pitches, zirconium acetyl acetonate was used as Zr-precursor. The mixture of petroleum residue with the titanium or zirconium compound was prepared in an ultrasonic bath. Pyrolysis of the mixtures was carried out in a laboratory pyrolysis plant using 350 g of mixture in the reactor. Pyrolysis was performed at 440°C, soak time 4 h and 1 MPa pressure. A reference sample without the titanium or zirconium compound was also prepared.

Self sintering mesophase powders containing Ti or Zr were prepared by extraction of the mesophase pitch, obtained in the pyrolysis, with tetrahydrofurane at its boiling point for 2 hours. This step was necessary to reduce the amount of volatile matter produced during heat-treatment of these self-sintering powders. In order to optimise the self-sintering abilities of the extracted materials and to improve the properties of the resulting graphites, a stabilisation stage has been carried out. All stabilisations were carried out at 175 °C, using dry air (100 mL/min), a heating rate of 2 °C/min and soaking time varying from 0 to 2 hours. All the stabilised samples were sieved to particle size smaller than 63 μm.

Table.1. Sample description.

Sample Code	Precursor	Dopant	Amount of metal in graphite (at%)	Open Porosity, $P_{O,im}$ (%)
C	none	none	0	4
C/Tin	TiC nanoparticles	Ti	2.4	2
C/Tib	Tetrabutyl-ortotitanate	Ti	2.2	6
C/Zr	Zr acetyl acetonate	Zr	1.4	6

Three grams of self sintering powder were uniaxially pressed at room temperature, at 175 MPa, to prepare green compacts of 5×10×50 mm. The compacts were heat treated under nitrogen flow of 60ml/min to 1500 °C, heating rate 1.0 °C/min, and further high-temperature treated at 2600 °C (10 °C/min) under inert atmosphere.

Table 1 describes the samples investigated in the current study and also the values of open porosity as measured by immersion of the bulk samples in the water are given.

2.2 SANS measurements

The theory of small-angle scattering is well described in the literature both generally [19, 20] and specifically for applications in porous materials [16, 21-22]. SANS with the utilization of the "contrast matching" technique has been used to determine both open and closed porosity in carbon and chars [14-17, 22-24].

In a SANS experiment the absolute neutron scattering cross section, $I(Q)$, is recorded as a function of the scattering vector $Q = 4\pi \sin\theta / \lambda$ where 2θ is the scattering angle, i.e. the angle between the incident and the scattered neutron beam and λ the neutron wavelength. SANS arises from the neutron scattering length density difference, $\Delta(SLD)$, between the matrix and inhomogeneities e.g. precipitates, pores. The neutron scattering length of the matrix, SLD_M, is given by

$$SLD_M = \frac{\rho_{sk} N_{Av}}{MB} \sum_i c_i b_i = C\rho_{sk} \quad (1)$$

where MB the average molecular mass, N_{Av} the Avogadro number, b_i is the neutron scattering length of element i with atomic concentration c_i and the sum is over all the elements present in the matrix. It should be noted that in equation (1) the constant C is known from the chemical composition of the matrix and the tabulated values of b_i, whereas the skeleton density ρ_{sk} has to be measured or assumed from the crystallographic structure. If we assume that the material density within the pores is very low then $\Delta(SLD) = SLD_M - SLD_{pore} \cong SLD_M$.

The SLD_M values as calculated vary between 5.99×10^{-6} (Zr doped sample) and 6.39×10^{-6} Å^{-2} (Ti doped sample using TiC nanoparticles). For the undoped sample the SLD_M value (6.70×10^{-6} Å^{-2}) is about 4% higher than that of D_2O. Therefore for the undoped sample 100% D_2O was used for the contrast matching.

Specimens for the SANS measurements were prepared by grinding the initially received samples to a powder of grain sizes larger than 40 μm. Two set of experiments were performed at room temperature, the called "dry" and "wet" sample measurements. In the "dry" sample measurements the neutron scattering measured from the powdered sample arises from both open and closed pores, whereas in the "wet" sample the neutron scattering arises from the closed pores. 1 mm thick Hellma cuvettes were used as sample holders. As a contrast matching liquid D_2O/H_2O was used and the concentration of D_2O/H_2O was calculated such as to match the SLD_M of the respective sample.

In the "wet" sample measurements, the sample powder was immersed in a H_2O/D_2O mixture which had the same neutron scattering length density as that of the solid ("contrast matching"). The pores accessible from the grain surface are filled with the liquid and since the neutron scattering length density difference of these filled pores with the matrix is zero, only SANS from the unfilled (closed) pores is received. The "wet" samples mixed with D_2O/H_2O were allowed for about 4 h to come to equilibrium before measurement. Some "wet" samples were re-measured after 24 h to see if there was any effect of the water wettability but no differences were observed in the values of the scattering at the different Q-values (i.e. scattering curves coincided within statistical errors).

The SANS scattering data were corrected for sample thickness, transmission and instrumental resolution and they were scaled to yield absolute cross sections.

The SANS measurements were carried out at the neutron double-crystal diffractometer MAUD (formerly DN-2) at NPI Řež (λ=0.209 nm) [25]. The measured intensity at a double-crystal diffractometer corresponds to the differential SANS cross section of the sample integrated in the vertical direction and smeared by the resolution function in the horizontal direction. The shape of the horizontal resolution function is equivalent to the shape of the curve measured without the sample. The MAUD instrument is tuned to operate over two ranges of the scattering vector: 2.6×10^{-3} to 0.022 nm^{-1} and 0.014 to 0.12 nm^{-1}. An absolute calibration of scattering cross-sections was performed by measuring the intensity of the direct beam.

The SANS data were fitted using SASProFit software [26] based on indirect Fourier transformation method [27]. In the data treatment it was assumed that the sample is composed of a polydisperse distribution of spherical scatterers. From the numerical processing of the measured cross sections [25] the pore volume distribution, $\varphi(R)$, the porosity, the pore mean size and the pores specific surface area are determined. From the "dry" sample measurements the total porosity parameters are determined whereas from the "wet" sample measurements only the closed porosity ones.

3. Results and Discussion

3.1 SANS results

Figure 1 presents the scattering factor from "dry" and "wet" samples. Two Q ranges have been measured with low and high resolution. The non-exact overlap of the two ranges is caused by the different resolution functions. For low resolution, much broader (angularly) instrumental curve is used than for the high resolution. The solid lines in Figure 1 are fitted curves to the data points and they have been derived using the procedure described in [26].

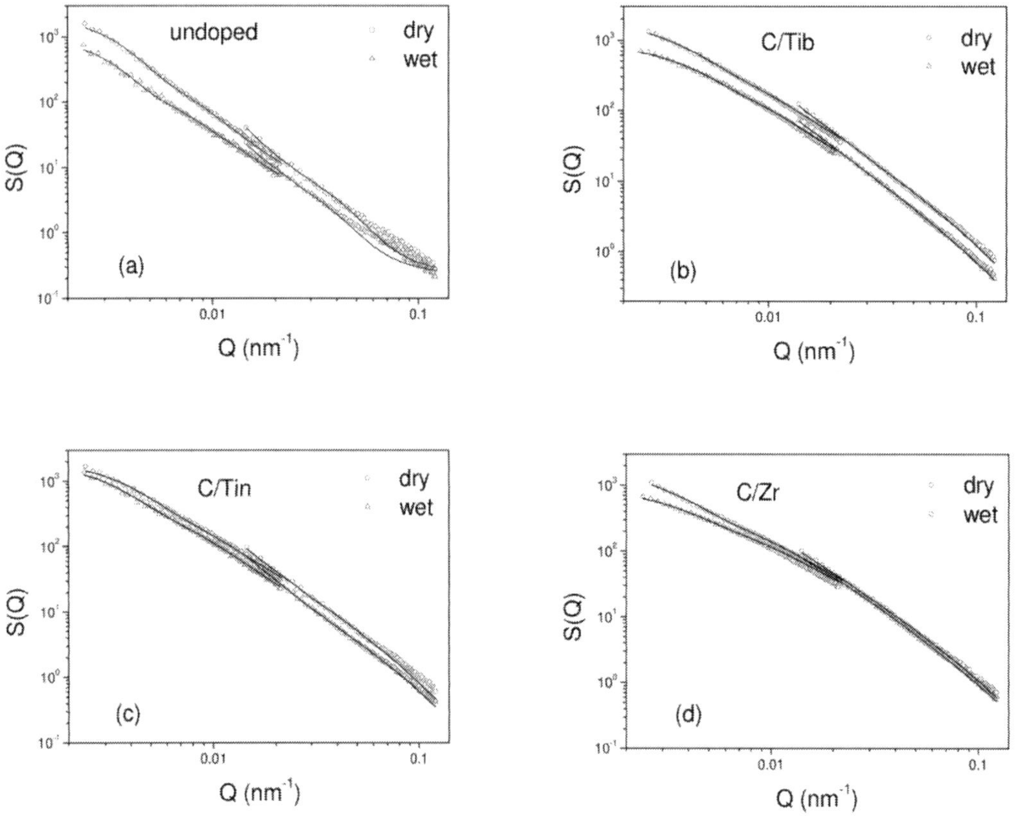

Figure 1. Scattering factors for the "dry" and "wet" samples: (a) undoped, (b) C/Tib, (c) C/Tin and (d) C/Zr. Solid lines are fitted curves to the data according to [26].

The scattering curves have very similar shapes for all the doped samples and in both "dry" and "wet" sample measurements and differ from those of the undoped sample. This observation demonstrates that the morphology and size distributions of open ("dry") and closed ("wet") pores are very similar and that the different dopants introduce similar changes with respect to the undoped sample. However, the absolute scattering cross section diversity among the different samples reflects the different values of either total or closed porosity. For the undoped samples the absolute intensity of the "wet" sample is about half that of the "dry" one which implies that of the total porosity half of it is closed. A similar reduction occurs for the C/Tib sample. On the contrary for the C/Tin sample the absolute intensity of the "wet" sample decreases by about 20% which implies that 20% of the porosity is open (figure 1c). For the Zr doped sample the absolute intensity decreases by about 40% (figure 1b).

From the numerical processing of the data the pore volume distribution, $\phi(R)$, was determined, for the "dry" and "wet" samples, which corresponds to total and closed porosity, respectively. The volume distribution for total and closed pores is depicted in figures 2 and 3, respectively. From these figures it is apparent that the doping does not affect the morphology of the volume distributions, but shifts the volume distribution to smaller sizes. In all samples the volume fraction distribution is trimodal. Doping generally increases the volume fraction of pores, shifts the mean size to lower values and makes the size distributions wider. For the total porosity the first size region is centered around 25 nm, the second around 160 nm and the third around 600 nm. Similar features are present in the size distribution of closed pores.

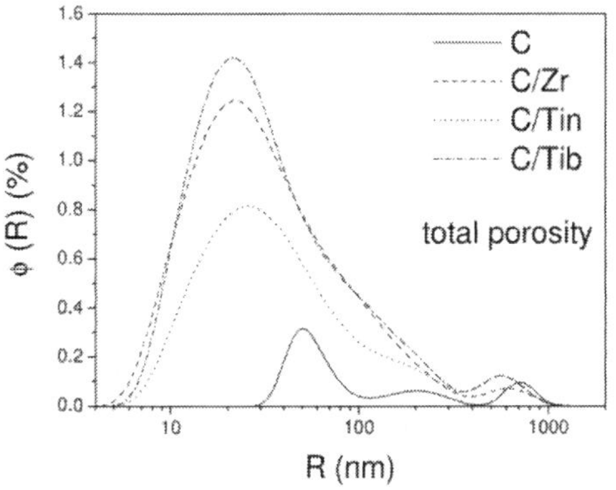

Figure 2. Volume distribution of total pores.

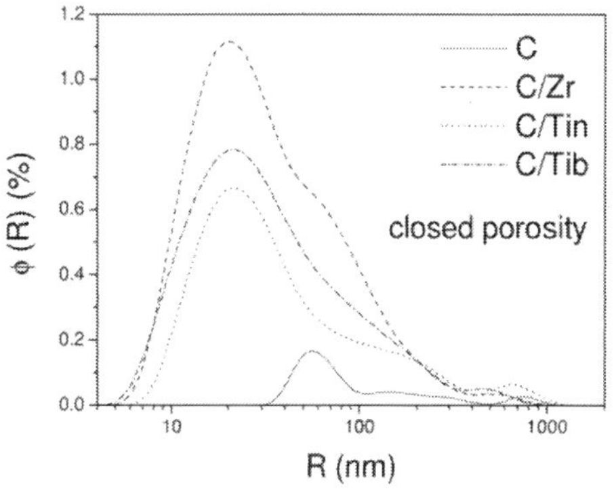

Figure 3. Volume distribution of closed pores.

In all the samples, except the C/Tin, the dopant was introduced chemically and thus the observed SANS arises solely from open or closed porosity. In the C/Tin sample the doping was made using TiC nanoparticles having an average size of 100 nm and volume fraction of about 4%. Therefore, it would be expected the presence of TiC nanoparticles to emerge in the volume distributions of C/Tin sample in both figures 2 and 3. Such a presence of the TiC is not at all obvious taking into account that the volume fraction of the incorporated nanoparticles of around 4% is much larger than the volume fraction at the peak of the distribution. Moreover, it is observed that the volume fraction of the C/Tin sample at around 10 nm is lower than that of the other doped samples. Therefore it is concluded that the TiC nanoparticles after the thermal processing have been dissolved in the matrix and no TiC agglomerates in a large extent remain.

From the size distribution the average radius of total and closed pores are determined (see table 2). As it can be seen, the doping reduces drastically the average size of the pores either open or closed with the maximum reduction observed for the Zr doped sample of about 70%.

Concerning the porosity values, it is observed that there is a drastic increase of the total porosity in the doped samples and this actually reflects the corresponding increase of the closed porosity. The closed porosity of the two Ti doped and the Zr doped samples compared to the undoped one is about three times larger. The sample doped with Tetrabutyl-ortotitanate has the largest total porosity. The dopant either Zr or Ti seems causes a four-fold increase of the closed porosity, whereas the open porosity remains almost the same as in the undoped sample except for the case of C/Tib.

Table 2. Pore size parameters determined from SANS data.

Sample	P_T (%)	Total porosity SSA ($\times 10^3$ cm^{-1})	$\langle R_T \rangle$ (open+closed) (nm)	P_C (%)	Closed porosity SSA ($\times 10^3$ cm^{-1})	$\langle R_C \rangle$ (close) (nm)	P_O (%)
C	5.4 ± 0.1	8.9 ± 0.4	500 ± 20	2.5 ± 0.1	8.9 ± 0.7	487 ± 38	2.9
C/Tin	13.0 ± 0.2	51.3 ± 1.4	300 ± 6	9.3 ± 0.2	37.3 ± 2.1	313 ± 9	3.7
C/Tib	17.1 ± 0.2	83.0 ± 3.2	244 ± 7	9.2 ± 0.1	49.6 ± 1.8	180 ± 4	7.9
C/Zr	14.7 ± 0.2	77.9 ± 3.0	221 ± 5	11.4 ± 0.2	67.0 ± 2.2	156 ± 3	3.3

The open porosity values differ from those determined by the immersion of the bulk samples in water from about 24% (C/Tib sample) up to 82%(Ti/Zr sample). This can be understood as follows. In a macroscopic sample (mm dimensions) there are two possibilities. Either a) all the network paths for each μm region (open grain porosity) are interconnected and terminate to the surface of the macroscopic sample (percolation) or b) only part of the network paths are interconnected and terminate to the surface. For the former case the open and consequently also the closed porosity values for the grain and the macroscopic sample will be the same. In the second case, if we grind the samples into grains, as it is done for the SANS measurements, more networks are exposed to the grain surface making the grain open porosity higher than that of the bulk sample [17]. The SSA for the total porosity increases with doping by a factor varying between 5.7 (C/Tin) and 9 (C/Tib, C/Zr) (table 2). For the closed porosity SSA is four (C/Tin) up to 7.5 times (C/Zr) higher than that of the undoped sample. The increase of the pore surface with doping is expected since the average size decreases and the volume fraction of pores increases with doping.

4. Conclusions

SANS measurements with contrast matching have been employed to determine the open and closed porosity in doped and undoped graphites. It is concluded that the volume distribution of the pores is very similar for all the doped samples and for either closed or open pores. In the volume distribution three distinct sizes of pores are present. Doping with either Zr or Ti decreases the average pore size but increases strongly the closed porosity values. Also in the doped samples the SSA of the closed pores increases drastically with respect to the undoped ones. In conclusion doping introduces microstructural changes which are manifested by the change in the pore volume distribution and closed porosity values.

Acknowledgements

This work has been carried out within the framework of the Integrated European Project "ExtreMat" (contract NMP-CT-2004-500253) with financial support by the European Community. This paper only reflects the views of the authors and the European Community is not liable for any use of the information contained therein. The support of the European Commission through the research project with Contract No. RII3-CT-2003-505925 (NMI3) is acknowledged.

References

[1] Fitzer E 1987 *Carbon* **25** 163

[2] Blazewicz S, Blocki J, Chlopek J, Godlewski J, Mickalowski J, Pakonski K, Piekarczyk J and Stodulski M 1996 *Carbon* **34** 1393
[3] Qian J P, Liu X, Li P Y and Guo Z 1992 *J. Nucl. Mater.* **191-194** 340
[4] Roth J, García-Rosales C 1996 *Nucl. Fus.* **36** (1996) 1647. Corrigendum 1997 *Nucl. Fus.* **37** 897
[5] Vietzke E and Haasz A A 1996 *Physical Processes of the Interaction of Fusion Plasmas with Solids* ed Hofer W O and Roth J (Academic Press) p 135
[6] Marinkovich S 1984 *Chemistry and Physics of Carbon* vol. 19 ed Thrower P A (New York: Marcel Decker) p 416
[7] Ogawa I, Kobayashi K and Nishikawa S 1988 *J. Mater. Sci.* **23** 1363
[8] García-Rosales C and Balden M 2001 *J. Nucl. Mater.* **290-293** 173
[9] Starke P, Fantz U and Balden M 2005 *J. Nucl. Mater.* **337–339** 1005
[10] Qiu H, Han L and Liu L 2005 *Carbon* **43** 1021
[11] Cregg S J and Sing K S W 1982 *Adsorption, Surface Area and Porosity* (New York: Academic Press)
[12] Zgrablich G, Mendioroz S, Daza L, Pajares J, Mayagoitia V, Rojas F and Conner W C, 1991 *Langmuir* **7** 779
[13] Huxham I M, Rowatt B, Sherrington D C and Tetley L 1992 *Polymer* **33** 2768
[14] Kostorz G 1979 Neutron scattering *Treatise on materials science and technology* ed Kostorz G (New York: Academic Press) vol 15 p 227
[15] Porod G 1982 *Small-angle X-ray scattering* ed Glatter O and Kratky O (New York: Academic Press) p 17-62
[16] Calo J M and Hall P J 2004 *Carbon* **42** 1299
[17] Mergia K, Stefanopoulos K L, Ordás N and García-Rosales C (2010) *Microporous and Mesoporous Materials* **134** 14
[18] Stuhrmann H B and Miller A 1978 *J. Appl. Cryst.* **11** 325
[19] Guinier A and Fournet G 1995 *Small-Angle Scattering of X-rays* (New York: John Wiley)
[20] Lindner P and Zemb T 1991 *Neutron, X-ray and Light Scattering* (Amsterdam: Elsevier)
[21] Schmidt P W 1991 *J. Appl. Cryst.* **24** 414
[22] Calo J M, Hall P J and Antxustegi M M 2001 *Colloids & Surfaces A: Physicochem. Eng. Aspects* **187-188** 219
[23] Gethner J S 1986 *J. Appl. Phys.* **59** 1068
[24] Hall P J, Antxustegi M M and Calo J M 1998 *Energy Fuels* **12** 542
[25] Strunz P, Šaroun J, Mikula P, Lukáš P and Eichhorn F *1997 J. Appl. Crystallogr.* **30** 844
[26] Šaroun J *2000 J. Appl. Cryst.* **33** 824
[27] Glatter O 1977 *Acta Phys. Austr.* **47** 83

Neutron scattering studies of Ti-Cr-V bcc alloy with the residual hydrogen and deuterium

K Mori[1], K Iwase[2], M Sugiyama[1], M Kofu[3], O Yamamuro[3], Y Onodera[1], T Otomo[4] and T Fukunaga[1]
[1]Research Reactor Institute, Kyoto University, 2-1010 Asashiro-Nishi, Kumatori, Sennan, Osaka 590-0494, Japan
[2]Frontier Research Center of Applied Nuclear Science, Ibaraki University, 162-1 Shirakata-Shirane, Tokai, Ibaraki 319-1106, Japan
[3]Neutron Science Laboratory, Institute for Solid State Physics, The University of Tokyo, 106-1 Shirakata, Tokai, Ibaraki 319-1106, Japan
[4]Institute of Materials Structure Science, High Energy Accelerator Research Organization, 1-1 Oho, Tsukuba, Ibaraki 305-0801, Japan

E-mail: kmori@rri.kyoto-u.ac.jp

Abstract. Hydrogen and deuterium pressure-composition (P-C) isotherm measurements were carried out for $Ti_{0.31}Cr_{0.49}V_{0.20}$ bcc alloy. As a result, the residual hydrogen-to-metal ratio $(H/M)_{res}$ was almost the same as the residual deuterium-to-metal one $(D/M)_{res}$; the $(H/M)_{res}$ and $(D/M)_{res}$ were approximately 0.5. Furthermore, the neutron scattering experiments were conducted with $Ti_{0.31}Cr_{0.49}V_{0.20}$ alloys including the residual hydrogen and/or deuterium. It was found that H atoms absorbed in the first absorption cycle up to H/M = 0.5 are hardly bound in the desorption cycle as the residual hydrogen, whereas D atoms are mainly bound in $Ti_{0.31}Cr_{0.49}V_{0.20}$ bcc alloy as the residual deuterium when D_2 gas is partially used in the H absorption reaction.

1. Introduction

Among the hydrogen absorbing materials, body centered cubic (bcc) alloys possess a large hydrogen capacity [1–3]. For example, the Ti-Cr-V bcc alloy transforms into $(Ti\text{-}Cr\text{-}V)H_2$, which has the CaF_2-type structure (i.e., fcc structure), by the hydrogenation reaction at room temperature (RT). The hydrogen-to-metal ratio (H/M) of $(Ti\text{-}Cr\text{-}V)H_2$ is 2.0, compared with 1.1 H/M for $LaNi_5H_7$ and 1.3 H/M for Mg_2NiH_4. Therefore, the Ti-Cr-V bcc alloy has attracted much attention as promising hydrogen storage applications. For the reversible hydrogen storage/release, however, we can only use ~50% to ~70% of the maximum hydrogen storage capacity. In other words, there is more than 30% of the residual hydrogen in the Ti-Cr-V bcc alloy. In order to increase the effective hydrogen storage capacity, it is very important to understand the state of the residual hydrogen in the Ti-Cr-V bcc alloy.

In this work, the hydrogen and deuterium pressure-composition (P-C) isotherm, X-ray powder diffraction, and neutron scattering experiments were carried out for hydrogenated and/or duterated $Ti_{0.31}Cr_{0.49}V_{0.20}$ alloys, using a quasi-elastic neutron scattering spectrometer. In particular, we qualitatively evaluated the ratio of the residual H to D for the samples, which were prepared by the different absorption processes: H → D and H → D, respectively.

2. Experimental Procedure

Seven kinds of samples were prepared as follows: (1) $Ti_{0.31}Cr_{0.49}V_{0.20}$, (2) $Ti_{0.31}Cr_{0.49}V_{0.20}H_{1.0}$, (3) $Ti_{0.31}Cr_{0.49}V_{0.20}D_{1.6}$, (4) $Ti_{0.31}Cr_{0.49}V_{0.20}H_{0.5}$, (5) $Ti_{0.31}Cr_{0.49}V_{0.20}H_x$, (6) $Ti_{0.31}Cr_{0.49}V_{0.20}(H\text{-}D)_x$, and (7) $Ti_{0.31}Cr_{0.49}V_{0.20}(D\text{-}H)_x$, where x is the H and/or D content. The (H-D) and (D-H) indicate the order in which high pressure H_2 and D_2 gases are given in the absorption process (i.e., H → D and D → H), as we will mention later. The $Ti_{0.31}Cr_{0.49}V_{0.20}$ alloy was synthesized by arc-melting of Ti (Furuuchi Chem. Co., 99.9%), Cr (99.99%), and V (99.9%), in the molar ratio. The obtained ingot was heated at 1473 K for 12 h in a vacuum furnace (~10^{-2} Pa), followed by quick cooling to 273 K. The sample was grinded to be powder of less than 212 μm prior to the hydrogenation (or deuteration) reaction. After heating at around 573 K in vacuum, it was reacted under hydrogen (or deuterium) pressure of several tens MPa at 298 K – we obtained (2) $Ti_{0.31}Cr_{0.49}V_{0.20}H_{1.0}$ and (3) $Ti_{0.31}Cr_{0.49}V_{0.20}D_{1.6}$ in the absorption cycle. The (4)-to-(7) samples were obtained in the desorption cycle, particularly (5) $Ti_{0.31}Cr_{0.49}V_{0.20}H_x$, (6) $Ti_{0.31}Cr_{0.49}V_{0.20}(H\text{-}D)_x$, and (7) $Ti_{0.31}Cr_{0.49}V_{0.20}(D\text{-}H)_x$ were synthesized by using a rotary pump.

The structural properties of the samples were characterized by the X-ray diffraction (Cu$K\alpha$ radiation) and the Rietveld method [4,5]. The hydrogen and deuterium P-C isotherm measurements were carried out at 298 K on the basis of the Sieverts method. Neutron scattering data were corrected at RT on a high-resolution pulse cold-neuron spectrometer, AGNES, of Institute for Solid State Physics (ISSP), The University of Tokyo, installed at C3-1-1 port in JRR-3M reactor of Japan Atomic Energy Agency (JAEA) [6,7]. The energy resolution of AGNES, ΔE, is 120 μeV (standard mode), and the incident neutron wavelength, λ, is 4.22 Å. The samples were put into a cylindrical aluminum can with 13 mm in diameter and 50 mm in height, and the thickness of sample was fixed to be 1.0 mm.

3. Results and discussion

Figure 1 shows the hydrogen and deuterium P-C isotherms at 298 K for $Ti_{0.31}Cr_{0.49}V_{0.20}$. Even though the plateau pressure of hydrogenation was higher than that of deuteration in the absorption and desorption cycles, the residual hydrogen-to-metal ratio $(H/M)_{res}$ was almost the same as the residual deuterium-to-metal one $(D/M)_{res}$; both of them were approximately 0.5. In the Rietveld analysis, the crystal structures of $Ti_{0.31}Cr_{0.49}V_{0.20}$ (without undergoing hydrogenation), $Ti_{0.31}Cr_{0.49}V_{0.20}H_{1.0}$ (in the absorption cycle), and $Ti_{0.31}Cr_{0.49}V_{0.20}H_{0.5}$ (in the desorption cycle) were refined, as shown in Fig. 2. Consequently, it was confirmed that $Ti_{0.31}Cr_{0.49}V_{0.20}$ has the bcc structure with the lattice parameter a = 3.049(1) Å, $Ti_{0.31}Cr_{0.49}V_{0.20}H_{1.0}$ the CaF_2-type structure with a = 4.339(4) Å (as a principal phase), and $Ti_{0.31}Cr_{0.49}V_{0.20}H_{0.5}$ the bcc structure with a = 3.111(2) Å. Note that the a of $Ti_{0.31}Cr_{0.49}V_{0.20}H_{0.5}$ is slightly larger than that of $Ti_{0.31}Cr_{0.49}V_{0.20}$ due to the existence of the residual hydrogen.

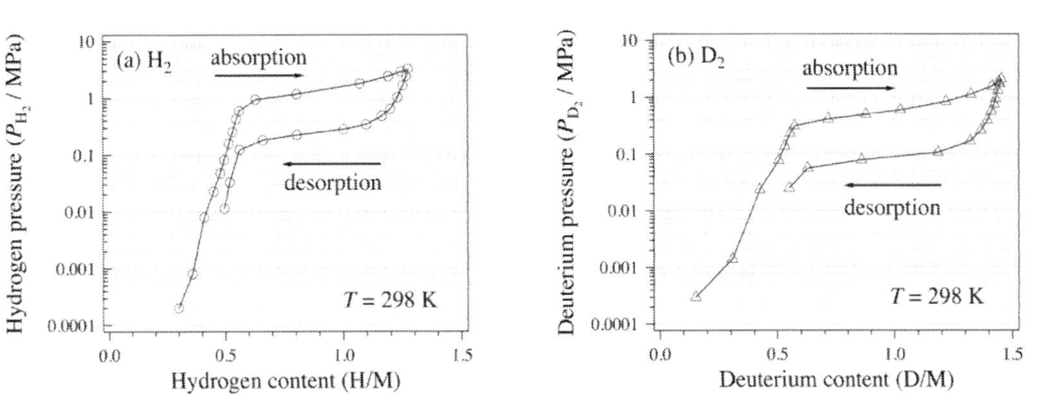

Figure 1. Hydrogen (a) and deuterium (b) pressure-composition isotherms at 298 K for $Ti_{0.31}Cr_{0.49}V_{0.20}$.

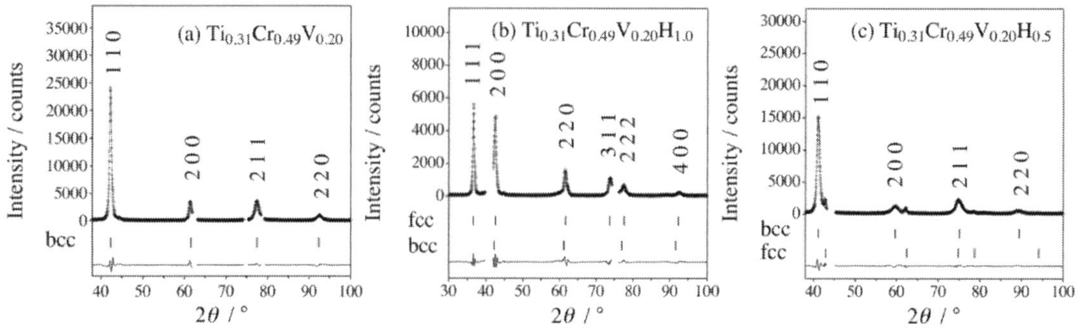

Figure 2. Full patterns of the Rietveld refinement for (a) $Ti_{0.31}Cr_{0.49}V_{0.20}$, (b) $Ti_{0.31}Cr_{0.49}V_{0.20}H_{1.0}$ (in absorption), and (c) $Ti_{0.31}Cr_{0.49}V_{0.20}H_{0.5}$ (in desorption). The plus marks are the observed X-ray diffraction intensities, and gray solid lines are the calculated ones. The vertical marks below the profiles indicate the positions of Bragg reflections for the bcc and/or CaF_2-type (fcc) structures. The curves at the bottom are the difference between the observed and calculated intensities.

Figure 3(a) shows the neutron scattering profile as a function of E, $I(E)$, at Q = 1.68 $Å^{-1}$ at RT for $Ti_{0.31}Cr_{0.49}V_{0.20}H_{0.5}$, together with the resolution function of AGNES, $R(E)$, on the basis of $I(E)$ for $Ti_{0.31}Cr_{0.49}V_{0.20}$, where E is the energy transfer, and Q is the magnitude of the scattering vector. It was found that the $I(E)$ of $Ti_{0.31}Cr_{0.49}V_{0.20}H_{0.5}$ is in good agreement with $R(E)$, meaning that it consists of only an elastic scattering. Furthermore, as shown in Fig. 2, no Bragg peak appears in the low-Q region ($Q \leq 2$ $Å^{-1}$), so that the elastic scattering can be regarded as an incoherent scattering. For $Ti_{0.31}Cr_{0.49}V_{0.20}$, $Ti_{0.31}Cr_{0.49}V_{0.20}H_{0.5}$, and $Ti_{0.31}Cr_{0.49}V_{0.20}D_{1.6}$, the intensities of the incoherent-elastic scattering in terms of Q, $I_{inc\_el}(Q)$, are shown in Fig. 3(b). Obviously, the $I_{inc\_el}(Q)$ of $Ti_{0.31}Cr_{0.49}V_{0.20}H_{0.5}$ is relatively high, because the intensity of the incoherent neutron scattering depends on the incoherent cross section, σ_{inc}, and the σ_{inc} of H (= 80.26 barn) is very large, compared with those of D (= 2.05 barn), Ti (= 2.87 barn), Cr (= 1.83 barn), and V (= 5.08 barn) [8]. In addition, this is the reason why the $I_{inc\_el}(Q)$ of $Ti_{0.31}Cr_{0.49}V_{0.20}D_{1.6}$ is quite low in spite of the large deuterium-to-metal ratio of 1.6 D/M.

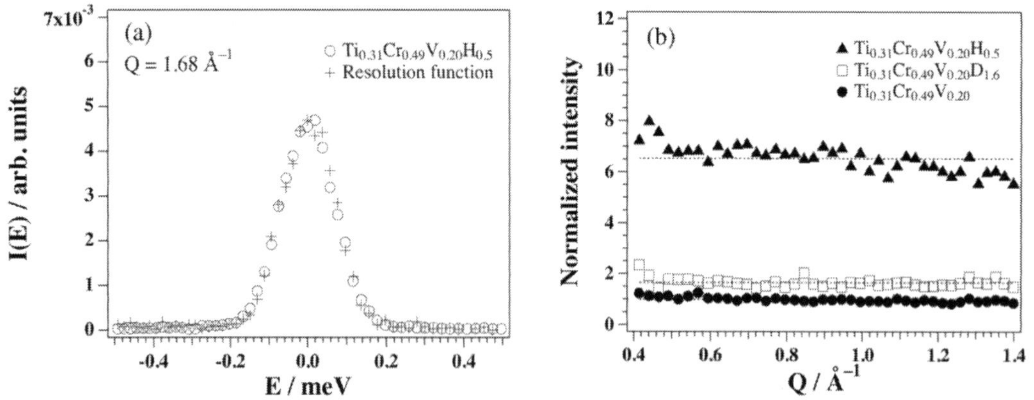

Figure 3. (a) Elastic neutron scattering profile as a function of E, $I(E)$, at Q = 1.68 $Å^{-1}$ at RT for $Ti_{0.31}Cr_{0.49}V_{0.20}H_{0.5}$ (○), together with the resolution function of AGNES (+). (b) Intensities of the incoherent-elastic scattering in terms of Q, $I_{inc\_el}(Q)$, for $Ti_{0.31}Cr_{0.49}V_{0.20}$ (●), $Ti_{0.31}Cr_{0.49}V_{0.20}D_{1.6}$ (□), and $Ti_{0.31}Cr_{0.49}V_{0.20}H_{0.5}$ (▲).

Journal of Physics: Conference Series **340** (2012) 012103 doi:10.1088/1742-6596/340/1/012103

As mentioned above, the P-C isotherm measurements gave us the following result: $(H/M)_{res}$ = $(D/M)_{res}$ = ~0.5. We assumed, therefore, that in the first absorption cycle up to H/M (or D/M) = 0.5 the H (or D) atoms are strongly bound to Ti or V ones to be the residual hydrogen (or deuterium). Note that Ti and V exhibit a high affinity to H. For the preparation of $Ti_{0.31}Cr_{0.49}V_{0.20}(H\text{-}D)_x$, the H absorption was carried out up to more than 0.5 H/M, and then D was additionally charged up to more than 1.0 (H-D)/M (i.e., H → D absorption), followed by quick desorption for 30 min using the rotary pump. Similarly, $Ti_{0.31}Cr_{0.49}V_{0.20}(D\text{-}H)_x$ was prepared by the D → H absorption and the quick desorption. Figure 4(a) shows the $I(E)$ profiles of $Ti_{0.31}Cr_{0.49}V_{0.20}(H\text{-}D)_x$ and $Ti_{0.31}Cr_{0.49}V_{0.20}(D\text{-}H)_x$, together with that of $Ti_{0.31}Cr_{0.49}V_{0.20}$. Since no quasi-elastic neutron scattering (QENS) appears, we regarded that the $I(E)$ profiles of $Ti_{0.31}Cr_{0.49}V_{0.20}(H\text{-}D)_x$ and $Ti_{0.31}Cr_{0.49}V_{0.20}(D\text{-}H)_x$ consist of only the incoherent-elastic scattering. Figure 4(b) shows the $I_{inc_el}(Q)$s of $Ti_{0.31}Cr_{0.49}V_{0.20}(H\text{-}D)_x$ and $Ti_{0.31}Cr_{0.49}V_{0.20}(D\text{-}H)_x$, compared with that of $Ti_{0.31}Cr_{0.49}V_{0.20}H_x$. Interestingly, the $I_{inc_el}(Q)$s of $Ti_{0.31}Cr_{0.49}V_{0.20}(H\text{-}D)_x$ and $Ti_{0.31}Cr_{0.49}V_{0.20}(D\text{-}H)_x$ were similar to that of $Ti_{0.31}Cr_{0.49}V_{0.20}D_{1.6}$, and the $I_{inc\_el}(Q)$ hardly changed by the order in which the high pressure H_2 and D_2 gases were given in the absorption cycle (i.e., H → D and D → H). This result leads us to the two most important conclusions: (1) H atoms absorbed in the first absorption cycle up to H/M = 0.5 are hardly bound in the desorption cycle as the residual hydrogen, whereas (2) D atoms are mainly bound in $Ti_{0.31}Cr_{0.49}V_{0.20}$ bcc alloy as the residual deuterium when D_2 gas is partially used in the H absorption reaction. It is most likely that this phenomenon can be understood by considering the different diffusion constant between H and D in the bcc alloy [9]. On the other hand, the further investigations concerning the state of the residual hydrogen are in progress. The results of the QENS experiments for $Ti_{0.31}Cr_{0.49}V_{0.20}$ bcc alloy including the residual hydrogen will be reported, elsewhere.

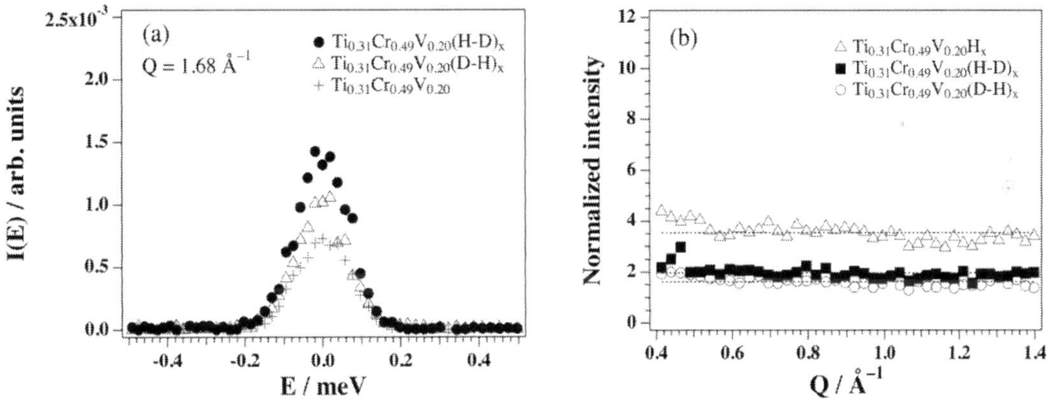

Figure 4. (a) Elastic neutron scattering profiles as a function of E, $I(E)$, at Q = 1.68 Å^{-1} at RT for $Ti_{0.31}Cr_{0.49}V_{0.20}(H\text{-}D)_x$ (●), $Ti_{0.31}Cr_{0.49}V_{0.20}(D\text{-}H)_x$ (Δ), and $Ti_{0.31}Cr_{0.49}V_{0.20}$ (+). (b) Intensities of the incoherent-elastic scattering in terms of Q, $I_{inc\_el}(Q)$, for $Ti_{0.31}Cr_{0.49}V_{0.20}H_x$ (Δ), $Ti_{0.31}Cr_{0.49}V_{0.20}(H\text{-}D)_x$ (■), and $Ti_{0.31}Cr_{0.49}V_{0.20}(D\text{-}H)_x$ (○).

4. Conclusions

Hydrogen and deuterium pressure-composition (P-C) isotherm, X-ray powder diffraction, and neutron scattering experiments were carried out for hydrogenated and/or duterated $Ti_{0.31}Cr_{0.49}V_{0.20}$ alloys. In the P-C isotherm measurements, it was found that the residual hydrogen-to-metal ratio $(H/M)_{res}$ is almost the same as the residual deuterium-to-metal one $(D/M)_{res}$; the $(H/M)_{res}$ and $(D/M)_{res}$ are approximately 0.5. In the Rietveld refinements for X-ray diffraction data, it was confirmed that $Ti_{0.31}Cr_{0.49}V_{0.20}$ has the bcc structure with the lattice parameter a = 3.049(1) Å, $Ti_{0.31}Cr_{0.49}V_{0.20}H_{1.0}$ the

CaF_2-type structure with a = 4.339(4) Å (as a principal phase), and $Ti_{0.31}Cr_{0.49}V_{0.20}H_{0.5}$ the bcc structure with a = 3.111(2) Å. Furthermore, the results of the neutron scattering experiments gave us the two most important conclusions: (1) H atoms absorbed in the first absorption cycle up to H/M = 0.5 are hardly bound in the desorption cycle as the residual hydrogen, whereas (2) D atoms are mainly bound in $Ti_{0.31}Cr_{0.49}V_{0.20}$ bcc alloy as the residual deuterium when D_2 gas is partially used in the H absorption reaction.

Acknowledgements
We wish to acknowledge the AGNES group of the ISSP, The University of Tokyo, for the help in neutron scattering experiments. This work is supported by a Grant-in-Aid for Yong Scientists (B) from the Ministry of Education, Science and Culture, Japan.

References
[1] Akiba E and Iba H 1998 *Intermetallics* **6** 461
[2] Nakamura Y, Oikawa K, Kamiyama T and Akiba E 2001 *Mater. J. Alloys Compd.* **316** 284
[3] Iwase K, Kamiyama T, Nakamura Y, Mori K, Yonemura M, Harjo S, Ishigaki T and Akiba E 2006 *Mater. Transations* **47** 271
[4] Rietveld H M 1969 *J. Appl. Crystallogr.* **2** 65
[5] Izumi F and Ikeda T 2000 *Mater. Sci. Forum* **321–324** 198
[6] Kajitani K, Shibata K, Ikeda S, Kohgi M, Yoshizawa H, Nemoto K and Suzuki K 1995 *Physica B* **213–214** 872
[7] Yamamuro O, Inamura Y, Kawamura Y, Watanabe S, Asami T and Yoshizawa H 2005 *Activity Report on Neutron Scattering Research, Neutron Science Laboratory, Institute for Solid State Physics, The University of Tokyo* **12** 12
[8] Dianoux A J and Lander G 2003 *Neutron data booklet, second ed., Institute Laue-Langevin*
[9] Fukai Y and Sugimoto H 1985 *Adv. in Physics* **34** 263

Absorption correction of peak positions for neutron strain measurements

Peter Staron

Institute of Materials Research, Helmholtz-Zentrum Geesthacht,
21502 Geesthacht, Germany

Peter.Staron@hzg.de

Abstract. In angle-dispersive neutron strain scanning the information about residual strain comes from the whole gauge volume that is defined by slits in the incoming and diffracted beams. Since the intensity of the neutron beam decreases with the amount of material it has travelled, neutrons diffracted from different locations within the gauge volume contribute with different intensities to the recorded diffraction peak. This can lead to peak shifts, and thus apparent strains. The magnitude of this peak shift depends mostly on the beam attenuation and the size of the gauge volume, but also on the sample geometry and position of the gauge volume within the sample. The peak shift plays a significant role when the size of the gauge volume becomes large because of peak broadening by the sample. An analytic expression for the peak shift was derived for a simple geometry to evaluate a numerical simulation. The numerical simulation was developed to quantify necessary corrections in detail. The attenuation-induced peak shift was demonstrated by measurements on a strain-free powder sample and the results were compared with the numerical predictions.

1. Introduction

Neutron diffraction is a well-established technique for the non-destructive characterization of residual stress states in the interior of crystalline material [1, 2]. The physical principal is the same as for X-rays from a laboratory source, with the important difference that the penetration depth of neutrons is about a thousand times larger than for such X-rays. This enables neutron strain measurements at any location within several cm thick samples for many materials. Typical spatial resolutions are between 1 and 3 mm, depending on parameters such stress gradients, grain size, and sample thickness. There are angle-dispersive and energy-dispersive neutron strains scanners, the former work with monochromatic neutrons and record one single peak at a time, the latter work with a polychromatic beam and a time-of-flight (TOF) setup recording several diffraction peaks at once. In the following, only angle-dispersive neutron diffractometers will be discussed.

In angle-dispersive neutron strain scanning a sample is shifted through a neutron beam while the position of a Bragg reflection is recorded with an area detector [3]. The information about residual strain comes from the whole gauge volume that is defined by slits in the incoming and diffracted beams. (The advantage of a radial collimator in the diffracted beam will be discussed at the end.) The divergence of the beam also affects the size and shape of the gauge volume. Since the intensity of the neutron beam decreases with the amount of material it has travelled due to absorption and scattering, neutrons diffracted from different locations within the gauge volume contribute with different intensities to the measured diffraction peak. This leads to peak shifts, and thus apparent strains, which have to be corrected in extreme cases to obtain true strains. The magnitude of the peak shift also depends on the sample geometry and, thus, can change as a function of sample position relative to the gauge volume. This means that a correction function is required for a whole strain scan along a line through a sample. However, this attenuation-induced peak shift is not included in standard data evaluation procedures for residual stress analysis, which is because it only plays a role when the size of the gauge volume becomes large. This is the case for sample-induced broadening of the diffracted beam. However, large peak broadening effects are not very often observed in engineering structural materials.

To be able to apply a correction for attenuation-induced peak shifts to measured strain data in such cases, a numerical simulation of the diffraction geometry including attenuation was developed. The numerical simulation was evaluated in two different ways: first, analytical expressions were derived for the peak shift for simple geometries, namely cylinder and plate samples. Second, a strain-free powder sample with a cylinder shape was measured and the results were directly compared with the predictions of the numerical simulation. The results show that the numerical simulation describes the attenuation-induced peak shift correctly and, thus, can be used for a correction of measured strain data.

2. Analytical expression

The effect of absorption on the peak position can be calculated analytically for simple geometries like e.g. the cylinder sample sketched in figure 1. It can also be easily formulated for plate samples. Incoming monochromatic neutrons travel along x. It is assumed that the width w_1 of the incoming beam is infinitesimally small and it has no divergence. The attenuation coefficient of the material is μ, corresponding to the total macroscopic cross section. Thus, the intensity of the incoming beam at a position x within the sample is

$$I(x) = I_0 \exp\left(-\mu(x - x_0)\right) \tag{1}$$

where the sample surface is at x_0, and I_0 is the intensity of the incoming beam. The width w_2 of the slit in the diffracted beam shall also be infinitesimally small. It is assumed that there are spreads in the lattice parameter of the sample material as well as in the wavelength of the incoming beam that lead to a Gaussian profile of the diffraction peak. The peak recorded on a linear detector in the diffracted beam is then described by

$$I(2\theta) = I_1 \exp\left(-\frac{(2\theta - 2\theta_0)^2}{2\sigma_a^2}\right) \tag{2}$$

where 2θ is the diffraction angle and I_1 is the maximum of the diffraction peak. The full width at half maximum (FWHM) of this diffraction peak is $\beta = 2\sqrt{2\ln 2}\,\sigma_a \approx 2.355\sigma_a$. With the distance d_2 of the secondary slit from the sample centre, the diffraction occurs along a line within the sample, where the

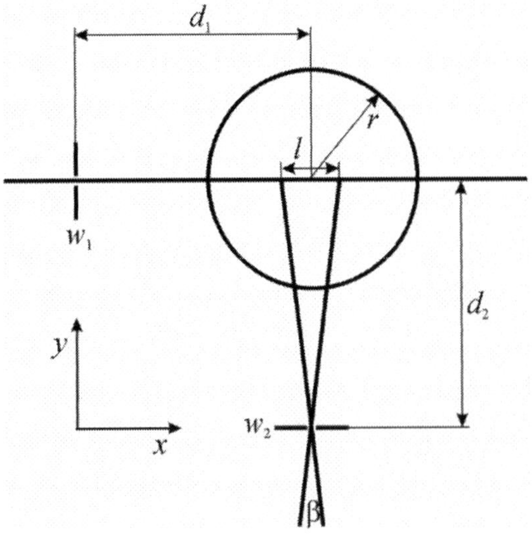

Figure 1. Geometry used for calculation of the analytical expression of the peak shift for cylinder sample.

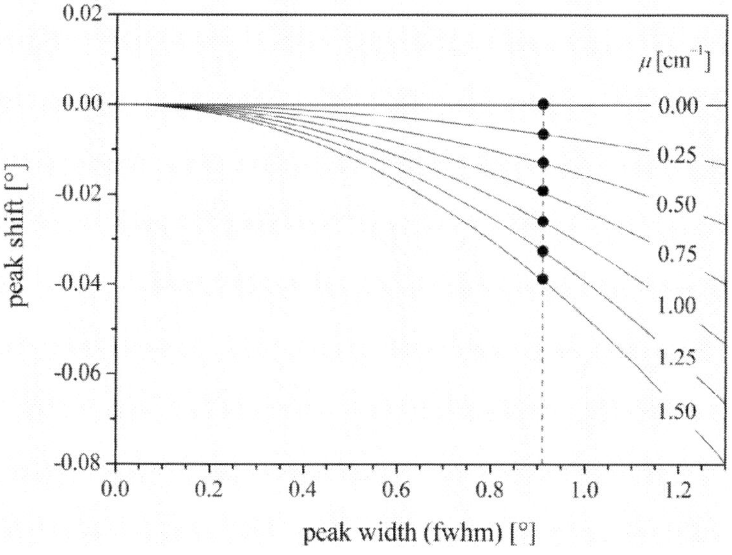

Figure 2. Calculated peak shift as a function of peak width for a cylinder sample and different attenuation coefficients. Lines are the results of the analytical expression, symbols are the results of numerical simulations for a peak width of 0.91° (d_1 = 50 mm, d_2 = 100 mm).

length of this line $l = d_2 \tan\beta$ (cf. figure 1) corresponds to the FWHM, assuming that β is small. A diffraction power can be defined along this line as

$$P(x) = P_0 \exp\left(-\frac{(x - x_1)^2}{2\sigma_x^2}\right) \tag{3}$$

where x_1 is the centre point and σ_x is the standard deviation with $\sigma_x = d_2\sigma_a$. This diffraction power has to be multiplied by the available intensity to give the diffracted intensity $I(x)\cdot P(x)$. By multiplying equations 1 and 3 and recollecting terms the peak shift due to absorption can be calculated as

$$\delta = -\frac{\mu\beta^2 d_2}{2.355^2} = -0.1803\mu\beta^2 d_2 \tag{4}$$

(details will be published elsewhere). The peak shift as a function of peak width for different attenuation coefficients μ as calculated with equation 4 were compared with peaks shifts obtained from the numerical simulation for the geometry shown in figure 1 and a peak width of 0.91° (figure 2). The analytical results agree well with the results of the numerical simulation.

3. Numerical simulation

A numerical simulation was used for the calculation of the peak shift for other positions of the sample relative to the gauge volume, where it is more difficult to obtain an analytic expression. The simulation was done in two dimensions, because in a first approximation only the geometry in the scattering plane is relevant. Cylinder and plate samples (reflection and transmission geometry) were taken into account so far (actually, a circle or rectangles in two dimensions). Neutrons in the primary beam with a mean wavelength of 1.64 Å were selected, the wavelength that was mainly used for the measurements at the strain scanner ARES-2 of the HZG neutron facility [4]. The monochromator with elastically bent perfect Si single crystals [5] was included in the simulation.

Behind the primary slit with width w_1 at a distance d_1 from the centre the beam enters the sample and the point of diffraction in the sample along the path of the beam is chosen by a random number. Diffraction is calculated by reflection of the neutron path at the diffracting lattice planes in a powder sample. The mean lattice parameter can be chosen such that the diffraction angle is close to 90°, which is normally used for strain measurements to obtain a near-rectangular gauge volume. A Gaussian distribution of lattice parameters was used with a given width, which can be adapted to the peak broadening observed in a measurement. This distribution of lattice parameters, together with the wavelength spread, gives a broadened peak. After the reflection at the lattice planes, it is tested if the

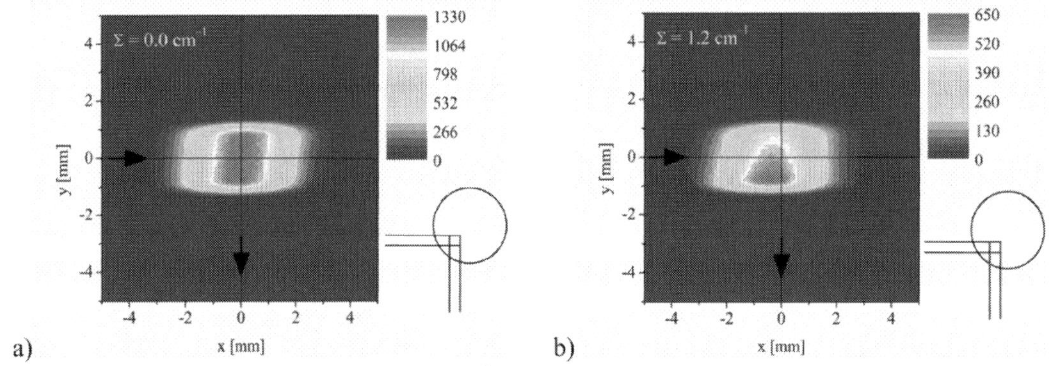

Figure 3. Calculated contribution to the peak intensity as a function of position within the gauge volume, which is located 7 mm from the centre of the steel cylinder. a) no absorption, b) attenuation coefficient μ = 1.2 cm^{-1}. The nominal location of the centre of the gauge volume at (0,0) is marked with a cross hair. ($w_1 = w_2 = 2$ mm, $d_1 = 50$ mm, $d_2 = 100$ mm, $2\theta = 101°$, $\beta = 1.5°$).

diffracted ray passes the secondary slit of width w_2 at distance d_2 from the centre. Those rays passing the slit are finally recorded on a virtual detector placed at a distance of 1 m from the sample.

The location of the sample can be changed with respect to the position of the gauge volume. In this way, scans of radial and tangential strains along a line through the sample can be simulated.

When the attenuation is set to zero, a gauge volume is obtained that is expected when attenuation plays only a minor role (figure 3a). However, when there is strong attenuation and the diffraction peaks are significantly broadened, the gauge volume is shifted towards the primary slit, leading to a peak shift (figure 3b). Moreover, there is also some weighting of the gauge volume towards the secondary slit. The location of the gauge volume shown in figure 3 is 7 mm from the centre of the cylinder as indicated in the insets. A large peak width of 1.5° and a large distance w_2 of the secondary slit of 100 mm were used for this exemplary calculation to get a strong visible effect. The scattering angle was 101° as for the powder measurement.

First, the simulation was tested by calculating the peak shift for the simple geometry that was used for deriving the analytical expression and a peak width of 0.91°, which was observed in the powder measurement. The simulation results match exactly with the analytical expression (figure 2). The second test involved the stress-free powder sample introduced in the following section.

4. Stress-free powder sample

A powder sample was used to evaluate the predictions of the numerical simulation. No peak shifts are expected in a scan of the powder sample because no residual strain gradients are present in the sample. The measurements were done at the strain scanner ARES-2 at the HZG neutron facility [4]. The powder container had a diameter of 20 mm and thin Al walls. Ni powder was used because it has a relatively large attenuation coefficient. The attenuation coefficient μ = 0.55 cm^{-1} for the powder sample was determined by a transmission measurement. The (311) reflection was used at $2\theta = 101°$ (λ = 1.64 Å). The radius of curvature of the monochromator crystal was optimized for $2\theta = 90°$; thus, at 101° the peak was slightly broadened to a width of 0.91°. Large attenuation and large peak width was desired in this case to obtain a large attenuation-induced peak shift. The slit width was 2 mm; the slit heights were 10 mm (slit 1) and 30 mm (slit 2). Both slit distances were 50 mm.

The peak shifts along sample diameters on scan lines for measuring radial and tangential strain directions calculated with the simulation were 0.12° and –0.07°, respectively (figure 4). This is the difference between the values at +10 mm and –10 mm. Since the peak positions can have a constant offset that depends on geometrical details, only the slopes of the curves shown in figure 4 were evaluated and compared with the measurements so far. In other words, the simulation results were shifted on the angular scale such that they fit to the measurements. With this restriction, calculated and measured peak shifts fit well within the statistics. Two things can be concluded from these measurements: *i*. there is a significant attenuation-induced peak shift with significant values for an

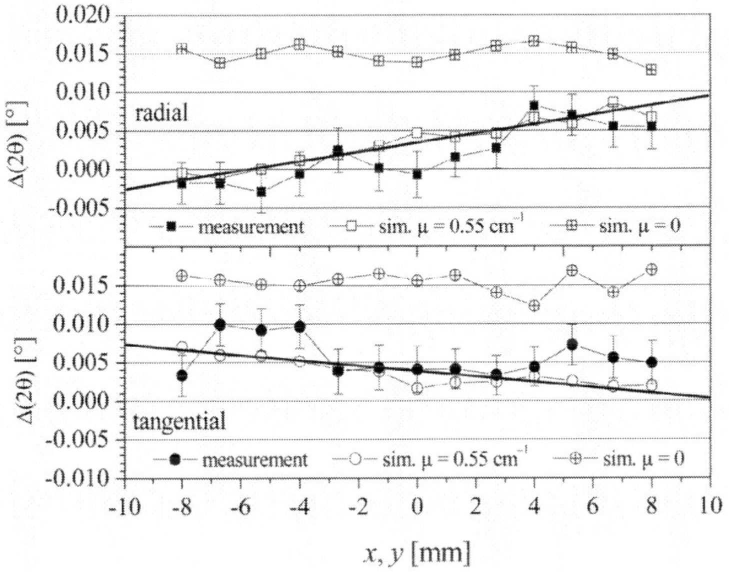

Figure 4. Calculated peak shift for radial and tangential scans for a cylinder with a diameter of 20 mm as a function of scan position compared with measurements on Ni powder ($w_1 = w_2 = 2$ mm, $d_1 = d_2 = 50$ mm, $2\theta = 101°$, $\beta = 0.91°$). The two solid lines interpolate the simulation results.

attenuation of 0.55 cm^{-1} and a peak width of 0.91°; *ii.* the developed numerical simulation is adequate for describing this attenuation-induced peak shift.

It should be noted that it is expected that significant values of the attenuation-induced peak shift are only observed when simple slits are used for strain scanning. With a radial collimator instead of the secondary slit, an increase of the gauge volume is avoided and, thus, an increase of the peak shift to significant values is expected to be suppressed. This was shown, e.g., by Pirling et al. [6]. A quantitative consideration of radial collimators will be subject of further work and published elsewhere.

5. Summary

The shift of a measured Bragg peak in a strain scanning experiment due to beam attenuation was analyzed. This shift can usually be neglected in residual stress analysis, but it becomes significant when there is strong peak broadening in the sample and simple slits are used for the measurement. A numerical simulation for the calculation of the peak shift was developed. A comparison with the results of an analytical expression for a simple geometry confirmed the correctness of the simulation. Moreover, the scanning results of a stress-free powder sample matched exactly the predictions of the numerical simulation. Thus, the numerical simulation can be used for correcting strain data that is influenced by attenuation-induced peak shifts.

References

[1] Allen A J, Hutchings M T, Windsor C G, Andreani C 1985 *Advances in Physics* **34** 445–473.
[2] Hutchings M T, Withers P J, Holden T M, Lorentzen T 2005 *Characterization of residual stress by neutron diffraction* (Boca Raton, FL: Taylor and Francis).
[3] Staron P 2008 Stress Analysis by Angle Dispersive Neutron Diffraction, in: *Application of Neutrons and Synchrotron Radiation in Materials Science* eds H Clemens, A Pyzalla, W Reimers, A Schreyer (Wiley).
[4] Staron P, Ruhnau H-U, Mikula P, Kampmann R 2000 *Physica B* **276–278** 158–159.
[5] Wagner V, Mikula P, Lukas P 1993 *Nucl. Instr. & Meth. A* **338** 53.
[6] Pirling T, Hughes D, Robinson J S 2010 *Materials Science Forum* **652** 80–85.

Residual stresses in a stainless steel – titanium alloy joint made with the explosive technique

Yu V Taran[1,*], A M Balagurov[1], B M Sabirov[2], A Evans[3], V Davydov[3] and A M Venter[4]

[1]Frank Laboratory of Neutron Physics, Joint Institute for Nuclear Research, Dubna, Russian Federation
[2]Dzhelepov Laboratory of Nuclear Problems, Joint Institute for Nuclear Research, Dubna, Russian Federation
[3]SINQ, Paul Scherrer Institute, Villigen, Switzerland
[4]Research and Development Division, NECSA Limited, Pretoria, South Africa

*E-mail: taran@nf.jinr.ru

Abstract. Joining of pipes from stainless steel (SS) and titanium (Ti) alloy still experience serious technical problems. Recently, reliable and hermetic joining of SS and Ti pipes has been achieved with the explosive bonding technique in the Russian Federal Nuclear Center. Such adapters are earmarked for use at the future International Linear Collider. The manufactured SS-Ti adapters have excellent mechanical behavior at room and liquid nitrogen temperatures, during high-pressure tests and thermal cycling. We here report the first neutron diffraction investigation of the residual stresses in a SS-Ti adapter on the POLDI instrument at the SINQ spallation source. The strain scanning across the adapter walls into the SS-SS and SS-Ti pipes sections encompassed measurement of the axial, radial and hoop strain components, which were transformed into residual stresses. The full stress information was successfully determined for the three steel pipes involved in the joint. The residual stresses do not exceed 300 MPa in magnitude. All stress components have tensile values close to the adapter internal surface, whilst they are compressive close to the outer surface. The strong incoherent and weak coherent neutron scattering cross-sections of Ti did not allow for the reliable determination of stresses inside the titanic pipe.

1. Introduction

At the end of 2008 one of the departments of the famous Siemens Company made and published the Siemens global calendar 2009 [1]. Each page of the calendar presents one of twelve outstanding achievements of industries from around the globe in 2008. On the first January page there is an art image of a cryomodule (Fig. 1) constructed at the Fermi National Accelerator Laboratory for the future electron-positron International Linear Collider (ILC). The cryomodule consists of a liquid helium cryostat and a superconducting microwave niobium cavity. One of the elements of this sophisticated structure is the transition element in the form of a two-layer tube (so-called adapter) (see Fig. 1), which connects the cryostat with a two-phase liquid helium supply line running along the entire 35 km length of the ILC.

Figure 1. Art image of the ILC cryomodule.

It was first planned to make the cryostat (as many as about 20 000) and the supply line from titanium. But in this case the cost of the supply line production would be too high. A group of specialists from JINR, which was engaged in the ILC project, proposed that the line be made from stainless steel (SS) and connected with the titanium cryostat through a SS-Ti adapter. Thus, it would be needed to join tightly the dissimilar materials, steel and titanium, which is a serious technical problem. To solve the problem, a group of specialists from the Russian Federal Nuclear Center (Sarov) did the research and development work that resulted in making a batch of SS-Ti adapters by the explosion bonding technique (EBT) [2]. The manufactured adapters have demonstrated excellent mechanical behavior at room and liquid nitrogen temperatures, during high-pressure tests and thermal cycling [3]. The measurements of the residual mechanical intrinsic stresses in the adapter are of large practical interest for the improvement of the EBT and minimization of the production cost. One of the tested adapters was used for the neutron diffraction investigations on the POLDI instrument at the SINQ spallation neutron source (Paul Scherrer Institute, Switzerland).

2. Experiment

The adapter is fabricated from AISI 316L SS and Gr 2 Ti-alloy pipes with the external diameter 59.6 mm and wall thickness 2.3 mm. The adapter joint was established by placing the pipe ends, precisely butted on a tube mandrel, symmetrically inside a loosely fitting external AISI 316L SS cover joint sleeve of wall thickness 2.7 mm. The constituents were welded by external explosion. After that the mandrel was removed from the adapter.

The incident beam was shaped with the sizes of 1.5 x 4 mm^2 by the motorized variable vertical and horizontal slits, respectively. The scattered beam at the angle 2θ=90° was shaped by the fixed multislit radial collimator with a space resolution of 1.5 mm in the horizontal plane. The strain scanning was performed across the adapter walls in the Ti-SS and SS-SS layers through cross-sections AB and CD symmetrically placed relatively to the butt joint of the SS and Ti pipes (Fig. 2). Three adapter orientations relative to the neutron scattering vector were used to measure the axial, radial and hoop (tangential) strain components. To measure the axial components, the adapter was placed horizontally on the strain scanner while the adapter was oriented vertically with the different azimuthal angles to measure the radial and hoop components.

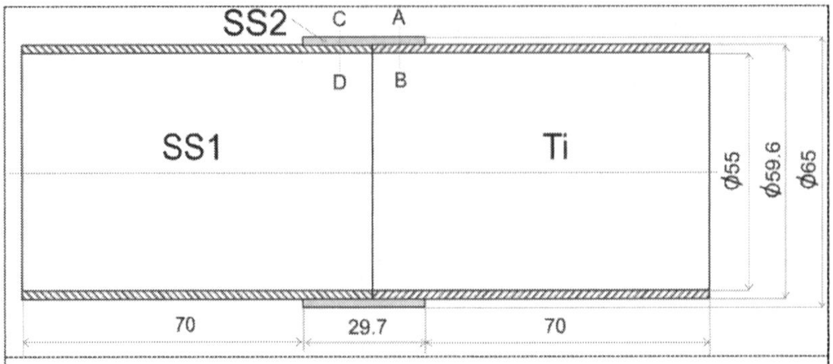

Figure 2. Sketch of the SS–Ti adapter. The cross-section AB is through the Ti and SS2 layers while the cross-section CD is through the SS1 and SS2 layers.

Note that when the gauge volume (GV) formed by collimators of direct and scattered beams is fully immerged in the sample material, the scattered volume (SV) is equal to GV and the geometric centre of GV coincides with the centre of gravity of SV. In the opposite case they do not coincide. Variation in the position of the SV centre of gravity due to the incomplete filling of GV in material will cause variation in the time of-flight path length L and the scattering angle 2θ and hence variation in the measured d-spacing of the lattice plane (hkl), independent of residual stress. It is the so-called pseudostrain shift of the diffraction (hkl)-peak. The evaluation of the pseudostrain effects (in the literature they are sometimes referred to as surface effects) was performed in a few papers, for example, [4-5]. The most trustworthy description of the pseudostrain effects can be obtained by the Monte Carlo simulation. To correct the measured radius $r_{measured}$ that coincides with the GV geometric centre to the true radius $r_{corrected}$ that coincides with the SV centre of gravity, we used a model of the one-slit scattered beam set developed for the scattering angle 90° [6]. Of course, it is too rough simplification but it permitted us to describe satisfactorily the experimental data obtained in [4] with a stress-free nickel sample. That is why this model was used to estimate the pseudostrain effects in the described experiment.

As an example, one of the experimental steel spectra obtained in the cross-section CD (the SS1-SS2 layers) is presented in Fig. 3. Note that the strong incoherent and weak coherent neutron scattering cross-sections of Ti did not allow reliable determination of strains inside the titanic pipe due to limitation on the allocated beam time. Each peak of the spectra was fitted individually by the Gauss function. All the steel layers exhibited strong texture different at different measurement points, which affected the peak intensity and, consequently, the accuracy of peak location in the d-spacing. After the diffraction spectra analysis only the (311) peak was observed in all spectra, the intensity of which did not vary more than three times.

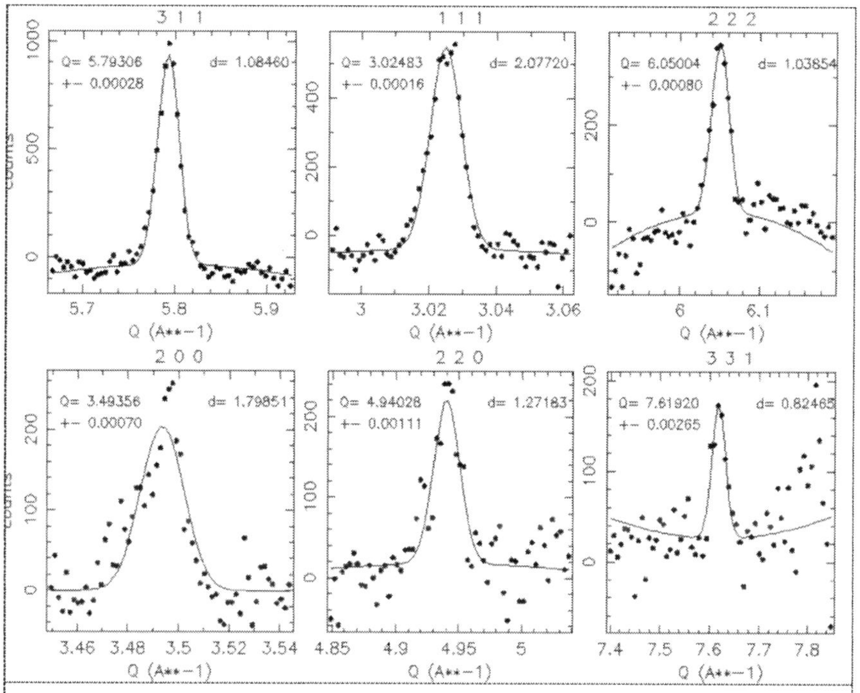

Figure 3. Diffraction spectrum obtained at the radius r=29 mm (SS2 layer).

Omitting the presentation of the d_{311}-spacing in all measured points as voluminous we passed to determination of a stress-free value $d_{0,311}$. Two ways were tested using: 1) small coupons from steel materials in the virgin state as delivered before application of EBT (we had in mind that the coupon did not have residual stresses) and 2) requirements of stress and momentum balance conditions and equilibrium relations (BC&ER) [7-9] on the diffraction experimental data. To realize the first way, each coupon was measured three times with its different orientation relative to the neutron scattering vector. The averaged values of $d_{0,311,SS1\text{-}coupon}$ and $d_{0,311,SS2\text{-}coupon}$ from these measurements were equal to 1.083934(33) and 1.083670(30) Å, respectively. The second way was more complicated. Omitting the BC&ER calculations of $d_{0,311,BC\&ER}$ and all admissions that were made during processing the diffraction experimental data for the cross-section CD, we present the result of the calculations as 1.084297 Å. All $d_{0,311}$-values differ from each other. This discrepancy will be discussed during the residual strain and stress calculations below.

3. Strain and stress results

3.1. Cross-section AB (SS sleeve layer)

There are two possibilities of calculating the residual strains in the SS sleeve layer from the experimental data using: 1) the value of $d_{0,311,BC\&ER}$ from the BC&ER calculations, and 2) the experimental result $d_{0,311,SS2\text{-}coupon}$=1.083670(30) Å obtained with the coupon from as-delivered steel used for the SS sleeve layer fabrication of the SS-Ti adapter. We chose the latter value, the substantiation of which will be done during the residual stress calculation. We omit the results of the strain calculation for lack of place and present the residual stress results in Fig. 4 obtained by using the generalized Hooke's law.

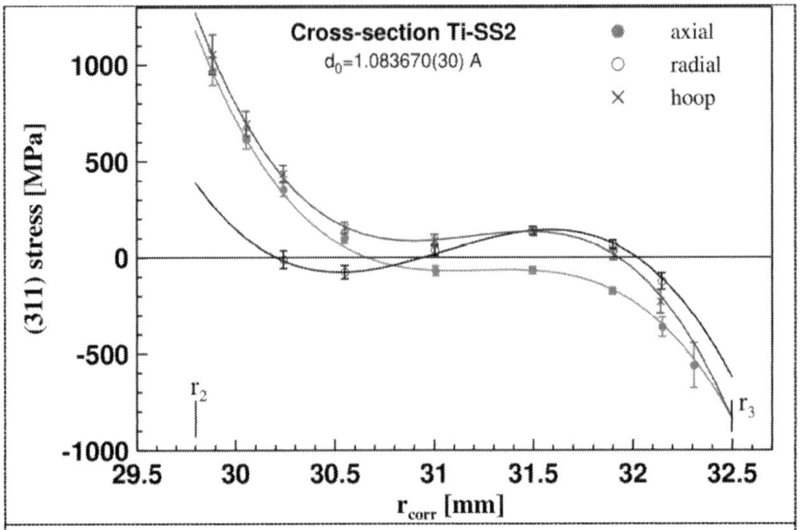

Figure 4. Measured (points) and fitted (lines) of the (311) peak residual stresses vs r_{corr} for the cross-section AB.

Here we come back to our promise to substantiate the choice of the experimental value $d_{0,311,SS2\text{-}coupon}$ for calculation of the strains and stress. Using the axial stress and momentum balance conditions for the two phase SS-Ti system

$$\int_{Ti}\sigma_a f(r)dr + \int_{SS}\sigma_a f(r)dr = 0 \tag{1}$$

and applying the generalized first theorem of the mean value [10]

$$\int_{Ti}\sigma_a f(r)dr = \mu\int_{Ti} f(r)dr, \tag{2}$$

where μ are the stress ($\overline{\sigma_a}$) or momentum ($\overline{\sigma_a^2}$) values averaged over the Ti layer, $f(r)$ are r or r^2, we can obtain σ_a and $\overline{\sigma_a^2}$ by following equations

$$\sigma_a = -\frac{2}{r_3^2 - r_2^2}\int_{SS}\sigma_a r dr, \tag{3}$$

$$\overline{\sigma_a^2} = -\frac{3}{r_3^3 - r_2^3}\int_{SS}\sigma_a r^2 dr. \tag{4}$$

Using the experimental data for the axial stresses presented in Fig. 4 and Eqs. (3-4) we calculated that σ_a=1.04 MPa, $\overline{\sigma_a^2}$=6.35 MPa. The values are very small as we expected from the stress and momentum balance consideration. If we use the value of $d_{0,311,BC\&ER}$ all curves in Fig. 4 would be shifted down by about 250 MPa, at that, we will have the values σ_a=177.2 MPa, $\overline{\sigma_a^2}$=176.9 MPa, which will be in conflict with the stress balance conditions. Thus, it becomes obvious that the value d_0 was taken as true for the SS sleeve layer.

3.2. Cross-section CD (SS-SS layers)

In this case, to calculate the residual strains in the SS-SS layers, we used the BC&ER value $d_{0,311,BC\&ER}$ =1.084297 Å. The results of the strain calculations were converted in the stresses, which are shown in Fig. 5. Now we can do a back check of the stress balance conditions $\int_{SS1+SS2}\sigma_a r dr = 0$ and $\int_{SS1+SS2}\frac{\sigma_h - \sigma_r}{r}dr = 0$ obtained in [9] for a sample of cylindrical symmetry (pipe). The results of integration were equal to –0.144 GPa·mm^2 and –1.286 MPa, respectively, from which the mean values were obtained to be equal to –0.03 GPa·mm and –0.26 MPa/mm. The first mean –0.03 GPa·mm must be compared with absolute difference of the means 2.56 and –2.69 GPa·mm obtained for only positive and only negative parts of the integrable function, respectively. The relative difference does not exceed 2 %. Thus, the back checking of the balance condition yielded a very satisfactory result. Note that such checking cannot be performed for the cross-section AB because the Ti layer was not investigated.

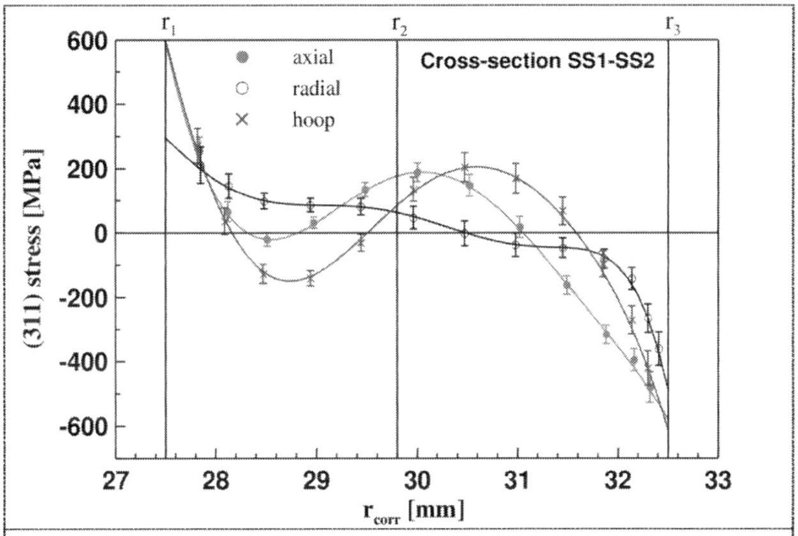

Figure 5. Measured (points) and fitted (lines) of the (311) peak residual stresses vs r_{corr} for the cross-section CD.

4. Equivalent Von Mises stress

Any hard manufacturing process creates a high residual mechanical intrinsic stress field in the material due to its plastic deformation. It was of great practical interest to investigate residual stresses in the Ti-SS adapter from the point of view of their effect on the hermiticity of the adapter at room temperature. The real operating conditions will be even more difficult because the adapters will be used at helium temperatures of 1-2K. In this case, thermostresses will arise from the mismatch of the steel and titanium thermal expansion coefficients.

Usually the thermal stress field in materials is calculated by the finite element method (FEM). The results of FEM are typically presented as the Von Mises equivalent stress

$$\sigma_e = \sqrt{\frac{1}{2}\left[(\sigma_a-\sigma_r)^2+(\sigma_r-\sigma_h)^2+(\sigma_h-\sigma_a)^2\right]}, \tag{5}$$

which includes all possible pairwise stress differences ($\sigma_i - \sigma_j$), where σ_i is the principal stress. Stress is in general a symmetric 3×3 tensor. The Von Mises stress reduces this to a scalar for the purposes of calculating yield criterion known as the maximum distortion energy criterion or the Von Mises criterion $\sigma_e \le R_{p0.2}$ where $R_{p0.2}$ is the yield stress.

The results of the equivalent residual stress calculation in the present experiment are shown in Fig. 6. As is seen in the figure, the equivalent stresses are less than 300 MPa in the cross-section CD, which does not exceed the yield stress $R_{p0.2}$=311 MPa measured on the as-delivered steel pipe at room temperature but is two times less than $R_{p0.2}$ in the middle of the SS sleeve layer of the cross-section AB. The stress dramatically increases up to 800 MPa close to the interface between the Ti and SS sleeve layers.

The calculations of the equivalent Von Mises thermal stresses in the most early variant of the adapter [11] showed that thermostresses in the SS and Ti layers amounted to about 420 and 1200 MPa, respectively, during cooling down to 4K. Taking in account last value and the residual stresses the Ti material may be passed deeply into the plastic region that may create a local microcrack. The adapter hermiticity will be under threat during operation. However, the real yield stresses of the adapter materials are unknown. The yield stresses of the materials after the explosion bonding process could probably be lower than in the origin-state materials, which would aggravate the situation. That is why the measurements of the residual stresses in the SS-Ti adapter were crucial for clarifying the situation with the total stresses. A more dramatic situation may be with the SS sleeve layer, where the edge stress effects are very high as a rule.

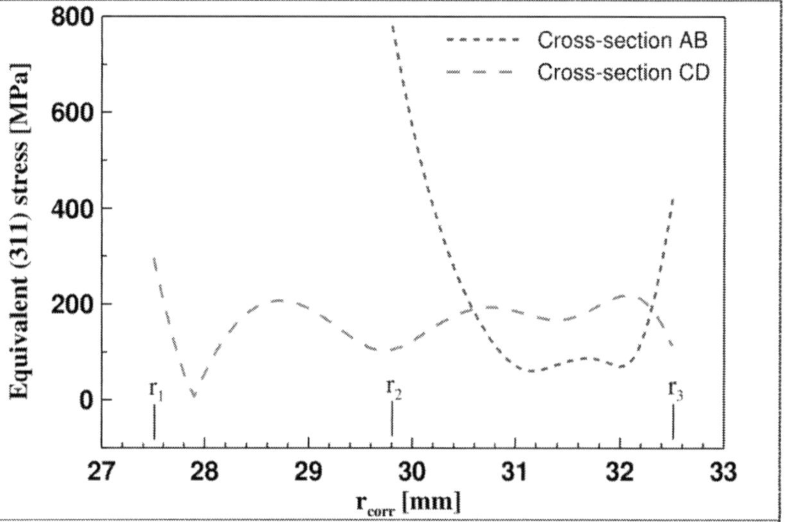

Figure 6. Equivalent stresses of (311) peak vs r_{corr}.

5. Conclusion

Note some important results of the described experiment:

- Neutron diffraction 3D-strain scanning of the SS-Ti adapter and residual stress tensor determination were carried out with the time-of-flight high resolution stress-diffractometer POLDI (PSI).
- The maximal value of all stress components in the cross-section AB through the SS-Ti layers is found to be almost two times larger than the same components in the cross-section CD. In other words, they are very high in comparison with the plastic constants of the austenite stainless steel AISI 316L.
- All the stress components close to the interface between the Ti and SS layers have a tension nature while close to the outer surface they were of a compression nature.
- All the stress components in the cross-section CD through the SS-SS layers close to the adapter intrinsic surface have a tensile nature while close to the outer surface they were of a compression nature. While no stress peculiarity around the interface between the SS-SS layers in the cross-section CD was observed, a rather great rise in the hoop stress component was found near the intrinsic and outer surface of the adapter.
- The radial stress component in the cross-section CD was found to be noticeably smaller than the other components that were observed before in the composite tube from the austenitic stainless steel with the welded ferritic steel cladding [12].

- No stress peculiarity was observed around the interface of the SS-SS layers in full contradiction with [12], where a very great rise in the hoop stress component was found in the ferritic steel layer close to the interface between the intrinsic austenitic steel layer and the outer ferritic steel layer.
- In general, the residual stresses in the steel part of the adapter did not exceed 1000 MPa, which is a rather large value in comparison with the plastic constants of the austenite stainless steel AISI 316L, such as the yield stress $R_{p0.2}$=170/311 MPa and the tensile strength R_m=485/559 MPa, where the first value before a slash corresponds to the generally accepted standard requirements on the material and the second value after a slash corresponds to the real tested starting material used to produce the adapter.
- The calculations of the equivalent stress in the adapter showed that they were everywhere lower than 800 MPa in the cross-section AB and lower than 200 MPa in the cross-section CD. However, real operation of the adapter at helium temperatures can be accompanied by higher stresses due to summation of the residual and thermal stresses. Thus, it is possible that the total stresses increase up to 2000 MPa in the SS layer of the cross-section AB and up to 620 MPa in the cross-section CD. In this case, the material may pass deeply into the plastic region that might create a local microcrack. The adapter hermiticity will be under threat during operation.

The authors are grateful to the SINQ facility (PSI) for the allocation of beam time at the POLDI instrument. The financial support from the JINR-SA Bilateral Scientific Cooperation Programme is gratefully acknowledged.

References

[1] Http://www.flickr.com/photos/31274959@N08/3214982315/sizes/o/in/set72157612782183019
[2] Malkov I et al. 2008 Investigations of the possibility of producing the bimetallic tube transition element by explosion welding for the cryomodule of the International Linear Collider *JINR Communication* E13-2008-109, Dubna
[3] Basti A et al. 2008 Leak rate measurements on bimetallic transition samples for ILC cryomodules *JINR Communication* E13-2008-110, Dubna
[4] Spooner S and Wang X L 1997 *J. Appl. Cryst.* **30** 449
[5] Bruno G et al. 2006 *Nucl. Instr. Meth. in Phys. Res.* **B 246** 425
[6] Taran Yu V and Balagurov A M 2011 Correction of diffraction peak shift due to partial immersion of a gauge volume in stressed sample *Int. Conf. "Stress and Texture Investigations by means of Neutron Diffraction"* (Dubna, Russia, 6-9 June 2011) Abstracts book JINR E14-2011-41, p 52
[7] Weidmann G, Lewis P and Reid N 1990 *Structural Materials* (London: Open University, Butterworth)
[8] Winholtz R A and Cohen J B 1989 *Adv. in X-ray Analysis* **32** 341
[9] Pintschovius L 1992 Macrostresses, microstresses and stress tensor *Measurement of Residual and Applied Stress using Neutron Diffraction* (The Netherlands: Klumer Academic Publishers) p 115
[10] Bronshtain N I and Semendyaev K A 1986 *Handbook on Mathematics* (Moscow: Nauka, 13th edition) p 239 (in Russian)
[11] Budagov J et al. 2008 Stress analysis of the Ti-SS transition joint *JINR Communication* P13-2008-58, Dubna
[12] Taran Yu V et al. 2007 *Zeit. fur Kristal.* Suppl. Issue no. **26** 355

In-situ investigation of hydrogen diffusion in Zircaloy-4 by means of neutron radiography

M Grosse[1], M van den Berg[1,2], C. Goulet[1,3], A Kaestner[4]

[1] Karlsruhe Institute of Technology, Institute for Applied Materials, P.O.Box 3640, D-76021 Karlsruhe, Germany
[2] Fontys Hogeschool, Rachelsmolen 1, NL-5612 MA Eindhoven, Netherlands
3 INSA University, 135 avenue de Rangueil, F-31077 Toulouse Cedex 4, Fance
[4] Department of Spallation Source, Paul Scherrer Institute (PSI), CH-5232, Villigen, Switzerland

E-mail: Mirco.Grosse@KIT.edu

Abstract. The hydrogen diffusion from the gas phase into Zircaloy-4 solid cylinders was investigated at various temperatures between 823 and 1473 K. Diffusion coefficients were fitted using axial hydrogen distributions for all temperatures investigated. The activation energy of the hydrogen diffusion was calculated for temperatures at which the bcc β phase is stable. The determined value of 48.3 kJ mol^{-1} K^{-1} is significantly higher than values published by other authors. At lower temperatures the hydrogen uptake is connected with a phase transformation from the hcp α- to the bcc β-phase. The spatial phase distribution and the hydrogen concentrations in the two phases were analysed.

1. Introduction

The progression of accidents in nuclear power reactors are studied experimentally at the Karlsruhe Institute of Technology (KIT, the former Research Centre Karlsruhe - FZK) in the framework of the QUENCH program. The investigations comprise loss of coolant accidents (LOCA) and severe accidents (accidents beyond LOCA). The processes occurring during such accidents are simulated on fuel rod bundle scale in the large scale QUENCH tests. The very expensive tests were supplemented by separate-effect tests on laboratory scale. In the framework of this program, the hydrogen diffusion in the β-phase of the classic nuclear fuel rod cladding material Zircaloy-4 (Zry-4) was studied in-situ by means of neutron radiography.

During loss of coolant accidents, free hydrogen is produced by the reaction between the hot zirconium alloy cladding material covering the fuel rods and steam. Very simplified the reaction can be described by:

$$Zr + 2\,H_2O = ZrO_2 + 4H \qquad (1)$$

Part of the produced hydrogen is released as gaseous H_2 into the reactor environment. The release can result in hydrogen detonations as happened in the Fukushima accident. The other part of the produced hydrogen is absorbed by the remaining metallic zirconium alloy. It degrades the mechanical properties and in particular the thermo-shock stability. An emergency cooling to terminate the accident can then result in a destruction of the cladding tube connected with release and redistribution of fuel and fission products.

Fig. 1 gives the zirconium-hydrogen binary phase diagram together with hydrogen partial pressure isobars according to Sieverts´ law [1]. In [1] the temperature correlations of the Sieverts´ coefficients were determined. For a given temperature and hydrogen partial pressure in the environment, the

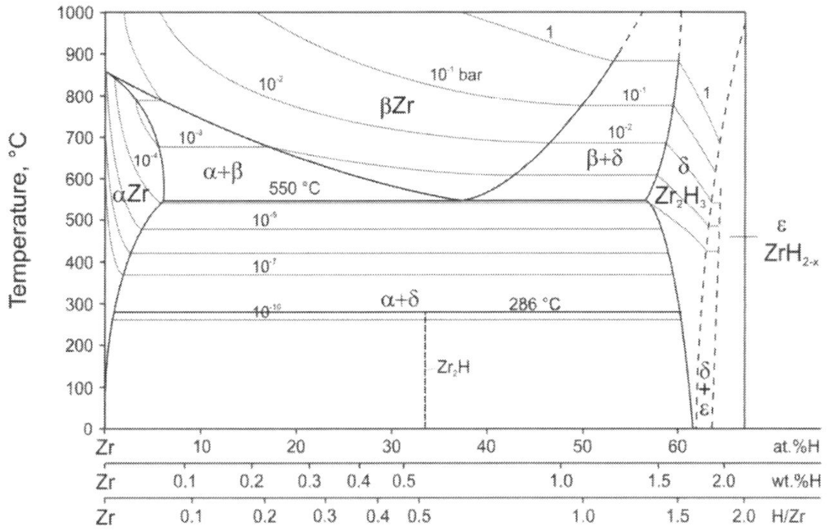

Fig. 1. Zr – H phase diagram with hydrogen partial pressure isobars.

hydrogen concentration in the metallic zirconium and the stable crystallographic phase can be examined. However, in order to understand the transient processes occurring during nuclear accidents, it is not sufficient to only know the thermodynamic equilibrium conditions. Kinetic parameters like the hydrogen transport have to be quantified as well.

Hydrogen diffusion in zirconium was studied in various works [2-8]. The published diffusion coefficients differ for some temperatures by more than two orders of magnitude. Fig. 2 shows the temperature correlations given in [2-4, 6] in an Arrhenius plot. The phase transformation between the hexagonal closest packed α- and the body centred cubic β-structure at about 1100 K ($10^4/T \sim 9\ K^{-1}$) cannot explain the discrepancy in these values completely.

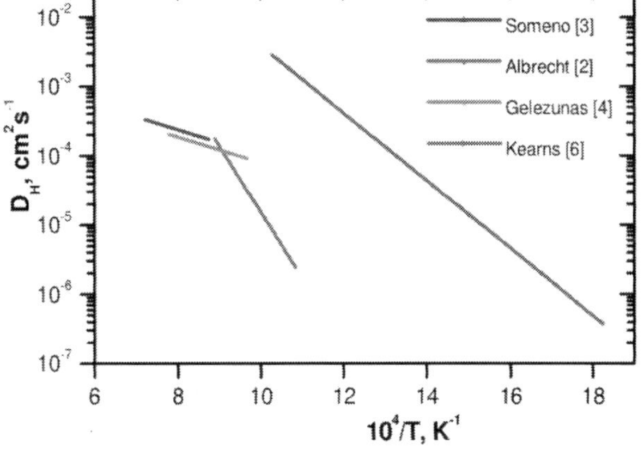

Fig. 2. Arrhenius plot of the temperature correlation of hydrogen diffusion coefficients in α- and β-zirconium published in [2-4, 6].

The aim of this work is the determination of the hydrogen diffusion coefficient in the β-phase of Zry-4 and the investigation of the diffusion processes in the two phase region between 800 and 1100 K. The investigations were performed by means of in-situ neutron radiography experiments. The advantage of neutron radiography in comparison to other methods like inelastic neutron scattering are the short measurement times, the high spatial resolution and the determination of macroscopic diffusion coefficients needed for nuclear engineering applications.

2. Material and Measurements

The investigations were performed with Zry-4 cylinders (1.5 % Sn, 0.21 % Fe, 0.10 % Cr, balance: Zr) having a length of 20 mm and a diameter of 12 mm. The cylinders were oxidized ex-situ for 2 h at 1473 K in Ar/O_2 atmosphere to produce a dense oxide layer of about 300 μm. The oxide layer suppresses the hydrogen uptake in significant manner. The oxide layer was removed at one base plane of the cylinder by mechanical grinding and polishing. Through this plane hydrogen can be absorbed.
The neutron radiography experiments were performed at the ICON beamline [9] at the Swiss neutron source SINQ (Paul Scherrer Institute Villigen, Switzerland). Experimental details are given in [10]. For the in-situ measurements a special reaction furnace, the INRRO furnace (see scheme in Fig. 3), with neutron transparent windows was constructed by the XERION company. A Bronkhorst system controls the gas flow through the furnace and allows establishing a well defined atmosphere in the furnace.
The specimen was put in the INRRO furnace at room temperature. It was heated up to test temperature in flowing argon (50 l/h under normal pressure) with a rate of 10 K/min. Recording of the radiograph was started when the test temperature was reached. A frame repetition rate of 0.05 Hz was applied. The first radiographs were used as reference. After about 180 s, in addition to the argon flow, the hydrogen flow (8 l/h under normal pressure) was switched on. After a predefined time the hydrogen was switched off and the release of hydrogen was measured. Finally, the specimen was cooled down in the furnace to room temperature.
The contrast of the radiography image is caused by differences in the neutron transmissions. Provided that scattered intensity does not contribute to the intensity measured at the detector (multiple scattering can be ignored), the neutron transmission of the specimen $T(x,y)$ detected at the point x,y of the radiograph is given by the ratio of the intensity behind and in front of the sample, $I(x,y)$ and $I_0(x,y)$, respectively, corrected by the background intensity $I_B(x,y)$.

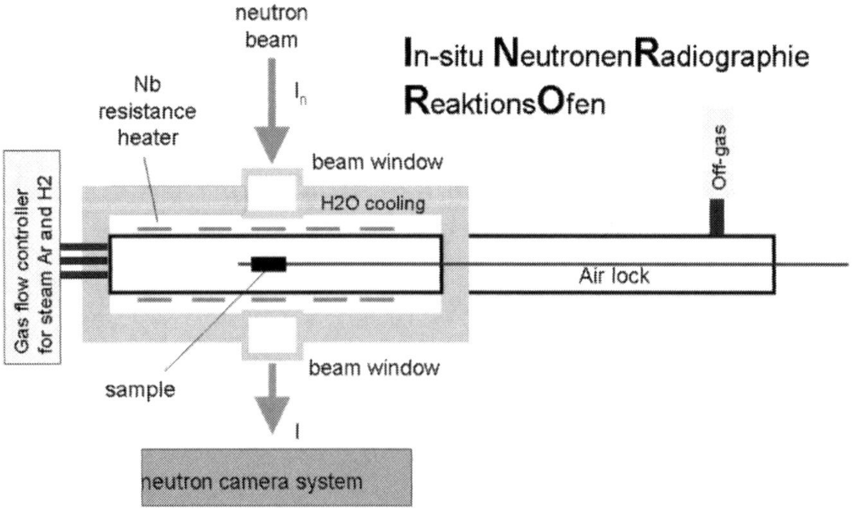

Fig. 3. Scheme of the INRRO furnace.

$$T(x,y) = \frac{I(x,y) - I_B(x,y)}{I_0(x,y) - I_B(x,y)} = \exp(-\Sigma_{total}(x,y) + s(x,y)) \qquad (2)$$

The transmission depends exponentially on the total macroscopic neutron cross section of the specimen $\Sigma_{total}(x,y)$ and on the path length of the neutron through the specimen $s(x,y)$. The total macroscopic neutron cross section can be described by the sum over all total microscopic neutron cross sections σ of the isotopes i in the specimens multiplied by their number density N:

$$\Sigma_{total} = \sum_i N_i \sigma_i \qquad (3)$$

For hydrogen diffusion in zirconium alloys Eq. (3) can be described by:

$$\Sigma_{total} = \underbrace{N_{Zr}\sigma_{Zr} + N_{Sn}\sigma_{Sn} + \ldots}_{\Sigma_{cladding\ material\ as\ received}} + N_H \sigma_H \qquad (3b)$$

The radiographs were referenced by images taken at test temperature but without hydrogen flow. The resulting image can be described by

$$\begin{aligned} T_R(x,y) &= \frac{T(x,y)}{T_{ref}(x,y)} = \frac{\exp\left(-\Sigma_{total}(x,y)\cdot s(x,y)\right)}{\exp\left(-\Sigma_{total_{ref}}(x,y)\cdot s(x,y)\right)} \\ &= \exp\left(-(\Sigma_{total}(x,y) - \Sigma_{total_{ref}}(x,y))\cdot s(x,y)\right) = \exp\left(-N_H\sigma_H \cdot s(x,y)\right) \end{aligned} \qquad (4)$$

Eq. (4) demonstrates that after the described referencing only the hydrogen containing parts are visible.

3. Results and Discussion

3.1. Hydrogen diffusion in single phase β-Zr

As an example, parts of the radiography images sequence obtained at 1273 K are given in Fig. 4. According to Eq. (4), the hydrogen containing parts appear darker. The propagation of the diffusion front is clearly visible. No significant hydrogen penetration through the oxide layer was observed.

From each of the images the axial distribution of the total macroscopic neutron cross section and with it the hydrogen distribution in the specimen can be obtained by quantitative image analysis. Fig. 5 gives as example of the axial distribution of the total macroscopic neutron cross section.

The diffusion can be described by the second Fick´s law:

$$\frac{\partial c}{\partial t} = \frac{\partial}{\partial x}\left(D\frac{\partial c}{\partial x}\right) \qquad (5)$$

The following boundary conditions were realized in the experiment:

1. The hydrogen concentration at the entrance plane is constant ($c(x=0,t) = c_0$)
2. The hydrogen concentration at the opposite side of the specimen is zero
3. The hydrogen diffusion coefficient D does not depend on the hydrogen concentration

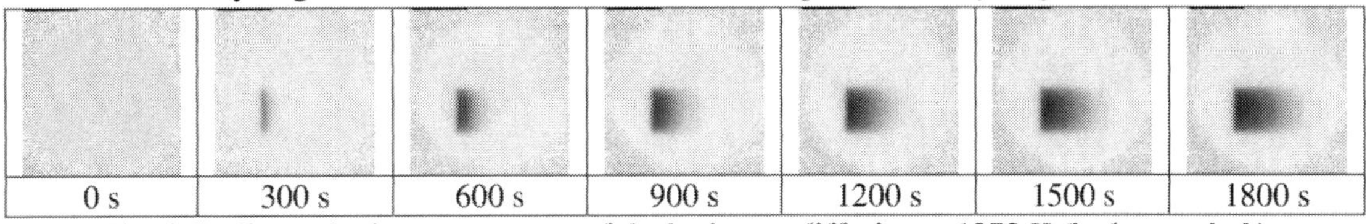

Fig. 4. Radiography image sequence of the hydrogen diffusion at 1273 K (hydrogen dark).

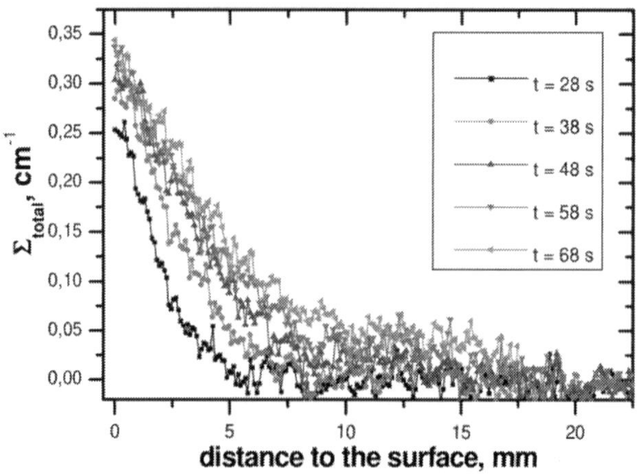

Fig. 5. Distribution of the total neutron cross section at various annealing times at 1473 K.

Boundary condition 1 is fulfilled because the experiments were performed in flowing gas atmosphere. Boundary condition 2 is fulfilled at the beginning of the experiment as long as the diffusion front has not reached the opposite side. On this basis Eq. (5) can be solved by:

$$c(x,t) = c_0\left(1 - erf\left(\frac{x}{2\sqrt{Dt}}\right)\right) \tag{6}$$

where *erf* is the Gaussian error function. Fig. 6 compares the fit according to Eq. (6) with the experimental data. The good agreement indicates that all three boundary conditions are fulfilled.

For each temperature investigated the diffusion coefficients were determined at different radiographs representing four to six different times. The temperature dependence of the diffusion coefficients determined is shown as an Arrhenius plot in Fig. 7. The temperature correlations given in [2-4] are compared. Whereas the data obtained for 1123 K and partially at 1173 K agree well with the correlation given in [4] and be not far away from [2], the slope of the curve and with it the activation energy determined is higher than the values given in [2, 3].

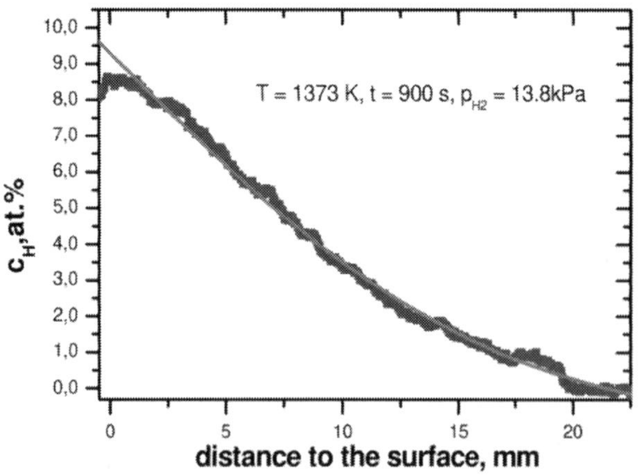

Fig. 6. Comparison between fit and experimental data of the hydrogen distribution after 900 s annealing in Ar/H_2 atmosphere at 1373 K.

The data obtained for 1273 and 1373 K and partially for 900 K agree better with the correlation given in [3]. At 1473 K the experimental data are significantly higher than the values expected from [3, 4]. The following correlation was determined from the experimental data:

$$D_H = 0.023 \frac{cm^2}{s} \exp\left(\frac{48.3 \frac{kJ}{mol\ K}}{R\,T} \right) \tag{7}$$

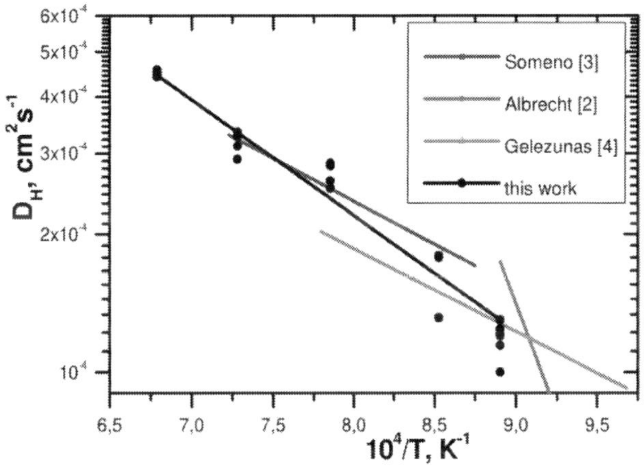

Fig. 7. Comparison of the results with temperature correlations from literature.

The strong texture of the applied specimens can be assumed as reason for the discrepancy between the data determined in this work and the correlations given in the references. The texture at high temperature was not yet measured. According to [11] a strong texture of the rod material at room temperature can be expected also for the β phase.

3.2. Hydrogen diffusion in the α-β two-phase range

The absorption of hydrogen extends the temperature range at which the bcc β phase is stable to lower temperatures (see Fig. 1). Therefore, the hydrogen diffusion at temperatures between 823 and 1123 K is connected with α- to β-phase transition.

Examples of the neutron radiograph sequence measured at 823 K are given in Fig. 8. The sample starts with hcp structure (α-phase). At the beginning, represented by the radiograph taken after 120 s, only a small region containing hydrogen is visible at the surface through which hydrogen can penetrate. The amount of hydrogen is low because the α-phase can absorb less hydrogen than the β-phase. The hydrogen is distributed more or less homogeneously over the whole sample cross section. After short time, represented by the radiograph taken at 1200 s, a local transformation to the β-phase starts at the upper left edge of the specimen. Because the β-region can absorb much more hydrogen than the former α-phase, it seems that the location transformed take hydrogen not only from the gas phase but also from the neighbouring α locations. This hinders reaching critical hydrogen concentration for the phase transformation at these locations. The further hydrogen uptake results only in a growth of the β-phase region. In no other region, the phase transformation occurs until the end of the experiment.

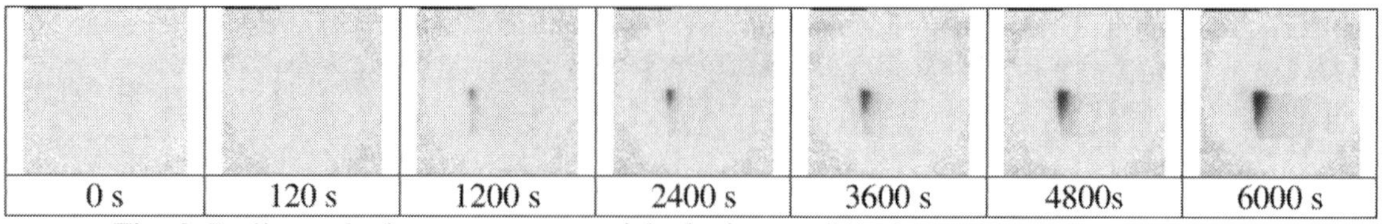

Fig. 8. Radiography image sequence of the hydrogen diffusion at 823 K (hydrogen dark).

Fig. 9 shows the hydrogen distribution perpendicular to the cylinder axis close to the surface through which the hydrogen penetrates. The different hydrogen concentrations in the two phases are clearly visible. Whereas in the β-region maximal concentrations of about 21.4 at.% were measured (corresponds with H/Zr = 27.4 % marked as solid green line in Fig. 9), a constant hydrogen concentration of 1.72 at.% was found in the α-region at the opposite side (H/Zr = 1.75 %, dotted blue line in Fig. 9). A monotonic decrease of the hydrogen concentration was observed between the two regions. At the left side of the diagram, the concentration is lower than found for the regions of the β-phase. It can be assumed that the pre-oxidation leads to a stabilization of the α-phase because of the oxygen absorption in the first 0.5 mm. According to the phase diagram given in Fig. 1, the maximal H/Zr atomic ratio in the α-phase at 823 K is about 7.5 %. The hydrogen concentration at the lower edge of the sample where the α-phase remains is significantly lower than this value marked by the black dashed line in Fig. 9. A slight kink in the hydrogen distribution gradient was found at this concentration value.

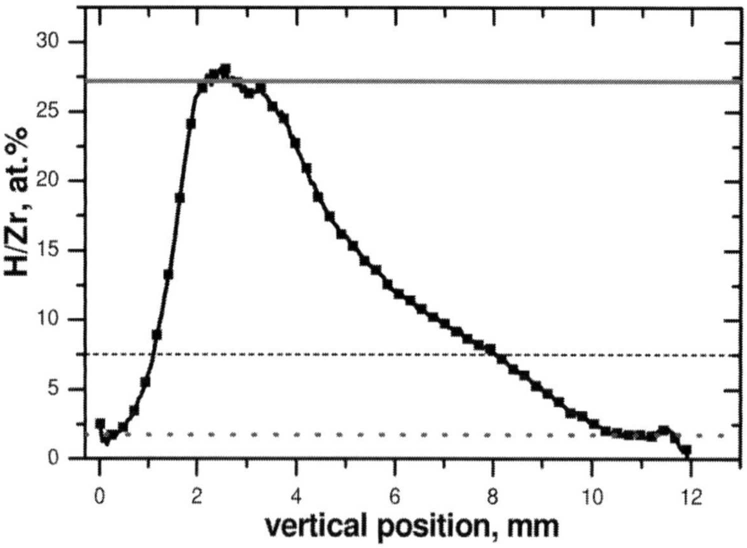

Fig. 9. Hydrogen distribution perpendicular to the cylinder axis close to the surface.

4. Summary and Conclusions

The hydrogen diffusion in Zry-4 was investigated by means of in-situ neutron radiography experiments.

At temperatures at which the bcc β-phase is stable, an activation energy of 48.3 kJ mol^{-1} K^{-1} was determined. It is significantly higher than formerly published data. However, the data presented here agree well with the temperature correlations published for various temperature ranges.

At temperatures between 823 and 1123 K the hydrogen uptake is connected with a stabilization of the β-phase to lower temperatures. The uptake starts homogeneously in the hcp α-phase. When a certain concentration level is reached, the phase transition occurs. Because the β-phase can absorb much more hydrogen than the α-phase, hydrogen diffuses from the α-region into the β-phase. The hydrogen depletion of the α-phase hinders further phase transitions at a second location. Hence, the further hydrogen uptake is connected with a growth of the location with bcc structure.

Acknowledgement
The authors thank the QUENCH team from KIT particularly U. Stegmaier, P. Severloh, J. Stuckert and M. Steinbrück for the support in the sample preparation and in construction and modification of the INRRO furnace. They are also grateful to the NIAG group of PSI for the support in the neutron radiography investigations.

References

[1] Steinbrück M 2004 *J. Nucl. Mater* **334,** 58.
[2] Albrecht W M and Goode W D 1960 The Diffusion of Hydrogen in Beta Zirconium *Report 1373*, Battelle Memorial Institute (referred in [3])
[3] Someno M (1960) *Nippon Kinzoku Gakkai-Shi* **24,** 249
[4] Gelezunas V L, Conn P G and Price R.H 1968 *J.Electr. Soc* **110,** 799
[5] Wagner C 1968 *Berichte der Bunsengesellschaft* **72,** 778
[6] Kearns J J 1972 J. Nucl. Mater. **43,** 330
[7] Sawatky A and Ells C E 2000 Understanding Hydrogen in Zirconium *Zirconium in the Nuclear Industry: 12th Intern. Symp.*, ASTM STP 1354, G.P. Sabol and G.D. Moan, Eds., American Society for Testing and Materials, West Conshocken, PA, 32
[8] Silva K R F, dos Santos D S, Ribeiro A F, and Almeida L H 2010 *Defect and Diffusion Forum* **297-301,** 722
[9] Kaestner A P, Hartmann S, Kühne G, Frei G, Grünzweig C, Josic L, Schmid F and Lehmann E H, The ICON beamline - a facility for cold neutron imaging at SINQ, Nuclear Instruments and Methods section A, 2011, doi:10.1016/j.nima.2011.08.022
[10] Grosse M, van den Ber M, Goulet C, Lehmann E and Schillinger B 2011 *Nucl. Instr. & Meths. In Physic. Res. A* (in press) doi: 10.1016/j.nima.2010.12.070
[11] Wenk, H R, Lonardelli I, and Williams D 2004 *Acta Mater.* **52,** 1899

Texture and Structure Analysis of Metagabbro by Neutron Diffraction

L Kalvoda[1], S Vratislav[1], M Dlouha[1], M Kruzelova[1], M Machek[2]
[1]Faculty of Nuclear Sciences and Physical Engineering, CTU in Prague, Brehova 7, 115 19 Prague 1, Czech Republic
[2]Institute of Geophysics AS CR, v.v.i, , Boční II/1401, 141 31 Prague 4, Czech Republic

E-mail: ladislav.kalvoda@fjfi.cvut.cz

Abstract. Crystallographic preferential orientation (CPO) of metagabbro mylonite from the eastern part of Stare Mesto belt, Bohemian Massif, is investigated by means of neutron diffraction method. The experiments were performed on the KSN-2 neutron diffractometer situated at the research reactor LVR-15 in the Nuclear Research Institute, plc. Rez, Czech Republic. Based on the collected diffraction patterns, the orientation distribution function of crystalline grains is determined by Rietveld harmonic method for the principal mineral phases - amphibole and plagioclase - and used to calculate (001), (020), (021), (110), (111) (plagioclase) and (001), ($11\bar{1}$), (020), (110), (200) (amphibole) pole figures. Lattice parameters of the phases have been obtained from diffraction patterns of a powder specimen. Main features of the observed CPO are compared with results of electron back-scatter diffraction measurements performed formerly on the same sample and with other known data.

1. Introduction

A substantial part of the lower continental and oceanic crust is formed by metabasic rocks, a heterogeneous and anisotropic microstructural nature of which is expected to have a pronounced effect on seismic properties. Consequently, the studies of crystallographic preferential orientation (CPO, texture) of such rock materials are of a high theoretical and practical importance because, for instance, the obtained texture results might be used in models of seismic wave propagation, the results calculated using such models can be used in prediction of seismic effects and further correlated with the laboratory data obtained by ultra-sound pulse transmission (UPT) method [1-3].

Since about three decades ago, neutron diffraction (ND) is increasingly applied to CPO analysis of rocks. Comparing with the other techniques, such as X-ray diffraction, electron back-scatter diffraction (EBSD) or special microscopic techniques, the ND method provides the unique possibility to integrate over a large specimen volume. The later feature is necessary for examination of coarse-grained, significantly heterogeneous materials and leads to statistically highly significant CPO characterization (cf. reviews [4-6]).

In our study, we present the results of investigation performed by ND on sample of metagabbro mylonite rock from the eastern part of the Stare Mesto belt, Bohemian Massif. The sample is part of a larger series used in complex studies of the seismic behaviour of meta-gabbros from the Stare Mesto belt [7, 8]. Methodically, the presented study links up to our former research devoted to CPO and structure of rocks [9].

2. Experimental

2.1. Samples
The investigated mylonite metagabbro rock was sampled from the eastern metagabbroic sheet of the Stare Mesto belt located in the eastern part of the Bohemian Massif. The rock underwent dynamic deformation under amphibolite facies conditions [7]. Two specimen types were prepared and investigated by the ND method: a powder specimen used to refine the lattice parameters of the constituting minerals and a spherical sample for CPO analysis (50 mm in diameter, further referred as GK3; the particular shape and size is used with aim to facilitate a consecutive juxtaposition of the results obtained by ND and UPT methods). The powder sample was prepared by grinding the raw rock material in a small ball mill to the final powder of average grain size ca 10 μm (the estimate obtained by means of optical microscopy).

2.2. ND measurements
The ND experiments were performed on the KSN-2 neutron diffractometer situated at the horizontal channel of the research reactor LVR-15 in the Nuclear Research Institute, plc. Rez, Czech Republic. The main experimental parameters were as follows: the wavelength of the monochromatic neutron beam $\lambda = 0.1362$ nm, the ultimate resolution $\Delta d/d = 7 \times 10^{-3}$ at the diffraction angle $2\theta = 19^{\circ}$, the diffraction patterns recorded within the 2θ range $7.5^{\circ} \div 80^{\circ}$ with $\Delta 2\theta = 0.1^{\circ}$. For the powder specimen measurement, the grinded rock material was placed in a cylindrical vanadium container. The bulk GK3 specimen employed in CPO characterization was fixed in a three axes texture goniometer and the latter was used to set the specimen gradually into 83 different positions covering uniformly one full hemisphere of the specimen orientations. At every specimen position, the full ND diagram was recorded. The coordination system of the GK3 specimen (which the texture analysis results are referred to) is defined by the normal to the plane of foliation (NF), the lineation direction (LD) and the third rectangular direction completing the right-handed set of axes.

3. Results
All the obtained ND patterns were corrected for non-linear background and then evaluated using the Rietveld method implemented in the software package GSAS [10]. At first, the powder specimen data were used to refine the structure parameters of the plagioclase (labradorite structure, triclinic space group C-1, further referred as Phase I) and the amphibole (monoclinic space group C2/m, referred as Phase II). Then, the GK3 specimen data were refined to reveal the CPO of the phases. The following procedure was applied in the latter case: At first, the scale factors, background coefficients and lattice parameters were determined for all diffraction patterns corresponding to the individual GK3 orientations. Then, the fixed average values of the lattice parameters were used to refine the orientation distribution functions (ODFs) of crystallites of both mineral phases. The ODF representation by spherical harmonics was employed (P1 sample symmetry, the harmonic expansion order limit $L = 8$).
The standard statistical descriptors related to the refinement process are given in Table 1. The calculated lattice parameters are summarized in Table 2. It is apparent that the refinement of the textured specimen GK3 converged to the final state statistically comparable with the fit of the powder specimen. There is also no significant difference between the lattice parameters obtained for both specimens. The calculated values of the lattice constants lie well within the common range expected upon the particular type of the mineral structures [11, 12].

Table 1. Statistical parameters of the Rietveld refinement.

Specimen	Rwp (%)	Rp (%)	χ^2
Powder	4.25	3.40	1.88
GK3	4.41	3.43	3.31

Table 2. Calculated lattice parameters.

Specimen	Phase	a (Å)	b (Å)	c (Å)	α (deg)	β (deg)	γ (deg)
Powder	Phase I	8.146(5)	12.83(6)	7.110(5)	93.76(6)	116.22(4)	89.89(4)
	Phase II	9.833(4)	17.99(5)	5.291(3)	90	105.01(6)	90
GK3	Phase I	8.154(3)	12.82(5)	7.104(2)	93.55(6)	116.18(7)	89.77(5)
	Phase II	9.837(4)	18.01(3)	5.296(6)	90	105.00(5)	90

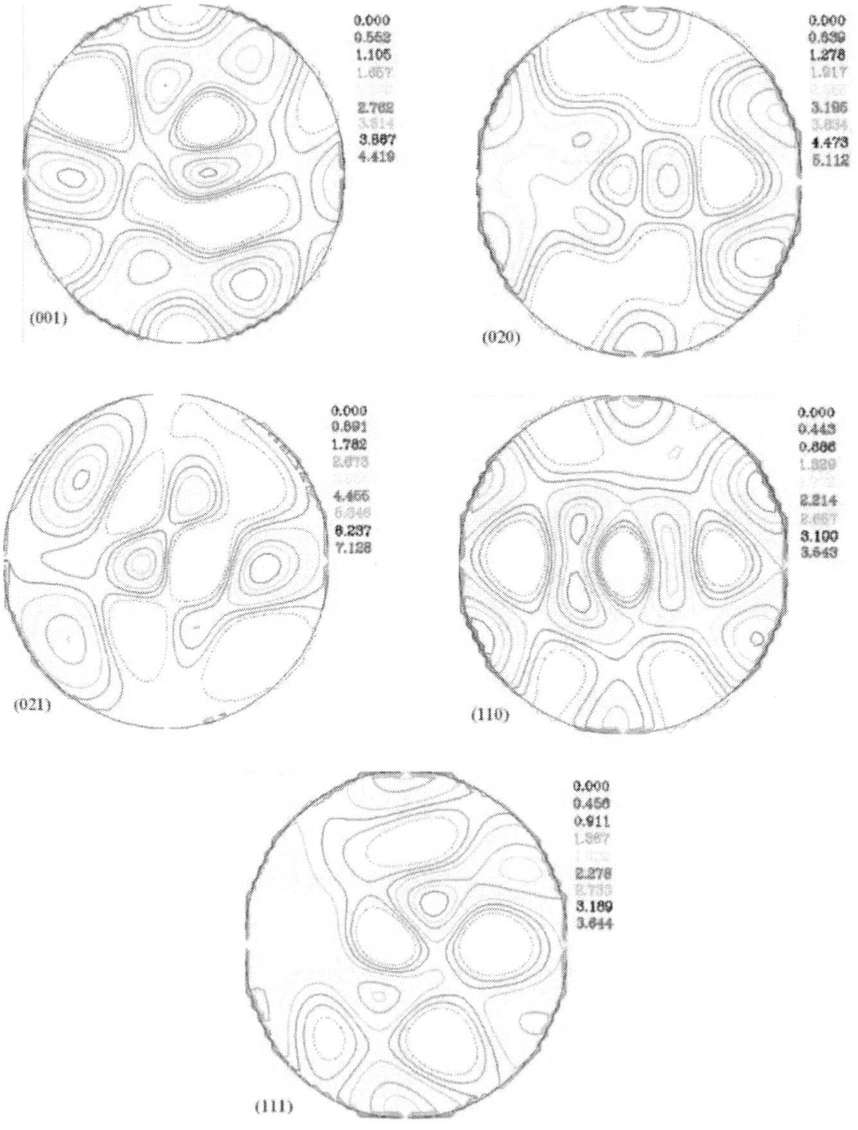

Figure 1. Calculated pole figures of the Phase I (plagioclase).

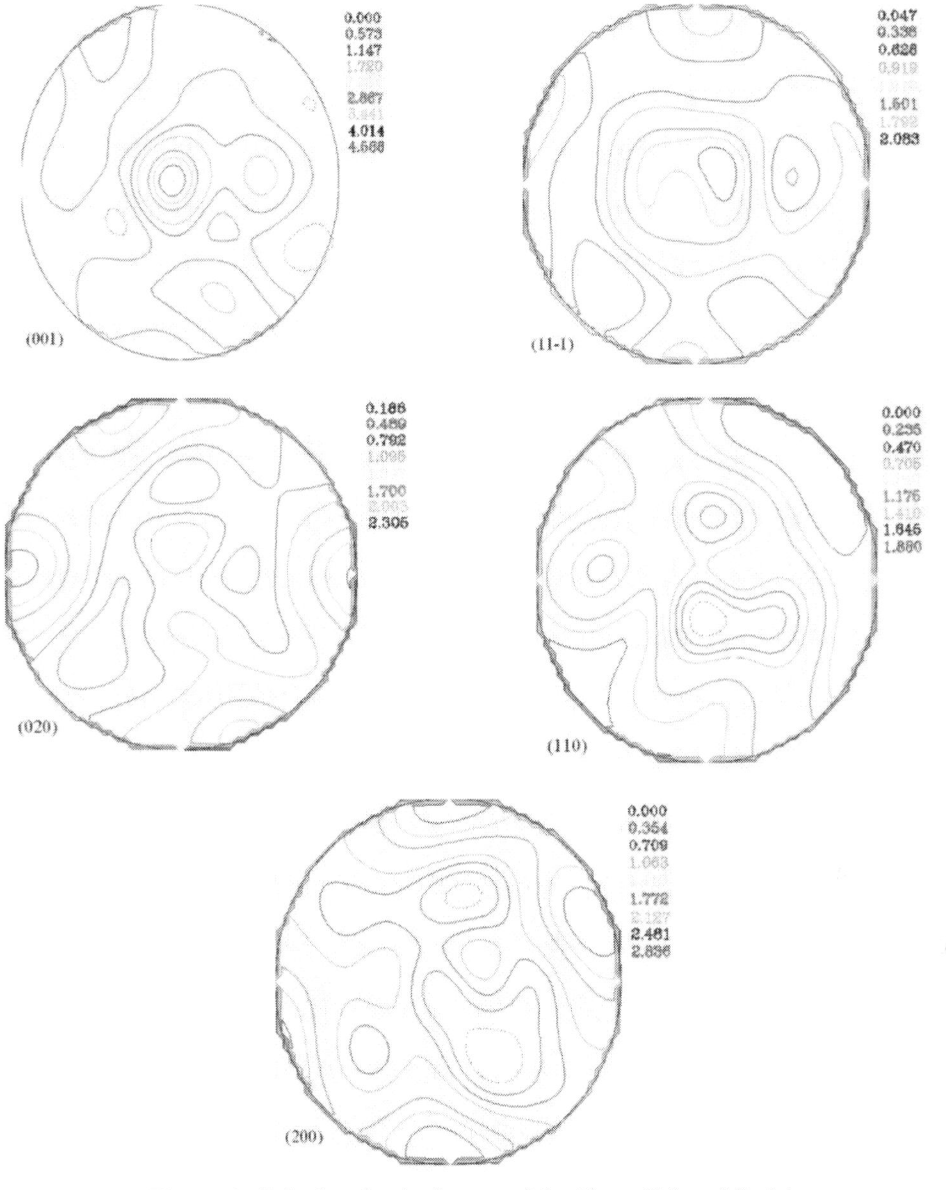

Figure 2. Calculated pole figures of the Phase II (amphibole).

Pole figures (PFs) of plagioclase (Phase I) and amphibole (Phase II) re-calculated from the corresponding ODFs are shown in figure 1 and figure 2, respectively. Scale of the contour lines is given in multiples of the random density (m. r. d.). Orientation of the stereographic projection is as follows: the NF points to the 'north' and the LD direction is in the 'zenith' of the projection. Thus, the foliation plane corresponds to the horizontal (equatorial) intersection of the projection.

In case of the Phase I (plagioclase), the PFs provide evidence about a complex multi-component CPO. Within the reconstructed PFs, the highest grain orientation population along LD and NF belongs to (021) and (001) poles, and (111) poles, respectively.

In case of the Phase II (amphibole), two distinguished texture components can be identified. The common feature of them is orientation of grains with the (001) pole parallel to LD, but they differ in orientation of the [100] and [010] directions. They can be considered as transient sub-types of the amphibole texture characterized by (001) poles parallel and (hk0) poles normal to LD observed with glaucophanite samples deformed under eclogite facies conditions and compatible with the (hk0)[001] slip system [13]. The more pronounced of the two components features orientation of the [010] direction within the plane of foliation and the [100] along the NF direction. It can be identified with the 'Type 1' texture introduced in [14] and agrees well with the results obtained by the EBSD method [7]. The second, weaker component shows [010] and [100] orientation out of the foliation plane, the directions inclined respectively ca. 30 and 60 degrees from NF and normal to LD. The second amphibole texture component was not observed in the upper mentioned EBSD measurements.

4. Conclusion

Neutron diffraction structure and texture analysis has been applied to investigate CPO of mylonite metagabbro sample. The obtained structural parameters agree well with the values expected for the particular forms of the plagioclase and amphibole minerals forming the rock. The results of the performed texture analysis suggest two-component nature of the CPO of the amphibole phase and a multi-component texture of the plagioclase phase. The amphibole CPO is in general features compatible with the published data and extends the results observed by EBSD method.
Taking in account that the amphibole CPO is known to control anisotropy of ultrasound transmission [15], presence of the additional second texture component can possibly influence on the calculated seismic wave velocities of the rock and might correct to the formerly found discrepancy between the measured and calculated UPT data [7]. Verification of the latter effect will be subject of further research.

Acknowledgment
This research has been supported by the Ministry of Education, Youth and Sports of the Czech Republic grants MSM68407721 and MSM6840770040, the Czech Science Foundation grant GA205/08/0767 and by the research plan AVOZ30120515 of the Institute of Geophysics AS CR..

5. References

[1] Pros Z, Babuska V 1968 *Studia geoph. et geod.* **12** 192
[2] Mainprice D 1990 *Computers and Geoscience* **16** 385
[3] Ullenmeyer K, Siegesmund S, Rasolofosaon P N J, Behrman J H 2006 Technophysics 414 97
[4] Leiss B et al 2000 *J. Struct. Geol.* **22** 1531
[5] Schäfer W 2002 *Eur. J. Mineral.* **14** 289
[6] Kockelmann W, Chapon L C, Radaelli P G 2006 *Physica B* **B385-386** 639
[7] Baratoux L 2004 *Petrology, Deformation Mechanisms, and Fabric Anisotropy of Metabasites Deformed at Natural Strain and Metamorphic Gradient - PhD Thesis* (Prague: Charles University)
[8] Machek M 2011 *Relation of Pore Space Geometry, Permeability and Microstructure in Low Porosity Rocks - PhD Thesis* (Prague: Charles University)
[9] Vratislav S, Dlouha M, Kalvoda L 2009 *Zeitschrift für Kristallographie Supl.* **30** 195
[10] Von Dreele R 1997 *J. Appl. Cryst.* **30** 517
[11] Ribbe P H 1975 Chemistry, Structure and Nomenclature of Feldspars *Feldspar Mineralogy (Reviews in Mineralogy* vol 2 2nd edition*)* ed P H Ribbe (Chantilly: Mineralogical Society of America) chapter 1 pp 1-12
[12] Hawthorn F C 1981 Crystal Chemistry of the Amphiboles *Amphiboles and Other Hydrous Pyriboles - Mineralogy (Reviews in Mineralogy* vol 9A*)* ed D R Veblen (Chantilly: Mineralogical Society of America) chapter 1 pp 91
[13] Zucali M, Chateigner D, Dugnani M, Lutterotti L, Ouladiaf B 2002 Quantitative texture Analysis of Glaucophanite Deformed under Eclogite Facies Conditions (Sesia-Lanzo Zone, Western Alps): Comparison between X-ray an Neutron Difraction Analysis *Deformation Mechanisms, Rheology and Tectonics: Current Status and Future Perspectives (Geological*

Society, London, Special Publications vol 200) ed S de Meer, M R Drury, J P H de Bresser, G M Pennock (The Geological Society of London) pp 239-253

[14] Leiss B, Groeger H R, Ullenmayer K, Lebit H 2002 Textures and Microstructures of Naturally Deformed Amphibolites from the Northern Cascades, NW USA: Methodology and Regional Aspects *Deformation Mechanisms, Rheology and Tectonics: Current Status and Future Perspectives* (*Geological Society, London, Special Publications* vol 200) ed S de Meer, M R Drury, J P H de Bresser, G M Pennock (The Geological Society of London) pp 219-238

[15] Barruol G, Kern H 1996 *Physics of the Earth and Planetary Interiors* **95** 175

Carbonation profiles in cement paste analyzed by neutron diffraction

I Galan[1], J Sanchez[1], C Andrade[1], A Evans[2]

[1] Eduardo Torroja Institute IETcc-CSIC, Serrano Galvache 4, 28033 Madrid, Spain
[2] Institut Laue Langevin ILL, 6, rue Jules Horowitz, 38042 Grenoble, France

isabelgalan@ietcc.csic.es

Abstract. The present work deals with the carbonation process in cement based materials such as concrete. In order to clarify the evolution of the two main phases involved in the process, portlandite and calcium carbonate as a function of depth, spatially resolved neutron diffraction experiments have been performed at SALSA diffractometer at ILL in carbonated cement paste samples. Specimens submitted to different carbonation processes, both natural and accelerated, have been analyzed with this non destructive technique. The evolution of the main diffraction peaks of portlandite and calcite has been followed by means of neutron diffraction patterns measured at different depths. The results indicate that, in specimens subjected to CO_2 atmospheres for 24 and 48 hours, the amount of calcite increases from the centre of the specimen to the surface. In both type of specimens calcite is formed at all depths analyzed, with higher quantities for the ones submitted to the longest carbonation period. Regarding the evolution of portlandite in these specimens, it almost completely disappeared, with only a low amount of the phase constant throughout the sample. In specimens subjected to air in a closed chamber for 21 months, higher amounts of portlandite were observed throughout the sample and little increase of calcite in the outer part, pointing out a much less severe reaction. The absorption effects are characterized by measuring in perpendicular directions and an absorption coefficient is calculated for portlandite.

1. Introduction

The term 'carbonation' in cement based materials applies to the reaction between acid components from the atmosphere, mainly carbon dioxide, and basic components from the hydrated cement, mainly portlandite (calcium hydroxide), to form calcium carbonate. The reaction begins at the exposed surface of the cement structure, continuing towards the inner part of the material. The diffusion rate and the amount of calcium carbonate formed in the pores depend on many variables: type of cement, water/cement ratio, environmental conditions, setting and curing, among others [1-5].

Considering steel reinforced concrete structures, carbonation may become a problem if the reaction front reaches the steel. The reason for this danger is the lowering of the pH that takes place in the concrete pores solution due to the calcium hydroxide carbonation. In safe conditions the steel inside the concrete is protected by a passive layer formed at pH values between 12.5 and 13.6. When the pH lowers below 8 the passive layer becomes unstable leading to a generalized corrosion on the steel surface [6-9].

The main technique used for preventing steel corrosion caused by concrete carbonation is the phenolphthalein indicator, which changes color at pH 8-9. Although this indicator shows two distinct zones separated in most cases by a sharp front, it has been already proved that between the carbonated and the non-carbonated zone there is a transitional region. In that region the portlandite decrease and the calcium carbonate increase are not abrupt, but gradual [10-14]. Most of these studies use Thermogravimetrical Analysis (TGA) for computing the amounts of portlandite and calcium carbonate present at different depths in the samples. For using this destructive technique the samples must be cut or sawn at the required depth and then milled, analyzing only some milligrams of the powder; the main disadvantages of the TGA are the unavoidable contact of the CO_2 with the cement paste powder and the difficulty to obtain representative samples. Other techniques used for studying carbonation, such as X-ray diffraction, Infrared spectroscopy and pH measurements, also work with little amounts of powder, having the same disadvantages as TGA. A more sophisticated technique used is the gammadensitometry, which is non-destructive, but requires uncarbonated references for interpreting results and does not give information about portlandite.

Neutron diffraction (ND), a non-destructive technique, has already been used for studying in-situ carbonation processes at constant relative humidity [15-16]. ND has been very useful for obtaining information about the crystalline phases involved in the process, mainly portlandite and calcite, in the bulk of the samples.

In the present work spatially resolved ND is used for analyzing carbonation profiles, that is, for following the evolution of the main phases with the depth.

2. Experimental

Cylindrical cement paste specimens, 1x4 cm, were fabricated using cement without additions (CEM I 42.5R) and deuterated water, with a water/cement ratio in weight of 0.5. The specimens were then cured for 28 days in their molds in a saturated humid chamber. After demoulding the specimens were placed in constant RH chambers, where the humidity was achieved and maintained at 65% by means of $NaNO_2$ saturated salt solutions. Some samples were kept there for 2 months and, after that, they were subjected to 100% CO_2 atmospheres for 24 or 48 hours, also in 65% RH atmospheres. Other samples were kept in the chambers with 65% RH and air, that is, 0.04% CO_2, for 21 months. During the time of exposure the chambers were opened three times, after 6, 12 and 18 months for few minutes, renewing the air.

The ND measurements were performed at SALSA instrument at Institut Laue Langevin (ILL). 2D diffraction patterns were collected every cubic millimeter from the sample center to the edge, that is, in five steps the whole radius was covered. As shown in Figures 1 and 2 the samples were placed vertically between the neutron source and the detector. The neutron wavelength used was 1.648 Å. With this lambda the main diffraction peak of portlandite appears at 36.5° and the one of calcite at 31.4°. The two phases were measured in the samples along the 'x' radial direction, which coincides with the bisection of the incident and diffracted beams. Each measurement took 1 hour, that is, 5 hours per phase and sample.

Measurements in the 'x' radial direction were performed in the 3 types of samples (carbonated at 100% CO_2 for 24 hours, at 100% CO_2 for 48 hours and at 0.04% CO_2 for 21 months).

In order to evaluate neutron absorption effects, measurements along the 'y' radial direction, perpendicular to 'x', were also performed in the samples carbonated in 0.04% CO_2. The main difference between the two directions is the distance the neutrons have to traverse inside the samples. In the 'x' direction this distance decreases with depth, while in the 'y' direction it is maintained almost constant in the five measurement volumes. The comparison of the patterns obtained in both perpendicular directions has allowed for the analysis of the absorption.

Figure 1. Experimental set-up in SALSA.

Figure 2. Specimens placed at the intersection of the primary neutron beam (defined by radial collimator) and diffracted beam (defined by radial collimator) in front of the detector.

3. Results and discussion

Figures 3 and 4 show the evolution with depth of the calcite main diffraction peak from the center of the specimen to the surface (from the left to the right), measured in the 'x' direction.

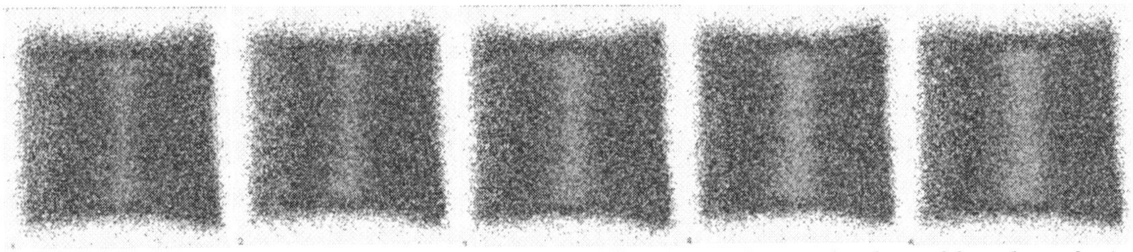

Figure 3. 2D diffraction patterns of the main calcite diffraction peak scanning in position along the 'x' direction in the sample exposed to 100% CO_2 24 hours. The horizontal axis is the diffraction angle two theta and the vertical is the over 5° around the diffraction ring.

Figure 4. 2D diffraction patterns of the main calcite diffraction peak scanning in position along the 'x' direction in the sample exposed to 100% CO_2 48 hours.

Figure 3 corresponds to the sample exposed to 100% CO_2 for 24 hours and Figure 4 to the one carbonated in the same atmosphere for 48 hours. The images were obtained from the Lamp data analysis program [17]. As it can be seen, in the samples submitted to accelerated carbonation, the evolution with depth is clear: calcite increases from the inside to the surface. In the samples exposed to air in a closed chamber for 21 months the calcite main diffraction peak is not very well defined, both in the 'x' direction as in the 'y'.

Considering portlandite evolution, in the samples exposed to accelerated carbonation the diffraction peaks are very weak and exhibit intensities almost half that of those corresponding to samples exposed to air.

Figures 5 and 6 represent the integration of the diffraction peak over the height of the 2D detector as a function of radial position in samples exposed to 0.04% CO_2 for 21 months measured in the 'x' and 'y' direction, respectively. The images were also obtained from Lamp. The lower part of the image (scan 66 in Figure 5) corresponds to the inner part of the sample and the top to the surface (scan 70). In Figure 5 it seems that portlandite increases from the inner part of the sample to the surface. In Figure 6 the evolution of the peak is very different: the intensity is almost maintained throughout the sample, decreasing slightly at the surface (scan -4.5). This fact may be due to the different paths covered by the neutrons inside the samples and the subsequent absorption. In order to analyze this effect, in Figure 7 a scheme of the neutron diffraction paths inside the samples is represented for the measurements in both directions.

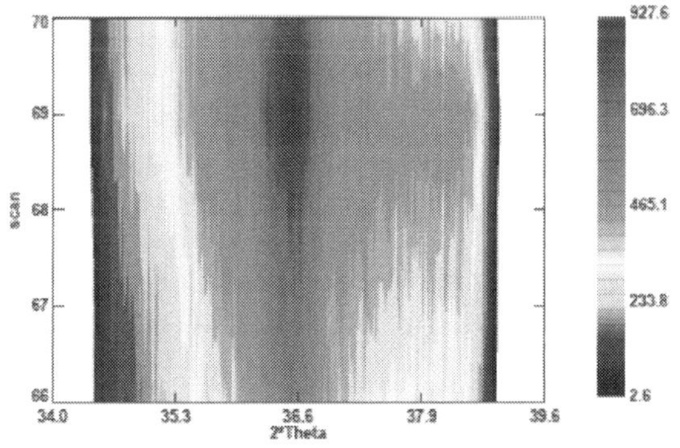

Figure 5. Portlandite main diffraction peak 'x' direction sample exposed to 0.04% CO_2 21 months. The centre of the sample is at the bottom on the Y axis (scan 66) and the surface at the top (scan 70).

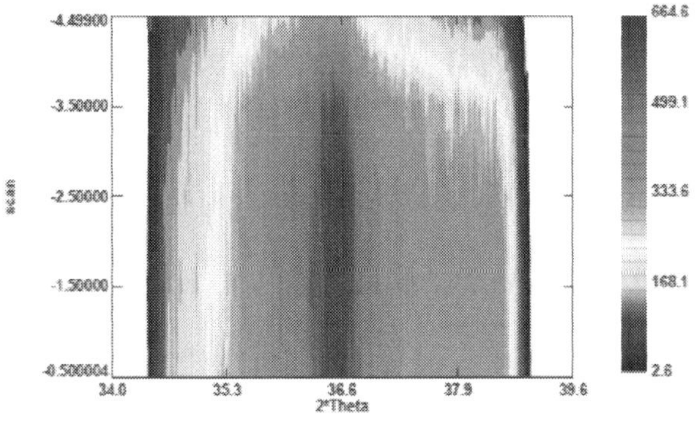

Figure 6. Portlandite main diffraction peak 'y' direction sample exposed to 0.04% CO_2 21 months. The centre of the sample is at the bottom on the Y axis (scan -0.5) and the surface at the top (scan -4.5).

The distances covered by the neutrons in the 'x' direction differ considerably for each measurement point, decreasing while 'going out'; in the 'y' direction the paths are all quite similar and almost independent from the measurement depth. The distance depends also slightly on the diffraction angle, that is, on the phase considered. For the calculations, the Lambert-Beer law is used:

$$I = I_0 \cdot \exp(-\alpha \cdot L \cdot C) \tag{1}$$

where I_0 is the incident intensity, I is the attenuated intensity when passing through a material, α is the absorption coefficient, L is the path covered and C is the concentration of 'absorbers' in the material. Relating the attenuated intensities with the paths in both directions, the value of $A = \alpha \cdot C$ can be calculated.

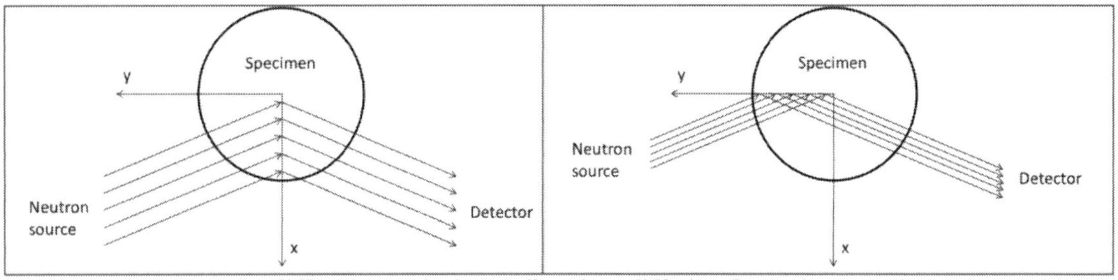

Figure 7. Neutron diffraction paths in 'x' direction, differing for each measuring point, and 'y' direction, being very similar for all measuring depths.

In this case, the intensities are the peaks' areas, calculated between 2θ=35.7° and 2θ=37.5° for the portlandite phase. Dividing the Lambert-Beer expression for 'y' by the corresponding for 'x' and taking logarithms, the following expression is obtained:

$$\ln\left(\frac{I_y}{I_x}\right) = A \cdot (L_x - L_y) \tag{2}$$

Figure 8 represents $\ln\left(\frac{I_y}{I_x}\right)$ versus $(L_x - L_y)$ and the corresponding linear fitting.

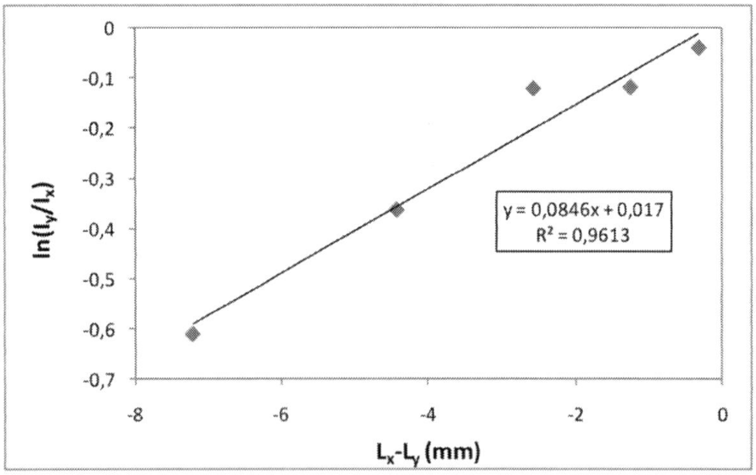

Figure 8. $\ln(I_y/I_x)$ versus (L_x-L_y), being the slope of the line the absorption parameter $A = \alpha \cdot C$.

The calculated value of the parameter $A = \alpha \cdot C$ is 0,085 mm^{-1}. This factor allows the conversion of the peak areas in the perpendicular directions.

Applying the A value to the samples submitted to accelerated carbonation, the portlandite peak areas obtained in the 'y' direction do not change throughout the sample and are approximately half of those in the natural carbonated sample.

Regarding the calcite peak in the sample submitted to air, the differences of intensity in both directions are not significant. As the amount of this phase in this sample is quite low and the peaks are not sharp, the corresponding absorption coefficient cannot be calculated from this set of data.

4. Conclusions

Neutron diffraction has allowed to measure carbonation profiles in cement paste samples. In the specimens subjected to CO_2 atmospheres for 24 and 48 hours the amount of calcite increases from the centre of the specimen to the surface. In both type of specimens calcite is formed at all depths analyzed, with higher quantities for the ones submitted to the longest carbonation period. Regarding the evolution of portlandite in these specimens, it almost completely disappeared, with only a low amount of the phase constant throughout the sample. In specimens subjected to air in a closed chamber for 21 months, higher amounts of portlandite are observed throughout the sample and little increase of calcite in the outer part, pointing out a much less severe reaction. The neutron absorption effects in this type of materials have been characterized by measuring in perpendicular directions and comparing the portlandite main peak areas.

Acknowledgments

The neutron diffraction experiment performed at SALSA was supported by the ILL, to whom the authors are grateful. We also thank the SALSA staff for their assistance.

References

[1] Venuat M 1977 *Rencontres CEFRACOR-77 JTBTP*
[2] Fattuhi N I 1988 *Cem. Concr. Res.* **18** 426-430.
[3] Papadakis V G, Vayenas C G and Fardis M N 1991 *ACI Mat. Jour.* ***88*** 186-196.
[4] Houst Y F and Wittmann F H 1994 *Cem. Concr. Res.* **24** 1165-1176.
[5] Gonen T and Yazicioglu S 2007 *Constr. Build. Mater.* **21** 1040-1045.
[6] Verbeck G J 1975 *Corr. Metals in Concr. ACI* 21-38.
[7] Gonzalez J A, Algaba S and Andrade C 1980 *Brit. Corr. Jour.* **15** 135-139.
[8] Alonso C and Andrade C 1987 *Mat. Constr.* **37** 5-16.
[9] Parrott L J 1994 *Mag. Concr. Res.* **46** 23-28.
[10] Parrott L J and Killoh D C 1989 *Cem. Concr. Res.* **19** 649-656.
[11] Rahman A A and Glasser F P 1989 *Adv. Cem. Res.* **2** 49-54.
[12] Houst Y F and Wittmann F H 2002 *Cem. Concr. Res.* **32** 1923-1930.
[13] Chang C F and Chen J W 2006 *Cem. Concr. Res.* **36** 1760-1767.
[14] Villain G, Thiery M and Platret G 2007 *Cem. Concr. Res.* **37** 1182-1192.
[15] Castellote M, Andrade C, Turrillas X, Campo J and Cuello G J 2008 *Cem. Concr. Res.* **38**, 1365-1373.
[16] Galan I, Andrade C, Castellote M, Rebolledo N, Sanchez J, Toro L, Puente I, Campo J and Fabelo O 2011 *J. Phys.: Conf. Ser.* **325**.
[17] LAMP, the Large Array Manipulation Program. http://www.ill.fr/data_treat/lamp/lamp.html.

Role of water on formation and structural features of Maya blue

C Mondelli[1], M Sánchez del Río[2], MA González[3], A. Magazzú[3], C. Cavallari[3], M. Suárez[4], E. García-Romero[5], and P. Romano[6]
[1]CNR-IOM-OGG, Institut Laue Langevin; 6, Rue Jules Horowitz; 38042 Grenoble Cedex 9, France
[2]European Synchrotron Radiation Facility, BP 220, 38043 Grenoble Cedex, France
[3]Institut Laue Langevin; 6, Rue Jules Horowitz; 38042 Grenoble Cedex, France
[4] Departamento de Geología, Universidad de Salamanca, 37008 Salamanca, Spain
[5]Departamento de Cristalografía y Mineralogía, Universidad Complutense de Madrid, 28040 Madrid, Spain, and Instituto de Geociencias, UCM-CSIC, Madris, Spain
[6] Diamond Light Source Ltd, Harwell Science and Innovation Campus, Didcot Oxfordshire OX11 0DE, England

E-mail: mondelli@ill.fr

Abstract. The Maya blue (MB) is an artificial pigment created between 500-800 A.D. and used in murals, pottery and sculptures by Mayas and other people in Mesoamerica. MB is resistant to age, acid, weathering, biodegradation and even modern chemical solvents, but the chemical reasons behind the resistance to chemical aggressions are still under debate. Water plays a fundamental role in the interactions between indigo and clay. The dynamics of the clay's zeolitic and structural water molecules during the formation of MB, usually stabilized by moderate heating, has been monitored by means of neutron inelastic scattering. Neutron incoherent scattering in these samples is only due to the hydrogen atoms, so the signal is very sensitive to the amount of released water, providing detailed information on the dehydration process. A simultaneous analysis of the coherent elastic scattering and the incoherent scattering allows observing and quantifying how the structure of the clay is affected by dehydration. Here we show that a quite resistant pigment can be obtained at room temperature simply by dehydrating a palygorskite-indigo mixture employing only vacuum, without any thermal treatment.

1. Introduction

Maya blue [1] is an artificial pigment fabricated by the Maya in the first millennium A.D. The pigment was used in Prehispanic times from the southern Maya region to most Mesoamerican cultures. In ancient times, the fabrication and distribution of this pigment probably required a sophisticated technical and commercial infrastructure. Maya blue is made from indigotin mixed with palygorskite. Palygorskite is a fibrous clay featuring channels in its structure, which are filled by zeolitic H_2O. Palygorskite has a great absorbing power, and can also be used as a carrier for organic molecules, as well as for drugs. MB can be easily fabricated by crushing a few percent of indigotin with palygorskite powder. After this, a mild heating (~200°C during a few hours) is necessary to make this light blue pigment surprisingly resistant to degradation in adverse environmental conditions, as it is unaffected by the attack of acids, alkalis, oxidants, reducing agents and organic solvents [2]. Scientists still do not fully agree in the mechanism of this organo-clay interaction [3]. It has been proposed that a partial or perhaps total evacuation of the zeolitic H_2O is essential to allow the indigo to either penetrate in the tunnels or to sit and block the entrance of the tunnels [4]. In this work we apply neutron scattering to monitor the dehydration of palygorskite during the formation of MB. In particular, we study here whether the dehydration of the clay, obtained only by vacuum pumping, is a sufficient condition for stabilizing MB, or if a subsequent heating is required.

2. Experimental details

Indigo was purchased from Sigma Aldrich. Palygorskite (theoretical molecular formula $Si_8(Mg_2Al_2)O_{20}(OH)_2(OH_2)_4 \cdot 4H_2O$) is a natural clay coming from Ticul, in Yucatán (Mexico). This palygorskite has been thoroughly studied [5]. The mixture of palygorskite equilibrated at ambient conditions and indigo was finely crushed in a mortar. For the present study, different samples of 0.7 g each have been prepared from the starting mixture having a content of indigo of 3% in weight. In situ inelastic neutron scattering measurements were performed on the Time-of-Flight spectrometer IN6 at the Institut Laue-Langevin (Grenoble). They were performed using an incident wavelength of 5.1 Å and the angular range covered was $10° \leq 2\theta \leq 115°$. Two different sets of measurements were done: One of the samples was measured only at room temperature, but applying a continuous pumping (vacuum $\sim 10^{-3}$ Torr) on it and acquiring data as a function of time. The second sample was subjected to the thermal treatment exemplified by the temperature ramp ($300 \leq T(K) \leq 520$) shown in Fig. 2. We applied to the data the standard corrections and calibration (subtraction of the empty cell signal and detector efficiency calibration). In all cases the transmission was higher than 90%, so no multiple scattering corrections has been applied.

Table 1. Coherent and incoherent cross sections for a unit formula of palygorskite ($Si_8(Mg_2Al_2)O_{20}(OH)_2(OH_2)_4{\cdot}4H_2O$) and indigo ($C_{16}H_{10}N_2O_2$). The numbers between parentheses indicate the number of atoms of each element in the sample.

Palygorskite	σ_{coh} (barn)	σ_{inc} (barn)
Si (8)	17.30	0.03
Mg (2)	7.26	0.16
Al (2)	2.99	0.16
O (30)	126.96	0.02
H (18)	31.62	1444.68
Total	**186.13**	**1445.05**
Indigo	σ_{coh} (barn)	σ_{inc} (barn)
C (16)	88.82	0.02
N (2)	22.04	1.00
O (2)	8.46	0
H (10)	17.57	802.60
Total	**136.89**	**803.62**

3. Results and discussion

In order to quantify the incoherent and coherent signals, related to the content of water and the atomic structure of the sample, respectively, we have evaluated the coherent and incoherent cross sections of the elements present in the samples (Table 1). The incoherent signal is mainly due to the atoms of hydrogen in the zeolitic and structural water molecules, together with the contribution of the structural OH groups in palygorskite and the H atoms in indigo. In addition, some hydration water present on the surface of the clay can contribute to it. This hydration water is rapidly lost when the vacuum or the thermal treatments are applied (Figures 1 and 2). The incoherent signal coming from the structural OH groups and the indigo is subtracted based on the chemical formulas of the constituents, so the remaining signal contains two equal contributions due to the four structural water molecules and the four zeolitic ones. This procedure allows following the formation of a pigment having the same characteristics of the ancient MB by monitoring the amount of structural and zeolitic waters in the samples under the hypothesis that the zeolitic water evacuates completely from the sample before the structural water starts to escape. In order to do this, we need to separate the incoherent and coherent contributions in the measured signal, so we integrated in energy the measured dynamical structure factor, $S(Q,\omega)$, and then we separated the elastic scattering $S(Q)_{coh}$ (assumed to correspond to the area below the Bragg peaks) from the incoherent flat signal using the XOP program [6].

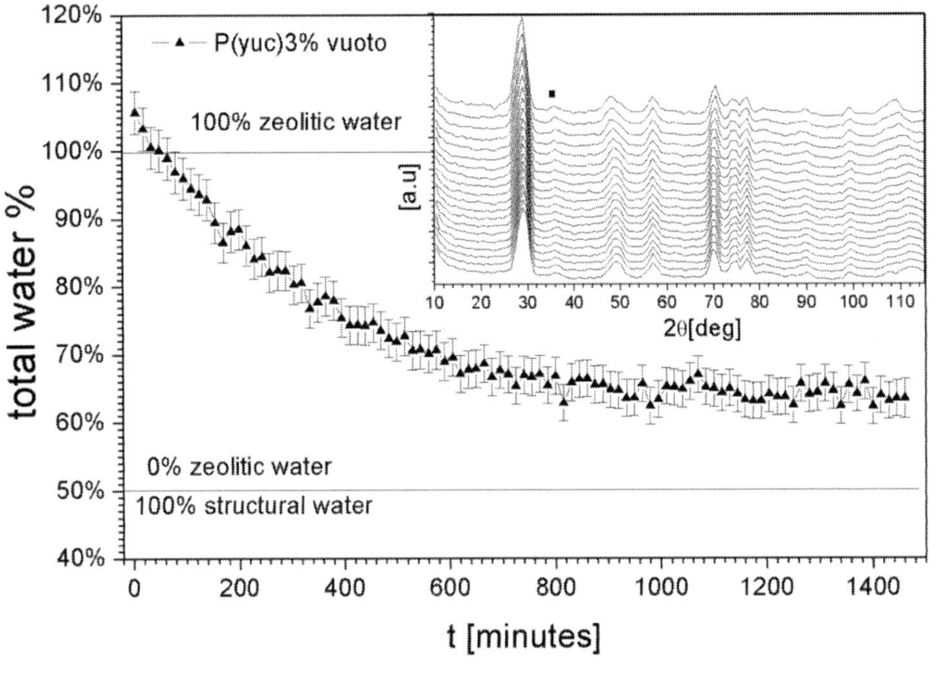

Figure 1. Percentage of water as a function of time in a mixture of palygorskite and 3% of indigo, when a continuous pumping is applied. A limiting value of 30% of zeolitic water remaining inside the palygorskite is reached after 800 minutes. The inset shows the elastic coherent signal measured at different times (time increasing from top to bottom). The peak marked with a square in the top curve corresponds to pure indigo that has not gone into the clay. The other peaks correspond to the structure of the palygorskite + indigo.

Figure 1 shows the evolution as a function of the time of the percentage of zeolitic and structural water, calculated as described in the previous paragraph. A continuous pumping is applied at time zero and the change of slope of the curve is related to the rate of water loss. It can be observed the contribution of the different types of water: a first quick loss corresponds to the hygroscopic water adsorbed to the clay and then the sample loses zeolitic H_2O progressively. In this case, all changes in water content are only due to the effect of vacuum, as no heating was applied. It is evident how the vacuum causes a partial loss of the total amount of water in the sample.

The hygroscopic H_2O is completely lost, plus an important part of the zeolitic H_2O. However the 30% of the total zeolitic water remains in the sample even after 8 hours of pumping. In the inset of figure 1 the peaks corresponding to the crystallographic structure of the mixture of palygorskite and indigo are reported. The crystallographic structure does not present relevant structural changes due to the vacuum, but we can observe a slow decrease of the intensity of the peak of un-reacted indigo (indicated with a square in the inset).

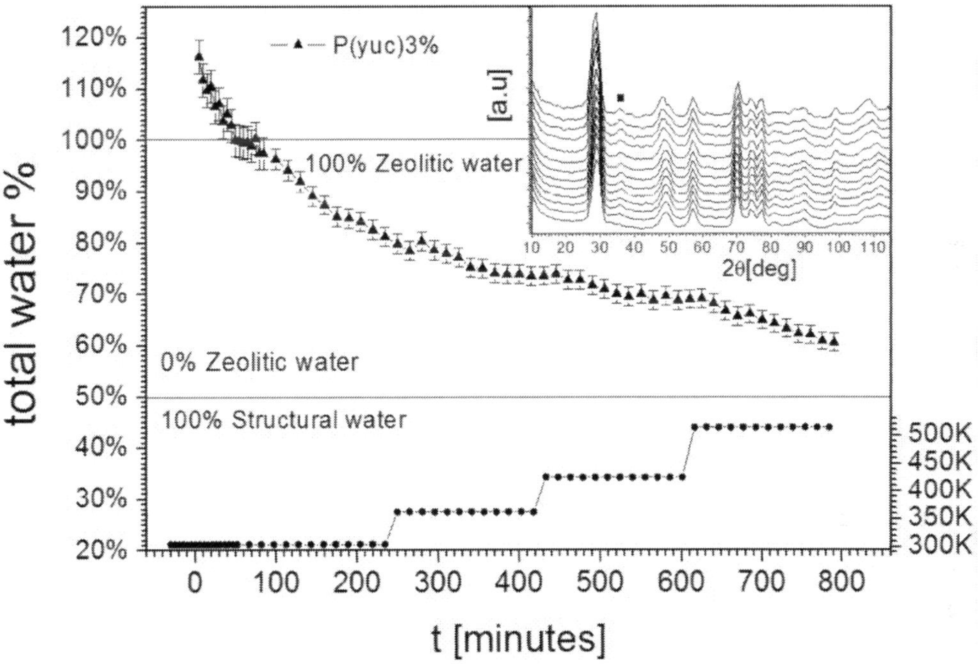

Figure 2. Change in the percentage of water as a function of time in a mixture of palygorskite and 3% of indigo when the thermal treatment indicated by the dots in the bottom of the picture is applied. As in the previous figure, the inset shows the elastic coherent signal measured at different times (time increasing from top to bottom).

A similar set of measurements was done using a second sample coming from the same bunch, but in this case applying a thermal treatment. The temperature ramp employed and the changes in the water content observed are reported in figure 2. The maximum temperature attained (520 K) corresponds to the temperature needed to produce a final pigment with the same stability as the original MB. Comparing with figure 1, we observe that the value of 30% of zeolitic water remaining in the sample is reached after only 650 minutes and that contrary to the previous case, the loss of water continues below this limit. Additionally we observe that the peak of un-reacted indigo disappears completely. We conclude that the thermal treatment anticipates the dehydration of the indigo-palygorskite mixture, but that the initial mechanism of water lost is equivalent down to 30% of zeolitic H_2O remaining in the sample.

In order to test the stability of the complexes formed by both methods, the final samples obtained from both experiments (with and without heating) were immersed in concentrated HNO_3 for at least 5 minutes. The high oxidant power of HNO_3 destroys the indigo molecules, a process that is accompanied by a complete lost of the blue colour. If indigo is just crushed with palygorskite, and the result is immersed in HNO_3, a complete decolouration is also observed because the indigo is not yet stabilized. However when the palygorskite stabilizes the indigo, e.g. after the heating process, the resulting complex becomes resistant and does not decolour in acid. After immersion in acid, the sample obtained by applying only vacuum shows a blue colour indicating that the MB pigment has been formed. However, a light decolouration is observed, indicating that not all the indigo was transformed into MB. On the contrary, the second sample (where heating was applied) did not show any decolouration, indicating a complete transformation of the indigo in MB. The resulting MB pigments are quite dark compared with the original Maya blue, which is much more turquoise, but this is surely due to the high content of indigo used in our samples (3%), while the quantity used by the Maya was probably around 1%. In figure 3 we compare the two pigments obtained in situ during the experiment. A higher content of indigo was employed to enhance the observed effects and facilitate the analysis, but the results with indigo contents less than approx 5% are certainly applicable to the ancient Maya blue with less indigo.

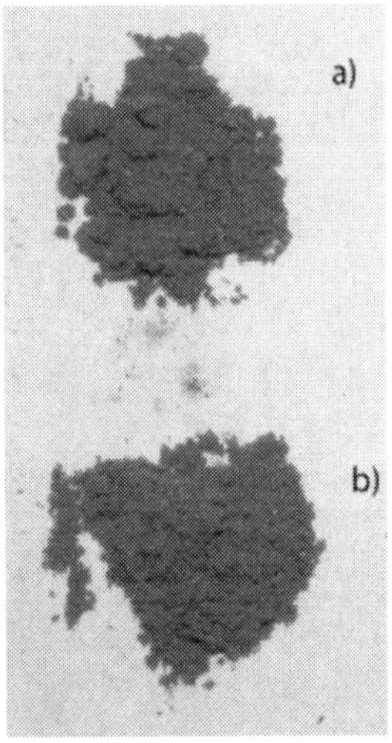

Figure 3. a) MB pigment obtained using only vacuum; b) MB pigment obtained using a thermal treatment. Both pigments are darker than the original MB due to the larger content of indigo employed in their preparation (3% instead of ≈ 1% used by the Mayas).

4. Conclusions

We have studied the lost of water during the formation of the final MB-type pigment by means of inelastic neutron scattering. As in our samples the incoherent scattering comes only from the hydrogen atoms, the measured signal is very sensitive to the amount of released water, providing quantitative information on the dehydration process. The simultaneous analysis of the incoherent and the coherent elastic scattering gave us a unique possibility to observe in parallel the dehydration process and the changes on the structure of the pigment. We show that the treatments applied are not strong enough to evacuate structural water and we quantify the amount of zeolitic water released during the MB formation. A quite resistant pigment is obtained from the zeolitic dehydration process of the palygorskite-indigo mixture only by means of vacuum, without any thermal treatment. This fact indicates that indigo bonds to palygorskite during the dehydration process, forming a stable pigment. This process is accelerated when heating is applied, but heating is not required to stabilize the pigment.

Acknowledgments

We acknowledge the ILL for the allocated beamtime and for financial support (ILL stage). Financial support from the Spanish CICYT (project CGL2006-09843) is also gratefully acknowledged.

References

[1] Reyes-Valerio C 1993 *De Bonampak al Templo Mayor. El azul maya en Mesoamerica: Colección America Nuestra* vol 40 (Mexico DF: Siglo XXI editores) p 157
[2] Van Olphen H 1966 *Science* **154** 645
[3] Sánchez del Río M, Doménech A, Doménech-Carbó M T, Vázquez de Agredos Pascual M L, Suárez M and García-Romero E 2011 *The Maya Blue Pigment* vol 3 Collection: Developments in Clay Science Chapter 18, ed A. Singer and E. Galan (Amsterdam, Elsevier)
[4] Giustetto R, Levy D, Chiari G 2006 *Eur. J. Mineralog.* **18** 629
[5] Sánchez del Río M, Suárez M, and García-Romero E 2009 *Archaeometry* **51** 214
[6] Sanchez del Rio M and Dejus R J 2001 XOP 2.1 *A new version of the x-ray optics software toolkit* American Institute of Physics Conference Proceedings **705** 784

Effect of Strontium doping on the Oxygen Diffusion in $La_{2-x}Sr_xCuO_{4\pm\delta}$ samples Investigated by Oxygen Isotope Back Exchange

R Sura[1, 2], M Ceretti[2], C Prestipino[2], W Paulus[2,4], J Schefer[1], L Keller[1], K Conder[3] and E Pomjakushina[3]

[1] Laboratory for Neutron Scattering (LNS), Paul Scherrer Institute, CH-5232 Villigen PSI, Switzerland.
[2] Sciences Chimiques de Rennes, UMR 6226, Campus de Beaulieu, Université de Rennes1, F-35042 Rennes, France.
[3] Laboratory for Developments and Methods (LDM), Paul Scherrer Institute, CH-5232 Villigen, PSI, Switzerland
[4] Chimie et cristallochimie des matériaux (C2M), ICG Montpellier, UMR 5253, Université de Montpellier2, F-34095, France.
E-mail: ravi.sura@psi.ch

Abstract Similar to $LaCoO_{4.14}$, $La_{2-x}Sr_xCuO_{4\pm\delta}$ shows oxygen diffusion at moderate temperatures. The Oxygen Isotope Back Exchange (OIBE) experiments are done in order to investigate the effect of strontium on the oxygen diffusion in as grown single crystals of $La_{2-x}Sr_xCuO_{4\pm\delta}$ (x =0, 0.05, 0.1, 0.15). Neutron diffraction data collected at TriCS@SINQ are in agreement with diffusion measurements achieved by oxygen back exchange. From $^{18}O/^{16}O$ isotope exchange between 40 ^{0}C to 1000 ^{0}C free oxygen mobility can be realized for x = 0, 0.05 below 500 ^{0}C but is depressed for x = 0.1 and 0.15. This gives evidence that low temperature oxygen mobility can be suppressed by replacing La with Sr. First crystal test using neutron single crystal diffraction proved the high quality of the samples.

1. Introduction

The understanding of oxygen ion conduction in solids at moderate temperatures is a key issue for the development of oxygen membranes in solid oxide fuel cells (SOFCs) [1]. Classically, oxygen ion conduction in solids is described in terms of a thermally activated hopping process with ions hopping from occupied to vacant lattice sites. As oxygen ions are doubly charged and have a large Van der Waals radius of about 1.52 Å, the energy barrier for the diffusion process is high, and normally requires fairly elevated temperatures (>1300K). As high temperature result in serious problems concerning thermal stability of the electrodes, suitable materials with operating at intermediate temperatures are searched by industry. On contrary two structure types are known showing oxygen bulk diffusion already at ambient temperature: Brownmillerites, *e.g.* $SrMO_{2.5}$ (M = Fe, Co) and Ruddlesden popper phases such as $La_2MO_{4+\delta}$ (M = Co, Ni, Cu).

A different view of the oxygen diffusion mechanism is necessary as the classical model fails in the explanation of the oxygen ion diffusion at ambient temperature. In the case of $SrFeO_{2.5}$ Brownmillerite-type structure [2] and $La_2CuO_{4+\delta}$ [3] oxygen mobility relies on the existence of specific, low energy lattice modes (phonons). These modes trigger and amplify oxygen mobility in solids. [2,3] The correlation between structure and diffusion has been demonstrated on $La_2CoO_{4.14}$ by the thesis work of Loïc Le Dréau [4], where oxygen diffusion sets in at 425 K and has been modelled by Density Functional Theory [5].

The oxygen diffusion in $La_{2-x}Sr_xCuO_{4-\delta}$ can be decreased by Sr doping as reported by Opila [6], going down from 2.0×10^{-5} cm^2/s for x=0.07 to 1.9×10^{-6} cm^2/s for x=0.12. Sr doping can therefore be used to investigate deeper how diffusion and structure [7] correlate. For this examining high quality single crystals are needed to avoid porosity and grain boundaries present in polycrystalline samples which can influence the equilibrium rate. Single crystals tested already on ORION@SINQ [8] are used to collect full data sets by using TriCS@SINQ [8] (x =0.15).

2. Experimental details

2.1 sample preparation

Polycrystalline $La_{2-x}Sr_xCuO_{4\pm\delta}$ (x = 0, 0.05, 0.10, 0.15) samples were prepared from pre annealed high purity (≥ 99.99%) La_2O_3, $SrCO_3$ and CuO powders by standard powder ceramic techniques. Samples were checked with powder X-ray diffraction for phase purity. These high quality polycrystalline samples were used to grow single crystals by traveling solvent floating zone technique at the Laboratory Development Methods at PSI. Typically, rods of about 6 mm diameter and several centimeters long were obtained. $La_{2-x}Sr_xCuO_{4-\delta}$ single grain crystals were grown by using a mirror furnace and tested them with X-ray Laue (Fig. 1) and neutron diffraction (Fig. 2).

2.2 Sample characterization with X-rays and Neutrons

The four as grown single crystals were annealed in order remove internal stress formed in the process of crystal growth. For phase purity X-ray powder diffraction on crushed powder from the single crystals were checked and showed pure tetragonal (x=0.15) and orthorhombic phases (x = 0, 0.05, 0.10) at room temperature. Additionally, the quality and growth axis of the all single crystals were checked with X-ray Laue back scattering technique (Fig. 1). Neutron measurements on ORION@SINQ, PSI showed that for x = 0.05 and 0.15 single grain crystals with mosaicity less than 0.5^0 were obtained, whereas x = 0, x = 0.10 were multiple sub-grain single crystals, not suitable for our diffusion investigations.

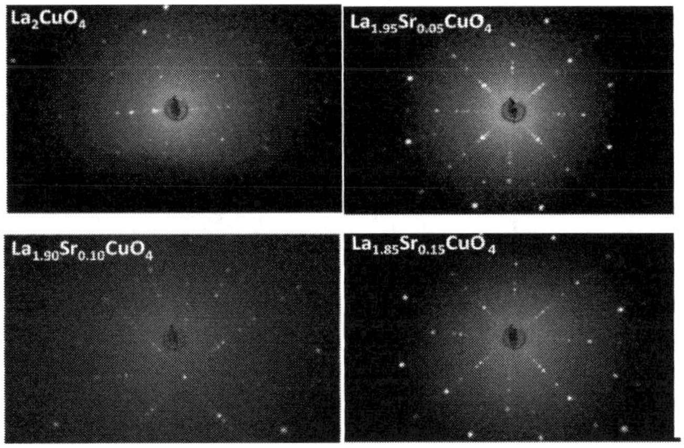

Figure 1: Back scattering Laue pictures of as grown single crystals. Data collected at LDM@PSI.

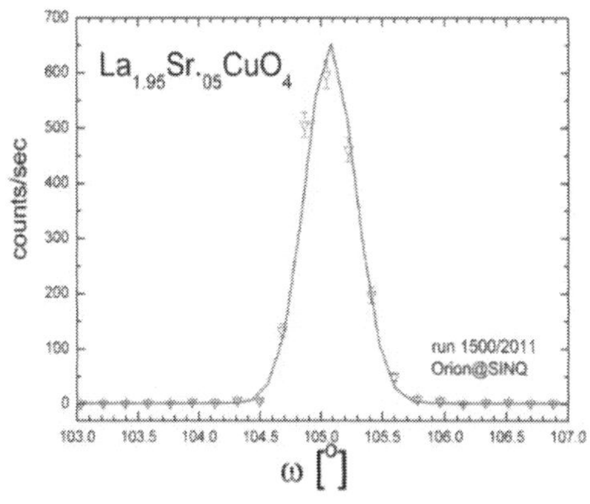

Figure 2: ω scan for (002) basic reflection measured on a high quality single crystal of $La_{1.95}Sr_{0.05}CuO_4$. Data collected at

2.3 Thermogravimetric Analysis (TGA)

The oxygen content in as grown $La_{2-x}Sr_xCuO_{4\pm\delta}$ (x = 0, 0.05, 0.10, 0.15) has been determined with TGA [9] hydrogen reduction (Fig. 3). The temperature was increased from 40 °C to 900 °C with a heating rate of 3°C/min. The gas flow was a mixture of helium (protective gas) with 5 volume % of hydrogen.

The total loss of the weight, attributed to the oxygen release, was measured for each crystal and is shown in Fig. 3. The oxygen content 4±δ value is close to 4 in all the cases. The measured $La_{2-x}Sr_xCuO_{4+\delta}$ (x/δ = 0/-0.004, 0.05/0.007, 0.1/-0.004 and 0.15/0.006) samples can be considered as oxygen stoichiometric.

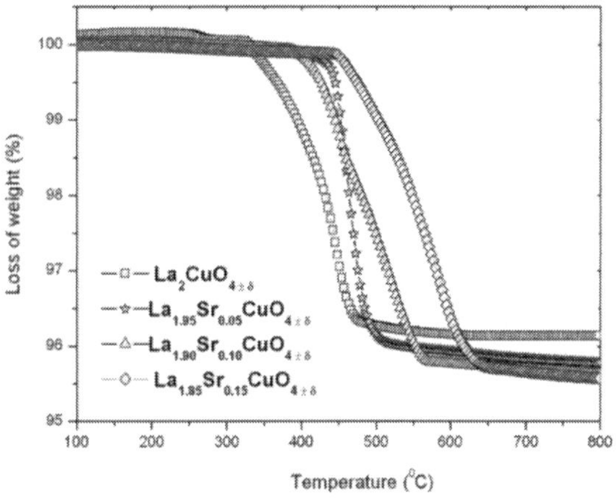

Figure 3: Thermogravimetric analysis (TGA) of the oxygen content in $La_{2-x}Sr_xCuO_{4+\delta}$.

Figure 4: Oxygen isotope exchange apparatus in LDM@PSI.

2.4 Oxygen isotope exchange

Oxygen isotope exchange has been proceed on $La_{2-x}Sr_xCuO_{4\pm\delta}$ (x = 0, 0.05, 0.10, 0.15) powders produced from crushed single crystals in order to study afterwards the oxygen self-diffusion.

The samples for the back exchange experiments were first enriched with ^{18}O in a closed quartz tube under controlled oxygen $^{18}O_2$ pressure slightly above one bar at 650°C for 50 hours and cooled down with a rate of 200°C/h. The exchange apparatus (Fig. 4) was equipped with a mass spectrometer (BALZERS Prisma), which allows to determine the progress of the exchange by measuring the isotopic composition of the gas inside the tube. During the isotope exchange between the gas (^{18}O) and the solid phase (containing ^{16}O at the beginning), the gas phase will be enriched by $^{16}O_2$ and $^{18}O^{16}O$ molecules. $La_{2-x}Sr_xCuO_{4\pm\delta}$ (x= 0, 0.05, 0.10, 0.15) samples with ^{18}O enrichment of approximately 82% have been obtained

3. Results and Discussion

OIBE was performed on crushed as grown single crystals of $La_{2-x}Sr_xCu^{18}O_{4\pm\delta}$ (x= 0, 0.05, 0.10, 0.15) under the same experimental conditions. Fig. 5 shows the loss of the weight of the sample from 40°C to 1000°C with heating rate of 3°C/min. The TG curve shows plateau at temperature higher than 700°C for x=0, therefore, a total oxygen isotope back exchange could be assumed for x=0 whereas for x=0.05 most of the oxygen isotope was back exchanged. The weight loss corresponds in the case of x=0 and 0.05 to a total amount of ^{18}O in the virgin sample.

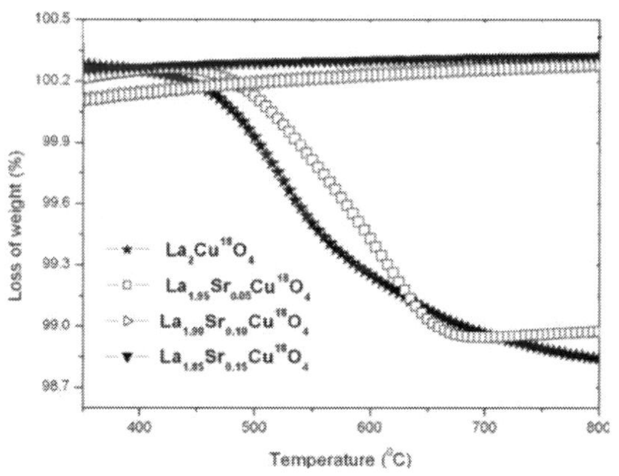

Figure 5: Oxygen Isotope Back Exchange (OIBE) in $La_{2-x}Sr_xCuO_{4+\delta}$ showing the suppression of oxygen diffusion for high Sr doping (x).

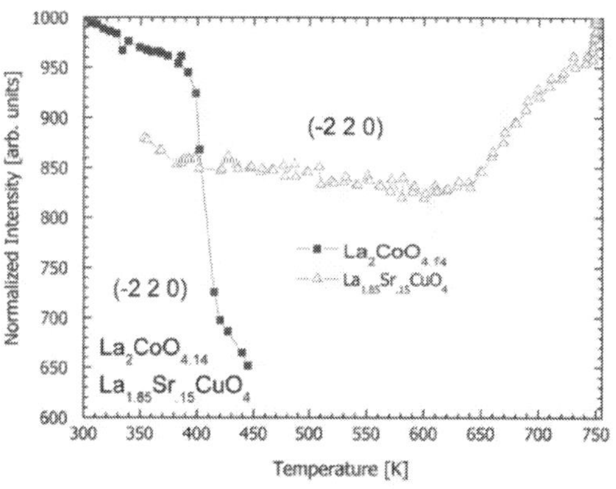

Figure 6: Temperature scan of (-2 2 0) reflection measured on a high quality single crystal of $La_{1.85}Sr_{0.15}CuO_4$. Data collected at TriCS@SINQ.

The results also show that the self diffusion starts at 450 ^{0}C for x=0 and at 470 ^{0}C for x=0.05, whereas the self diffusion of oxygen is not present in the case of x≥0.1.

Routbort *et al.* [10] measured ^{18}O tracer diffusivities by using SIMS analysis of gradients in polished and pre-annealed polycrystalline $La_{2-x}Sr_xCuO_{4\pm\delta}$ samples of compositions with x=0.10, 0.15 and 0.20. They found that the diffusivities decreased markedly with increasing strontium content and that the activation energy for oxygen tracer diffusion was highest for the x = 0.20 composition. They suggested that the immobilization of oxygen vacancies by strontium as $2Sr_{La}^{|} - V_o^{\cdot\cdot}$ clusters [10] may be responsible for the decrease in the oxygen diffusivity with increasing strontium content.

The effect of strontium doping on oxygen diffusivity is explained in terms of defect chemical models by Opila *et al.* [6]. The decreasing oxygen diffusivity with increasing strontium was attributed to the ordering of oxygen vacancies at large defect concentrations.

In order to correlate structure and oxygen diffusion as a function of temperature neutron measurements at TriCS@PSI on grain free 0.15 single crystal have been performed. Neutron diffraction data are in agreement with diffusion measurements achieved by oxygen back exchange. Around 650K the intensity of (-2 2 0) reflection was increased as shown in fig.6. This is the same temperature where the Oxygen diffusion was set in. This is higher temperature compared to $La_2CoO_{4.14}$. Later this data will be combined with SIMS (yields the diffusion constant and anisotropy on a macroscopic scale by observing the ^{18}O diffusion on oriented single crystals) data.

4. Conclusions

Free oxygen mobility in $La_{2-x}Sr_xCuO_{4+\delta}$ can be realized for x=0 and 0.05 already below 500 ^{0}C, but is depressed for x =0.1 and 0.15. This gives evidence that the low temperature oxygen mobility relies on Sr doping.

5. Acknowledgement This work is supported by CNRS, France and PSI, Switzerland.

References

[1] Tsujimoto Y, Tassel C, Hayashi N, Watanabe T, Kageyama H, Yoshimura K, Takano M, Ceretti M, Ritter C and Paulus W 2007 *Nature* **450** 1062
[2] Paulus W, Schober H, Eibl S, Johnson M, Berthier T, Hernandez O, Ceretti M, Plazanet M, Conder K and Lamberti C 2008 *J. Am. Chem. So.* **130** 16080
[3] Villesuzanne A, Paulus W, Cousson A, Hosoya S and Le Dréau L 2011 *J. Solid State Electrochem.* **15** 357
[4] Le Dréau L 2011 *Thesis* University of Rennes
[5] Kushima A, Parfitt D, Chroneos A, Yildiz B, Kilner J A and Grimes R W 2011 *Phys. Chem. Chem. Phys* **13** 2242
[6] Opila E J, Tuller H L, Wuensch B J and Maier J 1993 *J. Am. Ceram. Soc.* **6** 2363
[7] Böni P, Axe J D, Shirane G, Birgeneau J R, Gabbe D R, Jenssen H P, Kastner M A, Picone P J, Thurston T R and Sato M 1988 *Phys Rev B* **38** 185
[8] The Swiss spallation neutron source (SINQ) at the Paul Scherrer Institute (PSI) Switzerland
[9] NETZSCH, TG 449 *F3 Jupiter®* Thermo Microbalance
[10] Routbort J L, Rothman S J, Flandermeyer B K, Nowicki K J and Baker E 1988 *J.Mater. Res.* **3** 116

Towards room temperature, direct, solvent free synthesis of tetraborohydrides

A. Remhof[1], Y. Yan[1], O. Friedrichs[1], J. W. Kim[1,3,4], Ph. Mauron[1], A. Borgschulte[1], D. Wallacher[2], A. Buchsteiner[2], A. Hoser[2], K. H. Oh[4], Y. W. Cho[3], A. Züttel[1]

[1]Empa, Swiss Federal Laboratories for Materials Science and Technology, Hydrogen and Energy, 8600 Dübendorf, Switzerland

[2]Helmholtz Centre Berlin for Materials and Energy, 14109 Berlin, Germany

[3]Materials Science and Technology Research Division, Korea Institute of Science and Technology, Seoul 136-791, Republic of Korea

[4]Departement of Materials Science and Engineering, Seoul National University, Seoul 151-742, Republic of Korea

Arndt.Remhof@empa.ch

Abstract. Due to their high hydrogen content, tetraborohydrides are discussed as potential synthetic energy carriers. On the example of lithium borohydride $LiBH_4$, we discuss current approaches of direct, solvent free synthesis based on gas solid reactions of the elements or binary hydrides and/or borides with gaseous H_2 or B_2H_6. The direct synthesis from the elements requires high temperature and high pressure (700°C, 150bar D_2). Using LiB or AlB_2 as boron source reduces the required temperature by more than 300 K. Reactive milling of LiD with B_2H_6 leads to the formation of $LiBD_4$ already at room temperature. The reactive milling technique can also be applied to synthesize other borohydrides from their respective metal hydrides.

1. Introduction

Tetraborohydrides $M(BH_4)_n$ (n being the valence of the metal) are currently discussed as synthetic energy carriers [1]. Especially the lightweight alkali and alkaline earth metal borohydrides such as $LiBH_4$ which has a high gravimetric (18wt%) and volumetric hydrogen density (122kg/m^3), are promising candidates for mobile applications. They form ionic crystals composed of positively charged metal ions $[M]^{n+}$, and negatively charged borohydride $[BH_4]^-$ ions. To exploit the high hydrogen content, the material ought to have convenient working conditions in terms of desorption temperature, equilibrium pressure and sorption kinetics. Furthermore, it should be easy to regenerate: technical application requires an efficient way to synthesize the borohydrides either on-board or off-board.

The first borohydrides were isolated by Stock et al. in 1935 [2-3] from the reaction of potassium amalgam and sodium amalgam with diborane. The reaction products were mistaken to be "disodium-diborane" $Na_2B_2H_6$ and "dipotassium diborane" $K_2B_2H_6$. In 1949 Kasper et al. identified "disodium-diborane" to be sodium borohydride $NaBH_4$ by its X-ray diffraction pattern [4]. The synthesis of $LiBH_4$ from the reaction of gaseous diborane (B_2H_6) with ethyllithium (C_2H_5Li) was reported by Schlesinger and Brown in 1940 [5]. More conveniently, $LiBH_4$ can be produced either by the reaction of LiH and B_2H_6 (suspended in diethyl ether) or by the metathesis reaction of $NaBH_4$ and LiCl, respectively [6]. To obtain the pure borohydride, by-products of the reaction and the solvent have to be removed. A comprehensive review of the currently applied methods is given by Hagemann and R Černý [7].

Within the last years several methods for the solvent-free synthesis were applied. These methods comprise (a) the high temperature / high pressure direct synthesis from the elements, (b) high temperature / high pressure reactions involving binary metal hydrides and binary metal borides and finally (c) the reaction of gaseous B_2H_6 with the respective solid metal hydride. With the present paper we outline this development and give an overview over the current state of the art.

Diffraction techniques, especially neutron diffraction, are powerful tools to determine the crystalline structure of lightweight complex hydrides [8,9]. They are also used to recognize reaction pathways by the identification of intermediates and solid desorption products. The high penetration depth of neutrons allow to use massive sample environment, enabling in-situ experiments of high pressure – high temperature reactions in combination with volumetric measurements of the involved gases. The in-situ neutron diffraction experiments presented in this work were measured at the focusing diffractometer E6 at the 10 MW research reactor BERII of the Helmholtz Center for Materials and Energy in Berlin, Germany [10]. The neutron diffraction experiments were complimented by in-house X-ray diffraction and by imaging techniques.

2. Direct synthesis from the elements

Even though $LiBH_4$ possesses a high negative heat of formation (ΔH_f = -185kJ/mol) [11], the direct synthesis from the elements requires elevated temperature and high pressure. Goerrig observed the direct synthesis from lithium, boron and hydrogen at 650°C and 150 bar H_2 and claimed general applicability to group IA and IIA metals [12]. A significant activation energy barrier prevents the formation of $LiBH_4$ at lower temperatures. The activation barrier is attributed to the general chemical inertness of boron [13-15], which may be due to the strongly bound boron atoms ($\Delta H_{(B(s) \rightarrow B(g))}$ = 560 kJ/mol), compared to Li (($\Delta H_{(Li(s) \rightarrow Li(g))}$ = 159.3 kJ/mol) and H (($\Delta H_{(H2(g) \rightarrow 2H(g))}$ = 218 kJ/mol) [16] and/or a natural passivating oxide layer [17–19]. However, The results by Goerrig could not be reproduced so far [20,21].

We achieved the direct synthesis of $LiBH_4$ and $LiBD_4$ from the elements by exposing pre-milled boron powder and lithium to a hydrogen (deuterium) atmosphere of 150 bar at a temperature of 700 °C [22], which is slightly higher than the temperature observed by Goerrig and above the melting point of LiH (688°C). Above this temperature no passivating hydride layer is formed and all Li completely converts to LiH and liquid LiH intercalates the boron leading to an intermixing of Li, H and B on the atomic scale, facilitating the formation of $LiBH_4$.

The formation of $LiBH_4$ and $LiBD_4$ can be improved by Li intercalation into B [22-24] prior to a subsequent hydrogenation. At low Li-concentrations Li is dissolved into the interstitial sites of the boron lattice, leading to a stressed and expanded boron structure and causing the decrepitation of boron particles. At higher concentrations, stochiometric LiB forms. The reaction of LiB with D_2 to $LiBD_4$ involves LiD as an intermediate product. The diffraction data in figure 1 show that the nucleation of LiD is accompanied by the decomposition of the LiB compound. The absence of the boron diffraction peaks indicates that the boron resulting from the D-induced decomposition of LiB has no crystalline long-range order that is observed in the as-purchased boron. The crystalline order that is still observed in ball-milled boron is obviously destroyed after the formation of B from the decomposition of the binary LiB compound.

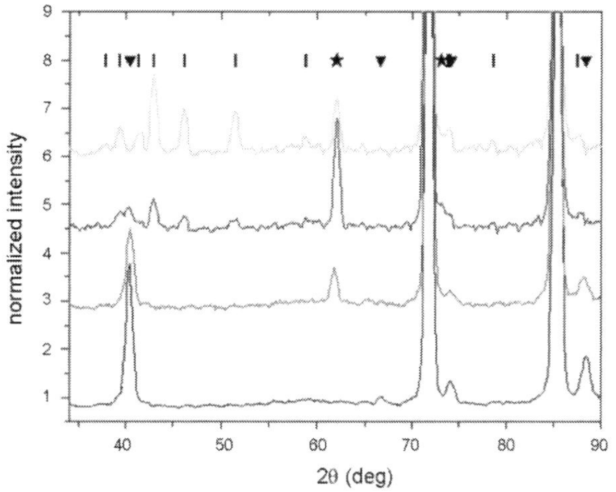

Figure 1: Neutron diffraction pattern recorded during the reaction of LiB with D_2 (180bar). From bottom to top: as prepared sample, after heating to 200°C, 350°C and 500°C. Ticks (|) mark the position of $LiBD_4$ reflections, down triangles (▼) mark the positions of the LiB reflections and the star (*) marks the reflection of LiD. The srong reflections at 72° and at 85° originate from the sample container. The curves are offset for clarity [24].

Similar temperatures can be achieved by exposing ball milled mixtures of LiH and B to H_2 [25, 26]. Chen at al. directly prepared $LiBH_4$ at 300 °C – 500 °C at 350 bar H_2 starting from high energy ball milled LiH/B mixtures under 100 bar H_2 atmosphere [26]. The milling reduces the particle size and induces mechanical alloying of LiH and B, reducing the solid diffusion pathway. The pre-treatment also deoxidizes the surface oxides. Chen et al., discuss several possible rate limiting steps that control the reaction kinetics, such as diffusion, nucleation and growth, or movement of phase boundaries. In a careful kinetic investigation, solid state diffusion was concluded to be the rate limiting step of the formation of $LiBH_4$ from LiH, B and H_2.

3. Synthesis paths involving metals, metal hydrides and metal borides

Binary compounds such as MgB_2 or AlB_2 can play an important role as destabilizing agents during hydrogen desorption. Vajo et al. [27] studied the hydrogen sorption behaviour of $LiBH_4$ using MgH_2 as additive. Including 2–3 mol% $TiCl_3$ as a catalyst, the reaction follows the equation:

$$2LiBH_4 + MgH_2 \rightarrow 2LiH + MgB_2 + 4H_2 \, (\Delta H = -45.8 kJ/molH_2)$$

The formation of MgB_2 upon dehydrogenation stabilizes the dehydrogenated state and, thereby, destabilizes the MgH_2 and the $LiBH_4$. The reduced decomposition temperature of the combined system as compared to the individual hydrides makes this system attractive for hydrogen storage. $LiBH_4 + MgH_2$ is thereby a prototype system for a whole class of pseudo-binary hydrides containing a complex hydride (as $LiBH_4$) and a binary metal hydride (as MgH_2) called *reactive hydride composites* (RHC).

Synthesis reactions according to

$$2MH+MgB_2+4H_2 \rightarrow 2MBH_4+MgH_2 \text{ (M=Li, Na) and } CaH_2+MgB_2+4H_2 \rightarrow Ca(BH_4)_2+MgH_2$$

have been realized at p_{H2}=350bar and at T= 400°C and by Barkhordarian et al. [28]. Closer investigations showed that the $Ca(BH_4)_2$ synthesis already takes place at p_{H2}=150 bar and at T= 350 °C [29]. Due to their reduced working temperature, enhanced kinetics and reversibility, RHC are currently in the focus of many research activities. Choosing the hydride and the boride from the same metal, pure borohydrides may be synthesized. $Ca(BH_4)_2$ prepared from CaB_6 and CaH_2 under 700 bar H_2 at about 400°C is an example for such a reaction [30].
The hydrogenation of metal hydride / metal boride mixtures may also result in the concurrent formation of the borohydride (from the respective metal hydride) and the elemental metal (from the metal boride). The most studied system in this respect is $AlB_2 + 2LiH + 3H_2 \rightarrow 2LiBH_4 + Al$. The reaction was observed to take place between 350 °C and 400 °C in applied hydrogen pressures between 70 bar and 150 bar [31–34]. Compared to the re-hydrogenation temperature of pure $LiBH_4$ these temperatures and pressures are more moderate and therefore a step towards technical application. Unfortunately, the system suffers from a severe capacity loss during cycling. A phase separation of Al and B during the decomposition of $LiBD_4$ was identified as the main reason for the hydrogen capacity loss and the decreasing kinetics upon subsequent cycling [35].

A common feature of the RCH is the prevention elemental boron as reactant, circumventing the breaking of B-B bonds during the synthesis. The boride serves as a source of boron during the hydrogen adsorption while the formation of the boride during hydrogen desorption binds the boron and avoids the formation of B-B bonds.

4. Synthesis paths involving diborane

The previous experiments suggest that the boron supply is crucial for the synthesis of $LiBD_4$. Breaking of the B-B or the B-metal bonds and the formation of B-H seem to be the limiting steps. Consequently, reaction paths involving diborane (B_2H_6), in which the B-H bonds are already established, should further facilitate the formation of $LiBD_4$ at lower temperatures and pressures.

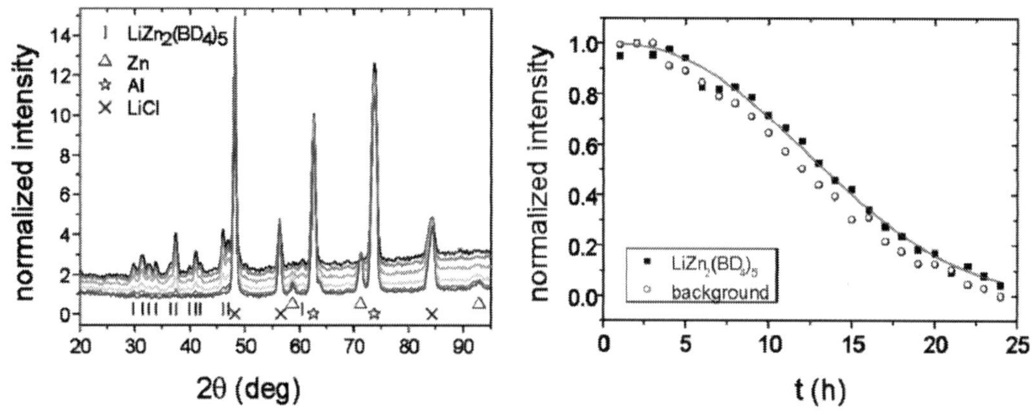

Figure 2a (left): *Top to bottom: Temporal evolution of the neutron powder diffraction pattern of a ball milled mixture of $LiBD_4$ and $ZnCl_2$, recorded in a sealed container at T=85°. Figure 2b (right): Temporal evolution of the integrated intensity of the $LiZn_2(BD_4)_5$ reflection (filled symbols) and of the integrated background intensity (open symbols).*

Diborane is a toxic, thermodynamically unstable gas. On the lab-scale small amounts of B_2H_6 are released, together with H_2, in a 1:1 ratio, from a ball milled mixture of $LiBH_4$ and $ZnCl_2$ [36]. Ravensbæk et al. identified the resulting compound to be the double cation lithium zincborohydride, formed according to $5LiBH_4 + 2ZnCl_2 \rightarrow LiZn_2(BH_4)_5 + 4LiCl$ [37].

Figure 2a (top curve) shows a series of neutron powder pattern during the decomposition of a deuterated $LiBD_4/ZnCl_2$ sample. Each individual diffractogram was recorded within one hour; the time lag between two depicted measurements equals five hours. Figure 1b shows the normalized temporal evolution of the integrated intensity of the strongest $LiZn_2(BD_4)_5$ reflection at $2\theta=37.47°$, together with the integrated background intensity. The strong initial background is obviously related to the presence of $LiZn_2(BD_4)_5$, most probably due to the presence of amorphous $LiZn_2(BD_4)_5$. In agreement with earlier studies, elemental zinc is observed within the solid residue [38].

Exposing LiH to a diborane/deuterium atmosphere of 10 bar at 100°C leads to the formation of $LiBD_4$. The reaction stops after about 50% of LiH is consumed for the formation of $LiBD_4$. The reaction proceeds in one step, no intermediate products are visible [36]. A core-shell structure of lithium hydride surrounded by lithium borohydride is observed, as shown in figure 3 [39].

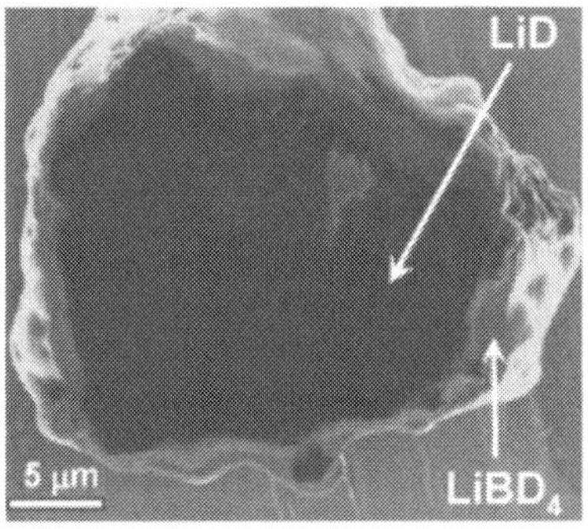

Figure 3: Secondary electron image (induced by 30 keV ion beam) of LiH after reaction with B_2H_6 showing a cross sectional view of a grain. The core shell structure is clearly visible. The chemical composition of the inner part of the grain and the outer layer were determined by electron loss spectroscopy [36].

The reaction stops due to kinetic constrains originating from the increased diffusion path of either B–H species into the grain or Li towards the exterior. The results are in agreement with the passivation layer proposed by Schlesinger et al., who synthesized different borohydrides in solvents in order to prevent the formation of the passivation layer [6]. Upon reactive ball milling the passivation layer is constantly broken and $LiBH_4$ forms already even at room temperature [39]. As long as sufficient gas is supplied, no influence on the pressure was detected. As long as fresh surface is provided, the reaction proceeds. Details of the reaction mechanism were studied by Gremaud et al. [40]. They found evidence for the heterolytic splitting of diborane on the alkali hydrides. The resulting $[BH_4]^-$ anion is subsequently exchanged with a $[H]^-$ ion of the underlying hydride. The alkali hydride surface is ionic and polarizes the B_2D_6 prior splitting, assisting the necessary charge transfer for binding of the negatively charged $[BD_4]^-$.

The synthesis of a borohydride by reactive ball milling is not limited to $LiBH_4$. The direct, solvent free method of synthesizing borohydrides from the respective binary hydride has been successfully applied to $Ca(BH_4)_2$, $Mg(BH_4)_2$ [39] and $Y(BH_4)_3$. Figure 4 shows the resulting XRD pattern of $LiBH_4$, $Ca(BH_4)_2$, $Mg(BH_4)_2$ and $Y(BH_4)_3$ (from top to bottom).

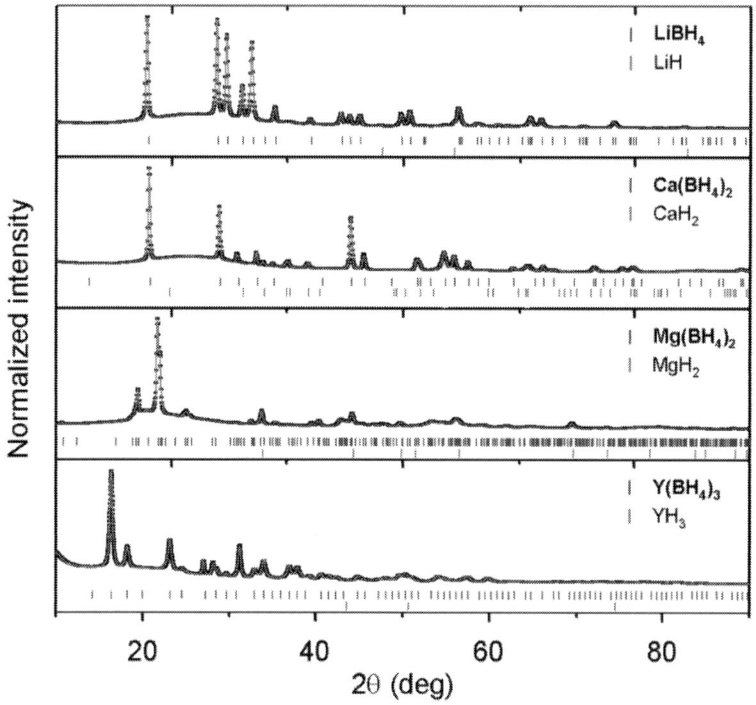

Figure 4: X-ray diffraction pattern of the products achieved by reactive ball milling of LiH, CaH_2, MgH_2 and YH_3 (from top to bottom) in a mixed B_2H_6/H_2 atmosphere. The main product is the respective borohydride, the educt is the only detectable solid contaminant.

The reactive ball milling is not generally applicable. The formation of the borohydride competes with the formation of the respective boride analogous to the reaction of TiH_2 and B_2H_6. The formation of a borohydride is generally accompanied by a reduction of the overall pressure in the milling vial due to the absorption of diborane. In the case of TiH_2 and B_2H_6 a pressure increase is observed, indicating the release of a gas. Figure 5 displays the pressure within the vial during the milling. Until saturation, the pressure increases from an initial value of 10.1 bar by 3.6 bar. The solid product has been identified as TiB_2, as shown in the inset to fig.5.

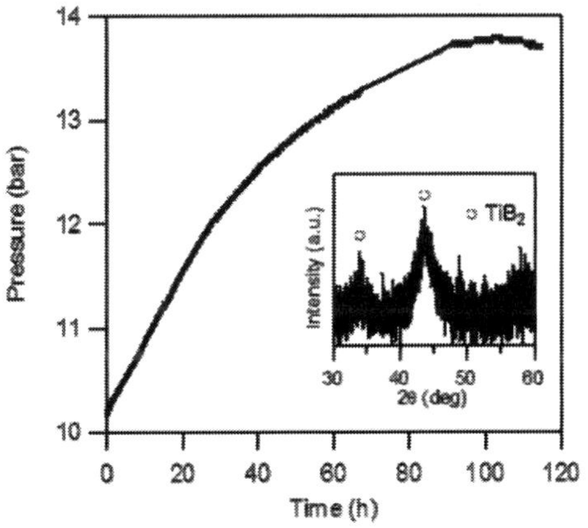

Figure 5: Pressure increase during the ball milling of TiH_2 in B_2H_6 atmosphere. The inset shows the XRD pattern of the reaction product, identified as TiB_2.

The applicability of the gas solid diborane synthesis is obviously limited by thermodynamics. The borohydride is only formed when it is more stable than the corresponding boride. Moreover, the increase of entropy originating from the net release of 3 formula units of gaseous products favours the formation of the boride instead of the borohydride.

5. Conclusion

Among the various methods to prepare borohydrides, the direct, mechanically assisted gas-solid reactions offer solvent free routes, circumventing the necessity to remove solvents or unwanted by-products that are unavoidable in metathesis reactions. Combining NPD, XRD, Raman spectroscopy and imaging techniques, the fundamental reaction mechanisms could be unveiled. The formation of the B-H bonds is the crucial step, mainly responsible for the activation energy needed for the direct synthesis from the elements. Lower barriers can be achieved by using additives that destabilize the stable borohydrides and that prevent the formation of B-B bonds during hydrogen desorption. Room temperature synthesis can be achieved starting from diborane, where the B-H bonds are already established.

Acknowledgements

Financial support from the Swiss National Science Foundation (SNF-Projekts 200021_119972/1, 200020-134442/1 and 200021_129603/1) and from the Hydrogen Energy R&D Center, one of 21st Century Frontier R&D Programs funded by the Ministry of Education, Science and Technology of Korea is gratefully acknowledged. We would also like to thank for the financial support by the European Commission through the Key Action: Strengthening the European Research Area, Research Infrastructure (contract number RII3-CT-2003-505925).

References

[1] Züttel A, Borgschulte A and Orimo S I 2007 *Scr. Mater.* **56** 823
[2] Stock A, Sütterlin W and Kurzen F 1935 *Z. Anorg. Allg. Chem.* **225** 225
[3] Stock A and Laudenklos H 1936 *Z. Anorg. Allg. Chem.* **228** 178
[4] Kasper J S, McCarty L V and Newkirk A E 1949 *J. Am. Chem. Soc.* **71** 2583
[5] Schlesinger H I and Brown H C 1940 *J. Am. Chem. Soc.* **75** 3429
[6] Schlesinger H I, Brown H C, Abraham B, Bond A C, Davidson N, Finholt A E, Gilbreath J R, Hoekstra H, Horvitz L, Hyde E K, Katz J J, Knight J, Lad R A, Mayfield D L, Rapp L, Ritter D M,

Schwartz A M, Sheft I, Tuck L D and Walker A O 1953 *J. Am. Chem. Soc.* **75** 186; Schlesinger H I, Brown H C, Hyde E K 1953 *J. Am. Chem. Soc.* **75** 20
[7] Hagemann H and Černý R 2010 *Dalton Trans.* **39** 6006
[8] Černý R 2008 *Z. Krist.* **223** 616
[9] Hauback B 2008 *Z. Krist.* **223** 636
[10] Stüsser N and Hofmann M 2002 *Nucl. Instrum. Methods Phys. Res. Sect. A* **482** 744
[11] Davis W D, Mason L S and Stegeman G 1949 *J. Am. Chem. Soc.* **71** 2775
[12] Goerrig D 1958 Ger. Pat. 1077644.
[13] Atkins P W 2001 *Physikalische Chemie* Wiley-VCH
[14] Greenwood N N and Earnshaw A 1984 *Chemistry of the elements*, Oxford: Pergamon Press
[15] Laubengayer A W, Hurd D T, Newkirk A E and Hoard L J, 1943 *J. Am. Chem. Soc.* **65** 1924
[16] Barin I 1989 *Thermochemical data of pure substances*, Weinheim: VCH Verlagsgemeinschaft
[17] Adams R M, 1964 *Boron Metallo-Boron. Compounds and boranes*, New York: Interscience Publishers
[18] Matkovich V I 1977 *Boron and refractory borides* Berlin: Springer-Verlag
[19] Mota J M, Abenojar J, Martinez M A, Velasco F, Criado A J 2004 *J. Solid State Chem.* **177** 619
[20] Elam C C ed. 2002 *IEA Annual Report: IEA Agreement on the Production and Utilization of Hydrogen* The International Energy Agency, Golden, CO, USA, http://www.ieahia.org/pdfs/2002_annual_report.pdf.
[21] Ohba N, Miwa K, Aoki M, Noritake T, Towata S, Nakamori Y, Orimo S and Züttel A 2004 *Phys. Rev. B* **69** 245120
[22] Friedrichs O, Buchter F, Zwicky C, Borgschulte A, Remhof A, Mauron P, Bielmann M and Züttel A 2008 *Acta Mater.* **56** 949.
[23] Çakanyıldırım Ç and Gürü M 2008 *Renew. Energy* **33** 2388
[24] Remhof A, Friedrichs O, Buchter F, Mauron P, Züttel A, Wallacher D 2008 *Phys. Chem. Chem. Phys.* **10** 5859
[25] Agresti F and Khandelwal A 2009 *Scripta Mater.* **60** 735
[26] Chen R, Wang X, Xu L, Li H, Chen C, Chen L and Pan H 2011 *J. Alloys Compd.* **509** 3481
[27] Vajo J J, Skeith S L and Mertens F 2005 *J. Phys. Chem. B* **109** 3719
[28] Barkhordarian G, Klassen T, Dornheim M and Bormann R 2007 *J. Alloys Compd.***440** L18
[29] Barkhordarian G, Jensen T R, Doppiu S, Boesenberg U, Borgschulte A, Gremaud R, Cerenius Y, Dornheim M, Klassen T and Bormann R 2008 *J. Phys. Chem. C***111** 2743
[30] Rönnebro E and Majzoub E H 2007 *J. Phys. Chem. B* **111** 12045
[31] Yang J, Sudik A and Wolverton C 2007 *J. Phys. Chem. C* **111** 19134
[32] Kang X D, Wang P, Ma L P and Cheng H M 2007 *Appl. Phys. A* **89** 963
[33] Jin S A, Shim J H, Cho Y W, Yi K W, Zabara O and Fichtner M 2008 *Scr. Mater.* **58** 963
[34] Friedrichs O, Kim J W, Remhof A, Buchter F, Borgschulte A, Wallacher D, Cho YW, Fichtner M, Oh K H and Züttel A 2009 *Phys. Chem. Chem. Phys.* **11** 1515
[35] Remhof A, Friedrichs O, Buchter F, Mauron P, Kim J W, Oh K H, Buchsteiner A, Wallacher D and Züttel A 2009 *J. Alloys Comp.* **484** 654
[36] Friedrichs O, Borgschulte A, Kato S, Buchter F, Gremaud R, Remhof A and Züttel A, 2009 *Chemistry Eur. J.* **15** 5531
[37] Ravnsbæk D, Filinchuk Y, Cerenius Y, Jakobsen H J, Besenbacher F, Skibsted J and Jensen T R 2009 *Angew. Chem. Int. Ed.* **48** 6659
[38] Kim J W, Friedrichs O, Ahn J P, Kim D H, Kim S C, Remhof A , Chung H S, Lee J, Shim J H, Cho Y W, Züttel A andOh K H 2009, *Scripta Mater.* **60** 1089
[39] Friedrichs O, Remhof A, Borgschulte A, Buchter F, Orimo S and Züttel A 2010 *Phys. Chem. Chem. Phys.***12** 10919
[40] Gremaud R, Borgschulte A, Friedrichs O and Züttel A 2011 *J. Phys. Chem. C* **115** 2489

Characterization of ancient Greek coins using non-destructive TOF neutron diffraction.

I.M. Siouris[1], S. Katsavounis[1] and W. Kockelmann[2]

[1] Democritus Univ. of Thrace, Dept. of Production Eng. and Management, 67 100 Xanthi, Greece.
[2] Rutherford Appleton Laboratory, Chilton, Didcot, Oxfordshire OX11 0QX, UK.

E-mail: jsiou@pme.duth.gr

Abstract. The characterization of ten ancient Greek silver coins from the Treasure collection of Gazoros (CH IX 61) found in Serres- North Greece; as well as two silver coin replicas, has been carried out using a polychromatic neutron beam of large cross-section to obtain diffraction patterns from the entire objects. The diffraction profiles indicate that there are three distinct categories of coins. The first one is a set of three coins consisting mainly of silver and copper alloy phases with high quantities of Cu_2O and CuCl . The second group is formed of coins with high silver/ copper alloy ratio and the third is a collection of five coins consisting of very high purity silver. The comparison between the diffraction profiles of the original coins to those of the replicas present distinct variations that may be used to differentiate the different groups.

1. Introduction

Information about long gone societies is logged in archeological or historic objects. The deciphering of such recordings requires apart from comparative ethnographic, visual, aesthetic, textual, stylistic, morphological or epigraphic investigations primarily performed by the archeologists, extended scientific approaches and multi-analytical techniques that can look into the chemical-physical properties , the microstructure and the elemental compositions of the materials used during the fabrication . To this purpose, analytical methods may be employed to collect data from the surface as well as the bulk of the object, however, this involves sampling which is either completely forbidden or in some very rare cases acceptable, but absolutely limited to certain concealed areas of the object and in very minute amounts. These facts demand the use of noninvasive procedures that on the one hand are well suited for near –surface analysis and can be used " in situ" e.g Laser induced break down spectroscopy LIBS [1-2], X-ray fluorescence (XRF) , Proton induced X- ray emission (PIXE) [3], 3D MicroXRF [4-7] , EDXRF - PIXE [8-9]. Although the portable methods are very useful to provide real time data at the site so that decisions for further investigation can be readily made, they are somewhat less specific as they are subjected to environmental conditions, absorption , X-ray production cross section etc., while the "ex situ" measurements are limited to several micro meters (~20-100 μm) of information depth. On the other hand, these constraints can be overcome by the use of High Energy PIXE [10] or neutron diffraction (ND) [11-17]. Neutron diffraction has been applied on metallic and ceramic artifacts and various objects at ISIS using a polychromatic pulsed neutron beam on ROTAX, ENGIN-X and GEM instruments, while later, the method has been extended in using monochromatic neutrons mainly on ceramics[18-19]. However, the time of flight (TOF) method is preferred to single wavelength neutrons for it requires smaller irradiation times. The crystal structure of each of the phases, the mineral and metal phase abundances, the grain sizes, orientations and, microstructures in materials can be examined. Surface anomalies and multifaceted articles can be overcome while the measuring time is cut down to minutes, reducing the irradiation time to a minimum, which in turn decreases the decay times of induced radioactivity levels on the isotopic

Table 1. Phase analysis (wt%) and data alloy composition (Sn wt%). Estimated error of ±0.5 wt%

Silver plated copper coins	CuSn	(Sn)	Ag	Cu_2O	CuCl	AgCl
01 Thasian stater triti	79.3	2.38	9.1	10.6	~1	+
02 Neapolis hemidrachmo	77.5	1.16	9.6	11.9	1	+
03 Thasian stater triti	79.7	1.04	10.8	8.8	0.7	+
Silver coins						
00 Replica			99.9			
04 Thasian hemiekto	-	-	96.9	-	-	3
05 Thasian hemiekto	-	-	96.9	-	-	3
06 Thasian hemiekto	-	-	99.8	-	-	0
07 Thasian hemiekto	-	-	98.7	-	-	1.3
09 Vergeo stater triti	2.4	0.72	96.6	-	+	~1
10 Iona hemiekto	-	-	98.9	-	-	1.1

compositions, to a few days for objects containing silver or copper. The (ND) technique can be employed to complement other long established analysis methods such as XRD, XRF, PNAA,EDX, PIXE in use by the archaeological science today as it can deliver information on the mineral phase abundances, by using both the Rietveld, or the two stage method of profile fitting, from the bulk rather than the surface of the object.

The present work focuses on finger printing selected silver/copper coins from the Treasure collection of Gazoros (CH IX 61) in Serres Greece, using a pulse neutron beam to obtain diffraction patterns from the entire sample.

2. Experimental

The selected coins shown in Figures 1 and 3(a-b), were discovered in a single grave at the archaeological site of Gazoros in Serres, North East Greece. They were considered to be representative of the coinage circulated in a rather wider area of ancient North East Greece and on the island of Thasos. Chronological estimation based on archaeological criteria in the style of the die-engraver, morphological details and other peripheral archaeological findings suggests that the samples were minted at the end of the 5th century B.C. Information on the archaeological classification of the objects is listed in Table 1. Given the archaeological importance of the coins, the surface of the coins were cleaned from all soil depositions and preserved with a silica gel cover in an attempt to prevent further environmental erosion. The two replica coins obtained from the Archaeological Museum of Kavalla were prepared using 950 high purity silver and a metallurgical mixture consisting of copper-tin-antimony-lead. These coins served as markers and standards.

The measurements were carried out on the ROTAX instrument, a medium resolution diffractometer, at the pulse neutron source at ISIS facility at the Rutherford Appleton Laboratory, England. The samples without any prior preparation were placed for support and protection in a pocket made of vanadium foil. The sample chamber was evacuated and the coins were exposed to a 15x25 mm^2 incident beam for 20-30 minutes. The overall measurements lasted 4 hours. Details on the experimental set-up are given in reference [11].The resulting multi-bank diffraction profiles were analysed using the public domain software GSAS [20] with the EXPGUI [21] interface and the program Amphorae [19] which employs profile decomposition analysis to specifically determine and evaluate multi- phase diffraction profiles.

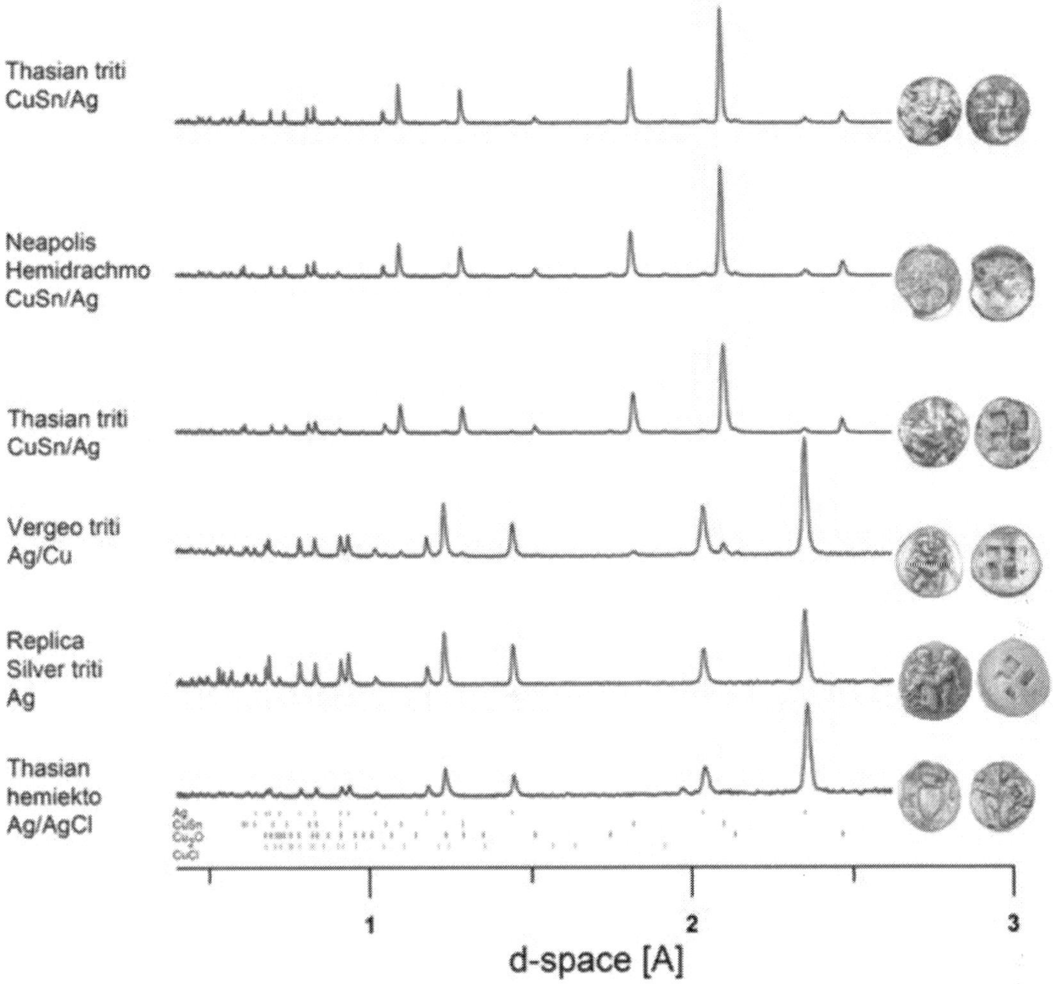

Figure 1: The coin images are shown on the right with their corresponding neutron diffraction profiles on the left. Bragg peaks for silver, Cu-alloy, cuprite (Cu_2O) and nantokite (CuCl) are displayed in the lower part of the diagram.

3. Results and discussion

The collected diffraction data for all the samples at room temperature is plotted in Figure 1, along with the corresponding calculated profiles derived from the theoretical analysis of the mineral fractions produced using the software. Measured data is represented by crosses (+), the final calculated spectra correspond to solid lines (-). The starting model for the analyses of the observed data consists of Cu, Ag, Pb, Au, CuSn, Cu_2O, CuCl and AgCl, based upon earlier EDS-XRD measurements and on comparable Greek coinage data [22-25]. The structural parameters of the starting phases were taken from the Inorganic Crystal Structure Database and CrystMet [26-27]. Phase analysis was carried out by refining one overall Debye–Waller parameter and one common absorption parameter for all patterns. Pattern fitting procedures and mineral phase refinement has been described in [19]. The sum of crystalline phases was normalized to one. The profile decomposition process and the resulting

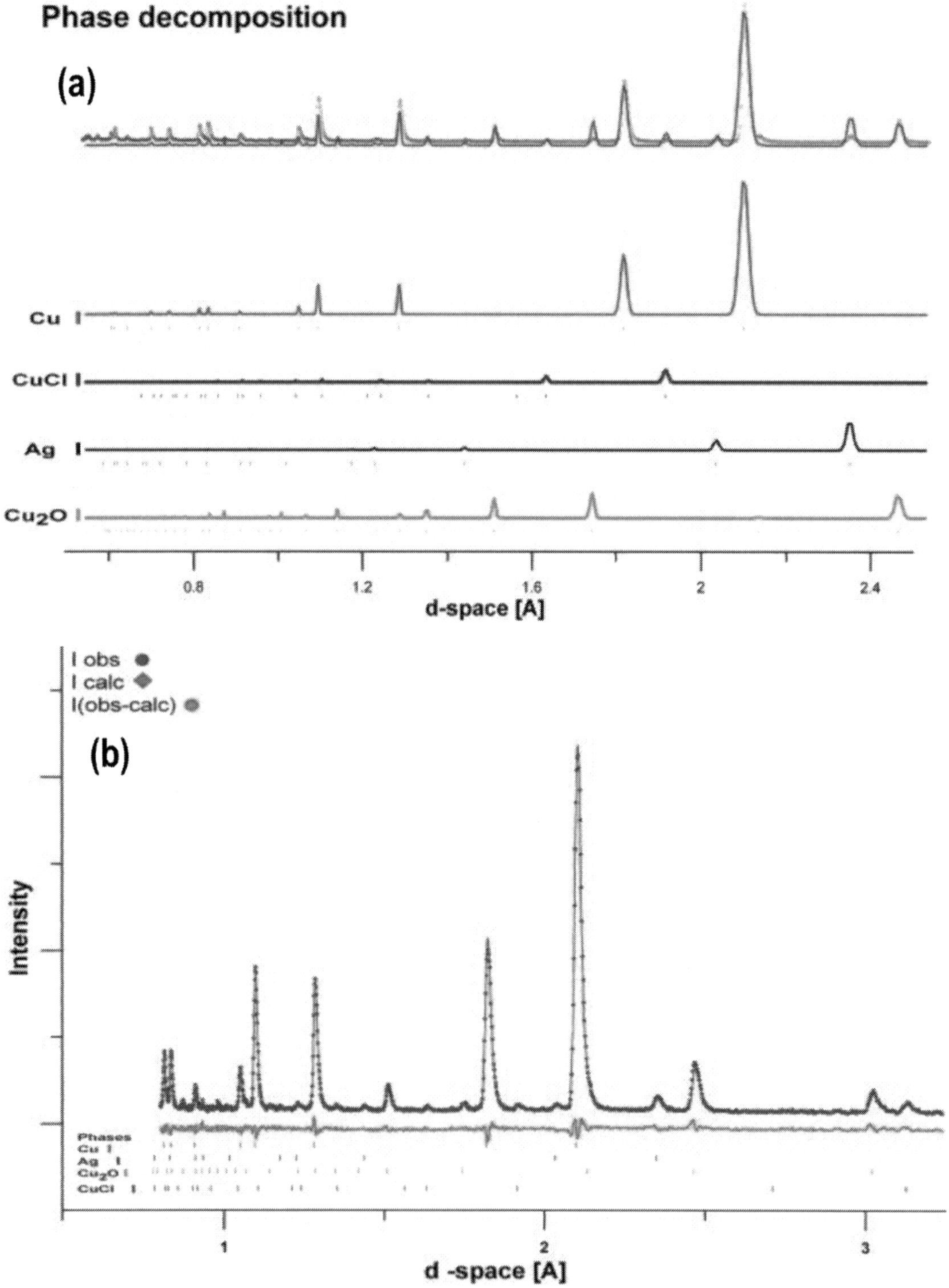

Figure 2: (a) Profile decomposition showing the individual calculated profiles of copper (Cu), nantokite (CuCl), silver (Ag) and cuprite (Cu_2O). (b) The neutron diffraction profile of the coin 01. Measured data is represented by crosses (+), the final calculated spectra correspond to solid lines (-). Bragg reflections for (Cu/Sn), (Ag), (Cu_2O) and (CuCl) are displayed in the lower part of the figure.

calculated pattern is illustrated in figure 2(a) and the measured and Rietveld refined profile of the coin No 01 is shown on figure 2(b). The reliability value of the refinement R_{Bragg} = 4.4 %.

From the diffraction profiles it is clearly observed that there are three distinct categories of coins. The first shown in Figure 3a, is a set of three items that present identical diffraction patterns consisting of two main phases, namely a copper-tin alloy and a pure silver and similar corrosion products phases. The pronounced copper alloy Bragg peaks originate from the core of the sample and are shifted with respect to pure copper Bragg reflections. These shifts can be linearly correlated with Sn wt% found in alloys containing up to 14 wt% Sn [28- 30]. The CuSn compositions were obtained by Rietveld refinement and the α-bronze phase lattice parameters are in good agreement with literature. Additional smaller silver peaks result from the surface of the coin while high quantities of copper oxide (Cu_2O) , small amounts of nantokite (CuCl) and silver- chlorine type compounds were also found. The presence of the various corrosion products developed on the coins is due to selective galvanic corrosion which is induced to metals of different electrochemical potential, namely copper (0.34) and silver (0.85), that are in contact or in very close proximity to each other [31-32]. In an electrolytic environment, the less noble metal (Cu) becomes anodic and consequently suffers a severe corrosion attack. The spread of corrosion is dependent on environmental factors, burial conditions, soil acidity or salinity and the micro-structure of the metals [33].

The second group consisting of coins with high silver to copper alloy ratio is illustrated in Figure 1 (***Vergeo triti Ag/Cu –coin 09***). These coins are very rich in silver and contain only a small fraction of copper- tin alloy with similar structural alloy characteristics as found in the first set. Copper chloride is also present, although in minute quantities. The third class group presented in Figure 3b, is a collection of five coins that display similar features on their diffraction profiles, presenting high amounts of silver content which are manifested by the prominent and slightly broad structured (111) and (200) Bragg reflections. Although these observations suggest the existence of additional phases such as gold (see ref. 26-27), Rietveld refinement carried out by assuming the presence of silver, gold and copper phases did not add to improving the theoretical profile or the reliability value of the refinement. The detected silver corrosion is identified as chlorargyrite (AgCl), and is considered among the most common alteration products of silver during burial in saline soils [29]. It is interesting to note (see figure 3b) that coins 04 and 05 are broken and are highly worn, coins 07 and 10 are pierced and slightly damaged, while coin 06 is in almost perfect condition.
The above observation may be correlated to the amount of AgCl found in each coin and reflects to the degree of damage and attrition due to corrosion. Although it is beyond the scope of this work, the archaeological and historic evidence suggest that the coins were issued during a period of Thasian domination and possession of the silver mining districts on the mainland, yielding enormous wealth to Thasians and their allies. The coins of the first group though referred to as being of Thasian origin, present a very careless style and differ significantly in silver / CuSn composition when compared to the genuine article such as found in group two, namely the coin 09. Therefore, these coins may be considered as poor imitations and they are more likely to be of local Thracian origin. The debasement of the stater triti and the Neapolis hemidrachmo further illustrate the deprivation of bullion in the particular local communities during the period the coins were minted. In contrast, the five Thasian silver hemiekto that chronologically coincide with the coins of group one, suggests an abundance of Thasian silver. Nonetheless, to avoid hasty deductions, one should try to analyze other similar coins.

The analysis of the replicas has shown that the inclusions of copper-tin-antimony-lead could not be detected, possibly due to overlapping with the main phase of silver and/or due to their small amount that was below the detection limit of the method. Additionally, the refined silver parameters obtained for the ancient and the replica coins differ from the pure silver parameters. This fact may signal the existence of metal inclusions (i.e. copper). At present a neutron diffraction investigation on laboratory prepared "silver and metal combinations" replicas are due, in order to complete a systematic calibration study.

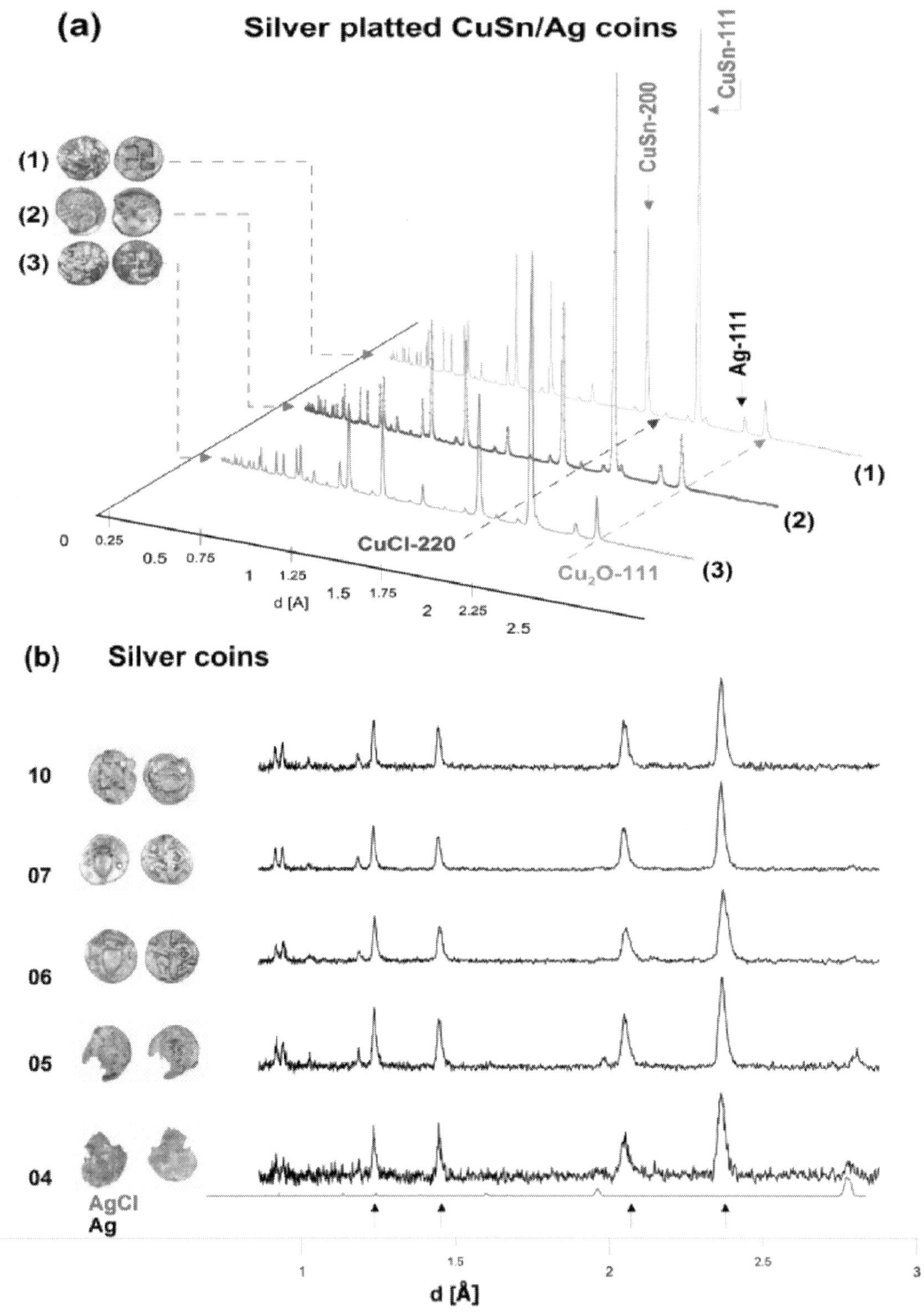

Figure 3: (a) 3D representaions of the neutron diffraction profiles of the coins 01, 02 and 03; b) Coins 04-10 and the corresponding measured data shown with solid lines (-).The prominet Bragg peaks are characteristic of the Cu-alloy in figure (a) and of the silver in (b). The corrosion products of cuprite, CuCl and AgCl are displayed in the lower part of each diagram.

4. Conclusions

Throughout this study ten ancient coins and two replicas have been examined using TOF -ND. The results of the analyses suggest that the coins can be classified into three categories according to the main metallic phases found, which confirms the initial hypothesis of the museum numismatists inferred from archeological sources. The fine style of coins constituting the third group along with the identical compositions, points to the same production technique. There is no apparent reason for a numismatist to suspect an imitation.

The comparison between the diffraction profiles of the original coins to those of the replicas, present distinct variations that may be used to differentiate the copies from the original items. As the results of this preliminary inquiry are encouraging, further analyses will probably help to further the comprehension of Thracian and Thasian coinage during the fifth century.

The above results also show the necessity in using complementary analysis techniques such as SR-XRD, High Energy PIXE and PGAA to separate the possible different phases, identify the elemental compositions, recognize additional trace elements and provide further information on minority phases which cannot be identified from the ND data, that has otherwise vanished into the background, or is lost due to the peak overlap.

References

[1] Anglos D, Georgiou S and Fotakis C 2009 *J. Nano Res.* **8** 47

[2] Corsi M Cristoforetti G Giuffrida M Legnaioli S Masotti L Palleschi V Salvetti A Tognoni E Vallebona C and Zanini A 2005 *Microchim. Acta* **152** 1-2 105 doi: 10.1007/s00604-005-0388-6

[3] Pappalardo G Costa E Marchetta C Pappalardo L Romano F P Zucchiatti A Prati P Mandò P. A Migliori A Palombo L and Vaccari M G 2004 *J. Cultural Heritage* **5** 2 183

[4] Kanngießer B Mantouvalou I Malzer W Wolff T and Hahn O 2008 *J. Anal. At. Spectrom.* **23** 814-819 doi: 10.1039/B717286A

[5] Mantouvalou I Lange K Wolff T Grötzsch D Lühl L Haschke M Hahn O and Kanngießer B 2010 *J. Anal. At. Spectrom* **25** 554-561 doi:10.1039/ B915912F

[6] Žitnik M Grlj N Vaupetič P Pelicon P Bŭcar K Sokaras D. Karydas A G and Kanngießer B 2010 *J. Anal. At. Spectrom.* **25** 28

[7] Sokaras D Karydas A G Malzer W Schütz R Kanngießer B Grlj N Pelicon P and Zitnik M 2009 *J. Anal. At. Spectrom.* **24** 611

[8] Santra S Mitra D, Sarkar M Bhattacharya D Denker A Opitz-Coutureau J and Rauschenberg J 2005 *Nucl. Instr. and Meth. Phys. Res.* B **229** 465

[9] Denker A and Maier K H 2000 *Nucl. Instr. and Meth.* B **161-163** 704

[10] Denker A Opitz- Coutureau J Griesser M Denk R and Winter H 2004 *Nucl. Instr. and Meth.* B **226** 163

[11] Kockelmann W Kirfel A Hähnel E 2001 *J. Archaeol. Sci.* **28** 213

[12] Siano S Kockelmann WA Bafile U Celli M Iozzo M Miccio M Moze O Pini R Salimbeni R and Zoppi M 2002 *Appl Phys* A **74** S1139

[13] Barone G Crupi V Majolino D Mazzoleni P Teixeira J and Venuti V 2009 *J. Appl. Phys.* **106** 5

[14] Imberti S Kockelmann W Celli M Grazzi F Zoppi M Botti A Sodo A Leo Imperiale M de Vries-Melein M Visser D and Postma H 2008 *Meas Sci Technol* **19** 03 40032008 doi: 10.1088/0957-0233/19/3/034003

[15] Barilaro D Crupi V Majolino D Venuti V, Barone G Tigano G Imberti S and Kockelmann W 2008 *J. of Phys. Cond. Mat.* **20** 10 doi: 10.1088/0953-8984/20/10/104254

[16] Barilaro D Crupi V Majolino D, Venuti V Barone G and Kockelmann W 2007 *J. Archaeol. Sci.* **34** 114 doi: 10.1016/j.jas.2006.10.024

[17] Huang W Kockelmann W Gordfrey E Scott D A and Wu X 2010 *Acta Scientiarum Naturalium Universitatis Pekinensis* **46** 2 245
[18] Siouris I M Walter J. 2006 *Physica* B **385** 225
[19] Siouris I M 2008 *J.Phys. Cond. Mat.* **200** 104252
[20] GSAS Larson A C and Von Dreele R B 1986 GSAS: General Structure Analysis System Report, LAUR 86-748 Los Alamos National Laboratories, USA. http://www.ccp14.ac.uk.
[21] Toby B H 2001 *J. Appl. Crystallogr.* **34** 210 .
[22] Flament C and Marchetti P 2004 Nucl. Instr. and Meth. B 226 179
[23] Ingo G M Balbi S De Caro T Fragalá I Angelini E and Bultrini G 2006 *Appl. Phys.* A **83** 493 ; Ingo G M Balbi S De Caro T Fragalá I Riccucci C and Bultrini G 2006 *Appl. Phys.* A **83** 623
[24] Giovannelli G Natali S Bozzini B Siciliano A Sarcinelli G and Vitale R 2005 *Archaeometry* **47** 4 817
[25] Pistofidis N Vourlias G Pavlidou E Dilo T Civici N Stamati F Gjongecaj S Prifti I Bilani O Stergioudis G and Polychroniadis E K 2006 *Appl. Phys.* **A 83** 637–642
[26] ICSD, Inorganic Crystal Structure Database Germany & National Institute of Standards and Technology USA, (2004). http:// www.fiz-informationsdienste.de/en/DB/icsd/index.html.
[27] CRYSTMET, Materials Toolkit 2.0.1 Toth Information Systems Inc. (2002). http://www.tothcanada.com/databases.html.
[28] Siano S Bartoli L Zoppi M Kockelmann W Dayamond M Dann J A Garagnani M G and Miccio 2003 *Proc. Int. Conf. on Archaeometallurgy in Europe 2*; (Milano: Associazione Italiana di Metallurgia) 19
[29] Sidot E Kahn-Harari A Cesari E and Robbiola L 2005 *Mater. Sci. Eng.* A **393** 147–56
[30] Grazzi F Bartoli L Siano S and Zoppi M 2010 *Anal Bioanal. Chem.* **397** 2501
[31] Oddy W A and Meeks ND 1982 *Proc. Conf. Sci in the Service of Conservation* (Washington) 119
[32] Clay T 1988 *Numis. Antichit. Class.* **17** 49
[33] Scott D A 1991 *Metallography and Microstructure of Ancient and Historic Metals.* Ed. The Getty Conservation Institute, CA, USA

AUTHOR INDEX

AUTHOR INDEX

AUTHOR INDEX

AUTHOR INDEX

AUTHOR INDEX

Institute of Physics
Dirac House, Temple Back
Bristol BS1 6BE UK

CURRAN ASSOCIATES INC
proceedings.com